PHILIP'S

WORLD REFERENCE
ATLAS

PICTURE ACKNOWLEDGEMENTS:

Page 9 (centre left) – NASA: Olympus Mons, *The Universe Revealed*, page 47 (top).
Page 10 – Science Photo Library/National Optical Astronomy Observationaries:
Sun's corona (blue) *The Universe Revealed*, page 21.
Page 12 – Royal Greenwich Observatory, Herstmonceaux: Sun maximum,
Joy of Knowledge Science and Technology (c 1976 pic), page 221.
Page 13 – NASA: UV shot of prominences *Joy of Knowledge Science and Technology*,
page 223 (c 1976).
Page 14 – NASA: Mercury from Mariner. *The Universe Revealed*, page 25.
Page 15 – NASA: Venus (Octopus Publishing Group Ltd).
Page 16 (centre) – NASA: *Joy of Knowledge Science and Technology*, page 192.
Page 16 (bottom) – NASA: *The Universe Revealed*, page 35.
Page 17 – NASA: *The Universe Revealed*, page 47 (bottom).
Page 18 – NASA: Jupiter and Io (Octopus Publishing Group Ltd).
Page 19 – NASA: Saturn (Octopus Publishing Group Ltd).
Page 20 – NASA/Science Photo Library: Uranus.
Page 21 (centre) – NASA: Triton, moon of Neptune. *The Universe Revealed*, page 74.
Page 21 (bottom) – Space Telescope Science Institute/NASA/Science Photo Library:
Hubble computer images of Pluto hemispheres.
Page 23 – NASA: comet Hale-Bopp. *The Universe Revealed*, page 78.
Page 24 – NASA: asteroid Ida. *The Universe Revealed*, page 80.

The Solar System and the Physical Earth compiled by Richard Widdows.

PHILIP'S

WORLD REFERENCE
ATLAS

Bounty
Books

This 2000 edition published
by Chancellor Press, an imprint of Bounty Books,
a division of Octopus Publishing Group Ltd,
2-4 Heron Quays, London E14 4JP.

Reprinted 2002, 2003, 2004(twice), 2005(twice), 2006

ISBN-13: 978-0-753714-59-1
ISBN-10: 0-753714-59-0

A CIP catalogue record for this book is available from the British Library

Produced by Toppan (HK) Ltd
Printed in Hong Kong

CONTENTS

THE SOLAR SYSTEM

THE PHYSICAL EARTH

THE MAP SECTION

INDEX

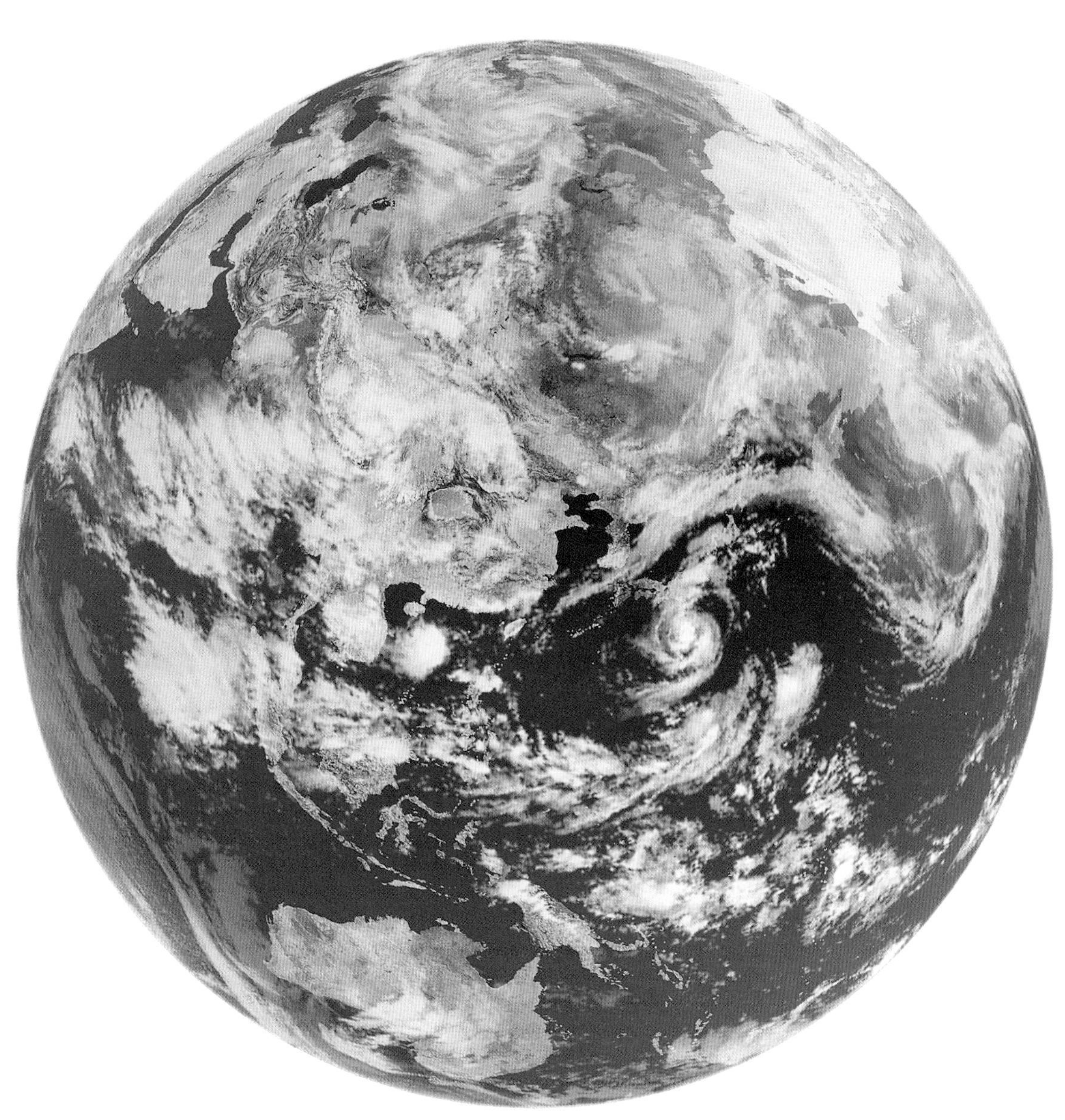

SOLAR SYSTEM: EVOLUTION

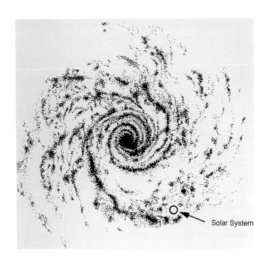

ABOUT 15 BILLION years ago, time and space began with the most colossal explosion in cosmic history: the "Big Bang" that initiated the Universe. According to current theory, in the first millionth of a second of its existence it expanded from a dimensionless point of infinite mass and density into a fireball about 30 billion km (18.6 billion miles) across – and has been expanding at a phenomenal rate ever since.

It took almost a million years for the primal fireball to cool enough for atoms to form. They were mostly hydrogen, still the most abundant material in the Universe. But the new matter was not evenly distributed around the young Universe, and a few billion years later atoms in relatively dense regions began to cling together under the influence of gravity, forming distinct masses of gas separated by vast expanses of empty space.

At the beginning these first proto-galaxies were dark places – the Universe had cooled – but gravitational attraction continued its work, condensing matter into coherent lumps inside the galactic gas clouds. About three billion years later, some of these masses had contracted so much that internal

pressure produced the high temperatures necessary to cause nuclear fusion: the first stars were born.

There were several generations of stars, each feeding on the wreckage of its extinct predecessors as well as the original galactic gas swirls. With each new generation, progressively larger atoms were forged in stellar furnaces and the galaxy's range of elements, once restricted to hydrogen, grew larger. About ten billion years after the Big Bang, a star formed on the outskirts of our galaxy with enough matter left over to create a retinue of planets. Some 4.6 billion years after that, a few planetary atoms had evolved into structures of complex molecules that lived, breathed and, eventually, pointed telescopes at the sky.

These early astronomers found that their Sun was just one of more than 100 billion stars in our home galaxy alone – the number of grains of rice it would take to fill a cathedral. Our galaxy, in turn, forms part of a local group of 25 or so similar structures, some much larger than ours. The most distant galaxy so far observed lies about 13.1 billion light-years away – and one light-year is some 9,461 million km (5,879 million miles).

ABOVE Our Solar System is located in one of the home galaxy's spiral arms, a little under 28,000 light-years away from the galactic centre and orbiting around it in a period of about some 200 million years. There are at least 100 million other galaxies in the Universe.

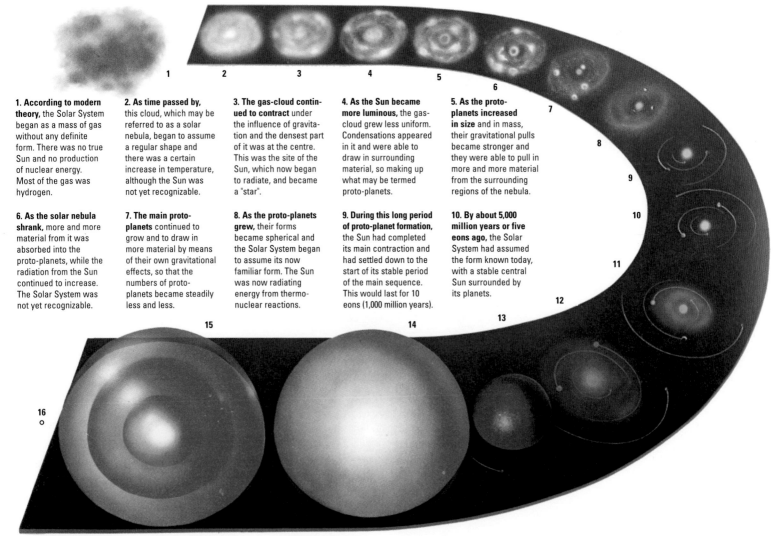

1. According to modern theory, the Solar System began as a mass of gas without any definite form. There was no true Sun and no production of nuclear energy. Most of the gas was hydrogen.

2. As time passed by, this cloud, which may be referred to as a solar nebula, began to assume a regular shape and there was a certain increase in temperature, although the Sun was not yet recognizable.

3. The gas-cloud continued to contract under the influence of gravitation and the densest part of it was at the centre. This was the site of the Sun, which now began to radiate, and became a "star".

4. As the Sun became more luminous, the gas-cloud grew less uniform. Condensations appeared in it and were able to draw in surrounding material, so making up what may be termed proto-planets.

5. As the proto-planets increased in size and in mass, their gravitational pulls became stronger and they were able to pull in more and more material from the surrounding regions of the nebula.

6. As the solar nebula shrank, more and more material from it was absorbed into the proto-planets, while the radiation from the Sun continued to increase. The Solar System was not yet recognizable.

7. The main proto-planets continued to grow and to draw in more material by means of their own gravitational effects, so that the numbers of proto-planets became steadily less and less.

8. As the proto-planets grew, their forms became spherical and the Solar System began to assume its now familiar form. The Sun was now radiating energy from thermo-nuclear reactions.

9. During this long period of proto-planet formation, the Sun had completed its main contraction and had settled down to the start of its stable period of the main sequence. This would last for 10 eons (1,000 million years).

10. By about 5,000 million years or five eons ago, the Solar System had assumed the form known today, with a stable central Sun surrounded by its planets.

11. In perhaps 5,000 million years from now the Sun will have exhausted its supply of available hydrogen and its structure will change. The core will shrink and the surface expand considerably, with a lower surface temperature.

12. The next stage of solar evolution will be expansion to the red giant stage, with luminosity increased by 100 times. The size of the globe will increase with the overall increase in energy output, and the inner planets will be destroyed.

13. With a further rise in core temperature, the Sun will begin to burn its helium, causing a rapid rise in temperature and increase in size. The Earth can hardly hope to survive this phase of evolution as the Sun expands to 50 times its size.

14. By now the Sun will be at its most unstable, with an intensely hot core and a rarefied atmosphere. The helium burning helium will give the so-called "helium flash". After a temporary contraction the Sun will be 400 times its present size.

15. Different kinds of reactions inside the Sun will lead to an even greater increase of core temperature. The system of planets will no longer exist in the form we know today, but the supply of nuclear energy will be almost exhausted.

16. When all the nuclear energy is used up, the Sun (as all stars eventually do) will collapse, very rapidly on the cosmic scale, into a small dense and very feeble white dwarf. It will continue to shine because it will still be contracting gravitationally.

Formation of the planets

The planets and larger satellites can be divided into two distinct classes. Mercury, Venus, Earth and Mars are all rocky "terrestrials", while Jupiter, Saturn, Uranus and Neptune are the large gaseous Jovian planets. Pluto can be classified, along with the large icy moons of the gas giants, as a third type. The terrestrial planets are closer to the Sun, have smaller masses and radii, and are more dense than the Jovian planets. These are big, low in density and have extensive satellite systems and rings.

The basic difference between the two families arose as a consequence of the temperature difference within the proto-solar cloud. This allowed icy material to condense well beyond the asteroid belt, producing cold proto-planets which effectively collected vast amounts of gas. The inner planets were too small and too hot to retain large amounts of original atmosphere in the face of the strong winds from the Sun.

Beyond the Solar System

Far beyond the gas giants, and outside the erratic orbit of Neptune, lie two regions of space that have intrigued astronomers since their discovery in the last half of the 20th century.

The Kuiper belt, named after one of the scientists who predicted its existence, is a disc of debris lying between about 35 and 100 astronomical units from the Sun; an astronomical unit (AU) is the average distance from the Earth to the Sun – 149,597,870km (92,958,350 miles). The first object was located there in 1992, so dim it was 10 million times fainter than the faintest stars seen by eye. It is now estimated that this belt may contain up to a billion comets, with a total mass just 1% of Earth.

Astronomers have now found over 60 Kuiper belt objects orbiting farther from the Sun than Neptune and Pluto, taking between 160 and 720 years to orbit the Sun. The smallest object seen is roughly 100km (60 miles) across, while the largest is 500 km (300 miles) in diameter, slightly smaller than Neptune's moon Triton. Indeed, Triton could be a body captured from the Kuiper belt, and Pluto and its moon Charon could be among its members.

Much further out in space is the Oort cloud, named in 1950 after the Dutch astronomer who identified it as a source of long-period comets. This is a rough sphere of rocky and icy debris left over from the solar nebula from which the Solar System formed. A vast size, it lies between 30,000 and 100,000AU from the Sun, a distance where gravity from passing stars could perurb it, sending comets in towards the Sun.

Future of the Solar System

We now know that dramatic consequences are in store for these terrestrial planets as a result of the dramatic changes that will happen to the Sun. Astronomers calculate that our star will be hot enough in 3 billion years to boil Earth's oceans away, leaving the planet a burned-out cinder, a dead and sterile place. Four billion years on, the Sun will balloon into a giant star, engulfing Mercury and becoming 2,000 times brighter than it is now. Its light will be intense enough to melt Earth's surface and turn the icy moons of the giant planets into globes of liquid.

Such events are in the almost inconceivably distant future, of course. For the present the Sun continues to provide us with an up-close laboratory of stellar astrophysics and evolution.

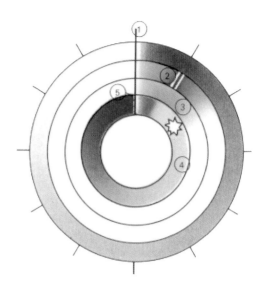

ABOVE The timescale of the Solar System can be represented on a 12-hour clock, tracing the lifespan of the Sun, the inner planets, Earth and the outer planets from the inner circle outwards. At the 12 o'clock position [1] the Solar System is created; after 4,000 million years, conditions on Earth are favourable for life [2]; as a red giant the Sun engulfs the inner planets [3] before collapsing as a white dwarf [4] and, possibly, end its long life as a brown dwarf [5].

BELOW The distance of the outermost planets – Jupiter, Saturn, Uranus, Neptune and Pluto – will save them from the Sun's helium burn, and each will continue its orbit. More precise predictions for their future are not possible.

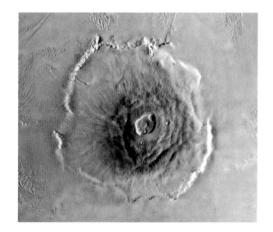

LEFT Olympus Mons is the largest volcano in the Solar System. Its peak rises to a staggering 27km (16.8 miles) above the mean surface level of Mars. More than three times as high as Earth's Mount Everest, it has a diameter of some 520km (323 miles). Olympus Mons is surrounded by a huge cliff up to 6km (3.7 miles) high, where the lower flanks appear to have fallen away in a gigantic landslide.

This collapse may have generated the peculiar blocky terrain of ridges separated by flat areas, the Olympus Mons aureole, that extends from the base of the cliff up to 1,000km (600 miles) from the volcano's summit. This contains a nested set of volcanic craters, the largest of them 80km (50 miles) across.

BELOW The lifespan of the Earth started from the material of the solar nebula [A] which at first had no regular form. When it reached its present size [B] the original hydrogen atmosphere had already been lost and replaced by a new one, caused by gases sent out from the interior. Life could begin and today the Earth is moving in a settled orbit round a stable star, so that it is habitable [C]. But this state of affairs will not persist indefinitely: long before the Sun enters the red giant stage, most scientists believe, the Earth will be overheated, the oceans will boil, and the atmosphere will be driven off [D]. Finally, the only planet known to have had life will be completely destroyed [E].

A

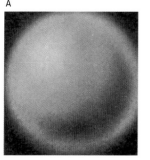

B

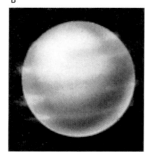

C

D

E

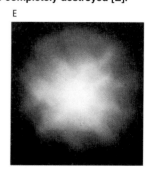

SOLAR SYSTEM: PROFILE

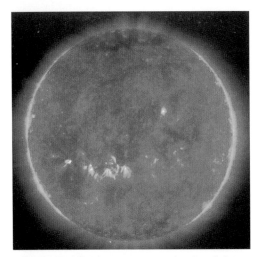

ABOVE The Sun's outer corona in ultraviolet light. The bright regions are areas of intense magnetism. This part of the corona is at a temperature of around 1,000,000°C. All the components of the Solar System are tethered by the immense gravitational pull of the Sun, the star whose thermonuclear furnaces provide them with virtually all their heat and light.

BELOW The planets of the Solar System shown to the same scale. On the right is a segment of the Sun [1]; from its surface rises a huge prominence [2]. Then come the inner planets: Mercury [3], Venus [4], Earth [5] with its Moon [6], and Mars [7]. Mars has two dwarf satellites Phobos [8] and Deimos [9], exaggerated here – if shown to the correct scale, they would be too small to be seen without a microscope.

A TINY PART of one of the millions of galaxies (collections of stars) that make up the known Universe, the Solar System orbits at a mean distance of 29,700 light-years from the centre of our own galaxy, the "Milky Way". The present distance is 27,700 light-years, and it will reach the minimum distance of 27,600 in around 15 million years' time. It comprises one star, which we call the Sun, nine principal planets, and various bodies of lesser importance, including the satellites that attend some of the planets and a range of cosmic debris, notably asteroids, meteors and comets.

The system is entirely dependent on the Sun, which is by far the most massive body and the only one to be self-luminous: the remaining members of the Solar System shine by reflected sunlight and appear so brilliant in our skies that it is not always easy to remember that in universal terms they are nowhere near as large or important as they appear.

The inner planets

The planets are divided into two well-defined groups. First come the four relatively small planets of Mercury, Venus, Earth and Mars, with diameters ranging from 12,756km (7,926 miles) for Earth down to only 4,878km (3,031 miles) for Mercury.

Then come the asteroids [10], of which even the largest is only about 913 (567 miles) in diameter. Beyond lie the giant planets: Jupiter [11], with its four largest satellites Io [12], Europa [13], Ganymede [14] and Callisto [15]; Saturn [16] with its retinue of satellites, of which the largest is Titan [17]; Uranus [18] with its many satellites; Neptune [19] with its large satellite Triton [20]; and finally misfit Pluto [21].

These planets have various factors in common. All, for example, have solid surfaces and are presumably made up of similar materials, although Earth and Mercury are more dense than Mars and Venus.

Although their orbits do not in general depart much from the circular, the paths of Mercury and Mars are considerably more eccentric than those of Earth and Venus. Mercury and Venus are known as the "inferior planets" because their orbits lie inside that of Earth; they show lunar-type phases from new to full and remain in the same region of the sky as the Sun. While Mercury and Venus are unattended by any satellites, Earth has one satellite (our familiar Moon) and Mars has two, Phobos and Deimos, both of which are very small and different in nature from the Moon.

Beyond Mars comes a wide gap, in which move thousands of small worlds known as the asteroids, or minor planets. Even Ceres, the largest, is only about 913km (567 miles) in diameter. This is much larger than was once thought, but still small by planetary standards. It is not therefore surprising that the asteroids remained hidden until relatively recent times, with Ceres discovered only in 1801. Just one of this new multitude, Vesta, is ever visible from Earth without the aid of a telescope.

The outer planets

Far beyond the main asteroid belt come the four giant planets of Jupiter, Saturn, Uranus and Neptune. These worlds are quite different from the terrestrial planets: they are fluid (that is, gas or liquid) rather than solid bodies with very dense atmospheres. Their masses are so great that they have been able to retain much of their original hydrogen; the escape velocity of Jupiter, for instance, is 60km (37 miles) per second as against

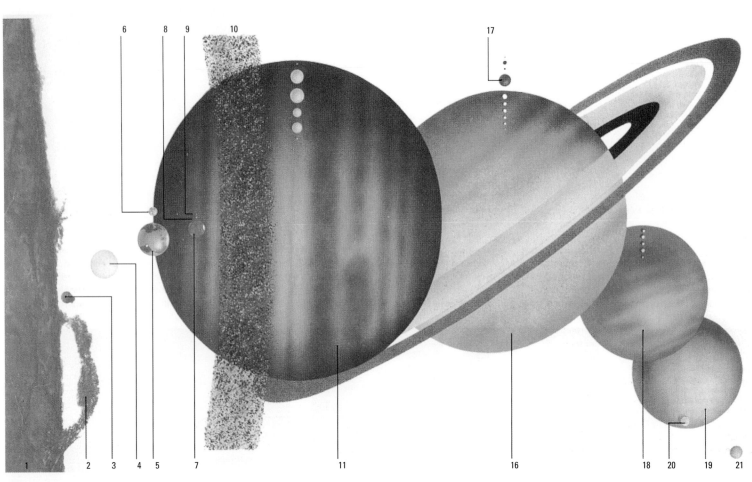

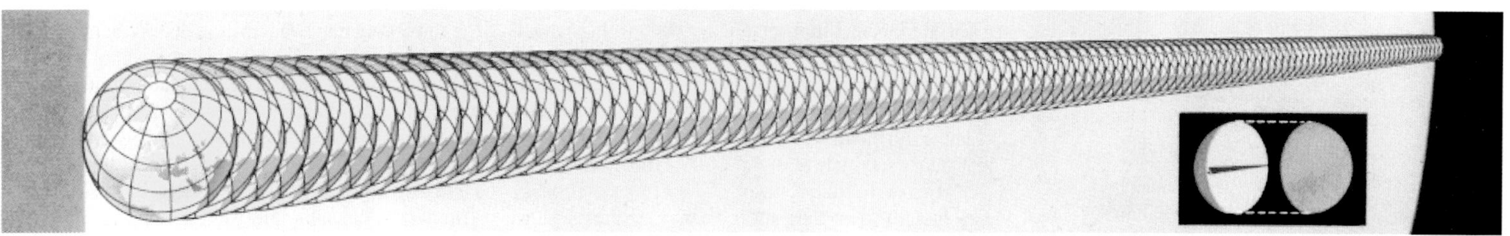

only 11.2km (7 miles) per second for Earth. Their mean distances from the Sun range from 778 million km (483 million miles) for Jupiter out to 4,497 million km (2,794 million miles) for Neptune. Conventional diagrams of the Solar System tend to be misleading as far as scale is concerned; it is tempting, for example, to assume that Saturn and Uranus are lying next to each other when in fact the distance of Uranus from the Earth's orbit is about twice that of Saturn.

The giant planets have various points in common, but differ markedly in detail. Their densities are comparatively low and the density of Saturn is actually less than that of water. Although Jupiter is seen solely by reflected sunlight, the planet does generate some heat of its own. However, even though the core temperature must be high, it is not nearly high enough for nuclear reactions to begin, so that Jupiter, though massive, cannot be compared to a star like the Sun.

Planetary discoveries

Five of the planets – Mercury, Venus, Mars, Jupiter and Saturn – have been known from ancient times, since all are prominent naked-eye objects. Uranus, just visible with the naked eye, was discovered fortuitously in 1781 by William Herschel and Neptune was added to the list of known planets in 1846 as a result of mathematical investigations carried out concerning movements of Uranus. All the giants are attended by satellites; Jupiter has 16 moons, Saturn 20, Uranus 15 and Neptune eight. Several of these are of planetary size, with diameters almost equal to Mercury's.

The outermost known planet is Pluto, discovered in 1930 by astronomers at the Lowell Observatory, Flagstaff, Arizona. It is far from another giant, being smaller than the Earth, and is usually ranked as a terrestrial-type planet, even though little is known about it.

Pluto's origin was long a mystery because of its size, rocky composition and highly unusual orbit. In recent years, however, it has become apparent that Pluto orbits within a "swarm" of tens of thousands of still smaller worlds orbiting well beyond the region of Neptune.

RIGHT The ecosphere is the region around the Sun in which a planet can be at a suitable temperature for life as we conceive it to exist – assuming that the planet is of Earth "type". The inner yellow zone [1] is way too hot, and beyond the ecosphere [orange, 2], temperatures will become too low [3]. Earth [4] lies in the middle of the ecosphere, enjoying a near-perfect set of balanced conditions for life. Inhospitable Venus [5] orbits at the very inner limit and barren Mars [6] at the outer, but recent probes have proved that neither has the prerequisites for evolution. The best hope of finding life as we know it seems now to rest with a similar ecosphere – in one of the billions of other solar systems in the Universe.

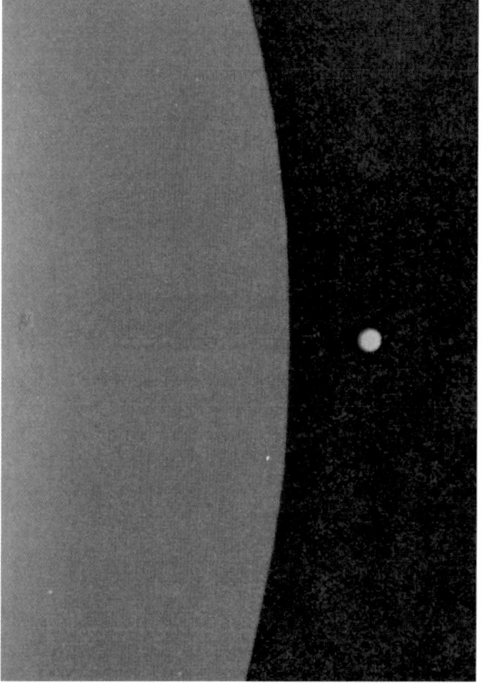

ABOVE Shown here in cross-section, the Sun has an equatorial diameter of 1,392,000km (865,000 miles), 109 times that of Earth. Despite the fact that its volume is more than a million times that of Earth, its mass is only 333,000 times greater because the density is lower: the mean specific gravity, on a scale where water = 1, is only 1.4.

LEFT While the Sun is the body on which the entire Solar System depends, and is more massive than all the planets combined, it is an ordinary main sequence star with a magnitude of +5 – small when compared with a giant star. The diagram shows the Sun alongside a segment of the red supergiant Betelgeuse, which marks Orion's right shoulder. Betelgeuse is of spectral class M2 – a very cool star – but has an absolute magnitude of –5.5. Its diameter is 300 to 400 times that of the Sun, and its globe is large enough to contain Earth's orbit. In 5 million years' time the Sun's life cycle will make it a modest red giant in its own right, and the solid inner planets of the Solar System will be destroyed by the heat and light that results from its phenomenal expansion.

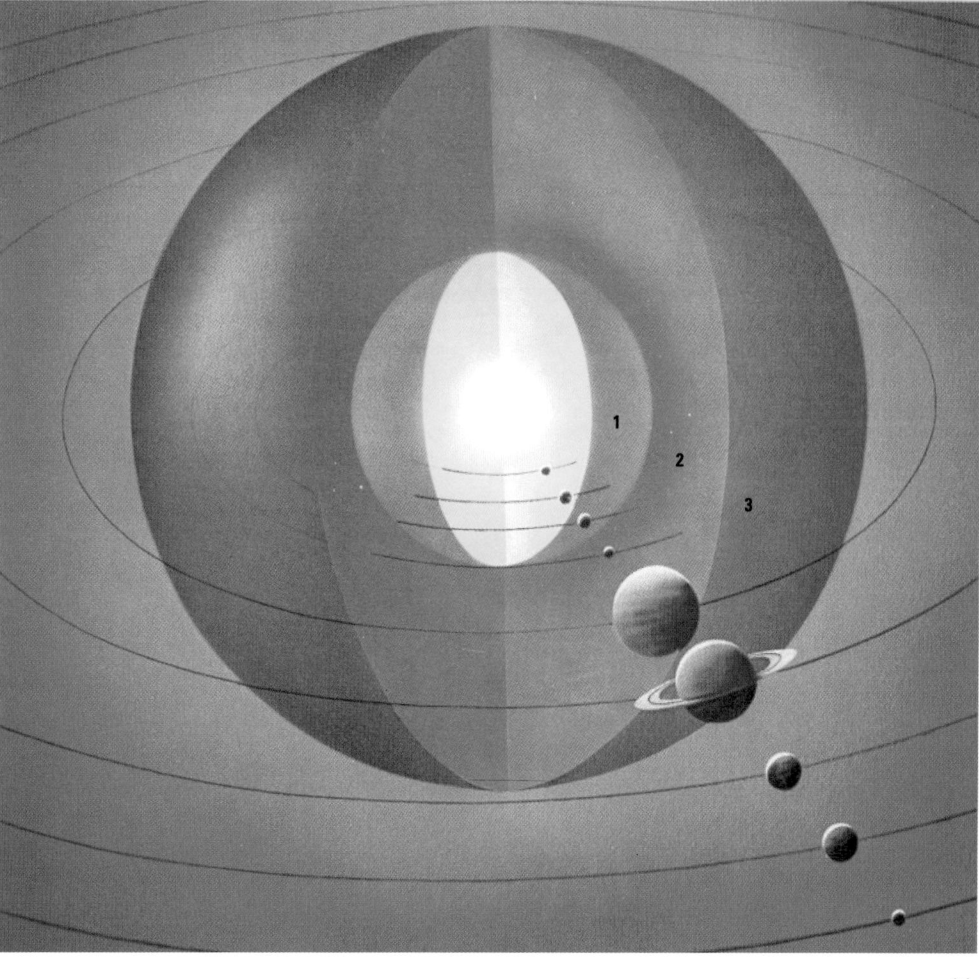

THE SUN

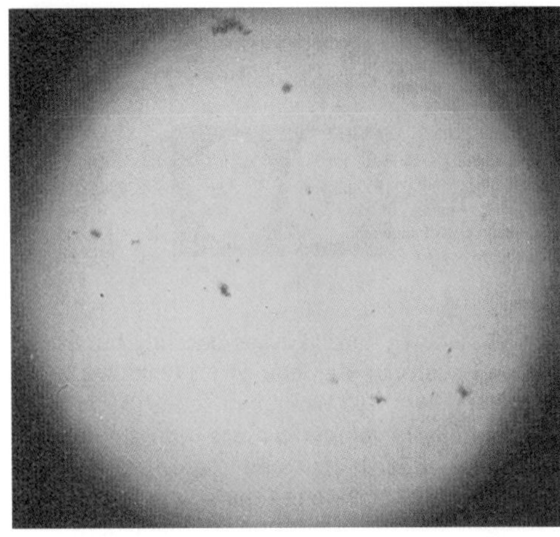

ABOVE The "solar maximum" of 1958, pictured here, was the most energetic phase of the solar cycle ever recorded, and sunspots are clearly visible. Occurring where there is a local strengthening of the Sun's magnetic field, sunspots are regions in the photosphere that are cooler than their surroundings and therefore appear darker. Varying in size from 1,000 to 50,000km (600 to 30,000 miles) and occasionally up to 200,000km (125,000 miles), they comprise a dark central region called the umbra, and a grey outer region, the penumbra. Their duration varies from a few hours to several weeks, or months for the biggest ones.

The number of spots visible depends on the stage of the solar cycle. This is fairly regular and lasts around 11 years and is part of a longer 22-year solar cycle, but at the intervening "spot minima" the disc may remain featureless for several days or even weeks. The exact cause of sunspots is not fully understood – and no theory has been able to explain their disappearance between 1645 and 1715.

Sunspots are seen to move across the face of the Sun as it rotates. Most appear in pairs, but often complex groups emerge. They can be seen if you project an image of the Sun onto a piece of white paper or card.

THE SUN is a star, one of 100,000 million stars in our galaxy. In the Universe as a whole it is insignificant – classed as a yellow dwarf star with a spectrum of type G – but in our planetary system it is the all-important controlling body.

Immensely larger than Earth, the Sun has a diameter of 1,392,000km (865,000 miles). Though big enough to contain more than a million bodies the volume of Earth, its mass is only 1.99×10^{30} kg – approximately 333,000 times that of Earth. The reason why it is not as massive as might be expected is that its density is lower than that of an Earth-type planet. The mean value for the specific gravity is 1.409 (that is, 1.409 times an equal volume of water), but the Sun is not homogenous and density, pressure and temperature all increase rapidly beneath the brilliant outer surface towards the centre. It consists of about 70% hydrogen (by weight) and some 28% helium, with the remainder mostly oxygen and carbon.

The Sun lies some 32,000 light-years from the centre of our galaxy and takes approximately 200 million years to complete one journey round the galactic nucleus. It has an axial rotation period of 25.4 days at its equator, but because the Sun does not rotate in the manner of a solid body, this period is considerably longer near the solar poles.

In ordinary light the Sun appears to have a clear edge. This is because only a 500-km (300-mile) layer of its atmosphere, the photosphere, is at the correct temperature to emit light at visible wavelengths – a very small layer in comparison to the star's vast diameter.

The Sun's magnetic field

Overall, the Sun's magnetic field is roughly the same strength as Earth's, but the mechanism is entirely different. The Sun is not a solid body but a plasma created by heat removing the electrons of hydrogen atoms to leave negatively charged electrons and, possibly, positively charged ions. Magnetic fields can be created by the motion of electrically charged particles, and the Sun's turbulence and rotation create localized fields.

As the Sun rotates, the magnetic field lines get "trapped" and move around with the rotation. As the top layers bubble with convection the field lines become twisted up, and this squashing together increases the strength of the magnetic field in those areas. These intense pockets cause many of the phenomena seen on the Sun, notably sunspots.

Prominences and flares

The part of the solar atmosphere lying immediately above the photosphere is called the chromosphere ("colour sphere") because it has a characteristically reddish appearance. This is also the region of the large and brilliant prominences. To observe the prominences, instruments based on the principle of the spectroscope are used. There are two main types of prominences: eruptive and quiescent. Eruptive prominences are in violent motion and have been observed extending to more than 50,000km (312,500 miles) above the Sun's surface; quiescent prominences are much more stable and may hang in the chromosphere for days before breaking up. Both are most common near the peak of the solar cycle of activity.

Prominences are often associated with major spot-groups. Active groups also produce "flares", which are not usually visible, although a few have been seen. The flares are short-lived and emit streams of particles as well as short-wave radiation. These emissions have marked effects on Earth, producing magnetic storms or disturbances of Earth's magnetic field that affect radio communications and compasses. They also produce the beautiful solar lights or aurorae.

The solar wind

Less dense areas of the corona, the outer layer of the Sun, called coronal holes by astronomers, appear where the Sun's magnetic field opens to interplanetary space rather than looping back down. These areas are believed to be the major source of the solar wind, where charged particles, mainly protons and electrons, stream out into the interplanetary medium.

It is this emission that has a strong effect on the tails of comets, forcing them to point away from the Sun. Even when it reaches Earth, the wind's velocity exceeds 950km (590 miles) per second.

BELOW The structure of the Sun cannot be drawn to an accurate scale, and attempts at full cross-sections are misleading. In the core, about 400,000km (250,000 miles) across, continual nuclear transformations create energy and the temperature is perhaps 15 million °C (27 million °F). Further out in the solar interior, the radiative zone [1], about 300,000km (200,000 miles) wide, diffuses radiation randomly, and temperatures range from 15 million to 1 million °C. In the convective layer [2] heat travels outward for 200,000km (125,000 miles) on convection currents, cooling from a million to 6,000°C (11,000°F). The relatively rarefied photosphere [3], the fairly well-defined "sphere of light" from which energy is radiated into space and where temperatures average 5,500°C (10,000°F), is surprisingly narrow – only 500km (300 miles) wide; because it is the layer of the Sun that radiates in visible wavelengths, this is the part of the Sun that we see, including the sunspots [4].

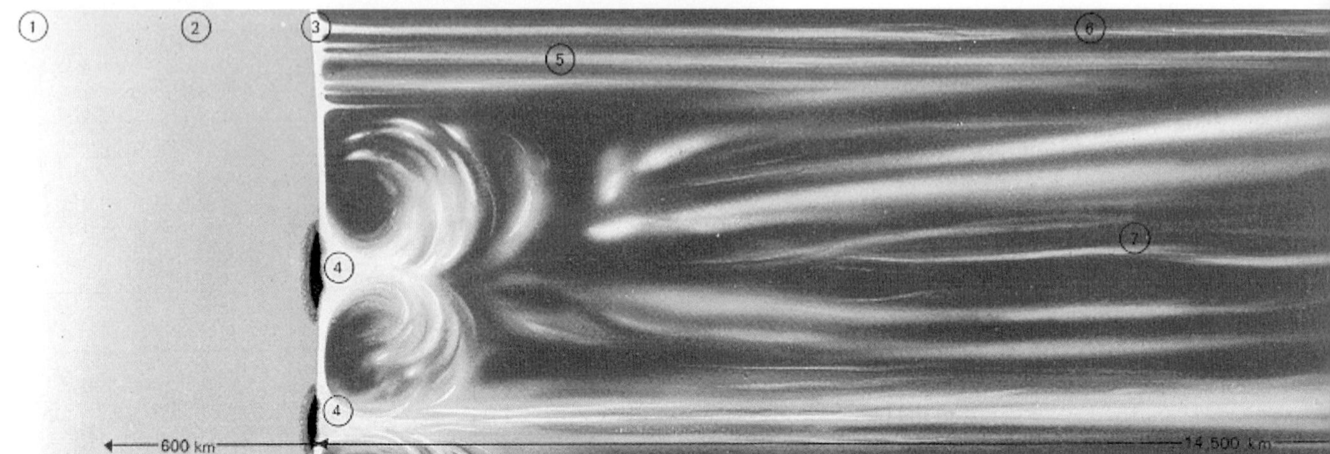

RIGHT Like all stars the Sun's energy is generated by nuclear reactions taking place under extreme conditions in the core. Here the Sun is continually converting four hydrogen atoms into one helium atom. The amount of energy produced in each individual reaction is tiny, but the Sun is converting 600,000 million kg (1,325,000 million lb) of hydrogen into helium every single second. The Sun's total power output, its luminosity, is 3.9×10^{26} watts (the equivalent of a million, million, million, million 100-watt light bulbs).

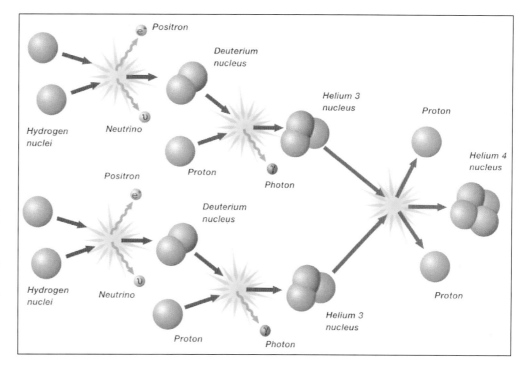

Powerhouse of a star

It is a mistake to think of the Sun burning in the same way that a fire burns. A star made up entirely of coal, and radiating as fiercely as the Sun does, would not last long on the cosmic scale, and astronomers believe that the Sun is at least 5,000 million years old.

The source of solar energy is to be found in nuclear transformations. Hydrogen is the main constituent and near the core, where the temperatures and pressures are so extreme that the second lightest element, helium, is formed from hydrogen nuclei by nuclear fusion. It takes four hydrogen nuclei to make one nucleus of helium; in the process a little mass is lost, being converted into a large amount of energy. The energy produced keeps the Sun radiating: the loss of mass amounts to four million tonnes per second. This may seem significant, but it is negligible compared with the total mass of the Sun – and there is enough hydrogen available to keep the Sun shining in its present form for at least another 5,000 million years.

Eventually the hydrogen will start to become exhausted and the Sun will change its structure drastically. According to current theory, it will pass through a red giant stage, when it will have a luminosity at least 100 times as great as it does today. Once all its nuclear fuel has been used up, it will start to collapse into a small dense star of the type known as a white dwarf. Earth will have long gone: it will not survive the heat of the Sun's red giant stage, and along with the other inner planets will be totally destroyed.

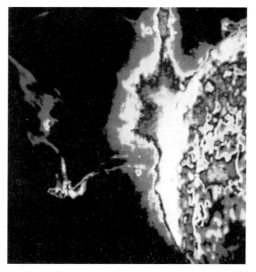

LEFT A solar prominence photographed by astronauts on board Skylab. In this extreme ultraviolet shot the colours are false: they represent the degree of radiation intensity from red, through yellow and blue, to purple and white, where the activity is most intense. This picture could only be taken with equipment carried above the layers of the Earth's atmosphere.

When viewed face-on against the bright photosphere, prominences are known as filaments. Narrow jets of gas called spicules can also be observed at the limb of the Sun. They move at around 20–30km (12–18 miles) a second from the lower chromosphere into the inner corona, and fall back or fade away after a few minutes. Flares, intense outpourings of energy, occur in complex sunspot groups, and can cause auroral activity and storms on Earth.

Above the photosphere lies the chromosphere [5], meaning "sphere of colour", and so-called because of its rosy tint when seen during a total solar eclipse. This is the region of flares and prominences [6], where the temperature rises from 6,000 to 50,000°C; temperature here is purely a measure of the speeds at which the

atomic particles are moving and does not necessarily indicate extra "heat". In the chromosphere there are spicules [7], masses of high-temperature gases shooting up into the immensely rarefied corona [8], where temperatures can reach 1 million °C (1,800,000°F) – possibly due to the action of the Sun's magnetic

field. Streamers [9] issue from the corona, which has no definite boundary and extends millions of kilometres out into space, eventually thinning to become the radiation we call the "solar wind". Together with the Sun's magnetic field, the solar wind dominates a vast indeterminate region of space called the heliosphere.

MERCURY

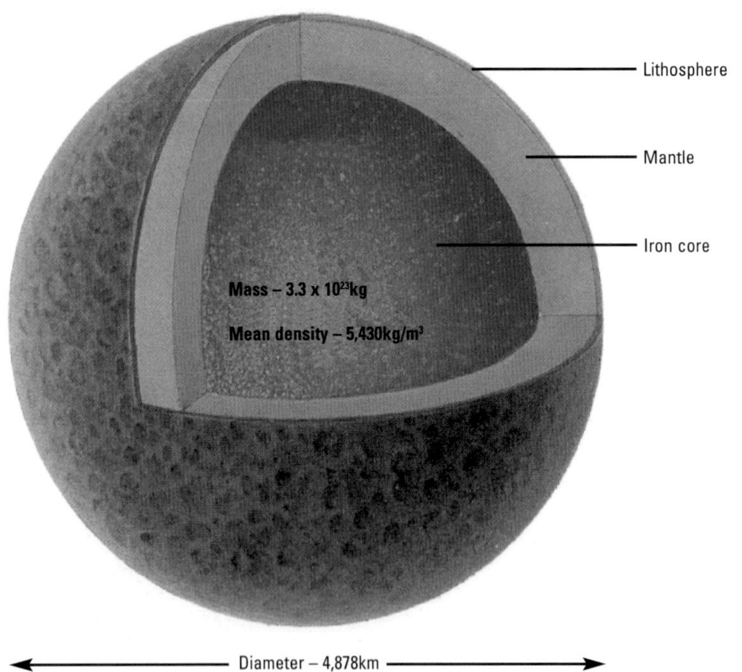

Lithosphere

Mantle

Iron core

Mass – 3.3 x 10²³kg

Mean density – 5,430kg/m³

Diameter – 4,878km

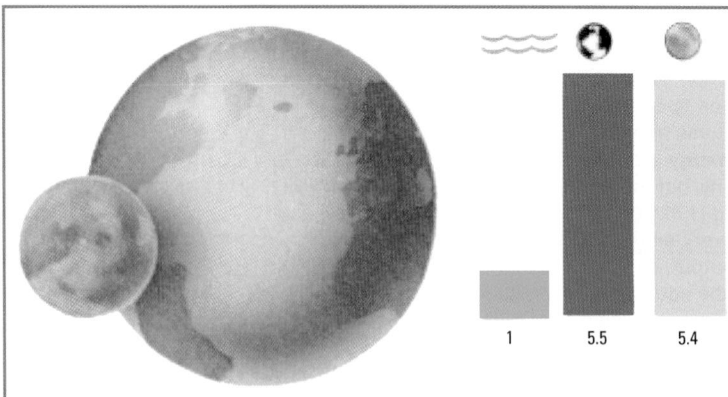

1   5.5   5.4

ABOVE With a diameter of 4,878km (3,031miles), Mercury is dwarfed by Earth and is the Solar System's smallest planet after Pluto. However, its mean density (5.4 times that of water) is similar to Earth's. A small planet must contain a lot of iron to have so high a density, and astronomers believe that Mercury has twice as much, by proportion, as any other planet. Its iron core, thought to extend out to three-quarters of its entire radius, is surrounded by a mantle of rock and a thick hard crust. Mercury has a very tenuous and thin atmosphere, mostly hydrogen and helium, with a ground pressure only two-trillionths that of Earth.

THE "FASTEST" planet, Mercury takes just 88 Earth days to orbit its massive close neighbour, the Sun – probably the reason the Romans named it after the fleet-footed messenger of the gods. It is the closest planet to the Sun and suffers the widest extremes of temperature: at noon, when the Sun is directly overhead, the temperature can soar to as high as 470°C (880°F), while during the long Mercurian night it can plunge to below –175°C (–283°F).

Mercury's orbit is elliptical: its aphelion (farthest point from the Sun) is 69,800,000km (43,400,000 miles), and its perihelion (closest point to the Sun) is 46,000,000km (28,600,000 miles).

The elusive planet

Although its existence has been known since the dawn of history, and it can appear to be brighter than the brightest star, Mercury is notoriously difficult to observe. This is because it is always too close to the Sun in the sky. The angle between Mercury and the Sun can never exceed 28°; this means that Mercury is lost in the Sun's glare because it sets no more than two hours after the Sun and rises no more than two hours before it. Once or twice a year, you may be able to see Mercury shining like a bright star close to the western horizon after sunset, or close to the eastern horizon before sunrise.

Mercury orbits the Sun in only 88 Earth days and undertakes the Earth at intervals of, on average, 115.88 days. On these occasions, Mercury lies between the Sun and Earth, but because of the tilt of its orbit (7°), usually passes above or below the Sun when viewed from Earth. Occasionally, when the alignment is right, Mercury passes directly in front of the Sun and can be seen as a small dot moving slowly across its face: such an event is called a transit. The alignments that allow transits of Mercury to take place occur only in the months of May or November, and the dates of early 21st-century transits are 7 May 2003, 8 November 2006, 9 May 2016, and 11 November 2019.

Until the 1960s, most astronomers believed that Mercury took exactly the same time to rotate on its axis as it took to orbit the Sun: one hemisphere would always face toward the Sun and constantly

suffer its boiling heat, while the other was in constant darkness. However, radar measurements carried out since then have shown that this is not the case: Mercury rotates every 58.65 Earth days, precisely two-thirds of its orbital period or year.

Mercury's magnetic puzzle

The strength of the magnetic field at Mercury's surface is very low: only about 1% that of the Earth's. This is only just strong enough to deflect most of the incoming solar wind and to form a magnetosphere around the planet. Nevertheless, Mariner

10's discovery of the magnetic field came as a surprise to most astronomers. According to conventional theory, a planet can only sustain a magnetic field if it has an electrically conductive liquid interior and rotates rapidly on its axis.

Although Mercury has a large iron core, this should in theory have cooled and solidified by now because of the planet's small size. The presence of a magnetic field suggests that at least part of the deep interior must still be liquid – but even if this is the case, Mercury's slow rotation still makes the presence of a magnetic field puzzling.

RIGHT A mosaic of Mercury created from images taken by Mariner 10, the first two-planet probe, on its outward journey in March 1974. The craft flew within 703km (437 miles) of the planet, and in three encounters during 1974–75 took more than 12,000 images covering over half its surface. Images returned by Mariner 10 revealed that most of Mercury's surface is heavily cratered from impacts by meteorites, asteroids and comets, with many over 200km (125 miles) wide. As on the Moon, some are surrounded by lighter-coloured ejecta – material splashed out by the impacts. The largest feature pictured by the probe was the Caloris Basin, measuring 1,300km (800 miles) across.

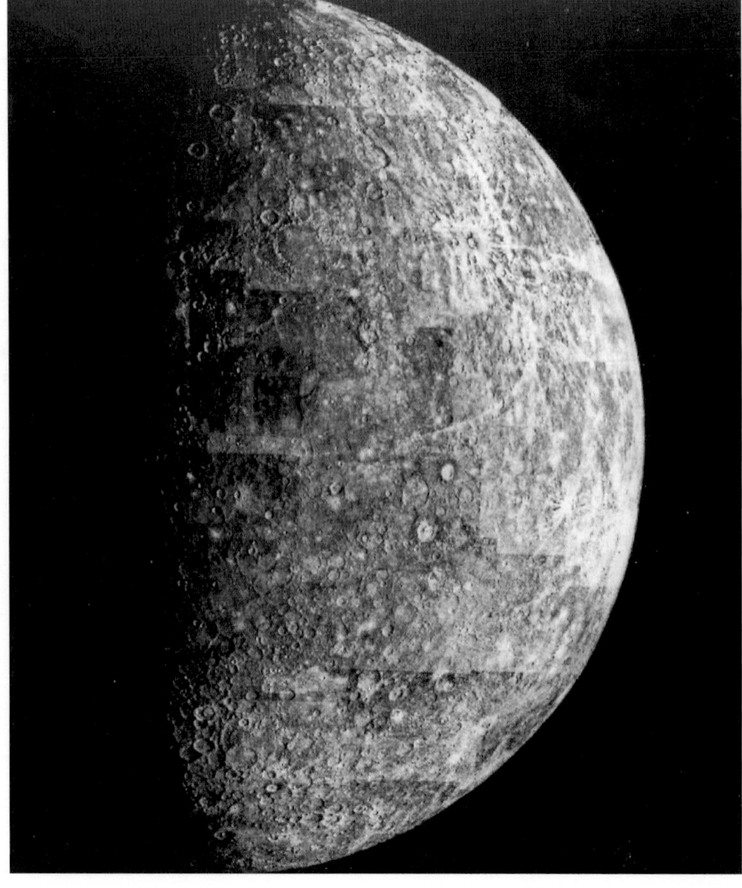

VENUS

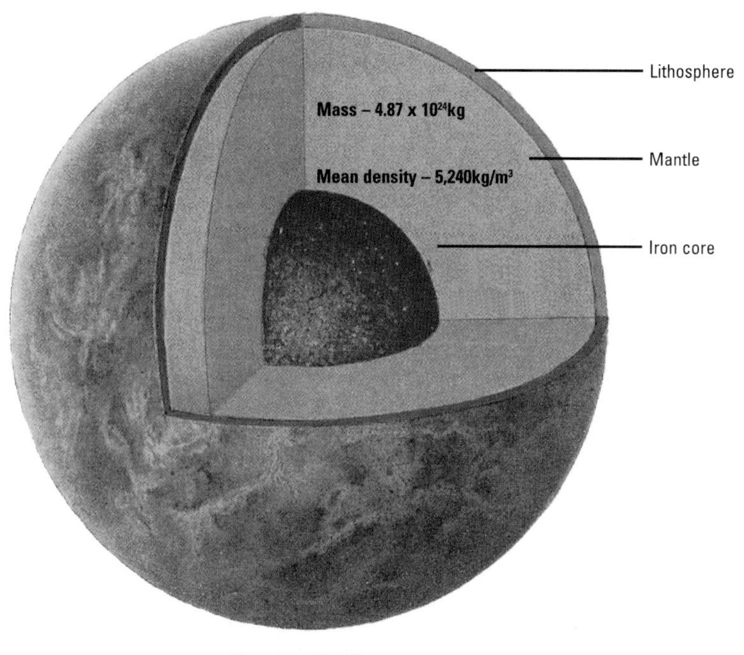

Lithosphere

Mass – 4.87 x 10²⁴kg

Mean density – 5,240kg/m³

Mantle

Iron core

Diameter – 12,104km

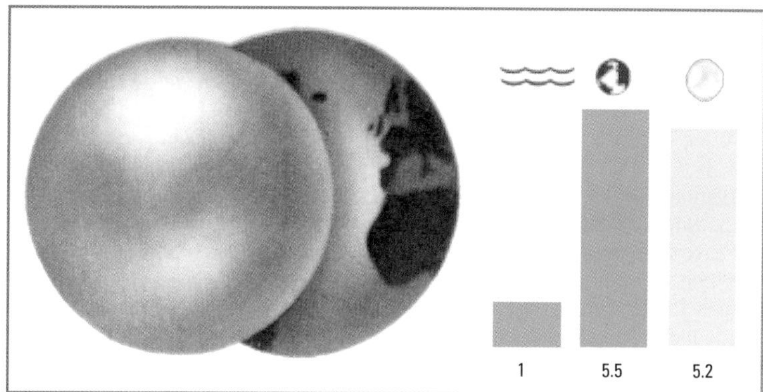

1   5.5   5.2

ABOVE The similarity in size, mass and density led to astronomers regarding Venus and Earth as "sister planets", formed at much the same time and part of space, but incredible heat and pressure make Venus an inhospitable body. Its internal structure, however, is probably much the same, with a nickel-iron core surrounded by a silicate mantle. The Mariner 2 probe of 1962 discovered it has a much weaker magnetic field than Earth, suggesting it may not have a liquid outer core. The lack of a strong magnetic field may also be a result of the planet's slow rotation of 243 days. Venus also rotates "backwards" (retrograde) as compared to other planets, but the reason remains a mystery.

TO THE NAKED EYE Venus is a splendid object and, as the evening and morning "star", is far brighter than any celestial object except the Sun and Moon – the reason it was named after the goddess of beauty. Telescopically, however, it has always been a disappointment, shrouded in cloud and, until very recently, mystery.

The orbit of Venus is nearer to circular than any planet, and the mean distance from the Sun – an average of 108,200,000km (67,200,000 miles) varies little. This revolution period is 224.7 Earth days, while the rotation takes 243 days.

Analysis of the sunlight reflected from the Venusian clouds revealed that the atmosphere was chiefly composed of carbon dioxide, and radio measurements suggested the surface was extremely hot. Space probe soundings of the atmosphere and surface later revealed a world completely devoid of all forms of water and confirmed searing surface temperatures that reached 480°C (895°F).

As if this were not enough, the dense atmosphere crushes down on the planet with a pressure 90 times that at the Earth's surface; a human being standing unprotected on the rock-strewn landscape of Venus would be simultaneously roasted, crushed and asphyxiated.

A dead planet

Liquid water is an essential ingredient for life as we know it, and without any water source it is extremely unlikely that any form of life ever existed on the planet. Venus's proximity to the Sun means that it probably started out with less water than the Earth – any water there would probably have existed as vapour rather than as liquid.

Even after the planets formed, the more intense solar radiation on Venus would have driven what little water remained from its atmosphere by breaking up the water molecules into their constituent parts of hydrogen and oxygen. Hydrogen is a very light gas and would have escaped off into space, while the oxygen would have been absorbed by the planet's surface. Even the rain of watery, icy comets that must have impacted Venus during its history was apparently insufficient to prevent the planet drying out.

The Venusian atmosphere

When the interiors of both Venus and Earth heated up from radioactivity, a great deal of volcanic activity occurred, causing vast amounts of carbon dioxide to be released. On Earth, the oceans dissolved some of this gas and carbonate rocks were formed, but on Venus there were no oceans and the carbon dioxide stayed in the atmosphere.

Findings in 1978 from the US Pioneer Venus 2 spacecraft, which parachuted through the atmosphere, established that Venus has sulphuric acid clouds concentrated in a layer at heights of 48–58km (30–36 miles) above the surface. Drops of the acid develop just like drops of water in our own clouds and when they are large enough, they fall as acid rain. However, this corrosive rain never reaches Venus' surface because the temperature difference, 13.3°C (56°F) at the top of the clouds and an oven-like 220°C (430°F) underneath them, causes them to evaporate at about 31km above the ground. Below this level, the Venera and Pioneer probes revealed that the atmosphere is remarkably clear, though the surface, subject to a fierce greenhouse effect, lies under a permanent overcast.

RIGHT This view of Venus was taken from 760,000km (450,000 miles) by Mariner 10's television cameras in 1974, en route to Mercury. Individual TV frames were computer-enhanced using invisible ultraviolet light: the blue appearance of the planet does not represent true colour, but is the result of darkroom processing of the images to enhance the UV markings on the clouds. It is this cloud cover that accounts for the brilliance of Venus. The picture is viewed with the predominant swirl at the South Pole. The clouds rotate 60 times faster than the planet's slow 243 days, taking only four days to go around Venus once, a rapid motion driven by the heating of the atmosphere by the nearby Sun.

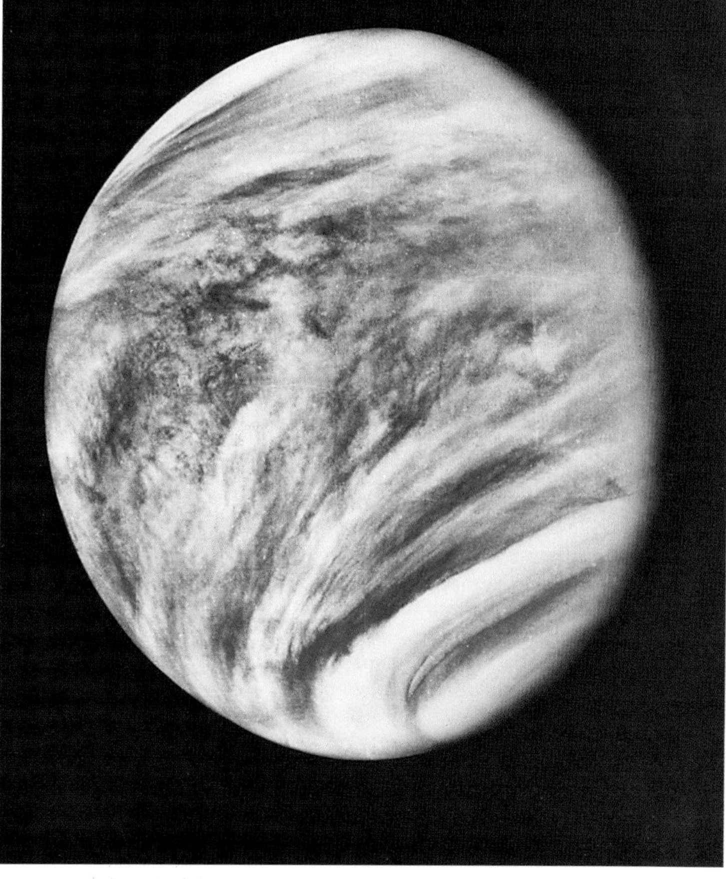

EARTH

THE "THIRD rock from the Sun" is the heaviest of the stony planets and the most dense of all planets. The difference in size and mass between Earth and Venus is slight but Mars is much smaller

What makes Earth unique, however, is the fact that it has the perfect physical and chemical credentials for the evolution of life; slightly closer to the Sun, or slightly farther away, and life could not have developed. The "ecosphere", or the region in which solar radiation will produce tolerable conditions for terrestrial-type life, extends from just inside the orbit of Venus out to that of Mars. Until about 1960, it was thought that such life might exist throughout the region, but spaceprobes have shown both Venus and Mars to be incapable of creating and sustaining any form of life.

Approximately equal in density as well as size and mass, Venus absorbs about the same amount of solar energy as Earth because of the high reflecting power of its cloud. It was not until 1967, when the surface temperature of Venus was shown to register up to 480°C (895°F), that it was commonly accepted

that advanced terrestrial life could develop only within a very limited zone.

Temperature depends not only on the distance of the planet from the Sun or the composition of its atmosphere; there is also the axial rotation period to be taken into account. Earth spins round once in approximately 24 hours, and the rotation period of Mars is only 37 minutes longer, but Mercury and Venus are very different – the periods are 58.7 days and 243 days respectively, leading to very peculiar "calendars". Were Earth a slow spinner, the climatic conditions would be both unfamiliar and hostile.

An atmosphere must not only enable living creatures to breathe, but also protect the planet from lethal short-wave radiations from space. There is no danger on the surface of Earth because the radiations are blocked out by layers in the upper atmosphere; had Earth been more massive, it might have been able to retain at least some of its original hydrogen (as the giants Jupiter and Saturn have done) and the resulting atmosphere might have been unsuitable for life.

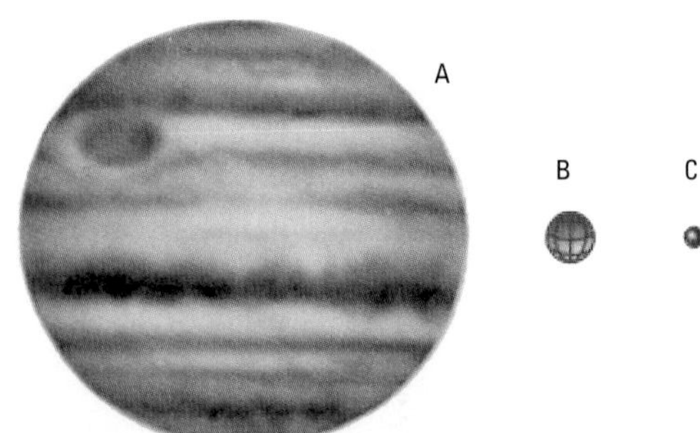

ABOVE The relative sizes of Jupiter [A], Earth [B] and Mercury [C]. Jupiter is the largest planet, Mercury the smallest (excluding the extraordinary misfit Pluto), and while Earth is intermediate in size, it is more nearly comparable with Mercury in the context of the Solar System. Earth is the largest of the so-called terrestrial planets – Mercury, Venus, Earth, Mars, Pluto – but far inferior in size even to the smallest of the four "gas giants", Neptune.

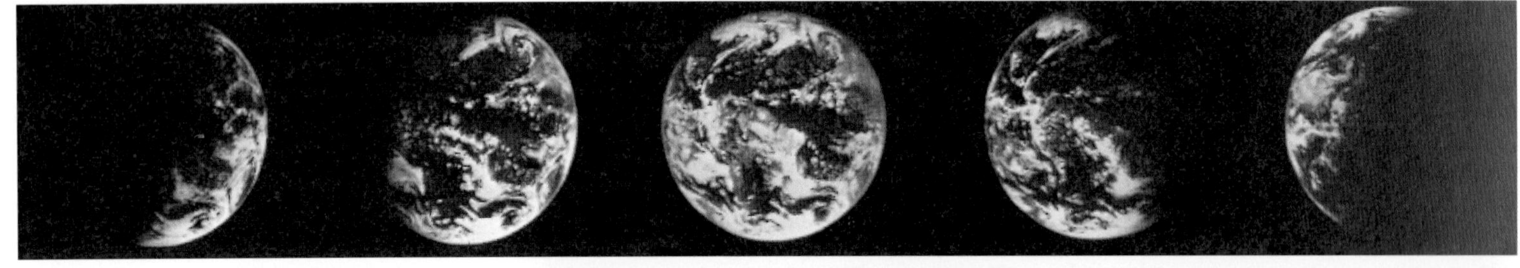

ABOVE Seen from space, Earth will show phases – just as the Moon does to us. These five photographs shown were taken from a satellite over a period of 12 hours.

RIGHT Earth as captured above the Moon's surface from an Apollo spacecraft. The contrast between the barren landscape of the Moon and the near-perfect balance of land, cloud and ocean on Earth is startling. Our planet is the only home of known life in the Solar System, though spheres in the same section of their ecospheres may well exist in the Universe.

Earth is unique in having a surface that is largely covered with water; thus although it is the largest of the four inner planets its land surface is much less than that of Venus and equal to that of Mars. There can be no oceans or even lakes on Mars, because of the low atmospheric pressure, and none on the Moon or Mercury, which are to all intents and purposes without atmosphere. On Venus the surface temperature is certainly too high for liquid water to exist, so that the old, intriguing picture of a "carboniferous" Venus, with luxuriant vegetation flourishing in a swampy and moist environment, has had to be given up.

Because Earth is so exceptional, it has been suggested that it was formed in a manner different from that of the other planets, but this is almost certainly not the case. The age of Earth, as measured by radioactive methods, is approximately 4,600 million years (4.6 eons) and studies of the lunar rocks show that the age of the Moon is the same; there is no reason to doubt that the Earth and all other members of the Solar System originated by the same process, and at about the same time, from the primeval solar nebula.

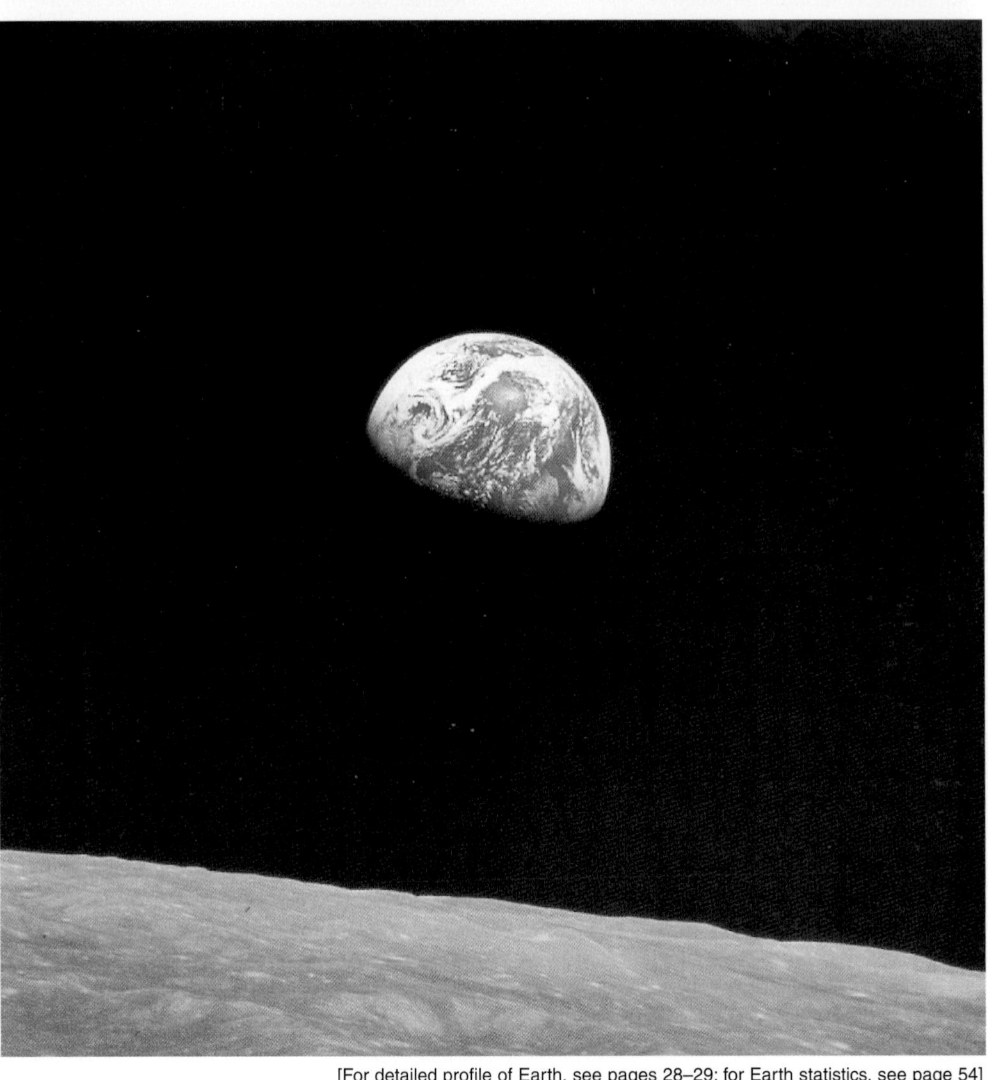

[For detailed profile of Earth, see pages 28–29; for Earth statistics, see page 54]

MARS

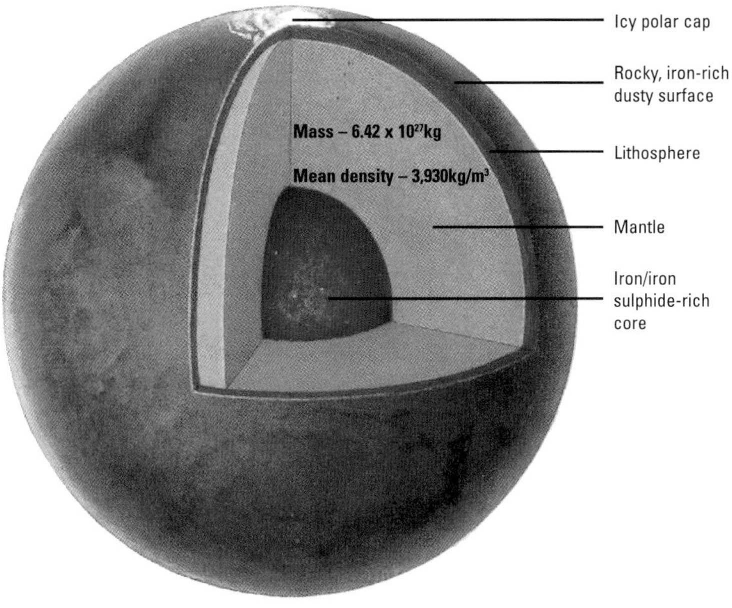

Icy polar cap

Rocky, iron-rich dusty surface

Lithosphere

Mantle

Iron/iron sulphide-rich core

Mass – 6.42 x 10²⁷kg

Mean density – 3,930kg/m³

Diameter – 6,787km

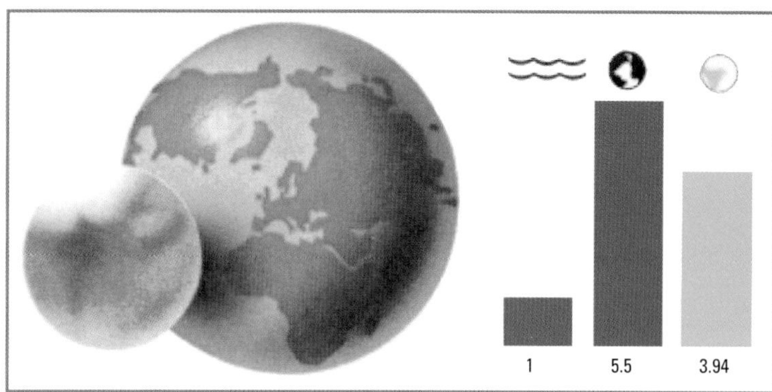

1   5.5   3.94

ABOVE The surface area of Mars is 28% that of Earth. Its diameter of 6,787km (4,217 miles) is a just over half that of Earth, and about twice that of the Moon. It has only a tenth of Earth's mass. Observations suggest that Mars contains an iron-rich core, about 1,700km (1,050 miles) in diameter. The low density of Mars compared to the other terrestrial planets hints that this core may also contain a significant amount of sulphur. Apparently, this core is not convecting enough to create as strong a magnetic field as Earth: indeed, it was not until 1997 that Mars Global Surveyor detected its weak and patchy magnetic field.

LIKE EARTH, Mars experiences seasons. Varying between 207 million km and 249 million km (129 and 158 million) miles from the Sun, its orbit is not circular and it is much closer to the Sun during the southern summer than in the northern summer, so that southern Martian summers are warmer than northern ones. But because the planet moves faster when it is closer to the Sun they are shorter and southern winters longer and colder than those in the north; one result is that the southern residual cap retains some frozen carbon dioxide (which melts at a lower temperature) as well as water.

In the late 1960s the Mariner 4, 6 and 7 spacecraft confirmed that the surface resides under only a thin atmosphere of carbon dioxide, with a pressure of only one hundredth of that at the Earth's surface at most, and in places even lower. They also revealed that Mars is cold, with mean annual temperatures ranging from –58°C (–72°F) at the equator to –123°C (–189°F) at the poles. At these temperatures and low pressures liquid water cannot currently exist on the Martian surface, although the Mariner and subsequent Viking pictures revealed evidence for the ancient action of flowing water.

RIGHT A mosaic image of the Schiaparelli hemisphere created from images taken by the Viking orbiter in 1980. Mars was once considered the likeliest of planets to share Earth's cargo of life, the seasonal expansion of dark patches strongly suggesting vegetation and the icecaps indicating the presence of water.

However, close inspection by spacecraft brought disappointment: some combination of chemical reactions, erosion and dark dust deposited by strong winds account for the "vegetation", and the "icecaps", though comprising a permanent layer of water ice, are covered from autumn to spring by a cover of carbon dioxide frost. Whatever oxygen the planet once possessed is now locked up in the iron-bearing rock that covers its cratered surface and gives it its characteristic red colour. The large crater near the centre is Schiaparelli, about 500km (370 miles) in diameter.

Mars is smaller and less "massive" than Earth or Venus, and so has a lower surface gravity and cannot hold on to a dense atmosphere. Mars' lower volume means that it could not generate and retain the same amount of internal heat as Venus or Earth, and does not maintain the same level of volcanic activity.

The core is surrounded by a molten rocky mantle denser and perhaps three times as rich in iron oxide as that of the Earth, overlain by a thin crust. The lack of plate tectonics and absence of current volcanic activity implies this mantle is also non-convecting – though one massive feature, the 4,500-km (2,800-mile) long Valles Marineris, may be a fracture in the crust caused by internal stresses.

Mars has two small moons, Phobos and Deimos, two potato-shape asteroids that were once captured by the planet's gravity.

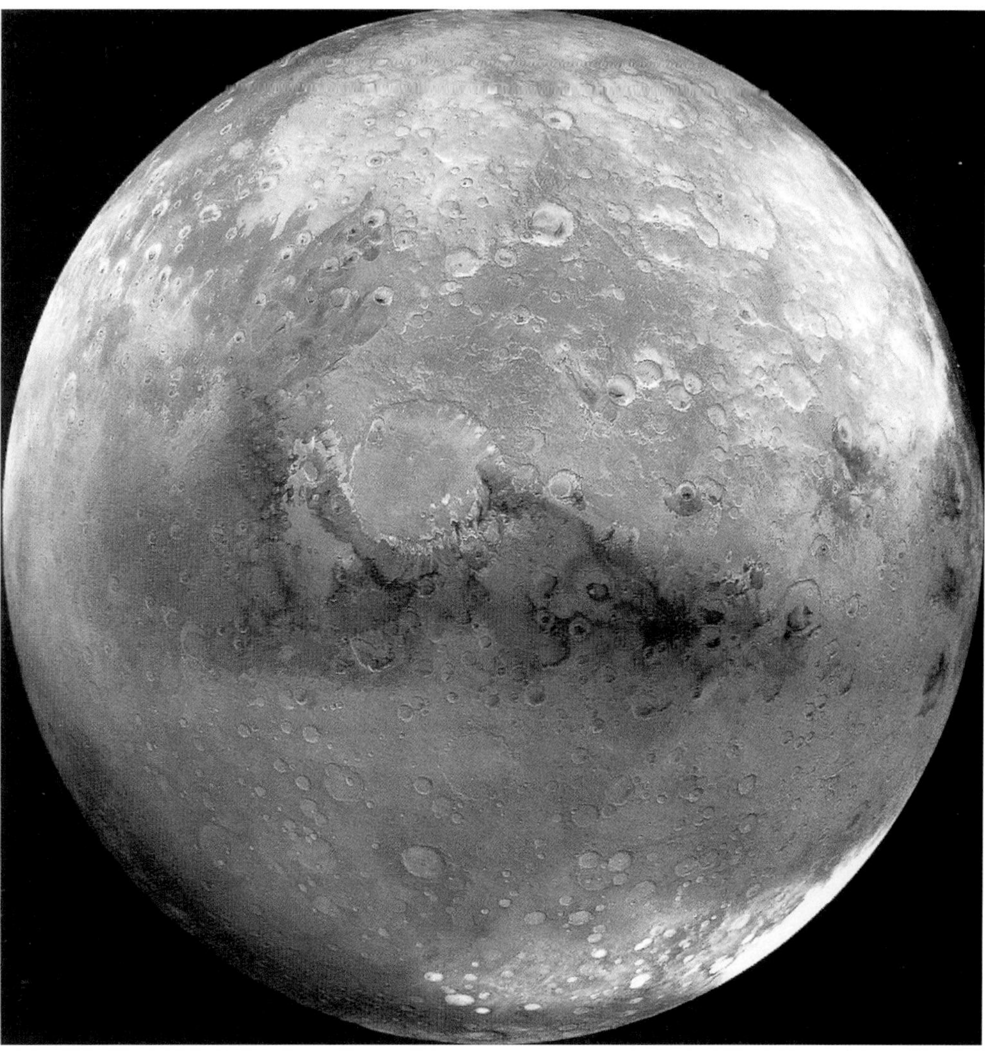

JUPITER

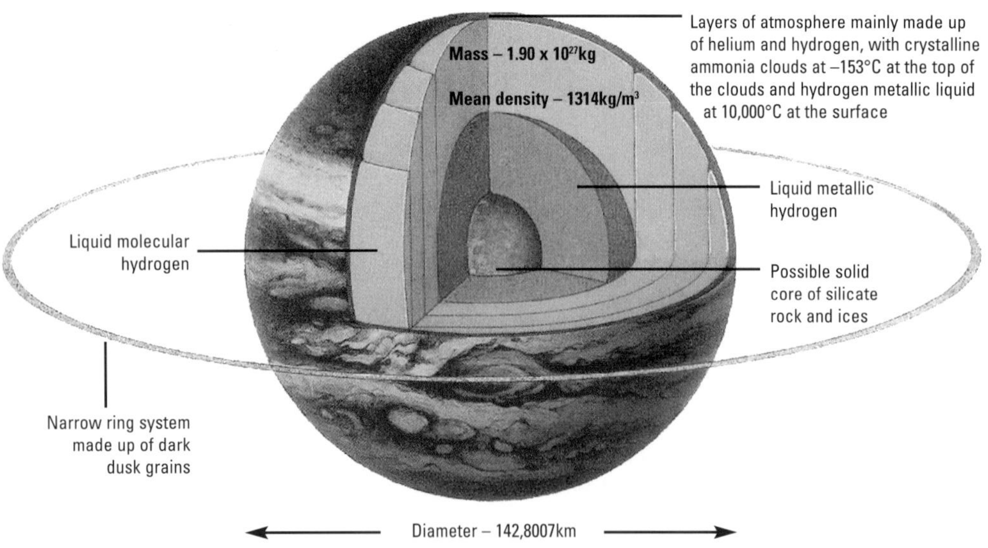

Mass – 1.90 x 10²⁷kg

Mean density – 1314kg/m³

Layers of atmosphere mainly made up of helium and hydrogen, with crystalline ammonia clouds at –153°C at the top of the clouds and hydrogen metallic liquid at 10,000°C at the surface

Liquid metallic hydrogen

Liquid molecular hydrogen

Possible solid core of silicate rock and ices

Narrow ring system made up of dark dusk grains

Diameter – 142,8007km

Jupiter's mean density is only 1.3 times that of water *(right)*, but the outer layers are tenuous and the core is far denser. The Earth's axis is tilted at an angle of 23¹/2° from the perpendicular to the plane of orbit, but Jupiter's is only just over 3° *(below)*.

1   5.5   1.3

23.5°   3.1°

FAR BEYOND the main asteroid belt, at a mean distance of 778,300,000km (483,600,000 miles) from the Sun, lies Jupiter, the largest of the planets. This huge globe could swallow up 1,300 bodies the volume of Earth but its mass – despite being nearly three times as much as the other planets combined – is only 318 times that Earth because Jupiter is much less dense.

The planet is mostly gas, under intense pressure in its lower atmosphere above a core of fiercely compressed hydrogen and helium. The upper layers form strikingly coloured rotating belts, outward signs of the intense storms created by its rapid rotation of less than ten hours. This also means that the equator tends to bulge, and like Saturn the planet is clearly flattened at the poles: Jupiter's equatorial diameter is 143,000km (89,000 miles), whereas the polar diameter is less than 135,000km (84,000 miles).

When viewing the planet, you can only see the outermost part of its very deep atmosphere, which has several layers of cloud of different composition and colour. Jupiter rotates so fast that it spins the clouds into bands in which various spots, waves and other dynamic weather systems occur. The banded patterns in Jupiter's clouds arise because of the existence of convection cells in the atmosphere. The giant spots between, and sometimes within,

the bands are giant eddies, or rotating masses of cloudy air, similar to enormous versions of our earthly hurricanes. Other weather systems, often of contrasting colour, appear embedded in the layers.

The Great Red Spot

While most of Jupiter's spots are short-lived, the Great Red Spot, by far the largest, is the notable exception. Under observation for over 300 years, it sometimes disappears but always returns, and has been prominent this time around since the mid-1960s. Occurring at a latitude of around 23° south, it is a huge, complex, cloudy vortex – variable in size but always far larger than the diameter of Earth – rotating in an anti-clockwise direction.

The "GRS" is believed to be a two-dimensional vortex which spirals outwards away from areas of high pressure, so although it appears like a hurricane it is a high- rather than low-pressure phenomenon. The reasons for its constant position and its characteristic colour, however, remain unclear.

Jupiter's rings

Recent investigations by spaceprobes have shown an orbiting ring system and discovered several previously unknown moons, and Jupiter has at least 16. The ring system is composed of three major

components. The main ring is some 7,000km (4,350 miles) wide and has an average radius of about 126,000km (80,000 miles). At its inner edge this merges into the halo, a faint doughnut-shaped ring about 20,000km (12,400 miles) across, which extends over half the distance to the planet itself. Just outside the main ring is a faint gossamer ring made of fine material, extending out past the orbit of the innermost satellite Amalthea. These rings are not only more tenuous than Saturn's but are also darker, probably comprising dust rather than ice.

The magnetic field

Jupiter has a strong magnetic field, caused by the planet still cooling from its time of formation and constantly collapsing in on itself under its own gravitational pull. This gives off heat, producing dynamic convection movements in the fluid metallic interior. Coupled with the spin of Jupiter's rapid rotation, it produces an extensive magnetic field about 20,000 times stronger than that of the Earth – one which constantly alters size and shape in response to changes in the solar wind.

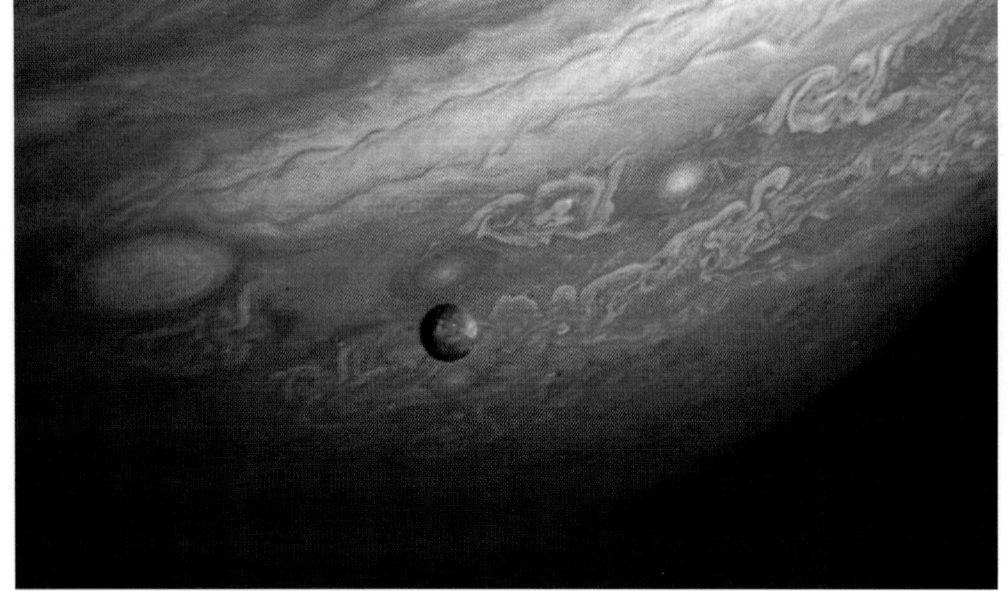

LEFT Voyager 1 took this photo of Jupiter and the innermost of its four Galilean satellites, Io, in 1979, with Io about 420,600km (260,000 miles) above Jupiter's Great Red Spot. The picture was taken about 20 million km (12,400,000 miles) from the planet. Slightly larger than our Moon, Io is the densest large object in the outer Solar System and most volcanically active, spewing material up to 300km (200 miles) into the air; in 1996, when Galileo detected an iron core and magnetic field, it found the moon's surface features had changed radically since the satellite was imaged by Voyager just 17 years before.

Jupiter's main moons are group-named after Galileo because it was his identification of their orbiting the planet that eventually led him to support Copernicus's revolutionary views that Earth revolved round the Sun. They are Europa, Callisto and Ganymede, the largest satellite in the Solar System and bigger than Mercury and Pluto, orbiting Jupiter at a distance of just over 1 million km (620,000 miles).

SATURN

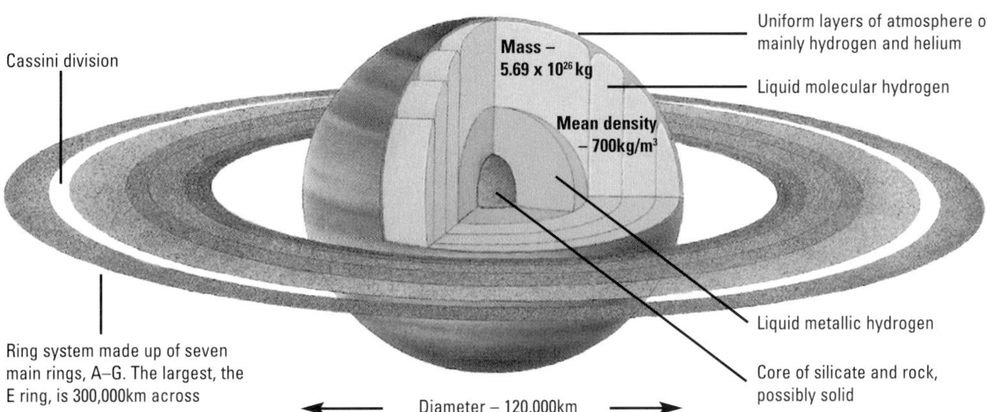

Cassini division

Mass – 5.69 x 10²⁶kg

Uniform layers of atmosphere of mainly hydrogen and helium

Liquid molecular hydrogen

Mean density – 700kg/m³

Liquid metallic hydrogen

Core of silicate and rock, possibly solid

Ring system made up of seven main rings, A–G. The largest, the E ring, is 300,000km across

← Diameter – 120,000km →

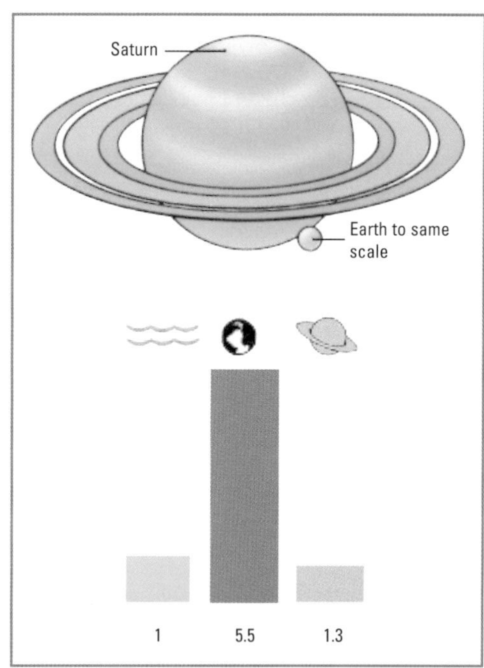

Saturn

Earth to same scale

1   5.5   1.3

OUTERMOST OF the planets known to ancient man and named after the Roman god of agriculture, Saturn lies at a mean distance of 1,427 million km (88 million miles) from the Sun and has a revolution period of 29.46 years. The second largest planet in the Solar System, its polar diameter of 120,000km (75,000 miles) is considerably less than its equatorial diameter.

Astronomers believe the temperatures in the core of Saturn exceed 11,700°C (21,000°F), and the atmosphere must be deeply convective since this is the only plausible way to transport the interior heat to levels where it can be radiated to space. Like its neighbours Jupiter and Uranus, Saturn radiates more energy into space than it receives from the Sun.

The atmosphere of Saturn broadly resembles that of Jupiter: it has 80–90% hydrogen, 10–20% helium and less than 1% traces of other gases, including methane and ammonia detected by Earth-based and Voyager spectroscopy. Because the cloud layers are cooler than those of Jupiter, they tend to be thicker and more uniform in shape, forming deeper in the atmosphere. Saturn's distinctive hazy yellow hue, plus the deeper orange-yellow of Titan, largest of its 18 moons, are thought to be caused by deep haze layers of condensed hydrocarbons.

Saturn's magnetosphere is smaller than that of Jupiter, though it still extends well beyond the orbits of the outer moons, while the field is about 30 times weaker than that of its huge neighbour.

ABOVE Though not as large as Jupiter, Saturn's globe is of impressive size – its volume is 1,000 times that of Earth. The mean density of Saturn is only 0.7 that of water, far less than any other planet, and it would float if it were dropped into an ocean. The low density is due to the preponderance of hydrogen.

RIGHT Voyager 2 returned this view of Saturn in 1981, when the spacecraft was approaching the large, gaseous planet at about 1 million km (620,000 miles) a day.

The so-called "ribbon-like" feature in the white cloud band marks a high-speed jet at about 47° north; there, the westerly wind speeds are about 530km/h (330mph). Although less pronounced than on Jupiter, the bands, storms, ovals and eddies are all evident here, too, caused by the same combination of rapid rotation (just under 10 hours and 14 minutes) and convective atmosphere.

Saturn's stunning ring system – hundreds or even thousands of narrow ringlets – are grouped, giving the impression of broad bands, each of which has been designated a letter. Brighter than those of other outer planets and no more than 1.5km thick, they comprise millions of small objects ranging in size from tiny stones to rocks several metres long, and are composed at least in part of water ice, possibly plus rocky particles with icy coatings.

The bright A and B rings and the fainter C ring are visible from Earth through a telescope. The space between the A and B rings is called the Cassini division, while the much narrower Encke division splits the A ring. The complex structure is due to the gravitational effects of the satellites, which orbit close to and within the rings.

Saturn has 18 named satellites, six of them icy (resembling the three outer moons of Jupiter) and the others small and rocky. The unique atmosphere of Titan, second largest moon in the Solar System after Ganymede, make it the odd one out. About every 15 years we see Saturn's rings edge-on because of the orbital geometry between Saturn and Earth.

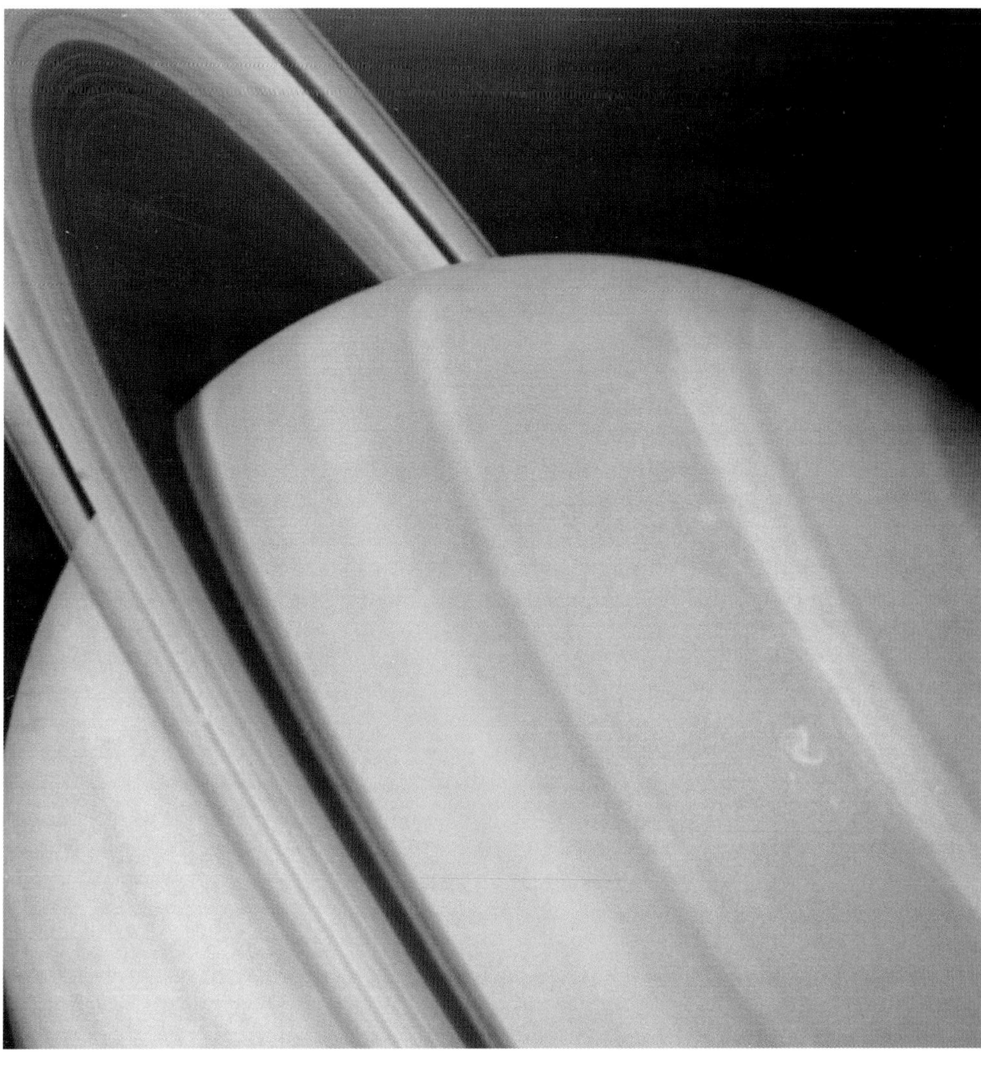

URANUS

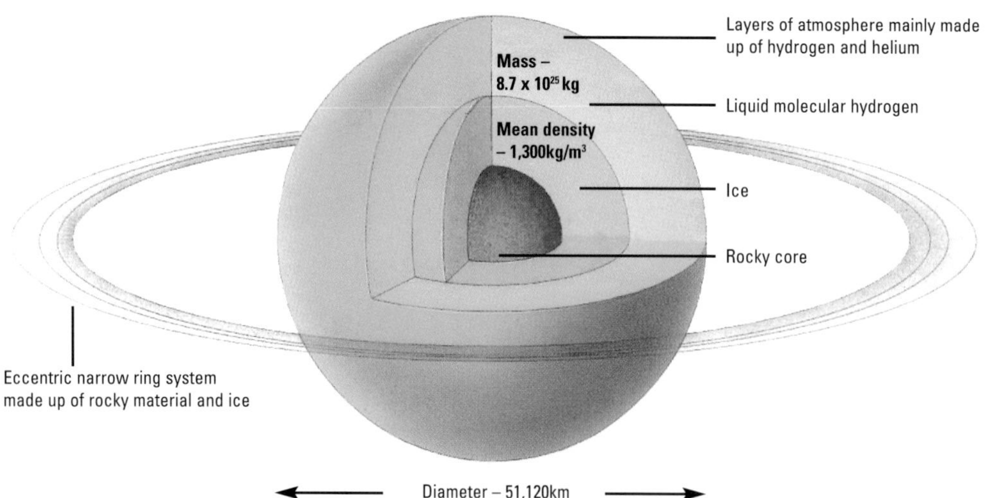

Mass –
8.7 x 10²⁵ kg

Mean density
– 1,300kg/m³

Layers of atmosphere mainly made up of hydrogen and helium

Liquid molecular hydrogen

Ice

Rocky core

Eccentric narrow ring system made up of rocky material and ice

Diameter – 51,120km

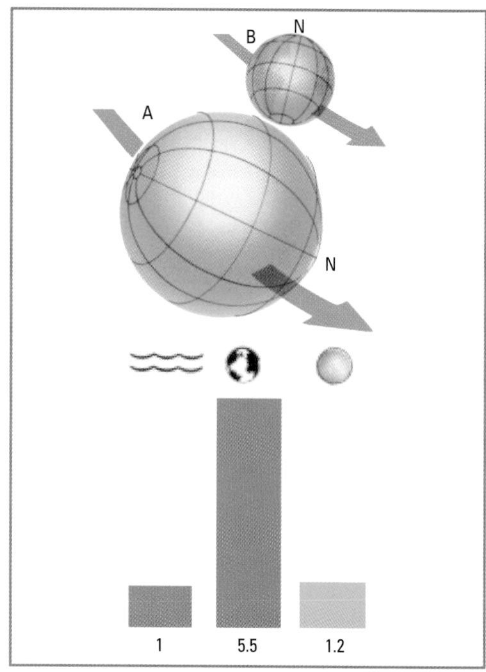

DISCOVERED BY William Herschel in 1781, Uranus appears as a smooth, aqua-coloured sphere with very subtle hints of bands, but this calm facade gives no hint of a history fraught with spectacular catastrophe: at some stage a mighty collision wrenched the young planet off its axis. As a result the planet is tipped over on its side so that its rotation axis lies almost in the plane of the planet's orbit, giving rise to the most striking seasonal changes. Another collision may have been responsible for the fantastic geology of its moon Miranda.

Uranus has a mean distance from the Sun of 2,869,600,000km (1,783 million miles) and a revolution period of just over 84 years. Its basic composition is the same as the other giant planets and similar to that of the Sun – predominantly hydrogen (about 80%) with some helium (15%), the remainder of the atmosphere being methane, hydrocarbons (molecular mixtures of carbon, nitrogen, hydrogen and oxygen) and other trace elements.

RIGHT The tilt of Uranus's axis [A] compared to Earth [B] is 98°, unique in the Solar System. Its density is 1.2 times that of water, more than Jupiter and Saturn but far less than Earth.

Uranus' colour is caused by the small amount of methane – probably less than 3% – that preferentially absorbs red light, meaning the reflected sunlight we see is greenish-blue.

Temperatures at the outer layers of the atmosphere are very cold, about –200°C (330°F), but pressures and temperatures rise with depth and the hydrogen and helium transform from gas to a liquid state. At still greater depth, a transition occurs to a thick, viscous, partly solidified layer of highly compressed liquid water, which may have traces of ammonia and methane. Deep within the centre of Uranus, at extremely high pressure, a core of rocky material is thought to exist, with a mass almost five times that of Earth.

BELOW A composite image of Uranus, the striking but featureless blue planet, and five of its 15 moons, made from photographs taken by Voyager 2 in 1986. The moons (clockwise from top left) are Umbriel, Oberon, Titania, Miranda and Ariel. While an unexplained jumble of huge geological features dominates Miranda, tectonic activity has given Ariel the youngest surface of the moons. Voyager 2's discovery of 10 moons tripled Uranus's known total, while in 1997 two unnamed satellites, probably captured asteroids, were found by the Palomar Observatory. In 1977, astronomers discovered that Uranus has a ring system: there are nine well-defined rings, plus a fainter one and a wider fuzzy ring.

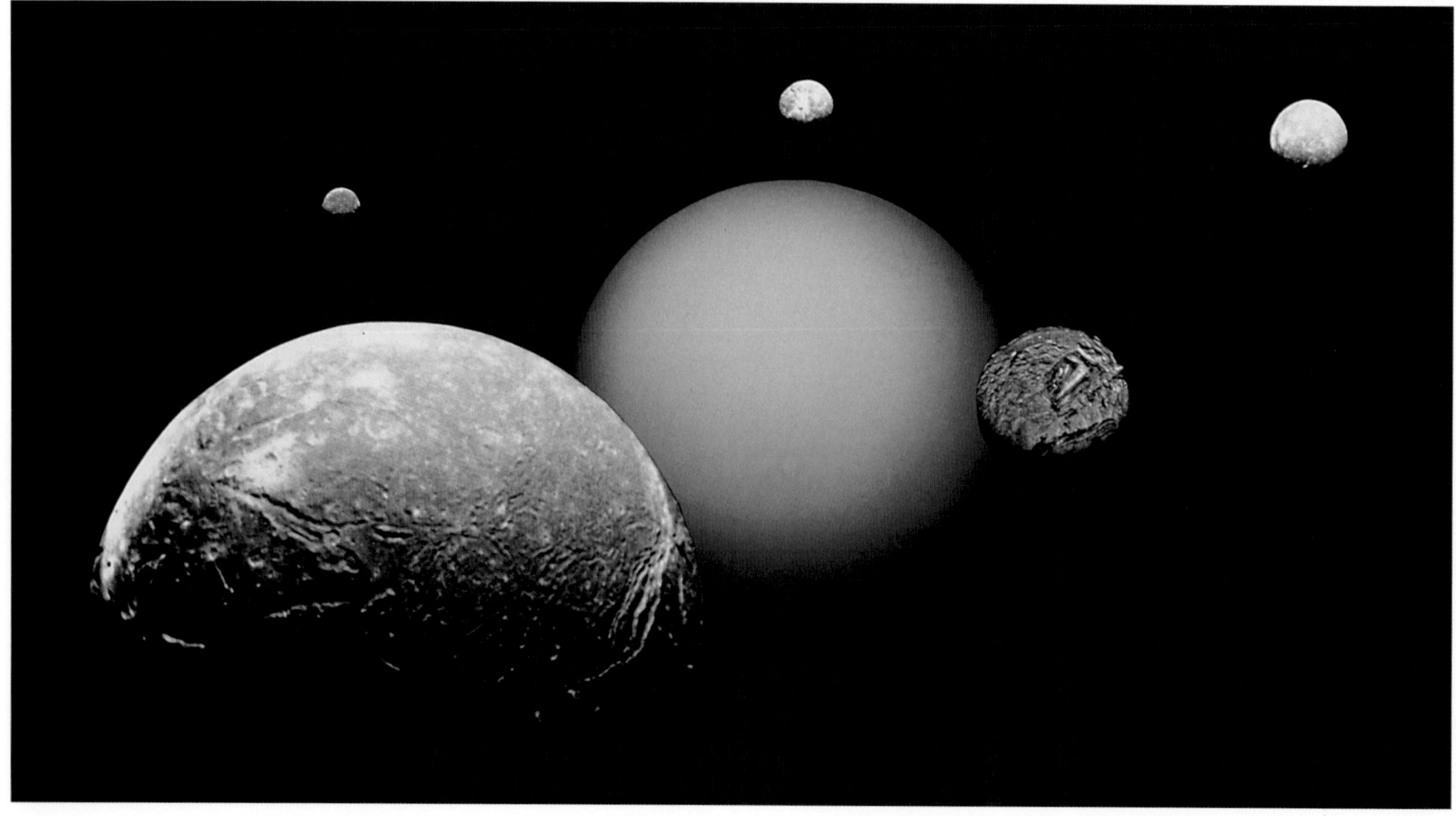

NEPTUNE

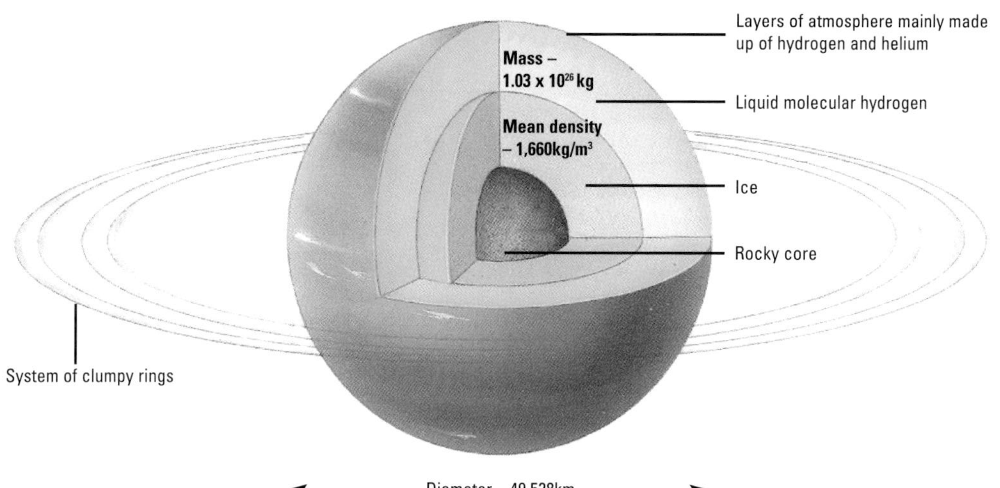

Mass – 1.03 x 10²⁶ kg

Mean density – 1,660kg/m³

Layers of atmosphere mainly made up of hydrogen and helium

Liquid molecular hydrogen

Ice

Rocky core

System of clumpy rings

Diameter – 49,528km

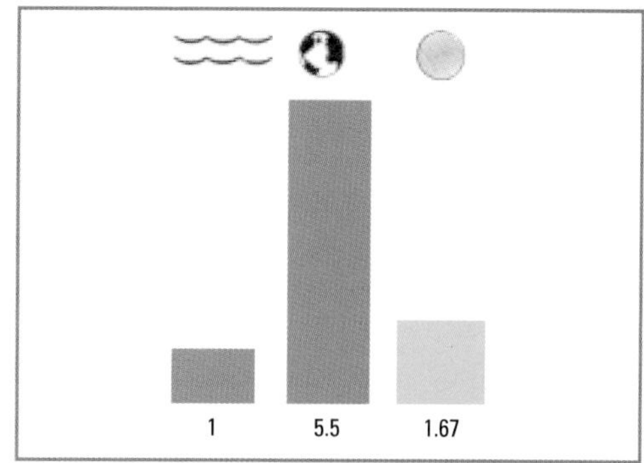

| | 1 | 5.5 | 1.67 |

ABOVE Densest of the four main outer planets, Neptune's mass is 17.1 times that of Earth. Its almost circular orbit is always more than 4 billion km (2.5 billion miles) from Earth. Its rotation period is just over 16 hours.

A NEAR TWIN to Uranus in size, Neptune has a similar atmospheric make-up and internal structure, though its magnetic field is 60% weaker. A gas giant surrounded by clumpy rings and eight moons, it takes 164.8 years to orbit the Sun.

Unlike Uranus, which has no detectable internal heat source, Neptune has the strongest internal heat source of all the giant planets. It radiates almost three times more heat than equilibrium conditions would predict, as opposed to Jupiter and Saturn, which radiate about twice as much energy as expected.

Clouds and storms are the main features of Neptune's dynamic atmosphere. Dominating all is the Great Dark Spot, a hurricane-like storm in the southern hemisphere about half the size of Earth. Like all Neptune's weather conditions, it is constantly and rapidly changing. Neptune's winds are among the fastest in the Solar System, dwarfed only by Saturn's high-speed equatorial jet.

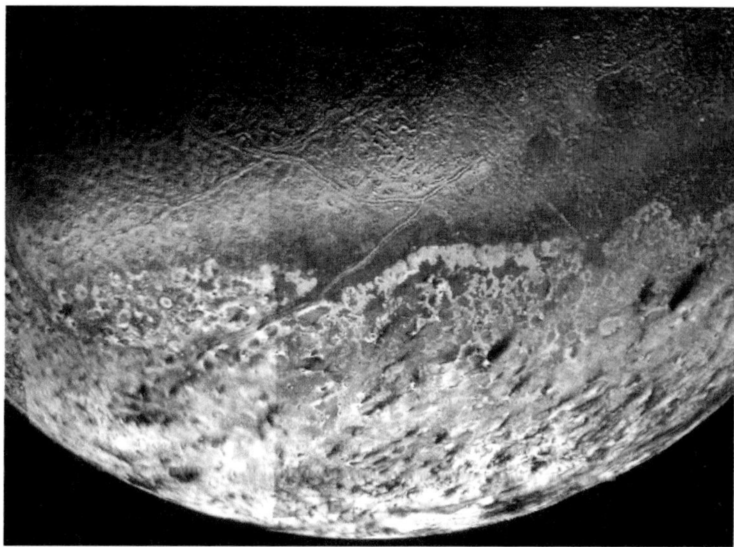

LEFT The southern hemisphere of Triton, largest of Neptune's eight moons, pictured from Voyager 2 in 1989. The large, lighter-coloured area is the polar icecap, probably nitrogen. Because of its retrograde and highly inclined orbit, it is thought Triton was captured by the gravitational pull of Neptune. Tiny Nereid was also known before 1989, when Voyager discovered six more satellites.

PLUTO

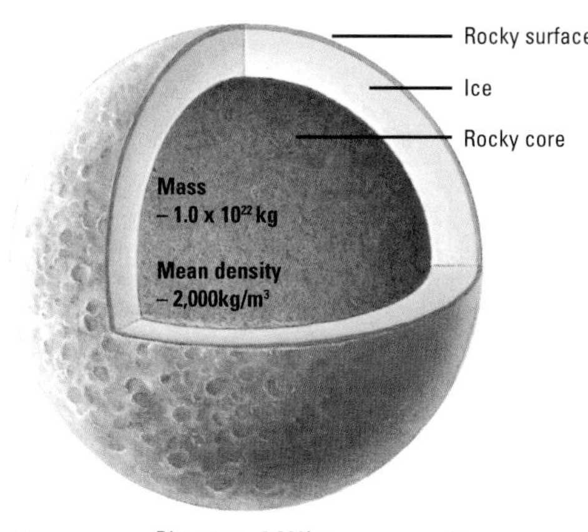

Rocky surface

Ice

Rocky core

Mass – 1.0 x 10²² kg

Mean density – 2,000kg/m³

Diameter – 2,300km

ABOVE Pluto has an average density about twice that of water ice, implying its interior is composed of rock; about 30% of its volume is thought to be water ice. Like Earth, Pluto's atmosphere is primarily nitrogen gas, but the changing surface pressure never exceeds about 10 millibars – around one hundred thousandth of the pressure on Earth at sea level.

FARTHEST PLANET from the Sun, Pluto's tiny size (smaller than Mercury) and rocky composition (like the terrestrial planets of the inner system) make it a real misfit among the gas giants of the outer system. In both size and surface constituents it is similar to Triton, a moon of Neptune, and many astronomers believe it is a former satellite of Neptune somehow separated from its parent.

Pluto has a long, elliptical and tilted orbit that takes over 248 Earth years to complete, of which about 20 years are inside the orbit of Neptune, the last occasion being from 1979 to 1999. Discovered only in 1930, its size and distance from Earth make it difficult to study, despite its high reflectivity: in our sky it is less than one 36,000th of a degree across – the equivalent of a walnut at a range of 50km (30 miles). It is thought that the surface of Pluto is largely nitrogen ice, with methane and carbon monoxide ices as impurities. At nearly half its size, the mysterious Charon (discovered in 1978) is the Solar System's largest moon in relation to its parent planet. Pluto did not form in isolation: it is simply the largest relic in space past Neptune left over from the formation of the Solar System.

RIGHT Hubble Space Telescope (HST) images from 1994 showing two hemispheres of Pluto. The two main images have been computer processed to show rotation and bring out the differences in brightness on the surface; the original "raw" images are at the top left of each panel. Twelve bright regions have been identified, including a large north polar cap.

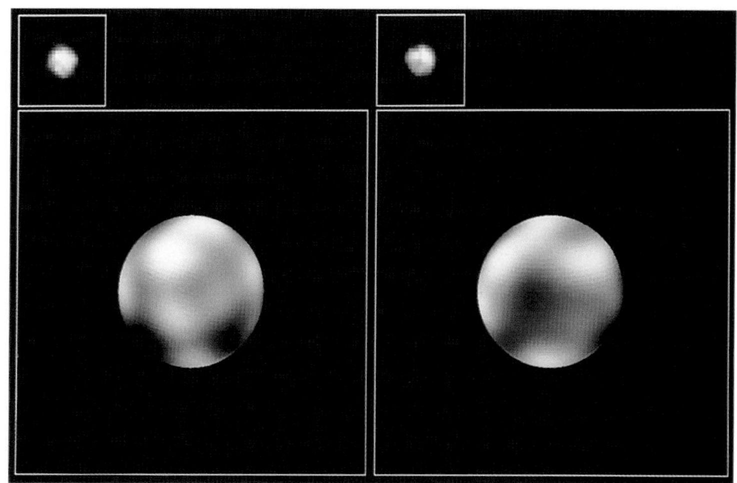

THE MOON

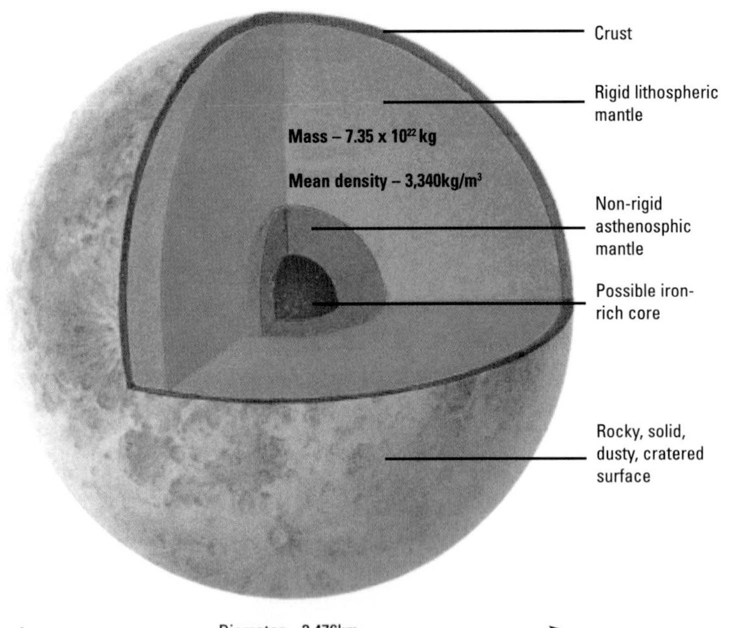

Crust

Rigid lithospheric mantle

Mass – 7.35 x 10²² kg

Mean density – 3,340kg/m³

Non-rigid asthenosphic mantle

Possible iron-rich core

Rocky, solid, dusty, cratered surface

Diameter – 3,476km

The average diameter of the Moon is 3,475km (2,159 miles), 0.27 times Earth. It has a mass of about $^1/_{81}$ that of Earth, its surface gravity is one-sixth of Earth, and its density is 3.344 times that of water. It orbits the Earth at a mean distance of 384,199 km (238,731 miles) at an average speed of 3,683km/h (2,289mph) in relation to the Earth. It orbits its parent in 27.3 days.

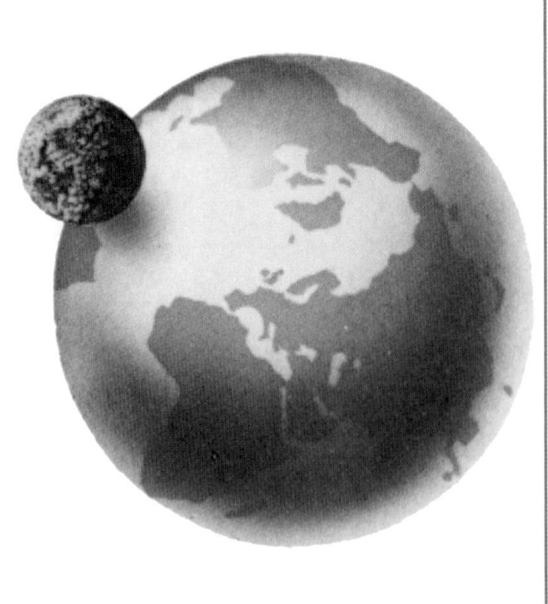

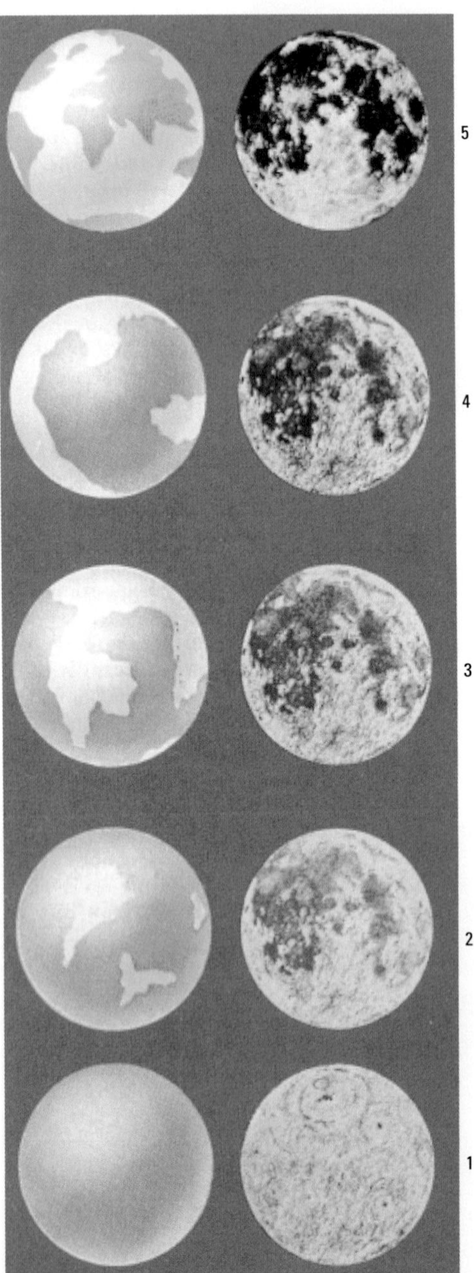

BECAUSE THE Moon is about 400 times smaller than the Sun, but about 400 times closer to Earth, humans have always seen them as roughly the same size. While the Moon is tiny in cosmic terms, however, its diameter is more than a quarter that of Earth, considerably more than Pluto and well over 70% that of Mercury. Despite its bright appearance it is a dark body, illuminated only by light reflected from the Sun.

Analysis of lunar samples suggests that the Moon was formed from the remnant of a Mars-sized body that collided with the juvenile planet Earth in a giant impact some 4,500 billion years ago. Any iron-rich core this body may have had appears to have been absorbed into the Earth's core, while the Moon grew out of the mostly rocky debris thrown into space by the crash. The lack of a large, fluid core explains the almost total lack of a magnetic field, which registers only one ten millionth of the strength of Earth's.

The Moon's characteristic dark patches – the face of the "Man in the Moon" – are low-lying regions once flooded by outpourings of basaltic lava, which scientists have dated as between 3 and 4 billion years ago. Known as the lunar seas (Latin *maria*, singular *mare*), they have appeared never to have contained any water, nor indeed any liquid.

Only about 59% of the Moon's surface is directly visible from Earth. Reflected light takes 1.25 seconds to reach us – compared to 8 minutes 27.3 seconds for light from the Sun. With the Sun overhead the temperature on the lunar equator can reach 117.2°C (243°F), and at night it can sink to –162.7°C (–261°C). An astronaut has only a sixth of his normal weight on the Moon, though his mass is unaltered. There is no local surface colour and the lunar sky is black, even when the Sun is above the horizon. There is no air or water – and there has never been any form of life.

LEFT The lunar "seas" were formed, either by internal accretion or by impact, at an early stage development of Moon and Earth surfaces around 4,000 million years ago [1]. The general aspect of the Moon then must have been similar to that of today, although the basins were not filled. The surface of both the Earth and the Moon then remained the same for a considerable period. Some 2,000 million years ago the basins on the Moon were filled in; 1,000 million years later [4] lunar activity was almost over, while Earth has since seen fabulous change.

BELOW After the dark patches of the lunar seas, huge craters are the most noticeable features on the Moon. Once presumed volcanic in origin, it is now accepted they were caused by the impact of asteroids and comets travelling at tens of kilometres a second. Around 30 times the size of the foreign bodies that created them, the craters are always roughly circular unless, rarely, the angle of impact was extremely oblique. The Moon's lack of any atmosphere means that its surface remains unprotected from any form of impactors – an atmospheric layer, as on Earth, helps to burn up any encroaching objects – and this, combined with the fact that it has no geological processes, means that no crater is ever worn away or changed . . . except by the arrival of another foreign body.

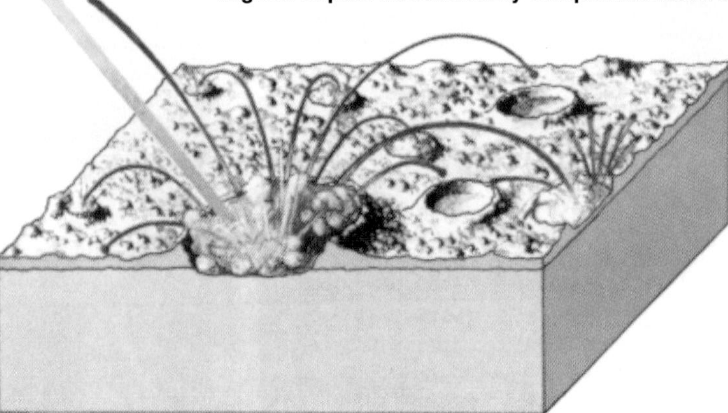

COMETS

A GREAT COMET, with a brilliant head and a tail stretching way across the sky, is a spectacular object – and it is easy to understand they caused such terror in ancient times. Comets have always been regarded as unlucky and fear of them is still not dead in some primitive societies.

Yet a comet is not nearly as important as it may look: it is made up of small rock and ice particles and tenuous gas. On several occasions Earth passed through a comet's tail without suffering the slightest damage. Since Edmund Halley first calculated the paths of several comets in 1695 – including Halley's, whose period is 76 years and which last appeared in 1986 – astronomers have found over 600 such bodies orbiting the Sun.

Analysis of a comet
At the heart of every comet is the nucleus, a solid mass of ice that also contains small solid particles of rock called "dust". Most nuclei are between 1 and 10km (0.6 and 6 miles) across, though they can reach 100km (60 miles). The dark thin crust of icy dust that covers them reflects only 4% of sunlight, making them difficult to detect when distant from the Sun. Over 80% of the ice is simple water ice – the nucleus of Halley's comet contains more than 300,000 tonnes of it – and another 10% or more is frozen carbon dioxide and carbon monoxide. The coma and tails appear only when the comet approaches the Sun, which can be from any angle; as the comet recedes the tail disappears.

Because a comet nucleus shrinks every time it passes the Sun – Halley's by perhaps about a metre (3ft) on each orbit – no comet can have been in its present orbit since the birth of the Solar System. It is now believed that while some comets come from the Kuiper belt beyond Pluto, far more spend most of their time in the Oort cloud much farther out in space. Collisions occur, too: in 1994 at least 21 fragments of Shoemaker-Levy 9 exploded in Jupiter's upper atrmosphere, and Jupuiter may well have swept up many comets from farther out in the Solar System in the past.

RIGHT There are three main classes of comet. The faint short-period comets [A] often have their aphelia (furthest points from the Sun) at approximately the distance of Jupiter's orbit [1], and their periods amount to a few years. Long-period comets [B] have aphelia near or beyond Neptune's orbit [2], though Halley's is the only conspicuous member of the class. Comets with very long periods [C] have such great orbital eccentricities that the paths are almost parabolic. Apart from Halley's, all the really brilliant comets are of this type. Half the known comets orbit almost entirely within the paths of Jupiter and Saturn, taking 20 or so years; the quickest, Encke, takes just 3.3 years. At the other extreme there are comets with huge orbits: Hyakutake, last seen in 1996, will not be near the Sun again for another 14,000 years.

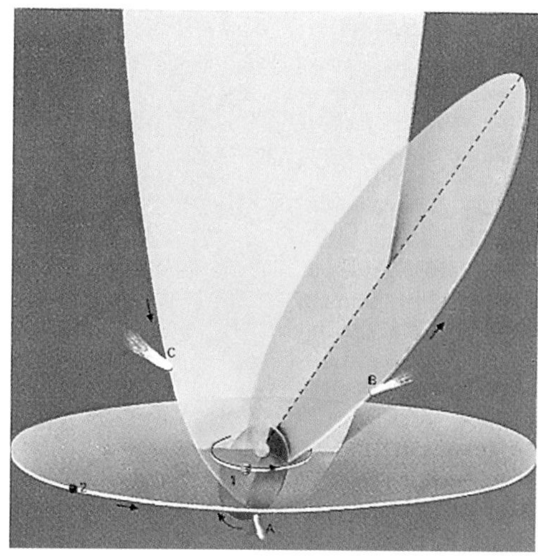

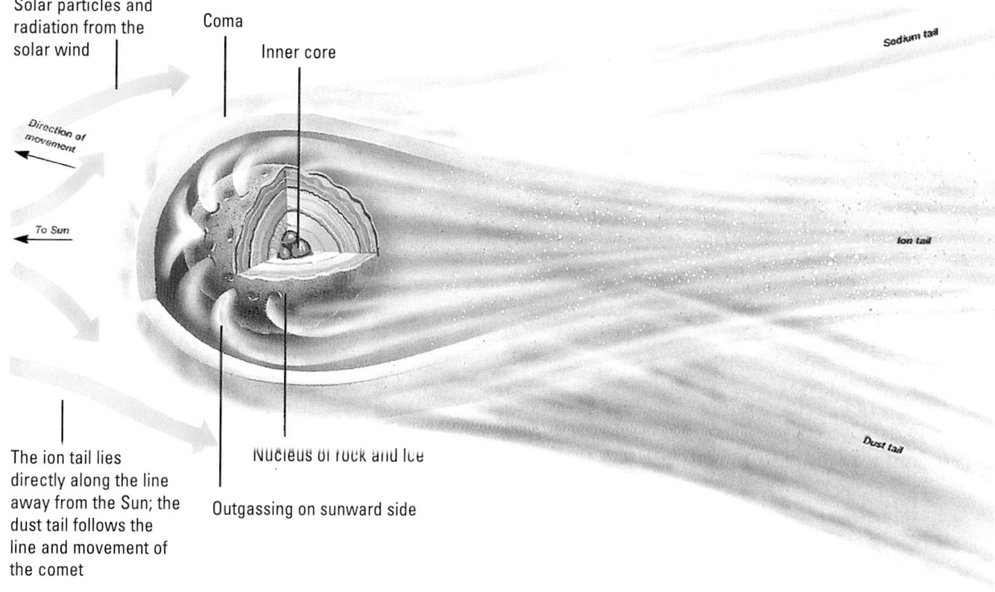

Solar particles and radiation from the solar wind

Coma

Inner core

Direction of movement

To Sun

Sodium tail

ion tail

Dust tail

The ion tail lies directly along the line away from the Sun; the dust tail follows the line and movement of the comet

Nucleus of rock and ice

Outgassing on sunward side

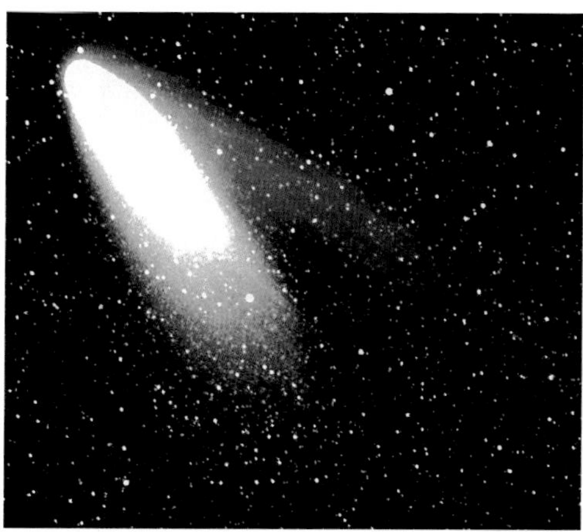

LEFT Hale-Bopp had its perihelion (closest approach to the Sun) in 1997, and it was easily seen with the naked eye from much of Earth for several days. Spectroscopy revealed 38 types of gas present in the comet's coma.

BELOW Following its visit in 1910, Halley's comet last returned to perihelion in 1986. Although not as bright as the great "non-periodical" comets, the increase and decline of the tail is clearly shown. As it approached perihelion the tail developed enormously; after the closest approach to the Sun the tail contracted, so that when the comet was last seen the tail had disappeared altogether. The seventh picture shows the tail shortly before perihelion.

ABOVE A comet has an irregular nucleus of rock and ice. As it nears the Sun, the ice vaporizes and combines with dust to produce the coma or head, hiding the nucleus from our view. The gas or ion tail, often blue and comprising charged electrons (ions) caught up in the solar wind, streams away from the direction of the Sun for up to 100 million km (60 million miles); the dust tail, often white in reflected sunlight, comprises tiny grains of rock and can stretch for up to 10 million km (6 million miles) behind the comet. When observing comet Hale-Bopp in 1997 astronomers also discovered a new type, the sodium tail, which accelerates a straight neutral gas tail up to 10 million km (6 million miles) behind the comet.

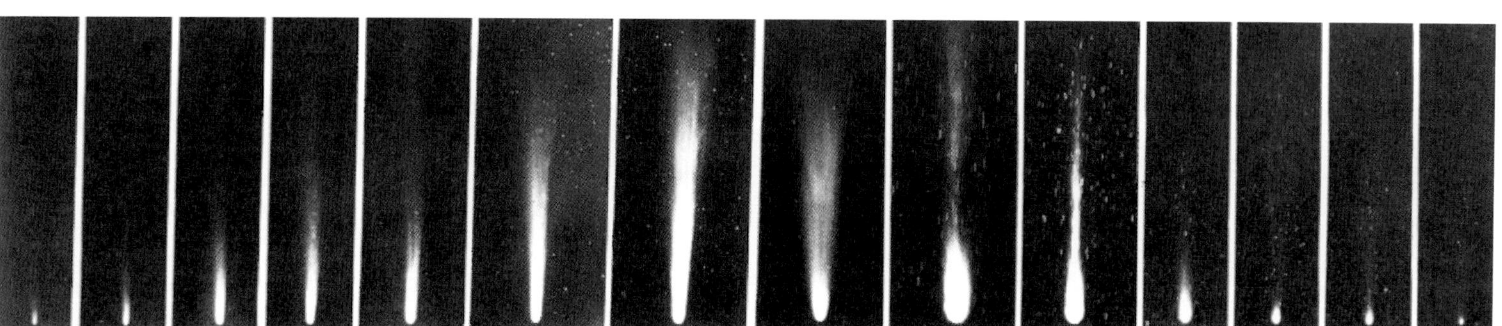

ASTEROIDS AND METEOROIDS

SINCE THE BEGINNING of the 19th century, astronomers have catalogued more than 8,000 asteroids orbiting our Sun, and at least 10,000 more have been observed. It is estimated that there are at least a million of these rocky bodies with diameters of over 1km (0.6 miles), their numbers making nonsense of their previous description the "minor planets". Along with comets and meteoroids, they are better described as "space debris", though 95% are found in the main asteroid belt between Mars and Jupiter. Some tiny examples find their way to Earth as meterorites, but our knowledge of their formation and composition remains limited.

Asteroid orbits

The main belt asteroids are not smoothly distributed in cloud between Mars and Jupiter. There are gaps in the main belt where very few asteroids exist. These were discovered by an American astronomer, Daniel Kirkwood, and are known as the Kirkwood gaps and mark places where the orbital period would be a simple fraction of Jupiter's. For example, an asteroid orbiting the Sun at a distance of 375 million km (233 million miles) would complete exactly three orbits while Jupiter orbited the Sun once. It would feel a gravitational tug from Jupiter, away from the Sun, every orbit, and quickly be moved out of that position.

However, in some places more asteroids are seen than expected: one such place is Jupiter's orbit. A swarm of a few hundred asteroids is found 60° ahead of and behind Jupiter in the same orbit. Known as the Trojans, they orbit the Sun at the same rate as Jupiter, but hardly ever come close enough to the planet for their orbits to be disturbed.

Swarms can also be found within the main belt of asteroids. These are known as asteroid families, and are formed when two larger asteroids collide. Astronomers then see the resulting fragments as many smaller asteroids sharing similar orbits around the Sun.

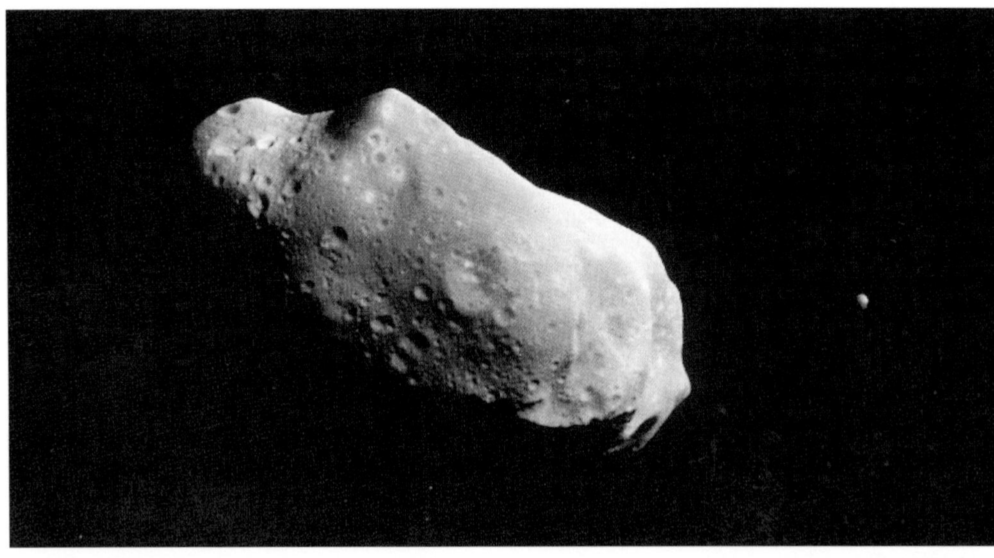

ABOVE The asteroid Ida as photographed by the Galileo spacecraft on its way to Jupiter in 1993. Galileo discovered a small moon, seen here on the right, orbiting at about 100km (60 miles); named Dactyl by surprised scientists, this irregularly shaped satellite measures only 1.7 cu. km (0.4 cu. miles). Galileo also passed the asteroids Gaspra and Mathilde, like Ida heavily cratered by small asteroids and meteorites: one crater on Mathilde was estimated at about 10km (6 miles) deep, huge in relation to the body's size.

Formation and composition of asteroids

For some time it was thought that the asteroids were the debris of a collision that destroyed a "missing" planet, but it now seems unlikely that a large planet ever formed between Mars and Jupiter – mainly because of the latter's gravitational field. Most of the mass present in that region during the early days of the Solar System was probably rotating in elliptical orbits and ended up colliding with the planets, their satellites or even the Sun.

No asteroid has ever been shown to have an atmosphere, so the light we see must be sunlight reflected from the surface. An asteroid's composition depends on its distance from the Sun. In the inner main belt nearest Mars, they are made of silicate rocks (minerals containing silicon and oxygen) similar to those found on Earth. These are called "S-type" asteroids. In the middle of the belt are mostly "C-type": these appear to have rocks containing carbon, similar to some types of meteorites landing on Earth.

The outer belt has asteroids that are so dark they only reflect 5% of the sunlight that reaches them, and are very red. Our best assumption about these "D-type" asteroids is that there is a large amount of ices such as water ice and frozen carbon monoxide

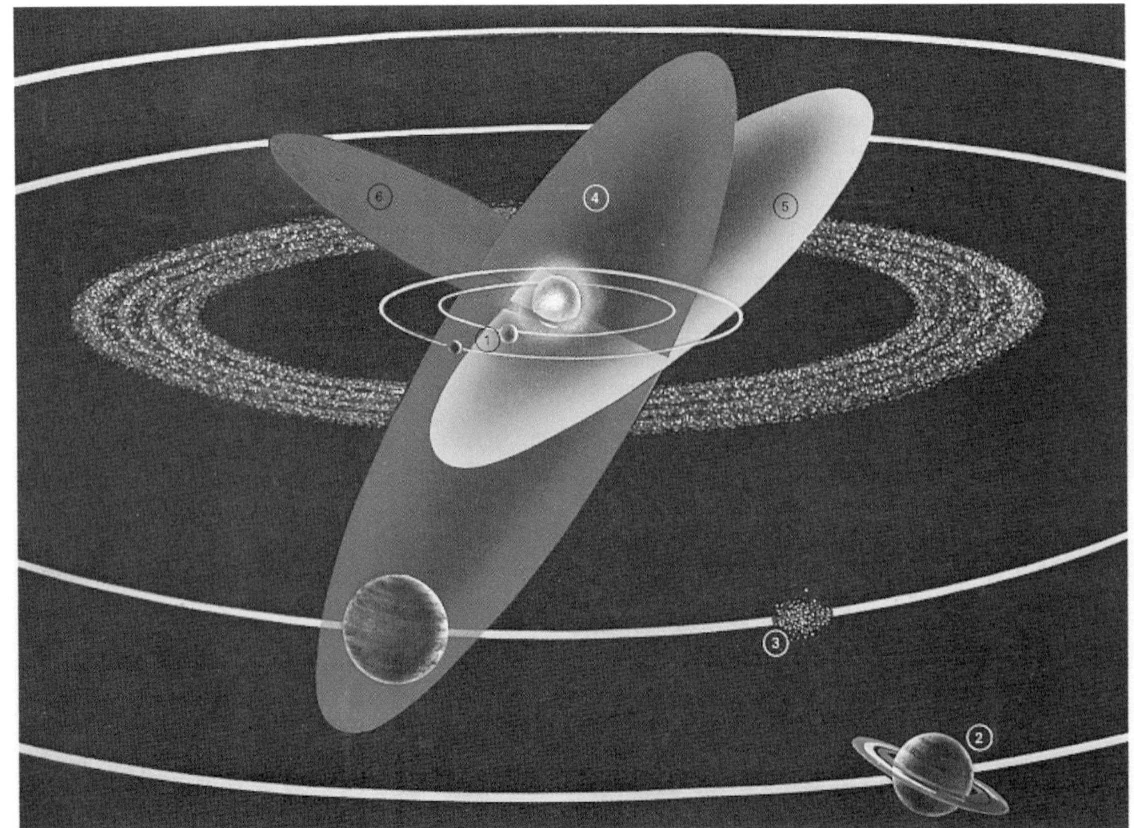

LEFT The orbits of the planets from Earth [1] out to Saturn [2], together with some notable asteroids (the illustration is not to scale). While most asteroids move in the region between the orbits of Mars and Jupiter, the so-called Trojan asteroids [3] move in the same orbit as Jupiter. They keep their distance, however, and collisions are unlikely to occur: one group moves 60° ahead of the planet and the other group 60° behind, though they move round for some distance to either side of their mean positions.

Hidalgo [4] has a path which is highly inclined and so eccentric – much like a comet – that its aphelion (farthest point from the Sun) is not far from the orbit of Saturn. Amor [5] and Apollo [6] belong to the so-called "Earth-grazing" asteroid group. All the Earth-grazers are very small: Amor has a diameter of 8km (5 miles) and Apollo only about 2km (1.25 miles). Both satellites of Mars, Phobos and Deimos, are asteroids captured by the gravitational pull of the planet.

mixed in with the rock, and that charged particles from the solar wind hitting them have created chemical reactions to form the dark red colour.

This change in asteroid make-up is logical if they were formed at the beginning of the Solar System, as this change in composition fits in with theories about how the planets formed. In addition, since their formation, asteroids nearer the Sun have been heated more than those farther out; this means that, over time, more ice melted and escaped. Farther away, lower temperatures mean that less of the ice has melted.

Asteroids undergo some of the most violent temperature changes in the Solar System. One asteroid, Icarus, actually approaches closer to the the Sun than the baked planet Mercury. At its perihelion (closest point), only 28 million km (17 million miles), its surface can reach more than 900°C (500°F); just 200 Earth days later it has reached its aphelion (farthest point), 295 million km (183 million miles) from the warmth of the Sun in the cool space beyond Mars.

Asteroids and Earth

In 1937 Hermes, a mere 1km (0.6 miles) in diameter, passed just 780,000km (485,000 miles) from Earth, less than twice the distance of the Moon. What would happen if such an object hit Earth? Besides the tremendous heat, enough rock and dust would be deposited in the atmosphere to change the climate all over the Earth, while if it landed near water, huge tsunamis would devastate cities on the edge of the ocean all over the globe. Indeed, it is now widely thought that the impact of an asteroid or comet 65 million years ago, producing a crater 180km (112 miles) wide in the Yucatan Peninsula of Mexico assisted in the extinction of the dinosaurs. Luckily for us, it's estimated that such devastating impacts are likely to happen only once every 100 million years or so.

Meteors

Commonly known as shooting stars, meteors are flashes of light caused by particles of rock entering the Earth's atmosphere at altitudes of around 100km (60 miles), most of them only the size of a grain of sand. As they travel at between 10 and 30km (6 to 18 miles) per second, friction with the air molecules rapidly heats them to thousands of degrees, and they vaporize in a flash of heat. Larger and therefore brighter meteors are known as fireballs, and can be anything from the size of a small pebble up to a large boulder. Before the rocks enter our atmosphere, they are following their own orbit about the Sun.

A "meteor shower" occurs when Earth passes through one of the meteor streams, belts of dust

RIGHT The sizes of the first four asteroids to be discovered, Ceres [C], Vesta [D], Pallas [E] and Juno [F], together with the irregularly-shaped Eros [B], are compared here with the Moon [A]. Being so small, the diameters are difficult to measure: earlier assessments of Ceres gave 685km (426 miles), but new methods show that it is much larger at 913km (567 miles). Still the largest known asteroid, Ceres is two-fifths the size of Pluto. It was the first asteroid to be identified, by the Italian astronomer Guiseppe Piazzi in 1801.

While Ceres and some other large asteroids are spherical, most are elongated and lumpy. All are pitted with craters – one on Mathilde is 10km (6 miles) deep. The average rotation period of an asteroid is eight hours, but while Florentina takes only three hours to spin once, Mathilde takes 17 days. They almost certainly originate from the time of the formation of the Solar System and are not remnants of a large planet that disintegrated, as was once thought. The largest asteroid is less than 1% of the mass of the Moon, and the known asteroids combined are less than 10% of Earth's mass. The first encounter by a space probe was made by the Galileo mission in 1989.

particles sharing their orbits with comets but too heavy to be swept out of the Solar System. For example, the Leonid shower shares the orbit of comet Temple-Tuttle. The best showers occur on 12 August and 13 December each year. They are called the Perseid meteor shower and the Geminid meteor shower because their radiant points appear to be in the constellations of Perseus and Gemini.

Meteorites

While meteoroids usually burn up in the atmosphere, some are big enough to make it through the atmosphere without being completely vaporized and reach the ground; they are then called meteorites. Scientists estimate that about 300,000 meteorites reach the surface of the Earth every year, though many fall in the oceans or remote forests,

deserts and mountains. Even those that fall near towns and cities can remain undiscovered, since many look like ordinary rocks to the untrained eye. Some meteorites are tiny particles, while others weigh up to 200 tonnes. Meteoroids weighing more than about 100 tonnes that don't break up are not decelerated as much as lighter bodies, and produce impressive impact craters.

When chemically analysed, there are many different types of meteorite. The most common finds are called chondrites, and appear to be the same type of iron- and silicon-bearing rock that S-type asteroids are made from. Much rarer are the carbonaceous chondrites, which have large amounts of carbon and appear to have come from the middle of the asteroid belt. Finally, about 10% of meteorites are the heavier stony-iron and iron-nickel type.

RIGHT Research suggests that craters such as the Barringer in Arizona, 1.6km (1 mile) wide and 180m (600ft) deep, was formed by nickel-iron meteorites up to about 50,000 years ago. Burning up as they plunged through the atmosphere, they shattered the Earth's outer layer of rock on impact (*top right*). Because of their high speed they burrowed into the ground, causing friction, heat, compression and shockwaves, culminating in a violent explosion that left the huge crater. More than 130 craters have so far been identified, though many more were created before being subsequently destroyed by geological activity and erosion.

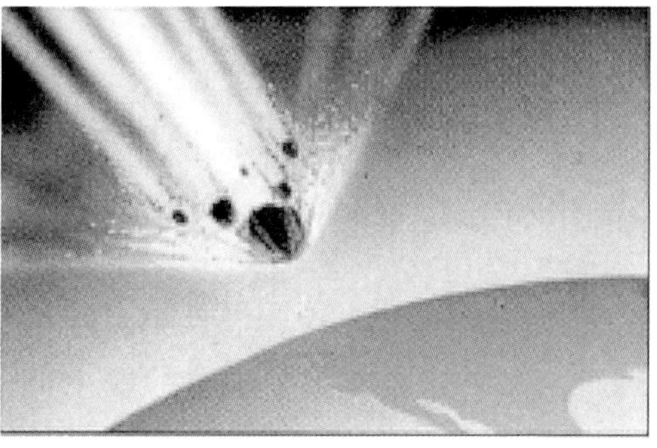
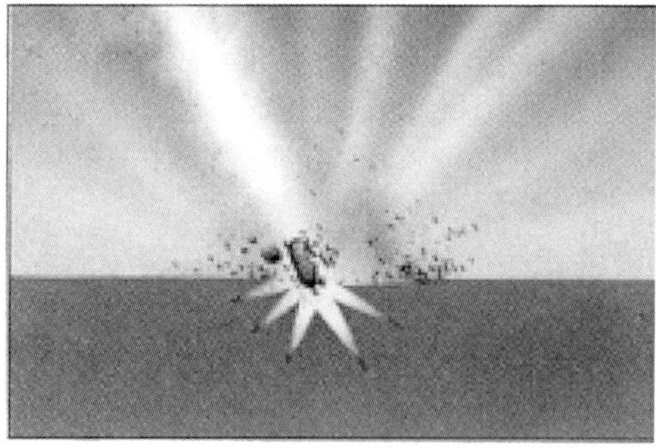
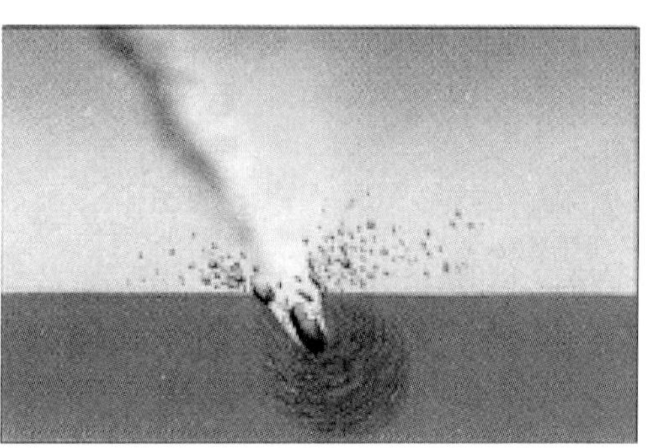
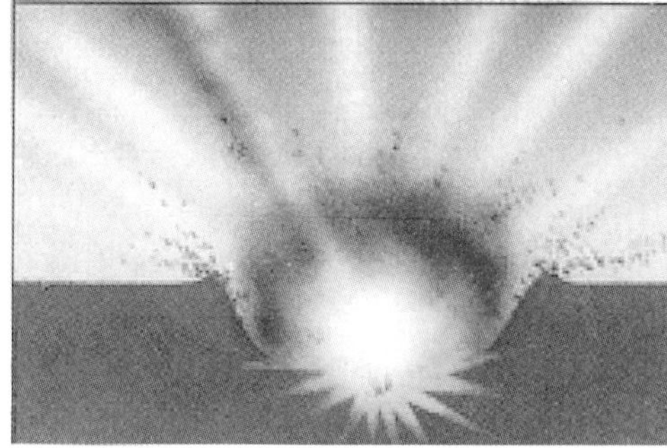

TIME AND MOTION

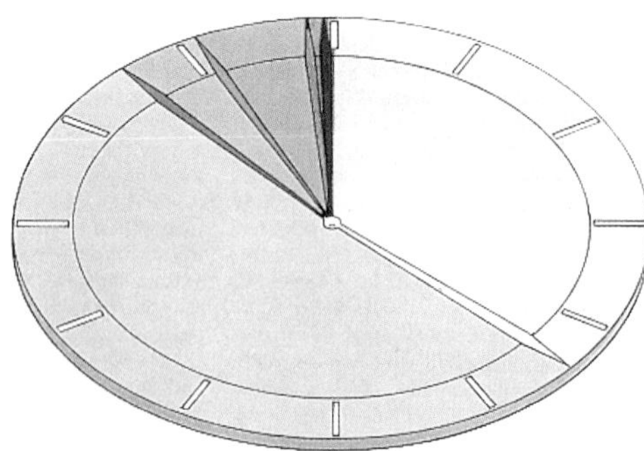

ABOVE A "clock" of the Earth's history, with 12 hours representing the 4,600 million years the world has been in existence. The first 2 hours and 52 minutes are still obscure, but the earliest rocks are then formed – though the planet remains a lifeless desert until 04.20, when bacterial organisms first appear.

Eons of time drag by until just after 10.30, when there is an explosion of invertebrate life in the oceans. Dinosaurs wander the land by 11.36, only to die out and replaced by birds and mammals 25 minutes later. Hominids arrive about 30 seconds before noon – and the last tenth of a second covers human civilization.

The oldest rocks of the great Precambrian shields of North America, Africa and Australia convey dates of up to about 3,500 million years ago. Only the past 570 million years show an abundance of plant and animal life. The most widely found fossil remains from any period are called index fossils and are used to correlate various rock formations of the same age.

T HE BASIC unit of time measurement is the day, one rotation of the Earth on its axis. The subdivision of the day into hours, minutes and seconds is simply for our convenience. The present Western calendar is based on the "solar year", the 365.24 days the Earth takes to orbit the Sun.

Calendars based on the movements of the Sun and Moon, however, have been used since ancient times. The average length of the year, fixed by the Julian Calendar introduced by Julius Caesar, was about 11 minutes too long, and the cumulative error was eventually rectified in 1582 by the Gregorian Calendar. Pope Gregory XIII decreed that the day following 4 October that year was in fact 15 October, and that century years do not count as leap years unless they are divisible by 400. Britain did not adopt the reformed calendar until 1752 – by which stage it was lagging 11 days behind the continent; the Gregorian Calendar was imposed on all its possessions, including the American colonies, with all dates preceding 2 September marked O.S., for Old Style.

The seasons are generated by a combination of the Earth's revolution around the Sun and the tilt of its axis of $23\frac{1}{2}°$. The solstices (from the Greek *sol*, sun, and *stitium*, standing) are the two times in the year when the Sun is overhead at one of the Tropics of Cancer and Capricorn, $23\frac{1}{2}°$ North and South, furthest from the Equator. The equinoxes (from the Greek *aequus*, equal, and *nox*, night) are the two times in the year when day and night are of equal length due to the Sun being overhead at the Equator. The longest and shortest days in each hemisphere fall on or around the solstices, and are opposites in each hemisphere.

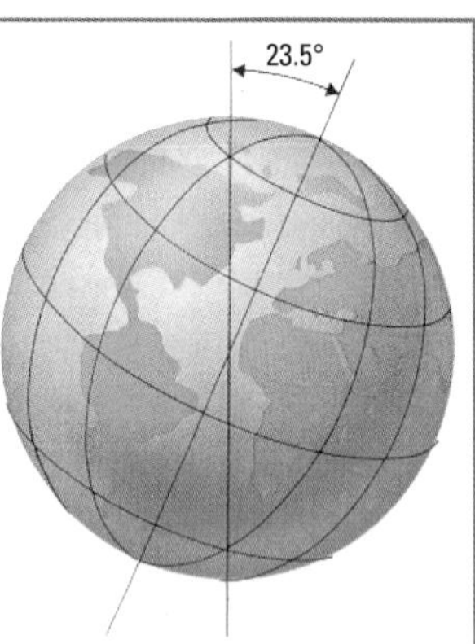

The Earth's axis is inclined at 23.5° to the perpendicular to the orbital plane. This angle accounts for the complex seasonal variations in climate, notably in mid-latitudes. The varying distance of the Earth from the Sun has only a minor effect.

DEFINITIONS OF TIME

Year: The time taken by the Earth to revolve around the Sun, or 365.24 days.
Month: The approximate time taken by the Moon to revolve around the Earth. The 12 months of the year in fact vary from 28 days (29 in a Leap Year – once every 4 years to offset the difference between the calendar and the solar year) to 31 days.
Week: An artificial period of 7 days. Unlike days, months and years – but like minutes and seconds – it is not based on astronomical time.
Day: The time taken by the Earth to complete one rotation (spin) on its axis.
Hour: A day comprises 24 hours, divided into hours a.m. (*ante meridiem*, before noon) and p.m. (*post meridiem*, after noon) – though timetables use the 24-hour system from midnight.

LEFT Seasons occur because the Earth's axis is tilted at a constant angle of $23\frac{1}{2}°$ as it spins. When the Northern Hemisphere is tilted to a maximum extent towards the Sun, on 21 June, the Sun is overhead at noon at the Tropic of Cancer ($23\frac{1}{2}°$ North): this is midsummer, or the summer solstice, in this hemisphere.

On 22 or 23 September the Sun is overhead at the Equator, and day and night are of equal length throughout the world: this is the autumn or fall equinox in the Northern Hemisphere. On 21 or 22 December, the Sun is overhead at the Tropic of Capricorn ($23\frac{1}{2}°$ South), the winter solstice in the Northern Hemisphere. The overhead Sun then tracks north, until on 21 March it is overhead at the Equator: this is the spring equinox in the Northern Hemisphere.

In the Southern Hemisphere the seasons are the reverse of those in the Northern Hemisphere: autumn corresponds to spring and winter to summer.

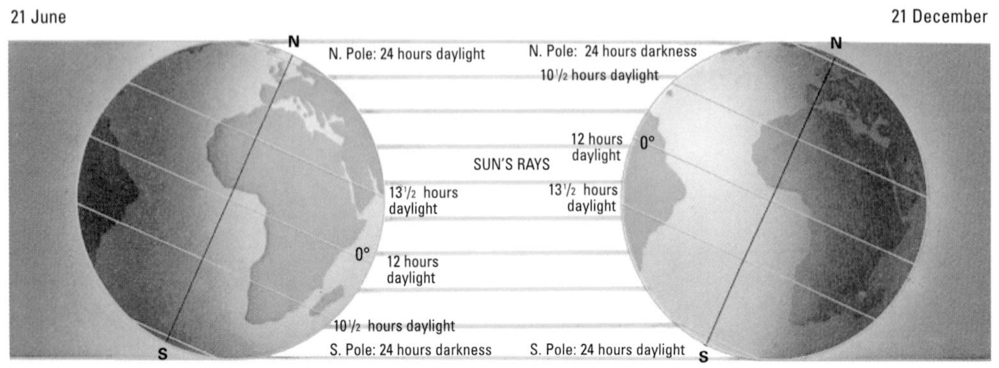

LEFT The Sun appears to "rise" in the east, reach its highest point at noon, and then "set" in the west, to be followed by night. In reality it is not the Sun that is moving but the Earth, rotating ("spinning" on its axis) from west to east. At the summer solstice in the Northern Hemisphere (21 June), the area inside the Arctic Circle has total daylight and the area inside the Antarctic Circle has total darkness. The opposite occurs at the winter solstice on 21 or 22 December. At the Equator, the length of day and night are almost equal all year round, with seasonal variations in between.

RIGHT The Moon rotates more slowly than the Earth, making one complete turn on its axis in just over 27 days. Since this corresponds to its period of revolution around the Earth, the Moon always presents the same hemisphere or face to us, and we never see its "dark side".

The interval between one full Moon and the next (and thus also between two new Moons) is about 29½ days – a lunar month. The apparent changes in the shape of the Moon are caused by its changing position in relation to the Earth; like the planets, the Moon produces no light of its own and shines only by reflecting the rays of the Sun.

BELOW The Earth rotates through 360° in 24 hours, and therefore moves 15° every hour. The world is divided into 24 standard time zones, each centred on lines of longitude at 15° intervals, 7½° on either side of its central meridian.

The prime or Greenwich meridian, based on the Royal Observatory in London, lies at the centre of the first zone. All places to the west of Greenwich are one hour behind for every 15° of longitude; places to the east are ahead by one hour for every 15°.

When it is 12 noon at the Greenwich meridian, at 180° east it is midnight of the same day – while at 180° west the day is only just beginning. To overcome this problem the International Dateline was established, approximately following the 180° meridian. If you travelled from Japan (140° east) to Samoa (170° west) you would pass from the night into the morning of the same day.

While some countries cope with several time zones (Russia experiences no fewer than 11), others "bend" the meridians to incorporate their territory in certain zones, and China, despite crossing five, follows just one. Others, including Iran and India, employ differences of half an hour.

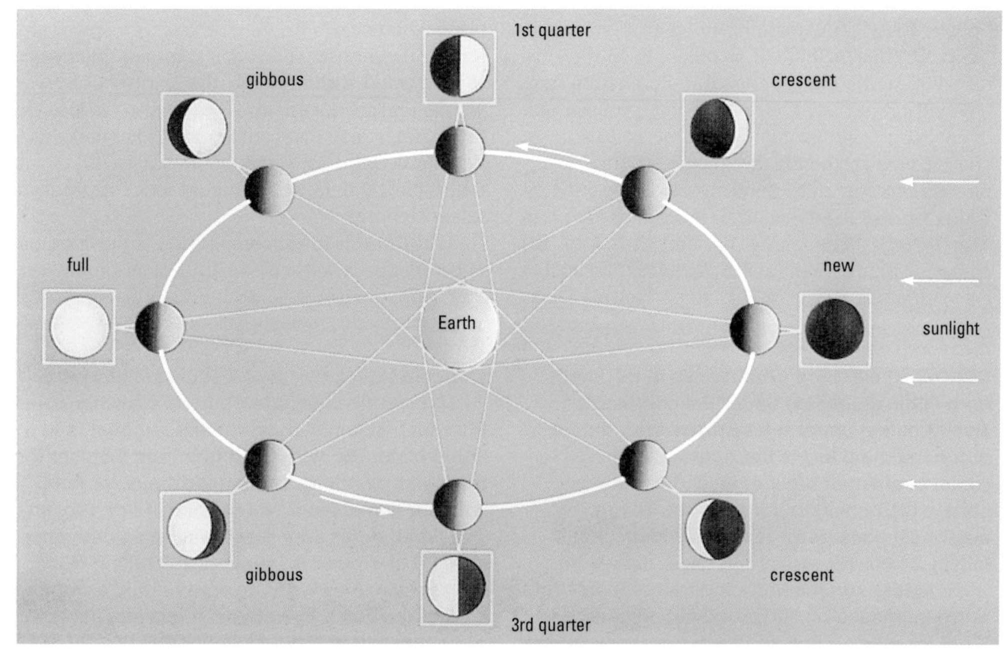

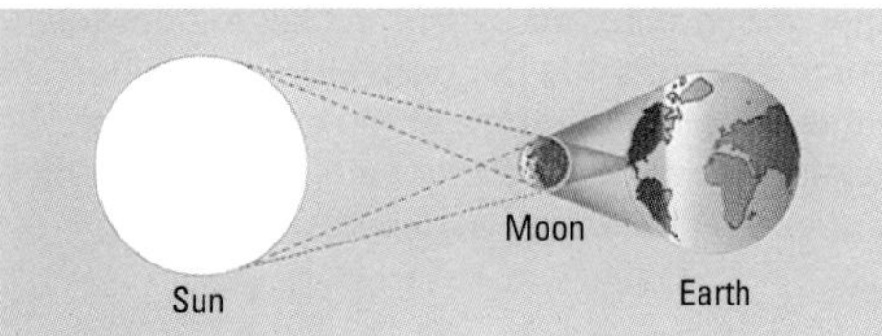

LEFT A solar eclipse occurs when the Moon passes between the Sun and the Earth. It will cause a partial eclipse of the Sun if the Earth passes through the Moon's outer shadow, or a total eclipse if the inner cone shadow crosses the Earth's surface. A total eclipse was visible in much of the Northern Hemisphere in 1999.

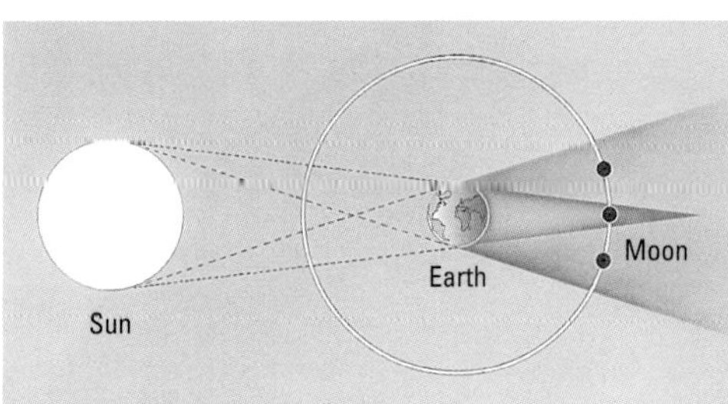

LEFT In a lunar eclipse the Earth's shadow crosses the Moon and, as with the solar version, provides either a partial or total eclipse. Eclipses do not occur every month because of the 5° difference between the plane of the Moon's orbit and the plane in which the Earth moves. In the 1990s, for example, only 14 eclipses were possible – seven partial and seven total – and each was visible only from certain and variable parts of the world.

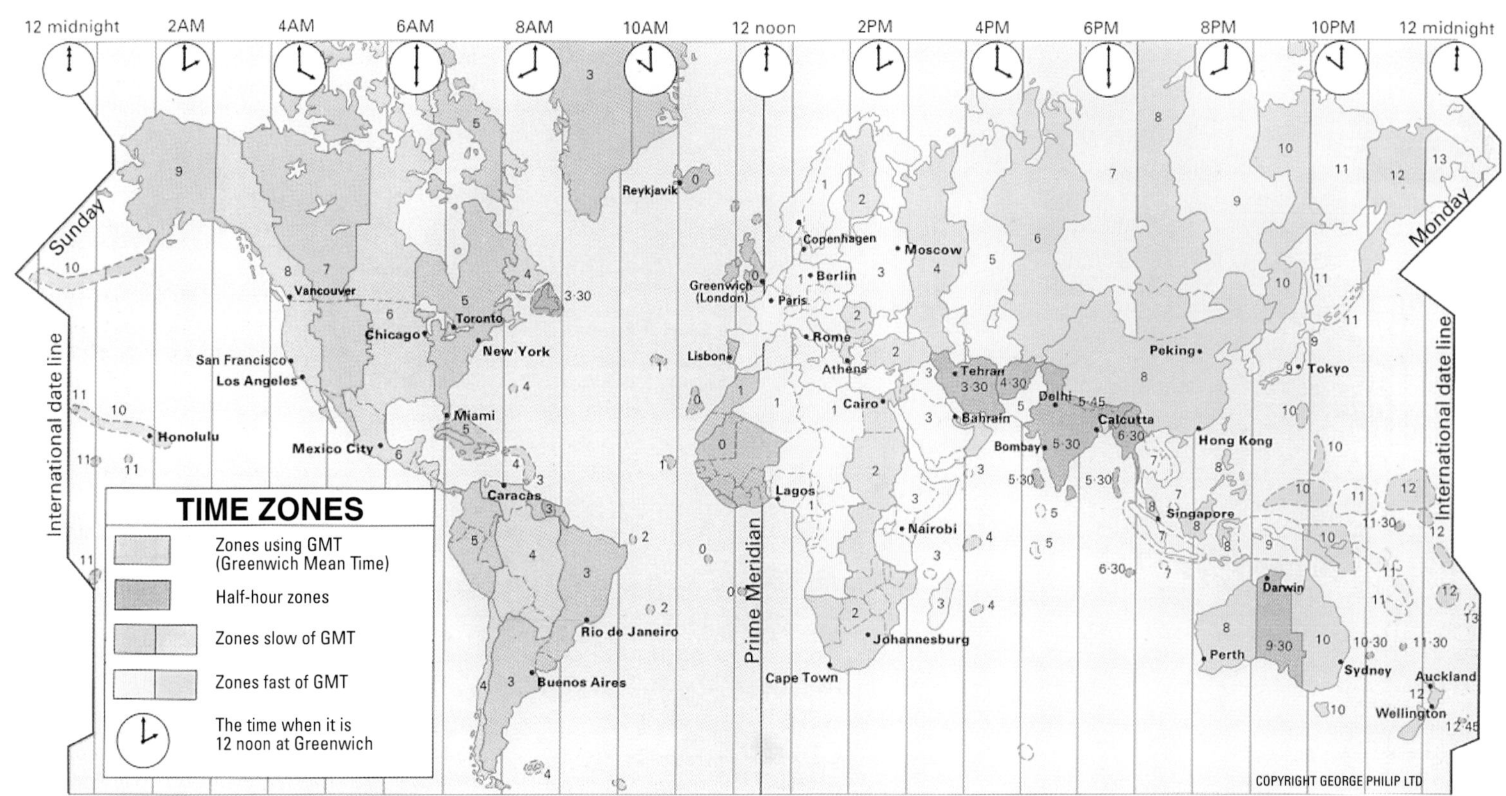

ANATOMY OF THE EARTH

THE EARTH is made up of several concentric shells, like the bulb of an onion. Each shell has its own particular chemical composition and physical properties. These layers are grouped into three main regions: the outermost is called the crust, which surrounds the mantle, and the innermost is the core. The solid, low-density crust on which we live is no thicker in relation to the Earth than an eggshell, taking up only 1.5% of the planet's volume. While the chemical distinction between crust and mantle is important, as far as

physical processes go they behave as a single unit termed the lithosphere. It is a common fallacy that if you could drill through the Earth's crust you would find a molten mass: even well below the brittle outer shell that forms part of the lithosphere, the convecting part of the the mantle is still essentially solid, and pockets of liquid rock (magma) are relatively rare.

While the chemical composition of the crust and upper mantle is well known, little is absolutely certain about the layers beneath.

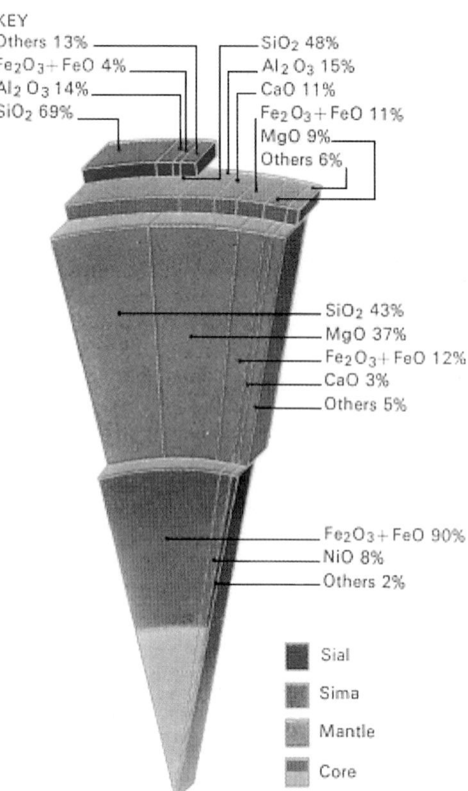

Others 13%
Fe$_2$O$_3$+FeO 4%
Al$_2$O$_3$ 14%
SiO$_2$ 69%

SiO$_2$ 48%
Al$_2$O$_3$ 15%
CaO 11%
Fe$_2$O$_3$+FeO 11%
MgO 9%
Others 6%

SiO$_2$ 43%
MgO 37%
Fe$_2$O$_3$+FeO 12%
CaO 3%
Others 5%

Fe$_2$O$_3$+FeO 90%
NiO 8%
Others 2%

▪ Sial
▪ Sima
▪ Mantle
▪ Core

BELOW The Earth's crust varies in thickness from 40km (25 miles) under the continents to 5km (3 miles) under the seafloor. With the top of the mantle it forms the rigid lithosphere [1], which overlies a "plastic" layer, the asthenosphere [2], on which it may move. The upper mantle [3] goes down to about 700km (430 miles), where it overlies the lower mantle [4].

From the surface the temperature inside the Earth increases by 30°C for every kilometre (85°F for every mile), so that the asthenosphere is close to melting point. At 50km (30 miles), in the upper mantle, it reaches 800°C (1,480°F). After around 100km (60 miles) the rate of increase slows dramatically, and scientists now estimate the temperature to be 2,500°C (4,600°F) at the boundary of the lower mantle

and core [5] – a depth of 2,900km (1,800 miles).

The mantle is separated from the outer core [6], which seismic observations suggest is in a liquid state. The density jumps from 5.5g/cm for the lower mantle to 10g/cm for the outer core, where it increases downwards to 12 or 13g/cm. The liquid outer core gives way to a solid inner core [7] at around 5,150km (3,200 miles] from the surface. Although the core is only around 16% of the Earth by volume, it represents 32% of its mass; it is thought to consist mostly of iron and some nickel, a hypothesis that fits the data and is inspired by iron-nickel meteorites which are probably the remnants of another planet. The temperature at the centre of the Earth (8) is estimated at least 3,000°C (5,400°F), and could be as high as 5,000°C (9,000°F).

ABOVE The Earth is composed of three main but unequal layers – the crust, mantle and core. The crust is subdivided into continental and oceanic material. The upper continental crust is mostly granite, abundant in silicon and aluminium – hence the term sial; over oceanic areas, and underlytng the continental sial, is a lighter material, essentially basalt and rich in silicon and magnesium – hence the term sima. The mantle comprises rock, rich in magnesium and iron silicates, and the dense core probably consists mainly of iron and nickel oxides, almost certainly in a molten condition. Heat is transferred to the surface by convection and conduction: in the solid layers it is probably transferred by conduction, and in the liquid layers it moves by convection.

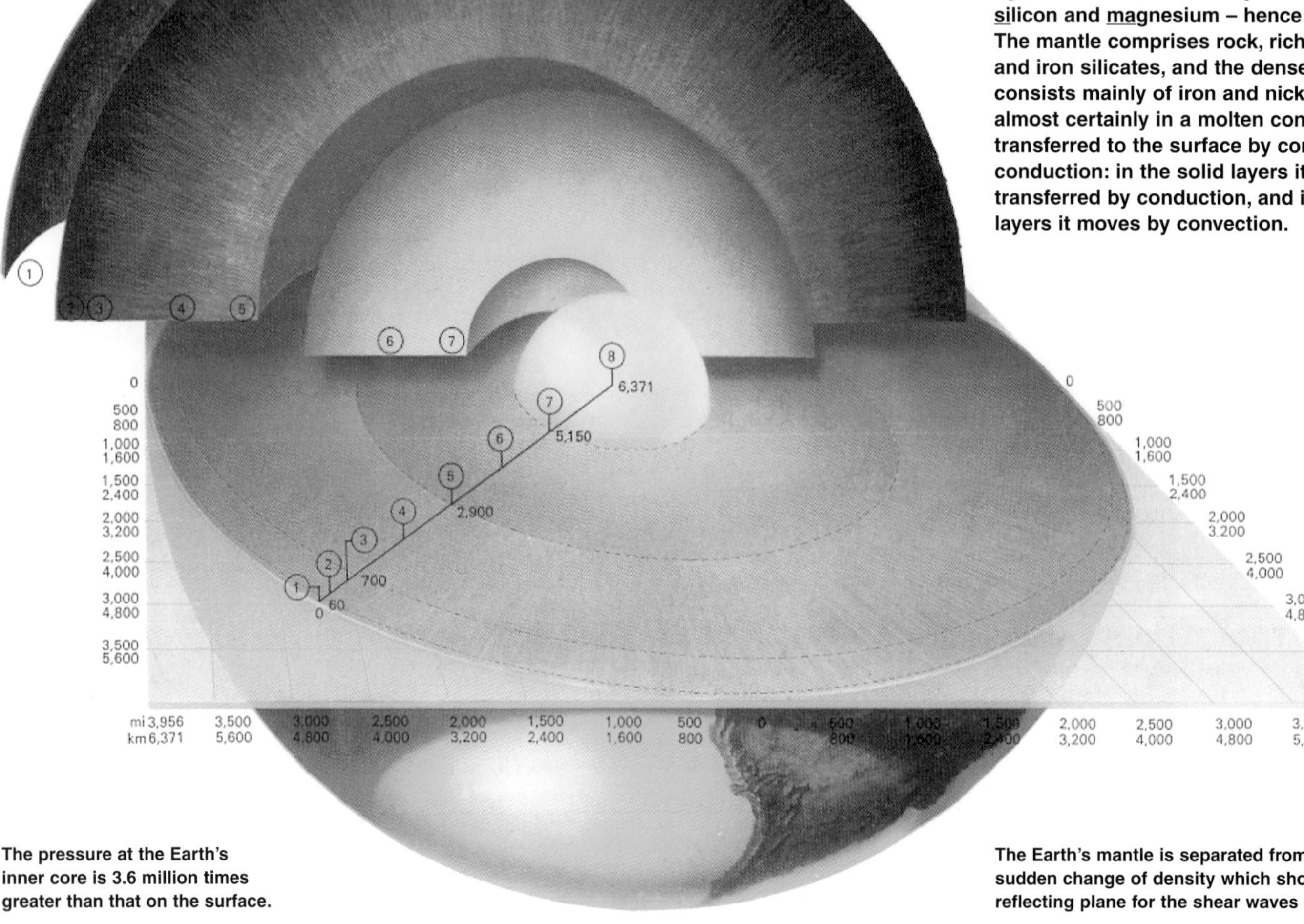

The pressure at the Earth's inner core is 3.6 million times greater than that on the surface.

The Earth's mantle is separated from the core by a sudden change of density which shows up as a reflecting plane for the shear waves of earthquakes.

THE MAGNETIC EARTH

As the Earth spins on its axis, the fluid layer of the outer core allows the mantle and solid crust to rotate relatively faster than the inner core. As a result, electrons in the core move relative to those in the mantle and crust. It is this electron movement that constitutes a natural dynamo and produces a magnetic field similar to that produced by an electric coil.

The Earth's magnetic axis is inclined to its geographical axis by about 11°, and the magnetic poles don't coincide with the geographic north and south poles. The Earth's magnetic axis is continually changing its angle in relation to the geographic axis, but over a long time – some tens of thousands of years – an average relative position is established.

A compass needle points to a position some distance away from the geographical north and south poles. The difference (the declination), varies from one geographical location to the next, with small-scale variations in the Earth's magnetism. The magnetosphere is the volume of space in which the Earth's magnetic field predominates.

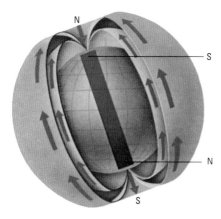

ABOVE The magnetic field originating inside the Earth makes up about 90% of the field observed at ground level: the remainder is due to currents of charged particles coming from the Sun and to the magnetism of rocks in the crust. The difference in rotation speed between the liquid outer core and the mantle creates a dynamo effect.

ABOVE The Earth's magnetic field is like that of a giant natural bar magnet placed inside the Earth, with its magnetic axis inclined at a small angle to the geographical axis. The poles of a compass needle are attracted by the magnetic poles of the Earth so that one end points to the north magnetic pole and the other to the south magnetic pole.

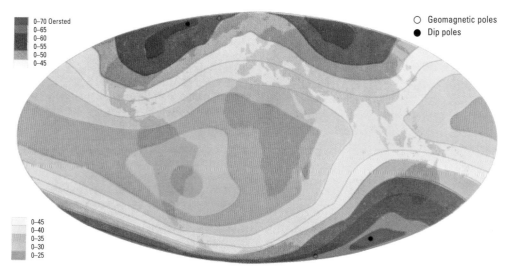

- ○ Geomagnetic poles
- ● Dip poles

LEFT The intensity of the Earth's magnetic field is strongest at the poles and weakest in the equatorial regions. If the field were purely that of a bar magnet in the centre of the Earth and parallel to the spin axis, the lines of equal intensity would follow the lines of latitude and the magnetic poles would coincide with the geographic poles. In reality, however, the "bar magnet" field is inclined at about 11° to the spin axis and so are its geomagnetic poles.

Neither is the real field purely that of a bar magnet. The "dip poles", where the field direction is vertical (downwards at the north pole and upwards at the south dip pole), are themselves offset in respect to the geomagnetic poles – each by a different amount so that the south dip pole is not exactly opposite the north dip pole. Oersted is a traditional unit of magnetic field strength.

BELOW Until recent times taking account of the difference between the magnetic and geographic poles was crucial in navigation. The needle of a ship's magnetic compass, for example, swings to a position where its ends point to north and south along a line of force of the Earth's magnetic field. In navigation today, the magnetic compass is often replaced by the motor-driven gyrocompass, which indicates true north.

RIGHT The magnetosphere is the region in which the Earth's magnetic field can be detected. It would be symmetrical were it not for the "solar wind", electrically-charged particles from the Sun [A], which distort it to a teardrop shape. The particles meet the Earth's magnetic field at the shock front [1]. Behind this is a region of turbulence and inside the turbulent region is the magnetopause [2], the boundary of the magnetic field. The Van Allen belts [3] are two zones of high radiation in the magnetopause. The inner belt consists of high-energy particles produced by cosmic rays, the outer comprises solar electrons.

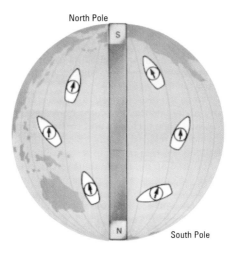

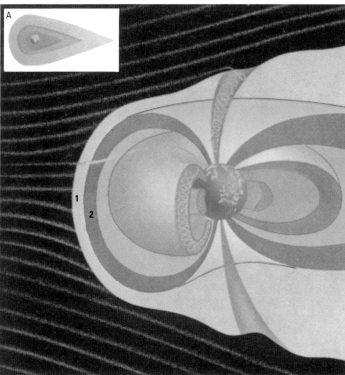

THE RESTLESS EARTH

THE THEORY of plate tectonics was advanced in the late 1960s and has had a revolutionary effect on the earth sciences. It is a unifying, all-embracing theory, offering a plausible and logical explanation for many of the Earth's varied structural phenomena, ranging from continental drift to earthquakes and mountain building.

The crust of the Earth, which together with the upper mantle forms the lithosphere, consists of rigid slabs called plates that are slowly but constantly moving their position in relation to each other. The plates are bounded by oceanic ridges, trenches and transform faults. Oceanic ridges are formed where two plates are moving apart, leaving a gap which is continuously filled by magma (molten rock) rising from the asthenosphere, on which the plates "float". As the magma cools, new crust is created on the ridges and becomes part of the oceanic plates.

This is the phenomenon known as seafloor spreading. Spreading rates, though slow, are not negligible: the North Atlantic is opening up by 4cm (1.6 in) a year, and the fastest rate is found at the East Pacific Rise, which creates 10cm (4 in) of new crust every year – 1,000km (620 miles) in the relatively short geological time of 10 million years.

Trenches as well as mountain ranges are formed where two plates converge. One of the plates slides steeply under the other and enters the mantle: the world's deepest trench, the Mariana, was formed when the Pacific plate was forced under the far smaller Philippine plate. Since the volume of the Earth does not change, the amount of crust created at the ridges is balanced by that destroyed at the trenches in an endless cycle of movement.

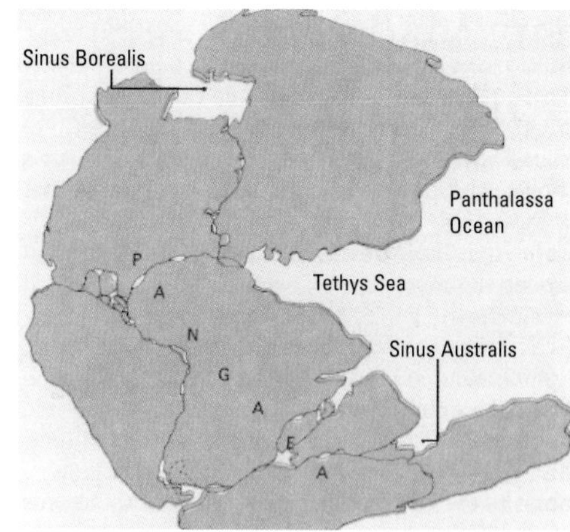

ABOVE First put forward by the German meteorologist Alfred Wegener in 1912, the theory of continental drift suggests the continents once formed a single land mass, Pangaea. The initial break-up created a northern mass, Laurasia, and a southern one, Gondwanaland, named after a province in India.

LEFT A map of Pangaea cannot be accurately constructed. The most suitable fit of the land masses is obtained by matching points midway down the continental slope, at about 200m (650ft). The easiest areas to fit together are the continents of Africa and South America, and while the linking of the northern lands is possible with a certain degree of accuracy, much remains to be learned of the complex fit of India, Antarctica and Australia with Africa and South America. The break-up of Pangaea began about 200 million years ago, and by the end of the Jurassic period, about 135 million years ago, the North Atlantic and Indian Oceans had become firmly established. The Tethys Sea was being diminished by the Asian land mass rotating in an anti-clockwise direction, and South America had begun to move away from Africa to form the South Atlantic.

LEFT By the end of the Cretaceous period, about 65 million years ago, the South Atlantic had grown, Madagascar had parted from Africa and India had continued northwards. Antarctica was moving away from the central land mass, though still linked with Australia. The North Atlantic rift forked at the north, starting to form the island of Greenland.

Geological evidence that the continents were once linked is provided by distinctive rock formations that can be assembled into continuous belts when South America and Africa are juxtaposed; by the processes of mountain building, notably India grinding into Asia and crumpling up sediments to form the Himalayas; and by the dovetailed distribution of many plants and animals.

Perhaps the most important impetus to the theory of continental drift came from the twin theories of plate tectonics and seafloor spreading, which developed rapidly from the 1960s. One of the weakest points in Wegener's argument centred on the tremendous forces needed to drive the continents apart. The new plate theories, which have been substantially proven, provide an explanation of the source of the necessary power. Even so, much has still to be learned about the original continent.

135 million years ago

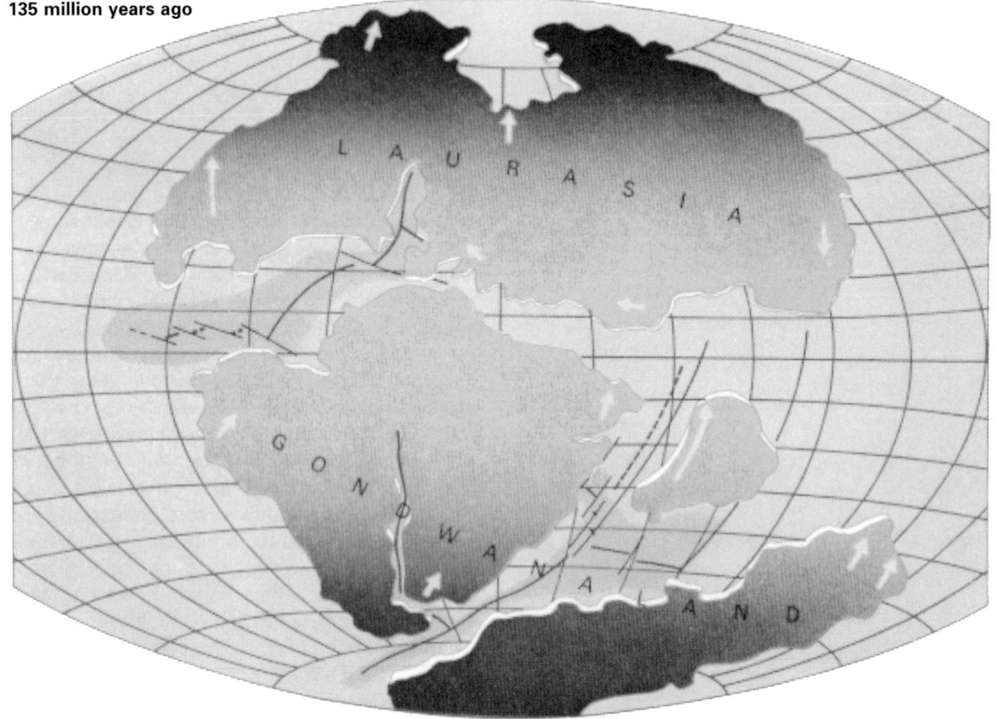

65 million years ago

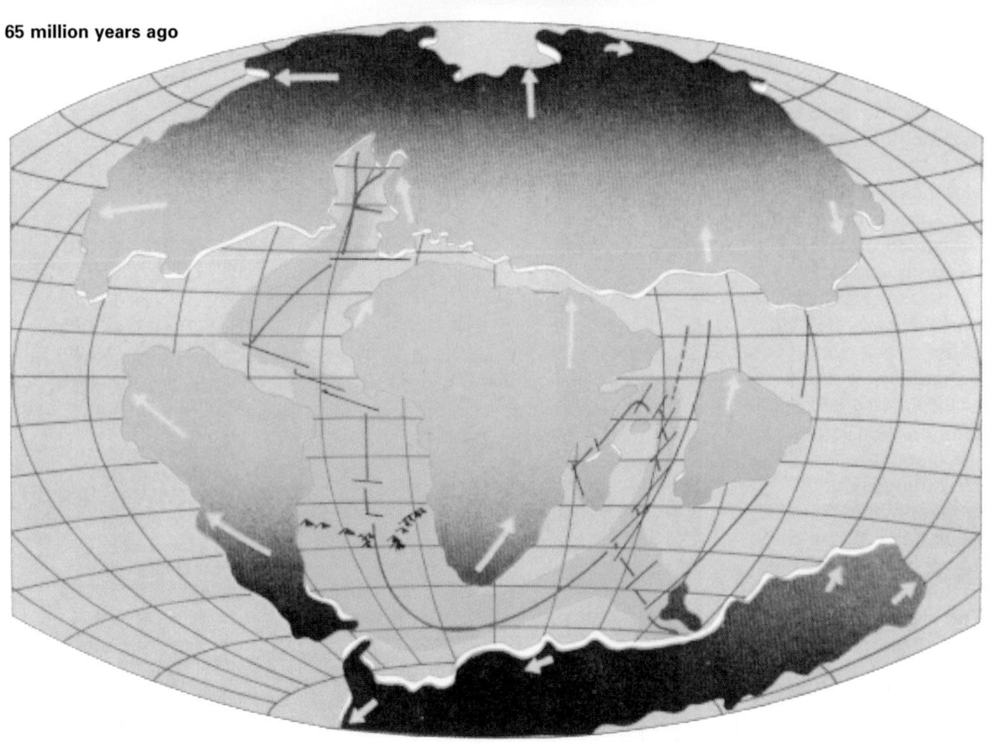

RIGHT The debate about continental drift was followed by a more radical idea: plate tectonics. The basic theory proposes that the Earth's crust comprises a series of rigid plates that "float" on a softer layer of the mantle, and are moved about by continental convection currents within the Earth's interior. These plates slowly converge and diverge along margins marked by seismic (earthquake) activity.

Converging plates form either trenches (where the oceanic plate sinks below the lighter continental rock), or mountain ranges. The theory not only supports the notion of continental drift: it also explains the paradox that while there have always been oceans, none of the present seabeds contain sediments more than 150 million years old.

The six major mobile plates (the American, Eurasian, African, Indo-Australian, Pacific and Antarctic) contain smaller plates such as the Arabian and West Indian plates which "absorb" the geometrical discrepancies between major plates by creating or destroying compensating amounts of crustal material.

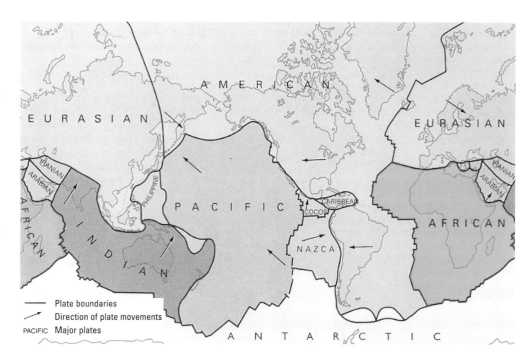

—— Plate boundaries
/ Direction of plate movements
PACIFIC Major plates

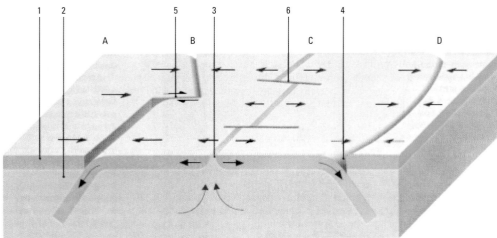

BELOW Collision zones are where two plates, each carrying a continental mass, meet. When one of the plates is forced beneath the other, the buoyant continental material is pushed upwards in a series of high overthrusts and folds, producing great mountain ranges. The Himalayas, formed when the northward-moving Indian plate crunched up against the Eurasian plate, were the result of such forces – as were other leading fold mountains such as the Alps in Europe (African/Eurasian plates), the Andes in South America (Nazca/American) and the western Rockies (Pacific/American).

ABOVE The plate tectonics theory sees the Earth's lithosphere [1] as a series of rigid but mobile slabs called plates [A,B,C,D]. The lithosphere floats on a "plastic" layer called the asthenosphere [2]. There are three types of boundaries. At the mid-oceanic ridges [3], upwelling of mantle material occurs and new seafloor is formed. A trench [4] is formed where

one plate of oceanic crust slides beneath the other, which may be oceanic or continental. The third type of boundary is where two plates slide past one another, creating a transform fault [5,6]. These link two segments of the same ridge [6], two ocean trenches [5] or a ridge to a trench. Plates move from ridges and travel like conveyor belts towards the trenches.

50 million years ahead

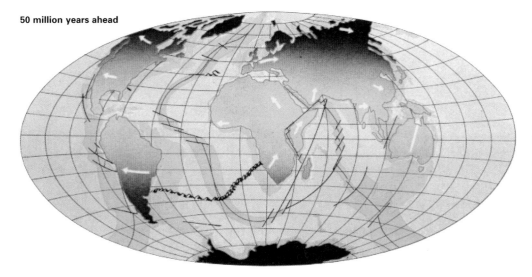

LEFT The continents are still drifting, and there is no reason to expect them to stop. This is how the world may look 50 million years from now if drift is maintained as predicted. The most striking changes in this "new world" are the joining of the Atlantic and Pacific Oceans; the splitting away from the USA of Baja California and the area west of the San Andreas fault line; the northward drift of Africa; the breaking away of that part of the African continent east of the present-day Great Rift Valley; and Australia's continued journey north towards Asia. The majority of the great continent of Antarctica, however, remains in its present southerly position. Plant fossils found in Antarctica's coal seams – remnants of its tropical past hundreds of millions of years ago – are among many examples of evidence supporting the tectonic theory of continent drift.

EARTHQUAKES

A N EARTHQUAKE is the sudden release of energy in the form of vibrations and tremors caused by compressed or stretched rock snapping along a fault in the Earth's surface. Rising lava under a volcano can also produce small tremors. It has been estimated that about a million earthquakes occur each year, but most of these are so minor that they pass unnoticed. While really violent earthquakes occur about once every two weeks, fortunately most of these take place under the oceans, and only rarely do they produce tsunamis.

Slippage along a fault is initially prevented by friction along the fault plane. This causes energy, which generates movement, to be stored up as elastic strain, similar to the effect created when a bow is drawn. Eventually the strain reaches a critical point, the friction is overcome and the rocks snap past each other, releasing the stored-up energy in the form of earthquakes by vibrating back and forth. Earthquakes can also occur when rock folds that can no longer support the elastic strain break to form a fault.

Shockwaves

Seismic or shockwaves spread outwards in all directions from the focus of an earthquake, much as sound waves do when a gun is fired. There are two main types of seismic wave: compressional and shear. Compressional waves cause the rock particles through which they pass to shake back and forth in the direction of the wave, and can be transmitted through both solids and liquids; they are therefore able to travel through the Earth's core. Shear waves make the particles vibrate at right-angles to the direction of their passage, and can

travel only through solids; at the boundary of lower mantle and liquid outer core, they are reflected back to the Earth's surface. Neither type of seismic wave physically moves the particles – it merely travels through them.

Compressional waves, which travel 1.7 times faster than shear waves, are the first ones to be distinguished at an earthquake recording station. Consequently seismologists refer to them as primary (P) waves and to the shear waves as secondary (S) waves. A third wave type is recognized by seismologists – the long (L) wave which travels slowly along the Earth's surface, vertically or horizontally. It is L waves that produce the most violent shocks.

Measuring earthquakes

The magnitude of earthquakes is usually rated according to either the Richter or the Modified Mercalli scales, both formulated in the 1930s. Developed by the US geologist Charles Richter, the Richter scale measures the total energy released by a quake with mathematical precision, each upward step representing a tenfold increase in shockwave power. A magnitude of 2 is hardly felt, while a magnitude of 7 is the lower limit of an earthquake that has a devastating effect over a large area. Theoretically there is no upper limit, but the largest measured have been rated at between 8.8 and 8.9. The 12-point Mercalli scale, named after the Italian seismologist Guiseppe Mercalli, is based on damage done and thus varies in different places. It ranges from I (noticed only by seismographs) to XII (total destruction); intermediate points include VII (collapse of substandard buildings) and IX (conspicuous cracks in the ground).

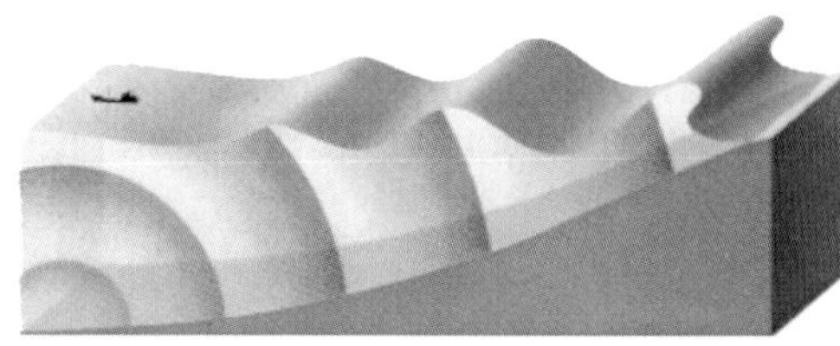

ABOVE The long wave length of tsunamis gives them tremendous speed. An earthquake in the Aleutian Trench in the far northern Pacific in 1946 triggered off a tsunami that devastated Honolulu; it took 4 hours 34 minutes to reach Hawaii, a distance of 3,220km (2,000 miles) – a speed of about 700km/h (440mph).

Tsunamis

Tsunami is the Japanese word for a seismic sea wave; they are often called tidal waves, though they have no connection with tides. Tsunamis are caused mainly by seismic disturbances below the seafloor (oceanic earthquakes), but also by submarine landslides and volcanic eruptions. Other tidal waves can be due to the surge of water when the barometric pressure is exceptionally low, such as in a hurricane. At sea, the height of the wave is seldom more than 60–90cm (2–3ft), but the wave length may be as long as 200km (120 miles), generating speeds of up to 750km/h (450mph).

Although the height of the crest is low out to sea, tsunamis have immense energy, which, as they lose speed in more shallow water, is converted into an increase in height. The waves, on reaching the shore, may be 40m (125ft) or more high. The most destructive tsunamis occur in the northern Pacific.

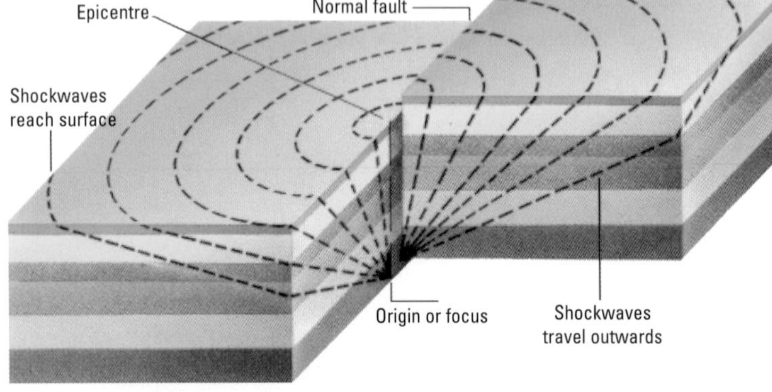

Epicentre · Normal fault
Shockwaves reach surface
Origin or focus
Shockwaves travel outwards

LEFT An earthquake takes place when two parts of the Earth's surface move suddenly in relation to each other along a crack called a fault. The point from which this movement originates is called the focus, usually located at depths between 8 and 30km (5 to 18 miles), and the point on the surface directly above this is called the epicentre. Shockwaves move outwards from the focus in a curved pattern; while speed varies with the density of rock, intensity decreases the farther the waves travel.

BELOW Earthquakes occur in geologically sensitive areas of the world such as mid-oceanic ridges and mountain-building regions, and can be classified according to the depth of their focus. Deep focus quakes occur at depths of between 300 and 650km (185-400 miles), intermediate focus quakes from 55 to 240km (35-150 miles), and shallow focus quakes from the surface down to a depth of 55km (35 miles).

EARTHQUAKE ZONES

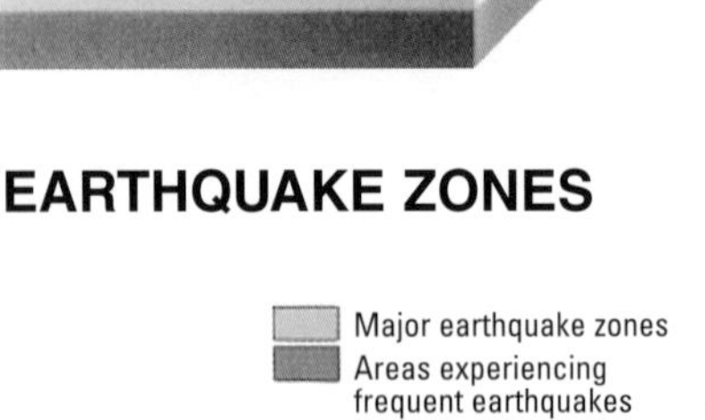

Major earthquake zones
Areas experiencing frequent earthquakes

The highest magnitude recorded on the Richter scale is 8.9, for a quake that killed 2,990 people in Japan on 2 March 1933. The most devastating earthquake ever affected three provinces of central China on 2 February 1556, when it is believed that about 830,000 people perished. The highest toll in modern times was at Tangshan, eastern China, on 28 July 1976: the original figure of over 655,000 deaths has since been twice revised by the Chinese government to stand at 242,000.

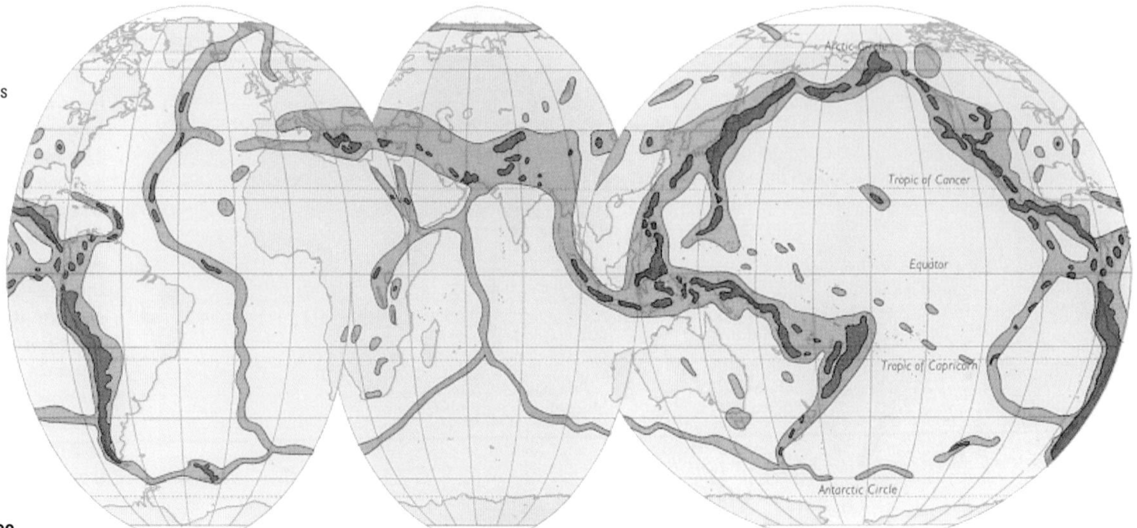

Tropic of Cancer
Equator
Tropic of Capricorn
Antarctic Circle

VOLCANOES

Fissure eruptions do not form volcanoes but release flows of fluid lava that can cover areas up to 500 sq km

Fluid rock in the magma chamber is released as ash and lava during eruptions

Lava flows can be released from side vents and gases can issue from crevices in the loose flanks

Stratified layers of volcanic rocks build up the main cone; each eruption adds at least one layer

Rainwater heated by the magma surfaces as geysers and hot springs

Geysers are fountains of water and steam created by the vaporising of ground waters.

Active or recent cones often form inside explosion craters or crater-shaped calderas

A laccolith is a giant lens-shaped intrusion that pushes up the strata above; it is fed from the magma chamber

Pressure in the main vent encourages the opening of side vents as alternative paths to the surface

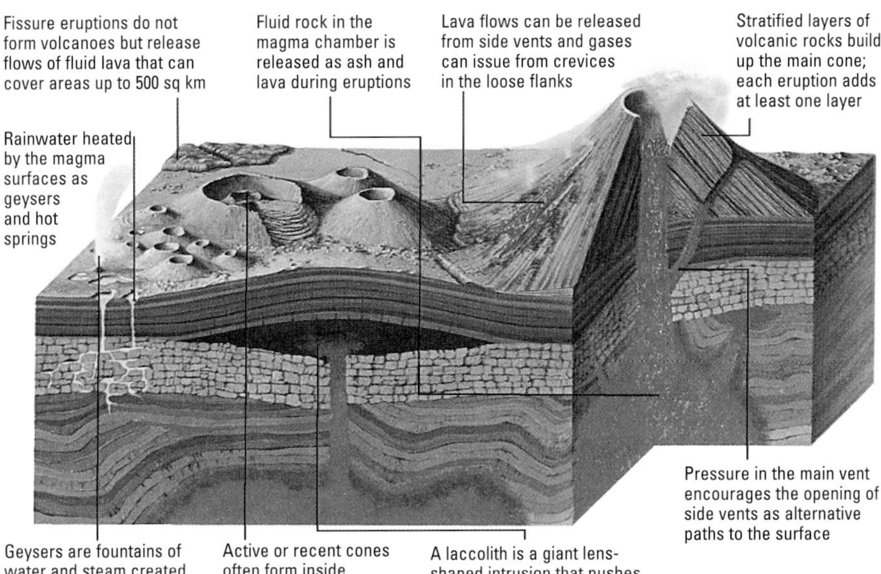

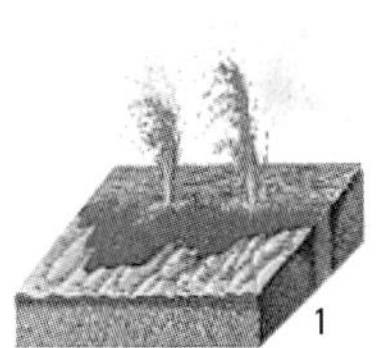

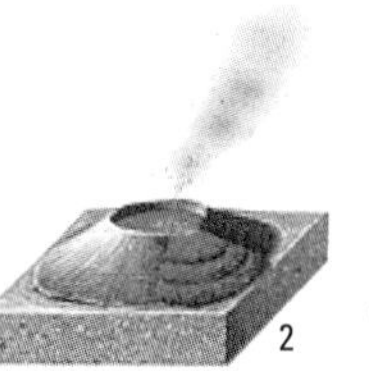

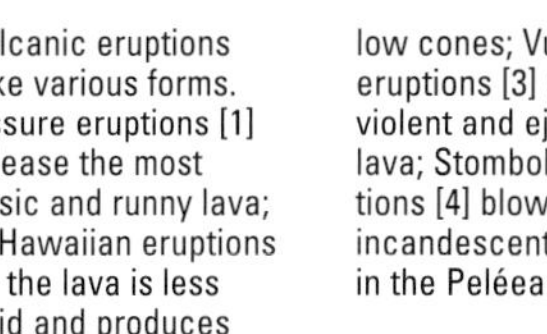

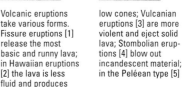

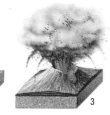

Volcanic eruptions take various forms. Fissure eruptions [1] release the most basic and runny lava; in Hawaiian eruptions [2] the lava is less fluid and produces

low cones; Vulcanian eruptions [3] are more violent and eject solid lava; Stombolian eruptions [4] blow out incandescent material; in the Peléean type [5]

a blocked vent is cleared explosively; and a Plinian eruption [6] is a continuous blast of gas that rises to immense heights.

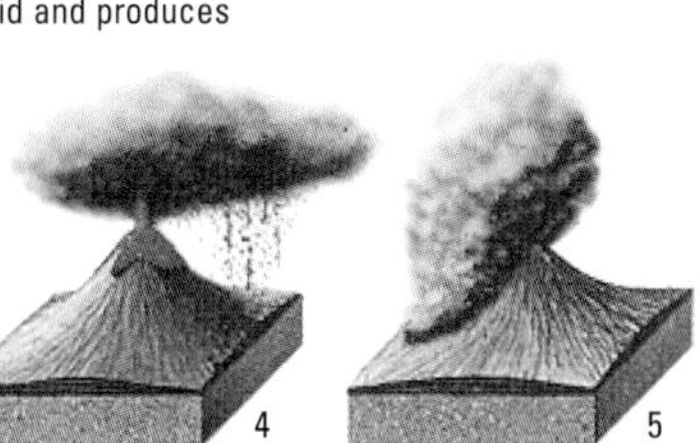

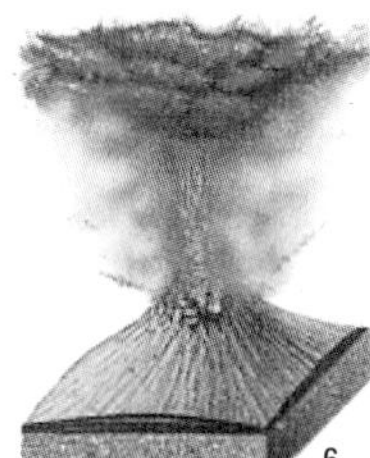

THE WORLD'S most spectacular natural displays of energy, volcanoes are responsible for forming large parts of the Earth's crust. Volcanoes occur when hot liquefied rock beneath the crust is pushed up by pressure to the surface as molten lava. They are found in places where the crust is weak – the mid-ocean ridges and their continental continuations, and along the collision edges of crustal plates. Some volcanoes erupt in an explosive way, throwing out rocks and ash, while others are effusive and lava flows out of the vent. Some, such as Mount Fuji in Japan, are both.

An accumulation of lava and cinders creates cones of various sizes and shapes. As a result of many eruptions over centuries Mount Etna in Sicily has a circumference of more than 120km (75 miles). Craters at rest are often filled by a lake – and the mudflow caused by an eruption can be as destructive as a lava flow and, because of its speed, even more lethal.

Despite the increasingly sophisticated technology available to geologists to monitor volcanoes, like earthquakes they remain both dramatic and unpredictable. For example, in 1991 Mount Pinatubo, located 100km (60 miles) north of the Philippines capital Manila, suddenly burst into life without any warning after lying dormant for over six centuries.

Most of the world's active volcanoes are located in a belt round the Pacific Ocean, on the edge of the Pacific crustal plate, called the "Ring of Fire" – a circle of fear that threatens over 400 million people. However, the soils formed by the weathering of volcanic rocks are usually exceptionally fertile, and despite the dangers large numbers of people have always lived in the shadows of volcanoes.

Climatologists believe that volcanic ash, if ejected high into the atmosphere, can influence temperature and weather conditions generally over a massive area and for several years afterwards. It has been estimated that the 1991 eruption of Mount Pinatubo in the Philippines threw up more than 20 million tonnes of dust and ash over 30km (18 miles) into the atmosphere, and it is widely believed that this accelerated the depletion of the ozone layer over large parts of the globe.

There are far more volcanoes on the seafloor than on the land, however. These "seamounts" exist because the oceanic crust is newer and thinner than continental crust and easily pierced by the underlying magma. The Pacific Ocean alone is thought to have more than 10,000 underwater volcanoes over 3,000m (9,850ft) high.

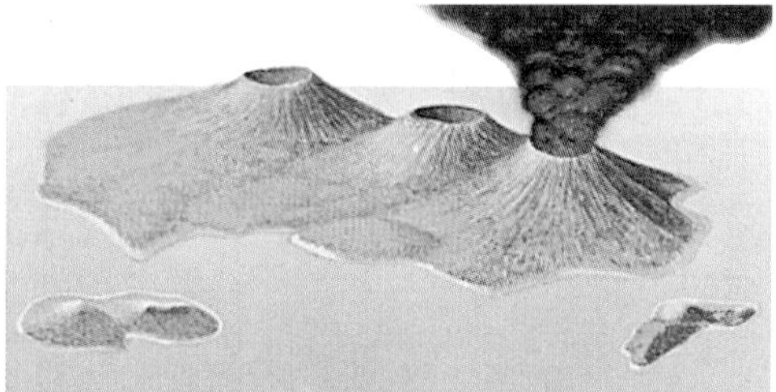

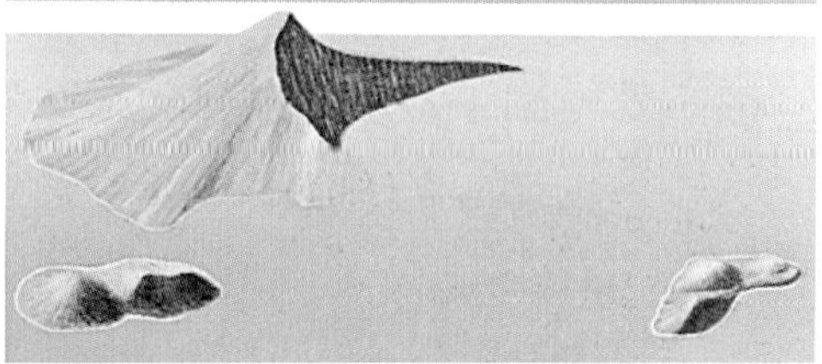

ABOVE Situated in the Sunda Strait of Indonesia, Krakatau was a small volcanic island inactive for over 200 years when, in August 1883, two-thirds of it was destroyed by a violent erruption. It was so powerful that the resulting tidal wave killed 36,000 people. Indonesia has the greatest concentration of volcanoes with 90, 12 of which are active.

VOLCANIC ZONES

. Volcanoes
— Seafloor spreading centre
◯ Ocean trench
Continental shelf

Structure

Pre-Cambrian
Caledonian folding
Hercynian folding
Tertiary folding
Great Rift Valley
// || Main trend lines

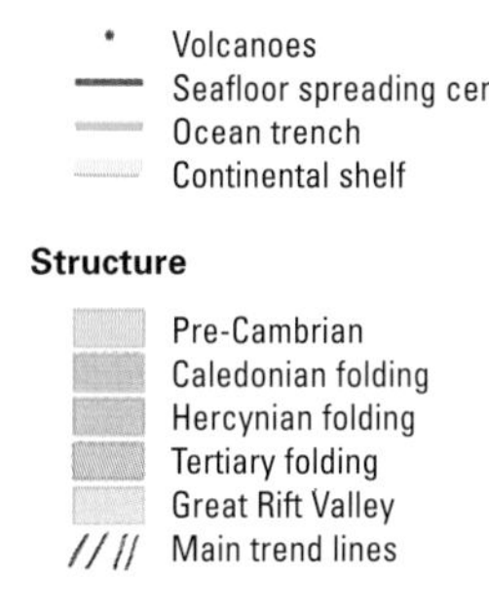

Of the 850 volcanoes to produce recorded eruptions, nearly three-quarters lie in the "Ring of Fire" that surrounds the Pacific Ocean on the edge of the Pacific plate.

SHAPING THE LANDSCAPE

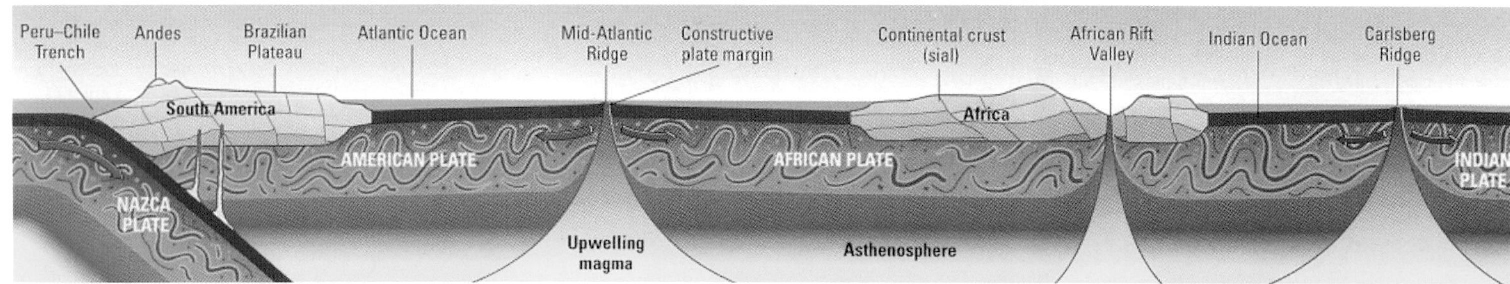

ABOVE A view of seafloor spreading along the Equator from the west coast of South America to the centre of the Indian Ocean. On the left, the Nazca plate has been subducted beneath the American plate to push up the Andes.

RIGHT A normal fault results when vertical movement causes the surface to break apart, while compression leads to a reverse fault. Horizontal movement causes shearing, known as a tear or strike-slip fault. When the rock breaks in two places, the central block may be pushed up in a "horst", or sink in a rift valley. Folds occur when rock strata are squeezed and compressed. Layers bending up form an anti-cline, those bending down form a syncline.

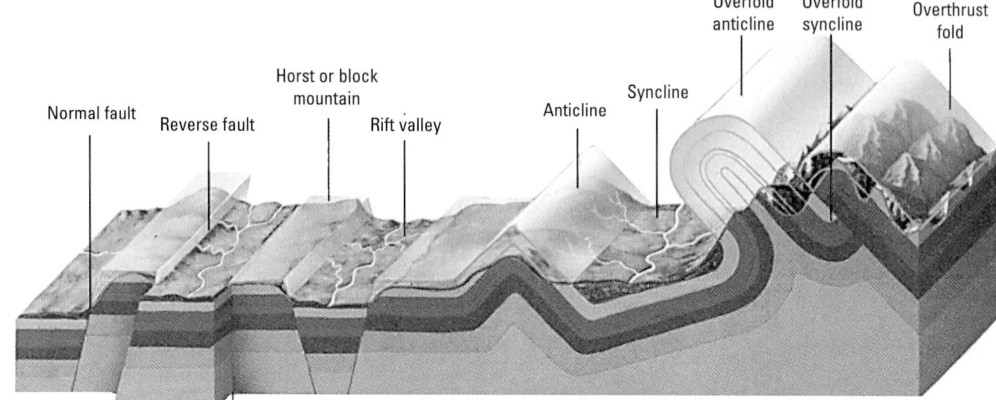

THE VAST ridges that divide the Earth beneath the world's oceans mark the boundaries between tectonic plates that are gradually moving in opposite directions. As the plates shift apart, molten magma rises from the mantle to seal the rift and the seafloor spreads towards the land masses. The rate of spreading has been calculated at about 40mm (1.6in) a year in the North Atlantic Ocean.

Near the ocean shore, underwater volcanoes mark the lines where the continental rise begins. As the plates meet, much of the denser oceanic crust dips beneath the continental plate at the "subduction zone" and falls back to the magma.

Mountains are formed when pressures on the Earth's crust caused by continental drift become so intense that the surface buckles or cracks. This happens where oceanic crust is subducted by continental crust, or where two tectonic plates collide: the Rockies, Andes, Alps, Urals and Himalayas all resulted from such impacts. These are known as fold mountains because they were formed by the compression of the sedimentary rocks, forcing the surface to bend and fold like a crumpled rug.

The other main mountain-building process occurs when the crust is being stretched or compressed so violently that the rock strata breaks to create faults, allowing rock to be forced upwards in large blocks; or when the pressure of magma inside the crust forces the surface to bulge into a dome, or erupts to form a volcano. Large and more complex mountain ranges may well reveal a combination of these features.

AGENTS OF EROSION

Destruction of the landscape, however, begins as soon as it is formed. Wind, ice, water and sea, the main agents of erosion, maintain a constant assault that even the hardest rocks cannot withstand. Mountain peaks may dwindle by only a few millimetres a year, but if they are not uplifted by further movements of the Earth's crust they will eventually disappear. Over millions of years, even great mountain ranges can be reduced to a low, rugged landscape.

Water is the most powerful destroyer: it has been estimated that 100 billion tonnes of rock are washed into the oceans each year. Three Asian rivers alone account for a fifth of this total – the Hwang Ho in China, and the Ganges and the Brahmaputra in Bangladesh.

When water freezes, its volume increases by about 9%, and no rock is strong enough to resist this pressure. Where water has penetrated fissures or seeped into softer rock, a freeze followed by a thaw may result in rockfalls or earthslides, creating major destruction in minutes.

Over much longer periods, acidity in rain water breaks down the chemical composition of porous rocks such as limestone, eating away the rock to form deep caves and tunnels. Chemical decomposition also occurs in river beds and glacier valleys, hastening the process of mechanical erosion.

Like the sea, rivers and glaciers generate much of their effect through abrasion, pounding or tearing the land with the debris they carry. Yet as well as destroying existing landforms they also create new ones, many of them spectacular. Prominent examples include the Grand Canyon, the vast deltas of the Mississippi and the Nile, the rock arches and stacks off the south coast of Australia, and the deep fjords cut by glaciers in British Columbia, Norway and New Zealand.

While landscapes evolve from a "young" mountainous stage, through a "mature" hilly stage to an "old age" of lowland plain, this long-term cycle of erosion is subject to interruption by a number of crucial factors, including the pronounced effects of plate tectonics and climate change.

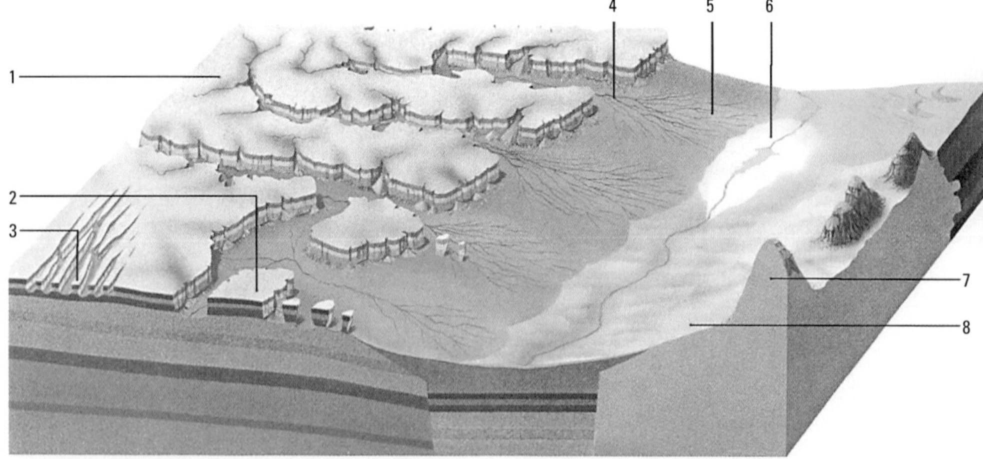

ABOVE The topography of a desert is characterized by the relative absence of the chemical weathering associated with water, and most erosion takes place mechanically through wind abrasion and the effect of heat – and cold.

Mesas [1] are large flat-topped areas with steep sides, while the butte [2] is a flat isolated hill, also with steep sides. Elongated in the direction of the prevailing wind, yardangs [3] comprise tabular masses of resistant rock resting on undercut pillars of softer material. Alluvial fans [5] are pebble-mounds deposited in desert deltas by flash floods, usually at the end of a wadi [4]. A saltpan [6] is a temporary lake of brackish water, also formed by flash floods. An inselberg [7] is an isolated hill rising from the plain, and a pediment [8] is a gently inclining rock surface.

Shaping forces: ice

Many of the world's most dramatic landscapes have been carved by icesheets and glaciers. During the Ice Ages of the Pleistocene epoch (over 10,000 years ago) up to a third of the land surface was glaciated; even today a tenth is still covered in ice – the vast majority locked up in the huge icesheets of Antarctica and Greenland.

Valley glaciers are found in mountainous regions throughout the world, except Australia. In the relatively short geological time scale of the recent Ice Ages, glaciers accomplished far more carving of the topography than rivers and wind.

They are formed from compressed snow, called névé, accumulating in a valley head or cirque. Slowly the glacier moves downhill, moving at rates of between a few millimetres and several metres a day, scraping away debris from the mountains and valleys through which it passes. The debris, or moraine, adds to the abrasive power of the ice. The sediments are transported by the ice to the edge of the glacier, where they are deposited or carried away by meltwater streams.

Shaping forces: rivers

From their origins as small upland rills and streams channelling rainfall, or as springs releasing water that has seeped into the ground, all rivers are incessantly at work cutting and shaping the landscape on their way to the sea.

In highland regions flow may be rapid and turbulent, pounding rocks to cut deep gorges and V-shaped valleys through softer rocks, or tumbling as waterfalls over harder ones.

As they reach more gentle slopes, rivers release some of the pebbles and heavier sediments they have carried downstream, flow more slowly and broaden out. Levées or ridges are raised along their banks by the deposition of mud and sand during floods. In lowland plains the river drifts into meanders, depositing layers of sediment, especially on the inside of bends where the flow is weakest. As the river reaches the sea it deposits its remaining load, and estuaries are formed where the tidal currents are strong enough to remove them; if not, the debris creates a delta.

Shaping forces: the sea

Under the constant assault from tides and currents, wind and waves, coastlines change faster than most landscape features, both by erosion and by the building up of sand and pebbles carried by the sea. In severe storms, giant waves pound the shoreline with rocks and boulders; but even in much quieter conditions, the sea steadily erodes cliffs and headlands, creating new features in the form of sand dunes, spits and salt marshes. Beaches, where sand and shingle have been deposited, form a buffer zone between the erosive power of the waves and the coast. Because it is composed of loose materials, a beach can rapidly adapt its shape to changes in wave energy.

Where the coastline is formed from soft rocks such as sandstones, debris may fall evenly and be carried away by currents from shelving beaches. In areas with harder rock, the waves may cut steep cliffs and wave-cut platforms; eroded debris is deposited as a terrace. Bays and smaller coves are formed when sections of soft rock are carved away between headlands of harder rock. These are then battered by waves from both sides, until the headlands are eventually reduced to rock arches, which as stacks are later separated from the mainland.

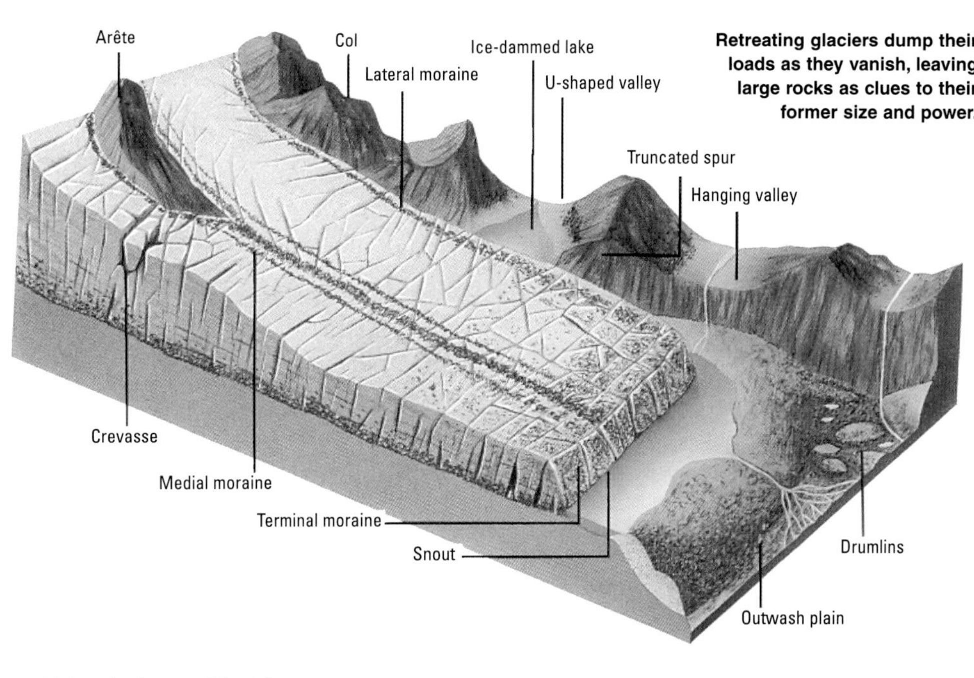

Retreating glaciers dump their loads as they vanish, leaving large rocks as clues to their former size and power.

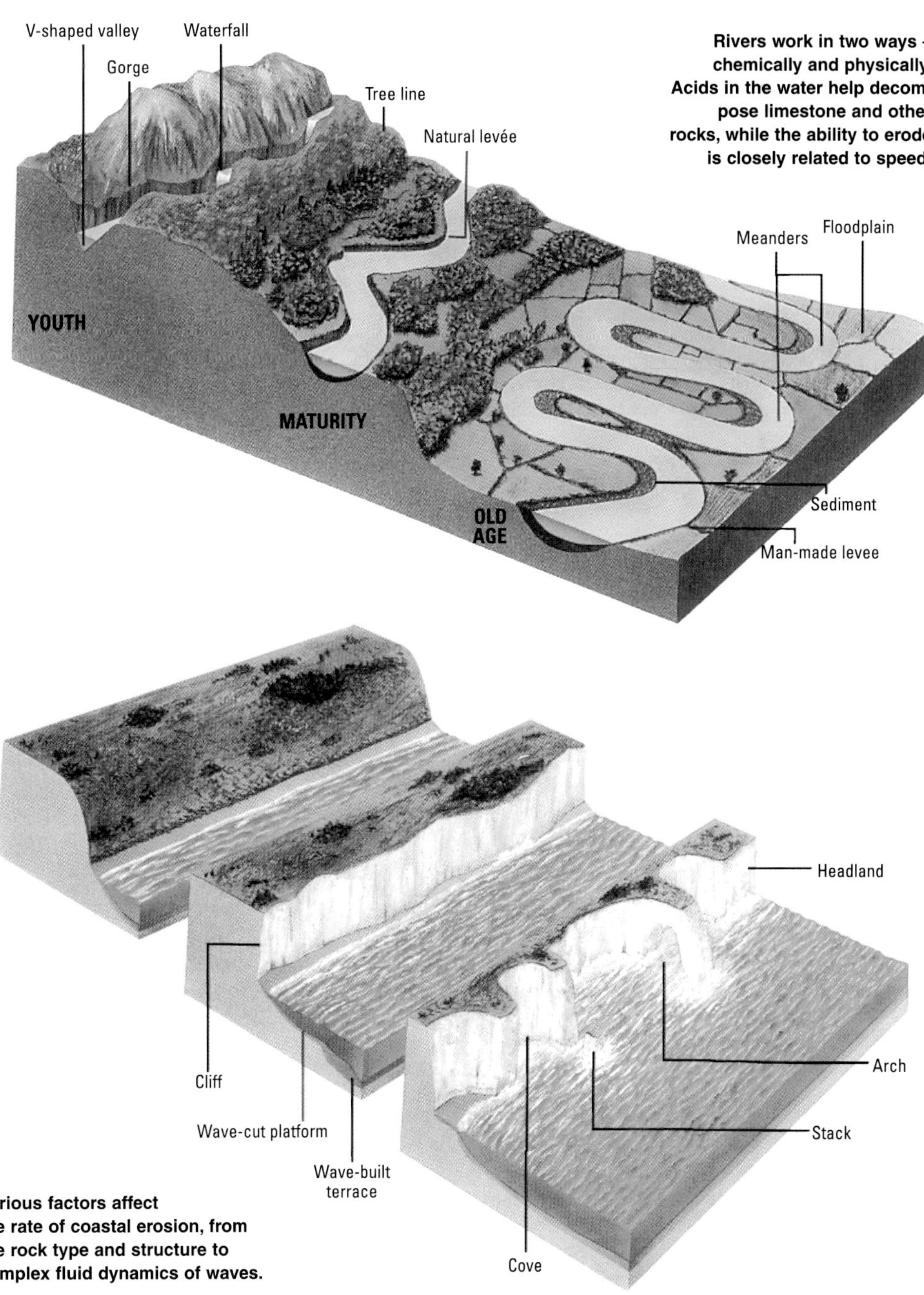

Rivers work in two ways – chemically and physically. Acids in the water help decompose limestone and other rocks, while the ability to erode is closely related to speed.

Various factors affect the rate of coastal erosion, from the rock type and structure to complex fluid dynamics of waves.

OCEANS: SEAWATER

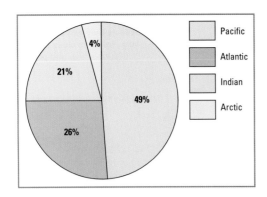

ABOVE In the strictest geographical sense the Earth has only three oceans – Atlantic, Indian and Pacific. The legendary "Seven Seas" would require these to be divided at the Equator and the addition of the smaller Arctic Ocean. Geographers do not recognize the Antarctic Ocean (much less the "Southern Ocean") as a separate entity.

The Earth is a watery planet: almost 71% of its surface is covered by its oceans and seas. This great liquid cloak gives our planet its characteristic and beautiful blue appearance from space, and is one of the two obvious differences between the Earth and its two near-neighbours, Venus and Mars. The other difference is the presence of life – and the two are closely linked.

ABOVE When sunlight strikes the surface of the ocean between 3% and 30% of it is immediately reflected. The amount reflected depends on the angle at which the light strikes – the smaller the angle the greater the reflection – which varies with latitude and the seasons.

Penetration of sunlight is selectively reduced according to wavelength. Radiation at the red or long-wave end of the visible spectrum is absorbed near the surface of the water, while the shorter blue wavelengths are scattered, giving the sea its characteristic blue colour.

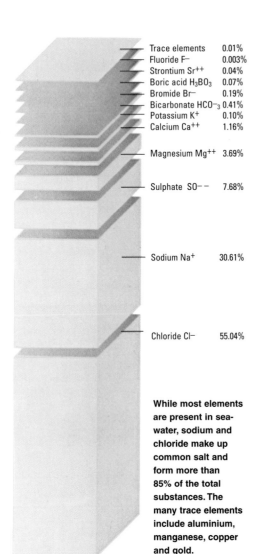

Trace elements	0.01%
Fluoride F–	0.003%
Strontium Sr++	0.04%
Boric acid H_3BO_3	0.07%
Bromide Br–	0.19%
Bicarbonate HCO_3^-	0.41%
Potassium K+	0.10%
Calcium Ca++	1.16%
Magnesium Mg++	3.69%
Sulphate SO– –	7.68%
Sodium Na+	30.61%
Chloride Cl–	55.04%

While most elements are present in seawater, sodium and chloride make up common salt and form more than 85% of the total substances. The many trace elements include aluminium, manganese, copper and gold.

EARTH IS something of a misnomer for our planet; "Ocean" would be a more suitable name, since the oceans and seas cover 70.8% of its total surface area. The oceans are not separate areas of water but form one continuous oceanic mass, and (as with some continental divisions) the boundaries between them are arbitrary lines drawn for convenience. The vast areas of interconnected oceans contain 97.2% of the world's total water supply.

The study of oceans, including their biology, chemistry, geology and physics, has now become a matter of urgency, because the future of humans on Earth may well depend on our knowledge of the ocean's potential resources not only of minerals and power but also of food.

Composition of seawater

The most obvious resource of the oceans is the water itself. But seawater is salty, containing sodium chloride (common salt), which makes it unsuitable for drinking or farming. One kilogramme (2.2lb) of seawater contains about 35g (1.2oz) of dissolved materials, of which chloride and sodium together make up nearly 30g (1oz) or about 85%.

Seawater is a highly complex substance in which 73 of the 93 natural chemical elements are present in measurable or detectable amounts. Apart from chloride and sodium it contains appreciable amounts of sulphate, magnesium, potassium and calcium, which together add up to over 13% of the total. The remainder, less than 1%, is made up of bicarbonate, bromide, boric acid, strontium, fluoride, silicon and various trace elements. Because the volume of the oceans is so great, there are substantial amounts of some trace elements: seawater contains more gold, for example, than there is on land, even though it's in a very low concentration of four-millionths of one part per million.

Also present in seawater are dissolved gases from the atmosphere, including nitrogen, oxygen and carbon dioxide. Of these, oxygen is vital to marine organisms. The amount of oxygen in seawater varies according to temperature. Cold water can contain more oxygen than warm water, but cold water in the ocean depths, which has been out of contact with the atmosphere for a long period, usually contains a much smaller amount of oxygen than surface water.

Other chemicals in seawater that are important to marine life include calcium, silicon and phosphates, all of which are used by marine creatures to form shells and skeletons. For building cells and tissue, marine organisms extract phosphates, certain nitrogen compounds, iron and silicon. The chief constituents of seawater – chloride, sodium, magnesium and sulphur – are hardly used by marine organisms.

Density, light and sound

The density of seawater is an important factor in causing ocean currents and is related to the interaction of salinity and temperature. The temperature of surface water varies between –2°C and 29°C (28°F and 85°F); ice will begin to form if the temperature drops below –2°C (28°F).

The properties of light passing through seawater determine the colour of the oceans. Radiation at the red or long-wave end of the spectrum is absorbed near the surface of the water, while the shorter blue wavelengths are scattered, giving the sea its characteristic colour.

The depth to which light can penetrate is important to marine life. In clear water light may reach to 110m (360ft), whereas in muddy coastal waters it may penetrate to only 15m (50ft). Below about 1,000m (3,300ft) there is virtually no light at all.

The most active zone in the oceans is the sunlit upper layer, falling to about 200m (650ft) at the edge of the continental shelf, where the water is moved around by windblown currents. This is the

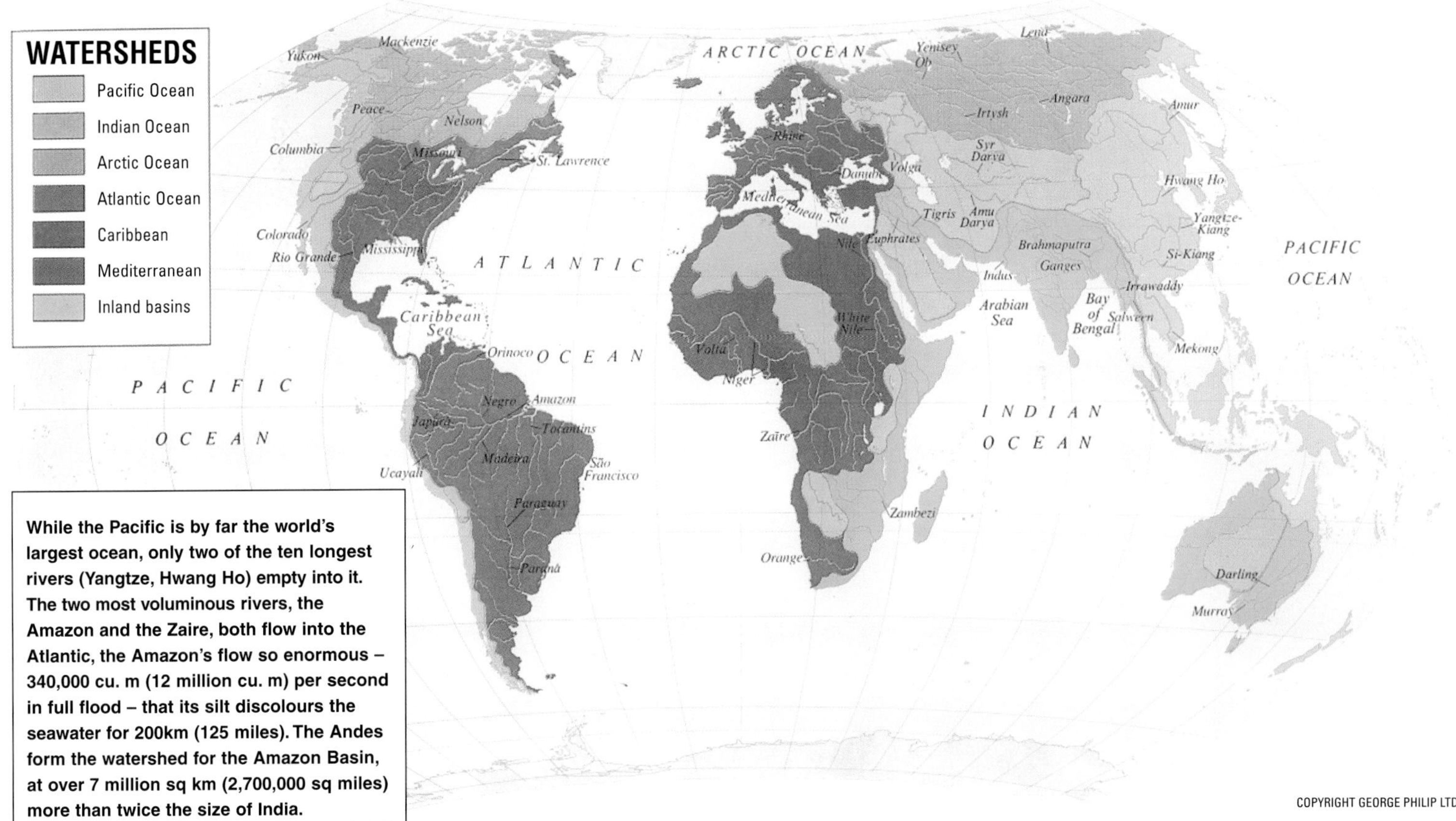

WATERSHEDS

- Pacific Ocean
- Indian Ocean
- Arctic Ocean
- Atlantic Ocean
- Caribbean
- Mediterranean
- Inland basins

While the Pacific is by far the world's largest ocean, only two of the ten longest rivers (Yangtze, Hwang Ho) empty into it. The two most voluminous rivers, the Amazon and the Zaire, both flow into the Atlantic, the Amazon's flow so enormous – 340,000 cu. m (12 million cu. m) per second in full flood – that its silt discolours the seawater for 200km (125 miles). The Andes form the watershed for the Amazon Basin, at over 7 million sq km (2,700,000 sq miles) more than twice the size of India.

home of most sealife and acts as a membrane through which the ocean breathes, absorbing great quantities of carbon dioxide and partly exchanging it for oxygen.

As the depth increases light fades and temperatures fall until just before around 950m (3,000ft), when there is a marked temperature change at the thermocline, the boundary between the warm surface zones and the cold deep zones.

Water is a good conductor of sound, which travels at about 1,507m (4,954ft) per second through seawater, compared with 331m (1,087ft) per second through air. Echo-sounding is based on the meas-

urement of the time taken for sound to travel from a ship to the seafloor and back again. However, temperature and pressure both affect the speed of sound, causing the speed to vary by about 100m (330ft) per second.

The salinity of the oceans

The volume of dissolved salts in seawater is called the salinity. The average salinity of seawater ranges between 33 and 37 parts of dissolved material per 1,000 parts of water. Oceanographers express these figures as 33 parts per thousand (33‰) to 37‰.

The salinity of ocean water varies with local

conditions. Large rivers or melting ice reduce salinity, for example, whereas it is increased in areas with little rainfall and high evaporation.

To produce fresh water from seawater the dissolved salts must be separated out. This desalination can be carried out by electrical, chemical and change of phase processes. Change of phase processes involve changing the water into steam and distilling it, or changing it into ice, a process that also expels the salt. Eskimos have used sea ice as a source of fresh water for centuries, while primitive coastal tribes still take salt from the sea by damming water in pools and letting it evaporate in the Sun.

RIGHT The average salinity of seawater ranges between 33 and 37 parts of dissolved material per 1,000 parts of water. While the most saline water is generally found in semi-enclosed seas in temperate and tropical areas such as the Gulf of Mexico, Mediterranean and the Red Sea (where high rates of evaporation can produce a figure of 41 parts per thousand), the Baltic Sea, which receives large quantities of freshwater from rivers and melting snow, has a remarkably low salinity of 7.2‰.

In the oceans themselves, the least saline waters occur in areas of high freshwater discharge such as the edge of the Antarctic icecap and the mouths of large rivers. The most pronounced regional example of this is Southeast Asia, where a string of massive rivers – including the Ganges, Brahmaputra, Irrawaddy, Salween, Mekong, Si Kiang, Yangtze and Hwang Ho – flow into the coastal area from the Bay of Bengal to the Yellow Sea.

If the salt in the oceans were precipitated, it would cover the Earth's land areas with a layer more than 150m (500ft) thick.

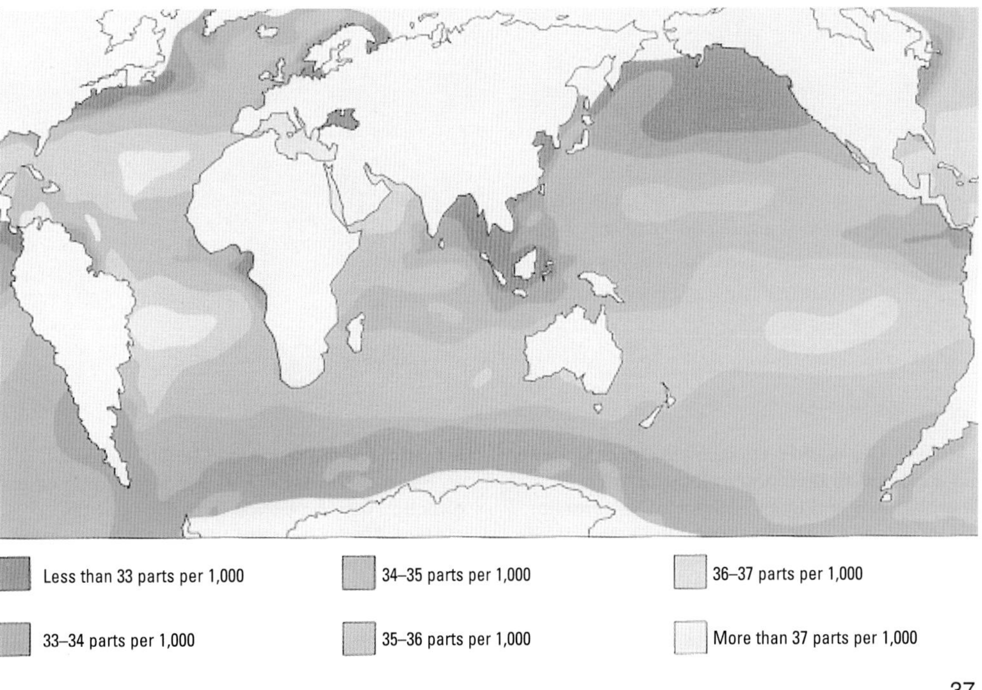

- Less than 33 parts per 1,000
- 33–34 parts per 1,000
- 34–35 parts per 1,000
- 35–36 parts per 1,000
- 36–37 parts per 1,000
- More than 37 parts per 1,000

THE OCEANS: CURRENTS

NO PART of the ocean is completely still – although, in the ocean depths, the movement of water is often extremely slow. Exploration of the deeper parts of the oceans has revealed the existence of marine life. If the water were not in motion, the oxygen on which all lifeforms depend would soon be used up and not replaced. No life would therefore be possible.

Prevailing winds sweep surface water along to form drift currents. These surface currents do not conform precisely with the direction of the prevailing wind because of the Coriolis effect caused by the rotation of the Earth. This effect, which increases away from the Equator, makes currents in the Northern Hemisphere veer to the right of the wind direction and currents in the Southern Hemisphere veer to the left. The result is a general clockwise circulation of water in the Northern Hemisphere

and an anticlockwise circulation in the Southern.

Other factors affecting currents are the configuration of the ocean bed and the shapes of land masses. For example, in the Atlantic Ocean the North Equatorial Current flows towards the West Indies. Most of this current is channelled into the Gulf of Mexico where it veers northeastwards, bursting into the Atlantic between Florida and Cuba as the Gulf Stream.

OCEAN CURRENTS

Winter in Northern Hemisphere

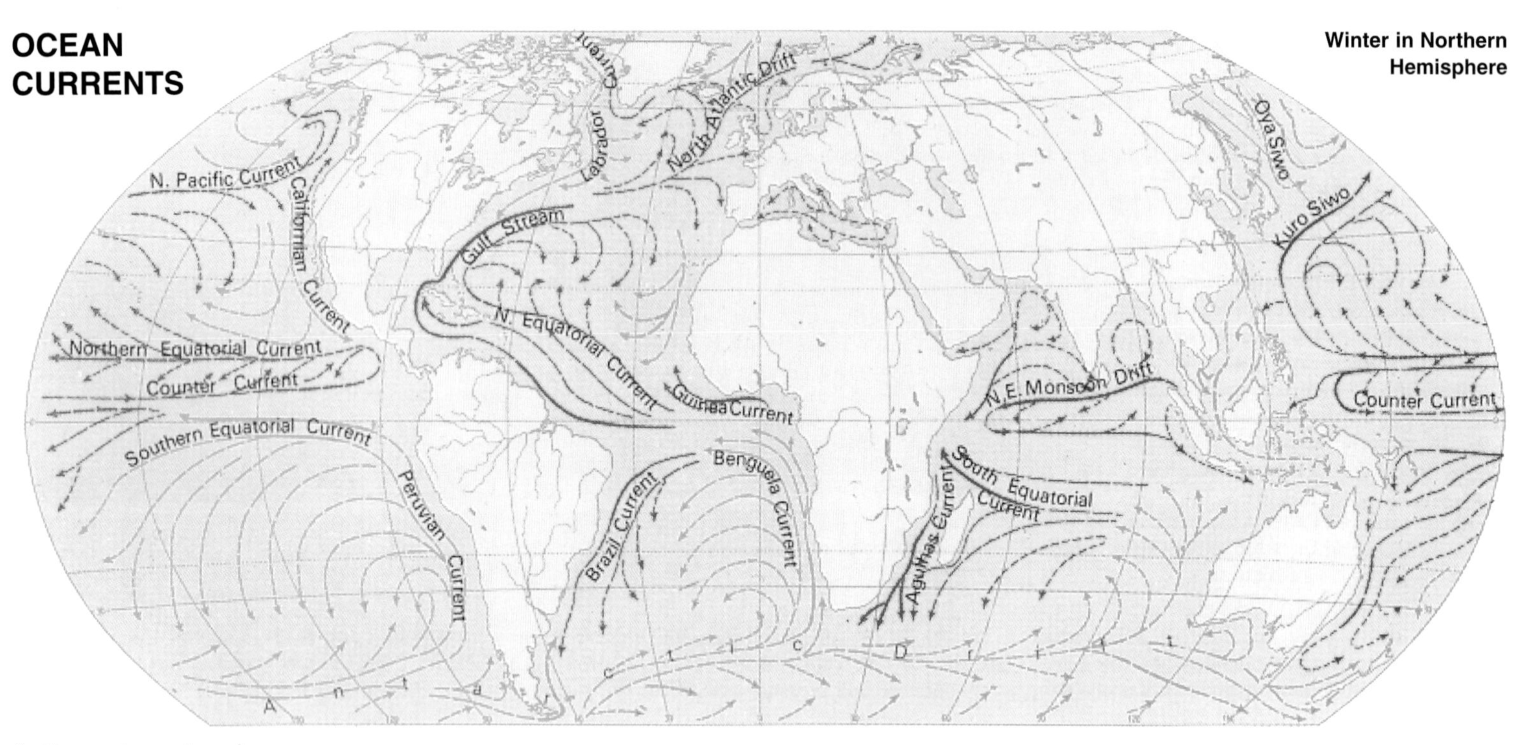

(cold currents are shown in blue, warm currents in red)

Summer in Northern Hemisphere

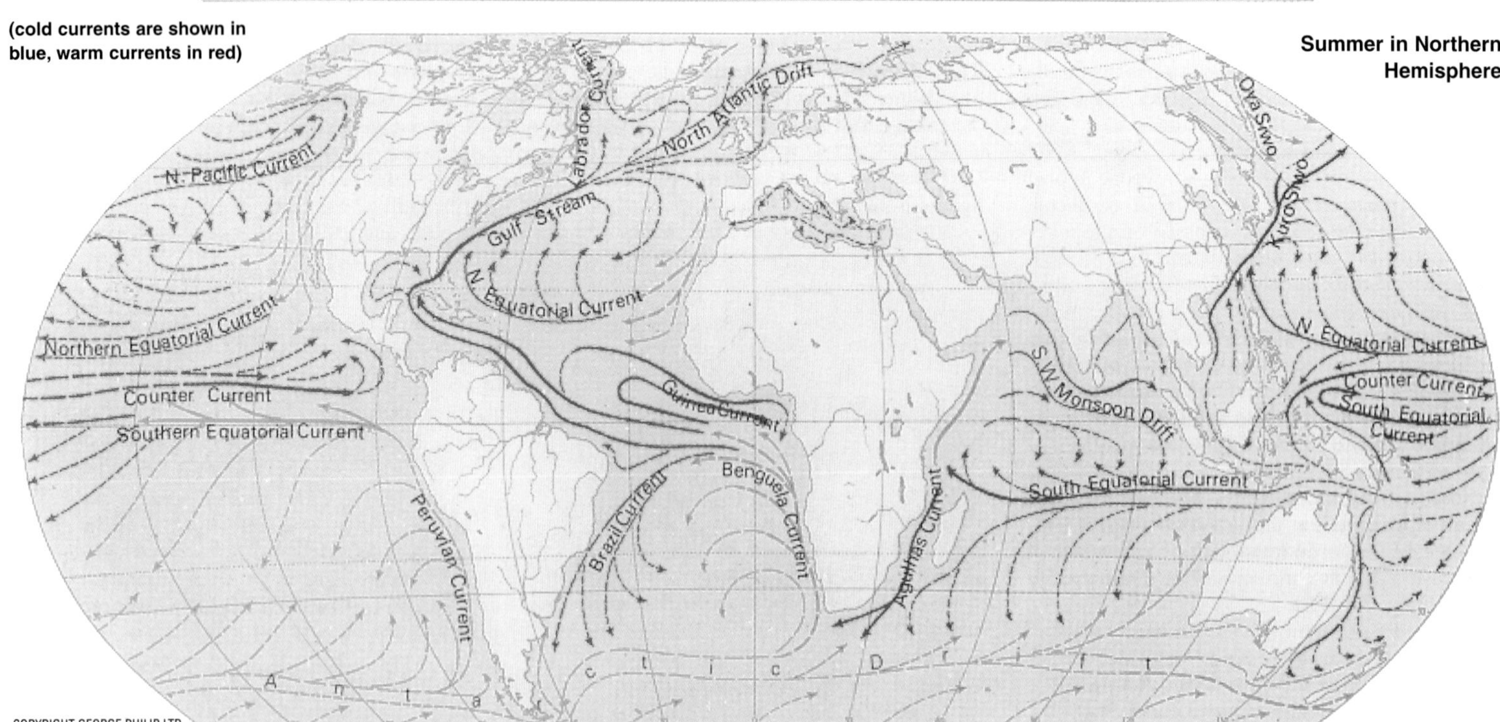

ABOVE The world's surface currents circulate in a clockwise direction in the Northern Hemisphere and in an anticlockwise direction in the Southern Hemisphere. These circulatory systems are called gyres. There are two large clockwise gyres in the Northern Hemisphere (North Atlantic and North Pacific) and three anticlockwise gyres in the Southern (South

Atlantic, South Pacific and Indian Ocean).

Beneath the surface are undercurrents whose direction may be opposite to those at the surface. Under the Gulf Stream off the eastern USA lies a large, cold current flowing south from the Arctic. The Gulf Stream finally splits: while the North Atlantic Drift branches past eastern Greenland and western Europe, part of

the current returns southwards to complete the gyre. Surface cold currents in the Northern Hemisphere generally flow southwards. In the Southern Hemisphere, cold water circulates around Antarctica, while offshoots flow northwards. The warm currents are very strong in tropical and subtropical regions, and include the various Equatorial currents.

The causes of currents that are not powered by winds are related to the density of ocean water, which varies according to temperature and salinity. Heating at the Equator causes the water to become less dense, while cooling round the poles has the opposite effect. Salinity is affected by the inflow of fresh water from rivers, melting ice and rainfall, and by evaporation. A high rate of evaporation increases the salinity and therefore the density.

Effects of ocean currents

One of the most important effects of ocean currents is that they mix ocean water and so affect directly the fertility of the sea. Mixing is especially important when subsurface water is mixed with surface water. The upwelling of subsurface water may be caused by strong coastal winds that push the surface water outwards, allowing subsurface water to rise up. Such upwelling occurs off the coasts of Peru, California and Mauritania, where subsurface water rich in nutrients (notably phosphorus and silicon) rises to the surface, stimulating the growth of plankton which provides food for great shoals of fish, such as Peruvian anchovies.

Water has a high heat capacity and can retain heat two and a half times as readily as land. The heat of the Sun absorbed by water around the Equator is transported north and south by currents. Part of the North Atlantic Drift flows past Norway, warming offshore winds and giving northwest Europe a winter temperature that is 11°C (20°F) above the average for those latitudes. The north-ward-flowing Peru and Benguela currents have a reverse effect, bringing cooler weather to the western coasts of South America and southern Africa.

In such ways, currents have a profound effect on climate. Currents from polar regions can also create hazards for shipping: the Labrador and East Greenland currents carry icebergs and pack ice into shipping lanes, and fog often occurs where cold and warm currents meet, most persistently off the coast of Newfoundland.

RIGHT Surface currents are caused largely by prevailing winds. The Coriolis effect caused by the rotation of the Earth results in the deflection of currents to the right of the wind direction in the Northern Hemisphere. In the same manner, the surface motion drags the subsurface layer at an angle to it, and so on.

Each layer moves at a slower speed than the one above it and at a greater angle from the wind. The spiral created has the overall effect of moving the water mass above the depth of frictional resistance at an angle of about 90° from the wind direction, while surface currents move at around 45°. The same effect reverses the direction of draining water from a bath in the Northern and Southern hemispheres.

Wind direction

Surface current

Net water mass transport

Depth of frictional resistance

4   2   1   3

18°C
16°C
14°C
12°C

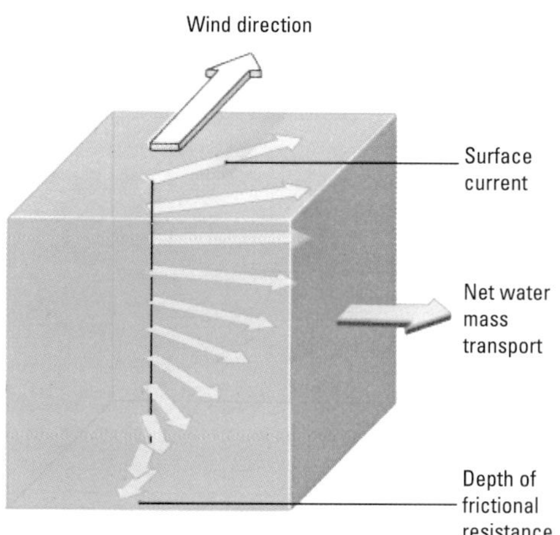

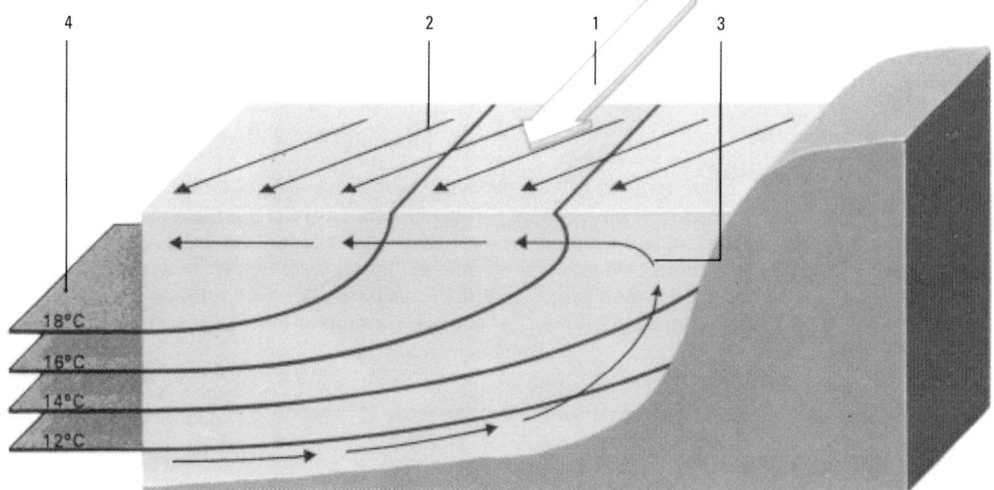

ABOVE Upwelling occurs when a longshore wind [1] pushes surface water away from a coast at an angle [2], allowing deeper water to rise [3]. The deeper water is not only colder [4] but usually rich in nutrients, and areas where upwelling occurs are often exceptional fishing grounds. A good example are the waters off the west coast of South America, the most productive in the world before the upwelling was suppressed by successive years of El Niño.

EL NIÑO

El Niño is the most dramatic and influential of current reversals, producing devastating effects. Its 1997–98 visit was the most damaging yet, triggering (among other things) floods and land-slides in northwest South America, storms in California, drought in southern Africa, monsoon failure in India and widespread rainforest fires in Southeast Asia. Estimates put the cost of the property damage alone as high as US$33 billion.

As the previous worst case in 1982–83 showed, the commercial cost is colossal and far-reaching. This includes declining fish stocks in the eastern Pacific (the anchovy catch dropped by over 90%), frost-wrecked orange groves in Florida, crop losses in Africa and brush fires in Australia.

While El Niño is unpredictable in both power and frequency – it used to appear every 2 to 7 years – the phenomenon now occurs more often, including five consecutive seasons from 1990 to 1994. It usually lasts about three months but can be far longer. The name was originally given by Peruvian fishermen to the warm but weak current that flowed south for a few weeks each year around Christmas – hence the name, which means "Christ child". Now the term is applied to a complex if irregular series of remarkable natural happenings.

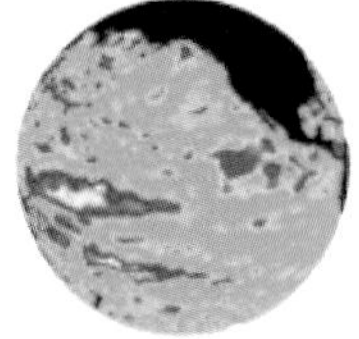

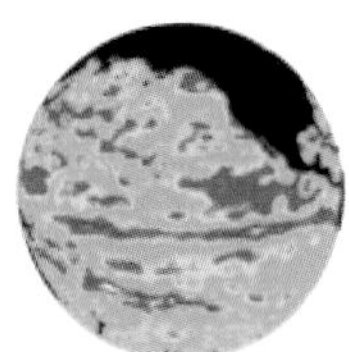

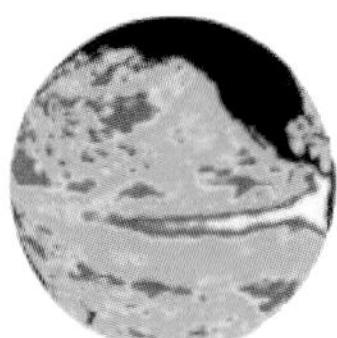

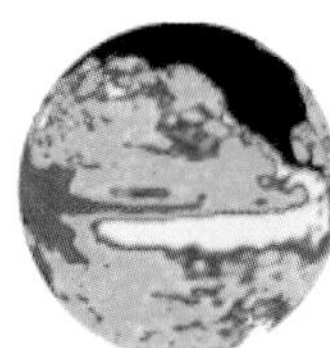

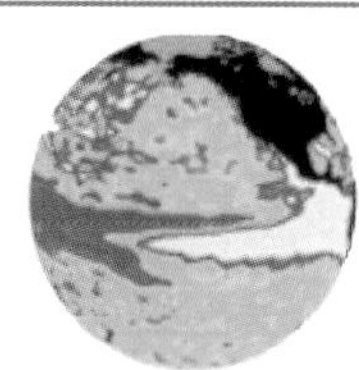

The El Niño sequence begins in the western Pacific. The mass of warm water (white in centre), 8°C higher than in the east and generally kept in check by the prevailing westerly trade winds, breaks free of its moorings as these winds subside and moves in an equatorial swell towards South America. There it raises both sea temperatures and sea levels, suppressing the normal upwelling of the cold and nutrient-rich Peruvian current. Meanwhile, the western Pacific waters cool (purple) as the warm water is displaced.

The movement of such vast amounts of warm water results in chaotic changes to wind patterns, in turn creating freak weather conditions well outside the Pacific tropics. The passage of El Niño's warm water is also tracked by rainfall, leading to droughts in Southeast Asia and Australia and excessive levels of precipitation in South America.

This sequence traces El Niño's passage from March 1997 *(above left)* to October 1998 *(below right)*, when normal oceanic conditions were finally resumed – until the next time. In 1983, 1987 and 1995 El Niño was followed by the cool current La Niña ("the little girl").

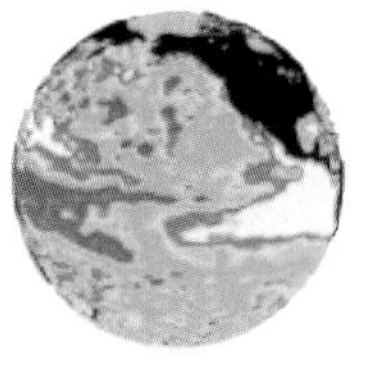

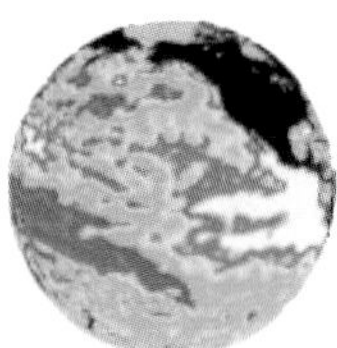

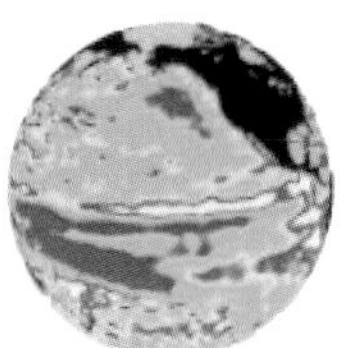

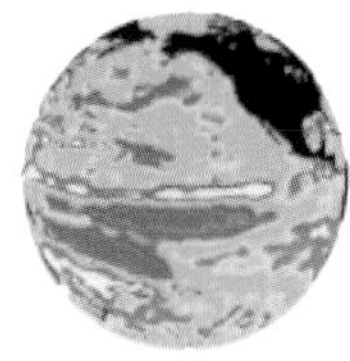

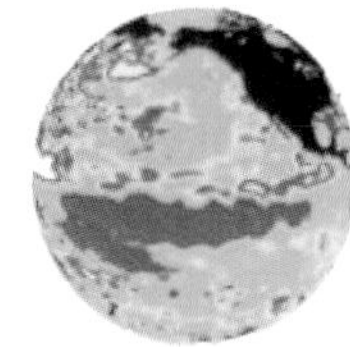

OCEANS: WAVES AND TIDES

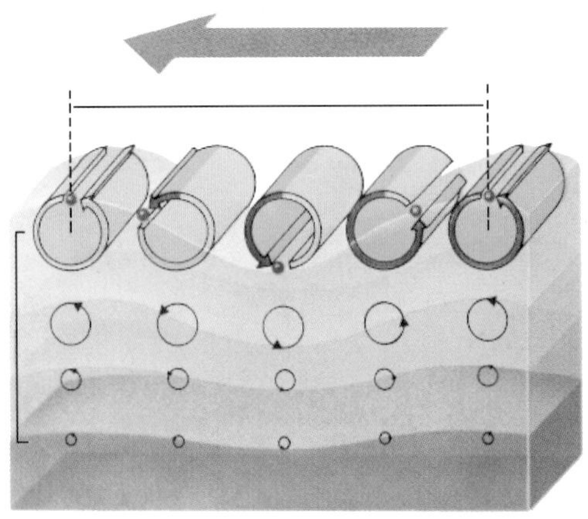

ABOVE Most waves are generated by the wind. As a wave travels in deep water, however, the water particles don't move up and down but rotate in circular orbits. As depth increases, the rotations of the water particles diminish rapidly – the reason why submarines escape the effects of severe storms at sea.

BECAUSE THEY affect coastal areas, waves and tides are the most familiar features of oceans and seas for most of us, but sometimes the energies of waves, tides and high winds combine with devastating effect. In January 1953 a high spring tide, storm waves and winds of 185km/h (115mph) combined to raise the level of the North Sea by 3m (10ft) higher than usual. This "surge" caused extensive flooding in eastern England, but in the Netherlands over 4% of the country was inundated: 1,800 people died and about 30,000 houses were destroyed or damaged by the seawater.

The motion of waves

While some wave motion occurs at great depth along the boundary of two opposing currents, most waves are caused by the wind blowing over an open stretch of water. This area where the wind blows is known as the "fetch". Waves there are confused and irregular and are referred to as a "sea". As they propagate beyond the fetch they combine into more orderly waves to form a "swell", which travels for long distances beyond

the fetch. Waves are movements of oscillation – that is, the shape of the wave moves across the water, but the water particles rotate in a circular orbit with hardly any lateral movement. As a result, if there is no wind or current, a corked bottle bobs up and down in the waves, but is more or less stationary.

At sea, waves seldom exceed 12m (40ft) in height, although one 34m (115ft) high was accurately measured in the Pacific Ocean in 1933. Such a wave requires a long fetch measuring thousands of kilometres and high-speed winds.

Waves that break along a seashore may have been generated by storms in mid-ocean or by local winds. As a wave approaches shallow water, which is defined as a depth of half a wave length, it "feels" the bottom, gradually slowing down, and the crests tend to crowd together. When the water in front of a wave is insufficient to fill the wave form, the rotating orbit – and hence the wave – breaks. There are two main kinds of breakers: spilling breakers occur on gently sloping beaches, when the crests spill over to form a mass of surf, while plunging breakers occur on steeper slopes.

BELOW Waves have dimensions of both length and height. The wave length [14] is the distance between one crest [5] and another – in this case a peaking wave [4] – and between two crests is a trough [11]. The wave height [6] is

the distance between the crest and the trough. If wave action ceased, the water would settle at the "still water level" [8]. Wave action extends to the wave base [7], where rotation becomes negligible. Wave distortion is caused by frictional

drag on the seabed: if waves pass over a sandbar [10], a spilling breaker [9] may form. Sometimes, waves in shallow water move the whole body of the water forward in translation waves [2] towards the shore [1].

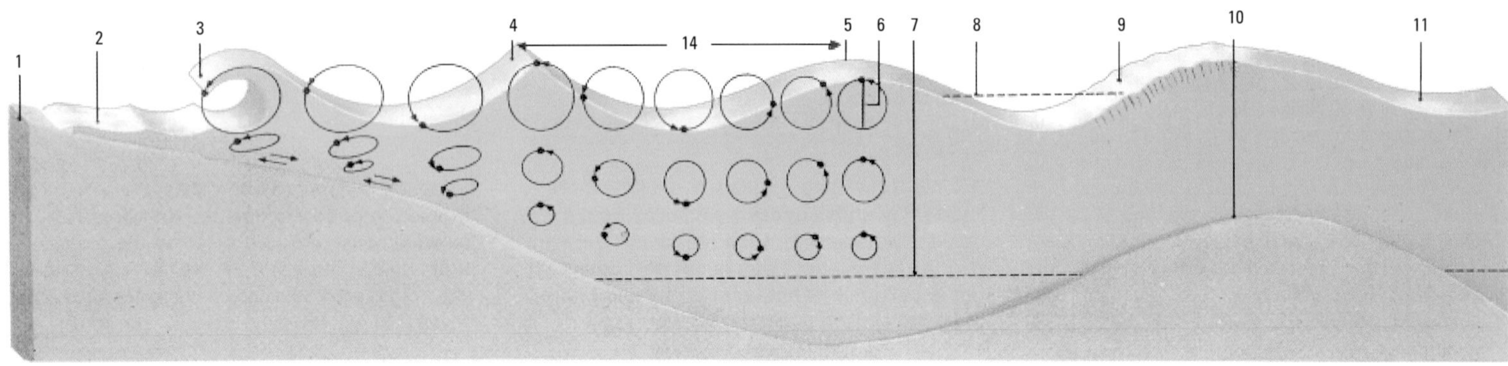

RIGHT Tides are the alternate rises and falls of the sea's surface level, caused by the gravitational pull of the Moon and the Sun. Although the Moon is much smaller than the Sun, it is much closer to Earth and its effect on the oceans is more than twice that of the Sun. The configurations of coasts and seafloors can accentuate these forces, while barometric pressure and wind effects can also superimpose an added "surge" element.

In the open sea the tidal range is small and in enclosed basins, such as the Mediterranean, it is little more than 30cm (12in). However, in shallow seas it may be more than 6m (20ft) and in tidal estuaries 12–15m (40–50ft). The highest tidal range recorded is about 16m (53ft) in the Bay of Fundy, which divides the peninsula of Nova Scotia from the Canadian mainland of New Brunswick.

In some 60 estuaries, such as Hangchow Bay in China and the Severn in England, tidal bores occur. These are bodies of water with a wall-like front that surge up rivers, formed because the estuaries act as funnels, leading to a rise in the height of the water. At spring tides the Hangchow bore attains heights of 7.5m (25ft) and speeds of 27km/h (17mph).

SPRING TIDE

Tidal bulge

Tidal bulge

Earth

Moon

Sun

When the Moon and Sun are roughly in the same direction (around the time of the new Moon), they each pull the oceans on the near side of the Earth towards them. They also pull the Earth towards them, away from the oceans on the far side of the Earth. The effect is to produce two bulges on opposite sides of the Earth. These will not rotate with the Earth but will stay with the forces that produced them, causing two high tides and two low tides a day.

NEAP TIDE

Moon

Tidal bulge

Earth

Tidal bulge

The effect is the same if the Moon and Sun are aligned on opposite sides of the Earth (at the time of full Moon). Thus the tides are greatest at the time of new Moon and full Moon (spring tides). Tides are less pronounced when the Sun, Moon and Earth are not aligned, and are least strong when the three are at right-angles to each other (near the Moon's first and third quarters). In this situation solar and lunar forces compete: the lunar tide wins, but the difference between high and low tides is much less (neap tides).

OCEANS: THE SEAFLOOR

THE DEEP ocean floor was once thought to be flat, but maps compiled from readings made by sonar equipment show that it's no more uniform than the surface of the continents. Here are not just the deepest trench – the Challenger Deep of the Pacific's Mariana Trench plunges 11,022m (36,161ft) – but also the Earth's longest mountain chains and its tallest peaks.

The vast underwater world starts in the shallows of the seaside. Surrounding the land masses is the shallow continental shelf, composed of rocks that less dense than the underlying oceanic crust. The shelf drops gently to around 200m (650ft), where the seafloor suddenly falls away at an angle of 3° to 6° via the continental slope. Submarine canyons such as the 1.5km (5,000ft) gorge off Monterey, California are found on the continental slopes. They can be caused either by river erosion before the land was submerged by the sea, or by turbidity currents – underwater avalanches that carry mud, pebbles and sand far out to sea, scouring gorges out of both slope rock and sediment.

The third stage, the continental rise, made up of sediments washed down from the shelves, is more gradual, with gradients varying from 1 in 100 to 1 in 700. At an average depth of 5,000m (9,000ft) there begins the aptly named abyssal plain, massive submarine depths where sunlight fails to penetrate and only creatures specially adapted to deal with the darkness and the pressure can survive.

Underwater highlands

While the abyss contains large plains it is broken by hills, volcanic seamounts and mid-ocean ridges. Here new rock is being continually formed as magma rises through the Earth's crust, pushing the tectonic plates on each side apart towards the continents in the process called seafloor spreading.

Taken from base to top, many of the seamounts which rise from these plains rival and even surpass the biggest of continental mountains in height. Mauna Kea, Hawaii's highest peak, reaches 10,203m (33,475ft), some 1,355m (4,380ft) more than Mount Everest, though only 4,205m (13,795ft) is above sea level. Nearby is Mauna Loa, the world's biggest active volcano, over 84% of which is hidden from view.

Life in the ocean depths

Manned submersibles have now established that life exists even in the deepest trenches, where the pressure reaches 1,000 "atmospheres" – the equivalent of the force of a tonne bearing down on every sq cm (6.5 tons per sq in).

Further exploration in the pitch-black environment of the oceanic ridges has revealed extraordinary forms of marine life around the scalding hot vents: creatures include giant tubeworms, blind shrimps, and bacteria, some of which are genetically different from any other known lifeforms.

In 1996 an analysis of one micro-organism revealed that at least half its 1,700 or so genes were hitherto unknown. Based on chemicals, not sunlight, this alien environment may well resemble the places where life on Earth first began.

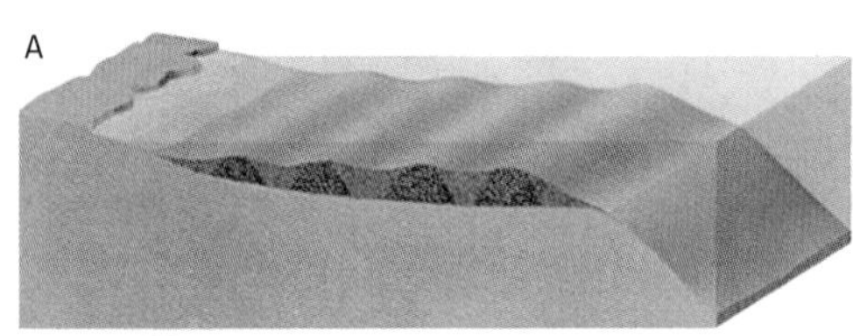
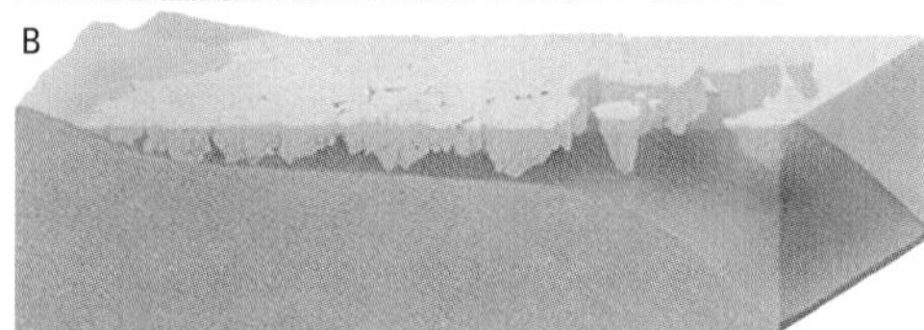
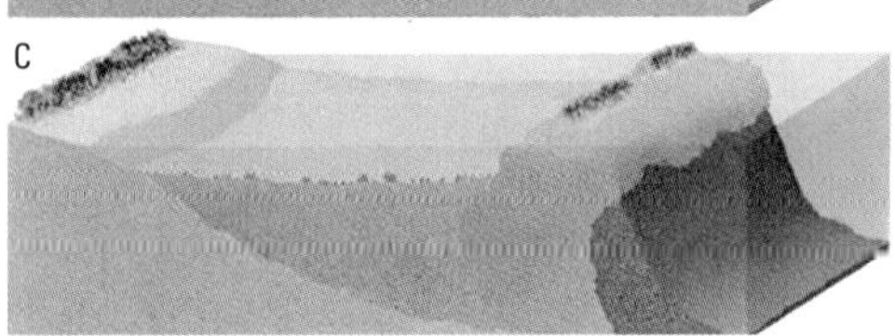

ABOVE Continental shelves are the regions immediately off the land masses, and there are several different types. Off Europe and North America the shelf has a gentle relief, often with sandy ridges and barriers [A]. In high latitudes, floating ice wears the shelf smooth [B], and in clear tropical seas a smooth shelf may be rimmed with a coral barrier such as the Great Barrier Reef off eastern Australia, leaving an inner lagoon area "dammed" by the reef [C].

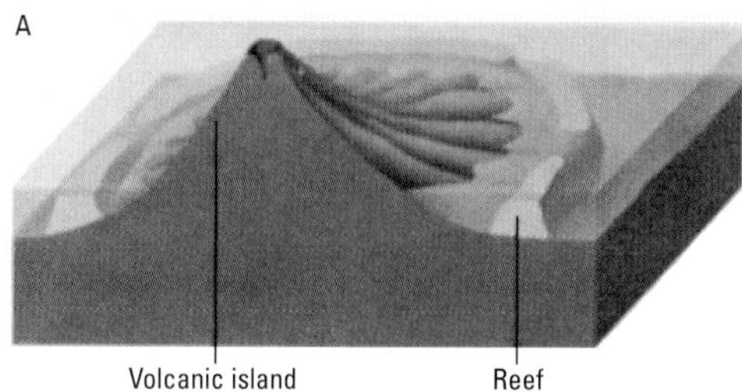

Volcanic island   Reef

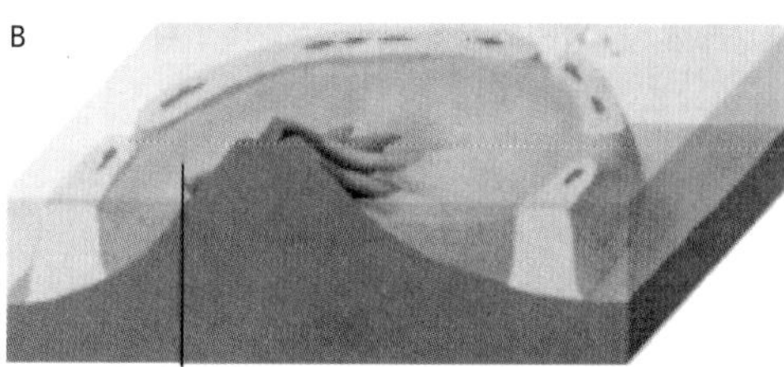

Reef lagoon

Low islands   Reef and detritus

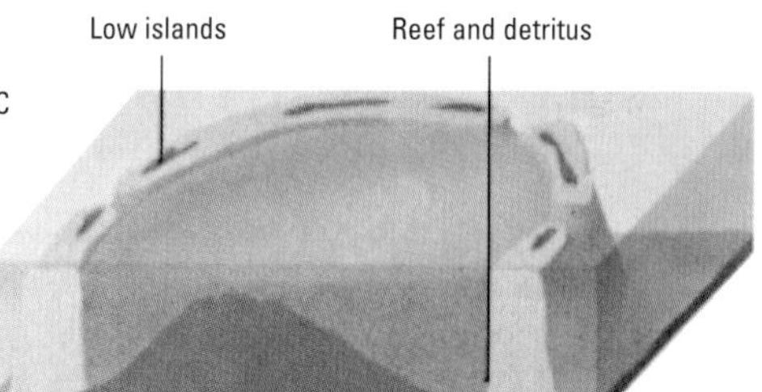

ABOVE The most intriguing of coral features, an atoll is a ring or horseshoe-shaped group of coral islands. Organisms with skeletons of calcium carbonate, corals grow in warm, fairly shallow water to depths of about 90m (300ft), but the depth of coral in many atolls is much greater than this.

The prevailing theory is that the coral began to form as a reef in the shallows of a volcanic island [A]. While the sea level began to rise and the island slowly sank [B], the coral growth kept pace with these gradual changes, leaving an atoll of hard limestone around its remnant [C]. In this way, coral can reach depths of up to 1,600m (5,250ft).

The world's largest atoll is Kwajalein in the Marshall Islands, in the central Pacific. Its slender 283-km (176-mile) coral reef encloses a lagoon of 2,850 sq km (1,100 sq miles).

RIGHT The seafloor consists of different zones, the most shallow being the continental shelf that lies between the coast and the 200m (650ft) depth contour. The shelf area occupies 7.5% of the seafloor and corresponds to the submerged portion of the continental crust. Beyond, the downward slope increases abruptly to form the continental slope (8.5%), an area that may be dissected by submarine canyons. The continental slope meets the abyssal basins at a more gentle incline (the continental rise). The basins lie at depths of 4,000m (13,200 ft) and feature mountain ranges and hills.

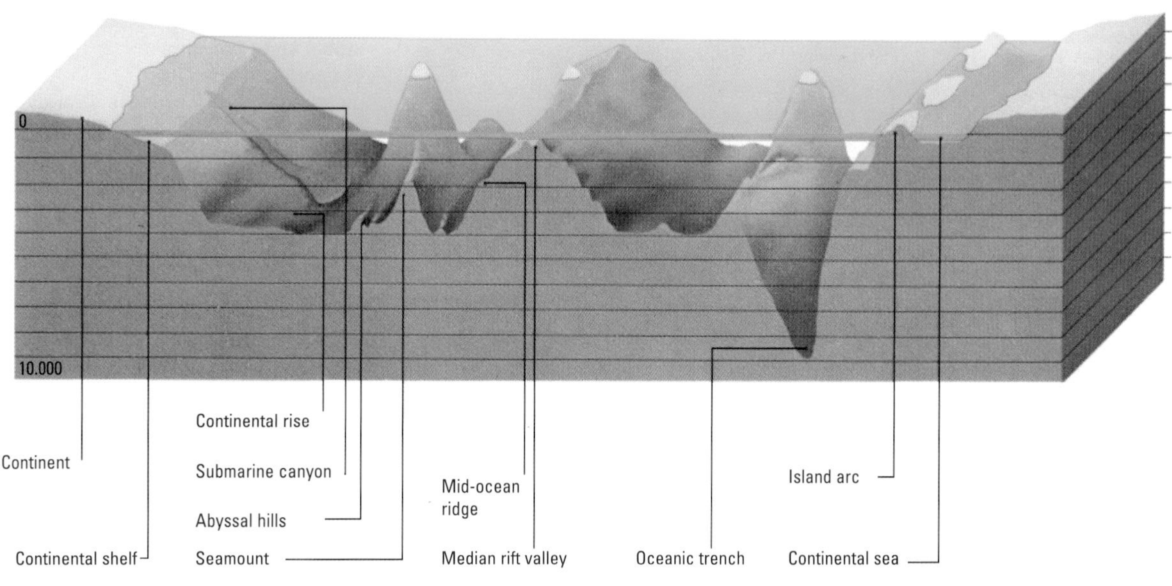

0

10.000

Continent

Continental shelf

Continental rise

Submarine canyon

Abyssal hills

Seamount

Mid-ocean ridge

Median rift valley

Oceanic trench

Island arc

Continental sea

THE ATMOSPHERE AND CLOUDS

THE ORIGIN of the atmosphere was closely associated with the origin of the Earth. When the Earth was still a molten ball, it was probably surrounded by a large atmosphere of cosmic gases, including hydrogen, that were gradually lost into space. As the Earth began to develop a solid crust over a molten core, gases such as carbon dioxide, nitrogen and water vapour were slowly released to form an atmosphere with a composition not unlike the present emissions from volcanoes. Further cooling probably led to massive precipitation of water vapour – so that today it occupies less than 4% by volume of the atmosphere. At a much later stage, the oxygen content of the atmosphere was created by green plants releasing oxygen.

Extending from the Earth's surface far into space, the atmosphere is a meteor shield, a radiation deflector, a thermal blanket and a source of chemical energy for the Earth's diverse lifeforms. Five-sixths of its total mass is located in the first 15 km (9 miles), the troposphere, which is no thicker in relative terms than the skin of an onion. Almost all the phenomena we call the weather occur in this narrow layer.

RIGHT Because air is easily compressed, the atmosphere becomes "squashed" by gravity. Thus the bulk of the atmosphere lies in the troposphere, occupying a volume of about 6 billion cu. km (1,560 million cu. miles). As air density decreases with altitude, the very much smaller amounts of air present in the stratosphere (19%) and the ionosphere and above (1%) occupy an increasingly greater volume.

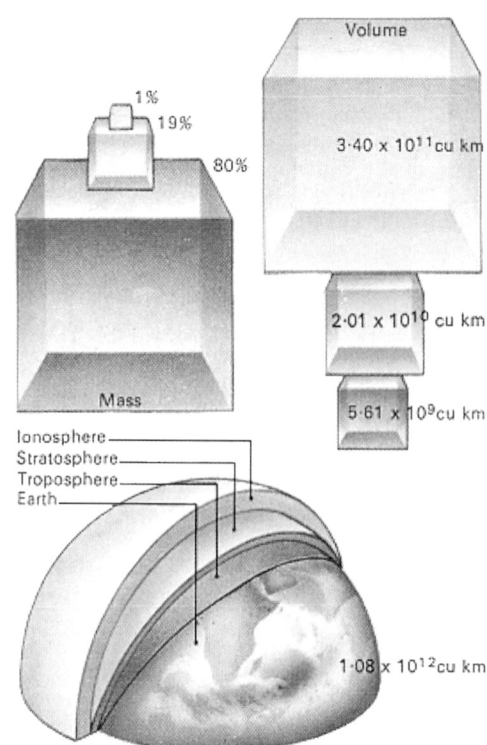

LEFT The discovery by British scientists of the hole in the ozone layer over Antarctica in 1985 triggered a growing interest in the structure of the atmosphere.

LAYERS OF THE ATMOSPHERE

1. EXOSHERE
The atmosphere's upper layer has no clear outer boundary, merging imperceptibly with interplanetary space. Its lower boundary, at an altitude of around 400km (250 miles), is almost equally vague. The exosphere is mainly composed of hydrogen and helium in changing proportions: helium vanishes with increasing altitude, and above 2,400km (1,500 miles) it is almost entirely hydrogen.

2. IONOSPHERE
Gas molecules in the ionosphere, mainly helium, oxygen and nitrogen, are ionized – electrically charged – by the Sun's radiation. Within the ionosphere's range of 50 to 400km (30 to 250 miles) they group themselves into four layers, known conventionally as D, E, F1 and F2, all of which can reflect radio waves of differing frequencies. The high energy of ionospheric gas gives it a notional temperature of more than 2000°C (3,600°F), although its density is negligible. The auroras – *aurora borealis* and its southern counterpart, *aurora australis* – occur in the ionosphere when charged particles from the Sun interact with the Earth's magnetic fields, at their strongest near the poles.

3. STRATOSPHERE
Separated at its upper and lower limits by the distinct thresholds of the stratopause and the tropopause, the stratosphere is a remarkably stable layer between about 15km and 50km (9 and 30 miles). Its temperature rises from –55°C (–67°F) at its lower extent to approximately 0°C (32°F) near the stratopause, where a thin layer of ozone – increasingly depleted with the acceleration in pollution by CFCs since the 1970s – absorbs ultraviolet radiation believed to cause skin cancer, cataracts and damage to the immune system in humans. Stratospheric air contains enough ozone to make it poisonous, although it is far too rarified to breathe. Overall, the stratopsher comprises 80% nitrogen, 18% oxygen, 1% argon and 1% ozone. "Mother-of-pearl" or nacreous cloud occurs at about 25km (15 miles).

4. TROPOSPHERE
The narrowest of all the atmospheric layers, the troposphere extends up to 15km (9 miles) at the Equator but only 8km (5 miles) at the poles. Since this thin region contains about 85% of the atmosphere's total mass and almost all of its water vapour, it is also the realm of the Earth's weather. Temperatures fall steadily with increasing height by about 1°C for every 100 metres (1.5°F for every 300 feet) above sea level. The main constituents are nitrogen (78%), oxygen (21%) and argon (1%).

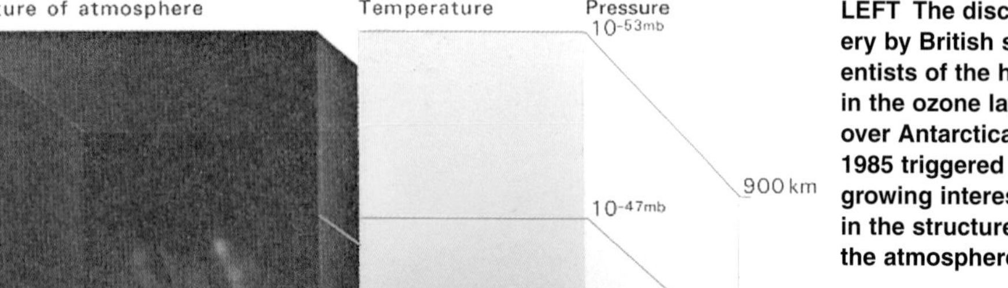

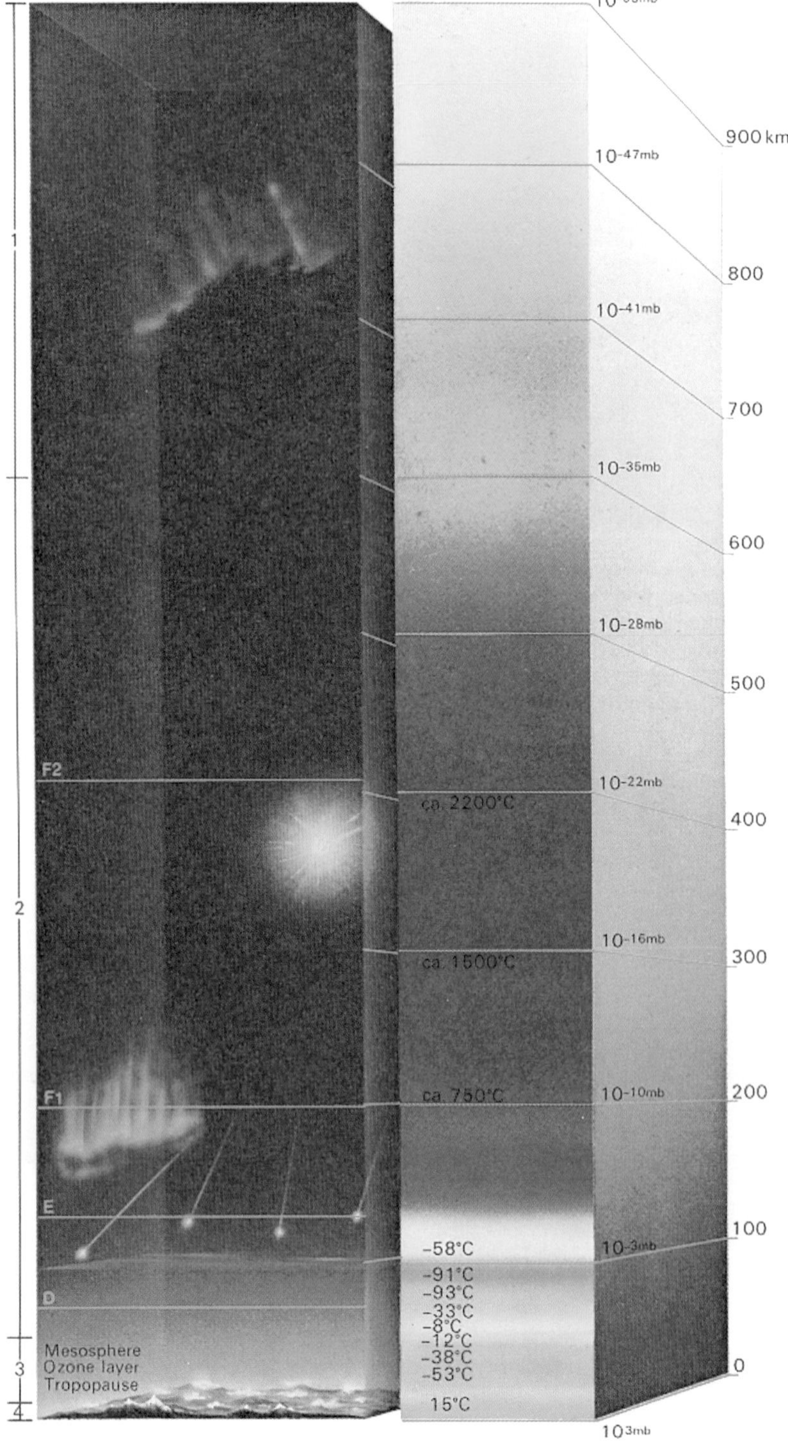

Structure of atmosphere	Temperature	Pressure

ca. 2200°C

ca. 1500°C

ca. 750°C

–58°C
–91°C
–93°C
–33°C
–8°C
–12°C
–38°C
–53°C

15°C

Mesosphere
Ozone layer
Tropopause

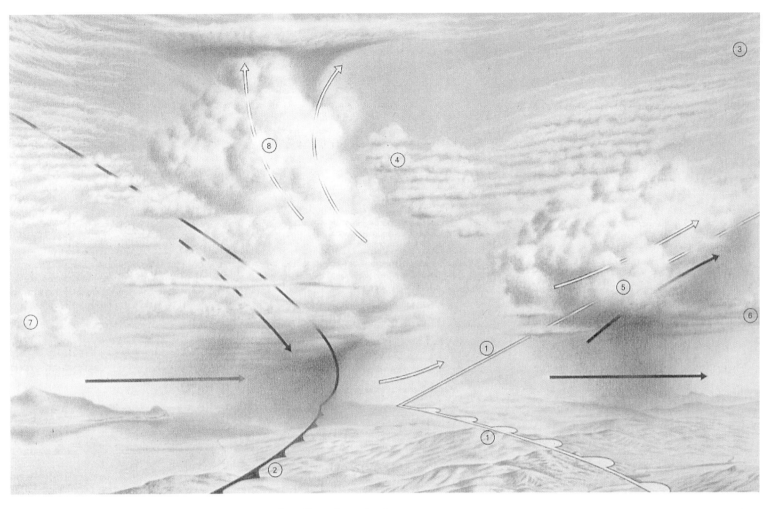

ABOVE The different cloud types are best illustrated within the context of the familiar mid-latitude frontal depression. Here a schematic, generalized Northern Hemisphere depression is viewed from the south as it moves eastwards, with both warm [1] and cold [2] fronts clearly visible. Over the warm front the air rises massively and slowly over the great depth of the atmosphere. This results in a fairly complete suite of layer-type clouds ranging from ice-crystal cirrus [3] and fluffy altocumulus [4] to grey-based nimbostratus [5].

The precipitation area often associated with such cloud types, and especially with nimbo-stratus, usually lies ahead of the surface warm front and roughly parallel to it [6]. Turbulence may cause some clouds to rise and produce heavy convective rainfall, as well as the generally lighter and more widespread classical warm front rainfall. Stratus often occupies the warm sector, but a marked change occurs at the cold front. Here the wind veers (blowing in a more clockwise direction) and cumulus clouds [7], brilliant white in sunlight, are often found in the cold air behind the front.

At the front itself the atmosphere is often unstable and cumulus clouds grow into dramatic cumulonimbus formations [8]. The canopy of cirrus clouds – of all types – may extend over the whole depression and is often juxtaposed with the anvil shape of the nimbus. These cloud changes are accompanied by changes in pressure, wind temperature and humidity as the fronts pass.

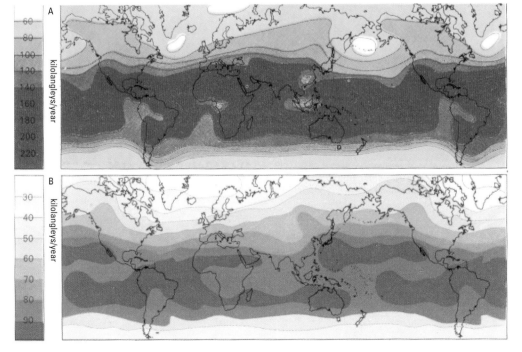

LEFT Temperatures in the atmosphere and on Earth result mainly from a balance of radiation inputs and outputs. Average annual solar radiation reaching the Earth, measured in kilolangleys – one calorie absorbed per sq cm (0.15 sq in) is highest in hot desert areas [A].

Comparison with the average annual long-wave radiation back from the Earth's surface [B] shows an overall surplus radiation for nearly all latitudes, but this is absorbed in the atmos-phere and then lost in space, ensuring an overall balance. The extreme imbalance of incoming radiation between equatorial and polar latitudes is somewhat equalized through heat transfers by atmosphere and oceans. This balancing transfer between surplus and deficit radiation is greatest in mid-latitudes, where most cyclones and anticyclones occur.

WINDS AND THE WEATHER

WIND IS the movement of air, and large-scale air movements, both horizontal and vertical, are crucial in shaping weather and climate. The chief forces affecting horizontal air movements are pressure gradients and the Coriolis effect.

Pressure gradients are caused by the unequal heating of the atmosphere by the Sun. Warm equatorial air is lighter and therefore has a lower pressure than cold, dense, polar air. The strength of air movement from areas of high to low pressure – known as the pressure gradient – is proportional to the difference in pressure.

Along the Equator is a region called the doldrums, where the Sun's heat warms the rising air. This air eventually spreads out and flows north and south away from the Equator. It finally sinks at about 30°N and 30°S, creating subtropical high-pressure belts (the horse latitudes), from which trade winds flow back towards the Equator and westerlies flow towards the mid-latitudes.

The Coriolis effect is the deflection of winds caused by the Earth's rotation, to the right in the Northern Hemisphere and to the left in the Southern. As a result, winds don't flow directly from the point of highest pressure to the lowest; those approaching a low-pressure system are deflected round it rather than flowing directly into it. This creates air systems, with high or low pressure, in which winds circulate round the centre. Horizontal air movements are important around cyclonic (low-pressure) and anticyclonic (high-pressure) systems. Horizontal and vertical movements combine to create a pattern of prevailing global winds.

Weather and depressions

To most of us "weather" means rain and sunshine, heat and cold, clouds and wind. Humidity and visibility might be added to the list. If not precise in terminology, this layman's catalogue comprises the six main elements which also comprise weather for meteorologists: in their language they are precipitation, air temperature, cloud cover, wind velocity, humidity and barometric pressure.

Depressions occur when warm air flows into waves in a polar front while cold air flows in behind it, creating rotating air systems that bring changeable weather. Along the warm front (the boundary on the ground between the warm and cold air), the warm air flows upwards over the cold air, producing a sequence of clouds that help forecasters predict a depression's advance.

Along the cold front the advancing cold air forces warm air to rise steeply, and towering cumulonimbus clouds form in the rising air. When the cold front overtakes the warm front, the warm air is pushed up to form an occluded front. Cloud and rain persist along occlusions until temperatures equalize, the air mixes, and the depression dies out.

BELOW The world's zones of high and low pressure are both areas of comparative calm, but between them lie the belts of prevailing winds. West of Africa, wind patterns are remarkably constant between summer and winter, but in much of the east variations are caused by monsoons (reversals of wind flows) stemming in part from the unequal heating of land masses and the sea.

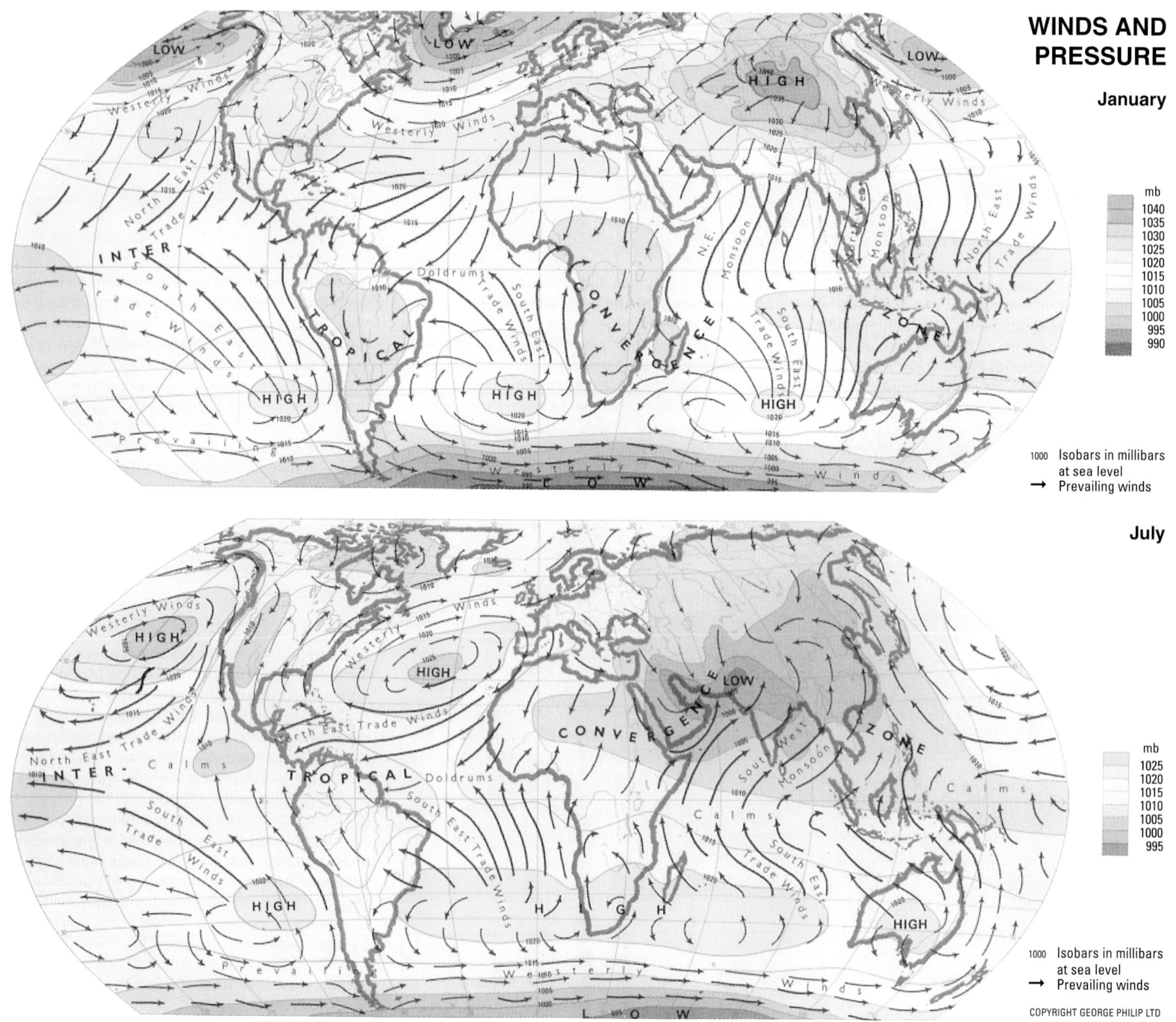

WINDS AND PRESSURE

January

mb
1040
1035
1030
1025
1020
1015
1010
1005
1000
995
990

1000 Isobars in millibars at sea level
→ Prevailing winds

July

mb
1025
1020
1015
1010
1005
1000
995

1000 Isobars in millibars at sea level
→ Prevailing winds

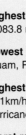

Legend

Warm air
Cool air
Cold air

▲▲▲▲ Warm front
△△△△ Cold front

H = High pressure
L = Low pressure

ABOVE The Earth's atmosphere acts as a giant heat engine. The temperature differences between the poles and the Equator provide the thermal energy to drive atmospheric circulation, both horizontal and vertical. In general, warm air at the Equator rises and moves towards the poles at high levels and cold polar air moves towards the Equator at low levels to replace it. Air also flows north and south from the high-pressure belts called the horse latitudes, and these airflows meet up with cold, dense air flowing from the poles along the polar front.

The basic global pattern of prevailing winds is complicated by the rotation of the Earth (which causes the Coriolis effect), by cells of high-pressure and low-pressure systems (depressions) and by the distribution and configuration of land and sea.

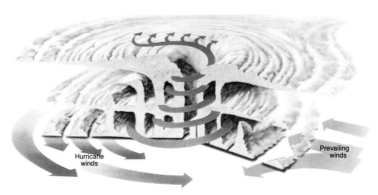

ABOVE Hurricanes consist of a huge swirl of clouds rotating round a calm centre – the "eye" – where warm air is sucked down. Hurricanes may be 400km (250 miles) in diameter and they extend through the troposphere, which is about 15-20km (9-12 miles) thick. Clouds, mainly cumulonimbus, are arranged in bands round the eye, the tallest forming the wall of the eye. Cirrus clouds usually cap the hurricane.

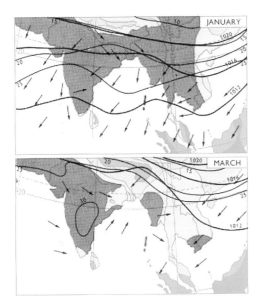

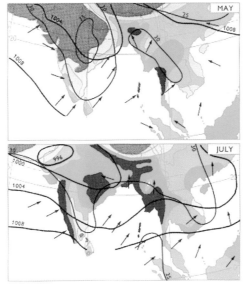

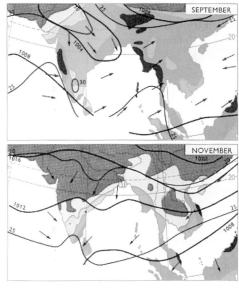

ABOVE "Monsoon" is the term given to the seasonal reversal of wind direction, most noticeably in South and Southeast Asia, where it results in very heavy rains. In January a weak anticylone in northern India gives the clear skies brought by northeasterly winds; in March temperatures increase and the anticyclone subsides, sea breezes bringing rain to coastal areas; by May the north is hot and a low pressure area begins to form, while the south is cooler with some rain; in July the low-pressure system over India caused by high temperatures brings the Southwest Monsoon from the high-pressure area in the south Indian Ocean; in September the Southwest Monsoon – with its strong winds, cloud cover, rain and cool temperatures – begins to retreat from the northwest; by the end of the cycle in November the subcontinent is cool and dry, though still wet in the southeast.

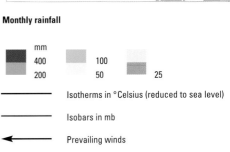

Monthly rainfall

mm
400
200
100
50
25

——— Isotherms in °Celsius (reduced to sea level)

——— Isobars in mb

← Prevailing winds

WORLD CLIMATE

CLIMATE IS weather in the longer term, the seasonal pattern of hot and cold, wet and dry, averaged over time. Its passage is marked by a ceaseless churning of the atmosphere and the oceans, further agitated by the Earth's rotation and the motion it imparts to moving air and water.

There are many classifications of world climate, but most are based on a system developed in the early 19th century by the Russian meteorologist Vladimir Köppen. Basing his divisions on two main features, temperature and precipitation, and using a code of letters, he identified five main climatic types: tropical (A), dry (B), warm temperate (C),

cool temperate (D) and cold (E). Each of these main regions was then further subdivided. (A highland mountain category was added later to account for the variety of climatic zones found in mountainous areas due to changes caused by altitude.)

Although latitude is a major factor in determining climate, other factors add to the complexity. These include the influence of ocean currents, different rates of heating and cooling of land and ocean, distance from the sea, and the effect of mountains on winds. New York, Naples and the Gobi Desert all share the same latitude, for example, but their climates are very different.

Climates are not stable indefinitely. Our planet regularly passes through cool periods – Ice Ages probably caused by the recurring long-term oscillations in the Earth's orbital path from almost circular to elliptical every 95,000 years, variations in the Earth's tilt from $21^1/2°$ to $24^1/2°$ every 42,000 years, and perhaps even fluctuations in the Sun's energy output. In the present era, the Earth is closest to the Sun in the middle of winter in the Northern Hemisphere and furthest away in summer; 12,000 years ago, at the height of the last Ice Age, northern winter fell with the Sun at its most distant.

Studies of these cycles suggest that we are now

CLIMATIC ZONES

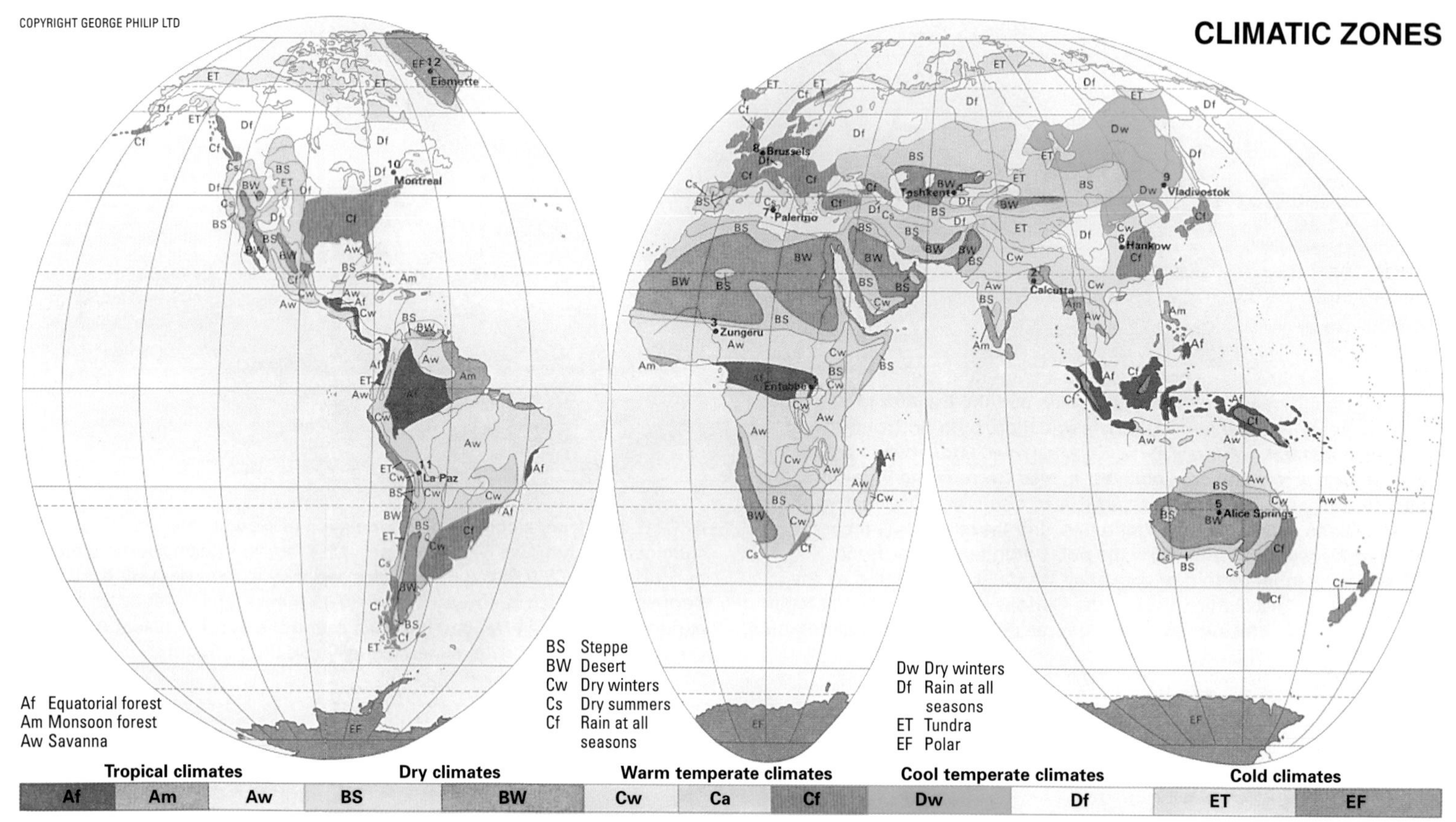

Af Equatorial forest
Am Monsoon forest
Aw Savanna

BS Steppe
BW Desert
Cw Dry winters
Cs Dry summers
Cf Rain at all seasons

Dw Dry winters
Df Rain at all seasons
ET Tundra
EF Polar

Tropical climates			Dry climates		Warm temperate climates			Cool temperate climates		Cold climates	
Af	Am	Aw	BS	BW	Cw	Ca	Cf	Dw	Df	ET	EF

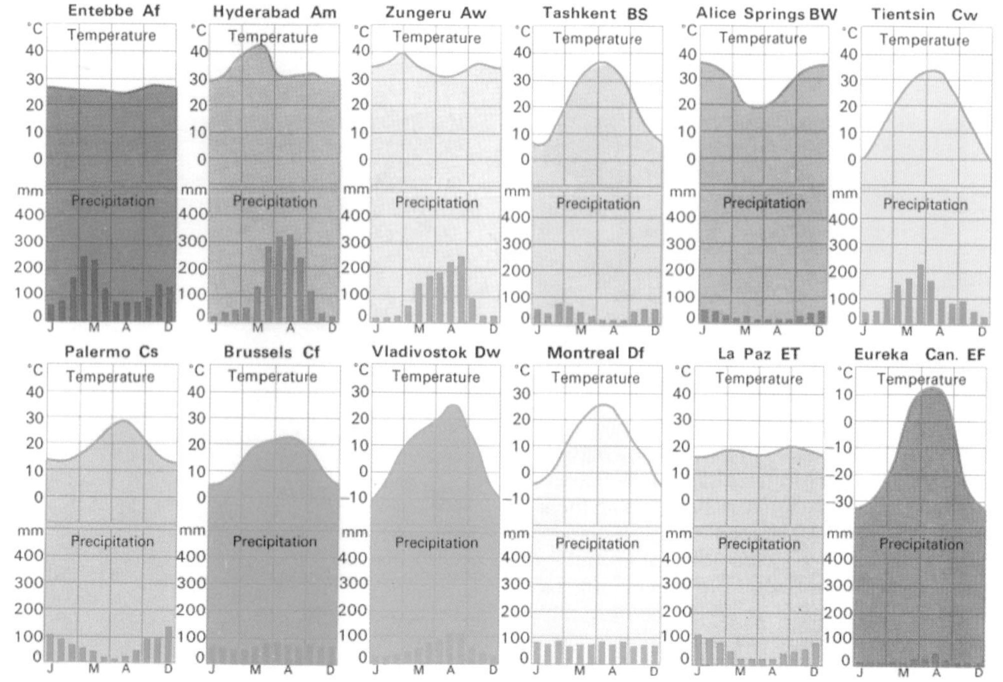

CLIMATE TERMS

Cyclone: Violent storm called a hurricane in N. America and a typhoon in the Far East.
Depression: Area of low pressure.
Frost: Dew when the air temperature falls below freezing point.
Hail: Frozen rain.
Humidity: Amount of moisture in the air.
Isobar: Line on a map connecting places of equal atmospheric pressure.
Isotherm: Line on a map connecting places of equal temperatutre.
Precipitation: Measurable amounts of rain, snow, sleet or hail.
Rain: Precipitation of liquid particles with diameter larger than 0.5mm (0.02 in); under this size is classified as drizzle.
Sleet: Partially melted snow.
Snow: Crystals formed when water vapour condenses below freezing point.
Tornado: Severe funnel-shaped storm that twists as hot air spins vertically; called a waterspout at sea.

in an interglacial period, but with a new glacial period on the way. For the forseeable future, however, the planet is likely to continue heating up because of global warming, caused largely by the burning of fossil fuels and deforestation. Figures show that average temperatures rose 1.7°C (0.9°F) in the 20th century, with most of that increase coming after about 1970, and despite attempts to stabilize the situation it's likely that the trend will continue. Such changes would not redraw Köppen's divisions, but they would make a significant difference to many local climates, with a dramatic effect on everything from agriculture to architecture.

CLIMATE RECORDS

TEMPERATURE

Highest recorded shade temperature: Al Aziziyah, Libya, 58°C (136.4°F), 13 Sep. 1922.

Highest mean annual temperature: Dallol, Ethiopia, 34.4°C (94°F), 1960-66.

Longest heatwave: Marble Bar, Western Australia, 162 days over 37.8°C (100°F), 23 October 1923 to 7 April 1924.

Lowest recorded temperature: Vostock Station, Eastern Antarctica, 21 July 1985, –89.2°C (–128.6°F)

(Lowest recorded temperature (outside poles): Verkhoyansk, Siberia, –68°C (–90°F), 6 February 1933.

Lowest mean annual temperature: Plateau Station, Antarctica, –56.6°C (–72.0°F).

PRECIPITATION

Longest drought: Calama, N. Chile – no recorded rainfall in 400 years to 1971.

Wettest place (12 months): Cherrapunji, Meghalaya, NE. India, 26,470mm (1,040 in), August 1860 to August 1861; Cherrapunji also holds the record for the most rainfall in a month: 2,930mm (115 in), July 1861.

Wettest place (average): Tututendo, Colombia, mean annual rainfall of 11,770mm (463.4 in).

Wettest place (24 hours): Cilaos, Réunion, Indian Ocean, 1,870mm (73.6 in), 15-16 March 1952.

Heaviest hailstones: Gopalganj, Bangladesh, up to 1.02kg (2.25lb), 14 April 1986 (92 people were killed).

Heaviest snowfall (continuous): Bessans, Savoie, France, 1,730mm (68 in) in 19 hours, 5-6 April 1969.

Heaviest snowfalls (season/year): Paradise Ranger Station, Mt Rainier, Washington, USA, 31,102mm (1,224.5 in), 19 February 1971 to 18 February 1972.

Conversions
°C = (°F -32) x 5/9; °F = (°C x 9/5) + 32; 0°C = 32°F
1 mm = 0.0394 in (100 mm = 3.94 in); 1 in = 25.4 mm

TEMPERATURE

Average temperature in January

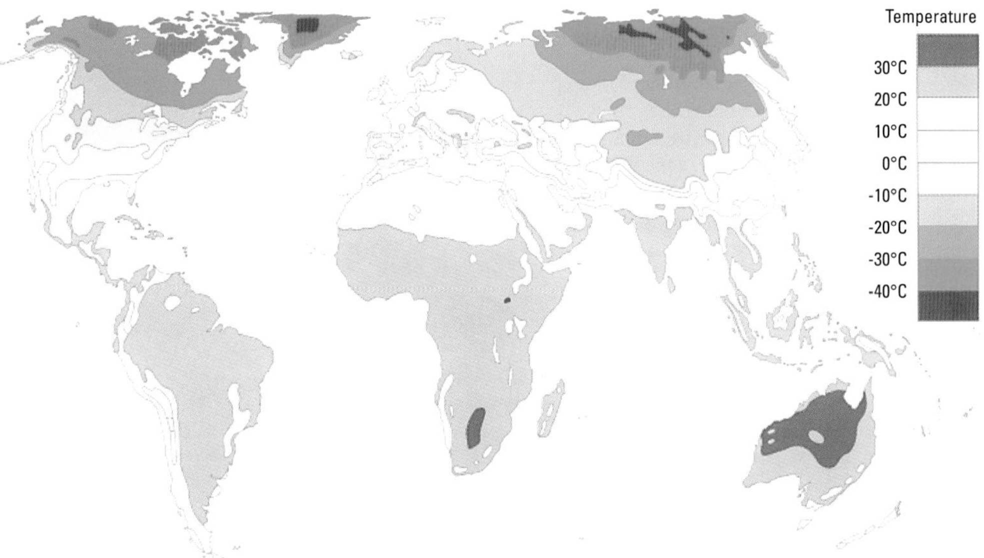

Temperature

30°C
20°C
10°C
0°C
-10°C
-20°C
-30°C
-40°C

Average temperature in July

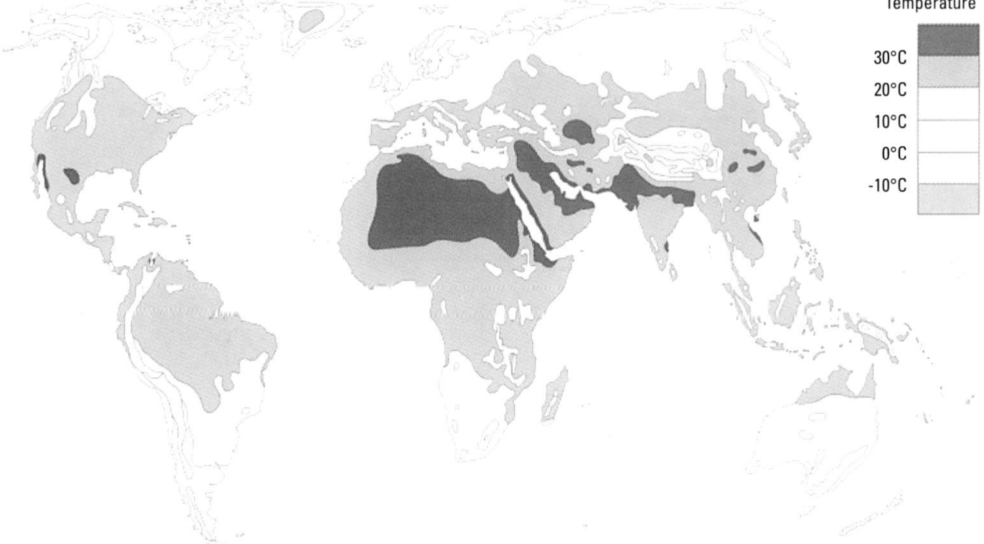

Temperature

30°C
20°C
10°C
0°C
-10°C

RAINFALL

Average annual precipitation

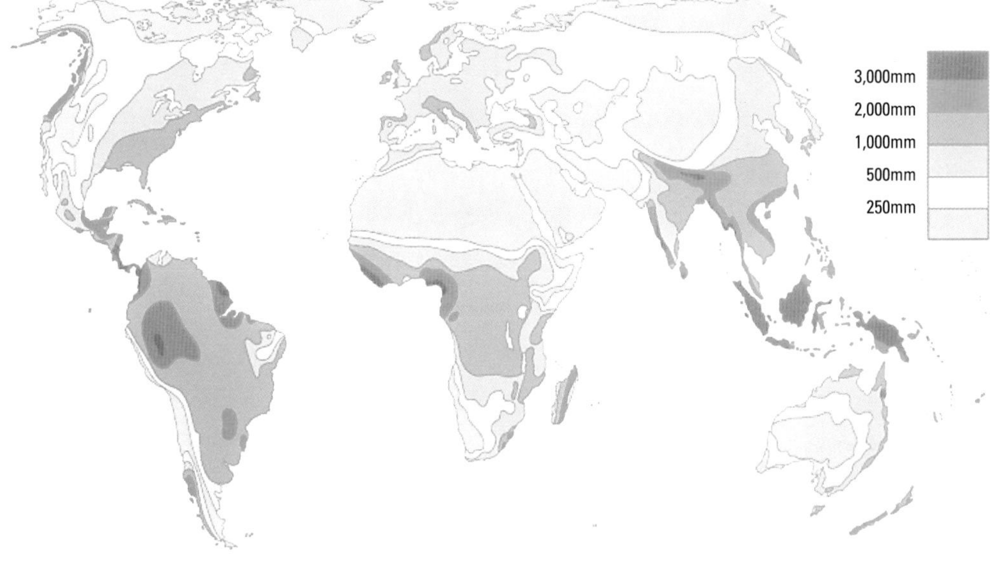

3,000mm
2,000mm
1,000mm
500mm
250mm

THE WORLD'S WATER

FRESH WATER is essential to all life on Earth, from the humblest bacterium to the most advanced technological society. Yet freshwater resources form a minute fraction of our 1.36 billion cu. km (326 million cu. miles) of water: most human needs must be met from the 2,000 cu. km (480 cu. miles) circulating in rivers.

Agriculture accounts for huge quantities: without large-scale irrigation, most of the world's people would starve. Since fresh water is just as essential for most industrial processes, the combination of growing population and advancing industry has put supplies under increasing strain.

Fortunately water is seldom used up: the planet's water cycle circulates it efficiently, at least on a global scale. More locally, however, human activity can cause severe shortages: water for industry and agriculture is being withdrawn from many river basins and underground aquifers faster than natural recirculation can replace it – a process exacerbated by global warming.

The demand for water has led to tensions between an increasing number of nations as supplies are diverted or hoarded. Both Iraq and Syria, for example, have protested at Turkey's dam-building programme, which they claim drastically reduces the flow of Tigris and Euphrates water to their land.

The water cycle

Oceanic water is salty and unsuitable for drinking or farming. In some desert regions, where fresh sources are in short supply, seawater is desalinated to make fresh water, but most of the world is constantly supplied with fresh water by the natural process of the water or hydrological cycle, which relies on the action of two factors: gravity and the Sun's heat.

Over the oceans, which cover almost 71% of the Earth's surface, the Sun's heat causes evaporation, and water vapour rises on air currents and winds. Some of this vapour condenses and returns directly to the oceans as rain, but because of the circulation of the atmosphere, air bearing large amounts is carried over land, where it falls as rain or snow.

Much of this precipitation is quickly re-evaporated by the Sun. Some soaks into the soil, where it is absorbed by plants and partly returned to the air through transpiration; some flows over the land surface as run-off, which flows into streams and rivers; and some rain and melted snow seeps through the soil into the rocks beneath to form ground water.

In polar and high mountainous regions most precipitation is in the form of snow. There it is compacted into ice, forming icesheets and glaciers. The force of gravity causes these bodies of ice to move downwards and outwards, and they may eventually return to the oceans where chunks of ice break off at the coastline to form icebergs. Thus all the water that does not return directly to the atmosphere gradually returns to the sea to complete the water cycle. This continual movement of water and ice plays a major part in the erosion of land areas.

Of the total water on land, more than 75% is frozen in icesheets and glaciers, and two-thirds of all the Earth's fresh water is held in Antarctica. Twice the size of Australia, this frozen continent contains ice to depths of 3,500m (11,500ft) and land is covered in ice to an average depth of more than 2,000m (6,500ft). However, Antarctica receives very little precipitation, not even in the form of snow. It is, effectively, a polar desert.

Most of the rest of the water on land (about 22%) is collected below the Earth's surface and is called ground water; comparatively small but crucially important quantities are in lakes, rivers and in the soil. Water that is held in the soil and that nourishes plant growth is called capillary water: it is retained in the upper few metres by molecular attraction between the water and soil particles.

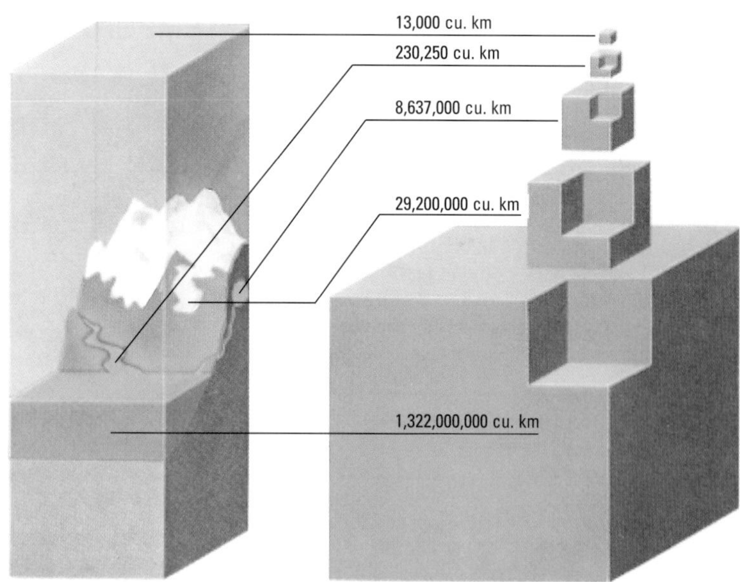

RIGHT Over 97% of the world's water is accounted for by the oceans. Of the total water on land more than 75% is frozen in icesheets and glaciers. Most of the rest, about 22%, is water collected below the Earth's surface (ground water). Relatively small quantities are in lakes and rivers (0.017% of the total), while water vapour represents only 0.001%. Without this, however, there would be no life on land.

13,000 cu. km
230,250 cu. km
8,637,000 cu. km
29,200,000 cu. km
1,322,000,000 cu. km

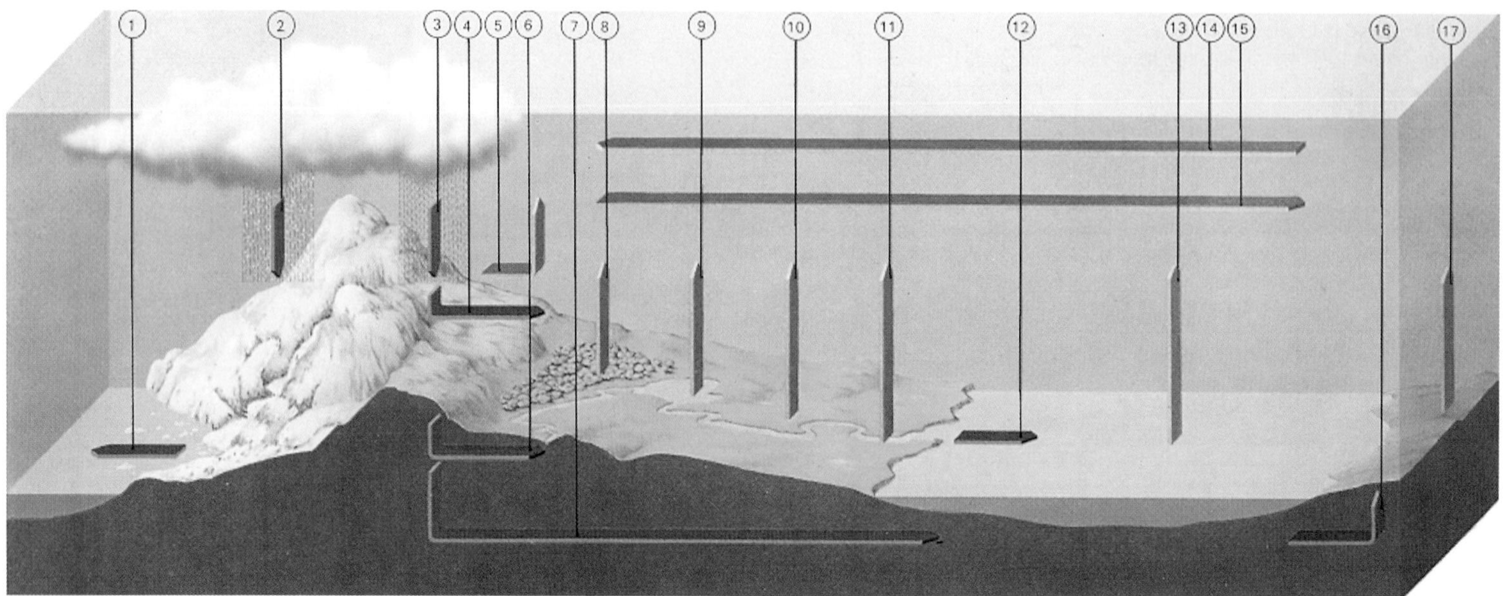

BELOW The water or hydrological cycle is the process whereby water, in its various forms, circulates from the oceans to land areas and back again. Fresh water is present on the Earth as water vapour in the atmosphere, as ice, and as liquid water.

The elements of the cycle are precipitation as rain [3], surface run-off [4], evaporation of rain in falling [5], ground water flow to rivers and streams [6], ground water flow to the oceans [7], transpiration from plants [8], evaporation from lakes and ponds [9], evaporation from the soil [10], evaporation from rivers and streams [11], evaporation from the oceans [13], flow of rivers and streams to the oceans [12], ground water flow from the oceans to arid land [16], intense evaporation from arid land [17], movement of moist air from and to the oceans [14,15], precipitation as snow [2], and ice-flow into the seas and oceans [1].

contain so much mineral substance in solution that their water is used for medicinal purposes and spa towns have grown up around them.

While sandstone is a highly porous rock through which water percolates easily, limestone is a permeable but non-porous rock. Ground water can seep through its maze of joints, fissures and caves, with apertures enlarged by the chemical action of rainwater containing dissolved carbon dioxide.

BELOW Limestone surfaces are often eroded into blocks called clints [1]. Surface streams flow into dissolved sink-holes [2] that lead to a deep chimney [3]; pot-holes [7] are dry chimneys. Gours [4] are ridges formed as carbonate is precipitated from turbulent water. Streams flow at the lowest level of the galleries [17], and abandoned galleries [13] are common. A siphon [12] occurs where the roof is below water level. Streams reappear at resurgences [20], and abandoned resurgences [19] may provide entrances to caves.

Stalactites [5] include macaroni stalactites [6], curtain stalactites or drapes [11] and "eccentric" stalactites [16], formed by water being blown sideways; stalagmites [14] sometimes have a fir-cone shape [15] caused by splashing, or resemble stacked plates [8]. Stalactites and stalagmites may also merge to form columns [10]. Signs of ancient humans [18] have been found in many caves, and they still harbour a variety of animal life adapted to the environment, including colourless shrimps and sightless newts – often called blind fish – which live in the dark pools [9].

Rain coming off the Atlantic Ocean and Mediterreanean Sea and falling on the Atlas Mountains of Morocco and Algeria then drains into porous rocks underlying the northern parts of the Sahara Desert. The water seeps through these rocks which, wherever they come to the surface, give rise to fertile oases.

BELOW As rain falls, it dissolves carbon dioxide from the atmosphere and becomes a weak carbonic acid that attacks carbonate rock (limestone and dolomite) by transforming it into the soluble bicarbonate. Carbonate rocks are crisscrossed by vertical cracks and horizontal breaks along bedding planes [A]. Some geologists believe the caves were formed when the rock was saturated by water; others reckon they formed gradually by solution [B] into a major cave network [C]. Limestone caves contain many features formed from calcium carbonate.

VEGETATION AND SOIL

THE DISTRIBUTION of natural resources over the Earth's surface is far from even. The whereabouts of mineral deposits depends on random events in a remote geological past, while patches of fertile soil depend on more recent events such as the flow of rivers or the movement of ice.

For agriculture, the activity that has been basic to the survival of humanity and our huge increase in population, about a fifth of the Earth's surface is barred by ice or perennially frozen soil; a fifth is arid or desert; and another fifth is composed of highlands too cold, rugged or barren for the cultivation of crops. Between 5% and 10% of the remainder has no soil, either because it has been scraped by ice or because it is permanently wet or flooded. This leaves only 30% to 35% of the land surface where food production is even possible.

The importance of soil

The whole structure of life on Earth, with its enormous diversity of plant and animal types, is dependent on a mantle of soil which is rich in moisture and nutrients.

Soil is a result of all the processes of physical and chemical weathering on the barren, underlying rock mass of the Earth that it covers, and varies in

BELOW The map illustrates the natural "climax" vegetation of a region, as dictated by its climate and topography. In the vast majority of cases, however, human agricultural activity has drastically altered the pattern of vegetation. Western Europe, for example, lost most of its broadleaf forest many centuries ago, and in many areas irrigation has gradually turned natural semi-desert into productive land.

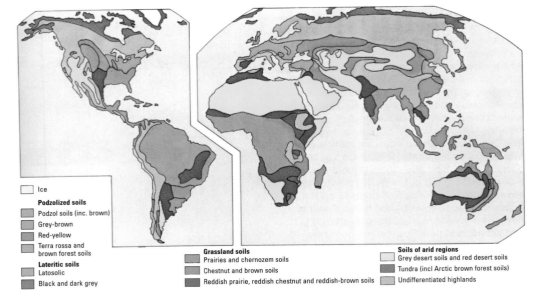

- Ice
- **Podzolized soils**
  - Podzol soils (inc. brown)
  - Grey-brown
  - Red-yellow
  - Terra rossa and brown forest soils
- **Lateritic soils**
  - Latosolic
  - Black and dark grey
- **Grassland soils**
  - Prairies and chernozem soils
  - Chestnut and brown soils
  - Reddish prairie, reddish chestnut and reddish-brown soils
- **Soils of arid regions**
  - Grey desert soils and red desert soils
  - Tundra (incl Arctic brown forest soils)
  - Undifferentiated highlands

depth from a few centimetres to several metres. The depth of soil is measured either by the distance to which plants send down their roots, or by the depth of soil directly influencing their systems. In some places only a very thin layer is necessary to support life.

Soil remains an unconsolidated mass of inorganic particles until it acquires a minimum organic content and plants take root and deposit their "litter". As the organic matter accumulates, fine humus builds up in the upper soil horizons, enriching them chemically and providing an environment for a wide variety of lifeforms. In the course of time

plants, fungi, bacteria, worms, insects and burrowing animals such as rodents and moles reproduce in the soil and thrive in the complex ecosystem of a mature soil.

Formation of soil is the result of the complex interaction of five major elements – the parent rock (the source of the vast bulk of soil material), land topography, time, climate and decay. However, by far the most single important factor in the development of soil is climate, with water essential to all chemical and biological change. As it percolates through, water both leaches the surface layers and deposits material in the subsoil.

NATURAL VEGETATION

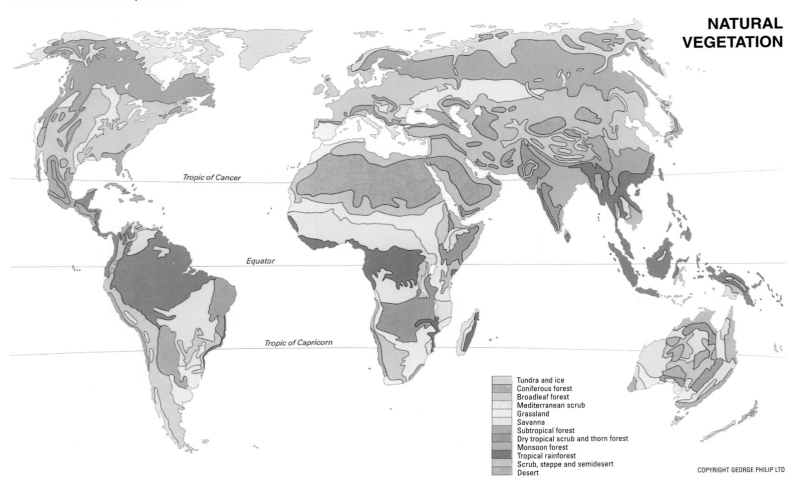

Tropic of Cancer

Equator

Tropic of Capricorn

- Tundra and ice
- Coniferous forest
- Broadleaf forest
- Mediterranean scrub
- Grassland
- Savanna
- Subtropical forest
- Dry tropical scrub and thorn forest
- Monsoon forest
- Tropical rainforest
- Scrub, steppe and semidesert
- Desert

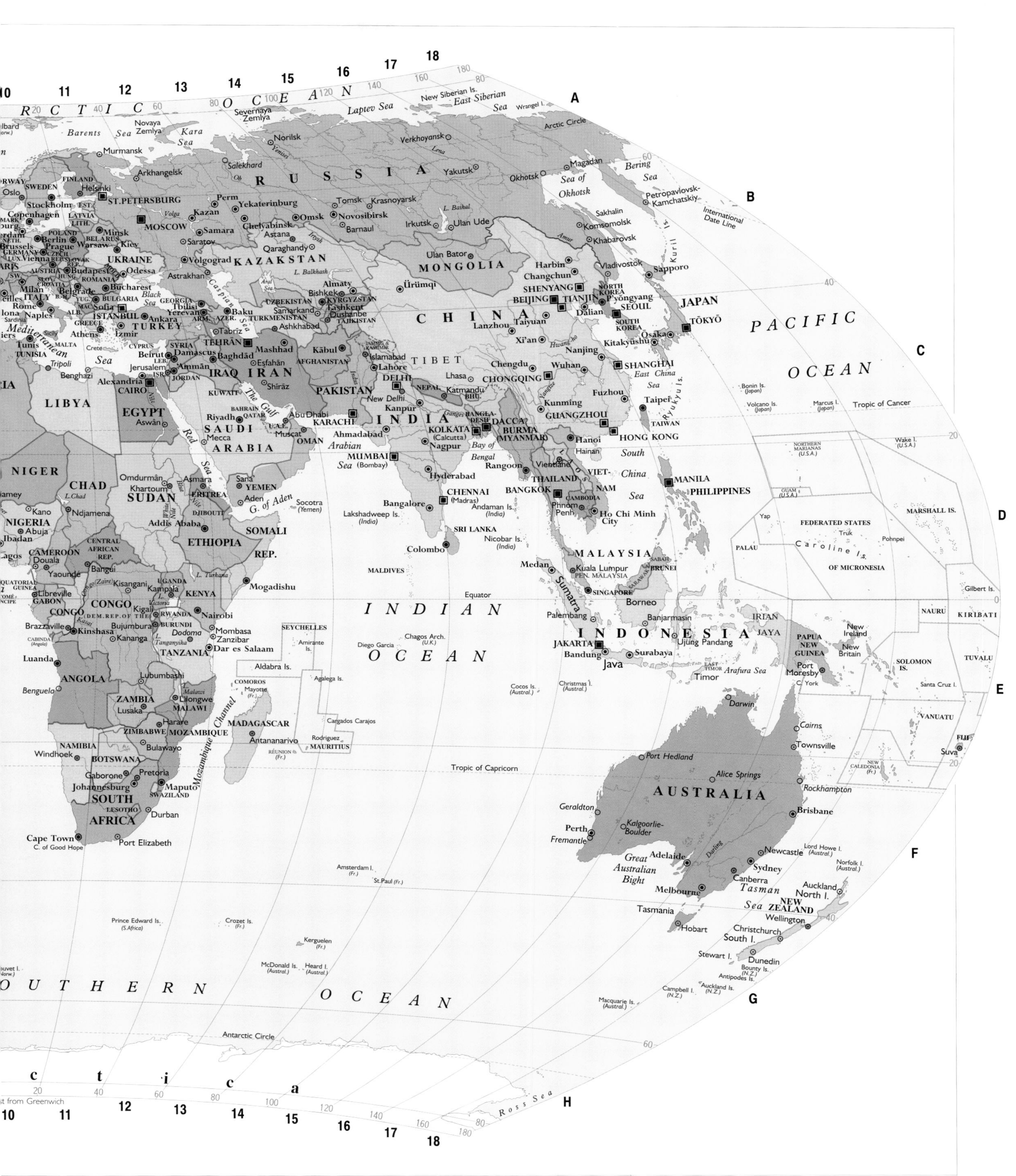

Hanoi ● Capital Cities

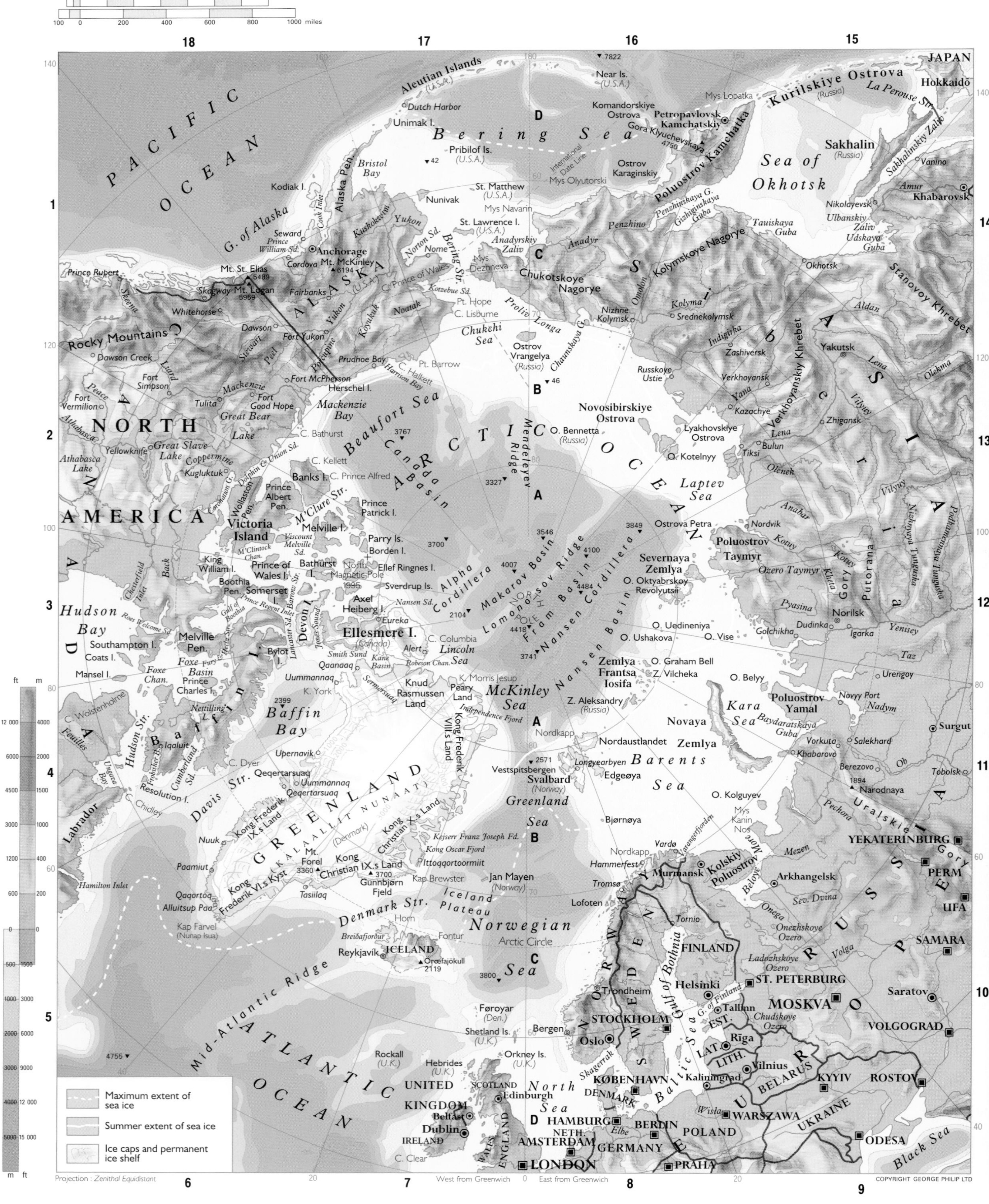

100 0 200 400 600 800 1000 1200 1400 km
100 0 200 400 600 800 1000 miles

Maximum extent of sea ice

Summer extent of sea ice

Ice caps and permanent ice shelf

Projection : Zenithal Equidistant

West from Greenwich   East from Greenwich

COPYRIGHT GEORGE PHILIP LTD

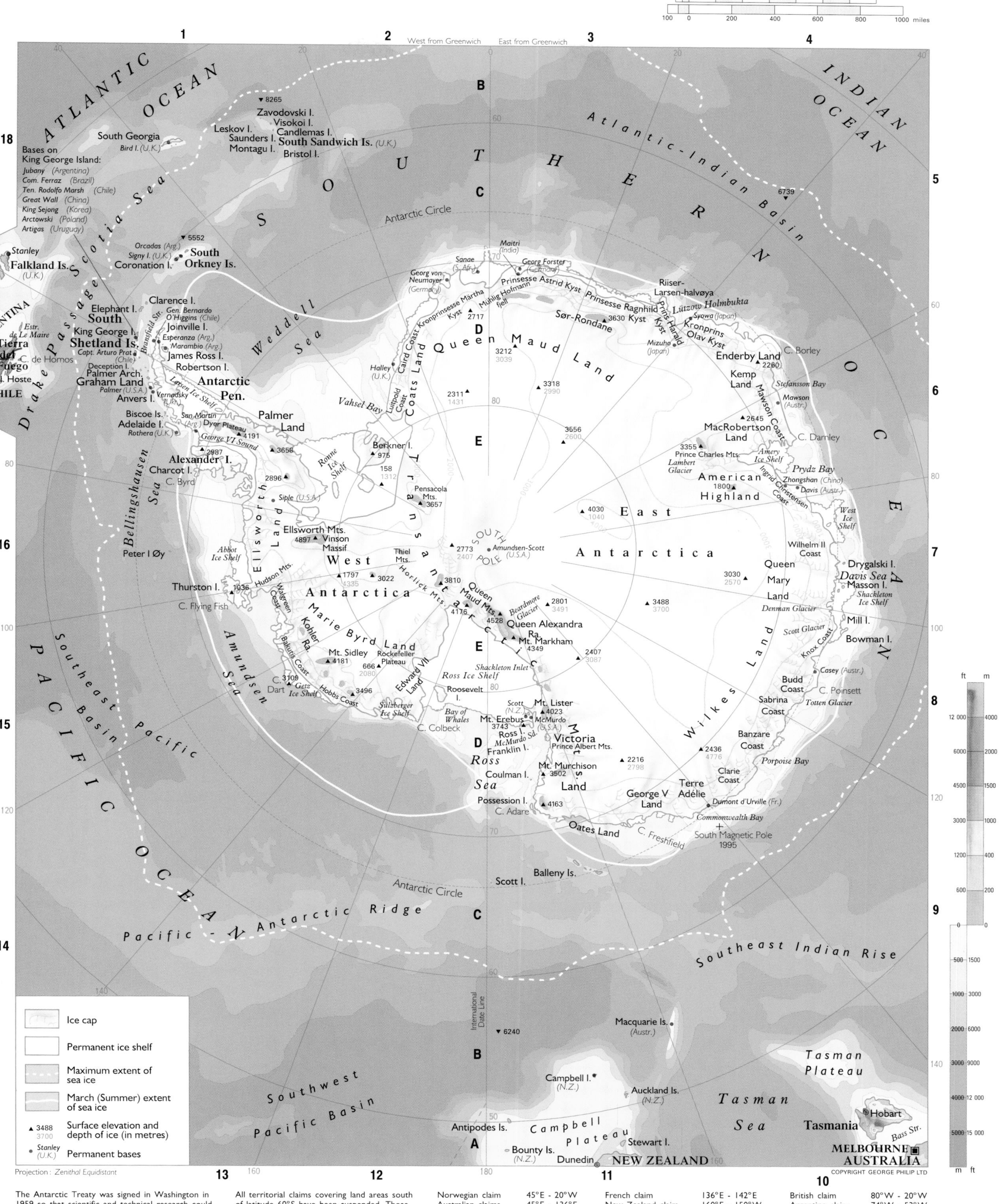

The Antarctic Treaty was signed in Washington in 1959 so that scientific and technical research could continue unhampered by international politics.

All territorial claims covering land areas south of latitude 60°S have been suspended. Those claims were:

Norwegian claim	45°E - 20°W
Australian claims	45°E - 136°E
	142°E - 160°E
French claim	136°E - 142°E
New Zealand claim	160°E - 150°W
Chilean claim	90°W - 53°W
British claim	80°W - 20°W
Argentine claim	74°W - 53°W

Legend:
- Ice cap
- Permanent ice shelf
- Maximum extent of sea ice
- March (Summer) extent of sea ice
- ▲ 3488 / 3700 Surface elevation and depth of ice (in metres)
- • Stanley (U.K.) Permanent bases

Projection: Zenithal Equidistant

COPYRIGHT GEORGE PHILIP LTD

Bases on King George Island:
- Jubany (Argentina)
- Com. Ferraz (Brazil)
- Ten. Rodolfo Marsh (Chile)
- Great Wall (China)
- King Sejong (Korea)
- Arctowski (Poland)
- Artigas (Uruguay)

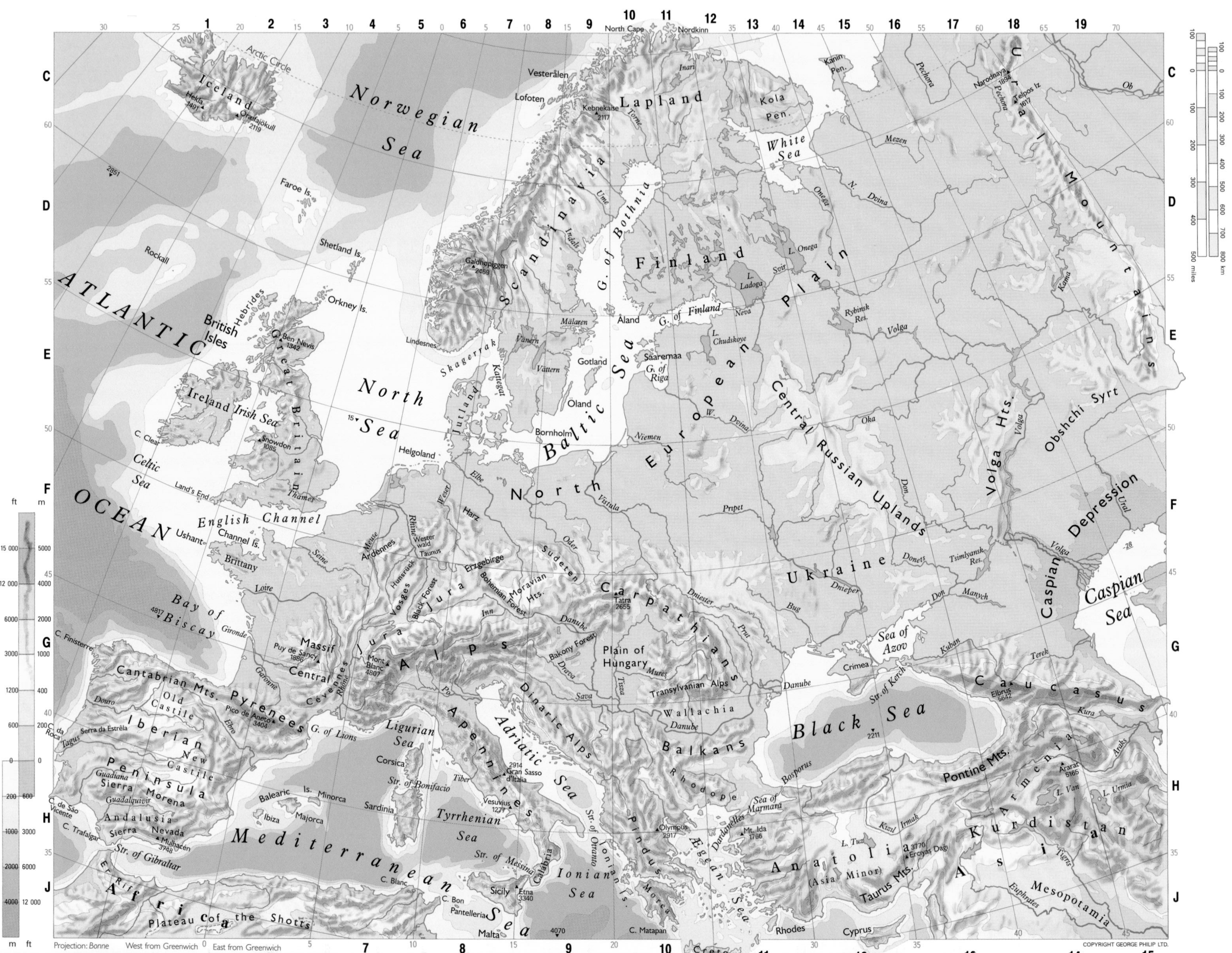

Projection: Bonne

EUROPE : Physical 1:20 000 000
100 0 100 200 300 400 500 600 700 800 km
100 0 100 200 300 400 500 miles

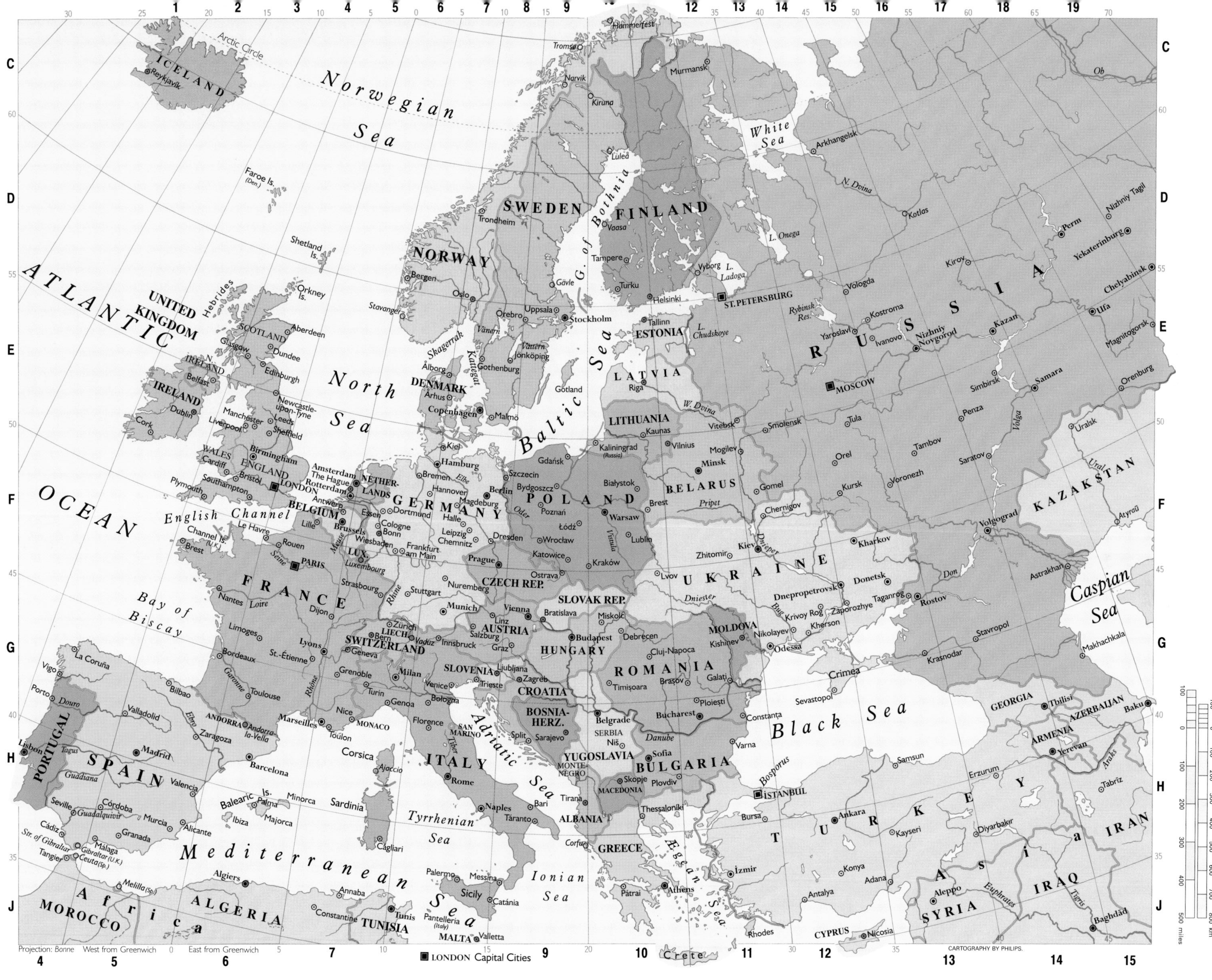
100 0 100 200 300 400 500 600 700 800 km
100 0 100 200 300 400 500 miles
ICELAND
Reykjavík
Arctic Circle
Norwegian Sea
Faroe Is. (Den.)
Shetland Is.
Orkney Is.
Hebrides
ATLANTIC OCEAN
UNITED KINGDOM
SCOTLAND
N. IRELAND
IRELAND
WALES
ENGLAND
LONDON
North Sea
NORWAY
SWEDEN
FINLAND
G. of Bothnia
White Sea
Baltic Sea
ESTONIA
LATVIA
LITHUANIA
BELARUS
UKRAINE
MOLDOVA
RUSSIA
MOSCOW
ST. PETERSBURG
KAZAKSTAN
Caspian Sea
DENMARK
NETHER-LANDS
BELGIUM
LUX.
GERMANY
POLAND
CZECH REP.
SLOVAK REP.
AUSTRIA
HUNGARY
ROMANIA
SWITZERLAND
LIECH.
SLOVENIA
CROATIA
BOSNIA-HERZ.
YUGOSLAVIA
SERBIA
MONTE-NEGRO
MACEDONIA
ALBANIA
BULGARIA
GREECE
FRANCE
PARIS
English Channel
Bay of Biscay
PORTUGAL
SPAIN
ANDORRA
MONACO
SAN MARINO
ITALY
Corsica
Sardinia
Sicily
MALTA
Adriatic Sea
Tyrrhenian Sea
Ionian Sea
Ægean Sea
Mediterranean Sea
Black Sea
Crimea
Crete
CYPRUS
GEORGIA
ARMENIA
AZERBAIJAN
TURKEY
ISTANBUL
SYRIA
IRAQ
IRAN
Asia
Africa
MOROCCO
ALGERIA
TUNISIA
Projection: Bonne West from Greenwich East from Greenwich
■ LONDON Capital Cities
CARTOGRAPHY BY PHILIPS.

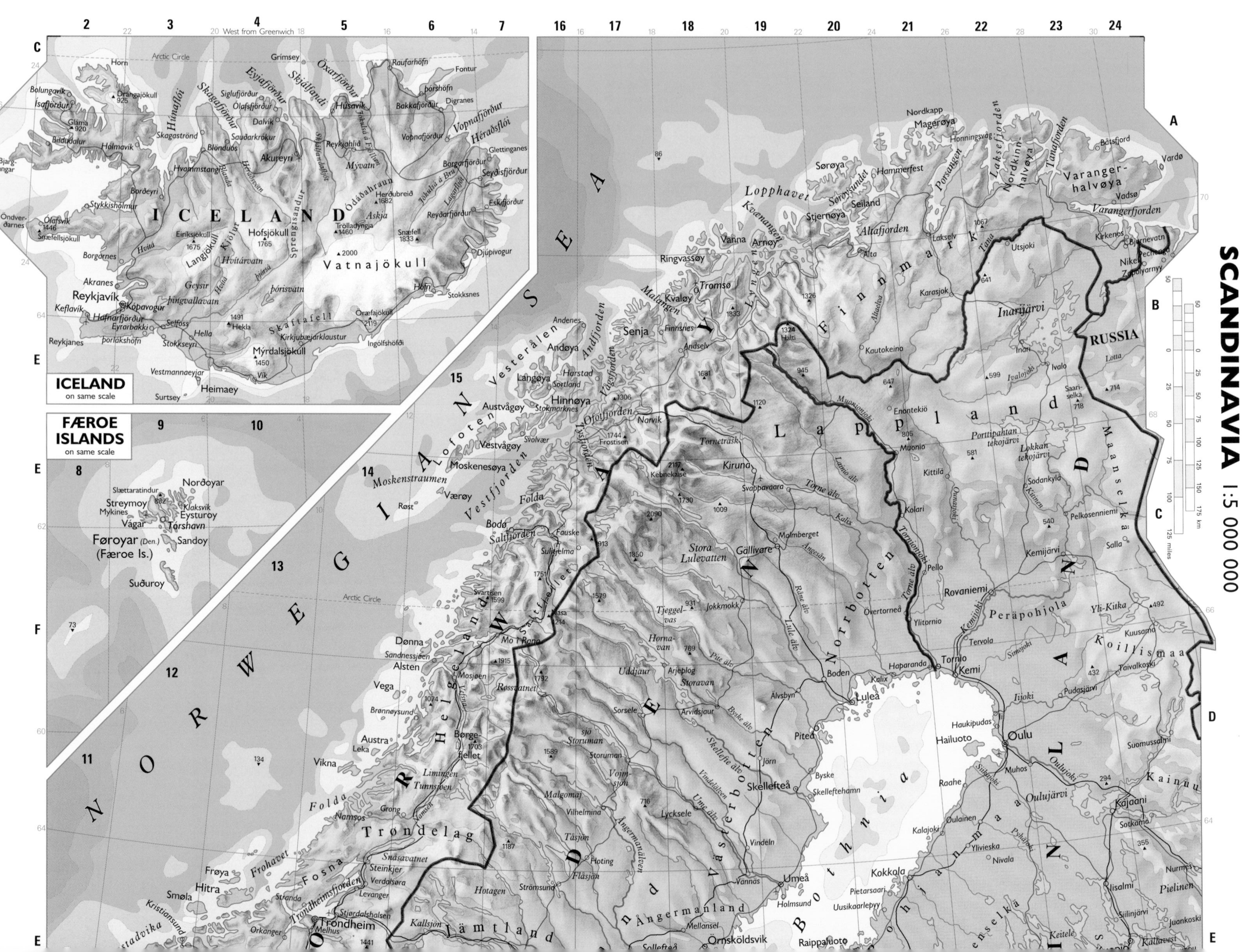

ICELAND
on same scale

FÆROE
ISLANDS
on same scale

Map of the Baltic Region

Countries and major features visible:

- FINLAND
- ESTONIA
- LATVIA
- LITHUANIA
- RUSSIA
- POLAND
- GERMANY
- DENMARK
- NORWAY
- SWEDEN

Seas / Gulfs: Gulf of Finland, Gulf of Riga, BALTIC SEA, Kattegat, Skagerrak, Ålands hav

Cities: Helsinki (Helsingfors), Tallinn, Tartu, Riga, Vilnius, Kaunas, Kaliningrad (Russia), Klaipėda, Gdańsk, Gdynia, København (Copenhagen), Malmö, STOCKHOLM, Göteborg (Gothenburg), Oslo, Bergen, Stavanger, Kiel, Lübeck, Rostock, Tampere, Turku (Åbo), Gävle, Uppsala, Norrköping, Linköping, Kalmar, Karlskrona

Islands: Gotland, Öland, Bornholm, Saaremaa (Ösel), Hiiumaa (Dago), Rügen, Fehmarn, Åland (Ahvenanmaa), Fårö

COPYRIGHT GEORGE PHILIP LTD.

East from Greenwich

Projection: Conical with two standard parallels

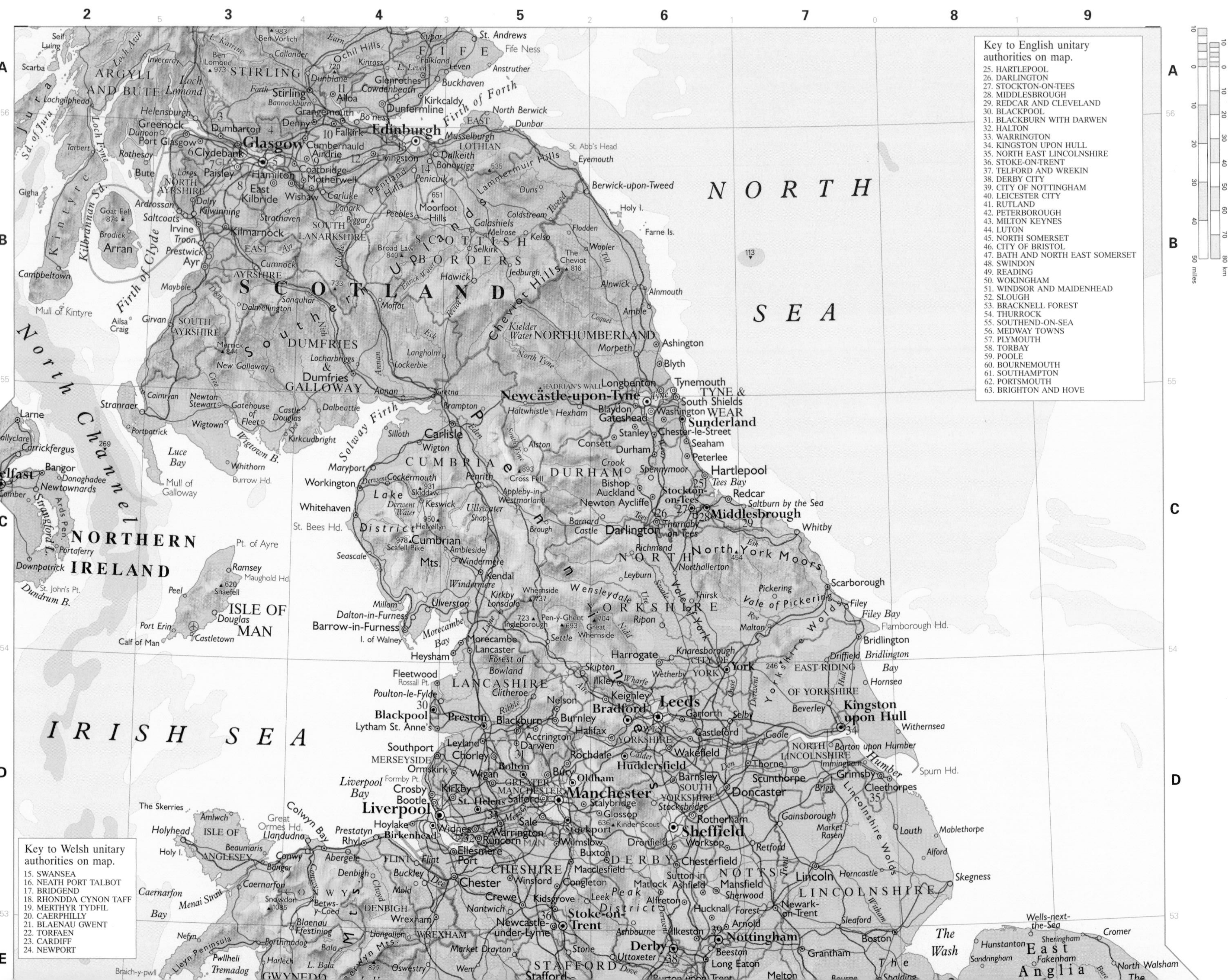

WORLD MAPS

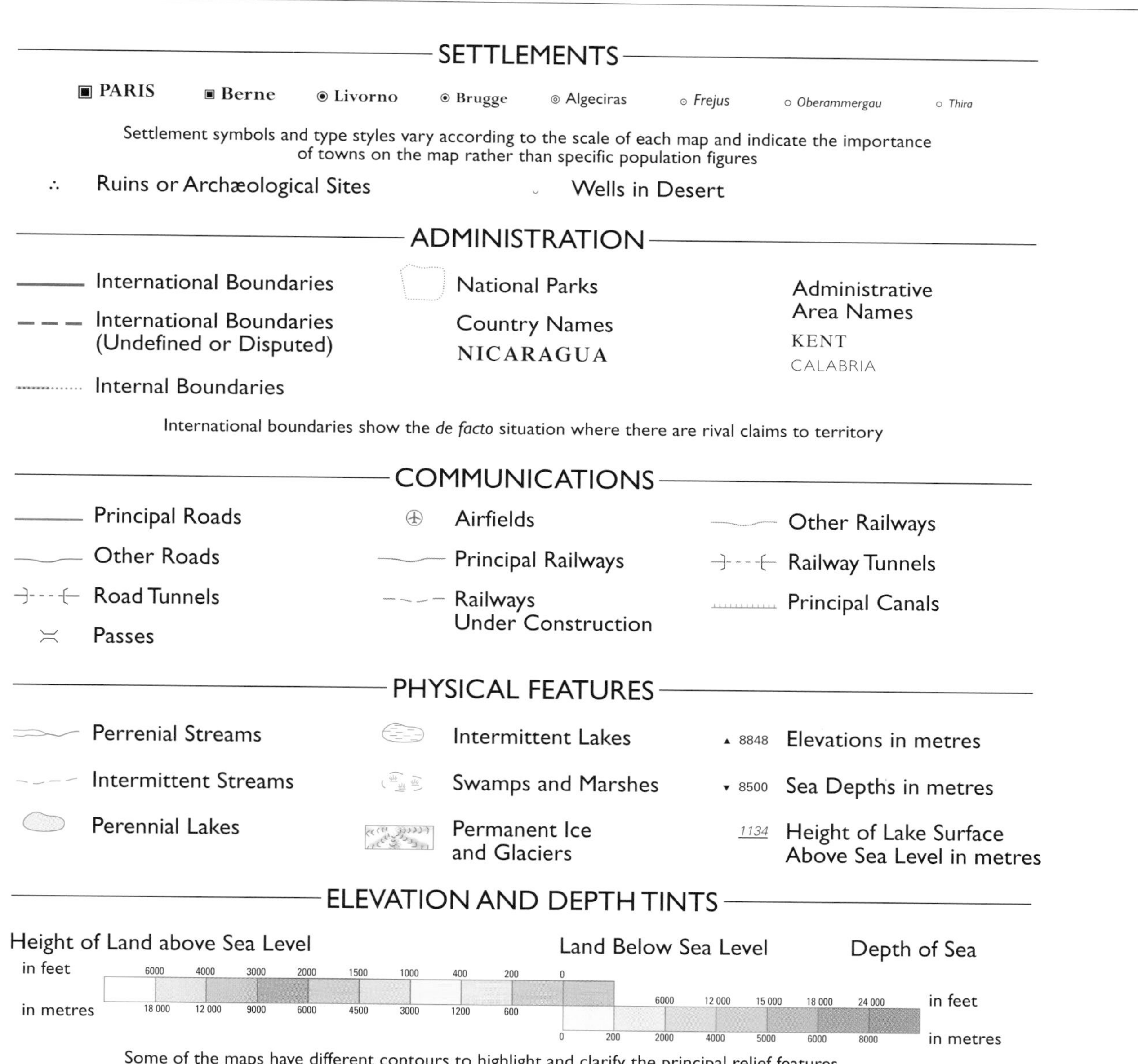

──────────────── SETTLEMENTS ────────────────

■ PARIS ■ Berne ⊙ Livorno ⊛ Brugge ◎ Algeciras ○ Frejus ○ Oberammergau ○ Thira

Settlement symbols and type styles vary according to the scale of each map and indicate the importance
of towns on the map rather than specific population figures

∴ Ruins or Archæological Sites ˅ Wells in Desert

──────────────── ADMINISTRATION ────────────────

────── International Boundaries

─ ─ ─ International Boundaries
(Undefined or Disputed)

·········· Internal Boundaries

National Parks

Country Names
NICARAGUA

Administrative
Area Names

KENT
CALABRIA

International boundaries show the *de facto* situation where there are rival claims to territory

──────────────── COMMUNICATIONS ────────────────

────── Principal Roads

────── Other Roads

⌐·-⌐ Road Tunnels

⌒ Passes

⊕ Airfields

────── Principal Railways

─ ─ ─ Railways
Under Construction

────── Other Railways

⌐·-⌐ Railway Tunnels

············ Principal Canals

──────────────── PHYSICAL FEATURES ────────────────

~~~ Perrenial Streams

─ ─ ─ Intermittent Streams

⬭ Perennial Lakes

⬭ Intermittent Lakes

Swamps and Marshes

Permanent Ice
and Glaciers

▲ 8848 Elevations in metres

▼ 8500 Sea Depths in metres

1134 Height of Lake Surface
Above Sea Level in metres

──────────────── ELEVATION AND DEPTH TINTS ────────────────

Height of Land above Sea Level

in feet

| 6000 | 4000 | 3000 | 2000 | 1500 | 1000 | 400 | 200 | 0 |

in metres

| 18 000 | 12 000 | 9000 | 6000 | 4500 | 3000 | 1200 | 600 |

Land Below Sea Level

| 6000 | 12 000 | 15 000 | 18 000 | 24 000 |

| 0 | 200 | 2000 | 4000 | 5000 | 6000 | 8000 |

Depth of Sea

in feet

in metres

Some of the maps have different contours to highlight and clarify the principal relief features

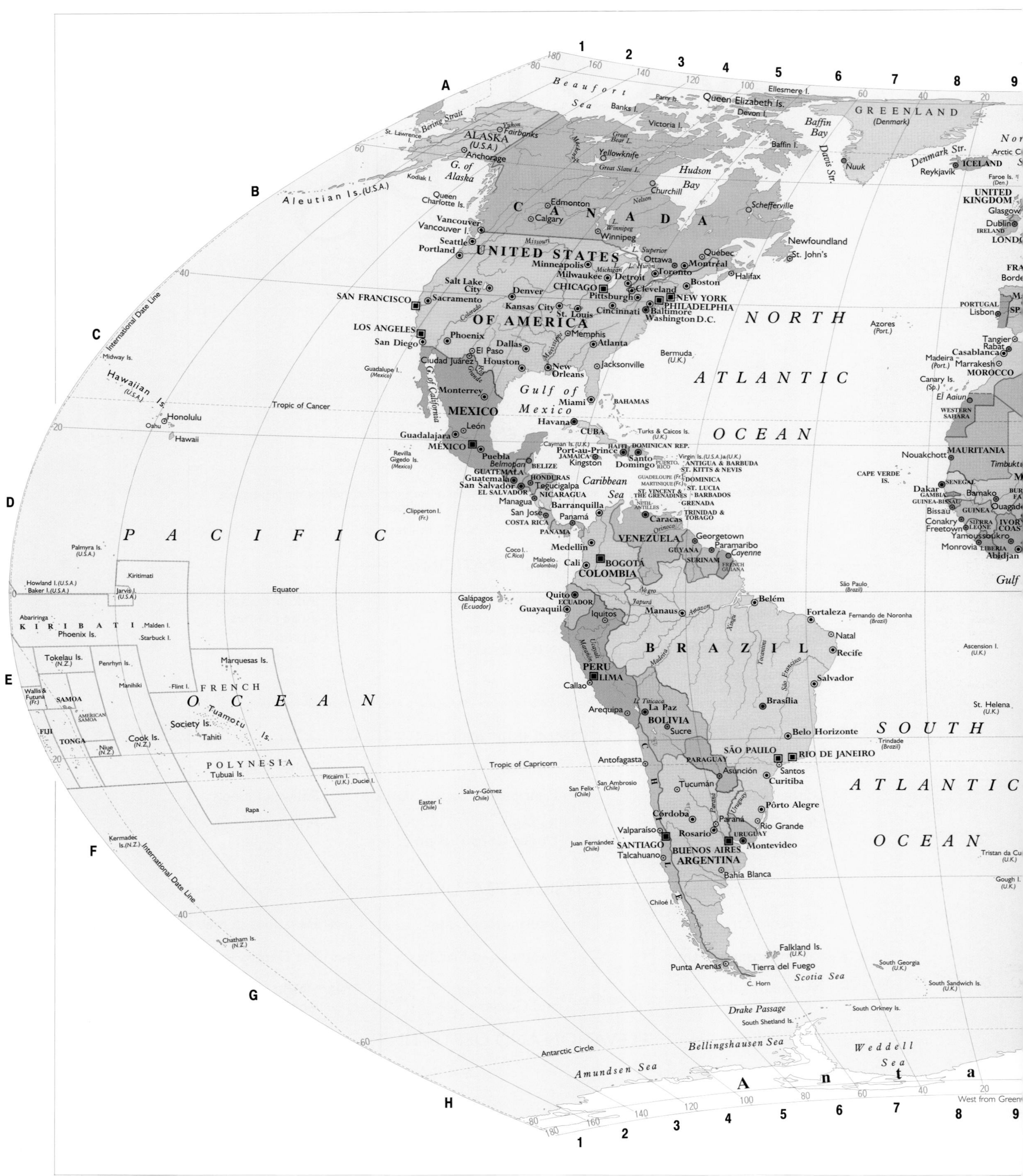

Projection: Hammer Equal Area

WALES
ENGLAND
FRANCE
NORMANDIE
HAUTE-
SEINE-MARITIME

ENGLISH CHANNEL

Strait of Dover

Bristol Channel

Cardigan Bay

London
Birmingham
Bristol
Cardiff
Swansea
Plymouth
Southampton
Portsmouth
Bournemouth
Brighton
Hove
Le Havre
Rouen

NORFOLK
SUFFOLK
CAMBRIDGE
NORTHANTS
BEDFORD
HERTS
BUCKS
BERKSHIRE
OXFORD
WEST
SURREY
WEST SUSSEX
EAST SUSSEX
KENT
ESSEX
HANTS
WILTSHIRE
SOMERSET
DORSET
DEVON
CORNWALL
GLOUCS
WORCESTER
SHROPSHIRE
HEREFORD
POWYS
CEREDIGION
PEMBROKESHIRE
CARMARTHENSHIRE
GLAMORGAN
VALE OF
CALVADOS
MANCHE
COTENTIN

Lowestoft
Southwold
Beccles
Bungay
Diss
Wymondham
Thetford
Downham Market
Wisbech
Peterborough
Stamford
Rutland Water
Leicester
Hinckley
Nuneaton
Wolverhampton
Walsall
Sutton Coldfield
West Bromwich
Dudley
Halesowen
Kidderminster
Bridgnorth
Telford
Ludlow
Leominster
Hereford
Knighton
Llandrindod Wells
Newtown
Machynlleth
Aberystwyth
Aberaeron
New Quay
Cardigan
Newcastle Emlyn
Carmarthen
Haverfordwest
Milford Haven
Pembroke
Tenby
Llanelli
Neath
Port Talbot
Bridgend
Barry
Penarth
Newport
Chepstow
Monmouth
Ross-on-Wye
Gloucester
Cheltenham
Tewkesbury
Malvern
Worcester
Droitwich
Redditch
Bromsgrove
Stourbridge
Solihull
Coventry
Rugby
Royal Leamington Spa
Warwick
Stratford-upon-Avon
Banbury
Daventry
Northampton
Wellingborough
Kettering
Corby
Market Harborough
Oundle
Huntingdon
St. Neots
Bedford
Biggleswade
Cambridge
Newmarket
Bury St. Edmunds
Sudbury
Stowmarket
Ipswich
Felixstowe
Harwich
Woodbridge
Aldeburgh
Saxmundham
Halstead
Braintree
Witham
Chelmsford
Maldon
Colchester
Clacton-on-Sea
Walton-on-the-Naze
Brentwood
Rayleigh
Canvey
Southend-on-Sea
Sheerness
Gillingham
Chatham
Rochester
Gravesend
Sittingbourne
Faversham
Whitstable
Herne Bay
Margate
Ramsgate
Deal
Dover
Folkestone
Hythe
New Romney
Dungeness
Hastings
Bexhill
Eastbourne
Newhaven
Seaford
Lewes
Worthing
Littlehampton
Bognor Regis
Chichester
Haywards Heath
East Grinstead
Crawley
Reigate
Croydon
Epsom
Leatherhead
Guildford
Aldershot
Farnham
Basingstoke
Alton
Petersfield
Havant
Gosport
Fareham
Ryde
Newport
Cowes
ISLE OF WIGHT
Ventnor
St. Catherine's Pt.
Lymington
Christchurch
Poole
Swanage
St. Alban's Head
Weymouth
Portland Bill
I. of Portland
Dorchester
Bridport
Lyme Regis
Sidmouth
Exmouth
Dawlish
Teignmouth
Torquay
Paignton
Brixham
Dartmouth
Kingsbridge
Salcombe
Start Pt.
Bolt Head
Totnes
Newton Abbot
Tavistock
Launceston
Bodmin
Bodmin Moor
Liskeard
Looe
Fowey
St. Austell
Truro
Falmouth
Helston
Penryn
Camborne
Redruth
Hayle
St. Ives
Newquay
Padstow
Wadebridge
Bude
Boscastle
Bideford
Barnstaple
Ilfracombe
Minehead
Lynton
Bridgwater
Burnham-on-Sea
Weston-super-Mare
Clevedon
Portishead
Bath
Trowbridge
Frome
Wells
Glastonbury
Bridgwater
Taunton
Wellington
Tiverton
Exeter
Crediton
Okehampton
Dartmoor
Exmoor
Wimborne
Blandford Forum
Shaftesbury
Sherborne
Yeovil
Chard
Honiton
Axminster
Salisbury
Warminster
Devizes
Calne
Chippenham
Cirencester
Stroud
Swindon
Marlborough
Newbury
Reading
Wokingham
Bracknell
Windsor
Maidenhead
High Wycombe
Aylesbury
Oxford
Abingdon
Witney
Woodstock
Chipping Norton
Evesham
Wantage
Henley
Slough
Staines
Bracknell
Esher
Kingston
Watford
Hemel Hempstead
St. Albans
Welwyn Garden City
Hatfield
Hertford
Harlow
Bishop's Stortford
Saffron Walden
Royston
Stevenage
Hitchin
Letchworth
Luton
Dunstable
Bletchley
Milton Keynes
Buckingham
Bicester
Newport Pagnell
Rushden
Ely
Mildenhall
Brandon
Breckland
March
Chatteris
Orford Ness
Orwell
Deben
Naze
Foulness
Thames
Estuary
Southend-on-Sea
Maidstone
Tonbridge
Tunbridge Wells
Ashford
Canterbury
Tenterden
Royal Tunbridge Wells
Horsham
Cuckfield
Midhurst
Selsey Bill
Hayling I.
Spithead
The Solent
The Needles
New Forest
South Downs
North Downs
Marlborough Downs
Salisbury Plain
Vale of White Horse
Cotswolds
Mendip Hills
Quantock Hills
Blackdown Hills
Cranborne Chase
Purbeck
I. of Purbeck
Romney Marsh
The Weald
Ashdown Forest

Channel Tunnel

CHANNEL ISLANDS
(U.K.)
Alderney
Guernsey
St. Peter Port
Herm
Sark
Jersey
St. Helier

Baie de la Seine
Baie de la Somme

Calais
Boulogne-sur-Mer
Le Touquet-Paris-Plage
Berck
Étaples
Dieppe
Le Tréport
Eu
Fécamp
St-Valery-en-Caux
Yport
Étretat
Sainte-Adresse
Honfleur
Deauville
Trouville
Houlgate
Cabourg
Ouistreham
Courseulles-sur-Mer
Arromanches-les-Bains
Port-en-Bessin
Bayeux
Caen
Bernay
Lisieux
Pont-l'Évêque
Évreux
Louviers
Elbeuf
Lillebonne
Bolbec
Yvetot
Neufchâtel-en-Bray
Montivilliers
Cherbourg
Octeville
Valognes
Barfleur
Pte. de Barfleur
Carentan
Coutances
Granville
St. Helier
Ste-Mère-Église
Isigny
Périers
Lessay

C. de la Hague
Nez de Jobourg

Seine
Risle
Orne
Vire
Dives
Touques

East from Greenwich
West from Greenwich

Projection : Lambert's Conformal Conic

Isles of Scilly
On same scale
Tresco
St. Mary's

Land's End
Land's
End
Newlyn
Penzance
St. Ives
Camborne
Hayle

ft   m
3000 1000
1500 500
600   200
300   100
0   0
        m ft
50-150
100-300
200-600

ft
m

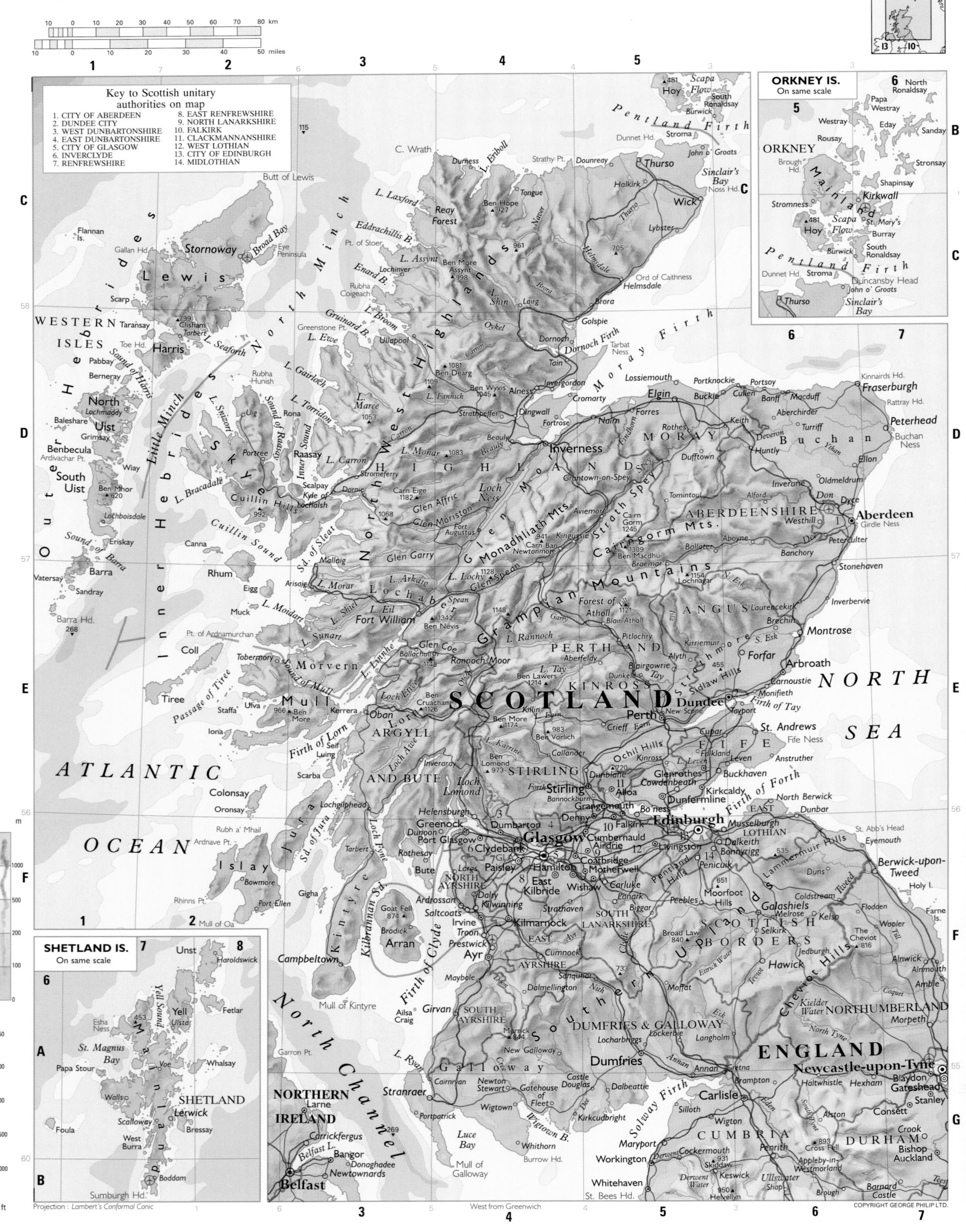

Key to Scottish unitary authorities on map
1. CITY OF ABERDEEN
2. DUNDEE CITY
3. WEST DUNBARTONSHIRE
4. EAST DUNBARTONSHIRE
5. CITY OF GLASGOW
6. INVERCLYDE
7. RENFREWSHIRE
8. EAST RENFREWSHIRE
9. NORTH LANARKSHIRE
10. FALKIRK
11. CLACKMANNANSHIRE
12. WEST LOTHIAN
13. CITY OF EDINBURGH
14. MIDLOTHIAN

ORKNEY IS.
On same scale

SHETLAND IS.
On same scale

Projection : Lambert's Conformal Conic

West from Greenwich

COPYRIGHT GEORGE PHILIP LTD.

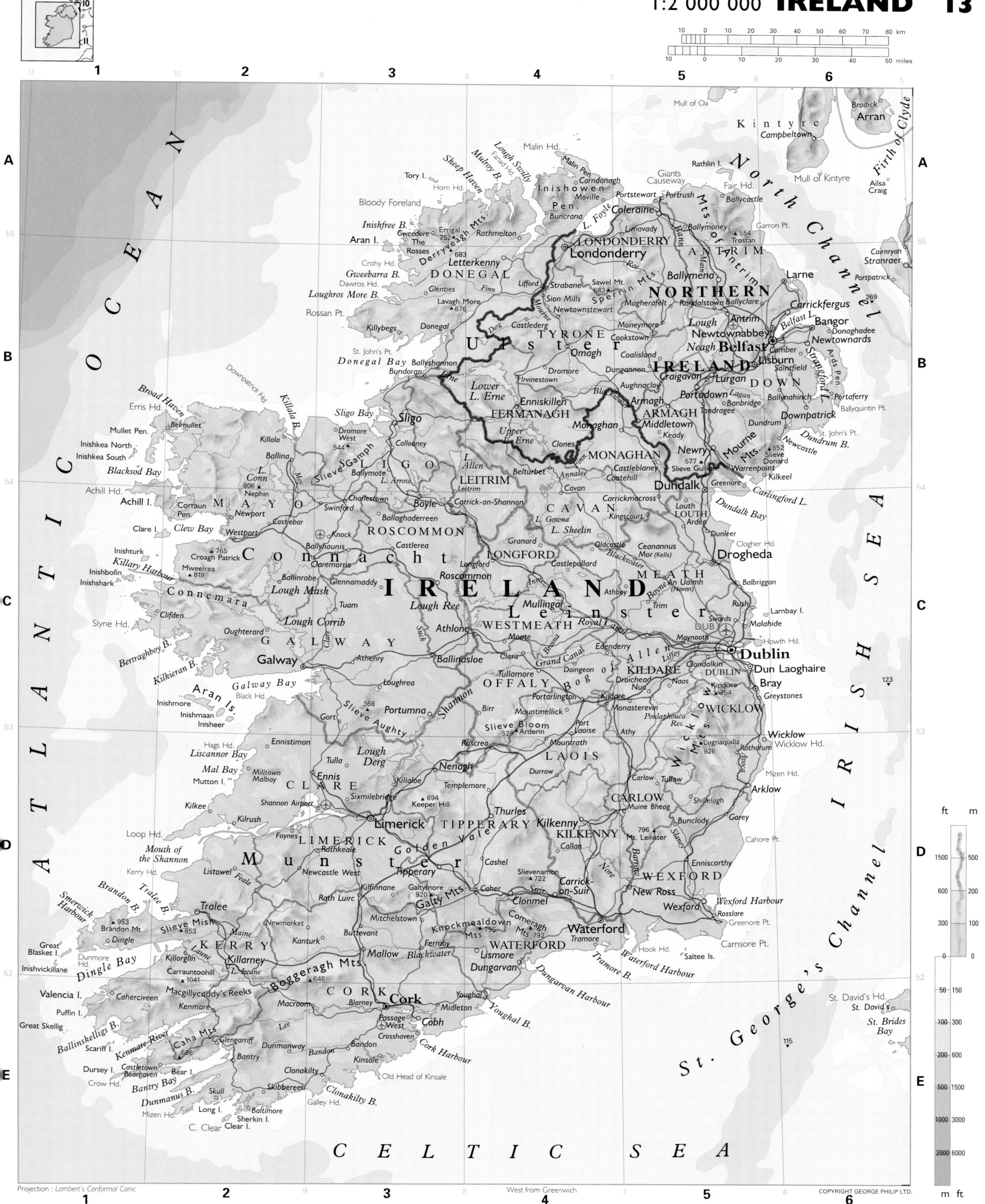

10   0   10   20   30   40   50   60   70   80 km
10   0   10   20   30   40   50 miles

ATLANTIC OCEAN
NORTHERN IRELAND
IRELAND
Ulster
Connacht
Leinster
Munster
IRISH SEA
CELTIC SEA
North Channel
St. George's Channel
Dublin
Belfast
Cork
Limerick
Galway
Waterford
Londonderry
Projection : Lambert's Conformal Conic
West from Greenwich

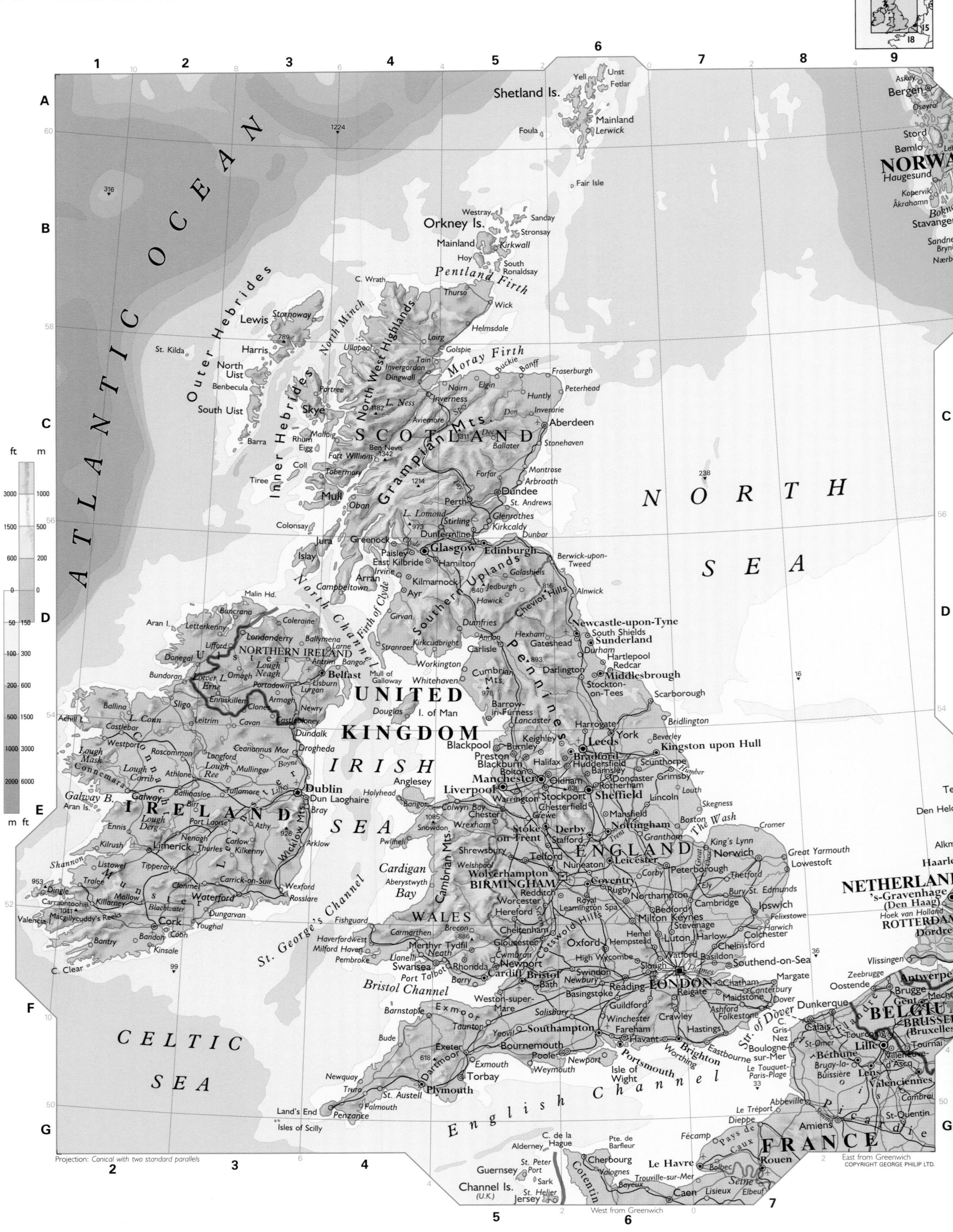

Projection: Conical with two standard parallels

East from Greenwich
COPYRIGHT GEORGE PHILIP LTD.
West from Greenwich

THE CONTINENTS

Continent	Area			Highest point above sea level				Lowest point below sea level			
	sq km	sq miles	%			metres	feet			metres	feet
Asia	44,500,000	17,179,000	29.8	Mt Everest (China/Nepal)		8,848	29,029	Dead Sea, Israel/Jordan		−396	−1,302
Africa	30,302,000	11,697,000	20.3	Mt Kilimanjaro, Tanzania		5,895	19,340	Lake Assal, Djibouti		−153	−502
North America	24,454,000	9,442,000	16.2	Mt McKinley, Alaska		6,194	20,321	Death Valley, California, USA		−86	−282
South America	17,793,000	6,868,000	11.9	Mt Aconcagua, Argentina		6,960	22,834	Peninsular Valdés, Argentina		−40	−131
Antarctica	14,100,000	5,443,000	9.4	Vinson Massif		4,897	16,066	*			
Europe	9,957,000	3,843.000	6.7	Mt Elbrus, Russia		5,633	18,481	Caspian Sea, W. Central Asia		−28	−92
Oceania	8,945,000	3,454,000	5.7	Puncak Jaya (Ngga Pulu), Indonesia		5,029	16,499	Lake Eyre (N), South Australia		−15	−50

*The Bentley trench (−2,540m/−8,333ft) is englacial and therefore not a surface point

THE OCEANS

Ocean	Area			Average depth		Greatest known depth			
	sq km	sq miles	%	metres	feet		metres	feet	
Pacific	179,679,000	69,356,000	49.9	4,300	14,100	Mariana Trench	11,022	36,161	
Atlantic	92,373,000	35,657,000	25.7	3,700	12,100	Puerto Rico Deep*	9,200	30,138	
Indian	73,917,000	28,532,000	20.5	3,900	12,800	Java Trench	7,450	24,442	
Arctic	14,090,000	5,439,000	3.9	1,330	4,300	Molloy Deep	5,608	18,399	

*7th deepest trench in the world; 8 of the deepest 10, including 1-6, are in the Pacific Ocean

LONGEST RIVERS

	Outflow	km	miles
Europe			
Volga	Caspian Sea	3,700	2,300
Danube	Black Sea	2,850	1,770
Ural*	Caspian Sea	2,535	1,575
Asia			
Yangtze [3]	Pacific Ocean	6,380	3,960
Yenisey-Angara [5]	Arctic Ocean	5,550	3,445
Hwang Ho [6]	Pacific Ocean	5,464	3,395
Ob-Irtysh [7]	Arctic Ocean	5,410	3,360
Mekong [9]	Pacific Ocean	4,500	2,795
Amur [10]	Pacific Ocean	4,400	2,730
Africa			
Nile [1]	Mediterranean	6,620	4,140
Zaire (Congo) [8]	Atlantic Ocean	4,670	2,900
Niger	Atlantic Ocean	4,180	2,595
Zambezi	Indian Ocean	3,540	2,200
North America			
Mississippi-Missouri[4]	Gulf of Mexico	6,020	3,740
Mackenzie	Arctic Ocean	4,240	2,630
Mississippi	Gulf of Mexico	3,780	2,350
Missouri	Mississippi	3,780	2,350
Yukon	Pacific Ocean	3,185	1,980
Rio Grande	Gulf of Mexico	3,030	1,880
Arkansas	Mississippi	2,840	1,450
Colorado	Pacific Ocean	2,330	1,445
South America			
Amazon [2]	Atlantic Ocean	6,450	4,010
Paraná-Plate	Atlantic Ocean	4,500	2,800
Purus	Amazon	3,350	2,080
Madeira	Amazon	3,200	1,990
Sao Francisco	Atlantic Ocean	2,900	1,800
Australia			
Murray-Darling	Southern Ocean	3,750	2,830
Darling	Murray	3,070	1,905
Murray	Southern Ocean	2,575	1,600
Murrumbidgee	Murray	1,690	1,050

* Flows through Europe and Asia

HIGHEST MOUNTAINS

	Location	metres	feet		Location	metres	feet
Europe				Ruwenzori	Uganda/Zaire	5,109	16,762
Elbrus*	Russia	5,642	18,510				
Mont Blanc[†][‡]	France/Italy	4,807	15,771	**North America**			
Monte Rosa[‡]	Italy/Switzerland	4,634	15,203	Mt McKinley (Denali)[‡]	USA (Alaska)	6,194	20,321
Also				Mt Logan	Canada	5,959	19,551
Matterhorn (Cervino)[‡]	Italy/Switzerland	4,478	14,691	Citlaltépetl (Orizaba)	Mexico	5,700	18,701
Jungfrau	Switzerland	4,158	13,642	Mt St Elias	USA/Canada	5,489	18,008
Grossglockner	Austria	3,797	12,457	Popocatépetl	Mexico	5,452	17,887
Mulhacen	Spain	3,478	11,411	*Also*			
Etna	Italy (Sicily)	3,340	10,958	Mt Whitney	USA	4,418	14,495
Zugspitze	Germany	2,962	9,718	Tajumulco	Guatemala	4,220	13,845
Olympus	Greece	2,917	9,570	Chirripo Grande	Costa Rica	3,837	12,589
Galdhopiggen	Norway	2,468	8,100	Pico Duarte	Dominican Rep.	3,175	10,417
Ben Nevis	UK (Scotland)	1,343	4,406				
				South America			
Asia[§]				Aconcagua[#]	Argentina	6,960	22,834
Everest	China/Nepal	8,848	29,029	Ojos del Salado	Argentina/Chile	6,863	22,516
K2 (Godwin Austen)	China/Kashmir	8,611	28,251	Pissis	Argentina	6,779	22,241
Kanchenjunga[‡]	India/Nepal	8,598	28,208	Mercedario	Argentina/Chile	6,770	22,211
Lhotse[‡]	China/Nepal	8,516	27,939	Huascarán[‡]	Peru	6,768	22,204
Makalu[‡]	China/Nepal	8,481	27,824				
Cho Oyu	China/Nepal	8,201	26,906	**Oceania**			
Dhaulagiri[‡]	Nepal	8,172	26,811	Puncak Jaya	Indonesia (W Irian)	5,029	16,499
Manaslu (Kutang)[‡]	Nepal	8,156	26,758	Puncak Trikora	Indonesia (W Irian)	4,750	15,584
Nanga Parbat	Kashmir	8,126	26,660	Puncak Mandala	Indonesia (W Irian)	4,702	15,427
Annapurna[‡]	Nepal	8,078	26,502	Mt Wilhelm	Papua New Guinea	4,508	14,790
Also				*Also*			
Pik Kommunizma	Tajikistan	7,495	24,590	Mauna Kea	USA (Hawaii)	4,205	13,796
Ararat	Turkey	5,165	16,945	Mauna Loa	USA (Hawaii)	4,170	13,681
Gunong Kinabalu	Malaysia (Borneo)	4,101	13,455	Mt Cook (Aorangi)	New Zealand	3,753	12,313
Fuji-san (Fujiyama)	Japan	3,776	12,388	Mt Kosciusko	Australia	2,237	7,339
Africa				**Antarctica**			
Kilimanjaro	Tanzania	5,895	19,340	Vinson Massif	—	4,897	16,066
Mt Kenya	Kenya	5,199	17,057	Mt Tyree	—	4,965	16,289

* The Caucasus Mountains include 14 other peaks higher than Mont Blanc, the highest point in non-Russian Europe
† The highest point is in France; the highest point wholly in Italian territory is 4,760m (15,616ft)
‡ Many mountains, especially in Asia, have two or more significant peaks; only the highest ones are listed here
§ The ranges of Central Asia have more than 100 peaks over 7,315m (24,000ft); thus the first 10 listed here constitute the world's 10 highest mountains   # Highest mountain outside Asia

HIGHEST WATERFALLS

Name	Total height		Location	River	Highest fall	
	m	ft			m	ft
Angel	979	3,212	Venezuela	Carrao	807	2,648
Tugela	947	3,110	Natal, South Africa	Tugela	410	1,350
Utigård	800	2,625	Nesdale, Norway	Jostedal Glacier	600	1,970
Mongefoseen	774	2,540	Mongebekk, Norway	Monge	—	—
Yosemite	739	2,425	California, USA	Yosemite Creek	739	2,425
Østre Mardøla Foss	656	2,154	Eikisdal, Norway	Mardals	296	974
Tyssestrengane	646	2,120	Hardanger, Norway	Tysso	289	948
Cuquenán	610	2,000	Venezuela	Arabopó	—	—
Sutherland	580	1,904	Otago, New Zealand	Arthur	248	815
Takkakaw	502	1,650	British Columbia, Canada	Daly Glacier	365	1,200
Ribbon	491	1,612	California, USA	Ribbon Fall Stream	491	1,612

The greatest falls by volume are the Boyoma (formerly Stanley) Falls on the Zaïre (formerly Congo), with a mean annual flow of 17,000 cu m/sec (600,000 cu ft/sec). The Niagara Falls come 4th and the Victoria Falls 9th in terms of volume, though both are relatively modest in height.

NOTABLE EARTHQUAKES*

Year	Location	Magnitude[†]	Deaths
1906	San Francisco, USA	8.3	503
1908	Messina, Italy	7.5	83,000
1920	Gansu (Kansu), China	8.6	180,000
1923	Yokohama, Japan	8.3	143,000
1927	Nan Xian, China	8.3	200,000
1932	Gansu (Kansu), China	7.6	70,000
1933	Sanriku, Japan	8.9[‡]	2,990
1935	Quetta, India[§]	7.5	60,000
1939	Chillan, Chile	8.3	28,000
1963	Skopje, Yugoslavia[#]	6.0	1,000
1964	Anchorage, Alaska	8.4	131
1970	N. Peru	7.7	86,794
1976	Guatemala	7.5	22,778
1976	Tangshan, China	8.2	242,000
1985	Mexico City, Mexico	8.1	4,200
1988	NW. Armenia	6.8	55,000
1990	N. Iran	7.7	36,000
1993	Maharastra, India	6.4	30,000
1995	Kobe, Japan	7.2	5,000
1995	Sakhalin Island, Russia	7.5	2,000
1997	NE Iran	7.1	2,400
1998	Takhar, Afghanistan	6.1	4,200
1999	NW Turkey	8.2	22,000
1999	Taiwan	7.6	4,600

* Since 1900   † On the Richter scale   ‡ Highest ever recorded   § Now Pakistan   # Now Macedonia

UNDERSTANDING MAPS

Mapmaking

While small areas can be mapped by plane (flat) surveying, larger areas must be done by geodesy, which takes into account the Earth's curvature. A variety of instruments and techniques is used to determine the position, height and extent of features – data essential to the cartographic process. Instruments such as graduated metal rods, chains, tapes and portable radar or radio transmitters are used for measuring distances, and the theodolite is used for angles. With measured distances and angles, further distances and angles as well as heights are calculated by triangulation.

Latitude and longitude

Accurate positioning of points on the Earth's surface is made possible by reference to latitude and longitude. Parallels of latitude are drawn west-east around the globe and numbered by degrees north and south of the Equator (0° of latitude). Meridians of longitude are drawn north-south and numbered by degrees east and west and the prime meridian (0° of longitude) which passes through the Royal Observatory at Greenwich in southeast London. Latitude and longitude are indicated by blue lines on the maps, and are straight or slightly curved according to the projection used.

Representing relief

Height and gradient can be represented on a map in many ways. Hachuring, in which fine lines follow the direction of the greatest slope, can give an excellent impression of the landscape but the lines may obscure other information. Hill shading, the representation of a landscape illuminated from one direction, is used alone or with colours. Contours can also be separated by colour and intermediate heights given as spot heights. These techniques are now often used in conjunction with sophisticated computerized technology, including digitalization.

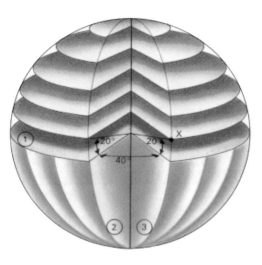

ABOVE Any point on the Earth's surface can be located in terms of longitude and latitude – in degrees, minutes and seconds east or west of the prime meridian for longitude, and north or south of the Equator for latitude. The latitude of X (the angle between X, the centre of the Earth and the plane of the Equator [1]) equals 20°, while its longitude (the angle between between the plane of the prime meridian [2] and that passing through X and the North and South Poles [3]) equals 40°.

RIGHT Reference to lines of longitude and latitude is the easiest and most common way of determining the relative positions of places on different maps, and for plotting compass directions.

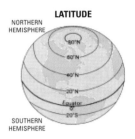

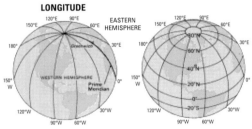

Projections

A map projection is the systematic depiction on a plane surface of the imaginary lines of latitude or longitude from a globe of the Earth. This network of lines is called the graticule and forms the framework on which an accurate depiction of the world is made. The basis of any map, the graticule, is constructed sometimes by graphical means but often by using mathematical formulas to give the intersections plotted as x and y co-ordinates.

The choice of projection is governed by the properties the cartographer wishes the map to possess, the map scale and also the extent of the area to be mapped. Since the globe is three-dimensional, it is not possible to depict its surface on a two-dimensional plane without distortion. Preservation of one of the basic properties involved – area, distance or shape – can only be secured at the expense of the others, and the choice of projection is often a compromise solution.

Map projections are constructions designed to maintain certain selected relationships of the Earth's surface. Most of the projections used for large-scale atlases, selected primarily to minimize distortion of size and distance, fall into one of three categories – conic [A], cylindrical [B] or azimuthal [C]. Each involves plotting the forms of the Earth's surface on a grid of lines of latitude and longitude, which may be shown as parallels, curved lines or radiating spokes (see below).

Conical projections use the projection of the graticule from the globe onto a cone which is tangential to a line of latitude (termed the standard parallel). This line is always an arc and scale is always true along it. Because of its method of construction it is used mainly for maps depicting the temperate latitudes around the standard parallel – that is, where there is least distortion.

Cylindrical projections are constructed by the projection of the graticule from the globe onto a cylinder tangential to the globe, and permit the whole of the Earth's surface to be depicted on one map. Though they can depict all the land masses, there is colossal exaggeration of area and shape towards the poles at the expense of equatorial regions: Greenland, for example, grows to almost the size of Africa. However, the best known example, named after the pioneering 16th-century cartographer Gerardus Mercator, has been invaluable to navigators because any straight line drawn on it is a line of constant bearing.

Azimuthal projections, sometimes called zenithal, are constructed by the projection of part of the graticule from the globe onto a plane tangential to any single point on it. This plane may be tangential to the equator (equatorial case), the poles (polar case) or any other point (oblique case). Any straight line drawn from the point where the plane touches the globe is the shortest distance from that point and is known as a great circle.

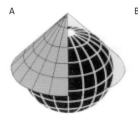

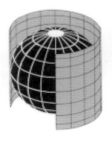

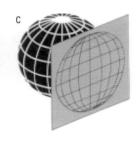

LEFT Most of the projections used for large-scale atlases, selected primarily to minimize distortion of size and distance, fall into one of three principal categories – conic [A], cylindrical [B] or azimuthal [C].

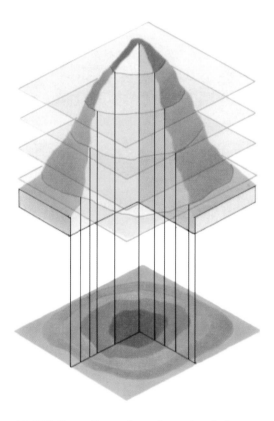

ABOVE Recording a three-dimensional shape on a flat surface can be achieved by contour scaling. Here the cross-sections of a hill at heights of 50, 100, and 150 metres (or feet) are projected onto a map of the hill. The topography of the hill can be visualized fairly well from such a map when graduated colour is employed – the closer the gradations the steeper the slope – though the crudeness of the contour intervals loses some finer detail.

10 0 10 20 30 40 50 60 70 80 90 km
10 0 10 20 30 40 50 60 miles

NORTH

SEA

Helgoland · Düne
Scharhörn · Neuwerk

Ostfriesische Inseln
Wangerooge · Langeoog · Spiekeroog · Baltrum · Norderney · Juist · Borkum
Alte Mellum

Waddeneilanden

UNITED
KINGDOM

Cromer
North Walsham
The Broads
Norwich · Great Yarmouth
Bungay · Beccles · Lowestoft
Southwold
Saxmundham · Aldeburgh
Woodbridge · Orford Ness
Felixstowe

Margate · North Foreland
Ramsgate
Deal
Dover

Calais
Sangatte · Wissant · C. Gris Nez
Marquise
Boulogne-sur-Mer · Étaples
Montreuil · Berck · Rue

Terschelling · Ameland · Schiermonnikoog
Vlieland · West-Terschelling · Holwerd · Uithuizen
Texel · Harlingen · Franeker · Leeuwarden · Kollum · Zuidhorn · Groningen · Winschoten
Den Helder · Bolsward · Sneek · FRIESLAND · Assen · Stadskanaal
Schagen · Medemblik · Heerenveen · Wolvega · Beilen · DRENTHE · Emmen
Hoorn · Enkhuizen · Urk · Meppel · Hoogeveen · Klazienaveen
Alkmaar · Edam · Lelystad · Zwolle · Hardenberg · Ommen
Haarlem · Amsterdam · Almere · Harderwijk · OVERIJSSEL · Almelo · Hengelo · Enschede
Zandvoort · Hilversum · Amersfoort · Apeldoorn · Deventer · Haaksbergen
Leiden · Utrecht · GELDERLAND · Arnhem · Winterswijk
's-Gravenhage (Den Haag) · Zoetermeer · Ede · Doesburg
Delft · Gouda · Wageningen · Doetinchem · Aalten
Hoek van Holland · Rotterdam · Tiel · Nijmegen · Emmerich
Vlaardingen · Schiedam · Dordrecht · Gorinchem · Oss · Cuijk
Hellevoetsluis · Waalwijk · 's-Hertogenbosch · Venray
ZEELAND · Bergen op Zoom · Tilburg · Helmond
Middelburg · Vlissingen · Goes · Breda · Eindhoven · Venlo
Knokke-Heist · Westerschelde · Roosendaal · BRABANT · Geldrop · Deurne
Zeebrugge · Blankenberge · Terneuzen · Brecht · LIMBURG · Weert · Roermond
Oostende · De Haan · Brugge · Antwerpen · Turnhout · Sittard
Nieuwpoort · Eeklo · Sint-Niklaas · Beveren · Lier · Geel · Bree · Maaseik
De Panne · Torhout · Gent (Gand) · Lokeren · Herentals · Hasselt · Genk
Diksmuide · Tielt · Dendermonde · Mechelen · Diest · VANDEREN
Veurne · Roeselare · Aalst · Vilvoorde · Aarschot · Leuven
Ieper · Kortrijk · Oudenaarde · Brussel (Bruxelles) · Tienen · Sint-Truiden · Maastricht
Poperinge · Menen · Ronse · Geraardsbergen · Wavre · Liège
Cassel · Tourcoing · Roubaix · Ath · Soignies · Waterloo · Namur

BELGIUM

HAINAUT
Lille · Tournai · Mons · Nivelles · Charleroi
Béthune · Lens · Seclin · Valenciennes · Binche · Thuin
NORD · Douai · Maubeuge · La Louvière · Dinant
PAS-DE-CALAIS · Cambrai · Avesnes · Philippeville · Rochefort

FRANCE

Amiens · St-Quentin · Laon · Charleville-Mézières · Sedan · Bouillon
Beauvais · Compiègne · Soissons · Reims · ARDENNES
SOMME · Montdidier · Noyon · PICARDIE
Abbeville · Péronne · Guise · THIÉRACHE · LUXEMBOURG
PARIS · Meaux · Château-Thierry · Épernay · Châlons-en-Champagne · LORRAINE
Versailles · Melun · Provins · Verdun · Metz · Nancy

GERMANY

Bremerhaven · Nordenham · Varel · Oldenburg
Wilhelmshaven · Emden · OSTFRIESLAND · Leer · Papenburg · Cloppenburg
Aurich · Norden · WESER-EMS · Quakenbrück · Lohne · Vechta
Meppen · Haselünne · Lingen · Nordhorn · Osnabrück · Bramsche · Wallenhorst
Rheine · Ibbenbüren · Münster · Warendorf · Gütersloh
Gronau · Ahaus · Coesfeld · NORDRHEIN · Hamm · Lippstadt · Soest
Bocholt · Dülmen · Recklinghausen · Herne · Dortmund · Unna · Arnsberg
Wesel · Oberhausen · Gelsenkirchen · Bochum · Witten · Hagen
Kleve · Duisburg · Essen · Wuppertal · Lüdenscheid
Krefeld · Düsseldorf · Solingen · Remscheid · WESTFALEN
Mönchengladbach · Neuss · Leverkusen · Bergisch Gladbach · Gummersbach
Viersen · Köln · Overath · Siegen
Erkelenz · Jülich · Düren · Bergheim · Troisdorf · Dillenburg
Heinsberg · Eschweiler · Bonn · Bad Honnef
Aachen · Stolberg · Euskirchen · Neuwied · Koblenz
Verviers · Eupen · Bad Münstereifel · RHEINLAND · Limburg · Lahnstein
Malmedy · Schleiden · Ahrweiler · Montabaur · Diez
St-Vith · Prüm · WESTERWALD · Wiesbaden · Mainz
Bitburg · Wittlich · Bernkastel-Kues · Bad Kreuznach · Bingen · Alzey
Trier · Idar-Oberstein · Kirchheimbolanden
LUXEMBOURG · Saarburg · Konz · Birkenfeld · Kusel · Kaiserslautern
Luxembourg · Mersch · Grevenmacher · PFALZ · Neunkirchen · Grünstadt
Differdange · Esch-sur-Alzette · Remich · SAARLAND · Homburg · Neustadt
Thionville · Merzig · Saarlouis · St. Wendel · RHEINHESSEN-PFALZ · Zweibrücken
Saarbrücken · Forbach · Pirmasens · Landau
Sarreguemines · Wissembourg · Wörth
Saverne · Strasbourg

East from Greenwich

Projection : Lambert's Conformal Conic

COPYRIGHT GEORGE PHILIP LTD.

Underlined towns give their name to the administrative area in which they stand.

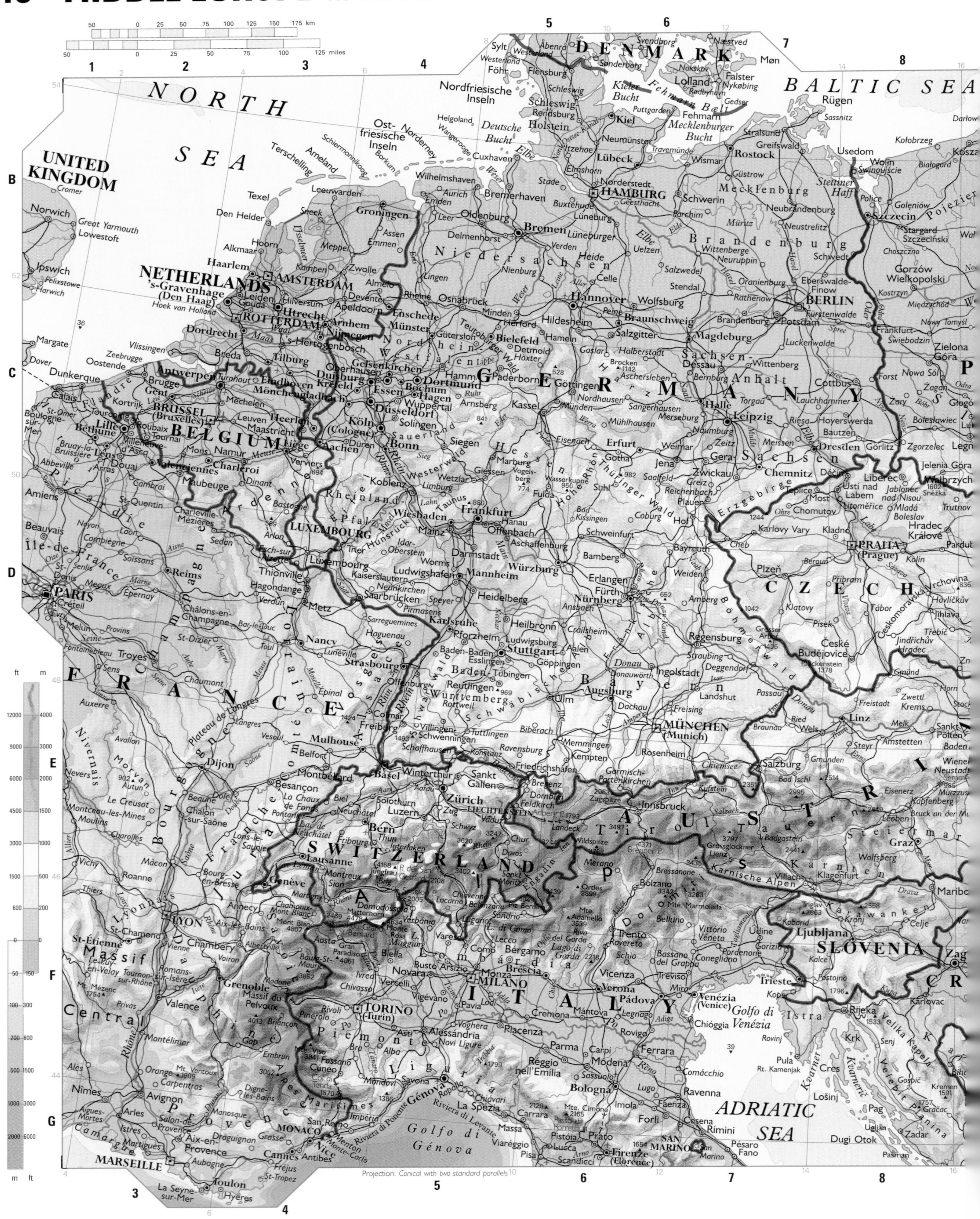

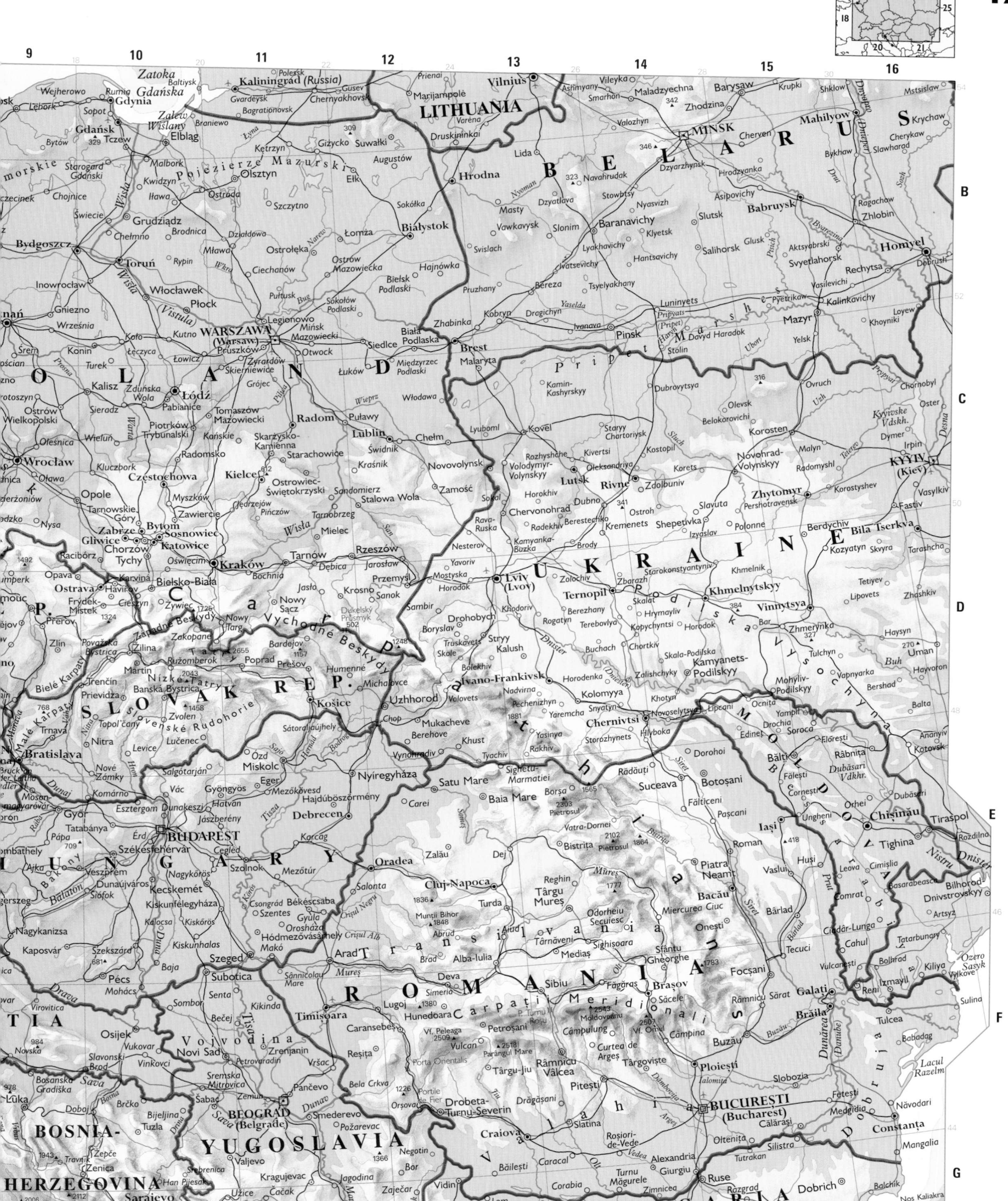
9
10
11
12
13
14
15
16
Zatoka Gdańska
Baltiysk
Polessk
Kaliningrad (Russia)
Prienai
Vilnius
Vileyka
Ashmyany
Smarhon
Maladzyechna
Barysaw
Krupki
Shklow
Mstsislaw
Wejherowo
Rumia
Gdynia
Lębork
Sopot
Gusev
Chernyakhovsk
Gvardeysk
Bagrationovsk
Marijampolė
LITHUANIA
Varėna
Zhodzina
Mahilyow
Krychaw
Cherykaw
Slawharad
Zalew Wiślany
Braniewo
Elbląg
Gdańsk
Tczew
Bytów
Kętrzyn
Giżycko
Suwałki
Druskininkai
Lida
MINSK
Cherven
Dzyarzhynsk
Valozhyn
Bykhaw
Starogard Gdański
Malbork
Kwidzyn
Pojezierze Mazurskie
Olsztyn
Augustów
Ełk
Hrodna
BELARUS
Navahrudak
Stowbtsy
Hrodzyanka
Asipovichy
Chojnice
Iława
Ostróda
Szczytno
Sokółka
Nyoman
Dzyatlava
Nyasvizh
Babruysk
Rahachow
Zhlobin
Świecie
Grudziądz
Brodnica
Działdowo
Białystok
Masty
Vawkavysk
Slonim
Baranavichy
Slutsk
Bydgoszcz
Chełmno
Toruń
Rypin
Mława
Ostrołęka
Łomża
Ostrów Mazowiecka
Hajnówka
Svislach
Lyakhavichy
Klyetsk
Hantsavichy
Salihorsk
Glusk
Aktsyabrski
Svyetlahorsk
Homyel
Dobrush
Inowrocław
Włocławek
Płock
Ciechanów
Bielsk Podlaski
Ivatsevichy
Rechytsa
Pułtusk
Bug
Sokołów Podlaski
Pruzhany
Bereza
Tsyelyakhany
Vasilevichi
Gniezno
Września
Wisła (Vistula)
Legionowo
WARSZAWA (Warsaw)
Mińsk Mazowiecki
Biała Podlaska
Zhabinka
Kobryn
Drogichyn
Yaselda
Ivanava
Pinsk
Luninyets
Pripyats (Pripet)
Pyetrikaw
Kalinkavichy
Mazyr
Khoyniki
Loyew
Konin
Koło
Kutno
Siedlce
Brest
Davyd Haradok
Stolin
Yelsk
Śrem
Łęczyca
Łowicz
Pruszków
Otwock
Żyrardów
Skierniewice
Łuków
Międzyrzec Podlaski
Malaryta
Pripet Marshes
Turek
Kalisz
Zduńska Wola
Łódź
Pabianice
Grójec
Dubrovytsya
Ovruch
Chornobyl
Kamin-Kashyrskyy
Sieradz
Tomaszów Mazowiecki
Radom
Puławy
Wieprz
Włodawa
Olevsk
Belokorovichi
Oster
Kyyivske Vdskh.
Wieluń
Piotrków Trybunalski
Końskie
Skarżysko-Kamienna
Lublin
Chełm
Lyuboml
Kovel
Staryy Chartoriysk
Korosten
Dymer
Irpin
KYYIV (Kiev)
Oleśnica
Wrocław
Radomsko
Starachowice
Świdnik
Kraśnik
Rozhyshche
Kivertsi
Kostopil
Sluch
Novohrad-Volynskyy
Malyn
Radomyshl
Kluczbork
Częstochowa
Kielce
Ostrowiec-Świętokrzyski
Novovolynsk
Volodymyr-Volynskyy
Lutsk
Oleksandriya
Rivne
Zdolbuniv
Korets
Zhytomyr
Korostyshev
Vasylkiv
Oława
Opole
Myszków
Jędrzejów
Sandomierz
Stalowa Wola
Zamość
Horokhiv
Dubno
Ostroh
Slavuta
Pershotravensk
Fastiv
Tarnowskie Góry
Zawiercie
Pińczów
Tarnobrzeg
Sokal
Chervonohrad
Radekhiv
Berestechko
Kremenets
Shepetivka
Polonne
Berdychiv
Bila Tserkva
Nysa
Zabrze
Bytom
Sosnowiec
Gliwice
Chorzów
Katowice
Mielec
Rava-Ruska
Kamyanka-Buzka
Brody
Izyaslav
Kozyatyn
Skvyra
Tarashcha
Racibórz
Tychy
Oświęcim
Kraków
Tarnów
Rzeszów
Jarosław
Nesterov
Yavoriv
UKRAINE
Starokonstyantyniv
Khmelnik
Opava
Karviná
Havířov
Bielsko-Biała
Bochnia
Dębica
Przemyśl
Mostyska
Lviv (Lvov)
Zolochiv
Zbarazh
Tetiyev
Ostrava
Frýdek-Místek
Český Těšín
Żywiec
Nowy Sącz
Jasło
Krosno
Sanok
Horodok
Ternopil
Khmelnytskyy
Lipovets
Zhashkiv
Přerov
Nowy Targ
Dukelský Průsmyk
Sambir
Drohobych
Khodoriv
Berezhany
Skalat
Hrymayliv
Vinnytsya
Zlín
Zakopane
Západné Beskydy
Východné Beskydy
Bardejov
Boryslav
Rogatyn
Terebovlya
Kopychyntsi
Horodok
Bar
Zhmerynka
Haysyn
Považská Bystrica
Žilina
Tatry
Ružomberok
Poprad
Prešov
Truskavets
Stryy
Kalush
Dnister
Buchach
Chortkiv
Skala-Podilska
Tulchyn
Uman
Martin
Nízke Tatry
Humenné
Bolekhiv
Ivano-Frankivsk
Horodenka
Zalishchyky
Kamyanets-Podilskyy
Vapnyarka
Buh
Hayvoron
Biele Karpaty
Trenčín
Banská Bystrica
SLOVAK REP.
Michalovce
Nadvirna
Kolomyya
Khotyn
Mohyliv-Podilskyy
Bershad
Prievidza
Košice
Uzhhorod
Volovets
Pechenizhyn
Snyatyn
Novoselytsya
Lipcani
Ocnița
Yampil
Balta
Malé Karpaty
Topoľčany
Zvolen
Slovenské Rudohorie
Chop
Mukacheve
Yaremcha
Chernivtsi
Drochia
Soroca
Trnava
Nitra
Lučenec
Sátoraljaújhely
Berehove
Khust
Yasinya
Storozhynets
Hlyboka
Edineț
Florești
Ananyiv
Bratislava
Levice
Sajó
Hernád
Bodrog
Rakhiv
Dorohoi
Bălți
Rîbnița
Kotovsk
Nové Zámky
Ózd
Miskolc
Vynohradiv
Tyachiv
Sighetu Marmației
Rădăuți
Botoșani
Fălești
Dubăsari Vdkhr.
Salgótarján
Eger
Nyíregyháza
Satu Mare
Suceava
Dunaj
Komárno
Vác
Gyöngyös
Mezőkövesd
Hajdúböszörmény
Baia Mare
Borșa
Pietrosul
Fălticeni
Cornești
Orhei
Dubăsari
Mosonmagyaróvár
Győr
Esztergom
Dunakeszi
Hatvan
Carei
Pașcani
Ungheni
Chișinău
Tiraspol
Tatabánya
Jászberény
Debrecen
Vatra Dornei
Iași
MOLDOVA
Pápa
BUDAPEST
Érd
Cegléd
Karcag
Bistrița
Roman
Tighina
Rozdilna
Székesfehérvár
Szolnok
Zalău
Dej
Hunedoara
Piatra Neamț
Huși
Vaslui
Leova
Cimișlia
Nistru
Dnister
Bakony
Veszprém
Balaton
Dunaújváros
Kecskemét
Nagykőrös
Mezőtúr
Oradea
Cluj-Napoca
Reghin
Mureș
Bacău
Basarabeasca
Bilhorod-Dnistrovskyy
Siófok
Kiskunfélegyháza
Salonta
Turda
Târgu Mureș
Miercurea Ciuc
Comrat
Artsyz
Csongrád
Békéscsaba
Szentes
Crișul Negru
Muntii Bihor
Odorheiu Secuiesc
Onești
Bârlad
Nagykanizsa
Kalocsa
Kiskőrös
Gyula
Hódmezővásárhely
Crișul Alb
Abrud
Aiud
Târnăveni
Sfântu Gheorghe
Ciadâr-Lunga
Cahul
Tatarbunary
Kaposvár
Szekszárd
Kiskunhalas
Szeged
Makó
Arad
Transilvania
Mediaș
Sighișoara
Tecuci
Bolhrad
Kiliya
Vylkove
Ozero Sasyk
Pécs
Baja
Subotica
Sânnicolau Mare
Brad
Alba-Iulia
Deva
Sibiu
Făgăraș
Brașov
Focșani
Vulcanești
Reni
Izmayil
Mohács
Senta
Simeria
ROMANIA
Galați
Dunărea (Danube)
Sulina
Virovitica
Dráva
Sombor
Kikinda
Lugoj
Carpații Meridionali
Moldoveanu
Săcele
Râmnicu Sărat
Brăila
Tulcea
Osijek
Bečej
Tisa
Timișoara
Caransebeș
Petroșani
Câmpulung
Vf. Omul
Câmpina
Buzău
Babadag
Vukovar
Vojvodina
Zrenjanin
Reșița
Vf. Peleaga
Vulcan
Parângul Mare
Curtea de Argeș
Târgoviște
Ploiești
Lacul Razelm
Slavonski Brod
Vinkovci
Novi Sad
Petrovaradin
Vršac
Porta Orientalis
Râmnicu Vâlcea
Târgu-Jiu
Pitești
Slobozia
Sava
Bosanska Gradiška
Sremska Mitrovica
Pančevo
Bela Crkva
Portile de Fier
Orșova
Drăgășani
Dâmbovița
Ialomița
Fetești
Năvodari
Luka
Brčko
Bijeljina
Šabac
Zemun
BEOGRAD (Belgrade)
Dunav
Smederevo
Drobeta-Turnu-Severin
Jiu
Argeș
BUCUREȘTI (Bucharest)
Medgidia
Doboj
Tuzla
Drina
Požarevac
Craiova
Slatina
Roșiori-de-Vede
Oltenița
Călărași
Constanța
BOSNIA-HERZEGOVINA
YUGOSLAVIA
Negotin
Bor
Olt
Vedea
Alexandria
Giurgiu
Silistra
Dobrogea
Mangalia
Travnik
Zepče
Zenica
Valjevo
Kragujevac
Jagodina
Zaječar
Vidin
Băilești
Caracal
Corabia
Turnu Măgurele
Zimnicea
Ruse
Tutrakan
Razgrad
Dobrich
Balchik
Srebrenica
Han Pijesak
Užice
Čačak
Kraljevo
Morava
Timok
Lom
Oryakhovo
BULGARIA
Varna
Nos Kaliakra
Cincar
Sarajevo
Višegrad
East from Greenwich
B
C
D
E
F
G
COPYRIGHT GEORGE PHILIP LTD.

Corse (Corsica)

MEDITERRANEAN SEA

Bay of Biscay

English Channel

Golfe de Gascogne

Golfe du Lion

UNITED KINGDOM

BELGIUM

LUXEMBOURG

GERMANY

SWITZERLAND

ITALY

AUSTRIA

SPAIN

ANDORRA

MONACO

FRANCE

Projection: Conical with two standard parallels

East from Greenwich

West from Greenwich

COPYRIGHT GEORGE PHILIP LTD

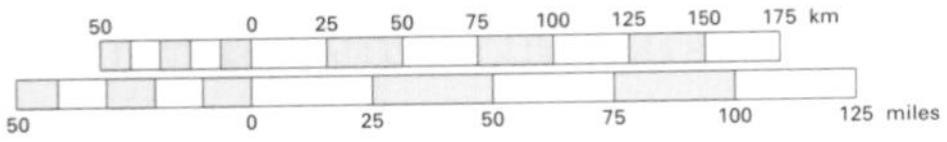

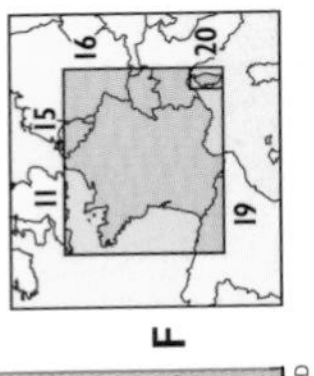

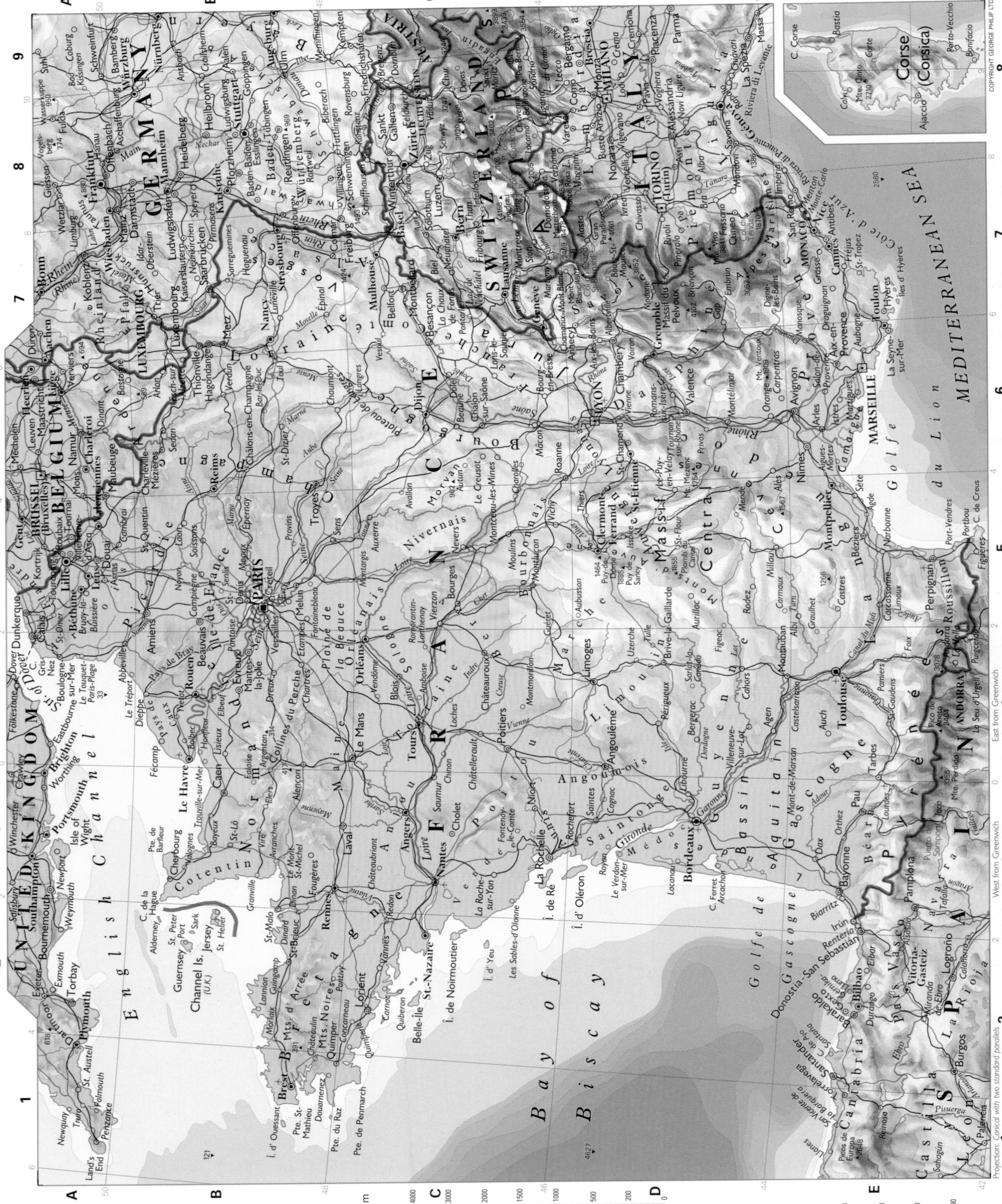

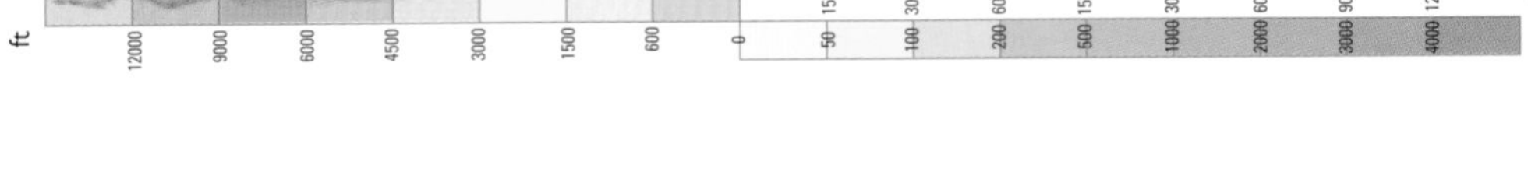

50 0 25 50 75 100 125 150 175 km
50 0 25 50 75 100 125 miles

SPAIN

PORTUGAL

FRANCE

ALGERIA

MOROCCO

MADRID · BARCELONA · Valencia · Sevilla · Zaragoza · Bilbao · Málaga · LISBOA · Porto · Toulouse · Montpellier · ALGER · Oran

MEDITERRANEAN SEA

ATLANTIC OCEAN

Golfe du Lion · Baleares · Mallorca · Menorca · Evissa (Ibiza) · Formentera

Pyrénées · ANDORRA · Cordillera Cantábrica · Sierra Nevada · Sierra Morena · Picos de Europa

Ebro · Duero · Tajo · Guadiana · Guadalquivir · Douro

Str. of Gibraltar · G. de Cádiz · Costa Brava · Costa Dorada · Costa Blanca · Costa del Sol · Golfo de Valencia · G. de San Jorge

Ceuta (Sp.) · Melilla (Sp.) · Gibraltar (U.K.)

Projection: Conical with two standard parallels

West from Greenwich 0 East from Greenwich

ft m 2000 · 1500 · 1000 · 500 · 200 · 0 · 150 · 300 · 600 · 1500 · 3000 · 6000 · 9000 · 12000 m ft

ITALY AND THE BALKAN STATES 1:5 000 000

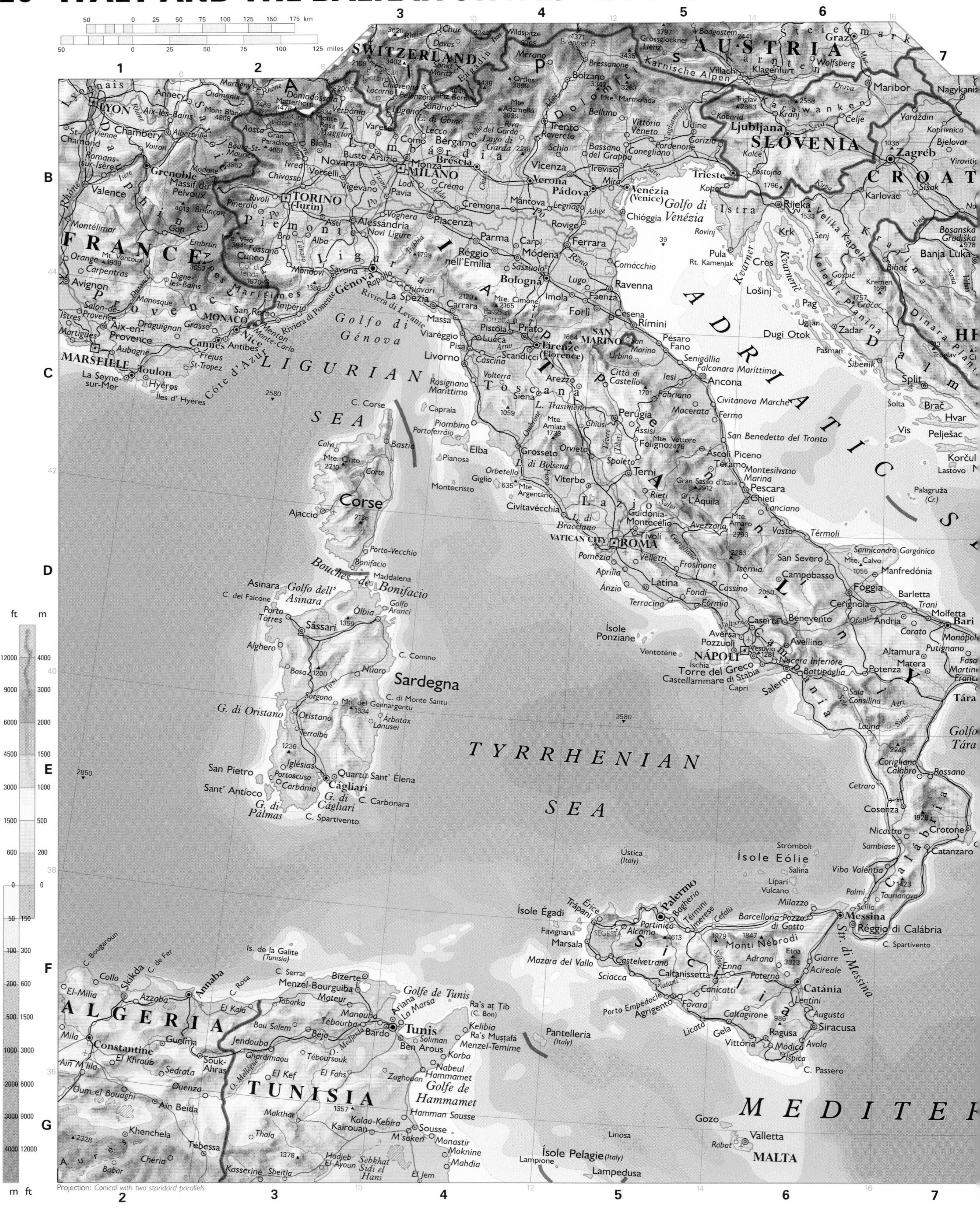

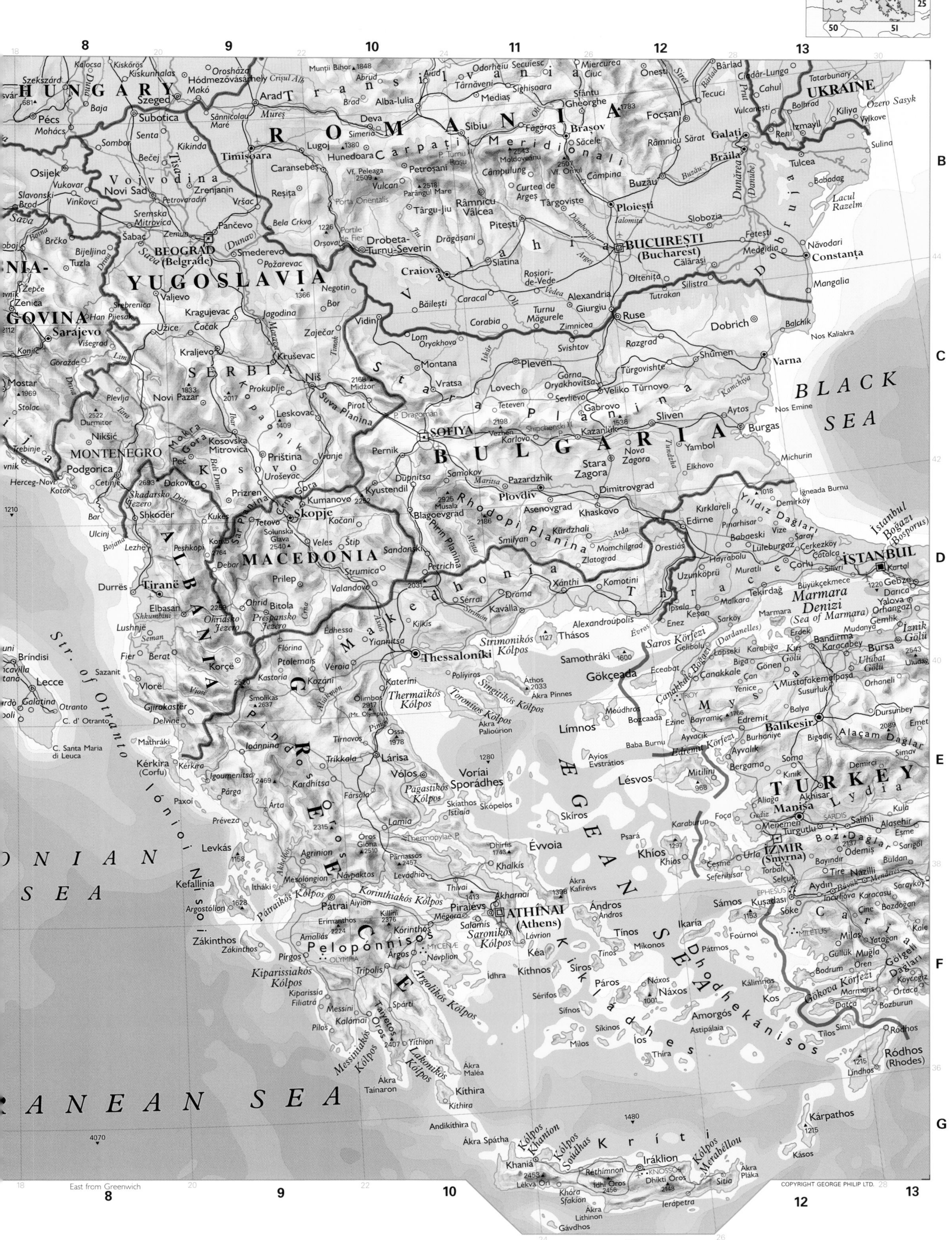

8
9
10
11
12
13
B
C
D
E
F
G
HUNGARY
ROMANIA
UKRAINE
YUGOSLAVIA
SERBIA
MONTENEGRO
MACEDONIA
ALBANIA
BULGARIA
GREECE
TURKEY
BLACK SEA
ÆGEAN SEA
IONIAN SEA
MEDITERRANEAN SEA
Str. of Otranto
Szekszárd
Kalocsa
Kiskőrös
Kiskunhalas
Orosháza
Hódmezővásárhely
Makó
Szeged
Pécs
Mohács
Baja
Subotica
Sombor
Senta
Kikinda
Bečej
Osijek
Vukovar
Vinkovci
Novi Sad
Petrovaradin
Zrenjanin
Vršac
Vojvodina
Tisza
Slavonski Brod
Sava
Brčko
Bijeljina
Tuzla
Sremska Mitrovica
Šabac
Zemun
Pančevo
BEOGRAD (Belgrade)
Smederevo
Požarevac
Dunav
Valjevo
Kragujevac
Jagodina
Užice
Čačak
Kraljevo
Kruševac
Niš
Prokuplje
Leskovac
Novi Pazar
Kosovska Mitrovica
Priština
Uroševac
Vranje
Prizren
Pec
Đakovica
Kosovo
Podgorica
Cetinje
Kotor
Herceg-Novi
Nikšić
Durmitor
Pljevlja
Sarajevo
Višegrad
Goražde
Mostar
Stolac
Konjic
Zenica
Žepče
Srebrenica
Han Pijesak
Skadarsko Jezero
Shkodër
Bar
Ulcinj
Kukës
Peshkopi
Lezhë
Durrës
Tiranë
Elbasan
Shkumbini
Lushnjë
Fier
Berat
Vlorë
Korçë
Gjirokastër
Delvinë
Sazanit
Ohrid
Ohridsko Jezero
Prespansko Jezero
Bitola
Prilep
Skopje
Tetovo
Kumanovo
Kočani
Veles
Štip
Strumica
Valandovo
Sandanski
Petrich
Kyustendil
Blagoevgrad
Rhodopi Planina
Pirin Planina
Musala
Arad
Timişoara
Lugoj
Caransebeş
Reşiţa
Hunedoara
Deva
Simeria
Alba-Iulia
Brad
Abrud
Munţii Bihor
Sânnicolau Mare
Mureş
Crişul Alb
Transilvania
Târnăveni
Odorheiu Secuiesc
Miercurea Ciuc
Sighişoara
Mediaş
Sibiu
Făgăraş
Braşov
Sfântu Gheorghe
Săcele
Carpaţii Meridionali
Petroşani
Vulcan
Târgu-Jiu
Râmnicu Vâlcea
Piteşti
Câmpulung
Curtea de Argeş
Târgovişte
Câmpina
Ploieşti
Buzău
Focşani
Oneşti
Râmnicu Sărat
Galaţi
Brăila
Tecuci
Bârlad
Ciadâr-Lunga
Cahul
Vulcaneşti
Reni
Izmayil
Kiliya
Vylkove
Bolhrad
Tatarbunary
Ozero Sasyk
Tulcea
Sulina
Babadag
Lacul Razelm
Dunărea (Danube)
Dobruja
Slobozia
Feteşti
Medgidia
Năvodari
Constanţa
Mangalia
BUCUREŞTI (Bucharest)
Călăraşi
Olteniţa
Silistra
Tutrakan
Alexandria
Giurgiu
Ruse
Turnu Măgurele
Zimnicea
Roşiori-de-Vede
Slatina
Craiova
Drobeta-Turnu-Severin
Orşova
Drăgăşani
Băileşti
Caracal
Corabia
Calafat
Vidin
Lom
Oryakhovo
Montana
Vratsa
Pleven
Lovech
Svishtov
Razgrad
Dobrich
Balchik
Nos Kaliakra
Shumen
Varna
Türgovishte
Veliko Türnovo
Gorna Oryakhovitsa
Sevlievo
Gabrovo
Stara Planina
Teteven
Kazanlŭk
Sliven
Aytos
Burgas
Nos Emine
Yambol
Elkhovo
Michurin
Nova Zagora
Stara Zagora
SOFIYA
Pernik
Dupnitsa
Samokov
Pazardzhik
Plovdiv
Asenovgrad
Khaskovo
Dimitrovgrad
Kŭrdzhali
Smilyan
Momchilgrad
Zlatograd
Zaječar
Negotin
Bor
Pirot
Suva Planina
Midžor
Edirne
Kırklareli
Yıldız Dağları
İğneada Burnu
Demirköy
Pınarhisar
Vize
Babaeski
Lüleburgaz
Saray
Çerkezköy
Çatalca
Çorlu
Silivri
İSTANBUL
İstanbul Boğazı (Bosporus)
Kartal
Gebze
Darıca
Hayrabolu
Muratlı
Uzunköprü
Büyükçekmece
Tekirdağ
Malkara
Keşan
İpsala
Enez
Şarköy
Marmara
Marmara Denizi (Sea of Marmara)
Yalova
Orhangazi
Gemlik
Mudanya
İznik Gölü
Bursa
Uludağ
Erdek
Bandırma
Karacabey
Gelibolu
Lapseki
Çanakkale Boğazı (Dardanelles)
Çanakkale
Biga
Gönen
Ulubat Gölü
Orhaneli
Yenice
Mustafakemalpaşa
Susurluk
Ezine
Bayramiç
Balya
Dursunbey
Edremit
Balıkesir
Bigadiç
Alaçam Dağlar
Simav
Emet
Ayvacık
Burhaniye
Edremit Körfezi
Baba Burnu
Ayvalık
Bergama
Soma
Kınık
Demirci
Akhisar
Manisa
Lydia
Kula
Salihli
Alaşehir
Eşme
Turgutlu
Menemen
Aliağa
Foça
Gediz
İZMİR (Smyrna)
Urla
Çeşme
Seferihisar
Bozdağlar
Ödemiş
Bayındır
Tire
Nazilli
Buldan
Torbalı
Selçuk
Aydın
Büyük Menderes
Sarayköy
Kuşadası
Söke
Karacasu
Çine
Bozdoğan
Milas
Yatağan
Muğla
Güllük
Bodrum
Ören
Gölgeli Dağları
Köyceğiz
Ortaca
Gökova Körfezi
Marmaris
Datça
Bozburun
Caria
Saros Körfezi
Alexandroúpolis
Évros
Komotini
Xánthi
Dráma
Kavála
Sérrai
Strimón
Kilkís
Thrace
Makedhonía
Thessaloníki
Yiannitsa
Édhessa
Flórina
Ptolemaís
Kastoría
Kozáni
Véroia
Kateríni
Thermaïkós Kólpos
Olimbos
Pindhos
Ioánnina
Igoumenitsa
Árta
Préveza
Párga
Kérkira (Corfu)
Mathráki
Paxoí
Levkás
Kefallinía
Argostólion
Itháki
Zákinthos
Agrinion
Mesolóngion
Návpaktos
Pátrai
Aiyion
Pátraikós Kólpos
Korinthiakós Kólpos
Kórinthos
Mégara
Piraiévs
ATHÍNAI (Athens)
Akharnai
Salamís
Saronikós Kólpos
Lávrion
Peloponnisos
Erimanthos
Amaliás
Pírgos
Olympia
Tripolis
Árgos
Návplion
Mycenae
Argolikós Kólpos
Kiparissiakós Kólpos
Kiparissia
Filiatrá
Messíni
Kalámai
Pílos
Spárti
Taíyetos Óros
Yíthion
Messiniakós Kólpos
Lakonikós Kólpos
Ákra Maléa
Ákra Tainaron
Kíthira
Andikíthira
Trikkala
Kardhítsa
Lárisa
Vólos
Pagastikós Kólpos
Fársala
Lamía
Thermopylae P.
Ossa
Tírnavos
Voríai Sporádhes
Skíathos
Skópelos
Skíros
Évvoia
Khalkís
Levádhia
Thívai
Párnassos
Áthos
Singitikós Kólpos
Toronaíos Kólpos
Políyiros
Strimonikós Kólpos
Thásos
Samothráki
Gökçeada
Límnos
Moúdhros
Bozcaada
Áyios Evstrátios
Lésvos
Mitilíni
Psará
Khíos
Sámos
Ikaría
Fournoi
Pátmos
Ándros
Tínos
Míkonos
Kéa
Kíthnos
Síros
Sérifos
Sífnos
Páros
Náxos
Mílos
Síkinos
Íos
Thíra
Amorgós
Astipálaia
Kálimnos
Kos
Tílos
Simi
Ródhos (Rhodes)
Líndhos
Kárpathos
Kásos
Kikládhes
Dhodhekánisos
Ídhra
Kríti
Khania
Kólpos Khanion
Kólpos Soúdhas
Réthimnon
Iráklion
Knossos
Ídhi Óros
Dhíkti Óros
Kólpos Merabéllou
Sitía
Ákra Pláka
Ierápetra
Khóra Sfakíon
Ákra Líthinon
Gávdhos
Lékva Óri
Ákra Spátha
Brindisi
Lecce
Galatina
Otranto
C. d' Otranto
C. Santa Maria di Leuca
East from Greenwich

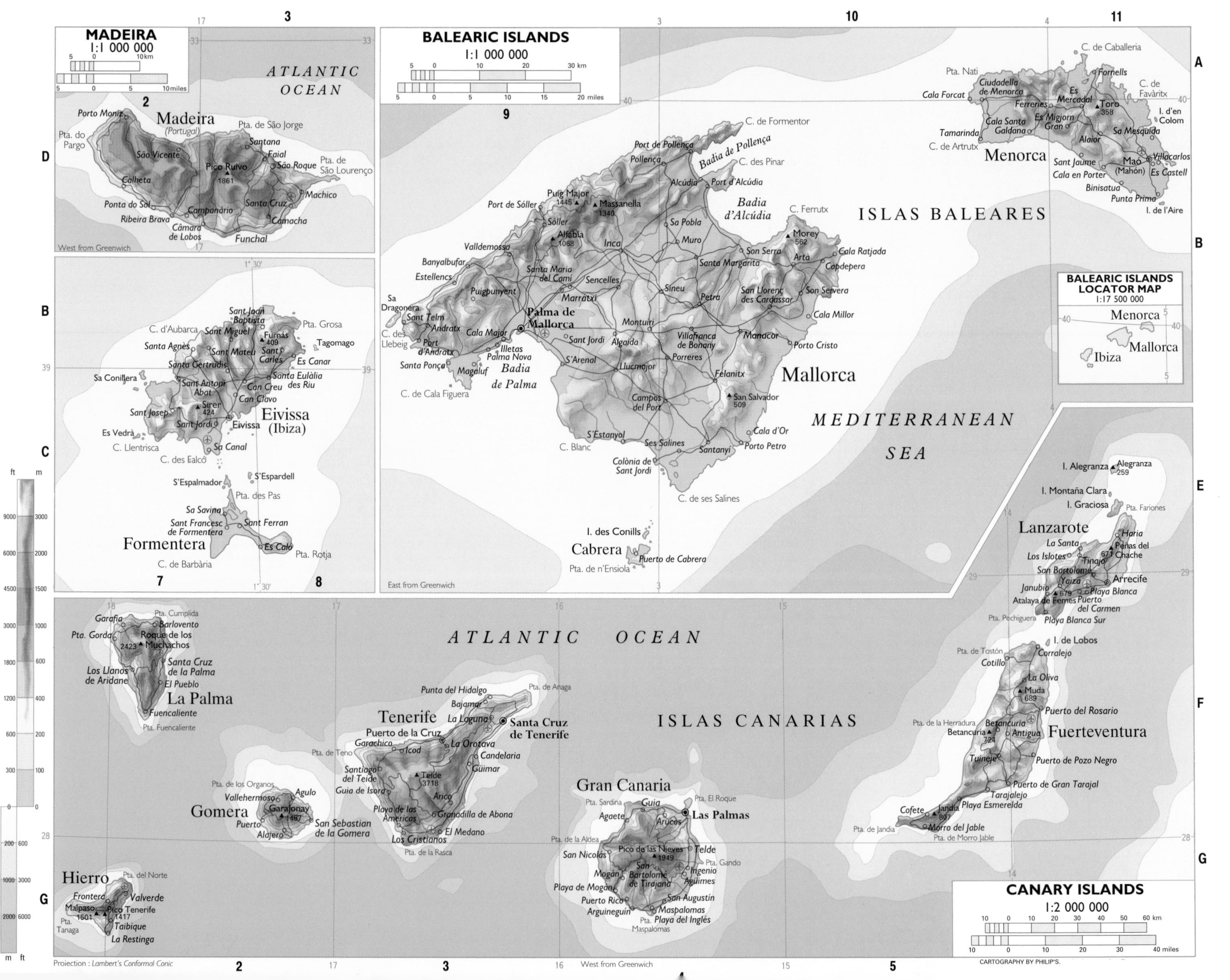

BALEARIC ISLANDS LOCATOR MAP
1:17 500 000

Menorca
Mallorca
Ibiza

CANARY ISLANDS
1:2 000 000

CARTOGRAPHY BY PHILIP'S.

BALEARIC ISLANDS
1:1 000 000

MADEIRA
1:1 000 000

Projection : Lambert's Conformal Conic

ISLAS BALEARES

MEDITERRANEAN SEA

Menorca

Mallorca

Cabrera

Eivissa (Ibiza)

Formentera

Madeira (Portugal)

ATLANTIC OCEAN

ISLAS CANARIAS

Lanzarote

Fuerteventura

Gran Canaria

Tenerife

Gomera

La Palma

Hierro

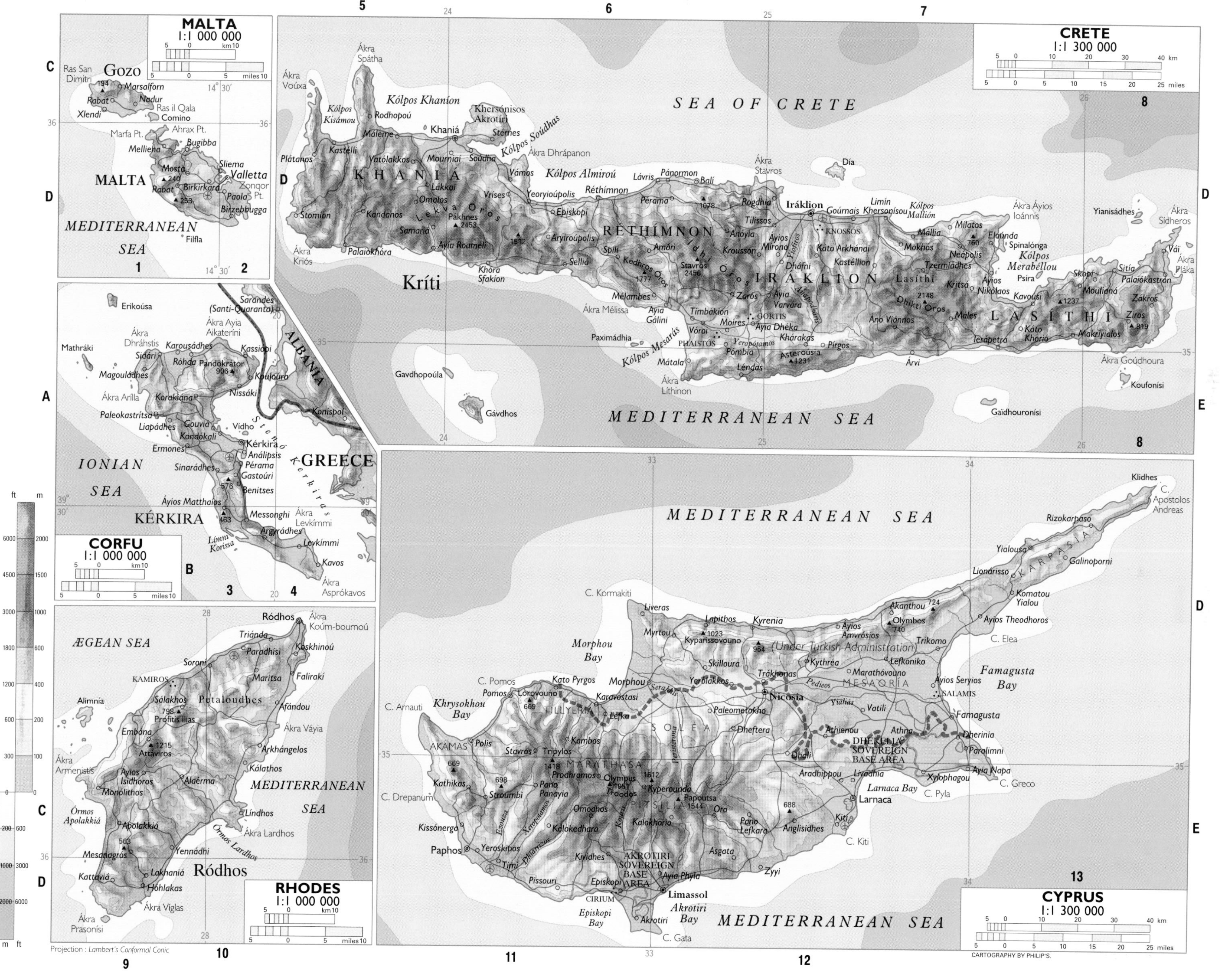

CRETE
1:1 300 000

MALTA
1:1 000 000

CORFU
1:1 000 000

RHODES
1:1 000 000

CYPRUS
1:1 300 000

CARTOGRAPHY BY PHILIP'S.

Projection: Lambert's Conformal Conic

24 EASTERN EUROPE AND TURKEY 1:10 000 000

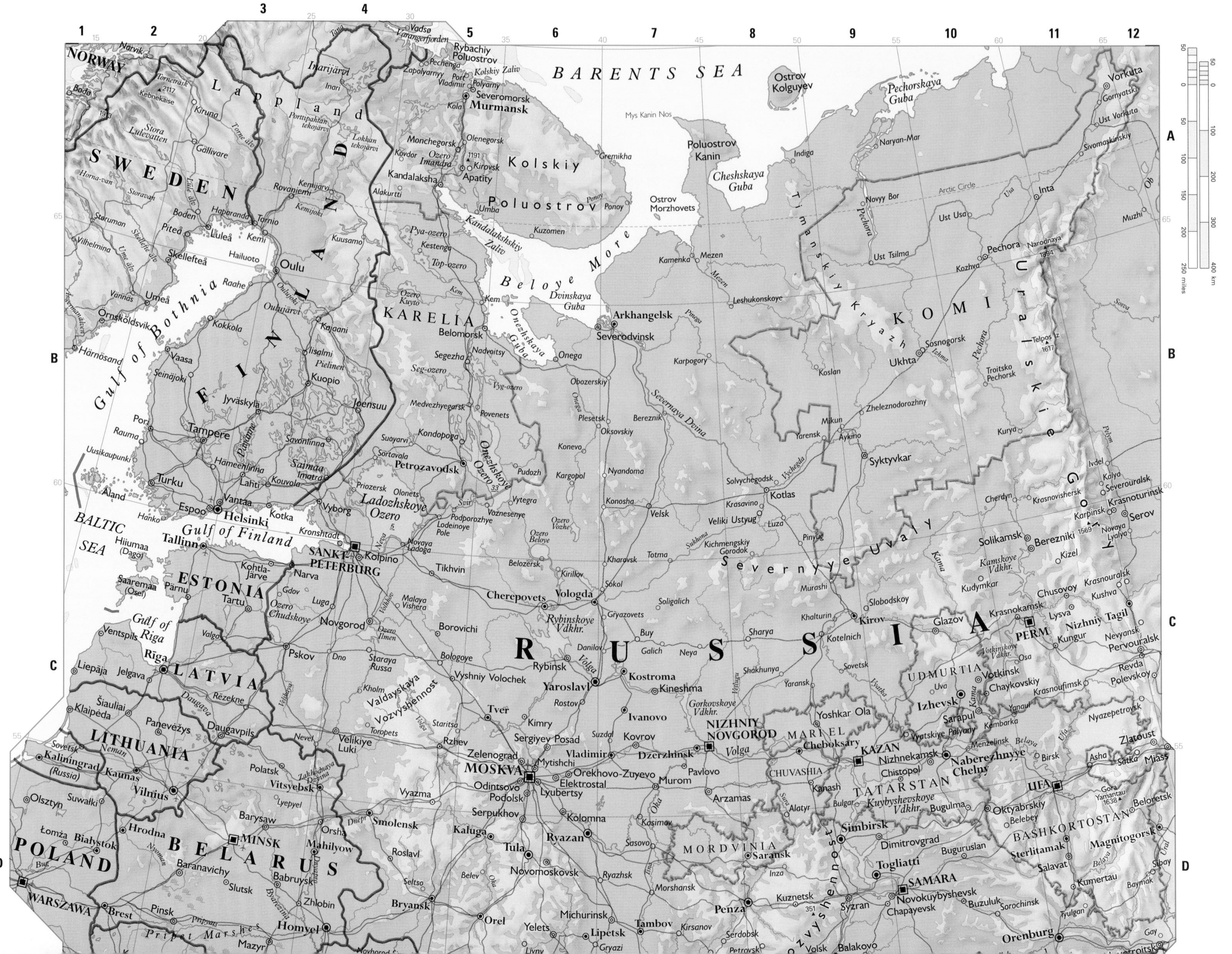

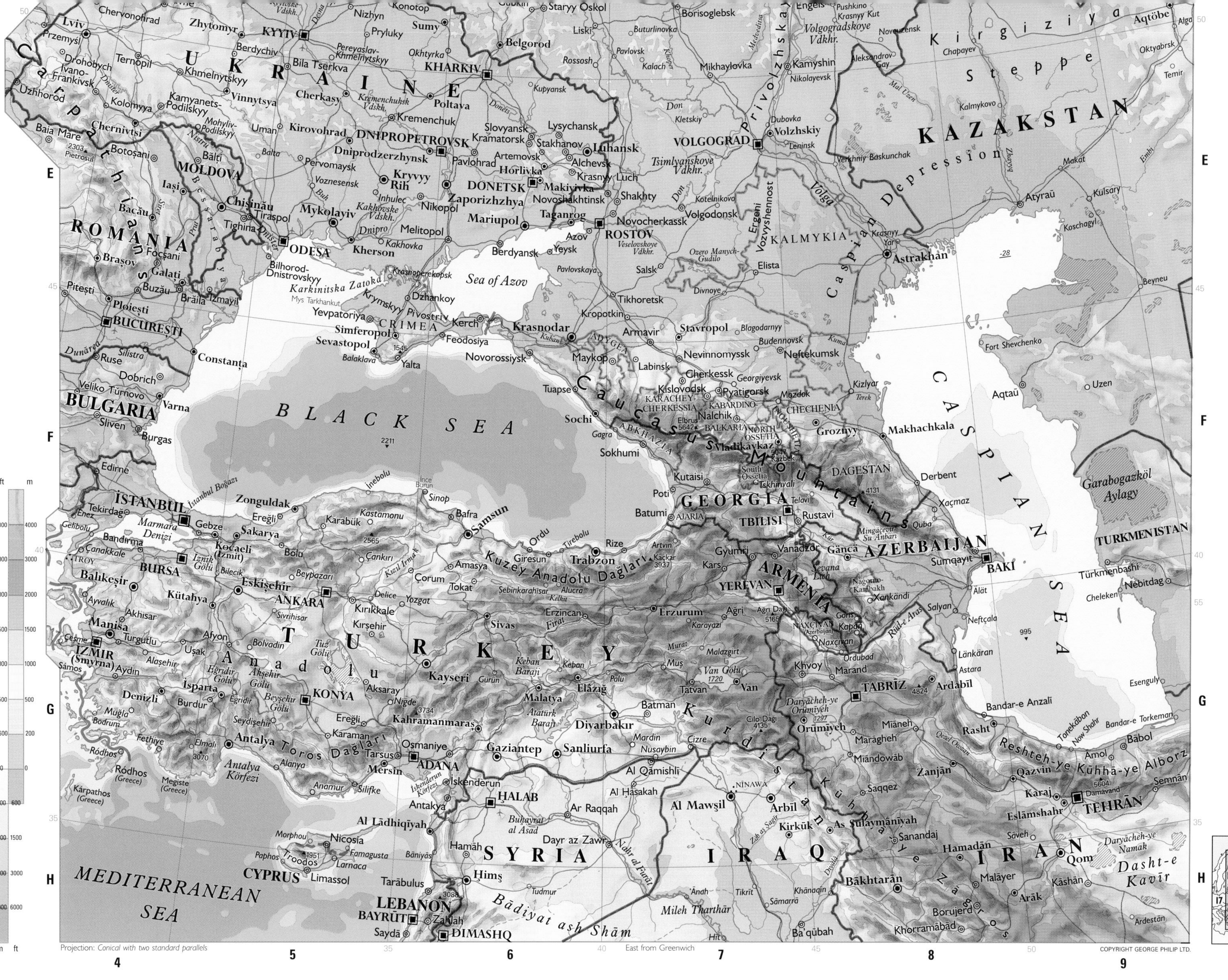

BLACK SEA

CASPIAN SEA

MEDITERRANEAN SEA

Sea of Azov

KAZAKSTAN

UKRAINE

ROMANIA

BULGARIA

MOLDOVA

GEORGIA

ARMENIA

AZERBAIJAN

TURKMENISTAN

IRAN

IRAQ

SYRIA

LEBANON

CYPRUS

TURKEY

Caucasus Mountains

Kurdistan

Anadolu

Kirgiz Steppe

Caspian Depression

Volga

Don

Projection: Conical with two standard parallels

COPYRIGHT GEORGE PHILIP LTD.

East from Greenwich

KYIV

KHARKIV

DNIPROPETROVSK

DONETSK

ODESA

VOLGOGRAD

ROSTOV

Astrakhan

Krasnodar

BAKI

TBILISI

YEREVAN

ANKARA

ISTANBUL

BURSA

İZMIR

KONYA

ADANA

HALAB

DIMASHQ

BAYRŪT

TEHRĀN

Tabrīz

Al Mawşil

Hamadān

BUCUREŞTI

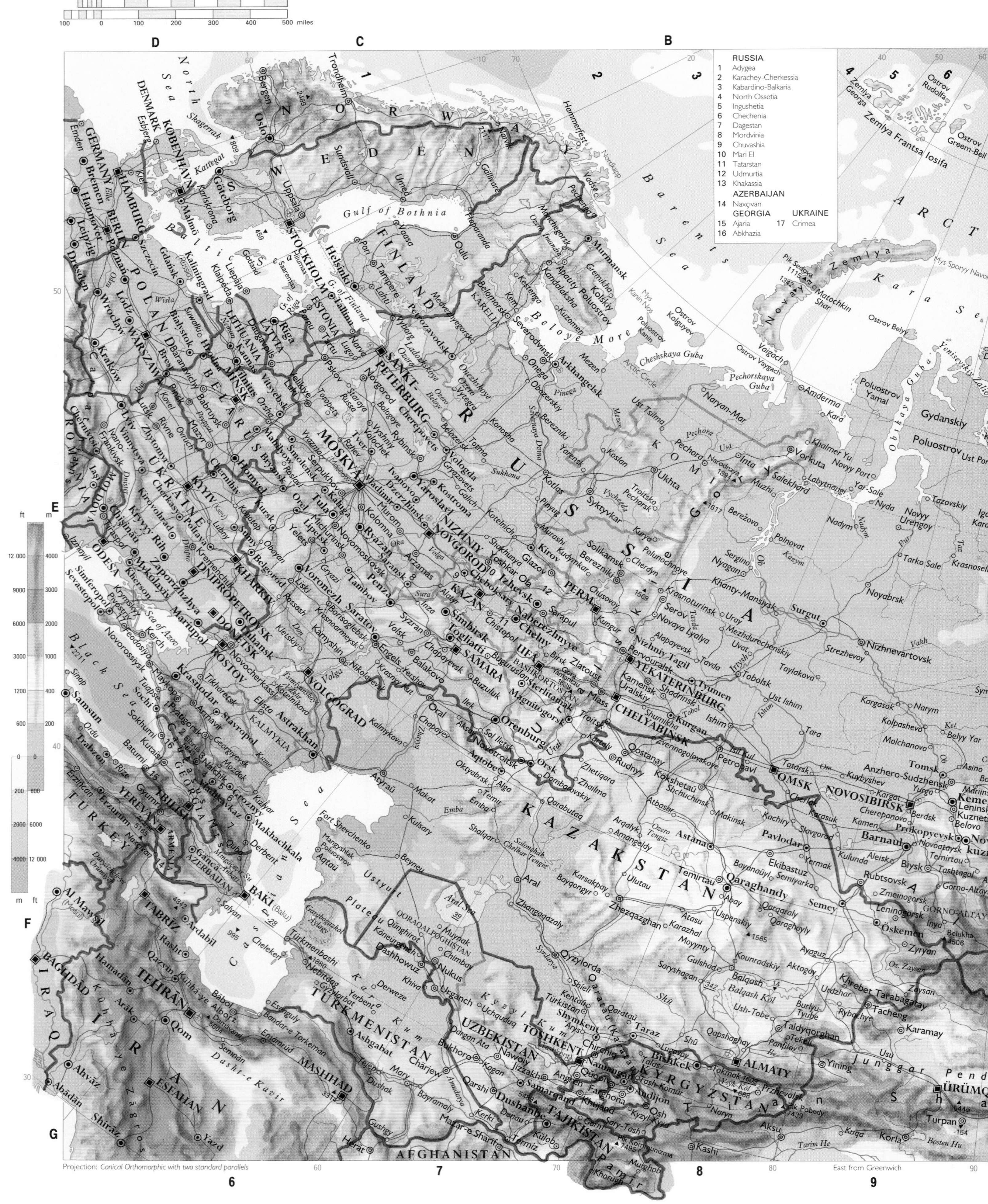
100 0 100 200 300 400 500 600 700 800 km
100 0 100 200 300 400 500 miles
RUSSIA
1 Adygea
2 Karachey-Cherkessia
3 Kabardino-Balkaria
4 North Ossetia
5 Ingushetia
6 Chechenia
7 Dagestan
8 Mordvinia
9 Chuvashia
10 Mari El
11 Tatarstan
12 Udmurtia
13 Khakassia
AZERBAIJAN
14 Naxçıvan
GEORGIA
15 Ajaria
16 Abkhazia
UKRAINE
17 Crimea
ft m
12 000 4000
9000 3000
6000 2000
3000 1000
1200 400
600 200
0 0
200 600
2000 6000
4000 12 000
m ft
NORWAY
SWEDEN
FINLAND
ESTONIA
LATVIA
LITHUANIA
BELARUS
POLAND
GERMANY
DENMARK
UKRAINE
MOLDOVA
ROMANIA
TURKEY
GEORGIA
ARMENIA
AZERBAIJAN
IRAN
IRAQ
TURKMENISTAN
UZBEKISTAN
KAZAKSTAN
KYRGYZSTAN
TAJIKISTAN
AFGHANISTAN
RUSSIA
Barents Sea
Kara Sea
Baltic Sea
Black Sea
Caspian Sea
Aral Sea
Gulf of Bothnia
North Sea
Beloye More
Zemlya Frantsa Iosifa
Novaya Zemlya
MOSKVA
SANKT-PETERBURG
NIZHNIY NOVGOROD
KAZAN
SAMARA
VOLGOGRAD
ROSTOV
YEKATERINBURG
CHELYABINSK
OMSK
NOVOSIBIRSK
KYIV
KHARKIV
DNIPROPETROVSK
DONETSK
ODESA
MINSK
WARSZAWA
STOCKHOLM
KØBENHAVN
HAMBURG
BERLIN
TBILISI
YEREVAN
BAKI
TEHRĀN
ESFAHAN
MASHHAD
TABRIZ
BAGHDĀD
TOSHKENT
ALMATY
ÜRÜMQI
Projection: Conical Orthomorphic with two standard parallels
East from Greenwich

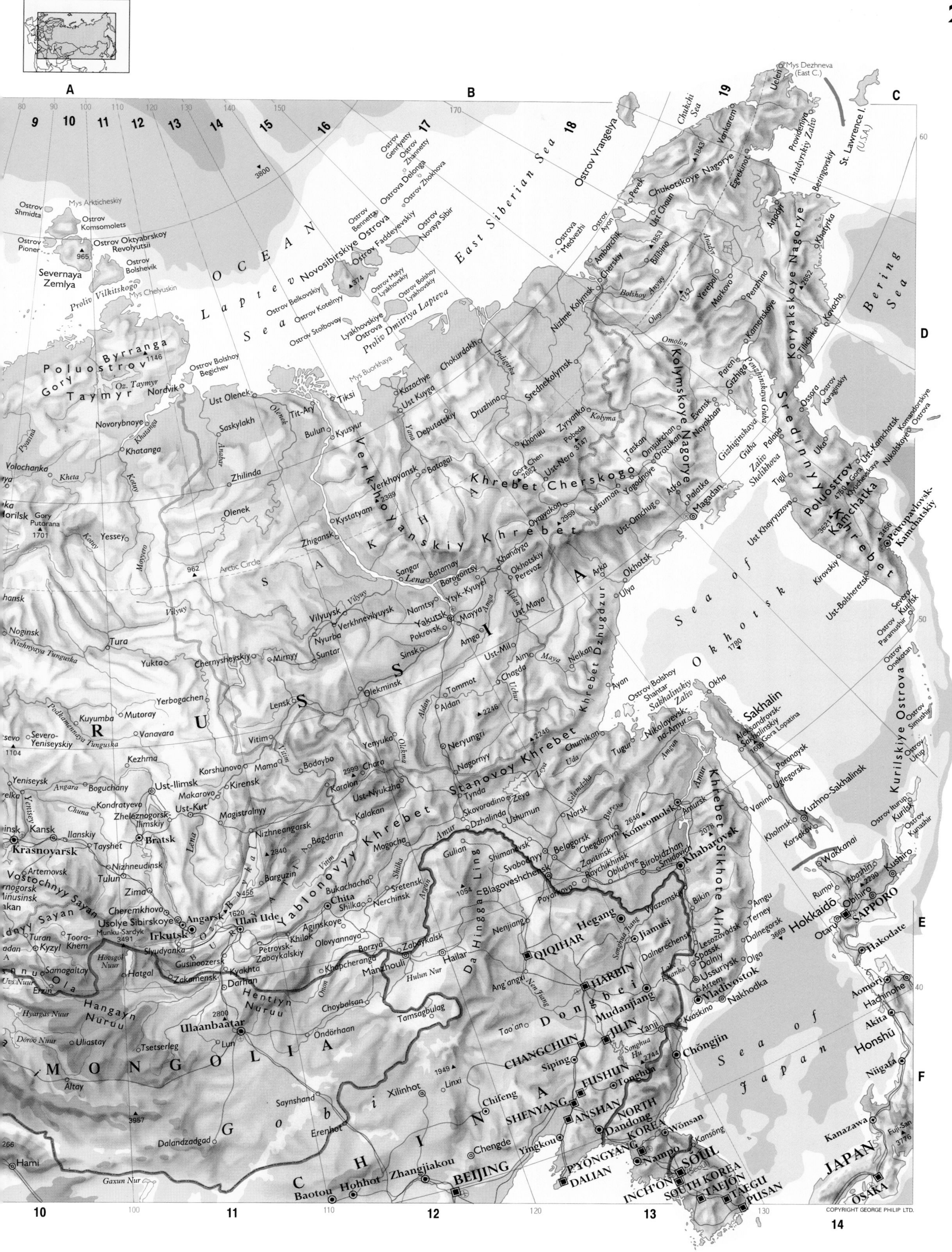

A   B   C

9 10 11 12 13 14 15 16 17 18 19

ARCTIC OCEAN

Ostrov Shmidta
Mys Arkticheskiy
Ostrov Komsomolets
Ostrov Oktyabrskoy Revolyutsii
965
Ostrov Pioner
Ostrov Bolshevik
Severnaya Zemlya
Proliv Vilkitskogo
Mys Chelyuskin

Laptev Sea

Ostrov Genriyetty
Ostrov Zhannetty
Ostrova Delonga
Ostrov Zhokhova
Novosibirskiye Ostrova
Ostrov Bennetta
Ostrov Faddeyevskiy
Ostrov Novaya Sibir
3800
Ostrov Malyy Lyakhovskiy
Ostrov Belkovskiy
374
Ostrov Kotelnyy
Lyakhovskiye Ostrova
Ostrov Bolshoy Lyakhovskiy
Ostrov Stolbovoy

East Siberian Sea

Ostrova Medvezhi
Ostrova Ayon
Ostrov Vrangelya

Chukchi Sea
Mys Dezhneva (East C.)
Uelen
Inchoun
Vankarem
Providenskiy Zaliv
St. Lawrence I. (U.S.A.)
60

Bering Sea

Poluostrov Gory Byrranga Taymyr
1146
Oz. Taymyr
Nordvik
Novorybnoye
Khatanga
Kheta
Volochanka
Pyasina
Norilsk
Gory Putorana
1701
Yessey
Kotuy
Moyero
Kuyumba
Mutoray
Vanavara
Tura
Podkamennaya Tunguska

Ust Olenek
Olenek
Tiksi
Ust Kuyga
Saskylakh
Zhilinda
Tit-Ary
Bulun
Kyusyur
Kazachye
Druzhina
Chokurdakh
Srednekolymsk
Nizhne Kolymsk
Cherskiy
Ambarchik
Bilibino
Pevek
Chukotskoye Nagorye
1853
Ust Chaun
Anadyr
Beringovskiy
Egvekinot
Koryakskoye Nagorye
1853
Enmelen

Zhigansk
Verkhoyansk
2389
Batagay
Khonuu
Zyryanka
Pobeda
3147
Gora Chen
2082
Omsukchon
Oratukan
Neyakhan
Omolon
Oloy
Markovo
1792
Penzhino
Kamenskoye
Gizhiginskaya
Gizhiga
Penzhinskaya Guba
Slautnoye
Ust-Kamchatsk
4750
Gora Ichinskaya
Klyuchi
Kozyrevsk
Esso

Vilyuy
Vilyuysk
Verkhnevilyuysk
Nyurba
Mirnyy
Suntar
Olekminsk
Lensk
Vitim
Bodaybo
Mama
Charo
2999
Karolon

Yakutsk
Pokrovsk
Sinsk
Tommot
Aldan
2246
Neryungri
Nagornyy
Tynda
Skovorodino
Zeya

Khandyga
Ytyk-Kyuyel
Borogontsy
Oymyakon
Ust-Nera
3147
2556
Susuman
Ust-Omchug
Magadan
Atka
Palatka
Ola
Arka

Sea of Okhotsk

Petropavlovsk-Kamchatskiy
3456
K a m c h a t k a
Kronotskaya
Poluostrov Kamchatka

MONGOLIA

CHINA

JAPAN

COPYRIGHT GEORGE PHILIP LTD.

10   100   110   12   120   13   130   14

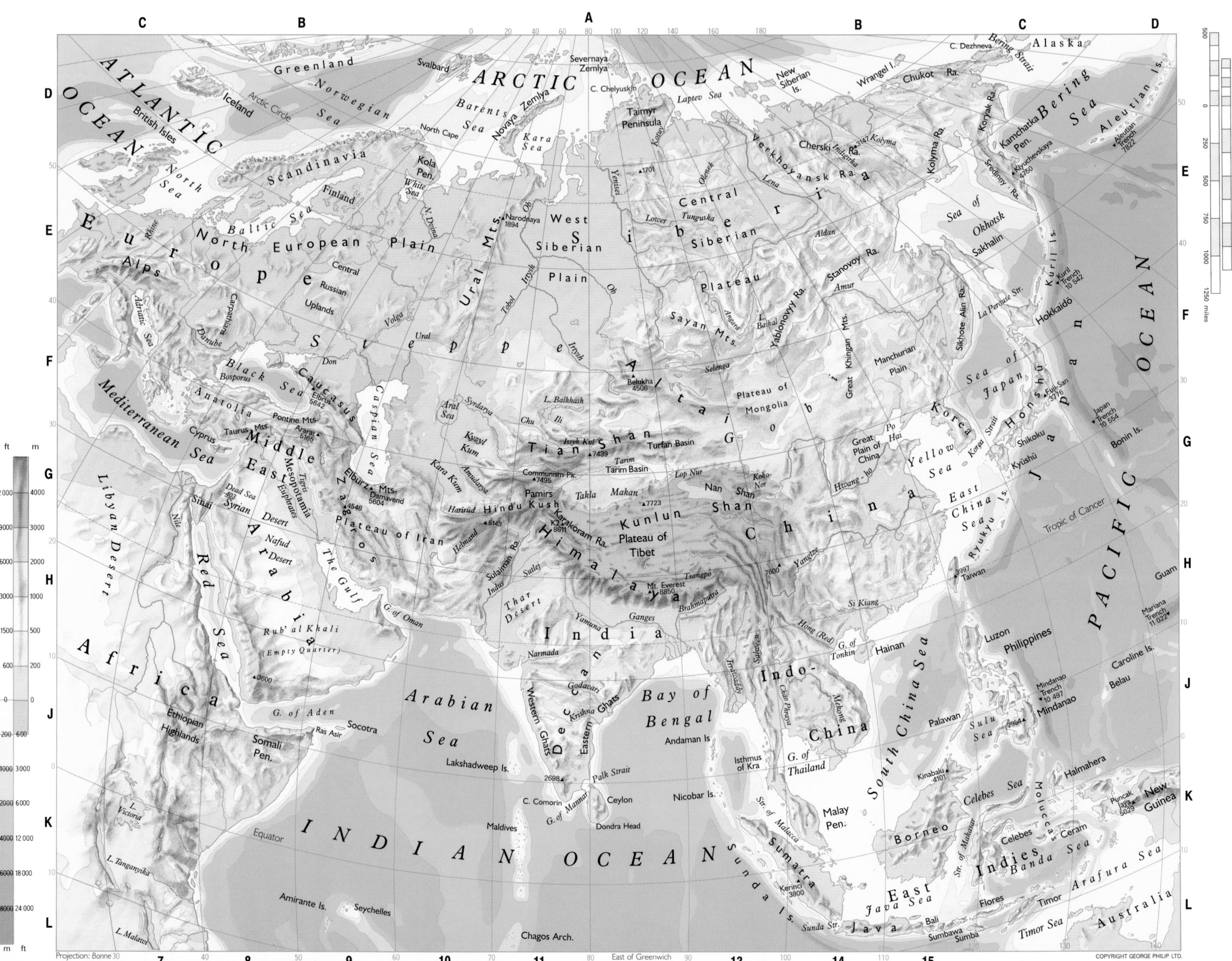

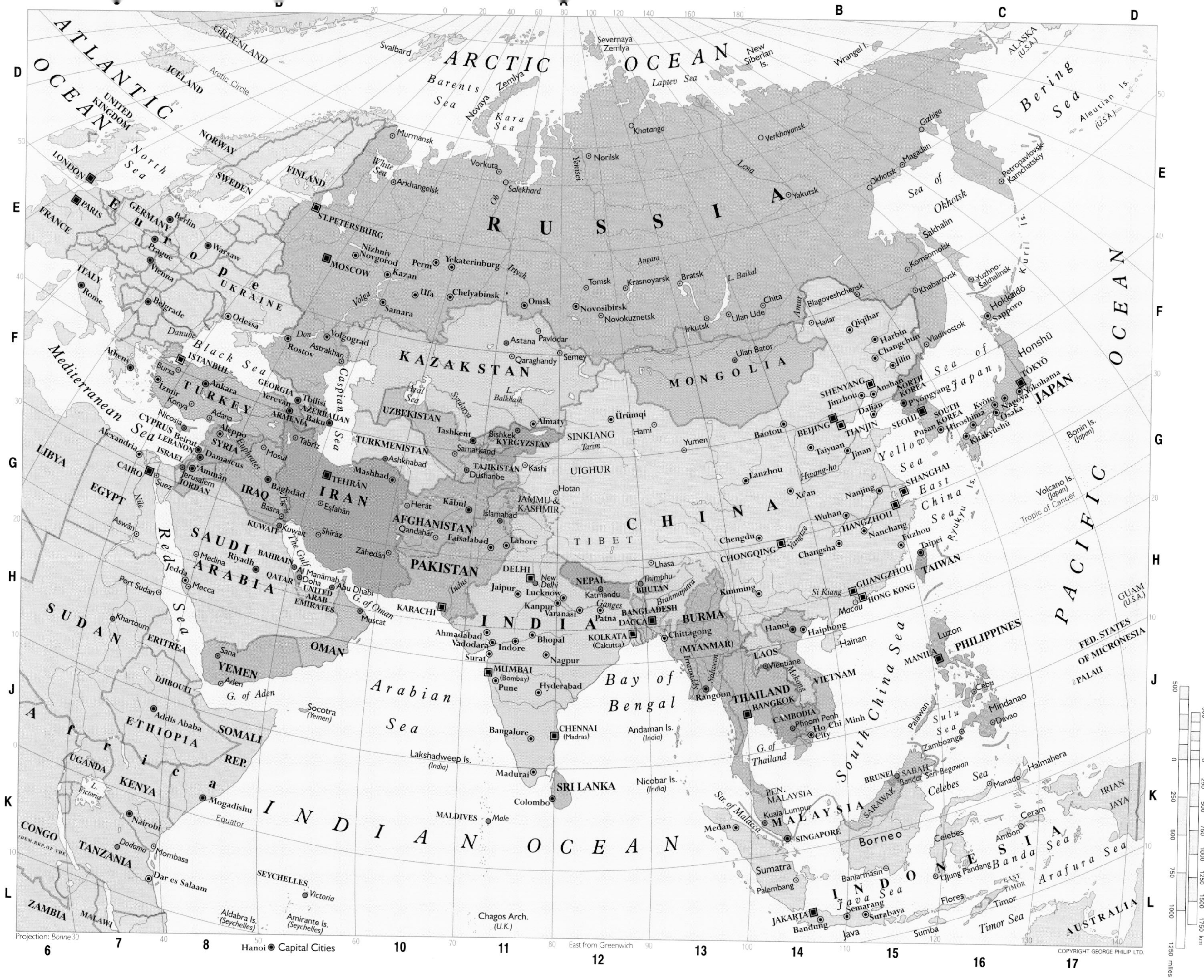
500 0 250 500 750 1000 1250 1500 1750 km
500 0 250 500 750 1000 1250 miles
ARCTIC OCEAN
PACIFIC OCEAN
INDIAN OCEAN
ATLANTIC OCEAN
RUSSIA
KAZAKSTAN
MONGOLIA
CHINA
TIBET
SINKIANG
UIGHUR
INDIA
PAKISTAN
AFGHANISTAN
IRAN
IRAQ
SAUDI ARABIA
OMAN
YEMEN
TURKEY
SYRIA
JORDAN
ISRAEL
LEBANON
CYPRUS
GEORGIA
ARMENIA
AZERBAIJAN
TURKMENISTAN
UZBEKISTAN
TAJIKISTAN
KYRGYZSTAN
JAMMU & KASHMIR
NEPAL
BHUTAN
BANGLADESH
BURMA (MYANMAR)
THAILAND
LAOS
VIETNAM
CAMBODIA
MALAYSIA
SINGAPORE
INDONESIA
PHILIPPINES
TAIWAN
JAPAN
NORTH KOREA
SOUTH KOREA
SRI LANKA
MALDIVES
Europe
Africa
EGYPT
SUDAN
ETHIOPIA
ERITREA
DJIBOUTI
SOMALI REP.
KENYA
UGANDA
TANZANIA
LIBYA
CONGO
ZAMBIA
MALAWI
KUWAIT
BAHRAIN
QATAR
UNITED ARAB EMIRATES
SEYCHELLES
MOSCOW
ST. PETERSBURG
BEIJING
TOKYO
SEOUL
SHANGHAI
HONG KONG
MANILA
BANGKOK
JAKARTA
DELHI
MUMBAI (Bombay)
KOLKATA (Calcutta)
CHENNAI (Madras)
KARACHI
TEHRAN
CAIRO
ISTANBUL
Caspian Sea
Black Sea
Red Sea
Mediterranean Sea
Arabian Sea
Bay of Bengal
South China Sea
East China Sea
Sea of Japan
Sea of Okhotsk
Bering Sea
Barents Sea
Kara Sea
Laptev Sea
Celebes Sea
Banda Sea
Arafura Sea
Timor Sea
Java Sea
Yellow Sea
The Gulf
G. of Aden
G. of Oman
G. of Thailand
Aral Sea
L. Baikal
L. Balkhash
Tropic of Cancer
Arctic Circle
Equator
East from Greenwich
Hanoi ● Capital Cities
Projection: Bonne 30
COPYRIGHT GEORGE PHILIP LTD.

JAPAN 1:5 000 000

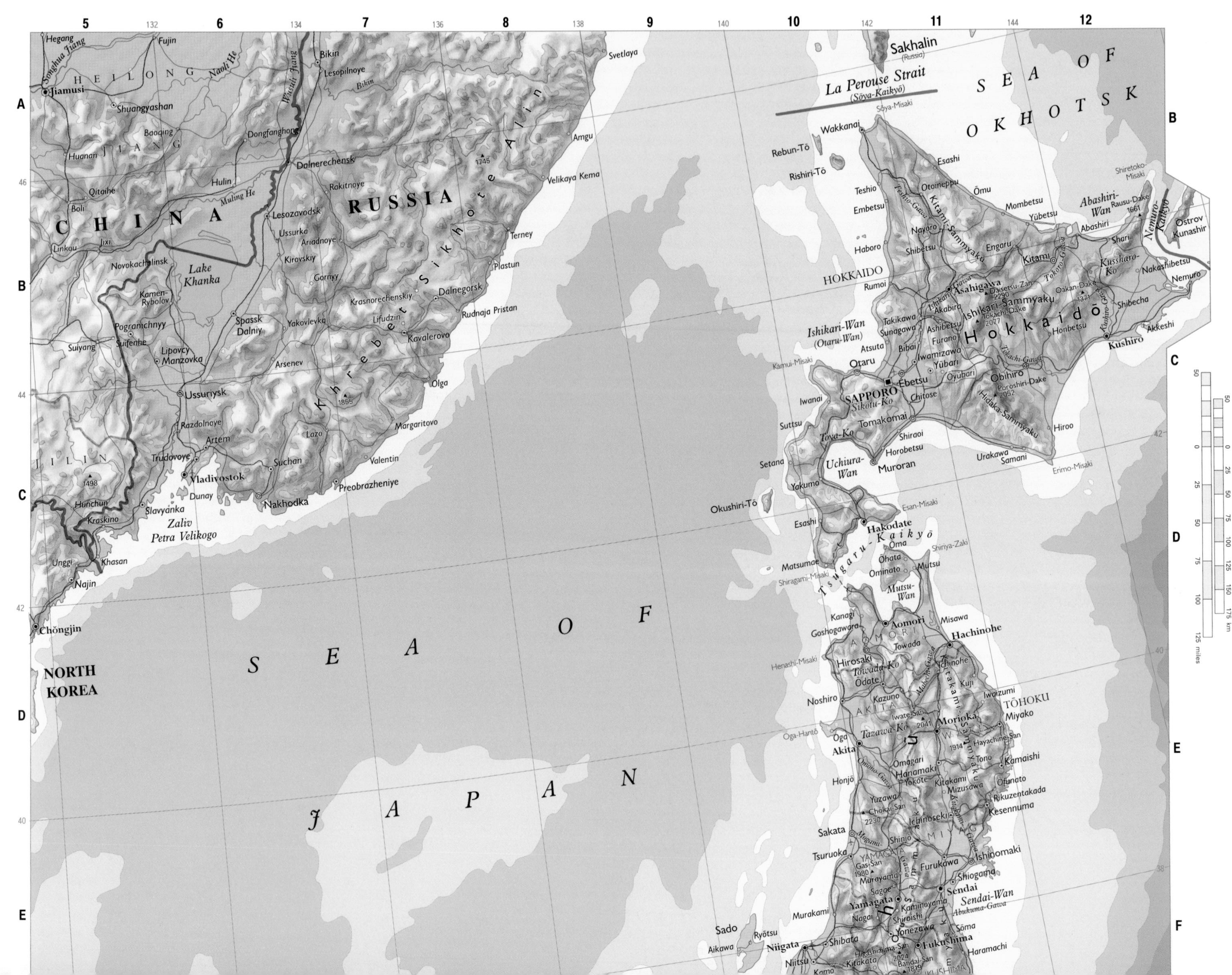

50 0 25 50 75 100 125 150 175 km

50 0 25 50 75 100 125 miles

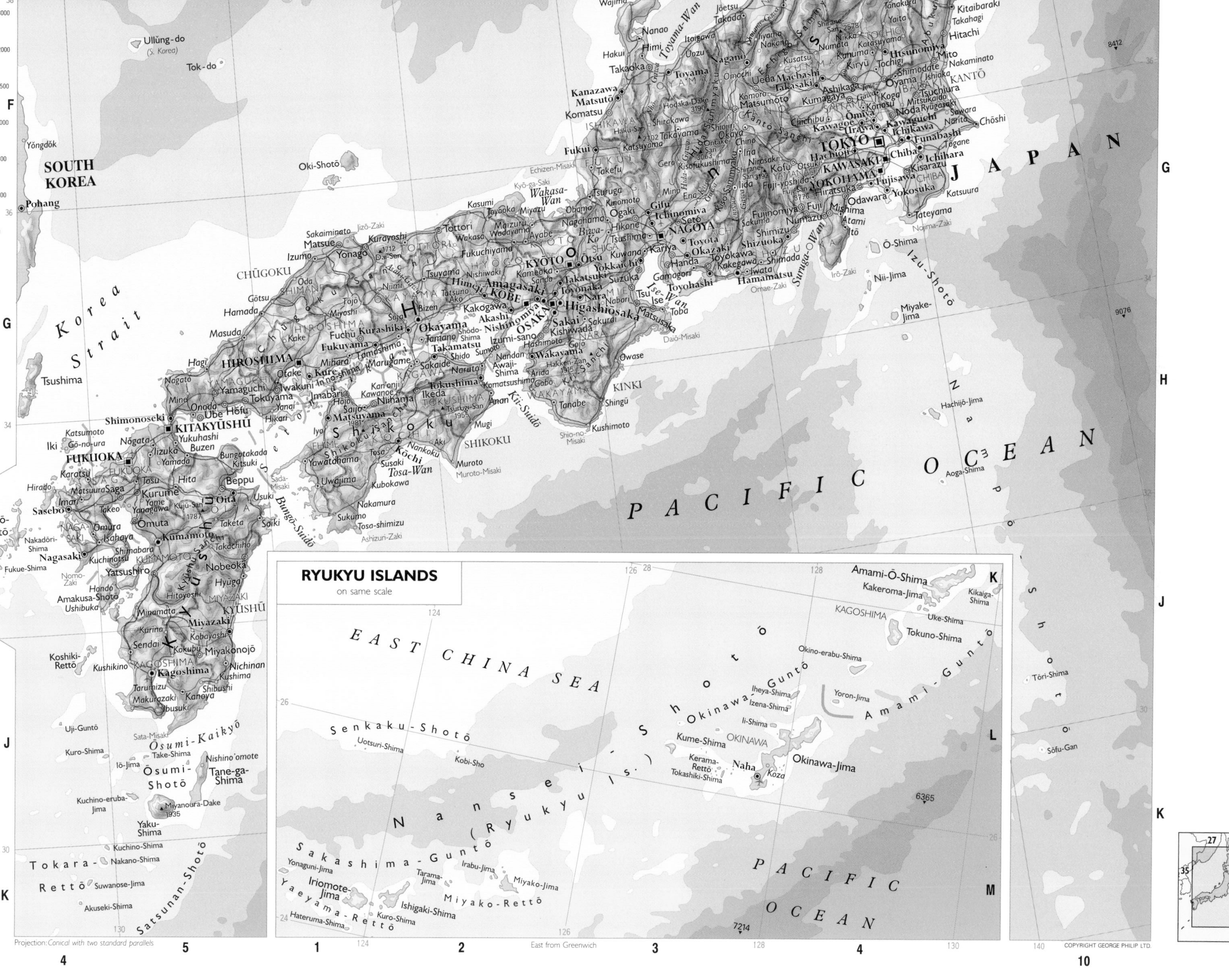

RYUKYU ISLANDS
on same scale

SOUTH KOREA

JAPAN

PACIFIC OCEAN

EAST CHINA SEA

PACIFIC OCEAN

Korea Strait

CHŪGOKU

SHIKOKU

KINKI

KYŪSHŪ

HIROSHIMA

KITAKYŪSHŪ

FUKUOKA

TŌKYŌ

YOKOHAMA

KAWASAKI

NAGOYA

KYŌTO

ŌSAKA

KŌBE

KANTŌ

CHIBA

East from Greenwich

Projection: Conical with two standard parallels

ft
3000
2000
1500
1000
400
200
0

m
9000
6000
4500
3000
1200
600
0
-200
-2000
-4000
-6000
-9000

6000
12 000
18 000
24 000

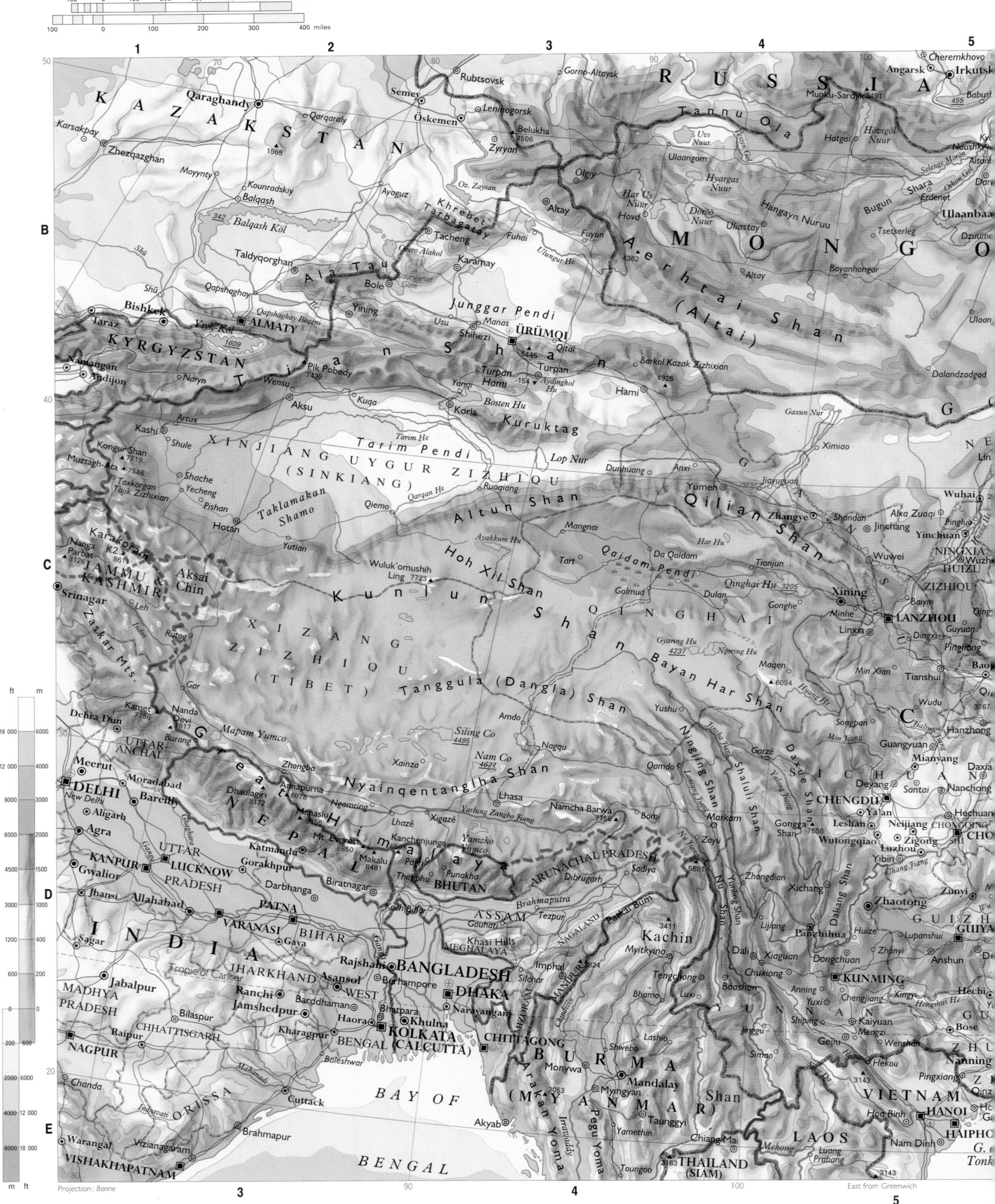
100 0 100 200 300 400 500 600 km
100 0 100 200 300 400 miles
1
2
3
4
5
B
C
D
E
K A Z A K S T A N
Qaraghandy
Karsakpay
Zhezqazghan
Qarqaraly
Moyynty
Kounradskiy
Balqash
Balqash Köl
342
Shū
Taldyqorghan
Qapshaghay
Qapshaghay Bögeni
Bishkek
ALMATY
Taraz
Ysyk-Köl
1609
KYRGYZSTAN
Namangan
Andijon
Naryn
Semey
Rubtsovsk
Gorno-Altaysk
Leninogorsk
Oskemen
Belukha
4506
Zyryan
Oz. Zaysan
Ayoguz
Khrebet Tarbagatay
Tacheng
Ozero Alakol
Ala Tau
Dzungarian Gate
Bole
Karamay
Fuhai
Ulungur He
Altay
Fuyun
Yining
Junggar Pendi
Usu
Manas
Shihezi
ÜRÜMQI
Qitai
5445
Turpan
Turpan Hami
-154
Aydingkol Hu
Pik Pobedy
7439
Wensu
Yanqi
Bosten Hu
Korla
Kuruktag
Aksu
Kuqa
Artux
Kashi
Shule
Kongur Shan
7719
Muztagh-Ata
7546
Taxkorgan Tajik Zizhixian
Shache
Yecheng
Pishan
Hotan
Tarim He
Tarim Pendi
X I N J I A N G U Y G U R Z I Z H I Q U
(S I N K I A N G)
Taklamakan Shamo
Qarqan He
Qiemo
Ruoqiang
Lop Nur
Yutian
Karakoram
Nanga Parbat
8126
K2
8611
JAMMU & KASHMIR
Aksai Chin
Srinagar
Leh
Indus
Rutog
Zaskar Mts.
Gar
Kamet
7756
Nanda Devi
7817
Burang
UTTAR-ANCHAL
Dehra Dun
Mapam Yumco
Wuluk'omushih Ling
7723
K u n l u n S h a n
X I Z A N G Z I Z H I Q U
(T I B E T)
Tanggula (Dangla) Shan
Siling Co
4495
Nam Co
4627
Nagqu
Amdo
Xainza
Zhongba
Nyainqentanglha Shan
Lhasa
Yarlung Zangbo Jiang
Namcha Barwa
7756
Xigazê
Lhazê
Annapurna
8078
Dhaulagiri
8172
Manaslu
8156
Ngamring
Mt. Everest
8850
Kanchenjunga
8598
Makalu
8481
Katmandu
NEPAL
G r e a t H i m a l a y a
Yamzho Yumco
Paro
Punakha
Thimphu
BHUTAN
ARUNACHAL PRADESH
Dibrugarh
Sadiya
Brahmaputra
Tezpur
ASSAM
Gauhati
Khasi Hills
MEGHALAYA
NAGALAND
Patkai Bum
Imphal
MANIPUR
2824
Silchar
MIZORAM
Meerut
DELHI
New Delhi
Moradabad
Bareilly
Aligarh
Agra
UTTAR PRADESH
LUCKNOW
KANPUR
Gwalior
Jhansi
Allahabad
Gorakhpur
Darbhanga
Biratnagar
PATNA
VARANASI
BIHAR
Gaya
Sagar
I N D I A
Tropic of Cancer
JHARKHAND
Asansol
Ranchi
Jamshedpur
WEST BENGAL
Barddhaman
Bhatpara
Haora
KOLKATA (CALCUTTA)
Kharagpur
Baleshwar
Rajshahi
BANGLADESH
Berhampore
DHAKA
Narayanganj
Khulna
CHITTAGONG
Jabalpur
MADHYA PRADESH
Bilaspur
CHHATTISGARH
Raipur
NAGPUR
Chanda
Mahanadi
ORISSA
Cuttack
Indravati
Brahmapur
Warangal
Vizianagaram
VISHAKHAPATNAM
BAY OF
BENGAL
Akyab
R U S S I A
Irkutsk
Angarsk
Cheremkhovo
Munku-Sardyk
3491
Babushkin
455
Tannu Ola
Uvs Nuur
Ulaangom
Hövsgöl Nuur
Hatgal
Ölgiy
Hyargas Nuur
Har Us Nuur
Hovd
Döröö Nuur
Uliastay
Hangayn Nuruu
Bugun
Erdenet
Ulaanbaatar
Tsetserleg
M O N G O L I A
4362
Aerhtai Shan (Altai)
Altay
Bayanhongor
Barkol Kazak Zizhixian
4925
Hami
Dalandzadgad
Gaxun Nur
Ximiao
Dunhuang
Anxi
Yumen
Jiayuguan
Altun Shan
Qilian Shan
Zhangye
Shandan
Jinchang
Alxa Zuoqi
Wuhai
Yinchuan
NINGXIA HUIZU ZIZHIQU
Wuwei
Mangnai
Ayakkum Hu
Hoh Xil Shan
Tart
Da Qaidam
Qaidam Pendi
Har Hu
Tianjun
Golmud
Dulan
Qinghai Hu
3205
Xining
Gonghe
Minhe
LANZHOU
Linxia
Q I N G H A I
Gyaring Hu
4237
Ngoring Hu
Maqên
6094
Bayan Har Shan
Min Xian
Tianshui
Yushu
Songpan
Wudu
Hanzhong
Min Jiang
Guangyuan
Mianyang
Deyang
CHENGDU
Ya'an
Leshan
Gongga Shan
7556
Garzê
Qamdo
Ningjing Shan
Shaluli Shan
Daxue Shan
Markam
Bomi
Zayü
5881
Nu Jiang
S I C H U A N
Santai
Nanchong
Neijiang
Zigong
Luzhou
Yibin
Chang Jiang
Wutongqiao
Xichang
Zhongdian
Dali
Xiaguan
Lijiang
Panzhihua
Huize
Zhaotong
Liupanshui
Dongchuan
KUNMING
Chuxiong
Baoshan
Anning
Yuxi
Chengjiang
Tengchong
Luxi
Bhamo
Myitkyina
Kachin
3411
Lashio
Shwebo
Monywa
Mandalay
Myingyan
3053
B U R M A (M Y A N M A R)
Arakan Yoma
Irrawaddy
Pegu Yoma
Yamethin
Taunggyi
Shan
Toungoo
Chiang Mai
THAILAND (SIAM)
2163
LAOS
Mekong
Luang Prabang
Y U N N A N
Jinggu
Simao
Gejiu
Mengzi
Kaiyuan
Hekou
Wenshan
3143
VIETNAM
Hoa Binh
HANOI
HAIPHONG
Nam Dinh
Nanning
Pingxiang
Bose
Projection: Bonne
East from Greenwich
ft m
18 000 6000
12 000 4000
9000 3000
6000 2000
4500 1500
3000 1000
1200 400
600 200
0 0
200 600
2000 6000
4000 12 000
6000 18 000
m ft

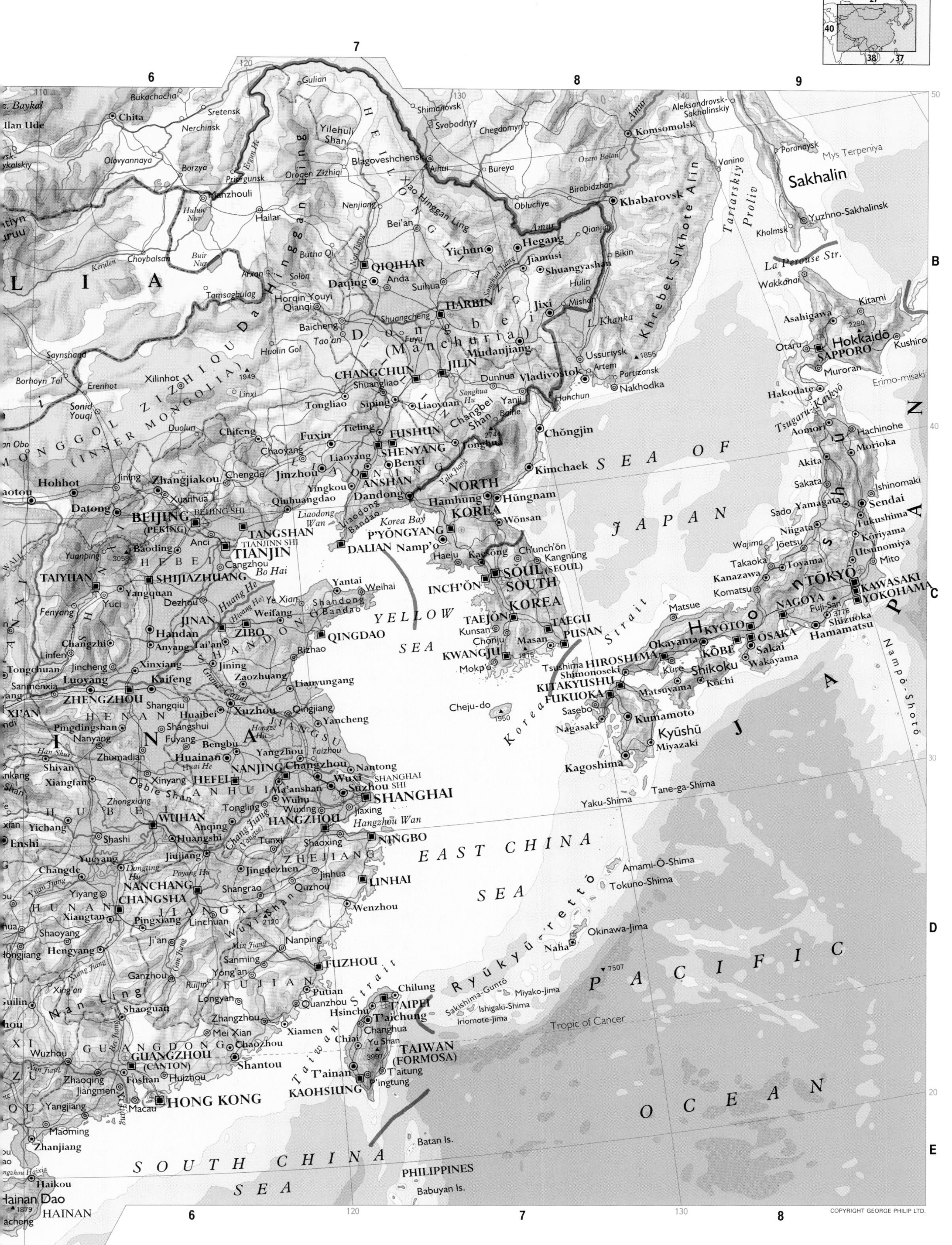
27
40
38
37
6
7
8
9
B
C
D
E
50
40
30
20
110
120
130
140
Bukachacha
Chita
Sretensk
Nerchinsk
Ulan Ude
Olovyannaya
Borzya
Priargunsk
Manzhouli
Hulun Nur
Hailar
Buir Nur
Kerulen
Choybalsan
Arxan
Solon
Tamsagbulag
Gulian
Yilehuli Shan
Oroqen Zizhiqi
Nenjiang
Bei'an
Butha Qi
Shimanovsk
Svobodnyy
Blagoveshchensk
Aihui
Bureya
Chegdomyn
Obluchye
Birobidzhan
Amur
Komsomolsk
Aleksandrovsk-Sakhalinskiy
Ozero Bolon'
Khabarovsk
Khrebet Sikhote Alin
Tartarskiy Proliv
Vanino
Poronaysk
Mys Terpeniya
Sakhalin
Kholmsk
Yuzhno-Sakhalinsk
La Perouse Str.
Wakkanai
Kitami
Asahigawa
2290
Hokkaido
SAPPORO
Otaru
Kushiro
Muroran
Erimo-misaki
Hakodate
Tsugaru-Kaikyo
Aomori
Hachinohe
Morioka
Akita
Ishinomaki
Sakata
Yamagata
Sendai
Sado
Niigata
Fukushima
Koriyama
Utsunomiya
Wajima
Joetsu
Toyama
Mito
Takaoka
Kanazawa
Komatsu
TOKYO
KAWASAKI
YOKOHAMA
NAGOYA
Fuji-San
3776
Shizuoka
Hamamatsu
Matsue
KYOTO
OSAKA
Okayama
KOBE
Sakai
Wakayama
HIROSHIMA
Kure
Shikoku
Kochi
Matsuyama
KITAKYUSHU
Shimonoseki
Tsushima
FUKUOKA
Sasebo
Nagasaki
Kumamoto
Kyushu
Miyazaki
Kagoshima
Yaku-Shima
Tane-ga-Shima
Amami-O-Shima
Tokuno-Shima
Okinawa-Jima
Naha
Ryukyu-retto
Miyako-Jima
Ishigaki-Shima
Iriomote-Jima
Sakishima-Gunto
Nampo-Shoto
7507
HEILONGJIANG
Xiao Hinggan Ling
Da Hinggan Ling
Yichun
Hegang
Jiamusi
Qianjin
Shuangyashan
Bikin
Hulin
Jixi
Mishan
L. Khanka
QIQIHAR
Daqing
Anda
Suihua
Nen Jiang
Songhua Jiang
HARBIN
Shuangcheng
Horqin Youyi Qianqi
Baicheng
Tao'an
Fuyu
Dongbei (Manchuria)
Mudanjiang
Ussuriysk
1855
Artem
Partizansk
Vladivostok
Nakhodka
Hunchun
Huolin Gol
CHANGCHUN
JILIN
Shuangliao
Dunhua
Songhua Hu
Yanji
Changbei Shan
Baitou
2744
Siping
Liaoyuan
Tongliao
Saynshand
Borhoyn Tal
Erenhot
Xilinhot
1949
Linxi
Sonid Youqi
NEI MONGOL ZIZHIQU (INNER MONGOLIA)
Duolun
Chifeng
Fuxin
Tieling
FUSHUN
SHENYANG
Benxi
Liaoyang
Chaoyang
Jinzhou
LIAONING
ANSHAN
Yingkou
Tonghua
Yalu Jiang
Dandong
Ch'ongjin
Kimchaek
Hohhot
Jining
Zhangjiakou
Chengde
Xuanhua
Datong
Qinhuangdao
BEIJING (PEKING)
BEIJING SHI
TANGSHAN
TIANJIN SHI
TIANJIN
Liaodong Wan
Liaodong Bandao
Korea Bay
Baoding
Anci
Yuanping
3058
HEBEI
Cangzhou
Bo Hai
TAIYUAN
SHIJIAZHUANG
Yangquan
Yuci
Dezhou
SHANXI
Fenyang
Huang He
Hwang Ho
JINAN
ZIBO
Weifang
Ye Xian
Yantai
Weihai
Shandong Bandao
SHANDONG
QINGDAO
Rizhao
Handan
Anyang
Tai'an
Changzhi
Linfen
Jincheng
Xinxiang
Jining
Zaozhuang
Tongchuan
Sanmenxia
Luoyang
Kaifeng
ZHENGZHOU
XI'AN
Lianyungang
Grand Canal
Shangqiu
Huaibei
Xuzhou
Qingjiang
Yancheng
HENAN
Pingdingshan
Shangshui
Nanyang
Fuyang
Hongze Hu
JIANGSU
Bengbu
Han Shui
Zhumadian
Huainan
Huai He
Yangzhou
Taizhou
Shiyan
Xiangfan
Xinyang
HEFEI
NANJING
Changzhou
Nantong
Wuxi
Suzhou
SHANGHAI SHI
SHANGHAI
Dabie Shan
ANHUI
Ma'anshan
Wuhu
Zhongxiang
HUBEI
Tongling
Wuxing
Jiaxing
Hangzhou Wan
WUHAN
HANGZHOU
Yichang
Anqing
Enshi
Shashi
Huangshi
Chang Jiang (Yangtse)
Tunxi
Shaoxing
NINGBO
ZHEJIANG
Yueyang
Jiujiang
Changde
Dongting Hu
Poyang Hu
Jingdezhen
Jinhua
LINHAI
Yuan Jiang
NANCHANG
Yiyang
CHANGSHA
Shangrao
Quzhou
HUNAN
Xiangtan
JIANGXI
Pingxiang
Linchuan
Wuyi Shan
2120
Wenzhou
Shaoyang
Ji'an
Gan Jiang
Min Jiang
Nanping
Hengyang
Xiang Jiang
Sanming
FUZHOU
Ganzhou
Yong'an
Ruijin
FUJIAN
Putian
Xing'an
Nan Ling
Shaoguan
Longyan
Quanzhou
Zhangzhou
Xiamen
Mei Xian
Bei Jiang
GUANGDONG
Chaozhou
Wuzhou
GUANGZHOU (CANTON)
Shantou
Xi Jiang
Foshan
Huizhou
Zhaoqing
Jiangmen
HONG KONG
Macau
Yangjiang
Maoming
Zhanjiang
Haikou
Hainan Dao
1879
HAINAN
Taiwan Strait
Chilung
T'AIPEI
Hsinchu
T'aichung
Changhua
Chiai
Yu Shan
3997
TAIWAN (FORMOSA)
T'ainan
T'aitung
KAOHSIUNG
P'ingtung
Batan Is.
PHILIPPINES
Babuyan Is.
Tropic of Cancer
NORTH KOREA
Hamhung
Hungnam
Wonsan
P'YONGYANG
Namp'o
DALIAN
Haeju
Kaesong
Ch'unch'on
Kangnung
SOUL (SEOUL)
SOUTH KOREA
INCH'ON
TAEJON
TAEGU
Kunsan
Chonju
KWANGJU
Masan
PUSAN
1915
Mokp'o
Cheju-do
1950
Korea Strait
SEA OF JAPAN
YELLOW SEA
EAST CHINA SEA
SOUTH CHINA SEA
PACIFIC OCEAN
JAPAN
CHINA
MONGOLIA

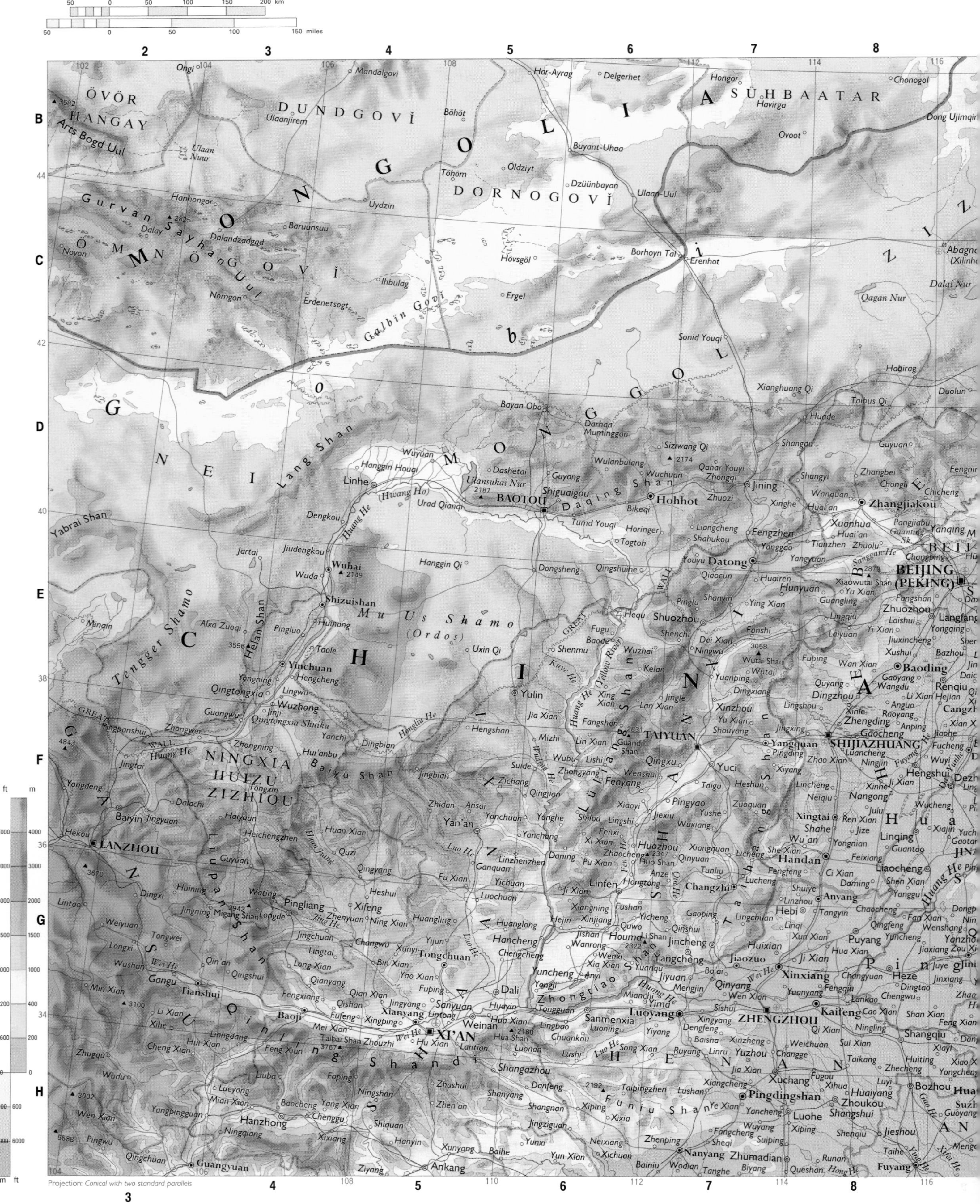

ÖVÖR HANGAY
Arts Bogd Uul
▲3582

O N G O L I A S Ü H B A A T A R

D U N D G O V Ĭ

Ongi Mandalgovĭ Har-Ayrag Delgerhet Hongor Chonogol
Ulaanjirem Böhöt Havirga Dong Ujimqin
Buyant-Uhaa Ovoot

D O R N O G O V Ĭ

Tôhöm Öldziyt Dzüünbayan Ulaan-Uul
Hanhongor Baruunsuu Ihbulag Hövsgöl Borhoyn Tal Erenhot Abagnar (Xilinhe)
Gurvan Sayhan Uul Dalandzadgad Ergel Qagan Nur Dalai Nur
Noyon Dalay Nömgön Erdenetsogt Galbin Govi b Sonid Youqi Hobirag Duolun
Xianghuang Qi Taibus Qi

G N E I Bayan Obo Darhan Muminggan Huade Shangdu Guyuan Fengnir
Wuyuan Siziwang Qi ▲2174 Zhangbei Chongli Chicheng
Lang Shan Hanggin Houqi Dashetai Guyang Wulanbulang Wuchuan Qahar Youyi Zhongqi Jining Shangyi Xinghe Wanquan Huai'an Zhangjiakou
Yabrai Shan Linhe Ulansuhai Nur ▲2187 Shiguaigou Hohhot Zhuozi Xuanhua Huai'an Yanqing
Dengkou BAOTOU Daqing Shan Bikeqi Liangcheng Shahukou Fengzhen Yanggao Tianzhen Zhuolu Changping BEIJING (PEKING)
Jartai Jiudengkou Urad Qianqi Tumd Youqi Horinger Togtoh Youyu Datong Qiaocun Huairen Hunyuan Xiaowutai Shan ▲2870 Yu Xian Zhuozhou Langfang
Wuhai ▲2149 Hanggin Qi Dongsheng Qingshuihe Guangling Zhuozhou Laishui
Wuda Shizuishan M u U s S h a m o Pingli Shanyin Ying Xian Linggiu Laiyuan Jiuxincheng Yongqing
Alxa Zuoqi Pingluo (Ordos) Uxin Qi Hequ Shuozhou Dai Xian Fanshi Wutai Shan ▲3058 Fuping Wan Xian Baoding
Minqin Huinong Fugu Baode Shenchi Kelan Yuanping Dingxiang Quyang Gaoyang Renqiu
Yinchuan Hengcheng Shenmu Wuzhai Jingle Lan Xian Xinzhou Lingshou Xinfe Hejian Cangzhou
Qingtongxia Lingwu Yulin Jia Xian Mizhi Lin Xian TAIYUAN Shouyang Jingxing Zhengding Anping Anguo
Yongning Wuzhong Hengliu He Hengshan Fangshan ▲2831 Qingxu Yuci Yangquan SHIJIAZHUANG Gaocheng Jiaohe
Guangwu Yanchi Suide Zhongyang Wenshui Taigu Heshun Lincheng Gaotan
Jinji Zhongwei Zhongning Qingtongxia Shuiku Dingbian Zichang Lishi Fenyang Pingyao Yushe Zuoquan Neiqiu Nangong Wucheng
Jingtai Baiyu Shan Ansai Yanchuan Yonghe Xiaoyi Lingshi Xiangyuan Xingtai Ren Xian Jize Xiajin
Yongdeng Huang He Hui'anbu Zhidan Yan'an Yanchang Daning Fenxi Xi Xian Huozhou Qinyuan Tunliu Wu'an Handan Linqing
Baiyin Haiyuan Heichengzhen Huan Xian Quzi Xiaoyi Zhaocheng ▲2347 Licheng She Xian Feixiang Daming
LANZHOU Guyuan Qingyang Yichuan Luochuan Ji Xian Pu Xian Hongtong Changzhi Ci Xian Shen Xian Liaocheng
Dingxi ▲3670 Huining Heshui Fu Xian Ganquan Xiangning Fushan Gaoping Pengfeng Anyang Tangyin Chaocheng Dongp
Weiyuan Pingliang ▲2942 Xifeng Ning Xian Huangling Hejin Xinjiang Quwo Qinshui Lingchuan Hebi Linqi Qingfeng Fan Xian Wenshang Yanzho
Longxi Tongwei Jingning Migang Shan Longde Zhenyuan Huanglong Jishan Houma ▲2322 Yangcheng Jiaozuo Ji Xian Huixian Xun Xian Puyang Yuncheng Yanzhou
Wushan Qin'an Jingchuan Changwu Huajing Hancheng Wanrong Wenxi Yangcheng Jiyuan Bo'ai Xinxiang Changyuan Heze Juye Jinxiang Jinxiang
Gangu Tianshui Lingtai Bin Xian Yao Xian Chengcheng Xia Xian Mianchi Qinyang Wen Xian Yuanyang Fengqiu Lankao Chengwu Dingtao
Min Xian ▲3100 Qianyang Fengxiang Qian Xian Fuping Dali Yuncheng Zhongtiao Shan Huang He Mengjin ZHENGZHOU Kaifeng Cao Xian Shan Xian Shangqiu
Li Xian Longdan Baoji Fufeng Xingping Huayin Tongguan Sanmenxia Luoyang Xingyang Yuanyang Qi Xian Ningling Zhao
Xihe Hui Xian Mei Xian Zhouzhi Weinan ▲2160 Lingbao Luoning Yima Yiyang Dengfeng Baisha Weichuan Sui Xian Xiayi
Zhugou Cheng Xian Taibai Shan ▲3767 Hu Xian XI'AN Lantian Hua Shan Chuankou Lushi Luo He Song Xian Ruyang Yuzhou Huiting Zhecheng Xiao X Yongcheng
Wudu ▲3002 Lueyang Liuba Foping Shangzhou Danfeng Xichuan Shangcheng Jia Xian Xuchang Fugou Luyi Bozhou Hua
Wen Xian Mian Xian Ningshan ▲2192 Zhashui Shanyang Shangnan Lushan Xiangcheng Xihua Huaiyang Zhoukou Suzh
Pingwu ▲5588 Hanzhong Baocheng Yang Xian Zhen'an Shiquan Jingziguan Xixia Pingdingshan Yancheng Taihe Guoyang
Qingchuan Ningqiang Chenggu Xunyang Hanyin Xichuan Zhenping Fangcheng Suiping Shenqiu Jieshou A N
Guangyuan Ziyang Baihe Yunxi Yun Xian Nanyang Zhumadian Runan Mengc
Ankang Yunyang Danjiang Tanghe Biyang Queshan Hong He Taihe Fuyang Yongche

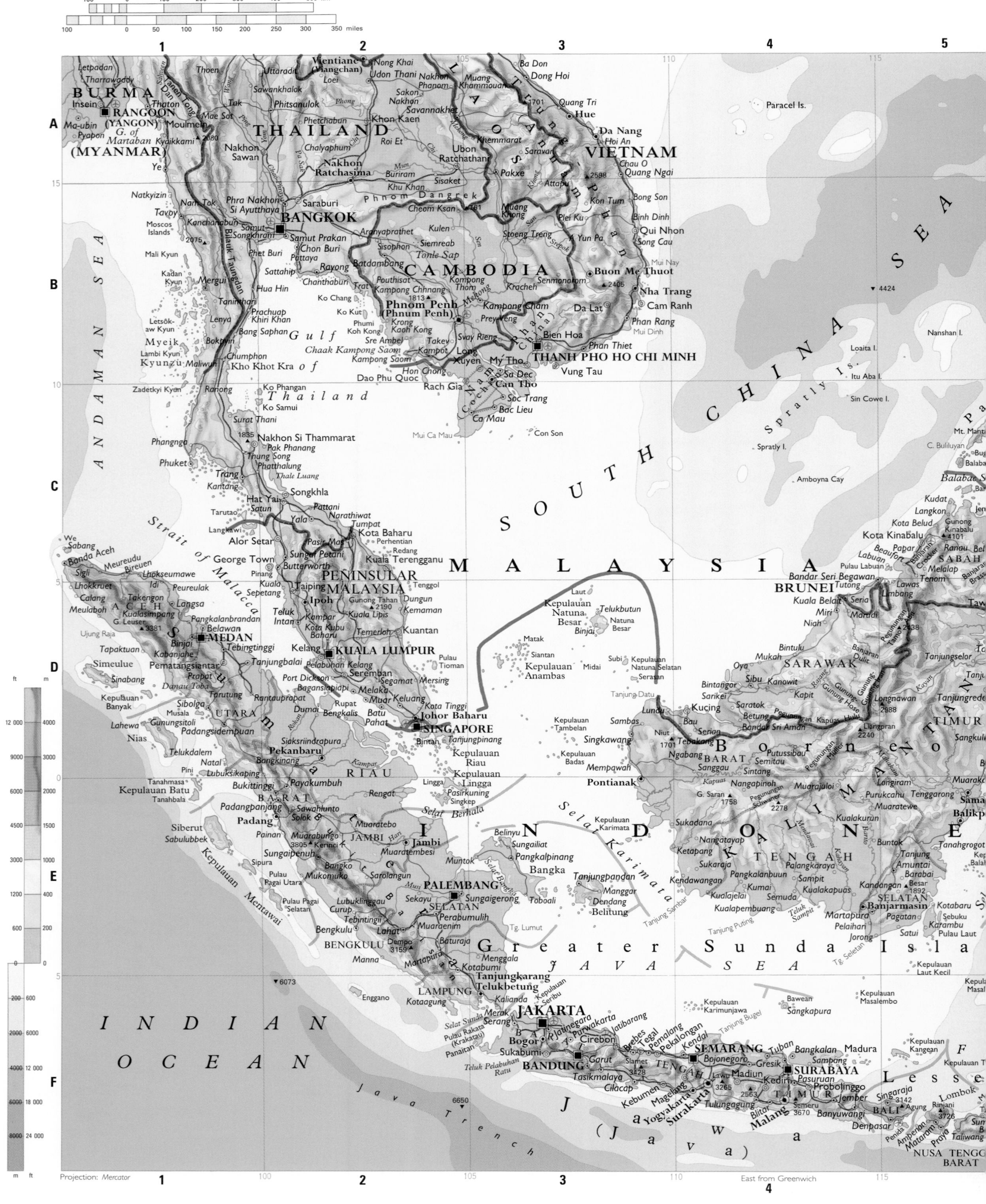

km
miles
1
2
3
4
5
A
B
C
D
E
F
BURMA
(MYANMAR)
RANGOON
(YANGON)
Moulmein
G. of Martaban
THAILAND
BANGKOK
Vientiane
(Viangchan)
LAOS
VIETNAM
Hue
Da Nang
CAMBODIA
Phnom Penh
(Phnum Penh)
THANH PHO HO CHI MINH
Vung Tau
Can Tho
Gulf of Thailand
ANDAMAN SEA
SOUTH CHINA SEA
Paracel Is.
Spratly Is.
Spratly I.
Strait of Malacca
MALAYSIA
PENINSULAR MALAYSIA
KUALA LUMPUR
George Town
Ipoh
Kota Baharu
Kuala Terengganu
Kuantan
Johor Baharu
SINGAPORE
BRUNEI
Bandar Seri Begawan
Kota Kinabalu
SABAH
SARAWAK
Kuching
Miri
Sibu
Kepulauan Natuna
Natuna Besar
Kepulauan Anambas
Borneo
KALIMANTAN
KALIMANTAN BARAT
KALIMANTAN TENGAH
KALIMANTAN SELATAN
KALIMANTAN TIMUR
Pontianak
Banjarmasin
Palangkaraya
INDONESIA
Sumatera
ACEH
Banda Aceh
MEDAN
Simeulue
Nias
SUMATERA UTARA
SUMATERA BARAT
Padang
Pekanbaru
RIAU
Kepulauan Riau
Kepulauan Lingga
JAMBI
Jambi
PALEMBANG
SUMATERA SELATAN
Bangka
Belitung
Selat Karimata
BENGKULU
Bengkulu
LAMPUNG
Tanjungkarang
Telukbetung
Siberut
Kepulauan Mentawai
Enggano
Greater Sunda Islands
JAVA SEA
JAKARTA
Bogor
BANDUNG
Cirebon
SEMARANG
Yogyakarta
Surakarta
Madiun
Kediri
Malang
SURABAYA
Madura
JAWA BARAT
JAWA TENGAH
JAWA TIMUR
Jawa
(Java)
BALI
Denpasar
Lombok
Mataram
NUSA TENGGARA BARAT
INDIAN OCEAN
Java Trench
Projection: Mercator
East from Greenwich
100
105
110
115
15
10
5
0
ft
m

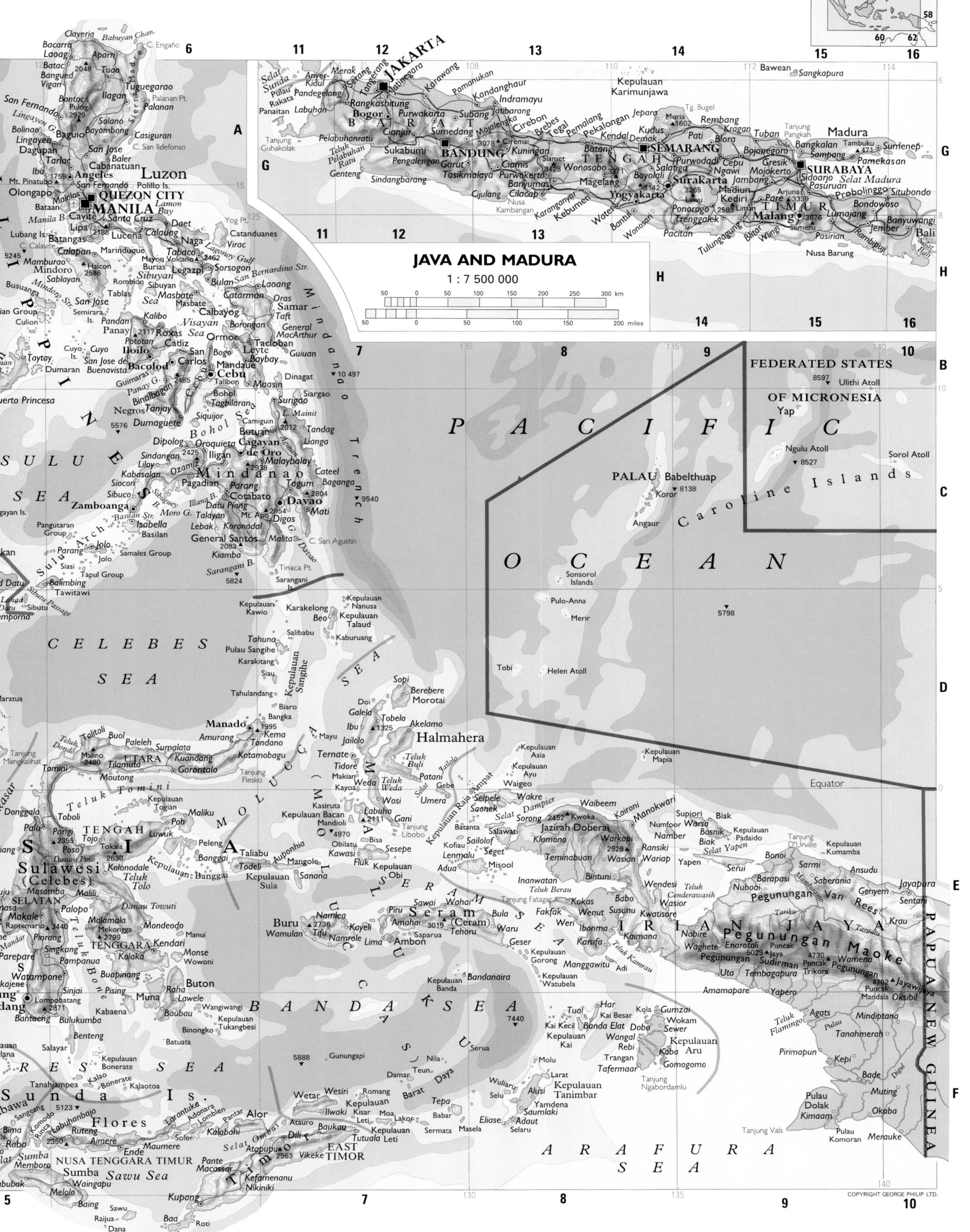

JAVA AND MADURA
1 : 7 500 000
JAKARTA
BANDUNG
SEMARANG
SURABAYA
Surakarta
Yogyakarta
Malang
Madura
Kepulauan Karimunjawa
Bali
Nusa Barung
PACIFIC OCEAN
FEDERATED STATES OF MICRONESIA
Yap
Ulithi Atoll
Ngulu Atoll
Sorol Atoll
Caroline Islands
PALAU
Babelthuap
Koror
Angaur
Sonsorol Islands
Pulo-Anna
Merir
Tobi
Helen Atoll
Equator
Luzon
QUEZON CITY
MANILA
Mindoro
Samar
Leyte
Cebu
Negros
Panay
Mindanao
Davao
Zamboanga
Cotabato
General Santos
Mindanao Trench
SULU SEA
CELEBES SEA
MOLUCCA SEA
Halmahera
Morotai
Manado
Sulawesi (Celebes)
Buru
Seram (Ceram)
Ambon
BANDA SEA
ARAFURA SEA
SERAM SEA
Jazirah Doberai
IRIAN JAYA
Pegunungan Maoke
Puncak Jaya
Jayapura
Merauke
Kepulauan Aru
Kepulauan Tanimbar
Kepulauan Kai
Flores
Sumba
Sawu Sea
Timor
EAST TIMOR
Kupang
Dili
Alor
Wetar
NUSA TENGGARA TIMUR
Sunda Is.
PAPUA NEW GUINEA
COPYRIGHT GEORGE PHILIP LTD.

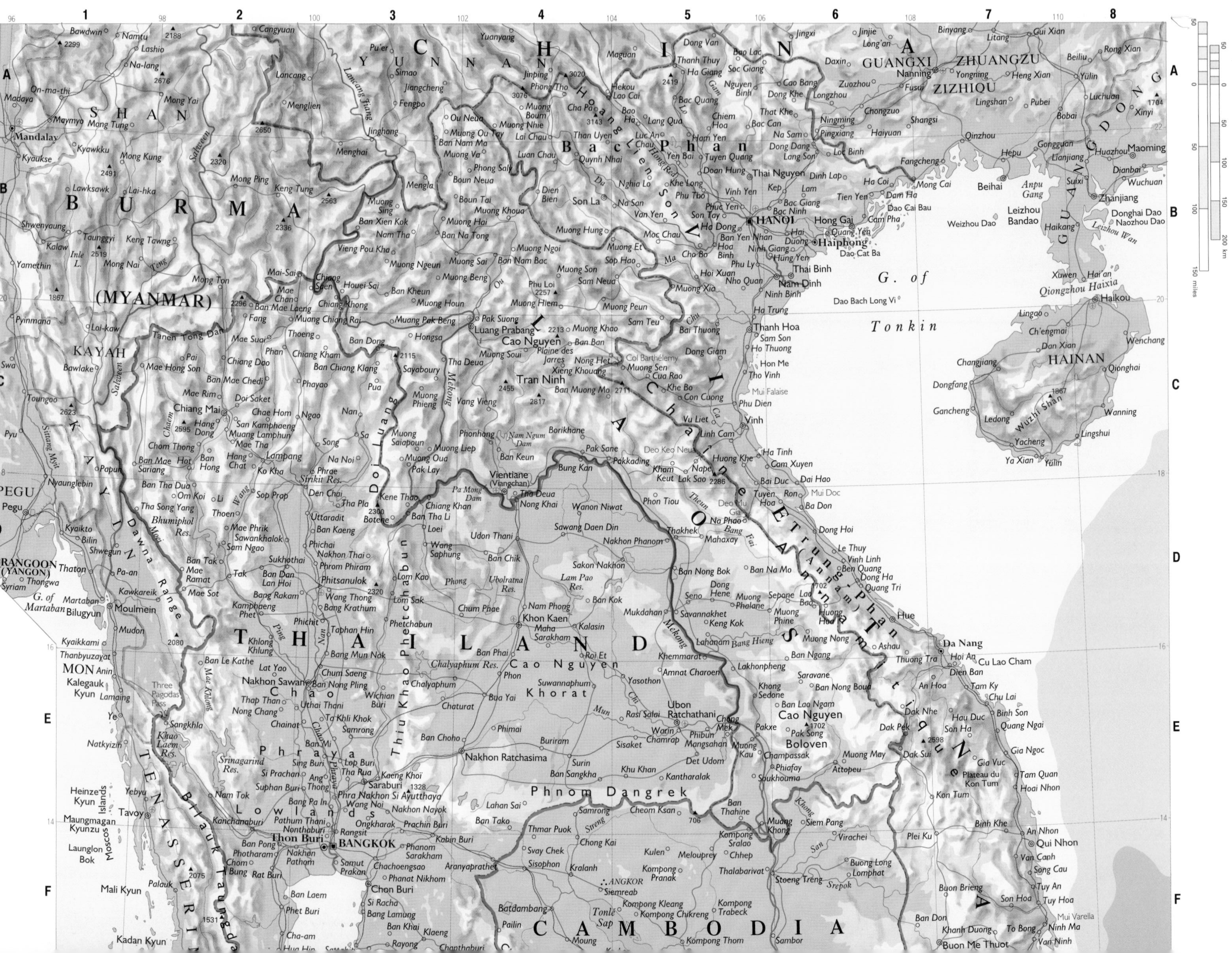

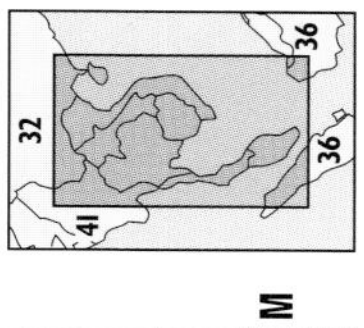

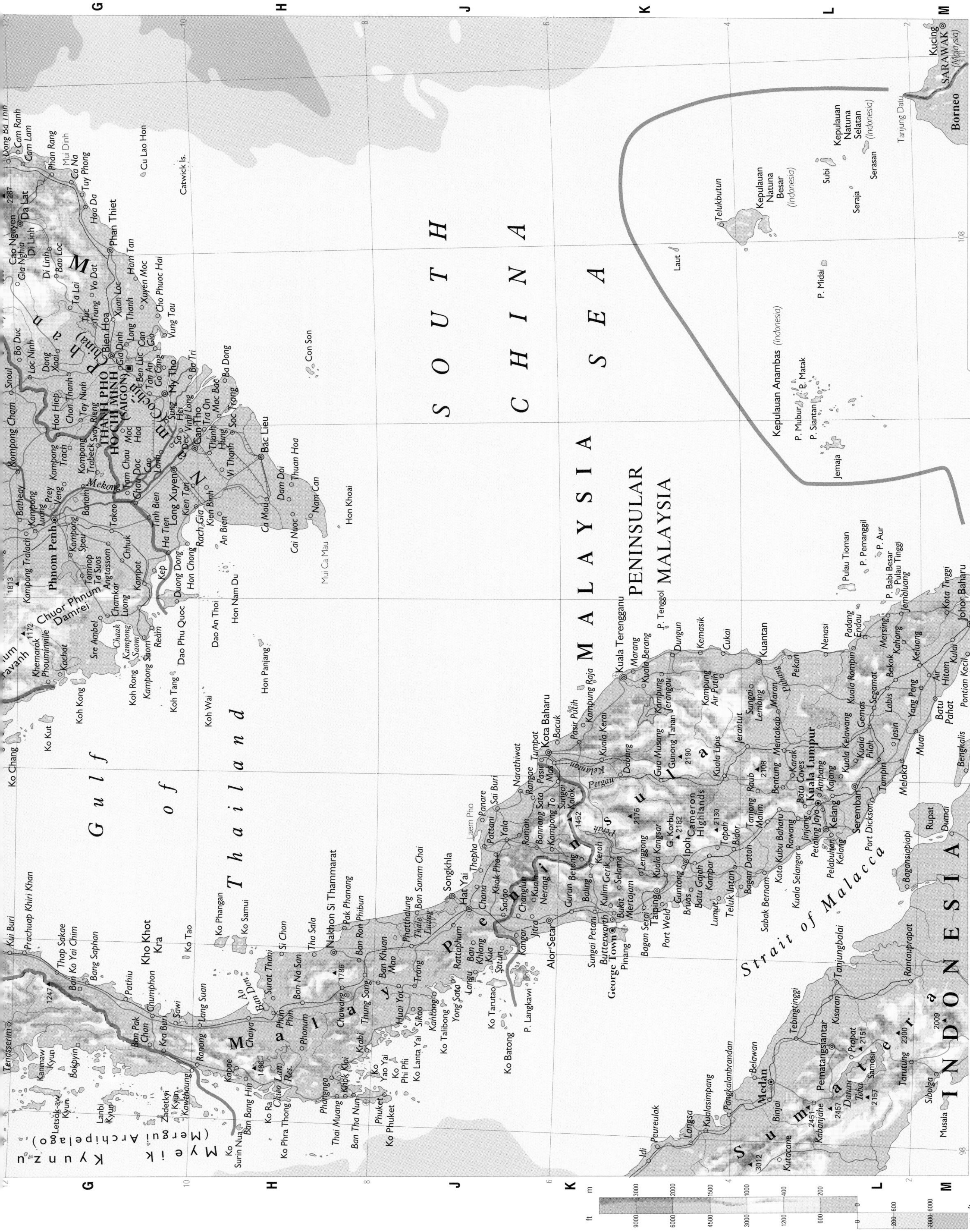

SOUTH CHINA SEA

Gulf of Thailand

Strait of Malacca

PENINSULAR MALAYSIA

M A L A Y S I A

I N D O N E S I A

Borneo

SARAWAK (Malaysia)

Kucing

Tanjung Datu

Kepulauan Natuna Besar (Indonesia)

Subi

Seraja

Serasan

Telukbutun

Kepulauan Anambas (Indonesia)

P. Midai

P. Mubur

P. Matak

P. Siantan

Jemaja

Laut

East from Greenwich

COPYRIGHT GEORGE PHILIP LTD.

Singapore

SINGAPORE

Strait of Singapore

Batam

Bintan

Tanjungpinang

Johor Baharu

Kota Tinggi

Kukup

Pontian Kecil

Pulau Tinggi

Pulau Besar

P. Babi Besar

P. Aur

P. Pemanggil

Pulau Tioman

Mersing

Endau

Kuala Rompin

Pekan

Kuantan

Cukai

Kemasik

Dungun

Kuala Berang

Marang

Kuala Terengganu

Kampung Raja

Pasir Putih

Kota Baharu

Tumpat

Bacuk

Pasir Mas

Sungai Kolok

Kota Baharu

Narathiwat

Rangae

Sai Buri

Pattani

Panare

Laem Pho

Thepha

Songkhla

Hat Yai

Sadao

Ban Sanam Chai

Phatthalung

Trang

Yong Sata

Ko Tarutao

Ko Talibong

Ko Batong

P. Langkawi

Satun

Kangar

Kua

Alor Setar

Jitra

Sungai Petani

Butterworth

George Town

Pinang

Bagan Serai

Port Weld

Taiping

Ipoh

Cameron Highlands

Kuala Lipis

Raub

Batu Caves

Kuala Lumpur

Kelang

Port Dickson

Seremban

Melaka

Muar

Batu Pahat

Pontian

Kulai

Keluang

Yong Peng

Labis

Segamat

Tampin

Bahau

Kuala Pilah

Jerantut

Gunung Tahan 2190

Kuala Krai

Kampung Kerai

Dabong

Gua Musang

Kuala Kangsar

Teluk Intan

Bidor

Tapah

Kuantan

Pekan

Kenasi

Padang Endau

Bengkalis

Rupat

Dumai

Bagansiapiapi

Tanjungbalai

Rantauprapat

Kisaran

Pematangsiantar

Prapat

Tebingtinggi

Belawan

Medan

Binjai

Langsa

Kualasimpang

Kutacane

Kabanjahe

Sibolga

Tarutung

Samosir

Toba

S u m a t e r a

Musala

East from Greenwich 98 100 102 104 106

VIETNAM

Ho Chi Minh

THANH PHO HO CHI MINH (SAIGON)

Phnom Penh

Mekong

Da Lat

Phan Rang

Cam Ranh

Phan Thiet

Vung Tau

My Tho

Can Tho

Long Xuyen

Rach Gia

Ca Mau

Mui Ca Mau

Con Son

Hon Khoai

Hon Nam Du

Dao Phu Quoc

Hon Chong

Kep

Kampot

Kampong Saom

Koh Kong

Koh Rong

Koh Kut

Ko Chang

K y u n z u

Myeik (Mergui Archipelago)

Kho Khot Kra

Chumphon

Lang Suan

Surat Thani

Nakhon Si Thammarat

Phangnga

Phuket

Ko Phuket

Krabi

Ko Lanta Yai

Trang

T h a i l a n d

M a l a y

Gulf of Thailand

Ko Samui

Ko Phangan

Ko Tao

Chaiya

Projection: Conical with two standard parallels

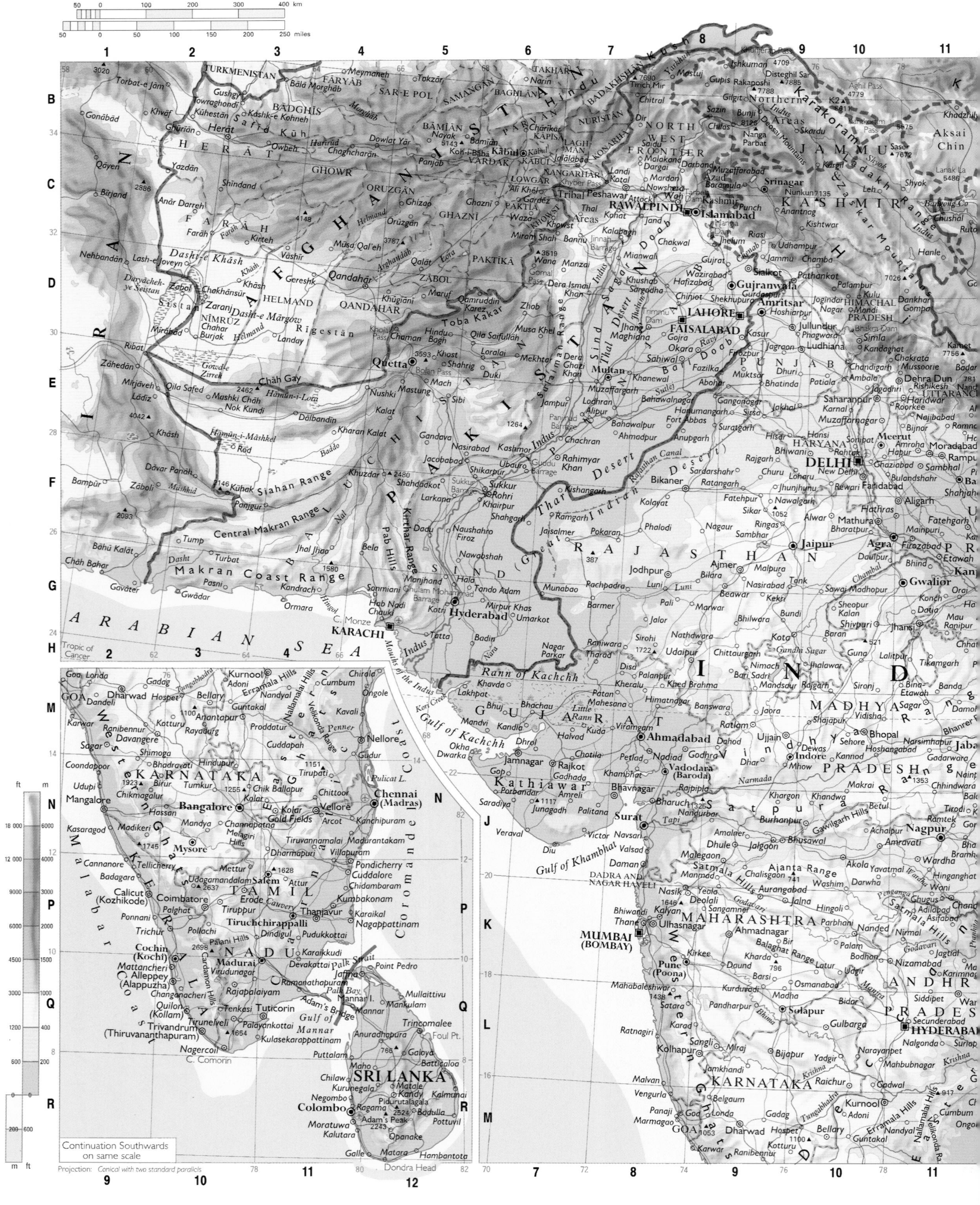

TURKMENISTAN
AFGHANISTAN
PAKISTAN
IRAN
INDIA
SRI LANKA
ARABIAN SEA
HERĀT
FARĀH
NĪMRŪZ
HELMAND
QANDAHĀR
ORUZGĀN
GHOWR
GHAZNĪ
ZĀBOL
PAKTĪKĀ
BĀMĪĀN
BAGHLĀN
PARVĀN
NURISTAN
BADAKHSHĀN
TAKHĀR
NORTH WEST FRONTIER
Northern Areas
JAMMU KASHMIR
Aksai Chin
HIMACHAL PRADESH
PUNJAB
HARYANA
DELHI
RAJASTHAN
SIND
GUJARAT
MADHYA PRADESH
MAHARASHTRA
KARNATAKA
KERALA
TAMIL NADU
ANDHRA PRADESH
Hindu Kush
Karakoram
Safīd Kūh
Dasht-e Khāsh
Dasht-e Mārgow
Rīgestān
Toba Kakar
Sulaiman Range
Kirthar Range
Pab Hills
Siahan Range
Central Makran Range
Makran Coast Range
Thar Desert
Great Indian Desert
Thal Desert
Rann of Kachchh
Gulf of Kachchh
Kathiawar
Gulf of Khambhat
Western Ghats
Eastern Ghats
Malabar Coast
Coromandel Coast
Gulf of Mannar
Palk Strait
Palk Bay
Adam's Bridge
Vindhya Range
Satpura Range
Ajanta Range
Satmala Hills
Balaghat Range
Kābul
Islamabad
RAWALPINDI
Peshawar
LAHORE
FAISALABAD
Multan
Quetta
KARACHI
Hyderabad
Srinagar
Amritsar
DELHI
New Delhi
Jaipur
Ahmadabad
Surat
MUMBAI (BOMBAY)
Pune (Poona)
Nagpur
Bhopal
Indore
Bangalore
Chennai (Madras)
Madurai
Colombo
Herāt
Qandahār
Zāhedān
Gwalior
Agra
K2
Nanga Parbat
Tirich Mir
Tropic of Cancer
Continuation Southwards on same scale
Projection: Conical with two standard parallels

C H I N A

XIZANG ZIZHIQU (TIBET)

QINGHAI

SICHUAN

YUNNAN

NEPAL

BHUTAN

BANGLADESH

BURMA (MYANMAR)

THAILAND

BAY OF BENGAL

INDIAN OCEAN

East from Greenwich

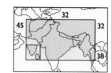

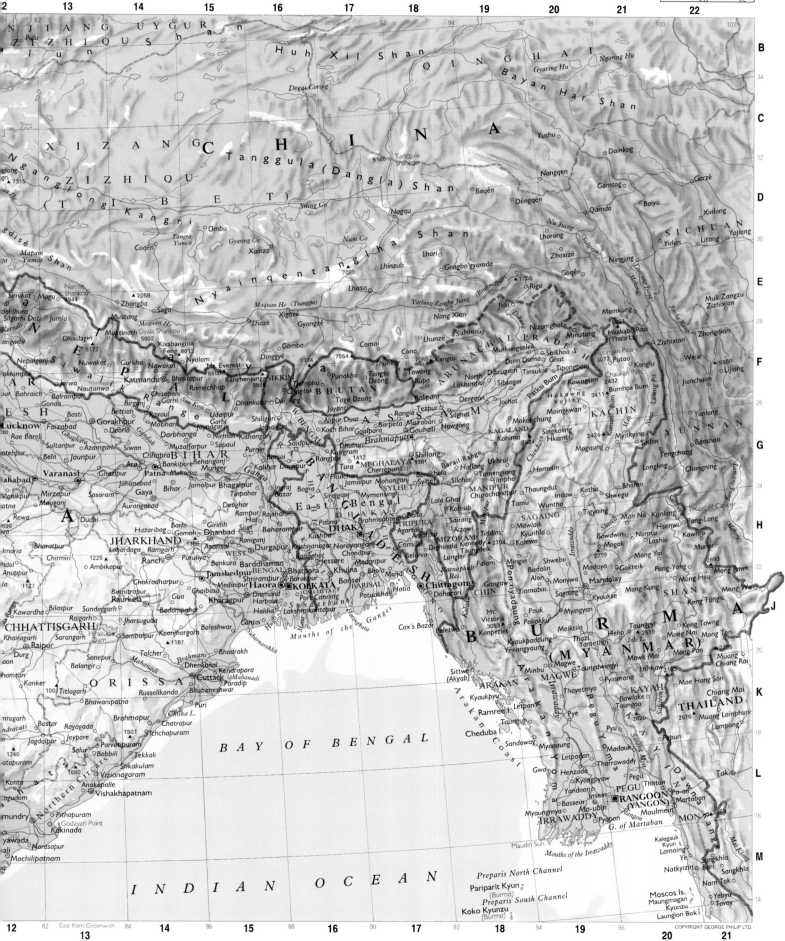

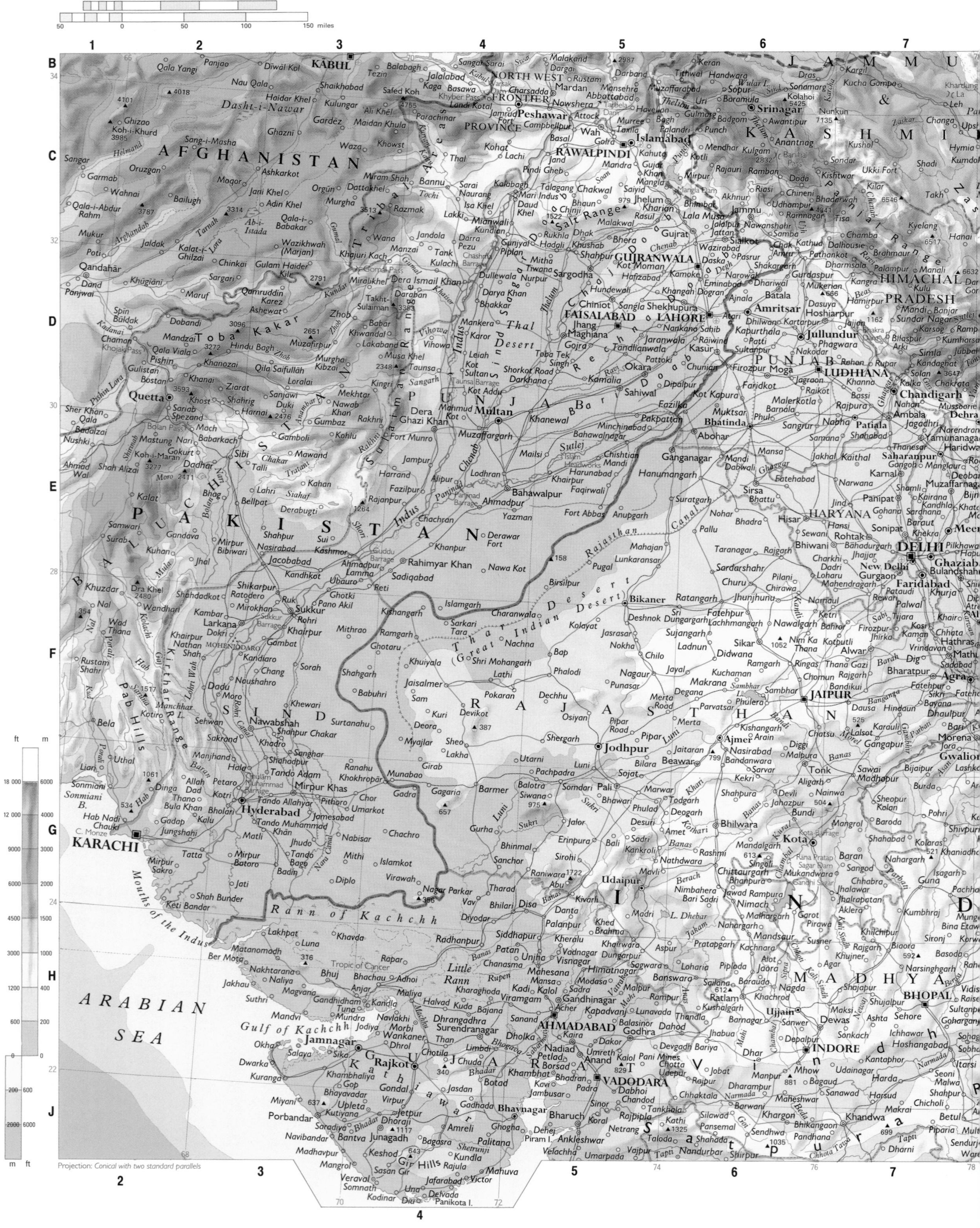

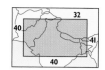

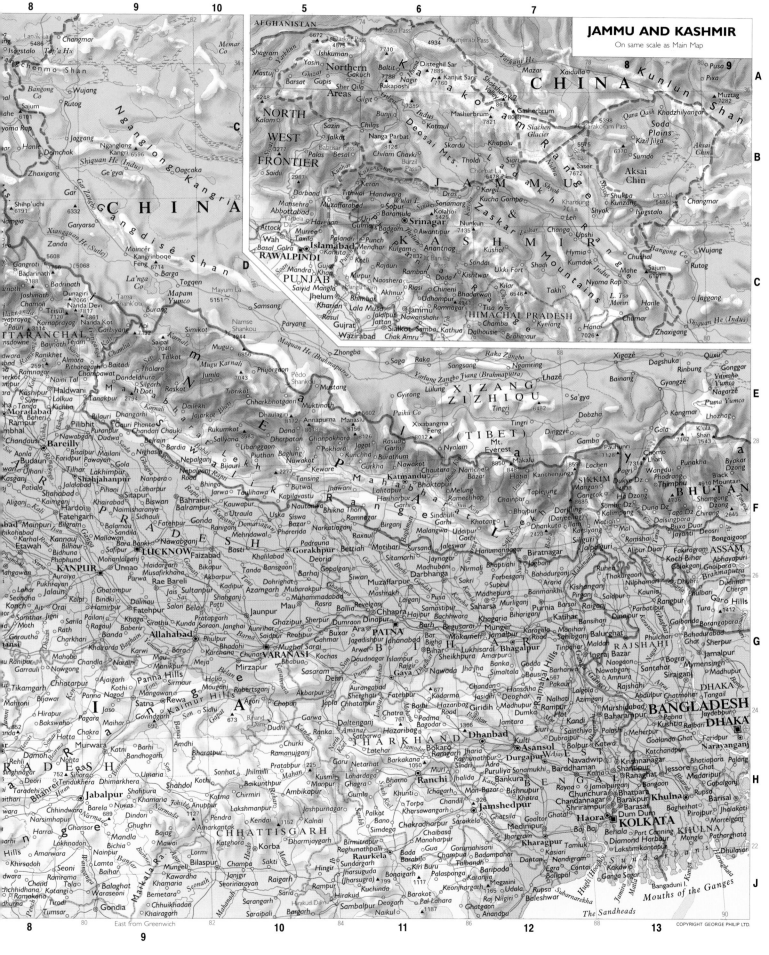

JAMMU AND KASHMIR
On same scale as Main Map

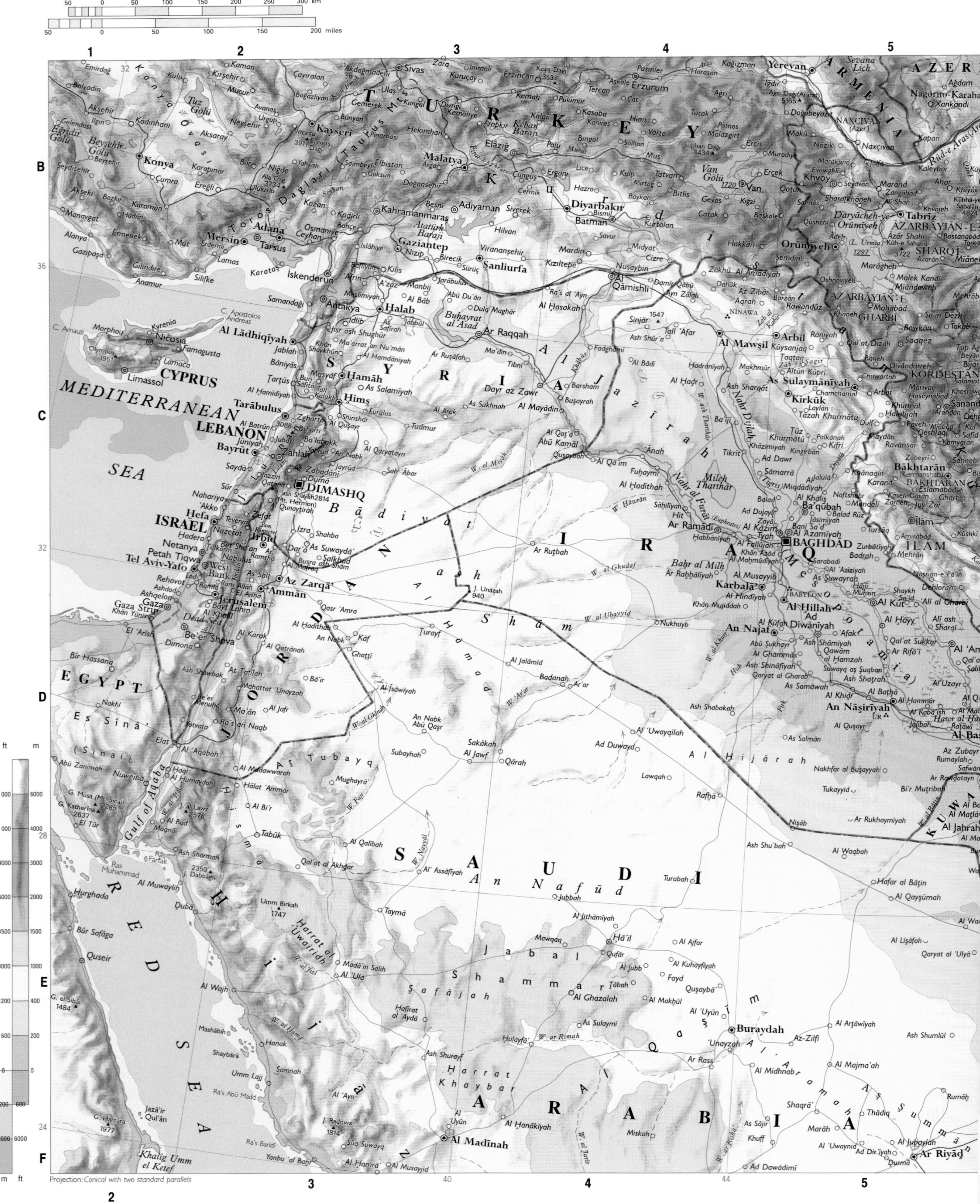

6 7 8 9 10

B C D E F

TURKMENISTAN
Kara Kum
Kopet Dagh
CASPIAN SEA
IRAN
AFGHANISTAN
PAKISTAN
OMAN
UNITED ARAB EMIRATES
QATAR
BAHRAIN
THE GULF
Gulf of Oman
Str. of Hormuz

Bakı, Türkmenbashi, Nebitdag, Gazanjyk, Gyzylarbat, Ashgabat, Tejen, Mary, Bayramaly, Yoloten, Chärjew, Amudarya, Sarakhs, Dashköpri, Gushgy
Cheleken Yarymadasy, Ostrov Ogurchinskiy, 26 Bakinskikh Komissarov, Gyzyletrek, Khrebet Bolshoy Balkhan 1880
Qazimämmäd, Älät, Neftçala, Länkäran, Ästärä, Ardabīl, Tālesh, Bandar-e Anzalī, Rasht, Āstāneh, Lāhījān, Fowman, Gīlān, Manjīl, Qazvīn, Rāmsar, Tonekābon, Now Shahr, Bābol Sar, Bābol, Sārī, Neka, Behshahr, Āmol, Qā'emshahr, Gorgān, Gonbad-e Kāvūs, Bandar-e Torkeman, Māzandarān
Reshteh-ye Kūhhā-ye Alborz
Karaj, Tajrīsh, Tehrān, Eslāmshahr, Rey, Damāvand 5604, Fīrūzkūh, Semnān, Semnān, Dāmghān, Shāhrūd, Markazī, Qom, Kāshān, Arāk, Malāyer, Borūjerd, Khomeyn
Bojnūrd, Shīrvān, Qūchān, Esfarāyen, Sabzevār, Neyshābūr, Mashhad, Kūh-e Bīnālūd 3314, Torbat-e Heydārīyeh, Torbat-e Jām, Kāshmar, Khorāsān, Gonābād, Ferdows, Qāyen, Bīrjand, Tabas, Kūh-e Sorkh 3020
Dasht-e Kavīr, Daryācheh-ye Namak, Dasht-e Lūt, Daryācheh-ye Namakzār, Namakzār-e Shahdād
Herāt, Harīrūd, Safīd Kūh, Shīndand, Farāh, Dowlatābād, Nīmrūz, Dasht-e Mārgow, Gowd-e Zirreh, Zaranj, Chakhānsūr
Eşfahān, Najafābād, Khomeynī Shahr, Shahr-e Kord, Chahār Mahāll va Bakhtīārī, Yazd, Ardakān, Bāfq, Mehrīz, Kermān, Rafsanjān, Sīrjān, Bam, Jīroft, Kūh-e Hazārān 4419, Kūh-e Jebāl Bārez 3962, Kūh-e Bazmān 3489, Kūh-e Taftān 4042
Zāhedān, Zābol, Daryācheh-ye Sīstān, Sīstān va Balūchestān, Īrānshahr, Khāsh, Chāh Bahār, Gwādar, Jīwani
Khūzestān, Ahvāz, Dezfūl, Shūshtar, Masjed Soleymān, Ābādān, Khorramshahr, Bandar-e Khomeynī, Bandar-e Māhshahr, Shatt al Arab, Kūh-e Dīnār 4431, Yāsūj, Kohkīlūyeh va Būyer Ahmadī, Būshehr, Kāzerūn, Shīrāz, Persepolis, Fārs, Fasā, Dārāb, Lār, Reshteh-ye Zāgros
Hormozgān, Bandar-e 'Abbās, Qeshm, Hormoz, Bandar-e Lengeh, Mīnāb, Kūhhā-ye Bashākerd, Jāsk, Makrān
Kuwayt, Būbiyān, Mīnā' al Aḥmadī, Al Khafjī, Ra's al Mish'āb, As Saffānīyah, Al Jubayl, Ad Dammām, Al Khubar, Az Zahrān, Al Hufūf, Al Manāmah, Al Muḥarraq, Ad Dawḥah, Dukhān, Al Jāfūrah
Abū Zaby, Dubayy, Ash Shāriqah, 'Ajmān, Umm al Qaywayn, Ra's al Khaymah, Al Fujayrah, Al 'Ayn, Ruways, Masandam, Ra's Masandam (Oman), Al Hajar al Gharbī, Ṣuḥār, 'Ibrī

East from Greenwich 52 56 60

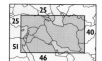

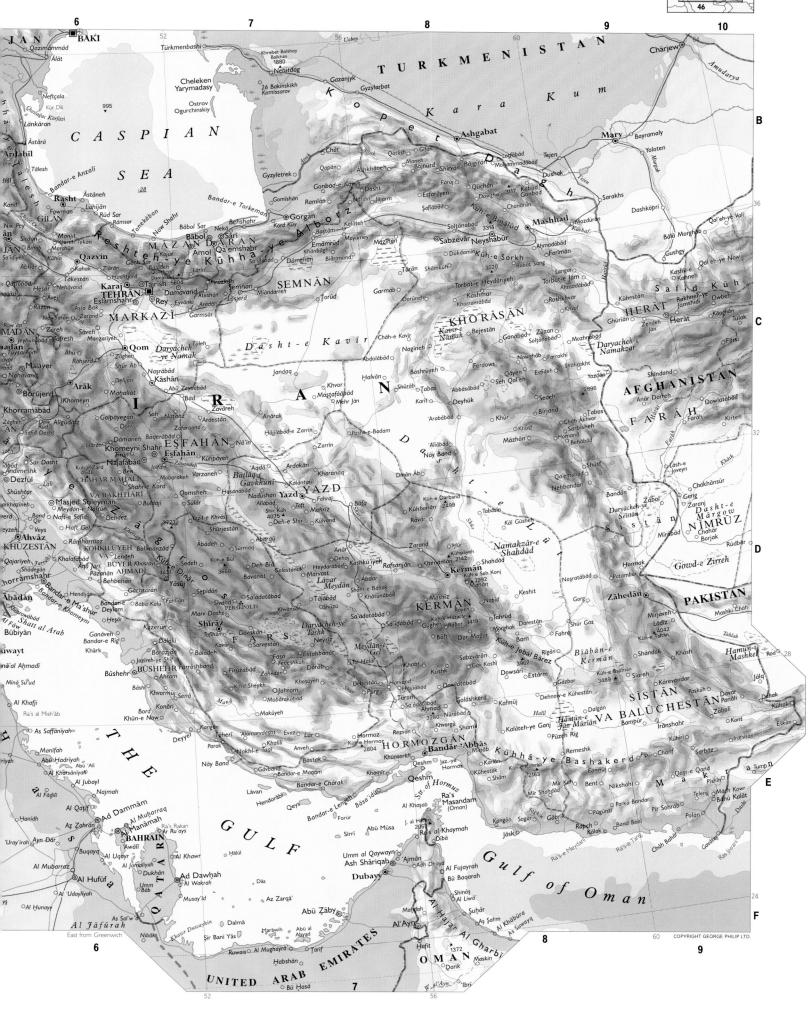

ARABIA AND THE HORN OF AFRICA 1:15 000 000

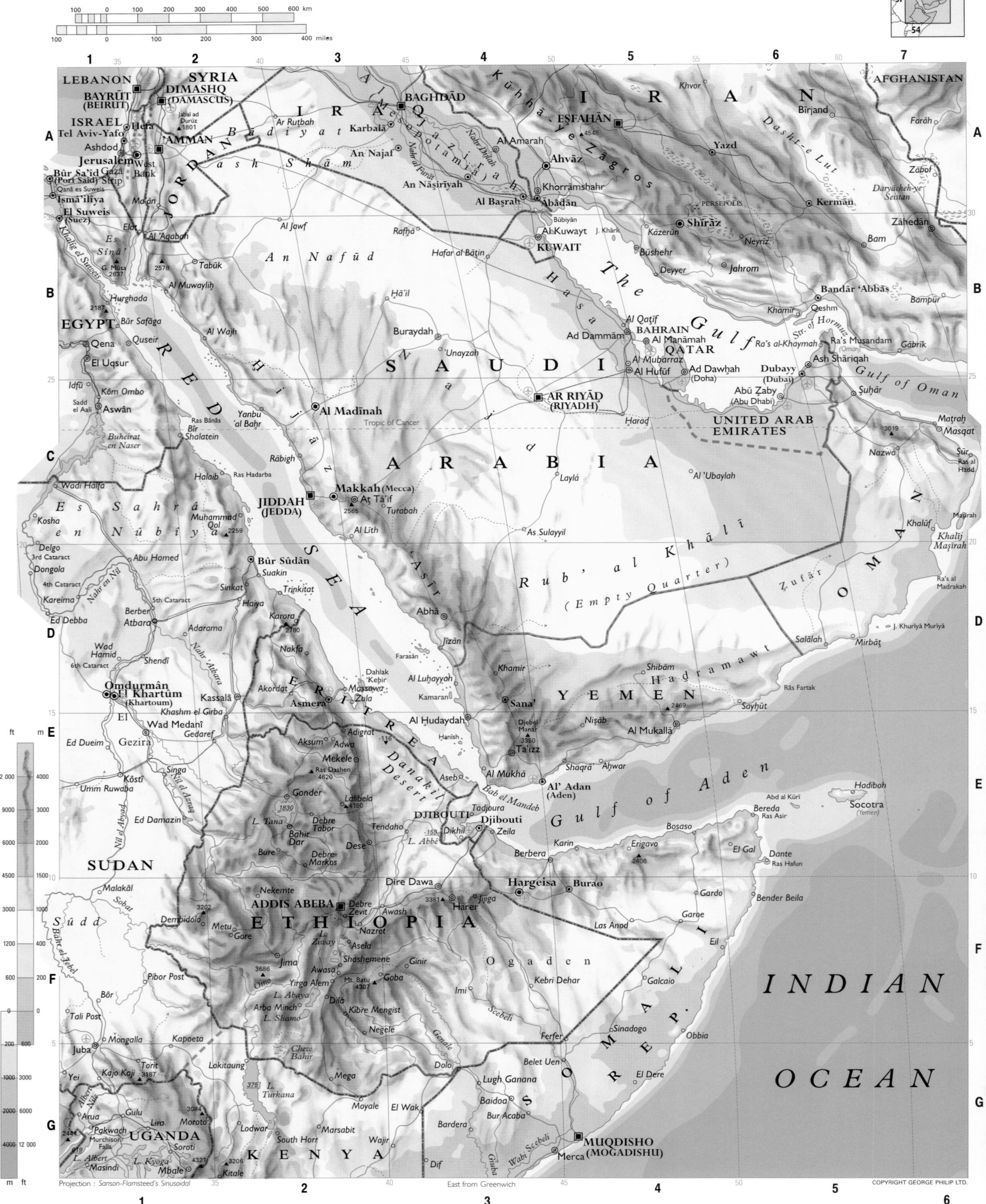

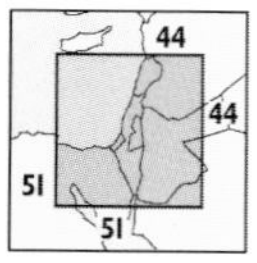

44
44
51
51

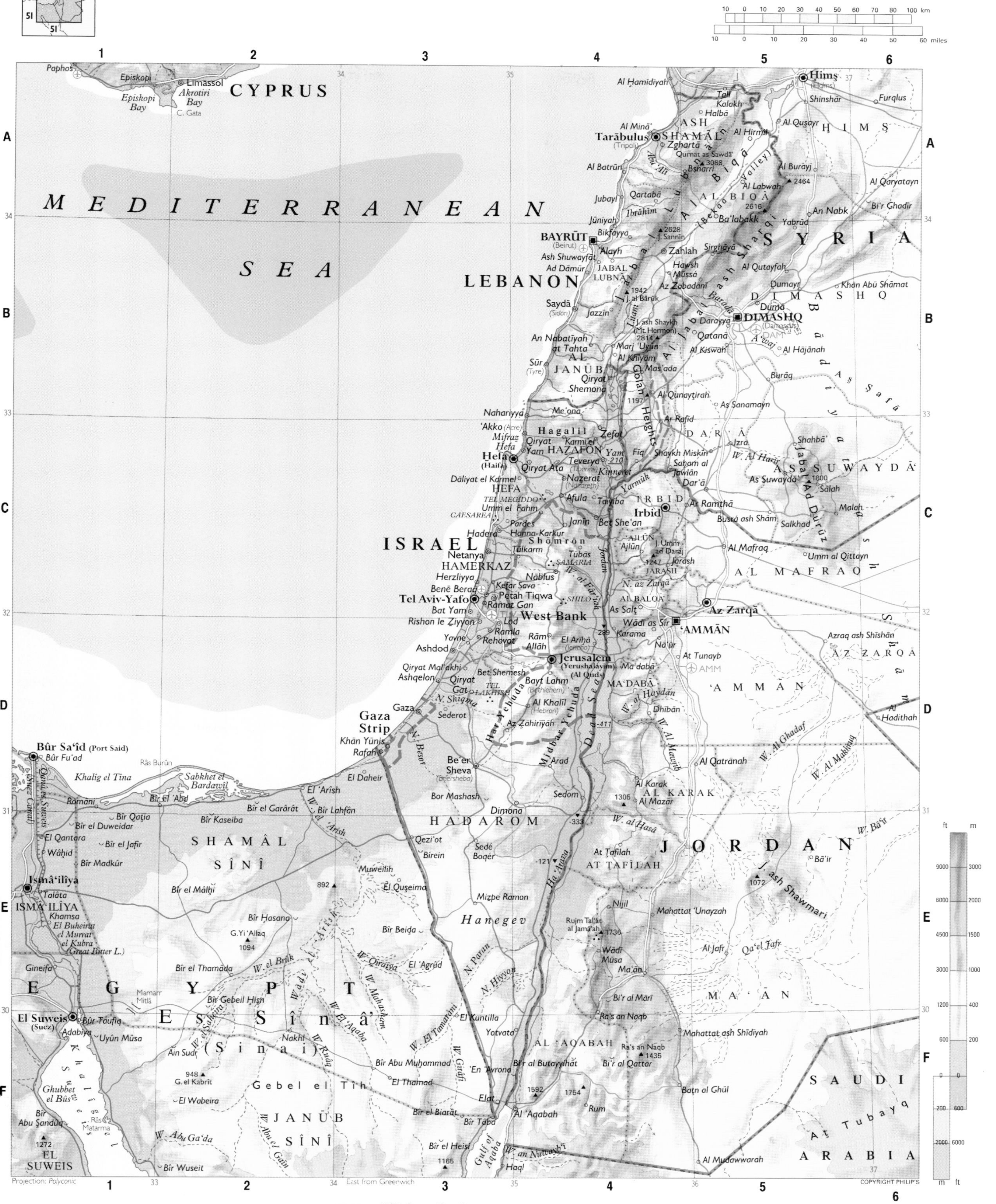

10 0 10 20 30 40 50 60 70 80 100 km
10 0 10 20 30 40 50 60 miles

1 2 3 4 5 6

CYPRUS
Paphos
Episkopi
Episkopi Bay
Limassol
Akrotiri
Akrotiri Bay
C. Gata

M E D I T E R R A N E A N

S E A

Hims
Shinshar
Furqlus
Al Hamidiyah
Tall Kalakh
Halbā
Al Qusayr
Al Qaryatayn
Al Hirmil
Al Minā'
Tarābulus (Tripoli)
ASH SHAMĀL
Zgharta
Al Batrūn
Qurnat as Sawdā 3088
2616
Al Labwah 2464
Bi'r Ghadir
Jubayl
Qartabā
Ba'labakk
Yabrūd
An Nabk
Ibrāhīm
Jūniyah
Bikfayya
2628 Sannīn
Ba'labakk
SYRIA
BAYRŪT (Beirut)
'Alayh
Ash Shuwayfāt
Ad Dāmūr
Zahlah
Sirghāyā
Al Qutayfah
Khān Abū Shāmat
LEBANON
1942
Jabal al Bārūk
Az Zabdānī
Dumayr
DIMASHQ
Saydā (Sidon)
Jazzīn
2814
Marj 'Uyūn
Al Khiyām
Qatanā
Al Kiswah
Al Hājānah
DIMASHQ (Damascus)
Dārayyā
An Nabatīyah at Tahta
AL JANŪB
Sūr (Tyre)
Qiryat Shemona
Mas'ada
Golan Heights
1197
Al Qunaytirah
As Sanamayn
Burāq
Nahariyya
Me'ona
Ar Rafid
DAR'Ā
'Akko (Acre)
Mifraz Hefa
Hagalil
Zefat
Fiq
Izra
Shahbā'
Qiryat Yam
Karmi'el
HAZAFON
Yam 210 Kinneret
Shaykh Miskin
As Suwaydā 1800
Hefa (Haifa)
Qiryat Ata
Teverya (Tiberias)
Saham al Jawlān
Dar'ā
Salah
HEFA
Nazerat (Nazareth)
Yarmūk
Ar Ramthā
AS SUWAYDĀ
Dāliyat el Karmel
Afula
IRBID
Busrá ash Shām
Salkhad
Malah
TEL MEGIDDO
Umm el Fahm
Taiyiba
Bet She'an
Irbid
Jabal ad Durūz
CAESAREA
Janin
AJLŪN
Al Mafraq
Umm al Qittayn
Hadera
Shōmrōn
Tūbās
Ajlūn
Umm ad Daraj
Hanna-Karkur
ISRAEL
Tulkarm
SAMARIA
1247 Jarash
JARASH
Al Mafraq
Umm al Qittayn
Netanya
HAMERKAZ
Nāblus
W. al Fār'ah
N. az Zarqā
AL MAFRAQ
Herzliyya
Kefar Sava
Benē Beraq
Petah Tiqwa
SHILO
As Salt
AL BALQA
Az Zarqā
Tel Aviv-Yafo
Ramat Gan
Wādī as Sīr
AMMĀN
Bat Yam
West Bank
Karama
Azraq ash Shishān
Rishon le Ziyyon
Lod
Rām Allāh
299
Na'ūr
AZ ZARQĀ
Ramla
Rehovot
El Arīhā (Jericho)
Yavne
Qiryat Mal'akhi
TEL LAKHISH
Jerusalem (Yerushalayim) (Al Quds)
Bet Shemesh
AMM
'AMMĀN
Ashdod
Ashqelon
Qiryat Gat
Bayt Lahm (Bethlehem)
Ma'daba
At Tunayb
Al Khalīl (Hebron)
MA'DĀBA
Haydān
Al Hadithah
Gaza
Sederot
Az Zāhirīyah
Dhībān
W. al Ghadaf
Gaza Strip
N. Shiqma
Midbar Yehuda
-411
W. al Mūjib
Khān Yūnis
Rafah
El Daheir
Be'er Sheva (Beersheba)
Arad
Būr Sa'īd (Port Said)
Būr Fu'ād
Rās Burūn
Sabkhet el Bardawil
El 'Arīsh
Bor Mashash
Sedom
1305
Al Mazār
AL KARAK
Al Qatrānah
W. al Mūjib
W. al Makhrūq
Khalig el Tīna
Bīr el Abd
W. el 'Arīsh
Bīr el Garārāt
Bīr Lahfān
Al Karak
Rūmāni
Bīr Qatia
Bīr el Duweidar
Bīr Kaseiba
Dimona
-333
Qezi'ot
Sedé Boqér
At Tafīlah
JORDAN
W. Bā'ir
El Qantara
Bīr el Jafir
892
SHAMĀL
SĪNĪ
Bīr Madkūr
Muweilih
El Qúseima
-121
AT TAFĪLAH
1072
Jabal ash Shawmari
Ismā'īlīya
Talāta
Bīr el Mālhi
Bīr Hasana
Bīr Beiqa
Hanegev
Mizpe Ramon
Nijil
Mahattat 'Unayzah
ISMĀ'ĪLĪYA
Khamsa
El Buheirat el Murrat el Kubra (Great Bitter L.)
G. Yi'Allaq 1094
1736 Rujm Talat al Jama'ah
Al Jafr
Qa'el Jafr
Gineifa
Bīr el Thamāda
W. el Brūk
W. Qiraiya
El 'Agrūd
N. Paran
MA'ĀN
E G Y P T
Mamarr Mitla
Bīr Gebeil Hisn
Wādi Mūsā
Ma'ān
Bi'r al Mārī
El Suweis (Suez)
Būr Taufīq
Adabiya
Uyūn Mūsa
W. el Sudira
W. el 'Aqaba
El Kuntilla
Yotvata
Ra's an Naqb
MA'ĀN
ES SĪNĀ' (Sinai)
Nakhl
W. el Ruqq
N. Hiyyon
Al 'Aqabah
1435
Mahattat ash Shīdiyah
Āin Sudr
948 G. el Kabrīt
Bīr Abu Muhammad
El Thamad
'En 'Avrona
Bi'r al Butayyihāt
Bi'r al Qattar
SAUDI
Ghubbet el Bûs
Gebel el Tîh
1592
1754
Batn al Ghūl
El Wabeira
Elat
Al 'Aqabah
Rum
1272
EL SUWEIS
Ras Matarma
Bīr Abu Sanduq
J A N Ū B
S Ī N Î
W. Abu Ga'da
W. Abu el Gain
Bīr el Biarat
Bīr Tāba
Gulf of Aqaba
Al Mudawwarah
At Tubayq
1165
Bīr el Heisi
Bīr Wuseit
W. an Nuttwe
Haql
A R A B I A

Projection: Polyconic
East from Greenwich
COPYRIGHT PHILIP'S

--- 1974 Cease Fire Lines

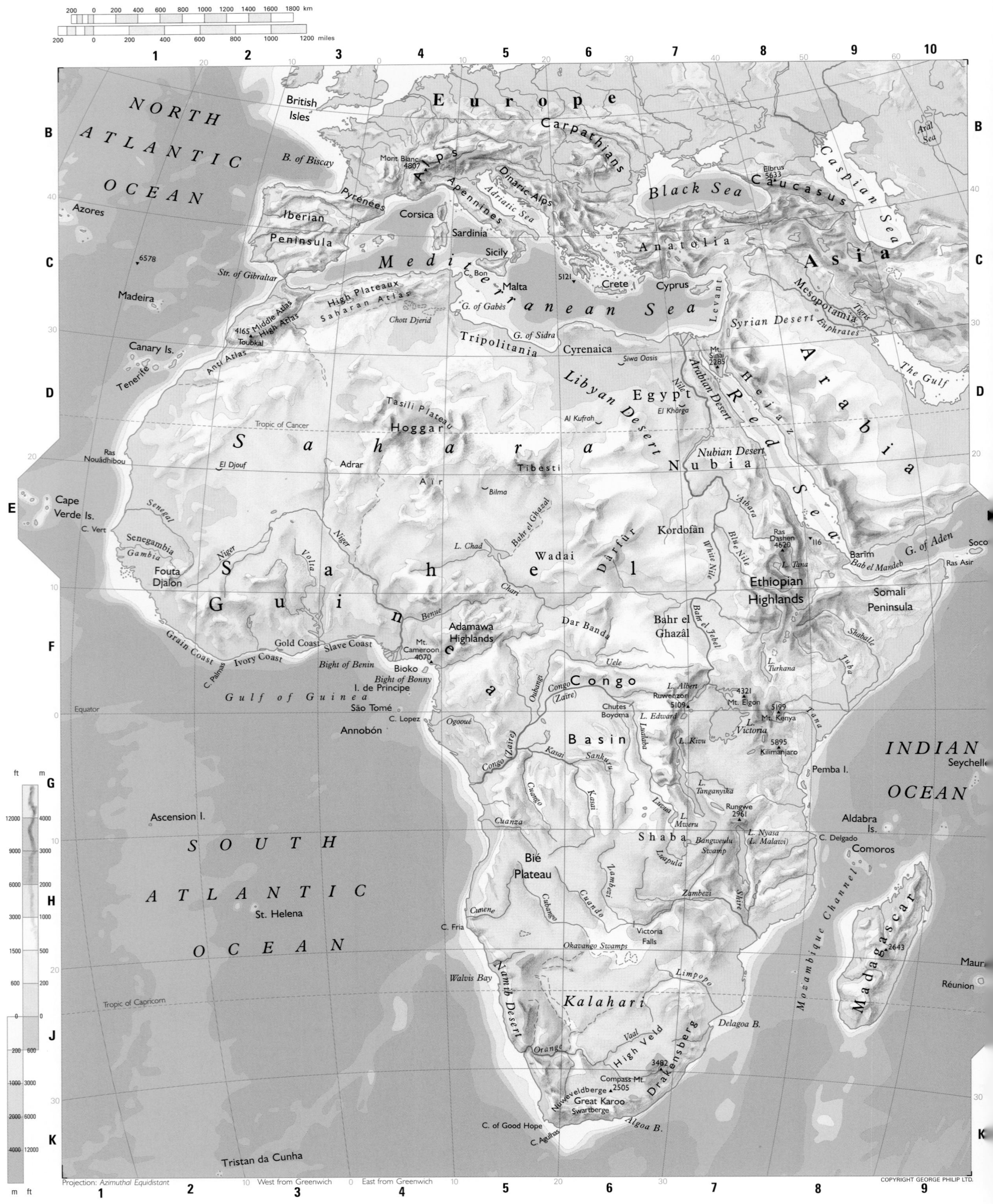

200 0 200 400 600 800 1000 1200 1400 1600 1800 km
200 0 200 400 600 800 1000 1200 miles

| | | | | | | | | | |
|1|2|3|4|5|6|7|8|9|10|

NORTH
ATLANTIC
OCEAN

British
Isles

E u r o p e

Carpathians

B. of Biscay

Mont Blanc
4807

Alps

Dinaric Alps

Apennines

Adriatic Sea

Black Sea

Caucasus

Elbrus
5633

Caspian Sea

Aral
Sea

Azores

Pyrénées

Corsica

Iberian
Peninsula

Sardinia

Sicily

Anatolia

Asia

6578

Madeira

Str. of Gibraltar

High Plateaux

Saharan Atlas

C. Bon

Malta

5121

Crete

Cyprus

Mediterranean Sea

Levant

Mesopotamia

Tigris

Euphrates

4165 Middle Atlas

Canary Is.

High Atlas

Toubkal

Chott Djerid

G. of Gabès

G. of Sidra

Tripolitania

Cyrenaica

Siwa Oasis

Syrian Desert

The Gulf

Tenerife

Anti Atlas

Tasili Plateau

Libyan Desert

Egypt

El Khârga

Al Kufrah

Arabian Desert

Mt.
Sinai
2285

Hejaz

Red Sea

Arabia

Ras
Nouâdhibou

El Djouf

Adrar

S a h a r a

Hoggar

Tibesti

Nubia

Nubian Desert

'Athara

Ras
Dashen
4620

116

G. of Aden

Tropic of Cancer

Aïr

Bilma

Nile

Socotra

Cape
Verde Is.

C. Vert

Senegal

Niger

L. Chad

Bahr el Ghazal

Kordofân

White Nile

Blue Nile

L. Tana

Barim

Bab el Mandeb

Ras Asir

Senegambia
Gambia

Volta

Niger

S a h e l

Wadai

Darfûr

Ethiopian
Highlands

Somali
Peninsula

Fouta
Djalon

G u i n e a

Benue

Chari

Dar Banda

Bahr el
Ghazâl

Bahr el Jebel

L.
Turkana

Juba

Shaballe

Grain Coast

Gold Coast

Slave Coast

Mt.
Cameroon
4070

Adamawa
Highlands

Uele

C. Palmas

Ivory Coast

Bight of Benin

Bioko

Bight of Bonny

I. de Principe

Ogooué

Oubangi

Congo
(Zaïre)

Congo

Ruwenzori
5109

L. Albert

4321

Mt. Elgon

5199

Mt. Kenya

Gulf of Guinea

São Tomé

C. Lopez

Chutes
Boyoma

L. Edward

L. Victoria

Tana

Equator

Annobón

Ogooué

Kasai

Sankuru

B a s i n

L. Kivu

5895

Kilimanjaro

INDIAN

Seychelle

Congo
(Zaïre)

Cuango

Kasai

L. Lualaba

L.
Tanganyika

OCEAN

Ascension I.

Cuanza

Lucua

Rungwe
2961

Pemba I.

SOUTH

Shaba

L.
Mweru

Bangweulu
Swamp

L. Nyasa
(L. Malawi)

Aldabra
Is.

C. Delgado

ATLANTIC

Bié
Plateau

Luapula

Comoros

St. Helena

Zambezi

Zambezi

Shire

OCEAN

Cunene

Cubango

Cuando

Victoria
Falls

Madagascar

2643

C. Fria

Okavango Swamps

Mozambique Channel

Mauri

Tropic of Capricorn

Walvis Bay

Namib Desert

Kalahari

Limpopo

Réunion

Vaal

High Veld

Delagoa B.

Orange

3482

Drakensberg

Compass Mt.
2505

Nuweveldberge

Great Karoo

Swartberge

Algoa B.

C. of Good Hope

C. Agulhas

Tristan da Cunha

ft   m

12000 4000
9000   3000
6000   2000
3000   1000
1500   500
600   200
0   0
200   600
1000   3000
2000   6000
4000 12000
m   ft

Projection: Azimuthal Equidistant

10 West from Greenwich | 0 East from Greenwich | 10

COPYRIGHT GEORGE PHILIP LTD.

| | | | | | | | |
|1|2|3|4|5|6|7|8|9|

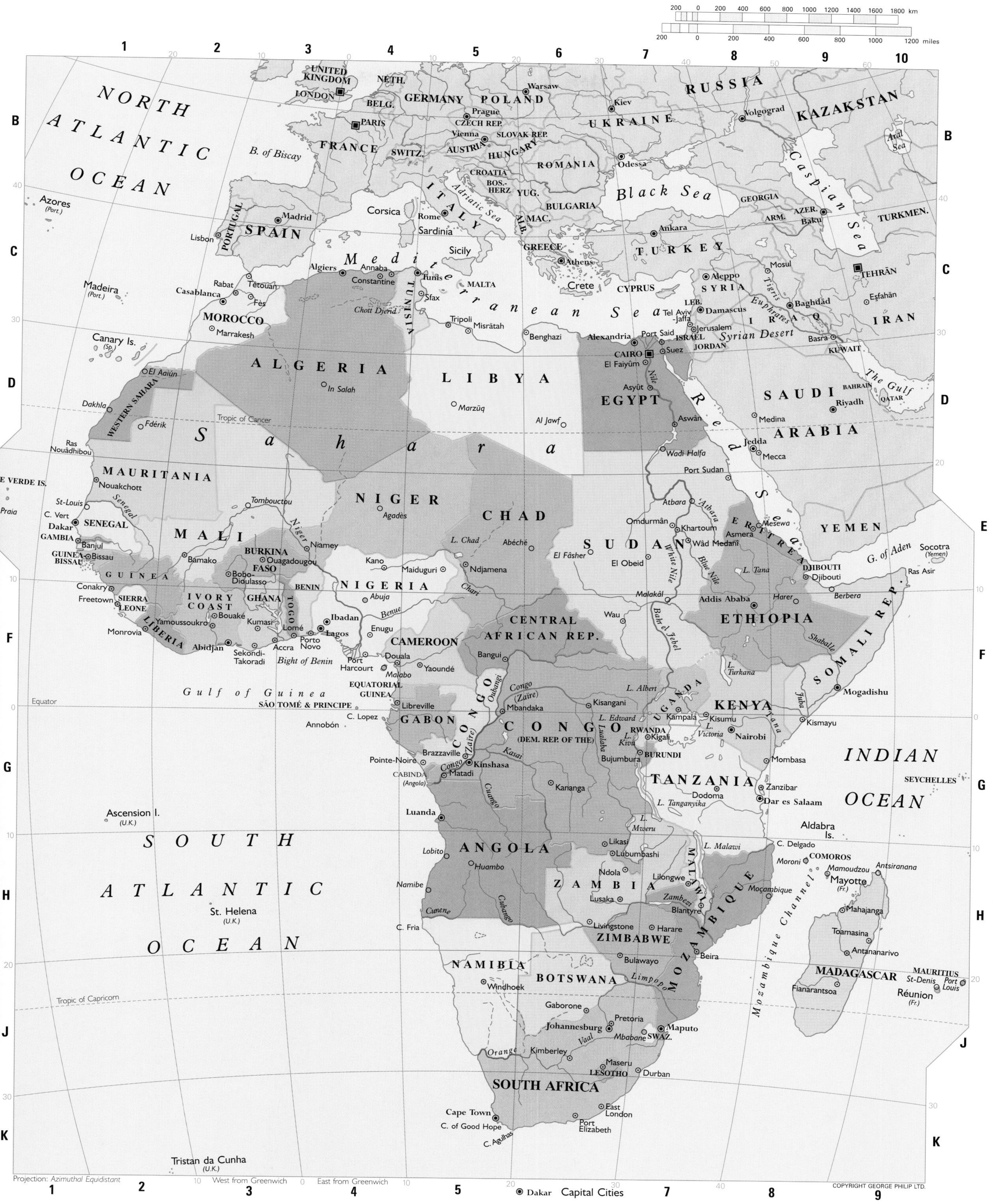

◉ Dakar Capital Cities

COPYRIGHT GEORGE PHILIP LTD.

NORTHERN AFRICA

1:15 000 000

100 0 100 200 300 400 500 600 km

100 0 100 200 300 400 miles

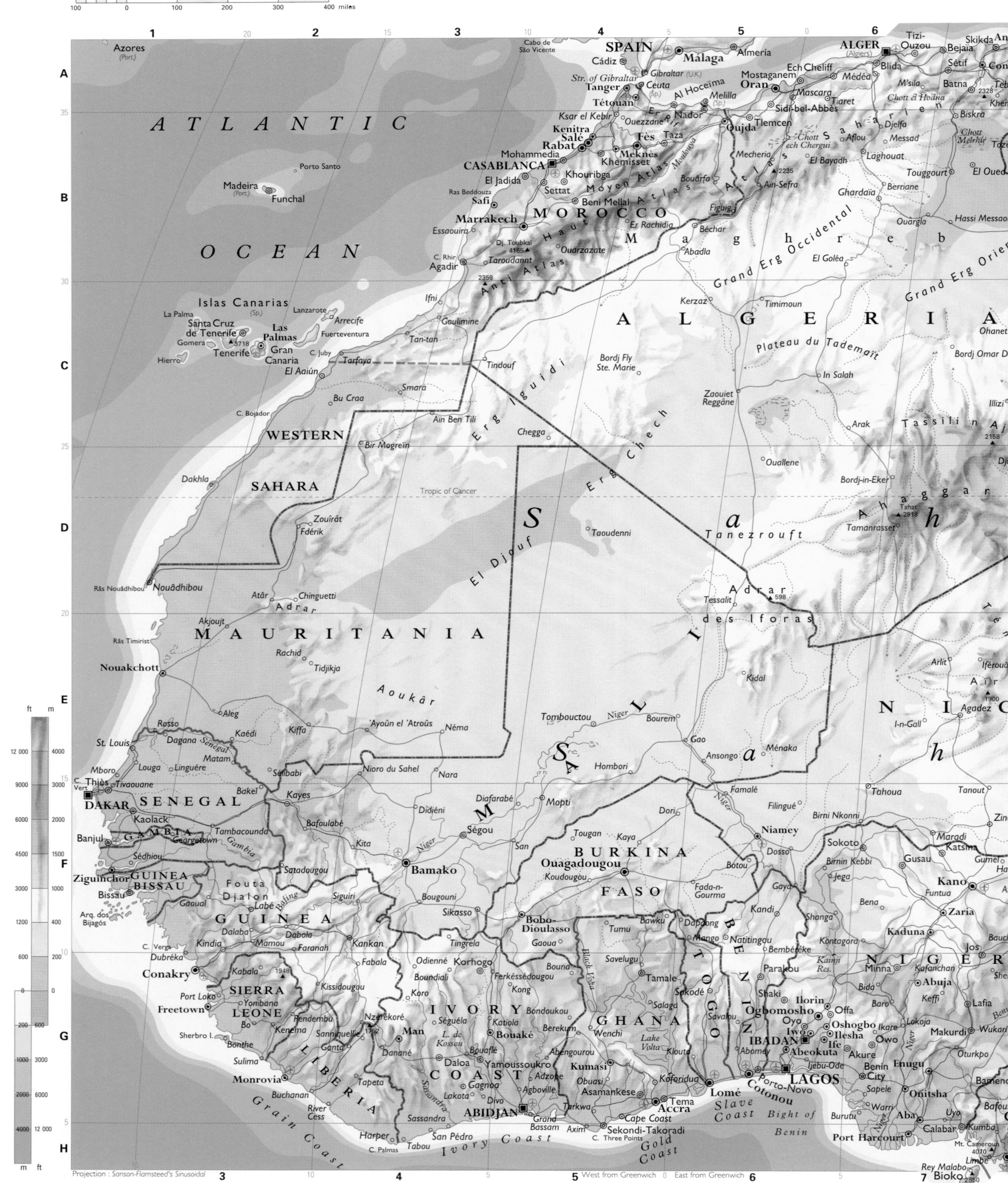

ft m

12 000 4000

9000 3000

6000 2000

4500 1500

3000 1000

1200 400

600 200

0 0

200 600

1000 3000

2000 6000

4000 12 000

m ft

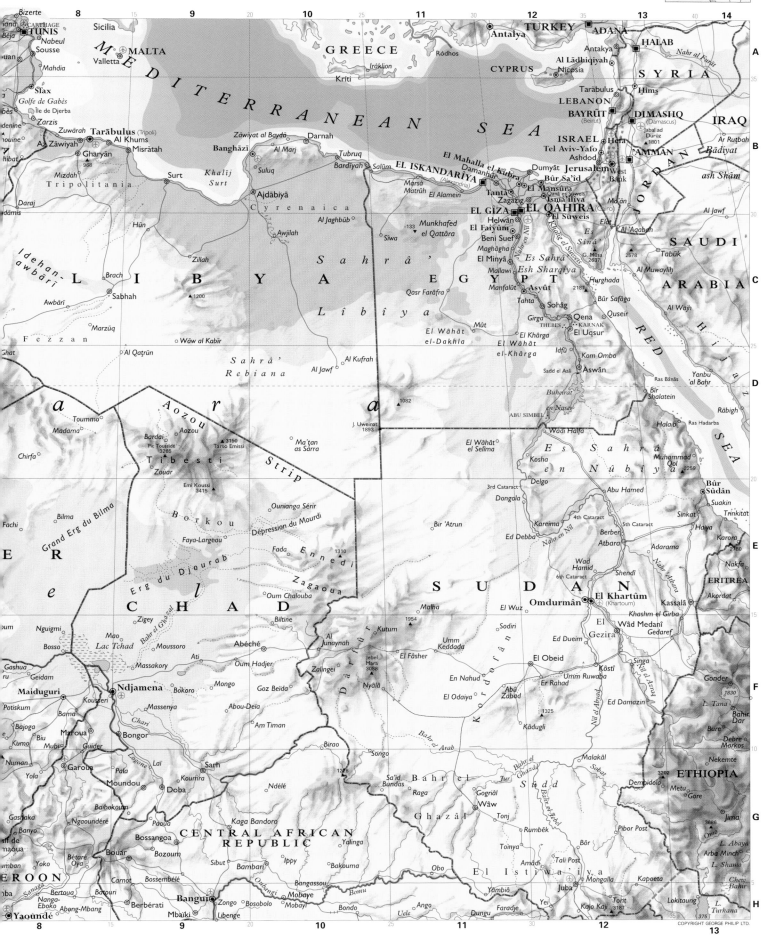

8 9 10 11 12 13 14
15 20 25 30 35 40

A B C D E F G H

MEDITERRANEAN SEA
Bizerte
CARTHAGE
TUNIS
Béja
Nabeul
Sousse
Mahdia
Sfax
Golfe de Gabès
Île de Djerba
Zarzis
Sicilia
MALTA
Valletta
Zuwārah
Tarābulus (Tripoli)
Az Zāwiyah
Al Khums
Gharyān
968
Misrātah
Mizdah
Tripolitania
Daraj
Surt
Khalīj Surt
Banghāzī
Zāwiyat al Baydā
Al Marj
Darnah
Tubruq
Bardīyah
Suluq
Ajdābiyā
Cyrenaica
Al Jaghbūb
Awjilah
Hūn
Zillah
Idehan-Awbārī
Brach
Sabhah
Awbārī
1200
Marzūq
Fezzan
Wāw al Kabīr
Al Qaţrūn
LIBYA
Sahrâ' Lîbîya
Sahrâ' Rebiana
Al Jawf
Al Kufrah
GREECE
Iráklion
Kríti
Ródhos
TURKEY
Antalya
ADANA
Antakya
HALAB
Nahr al Furāt
CYPRUS
Nicosia
Al Lādhiqīyah
SYRIA
Ḩimş
Tarābulus
LEBANON
BAYRŪT (Beirut)
DIMASHQ (Damascus)
Jabal ad Durūz
1801
IRAQ
Ar Ruţbah
Bādiyat ash Shām
ISRAEL
Haifa
Tel Aviv-Yafo
Ashdod
'AMMĀN
JORDAN
Jerusalem
West Bank
Ma'ān
Al Jawf
El Mahalla el Kubra
Damanhûr
EL ISKANDARÎYA (Alexandria)
Salûm
Marsâ Matrûh
El Alamein
Dumyât
Bûr Sa'id
El Mansûra
Qana es Suweis
Tanta
Zagazig
Ismâ'ilîya
EL GÎZA
EL QAHIRA (Cairo)
El Suweis
Helwân
El Faiyûm
Beni Suef
Maghâgha
El Minyâ
Mallawi
Manfalût
Asyût
Siwa
-133
Munkhafed el Qattâra
Es Sahrâ' Esh Sharqîya
Khalîg el Suweis
Es Sinâ'
Elat
Al 'Aqabah
G. Mûsa 2637
2578
Tabūk
SAUDI ARABIA
Al Muwaylih
Hurghada
2187
Bûr Safâga
Quseir
Al Wajh
Hijaz
EGYPT
Qasr Farâfra
Tahta
Sohâg
Girga
THEBES
KARNAK
Qena
El Uqsur
Mût
El Khârga
El Wâhât el-Dakhla
El Wâhât el-Khârga
Idfû
Kom Ombo
Aswân
Sadd el Aali
RED SEA
Ras Bânâs
Yanbu 'al Baḩr
Buheirat en Naser
Bîr Shalatein
Râbigh
ABU SIMBEL
Halaib
Ras Hadarba
1082
J. Uweinat 1893
Wadi Halfa
Sahara
Aozou Strip
Aozou
Toummo
Madama
Bardai
Pic Toussidé 3265
Tarso Emissi 3150
Tibesti
Zouar
Chirfa
Emi Koussi 3415
Ma'tan as Sarra
El Wâhât el Selîma
Es Sahrâ en Nûbîya
Kosha
Delgo
Muhammad Qol
2259
Bûr Sûdân
Suakin
Trinkitat
3rd Cataract
Dongola
Abu Hamed
Sinkat
Bilma
Grand Erg du Bilma
Fachi
Borkou
Ouniangua Sérir
Dépression du Mourdi
Faya-Largeau
Bir 'Atrun
Kareima
4th Cataract
5th Cataract
Berber
Atbara
Haiya
Karora
Ed Debba
Nahr en Nil
Adarama
Nahr 'Atbara
2780
Nakfa
ERITREA
Fada
Ennedi
1310
Erg du Djourab
Zagaoua
Oum Chalouba
NIGER
CHAD
SUDAN
Wad Hamid
Shendi
6th Cataract
Omdurmân
El Khartûm (Khartoum)
Kassalâ
Akordat
Malha
El Wuz
Khashm el Girba
Zigey
Bahr el Ghazal
Biltine
1954
Kutum
Sodiri
El Gezira
Wâd Medanî
Gedaref
Nguigmi
Mao
Lac Tchad
Bosso
Moussoro
Abéché
Al Junaynah
Umm Keddada
Ed Dueim
Gashua
Geidam
Ati
Massakory
Oum Hadjer
Zalingei
Jebel Marra 3088
El Fâsher
El Obeid
Kôstî
Singa
Nîl el Azraq
Maiduguri
Ndjamena
Bokoro
Mongo
Goz Beïda
Dârfûr
Nyâlâ
En Nahud
Umm Ruwaba
Er Rahad
Gonder
1830
Potiskum
Kousseri
Massenya
El Odaiya
Abû Zabad
Kordofân
Nîl el Abyad
Ed Damazin
L. Tana
Bahir Dar
Bama
Abou-Deïa
Am Timan
1325
Kâdugli
Bure
Bajoga
Chari
Debre Markos
Kumo
Biu
Maroua
Bongor
Mubi
Guider
Birao
Bahr el Arab
Songo
Malakâl
Nekemte
Numan
Garoua
Logone
Laï
Sarh
1276
Bahr el Jebel
Sobat
Yola
Pala
Moundou
Koumra
Doba
Ndélé
Sa'id Bundas
Bahr el Ghazâl
Raga
Sudd
Bahr el Jur
Dembidolo
3202
ETHIOPIA
Metu
Gore
Gashaka
Baïbokoum
Gogriâl
Wâw
Tonj
Rumbêk
Jima
Banyo
Ngaoundéré
Paoua
Kaga Bandoro
Pibor Post
3686
Omo
Bossangoa
CENTRAL AFRICAN REPUBLIC
Yalinga
Bôr
L. Abaya
Arba Minch
Bouar
Bozoum
Bétaré Oya
Toinya
Yokko
Sibut
Bambari
Ippy
Bakouma
Obo
Amadi
Tali Post
L. Shamo
CAMEROON
Carnot
Bossembélé
El Istiwâ'îya
Mongalla
Kapoeta
Chew Bahir
Sanaga
Bertoua
Batouri
Bangassou
Bomu
Yambio
Juba
Nanga-Eboko
Abong-Mbang
Berbérati
Bangui
Zongo
Bosobolo
Oubangi
Mobaye
Mobayi
Bondo
Ango
Uele
Dungu
Faradje
Yei
Kajo Kaji
Torit 3187
Lokitaung
L. Turkana
375
Yaoundé
Mbaïki
Libenge

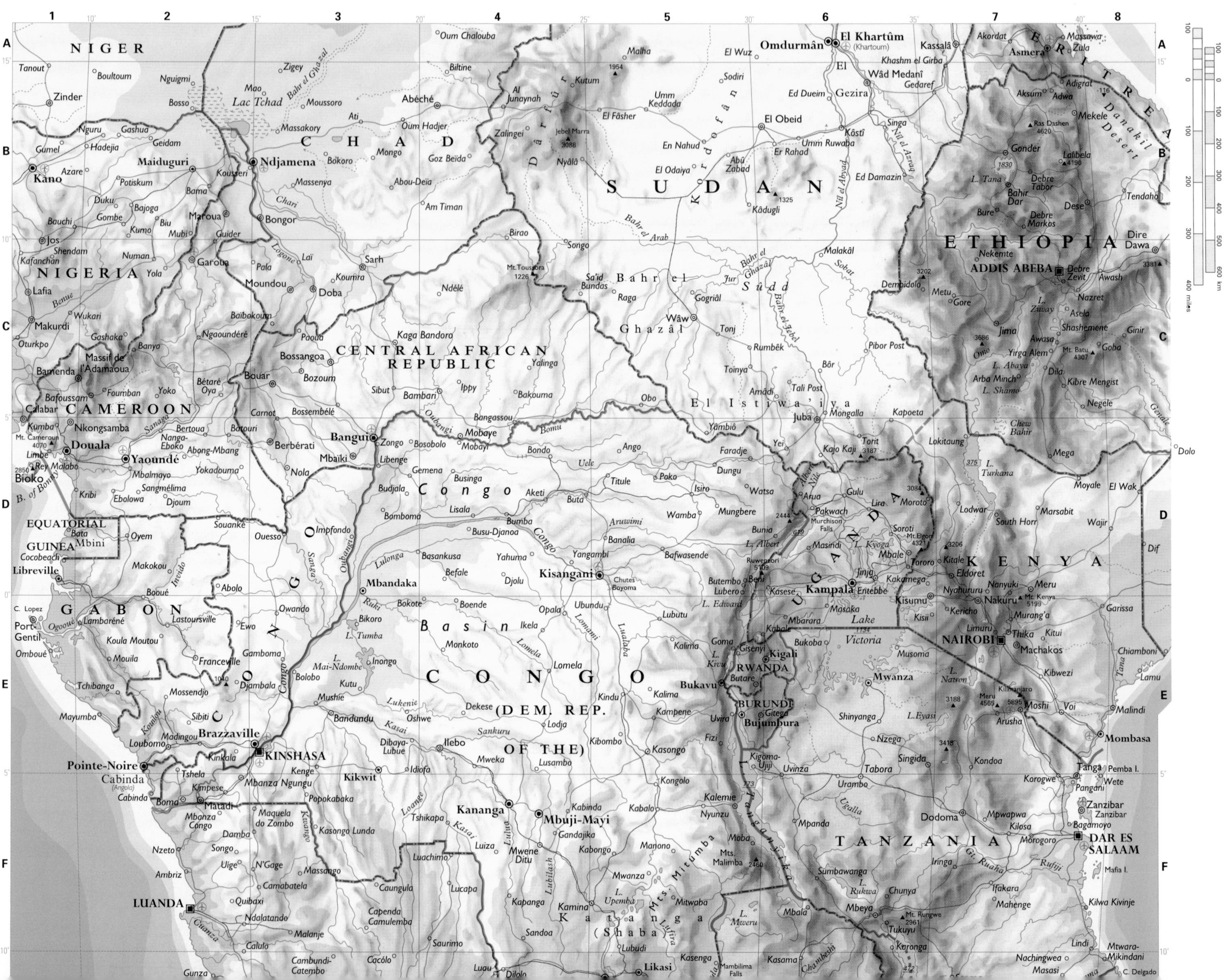

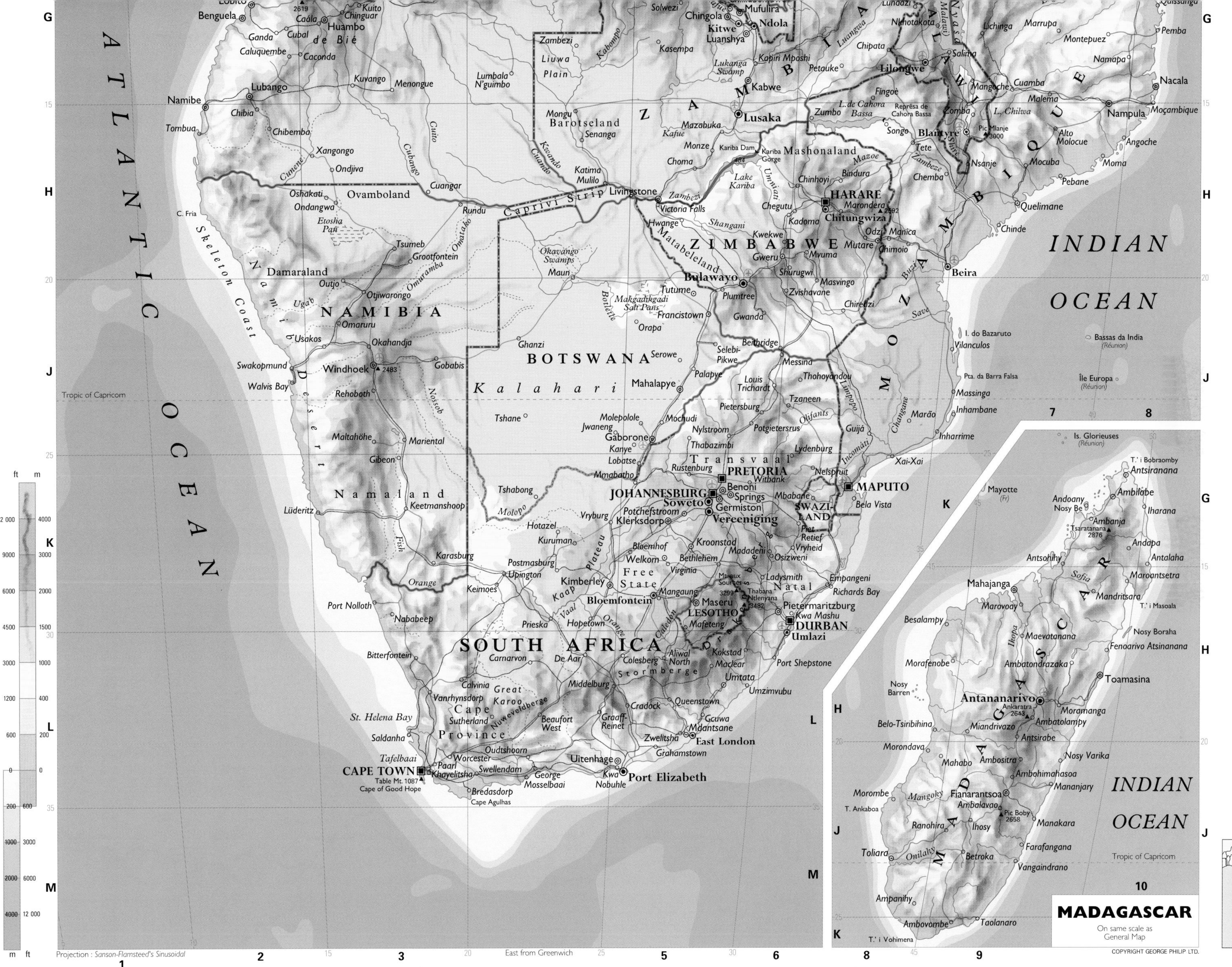

MADAGASCAR

On same scale as
General Map

COPYRIGHT GEORGE PHILIP LTD.

ATLANTIC OCEAN

INDIAN OCEAN

INDIAN OCEAN

Projection: Sanson-Flamsteed's Sinusoidal

East from Greenwich

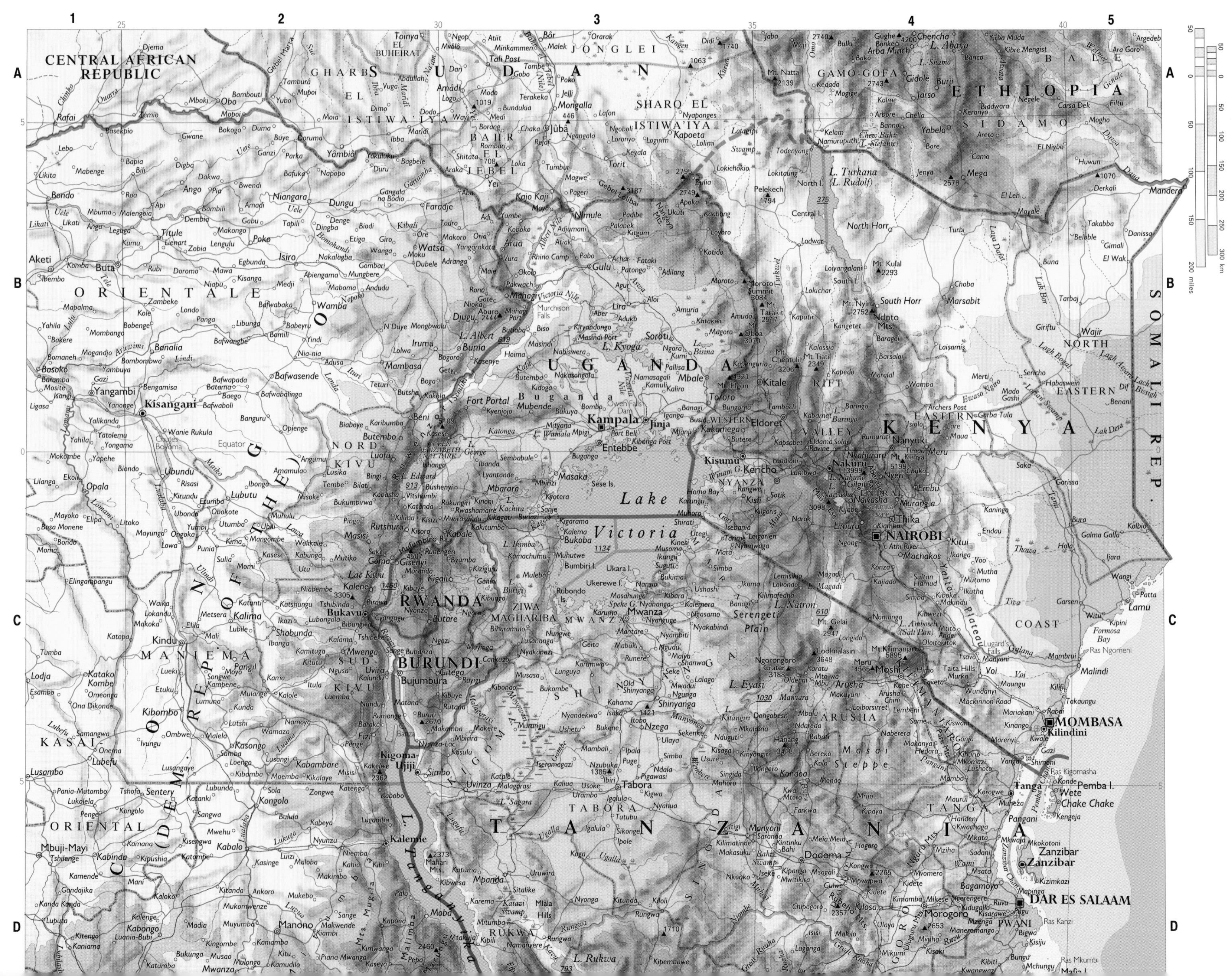

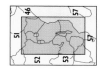

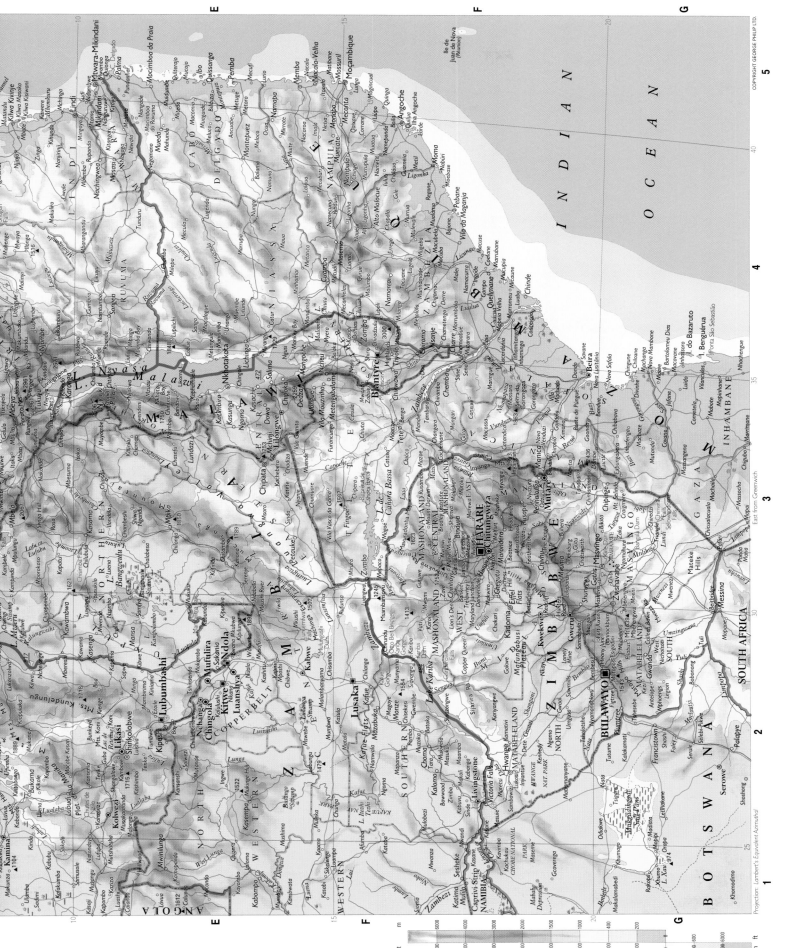

E

F

G

5

4

3

2

1

I N D I A N

O C E A N

M O Z A M B I Q U E

Z A M B I A

Z I M B A B W E

B O T S W A N A

SOUTH AFRICA

A N G O L A

MALAWI

L. Nyasa

Harare

Lusaka

Lilongwe

Blantyre

Bulawayo

Lubumbashi

Beira

Projection: Lambert's Equivalent Azimuthal

East from Greenwich

ft / m

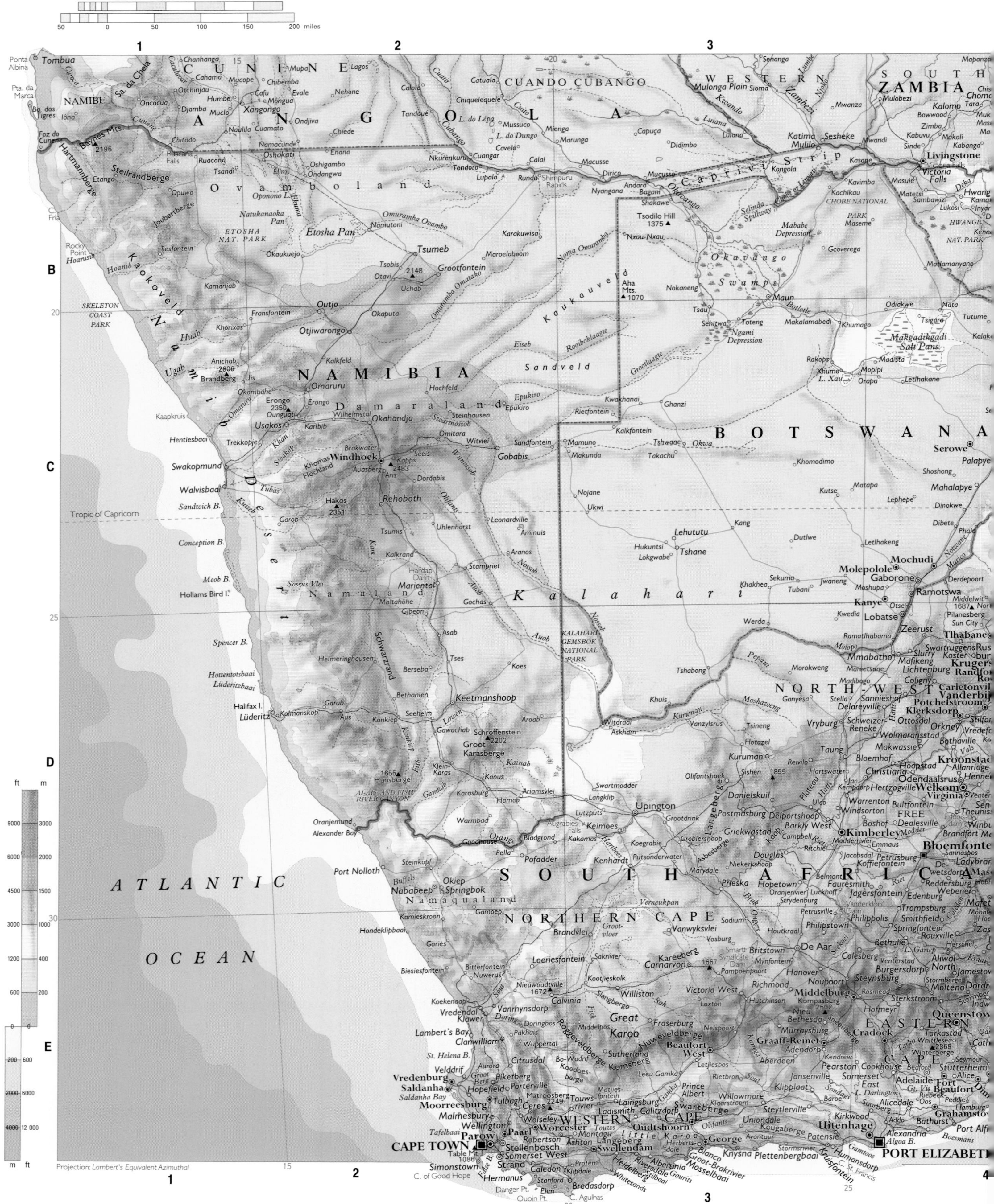

50 0 50 100 150 200 250 300 km
50 0 50 100 150 200 miles
1
2
3
4
B
C
D
E
ANGOLA
CUNENE
CUANDO CUBANGO
WESTERN
ZAMBIA
SOUTH
NAMIBIA
BOTSWANA
SOUTH AFRICA
NORTHERN CAPE
WESTERN CAPE
EASTERN CAPE
NORTH-WEST
FREE
ATLANTIC
OCEAN
Tropic of Capricorn
Ovamboland
Damaraland
Namaland
Namaqualand
Kaokoveld
Namib Desert
Kalahari
Sandveld
Kaukauveld
Caprivi Strip
Okavango Swamps
Etosha Pan
ETOSHA NAT. PARK
SKELETON COAST PARK
KALAHARI GEMSBOK NATIONAL PARK
CHOBE NATIONAL PARK
HWANGE NAT. PARK
AI-AIS AND FISH RIVER CANYON
Makgadikgadi Salt Pans
Ngami Depression
Great Karoo
Little Karoo
Tombua
NAMIBE
Baynes Mts
2195
Xangongo
Ondjiva
Ruacana Falls
Oshakati
Ondangwa
Tsumeb
Grootfontein
Otavi
2148
Outjo
Otjiwarongo
Omaruru
Brandberg
2606
Erongo
2350
Usakos
Okahandja
Swakopmund
Walvisbaai
Windhoek
2483
Rehoboth
Hakos
2351
Gobabis
Mariental
Keetmanshoop
Lüderitz
Schroffenstein
2202
Groot Karasberge
Oranjemund
Alexander Bay
Rundu
Katima Mulilo
Sesheke
Livingstone
Victoria Falls
Tsodilo Hill
1375
Aha Mts.
1070
Maun
Ghanzi
Serowe
Palapye
Mahalapye
Mochudi
Molepolole
Gaborone
Kanye
Lobatse
Ramotswa
Zeerust
Mmabatho
Mafikeng
Lichtenburg
Vryburg
Kuruman
Upington
Kimberley
Bloemfontein
Welkom
Kroonstad
Port Nolloth
Springbok
Calvinia
Carnarvon
De Aar
Colesberg
Beaufort West
Graaff-Reinet
Cradock
Queenstown
Oudtshoorn
George
Knysna
Plettenbergbaai
Mosselbaai
Uitenhage
PORT ELIZABETH
Algoa B.
Saldanha
Vredenburg
Moorreesburg
Malmesbury
Paarl
Worcester
Stellenbosch
Somerset West
Parow
CAPE TOWN
Table Mt
1086
Simonstown
C. of Good Hope
Hermanus
Swellendam
Bredasdorp
C. Agulhas
Danger Pt.
Lambert's Bay
Clanwilliam
Vredendal
Vanrhynsdorp
Nieuwoudtville
1672
2249
2502
2369
1667
1855
1687
Projection: Lambert's Equivalent Azimuthal
15
20
25
30

MOZAMBIQUE CHANNEL

INDIAN OCEAN

INDIAN OCEAN

Tropic of Capricorn

MADAGASCAR

On same scale as General Map

COPYRIGHT GEORGE PHILIP LTD.

East from Greenwich

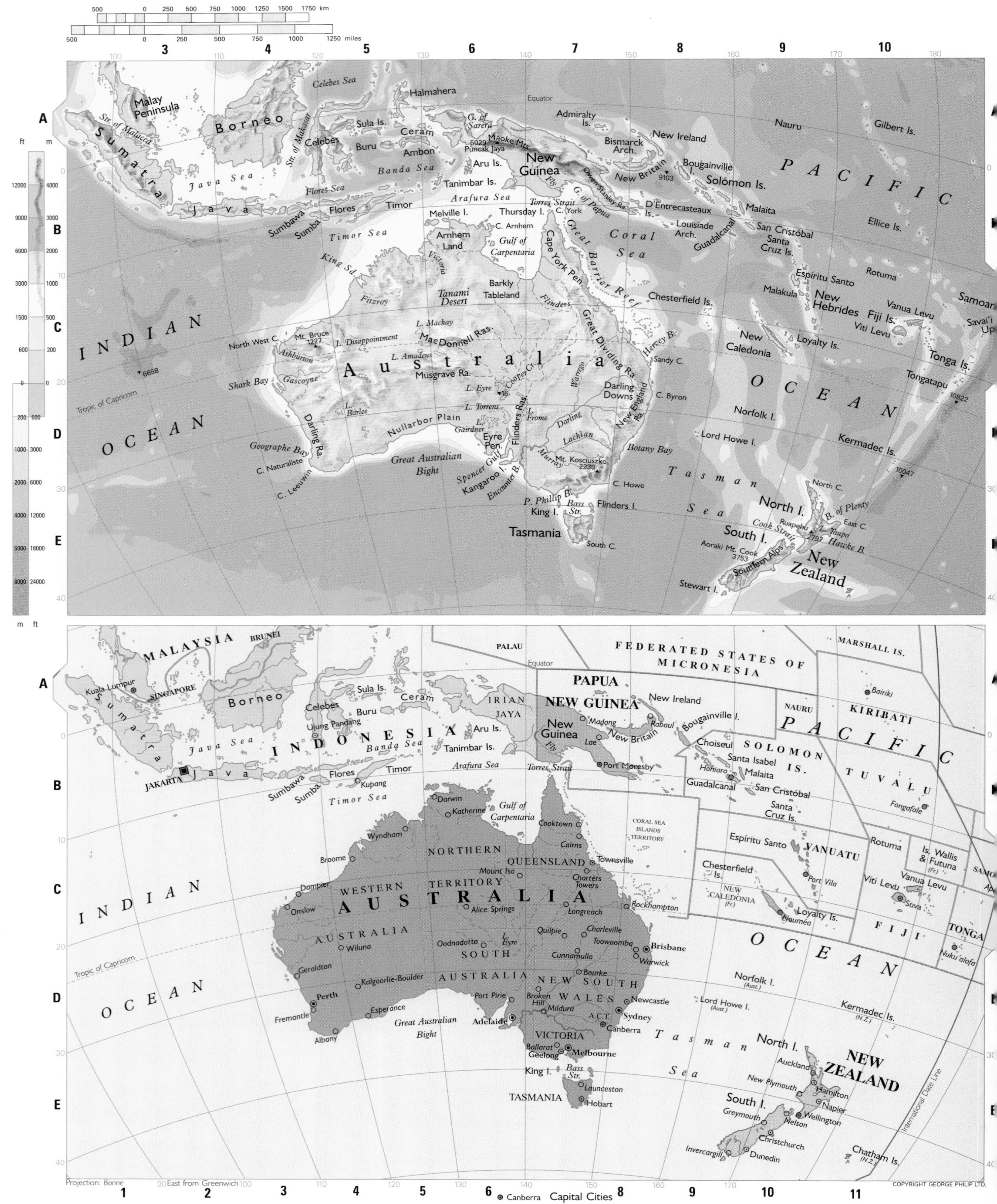

Projection: Bonne   90 East from Greenwich 100

● Canberra Capital Cities

COPYRIGHT GEORGE PHILIP LTD.

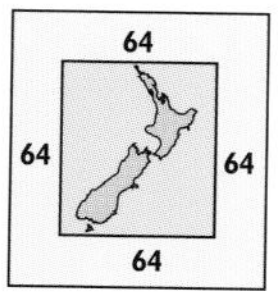

64
64 64 64
64

50 0 50 100 150 200 km
50 0 50 100 150 miles

PACIFIC

OCEAN

North C.
C. Reinga
C. Maria van Diemen
Rangaunu B.
Houhora Heads
Doubtless B.
Mangonui
Whangaroa Harb.
Ahipara B.
Kaitaia
Tauroa Pt.
Okahau
B. of Islands
Raweae
C. Brett
Opua
Hokianga Harbour
Kaikohe
Hikurangi
Whangarei
Donnelly's Crossing
Whangarei Harb.
Bream Hd.
Dargaville
Waipu
Bream B.
Little Barrier I.
Great Barrier I.
Warkworth
C. Rodney
C. Colville
Cuvier I.
Helensville
Hauraki Gulf
Coromandel
Whitianga
Takapuna
Devonport
AUCKLAND
Manukau
Papakura
Thames
Waiuku
Pukekohe
Mercer
Paeroa
Waihi
Mayor I.
Waikato
Huntly
Te Aroha
Tauranga Harb.
Morrinsville
Mount Maunganui
White I.
C. Runaway
Bay of Plenty
Hamilton
Cambridge
Tauranga
Te Puke
Whakatane
Raglan
Te Awamutu
Kawerau
Opotiki
Kawhia Harbour
Putaruru
Rotorua
Rotorua
L. Tarawera
Murupara
Raukumara Ra.
Hikurangi 1763
Waipiro
Otorohanga
Tokoroa
Te Kuiti
Kinleith
Motu
Tolaga Bay
Mokau
Mokai
Taupo
Taneatua
Waikaremoana
Ongarue L.
Whirinaki
Ormond
North Taranaki Bight
Waitara
Whangamomona
Turangi
Ruapehu
Kaimanawa Mts.
Tarawera
Nuhaka
Waikokopu
Mahia Pen.
New Plymouth
Inglewood
Mt. Taranaki (Mt. Egmont)
C. Egmont 2518
Stratford
Ohakune 2797
Raetihi
Waiouru
Taihape
Waipawa
Hawke Bay
Napier
C. Kidnappers
Hastings
Opunake
Kapuni
Eltham
Hawera
Waverley
South Taranaki Bight
Patea
Wanganui
Marton
Mangaweka
Hunterville
Halcombe
Feilding
Dannevirke
Waipukurau
C. Turnagain
Palmerston North
Foxton
Shannon
Levin
Woodville
Pahiatua
Eketahuna
Otaki
Masterton
Paraparaumu
Kapiti I.
Carterton
Greytown
Martinborough
Wairarapa
Upper Hutt
Featherston
Eastbourne
Petone
WELLINGTON
Lower Hutt
Cook Strait

North Island

TASMAN

SEA

C. Farewell
Collingwood
Golden B.
D'Urville I.
Takaka
Tasman Mts.
Tasman B.
Motueka
Pelorus Sd.
Karamea
Karamea Bight
Tadmor
Nelson
Havelock
Picton
Seddonville
Granity
Richmond
Wakefield
Blenheim
Martin Ra.
Westport
Lyell
Murchison
Seddon
Ward
Reefton
Inangahua
Rotoroa
Mt. Travers 2338
2885 Tapuaenuku
Blackball
Runanga
Spenser Mts.
Kaikoura
Greymouth
Stillwater
Hanmer Springs
Kumara
L. Bruner
Jacksons
Waiau
Kaikoura
Hokitika
Arthur's Pass
Waikari
Hurunui
Ross
Culverden
Amberley
Waipara
Abut Hd.
Coleridge
Oxford
Rangiora
Kaiapoi
Pegasus Bay
Springfield
Whitecliffs
New Brighton
Methven
Christchurch
Aoraki Mt. Cook 3753
Staveley
Riccarton
Lyttelton
Lincoln
Banks Pen.
Mount Cook
L. Tekapo
Fairlie
Akaroa
Rakaia
L. Ellesmere
Little River
Jackson B.
Mt. Aspiring 3027
Ashburton
Canterbury Bight
Okuru
Haast
Mt. Earnslaw 2818
Wanaka
Ohai
Temuka
Timaru
St. Andrews
Milford Sd.
Sutherland Falls
Bligh Sound
Milford Sound
George Sound
Arrowtown
Cromwell
Kurow
Waimate
Queenstown
Clyde
Naseby
Tokarahi
Ngapara
Secretary I.
Doubtful Sd.
Wakatipu
Alexandra
Kokanui
Oamaru
Maheno
Breaksea Sd.
Te Anau
Kingston
Roxburgh
Hampden
Dunback
Dusky Sd.
Manapouri
Waikouaiti
Palmerston
Resolution I.
Mossburn
Otago
Port Chalmers
Otago Harbour
Lumsden
Clutha
Mosgiel
Saunders C.
Ohai
Edievale
Kelso
Lawrence
Dunedin
Fairfield
Milton
Nightcaps
Clifden
Tuatapere
Hedgehope
Winton
Clinton
Balclutha
Orepuki
Riverton
Gore
Mataura
Kaitangata
Te Waewae B.
Invercargill
Wyndham
Nugget Pt.
Owaka
Bluff
Invercargill
Tahakopa
Foveaux Str.
Ruapuke I.
Halfmoon Bay
Stewart I.
Southwest C.
Port Pegasus

South Island
Westland Bight
Southern Alps
Eyre Mts.
Garvie Mts.

Projection: Conical with two standard parallels
East from Greenwich

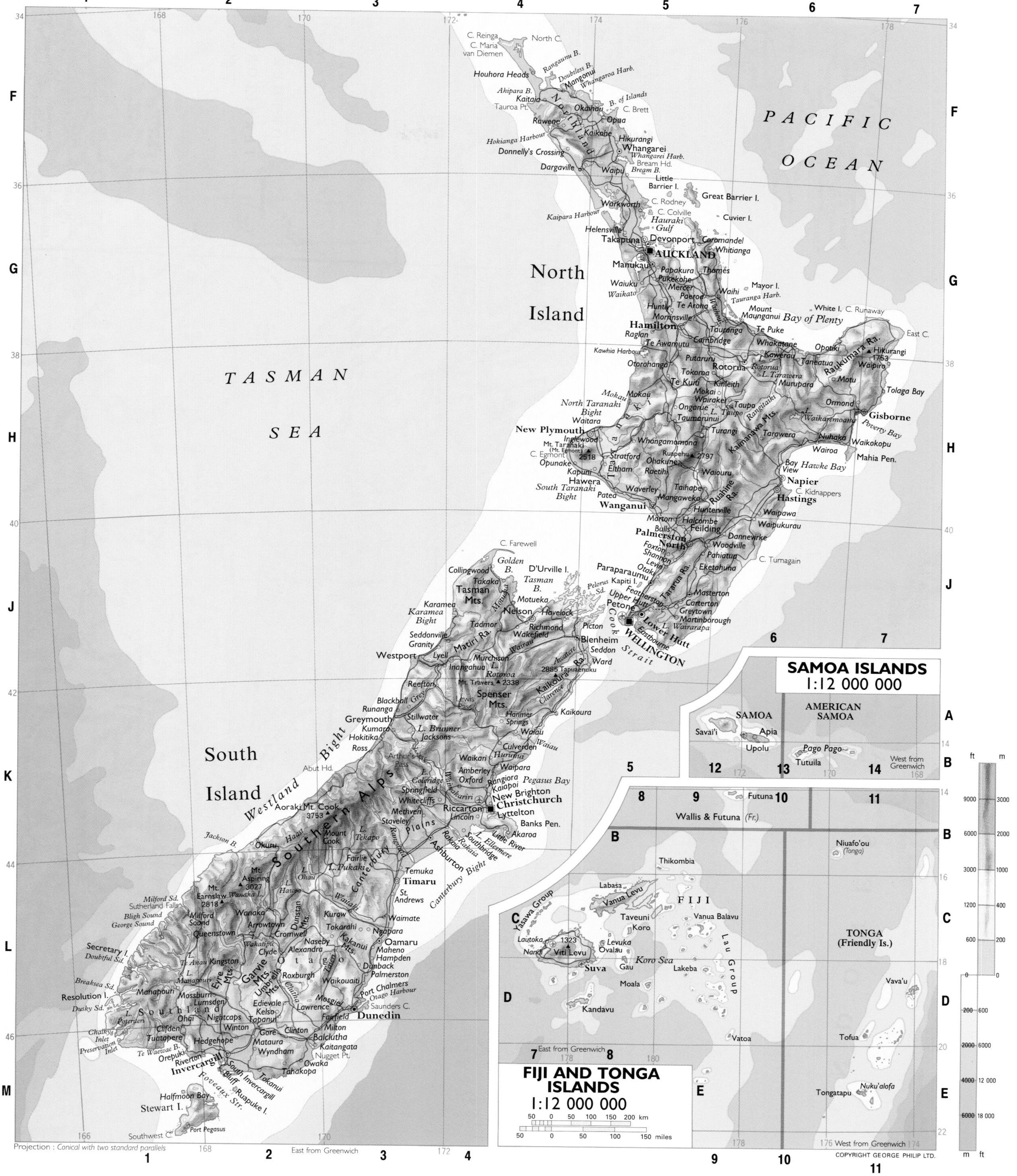

COPYRIGHT GEORGE PHILIP LTD.

SAMOA ISLANDS
1:12 000 000

SAMOA
AMERICAN SAMOA
Savai'i
Apia
Upolu
Pago Pago
Tutuila
West from Greenwich

Wallis & Futuna (Fr.)
Futuna

Niuafo'ou (Tonga)

Thikombia
Labasa
Vanua Levu
FIJI
Vanua Balavu
Yasawa Group
Taveuni
Koro
Lau Group
Lautoka 1323
Levuka
Ovalau
Nandi
Viti Levu
Gau
Koro Sea
Lakeba
TONGA (Friendly Is.)
Suva
Moala
Vava'u
Kandavu
Moala
Vatoa
Tofua
Tongatapu
Nuku'alofa

FIJI AND TONGA ISLANDS
1:12 000 000

50 0 50 100 150 200 km
50 0 50 100 150 miles

East from Greenwich
West from Greenwich

ft m
9000 3000
6000 2000
3000 1000
1200 400
600 200
0 0
200 600
2000 6000
4000 12 000
6000 18 000
m ft

PACIFIC OCEAN Equatorial Scale 1:54 000 000

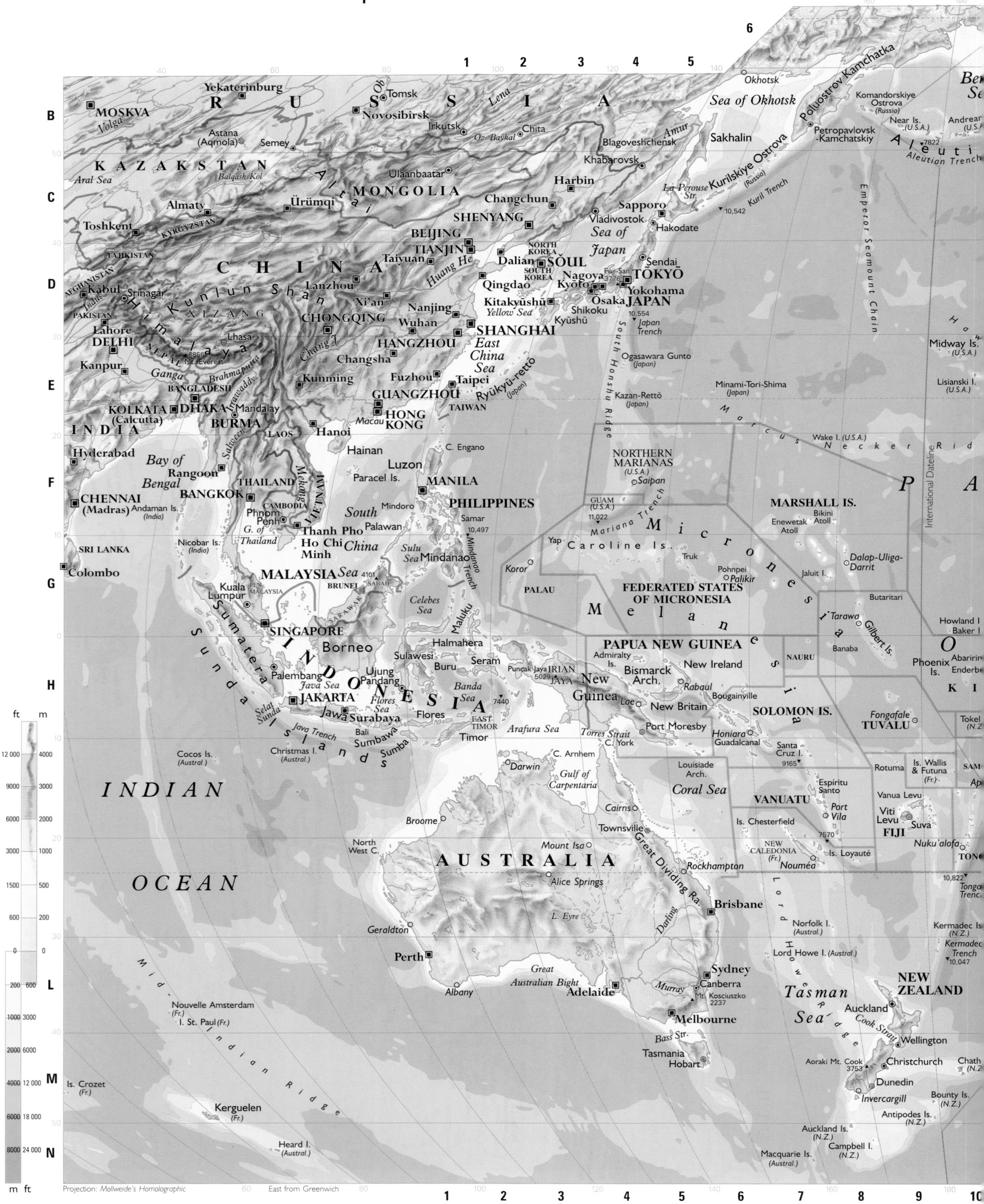

RUSSIA
MOSKVA
Volga
Yekaterinburg
Tomsk
Ob
Novosibirsk
Irkutsk
Oz. Baykal
Chita
Lena
Okhotsk
Sea of Okhotsk
Poluostrov Kamchatka
Komandorskiye Ostrova (Russia)
Petropavlovsk-Kamchatskiy
Near Is. (U.S.A.)
Andreanof Is. (U.S.A.)
7822
Aleutian Trench
Bering Sea
Sakhalin
Amur
Blagoveshchensk
Khabarovsk
Kurilskiye Ostrova (Russia)
Kuril Trench
10,542
La Perouse Str.
Emperor Seamount Chain
Astana (Aqmola)
Semey
KAZAKSTAN
Aral Sea
Balqash Köl
Altai
Ulaanbaatar
MONGOLIA
Harbin
Changchun
Sapporo
Hakodate
Vladivostok
Sea of Japan
Almaty
Ürümqi
Toshkent
KYRGYZSTAN
TAJIKISTAN
SHENYANG
BEIJING
TIANJIN
NORTH KOREA
SOUTH KOREA
Dalian
SÖUL
Sendai
CHINA
Taiyuan
Huang He
AFGHANISTAN
Kabul
Srinagar
Indus
Kunlun Shan
Lanzhou
Xi'an
Qingdao
Nagoya
Fuji-San 3776
TŌKYŌ
Yokohama
Kyoto
Osaka
JAPAN
Kitakyūshū
Shikoku
Kyūshū
Yellow Sea
PAKISTAN
XIZANG
Nanjing
CHONGQING
Wuhan
Lahore
DELHI
Himalaya
Lhasa
NEPAL
8850 Mt. Everest
Chang Jiang
HANGZHOU
SHANGHAI
East China Sea
10,554 Japan Trench
South Honshu Ridge
Ogasawara Gunto (Japan)
Midway Is. (U.S.A.)
Kanpur
Ganga
Brahmaputra
BANGLADESH
Changsha
Kunming
Fuzhou
Taipei
Ryūkyū-rettō (Japan)
Minami-Tori-Shima (Japan)
Lisianski I. (U.S.A.)
KOLKATA (Calcutta)
DHAKA
Irrawaddy
Mandalay
GUANGZHOU
Macau
HONG KONG
TAIWAN
Kazan-Rettō (Japan)
Marcus Necker Ridge
INDIA
BURMA
LAOS
Hanoi
Hainan
C. Engano
NORTHERN MARIANAS (U.S.A.)
Saipan
Wake I. (U.S.A.)
International Dateline
Hyderabad
Bay of Bengal
Salween
Rangoon
Mekong
VIETNAM
Luzon
Paracel Is.
MANILA
PHILIPPINES
GUAM (U.S.A.)
11,022
Mariana Trench
MARSHALL IS.
PACIFIC
CHENNAI (Madras)
Andaman Is. (India)
THAILAND
BANGKOK
CAMBODIA
Phnom Penh
Mindoro
South China Sea
Samar
10,497
Mindanao Trench
Bikini Atoll
Enewetak Atoll
Micronesia
G. of Thailand
Thanh Pho Ho Chi Minh
Palawan
Sulu Sea
Mindanao
Yap
Caroline Is.
Truk
Pohnpei
Palikir
Jaluit I.
Dalap-Uliga-Darrit
SRI LANKA
Colombo
Nicobar Is. (India)
MALAYSIA
4101
BRUNEI
SABAH
Koror
PALAU
FEDERATED STATES OF MICRONESIA
Butaritari
Kuala Lumpur
PEN. MALAYSIA
SARAWAK
Celebes Sea
Maluku
Melanesia
Tarawa
Gilbert Is.
Howland I.
Baker I.
Sumatera
SINGAPORE
Borneo
INDONESIA
Halmahera
PAPUA NEW GUINEA
Admiralty Is.
NAURU
Banaba
Sulawesi
Buru
Seram
Puncak Jaya 5029
IRIAN JAYA
Bismarck Arch.
New Ireland
Phoenix Is.
Abariringa
Enderbury
Sunda Islands
Palembang
Java Sea
Ujung Pandang
Banda Sea
New Guinea
Rabaul
JAKARTA
Flores Sea
7440
Lae
Bougainville
Selat Sunda
Jawa
Surabaya
Flores
EAST TIMOR
Arafura Sea
New Britain
SOLOMON IS.
Fongafale
TUVALU
Tokelau (N.Z.)
Java Trench
Bali
Sumbawa
Sumba
Timor
Torres Strait
C. York
Port Moresby
Honiara
Guadalcanal
Santa Cruz I.
9165
Christmas I. (Austral.)
Cocos Is. (Austral.)
C. Arnhem
Darwin
Gulf of Carpentaria
Louisiade Arch.
Coral Sea
Rotuma
Is. Wallis & Futuna (Fr.)
SAMOA
Apia
INDIAN OCEAN
Espíritu Santo
VANUATU
Port Vila
Vanua Levu
Viti Levu
Suva
FIJI
Broome
Cairns
Townsville
Is. Chesterfield
7570
North West C.
Mount Isa
Great Dividing Ra.
NEW CALEDONIA (Fr.)
Is. Loyauté
Nouméa
Nuku'alofa
TONGA
AUSTRALIA
Rockhampton
Alice Springs
10,822
Tonga Trench
Brisbane
L. Eyre
Lord Howe Ridge
Darling
Norfolk I. (Austral.)
Kermadec Is. (N.Z.)
Kermadec Trench
10,047
Geraldton
Lord Howe I. (Austral.)
Perth
Sydney
Canberra
Great Australian Bight
Albany
Adelaide
Murray
Mt. Kosciuszko 2237
Tasman Sea
NEW ZEALAND
Auckland
Nouvelle Amsterdam (Fr.)
I. St. Paul (Fr.)
Melbourne
Bass Str.
Cook Strait
Wellington
Mid-Indian Ridge
Tasmania
Hobart
Aoraki Mt. Cook 3753
Christchurch
Chatham Is. (N.Z.)
Is. Crozet (Fr.)
Dunedin
Invercargill
Bounty Is. (N.Z.)
Kerguelen (Fr.)
Antipodes Is. (N.Z.)
Auckland Is. (N.Z.)
Campbell I. (N.Z.)
Heard I. (Austral.)
Macquarie Is. (Austral.)
ft m
12 000 4000
9000 3000
6000 2000
3000 1000
1500 500
600 200
0 0
200 600
1000 3000
2000 6000
4000 12 000
6000 18 000
8000 24 000
m ft
Projection: Mollweide's Homolographic
East from Greenwich

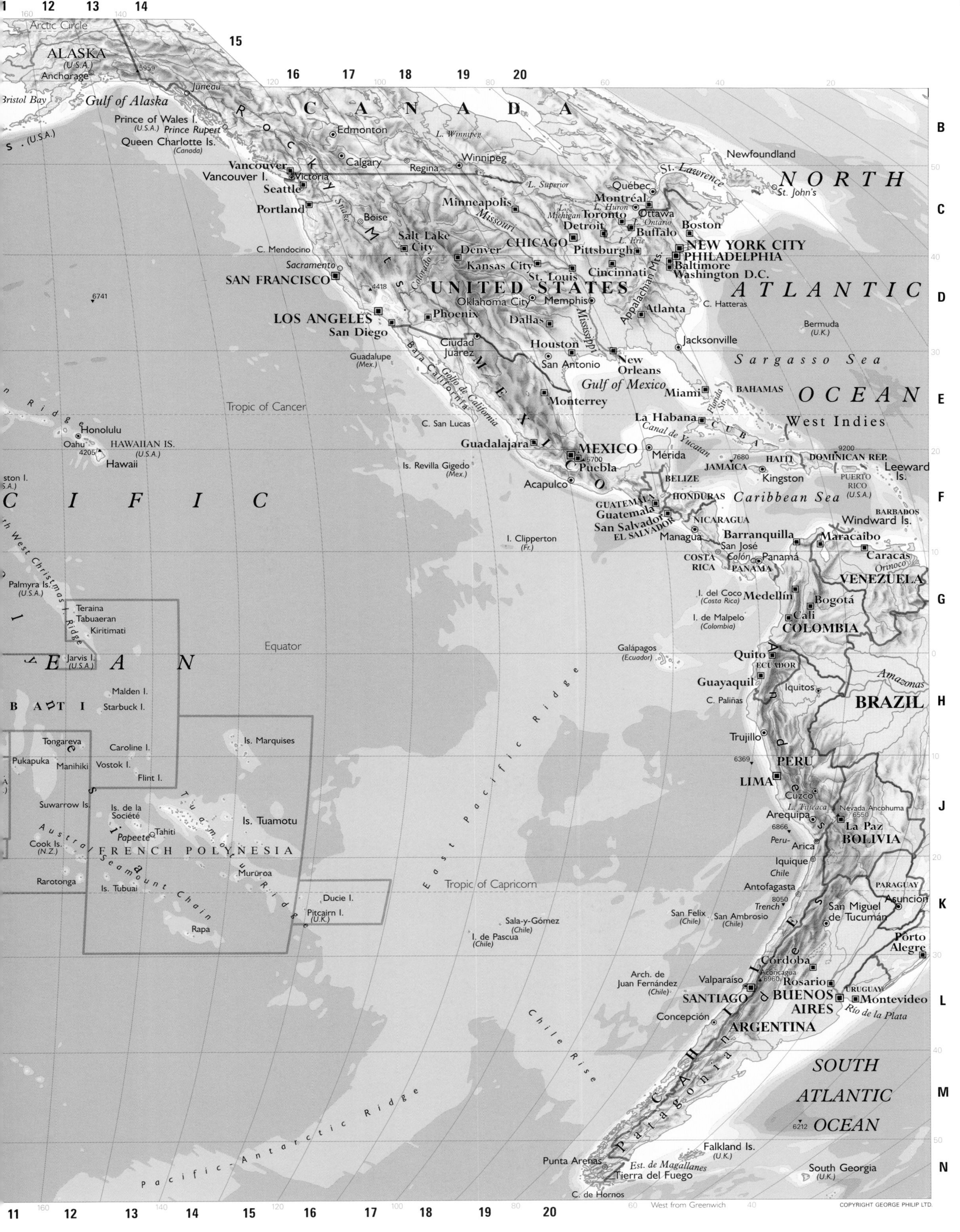

1 12 13 14

Arctic Circle
ALASKA
(U.S.A.)
Anchorage
Bristol Bay
Gulf of Alaska
Prince of Wales I.
(U.S.A.) Prince Rupert
Queen Charlotte Is.
(Canada)
Juneau

15

16 17 18 19 20

ROCKY C A N A D A
Edmonton
L. Winnipeg
Calgary
Regina Winnipeg
Newfoundland
Vancouver
Vancouver I. Victoria
Seattle
Portland
Boise
St. Lawrence
Québec St. John's
Montréal
NORTH
L. Superior
Minneapolis
Missouri
L. Michigan Toronto Ottawa
Detroit L. Ontario Boston
Buffalo L. Erie
CHICAGO Pittsburgh NEW YORK CITY
Salt Lake City
Denver Cincinnati PHILADELPHIA
Kansas City St. Louis Washington D.C. Baltimore
C. Mendocino
Sacramento
SAN FRANCISCO
UNITED STATES
Oklahoma City Memphis
Atlanta ATLANTIC
C. Hatteras
LOS ANGELES
San Diego
Phoenix Dallas
Ciudad Juárez Houston Jacksonville Bermuda (U.K.)
San Antonio New Orleans
Guadalupe (Mex.) Monterrey Miami Sargasso Sea
Tropic of Cancer BAHAMAS
Gulf of Mexico
OCEAN
Honolulu La Habana West Indies
Oahu HAWAIIAN IS. (U.S.A.) C. San Lucas CUBA
Hawaii Guadalajara Mérida HAITI DOMINICAN REP.
Is. Revilla Gigedo (Mex.) MEXICO Puebla JAMAICA Kingston PUERTO RICO (U.S.A.) Leeward Is.
Acapulco BELIZE BARBADOS
GUATEMALA HONDURAS Caribbean Sea Windward Is.
Guatemala NICARAGUA
San Salvador Managua Barranquilla Maracaibo
EL SALVADOR San José Caracas
COSTA RICA Colón Panamá VENEZUELA
I. del Coco (Costa Rica) PANAMA
Medellín Bogotá
I. de Malpelo (Colombia) Cali COLOMBIA
Galápagos (Ecuador) Quito ECUADOR
Equator Guayaquil Iquitos
C. Paliñas BRAZIL
Trujillo
PERU
LIMA
Cuzco
Arequipa La Paz
Iquique BOLIVIA
Antofagasta PARAGUAY
Asunción
San Miguel de Tucumán
Córdoba Porto Alegre
Valparaíso Rosario
SANTIAGO BUENOS AIRES URUGUAY Montevideo
Concepción Río de la Plata
ARGENTINA
SOUTH
ATLANTIC
OCEAN
Falkland Is. (U.K.)
Punta Arenas South Georgia (U.K.)
Tierra del Fuego
C. de Hornos

11 12 13 14 15 16 17 18 19 20

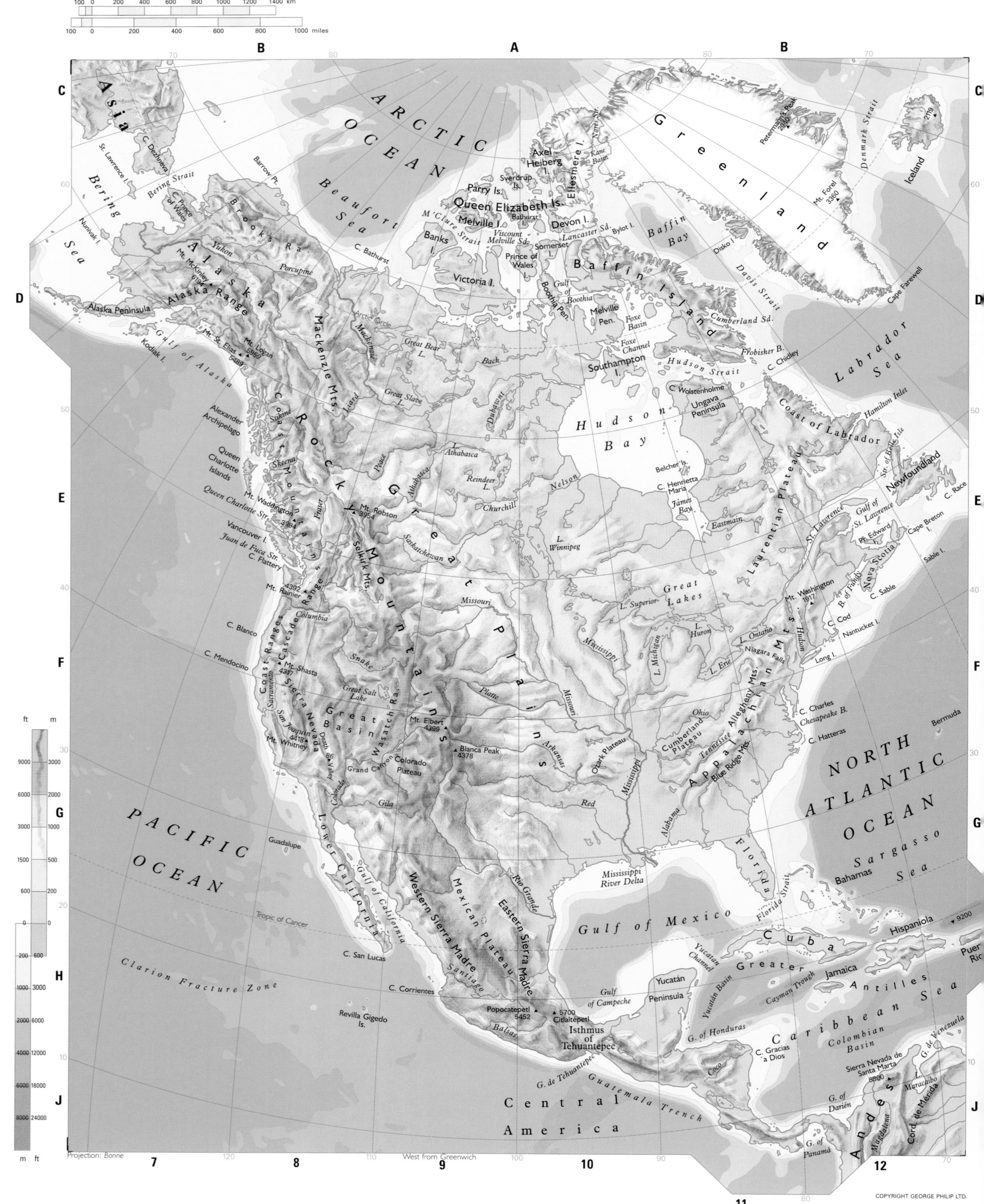

100 0 200 400 600 800 1000 1200 1400 km

100 0 200 400 600 800 1000 miles

ft m

9000 3000

6000 2000

3000 1000

1500 500

600 200

0 0

200 600

1000 3000

2000 6000

4000 12000

6000 18000

8000 24000

m ft

Projection: Bonne

A S I A

ARCTIC OCEAN

G r e e n l a n d

Petermanns Peak

Mt. Forel 3360

Denmark Strait

Iceland

C. Dezhnev

St. Lawrence I.

C. Prince of Wales

Bering Strait

Barrow Pt.

Beaufort Sea

Axel Heiberg I.

Ellesmere I.

Kane Basin

Kennedy Str.

Sverdrup Is.

Parry Is.

Queen Elizabeth Is.

Melville I.

Bathurst

Devon I.

Bylot I.

Lancaster Sd.

Disko I.

Davis Strait

Baffin Bay

Numak I.

Bering Sea

Brooks Ra.

Yukon

Mt. McKinley 6194

Porcupine

M'Clure Strait

Banks I.

Viscount Melville Sd.

Prince of Wales

Somerset

Gulf of Boothia

Boothia Pen.

Cape Farewell

A l a s k a

Alaska Range

Mt. St. Elias 5489

Mt. Logan 5960

Gulf of Alaska

Kodiak I.

Mackenzie Mts.

Mackenzie

Arctic Circle

Great Bear L.

Victoria I.

Melville Pen.

Foxe Basin

Foxe Channel

Cumberland Sd.

L a b r a d o r S e a

Alexander Archipelago

C O A S T

R O C K Y

Stikine

Liard

Back

Dubawnt

Great Slave L.

Southampton I.

Hudson Strait

Frobisher B.

C. Chidley

C. Wolstenholme

Ungava Peninsula

Hamilton Inlet

Queen Charlotte Islands

Mt. Waddington 4042

Queen Charlotte Str.

Peace

Athabasca

Mt. Robson 3954

Athabasca

Reindeer L.

L. Athabasca

Nelson

Churchill

Belcher Is.

C. Henrietta Maria

James Bay

Eastmain

Coast of Labrador

Laurentian Plateau

St. Lawrence

Gulf of St. Lawrence

Newfoundland

Str. of Belle Isle

C. Race

Vancouver I.

Juan de Fuca Str.

C. Flattery

Fraser

Selkirk Mts.

Saskatchewan

L. Winnipeg

Pt. Edward

Cape Breton

Nova Scotia

Sable I.

M O U N T A I N S

Cascade Ranges

Mt. Rainier 4392

Columbia

Snake

Missouri

L. Superior

Great Lakes

L. Michigan

L. Huron

Mt. Washington 1917

B. of Fundy

C. Sable

C. Blanco

C. Mendocino

Coast Ranges

Mt. Shasta 4317

Great Salt Lake

G r e a t

P l a i n s

Missouri

Mississippi

L. Ontario

L. Erie

Niagara Falls

Hudson

Long I.

C. Cod

Nantucket I.

Sacramento

Sierra Nevada

San Joaquin

Mt. Whitney 4418

Death Valley 86

Great Basin

Wasatch Ra.

Mt. Elbert 4399

Blanca Peak 4378

Platte

Arkansas

Ohio

Ozark Plateau

Allegheny Mts.

Tennessee

Cumberland Plateau

Appalachian Mts.

Blue Ridge Mts.

C. Charles

Chesapeake B.

C. Hatteras

Grand Canyon

Colorado Plateau

Colorado

Gila

Red

Mississippi

Alabama

Florida

PACIFIC OCEAN

Guadalupe

Lower California

Western Sierra Madre

Mexican Plateau

Eastern Sierra Madre

Rio Grande

Mississippi River Delta

Gulf of Mexico

Florida Strait

Bahamas

NORTH ATLANTIC OCEAN

Sargasso Sea

Bermuda

Tropic of Cancer

Gulf of California

C. San Lucas

C. Corrientes

Santiago

Balsas

Popocatepetl 5452

Citlaltepetl 5700

Isthmus of Tehuantepec

G. of Campeche

Yucatán Peninsula

Yucatán Channel

Yucatán Basin

C u b a

G r e a t e r

Jamaica

A n t i l l e s

Cayman Trough

Hispaniola

Puerto Rico

Clarion Fracture Zone

Revilla Gigedo Is.

G. de Tehuantepec

Guatemala Trench

C e n t r a l

A m e r i c a

G. of Honduras

C. Gracias a Dios

Coco

Colombian Basin

C a r i b b e a n S e a

Sierra Nevada de Santa Marta 5800

G. de Venezuela

G. of Darién

G. of Panamá

Magdalena

Maracaibo

Cord. de Mérida

A n d e s

West from Greenwich

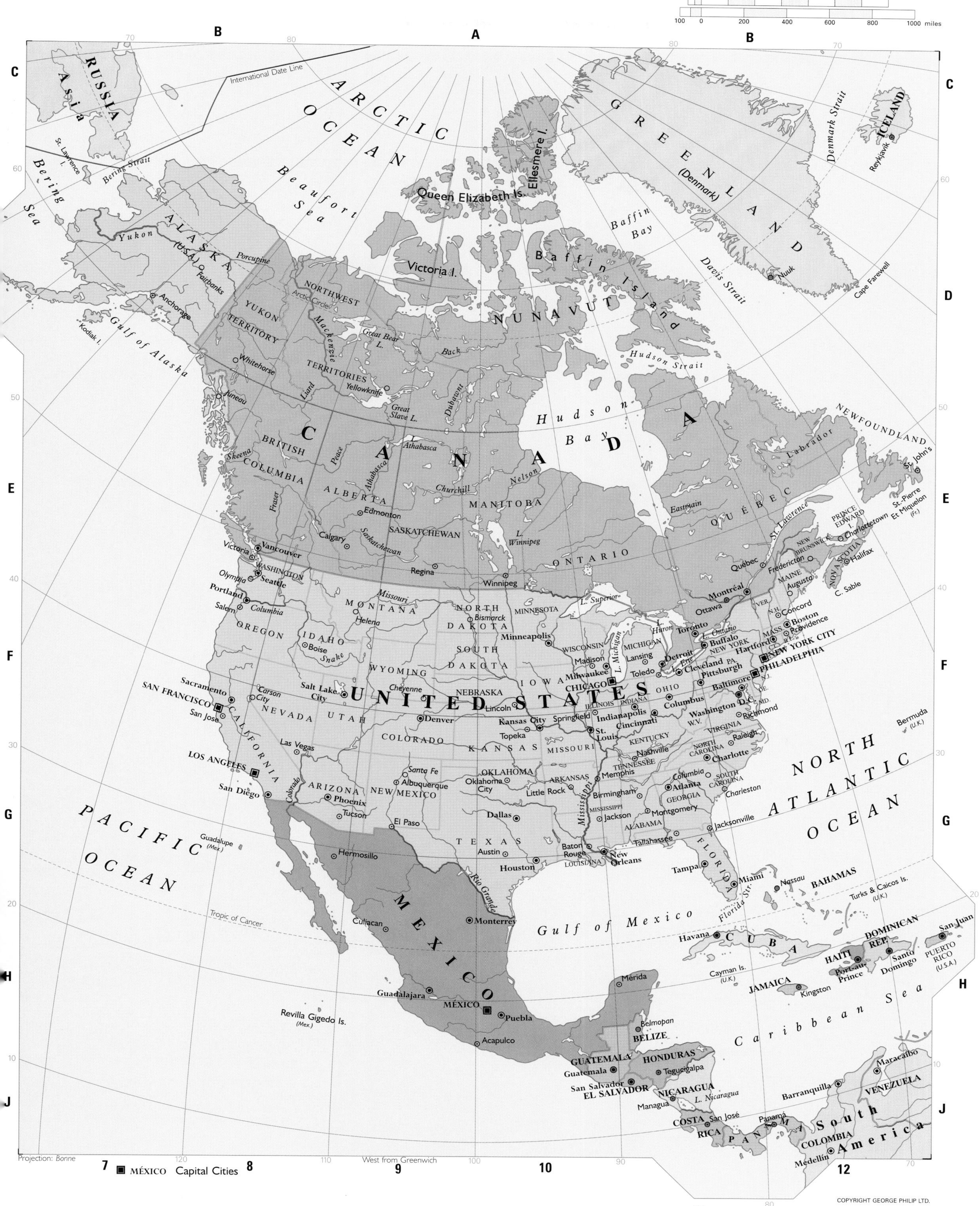

Projection: Bonne

7 ■ MÉXICO Capital Cities 8

COPYRIGHT GEORGE PHILIP LTD.

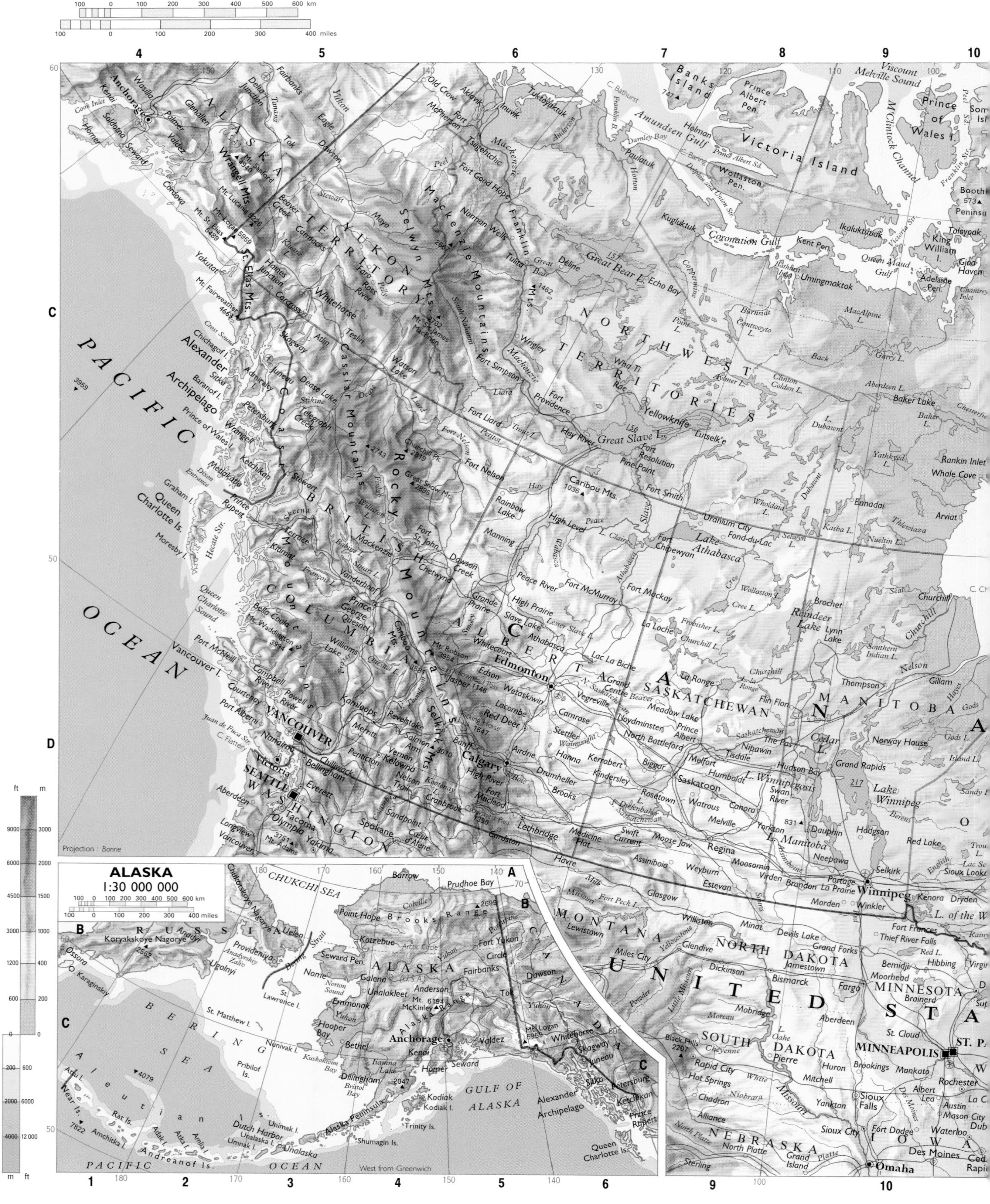

100 0 100 200 300 400 500 600 km
100 0 100 200 300 400 miles
PACIFIC OCEAN
ALASKA
YUKON TERRITORY
NORTHWEST TERRITORIES
BRITISH COLUMBIA
ALBERTA
SASKATCHEWAN
MANITOBA
WASHINGTON
MONTANA
NORTH DAKOTA
SOUTH DAKOTA
MINNESOTA
NEBRASKA
IOWA
UNITED STATES
Victoria Island
Banks Island
Prince of Wales I.
Amundsen Gulf
Coronation Gulf
Great Bear L.
Great Slave L.
Lake Athabasca
Lake Winnipeg
Mackenzie Mountains
Selwyn Mts.
Rocky Mountains
Coast Mountains
Cassiar Mountains
Columbia Mountains
Selkirk Mts.
St. Elias Mts.
Wrangell Mts.
Alexander Archipelago
Queen Charlotte Is.
Vancouver I.
Anchorage
Whitehorse
Yellowknife
Inuvik
Edmonton
Calgary
VANCOUVER
Victoria
SEATTLE
Saskatoon
Regina
Winnipeg
MINNEAPOLIS
Omaha
Projection : Bonne
ALASKA
1:30 000 000
100 0 100 200 300 400 500 600 km
100 0 100 200 300 400 miles
CHUKCHI SEA
BERING SEA
GULF OF ALASKA
PACIFIC OCEAN
RUSSIA
Brooks Range
Alaska Range
Aleutian Is.
Mt. McKinley 6194
Anchorage
Fairbanks
Barrow
Nome
West from Greenwich
ft m
9000 3000
6000 2000
4500 1500
3000 1000
1200 400
600 200
0 0
200 600
2000 6000
4000 12 000
m ft

B

11   12   13   14   15   16

Devon I.
Lancaster Sound

Baffin Bay

2136

Nunavik

Uummannaq

Qeqertarsuaq

G R E E N L A N D
(KALAALLIT NUNAAT)
(Denmark)

Kong Frederik VI's Kyst

2850

Tasiilak

Arctic Bay
Nanisivik
Borden
Pen.

Bylot I.

Pond Inlet

Clyde River

C. Adair

1890

rodeur
eninsula

Eclipse

Sisimiut

Qeqertarsuaq

Ataniganguit

Ilulissat

Kangerlussuaq

Arsuk

Maniitsoq

Qasigianguit

Home B.

Davis Strait

C. Raper

Qikiqtarjuaq

2591

Cumberland
Peninsula

C. Dyer

Nuuk

Qeqertarsuatsiaat

Paamiut

Qaqortoq
Nanortalik

Muitsup Paa

Uummannarsuaq

mpson
Pen.

Pelly
Bay

Commute B.

Fury and Hecla Str.

Igloolik

Sanirajak

Prince
Charles
I.

Melville

Peninsula

Foxe

Basin

Air
Force I.

Pangnirtung

Hoare B.

Mercy C.

Cumberland Sd.

Rae Isthmus
Repulse
Bay

C. Dorchester

Meta

Iqaluit

Hall Peninsula

Amadjuak

Foxe
Pen.

Salluit

Bell
Pen.

Coats
I.

Mansel
I.

N U N A V U T

Foxe Channel

Salisbury
I.

Nottingham I.

Incognita
Peninsula

Kimmirut

Frobisher Bay

Resolution I.

Cape Dorset

Nattilling L.

Roes Welcome Sd.

Southampton
I.

H u d s o n   S t r a i t

Ivujivik

Salluit

Kangiqsujuaq

Quaqtaq

Akpatok I.

C. Chidley

Labrador

Sea

3809

gaarjuk

Hudson

Bay

257

Ottawa Is.

Péninsule

d'Ungava

Puvirnituq

Arnaud

L. Payne

Feuilles

Kangirsuk

Ungava Bay

Kangiqsualujjuaq

Hebron

Nain

Hopedale

C. Harrison

Severn

Peawanuck

tnam

Winisk

King George Is.

Baker's
Dozen
Is.

Sleeper Is.

Saniikiluaq

Belcher Is.

C. Henrietta
Maria

Inukjuak

Kuujjuarapik

Pte. Louis
XIV

L. à l'Eau
Claire

L. Minto

Mézels

L. Bienville

Kanaaupscow

La Grande

Camapiscau

Schefferville

N E W F O U N D L A N D

Smallwood
Res.

North West River

Happy Valley
Goose Bay

Rigolet

Cartwright

Port Hope Simpson

Belle Isle

St. Anthony

C. Bauld

Big
Trout L.

Chisasibi

Wemindji

James Bay

Akimiski I.

Charlton
I.

Grande
Baleine

Labrador
City

Fermont

Esker

Ashuanipi

Petitsikapau

Churchill
Falls

Churchill

Labrador

Gagnon

St-Augustin

Romaine

C. Mecatina

Natashquan

Deer
Lake

Baie
Verte

Grand
Falls

Corner Brook

Twillingate

Gander

Notre Dame B.

Bonavista

Carbonear

St. John's

O N T A R I O

Albany

Fort Albany

Waskaganish

Eastmain

Rupert

Moosonee

Attawapiskat

Attawapiskat

1135

Q U É B E C

Rés. de
Caniapiscau

Mistassini

L. Albanel

Chibougamau

Manicouagan

Sept-Îles

Port-Cartier

Havre-
St-Pierre

Î. d'Anticosti

Gulf of

St. Lawrence

Cabot Str.

Stephenville

Channel-Port
aux Basques

814

Ray

North C.

Marystown

Placentia

St-PIERRE
et MIQUELON
(Fr)

Placentia B.

C. Race

St. Joseph

Nakina

Kenogami

Missinaibi

Hearst

Geraldton

Marathon

Oba

Kapuskasing

Cochrane

Timmins

Val-d'Or

Matagami

Abitibi L.

Rés. Gouin

Dolbeau

Roberval

St-Jean

Chicoutimi

Jonquière

La Tuque

1190

Québec

Lévis

Matane

Rimouski

Rivière-du-Loup

Edmundston

Pén. de Gaspé

Gaspé

Campbellton

Bathurst

Miramichi

Grand Falls

Woodstock

N E W
B R U N S W I C K

Moncton

Sydney

Glace Bay

Cape Breton I.

Port Hawkesbury

Antigonish

New Glasgow

lipigon

Thunder Bay

Houghton

183

Marquette

Wawa

Chapleau

Kirkland
Lake

New
Liskeard

Rouyn-
Noranda

Rés.
Cabonga

Mont-
Laurier

Shawinigan

Trois-Rivières

Sherbrooke

Thetford
Mines

Fredericton

Saint
John

B. of Fundy

Amherst

Kentville

Truro

N O V A   S C O T I A

Dartmouth

Halifax

Bridgewater

Liverpool

Sable I.
(Nova Scotia)

6309

M I C H I G A N

onwood

Ironwood

Sault Ste.
Marie

Elliot
Lake

Sudbury

North
Bay

Parry
Sound

Pembroke

Huntsville

Nipissing

Ottawa

Hull

MONTRÉAL

St-Hyacinthe

Granby

St-Jean

Drummondville

Joliette

Champlain

M A I N E

Bangor

Augusta

Lewiston

Portland

V E R M O N T

N E W
H A M P S H I R E

Concord

Manchester

P R. EDWARD I.
Summerside
Charlottetown
Northumberland Str.

B O S T O N

C. Cod

Manistique

Escanaba

Menominee

Petoskey

Traverse City

Cadillac

Lake
Huron

Georgian
Bay

Manitoulin
I.

Owen Sound

Barrie

Peterborough

Oshawa

TORONTO

Hamilton

Belleville

Kingston

Cornwall

Burlington

Montpelier

L. Champlain

Springfield

Albany

HARTFORD

CONN.

Providence

R.I.

New Haven

MASS.

elander

Rhinelander

Wausau

Green
Bay

W I S C O N S I N

ppleton

Appleton

Sheboygan

MILWAUKEE

Racine

Kenosha

ford

Rockford

Lansing

CHICAGO

Gary

South Bend

Toledo

I L L I N O I S

INDIANA

OHIO

L. Michigan

Grand
Rapids

Flint

Saginaw

DETROIT

Windsor

174

Sarnia

London

Kitchener

BUFFALO

Niagara
Falls

N E W   Y O R K

Rochester

Syracuse

Elmira

Binghamton

Scranton

Jamestown

Erie

CLEVELAND

P E N N S Y L V A N I A

Allentown

Trenton

Newark

N E W   Y O R K

N.J.

Bridgeport

A T L A N T I C

O C E A N

C

D

E

11   80   12   70   West from Greenwich   60   COPYRIGHT GEORGE PHILIP LTD.

13   14

60

50

40

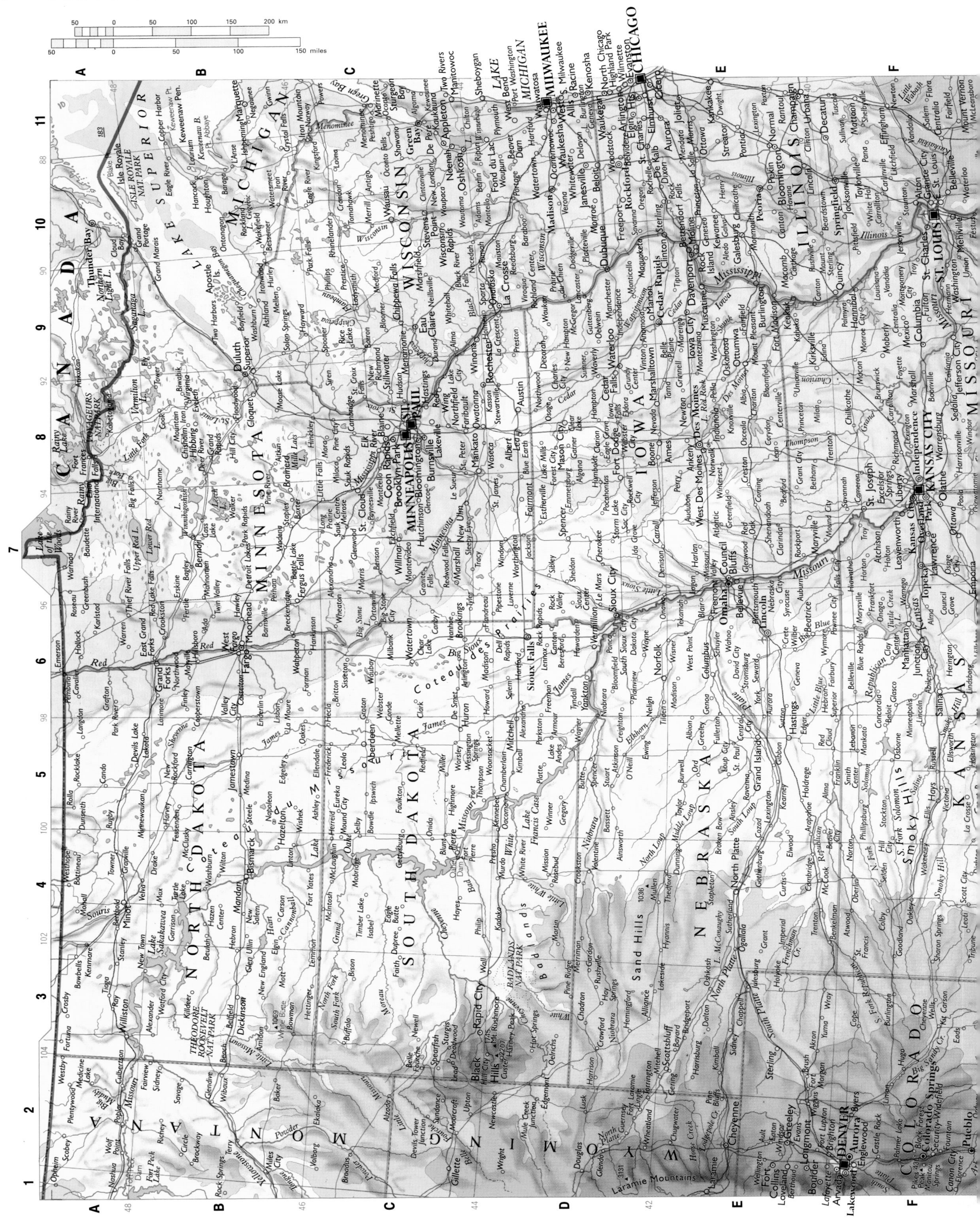

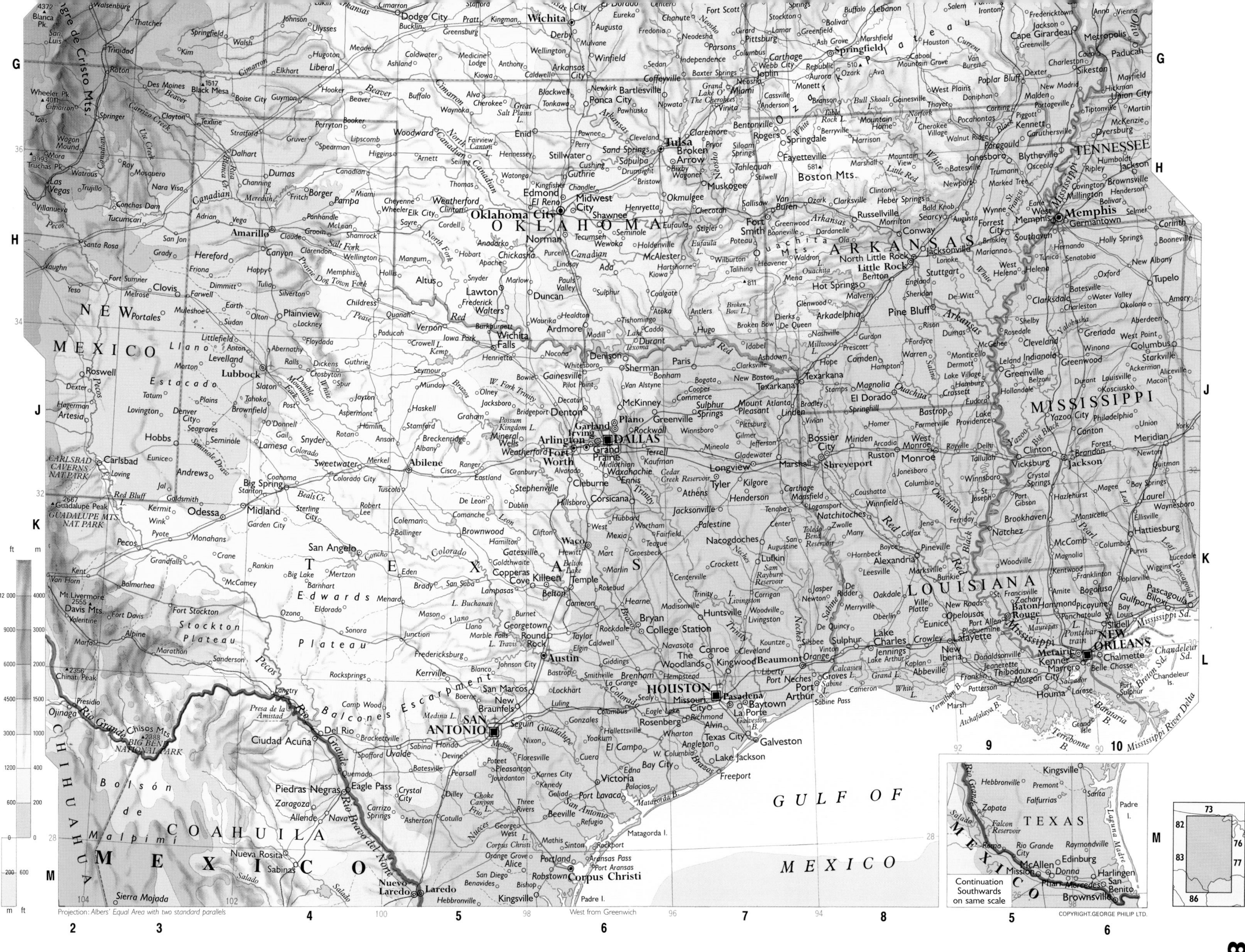

OKLAHOMA
ARKANSAS
MISSISSIPPI
TENNESSEE
LOUISIANA
TEXAS
NEW MEXICO
MEXICO
COAHUILA
CHIHUAHUA
GULF OF MEXICO
Sangre de Cristo Mts.
Llano Estacado
Edwards Plateau
Stockton Plateau
Balcones Escarpment
Bolsón de Mapimí
Boston Mts.
Ouachita Mts.
Ozark Plateau
Rio Grande
Rio Bravo del Norte
Mississippi
Red
Arkansas
Canadian
Brazos
Colorado
Pecos
Oklahoma City
Tulsa
Wichita
Dallas
Fort Worth
Houston
San Antonio
Austin
Corpus Christi
Galveston
Memphis
Little Rock
Jackson
New Orleans
Baton Rouge
Shreveport
Amarillo
Lubbock
El Paso
Laredo
Nuevo Laredo
Big Bend National Park
Carlsbad Caverns Nat. Park
Guadalupe Mts. Nat. Park
Continuation Southwards on same scale
Projection: Albers' Equal Area with two standard parallels
West from Greenwich
COPYRIGHT GEORGE PHILIP LTD.
ft m
12 000 4000
9000 3000
6000 2000
4500 1500
3000 1000
1200 400
600 200
0 0
200 600
m ft
18

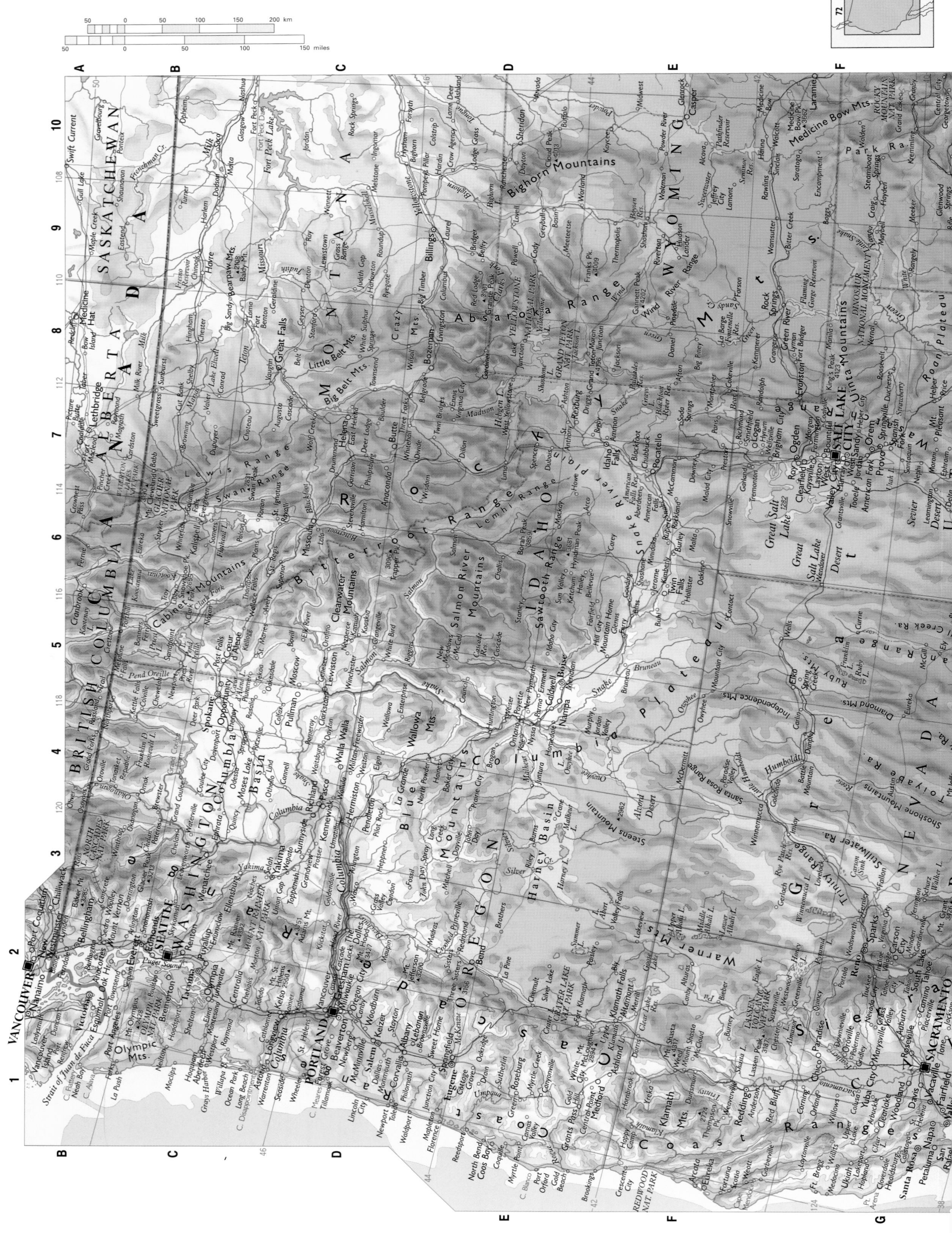

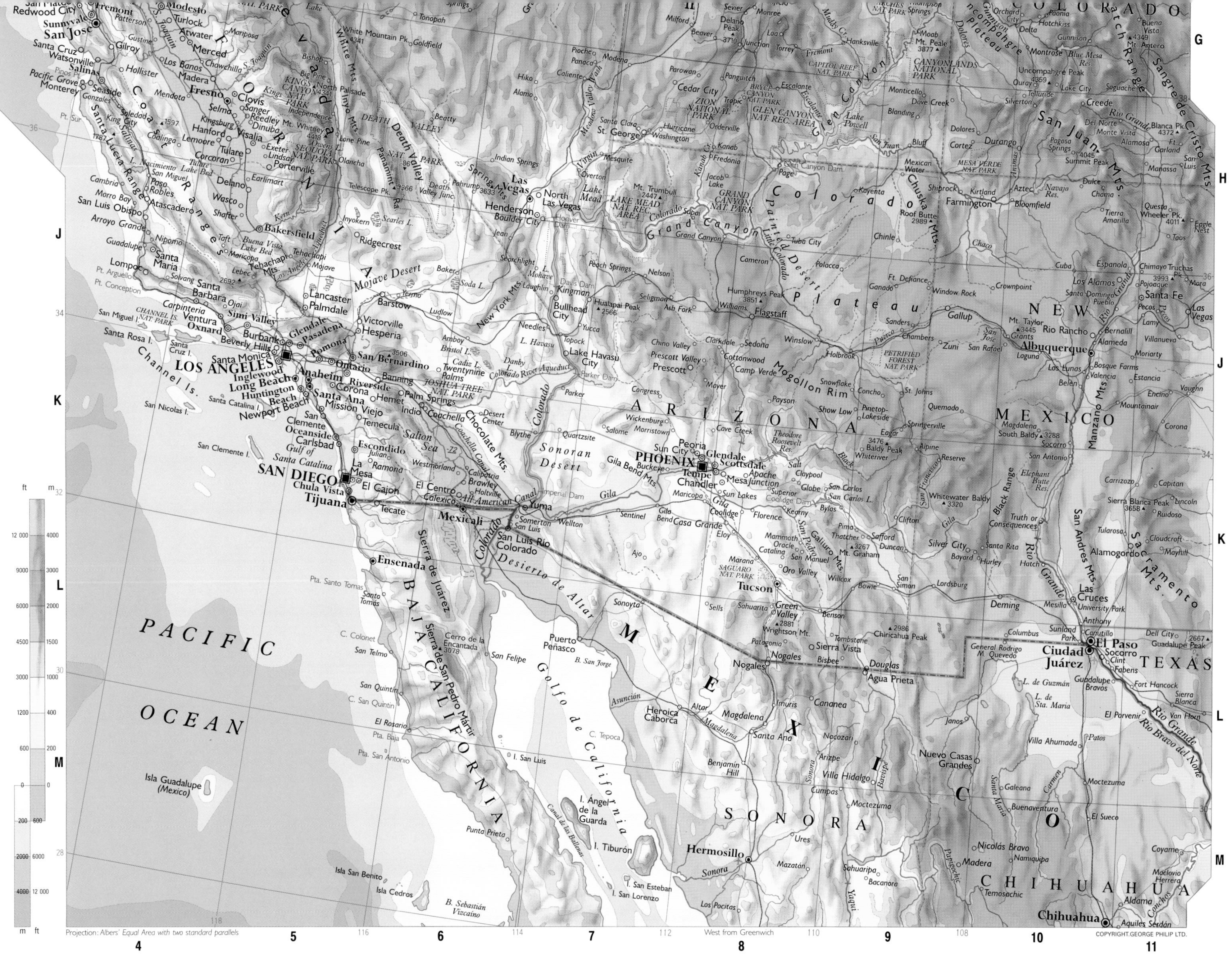

COPYRIGHT GEORGE PHILIP LTD.

Projection: Albers' Equal Area with two standard parallels

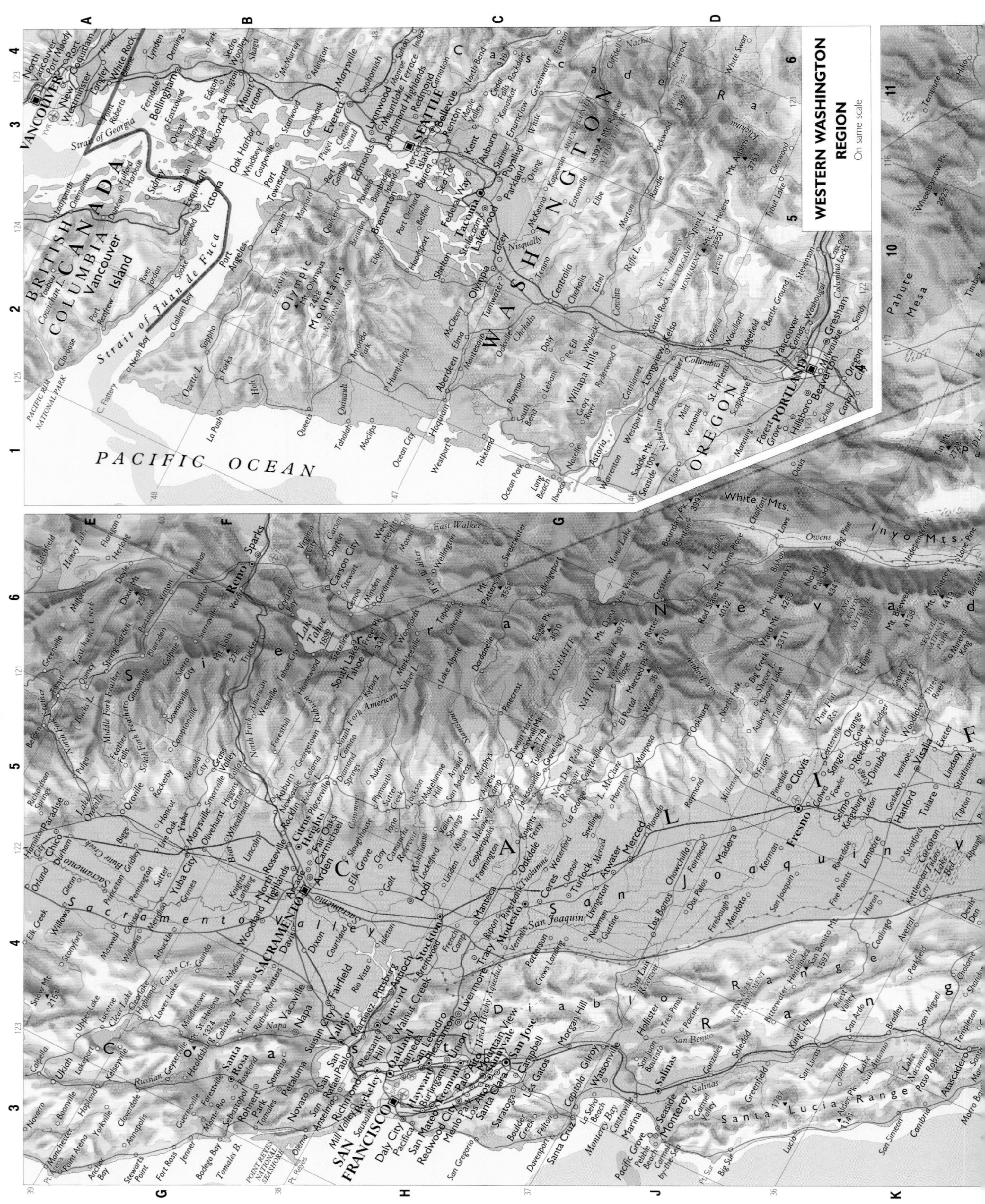

WESTERN WASHINGTON REGION
On same scale

PACIFIC OCEAN

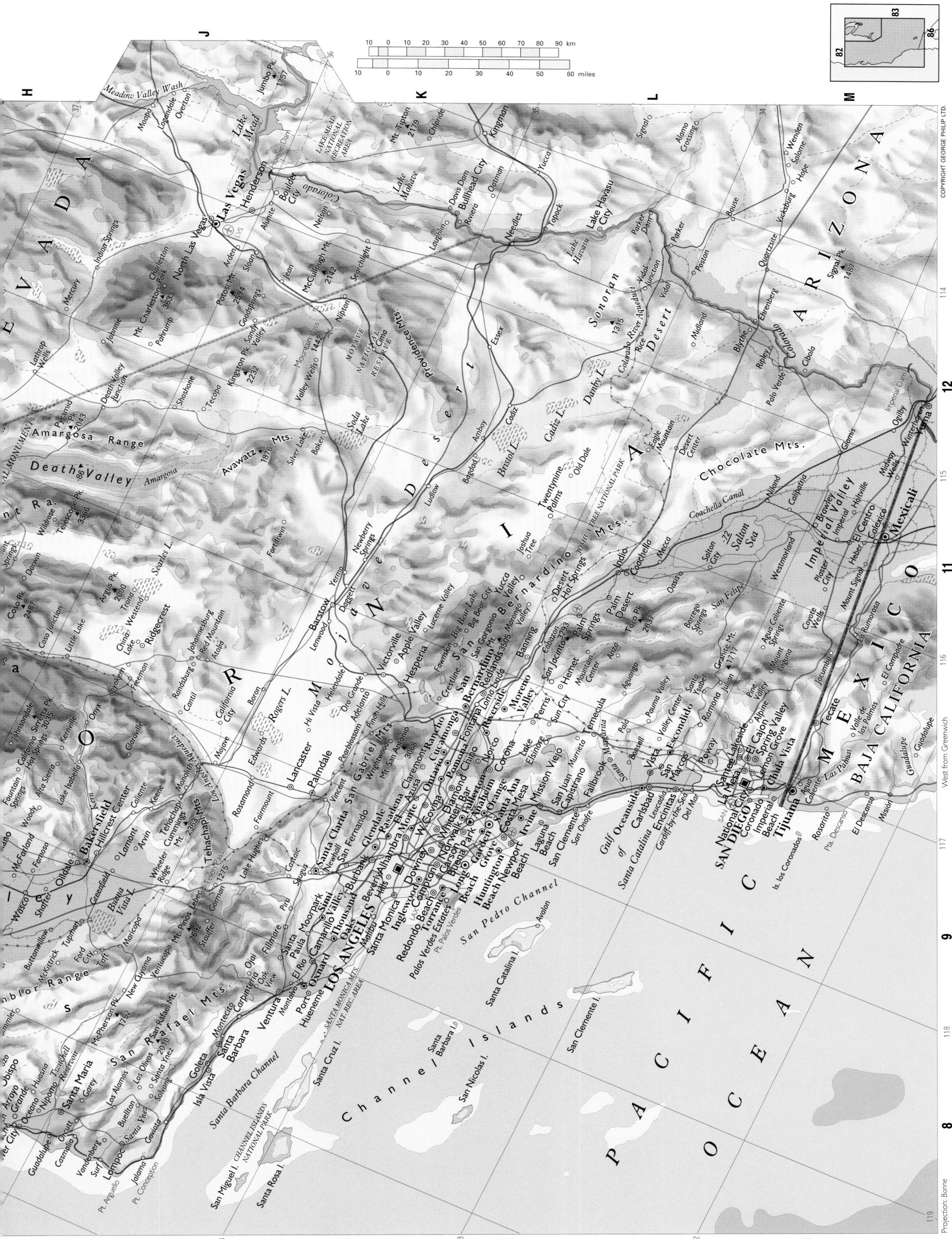

10 0 10 20 30 40 50 60 70 80 90 km

10 0 10 20 30 40 50 60 miles

COPYRIGHT GEORGE PHILIP LTD.

Projection: Bonne

West from Greenwich

NEVADA

ARIZONA

CALIFORNIA

BAJA CALIFORNIA

MEXICO

PACIFIC OCEAN

Channel Islands

Meadow Valley Wash
Overton
Moapa
Logandale
Lake Mead
LAKE MEAD NATIONAL RECREATION AREA
Jumbo Pk. 1357
Mt. Tipton 2179
Chloride
Kingman
Yucca
Hope
Solome
Wenden
Alamo Crossing
Signal
Bouse
Vicksburg
Quartzsite
Ehrenberg
Blythe
Parker
Poston
Midland
Desert Center
Eagle Mountain
Cadiz L.
Amboy
Bagdad
Ludlow
Essex
Cima
Goffs
Needles
Topock
Oatman
Bullhead City
Davis Dam
Lake Mohave
Searchlight
Nelson
Boulder City
Boulder Dam
Henderson
Las Vegas
North Las Vegas
Indian Springs
Mercury
Mt. Charleston 3633
Charleston Pk.
Potosi Mt. 2194
Goodsprings
Sloan
Jean
McCullough Mt. 2142
Nipton
Mountain Pass
Kingston Pk. 2232
Tecopa
Shoshone
Death Valley Junction
Pahrump
Johnnie
Indian Springs
Lathrop Wells
Amargosa
Death Valley
Telescope Pk. 3366
Wildrose
Amargosa Range
Silver Lake
Soda Lake
Baker
Mojave Natl. Reserve
Providence Mts.
Kelso
Cadiz
Bristol L.
Twentynine Palms
Old Dole
JOSHUA TREE NATIONAL PARK
Joshua Tree
Yucca Valley
Desert Hot Springs
Indio
Coachella
Mecca
Thermal
Salton City
Salton Sea
Westmorland
Brawley
Calipatria
Niland
Imperial
El Centro
Calexico
Mexicali
Coachella Canal
Colorado River Aqueduct
Imperial Dam
Yuma
Winterhaven
Ogilby
Midway Wells
Calexico
Heber
El Compadre
Tecate
Tijuana
Rosarito
Pta. Descanso
El Descanso
Misión
Guadalupe
Las Palmas
Valle de las Palmas
Palm Springs
Palm Desert
San Jacinto 3293
Toro Pk. 2657
Borrego Springs
San Felipe
Aqua Caliente Springs
Mount Laguna
Campo
Pine Valley
Julian
Alpine
El Cajon
La Mesa
National City
Chula Vista
Coronado
Imperial Beach
SAN DIEGO
Lemon Grove
Santee
Lakeside
Poway
Ramona
Escondido
San Marcos
Vista
Fallbrook
Temecula
Murrieta
Sun City
Hemet
Perris
Elsinore
Lake Elsinore
San Juan Capistrano
San Clemente
San Onofre
Oceanside
Carlsbad
Encinitas
Leucadia
Cardiff-by-the-Sea
Del Mar
Solana Beach
Laguna Beach
Irvine
Mission Viejo
Newport Beach
Huntington Beach
Costa Mesa
Santa Ana
Garden Grove
Orange
Anaheim
Buena Park
Fullerton
La Habra
Whittier
Norwalk
Downey
Compton
Long Beach
Torrance
Redondo Beach
Palos Verdes Estates
Pt. Palos Verdes
Inglewood
Santa Monica
Malibu
LOS ANGELES
Beverly Hills
Burbank
Glendale
Pasadena
Alhambra
Monterey Park
W. Covina
Pomona
Diamond Bar
Chino
Ontario
Upland
Claremont
Fontana
San Bernardino
Rialto
Colton
Redlands
Loma Linda
Moreno Valley
Riverside
Corona
Norco
San Gorgonio Mt. 3506
Banning
Beaumont
Cabazon
Cucamonga
Rancho Cucamonga
San Gabriel Mts.
Mt. San Antonio 3068
Wrightwood
Crestline
Big Bear Lake
Big Bear City
Lake Arrowhead
San Bernardino Mts.
Lucerne Valley
Morongo Valley
Landers
Fawnskin
Victorville
Hesperia
Apple Valley
Adelanto
Hi Vista
Oro Grande
Helendale
Lenwood
Barstow
Daggett
Yermo
Newberry Springs
Fort Irwin
Rogers L.
Edwards
Mojave
Boron
California City
Rosamond
Lancaster
Palmdale
Fairmont
Vincent
Santa Clarita
Newhall
Saugus
San Fernando
Simi
Thousand Oaks
Moorpark
Fillmore
Santa Paula
Ojai
Ventura
Oxnard
Port Hueneme
SANTA MONICA MTS. NAT. REC. AREA
Camarillo
Somis
El Rio
Montalvo
Santa Barbara
Goleta
Isla Vista
Carpinteria
Montecito
Summerland
San Rafael Mts.
2010
San Emigdio Mts.
Gorman
Mt. Pinos 2692
Cuyama
Mt. Abel
Frazier Park
Lebec
Tejon
Grapevine
Tehachapi
Tehachapi Mts.
2331
Cummings
Caliente
Bodfish
Kernville
Isabella
Lake Isabella
Wofford Heights
Woody
Glennville
Havilah
Weldon
Bakersfield
Oildale
Lamont
Arvin
Wheeler Ridge
Maricopa
McKittrick
Taft
Fellows
Ford City
Buttonwillow
Wasco
Shafter
Greenfield
Delano
McFarland
Oil City
Coalinga
New Cuyama
Cuyama
Temblor Range
Santa Maria
Orcutt
Nipomo
Guadalupe
Betteravia
Los Alamos
Los Olivos
Santa Ynez
Solvang
Buellton
Lompoc
Vandenberg
Surf
Jalama
Pt. Arguello
Pt. Conception
Gaviota
Goleta
Santa Ynez Mts.
Arroyo Grande
Oceano
Grover City
Pismo Beach
Morro Bay
Obispo
San Luis Obispo
Santa Barbara Channel
San Miguel I.
Santa Rosa I.
Santa Cruz I.
Santa Barbara I.
San Nicolas I.
San Clemente I.
Santa Catalina I.
Avalon
Is. los Coronados
Gulf of Santa Catalina
San Pedro Channel
CHANNEL ISLANDS NATIONAL PARK
Chocolate Mts.
Santa Rosa Mts.
Sonoran Desert
Colorado Desert
Lake Havasu City
Lake Havasu
Havasu L.
Parker Dam
Danby L.
Chuckwalla Mts.
Eagle Mt.
Graphite Mt. 1717
Sierra Pk.
Dome Rock Mts.
Signal Pk. 1451
Palomar Mtn.
Valley Center
Pala
Bonsall
Warner Springs
Aguanga
Anza
Coyote Wells

m ft
12 000
9000
6000
4500
3000
1500
600
200
0
200
2000 6000
m ft

4000
3000
2000
1500
1000
400
200
0

37
36
35
34
33
32
119
118
117
116
115
114
113

H J K L M
L M N P

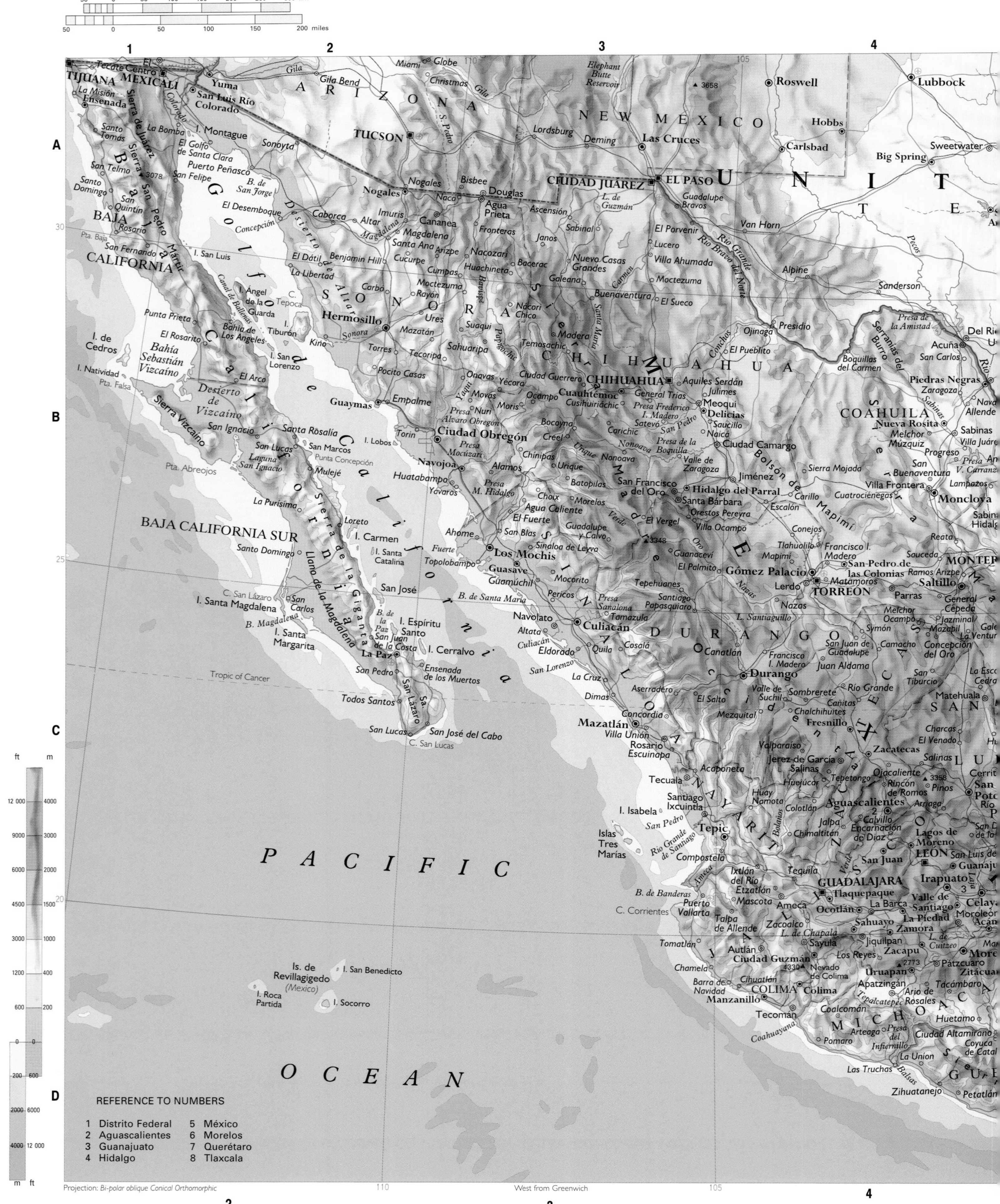

REFERENCE TO NUMBERS

1 Distrito Federal   5 México
2 Aguascalientes   6 Morelos
3 Guanajuato   7 Querétaro
4 Hidalgo   8 Tlaxcala

Projection: Bi-polar oblique Conical Orthomorphic

West from Greenwich

5 6 7 8

A B C D E

UNITED STATES

Wichita Falls, Denison, Sherman, Paris, Red, Hope, Camden, ARKANSAS, Texarkana, El Dorado, Greenville, MISSISSIPPI, Tuscaloosa, Opelika, Columbus, McRae, Brazos, Possum Kingdom Res., Denton, Greenville, FORT WORTH, DALLAS, Marshall, Monroe, Vicksburg, Meridian, ALABAMA, Selma, Montgomery, Phenix City, Americus, Cordele, GEORGIA, Ranger, Longview, Tyler, Shreveport, Tallulah, Jackson, Troy, Albany, Tifton, Cleburne, LOUISIANA, STATES, Waycross, Hillsboro, Corsicana, Palestine, Toledo Bend Res., Natchez, Laurel, Dothan, Valdosta, Brownwood, Waco, Nacogdoches, Pearl, Hattiesburg, Flomaton, Jim Woodruff Res., Chattahoochee, Tallahassee, Lufkin, Sam Rayburn Reservoir, Jewett, Alexandria, McComb, MOBILE, FLORIDA, Lake City, Temple, Bogalusa, Biloxi, Pensacola, Panama City, Baton Rouge, Hammond, Colorado, Huntsville, Trinity, Sabine, Gulfport, Mobile Bay, Apalachee Bay, Bryan, L. Pontchartrain, C. San Blas, Navasota, Lake Charles, Lafayette, NEW ORLEANS, Suwannee, Austin, Beaumont, Breton Sd., HOUSTON, Port Arthur, Atchafalaya Bay, Terrebonne Bay, Mississippi River Delta, SAN ANTONIO, Rosenberg, Galveston, Guadalupe, Victoria, Clearwater, Dilley, Nueces, Alice, Corpus Christi, Laredo, Kingsville

GULF OF MEXICO

Nuevo Laredo, Zapata, Laguna Madre, Camargo, McAllen, Harlingen, Brownsville, Presa M. R. Gomez, China, Reynosa, Matamoros, Valle Hermoso, Santa Teresa, Montemorelos, Conchos, Mendez, Laguna Madre, Linares, San Fernando, TAMAULIPAS, Villagrán, Hidalgo, Santander Jiménez, Zaragoza, La Pesca, Soto la Marina, Ciudad Victoria, Sierra de Tamaulipas, Llera, Aldama, Pta. Jerez, Ocampo, Ciudad Mante, Altamira, Ciudad Madero, Tampico, Ciudad Cárdenas, Ciudad de Valles, Pánuco, SAN LUIS POTOSÍ, Ozuluama, L. de Tamiahua, Tempoal, Magozal, C. Rojo, Tantoyuca, Tamazunchale, Chicontepec, Tuxpan, Zimapán, Zacualtipán, Poza Rica, Papantla, Juan del Río, Huichapan, Huauchinango, Nautla, Pachuca, Tulancingo, Misantla, Tula, Zumpango, Teziutlán, Jalapa Enríquez, MÉXICO, VERACRUZ, Zempoala, Tlaxcala, Apizaco, Coatepec, Veracruz, Amecameca, PUEBLA, Citlaltépetl 5700, Popocatépetl 5452, Tenango, Córdoba, Alvarado, Tlacotalpan, Cuernavaca, Orizaba, San Andrés Tuxtla 1879, Tehuacán, Cosamaloapan, Jojutla, Izúcar de Matamoros, Chilac, Ajalpan, Tres Valles, Coatzacoalcos, Iguala, Chiautla, Acatlán, Huautla de Jiménez, Presa Miguel Alemán, Acayucan, Minatitlán, Huajuapan de León, Asunción Nochixtlán, San Juan Bautista Valle Nacional, GUERRERO, Chilapa, Chilpancingo, Silacayoapan, Tlaxiaco, Oaxaca, Tlacolula, Sierra Madre del Sur, Tierra Colorada, Monte Albán, OAXACA, Ayutla, Ometepec, Ocotlán, San Jerónimo Taviche, Ixtepec, Acapulco, Pinotepa Nacional, Ejutla, Tehuantepec, Miahuatlán, Juchitán, Papagayo, Punta Maldonado, Jamiltepec, Tututepec, Salina Cruz, San Pedro Mixtepec, 3139, Pochutla, Verde, Puerto Escondido, Puerto Ángel, Golfo de Tehuantepec, Mar Muerto

Golfo de Campeche

Frontera, Paraíso, Comalcalco, TABASCO, Villahermosa, Cárdenas, La Venta, Macuspana, Teapa, Jesús Carranza, Istmo de Tehuantepec, Presa Malpaso, Presa Netzahualcóyotl, Simojovel, Ocosingo, Copainalá, Chiapa de Corzo, Tuxtla Gutiérrez, San Cristóbal de las Casas, La Independencia, Matías Romero, CHIAPAS, Comitán, Sierra Madre, Arriaga, Tonalá, La Concordia, Puerto Arista, Pijijiapan, Mapastepec, Motozintla, Huixtla, Tapachula, Palenque, Usumacinta, Lacantún, Tenosique, Balancán, Palizada, Ciudad del Carmen, L. de Términos, CAMPECHE, Concepción, Champotón, Chenkán, Campeche, Edzná, Hopelchén, Tenabo, Maxcanú, Uxmal, Ticul, Tekax, Bolonchenticul, Peto, Progreso, Dzibilchaltún, Mérida, YUCATÁN, Mayapán, Sotuta, Chichén Itzá, Izamal, Motul, Temax, Dzilam de Bravo, Pta. Yalkubul, Río Lagartos, El Cuyo, Tizimín, Espita, Valladolid, C. Catoche, Cancún, Puerto Juárez, Puerto Morelos, Cozumel, Isla Cozumel, Vigía Chico, B. de la Ascensión, B. del Espíritu Santo, Felipe Carrillo Puerto, QUINTANA ROO, Bacalar, Chetumal, B. de Chetumal, Banco Chinchorro, Corozal, Orange Walk, Hondo, Ambergris Cay, Matamoros, I. Desterrada, I. Pérez (Mexico), Tropic of Cancer, Canal de Yucatán

CUBA, La Esperanza, Guane, La Fe, C. San Antonio, C. Corrientes

Uaxactún, Tikal, L. Petén Itzá, La Libertad, Flores, San Ignacio, Belmopan, Belize City, BELIZE, Turneffe Is., Benque Viejo, Dangriga, Maya Mts., Monkey River, Golfo de Honduras, San Luis, Punta Gorda, San Antonio, Salinas, Puerto Barrios, Livingston, Sebol, GUATEMALA, L. de Izabal, Cobán, Cuchumatanes, Huehuetenango, Cuilco, 3993, 4220, San Marcos, Totonicapán, Utatlán, Sololá, Sierra de las Minas, Motagua, Gualán, Zacapa, Copán, Chiquimula, Jalapa, Antigua, Mazatenango, Retalhuleu, Coatepeque, Amatitlán, GUATEMALA, La Esperanza, Santa Bárbara, Santa Rosa de Copán, L. de Yojoa, San Pedro Sula, El Progreso, Arenal, HONDURAS, Tela, La Ceiba, Balfate, Savá, Olanchito, Trujillo, Yoro, Sulaco, Juticalpa, Catacamas, Patuca, Comayagua, Tegucigalpa, La Paz, Danlí, Yuscarán, Is. de la Bahía, Roatán, Puerto Castilla, Iriona

95 90 85 30 25 20 15

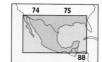

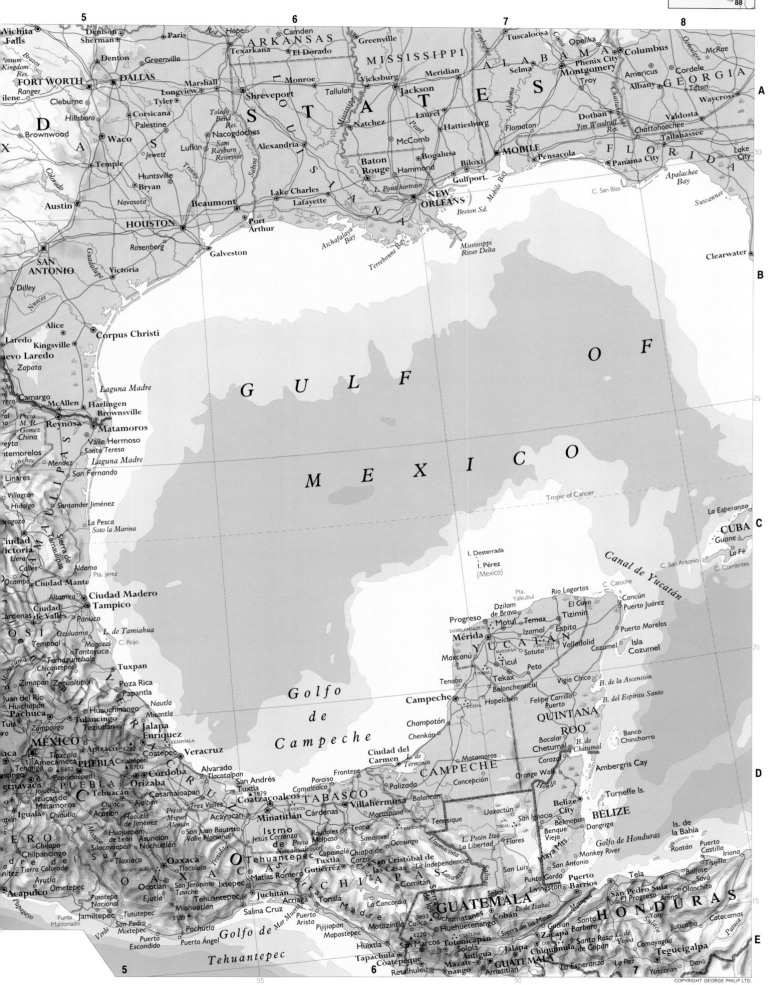

1:8 000 000

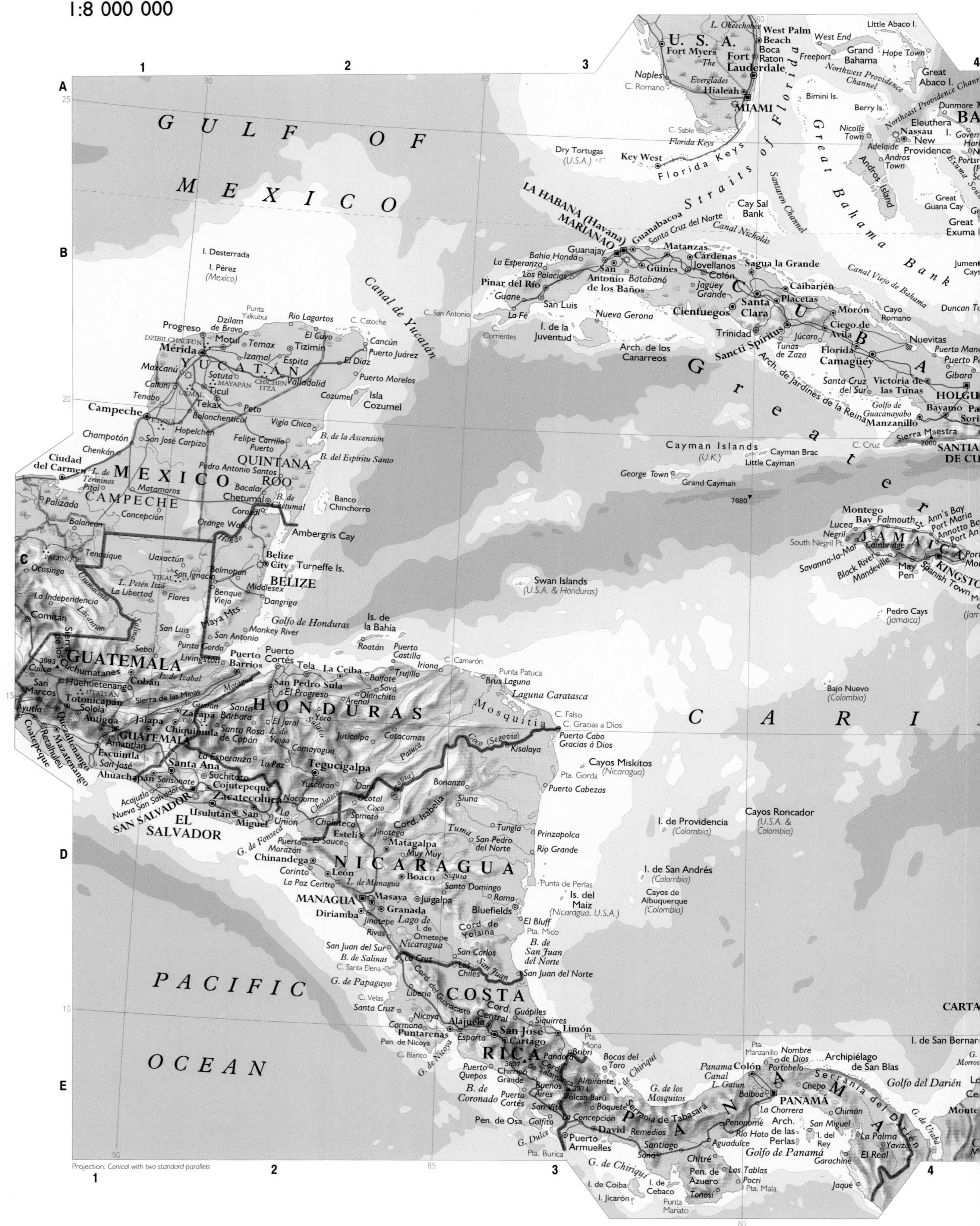

Projection: Conical with two standard parallels

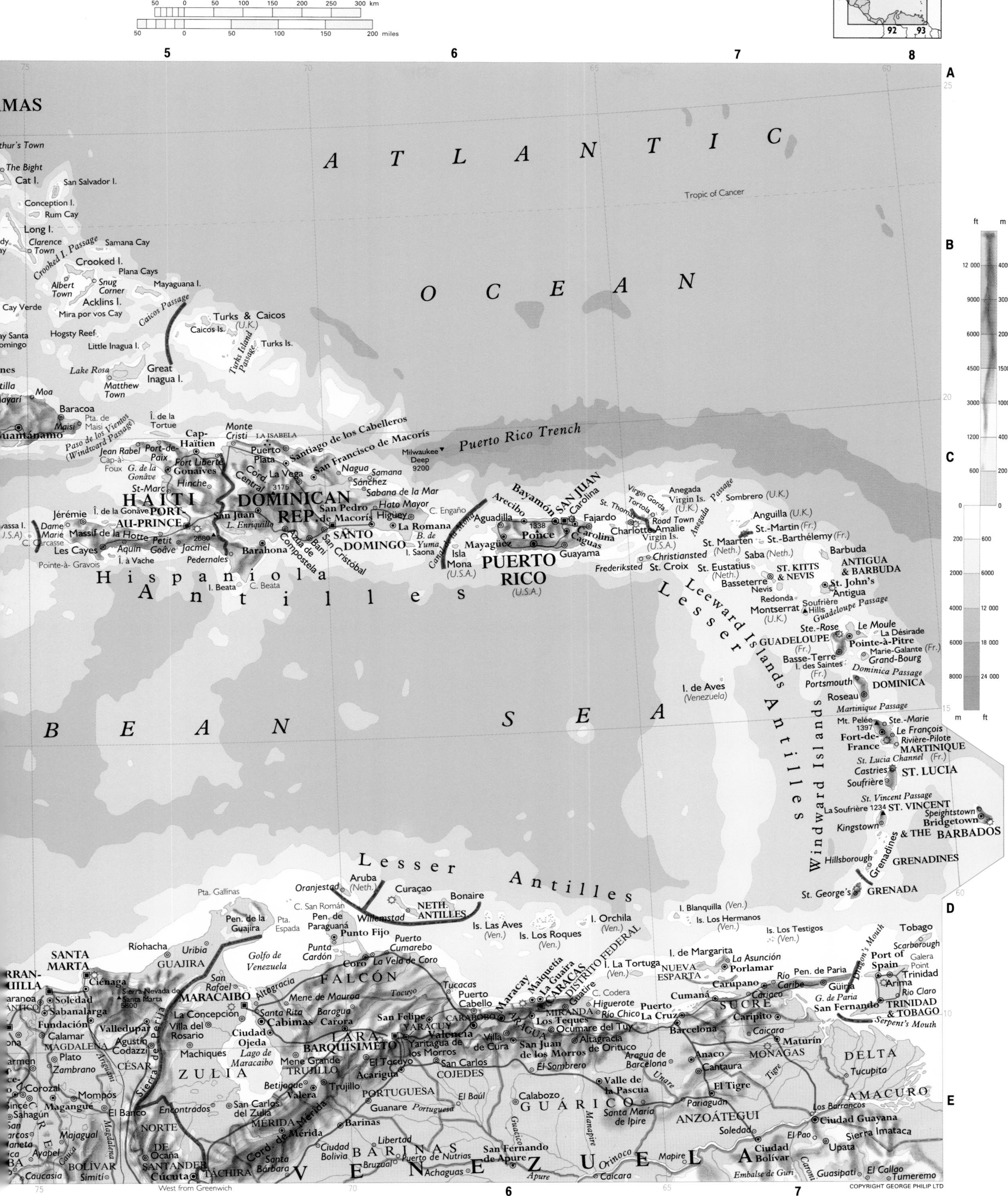
50 0 50 100 150 200 250 300 km
50 0 50 100 150 200 miles
5
6
7
8
A
B
C
D
E
87
92
93
ATLANTIC
OCEAN
Tropic of Cancer
Puerto Rico Trench
Milwaukee Deep 9200
MAS
Arthur's Town
The Bight
Cat I.
San Salvador I.
Conception I.
Rum Cay
Long I.
Clarence Town
Crooked I. Passage
Samana Cay
Crooked I.
Plana Cays
Albert Town
Snug Corner
Mayaguana I.
Acklins I.
Cay Verde
Mira por vos Cay
Caicos Passage
Turks & Caicos
(U.K.)
Caicos Is.
Turks Is.
Turks Island Passage
Hogsty Reef
Little Inagua I.
Lake Rosa
Great Inagua I.
Matthew Town
Moa
Baracoa
Maisí
Pta. de Maisí
Guantánamo
Î. de la Tortue
Paso de los Vientos (Windward Passage)
Cap-Haïtien
Port-de-Paix
Monte Cristi
LA ISABELA
Santiago de los Caballeros
San Francisco de Macorís
Jean Rabel
Fort Liberté
Puerto Plata
Cap-à-Foux
G. de la Gonâve
Gonaïves
Hinche
Cord. Central
La Vega
3175
Nagua
Samana
Sánchez
Sabana de la Mar
St-Marc
HAITI
DOMINICAN REP.
Jérémie
Î. de la Gonâve
PORT-AU-PRINCE
San Juan
L. Enriquillo
San Pedro de Macorís
Hato Mayor
Higüey
C. Engaño
Dame Marie
Massif de la Hotte
Petit Goâve
Jacmel
2680
Azua de Compostela
Baní
San Cristóbal
SANTO DOMINGO
La Romana
B. de Yuma
I. Saona
Les Cayes
Aquin
Î. à Vache
Barahona
Pedernales
C. Carcasse
Pointe-à-Gravois
Hispaniola
Antilles
I. Beata
C. Beata
Canal de la Mona
Isla Mona (U.S.A.)
Aguadilla
Arecibo
Bayamón
SAN JUAN
Carolina
1338
Mayagüez
Ponce
Caguas
Guayama
Fajardo
PUERTO RICO
(U.S.A.)
Virgin Gorda
Tortola
St. Thomas
Road Town
Charlotte Amalie
Virgin Is. (U.S.A.)
Anegada
Virgin Is. (U.K.)
Anegada Passage
Sombrero (U.K.)
Christiansted
Frederiksted
St. Croix
Anguilla (U.K.)
St.-Martin (Fr.)
St. Maarten (Neth.)
St.-Barthélemy (Fr.)
Saba (Neth.)
St. Eustatius (Neth.)
Barbuda
ST. KITTS & NEVIS
Basseterre
Nevis
ANTIGUA & BARBUDA
St. John's
Antigua
Redonda
Montserrat (U.K.)
Soufrière Hills
Guadeloupe Passage
Leeward Islands
Lesser Antilles
Ste.-Rose
Le Moule
La Désirade
GUADELOUPE (Fr.)
Pointe-à-Pitre
Marie-Galante (Fr.)
Basse-Terre
Grand-Bourg
I. des Saintes (Fr.)
Dominica Passage
Portsmouth
DOMINICA
Roseau
I. de Aves (Venezuela)
Martinique Passage
Mt. Pelée 1397
Ste.-Marie
Le François
Fort-de-France
Rivière-Pilote
MARTINIQUE (Fr.)
St. Lucia Channel
Castries
ST. LUCIA
Soufrière
St. Vincent Passage
La Soufrière 1234
ST. VINCENT & THE GRENADINES
Speightstown
Bridgetown
BARBADOS
Kingstown
Windward Islands
Grenadines
Hillsborough
GRENADINES
St. George's
GRENADA
BEAN
SEA
Lesser Antilles
Aruba (Neth.)
Oranjestad
Curaçao
Bonaire
Willemstad
NETH. ANTILLES
Pta. Gallinas
C. San Román
Pen. de la Guajira
Pta. Espada
Pen. de Paraguaná
Punto Fijo
Puerto Cumarebo
Punta Cardón
Coro
La Vela de Coro
Is. Las Aves (Ven.)
Is. Los Roques (Ven.)
I. Orchila (Ven.)
I. Blanquilla (Ven.)
Is. Los Hermanos (Ven.)
Is. Los Testigos (Ven.)
Tobago
Scarborough
Galera Point
Ríohacha
Uribia
GUAJIRA
Golfo de Venezuela
FALCÓN
SANTA MARTA
Ciénaga
Sierra Nevada de Santa Marta 5800
MARACAIBO
Altagracia
Mene de Mauroa
Tocuyo
Tucacas
Puerto Cabello
Maracay
Maiquetía
La Guaira
CARACAS
DISTRITO FEDERAL
I. La Tortuga (Ven.)
I. de Margarita
La Asunción
Porlamar
NUEVA ESPARTA
Pen. de Paria
Dragon's Mouth
Port of Spain
Trinidad
Arima
Río Claro
Carúpano
Río Caribe
Güiria
G. de Paria
San Fernando
TRINIDAD & TOBAGO
Serpent's Mouth
Soledad
Sabanalarga
Fundación
Valledupar
La Concepción
Villa del Rosario
Santa Rita
Cabimas
Baragua
Carora
San Felipe
YARACUY
CARABOBO
Valencia
ARAGUA
MIRANDA
Los Teques
Ocumare del Tuy
C. Codera
Higuerote
Río Chico
Guatire
Cumaná
Puerto La Cruz
Barcelona
SUCRE
Cariaco
Caripito
Caicara
Maturín
MONAGAS
DELTA AMACURO
Tucupita
Calamar
MAGDALENA
Agustín Codazzi
Plato
Zambrano
CÉSAR
Ciudad Ojeda
Lago de Maracaibo
Machiques
ZULIA
LARA
BARQUISIMETO
Yaritagua
Mene Grande
TRUJILLO
El Tocuyo
Acarigua
Villa de Cura
San Juan de los Morros
Altagracia de Orituco
Aragua de Barcelona
Anaco
Cantaura
San Carlos
COJEDES
El Sombrero
Unare
Tigre
El Tigre
Valle de la Pascua
Corozal
Mompós
Magangué
Sincé
Sahagún
El Banco
Sierra de Perijá
Encontrados
San Carlos del Zulia
Betijoque
Valera
Trujillo
PORTUGUESA
Guanare
Portuguesa
El Baúl
Calabozo
GUÁRICO
Santa María de Ipire
Pariaguán
ANZOÁTEGUI
Los Barrancos
Ciudad Guayana
Sierra Imataca
Majagual
Magdalena
Cauca
Ayapel
BOLÍVAR
Simití
Caucasia
NORTE DE SANTANDER
Ocaña
Cúcuta
TÁCHIRA
MÉRIDA
Cord. de Mérida
Barinas
Santa Bárbara
Ciudad Bolivia
BARINAS
Libertad
Puerto de Nutrias
Bruzual
Achaguas
San Fernando de Apure
Apure
Guárico
Manapire
Orinoco
Mapire
Soledad
Ciudad Bolívar
El Pao
Upata
Caroní
Embalse de Guri
Guasipati
El Callao
Tumeremo
VENEZUELA
Caicara
West from Greenwich
75
70
65
60
25
20
15
10
ft m
12 000 4000
9000 3000
6000 2000
4500 1500
3000 1000
1200 400
600 200
0 0
200 600
2000 6000
4000 12 000
6000 18 000
8000 24 000
m ft

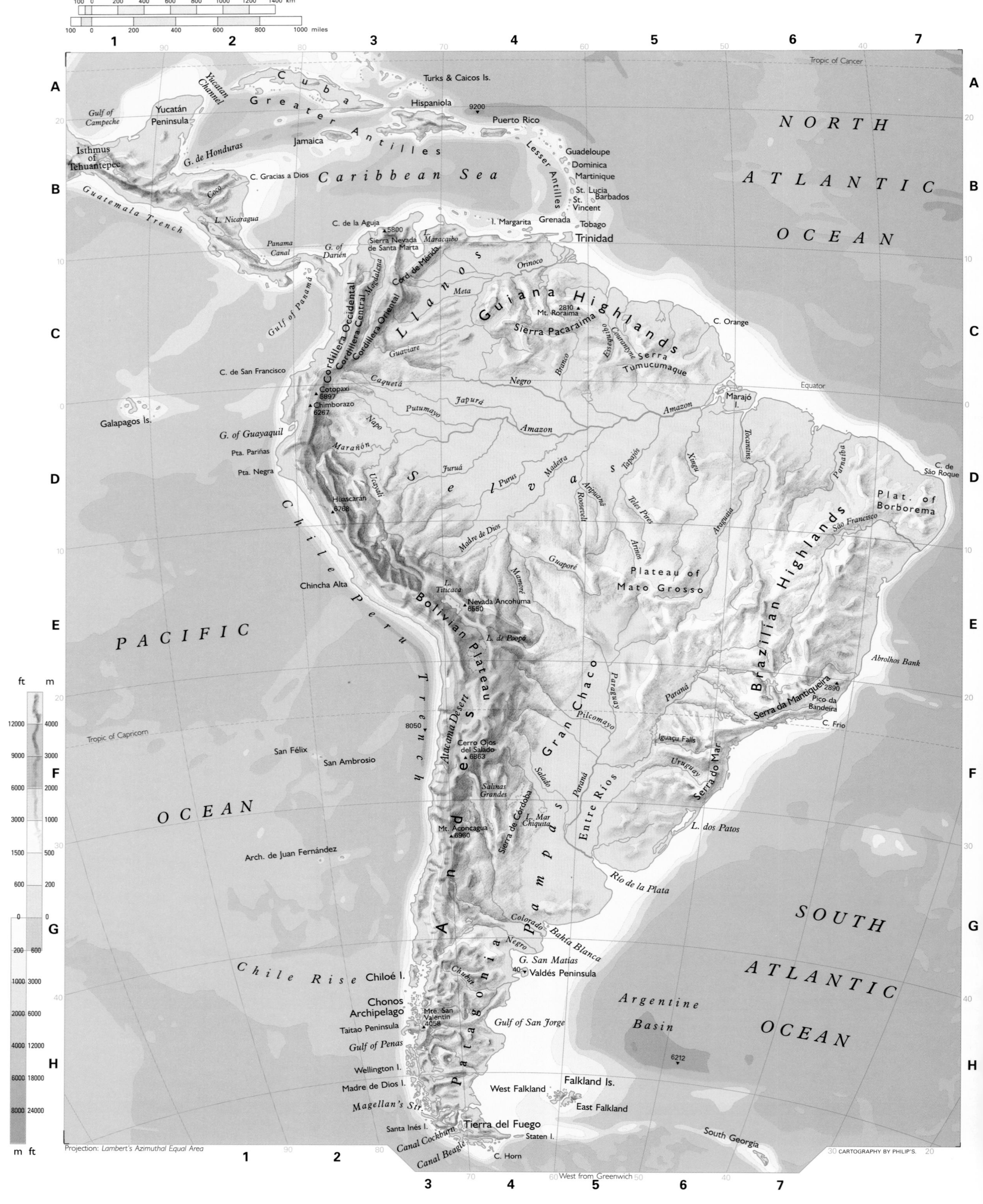

100 0 200 400 600 800 1000 1200 1400 km
100 0 200 400 600 800 1000 miles

NORTH ATLANTIC OCEAN

Tropic of Cancer

Yucatán Channel
Cuba
Greater Antilles
Turks & Caicos Is.
Gulf of Campeche
Yucatán Peninsula
Hispaniola
9200
Puerto Rico
Isthmus of Tehuantepec
G. de Honduras
Jamaica
Lesser Antilles
Guadeloupe
Dominica
Martinique
St. Lucia
St. Vincent
Barbados

Guatemala Trench
Coco
C. Gracias a Dios
Caribbean Sea
Grenada
Tobago
Trinidad

L. Nicaragua
Panama Canal
C. de la Aguja
5800
Sierra Nevada de Santa Marta
Maracaibo
Grenada

Gulf of Panamá
G. of Darién
Cord. de Mérida
Llanos
Orinoco
Meta
Guiana Highlands
2810 Mt. Roraima
Sierra Pacaraima
C. Orange

C. de San Francisco
Cordillera Occidental
Cordillera Central
Cordillera Oriental
Guaviare
Caquetá
Negro
Branco
Caroni
Serra Tumucumaque

Cotopaxi 5897
Chimborazo 6267
Putumayo
Japurá
Equator

Galapagos Is.
G. of Guayaquil
Napo
Marañón
S   e   l   v   a   s
Amazon
Amazon
Marajó I.

Pta. Pariñas
Pta. Negra
Ucayali
Juruá
Purus
Madeira
Tapajós
Xingu
Tocantins
Parnaíba
C. de São Roque

Huascarán 6768
Madre de Dios
Roosevelt
Aripuaná
Teles Pires
Araguaia
São Francisco
Plat. of Borborema

Chile   Peru   Trench
Chincha Alta
L. Titicaca
Guaporé
Mamoré
Arinos
Plateau of Mato Grosso
Brazilian Highlands

PACIFIC
Nevada Ancohuma 6550
Bolivian Plateau
L. de Poopó
2890
Pico da Bandeira
Serra da Mantiqueira

Abrolhos Bank

Tropic of Capricorn
San Félix
San Ambrosio
8050
Atacama Desert
A   n   d   e   s
Paraguay
Paraná
Gran Chaco
Uruguay
Serra do Mar
Iguaçu Falls
C. Frio

OCEAN
Cerro Ojos del Salado 6863
Salinas Grandes
Salado
Pilcomayo
Paraná
Entre Ríos

Arch. de Juan Fernández
Mt. Aconcagua 6960
Sierra de Córdoba
L. Mar Chiquita
L. dos Patos

P   a   m   p   a   s
Río de la Plata

SOUTH ATLANTIC OCEAN

Colorado
Bahía Blanca
Negro
G. San Matías
Valdés Peninsula

Chile Rise
Chiloé I.
Chonos Archipelago
Mte. San Valentín 4058
Taitao Peninsula
Gulf of Penas
P   a   t   a   g   o   n   i   a
Chubut
Gulf of San Jorge
Argentine Basin
6212

Wellington I.
Madre de Dios I.
West Falkland
East Falkland
Falkland Is.

Magellan's Str.
Santa Inés I.
Canal Cockburn
Tierra del Fuego
Staten I.
South Georgia
Canal Beagle
C. Horn

Projection: Lambert's Azimuthal Equal Area

ft / m scale:
12000 / 4000
9000 / 3000
6000 / 2000
3000 / 1000
1500 / 500
600 / 200
0 / 0
200 / 600
1000 / 3000
2000 / 6000
4000 / 12000
6000 / 18000
8000 / 24000
m ft

West from Greenwich

30 CARTOGRAPHY BY PHILIP'S. 20

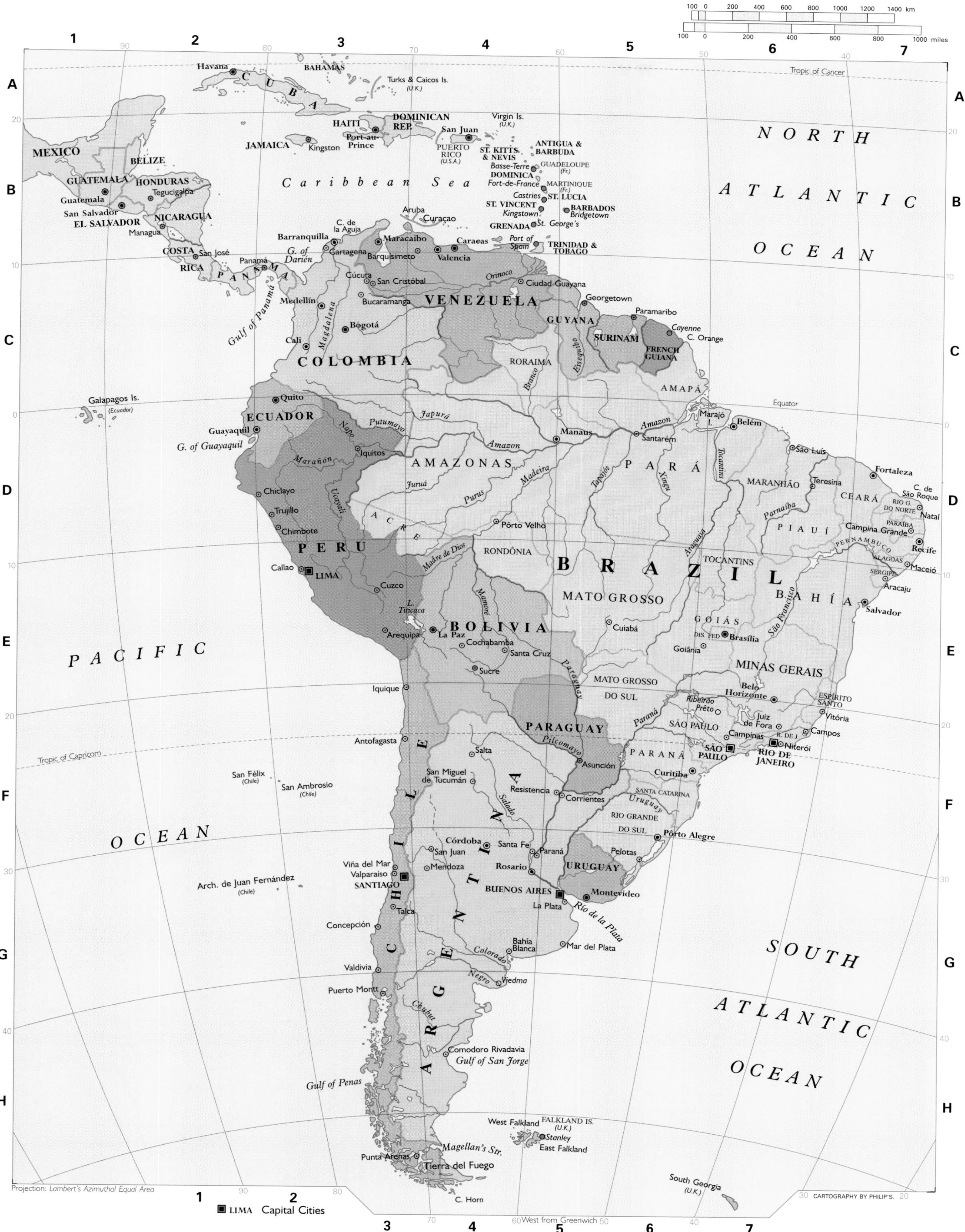
NORTH ATLANTIC OCEAN
PACIFIC OCEAN
SOUTH ATLANTIC OCEAN
Caribbean Sea
BRAZIL
ARGENTINA
CHILE
PERU
BOLIVIA
COLOMBIA
VENEZUELA
ECUADOR
PARAGUAY
URUGUAY
GUYANA
SURINAM
FRENCH GUIANA
CUBA
MEXICO
BELIZE
GUATEMALA
HONDURAS
EL SALVADOR
NICARAGUA
COSTA RICA
PANAMA
JAMAICA
HAITI
DOMINICAN REP.
BAHAMAS
PUERTO RICO (U.S.A.)
ST. KITTS & NEVIS
ANTIGUA & BARBUDA
DOMINICA
ST. LUCIA
ST. VINCENT
BARBADOS
GRENADA
TRINIDAD & TOBAGO
Turks & Caicos Is. (U.K.)
Virgin Is. (U.K.)
GUADELOUPE (Fr.)
MARTINIQUE (Fr.)
Galapagos Is. (Ecuador)
San Félix (Chile)
San Ambrosio (Chile)
Arch. de Juan Fernández (Chile)
FALKLAND IS. (U.K.)
South Georgia (U.K.)
Tropic of Cancer
Equator
Tropic of Capricorn
Amazon
Orinoco
Magdalena
Paraná
Paraguay
Pilcomayo
Salado
Colorado
Negro
Chubut
Uruguay
Río de la Plata
São Francisco
Tocantins
Araguaia
Xingu
Tapajós
Madeira
Purus
Juruá
Japurá
Putumayo
Napo
Marañón
Ucayali
Madre de Dios
Mamoré
Branco
Essequibo
Parnaíba
L. Titicaca
Gulf of Panama
G. of Darién
G. of Guayaquil
Gulf of Penas
Gulf of San Jorge
Magellan's Str.
C. Horn
C. Orange
C. de São Roque
C. de la Aguja
Marajó I.
Tierra del Fuego
AMAZONAS
PARÁ
AMAPÁ
RORAIMA
ACRE
RONDÔNIA
MATO GROSSO
MATO GROSSO DO SUL
TOCANTINS
GOIÁS
DIS. FED
MARANHÃO
PIAUÍ
CEARÁ
RIO G. DO NORTE
PARAÍBA
PERNAMBUCO
ALAGOAS
SERGIPE
BAHÍA
MINAS GERAIS
ESPÍRITO SANTO
R. DE J.
SÃO PAULO
PARANÁ
SANTA CATARINA
RIO GRANDE DO SUL
Havana
Kingston
Port-au-Prince
San Juan
Basse-Terre
Fort-de-France
Castries
Kingstown
Bridgetown
St. George's
Port of Spain
Aruba
Curaçao
Guatemala
Tegucigalpa
San Salvador
Managua
San José
Panamá
Barranquilla
Cartagena
Maracaibo
Barquisimeto
Caracas
Valencia
Cúcuta
San Cristóbal
Ciudad Guayana
Medellín
Bucaramanga
Bogotá
Cali
Georgetown
Paramaribo
Cayenne
Quito
Guayaquil
Iquitos
Manaus
Santarém
Belém
São Luís
Fortaleza
Teresina
Natal
Campina Grande
Recife
Maceió
Aracaju
Salvador
Chiclayo
Trujillo
Chimbote
Pôrto Velho
Callao
LIMA
Cuzco
Arequipa
La Paz
Cochabamba
Santa Cruz
Sucre
Cuiabá
Brasília
Goiânia
Belo Horizonte
Vitória
Ribeirão Prêto
Juiz de Fora
Campinas
Campos
SÃO PAULO
Niterói
RIO DE JANEIRO
Curitiba
Iquique
Antofagasta
Salta
San Miguel de Tucumán
Asunción
Resistencia
Corrientes
Pôrto Alegre
Pelotas
Córdoba
Santa Fe
San Juan
Paraná
Mendoza
Rosario
Viña del Mar
Valparaíso
SANTIAGO
BUENOS AIRES
La Plata
Montevideo
Talca
Concepción
Bahía Blanca
Mar del Plata
Valdivia
Puerto Montt
Viedma
Comodoro Rivadavia
West Falkland
Stanley
East Falkland
Punta Arenas
West from Greenwich
Projection: Lambert's Azimuthal Equal Area
CARTOGRAPHY BY PHILIP'S.
LIMA Capital Cities
100 0 200 400 600 800 1000 1200 1400 km
100 0 200 400 600 800 1000 miles

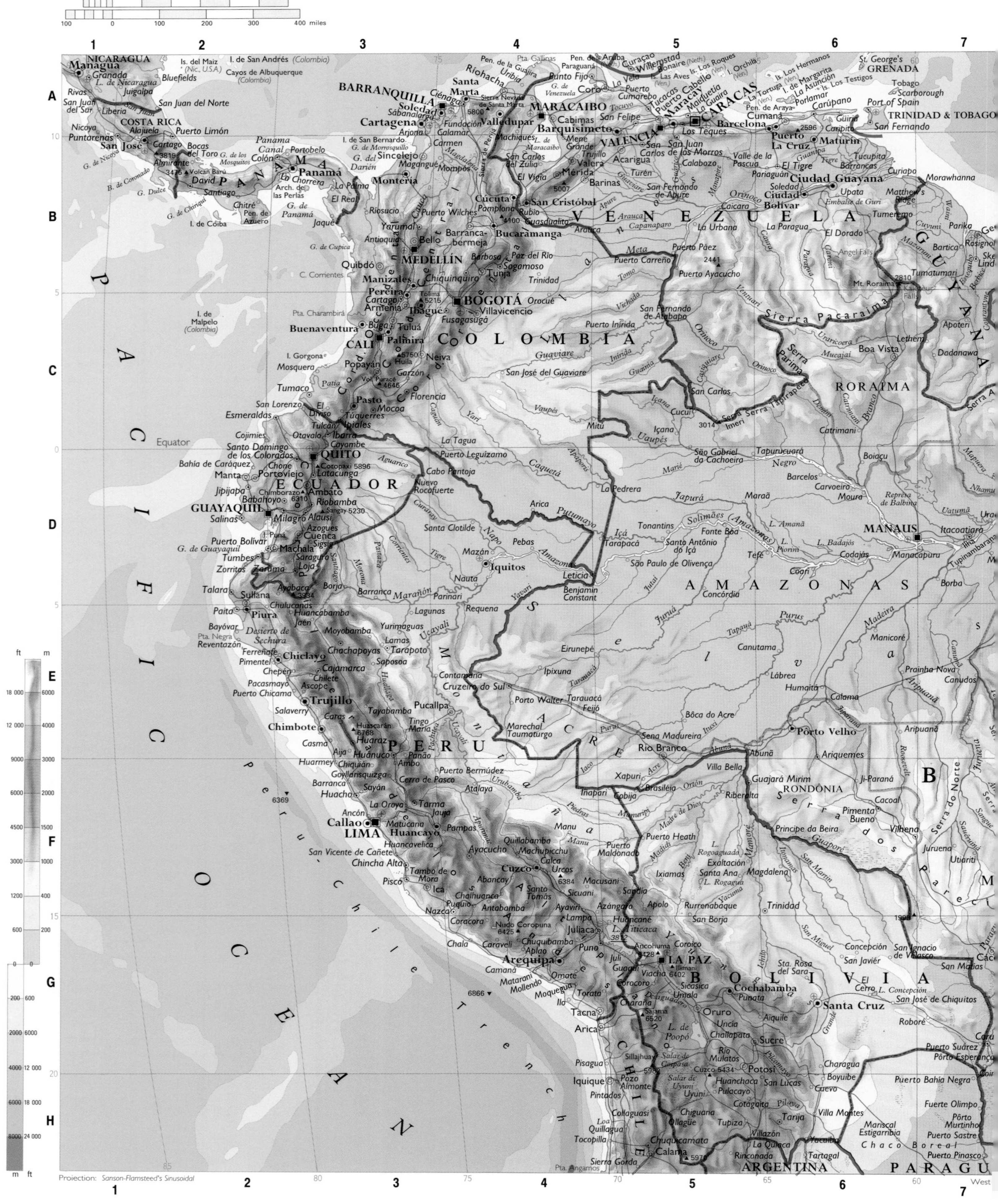

A

B

_A T L A N T I C_

_O C E A N_

C

São Paulo
(Braz.)

Equator

D

FORTALEZA
Rocas
Fernando de Noronha
(Braz.)

E

RECIFE
Olinda

MARANHÃO
Teresina
CEARÁ
PARAÍBA
PERNAMBUCO
Maceió

F

TOCANTINS
B A H I A
SERGIPE
Aracaju
SALVADOR

G

GOIÁS
BRASÍLIA
Goiânia
MINAS GERAIS
Trindade
(Braz.)

H

BELO HORIZONTE
Vitória
Campo
Grande
SÃO
PAULO
Campinas
RIO DE JANEIRO

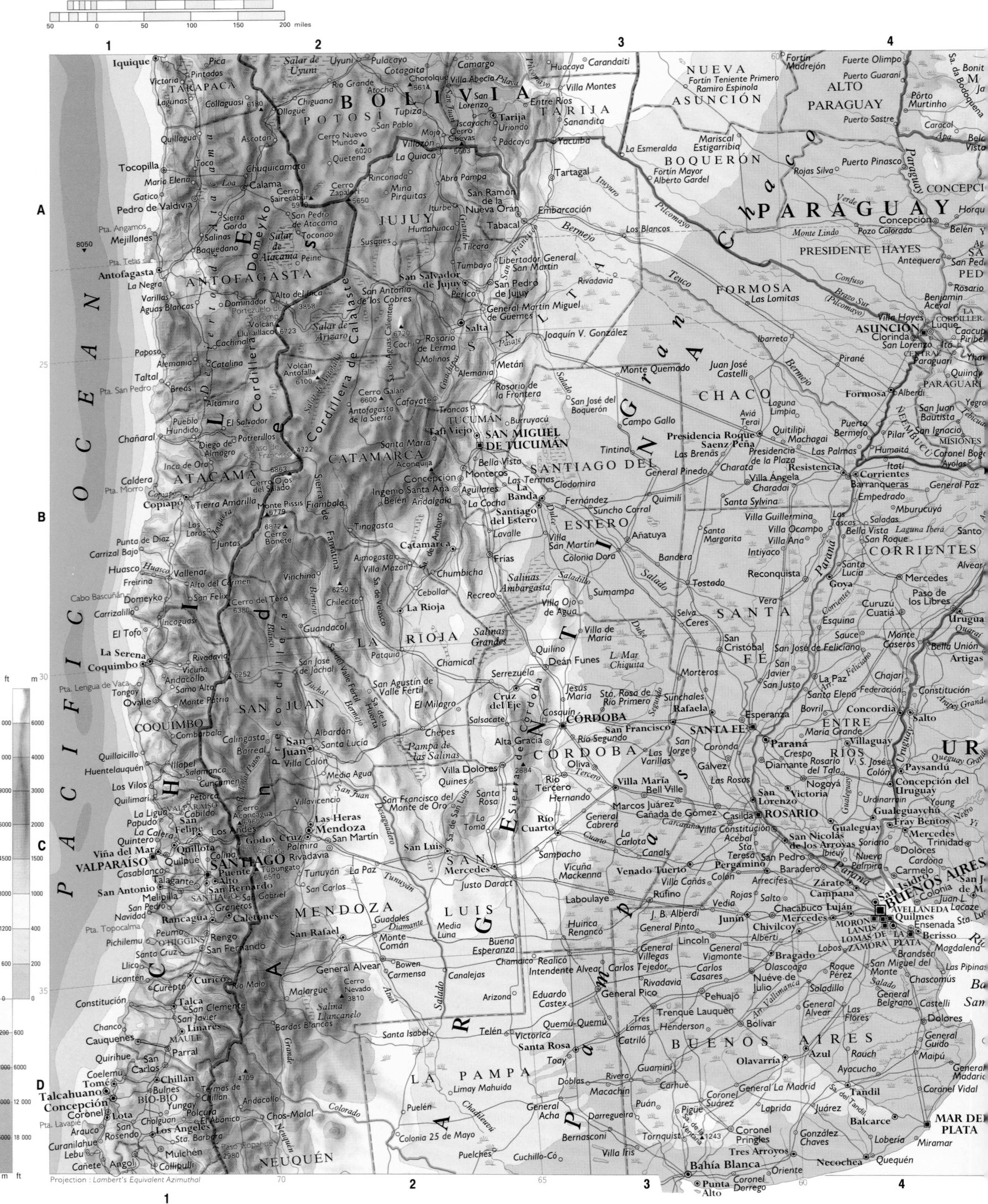

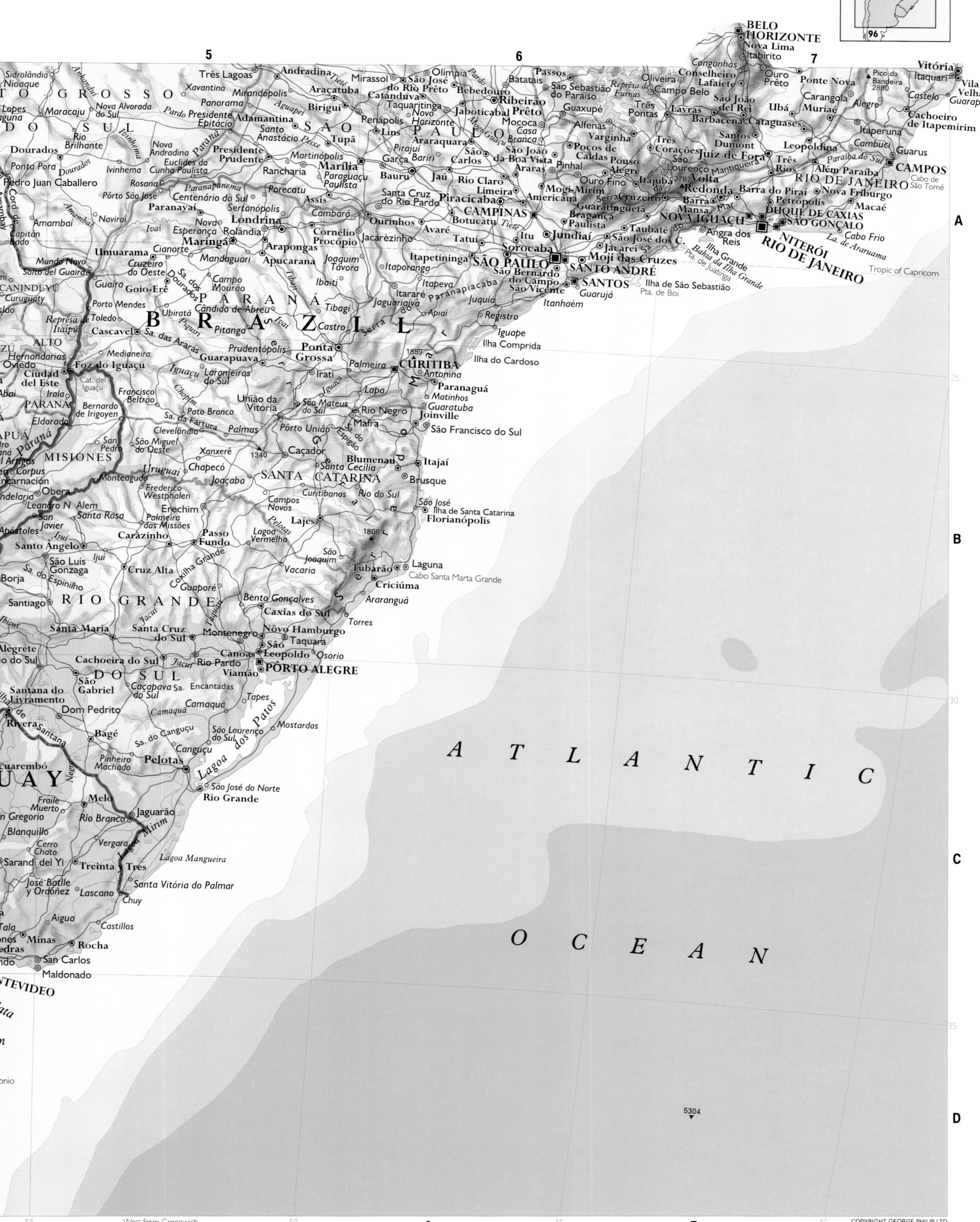

BELO
HORIZONTE
Nova Lima
Itabirito
Congonhas
Conselheiro
Lafaiete
Oliveira
Ouro
Prêto
Campo Belo
Ponte Nova
Vitória
Itaquari
Vila
Velha
Guarapari
Pico da
Bandeira
2890
Castelo
Cachoeiro
de Itapemirim

Três Lagoas
Andradina
Mirassol
São José
do Rio Prêto
Olímpia
Passos
São Sebastião
do Paraíso
Sebastião
Cruzeiro
São João
del Rei
Barbacena
Cataguases
Carangola
Muriaé
Ponte Nova

TO GROSSO
DO SUL
Sidrolândia
Nioaque
Xavantina
Panorama
Mirandópolis
Araçatuba
Catanduva
Bebedouro
Ribeirão
Prêto
Guaxupé
Oliveira
Lavras
Santos
Juiz de Fora
Leopoldina
Cambuci
Guarus

Maracaju
Nova Alvorada
do Sul
Adamantina
Penápolis
Jaboticabal
Mococa
Casa
Branca
Alfenas
Varginha
Três
Corações
Pouso
Juiz de Fora
Três
Rios
Além Paraíba
CAMPOS

Dourados
Presidente
Epitácio
Santo
Anastácio
Tupã
Lins
Araraquara
São
Carlos
Rio Claro
Pinhal
São
Lourenço
Redonda
Volta
Barra do Piraí
Nova Friburgo
Cabo de
São Tomé

Ponta Porã
Dourados
Presidente
Prudente
Marília
Paraguaçu
Paulista
Bauru
Santa Cruz
do Rio Pardo
Limeira
Americana
Guaratinguetá
Taubaté
Barra
Mansa
RIO DE JANEIRO
Macaé

Pedro Juan Caballero
Ivinhema
Euclides da
Cunha Paulista
Rancharia
Assis
Jaú
Piracicaba
CAMPINAS
Botucatu
Bragança
Paulista
São José dos C.
NOVA IGUAÇU
DUQUE DE CAXIAS
SÃO GONÇALO

CANINDEYÚ
Curuguaty
Mundo Novo
Salto del Guairá
Nova
Esperança
Londrina
Rolândia
Cornélio
Procópio
Jacarèzinho
Avaré
Tatuí
Itapetininga
SÃO PAULO
SANTO ANDRÉ
Moji das Cruzes
NITERÓI
RIO DE JANEIRO

Maringá
Apucarana
Joaquim
Távora
Ibaiti
Itaporanga
São Bernardo
do Campo
SANTOS
Ilha de São Sebastião
Tropic of Capricorn

BRAZIL
Umuarama
Cianorte
Mandaguari
Campo
Mourão
Itaporanga
Itapeva
Itapetininga
São Vicente
Guarujá
Ilha Grande
Angra dos
Reis
La. de Araruama
Cabo Frio

ABAI
Guaira
Porto Mendes
Goio-Erê
PARANÁ
Itararé
Paranapiacaba
Juquiá
Registro
Iguape
Ilha Comprida
Ilha de São Sebastião
Pta. de Boi

Ciudad
del Este
Cascavel
Medianeira
Sa. das Araras
Guarapuava
Prudentópolis
Ponta
Grossa
Castro
Palmeira
CURITIBA
Antonina
Paranaguá
Ilha do Cardoso

Foz do Iguaçu
Toledo
Cândido de Abreu
Pitanga
Irati
Lapa
Guaratuba
Joinville
São Francisco do Sul

PARANÁ
Bernardo
de Irigoyen
Francisco
Beltrão
União da
Vitória
Pôrto União
São Mateus
do Sul
Rio Negro
Mafra

Eldorado
MISIONES
Pato Branco
Palmas
Clevelândia
Sa. da Fartura
São
Miguel
do Oeste
Xanxerê
1340
Caçador
Blumenau
Santa Cecília
Itajaí
São Francisco do Sul

APÚA
Bernardo
de Irigoyen
Irala
São Pedro
Frederico
Westphalen
Chapecó
Joaçaba
SANTA
CATARINA
Rio do Sul
Brusque

Obera
Leandro N. Alem
Santa Rosa
Enechim
Palmeira
das Missões
Campos
Novos
Curitibanos
São José
Ilha de Santa Catarina
Florianópolis

Apóstoles
San
Javier
Montenegro
Carazinho
Passo
Fundo
Lajes
1808

Santo
Angelo
Ijuí
Cruz Alta
Coxilha Grande
Vacaria
São
Joaquim
Tubarão
Laguna

Borja
São Luís
Gonzaga
Sa. do Espinilho
Guaporé
Bento Gonçalves
Criciúma
Cabo Santa Marta Grande

Santiago
RIO GRANDE
Caxias do Sul
Araranguá

Santa Maria
Santa Cruz
do Sul
Nôvo Hamburgo
Taquara
Torres

Alegrete
Cachoeira do Sul
Rio Pardo
Canoas
São
Leopoldo
Osorio

Rivera
Santana do
Livramento
DO SUL
São
Gabriel
Caçapava
do Sul
Encantadas
Viamão
PÔRTO ALEGRE

Dom Pedrito
Camaquã
Tapes

UAY
Bagé
Sa. do Canguçu
São Lourenço
do Sul
Mostardas

Fraile
Muerto
Pinheiro
Machado
Canguçu
Pelotas
Lagoa
dos
Patos

Melo
Rio Branco
Jaguarão
São José do Norte
Rio Grande

Vergara
Mirim
Lagoa Mangueira

Sarandi del Yi
Treinta y Tres
Santa Vitória do Palmar

José Batlle
y Ordóñez
Lascano
Chuy

Aigua
Castillos

Minas
Rocha

San Carlos

MONTEVIDEO
Maldonado

A T L A N T I C

O C E A N
5304

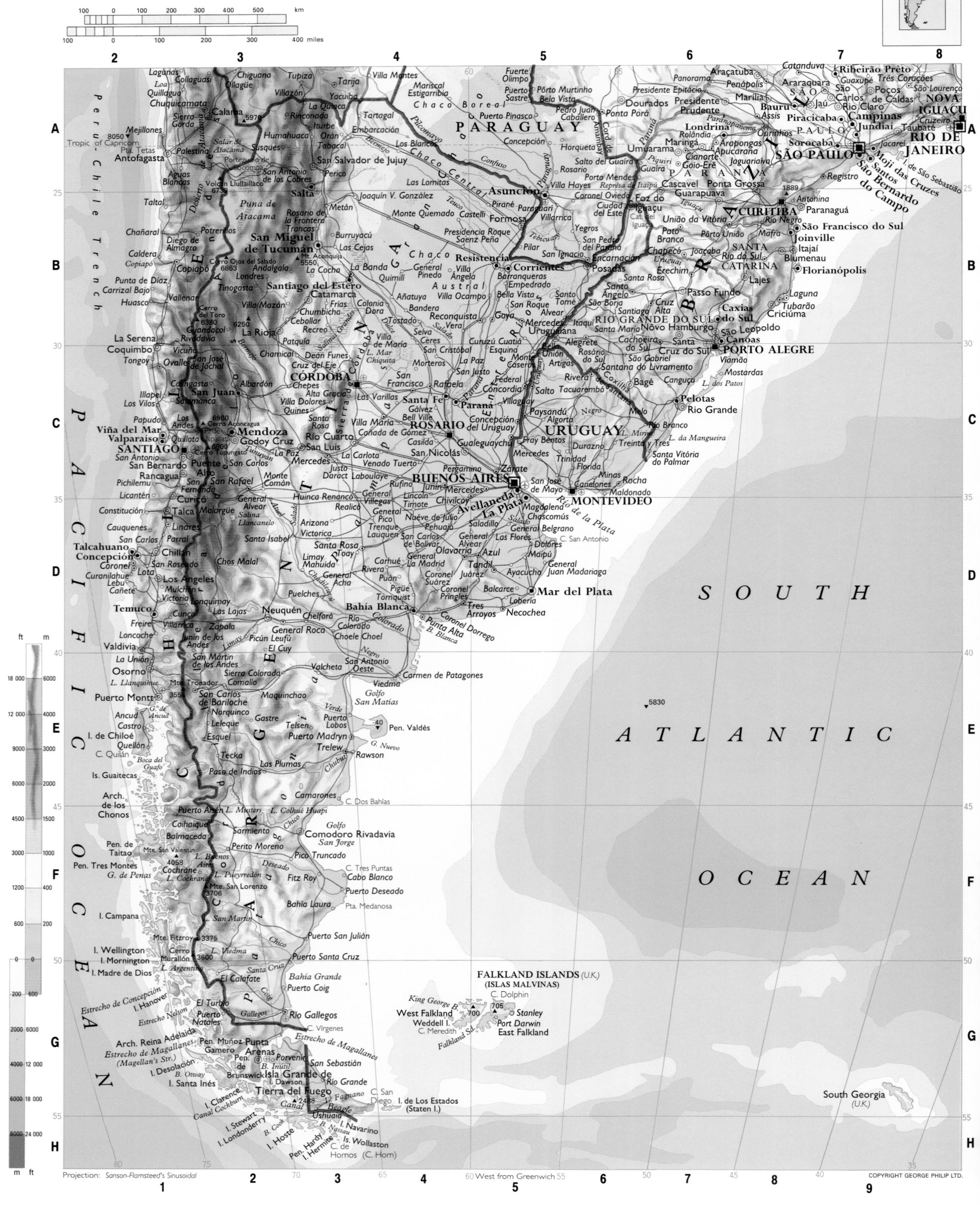
km
miles
PARAGUAY
URUGUAY
BRAZIL
SOUTH ATLANTIC OCEAN
PACIFIC OCEAN
Peru-Chile Trench
Tropic of Capricorn
Chaco Boreal
Chaco Central
Chaco Austral
Puna de Atacama
Antofagasta
Calama
Chuquicamata
Salta
San Salvador de Jujuy
San Miguel de Tucumán
Santiago del Estero
Catamarca
La Rioja
Copiapó
La Serena
Coquimbo
San Juan
Mendoza
Valparaíso
Viña del Mar
SANTIAGO
Rancagua
Talca
Chillán
Concepción
Talcahuano
Los Angeles
Temuco
Valdivia
Osorno
Puerto Montt
I. de Chiloé
Is. Guaitecas
Arch. de los Chonos
Puerto Aisén
Coihaique
Pen. de Taitao
Pen. Tres Montes
G. de Penas
I. Campana
I. Wellington
I. Madre de Dios
Arch. Reina Adelaida
Punta Arenas
Puerto Natales
Río Gallegos
Isla Grande de Tierra del Fuego
Ushuaia
Río Grande
C. de Hornos (C. Horn)
I. de Los Estados (Staten I.)
Estrecho de Magallanes
FALKLAND ISLANDS (U.K.)
(ISLAS MALVINAS)
West Falkland
East Falkland
Stanley
South Georgia (U.K.)
Comodoro Rivadavia
Golfo San Jorge
Trelew
Rawson
Puerto Madryn
Pen. Valdés
Golfo San Matías
Viedma
Carmen de Patagones
San Carlos de Bariloche
Neuquén
Bahía Blanca
Mar del Plata
Necochea
Tandil
Santa Rosa
CÓRDOBA
Río Cuarto
San Luis
ROSARIO
Santa Fe
Paraná
BUENOS AIRES
La Plata
Avellaneda
Río de la Plata
MONTEVIDEO
Salto
Paysandú
Rivera
Pelotas
Rio Grande
PORTO ALEGRE
Canoas
Caxias do Sul
Florianópolis
Blumenau
Joinville
CURITIBA
Ponta Grossa
Londrina
Maringá
SÃO PAULO
Santos
Campinas
RIO DE JANEIRO
NOVA IGUAÇU
Ribeirão Prêto
Asunción
Resistencia
Corrientes
Posadas
Encarnación
Formosa
Concepción
Foz do Iguaçu
Cerro Aconcagua 6960
Cerro Ojos del Salado 6863
Mte. Fitzroy 3375
5830
Projection: Sanson-Flamsteed's Sinusoidal
West from Greenwich
COPYRIGHT GEORGE PHILIP LTD.

INDEX

The index contains the names of all the principal places and features shown on the World Maps. Each name is followed by an additional entry in italics giving the country or region within which it is located. The alphabetical order of names composed of two or more words is governed primarily by the first word and then by the second. This is an example of the rule:

Mīr Kūh, *Iran*	**45 E8**	26 22N	58 55 E
Mīr Shahdād, *Iran*	**45 E8**	26 15N	58 29 E
Mira, *Italy*	**20 B5**	45 26N	12 8 E
Mira por vos Cay, *Bahamas*	**89 B5**	22 9N	74 30W
Miraj, *India*	**40 L9**	16 50N	74 45 E

Physical features composed of a proper name (Erie) and a description (Lake) are positioned alphabetically by the proper name. The description is positioned after the proper name and is usually abbreviated:

Erie, L., *N. Amer.*	**78 D4**	42 15N	81 0W

Where a description forms part of a settlement or administrative name however, it is always written in full and put in its true alphabetic position:

Mount Morris, *U.S.A.*	**78 D7**	42 44N	77 52W

Names beginning with M' and Mc are indexed as if they were spelled Mac. Names beginning St. are alphabetised under Saint, but Sankt, Sint, Sant', Santa and San are all spelt in full and are alphabetised accordingly. If the same place name occurs two or more times in the index and all are in the same country, each is followed by the name of the administrative subdivision in which it is located. The names are placed in the alphabetical order of the subdivisions. For example:

Jackson, *Ky., U.S.A.*	**76 G4**	37 33N	83 23W
Jackson, *Mich., U.S.A.*	**76 D3**	42 15N	84 24W
Jackson, *Minn., U.S.A.*	**80 D7**	43 37N	95 1W

The number in bold type which follows each name in the index refers to the number of the map page where that feature or place will be found. This is usually the largest scale at which the place or feature appears.

The letter and figure which are in bold type immediately after the page number give the grid square on the map page, within which the feature is situated. The letter represents the latitude and the figure the longitude.

In some cases the feature itself may fall within the specified square, while the name is outside. This is usually the case only with features which are larger than a grid square.

For a more precise location the geographical coordinates which follow the letter/figure references give the latitude and the longitude of each place. The first set of figures represent the latitude which is the distance north or south of the Equator measured as an angle at the centre of the earth. The Equator is latitude 0°, the North Pole is 90°N, and the South Pole 90°S.

The second set of figures represent the longitude, which is the distance East or West of the prime meridian, which runs through Greenwich, England. Longitude is also measured as an angle at the centre of the earth and is given East or West of the prime meridian, from 0° to 180° in either direction.

The unit of measurement for latitude and longitude is the degree, which is subdivided into 60 minutes. Each index entry states the position of a place in degrees and minutes, a space being left between the degrees and the minutes.

The latitude is followed by N(orth) or S(outh) and the longitude by E(ast) or W(est).

Rivers are indexed to their mouths or confluences, and carry the symbol ➔ after their names. A solid square ■ follows the name of a country, while an open square □ refers to a first order administrative area.

Abbreviations used in the index

A.C.T. – Australian Capital Territory
Afghan. – Afghanistan
Ala. – Alabama
Alta. – Alberta
Amer. – America(n)
Arch. – Archipelago
Ariz. – Arizona
Ark. – Arkansas
Atl. Oc. – Atlantic Ocean
B. – Baie, Bahía, Bay, Bucht, Bugt
B.C. – British Columbia
Bangla. – Bangladesh
Barr. – Barrage
Bos.-H. – Bosnia-Herzegovina
C. – Cabo, Cap, Cape, Coast
C.A.R. – Central African Republic
C. Prov. – Cape Province
Calif. – California
Cent. – Central
Chan. – Channel
Colo. – Colorado
Conn. – Connecticut
Cord. – Cordillera
Cr. – Creek
Czech. – Czech Republic
D.C. – District of Columbia
Del. – Delaware
Dep. – Dependency
Des. – Desert
Dist. – District
Dj. – Djebel
Domin. – Dominica
Dom. Rep. – Dominican Republic
E. – East

E. Salv. – El Salvador
Eq. Guin. – Equatorial Guinea
Fla. – Florida
Falk. Is. – Falkland Is.
G. – Golfe, Golfo, Gulf, Guba, Gebel
Ga. – Georgia
Gt. – Great, Greater
Guinea-Biss. – Guinea-Bissau
H.K. – Hong Kong
H.P. – Himachal Pradesh
Hants. – Hampshire
Harb. – Harbor, Harbour
Hd. – Head
Hts. – Heights
I.(s). – Île, Ilha, Insel, Isla, Island, Isle
Ill. – Illinois
Ind. – Indiana
Ind. Oc. – Indian Ocean
Ivory C. – Ivory Coast
J. – Jabal, Jebel, Jazira
Junc. – Junction
K. – Kap, Kapp
Kans. – Kansas
Kep. – Kepulauan
Ky. – Kentucky
L. – Lac, Lacul, Lago, Lagoa, Lake, Limni, Loch, Lough
La. – Louisiana
Liech. – Liechtenstein
Lux. – Luxembourg
Mad. P. – Madhya Pradesh
Madag. – Madagascar
Man. – Manitoba
Mass. – Massachusetts

Md. – Maryland
Me. – Maine
Medit. S. – Mediterranean Sea
Mich. – Michigan
Minn. – Minnesota
Miss. – Mississippi
Mo. – Missouri
Mont. – Montana
Mozam. – Mozambique
Mt.(e) – Mont, Monte, Monti, Montaña, Mountain
N. – Nord, Norte, North, Northern, Nouveau
N.B. – New Brunswick
N.C. – North Carolina
N. Cal. – New Caledonia
N. Dak. – North Dakota
N.H. – New Hampshire
N.I. – North Island
N.J. – New Jersey
N. Mex. – New Mexico
N.S. – Nova Scotia
N.S.W. – New South Wales
N.W.T. – North West Territory
N.Y. – New York
N.Z. – New Zealand
Nebr. – Nebraska
Neths. – Netherlands
Nev. – Nevada
Nfld. – Newfoundland
Nic. – Nicaragua
O. – Oued, Ouadi
Occ. – Occidentale
Okla. – Oklahoma
Ont. – Ontario
Or. – Orientale

Oreg. – Oregon
Os. – Ostrov
Oz. – Ozero
P. – Pass, Passo, Pasul, Pulau
P.E.I. – Prince Edward Island
Pa. – Pennsylvania
Pac. Oc. – Pacific Ocean
Papua N.G. – Papua New Guinea
Pass. – Passage
Pen. – Peninsula, Péninsule
Phil. – Philippines
Pk. – Park, Peak
Plat. – Plateau
Prov. – Province, Provincial
Pt. – Point
Pta. – Ponta, Punta
Pte. – Pointe
Qué. – Québec
Queens. – Queensland
R. – Rio, River
R.I. – Rhode Island
Ra.(s). – Range(s)
Raj. – Rajasthan
Reg. – Region
Rep. – Republic
Res. – Reserve, Reservoir
S. – San, South, Sea
Si. Arabia – Saudi Arabia
S.C. – South Carolina
S. Dak. – South Dakota
S.I. – South Island
S. Leone – Sierra Leone
Sa. – Serra, Sierra
Sask. – Saskatchewan
Scot. – Scotland
Sd. – Sound

Sev. – Severnaya
Sib. – Siberia
Sprs. – Springs
St. – Saint
Sta. – Santa, Station
Ste. – Sainte
Sto. – Santo
Str. – Strait, Stretto
Switz. – Switzerland
Tas. – Tasmania
Tenn. – Tennessee
Tex. – Texas
Tg. – Tanjung
Trin. & Tob. – Trinidad & Tobago
U.A.E. – United Arab Emirates
U.K. – United Kingdom
U.S.A. – United States of America
Ut. P. – Uttar Pradesh
Va. – Virginia
Vdkhr. – Vodokhranilishche
Vf. – Vírful
Vic. – Victoria
Vol. – Volcano
Vt. – Vermont
W. – Wadi, West
W. Va. – West Virginia
Wash. – Washington
Wis. – Wisconsin
Wlkp. – Wielkopolski
Wyo. – Wyoming
Yorks. – Yorkshire
Yug. – Yugoslavia

A

A Coruña, Spain 19 A1 43 20N 8 25W
A Estrada, Spain 19 A1 42 43N 8 27W
A Fonsagrada, Spain 19 A2 43 8N 7 4W
Aachen, Germany 16 C4 50 45N 6 6 E
Aalborg = Ålborg, Denmark 9 H13 57 2N 9 54 E
Aalen, Germany 16 D6 48 51N 10 6 E
Aalst, Belgium 15 D4 50 56N 4 2 E
Aalten, Neths. 15 C6 51 56N 6 35 E
Aalter, Belgium 15 C3 51 5N 3 28 E
Äänekoski, Finland 9 E21 62 36N 25 44 E
Aarau, Switz. 18 C8 47 23N 8 4 E
Aare →, Switz. 18 C8 47 33N 8 14 E
Aarhus = Århus, Denmark 9 H14 56 8N 10 11 E
Aarschot, Belgium 15 D4 50 59N 4 49 E
Aba, Dem. Rep. of the Congo 54 B3 3 58N 30 17 E
Aba, Nigeria 50 G7 5 10N 7 19 E
Ābādān, Iran 45 D6 30 22N 48 20 E
Ābādeh, Iran 45 D7 31 8N 52 40 E
Abadla, Algeria 50 B5 31 2N 2 45W
Abaetetuba, Brazil 93 D9 1 40S 48 50W
Abagnar Qi, China 34 C9 43 52N 116 2 E
Abai, Paraguay 95 B4 25 58S 55 54W
Abakan, Russia 27 D10 53 40N 91 10 E
Abancay, Peru 92 F4 13 35S 72 55W
Abariringa, Kiribati 64 H10 2 50S 171 40W
Abarqū, Iran 45 D7 31 10N 53 20 E
Abashiri, Japan 30 B12 44 0N 144 15 E
Abashiri-Wan, Japan 30 C12 44 0N 144 30 E
Ābay = Nîl el Azraq →, Sudan 51 E12 15 38N 32 31 E
Abay, Kazakstan 26 E8 49 38N 72 53 E
Abaya, L., Ethiopia 46 F2 6 30N 37 50 E
Abaza, Russia 26 D9 52 39N 90 6 E
'Abbāsābād, Iran 45 C8 33 34N 58 23 E
Abbay = Nîl el Azraq →, Sudan 51 E12 15 38N 32 31 E
Abbaye, Pt., U.S.A. 76 B1 46 58N 88 8W
Abbé, L., Ethiopia 46 E3 11 8N 41 47 E
Abbeville, France 18 A4 50 6N 1 49 E
Abbeville, Ala., U.S.A. 77 K3 31 34N 85 15W
Abbeville, La., U.S.A. 81 L8 29 58N 92 8W
Abbeville, S.C., U.S.A. 77 H4 34 11N 82 23W
Abbot Ice Shelf, Antarctica 5 D16 73 0S 92 0W
Abbottabad, Pakistan 42 B5 34 10N 73 15 E
Abd al Kūrī, Yemen 46 E5 12 5N 52 20 E
Ābdar, Iran 45 D7 30 16N 55 19 E
'Abdolābād, Iran 45 C8 34 12N 56 30 E
Abdulpur, Bangla. 43 G13 24 15N 88 59 E
Abéché, Chad 51 F10 13 50N 20 35 E
Abengourou, Ivory C. 50 G5 6 42N 3 27W
Åbenrå, Denmark 9 J13 55 3N 9 25 E
Abeokuta, Nigeria 50 G6 7 3N 3 19 E
Aber, Uganda 54 B3 2 12N 32 25 E
Aberaeron, U.K. 11 E3 52 15N 4 15W
Aberayron = Aberaeron, U.K. 11 E3 52 15N 4 15W
Aberchirder, U.K. 12 D6 57 34N 2 37W
Abercorn = Mbala, Zambia 55 D3 8 46S 31 24 E
Abercorn, Australia 63 D5 25 12S 151 5 E
Aberdare, U.K. 11 F4 51 43N 3 27W
Aberdare Ra., Kenya 54 C4 0 15S 36 50 E
Aberdeen, Australia 63 E5 32 9S 150 56 E
Aberdeen, Canada 73 C7 52 20N 106 8W
Aberdeen, S. Africa 56 E3 32 28S 24 2 E
Aberdeen, U.K. 12 D6 57 9N 2 5W
Aberdeen, Ala., U.S.A. 77 J1 33 49N 88 33W
Aberdeen, Idaho, U.S.A. 82 E7 42 57N 112 50W
Aberdeen, Md., U.S.A. 76 F7 39 31N 76 10W
Aberdeen, S. Dak., U.S.A. 80 C5 45 28N 98 29W
Aberdeen, Wash., U.S.A. 84 D3 46 59N 123 50W
Aberdeen, City of □, U.K. 12 D6 57 10N 2 10W
Aberdeenshire □, U.K. 12 D6 57 17N 2 36W
Aberdovey = Aberdyfi, U.K. 11 E3 52 33N 4 3W
Aberdyfi, U.K. 11 E3 52 33N 4 3W
Aberfeldy, U.K. 12 E5 56 37N 3 51W
Abergavenny, U.K. 11 F4 51 49N 3 1W
Abergele, U.K. 10 D4 53 17N 3 35W
Abernathy, U.S.A. 81 J4 33 50N 101 51W
Abert, L., U.S.A. 82 E3 42 38N 120 14W
Aberystwyth, U.K. 11 E3 52 25N 4 5W
Abhā, Si. Arabia 46 D3 18 0N 42 34 E
Abhar, Iran 45 B6 36 9N 49 13 E
Abhayapuri, India 43 F14 26 24N 90 38 E
Abidjan, Ivory C. 50 G5 5 26N 3 58W
Abilene, Kans., U.S.A. 80 F6 38 55N 97 13W
Abilene, Tex., U.S.A. 81 J5 32 28N 99 43W
Abingdon, U.K. 11 F6 51 40N 1 17W
Abingdon, U.S.A. 77 G5 36 43N 81 59W
Abington Reef, Australia 62 B4 18 0S 149 35 E
Abitau →, Canada 73 B7 59 53N 109 3W
Abitibi →, Canada 70 B3 51 3N 80 55W
Abitibi, L., Canada 70 C4 48 40N 79 40W
Abkhaz Republic = Abkhazia □, Georgia 25 F7 43 12N 41 5 E
Abkhazia □, Georgia 25 F7 43 12N 41 5 E
Abminga, Australia 63 D1 26 8S 134 51 E
Åbo = Turku, Finland 9 F20 60 30N 22 19 E
Abohar, India 42 D6 30 10N 74 10 E
Abomey, Benin 50 G6 7 10N 2 5 E
Abong-Mbang, Cameroon 52 D2 4 0N 13 8 E
Abou-Deïa, Chad 51 F9 11 20N 19 20 E
Aboyne, U.K. 12 D6 57 4N 2 47W
Abra Pampa, Argentina 94 A2 22 43S 65 42W
Abraham L., Canada 72 C5 52 15N 116 35W
Abreojos, Pta., Mexico 86 B2 26 50N 113 40W
Abrud, Romania 17 E12 46 19N 23 5 E
Absaroka Range, U.S.A. 82 D9 44 45N 109 50W
Abu, India 42 G5 24 41N 72 50 E
Abū al Abyad, U.A.E. 45 E7 24 11N 53 50 E
Abū al Khaṣīb, Iraq 45 D6 30 25N 48 0 E
Abū 'Alī, Si. Arabia 45 E6 27 20N 49 27 E
Abū 'Alī →, Lebanon 47 A4 34 25N 35 50 E
Abu Dhabi = Abū Ẓāby, U.A.E. 46 C5 24 28N 54 22 E
Abū Du'ān, Syria 44 B3 36 25N 38 15 E
Abu el Gaīri, W. →, Egypt 47 F2 29 35N 33 30 E
Abu Ga'da, W. →, Egypt 47 F1 29 15N 32 53 E
Abū Ḥadrīyah, Si. Arabia 45 E6 27 20N 48 58 E
Abu Hamed, Sudan 51 E12 19 32N 33 13 E
Abū Kamāl, Syria 44 C4 34 30N 41 0 E
Abū Madd, Ra's, Si. Arabia 44 E3 24 50N 37 7 E
Abū Mūsā, U.A.E. 45 E7 25 52N 55 3 E
Abū Qaṣr, Si. Arabia 44 D3 30 21N 38 34 E
Abu Ṣafāt, W. →, Jordan 47 E5 30 24N 36 7 E
Abu Simbel, Egypt 51 D12 22 18N 31 40 E
Abū Şukhayr, Iraq 44 D5 31 54N 44 30 E
Abu Zabad, Sudan 51 F11 12 25N 29 10 E
Abū Ẓāby, U.A.E. 46 C5 24 28N 54 22 E
Abū Zeydābād, Iran 45 C6 33 54N 51 45 E
Abuja, Nigeria 50 G7 9 5N 7 32 E
Abukuma-Gawa →, Japan 30 E10 38 6N 140 52 E
Abukuma-Sammyaku, Japan 30 F10 37 30N 140 45 E
Abunã, Brazil 92 E5 9 40S 65 20W
Abunã →, Brazil 92 E5 9 41S 65 20W
Aburo, Dem. Rep. of the Congo 54 B3 2 4N 30 53 E
Abut Hd., N.Z. 59 K3 43 7S 170 15 E
Acadia Nat. Park, U.S.A. 77 C11 44 20N 68 13W
Açailândia, Brazil 93 D9 4 57S 47 0W
Acajutla, El Salv. 88 D2 13 36N 89 50W
Acámbaro, Mexico 86 D4 20 0N 100 40W
Acaponeta, Mexico 86 C3 22 30N 105 20W
Acapulco, Mexico 87 D5 16 51N 99 56W
Acarai, Serra, Brazil 92 C7 1 50N 57 50W
Acarigua, Venezuela 92 B5 9 33N 69 12W
Acatlán, Mexico 87 D5 18 10N 98 3W
Acayucan, Mexico 87 D6 17 59N 94 58W
Accomac, U.S.A. 76 G8 37 43N 75 40W
Accra, Ghana 50 G5 5 35N 0 6W
Accrington, U.K. 10 D5 53 45N 2 22W
Acebal, Argentina 94 C3 33 20S 60 50W
Aceh □, Indonesia 36 D1 4 15N 97 30 E
Achalpur, India 40 J10 21 22N 77 32 E
Acheng, China 35 B14 45 30N 126 58 E
Acher, India 42 H5 23 10N 72 32 E
Achill Hd., Ireland 13 C1 53 58N 10 15W
Achill I., Ireland 13 C1 53 58N 10 1W
Achinsk, Russia 27 D10 56 20N 90 20 E
Acireale, Italy 20 F6 37 37N 15 10 E
Ackerman, U.S.A. 81 J10 33 19N 89 11W
Acklins I., Bahamas 89 B5 22 30N 74 0W
Acme, Canada 72 C6 51 33N 113 30W
Acme, U.S.A. 78 F5 40 8N 79 26W
Aconcagua, Cerro, Argentina 94 C2 32 39S 70 0W
Aconquija, Mt., Argentina 94 B2 27 0S 66 0W
Açores, Is. dos, Atl. Oc. 50 A1 38 0N 27 0W
Acornhoek, S. Africa 57 C5 24 37S 31 2 E
Acraman, L., Australia 63 E2 32 2S 135 23 E
Acre = 'Akko, Israel 47 C4 32 55N 35 4 E
Acre □, Brazil 92 E4 9 1S 71 0W
Acre →, Brazil 92 E5 8 45S 67 22W
Acton, Canada 78 C4 43 38N 80 3W
Acuña, Mexico 86 B4 29 18N 100 55W
Ad Dammām, Si. Arabia 45 E6 26 20N 50 5 E
Ad Dāmūr, Lebanon 47 B4 33 44N 35 27 E
Ad Dawādimī, Si. Arabia 44 E5 24 35N 44 15 E
Ad Dawḥah, Qatar 46 B5 25 15N 51 35 E
Ad Dawr, Iraq 44 C4 34 27N 43 47 E
Ad Dir'iyah, Si. Arabia 44 E5 24 44N 46 35 E
Ad Dīwānīyah, Iraq 44 D5 32 0N 45 0 E
Ad Dujayl, Iraq 44 C5 33 51N 44 14 E
Ad Duwayd, Si. Arabia 44 D4 30 15N 42 17 E
Ada, Minn., U.S.A. 80 B6 47 18N 96 31W
Ada, Okla., U.S.A. 81 H6 34 46N 96 41W
Adabiya, Egypt 47 F1 29 53N 32 28 E
Adair, C., Canada 69 A12 71 31N 71 24W
Adaja →, Spain 19 B3 41 32N 4 52W
Adak I., U.S.A. 68 C2 51 45N 176 45W
Adamaoua, Massif de l', Cameroon 51 G7 7 20N 12 20 E
Adamawa Highlands = Adamaoua, Massif de l', Cameroon 51 G7 7 20N 12 20 E
Adamello, Mte., Italy 18 C9 46 9N 10 30 E
Adaminaby, Australia 63 F4 36 0S 148 45 E
Adams, Mass., U.S.A. 79 D11 42 38N 73 7W
Adams, N.Y., U.S.A. 79 C8 43 49N 76 1W
Adams, Wis., U.S.A. 80 D10 43 57N 89 49W
Adam's Bridge, Sri Lanka 40 Q11 9 15N 79 40 E
Adams L., Canada 72 C5 51 10N 119 40W
Adams Mt., U.S.A. 84 D5 46 12N 121 30W
Adam's Peak, Sri Lanka 40 R12 6 48N 80 30 E
Adana, Turkey 25 G6 37 0N 35 16 E
Adapazarı = Sakarya, Turkey 25 F5 40 48N 30 25 E
Adarama, Sudan 51 E12 17 10N 34 52 E
Adare, C., Antarctica 5 D11 71 0S 171 0 E
Adaut, Indonesia 37 F8 8 8S 131 7 E
Adavale, Australia 63 D3 25 52S 144 32 E
Adda →, Italy 18 D8 45 8N 9 53 E
Addis Ababa = Addis Abeba, Ethiopia 46 F2 9 2N 38 42 E
Addis Abeba, Ethiopia 46 F2 9 2N 38 42 E
Addison, U.S.A. 78 D7 42 1N 77 14W
Addo, S. Africa 56 E4 33 32S 25 45 E
Ādeh, Iran 44 B5 37 42N 45 11 E
Adel, U.S.A. 77 K4 31 8N 83 25W
Adelaide, Australia 63 E2 34 52S 138 30 E
Adelaide, Bahamas 88 A4 25 4N 77 31W
Adelaide, S. Africa 56 E4 32 42S 26 20 E
Adelaide I., Antarctica 5 C17 67 15S 68 30W
Adelaide Pen., Canada 68 B10 68 15N 97 30W
Adelaide River, Australia 60 B5 13 15S 131 7 E
Adelanto, U.S.A. 85 L9 34 35N 117 22W
Adele I., Australia 60 C3 15 32S 123 9 E
Adélie, Terre, Antarctica 5 C10 68 0S 140 0 E
Adélie Land = Adélie, Terre, Antarctica 5 C10 68 0S 140 0 E
Aden = Al 'Adan, Yemen 46 E4 12 45N 45 0 E
Aden, G. of, Asia 46 E4 12 30N 47 30 E
Adendorp, S. Africa 56 E3 32 15S 24 30 E
Adh Dhayd, U.A.E. 45 E7 25 17N 55 53 E
Adhoi, India 42 H4 23 26N 70 32 E
Adi, Indonesia 37 E8 4 15S 133 30 E
Adieu, C., Australia 61 F5 32 0S 132 10 E
Adieu Pt., Australia 60 C3 15 14S 124 35 E
Adige →, Italy 20 B5 45 9N 12 20 E
Adigrat, Ethiopia 46 E2 14 20N 39 26 E
Adilabad, India 40 K11 19 33N 78 20 E
Adirondack Mts., U.S.A. 79 C10 44 0N 74 0W
Adjumani, Uganda 54 B3 3 20N 31 50 E
Adlavik Is., Canada 71 B8 55 0N 58 40W
Admiralty G., Australia 60 B4 14 20S 125 55 E
Admiralty I., U.S.A. 72 B2 57 30N 134 30W
Admiralty Is., Papua N. G. 64 H6 2 0S 147 0 E
Adonara, Indonesia 37 F6 8 15S 123 5 E
Adoni, India 40 M10 15 33N 77 18 E
Adour →, France 18 E3 43 32N 1 32W
Adra, India 43 H12 23 30N 86 42 E
Adra, Spain 19 D4 36 43N 3 3W
Adrano, Italy 20 F6 37 40N 14 50 E
Adrar, Mauritania 50 D3 20 30N 7 30 E
Adrar des Iforas, Algeria 50 C5 27 51N 0 11 E
Adrian, Mich., U.S.A. 76 E3 41 54N 84 2W
Adrian, Tex., U.S.A. 81 H3 35 16N 102 40W
Adriatic Sea, Medit. S. 20 C6 43 0N 16 0 E
Adua, Indonesia 37 E7 1 45S 129 50 E
Adwa, Ethiopia 46 E2 14 15N 38 52 E
Adygea □, Russia 25 F7 45 0N 40 0 E
Adzhar Republic = Ajaria □, Georgia 25 F7 41 30N 42 0 E
Adzopé, Ivory C. 50 G5 6 7N 3 49W
Ægean Sea, Medit. S. 21 E11 38 30N 25 0 E
Aerhtai Shan, Mongolia 32 B4 46 40N 92 45 E
'Afak, Iraq 44 C5 32 4N 45 15 E
Afándou, Greece 23 C10 36 18N 28 12 E
Afghanistan ■, Asia 40 C4 33 0N 65 0 E
Aflou, Algeria 50 B6 34 7N 2 3 E
Africa 48 E6 10 0N 20 0 E
'Afrīn, Syria 44 B3 36 32N 36 50 E
Afton, N.Y., U.S.A. 79 D9 42 14N 75 32W
Afton, Wyo., U.S.A. 82 E8 42 44N 110 56W
Afuá, Brazil 93 D8 0 15S 50 20W
'Afula, Israel 47 C4 32 37N 35 17 E
Afyon, Turkey 25 G5 38 45N 30 33 E
Afyonkarahisar = Afyon, Turkey 25 G5 38 45N 30 33 E
Agadès = Agadez, Niger 50 E7 16 58N 7 59 E
Agadez, Niger 50 E7 16 58N 7 59 E
Agadir, Morocco 50 B4 30 28N 9 55W
Agaete, Canary Is. 22 F4 28 6N 15 43W
Agar, India 42 H7 23 40N 76 2 E
Agartala, India 41 H17 23 50N 91 23 E
Agassiz, Canada 72 D4 49 14N 121 46W
Agats, Indonesia 37 F9 5 33S 138 0 E
Agawam, U.S.A. 79 D12 42 5N 72 37W
Agboville, Ivory C. 50 G5 5 55N 4 15W
Ağdam, Azerbaijan 44 B5 40 0N 46 58 E
Agde, France 18 E5 43 19N 3 28 E
Agen, France 18 D4 44 12N 0 38 E
Āgh Kand, Iran 45 B6 37 15N 48 4 E
Aginskoye, Russia 27 D12 51 6N 114 32 E
Agnew, Australia 61 E3 28 1S 120 31 E
Agori, India 43 G10 24 33N 82 57 E
Agra, India 42 F7 27 17N 77 58 E
Ağri, Turkey 25 G7 39 44N 43 3 E
Agri →, Italy 20 D7 40 13N 16 44 E
Ağrı Dağı, Turkey 25 G7 39 50N 44 15 E
Ağri Karakose = Ağri, Turkey 25 G7 39 44N 43 3 E
Agrigento, Italy 20 F5 37 19N 13 34 E
Agrínion, Greece 21 E9 38 37N 21 27 E
Agua Caliente, Baja Calif., Mexico 85 N10 32 29N 116 59W
Agua Caliente, Sinaloa, Mexico 86 B3 26 30N 108 20W
Agua Caliente Springs, U.S.A. 85 N10 32 56N 116 19W
Água Clara, Brazil 93 H8 20 25S 52 45W
Agua Hechicero, Mexico 85 N10 32 26N 116 14W
Agua Prieta, Mexico 86 A3 31 20N 109 32W
Aguadilla, Puerto Rico 89 C6 18 26N 67 10W
Aguadulce, Panama 88 E3 8 15N 80 32W
Aguanga, U.S.A. 85 M10 33 27N 116 51W
Aguanish, Canada 71 B7 50 14N 62 2W
Aguanus →, Canada 71 B7 50 13N 62 5W
Aguapey →, Argentina 94 B4 29 7S 56 36W
Aguaray Guazú →, Paraguay 94 A4 24 47S 57 19W
Aguarico →, Ecuador 92 D3 0 59S 75 11W
Aguas Blancas, Chile 94 A2 24 15S 69 55W
Aguas Calientes, Sierra de, Argentina 94 B2 25 26S 66 40W
Aguascalientes, Mexico 86 C4 21 53N 102 12W
Aguascalientes □, Mexico 86 C4 22 0N 102 20W
Aguilares, Argentina 94 B2 27 26S 65 35W
Aguilas, Spain 19 D5 37 23N 1 35W
Agüimes, Canary Is. 22 G4 27 58N 15 27W
Aguja, C. de la, Colombia 90 B3 11 18N 74 12W
Agulhas, C., S. Africa 56 E3 34 52S 20 0 E
Agulo, Canary Is. 22 F2 28 11N 17 12W
Agung, Gunung, Indonesia 36 F5 8 20S 115 28 E
Agur, Uganda 54 B3 2 28N 32 55 E
Agusan →, Phil. 37 C7 9 0N 125 30 E
Aha Mts., Botswana 56 B3 19 45S 21 0 E
Ahaggar, Algeria 50 D7 23 0N 6 30 E
Ahar, Iran 44 B5 38 35N 47 0 E
Ahipara B., N.Z. 59 F4 35 5S 173 5 E
Ahiri, India 40 K12 19 30N 80 0 E
Ahmad Wal, Pakistan 42 E1 29 18N 65 58 E
Ahmadabad, India 42 H5 23 0N 72 40 E
Aḥmadābād, Khorāsān, Iran 45 C9 35 3N 60 50 E
Aḥmadābād, Khorāsān, Iran 45 C8 35 49N 59 42 E
Aḥmadī, Iran 45 E8 27 56N 56 42 E
Ahmadnagar, India 40 K9 19 7N 74 46 E
Ahmadpur, Pakistan 42 E4 29 12N 71 10 E
Ahmadpur Lamma, Pakistan 42 E4 28 19N 70 3 E
Ahmedabad = Ahmadabad, India 42 H5 23 0N 72 40 E
Ahmednagar = Ahmadnagar, India 40 K9 19 7N 74 46 E
Ahome, Mexico 86 B3 25 55N 109 11W
Ahoskie, U.S.A. 77 G7 36 17N 76 59W
Ahram, Iran 45 D6 28 52N 51 16 E
Ahrax Pt., Malta 23 D1 36 0N 14 22 E
Āhū, Iran 45 C6 34 33N 50 2 E
Ahuachapán, El Salv. 88 D2 13 54N 89 52W
Ahvāz, Iran 45 D6 31 20N 48 40 E
Ahvenanmaa = Åland, Finland 9 F19 60 15N 20 0 E
Aḥwar, Yemen 46 E4 13 30N 46 40 E
Ai →, India 43 F14 26 26N 90 44 E
Ai-Ais, Namibia 56 D2 27 54S 17 59 E
Aichi □, Japan 31 G8 35 0N 137 15 E
Aigua, Uruguay 95 C5 34 13S 54 46W
Aigues-Mortes, France 18 E6 43 35N 4 12 E
Aihui, China 33 A7 50 10N 127 30 E
Aija, Peru 92 E3 9 50S 77 45W
Aikawa, Japan 30 E9 38 2N 138 15 E
Aiken, U.S.A. 77 J5 33 34N 81 43W
Aileron, Australia 62 C1 22 39S 133 20 E
Aillik, Canada 71 A8 55 11N 59 18W
Ailsa Craig, U.K. 12 F3 55 15N 5 6W
'Ailūn, Jordan 47 C4 32 18N 35 47 E
Aim, Russia 27 D14 59 0N 133 55 E
Aimere, Indonesia 37 F6 8 45S 121 3 E
Aimogasta, Argentina 94 B2 28 33S 66 50W
Aïn Ben Tili, Mauritania 50 C4 25 59N 9 27W
Aïn Sefra, Algeria 50 B5 32 47N 0 37W
Ain Sudr, Egypt 47 F2 29 50N 33 6 E
Ainaži, Latvia 9 H21 57 50N 24 24 E
Ainsworth, U.S.A. 80 D5 42 33N 99 52W
Aiquile, Bolivia 92 G5 18 10S 65 10W
Aïr, Niger 50 E7 18 30N 8 0 E
Air Force I., Canada 69 B12 67 58N 74 5W
Air Hitam, Malaysia 39 M4 1 55N 103 11 E
Airdrie, Canada 72 C6 51 18N 114 2W
Airdrie, U.K. 12 F5 55 52N 3 57W
Aire →, U.K. 10 D7 53 43N 0 55W
Aire, I. de l', Spain 22 B11 39 48N 4 16 E
Airlie Beach, Australia 62 C4 20 16S 148 43 E
Aisne →, France 18 B5 49 26N 2 50 E
Ait, India 43 G8 25 54N 79 14 E
Aitkin, U.S.A. 80 B8 46 32N 93 42W
Aiud, Romania 17 E12 46 19N 23 44 E
Aix-en-Provence, France 18 E6 43 32N 5 27 E
Aix-la-Chapelle = Aachen, Germany 16 C4 50 45N 6 6 E
Aix-les-Bains, France 18 D6 45 41N 5 53 E
Aíyion, Greece 21 E10 38 15N 22 5 E
Aizawl, India 41 H18 23 40N 92 44 E
Aizkraukle, Latvia 9 H21 56 36N 25 11 E
Aizpute, Latvia 9 H19 56 43N 21 40 E
Aizuwakamatsu, Japan 30 F9 37 30N 139 56 E
Ajaccio, France 18 F8 41 55N 8 40 E
Ajaigarh, India 43 G9 24 52N 80 16 E
Ajalpan, Mexico 87 D5 18 22N 97 15W
Ajanta Ra., India 40 J9 20 28N 75 50 E
Ajari Rep. = Ajaria □, Georgia 25 F7 41 30N 42 0 E
Ajaria □, Georgia 25 F7 41 30N 42 0 E
Ajax, Canada 78 C5 43 50N 79 1W
Ajdābiyā, Libya 51 B10 30 54N 20 4 E
Ajka, Hungary 17 E9 47 4N 17 31 E
'Ajmān, U.A.E. 45 E7 25 25N 55 30 E
Ajmer, India 42 F6 26 28N 74 37 E
Ajnala, India 42 D6 31 50N 74 48 E
Ajo, U.S.A. 83 K7 32 22N 112 52W
Ajo, C. de, Spain 19 A4 43 31N 3 35W
Akabira, Japan 30 C11 43 33N 142 5 E
Akamas, Cyprus 23 D11 35 3N 32 18 E
Akanthou, Cyprus 23 D12 35 22N 33 45 E
Akaroa, N.Z. 59 K4 43 49S 172 59 E
Akashi, Japan 31 G7 34 45N 134 58 E
Akbarpur, Bihar, India 43 G10 24 39N 83 58 E
Akbarpur, Ut. P., India 43 F10 26 25N 82 32 E
Akelamo, Indonesia 37 D7 1 35N 129 40 E
Aketi, Dem. Rep. of the Congo 52 D4 2 38N 23 47 E
Akharnaí, Greece 21 E10 38 5N 23 44 E
Akhelóös →, Greece 21 E9 38 19N 21 7 E
Akhisar, Turkey 21 E12 38 56N 27 48 E
Akhnur, India 43 C6 32 52N 74 45 E
Akhtyrka = Okhtyrka, Ukraine 25 D5 50 25N 35 0 E
Aki, Japan 31 H6 33 30N 133 54 E
Akimiski I., Canada 70 B3 52 50N 81 30W
Akita, Japan 30 E10 39 45N 140 7 E
Akita □, Japan 30 E10 39 40N 140 30 E
Akjoujt, Mauritania 50 E3 19 45N 14 15W
Akkeshi, Japan 30 C12 43 2N 144 51 E
'Akko, Israel 47 C4 32 55N 35 4 E
Aklavik, Canada 68 B6 68 12N 135 0W
Aklera, India 42 G7 24 26N 76 32 E
Akmolinsk = Astana, Kazakstan 26 D8 51 10N 71 30 E
Akō, Japan 31 G7 34 45N 134 24 E
Akola, India 40 J10 20 42N 77 2 E
Akordat, Eritrea 46 D2 15 30N 37 40 E
Akpatok I., Canada 69 B13 60 25N 68 8W
Åkrahamn, Norway 9 G11 59 15N 5 10 E
Akranes, Iceland 8 D2 64 19N 22 5W
Akron, Colo., U.S.A. 80 E3 40 10N 103 13W
Akron, Ohio, U.S.A. 78 E3 41 5N 81 31W
Akrotiri, Cyprus 23 E11 34 36N 32 57 E
Akrotiri Bay, Cyprus 23 E12 34 35N 33 10 E
Aksai Chin, China 43 B8 35 15N 79 55 E
Aksaray, Turkey 25 G5 38 25N 34 2 E
Aksay, Kazakstan 25 D9 51 11N 53 0 E
Akşehir, Turkey 44 B1 38 18N 31 30 E
Akşehir Gölü, Turkey 25 G5 38 30N 31 25 E
Aksu, China 32 B3 41 5N 80 10 E
Aksum, Ethiopia 46 E2 14 5N 38 40 E
Aktogay, Kazakstan 26 E8 46 57N 79 40 E
Aktsyabrski, Belarus 17 B15 52 38N 28 53 E
Aktyubinsk = Aqtöbe, Kazakstan 25 D10 50 17N 57 10 E
Akure, Nigeria 50 G7 7 15N 5 5 E
Akureyri, Iceland 8 D4 65 40N 18 6W
Akuseki-Shima, Japan 31 K4 29 27N 129 37 E
Akyab = Sittwe, Burma 41 J18 20 18N 92 45 E
Al 'Adan, Yemen 46 E4 12 45N 45 0 E
Al Aḥsā = Hasa □, Si. Arabia 45 E6 25 50N 49 0 E
Al Ajfar, Si. Arabia 44 E4 27 26N 43 0 E
Al Amādīyah, Iraq 44 B4 37 5N 43 30 E
Al 'Amārah, Iraq 44 D5 31 55N 47 15 E
Al 'Aqabah, Jordan 47 F4 29 31N 35 0 E
Al Arak, Syria 44 C3 34 38N 38 35 E
Al 'Aramah, Si. Arabia 44 E5 25 30N 46 0 E
Al Arṭāwīyah, Si. Arabia 44 E5 26 31N 45 20 E
Al 'Āṣimah = 'Ammān □, Jordan 47 D5 31 40N 36 30 E
Al 'Assāfīyah, Si. Arabia 44 D3 28 17N 38 59 E
Al 'Ayn, Oman 45 E7 24 15N 55 45 E
Al 'Ayn, Si. Arabia 44 E3 25 4N 38 6 E
Al 'Aẓamīyah, Iraq 44 C5 33 22N 44 22 E
Al 'Azīzīyah, Iraq 44 C5 32 54N 45 4 E
Al Bāb, Syria 44 B3 36 23N 37 29 E
Al Bad', Si. Arabia 44 D2 28 28N 35 1 E
Al Bādī, Iraq 44 C4 35 56N 41 32 E
Al Baḥrah, Kuwait 44 D5 29 40N 47 52 E
Al Baḥral Mayyit = Dead Sea, Asia 47 D4 31 30N 35 30 E
Al Balqā' □, Jordan 47 C4 32 5N 35 45 E
Al Bārūk, J., Lebanon 47 B4 33 39N 35 40 E
Al Başrah, Iraq 44 D5 30 30N 47 50 E
Al Baṭḥā, Iraq 44 D5 31 6N 45 53 E
Al Batrūn, Lebanon 47 A4 34 15N 35 40 E
Al Bayḍā, Libya 51 B10 32 50N 21 44 E
Al Biqā, Lebanon 47 A5 34 10N 36 10 E
Al Bi'r, Si. Arabia 44 D3 28 51N 36 16 E
Al Burayj, Syria 47 A5 34 15N 36 46 E
Al Faḍilī, Si. Arabia 45 E6 26 58N 49 10 E
Al Fallūjah, Iraq 44 C4 33 20N 43 55 E
Al Fāw, Iraq 45 D6 30 0N 48 30 E
Al Fujayrah, U.A.E. 45 E8 25 7N 56 18 E
Al Ghadaf, W. →, Jordan 47 D5 31 26N 36 43 E
Al Ghammās, Iraq 44 D5 31 45N 44 37 E

Amakusa-Shotō

Al Ghazālah, *Si. Arabia* **44 E4** 26 48N 41 19 E
Al Ḥadīthah, *Iraq* **44 C4** 34 0N 41 13 E
Al Ḥadīthah, *Si. Arabia* **47 D6** 31 28N 37 8 E
Al Ḥaḍr, *Iraq* **44 C4** 35 35N 42 44 E
Al Hājānah, *Syria* **47 B5** 33 20N 36 33 E
Al Ḥajar al Gharbī, *Oman* **45 E8** 24 10N 56 15 E
Al Hāmad, *Si. Arabia* **44 D3** 31 30N 39 30 E
Al Hamdāniyah, *Syria* **44 C3** 35 25N 36 50 E
Al Ḥamīdīyah, *Syria* **47 A4** 34 42N 35 57 E
Al Ḥammār, *Iraq* **44 D5** 30 57N 46 51 E
Al Ḥamrā', *Si. Arabia* **44 E3** 24 2N 38 55 E
Al Ḥanākīyah, *Si. Arabia* **44 E4** 24 51N 40 31 E
Al Harir, W. →, *Syria* **47 C4** 32 44N 35 59 E
Al Ḥasā, W. →, *Jordan* **47 D4** 31 4N 35 29 E
Al Ḥasakah, *Syria* **44 B4** 36 35N 40 45 E
Al Haydān, W. →, *Jordan* **47 D4** 31 29N 35 34 E
Al Ḥayy, *Iraq* **44 C5** 32 5N 46 5 E
Al Ḥijarah, *Asia* **44 D4** 30 0N 44 0 E
Al Ḥillah, *Iraq* **44 C5** 32 30N 44 25 E
Al Ḥillah, *Si. Arabia* **46 B4** 23 35N 46 50 E
Al Hindīyah, *Iraq* **44 C5** 32 30N 44 10 E
Al Hirmil, *Lebanon* **47 A5** 34 26N 36 24 E
Al Hoceïma, *Morocco* **50 A5** 35 8N 3 58W
Al Ḥudaydah, *Yemen* **46 E3** 14 50N 43 0 E
Al Hufūf, *Si. Arabia* **46 B4** 25 25N 49 45 E
Al Ḥumayḍah, *Si. Arabia* **44 D2** 29 14N 34 56 E
Al Ḥunayy, *Si. Arabia* **45 E6** 25 58N 48 45 E
Al Īsāwīyah, *Si. Arabia* **44 D3** 30 43N 37 59 E
Al Jafr, *Jordan* **47 E5** 30 18N 36 14 E
Al Jāfūrah, *Si. Arabia* **45 E7** 25 0N 50 15 E
Al Jaghbūb, *Libya* **51 C10** 29 42N 24 38 E
Al Jahrah, *Kuwait* **44 D5** 29 25N 47 40 E
Al Jalāmīd, *Si. Arabia* **44 D3** 31 20N 40 6 E
Al Jamalīyah, *Qatar* **45 E6** 25 37N 51 5 E
Al Janūb □, *Lebanon* **47 B4** 33 20N 35 20 E
Al Jawf, *Libya* **51 D10** 24 10N 23 24 E
Al Jawf, *Si. Arabia* **44 D3** 29 55N 39 40 E
Al Jazirah, *Iraq* **44 C5** 33 30N 44 0 E
Al Jithāmīyah, *Si. Arabia* **44 E4** 27 41N 41 43 E
Al Jubayl, *Si. Arabia* **45 E6** 27 0N 49 50 E
Al Jubaylah, *Si. Arabia* **44 E5** 24 55N 46 25 E
Al Jubb, *Si. Arabia* **44 E4** 27 11N 42 17 E
Al Junaynah, *Sudan* **51 F10** 13 27N 22 45 E
Al Kabā'ish, *Iraq* **44 D5** 30 58N 47 0 E
Al Karak, *Jordan* **47 D4** 31 11N 35 42 E
Al Karak □, *Jordan* **47 E5** 31 0N 36 0 E
Al Kāzim Tyah, *Iraq* **44 C5** 33 22N 44 12 E
Al Khābūra, *Oman* **45 F8** 23 57N 57 5 E
Al Khafji, *Si. Arabia* **45 E6** 28 24N 48 29 E
Al Khalīl, *West Bank* **47 D4** 31 32N 35 6 E
Al Khāliş, *Iraq* **44 C5** 33 49N 44 32 E
Al Kharsānīyah, *Si. Arabia* **45 E6** 27 13N 49 18 E
Al Khaşab, *Oman* **45 E8** 26 14N 56 15 E
Al Khawr, *Qatar* **45 E6** 25 41N 51 30 E
Al Khiḍr, *Iraq* **44 D5** 31 12N 45 33 E
Al Khiyām, *Lebanon* **47 B4** 33 20N 35 36 E
Al Kiswah, *Syria* **47 B5** 33 23N 36 14 E
Al Kūfah, *Iraq* **44 C5** 32 2N 44 24 E
Al Kufrah, *Libya* **51 D10** 24 17N 23 15 E
Al Kuhayfiyah, *Si. Arabia* **44 E4** 27 12N 43 3 E
Al Kūt, *Iraq* **44 C5** 32 30N 46 0 E
Al Kuwayt, *Kuwait* **46 B4** 29 30N 48 0 E
Al Labwah, *Lebanon* **47 A5** 34 11N 36 20 E
Al Lādhiqīyah, *Syria* **44 C2** 35 30N 35 45 E
Al Līth, *Si. Arabia* **46 C3** 20 9N 40 15 E
Al Liwā', *Oman* **45 E8** 24 31N 56 36 E
Al Luḥayyah, *Yemen* **46 D3** 15 45N 42 40 E
Al Madīnah, *Iraq* **44 D5** 30 57N 47 16 E
Al Madīnah, *Si. Arabia* **46 C2** 24 35N 39 52 E
Al Mafraq, *Jordan* **47 C5** 32 17N 36 14 E
Al Maḥmūdīyah, *Iraq* **44 C5** 33 3N 44 21 E
Al Majma'ah, *Si. Arabia* **44 E5** 25 57N 45 22 E
Al Makhruq, W. →, *Jordan* **47 D6** 31 28N 37 0 E
Al Makḥūl, *Si. Arabia* **44 E4** 26 37N 42 39 E
Al Manāmah, *Bahrain* **46 B5** 26 10N 50 30 E
Al Maqwa', *Kuwait* **44 D5** 29 10N 47 59 E
Al Marj, *Libya* **51 B10** 32 25N 20 30 E
Al Maṭlā, *Kuwait* **44 D5** 29 24N 47 40 E
Al Mawjib, W. →, *Jordan* **47 D4** 31 28N 35 36 E
Al Mawşil, *Iraq* **44 B4** 36 15N 43 5 E
Al Mayādin, *Syria* **44 C4** 35 1N 40 27 E
Al Mazār, *Jordan* **47 D4** 31 4N 35 41 E
Al Midhnab, *Si. Arabia* **44 E5** 25 50N 44 18 E
Al Minā', *Lebanon* **47 A4** 34 24N 35 49 E
Al Miqdādīyah, *Iraq* **44 C5** 34 0N 45 0 E
Al Mubarraz, *Si. Arabia* **45 E6** 25 30N 49 40 E
Al Mudawwarah, *Jordan* **47 F5** 29 19N 36 0 E
Al Mughayrā', *U.A.E.* **45 E7** 24 5N 53 32 E
Al Muḥarraq, *Bahrain* **45 E6** 26 15N 50 40 E
Al Mukallā, *Yemen* **46 E4** 14 33N 49 2 E
Al Mukhā, *Yemen* **46 E3** 13 18N 43 15 E
Al Musayjīd, *Si. Arabia* **44 E3** 24 5N 39 5 E
Al Musayyib, *Iraq* **44 C5** 32 49N 44 20 E
Al Muwayliḥ, *Si. Arabia* **44 E2** 27 40N 35 30 E
Al Qā'im, *Iraq* **44 C4** 34 21N 41 7 E
Al Qalībah, *Si. Arabia* **44 D3** 28 24N 37 42 E
Al Qāmishlī, *Syria* **44 B4** 37 2N 41 14 E
Al Qaryatayn, *Syria* **47 A6** 34 12N 37 13 E
Al Qaşīm, *Si. Arabia* **44 E4** 26 0N 43 0 E
Al Qaţ'ā, *Syria* **44 C4** 34 40N 40 48 E
Al Qaţīf, *Si. Arabia* **45 E6** 26 35N 50 0 E
Al Qatrānah, *Jordan* **47 D5** 31 12N 36 6 E
Al Qaṭrūn, *Libya* **51 D9** 24 56N 15 3 E
Al Qayşūmah, *Si. Arabia* **44 D5** 28 20N 46 7 E
Al Quds = Jerusalem, *Israel* **47 D4** 31 47N 35 10 E
Al Qunayṭirah, *Syria* **47 C4** 32 55N 35 45 E
Al Qurnah, *Iraq* **44 D5** 31 1N 47 25 E
Al Quşayr, *Iraq* **44 D5** 30 39N 45 50 E
Al Quşayr, *Syria* **47 A5** 34 31N 36 34 E
Al Quṭayfah, *Syria* **47 B5** 33 44N 36 36 E
Al 'Ubaylah, *Si. Arabia* **46 C5** 21 59N 50 57 E
Al 'Uḍaylīyah, *Si. Arabia* **45 E6** 25 8N 49 18 E
Al 'Ulā, *Si. Arabia* **44 E3** 26 35N 38 0 E
Al 'Uqayr, *Si. Arabia* **45 E6** 25 40N 50 15 E
Al 'Uwaynid, *Si. Arabia* **44 E5** 24 50N 46 0 E
Al 'Uwayqīlah, *Si. Arabia* **44 D4** 30 30N 42 10 E
Al 'Uyūn, *Ḥijāz, Si. Arabia* **44 E3** 24 33N 39 35 E
Al 'Uyūn, *Najd, Si. Arabia* **44 E4** 26 30N 43 50 E
Al 'Uzayr, *Iraq* **44 D5** 31 19N 47 25 E
Al Wajh, *Si. Arabia* **44 E3** 26 10N 36 30 E
Al Wakrah, *Qatar* **45 E6** 25 10N 51 40 E
Al Waqbah, *Si. Arabia* **44 D5** 28 48N 45 33 E
Al Wari'āh, *Si. Arabia* **44 E5** 27 51N 47 25 E
Ala Dağ, *Turkey* **44 B2** 37 44N 35 9 E
Ala Tau Shankou = Dzungarian Gates, *Asia* **32 B3** 45 0N 82 0 E
Alabama □, *U.S.A.* **77 J2** 33 0N 87 0W
Alabama →, *U.S.A.* **77 K2** 31 8N 87 57W
Alabaster, *U.S.A.* **77 J2** 33 15N 86 49W
Alaçam Dağları, *Turkey* **21 E13** 39 18N 28 49 E
Alachua, *U.S.A.* **77 L4** 29 47N 82 30W
Alaérma, *Greece* **23 C9** 36 9N 27 57 E
Alagoa Grande, *Brazil* **93 E11** 7 3S 35 35W
Alagoas □, *Brazil* **93 E11** 9 0S 36 0W
Alagoinhas, *Brazil* **93 F11** 12 7S 38 20W
Alaior, *Spain* **22 B11** 39 57N 4 8 E
Alajero, *Canary Is.* **22 F2** 28 3N 17 13W
Alajuela, *Costa Rica* **88 D3** 10 2N 84 8W
Alakamisy, *Madag.* **57 C8** 21 19S 47 14 E
Alaknanda →, *India* **43 D8** 30 8N 78 36 E
Alakurtti, *Russia* **24 A5** 67 0N 30 30 E
Alamarvdasht, *Iran* **45 E7** 27 37N 52 59 E
Alameda, *Calif., U.S.A.* **84 H4** 37 46N 122 15W
Alameda, *N. Mex., U.S.A.* **83 J10** 35 11N 106 37W
Alamo, *U.S.A.* **85 J11** 37 22N 115 10W
Alamo Crossing, *U.S.A.* **85 L13** 34 16N 113 33W
Alamogordo, *U.S.A.* **83 K11** 32 54N 105 57W
Alamos, *Mexico* **86 B3** 27 0N 109 0W
Alamosa, *U.S.A.* **83 H11** 37 28N 105 52W
Åland, *Finland* **9 F19** 60 15N 20 0 E
Ålands hav, *Sweden* **9 F18** 60 0N 19 30 E
Alania = North Ossetia □, *Russia* **25 F7** 43 30N 44 30 E
Alanya, *Turkey* **25 G5** 36 38N 32 0 E
Alaotra, Farihin', *Madag.* **57 B8** 17 30S 48 30 E
Alapayevsk, *Russia* **26 D7** 57 52N 61 42 E
Alappuzha = Alleppey, *India* **40 Q10** 9 30N 76 28 E
Alarobia-Vohiposa, *Madag.* **57 C8** 20 59S 47 9 E
Alaşehir, *Turkey* **21 E13** 38 23N 28 30 E
Alaska □, *U.S.A.* **68 B5** 64 0N 154 0W
Alaska, G. of, *Pac. Oc.* **68 C5** 58 0N 145 0W
Alaska Peninsula, *U.S.A.* **68 C4** 56 0N 159 0W
Alaska Range, *U.S.A.* **68 B4** 62 50N 151 0W
Älät, *Azerbaijan* **25 G8** 39 58N 49 25 E
Alatyr, *Russia* **24 D8** 54 55N 46 35 E
Alausi, *Ecuador* **92 D3** 2 0S 78 50W
Alava, C., *U.S.A.* **82 B1** 48 10N 124 44W
Alavus, *Finland* **9 E20** 62 35N 23 36 E
Alawoona, *Australia* **63 E3** 34 45S 140 30 E
'Alayh, *Lebanon* **47 B4** 33 46N 35 33 E
Alba, *Italy* **18 D8** 44 42N 8 2 E
Alba-Iulia, *Romania* **17 E12** 46 8N 23 39 E
Albacete, *Spain* **19 C5** 39 0N 1 50W
Albacutya, L., *Australia* **63 F3** 35 45S 141 58 E
Albanel, L., *Canada* **70 B5** 50 55N 73 12W
Albania ■, *Europe* **21 D9** 41 0N 20 0 E
Albany, *Australia* **61 G2** 35 1S 117 58 E
Albany, *Ga., U.S.A.* **77 K3** 31 35N 84 10W
Albany, *N.Y., U.S.A.* **79 D11** 42 39N 73 45W
Albany, *Oreg., U.S.A.* **82 D2** 44 38N 123 6W
Albany, *Tex., U.S.A.* **81 J5** 32 44N 99 18W
Albany →, *Canada* **70 B3** 52 17N 81 31W
Albardón, *Argentina* **94 C2** 31 20S 68 30W
Albatross B., *Australia* **62 A3** 12 45S 141 30 E
Albemarle, *U.S.A.* **77 H5** 35 21N 80 11W
Albemarle Sd., *U.S.A.* **77 H7** 36 5N 76 0W
Alberche →, *Spain* **19 C3** 39 58N 4 46W
Alberdi, *Paraguay* **94 B4** 26 14S 58 20W
Albert, L., *Africa* **54 B3** 1 30N 31 0 E
Albert, L., *Australia* **63 F2** 35 30S 139 10 E
Albert Edward Ra., *Australia* **60 C4** 18 17S 127 57 E
Albert Lea, *U.S.A.* **80 D8** 43 39N 93 22W
Albert Nile →, *Uganda* **54 B3** 3 36N 32 2 E
Albert Town, *Bahamas* **89 B5** 22 37N 74 33W
Alberta □, *Canada* **72 C6** 54 40N 115 0W
Alberti, *Argentina* **94 D3** 35 1S 60 16W
Albertinia, *S. Africa* **56 E3** 34 11S 21 34 E
Alberton, *Canada* **71 C7** 46 50N 64 0W
Albertville = Kalemie, *Dem. Rep. of the Congo* **54 D2** 5 55S 29 9 E
Albertville, *France* **18 D7** 45 40N 6 22 E
Albertville, *U.S.A.* **77 H2** 34 16N 86 13W
Albi, *France* **18 E5** 43 56N 2 9 E
Albia, *U.S.A.* **80 E8** 41 2N 92 48W
Albina, *Surinam* **93 B8** 5 37N 54 15W
Albina, Ponta, *Angola* **56 B1** 15 52S 11 44 E
Albion, *Mich., U.S.A.* **76 D3** 42 15N 84 45W
Albion, *Nebr., U.S.A.* **80 E6** 41 42N 98 0W
Albion, *Pa., U.S.A.* **78 E4** 41 53N 80 22W
Alborán, *Medit. S.* **19 E4** 35 57N 3 0W
Ålborg, *Denmark* **9 H13** 57 2N 9 54 E
Alborz, Reshteh-ye Kūhhā-ye, *Iran* **45 C7** 36 0N 52 0 E
Albuquerque, *U.S.A.* **83 J10** 35 5N 106 39W
Albuquerque, Cayos de, *Caribbean* **88 D3** 12 10N 81 50W
Alburg, *U.S.A.* **79 B11** 44 59N 73 18W
Albury = Albury-Wodonga, *Australia* **63 F4** 36 3S 146 56 E
Albury-Wodonga, *Australia* **63 F4** 36 3S 146 56 E
Alcalá de Henares, *Spain* **19 B4** 40 28N 3 22W
Alcalá la Real, *Spain* **19 D4** 37 27N 3 57W
Álcamo, *Italy* **20 F5** 37 59N 12 55 E
Alcaniz, *Spain* **19 B5** 41 2N 0 8W
Alcântara, *Brazil* **93 D10** 2 20S 44 30W
Alcántara, Embalse de, *Spain* **19 C2** 39 44N 6 50W
Alcantarilla, *Spain* **19 D5** 37 59N 1 12W
Alcaraz, Sierra de, *Spain* **19 C4** 38 40N 2 20W
Alcaudete, *Spain* **19 D3** 37 35N 4 5W
Alcázar de San Juan, *Spain* **19 C4** 39 24N 3 12W
Alchevsk, *Ukraine* **25 E6** 48 30N 38 45 E
Alcira = Alzira, *Spain* **19 C5** 39 9N 0 30W
Alcova, *U.S.A.* **82 E10** 42 34N 106 43W
Alcoy, *Spain* **19 C5** 38 43N 0 30W
Alcúdia, *Spain* **22 B10** 39 51N 3 7 E
Alcúdia, B. d', *Spain* **22 B10** 39 47N 3 15 E
Aldabra Is., *Seychelles* **49 G8** 9 22S 46 28 E
Aldama, *Mexico* **87 C5** 23 0N 98 4W
Aldan, *Russia* **27 D13** 58 40N 125 30 E
Aldan →, *Russia* **27 C13** 63 28N 129 35 E
Aldea, Pta. de la, *Canary Is.* **22 G4** 28 0N 15 50W
Aldeburgh, *U.K.* **11 E9** 52 10N 1 37 E
Alder Pk., *U.S.A.* **84 K5** 35 53N 121 22W
Alderney, *U.K.* **11 H5** 49 42N 2 11W
Aldershot, *U.K.* **11 F7** 51 15N 0 44W
Aledo, *U.S.A.* **80 E9** 41 12N 90 45W
Aleg, *Mauritania* **50 E3** 17 3N 13 55W
Alegranza, *Canary Is.* **22 E6** 29 23N 13 32W
Alegranza, I., *Canary Is.* **22 E6** 29 23N 13 32W
Alegre, *Brazil* **95 A7** 20 50S 41 30W
Alegrete, *Brazil* **95 B4** 29 40S 56 0W
Aleisk, *Russia* **26 D9** 52 40N 83 0 E
Aleksandriya = Oleksandriya, *Ukraine* **17 C14** 50 37N 26 19 E
Aleksandrov Gay, *Russia* **25 D8** 50 9N 48 34 E
Aleksandrovsk-Sakhalinskiy, *Russia* **27 D15** 50 50N 142 20 E
Além Paraíba, *Brazil* **95 A7** 21 52S 42 41W
Alemania, *Argentina* **94 B2** 25 40S 65 30W
Alemania, *Chile* **94 B2** 25 10S 69 55W
Alençon, *France* **18 B4** 48 27N 0 4 E
Alenquer, *Brazil* **93 D8** 1 56S 54 46W
Alenuihaha Channel, *U.S.A.* **74 H17** 20 30N 156 0W
Aleppo = Ḥalab, *Syria* **44 B3** 36 10N 37 15 E
Alès, *France* **18 D6** 44 9N 4 5 E
Alessándria, *Italy* **18 D8** 44 54N 8 37 E
Ålesund, *Norway* **9 E12** 62 28N 6 12 E
Aleutian Is., *Pac. Oc.* **68 C2** 52 0N 175 0W
Aleutian Trench, *Pac. Oc.* **64 C10** 48 0N 180 0 E
Alexander, *U.S.A.* **80 B3** 47 51N 103 39W
Alexander, Mt., *Australia* **61 E3** 28 58S 120 16 E
Alexander Arch., *U.S.A.* **68 C6** 56 0N 136 0W
Alexander Bay, *S. Africa* **56 D2** 28 40S 16 30 E
Alexander City, *U.S.A.* **77 J3** 32 56N 85 58W
Alexander I., *Antarctica* **5 C17** 69 0S 70 0W
Alexandra, *Australia* **63 F4** 37 8S 145 40 E
Alexandra, *N.Z.* **59 L2** 45 14S 169 25 E
Alexandra Falls, *Canada* **72 A5** 60 29N 116 18W
Alexandria = El Iskandarîya, *Egypt* **51 B11** 31 13N 29 58 E
Alexandria, *B.C., Canada* **72 C4** 52 35N 122 27W
Alexandria, *Ont., Canada* **79 A10** 45 19N 74 38W
Alexandria, *Romania* **17 G13** 43 57N 25 24 E
Alexandria, *S. Africa* **56 E4** 33 38S 26 28 E
Alexandria, *U.K.* **12 F4** 55 59N 4 35W
Alexandria, *La., U.S.A.* **81 K8** 31 18N 92 27W
Alexandria, *Minn., U.S.A.* **80 C7** 45 53N 95 22W
Alexandria, *S. Dak., U.S.A.* **80 D6** 43 39N 97 47W
Alexandria, *Va., U.S.A.* **76 F7** 38 48N 77 3W
Alexandria Bay, *U.S.A.* **79 B9** 44 20N 75 55W
Alexandrina, L., *Australia* **63 F2** 35 25S 139 10 E
Alexandroúpolis, *Greece* **21 D11** 40 50N 25 54 E
Alexis →, *Canada* **71 B8** 52 33N 56 8W
Alexis Creek, *Canada* **72 C4** 52 10N 123 20W
Alfabia, *Spain* **22 B9** 39 44N 2 44 E
Alfenas, *Brazil* **95 A6** 21 20S 46 10W
Alford, *Aberds., U.K.* **12 D6** 57 14N 2 41W
Alford, *Lincs., U.K.* **10 D8** 53 15N 0 10 E
Alfred, *Maine, U.S.A.* **79 C14** 43 29N 70 43W
Alfred, *N.Y., U.S.A.* **78 D7** 42 16N 77 48W
Alfreton, *U.K.* **10 D6** 53 6N 1 24W
Alga, *Kazakstan* **25 E10** 49 53N 57 20 E
Algaida, *Spain* **22 B9** 39 33N 2 53 E
Ålgård, *Norway* **9 G11** 58 46N 5 53 E
Algarve, *Portugal* **19 D1** 36 58N 8 20W
Algeciras, *Spain* **19 D3** 36 9N 5 28W
Algemesi, *Spain* **19 C5** 39 11N 0 27W
Alger, *Algeria* **50 A6** 36 42N 3 8 E
Algeria ■, *Africa* **50 C6** 28 30N 2 0 E
Alghero, *Italy* **20 D3** 40 33N 8 19 E
Algiers = Alger, *Algeria* **50 A6** 36 42N 3 8 E
Algoa B., *S. Africa* **56 E4** 33 50S 25 45 E
Algoma, *U.S.A.* **76 C2** 44 36N 87 26W
Algona, *U.S.A.* **80 D7** 43 4N 94 14W
Algonac, *U.S.A.* **78 D2** 42 37N 82 32W
Algonquin Prov. Park, *Canada* **70 C4** 45 50N 78 30W
Algorta, *Uruguay* **96 C5** 32 25S 57 23W
Alhambra, *U.S.A.* **85 L8** 34 8N 118 6W
Alhucemas = Al Hoceïma, *Morocco* **50 A5** 35 8N 3 58W
'Alī al Gharbī, *Iraq* **44 C5** 32 30N 46 45 E
'Alī ash Sharqī, *Iraq* **44 C5** 32 7N 46 44 E
'Alī Khēl, *Afghan.* **42 C3** 33 57N 69 43 E
Alī Shāh, *Iran* **44 B5** 38 9N 45 50 E
'Alīābād, *Khorāsān, Iran* **45 C8** 32 30N 57 30 E
'Alīābād, *Kordestān, Iran* **44 C5** 35 4N 46 58 E
'Alīābād, *Yazd, Iran* **45 D7** 31 41N 53 49 E
Aliağa, *Turkey* **21 E12** 38 47N 26 59 E
Aliákmon →, *Greece* **21 D10** 40 30N 22 36 E
Alicante, *Spain* **19 C5** 38 23N 0 30W
Alice, *S. Africa* **56 E4** 32 48S 26 55 E
Alice, *U.S.A.* **81 M5** 27 45N 98 5W
Alice →, *Queens., Australia* **62 C3** 24 2S 144 50 E
Alice →, *Queens., Australia* **62 B3** 15 35S 142 20 E
Alice Arm, *Canada* **72 B3** 55 29N 129 31W
Alice Springs, *Australia* **62 C1** 23 40S 133 50 E
Alicedale, *S. Africa* **56 E4** 33 15S 26 4 E
Aliceville, *U.S.A.* **77 J1** 33 8N 88 9W
Aliganj, *India* **43 F8** 27 30N 79 10 E
Aligarh, *Raj., India* **42 G7** 25 55N 76 15 E
Aligarh, *Ut. P., India* **42 F8** 27 55N 78 10 E
Alīgūdarz, *Iran* **45 C6** 33 25N 49 45 E
Alimnía, *Greece* **23 C9** 36 16N 27 43 E
Alingsås, *Sweden* **9 H15** 57 56N 12 31 E
Alipur, *Pakistan* **42 E4** 29 25N 70 55 E
Alipur Duar, *India* **41 F16** 26 30N 89 35 E
Aliquippa, *U.S.A.* **78 F4** 40 37N 80 15W
Alitus = Alytus, *Lithuania* **9 J21** 54 24N 24 3 E
Aliwal North, *S. Africa* **56 E4** 30 45S 26 45 E
Alix, *Canada* **72 C6** 52 24N 113 11W
Aljustrel, *Portugal* **19 D1** 37 55N 8 10W
Alkmaar, *Neths.* **15 B4** 52 37N 4 45 E
All American Canal, *U.S.A.* **83 K6** 32 45N 115 15W
Allagash →, *U.S.A.* **77 B11** 47 5N 69 3W
Allah Dad, *Pakistan* **42 G2** 25 38N 67 34 E
Allahabad, *India* **43 G9** 25 25N 81 58 E
Allan, *Canada* **73 C7** 51 53N 106 4W
Allanridge, *S. Africa* **56 D4** 27 45S 26 40 E
Allegany, *U.S.A.* **78 D6** 42 6N 78 30W
Allegheny →, *U.S.A.* **78 F5** 40 27N 80 1W
Allegheny Mts., *U.S.A.* **76 G6** 38 15N 80 10W
Allegheny Reservoir, *U.S.A.* **78 E6** 41 50N 79 0W
Allen, Bog of, *Ireland* **13 C5** 53 15N 7 0W
Allen, L., *Ireland* **13 B3** 54 8N 8 4W
Allendale, *U.S.A.* **77 J5** 33 1N 81 18W
Allende, *Mexico* **86 B4** 28 20N 100 50W
Allentown, *U.S.A.* **79 F9** 40 37N 75 29W
Alleppey, *India* **40 Q10** 9 30N 76 28 E
Aller →, *Germany* **16 B5** 52 56N 9 12 E
Alliance, *Nebr., U.S.A.* **80 D3** 42 6N 102 52W
Alliance, *Ohio, U.S.A.* **78 F3** 40 55N 81 6W
Allier →, *France* **18 C5** 46 57N 3 4 E
Alliford Bay, *Canada* **72 C2** 53 12N 131 58W
Alliston, *Canada* **78 B5** 44 9N 79 52W
Alloa, *U.K.* **12 E5** 56 7N 3 47W
Allora, *Australia* **63 D5** 28 2S 152 0 E
Alluitsup Paa, *Greenland* **4 C5** 60 30N 45 35W
Alma, *Canada* **71 C5** 48 35N 71 40W
Alma, *Ga., U.S.A.* **77 K4** 31 33N 82 28W
Alma, *Kans., U.S.A.* **80 F6** 39 1N 96 17W
Alma, *Mich., U.S.A.* **76 D3** 43 23N 84 39W
Alma, *Nebr., U.S.A.* **80 E5** 40 6N 99 22W
Alma Ata = Almaty, *Kazakstan* **26 E8** 43 15N 76 57 E
Almada, *Portugal* **19 C1** 38 40N 9 9W
Almaden, *Australia* **62 B3** 17 22S 144 40 E
Almadén, *Spain* **19 C3** 38 49N 4 52W
Almanor, L., *U.S.A.* **82 F3** 40 14N 121 9W
Almansa, *Spain* **19 C5** 38 51N 1 5W
Almanzor, Pico, *Spain* **19 B3** 40 15N 5 18W
Almanzora →, *Spain* **19 D5** 37 14N 1 46W
Almaty, *Kazakstan* **26 E8** 43 15N 76 57 E
Almazán, *Spain* **19 B4** 41 30N 2 30W
Almeirim, *Brazil* **93 D8** 1 30S 52 34W
Almelo, *Neths.* **15 B6** 52 22N 6 42 E
Almendralejo, *Spain* **19 C2** 38 41N 6 26W
Almere-Stad, *Neths.* **15 B5** 52 20N 5 15 E
Almería, *Spain* **19 D4** 36 52N 2 27W
Almirante, *Panama* **88 E3** 9 10N 82 30W
Almiroú, Kólpos, *Greece* **23 D6** 35 23N 24 20 E
Almond, *U.S.A.* **78 D7** 42 19N 77 44W
Almont, *U.S.A.* **78 D1** 42 55N 83 3W
Almonte, *Canada* **79 A8** 45 14N 76 12W
Almora, *India* **43 E8** 29 38N 79 40 E
Alness, *U.K.* **12 D4** 57 41N 4 16W
Alnmouth, *U.K.* **10 B6** 55 24N 1 37W
Alnwick, *U.K.* **10 B6** 55 24N 1 42W
Aloi, *Uganda* **54 B3** 2 16N 33 10 E
Alon, *Burma* **41 H19** 22 12N 95 5 E
Alor, *Indonesia* **37 F6** 8 15S 124 30 E
Alor Setar, *Malaysia* **39 J3** 6 7N 100 22 E
Alot, *India* **42 H6** 23 56N 75 40 E
Aloysius, Mt., *Australia* **61 E4** 26 0S 128 38 E
Alpaugh, *U.S.A.* **84 K7** 35 53N 119 29W
Alpena, *U.S.A.* **76 C4** 45 4N 83 27W
Alpha, *Australia* **62 C4** 23 39S 146 37 E
Alphen aan den Rijn, *Neths.* **15 B4** 52 7N 4 40 E
Alpine, *Ariz., U.S.A.* **83 K9** 33 51N 109 9W
Alpine, *Calif., U.S.A.* **85 N10** 32 50N 116 46W
Alpine, *Tex., U.S.A.* **81 K3** 30 22N 103 40W
Alps, *Europe* **18 C8** 46 30N 9 30 E
Alsace, *France* **18 B7** 48 15N 7 25 E
Alsask, *Canada* **73 C7** 51 21N 109 59W
Alsasua, *Spain* **19 A4** 42 54N 2 10W
Alsek →, *U.S.A.* **72 B1** 59 10N 138 12W
Alsten, *Norway* **8 D15** 65 58N 12 40 E
Alston, *U.K.* **10 C5** 54 49N 2 25W
Alta, *Norway* **8 B20** 69 57N 23 10 E
Alta Gracia, *Argentina* **94 C3** 31 40S 64 30W
Alta Sierra, *U.S.A.* **85 K8** 35 42N 118 33W
Altaelva →, *Norway* **8 B20** 69 54N 23 17 E
Altafjorden, *Norway* **8 A20** 70 5N 23 5 E
Altai = Aerhtai Shan, *Mongolia* **32 B4** 46 40N 92 45 E
Altamaha →, *U.S.A.* **77 K5** 31 20N 81 20W
Altamira, *Brazil* **93 D8** 3 12S 52 10W
Altamira, *Chile* **94 B2** 25 47S 69 51W
Altamira, *Mexico* **87 C5** 22 24N 97 55W
Altamont, *U.S.A.* **79 D10** 42 43N 74 3W
Altamura, *Italy* **20 D7** 40 49N 16 33 E
Altanbulag, *Mongolia* **32 A5** 50 16N 106 30 E
Altar, *Mexico* **86 A2** 30 40N 111 50W
Altar, Desierto de, *Mexico* **86 B2** 30 10N 112 0W
Altata, *Mexico* **86 C3** 24 30N 108 0W
Altavista, *U.S.A.* **76 G6** 37 6N 79 17W
Altay, *China* **32 B3** 47 48N 88 10 E
Altea, *Spain* **19 C5** 38 38N 0 2W
Altiplano = Bolivian Plateau, *S. Amer.* **90 E4** 20 0S 67 30W
Alto Araguaia, *Brazil* **93 G8** 17 15S 53 20W
Alto Cuchumatanes = Cuchumatanes, Sierra de los, *Guatemala* **88 C1** 15 35N 91 25W
Alto del Carmen, *Chile* **94 B1** 28 46S 70 30W
Alto del Inca, *Chile* **94 A2** 24 10S 68 10W
Alto Ligonha, *Mozam.* **55 F4** 15 30S 38 11 E
Alto Molocue, *Mozam.* **55 F4** 15 50S 37 35 E
Alto Paraguay □, *Paraguay* **94 A4** 21 0S 58 30W
Alto Paraná □, *Paraguay* **95 B5** 25 30S 54 50W
Alton, *Canada* **78 C4** 43 54N 80 5W
Alton, *U.K.* **11 F7** 51 9N 0 59W
Alton, *Ill., U.S.A.* **80 F9** 38 53N 90 11W
Alton, *N.H., U.S.A.* **79 C13** 43 27N 71 13W
Altoona, *U.S.A.* **78 F6** 40 31N 78 24W
Altūn Kūprī, *Iraq* **44 C5** 35 45N 44 9 E
Altun Shan, *China* **32 C3** 38 30N 88 0 E
Alturas, *U.S.A.* **82 F3** 41 29N 120 32W
Altus, *U.S.A.* **81 H5** 34 38N 99 20W
Alucra, *Turkey* **25 F6** 40 22N 38 47 E
Alūksne, *Latvia* **9 H22** 57 24N 27 3 E
Alunite, *U.S.A.* **85 K12** 35 59N 114 55W
Alusi, *Indonesia* **37 F8** 7 35S 131 40 E
Alva, *U.S.A.* **81 G5** 36 48N 98 40W
Alvarado, *Mexico* **87 D5** 18 40N 95 50W
Alvarado, *U.S.A.* **81 J6** 32 24N 97 13W
Alvaro Obregón, Presa, *Mexico* **86 B3** 27 55N 109 52W
Alvear, *Argentina* **94 B4** 29 5S 56 30W
Alvesta, *Sweden* **9 H16** 56 54N 14 35 E
Alvin, *U.S.A.* **81 L7** 29 26N 95 15W
Alvinston, *Canada* **78 D3** 42 49N 81 52W
Älvkarleby, *Sweden* **9 F17** 60 34N 17 26 E
Alvord Desert, *U.S.A.* **82 E4** 42 30N 118 25W
Älvsbyn, *Sweden* **8 D19** 65 40N 21 0 E
Alwar, *India* **42 F7** 27 38N 76 34 E
Alxa Zuoqi, *China* **34 E3** 38 50N 105 40 E
Alyangula, *Australia* **62 A2** 13 55S 136 30 E
Alyata = Älät, *Azerbaijan* **25 G8** 39 58N 49 25 E
Alyth, *U.K.* **12 E5** 56 38N 3 13W
Alytus, *Lithuania* **9 J21** 54 24N 24 3 E
Alzada, *U.S.A.* **80 C2** 45 2N 104 25W
Alzira, *Spain* **19 C5** 39 9N 0 30W
Am Timan, *Chad* **51 F10** 11 0N 20 10 E
Amadeus, L., *Australia* **61 D5** 24 54S 131 0 E
Amadi, *Dem. Rep. of the Congo* **54 B2** 3 40N 26 40 E
Amâdi, *Sudan* **51 G12** 5 29N 30 25 E
Amadjuak L., *Canada* **69 B12** 65 0N 71 8W
Amagansett, *U.S.A.* **79 F12** 40 59N 72 9W
Amagasaki, *Japan* **31 G7** 34 42N 135 20 E
Amahai, *Indonesia* **37 E7** 3 20S 128 55 E
Amakusa-Shotō, *Japan* **31 H5** 32 15N 130 10 E

Ashkhabad

Antsenavolo, *Madag.* **57 C8** 21 24S 48 3 E
Antsiafabositra, *Madag.* **57 B8** 17 18S 46 57 E
Antsirabe, *Antananarivo, Madag.* **57 B8** 19 55S 47 2 E
Antsirabe, *Antsiranana, Madag.* **57 A8** 14 0S 49 59 E
Antsirabe, *Mahajanga, Madag.* **57 B8** 15 57S 48 58 E
Antsiranana, *Madag.* **57 A8** 12 25S 49 20 E
Antsiranana □, *Madag.* **57 A8** 12 16S 49 17 E
Antsohihy, *Madag.* **57 A8** 14 50S 47 59 E
Antsohimbondrona Seranana, *Madag.* **57 A8** 13 7S 48 48 E
Antu, *China* **35 C15** 42 30N 128 20 E
Antwerp = Antwerpen, *Belgium* **15 C4** 51 13N 4 25 E
Antwerp, *U.S.A.* **79 B9** 44 12N 75 37W
Antwerpen, *Belgium* **15 C4** 51 13N 4 25 E
Antwerpen □, *Belgium* **15 C4** 51 15N 4 40 E
Anupgarh, *India* **42 E5** 29 10N 73 10 E
Anuppur, *India* **43 H9** 23 6N 81 41 E
Anuradhapura, *Sri Lanka* **40 Q12** 8 22N 80 28 E
Anveh, *Iran* **45 E7** 27 23N 54 11 E
Anvers = Antwerpen, *Belgium* **15 C4** 51 13N 4 25 E
Anvers I., *Antarctica* **5 C17** 64 30S 63 40W
Anxi, *China* **32 B4** 40 30N 95 43 E
Anxious B., *Australia* **63 E1** 33 24S 134 45 E
Anyang, *China* **34 F8** 36 5N 114 21 E
Anyer-Kidul, *Indonesia* **37 G11** 6 4S 105 53 E
Anyi, *China* **34 G6** 35 2N 111 2 E
Anza, *U.S.A.* **85 M10** 33 35N 116 39W
Anze, *China* **34 F7** 36 10N 112 12 E
Anzhero-Sudzhensk, *Russia* **26 D9** 56 10N 86 0 E
Ánzio, *Italy* **20 D5** 41 27N 12 37 E
Aoga-Shima, *Japan* **31 H9** 32 28N 139 46 E
Aomen = Macau, *China* **33 D6** 22 12N 113 33 E
Aomori, *Japan* **30 D10** 40 45N 140 45 E
Aomori □, *Japan* **30 D10** 40 45N 140 40 E
Aonla, *India* **43 E8** 28 16N 79 11 E
Aoraki Mount Cook, *N.Z.* **59 K3** 43 36S 170 9 E
Aosta, *Italy* **18 D7** 45 45N 7 20 E
Aoukâr, *Mauritania* **50 E4** 17 40N 10 0W
Apa →, *S. Amer.* **94 A4** 22 6S 58 2W
Apache, *U.S.A.* **81 H5** 34 54N 98 22W
Apache Junction, *U.S.A.* **83 K8** 33 25N 111 33W
Apalachee B., *U.S.A.* **77 L4** 30 0N 84 0W
Apalachicola, *U.S.A.* **77 L3** 29 43N 84 59W
Apalachicola →, *U.S.A.* **77 L3** 29 43N 84 58W
Apaporis →, *Colombia* **92 D5** 1 23S 69 25W
Aparri, *Phil.* **37 A6** 18 22N 121 38 E
Apatity, *Russia* **24 A5** 67 34N 33 22 E
Apatzingán, *Mexico* **86 D4** 19 0N 102 20W
Apeldoorn, *Neths.* **15 B5** 52 13N 5 57 E
Apennines = Appennini, *Italy* **20 B4** 44 0N 10 0 E
Apia, *Samoa* **59 A13** 13 50S 171 50W
Apiacás, Serra dos, *Brazil* **92 E7** 9 50S 57 0W
Apies →, *S. Africa* **57 D4** 25 15S 28 8 E
Apizaco, *Mexico* **87 D5** 19 26N 98 9W
Aplao, *Peru* **92 G4** 16 0S 72 40W
Apo, Mt., *Phil.* **37 C7** 6 53N 125 14 E
Apolakkiá, *Greece* **23 C9** 36 5N 27 48 E
Apolakkiá, Órmos, *Greece* **23 C9** 36 5N 27 45 E
Apolo, *Bolivia* **92 F5** 14 30S 68 30W
Aporé →, *Brazil* **93 G8** 19 27S 50 57W
Apostle Is., *U.S.A.* **80 B9** 47 0N 90 40W
Apóstoles, *Argentina* **95 B4** 28 0S 56 0W
Apostolos Andreas, C., *Cyprus* **23 D13** 35 42N 34 35 E
Apoteri, *Guyana* **92 C7** 4 2N 58 32W
Appalachian Mts., *U.S.A.* **76 G6** 38 0N 80 0W
Appennini, *Italy* **20 B4** 44 0N 10 0 E
Apple Hill, *Canada* **79 A10** 45 13N 74 46W
Apple Valley, *U.S.A.* **85 L9** 34 32N 117 14W
Appleby-in-Westmorland, *U.K.* **10 C5** 54 35N 2 29W
Appleton, *U.S.A.* **76 C1** 44 16N 88 25W
Approuague →, *Fr. Guiana* **93 C8** 4 30N 51 57W
Aprília, *Italy* **20 D5** 41 36N 12 39 E
Apsley, *Canada* **78 B6** 44 45N 78 6W
Apucarana, *Brazil* **95 A5** 23 55S 51 33W
Apure →, *Venezuela* **92 B5** 7 37N 66 25W
Apurimac →, *Peru* **92 F4** 12 17S 73 56W
Āqā Jarī, *Iran* **45 D6** 30 42N 49 50 E
Aqaba = Al 'Aqabah, *Jordan* **47 F4** 29 31N 35 0 E
Aqaba, G. of, *Red Sea* **44 D2** 28 15N 33 20 E
'Aqabah, Khalīj al = Aqaba, G. of, *Red Sea* **44 D2** 28 15N 33 20 E
'Aqdā, *Iran* **45 C7** 32 26N 53 37 E
Aqmola = Astana, *Kazakstan* **26 D8** 51 10N 71 30 E
'Aqrah, *Iraq* **44 B4** 36 46N 43 45 E
Aqtaū, *Kazakstan* **26 E6** 43 39N 51 12 E
Aqtöbe, *Kazakstan* **25 D10** 50 17N 57 10 E
Aquidauana, *Brazil* **93 H7** 20 30S 55 50W
Aquiles Serdán, *Mexico* **86 B3** 28 37N 105 54W
Aquin, *Haiti* **89 C5** 18 16N 73 24W
Aquitain, Bassin, *France* **18 D3** 44 0N 0 30W
Aqviligjuaq = Pelly Bay, *Canada* **69 B11** 68 38N 89 50W
Ar Rachidiya = Er Rachidia, *Morocco* **50 B5** 31 58N 4 20W
Ar Rafīd, *Syria* **47 C4** 32 57N 35 52 E
Ar Raḩḩālīyah, *Iraq* **44 C4** 32 44N 43 23 E
Ar Ramādī, *Iraq* **44 C4** 33 25N 43 20 E
Ar Ramthā, *Jordan* **47 C5** 32 34N 36 0 E
Ar Raqqah, *Syria* **44 C3** 35 59N 39 8 E
Ar Rass, *Si. Arabia* **44 E4** 25 50N 43 40 E
Ar Rifā'ī, *Iraq* **44 D5** 31 50N 46 10 E
Ar Riyāḑ, *Si. Arabia* **46 C4** 24 41N 46 42 E
Ar Ru'ays, *Qatar* **45 E6** 26 8N 51 12 E
Ar Rukhaymīyah, *Iraq* **44 D5** 29 22N 45 38 E
Ar Ruṣāfah, *Syria* **44 C3** 35 45N 38 49 E
Ar Ruṭbah, *Iraq* **44 C4** 33 0N 40 15 E
Ara, *India* **43 G11** 25 35N 84 32 E
Arab, *U.S.A.* **77 H2** 34 19N 86 30W
'Arab, Bahr el →, *Sudan* **51 G11** 9 0N 29 30 E
Arab, Shatt al →, *Asia* **45 D6** 30 0N 48 31 E
'Arabābād, *Iran* **45 C8** 33 2N 57 41 E
Arabia, *Asia* **28 G8** 25 0N 45 0 E
Arabian Desert = Es Sahrâ' Esh Sharqîya, *Egypt* **51 C12** 27 30N 32 30 E
Arabian Gulf = Gulf, The, *Asia* **45 E6** 27 0N 50 0 E
Arabian Sea, *Ind. Oc.* **29 H10** 16 0N 65 0 E
Aracaju, *Brazil* **93 F11** 10 55S 37 4W
Aracati, *Brazil* **93 D11** 4 30S 37 44W
Araçatuba, *Brazil* **95 A5** 21 10S 50 30W
Aracena, *Spain* **19 D2** 37 53N 6 38W
Araçuaí, *Brazil* **93 G10** 16 52S 42 4W
'Arad, *Israel* **47 D4** 31 15N 35 12 E
Arad, *Romania* **17 E11** 46 10N 21 20 E
Arādān, *Iran* **45 C7** 35 21N 52 30 E
Aradhippou, *Cyprus* **23 E12** 34 57N 33 36 E
Arafura Sea, *E. Indies* **28 K17** 9 0S 135 0 E
Aragón □, *Spain* **19 B5** 41 25N 0 40W
Aragón →, *Spain* **19 A5** 42 13N 1 44W
Araguacema, *Brazil* **93 E9** 8 50S 49 20W
Araguaia →, *Brazil* **93 E9** 5 21S 48 41W
Araguaína, *Brazil* **93 E9** 7 12S 48 12W
Araguari, *Brazil* **93 G9** 18 38S 48 11W
Araguari →, *Brazil* **93 C9** 1 15N 49 55W
Arain, *India* **42 F6** 26 27N 75 2 E
Arak, *Algeria* **50 C6** 25 20N 3 45 E
Arāk, *Iran* **45 C6** 34 0N 49 40 E
Arakan Coast, *Burma* **41 K19** 19 0N 94 0 E
Arakan Yoma, *Burma* **41 K19** 20 0N 94 40 E
Araks = Aras, Rūd-e →, *Asia* **44 B5** 40 5N 48 29 E
Aral, *Kazakstan* **26 E7** 46 41N 61 45 E
Aral Sea, *Asia* **26 E7** 44 30N 60 0 E
Aral Tengizi = Aral Sea, *Asia* **26 E7** 44 30N 60 0 E
Aralsk = Aral, *Kazakstan* **26 E7** 46 41N 61 45 E
Aralskoye More = Aral Sea, *Asia* **26 E7** 44 30N 60 0 E
Aramac, *Australia* **62 C4** 22 58S 145 14 E
Aran I., *Ireland* **13 A3** 55 0N 8 30W
Aran Is., *Ireland* **13 C2** 53 6N 9 38W
Aranda de Duero, *Spain* **19 B4** 41 39N 3 42W
Arandān, *Iran* **44 C5** 35 23N 46 55 E
Aranjuez, *Spain* **19 B4** 40 1N 3 40W
Aranos, *Namibia* **56 C2** 24 9S 19 7 E
Aransas Pass, *U.S.A.* **81 M6** 27 55N 97 9W
Aranyaprathet, *Thailand* **38 F4** 13 41N 102 30 E
Arapahoe, *U.S.A.* **80 E5** 40 18N 99 54W
Arapey Grande →, *Uruguay* **94 C4** 30 55S 57 49W
Arapgir, *Turkey* **44 B3** 39 5N 38 30 E
Arapiraca, *Brazil* **93 E11** 9 45S 36 39W
Arapongas, *Brazil* **95 A5** 23 29S 51 28W
Ar'ar, *Si. Arabia* **44 D4** 30 59N 41 2 E
Araranguá, *Brazil* **95 B6** 29 0S 49 30W
Araraquara, *Brazil* **93 H9** 21 50S 48 0W
Ararás, Serra das, *Brazil* **95 B5** 25 0S 53 10W
Ararat, *Australia* **63 F3** 37 16S 143 0 E
Ararat, Mt. = Ağrı Dağı, *Turkey* **25 G7** 39 50N 44 15 E
Araria, *India* **43 F12** 26 9N 87 33 E
Araripe, Chapada do, *Brazil* **93 E11** 7 20S 40 0W
Araruama, L. de, *Brazil* **95 A7** 22 53S 42 12W
Aras, Rūd-e →, *Asia* **44 B5** 40 5N 48 29 E
Arauca, *Colombia* **92 B4** 7 0N 70 40W
Arauca →, *Venezuela* **92 B5** 7 24N 66 35W
Arauco, *Chile* **94 D1** 37 16S 73 25W
Araxá, *Brazil* **93 G9** 19 35S 46 55W
Araya, Pen. de, *Venezuela* **92 A6** 10 40N 64 0W
Arba Minch, *Ethiopia* **46 F2** 6 0N 37 30 E
Arbat, *Iraq* **44 C5** 35 25N 45 35 E
Árbatax, *Italy* **20 E3** 39 56N 9 42 E
Arbīl, *Iraq* **44 B5** 36 15N 44 5 E
Arborfield, *Canada* **73 C8** 53 6N 103 39W
Arborg, *Canada* **73 C9** 50 54N 97 13W
Arbroath, *U.K.* **12 E6** 56 34N 2 35W
Arbuckle, *U.S.A.* **84 F4** 39 1N 122 3W
Arcachon, *France* **18 D3** 44 40N 1 10W
Arcade, *Calif., U.S.A.* **85 L8** 34 2N 118 15W
Arcade, *N.Y., U.S.A.* **78 D6** 42 32N 78 25W
Arcadia, *Fla., U.S.A.* **77 M5** 27 13N 81 52W
Arcadia, *La., U.S.A.* **81 J8** 32 33N 92 55W
Arcadia, *Pa., U.S.A.* **78 F6** 40 47N 78 51W
Arcata, *U.S.A.* **82 F1** 40 52N 124 5W
Archangel = Arkhangelsk, *Russia* **24 B7** 64 38N 40 36 E
Archbald, *U.S.A.* **79 E9** 41 30N 75 32W
Archer →, *Australia* **62 A3** 13 28S 141 41 E
Archer B., *Australia* **62 A3** 13 20S 141 30 E
Archers Post, *Kenya* **54 B4** 0 35N 37 35 E
Arches Nat. Park, *U.S.A.* **83 G9** 38 45N 109 25W
Arckaringa Cr. →, *Australia* **63 D2** 28 10S 135 22 E
Arco, *U.S.A.* **82 E7** 43 38N 113 18W
Arcos de la Frontera, *Spain* **19 D3** 36 45N 5 49W
Arcot, *India* **40 N11** 12 53N 79 20 E
Arctic Bay, *Canada* **69 A11** 73 1N 85 7W
Arctic Ocean, *Arctic* **4 B18** 78 0N 160 0W
Arctic Red River = Tsiigehtchic, *Canada* **68 B6** 67 15N 134 0W
Arda →, *Bulgaria* **21 D12** 41 40N 26 30 E
Ardabīl, *Iran* **45 B6** 38 15N 48 18 E
Ardakān = Sepīdān, *Iran* **45 D7** 30 20N 52 5 E
Ardakān, *Iran* **45 C7** 32 19N 53 59 E
Ardee, *Ireland* **13 C5** 53 52N 6 33W
Arden, *Canada* **78 B8** 44 43N 76 56W
Arden, *Calif., U.S.A.* **84 G5** 38 36N 121 33W
Arden, *Nev., U.S.A.* **85 J11** 36 1N 115 14W
Ardenne, *Belgium* **16 D3** 49 50N 5 5 E
Ardennes = Ardenne, *Belgium* **16 D3** 49 50N 5 5 E
Arderin, *Ireland* **13 C4** 53 2N 7 39W
Ardestān, *Iran* **45 C7** 33 20N 52 25 E
Ardivachar Pt., *U.K.* **12 D1** 57 23N 7 26W
Ardlethan, *Australia* **63 E4** 34 22S 146 53 E
Ardmore, *Okla., U.S.A.* **81 H6** 34 10N 97 8W
Ardmore, *Pa., U.S.A.* **79 G9** 39 58N 75 18W
Ardnamurchan, Pt. of, *U.K.* **12 E2** 56 43N 6 14W
Ardnave Pt., *U.K.* **12 F2** 55 53N 6 20W
Ardrossan, *Australia* **63 E2** 34 26S 137 53 E
Ardrossan, *U.K.* **12 F4** 55 39N 4 49W
Ards Pen., *U.K.* **13 B6** 54 33N 5 34W
Arecibo, *Puerto Rico* **89 C6** 18 29N 66 43W
Areia Branca, *Brazil* **93 E11** 5 0S 37 0W
Arena, Pt., *U.S.A.* **84 G3** 38 57N 123 44W
Arenal, *Honduras* **88 C2** 15 21N 86 50W
Arendal, *Norway* **9 G13** 58 28N 8 46 E
Arequipa, *Peru* **92 G4** 16 20S 71 30W
Arévalo, *Spain* **19 B3** 41 3N 4 43W
Arezzo, *Italy* **20 C4** 43 25N 11 53 E
Arga, *Turkey* **44 B3** 38 21N 37 59 E
Arganda, *Spain* **19 B4** 40 19N 3 26W
Argenta, *Canada* **72 C5** 50 11N 116 56W
Argentan, *France* **18 B3** 48 45N 0 1W
Argentário, Mte., *Italy* **20 C4** 42 24N 11 9 E
Argentia, *Canada* **71 C9** 47 18N 53 58W
Argentina ■, *S. Amer.* **96 D3** 35 0S 66 0W
Argentina Is., *Antarctica* **5 C17** 66 0S 64 0W
Argentino, L., *Argentina* **96 G2** 50 10S 73 0W
Argeş →, *Romania* **17 F14** 44 5N 26 38 E
Arghandab →, *Afghan.* **42 D1** 31 30N 64 15 E
Argolikós Kólpos, *Greece* **21 F10** 37 20N 22 52 E
Árgos, *Greece* **21 F10** 37 40N 22 43 E
Argostólion, *Greece* **21 E9** 38 11N 20 29 E
Arguello, Pt., *U.S.A.* **85 L6** 34 35N 120 39W
Arguineguín, *Canary Is.* **22 G4** 27 46N 15 41W
Argun →, *Russia* **27 D13** 53 20N 121 28 E
Argus Pk., *U.S.A.* **85 K9** 35 52N 117 26W
Argyle, L., *Australia* **60 C4** 16 20S 128 40 E
Argyll & Bute □, *U.K.* **12 E3** 56 13N 5 28W
Århus, *Denmark* **9 H14** 56 8N 10 11 E
Ariadnoye, *Russia* **30 B7** 45 8N 134 25 E
Ariamsvlei, *Namibia* **56 D2** 28 9S 19 51 E
Arica, *Chile* **92 G4** 18 32S 70 20W
Arica, *Colombia* **92 D4** 2 0S 71 50W
Arico, *Canary Is.* **22 F3** 28 9N 16 29W
Arid, C., *Australia* **61 F3** 34 1S 123 10 E
Arida, *Japan* **31 G7** 34 5N 135 8 E
Arílla, Ákra, *Greece* **23 A3** 39 43N 19 39 E
Arima, *Trin. & Tob.* **89 D7** 10 38N 61 17W
Arinos →, *Brazil* **92 F7** 10 25S 58 20W
Ario de Rosales, *Mexico* **86 D4** 19 12N 102 0W
Aripuanã, *Brazil* **92 E6** 9 25S 60 30W
Aripuanã →, *Brazil* **92 E6** 5 7S 60 25W
Ariquemes, *Brazil* **92 E6** 9 55S 63 6W
Arisaig, *U.K.* **12 E3** 56 55N 5 51W
Aristazabal I., *Canada* **72 C3** 52 40N 129 10W
Arivonimamo, *Madag.* **57 B8** 19 1S 47 11 E
Arizaro, Salar de, *Argentina* **94 A2** 24 40S 67 50W
Arizona, *Argentina* **94 D2** 35 45S 65 25W
Arizona □, *U.S.A.* **83 J8** 34 0N 112 0W
Arizpe, *Mexico* **86 A2** 30 20N 110 11W
Arjeplog, *Sweden* **8 D18** 66 3N 18 2 E
Arjona, *Colombia* **92 A3** 10 14N 75 22W
Arjuna, *Indonesia* **37 G15** 7 49S 112 34 E
Arka, *Russia* **27 C15** 60 15N 142 0 E
Arkadelphia, *U.S.A.* **81 H8** 34 7N 93 4W
Arkaig, L., *U.K.* **12 E3** 56 59N 5 10W
Arkalyk = Arqalyk, *Kazakstan* **26 D7** 50 13N 66 50 E
Arkansas □, *U.S.A.* **81 H8** 35 0N 92 30W
Arkansas →, *U.S.A.* **81 J9** 33 47N 91 4W
Arkansas City, *U.S.A.* **81 G6** 37 4N 97 2W
Arkaroola, *Australia* **63 E2** 30 20S 139 22 E
Arkhángelos, *Greece* **23 C10** 36 13N 28 7 E
Arkhangelsk, *Russia* **24 B7** 64 38N 40 36 E
Arki, *India* **42 D7** 31 9N 76 58 E
Arklow, *Ireland* **13 D5** 52 48N 6 10W
Arkport, *U.S.A.* **78 D7** 42 24N 77 42W
Arkticheskiy, Mys, *Russia* **27 A10** 81 10N 95 0 E
Arkville, *U.S.A.* **79 D10** 42 9N 74 37W
Arlanzón →, *Spain* **19 A3** 42 3N 4 17W
Arlbergpass, *Austria* **16 E6** 47 9N 10 12 E
Arles, *France* **18 E6** 43 41N 4 40 E
Arlington, *S. Africa* **57 D4** 28 1S 27 53 E
Arlington, *N.Y., U.S.A.* **79 E11** 41 42N 73 54W
Arlington, *Oreg., U.S.A.* **82 D3** 45 43N 120 12W
Arlington, *S. Dak., U.S.A.* **80 C6** 44 22N 97 8W
Arlington, *Tex., U.S.A.* **81 J6** 32 44N 97 7W
Arlington, *Va., U.S.A.* **76 F7** 38 53N 77 7W
Arlington, *Vt., U.S.A.* **79 C11** 43 5N 73 9W
Arlington, *Wash., U.S.A.* **84 B4** 48 12N 122 8W
Arlington Heights, *U.S.A.* **76 D2** 42 5N 87 59W
Arlit, *Niger* **50 E7** 19 0N 7 38 E
Arlon, *Belgium* **15 E5** 49 42N 5 49 E
Arltunga, *Australia* **62 C1** 23 26S 134 41 E
Armagh, *U.K.* **13 B5** 54 21N 6 39W
Armagh □, *U.K.* **13 B5** 54 18N 6 37W
Armavir, *Russia* **25 E7** 45 2N 41 7 E
Armenia, *Colombia* **92 C3** 4 35N 75 45W
Armenia ■, *Asia* **25 F7** 40 20N 45 0 E
Armenistís, Ákra, *Greece* **23 C9** 36 8N 27 42 E
Armidale, *Australia* **63 E5** 30 30S 151 40 E
Armour, *U.S.A.* **80 D5** 43 19N 98 21W
Armstrong, *B.C., Canada* **72 C5** 50 25N 119 10W
Armstrong, *Ont., Canada* **70 B2** 50 18N 89 4W
Arnarfjörður, *Iceland* **8 D2** 65 48N 23 40W
Arnaud →, *Canada* **69 B12** 60 0N 70 0W
Arnauti, C., *Cyprus* **23 D11** 35 6N 32 17 E
Arnett, *U.S.A.* **81 G5** 36 8N 99 46W
Arnhem, *Neths.* **15 C5** 51 58N 5 55 E
Arnhem, C., *Australia* **62 A2** 12 20S 137 30 E
Arnhem B., *Australia* **62 A2** 12 20S 136 10 E
Arnhem Land, *Australia* **62 A1** 13 10S 134 30 E
Arno →, *Italy* **20 C4** 43 41N 10 17 E
Arno Bay, *Australia* **63 E2** 33 54S 136 34 E
Arnold, *U.K.* **10 D6** 53 1N 1 7W
Arnold, *U.S.A.* **84 G6** 38 15N 120 20W
Arnot, *Canada* **73 B9** 55 56N 96 41W
Arnøy, *Norway* **8 A19** 70 9N 20 40 E
Arnprior, *Canada* **79 A8** 45 26N 76 21W
Arnsberg, *Germany* **16 C5** 51 24N 8 5 E
Aroab, *Namibia* **56 D2** 26 41S 19 39 E
Aron, *India* **42 G6** 25 57N 77 56 E
Arqalyk, *Kazakstan* **26 D7** 50 13N 66 50 E
Arrah = Ara, *India* **43 G11** 25 35N 84 32 E
Arran, *U.K.* **12 F3** 55 34N 5 12W
Arras, *France* **18 A5** 50 17N 2 46 E
Arrecife, *Canary Is.* **22 F6** 28 57N 13 37W
Arrecifes, *Argentina* **94 C3** 34 6S 60 9W
Arrée, Mts. d', *France* **18 B2** 48 26N 3 55W
Arriaga, *Chiapas, Mexico* **87 D6** 16 15N 93 52W
Arriaga, *San Luis Potosí, Mexico* **86 C4** 21 55N 101 23W
Arrilalah, *Australia* **62 C3** 23 43S 143 54 E
Arrino, *Australia* **61 E2** 29 30S 115 40 E
Arrow, L., *Ireland* **13 B3** 54 3N 8 19W
Arrowhead, L., *U.S.A.* **85 L9** 34 16N 117 10W
Arrowtown, *N.Z.* **59 L2** 44 57S 168 50 E
Arroyo Grande, *U.S.A.* **85 K6** 35 7N 120 35W
Ars, *Iran* **44 B5** 37 9N 47 46 E
Arsenault L., *Canada* **73 B7** 55 6N 108 32W
Arsenev, *Russia* **30 B6** 44 10N 133 15 E
Árta, *Greece* **21 E9** 39 8N 21 2 E
Artà, *Spain* **22 B10** 39 41N 3 21 E
Arteaga, *Mexico* **86 D4** 18 50N 102 20W
Artem, *Russia* **30 C6** 43 22N 132 13 E
Artemovsk, *Russia* **27 D10** 54 45N 93 35 E
Artemovsk, *Ukraine* **25 E6** 48 35N 38 0 E
Artesia = Mosomane, *Botswana* **56 C4** 24 2S 26 19 E
Artesia, *U.S.A.* **81 J2** 32 51N 104 24W
Arthur, *Canada* **78 C4** 43 50N 80 32W
Arthur →, *Australia* **62 G3** 41 2S 144 40 E
Arthur Cr. →, *Australia* **62 C2** 22 30S 136 25 E
Arthur Pt., *Australia* **62 C5** 22 7S 150 3 E
Arthur River, *Australia* **61 F2** 33 20S 117 2 E
Arthur's Pass, *N.Z.* **59 K3** 42 54S 171 35 E
Arthur's Town, *Bahamas* **89 B4** 24 38N 75 42W
Artigas, *Uruguay* **94 C4** 30 20S 56 30W
Artillery L., *Canada* **73 A7** 63 9N 107 52W
Artois, *France* **18 A5** 50 20N 2 30 E
Artrutx, C. de, *Spain* **22 B10** 39 55N 3 49 E
Artsyz, *Ukraine* **17 E15** 46 4N 29 26 E
Artvin, *Turkey* **25 F7** 41 14N 41 44 E
Aru, Kepulauan, *Indonesia* **37 F8** 6 0S 134 30 E
Aru Is. = Aru, Kepulauan, *Indonesia* **37 F8** 6 0S 134 30 E
Arua, *Uganda* **54 B3** 3 1N 30 58 E
Aruanã, *Brazil* **93 F8** 14 54S 51 10W
Aruba ■, *W. Indies* **89 D6** 12 30N 70 0W
Arucas, *Canary Is.* **22 F4** 28 7N 15 32W
Arun →, *Nepal* **43 F12** 26 55N 87 10 E
Arun →, *U.K.* **11 G7** 50 49N 0 33W
Arunachal Pradesh □, *India* **41 F19** 28 0N 95 0 E
Arusha, *Tanzania* **54 C4** 3 20S 36 40 E
Arusha □, *Tanzania* **54 C4** 4 0S 36 30 E
Arusha Chini, *Tanzania* **54 C4** 3 32S 37 20 E
Aruwimi →, *Dem. Rep. of the Congo* **54 B1** 1 13N 23 36 E
Arvada, *Colo., U.S.A.* **80 F2** 39 48N 105 5W
Arvada, *Wyo., U.S.A.* **82 D10** 44 39N 106 8W
Árvi, *Greece* **23 E7** 34 59N 25 28 E
Arviat, *Canada* **73 A10** 61 6N 93 59W
Arvidsjaur, *Sweden* **8 D18** 65 35N 19 10 E
Arvika, *Sweden* **9 G15** 59 40N 12 36 E
Arvin, *U.S.A.* **85 K8** 35 12N 118 50W
Arwal, *India* **43 G11** 25 15N 84 41 E
Arxan, *China* **33 B6** 47 11N 119 57 E
Aryirádhes, *Greece* **23 B3** 39 27N 19 58 E
Aryiroúpolis, *Greece* **23 D6** 35 17N 24 20 E
Arys, *Kazakstan* **26 E7** 42 26N 68 48 E
Arzamas, *Russia* **24 C7** 55 27N 43 55 E
Aş Şafā, *Syria* **47 B6** 33 10N 37 0 E
As Saffānīyah, *Si. Arabia* **45 E6** 27 55N 48 50 E
As Safīrah, *Syria* **44 B3** 36 5N 37 21 E
Aş Şahm, *Oman* **45 E8** 24 10N 56 53 E
As Sājir, *Si. Arabia* **44 E5** 25 11N 44 36 E
As Salamīyah, *Syria* **44 C3** 35 1N 37 2 E
As Salmān, *Iraq* **44 D5** 30 30N 44 32 E
As Salṭ, *Jordan* **47 C4** 32 2N 35 43 E
As Sal'w'a, *Qatar* **45 E6** 24 23N 50 50 E
As Samāwah, *Iraq* **44 D5** 31 15N 45 15 E
As Sanamayn, *Syria* **47 B5** 33 3N 36 10 E
As Sohar = Şuḩār, *Oman* **46 C6** 24 20N 56 40 E
As Sukhnah, *Syria* **44 C3** 34 52N 38 52 E
As Sulaymānīyah, *Iraq* **44 C5** 35 35N 45 29 E
As Sulaymī, *Si. Arabia* **44 E4** 26 17N 41 21 E
As Sulayyil, *Si. Arabia* **46 C4** 20 27N 45 34 E
As Summān, *Si. Arabia* **44 E5** 25 0N 47 0 E
As Suwaydā', *Syria* **47 C5** 32 40N 36 30 E
As Suwaydā' □, *Syria* **47 C5** 32 45N 36 45 E
As Suwayq, *Oman* **45 F8** 23 51N 57 26 E
As Şuwayrah, *Iraq* **44 C5** 32 55N 45 0 E
Asab, *Namibia* **56 D2** 25 30S 18 0 E
Asad, Buḩayrat al, *Syria* **44 C3** 36 0N 38 15 E
Asahi-Gawa →, *Japan* **31 G6** 34 36N 133 58 E
Asahigawa, *Japan* **30 C11** 43 46N 142 22 E
Asamankese, *Ghana* **50 G5** 5 50N 0 40W
Asan →, *India* **43 F8** 26 37N 78 24 E
Asansol, *India* **43 H12** 23 40N 87 1 E
Asbesberge, *S. Africa* **56 D3** 29 0S 23 0 E
Asbestos, *Canada* **71 C5** 45 47N 71 58W
Asbury Park, *U.S.A.* **79 F10** 40 13N 74 1W
Ascensión, *Mexico* **86 A3** 31 6N 107 59W
Ascensión, B. de la, *Mexico* **87 D7** 19 50N 87 20W
Ascension I., *Atl. Oc.* **49 G2** 7 57S 14 23W
Aschaffenburg, *Germany* **16 D5** 49 58N 9 6 E
Aschersleben, *Germany* **16 C6** 51 45N 11 29 E
Ascoli Piceno, *Italy* **20 C5** 42 51N 13 34 E
Ascope, *Peru* **92 E3** 7 46S 79 8W
Ascotán, *Chile* **94 A2** 21 45S 68 17W
Aseb, *Eritrea* **46 E3** 13 0N 42 40 E
Asela, *Ethiopia* **46 F2** 8 0N 39 0 E
Asenovgrad, *Bulgaria* **21 C11** 42 1N 24 51 E
Aserradero, *Mexico* **86 C3** 23 40N 105 43W
Asgata, *Cyprus* **23 E12** 34 46N 33 15 E
Ash Fork, *U.S.A.* **83 J7** 35 13N 112 29W
Ash Grove, *U.S.A.* **81 G8** 37 19N 93 35W
Ash Shabakah, *Iraq* **44 D4** 30 49N 43 39 E
Ash Shamāl □, *Lebanon* **47 A5** 34 25N 36 0 E
Ash Shāmīyah, *Iraq* **44 D5** 31 55N 44 35 E
Ash Shāriqah, *U.A.E.* **46 B6** 25 23N 55 26 E
Ash Sharmah, *Si. Arabia* **44 D2** 28 1N 35 16 E
Ash Sharqāt, *Iraq* **44 C4** 35 27N 43 16 E
Ash Sharqi, Al Jabal, *Lebanon* **47 B5** 33 40N 36 10 E
Ash Shaṭrah, *Iraq* **44 D5** 31 30N 46 10 E
Ash Shawbak, *Jordan* **44 D2** 30 32N 35 34 E
Ash Shawmari, J., *Jordan* **47 E5** 30 35N 36 35 E
Ash Shināfīyah, *Iraq* **44 D5** 31 35N 44 39 E
Ash Shu'bah, *Si. Arabia* **44 D5** 28 54N 44 44 E
Ash Shumlūl, *Si. Arabia* **44 E5** 26 31N 47 20 E
Ash Shūr'a, *Iraq* **44 C4** 35 58N 43 13 E
Ash Shurayf, *Si. Arabia* **44 E3** 25 43N 39 14 E
Ash Shuwayfāt, *Lebanon* **47 B4** 33 45N 35 30 E
Asha, *Russia* **24 D10** 55 0N 57 16 E
Ashau, *Vietnam* **38 D6** 16 6N 107 22 E
Ashbourne, *U.K.* **10 D6** 53 2N 1 43W
Ashburn, *U.S.A.* **77 K4** 31 43N 83 39W
Ashburton, *N.Z.* **59 K3** 43 53S 171 48 E
Ashburton →, *Australia* **60 D1** 21 40S 114 56 E
Ashcroft, *Canada* **72 C4** 50 40N 121 20W
Ashdod, *Israel* **47 D3** 31 49N 34 35 E
Ashdown, *U.S.A.* **81 J7** 33 40N 94 8W
Asheboro, *U.S.A.* **77 H6** 35 43N 79 49W
Ashern, *Canada* **73 C9** 51 11N 98 21W
Asherton, *U.S.A.* **81 L5** 28 27N 99 46W
Asheville, *U.S.A.* **77 H4** 35 36N 82 33W
Ashewat, *Pakistan* **42 D3** 31 22N 68 32 E
Asheweig →, *Canada* **70 B2** 54 17N 87 12W
Ashford, *Australia* **63 D5** 29 15S 151 3 E
Ashford, *U.K.* **11 F8** 51 8N 0 53 E
Ashgabat, *Turkmenistan* **26 F6** 38 0N 57 50 E
Ashibetsu, *Japan* **30 C11** 43 31N 142 11 E
Ashikaga, *Japan* **31 F9** 36 28N 139 29 E
Ashington, *U.K.* **10 B6** 55 11N 1 33W
Ashizuri-Zaki, *Japan* **31 H6** 32 44N 133 0 E
Ashkarkot, *Afghan.* **42 C2** 33 3N 67 58 E
Ashkhabad = Ashgabat, *Turkmenistan* **26 F6** 38 0N 57 50 E

Barakot

Bagamoyo, Tanzania **54 D4** 6 28S 38 55 E
Bagan Datoh, Malaysia **39 L3** 3 59N 100 47 E
Bagan Serai, Malaysia **39 K3** 5 1N 100 32 E
Baganga, Phil. **37 C7** 7 34N 126 33 E
Bagani, Namibia **56 B3** 18 7S 21 41 E
Bagansiapiapi, Indonesia **36 D2** 2 12N 100 50 E
Bagasra, India **42 J4** 21 30N 71 0 E
Bagaud, India **42 H6** 22 19N 75 53 E
Bagdad, U.S.A. **85 L11** 34 35N 115 53W
Bagdarin, Russia **27 D12** 54 26N 113 36 E
Bagé, Brazil **95 C5** 31 20S 54 15W
Bagenalstown = Muine Bheag, Ireland **13 D5** 52 42N 6 58W
Baggs, U.S.A. **82 F10** 41 2N 107 39W
Bagh, Pakistan **43 C5** 33 59N 73 45 E
Baghain →, India **43 G9** 25 32N 81 1 E
Baghdād, Iraq **44 C5** 33 20N 44 30 E
Bagheria, Italy **20 E5** 38 5N 13 30 E
Baghlān, Afghan. **40 A6** 32 12N 68 46 E
Baghlān □, Afghan. **40 B6** 36 0N 68 30 E
Bagley, U.S.A. **80 B7** 47 32N 95 24W
Bago = Pegu, Burma **41 L20** 17 20N 96 29 E
Bagodar, India **43 G11** 24 5N 85 52 E
Bagrationovsk, Russia **9 J19** 54 23N 20 39 E
Baguio, Phil. **37 A6** 16 26N 120 34 E
Bah, India **43 F8** 26 53N 78 36 E
Bahadurganj, India **43 F12** 26 16N 87 49 E
Bahadurgarh, India **42 E7** 28 40N 76 57 E
Bahama, Canal Viejo de, W. Indies **88 B4** 22 10N 77 30W
Bahamas ■, N. Amer. **89 B5** 24 0N 75 0W
Baharampur, India **43 G13** 24 2N 88 27 E
Bahawalnagar, Pakistan **42 E5** 30 0N 73 15 E
Bahawalpur, Pakistan **42 E4** 29 24N 71 40 E
Baheri, India **43 E8** 28 45N 79 34 E
Bahgul →, India **43 F8** 27 45N 79 36 E
Bahi, Tanzania **54 D4** 5 58S 35 21 E
Bahi Swamp, Tanzania **54 D4** 6 10S 35 0 E
Bahía = Salvador, Brazil **93 F11** 13 0S 38 30W
Bahía □, Brazil **93 F10** 12 0S 42 0W
Bahía, Is. de la, Honduras **88 C2** 16 45N 86 15W
Bahía Blanca, Argentina **94 D3** 38 35S 62 13W
Bahía de Caráquez, Ecuador **92 D2** 0 40S 80 27W
Bahía Honda, Cuba **88 B3** 22 54N 83 10W
Bahía Laura, Argentina **96 F3** 48 10S 66 30W
Bahía Negra, Paraguay **92 H7** 20 5S 58 5W
Bahir Dar, Ethiopia **46 E2** 11 37N 37 10 E
Bahmanzād, Iran **45 D6** 31 15N 51 47 E
Bahr el Ghazâl □, Sudan **51 G11** 7 0N 28 0 E
Bahraich, India **43 F9** 27 38N 81 37 E
Bahrain ■, Asia **46 B5** 26 0N 50 35 E
Bahror, India **42 F7** 27 51N 76 20 E
Bāhū Kalāt, Iran **45 E9** 25 43N 61 25 E
Bai Bung, Mui = Ca Mau, Mui, Vietnam **39 H5** 8 38N 104 44 E
Bai Duc, Vietnam **38 C5** 18 3N 105 49 E
Bai Thuong, Vietnam **38 C5** 19 54N 105 23 E
Baia Mare, Romania **17 E12** 47 40N 23 35 E
Baião, Brazil **93 D9** 2 40S 49 40W
Baïbokoum, Chad **51 G9** 7 46N 15 43 E
Baicheng, China **35 B12** 45 38N 122 42 E
Baidoa, Somali Rep. **46 G3** 3 8N 43 30 E
Baie Comeau, Canada **71 C6** 49 12N 68 10W
Baie-St-Paul, Canada **71 C5** 47 28N 70 32W
Baie Trinité, Canada **71 C6** 49 25N 67 20W
Baie Verte, Canada **71 C8** 49 55N 56 12W
Baihar, India **43 H9** 22 6N 80 33 E
Baihe, China **34 H6** 32 50N 110 5 E
Ba'ījī, Iraq **44 C4** 35 0N 43 30 E
Baijnath, India **43 E8** 29 55N 79 37 E
Baikal, L. = Baykal, Oz., Russia **27 D11** 53 0N 108 0 E
Baikunthpur, India **43 H10** 23 15N 82 33 E
Baile Atha Cliath = Dublin, Ireland **13 C5** 53 21N 6 15W
Băileşti, Romania **17 F12** 44 1N 23 20 E
Bainbridge, Ga., U.S.A. **77 K3** 30 55N 84 35W
Bainbridge, N.Y., U.S.A. **79 D9** 42 18N 75 29W
Baing, Indonesia **37 F6** 10 14S 120 34 E
Bainiu, China **34 H7** 32 50N 112 15 E
Bā'ir, Jordan **47 E5** 30 45N 36 55 E
Bairin Youqi, China **35 C10** 43 30N 118 35 E
Bairin Zuoqi, China **35 C10** 43 58N 119 15 E
Bairnsdale, Australia **63 F4** 37 48S 147 36 E
Baisha, China **34 G7** 34 20N 112 32 E
Baitadi, Nepal **43 E9** 29 35N 80 25 E
Baiyin, China **34 F3** 36 45N 104 14 E
Baiyu Shan, China **34 F4** 37 15N 107 30 E
Baj Baj, India **43 H13** 22 30N 88 5 E
Baja, Hungary **17 E10** 46 12N 18 59 E
Baja, Pta., Mexico **86 B1** 29 50N 116 0W
Baja California, Mexico **86 A1** 31 10N 115 12W
Baja California □, Mexico **86 B2** 30 0N 115 0W
Baja California Sur □, Mexico **86 B2** 25 50N 111 50W
Bajag, India **43 H9** 22 40N 81 21 E
Bajamar, Canary Is. **22 F3** 28 33N 16 20W
Bajana, India **42 H4** 23 7N 71 49 E
Bājgīrān, Iran **45 B8** 37 36N 58 24 E
Bajimba, Mt., Australia **63 D5** 29 17S 152 6 E
Bajo Nuevo, Caribbean **88 C4** 15 40N 78 50W
Bajoga, Nigeria **51 F8** 10 57N 11 20 E
Bajool, Australia **62 C5** 23 40S 150 35 E
Bakel, Senegal **50 F3** 14 56N 12 20W
Baker, Calif., U.S.A. **85 K10** 35 16N 116 4W
Baker, Mont., U.S.A. **80 B2** 46 22N 104 17W
Baker, L., Canada **68 B10** 64 0N 96 0W
Baker City, U.S.A. **82 D5** 44 47N 117 50W
Baker I., Pac. Oc. **64 G10** 0 10N 176 35W
Baker I., U.S.A. **72 B2** 55 20N 133 40W
Baker L., Australia **61 E4** 26 54S 126 5 E
Baker Lake, Canada **68 B10** 64 20N 96 3W
Baker Mt., U.S.A. **82 B3** 48 50N 121 49W
Bakers Creek, Australia **62 C4** 21 13S 149 7 E
Baker's Dozen Is., Canada **70 A4** 56 45N 78 45W
Bakersfield, Calif., U.S.A. **85 K8** 35 23N 119 1W
Bakersfield, Vt., U.S.A. **79 B12** 44 45N 72 48W
Bākhtarān, Iran **44 C5** 34 23N 47 0 E
Bākhtarān □, Iran **44 C5** 34 0N 46 30 E
Bakı, Azerbaijan **25 F8** 40 29N 49 56 E
Bakkafjörður, Iceland **8 C6** 66 2N 14 48W
Bakony, Hungary **17 E9** 47 10N 17 30 E
Bakony Forest = Bakony, Hungary **17 E9** 47 10N 17 30 E
Bakouma, C.A.R. **52 C4** 5 40N 22 56 E
Bakswaho, India **43 G8** 24 15N 79 18 E
Baku = Bakı, Azerbaijan **25 F8** 40 29N 49 56 E
Bakutis Coast, Antarctica **5 D15** 74 0S 120 0W
Baky = Bakı, Azerbaijan **25 F8** 40 29N 49 56 E
Bala, Canada **78 A5** 45 1N 79 37W
Bala, U.K. **10 E4** 52 54N 3 36W
Bala, L., U.K. **10 E4** 52 53N 3 37W
Balabac I., Phil. **36 C5** 8 0N 117 0 E
Balabac Str., E. Indies **36 C5** 7 53N 117 5 E
Balabagh, Afghan. **42 B4** 34 25N 70 12 E
Ba'labakk, Lebanon **47 B5** 34 0N 36 10 E
Balabalangan, Kepulauan, Indonesia **36 E5** 2 20S 117 30 E
Balad, Iraq **44 C5** 34 1N 44 9 E
Balad Rūz, Iraq **44 C5** 33 42N 45 5 E
Bālādeh, Fārs, Iran **45 D6** 29 17N 51 56 E
Bālādeh, Māzandaran, Iran **45 B6** 36 12N 51 48 E
Balaghat, India **40 J12** 21 49N 80 12 E
Balaghat Ra., India **40 K10** 18 50N 76 30 E
Balaguer, Spain **19 B6** 41 50N 0 50 E
Balaklava, Ukraine **25 F5** 44 30N 33 30 E
Balakovo, Russia **24 D8** 52 4N 47 55 E
Balamau, India **43 F9** 27 10N 80 21 E
Balancán, Mexico **87 D6** 17 48N 91 32W
Balashov, Russia **25 D7** 51 30N 43 10 E
Balasinor, India **42 H5** 22 57N 73 23 E
Balasore = Baleshwar, India **41 J15** 21 35N 87 3 E
Balaton, Hungary **17 E9** 46 50N 17 40 E
Balbina, Reprêsa de, Brazil **92 D7** 2 0S 59 30W
Balboa, Panama **88 E4** 8 57N 79 34W
Balbriggan, Ireland **13 C5** 53 37N 6 11W
Balcarce, Argentina **94 D4** 38 0S 58 10W
Balcarres, Canada **73 C8** 50 50N 103 35W
Balchik, Bulgaria **21 C13** 43 28N 28 11 E
Balclutha, N.Z. **59 M2** 46 15S 169 45 E
Balcones Escarpment, U.S.A. **81 L5** 29 30N 99 15W
Bald Hd., Australia **61 G2** 35 6S 118 1 E
Bald I., Australia **61 F2** 34 57S 118 27 E
Bald Knob, U.S.A. **81 H9** 35 19N 91 34W
Baldock L., Canada **73 B9** 56 33N 97 57W
Baldwin, Mich., U.S.A. **76 D3** 43 54N 85 51W
Baldwin, Pa., U.S.A. **78 F5** 40 23N 79 59W
Baldwinsville, U.S.A. **79 C8** 43 10N 76 20W
Baldy Mt., U.S.A. **82 B9** 48 9N 109 39W
Baldy Peak, U.S.A. **83 K9** 33 54N 109 34W
Baleares, Is., Spain **22 B10** 39 30N 3 0 E
Balearic Is. = Baleares, Is., Spain **22 B10** 39 30N 3 0 E
Baleine = Whale →, Canada **71 A6** 58 15N 67 40W
Baler, Phil. **37 A6** 15 46N 121 34 E
Baleshare, U.K. **12 D1** 57 31N 7 22W
Baleshwar, India **41 J15** 21 35N 87 3 E
Balfate, Honduras **88 C2** 15 48N 86 25W
Bali, Greece **23 D6** 35 25N 24 47 E
Bali, India **42 G5** 25 11N 73 17 E
Bali, Indonesia **36 F4** 8 20S 115 0 E
Bali □, Indonesia **36 F5** 8 20S 115 0 E
Bali, Selat, Indonesia **37 H16** 8 18S 114 25 E
Baliapal, India **43 J12** 21 40N 87 17 E
Balikeşir, Turkey **21 E12** 39 39N 27 53 E
Balikpapan, Indonesia **36 E5** 1 10S 116 55 E
Balimbing, Phil. **37 C5** 5 5N 119 58 E
Baling, Malaysia **39 K3** 5 41N 100 55 E
Balipara, India **41 F18** 26 50N 92 45 E
Balkan Mts. = Stara Planina, Bulgaria **21 C10** 43 15N 23 0 E
Balkhash = Balqash, Kazakstan **26 E8** 46 50N 74 50 E
Balkhash, Ozero = Balqash Köl, Kazakstan **26 E8** 46 0N 74 50 E
Balla, Bangla. **41 G17** 24 10N 91 35 E
Ballachulish, U.K. **12 E3** 56 41N 5 8W
Balladonia, Australia **61 F3** 32 27S 123 51 E
Ballaghaderreen, Ireland **13 C3** 53 55N 8 34W
Ballarat, Australia **63 F3** 37 33S 143 50 E
Ballard, L., Australia **61 E3** 29 20S 120 40 E
Ballater, U.K. **12 D5** 57 3N 3 3W
Ballenas, Canal de, Mexico **86 B2** 29 10N 113 45W
Balleny Is., Antarctica **5 C11** 66 30S 163 0 E
Ballia, India **43 G11** 25 46N 84 12 E
Ballina, Australia **63 D5** 28 50S 153 31 E
Ballina, Ireland **13 B2** 54 7N 9 9W
Ballinasloe, Ireland **13 C3** 53 20N 8 13W
Ballinger, U.S.A. **81 K5** 31 45N 99 57W
Ballinrobe, Ireland **13 C2** 53 38N 9 13W
Ballinskelligs B., Ireland **13 E1** 51 48N 10 13W
Ballston Spa, U.S.A. **79 D11** 43 0N 73 51W
Ballycastle, U.K. **13 A5** 55 12N 6 15W
Ballyclare, U.K. **13 B5** 54 46N 6 0W
Ballyhaunis, Ireland **13 C3** 53 46N 8 46W
Ballymena, U.K. **13 B5** 54 52N 6 17W
Ballymoney, U.K. **13 A5** 55 5N 6 31W
Ballymote, Ireland **13 B3** 54 5N 8 31W
Ballynahinch, U.K. **13 B6** 54 24N 5 54W
Ballyquintin Pt., U.K. **13 B6** 54 20N 5 30W
Ballyshannon, Ireland **13 B3** 54 30N 8 11W
Balmaceda, Chile **96 F2** 46 0S 71 50W
Balmertown, Canada **73 C10** 51 4N 93 41W
Balmoral, Australia **63 F3** 37 15S 141 48 E
Balmorhea, U.S.A. **81 K3** 30 59N 103 45W
Balochistan = Baluchistan □, Pakistan **40 F4** 27 30N 65 0 E
Balonne →, Australia **63 D4** 28 47S 147 56 E
Balotra, India **42 G5** 25 50N 72 14 E
Balqash, Kazakstan **26 E8** 46 50N 74 50 E
Balqash Köl, Kazakstan **26 E8** 46 0N 74 50 E
Balrampur, India **43 F10** 27 30N 82 20 E
Balranald, Australia **63 E3** 34 38S 143 33 E
Balsas, Mexico **87 D5** 18 0N 99 40W
Balsas →, Brazil **93 E9** 7 15S 44 35W
Balsas →, Mexico **86 D4** 17 55N 102 10W
Balston Spa, U.S.A. **79 D11** 43 0N 73 52W
Balta, Ukraine **17 D15** 48 2N 29 45 E
Bălţi, Moldova **17 E14** 47 48N 27 58 E
Baltic Sea, Europe **9 H18** 57 0N 19 0 E
Baltimore, Ireland **13 E2** 51 29N 9 22W
Baltimore, Md., U.S.A. **76 F7** 39 17N 76 37W
Baltimore, Ohio, U.S.A. **78 G2** 39 51N 82 36W
Baltit, Pakistan **43 A6** 36 15N 74 40 E
Baltiysk, Russia **9 J18** 54 41N 19 58 E
Baluchistan □, Pakistan **40 F4** 27 30N 65 0 E
Balurghat, India **43 G13** 25 15N 88 44 E
Balvi, Latvia **9 H22** 57 8N 27 15 E
Balya, Turkey **21 E12** 39 44N 27 35 E
Bam, Iran **45 D8** 29 7N 58 14 E
Bama, Nigeria **51 F8** 11 33N 13 41 E
Bamaga, Australia **62 A3** 10 50S 142 25 E
Bamaji L., Canada **70 B1** 51 9N 91 25W
Bamako, Mali **50 F4** 12 34N 7 55W
Bambari, C.A.R. **52 C4** 5 40N 20 35 E
Bambaroo, Australia **62 B4** 18 50S 146 10 E
Bamberg, Germany **16 D6** 49 54N 10 54 E
Bamberg, U.S.A. **77 J5** 33 18N 81 2W
Bambili, Dem. Rep. of the Congo **54 B2** 3 40N 26 0 E
Bamfield, Canada **72 D3** 48 45N 125 10W
Bāmīān □, Afghan. **40 B5** 35 0N 67 0 E
Bamiancheng, China **35 C13** 43 15N 124 2 E
Bampūr, Iran **45 E9** 27 15N 60 21 E
Ban Ban, Laos **38 C4** 19 31N 103 30 E
Ban Bang Hin, Thailand **39 H2** 9 32N 98 35 E
Ban Chiang Klang, Thailand **38 C3** 19 25N 100 55 E
Ban Chik, Laos **38 D4** 17 15N 102 22 E
Ban Choho, Thailand **38 E4** 15 2N 102 9 E
Ban Dan Lan Hoi, Thailand **38 D2** 17 0N 99 35 E
Ban Don = Surat Thani, Thailand **39 H2** 9 6N 99 20 E
Ban Don, Vietnam **38 F6** 12 53N 107 48 E
Ban Don, Ao →, Thailand **39 H2** 9 20N 99 25 E
Ban Dong, Thailand **38 C3** 19 30N 100 59 E
Ban Hong, Thailand **38 C2** 18 18N 98 50 E
Ban Kaeng, Thailand **38 D3** 17 29N 100 7 E
Ban Kantang, Thailand **39 J2** 7 25N 99 31 E
Ban Keun, Laos **38 C4** 18 22N 102 35 E
Ban Khai, Thailand **38 F3** 12 46N 101 18 E
Ban Kheun, Laos **38 B3** 20 13N 101 7 E
Ban Khlong Kua, Thailand **39 J3** 6 57N 100 8 E
Ban Khuan Mao, Thailand **39 J2** 7 50N 99 37 E
Ban Ko Yai Chim, Thailand **39 G2** 11 17N 99 26 E
Ban Kok, Thailand **38 D4** 16 40N 103 40 E
Ban Laem, Thailand **38 F2** 13 13N 99 59 E
Ban Lao Ngam, Laos **38 E6** 15 28N 106 10 E
Ban Le Kathe, Thailand **38 E2** 15 49N 98 53 E
Ban Mae Chedi, Thailand **38 C2** 19 11N 99 31 E
Ban Mae Laeng, Thailand **38 B2** 20 1N 99 17 E
Ban Mae Sariang, Thailand **38 C1** 18 10N 97 56 E
Ban Mê Thuôt = Buon Ma Thuot, Vietnam **38 F7** 12 40N 108 3 E
Ban Mi, Thailand **38 E3** 15 3N 100 32 E
Ban Muong Mo, Laos **38 C4** 19 4N 103 58 E
Ban Na Mo, Laos **38 D5** 17 7N 105 40 E
Ban Na San, Thailand **39 H2** 8 53N 99 52 E
Ban Na Tong, Laos **38 B3** 20 56N 101 47 E
Ban Nam Bac, Laos **38 B4** 20 38N 102 20 E
Ban Nam Ma, Laos **38 A3** 22 2N 101 37 E
Ban Ngang, Laos **38 E6** 15 59N 106 11 E
Ban Nong Bok, Laos **38 D5** 17 5N 104 48 E
Ban Nong Boua, Laos **38 E6** 15 40N 106 33 E
Ban Nong Pling, Thailand **38 E3** 15 40N 100 10 E
Ban Pak Chan, Thailand **39 G2** 10 32N 98 51 E
Ban Phai, Thailand **38 D4** 16 4N 102 44 E
Ban Pong, Thailand **38 F2** 13 50N 99 55 E
Ban Ron Phibun, Thailand **39 H2** 8 9N 99 51 E
Ban Sanam Chai, Thailand **39 J3** 7 33N 100 25 E
Ban Sangkha, Thailand **38 E4** 14 37N 103 52 E
Ban Tak, Thailand **38 D2** 17 2N 99 4 E
Ban Tako, Thailand **38 E4** 14 5N 102 40 E
Ban Tha Dua, Thailand **38 D2** 17 59N 98 39 E
Ban Tha Li, Thailand **38 D3** 17 37N 101 25 E
Ban Tha Nun, Thailand **39 H2** 8 12N 98 18 E
Ban Thahine, Laos **38 E5** 14 12N 105 33 E
Ban Xien Kok, Laos **38 B3** 20 54N 100 39 E
Ban Yen Nhan, Vietnam **38 B6** 20 57N 106 2 E
Banaba, Kiribati **64 H8** 0 45S 169 50 E
Banalia, Dem. Rep. of the Congo **54 B2** 1 32N 25 5 E
Banam, Cambodia **39 G5** 11 20N 105 17 E
Bananal, I. do, Brazil **93 F8** 11 30S 50 30W
Banaras = Varanasi, India **43 G10** 25 22N 83 0 E
Banas →, Gujarat, India **42 H4** 23 45N 71 25 E
Banas →, Mad. P., India **43 G9** 24 15N 81 30 E
Bânâs, Ras, Egypt **51 D13** 23 57N 35 59 E
Banbridge, U.K. **13 B5** 54 22N 6 16W
Banbury, U.K. **11 E6** 52 4N 1 20W
Banchory, U.K. **12 D6** 57 3N 2 29W
Bancroft, Canada **78 A7** 45 3N 77 51W
Band Bonī, Iran **45 E8** 25 30N 59 33 E
Band Qīr, Iran **45 D6** 31 39N 48 53 E
Banda, Mad. P., India **43 G8** 24 3N 78 57 E
Banda, Ut. P., India **43 G9** 25 30N 80 26 E
Banda, Kepulauan, Indonesia **37 E7** 4 37S 129 50 E
Banda Aceh, Indonesia **36 C1** 5 35N 95 20 E
Banda Banda, Mt., Australia **63 E5** 31 10S 152 28 E
Banda Elat, Indonesia **37 F8** 5 40S 133 5 E
Banda Is. = Banda, Kepulauan, Indonesia **37 E7** 4 37S 129 50 E
Banda Sea, Indonesia **37 F7** 6 0S 130 0 E
Bandai-San, Japan **30 F10** 37 36N 140 4 E
Bandān, Iran **45 D9** 31 23N 60 44 E
Bandanaira, Indonesia **37 E7** 4 32S 129 54 E
Bandanwara, India **42 F6** 26 9N 74 38 E
Bandar = Machilipatnam, India **41 L12** 16 12N 81 8 E
Bandār 'Abbās, Iran **45 E8** 27 15N 56 15 E
Bandar-e Anzalī, Iran **45 B6** 37 30N 49 30 E
Bandar-e Bushehr = Būshehr, Iran **45 D6** 28 55N 50 55 E
Bandar-e Chārak, Iran **45 E7** 26 45N 54 20 E
Bandar-e Deylam, Iran **45 D6** 30 5N 50 10 E
Bandar-e Khomeynı, Iran **45 D6** 30 30N 49 5 E
Bandar-e Lengeh, Iran **45 E7** 26 35N 54 58 E
Bandar-e Maqām, Iran **45 E7** 26 56N 53 29 E
Bandar-e Ma'shur, Iran **45 D6** 30 35N 49 10 E
Bandar-e Rīg, Iran **45 D6** 29 29N 50 38 E
Bandar-e Torkeman, Iran **45 B7** 37 0N 54 10 E
Bandar Maharani = Muar, Malaysia **39 L4** 2 3N 102 34 E
Bandar Penggaram = Batu Pahat, Malaysia **39 M4** 1 50N 102 56 E
Bandar Seri Begawan, Brunei **36 C4** 4 52N 115 0 E
Bandar Sri Aman, Malaysia **36 D4** 1 15N 111 32 E
Bandawe, Malawi **55 E3** 11 58S 34 5 E
Bandeira, Pico da, Brazil **95 A7** 20 26S 41 47W
Bandera, Argentina **94 B3** 28 55S 62 20W
Banderas, B. de, Mexico **86 C3** 20 40N 105 30W
Bandhogarh, India **43 H9** 23 40N 81 2 E
Bandi →, India **42 F6** 26 12N 75 47 E
Bandikui, India **42 F7** 27 3N 76 34 E
Bandırma, Turkey **21 D13** 40 20N 28 0 E
Bandon, Ireland **13 E3** 51 44N 8 44W
Bandon →, Ireland **13 E3** 51 43N 8 37W
Bandula, Mozam. **55 F3** 19 0S 33 7 E
Bandundu, Dem. Rep. of the Congo **52 E3** 3 15S 17 22 E
Bandung, Indonesia **36 F3** 6 54S 107 36 E
Bāneh, Iran **44 C5** 35 59N 45 53 E
Banes, Cuba **89 B4** 21 0N 75 42W
Banff, Canada **72 C5** 51 10N 115 34W
Banff, U.K. **12 D6** 57 40N 2 33W
Banff Nat. Park, Canada **72 C5** 51 30N 116 15W
Bang Fai →, Laos **38 D5** 16 57N 104 45 E
Bang Hieng →, Laos **38 D5** 16 10N 105 10 E
Bang Krathum, Thailand **38 D3** 16 34N 100 18 E
Bang Lamung, Thailand **38 F3** 13 3N 100 56 E
Bang Mun Nak, Thailand **38 D3** 16 2N 100 23 E
Bang Pa In, Thailand **38 E3** 14 14N 100 35 E
Bang Rakam, Thailand **38 D3** 16 45N 100 7 E
Bang Saphan, Thailand **39 G2** 11 14N 99 28 E
Bangaduni I., India **43 J13** 21 34N 88 52 E
Bangala Dam, Zimbabwe **55 G3** 21 7S 31 25 E
Bangalore, India **40 N10** 12 59N 77 40 E
Banganga →, India **42 F6** 27 6N 77 25 E
Bangaon, India **43 H13** 23 0N 88 47 E
Bangassou, C.A.R. **52 D4** 4 55N 23 7 E
Banggai, Indonesia **37 E6** 1 34S 123 30 E
Banggai, Kepulauan, Indonesia **37 E6** 1 40S 123 30 E
Banggai Arch. = Banggai, Kepulauan, Indonesia **37 E6** 1 40S 123 30 E
Banggi, Malaysia **36 C5** 7 17N 117 12 E
Banghāzī, Libya **51 B10** 32 11N 20 3 E
Bangka, Sulawesi, Indonesia **37 D7** 1 50N 125 5 E
Bangka, Sumatera, Indonesia **36 E3** 2 0S 105 50 E
Bangka, Selat, Indonesia **36 E3** 2 30S 105 30 E
Bangkalan, Indonesia **37 G15** 7 2S 112 46 E
Bangkinang, Indonesia **36 D2** 0 18N 101 5 E
Bangko, Indonesia **36 E2** 2 5S 102 9 E
Bangkok, Thailand **38 F3** 13 45N 100 35 E
Bangladesh ■, Asia **41 H17** 24 0N 90 0 E
Bangong Co, India **43 B8** 35 50N 79 20 E
Bangor, Down, U.K. **13 B6** 54 40N 5 40W
Bangor, Gwynedd, U.K. **10 D3** 53 14N 4 8W
Bangor, Maine, U.S.A. **69 D13** 44 48N 68 46W
Bangor, Pa., U.S.A. **79 F9** 40 52N 75 13W
Bangued, Phil. **37 A6** 17 40N 120 37 E
Bangui, C.A.R. **52 D3** 4 23N 18 35 E
Banguru, Dem. Rep. of the Congo **54 B2** 0 30N 27 10 E
Bangweulu, L., Zambia **55 E3** 11 0S 30 0 E
Bangweulu Swamp, Zambia **55 E3** 11 20S 30 15 E
Bani, Dom. Rep. **89 C5** 18 16N 70 22W
Banī Sa'd, Iraq **44 C5** 33 34N 44 32 E
Banihal Pass, India **43 C6** 33 30N 75 12 E
Bāniyās, Syria **44 C3** 35 10N 36 0 E
Banja Luka, Bos.-H. **20 B7** 44 49N 17 11 E
Banjar, India **42 D7** 31 38N 77 21 E
Banjar →, India **43 H9** 22 36N 80 22 E
Banjarmasin, Indonesia **36 E4** 3 20S 114 35 E
Banjul, Gambia **50 F2** 13 28N 16 40W
Banka, India **43 G12** 24 53N 86 55 E
Banket, Zimbabwe **55 F3** 17 27S 30 19 E
Bankipore, India **41 G14** 25 35N 85 10 E
Banks I., B.C., Canada **72 C3** 53 20N 130 0W
Banks I., N.W.T., Canada **68 A7** 73 15N 121 30W
Banks Pen., N.Z. **59 K4** 43 45S 173 15 E
Banks Str., Australia **62 G4** 40 40S 148 10 E
Bankura, India **43 H12** 23 11N 87 18 E
Banmankhi, India **43 G12** 25 53N 87 11 E
Bann →, Arm., U.K. **13 B5** 54 30N 6 31W
Bann →, L'derry., U.K. **13 A5** 55 8N 6 41W
Bannang Sata, Thailand **39 J3** 6 16N 101 16 E
Banning, U.S.A. **85 M10** 33 56N 116 53W
Banningville = Bandundu, Dem. Rep. of the Congo **52 E3** 3 15S 17 22 E
Bannockburn, Canada **78 B7** 44 39N 77 33W
Bannockburn, U.K. **12 E5** 56 5N 3 55W
Bannockburn, Zimbabwe **55 G2** 20 17S 29 48 E
Bannu, Pakistan **40 C7** 33 0N 70 18 E
Bano, India **43 H11** 22 40N 84 55 E
Bansgaon, India **43 F10** 26 33N 83 21 E
Banská Bystrica, Slovak Rep. **17 D10** 48 46N 19 14 E
Banswara, India **42 H6** 23 32N 74 24 E
Bantaeng, Indonesia **37 F5** 5 32S 119 56 E
Bantry, Ireland **13 E2** 51 41N 9 27W
Bantry B., Ireland **13 E2** 51 37N 9 44W
Bantul, Indonesia **37 G14** 7 55S 110 19 E
Bantva, India **42 J4** 21 29N 70 12 E
Banyak, Kepulauan, Indonesia **36 D1** 2 10N 97 10 E
Banyalbufar, Spain **22 B9** 39 42N 2 31 E
Banyo, Cameroon **52 C2** 6 52N 11 45 E
Banyumas, Indonesia **37 G13** 7 32S 109 18 E
Banyuwangi, Indonesia **37 H16** 8 13S 114 21 E
Banzare Coast, Antarctica **5 C9** 68 0S 125 0 E
Bao Ha, Vietnam **38 A5** 22 11N 104 21 E
Bao Lac, Vietnam **38 A5** 22 57N 105 40 E
Bao Loc, Vietnam **39 G6** 11 32N 107 48 E
Baocheng, China **34 H4** 33 12N 106 56 E
Baode, China **34 E6** 39 1N 111 5 E
Baodi, China **35 E9** 39 38N 117 20 E
Baoding, China **34 E8** 38 50N 115 28 E
Baoji, China **34 G4** 34 20N 107 5 E
Baoshan, China **32 D4** 25 10N 99 5 E
Baotou, China **34 D6** 40 32N 110 2 E
Baoying, China **35 H10** 33 17N 119 20 E
Bap, India **42 F5** 27 23N 72 18 E
Bapatla, India **41 M12** 15 55N 80 30 E
Bāqerābād, Iran **45 C6** 33 2N 51 58 E
Ba'qūbah, Iraq **44 C5** 33 45N 44 50 E
Baquedano, Chile **94 A2** 23 20S 69 52W
Bar, Montenegro, Yug. **21 C8** 42 8N 19 6 E
Bar, Ukraine **17 D14** 49 4N 27 40 E
Bar Bigha, India **43 G11** 25 21N 85 47 E
Bar Harbor, U.S.A. **77 C11** 44 23N 68 13W
Bar-le-Duc, France **18 B6** 48 47N 5 10 E
Bara, India **43 G9** 25 16N 81 43 E
Bara Banki, India **43 F9** 26 55N 81 12 E
Barabai, Indonesia **36 E5** 2 32S 115 34 E
Baraboo, U.S.A. **80 D10** 43 28N 89 45W
Baracoa, Cuba **89 B5** 20 20N 74 30W
Baradā →, Syria **47 B5** 33 33N 36 34 E
Baradero, Argentina **94 C4** 33 52S 59 29W
Baraga, U.S.A. **80 B10** 46 47N 88 30W
Barah →, India **42 F6** 27 42N 77 5 E
Barahona, Dom. Rep. **89 C5** 18 13N 71 7W
Barail Range, India **41 G18** 25 15N 93 20 E
Barakaldo, Spain **19 A4** 43 18N 2 59W
Barakar →, India **43 G12** 24 7N 86 14 E
Barakhola, India **41 G18** 25 0N 92 45 E
Barakot, India **43 J11** 21 33N 84 59 E

Barakpur

Barakpur, India 43 H13 22 44N 88 30 E
Baralaba, Australia 62 C4 24 13S 149 50 E
Baralzon L., Canada 73 B9 60 0N 98 3W
Baramula, India 43 B6 34 15N 74 20 E
Baran, India 42 G7 25 9N 76 40 E
Baran →, Pakistan 42 G3 25 13N 68 17 E
Baranavichy, Belarus 17 B14 53 10N 26 0 E
Baranof, U.S.A. 72 B2 57 5N 134 50W
Baranof I., U.S.A. 68 C6 57 0N 135 0W
Barapasi, Indonesia 37 E9 2 15S 137 5 E
Barasat, India 43 H13 22 46N 88 31 E
Barat Daya, Kepulauan, Indonesia 37 F7 7 30S 128 0 E
Barataria B., U.S.A. 81 L10 29 20N 89 55W
Barauda, India 42 H6 23 33N 75 15 E
Baraut, India 42 E7 29 13N 77 7 E
Barbacena, Brazil 95 A7 21 15S 43 56W
Barbados ■, W. Indies 89 D8 13 10N 59 30W
Barbària, C. de, Spain 22 C7 38 39N 1 24 E
Barbastro, Spain 19 A6 42 2N 0 5 E
Barberton, S. Africa 57 D5 25 42S 31 2 E
Barberton, U.S.A. 78 E3 41 0N 81 39W
Barbosa, Colombia 92 B4 5 57N 73 37W
Barbourville, U.S.A. 77 G4 36 52N 83 53W
Barbuda, W. Indies 89 C7 17 30N 61 40W
Barcaldine, Australia 62 C4 23 43S 145 6 E
Barcellona Pozzo di Gotto, Italy 20 E6 38 9N 15 13 E
Barcelona, Spain 19 B7 41 21N 2 10 E
Barcelona, Venezuela 92 A6 10 10N 64 40W
Barcelos, Brazil 92 D6 1 0S 63 0W
Barcoo →, Australia 62 D3 25 30S 142 50 E
Bardaï, Chad 51 D9 21 25N 17 0 E
Bardas Blancas, Argentina 94 D2 35 49S 69 45W
Bardawîl, Sabkhet el, Egypt 47 D2 31 10N 33 15 E
Barddhaman, India 43 H12 23 14N 87 39 E
Bardejov, Slovak Rep. 17 D11 49 18N 21 15 E
Bardera, Somali Rep. 46 G3 2 20N 42 27 E
Bardīyah, Libya 51 B10 31 45N 25 5 E
Bardsey I., U.K. 10 E3 52 45N 4 47W
Bardstown, U.S.A. 76 G3 37 49N 85 28W
Bareilly, India 43 E8 28 22N 79 27 E
Barela, India 43 H9 23 6N 80 3 E
Barents Sea, Arctic 4 B9 73 0N 39 0 E
Barfleur, Pte. de, France 18 B3 49 42N 1 16W
Bargara, Australia 62 C5 24 50S 152 25 E
Barguzin, Russia 27 D11 53 37N 109 37 E
Barh, India 43 G11 25 29N 85 46 E
Barhaj, India 43 F10 26 18N 83 44 E
Barharwa, India 43 G12 24 52N 87 47 E
Barhi, India 43 G11 24 15N 85 25 E
Bari, India 42 F7 26 39N 77 39 E
Bari, Italy 20 D7 41 8N 16 51 E
Bari Doab, Pakistan 42 D5 30 20N 73 0 E
Bari Sadri, India 42 G6 24 28N 74 30 E
Barīdī, Ra's, Si. Arabia 44 E3 24 17N 37 31 E
Barīm, Yemen 48 E8 12 39N 43 25 E
Barinas, Venezuela 92 B4 8 36N 70 15W
Baring, C., Canada 68 B8 70 0N 117 30W
Baringo, Kenya 54 B4 0 47N 36 16 E
Baringo, L., Kenya 54 B4 0 47N 36 16 E
Barisal, Bangla. 41 H17 22 45N 90 20 E
Barisan, Bukit, Indonesia 36 E2 3 30S 102 15 E
Barito →, Indonesia 36 E4 4 0S 114 50 E
Bark L., Canada 78 A7 45 27N 77 51W
Barkakana, India 43 H11 23 37N 85 29 E
Barker, U.S.A. 78 C6 43 20N 78 33W
Barkley, L., U.S.A. 77 G2 37 1N 88 14W
Barkley Sound, Canada 72 D3 48 50N 125 10W
Barkly East, S. Africa 56 E4 30 58S 27 33 E
Barkly Roadhouse, Australia 62 B2 19 52S 135 50 E
Barkly Tableland, Australia 62 B2 17 50S 136 40 E
Barkly West, S. Africa 56 D3 28 5S 24 31 E
Barkol Kazak Zizhixian, China 32 B4 43 37N 93 2 E
Bârlad, Romania 17 E14 46 15N 27 38 E
Bârlad →, Romania 17 F14 45 38N 27 32 E
Barlee, L., Australia 61 E2 29 15S 119 30 E
Barlee, Mt., Australia 61 D4 24 38S 128 13 E
Barletta, Italy 20 D7 41 19N 16 17 E
Barlovento, Canary Is. 22 F2 28 48N 17 48W
Barlow L., Canada 73 A8 62 0N 103 0W
Barmedman, Australia 63 E4 34 9S 147 21 E
Barmer, India 42 G4 25 45N 71 20 E
Barmera, Australia 63 E3 34 15S 140 28 E
Barmouth, U.K. 10 E3 52 44N 4 4W
Barna →, India 43 G10 25 21N 83 3 E
Barnagar, India 42 H6 23 7N 75 19 E
Barnala, India 42 D6 30 23N 75 33 E
Barnard Castle, U.K. 10 C6 54 33N 1 55W
Barnaul, Russia 26 D9 53 20N 83 40 E
Barnesville, U.S.A. 77 J3 33 3N 84 9W
Barnet □, U.K. 11 F7 51 38N 0 9W
Barneveld, Neths. 15 B5 52 7N 5 36 E
Barneveld, U.S.A. 79 C9 43 16N 75 14W
Barnhart, U.S.A. 81 K4 31 8N 101 10W
Barnsley, U.K. 10 D6 53 34N 1 27W
Barnstaple, U.K. 11 F3 51 5N 4 4W
Barnstaple Bay = Bideford Bay, U.K. 11 F3 51 5N 4 20W
Barnsville, U.S.A. 80 B6 46 43N 96 28W
Barnwell, U.S.A. 77 J5 33 15N 81 23W
Baro, Nigeria 50 G7 8 35N 6 18 E
Baroda = Vadodara, India 42 H5 22 20N 73 10 E
Baroda, India 42 G7 25 29N 76 35 E
Baroe, S. Africa 56 E3 33 13S 24 33 E
Baron Ra., Australia 60 D4 23 30S 127 45 E
Barotseland, Zambia 53 H4 15 0S 24 0 E
Barpeta, India 41 F17 26 20N 91 10 E
Barques, Pt. Aux, U.S.A. 78 B2 44 4N 82 58W
Barquísimeto, Venezuela 92 A5 10 4N 69 19W
Barr Smith Range, Australia 61 E3 27 4S 120 20 E
Barra, Brazil 93 F10 11 5S 43 10W
Barra, U.K. 12 E1 57 0N 7 29W
Barra, Sd. of, U.K. 12 D1 57 4N 7 25W
Barra de Navidad, Mexico 86 D4 19 12N 104 41W
Barra do Corda, Brazil 93 E9 5 30S 45 10W
Barra do Piraí, Brazil 95 A7 22 30S 43 50W
Barra Falsa, Pta. da, Mozam. 57 C6 22 58S 35 37 E
Barra Hd., U.K. 12 E1 56 47N 7 40W
Barra Mansa, Brazil 95 A7 22 35S 44 12W
Barraba, Australia 63 E5 30 21S 150 35 E
Barrackpur = Barakpur, India 43 H13 22 44N 88 30 E
Barradale Roadhouse, Australia 60 D1 22 42S 114 58 E
Barraigh = Barra, U.K. 12 E1 57 0N 7 29W
Barranca, Lima, Peru 92 F3 10 45S 77 50W
Barranca, Loreto, Peru 92 D3 4 50S 76 50W
Barrancabermeja, Colombia 92 B4 7 0N 73 50W
Barrancas, Venezuela 92 B6 8 55N 62 5W
Barrancos, Portugal 19 C2 38 10N 6 58W
Barranqueras, Argentina 94 B4 27 30S 59 0W
Barranquilla, Colombia 92 A4 11 0N 74 50W
Barraute, Canada 70 C4 48 26N 77 38W
Barre, Mass., U.S.A. 79 D12 42 25N 72 6W
Barre, Vt., U.S.A. 79 B12 44 12N 72 30W
Barreal, Argentina 94 C2 31 33S 69 28W
Barreiras, Brazil 93 F10 12 8S 45 0W
Barreirinhas, Brazil 93 D10 2 30S 42 50W
Barreiro, Portugal 19 C1 38 40N 9 6W
Barren, Nosy, Madag. 57 B7 18 25S 43 40 E
Barretos, Brazil 93 H9 20 30S 48 35W
Barrhead, Canada 72 C6 54 10N 114 24W
Barrie, Canada 78 B5 44 24N 79 40W
Barrier Ra., Australia 63 E3 31 0S 141 30 E
Barrière, Canada 72 C4 51 12N 120 7W
Barrington, U.S.A. 79 E13 41 44N 71 18W
Barrington L., Canada 73 B8 56 55N 100 15W
Barrington Tops, Australia 63 E5 32 6S 151 28 E
Barringun, Australia 63 D4 29 1S 145 41 E
Barro do Garças, Brazil 93 G8 15 54S 52 16W
Barron, U.S.A. 80 C9 45 24N 91 51W
Barrow, U.S.A. 68 A4 71 18N 156 47W
Barrow →, Ireland 13 D5 52 25N 6 58W
Barrow Creek, Australia 62 C1 21 30S 133 55 E
Barrow I., Australia 60 D2 20 45S 115 20 E
Barrow-in-Furness, U.K. 10 C4 54 7N 3 14W
Barrow Pt., Australia 62 A3 14 20S 144 40 E
Barrow Pt., U.S.A. 66 B4 71 10N 156 20W
Barrow Ra., Australia 61 E4 26 0S 127 40 E
Barrow Str., Canada 4 B3 74 20N 95 0W
Barry, U.K. 11 F4 51 24N 3 16W
Barry's Bay, Canada 78 A7 45 29N 77 41W
Barsat, Pakistan 43 A5 36 10N 72 45 E
Barsham, Syria 44 C4 35 21N 40 33 E
Barsi, India 40 K9 18 10N 75 50 E
Barsoi, India 41 G15 25 48N 87 57 E
Barstow, U.S.A. 85 L9 34 54N 117 1W
Barthélemy, Col, Vietnam 38 C5 19 26N 104 6 E
Bartica, Guyana 92 B7 6 25N 58 40W
Bartlesville, U.S.A. 81 G7 36 45N 95 59W
Bartlett, U.S.A. 84 J8 36 29N 118 2W
Bartlett, L., Canada 72 A5 63 5N 118 20W
Bartolomeu Dias, Mozam. 55 G4 21 10S 35 8 E
Barton, U.S.A. 79 B12 44 45N 72 11W
Barton upon Humber, U.K. 10 D7 53 41N 0 25W
Bartow, U.S.A. 77 M5 27 54N 81 50W
Barú, Volcan, Panama 88 E3 8 55N 82 35W
Barumba, Dem. Rep. of the Congo 54 B1 1 3N 23 37 E
Baruunsuu, Mongolia 34 C3 43 43N 105 35 E
Barwani, India 42 H6 22 2N 74 57 E
Barysaw, Belarus 17 A15 54 17N 28 28 E
Barzān, Iraq 44 B5 36 55N 44 3 E
Bāsa'idū, Iran 45 E7 26 35N 55 20 E
Basal, Pakistan 42 C5 33 33N 72 13 E
Basankusa, Dem. Rep. of the Congo 52 D3 1 5N 19 50 E
Basarabeasca, Moldova 17 E15 46 21N 28 58 E
Basarabia = Bessarabiya, Moldova 17 E15 47 0N 28 10 E
Basawa, Afghan. 42 B4 34 15N 70 50 E
Bascuñán, C., Chile 94 B1 28 52S 71 35W
Basel, Switz. 18 C7 47 35N 7 35 E
Bashākerd, Kūhhā-ye, Iran 45 E8 26 42N 58 35 E
Bashaw, Canada 72 C6 52 35N 112 58W
Bāshī, Iran 45 D6 28 41N 51 4 E
Bashkir Republic = Bashkortostan □, Russia 24 D10 54 0N 57 0 E
Bashkortostan □, Russia 24 D10 54 0N 57 0 E
Basibasy, Madag. 57 C7 22 10S 43 40 E
Basilan I., Phil. 37 C6 6 35N 122 0 E
Basilan Str., Phil. 37 C6 6 50N 122 0 E
Basildon, U.K. 11 F8 51 34N 0 28 E
Basim = Washim, India 40 J10 20 3N 77 0 E
Basin, U.S.A. 82 D9 44 23N 108 2W
Basingstoke, U.K. 11 F6 51 15N 1 5W
Baskatong, Rés., Canada 70 C4 46 46N 75 50W
Basle = Basel, Switz. 18 C7 47 35N 7 35 E
Basoda, India 42 H7 23 52N 77 54 E
Basoko, Dem. Rep. of the Congo 54 B1 1 16N 23 40 E
Basque Provinces = País Vasco □, Spain 19 A4 42 50N 2 45W
Basra = Al Başrah, Iraq 44 D5 30 30N 47 50 E
Bass Str., Australia 62 F4 39 15S 146 30 E
Bassano, Canada 72 C6 50 48N 112 20W
Bassano del Grappa, Italy 20 B4 45 46N 11 44 E
Bassas da India, Ind. Oc. 53 J7 22 0S 39 0 E
Basse-Terre, Guadeloupe 89 C7 16 0N 61 44W
Bassein, Burma 41 L19 16 45N 94 30 E
Basseterre, St. Kitts & Nevis 89 C7 17 17N 62 43W
Bassett, U.S.A. 80 D5 42 35N 99 32W
Bassi, India 42 D7 30 44N 76 21 E
Bastak, Iran 45 E7 27 15N 54 25 E
Baştām, Iran 45 B7 36 29N 55 4 E
Bastar, India 41 K12 19 15N 81 40 E
Basti, India 43 F10 26 52N 82 55 E
Bastia, France 18 E8 42 40N 9 30 E
Bastogne, Belgium 15 D5 50 1N 5 43 E
Bastrop, La., U.S.A. 81 J9 32 47N 91 55W
Bastrop, Tex., U.S.A. 81 K6 30 7N 97 19W
Bat Yam, Israel 47 C3 32 2N 34 44 E
Bata, Eq. Guin. 52 D1 1 57N 9 50 E
Bataan □, Phil. 37 B6 14 40N 120 25 E
Batabanó, Cuba 88 B3 22 40N 82 20W
Batabanó, G. de, Cuba 88 B3 22 30N 82 30W
Batac, Phil. 37 A6 18 3N 120 34 E
Batagai, Russia 27 C14 67 38N 134 38 E
Batala, India 42 D6 31 48N 75 12 E
Batama, Dem. Rep. of the Congo 54 B2 0 58N 26 33 E
Batamay, Russia 27 C13 63 30N 129 15 E
Batang, Indonesia 37 G13 6 55S 109 45 E
Batangas, Phil. 37 B6 13 35N 121 10 E
Batanta, Indonesia 37 E8 0 55S 130 40 E
Batatais, Brazil 95 A6 20 54S 47 37W
Batavia, U.S.A. 78 D6 43 0N 78 11W
Batchelor, Australia 60 B5 13 4S 131 1 E
Batdambang, Cambodia 38 F4 13 7N 103 12 E
Batemans B., Australia 63 F5 35 40S 150 12 E
Batemans Bay, Australia 63 F5 35 44S 150 11 E
Bates Ra., Australia 61 E3 27 27S 121 5 E
Batesburg-Leesville, U.S.A. 77 J5 33 54N 81 33W
Batesville, Ark., U.S.A. 81 H9 35 46N 91 39W
Batesville, Miss., U.S.A. 81 H10 34 19N 89 57W
Batesville, Tex., U.S.A. 81 L5 28 58N 99 37W
Bath, Canada 79 B8 44 11N 76 47W
Bath, U.K. 11 F5 51 23N 2 22W
Bath, Maine, U.S.A. 77 D11 43 55N 69 49W
Bath, N.Y., U.S.A. 78 D7 42 20N 77 19W
Bath & North East Somerset □, U.K. 11 F5 51 21N 2 27W
Batheay, Cambodia 39 G5 11 59N 104 57 E
Bathurst = Banjul, Gambia 50 F2 13 28N 16 40W
Bathurst, Australia 63 E4 33 25S 149 31 E
Bathurst, Canada 71 C6 47 37N 65 43W
Bathurst, S. Africa 56 E4 33 30S 26 50 E
Bathurst, C., Canada 68 A7 70 34N 128 0W
Bathurst B., Australia 62 A3 14 16S 144 25 E
Bathurst Harb., Australia 62 G4 43 15S 146 10 E
Bathurst I., Australia 60 B5 11 30S 130 10 E
Bathurst I., Canada 4 B2 76 0N 100 30W
Bathurst Inlet, Canada 68 B9 66 50N 108 1W
Batlow, Australia 63 F4 35 31S 148 9 E
Batman, Turkey 25 G7 37 55N 41 5 E
Baţn al Ghūl, Jordan 47 F4 29 36N 35 56 E
Batna, Algeria 50 A7 35 34N 6 15 E
Batoka, Zambia 55 F2 16 45S 27 15 E
Baton Rouge, U.S.A. 81 K9 30 27N 91 11W
Batong, Ko, Thailand 39 J2 6 32N 99 12 E
Batopilas, Mexico 86 B3 27 0N 107 45W
Batouri, Cameroon 52 D2 4 30N 14 25 E
Båtsfjord, Norway 8 A23 70 38N 29 39 E
Battambang = Batdambang, Cambodia 38 F4 13 7N 103 12 E
Batticaloa, Sri Lanka 40 R12 7 43N 81 45 E
Battipáglia, Italy 20 D6 40 37N 14 58 E
Battle, U.K. 11 G8 50 55N 0 30 E
Battle →, Canada 73 C7 52 43N 108 15W
Battle Creek, U.S.A. 76 D3 42 19N 85 11W
Battle Ground, U.S.A. 84 E4 45 47N 122 32W
Battle Harbour, Canada 71 B8 52 16N 55 35W
Battle Lake, U.S.A. 80 B7 46 17N 95 43W
Battle Mountain, U.S.A. 82 F5 40 38N 116 56W
Battlefields, Zimbabwe 55 F2 18 37S 29 47 E
Battleford, Canada 73 C7 52 45N 108 15W
Batu, Ethiopia 46 F2 6 55N 39 45 E
Batu, Kepulauan, Indonesia 36 E1 0 30S 98 25 E
Batu Caves, Malaysia 39 L3 3 15N 101 40 E
Batu Gajah, Malaysia 39 K3 4 28N 101 3 E
Batu Is. = Batu, Kepulauan, Indonesia 36 E1 0 30S 98 25 E
Batu Pahat, Malaysia 39 M4 1 50N 102 56 E
Batuata, Indonesia 37 F6 6 12S 122 42 E
Batumi, Georgia 25 F7 41 39N 41 44 E
Baturaja, Indonesia 36 E2 4 11S 104 15 E
Baturité, Brazil 93 D11 4 28S 38 45W
Bau, Malaysia 36 D4 1 25N 110 9 E
Baubau, Indonesia 37 F6 5 25S 122 38 E
Baucau, E. Timor 37 F7 8 27S 126 27 E
Bauchi, Nigeria 50 F7 10 22N 9 48 E
Baudette, U.S.A. 80 A7 48 43N 94 36W
Bauer, C., Australia 63 E1 32 44S 134 4 E
Bauhinia, Australia 62 C4 24 35S 149 18 E
Baukau = Baucau, E. Timor 37 F7 8 27S 126 27 E
Bauld, C., Canada 69 C14 51 38N 55 26W
Bauru, Brazil 95 A6 22 10S 49 0W
Bausi, India 43 G12 24 48N 87 1 E
Bauska, Latvia 9 H21 56 24N 24 15 E
Bautzen, Germany 16 C8 51 10N 14 26 E
Bavānāt, Iran 45 D7 30 28N 53 27 E
Bavaria = Bayern □, Germany 16 D6 48 50N 12 0 E
Bavispe →, Mexico 86 B3 29 30N 109 11W
Bawdwin, Burma 41 H20 23 5N 97 20 E
Bawean, Indonesia 36 F4 5 46S 112 35 E
Bawku, Ghana 50 F5 11 3N 0 19W
Bawlake, Burma 41 K20 19 11N 97 21 E
Baxley, U.S.A. 77 K4 31 47N 82 21W
Baxter, U.S.A. 80 B7 46 21N 94 17W
Baxter Springs, U.S.A. 81 G7 37 2N 94 44W
Bay City, Mich., U.S.A. 76 D4 43 36N 83 54W
Bay City, Tex., U.S.A. 81 L7 28 59N 95 58W
Bay Minette, U.S.A. 77 K2 30 53N 87 46W
Bay Roberts, Canada 71 C9 47 36N 53 16W
Bay St. Louis, U.S.A. 81 K10 30 19N 89 20W
Bay Springs, U.S.A. 81 K10 31 59N 89 17W
Bay View, N.Z. 59 H6 39 25S 176 50 E
Baya, Dem. Rep. of the Congo 55 E2 11 53S 27 25 E
Bayamo, Cuba 88 B4 20 20N 76 40W
Bayamón, Puerto Rico 89 C6 18 24N 66 10W
Bayan Har Shan, China 32 C4 34 0N 98 0 E
Bayan Hot = Alxa Zuoqi, China 34 E3 38 50N 105 40 E
Bayan Obo, China 34 D5 41 52N 109 59 E
Bayan-Ovoo = Erdenetsogt, Mongolia 34 C4 42 55N 106 5 E
Bayana, India 42 F7 26 55N 77 18 E
Bayanaūyl, Kazakstan 26 D8 50 45N 75 45 E
Bayandalay, Mongolia 34 C2 43 30N 103 29 E
Bayanhongor, Mongolia 32 B5 46 8N 102 43 E
Bayard, N. Mex., U.S.A. 83 K9 32 46N 108 8W
Bayard, Nebr., U.S.A. 80 E3 41 45N 103 20W
Baybay, Phil. 37 B6 10 40N 124 55 E
Baydhabo = Baidoa, Somali Rep. 46 G3 3 8N 43 30 E
Bayern □, Germany 16 D6 48 50N 12 0 E
Bayeux, France 18 B3 49 17N 0 42W
Bayfield, Canada 78 C3 43 34N 81 42W
Bayfield, U.S.A. 80 B9 46 49N 90 49W
Bayındır, Turkey 21 E12 38 13N 27 39 E
Baykal, Oz., Russia 27 D11 53 0N 108 0 E
Baykan, Turkey 44 B4 38 7N 41 44 E
Baykonur = Bayqongyr, Kazakstan 26 E7 47 48N 65 50 E
Baymak, Russia 24 D10 52 36N 58 19 E
Baynes Mts., Namibia 56 B1 17 15S 13 0 E
Bayombong, Phil. 37 A6 16 30N 121 10 E
Bayonne, France 18 E3 43 30N 1 28W
Bayonne, U.S.A. 79 F10 40 40N 74 7W
Bayovar, Peru 92 E2 5 50S 81 0W
Bayqongyr, Kazakstan 26 E7 47 48N 65 50 E
Bayram-Ali = Bayramaly, Turkmenistan 26 F7 37 37N 62 10 E
Bayramaly, Turkmenistan 26 F7 37 37N 62 10 E
Bayramiç, Turkey 21 E12 39 48N 26 36 E
Bayreuth, Germany 16 D6 49 56N 11 35 E
Bayrūt, Lebanon 47 B4 33 53N 35 31 E
Bays, L. of, Canada 78 A5 45 15N 79 4W
Baysville, Canada 78 A5 45 9N 79 7W
Bayt Lahm, West Bank 47 D4 31 43N 35 12 E
Baytown, U.S.A. 81 L7 29 43N 94 59W
Baza, Spain 19 D4 37 30N 2 47W
Bazaruto, I. do, Mozam. 57 C6 21 40S 35 28 E
Bazhou, China 34 E9 39 8N 116 22 E
Bazmān, Kūh-e, Iran 45 D9 28 4N 60 1 E
Beach, U.S.A. 80 B3 46 58N 104 0W
Beach City, U.S.A. 78 F3 40 39N 81 35W
Beachport, Australia 63 F3 37 29S 140 0 E
Beachy Hd., U.K. 11 G8 50 44N 0 15 E
Beacon, Australia 61 F2 30 26S 117 52 E
Beacon, U.S.A. 79 E11 41 30N 73 58W
Beaconsfield, Australia 62 G4 41 11S 146 48 E
Beagle, Canal, S. Amer. 96 H3 55 0S 68 30W
Beagle Bay, Australia 60 C3 16 58S 122 40 E
Bealanana, Madag. 57 A8 14 33S 48 44 E
Beals Cr. →, U.S.A. 81 J4 32 10N 100 51W
Beamsville, Canada 78 C5 43 12N 79 28W
Bear →, Calif., U.S.A. 84 G5 38 56N 121 36W
Bear →, Utah, U.S.A. 74 B4 41 30N 112 8W
Bear I., Ireland 13 E2 51 38N 9 50W
Bear L., Canada 73 B9 55 8N 96 0W
Bear L., U.S.A. 82 F8 41 59N 111 21W
Beardmore, Canada 70 C2 49 36N 87 57W
Beardmore Glacier, Antarctica 5 E11 84 30S 170 0 E
Beardstown, U.S.A. 80 F9 40 1N 90 26W
Bearma →, India 43 G8 24 20N 79 51 E
Béarn, France 18 E3 43 20N 0 30W
Bearpaw Mts., U.S.A. 82 B9 48 12N 109 30W
Bearskin Lake, Canada 70 B1 53 58N 91 2W
Beas →, India 42 D6 31 10N 74 59 E
Beata, C., Dom. Rep. 89 C5 17 40N 71 30W
Beata, I., Dom. Rep. 89 C5 17 34N 71 31W
Beatrice, U.S.A. 80 E6 40 16N 96 45W
Beatrice, Zimbabwe 55 F3 18 15S 30 55 E
Beatrice, C., Australia 62 A2 14 20S 136 55 E
Beatton →, Canada 72 B4 56 15N 120 45W
Beatton River, Canada 72 B4 57 26N 121 20W
Beatty, U.S.A. 84 J10 36 54N 116 46W
Beauce, Plaine de la, France 18 B4 48 10N 1 45 E
Beauceville, Canada 71 C5 46 13N 70 46W
Beaudesert, Australia 63 D5 27 59S 153 0 E
Beaufort, Malaysia 36 C5 5 30N 115 40 E
Beaufort, N.C., U.S.A. 77 H7 34 43N 76 40W
Beaufort, S.C., U.S.A. 77 J5 32 26N 80 40W
Beaufort Sea, Arctic 4 B1 72 0N 140 0W
Beaufort West, S. Africa 56 E3 32 18S 22 36 E
Beauharnois, Canada 79 A11 45 20N 73 52W
Beaulieu →, Canada 72 A6 62 3N 113 11W
Beauly, U.K. 12 D4 57 30N 4 28W
Beauly →, U.K. 12 D4 57 29N 4 27W
Beaumaris, U.K. 10 D3 53 16N 4 6W
Beaumont, Belgium 15 D4 50 15N 4 14 E
Beaumont, U.S.A. 81 K7 30 5N 94 6W
Beaune, France 18 C6 47 2N 4 50 E
Beaupré, Canada 71 C5 47 3N 70 54W
Beauraing, Belgium 15 D4 50 7N 4 57 E
Beauséjour, Canada 73 C9 50 5N 96 35W
Beauvais, France 18 B5 49 25N 2 8 E
Beauval, Canada 73 B7 55 9N 107 37W
Beaver, Okla., U.S.A. 81 G4 36 49N 100 31W
Beaver, Pa., U.S.A. 78 F4 40 42N 80 19W
Beaver, Utah, U.S.A. 83 G7 38 17N 112 38W
Beaver →, B.C., Canada 72 B4 59 52N 124 20W
Beaver →, Ont., Canada 70 A2 55 55N 87 48W
Beaver →, Sask., Canada 73 B7 55 26N 107 45W
Beaver →, U.S.A. 81 G5 36 35N 99 30W
Beaver City, U.S.A. 80 E5 40 8N 99 50W
Beaver Creek, Canada 68 B5 63 0N 141 0W
Beaver Dam, U.S.A. 80 D10 43 28N 88 50W
Beaver Falls, U.S.A. 78 F4 40 46N 80 20W
Beaver Hill L., Canada 73 C10 54 5N 94 50W
Beaver I., U.S.A. 76 C3 45 40N 85 33W
Beaverhill L., Canada 72 C6 53 27N 112 32W
Beaverlodge, Canada 72 B5 55 11N 119 29W
Beaverstone →, Canada 70 B2 54 59N 89 25W
Beaverton, Canada 78 B5 44 26N 79 9W
Beaverton, U.S.A. 84 E4 45 29N 122 48W
Beawar, India 42 F6 26 3N 74 18 E
Bebedouro, Brazil 95 A6 21 0S 48 25W
Beboa, Madag. 57 B7 17 22S 44 33 E
Beccles, U.K. 11 E9 52 27N 1 35 E
Bečej, Serbia, Yug. 21 B9 45 36N 20 3 E
Béchar, Algeria 50 B5 31 38N 2 18W
Beckley, U.S.A. 76 G5 37 47N 81 11W
Beddouza, Ras, Morocco 50 B4 32 33N 9 9W
Bedford, Canada 79 A12 45 7N 72 59W
Bedford, S. Africa 56 E4 32 40S 26 10 E
Bedford, U.K. 11 E7 52 8N 0 28W
Bedford, Ind., U.S.A. 76 F2 38 52N 86 29W
Bedford, Iowa, U.S.A. 80 E7 40 40N 94 44W
Bedford, Ohio, U.S.A. 78 E3 41 23N 81 32W
Bedford, Pa., U.S.A. 78 F6 40 1N 78 30W
Bedford, Va., U.S.A. 76 G6 37 20N 79 31W
Bedford, C., Australia 62 B4 15 14S 145 21 E
Bedfordshire □, U.K. 11 E7 52 4N 0 28W
Bedourie, Australia 62 C2 24 30S 139 30 E
Bedum, Neths. 15 A6 53 18N 6 36 E
Beebe Plain, Canada 79 A12 45 1N 72 9W
Beech Creek, U.S.A. 78 E7 41 5N 77 36W
Beenleigh, Australia 63 D5 27 43S 153 10 E
Be'er Menuha, Israel 44 D2 30 19N 35 8 E
Be'er Sheva, Israel 47 D3 31 15N 34 48 E
Beersheba = Be'er Sheva, Israel 47 D3 31 15N 34 48 E
Beestekraal, S. Africa 57 D4 25 23S 27 38 E
Beeston, U.K. 10 E6 52 56N 1 14W
Beeville, U.S.A. 81 L6 28 24N 97 45W
Befale, Dem. Rep. of the Congo 52 D4 0 25N 20 45 E
Befandriana, Mahajanga, Madag. 57 B8 15 16S 48 32 E
Befandriana, Toliara, Madag. 57 C7 21 55S 44 0 E
Befasy, Madag. 57 C7 20 33S 44 23 E
Befotaka, Antsiranana, Madag. 57 A8 13 15S 48 16 E
Befotaka, Fianarantsoa, Madag. 57 C8 23 49S 47 0 E
Bega, Australia 63 F4 36 41S 149 51 E
Begusarai, India 43 G12 25 24N 86 9 E
Behābād, Iran 45 C8 32 24N 59 47 E

Biaora

Behala, India 43 H13 22 30N 88 20 E
Behara, Madag. 57 C8 24 55S 46 20 E
Behbehān, Iran 45 D6 30 30N 50 15 E
Behm Canal, U.S.A. 72 B2 55 10N 131 0W
Behshahr, Iran 45 B7 36 45N 53 35 E
Bei Jiang →, China 33 D6 23 2N 112 58 E
Bei'an, China 33 B7 48 10N 126 20 E
Beihai, China 33 D5 21 28N 109 6 E
Beijing, China 34 E9 39 55N 116 20 E
Beijing □, China 34 E9 39 55N 116 20 E
Beilen, Neths. 15 B6 52 52N 6 27 E
Beilpajah, Australia 63 E3 32 54S 143 52 E
Beinn na Faoghla = Benbecula, U.K. 12 D1 57 26N 7 21W
Beipiao, China 35 D11 41 52N 120 32 E
Beira, Mozam. 55 F3 19 50S 34 52 E
Beirut = Bayrūt, Lebanon 47 B4 33 53N 35 31 E
Beiseker, Canada 72 C6 51 23N 113 32W
Beitaolaizhao, China 35 B13 44 58N 125 58 E
Beitbridge, Zimbabwe 55 G3 22 12S 30 0 E
Beizhen = Binzhou, China 35 F10 37 20N 118 2 E
Beizhen, China 35 D11 41 38N 121 54 E
Beizhengzhen, China 35 B12 44 31N 123 30 E
Beja, Portugal 19 C2 38 2N 7 53W
Béja, Tunisia 51 A7 36 43N 9 12 E
Bejaïa, Algeria 50 A7 36 42N 5 2 E
Béjar, Spain 19 B3 40 23N 5 46W
Bejestān, Iran 45 C8 34 30N 58 5 E
Békéscsaba, Hungary 17 E11 46 40N 21 5 E
Bekily, Madag. 57 C8 24 13S 45 19 E
Bekisopa, Madag. 57 C8 21 40S 45 54 E
Bekitro, Madag. 57 C8 24 33S 45 18 E
Bekodoka, Madag. 57 B8 16 58S 45 7 E
Bekok, Malaysia 39 L4 2 20N 103 7 E
Bekopaka, Madag. 57 B7 19 9S 44 48 E
Bela, India 43 G10 25 50N 82 0 E
Bela, Pakistan 42 F2 26 12N 66 20 E
Bela Crkva, Serbia, Yug. 21 B9 44 55N 21 27 E
Bela Vista, Brazil 94 A4 22 12S 56 20W
Bela Vista, Mozam. 57 D5 26 10S 32 44 E
Belan →, India 43 G9 24 2N 81 45 E
Belarus ■, Europe 17 B14 53 30N 27 0 E
Belau = Palau ■, Pac. Oc. 28 J17 7 30N 134 30 E
Belavenona, Madag. 57 C8 24 50S 47 4 E
Belawan, Indonesia 36 D1 3 33N 98 32 E
Belaya →, Russia 24 C9 54 40N 56 0 E
Belaya Tserkov = Bila Tserkva, Ukraine 17 D16 49 45N 30 10 E
Belcher Is., Canada 70 A3 56 15N 78 45W
Belden, U.S.A. 84 E5 40 2N 121 17W
Belebey, Russia 24 D9 54 7N 54 7 E
Beled Weyne = Belet Uen, Somali Rep. 46 G4 4 30N 45 5 E
Belém, Brazil 93 D9 1 20S 48 30W
Belén, Argentina 94 B2 27 40S 67 5W
Belén, Paraguay 94 A4 23 30S 57 6W
Belen, U.S.A. 83 J10 34 40N 106 46W
Belet Uen, Somali Rep. 46 G4 4 30N 45 5 E
Belev, Russia 24 D6 53 50N 36 5 E
Belfair, U.S.A. 84 C4 47 27N 122 50W
Belfast, S. Africa 57 D5 25 42S 30 2 E
Belfast, U.K. 13 B6 54 37N 5 56W
Belfast, Maine, U.S.A. 77 C11 44 26N 69 1W
Belfast, N.Y., U.S.A. 78 D6 42 21N 78 7W
Belfast L., U.K. 13 B6 54 40N 5 50W
Belfield, U.S.A. 80 B3 46 53N 103 12W
Belfort, France 18 C7 47 38N 6 50 E
Belfry, U.S.A. 82 D9 45 9N 109 1W
Belgaum, India 40 M9 15 55N 74 35 E
Belgium ■, Europe 15 D4 50 30N 5 0 E
Belgorod, Russia 25 D6 50 35N 36 35 E
Belgorod-Dnestrovskiy = Bilhorod-Dnistrovskyy, Ukraine 25 E5 46 11N 30 23 E
Belgrade = Beograd, Serbia, Yug. 21 B9 44 50N 20 37 E
Belgrade, U.S.A. 82 D8 45 47N 111 11W
Belhaven, U.S.A. 77 H7 35 33N 76 37W
Beli Drim →, Europe 21 C9 42 6N 20 25 E
Belinyu, Indonesia 36 E3 1 35S 105 50 E
Beliton Is. = Belitung, Indonesia 36 E3 3 10S 107 50 E
Belitung, Indonesia 36 E3 3 10S 107 50 E
Belize ■, Cent. Amer. 87 D7 17 0N 88 30W
Belize City, Belize 87 D7 17 25N 88 0W
Belkovskiy, Ostrov, Russia 27 B14 75 32N 135 44 E
Bell →, Canada 70 C4 49 48N 77 38W
Bell I., Canada 71 B8 50 46N 55 35W
Bell-Irving →, Canada 72 B3 56 12N 129 5W
Bell Peninsula, Canada 69 B11 63 50N 82 0W
Bell Ville, Argentina 94 C3 32 40S 62 40W
Bella Bella, Canada 72 C3 52 10N 128 10W
Bella Coola, Canada 72 C3 52 25N 126 40W
Bella Unión, Uruguay 94 C4 30 15S 57 40W
Bella Vista, Corrientes, Argentina 94 B4 28 33S 59 0W
Bella Vista, Tucuman, Argentina 94 B2 27 10S 65 25W
Bellaire, U.S.A. 78 F4 40 1N 80 45W
Bellary, India 40 M10 15 10N 76 56 E
Bellata, Australia 63 D4 29 53S 149 46 E
Belle-Chasse, U.S.A. 81 L10 29 51N 89 59W
Belle Fourche, U.S.A. 80 C3 44 40N 103 51W
Belle Fourche →, U.S.A. 80 C3 44 26N 102 18W
Belle Glade, U.S.A. 77 M5 26 41N 80 40W
Belle-Île, France 18 C2 47 20N 3 10W
Belle Isle, Canada 71 B8 51 57N 55 25W
Belle Isle, Str. of, Canada 71 B8 51 30N 56 30W
Belle Plaine, U.S.A. 80 E8 41 54N 92 17W
Bellefontaine, U.S.A. 76 E4 40 22N 83 46W
Bellefonte, U.S.A. 78 F7 40 55N 77 47W
Belleoram, Canada 71 C8 47 31N 55 25W
Belleville, Canada 78 B7 44 10N 77 23W
Belleville, Ill., U.S.A. 80 F10 38 31N 89 59W
Belleville, Kans., U.S.A. 80 F6 39 50N 97 38W
Belleville, N.Y., U.S.A. 79 C8 43 46N 76 10W
Bellevue, Canada 72 D6 49 35N 114 22W
Bellevue, Idaho, U.S.A. 82 E6 43 28N 114 16W
Bellevue, Nebr., U.S.A. 80 E7 41 8N 95 53W
Bellevue, Ohio, U.S.A. 78 E2 41 17N 82 51W
Bellevue, Wash., U.S.A. 84 C4 47 37N 122 12W
Bellin = Kangirsuk, Canada 69 B13 60 0N 70 0W
Bellingen, Australia 63 E5 30 25S 152 50 E
Bellingham, U.S.A. 68 D7 48 46N 122 29W
Bellingshausen Sea, Antarctica 5 C17 66 0S 80 0W
Bellinzona, Switz. 18 C8 46 11N 9 1 E
Bello, Colombia 92 B3 6 20N 75 33W
Bellows Falls, U.S.A. 79 C12 43 8N 72 27W
Bellpat, Pakistan 42 E3 29 0N 68 5 E
Belluno, Italy 20 A5 46 9N 12 13 E
Bellwood, U.S.A. 78 F6 40 36N 78 20W
Belmont, Canada 78 D3 42 53N 81 5W
Belmont, S. Africa 56 D3 29 28S 24 22 E
Belmont, U.S.A. 78 D6 42 14N 78 2W
Belmonte, Brazil 93 G11 16 0S 39 0W
Belmopan, Belize 87 D7 17 18N 88 30W
Belmullet, Ireland 13 B2 54 14N 9 58W
Belo Horizonte, Brazil 93 G10 19 55S 43 56W
Belo-sur-Mer, Madag. 57 C7 20 42S 44 0 E
Belo-Tsiribihina, Madag. 57 B7 19 40S 44 30 E
Belogorsk, Russia 27 D13 51 0N 128 20 E
Beloha, Madag. 57 D8 25 10S 45 3 E
Beloit, Kans., U.S.A. 80 F5 39 28N 98 6W
Beloit, Wis., U.S.A. 80 D10 42 31N 89 2W
Belokorovichi, Ukraine 17 C15 51 7N 28 2 E
Belomorsk, Russia 24 B5 64 35N 34 54 E
Belonia, India 41 H17 23 15N 91 30 E
Beloretsk, Russia 24 D10 53 58N 58 24 E
Belorussia = Belarus ■, Europe 17 B14 53 30N 27 0 E
Belovo, Russia 26 D9 54 30N 86 0 E
Beloye, Ozero, Russia 24 B6 60 10N 37 35 E
Beloye More, Russia 24 A6 66 30N 38 0 E
Belozersk, Russia 24 B6 60 1N 37 45 E
Belpre, U.S.A. 76 F5 39 17N 81 34W
Belrain, India 43 E9 28 23N 80 55 E
Belt, U.S.A. 82 C8 47 23N 110 55W
Beltana, Australia 63 E2 30 48S 138 25 E
Belterra, Brazil 93 D8 2 45S 55 0W
Belton, U.S.A. 81 K6 31 3N 97 28W
Belton L., U.S.A. 81 K6 31 8N 97 32W
Beltsy = Bălţi, Moldova 17 E14 47 48N 27 58 E
Belturbet, Ireland 13 B4 54 6N 7 26W
Belukha, Russia 26 E9 49 50N 86 50 E
Beluran, Malaysia 36 C5 5 48N 117 35 E
Belvidere, Ill., U.S.A. 80 D10 42 15N 88 50W
Belvidere, N.J., U.S.A. 79 F9 40 50N 75 5W
Belyando →, Australia 62 C4 21 38S 146 50 E
Belyy, Ostrov, Russia 26 B8 73 30N 71 0 E
Belyy Yar, Russia 26 D9 58 26N 84 39 E
Belzoni, U.S.A. 81 J9 33 11N 90 29W
Bemaraha, Lembalemban' i, Madag. 57 B7 18 40S 44 45 E
Bemarivo, Madag. 57 C7 21 45S 44 45 E
Bemarivo →, Antsiranana, Madag. 57 A9 14 9S 50 9 E
Bemarivo →, Mahajanga, Madag. 57 B8 15 27S 47 40 E
Bemavo, Madag. 57 C8 21 33S 45 25 E
Bembéréke, Benin 50 F6 10 11N 2 43 E
Bembesi, Zimbabwe 55 G2 20 0S 28 58 E
Bembesi →, Zimbabwe 55 F2 18 57S 27 47 E
Bemetara, India 43 J9 21 42N 81 32 E
Bemidji, U.S.A. 80 B7 47 28N 94 53W
Bemolanga, Madag. 57 B8 17 44S 45 6 E
Ben, Iran 45 C6 32 32N 50 45 E
Ben Cruachan, U.K. 12 E3 56 26N 5 8W
Ben Dearg, U.K. 12 D4 57 47N 4 56W
Ben Hope, U.K. 12 C4 58 25N 4 36W
Ben Lawers, U.K. 12 E4 56 32N 4 14W
Ben Lomond, N.S.W., Australia 63 E5 30 1S 151 43 E
Ben Lomond, Tas., Australia 62 G4 41 38S 147 42 E
Ben Lomond, U.K. 12 E4 56 11N 4 38W
Ben Luc, Vietnam 39 G6 10 39N 106 29 E
Ben Macdhui, U.K. 12 D5 57 4N 3 40W
Ben Mhor, U.K. 12 D1 57 15N 7 18W
Ben More, Arg. & Bute, U.K. 12 E2 56 26N 6 1W
Ben More, Stirl., U.K. 12 E4 56 23N 4 32W
Ben More Assynt, U.K. 12 C4 58 8N 4 52W
Ben Nevis, U.K. 12 E3 56 48N 5 1W
Ben Quang, Vietnam 38 D6 17 3N 106 55 E
Ben Vorlich, U.K. 12 E4 56 21N 4 14W
Ben Wyvis, U.K. 12 D4 57 40N 4 35W
Bena, Nigeria 50 F7 11 20N 5 50 E
Benalla, Australia 63 F4 36 30S 146 0 E
Benares = Varanasi, India 43 G10 25 22N 83 0 E
Benavente, Spain 19 A3 42 2N 5 43W
Benavides, U.S.A. 81 M5 27 36N 98 25W
Benbecula, U.K. 12 D1 57 26N 7 21W
Benbonyathe, Australia 63 E2 30 25S 139 11 E
Bend, U.S.A. 82 D3 44 4N 121 19W
Bender Beila, Somali Rep. 46 F5 9 30N 50 48 E
Bendery = Tighina, Moldova 17 E15 46 50N 29 30 E
Bendigo, Australia 63 F3 36 40S 144 15 E
Benë Beraq, Israel 47 C3 32 6N 34 51 E
Benenitra, Madag. 57 C8 23 27S 45 5 E
Benevento, Italy 20 D6 41 8N 14 45 E
Benga, Mozam. 55 F3 16 11S 33 40 E
Bengal, Bay of, Ind. Oc. 41 M17 15 0N 90 0 E
Bengbu, China 35 H9 32 58N 117 20 E
Benghazi = Banghāzī, Libya 51 B10 32 11N 20 3 E
Bengkalis, Indonesia 36 D2 1 30N 102 10 E
Bengkulu, Indonesia 36 E2 3 50S 102 12 E
Bengkulu □, Indonesia 36 E2 3 48S 102 16 E
Bengough, Canada 73 D7 49 25N 105 10W
Benguela, Angola 53 G2 12 37S 13 25 E
Benguérua, I., Mozam. 57 C6 21 58S 35 28 E
Beni, Dem. Rep. of the Congo 54 B2 0 30N 29 27 E
Beni →, Bolivia 92 F5 10 23S 65 24W
Beni Mellal, Morocco 50 B4 32 21N 6 21W
Beni Suef, Egypt 51 C12 29 5N 31 6 E
Beniah L., Canada 72 A6 63 23N 112 17W
Benicia, U.S.A. 84 G4 38 3N 122 9W
Benidorm, Spain 19 C5 38 33N 0 9W
Benin ■, Africa 50 G6 10 0N 2 0 E
Benin, Bight of, W. Afr. 50 H6 5 0N 3 0 E
Benin City, Nigeria 50 G7 6 20N 5 31 E
Benitses, Greece 23 A3 39 32N 19 55 E
Benjamin Aceval, Paraguay 94 A4 24 58S 57 34W
Benjamin Constant, Brazil 92 D4 4 40S 70 15W
Benjamin Hill, Mexico 86 A2 30 10N 111 10W
Benkelman, U.S.A. 80 E4 40 3N 101 32W
Bennett, Canada 72 B2 59 56N 134 53W
Bennett, L., Australia 60 D5 22 50S 131 2 E
Bennetta, Ostrov, Russia 27 B15 76 21N 148 56 E
Bennettsville, U.S.A. 77 H6 34 37N 79 41W
Bennington, N.H., U.S.A. 79 D11 43 0N 71 55W
Bennington, Vt., U.S.A. 79 D11 42 53N 73 12W
Benoni, S. Africa 57 D4 26 11S 28 18 E
Benque Viejo, Belize 87 D7 17 5N 89 8W
Benson, Ariz., U.S.A. 83 L8 31 58N 110 18W
Benson, Minn., U.S.A. 80 C7 45 19N 95 36W
Bent, Iran 45 E8 26 20N 59 31 E
Benteng, Indonesia 37 F6 6 10S 120 30 E
Bentinck I., Australia 62 B2 17 3S 139 35 E
Bento Gonçalves, Brazil 95 B5 29 10S 51 31W
Benton, Ark., U.S.A. 81 H8 34 34N 92 35W
Benton, Calif., U.S.A. 84 H8 37 48N 118 32W
Benton, Ill., U.S.A. 80 G10 38 0N 88 55W
Benton, Pa., U.S.A. 79 E8 41 12N 76 23W
Benton Harbor, U.S.A. 76 D2 42 6N 86 27W
Bentonville, U.S.A. 81 G7 36 22N 94 13W
Bentung, Malaysia 39 L3 3 31N 101 55 E
Benue →, Nigeria 50 G7 7 48N 6 46 E
Benxi, China 35 D12 41 20N 123 48 E
Beo, Indonesia 37 D7 4 25N 126 50 E
Beograd, Serbia, Yug. 21 B9 44 50N 20 37 E
Beppu, Japan 31 H5 33 15N 131 30 E
Beqaa Valley = Al Biqā, Lebanon 47 A5 34 10N 36 10 E
Ber Mota, India 42 H3 23 27N 68 34 E
Berach →, India 42 G6 25 15N 75 2 E
Beraketa, Madag. 57 C7 23 7S 44 25 E
Berat, Albania 21 D8 40 43N 19 59 E
Berau, Teluk, Indonesia 37 E8 2 30S 132 30 E
Beravina, Madag. 57 B8 18 10S 45 14 E
Berber, Sudan 51 E12 18 0N 34 0 E
Berbera, Somali Rep. 46 E4 10 30N 45 2 E
Berbérati, C.A.R. 52 D3 4 15N 15 40 E
Berbice →, Guyana 92 B7 6 20N 57 32W
Berdichev = Berdychiv, Ukraine 17 D15 49 57N 28 30 E
Berdsk, Russia 26 D9 54 47N 83 2 E
Berdyansk, Ukraine 25 E6 46 45N 36 50 E
Berdychiv, Ukraine 17 D15 49 57N 28 30 E
Berea, U.S.A. 76 G3 37 34N 84 17W
Berebere, Indonesia 37 D7 2 25N 128 45 E
Bereda, Somali Rep. 46 E5 11 45N 51 0 E
Berehove, Ukraine 17 D12 48 15N 22 35 E
Berekum, Ghana 50 G5 7 29N 2 34W
Berens →, Canada 73 C9 52 25N 97 2W
Berens I., Canada 73 C9 52 18N 97 18W
Berens River, Canada 73 C9 52 25N 97 0W
Beresford, U.S.A. 80 D6 43 5N 96 47W
Berestechko, Ukraine 17 C13 50 22N 25 5 E
Berevo, Mahajanga, Madag. 57 B7 17 14S 44 17 E
Berevo, Toliara, Madag. 57 B7 19 44S 44 58 E
Bereza = Byaroza, Belarus 17 B13 52 31N 24 51 E
Berezhany, Ukraine 17 D13 49 26N 24 58 E
Berezina = Byarezina →, Belarus 17 B16 52 33N 30 14 E
Bereznik, Russia 24 B7 62 51N 42 40 E
Berezniki, Russia 24 C10 59 24N 56 46 E
Berezovo, Russia 26 C7 64 0N 65 0 E
Berga, Spain 19 A6 42 6N 1 48 E
Bergama, Turkey 21 E12 39 8N 27 11 E
Bérgamo, Italy 18 D8 45 41N 9 43 E
Bergen, Neths. 15 B4 52 40N 4 43 E
Bergen, Norway 9 F11 60 20N 5 20 E
Bergen, U.S.A. 78 C7 43 5N 77 57W
Bergen op Zoom, Neths. 15 C4 51 28N 4 18 E
Bergerac, France 18 D4 44 51N 0 30 E
Bergholz, U.S.A. 78 F4 40 31N 80 53W
Bergisch Gladbach, Germany 15 D7 50 59N 7 8 E
Bergville, S. Africa 57 D4 28 52S 29 18 E
Berhala, Selat, Indonesia 36 E2 1 0S 104 15 E
Berhampore = Baharampur, India 43 G13 24 2N 88 27 E
Berhampur = Brahmapur, India 41 K14 19 15N 84 54 E
Bering Sea, Pac. Oc. 68 C1 58 0N 171 0 E
Bering Strait, Pac. Oc. 68 B3 65 30N 169 0W
Beringovskiy, Russia 27 C18 63 3N 179 19 E
Berisso, Argentina 94 C4 34 56S 57 50W
Berja, Spain 19 D4 36 50N 2 56W
Berkeley, U.S.A. 84 H4 37 52N 122 16W
Berkner I., Antarctica 5 D18 79 30S 50 0W
Berkshire, U.S.A. 79 D8 42 19N 76 11W
Berkshire Downs, U.K. 11 F6 51 33N 1 29W
Berlin, Germany 16 B7 52 30N 13 25 E
Berlin, Md., U.S.A. 76 F8 38 20N 75 13W
Berlin, N.H., U.S.A. 79 B13 44 28N 71 11W
Berlin, N.Y., U.S.A. 79 D11 42 42N 73 23W
Berlin, Wis., U.S.A. 76 D1 43 58N 88 57W
Berlin L., U.S.A. 78 E4 41 3N 81 0W
Bermejo →, Formosa, Argentina 94 B4 26 51S 58 23W
Bermejo →, San Juan, Argentina 94 C2 32 30S 67 30W
Bermen, L., Canada 71 B6 53 35N 68 55W
Bermuda ■, Atl. Oc. 66 F13 32 45N 65 0W
Bern, Switz. 18 C7 46 57N 7 28 E
Bernalillo, U.S.A. 83 J10 35 18N 106 33W
Bernardo de Irigoyen, Argentina 95 B5 26 15S 53 40W
Bernardo O'Higgins □, Chile 94 C1 34 15S 70 45W
Bernardsville, U.S.A. 79 F10 40 43N 74 34W
Bernasconi, Argentina 94 D3 37 55S 63 44W
Bernburg, Germany 16 C6 51 47N 11 44 E
Berne = Bern, Switz. 18 C7 46 57N 7 28 E
Berneray, U.K. 12 D1 57 43N 7 11W
Bernier I., Australia 61 D1 24 50S 113 12 E
Bernina, Piz, Switz. 18 C8 46 20N 9 54 E
Beroroha, Madag. 57 C8 21 40S 45 10 E
Beroun, Czech Rep. 16 D8 49 57N 14 5 E
Berri, Australia 63 E3 34 14S 140 35 E
Berriane, Algeria 50 B6 32 50N 3 46 E
Berry, Australia 63 E5 34 46S 150 43 E
Berry, France 18 C5 46 50N 2 0 E
Berry Is., Bahamas 88 A4 25 40N 77 50W
Berryessa L., U.S.A. 84 G4 38 31N 122 6W
Berryville, U.S.A. 81 G8 36 22N 93 34W
Berseba, Namibia 56 D2 26 0S 17 46 E
Bershad, Ukraine 17 D15 48 22N 29 31 E
Berthold, U.S.A. 80 A4 48 19N 101 44W
Berthoud, U.S.A. 80 E2 40 19N 105 5W
Bertoua, Cameroon 52 D2 4 30N 13 45 E
Bertraghboy B., Ireland 13 C2 53 22N 9 54W
Berwick, U.S.A. 79 E8 41 3N 76 14W
Berwick-upon-Tweed, U.K. 10 B6 55 46N 2 0W
Berwyn Mts., U.K. 10 E4 52 54N 3 26W
Besal, Pakistan 43 B5 35 4N 73 56 E
Besalampy, Madag. 57 B7 16 43S 44 29 E
Besançon, France 18 C7 47 15N 6 2 E
Besar, Indonesia 36 E5 2 40S 116 0 E
Besnard L., Canada 73 B7 55 25N 106 0W
Besni, Turkey 44 B3 37 41N 37 52 E
Besor, N. →, Egypt 47 D3 31 28N 34 22 E
Bessarabiya, Moldova 17 E15 47 0N 28 10 E
Bessarabka = Basarabeasca, Moldova 17 E15 46 21N 28 58 E
Bessemer, Ala., U.S.A. 77 J2 33 24N 86 58W
Bessemer, Mich., U.S.A. 80 B9 46 29N 90 3W
Bessemer, Pa., U.S.A. 78 F4 40 59N 80 30W
Beswick, Australia 60 B5 14 34S 132 53 E
Bet She'an, Israel 47 C4 32 30N 35 30 E
Bet Shemesh, Israel 47 D4 31 44N 35 0 E
Betafo, Madag. 57 B8 19 50S 46 51 E
Betancuria, Canary Is. 22 F5 28 25N 14 3W
Betanzos, Spain 19 A1 43 15N 8 12W
Bétaré Oya, Cameroon 52 C2 5 40N 14 5 E
Betatao, Madag. 57 B8 18 11S 47 52 E
Bethal, S. Africa 57 D4 26 27S 29 28 E
Bethanien, Namibia 56 D2 26 31S 17 8 E
Bethany, Canada 78 B6 44 11N 78 34W
Bethany, U.S.A. 80 E7 40 16N 94 2W
Bethel, Alaska, U.S.A. 68 B3 60 48N 161 45W
Bethel, Conn., U.S.A. 79 E11 41 22N 73 25W
Bethel, Maine, U.S.A. 79 B14 44 25N 70 47W
Bethel, Vt., U.S.A. 79 C12 43 50N 72 38W
Bethel Park, U.S.A. 78 F4 40 20N 80 1W
Bethlehem = Bayt Lahm, West Bank 47 D4 31 43N 35 12 E
Bethlehem, S. Africa 57 D4 28 14S 28 18 E
Bethlehem, U.S.A. 79 F9 40 37N 75 23W
Bethulie, S. Africa 56 E4 30 30S 25 59 E
Béthune, France 18 A5 50 30N 2 38 E
Betioky, Madag. 57 C7 23 48S 44 20 E
Betong, Thailand 39 K3 5 45N 101 5 E
Betoota, Australia 62 D3 25 45S 140 42 E
Betroka, Madag. 57 C8 23 16S 46 0 E
Betsiamites, Canada 71 C6 48 56N 68 40W
Betsiamites →, Canada 71 C6 48 56N 68 38W
Betsiboka →, Madag. 57 B8 16 3S 46 36 E
Bettendorf, U.S.A. 80 E9 41 32N 90 30W
Bettiah, India 43 F11 26 48N 84 33 E
Betul, India 40 J10 21 58N 77 59 E
Betung, Malaysia 36 D4 1 24N 111 31 E
Betws-y-Coed, U.K. 10 D4 53 5N 3 48W
Beulah, Mich., U.S.A. 76 C2 44 38N 86 6W
Beulah, N. Dak., U.S.A. 80 B4 47 16N 101 47W
Beveren, Belgium 15 C4 51 12N 4 16 E
Beverley, Australia 61 F2 32 9S 116 56 E
Beverley, U.K. 10 D7 53 51N 0 26W
Beverley Hills, U.S.A. 77 L4 28 56N 82 28W
Beverly, U.S.A. 79 D14 42 33N 70 53W
Beverly Hills, U.S.A. 85 L8 34 4N 118 25W
Bevoalavo, Madag. 57 D7 25 13S 45 26 E
Bewas →, India 43 H8 23 59N 79 21 E
Bexhill, U.K. 11 G8 50 51N 0 29 E
Beyānlū, Iran 44 C5 36 0N 47 51 E
Beyneu, Kazakstan 25 E10 45 18N 55 9 E
Beypazarı, Turkey 25 F5 40 10N 31 56 E
Beyşehir Gölü, Turkey 25 G5 37 41N 31 33 E
Béziers, France 18 E5 43 20N 3 12 E
Bezwada = Vijayawada, India 41 L12 16 31N 80 39 E
Bhabua, India 43 G10 25 3N 83 37 E
Bhachau, India 40 H7 23 20N 70 16 E
Bhadar →, Gujarat, India 42 H5 22 17N 72 20 E
Bhadar →, Gujarat, India 42 J3 21 27N 69 47 E
Bhadarwah, India 43 C6 32 58N 75 46 E
Bhadohi, India 43 G10 25 25N 82 34 E
Bhadra, India 42 E6 29 8N 75 14 E
Bhadrakh, India 41 J15 21 10N 86 30 E
Bhadran, India 42 H5 22 19N 72 6 E
Bhadravati, India 40 N9 13 49N 75 40 E
Bhag, Pakistan 42 E2 29 2N 67 49 E
Bhagalpur, India 43 G12 25 10N 87 0 E
Bhagirathi →, Uttaranchal, India 43 D8 30 8N 78 35 E
Bhagirathi →, W. Bengal, India 43 H13 23 25N 88 23 E
Bhakkar, Pakistan 42 D4 31 40N 71 5 E
Bhakra Dam, India 42 D7 31 30N 76 45 E
Bhaktapur, Nepal 43 F11 27 38N 85 24 E
Bhamo, Burma 41 G20 24 15N 97 15 E
Bhandara, India 40 J11 21 5N 79 42 E
Bhanpura, India 42 G6 24 31N 75 44 E
Bhanrer Ra., India 43 H8 23 40N 79 45 E
Bhaptiahi, India 43 F12 26 19N 86 44 E
Bharat = India ■, Asia 40 K11 20 0N 78 0 E
Bharatpur, Chhattisgarh, India 43 H9 23 44N 81 46 E
Bharatpur, Raj., India 42 F7 27 15N 77 30 E
Bharno, India 43 H11 23 14N 84 53 E
Bhatinda, India 42 D6 30 15N 74 57 E
Bhatpara, India 43 H13 22 50N 88 25 E
Bhattu, India 42 E6 29 36N 75 19 E
Bhaun, Pakistan 42 C5 32 55N 72 40 E
Bhaunagar = Bhavnagar, India 40 J8 21 45N 72 10 E
Bhavnagar, India 40 J8 21 45N 72 10 E
Bhawari, India 42 G5 25 42N 73 4 E
Bhayavadar, India 42 J4 21 51N 70 15 E
Bhera, Pakistan 42 C5 32 29N 72 57 E
Bhikangaon, India 42 J6 21 52N 75 57 E
Bhilsa = Vidisha, India 42 H7 23 28N 77 53 E
Bhilwara, India 42 G6 25 25N 74 38 E
Bhima →, India 40 L10 16 25N 77 17 E
Bhimbar, Pakistan 43 C6 32 59N 74 3 E
Bhind, India 43 F8 26 30N 78 46 E
Bhinga, India 43 F9 27 43N 81 56 E
Bhinmal, India 42 G5 25 0N 72 15 E
Bhiwandi, India 40 K8 19 20N 73 0 E
Bhiwani, India 42 E7 28 50N 76 9 E
Bhogava →, India 42 H5 22 26N 72 20 E
Bhola, Bangla. 41 H17 22 45N 90 35 E
Bholari, Pakistan 42 G3 25 19N 68 13 E
Bhopal, India 42 H7 23 20N 77 30 E
Bhubaneshwar, India 41 J14 20 15N 85 50 E
Bhuj, India 42 H3 23 15N 69 49 E
Bhusaval, India 40 J9 21 3N 75 46 E
Bhutan ■, Asia 41 F17 27 25N 90 30 E
Biafra, B. of = Bonny, Bight of, Africa 52 D1 3 30N 9 20 E
Biak, Indonesia 37 E9 1 10S 136 6 E
Biała Podlaska, Poland 17 B12 52 4N 23 6 E
Białogard, Poland 16 A8 54 2N 15 58 E
Białystok, Poland 17 B12 53 10N 23 10 E
Biaora, India 42 H7 23 56N 76 56 E

Brewster

Boggabri, *Australia* 63 E5 30 45S 150 5 E
Boggeragh Mts., *Ireland* 13 D3 52 2N 8 55W
Boglan = Solhan, *Turkey* 44 B4 38 57N 41 3 E
Bognor Regis, *U.K.* 11 G7 50 47N 0 40W
Bogo, *Phil.* 37 B6 11 3N 124 0 E
Bogong, Mt., *Australia* 63 F4 36 47S 147 17 E
Bogor, *Indonesia* 36 F3 6 36S 106 48 E
Bogotá, *Colombia* 92 C4 4 34N 74 0W
Bogotol, *Russia* 26 D9 56 15N 89 50 E
Bogra, *Bangla.* 41 G16 24 51N 89 22 E
Boguchany, *Russia* 27 D10 58 40N 97 30 E
Bohemian Forest =
Böhmerwald, *Germany* 16 D7 49 8N 13 14 E
Böhmerwald, *Germany* 16 D7 49 8N 13 14 E
Bohol □, *Phil.* 37 C6 9 50N 124 10 E
Bohol Sea, *Phil.* 37 C6 9 0N 124 0 E
Bohuslän, *Sweden* 9 G14 58 25N 12 0 E
Boi, Pta. de, *Brazil* 95 A6 23 55S 45 15W
Boiaçu, *Brazil* 92 D6 0 27S 61 46W
Boileau, C., *Australia* 60 C3 17 40S 122 7 E
Boise, *U.S.A.* 82 E5 43 37N 116 13W
Boise City, *U.S.A.* 81 G3 36 44N 102 31W
Boissevain, *Canada* 73 D8 49 15N 100 5W
Bojador, C., *W. Sahara* 50 C3 26 0N 14 30W
Bojana →, *Albania* 21 D8 41 52N 19 22 E
Bojnūrd, *Iran* 45 B8 37 30N 57 20 E
Bojonegoro, *Indonesia* 37 G14 7 11S 111 54 E
Bokaro, *India* 43 H11 23 46N 85 55 E
Bokhara →, *Australia* 63 D4 29 55S 146 42 E
Boknafjorden, *Norway* 9 G11 59 14N 5 40 E
Bokoro, *Chad* 51 F9 12 25N 17 14 E
Bokpyin, *Burma* 39 G2 11 18N 98 42 E
Bolan →, *Pakistan* 42 E2 28 38N 67 42 E
Bolan Pass, *Pakistan* 40 E5 29 50N 67 20 E
Bolaños →, *Mexico* 86 C4 21 14N 104 8W
Bolbec, *France* 18 B4 49 30N 0 30 E
Boldāji, *Iran* 45 D6 31 56N 51 3 E
Bole, *China* 32 B3 45 11N 81 37 E
Bolekhiv, *Ukraine* 17 D12 49 0N 23 57 E
Bolesławiec, *Poland* 16 C8 51 17N 15 37 E
Bolgrad = Bolhrad, *Ukraine* 17 F15 45 40N 28 32 E
Bolhrad, *Ukraine* 17 F15 45 40N 28 32 E
Bolívar, *Argentina* 94 D3 36 15S 60 53W
Bolivar, *Mo., U.S.A.* 81 G8 37 37N 93 25W
Bolivar, *N.Y., U.S.A.* 78 D6 42 4N 78 10W
Bolivar, *Tenn., U.S.A.* 81 H10 35 12N 89 0W
Bolivia ■, *S. Amer.* 92 G6 17 6S 64 0W
Bolivian Plateau, *S. Amer.* 90 E4 20 0S 67 30W
Bollnäs, *Sweden* 9 F17 61 21N 16 24 E
Bollon, *Australia* 63 D4 28 2S 147 29 E
Bolmen, *Sweden* 9 H15 56 55N 13 40 E
Bolobo, *Dem. Rep. of the Congo* 52 E3 2 6S 16 20 E
Bologna, *Italy* 20 B4 44 29N 11 20 E
Bologoye, *Russia* 24 C5 57 55N 34 5 E
Bolonchenticul, *Mexico* 87 D7 20 0N 89 49W
Boloven, Cao Nguyen, *Laos* 38 E6 15 10N 106 30 E
Bolpur, *India* 43 H12 23 40N 87 45 E
Bolsena, L. di, *Italy* 20 C4 42 36N 11 56 E
Bolshevik, Ostrov, *Russia* 27 B11 78 30N 102 0 E
Bolshoi Kavkas = Caucasus Mountains, *Eurasia* 25 F7 42 50N 44 0 E
Bolshoy Anyuy →, *Russia* 27 C17 68 30N 160 49 E
Bolshoy Begichev, Ostrov, *Russia* 27 B12 74 20N 112 30 E
Bolshoy Lyakhovskiy, Ostrov, *Russia* 27 B15 73 35N 142 0 E
Bolshoy Tyuters, Ostrov, *Russia* 9 G22 59 51N 27 13 E
Bolsward, *Neths.* 15 A5 53 3N 5 32 E
Bolt Head, *U.K.* 11 G4 50 12N 3 48W
Bolton, *Canada* 78 C5 43 54N 79 45W
Bolton, *U.K.* 10 D5 53 35N 2 26W
Bolton Landing, *U.S.A.* 79 C11 43 32N 73 35W
Bolu, *Turkey* 25 F5 40 45N 31 35 E
Bolungavík, *Iceland* 8 C2 66 9N 23 15W
Bolvadin, *Turkey* 25 G5 38 45N 31 4 E
Bolzano, *Italy* 20 A4 46 31N 11 22 E
Bom Jesus da Lapa, *Brazil* 93 F10 13 15S 43 25W
Boma, *Dem. Rep. of the Congo* 52 F2 5 50S 13 4 E
Bombala, *Australia* 63 F4 36 56S 149 15 E
Bombay = Mumbai, *India* 40 K8 18 55N 72 50 E
Bomboma, *Dem. Rep. of the Congo* 52 D3 2 25N 18 55 E
Bombombwa, *Dem. Rep. of the Congo* 54 B2 1 40N 25 40 E
Bomili, *Dem. Rep. of the Congo* 54 B2 1 45N 27 5 E
Bømlo, *Norway* 9 G11 59 37N 5 13 E
Bomokandi →, *Dem. Rep. of the Congo* 54 B2 3 39N 26 8 E
Bomu →, *C.A.R.* 52 D4 4 40N 22 30 E
Bon, C., *Tunisia* 48 C5 37 1N 11 2 E
Bon Sar Pa, *Vietnam* 38 F6 12 24N 107 35 E
Bonaigarh, *India* 43 J11 21 50N 84 57 E
Bonaire, *Neth. Ant.* 89 D6 12 10N 68 15W
Bonang, *Australia* 63 F4 37 11S 148 41 E
Bonanza, *Nic.* 88 D3 13 54N 84 35W
Bonaparte Arch., *Australia* 60 B3 14 0S 124 30 E
Bonaventure, *Canada* 71 C6 48 5N 65 32W
Bonavista, *Canada* 71 C9 48 40N 53 5W
Bonavista, C., *Canada* 71 C9 48 42N 53 5W
Bonavista B., *Canada* 71 C9 48 45N 53 25W
Bondo, *Dem. Rep. of the Congo* 54 B1 3 55N 23 53 E
Bondoukou, *Ivory C.* 50 G5 8 2N 2 47W
Bondowoso, *Indonesia* 37 G15 7 55S 113 49 E
Bone, Teluk, *Indonesia* 37 E6 4 10S 120 50 E
Bonerate, *Indonesia* 37 F6 7 25S 121 5 E
Bonerate, Kepulauan, *Indonesia* 37 F6 6 30S 121 10 E
Bo'ness, *U.K.* 12 E5 56 1N 3 37W
Bonete, Cerro, *Argentina* 94 B2 27 55S 68 40W
Bong Son = Hoai Nhon, *Vietnam* 38 E7 14 28N 109 1 E
Bongor, *Chad* 51 F9 10 35N 15 20 E
Bonham, *U.S.A.* 81 J6 33 35N 96 11W
Bonifacio, *France* 18 F8 41 24N 9 10 E
Bonifacio, Bouches de, *Medit. S.* 20 D3 41 12N 9 15 E
Bonin Is. = Ogasawara Gunto, *Pac. Oc.* 28 G18 27 0N 142 0 E
Bonn, *Germany* 16 C4 50 46N 7 6 E
Bonne Terre, *U.S.A.* 81 G9 37 55N 90 33W
Bonners Ferry, *U.S.A.* 82 B5 48 42N 116 19W
Bonney, L., *Australia* 63 F3 37 50S 140 20 E
Bonnie Rock, *Australia* 61 F2 30 29S 118 22 E
Bonny, Bight of, *Africa* 52 D1 3 30N 9 20 E
Bonnyrigg, *U.K.* 12 F5 55 53N 3 6W
Bonnyville, *Canada* 73 C6 54 20N 110 45W
Bonoi, *Indonesia* 37 E9 1 45S 137 41 E
Bonsall, *U.S.A.* 85 M9 33 16N 117 14W
Bontang, *Indonesia* 36 D5 0 10N 117 30 E
Bonthe, *S. Leone* 50 G3 7 30N 12 33W
Bontoc, *Phil.* 37 A6 17 7N 120 58 E
Bonython Ra., *Australia* 60 D4 23 40S 128 45 E
Bookabie, *Australia* 61 F5 31 50S 132 41 E
Booker, *U.S.A.* 81 G4 36 27N 100 32W
Boolaboolka L., *Australia* 63 E3 32 38S 143 10 E
Booligal, *Australia* 63 E3 33 58S 144 53 E
Boonah, *Australia* 63 D5 27 58S 152 41 E
Boone, *Iowa, U.S.A.* 80 D8 42 4N 93 53W
Boone, *N.C., U.S.A.* 77 G5 36 13N 81 41W
Booneville, *Ark., U.S.A.* 81 H8 35 8N 93 55W
Booneville, *Miss., U.S.A.* 77 H1 34 39N 88 34W
Boonville, *Calif., U.S.A.* 84 F3 39 1N 123 22W
Boonville, *Ind., U.S.A.* 76 F2 38 3N 87 16W
Boonville, *Mo., U.S.A.* 80 F8 38 58N 92 44W
Boonville, *N.Y., U.S.A.* 79 C9 43 29N 75 20W
Boorindal, *Australia* 63 E4 30 22S 146 11 E
Boorowa, *Australia* 63 E4 34 28S 148 44 E
Boosaaso = Bosaso, *Somali Rep.* 46 E4 11 12N 49 18 E
Boothia, Gulf of, *Canada* 69 A11 71 0N 90 0W
Boothia Pen., *Canada* 68 A10 71 0N 94 0W
Bootle, *U.K.* 10 D4 53 28N 3 1W
Booué, *Gabon* 52 E2 0 5S 11 55 E
Boquete, *Panama* 88 E3 8 46N 82 27W
Boquilla, Presa de la, *Mexico* 86 B3 27 40N 105 30W
Boquillas del Carmen, *Mexico* 86 B4 29 17N 102 53W
Bor, *Serbia, Yug.* 21 B10 44 5N 22 7 E
Bôr, *Sudan* 51 G12 6 10N 31 40 E
Bor Mashash, *Israel* 47 D3 31 7N 34 50 E
Borah Peak, *U.S.A.* 82 D7 44 8N 113 47W
Borås, *Sweden* 9 H15 57 43N 12 56 E
Borāzjān, *Iran* 45 D6 29 22N 51 10 E
Borba, *Brazil* 92 D7 4 12S 59 34W
Borborema, Planalto da, *Brazil* 90 D7 7 0S 37 0W
Bord Khūn-e Now, *Iran* 45 D6 28 3N 51 28 E
Borda, C., *Australia* 63 F2 35 45S 136 34 E
Bordeaux, *France* 18 D3 44 50N 0 36W
Borden, *Australia* 61 F2 34 3S 118 12 E
Borden, *Canada* 71 C7 46 18N 63 47W
Borden I., *Canada* 4 B2 78 30N 111 30W
Borden Pen., *Canada* 69 A11 73 0N 83 0W
Borders = Scottish Borders □, *U.K.* 12 F6 55 35N 2 50W
Bordertown, *Australia* 63 F3 36 19S 140 45 E
Borðeyri, *Iceland* 8 D3 65 12N 21 6W
Bordj Fly Ste. Marie, *Algeria* 50 C5 27 19N 2 32W
Bordj-in-Eker, *Algeria* 50 D7 24 9N 5 3 E
Bordj Omar Driss, *Algeria* 50 C7 28 10N 6 40 E
Borehamwood, *U.K.* 11 F7 51 40N 0 15W
Borgå = Porvoo, *Finland* 9 F21 60 24N 25 40 E
Borgarfjörður, *Iceland* 8 D7 65 31N 13 49W
Borgarnes, *Iceland* 8 D3 64 32N 21 55W
Børgefjellet, *Norway* 8 D15 65 20N 13 45 E
Borger, *Neths.* 15 B6 52 54N 6 44 E
Borger, *U.S.A.* 81 H4 35 39N 101 24W
Borgholm, *Sweden* 9 H17 56 52N 16 39 E
Borhoyn Tal, *Mongolia* 34 C6 43 50N 111 58 E
Borikhane, *Laos* 38 C4 18 33N 103 43 E
Borisoglebsk, *Russia* 25 D7 51 27N 42 5 E
Borisov = Barysaw, *Belarus* 17 A15 54 17N 28 28 E
Borja, *Peru* 92 D3 4 20S 77 40W
Borkou, *Chad* 51 E9 18 15N 18 50 E
Borkum, *Germany* 16 B4 53 34N 6 40 E
Borlänge, *Sweden* 9 F16 60 29N 15 26 E
Borley, C., *Antarctica* 5 C5 66 15S 52 30 E
Borneo, *E. Indies* 36 D5 1 0N 115 0 E
Bornholm, *Denmark* 9 J16 55 10N 15 0 E
Borogontsy, *Russia* 27 C14 62 42N 131 8 E
Boron, *U.S.A.* 85 L9 35 0N 117 39W
Borongan, *Phil.* 37 B7 11 37N 125 26 E
Borovichi, *Russia* 24 C5 58 25N 33 55 E
Borrego Springs, *U.S.A.* 85 M10 33 15N 116 23W
Borroloola, *Australia* 62 B2 16 4S 136 17 E
Borşa, *Romania* 17 E13 47 41N 24 50 E
Borsad, *India* 42 H5 22 25N 72 54 E
Borth, *U.K.* 11 E3 52 29N 4 2W
Borūjerd, *Iran* 45 C6 33 55N 48 50 E
Boryslav, *Ukraine* 17 D12 49 18N 23 28 E
Borzya, *Russia* 27 D12 50 24N 116 31 E
Bosa, *Italy* 20 D3 40 18N 8 30 E
Bosanska Gradiška, *Bos.-H.* 20 B7 45 10N 17 15 E
Bosaso, *Somali Rep.* 46 E4 11 12N 49 18 E
Boscastle, *U.K.* 11 G3 50 41N 4 42W
Boshan, *China* 35 F9 36 28N 117 49 E
Boshof, *S. Africa* 56 D4 28 31S 25 13 E
Boshrūyeh, *Iran* 45 C8 33 50N 57 30 E
Bosna →, *Bos.-H.* 21 B8 45 4N 18 29 E
Bosna i Hercegovina = Bosnia-Herzegovina ■, *Europe* 20 B7 44 0N 18 0 E
Bosnia-Herzegovina ■, *Europe* 20 B7 44 0N 18 0 E
Bosnik, *Indonesia* 37 E9 1 5S 136 10 E
Bosobolo, *Dem. Rep. of the Congo* 52 D3 4 15N 19 50 E
Bosporus = İstanbul Boğazı, *Turkey* 21 D13 41 10N 29 10 E
Bosque Farms, *U.S.A.* 83 J10 34 53N 106 40W
Bossangoa, *C.A.R.* 52 C3 6 35N 17 30 E
Bossier City, *U.S.A.* 81 J8 32 31N 93 44W
Bosso, *Niger* 51 F8 13 43N 13 19 E
Bostan, *Pakistan* 42 D2 30 26N 67 2 E
Bostānābād, *Iran* 44 B5 37 50N 46 50 E
Bosten Hu, *China* 32 B3 41 55N 87 40 E
Boston, *U.K.* 10 E7 52 59N 0 2W
Boston, *U.S.A.* 79 D13 42 22N 71 4W
Boston Bar, *Canada* 72 D4 49 52N 121 30W
Boston Mts., *U.S.A.* 81 H8 35 42N 93 15W
Boswell, *Canada* 72 D5 49 28N 116 45W
Boswell, *U.S.A.* 78 F5 40 10N 79 2W
Botad, *India* 42 H4 22 15N 71 40 E
Botene, *Laos* 38 D3 17 35N 101 12 E
Bothaville, *S. Africa* 56 D4 27 23S 26 34 E
Bothnia, G. of, *Europe* 8 E19 63 0N 20 15 E
Bothwell, *Australia* 62 G4 42 20S 147 1 E
Bothwell, *Canada* 78 D3 42 38N 81 52W
Botletle →, *Botswana* 56 C3 20 10S 23 15 E
Botoşani, *Romania* 17 E14 47 42N 26 41 E
Botou, *Burkina Faso* 50 F6 12 42N 1 59 E
Botswana ■, *Africa* 56 C3 22 0S 24 0 E
Bottineau, *U.S.A.* 80 A4 48 50N 100 27W
Bottrop, *Germany* 15 C6 51 31N 6 58 E
Botucatu, *Brazil* 95 A6 22 55S 48 30W
Botwood, *Canada* 71 C8 49 6N 55 23W
Bouaflé, *Ivory C.* 50 G4 7 1N 5 47W
Bouaké, *Ivory C.* 50 G4 7 40N 5 2W
Bouar, *C.A.R.* 52 C3 6 0N 15 40 E
Bouârfa, *Morocco* 50 B5 32 32N 1 58W
Boucaut B., *Australia* 62 A1 12 0S 134 25 E
Bougainville, C., *Australia* 60 B4 13 57S 126 4 E
Bougainville I., *Papua N. G.* 64 H7 6 0S 155 0 E
Bougainville Reef, *Australia* 62 B4 15 30S 147 5 E
Bougie = Bejaïa, *Algeria* 50 A7 36 42N 5 2 E
Bougouni, *Mali* 50 F4 11 30N 7 20W
Bouillon, *Belgium* 15 E5 49 44N 5 3 E
Boulder, *Colo., U.S.A.* 80 E2 40 1N 105 17W
Boulder, *Mont., U.S.A.* 82 C7 46 14N 112 7W
Boulder City, *U.S.A.* 85 K12 35 59N 114 50W
Boulder Creek, *U.S.A.* 84 H4 37 7N 122 7W
Boulder Dam = Hoover Dam, *U.S.A.* 85 K12 36 1N 114 44W
Boulia, *Australia* 62 C2 22 52S 139 51 E
Boulogne-sur-Mer, *France* 18 A4 50 42N 1 36 E
Boultoum, *Niger* 51 F8 14 45N 10 25 E
Boun Neua, *Laos* 38 B3 21 38N 101 54 E
Boun Tai, *Laos* 38 B3 21 23N 101 58 E
Bouna, *Ivory C.* 50 G5 9 10N 3 0W
Boundary Peak, *U.S.A.* 84 H8 37 51N 118 21W
Boundiali, *Ivory C.* 50 G4 9 30N 6 20W
Bountiful, *U.S.A.* 82 F8 40 53N 111 53W
Bounty Is., *Pac. Oc.* 64 M9 48 0S 178 30 E
Bourbonnais, *France* 18 C5 46 28N 3 0 E
Bourdel L., *Canada* 70 A5 56 43N 74 10W
Bourem, *Mali* 50 E5 17 0N 0 24W
Bourg-en-Bresse, *France* 18 C6 46 13N 5 12 E
Bourg-St-Maurice, *France* 18 D7 45 35N 6 46 E
Bourges, *France* 18 C5 47 9N 2 25 E
Bourget, *Canada* 79 A9 45 26N 75 9W
Bourgogne, *France* 18 C6 47 0N 4 50 E
Bourke, *Australia* 63 E4 30 8S 145 55 E
Bourne, *U.K.* 10 E7 52 47N 0 22W
Bournemouth, *U.K.* 11 G6 50 43N 1 52W
Bournemouth □, *U.K.* 11 G6 50 43N 1 52W
Bouse, *U.S.A.* 85 M13 33 56N 114 0W
Bouvet I. = Bouvetøya, *Antarctica* 3 G10 54 26S 3 24 E
Bouvetøya, *Antarctica* 3 G10 54 26S 3 24 E
Bovill, *U.S.A.* 82 C5 46 51N 116 24W
Bovril, *Argentina* 94 C4 31 21S 59 26W
Bow →, *Canada* 72 C6 49 57N 111 41W
Bow Island, *Canada* 72 D6 49 50N 111 23W
Bowbells, *U.S.A.* 80 A3 48 48N 102 15W
Bowdle, *U.S.A.* 80 C5 45 27N 99 39W
Bowelling, *Australia* 61 F2 33 25S 116 30 E
Bowen, *Argentina* 94 D2 35 0S 67 31W
Bowen, *Australia* 62 C4 20 0S 148 16 E
Bowen Mts., *Australia* 63 F4 37 0S 147 50 E
Bowie, *Ariz., U.S.A.* 83 K9 32 19N 109 29W
Bowie, *Tex., U.S.A.* 81 J6 33 34N 97 51W
Bowkān, *Iran* 44 B5 36 31N 46 12 E
Bowland, Forest of, *U.K.* 10 D5 54 0N 2 30W
Bowling Green, *Ky., U.S.A.* 76 G2 36 59N 86 27W
Bowling Green, *Ohio, U.S.A.* 76 E4 41 23N 83 39W
Bowling Green, C., *Australia* 62 B4 19 19S 147 25 E
Bowman, *U.S.A.* 80 B3 46 11N 103 24W
Bowman I., *Antarctica* 5 C8 65 0S 104 0 E
Bowmanville, *Canada* 78 C6 43 55N 78 41W
Bowmore, *U.K.* 12 F2 55 45N 6 17W
Bowral, *Australia* 63 E5 34 26S 150 27 E
Bowraville, *Australia* 63 E5 30 37S 152 52 E
Bowron →, *Canada* 72 C4 54 3N 121 50W
Bowron Lake Prov. Park, *Canada* 72 C4 53 10N 121 5W
Bowser L., *Canada* 72 B3 56 30N 129 30W
Bowsman, *Canada* 73 C8 52 14N 101 12W
Bowwood, *Zambia* 55 F2 17 5S 26 20 E
Box Cr. →, *Australia* 63 E3 34 10S 143 50 E
Boxmeer, *Neths.* 15 C5 51 38N 5 56 E
Boxtel, *Neths.* 15 C5 51 36N 5 20 E
Boyce, *U.S.A.* 81 K8 31 23N 92 40W
Boyd L., *Canada* 70 B4 52 46N 76 42W
Boyle, *Canada* 72 C6 54 35N 112 49W
Boyle, *Ireland* 13 C3 53 59N 8 18W
Boyne →, *Ireland* 13 C5 53 43N 6 15W
Boyne City, *U.S.A.* 76 C3 45 13N 85 1W
Boynton Beach, *U.S.A.* 77 M5 26 32N 80 4W
Boyolali, *Indonesia* 37 G14 7 32S 110 35 E
Boyoma, Chutes, *Dem. Rep. of the Congo* 54 B2 0 35N 25 23 E
Boysen Reservoir, *U.S.A.* 82 E9 43 25N 108 11W
Boyuibe, *Bolivia* 92 G6 20 25S 63 17W
Boyup Brook, *Australia* 61 F2 33 50S 116 23 E
Boz Dağları, *Turkey* 21 E13 38 20N 28 0 E
Bozburun, *Turkey* 21 F13 36 43N 28 4 E
Bozcaada, *Turkey* 21 E12 39 49N 26 3 E
Bozdoğan, *Turkey* 21 F13 37 40N 28 17 E
Bozeman, *U.S.A.* 82 D8 45 41N 111 2W
Bozen = Bolzano, *Italy* 20 A4 46 31N 11 22 E
Bozhou, *China* 34 H8 33 55N 115 41 E
Bozoum, *C.A.R.* 52 C3 6 25N 16 35 E
Bra, *Italy* 18 D7 44 42N 7 51 E
Brabant □, *Belgium* 15 D4 50 46N 4 30 E
Brabant L., *Canada* 73 B8 55 58N 103 43W
Brač, *Croatia* 20 C7 43 20N 16 40 E
Bracadale, L., *U.K.* 12 D2 57 20N 6 30W
Bracciano, L. di, *Italy* 20 C5 42 7N 12 14 E
Bracebridge, *Canada* 78 A5 45 2N 79 19W
Brach, *Libya* 51 C8 27 31N 14 20 E
Bräcke, *Sweden* 9 E16 62 45N 15 26 E
Brackettville, *U.S.A.* 81 L4 29 19N 100 25W
Bracknell, *U.K.* 11 F7 51 25N 0 43W
Bracknell Forest □, *U.K.* 11 F7 51 25N 0 44W
Brad, *Romania* 17 E12 46 10N 22 50 E
Bradenton, *U.S.A.* 77 M4 27 30N 82 34W
Bradford, *Canada* 78 B5 44 7N 79 34W
Bradford, *U.K.* 10 D6 53 47N 1 45W
Bradford, *Pa., U.S.A.* 78 E6 41 58N 78 38W
Bradford, *Vt., U.S.A.* 79 C12 43 59N 72 9W
Bradley, *Ark., U.S.A.* 81 J8 33 6N 93 39W
Bradley, *Calif., U.S.A.* 84 K6 35 52N 120 48W
Bradley Institute, *Zimbabwe* 55 F3 17 7S 31 25 E
Brady, *U.S.A.* 81 K5 31 9N 99 20W
Braeside, *Canada* 79 A8 45 28N 76 24W
Braga, *Portugal* 19 B1 41 35N 8 25W
Bragado, *Argentina* 94 D3 35 2S 60 27W
Bragança, *Brazil* 93 D9 1 0S 47 2W
Bragança, *Portugal* 19 B2 41 48N 6 50W
Bragança Paulista, *Brazil* 95 A6 22 55S 46 32W
Brahmanbaria, *Bangla.* 41 H17 23 58N 91 15 E
Brahmani →, *India* 41 J15 20 39N 86 46 E
Brahmapur, *India* 41 K14 19 15N 84 54 E
Brahmaputra →, *India* 41 F19 27 48N 95 30 E
Braich-y-pwll, *U.K.* 10 E3 52 47N 4 46W
Braidwood, *Australia* 63 F4 35 27S 149 49 E
Brăila, *Romania* 17 F14 45 19N 27 59 E
Brainerd, *U.S.A.* 80 B7 46 22N 94 12W
Braintree, *U.K.* 11 F8 51 53N 0 34 E
Braintree, *U.S.A.* 79 D14 42 13N 71 0W
Brak →, *S. Africa* 56 D3 29 35S 22 55 E
Brakwater, *Namibia* 56 C2 22 28S 17 3 E
Brampton, *Canada* 78 C5 43 45N 79 45W
Brampton, *U.K.* 10 C5 54 57N 2 44W
Branco →, *Brazil* 92 D6 1 20S 61 50W
Brandberg, *Namibia* 56 B2 21 10S 14 33 E
Brandenburg = Neubrandenburg, *Germany* 16 B7 53 33N 13 15 E
Brandenburg, *Germany* 16 B7 52 25N 12 33 E
Brandenburg □, *Germany* 16 B6 52 50N 13 0 E
Brandfort, *S. Africa* 56 D4 28 40S 26 30 E
Brandon, *Canada* 73 D9 49 50N 99 57W
Brandon, *U.S.A.* 79 C11 43 48N 73 4W
Brandon B., *Ireland* 13 D1 52 17N 10 8W
Brandon Mt., *Ireland* 13 D1 52 15N 10 15W
Brandsen, *Argentina* 94 D4 35 10S 58 15W
Brandvlei, *S. Africa* 56 E3 30 25S 20 30 E
Branford, *U.S.A.* 79 E12 41 17N 72 49W
Braniewo, *Poland* 17 A10 54 25N 19 50 E
Bransfield Str., *Antarctica* 5 C18 63 0S 59 0W
Branson, *U.S.A.* 81 G8 36 39N 93 13W
Brantford, *Canada* 78 C4 43 10N 80 15W
Bras d'Or L., *Canada* 71 C7 45 50N 60 50W
Brasher Falls, *U.S.A.* 79 B10 44 49N 74 47W
Brasil, Planalto, *Brazil* 90 E6 18 0S 46 30W
Brasiléia, *Brazil* 92 F5 11 0S 68 45W
Brasília, *Brazil* 93 G9 15 47S 47 55W
Brasília Legal, *Brazil* 93 D7 3 49S 55 36W
Braslaw, *Belarus* 9 J22 55 38N 27 0 E
Braşov, *Romania* 17 F13 45 38N 25 35 E
Brasschaat, *Belgium* 15 C4 51 19N 4 27 E
Brassey, Banjaran, *Malaysia* 36 D5 5 0N 117 15 E
Brassey Ra., *Australia* 61 E3 25 8S 122 15 E
Brasstown Bald, *U.S.A.* 77 H4 34 53N 83 49W
Brastad, *Sweden* 9 G14 58 23N 11 30 E
Bratislava, *Slovak Rep.* 17 D9 48 10N 17 7 E
Bratsk, *Russia* 27 D11 56 10N 101 30 E
Brattleboro, *U.S.A.* 79 D12 42 51N 72 34W
Braunau, *Austria* 16 D7 48 15N 13 3 E
Braunschweig, *Germany* 16 B6 52 15N 10 31 E
Braunton, *U.K.* 11 F3 51 7N 4 10W
Bravo del Norte, Rio = Grande, Rio →, *U.S.A.* 81 N6 25 58N 97 9W
Brawley, *U.S.A.* 85 N11 32 59N 115 31W
Bray, *Ireland* 13 C5 53 13N 6 7W
Bray, Mt., *Australia* 62 A1 14 0S 134 30 E
Bray, Pays de, *France* 18 B4 49 46N 1 26 E
Brazeau →, *Canada* 72 C5 52 55N 115 14W
Brazil, *U.S.A.* 76 F2 39 32N 87 8W
Brazil ■, *S. Amer.* 93 F9 12 0S 50 0W
Brazilian Highlands = Brasil, Planalto, *Brazil* 90 E6 18 0S 46 30W
Brazo Sur →, *S. Amer.* 94 B4 25 21S 57 42W
Brazos →, *U.S.A.* 81 L7 28 53N 95 23W
Brazzaville, *Congo* 52 E3 4 9S 15 12 E
Brčko, *Bos.-H.* 21 B8 44 54N 18 46 E
Breaden, L., *Australia* 61 E4 25 51S 125 28 E
Breaksea Sd., *N.Z.* 59 L1 45 35S 166 35 E
Bream B., *N.Z.* 59 F5 35 56S 174 28 E
Bream Hd., *N.Z.* 59 F5 35 51S 174 36 E
Breas, *Chile* 94 B1 25 29S 70 24W
Brebes, *Indonesia* 37 G13 6 52S 109 3 E
Brechin, *Canada* 78 B5 44 32N 79 10W
Brechin, *U.K.* 12 E6 56 44N 2 39W
Brecht, *Belgium* 15 C4 51 21N 4 38 E
Breckenridge, *Colo., U.S.A.* 82 G10 39 29N 106 3W
Breckenridge, *Minn., U.S.A.* 80 B6 46 16N 96 35W
Breckenridge, *Tex., U.S.A.* 81 J5 32 45N 98 54W
Breckland, *U.K.* 11 E8 52 30N 0 40 E
Brecon, *U.K.* 11 F4 51 57N 3 23W
Brecon Beacons, *U.K.* 11 F4 51 53N 3 26W
Breda, *Neths.* 15 C4 51 35N 4 45 E
Bredasdorp, *S. Africa* 56 E3 34 33S 20 2 E
Bree, *Belgium* 15 C5 51 8N 5 35 E
Bregenz, *Austria* 16 E5 47 30N 9 45 E
Breiðafjörður, *Iceland* 8 D2 65 15N 23 15W
Brejo, *Brazil* 93 D10 3 41S 42 47W
Bremen, *Germany* 16 B5 53 4N 8 47 E
Bremer Bay, *Australia* 61 F2 34 21S 119 20 E
Bremer I., *Australia* 62 A2 12 5S 136 45 E
Bremerhaven, *Germany* 16 B5 53 33N 8 36 E
Bremerton, *U.S.A.* 84 C4 47 34N 122 38W
Brenham, *U.S.A.* 81 K6 30 10N 96 24W
Brennerpass, *Austria* 16 E6 47 2N 11 30 E
Brent, *U.S.A.* 77 J2 32 56N 87 10W
Brentwood, *U.K.* 11 F8 51 37N 0 19 E
Brentwood, *Calif., U.S.A.* 84 H5 37 56N 121 42W
Brentwood, *N.Y., U.S.A.* 79 F11 40 47N 73 15W
Bréscia, *Italy* 18 D9 45 33N 10 15 E
Breskens, *Neths.* 15 C3 51 23N 3 33 E
Breslau = Wrocław, *Poland* 17 C9 51 5N 17 5 E
Bressanone, *Italy* 20 A4 46 43N 11 39 E
Bressay, *U.K.* 12 A7 60 9N 1 6W
Brest, *Belarus* 17 B12 52 10N 23 40 E
Brest, *France* 18 B1 48 24N 4 31W
Brest-Litovsk = Brest, *Belarus* 17 B12 52 10N 23 40 E
Bretagne, *France* 18 B2 48 10N 3 0W
Breton, *Canada* 72 C6 53 7N 114 28W
Breton Sd., *U.S.A.* 81 L10 29 35N 89 15W
Brett, C., *N.Z.* 59 F5 35 10S 174 20 E
Brevard, *U.S.A.* 77 H4 35 14N 82 44W
Breves, *Brazil* 93 D8 1 40S 50 29W
Brewarrina, *Australia* 63 E4 30 0S 146 51 E
Brewer, *U.S.A.* 77 C11 44 48N 68 46W
Brewer, Mt., *U.S.A.* 84 J8 36 44N 118 28W
Brewster, *N.Y., U.S.A.* 79 E11 41 23N 73 37W

Brewster

Brewster, Ohio, U.S.A. 78 F3 40 43N 81 36W
Brewster, Wash., U.S.A. 82 B4 48 6N 119 47W
Brewster, Kap = Kangikajik, Greenland 4 B6 70 7N 22 0W
Brewton, U.S.A. 77 K2 31 7N 87 4W
Breyten, S. Africa 57 D5 26 16S 30 0 E
Brezhnev = Naberezhnyye Chelny, Russia 24 C9 55 42N 52 19 E
Briançon, France 18 D7 44 54N 6 39 E
Bribie I., Australia 63 D5 27 0S 153 10 E
Bribri, Costa Rica 88 E3 9 38N 82 50W
Bridgehampton, U.S.A. 79 F12 40 56N 72 19W
Bridgend, U.K. 11 F4 51 30N 3 34W
Bridgend □, U.K. 11 F4 51 36N 3 36W
Bridgeport, Calif., U.S.A. 84 G7 38 15N 119 14W
Bridgeport, Conn., U.S.A. 79 E11 41 11N 73 12W
Bridgeport, Nebr., U.S.A. 80 E3 41 40N 103 6W
Bridgeport, Tex., U.S.A. 81 J6 33 13N 97 45W
Bridger, U.S.A. 82 D9 45 18N 108 55W
Bridgeton, U.S.A. 76 F8 39 26N 75 14W
Bridgetown, Australia 61 F2 33 58S 116 7 E
Bridgetown, Barbados 89 D8 13 5N 59 30W
Bridgetown, Canada 71 D6 44 55N 65 18W
Bridgewater, Canada 71 D7 44 25N 64 31W
Bridgewater, Mass., U.S.A. 79 E14 41 59N 70 58W
Bridgewater, N.Y., U.S.A. 79 D9 42 53N 75 15W
Bridgewater, C., Australia 63 F3 38 23S 141 23 E
Bridgewater-Gagebrook, Australia 62 G4 42 44S 147 14 E
Bridgnorth, U.K. 11 E5 52 32N 2 25W
Bridgton, U.S.A. 79 B14 44 3N 70 42W
Bridgwater, U.K. 11 F5 51 8N 2 59W
Bridgwater B., U.K. 11 F4 51 15N 3 15W
Bridlington, U.K. 10 C7 54 5N 0 12W
Bridlington B., U.K. 10 C7 54 4N 0 10W
Bridport, Australia 62 G4 40 59S 147 23 E
Bridport, U.K. 11 G5 50 44N 2 45W
Brig, Switz. 18 C7 46 18N 7 59 E
Brigg, U.K. 10 D7 53 34N 0 28W
Brigham City, U.S.A. 82 F7 41 31N 112 1W
Bright, Australia 63 F4 36 42S 146 56 E
Brighton, Australia 63 F2 35 5S 138 30 E
Brighton, Canada 78 B7 44 2N 77 44W
Brighton, U.K. 11 G7 50 49N 0 7W
Brighton, Colo., U.S.A. 80 F2 39 59N 104 49W
Brighton, N.Y., U.S.A. 78 C7 43 8N 77 34W
Brilliant, U.S.A. 78 F4 40 15N 80 39W
Bríndisi, Italy 21 D7 40 39N 17 55 E
Brinkley, U.S.A. 81 H9 34 53N 91 12W
Brinnon, U.S.A. 84 C4 47 41N 122 54W
Brion, I., Canada 71 C7 47 46N 61 26W
Brisbane, Australia 63 D5 27 25S 153 2 E
Brisbane →, Australia 63 D5 27 24S 153 9 E
Bristol, U.K. 11 F5 51 26N 2 35W
Bristol, Conn., U.S.A. 79 E12 41 40N 72 57W
Bristol, Pa., U.S.A. 79 F10 40 6N 74 51W
Bristol, R.I., U.S.A. 79 E13 41 40N 71 16W
Bristol, Tenn., U.S.A. 77 G4 36 36N 82 11W
Bristol, City of □, U.K. 11 F5 51 27N 2 36W
Bristol B., U.S.A. 68 C4 58 0N 160 0W
Bristol Channel, U.K. 11 F3 51 18N 4 30W
Bristol I., Antarctica 5 B1 58 45S 28 0W
Bristol L., U.S.A. 83 J5 34 23N 116 50W
Bristow, U.S.A. 81 H6 35 50N 96 23W
Britain = Great Britain, Europe 6 E5 54 0N 2 15W
British Columbia □, Canada 72 C3 55 0N 125 15W
British Indian Ocean Terr. = Chagos Arch., Ind. Oc. 29 K11 6 0S 72 0 E
British Isles, Europe 6 E5 54 0N 4 0W
Brits, S. Africa 57 D4 25 37S 27 48 E
Britstown, S. Africa 56 E3 30 37S 23 30 E
Britt, Canada 70 C3 45 46N 80 34W
Brittany = Bretagne, France 18 B2 48 10N 3 0W
Britton, U.S.A. 80 C6 45 48N 97 45W
Brive-la-Gaillarde, France 18 D4 45 10N 1 32 E
Brixen = Bressanone, Italy 20 A4 46 43N 11 39 E
Brixham, U.K. 11 G4 50 23N 3 31W
Brno, Czech Rep. 17 D9 49 10N 16 35 E
Broad →, U.S.A. 77 J5 34 1N 81 4W
Broad Arrow, Australia 61 F3 30 23S 121 15 E
Broad B., U.K. 12 C2 58 14N 6 18W
Broad Haven, Ireland 13 B2 54 20N 9 55W
Broad Law, U.K. 12 F5 55 30N 3 21W
Broad Sd., Australia 62 C4 22 0S 149 45 E
Broadalbin, U.S.A. 79 C10 43 4N 74 12W
Broadback →, Canada 70 B4 51 21N 78 52W
Broadhurst Ra., Australia 60 D3 22 30S 122 30 E
Broads, The, U.K. 10 E9 52 45N 1 30 E
Broadus, U.S.A. 80 C2 45 27N 105 25W
Brochet, Canada 73 B8 57 53N 101 40W
Brochet, L., Canada 73 B8 58 36N 101 35W
Brocken, Germany 16 C6 51 47N 10 37 E
Brockport, U.S.A. 78 C7 43 13N 77 56W
Brockton, U.S.A. 79 D13 42 5N 71 1W
Brockville, Canada 79 B9 44 35N 75 41W
Brockway, Mont., U.S.A. 80 B2 47 18N 105 45W
Brockway, Pa., U.S.A. 78 E6 41 15N 78 47W
Brocton, U.S.A. 78 D5 42 23N 79 26W
Brodeur Pen., Canada 69 A11 72 30N 88 10W
Brodhead, Mt., U.S.A. 78 E7 41 39N 77 47W
Brodick, U.K. 12 F3 55 35N 5 9W
Brodnica, Poland 17 B10 53 15N 19 25 E
Brody, Ukraine 17 C13 50 5N 25 10 E
Brogan, U.S.A. 82 D5 44 15N 117 31W
Broken Arrow, U.S.A. 81 G7 36 3N 95 48W
Broken Bow, Nebr., U.S.A. 80 E5 41 24N 99 38W
Broken Bow, Okla., U.S.A. 81 H7 34 2N 94 44W
Broken Bow Lake, U.S.A. 81 H7 34 9N 94 40W
Broken Hill = Kabwe, Zambia 55 E2 14 30S 28 29 E
Broken Hill, Australia 63 E3 31 58S 141 29 E
Bromley □, U.K. 11 F8 51 24N 0 2 E
Bromsgrove, U.K. 11 E5 52 21N 2 2W
Brønderslev, Denmark 9 H13 57 16N 9 57 E
Bronkhorstspruit, S. Africa 57 D4 25 46S 28 45 E
Brønnøysund, Norway 8 D15 65 28N 12 14 E
Brook Park, U.S.A. 78 E4 41 24N 81 51W
Brookhaven, U.S.A. 81 K9 31 35N 90 26W
Brookings, Oreg., U.S.A. 82 E1 42 3N 124 17W
Brookings, S. Dak., U.S.A. 80 C6 44 19N 96 48W
Brooklin, Canada 78 C6 43 55N 78 55W
Brooklyn Park, U.S.A. 80 C8 45 6N 93 23W
Brooks, Canada 72 C6 50 35N 111 55W
Brooks Range, U.S.A. 68 B5 68 0N 152 0W
Brooksville, U.S.A. 77 L4 28 33N 82 23W
Brookton, Australia 61 F2 32 22S 117 0 E
Brookville, U.S.A. 78 E5 41 10N 79 5W
Broom, L., U.K. 12 D3 57 55N 5 15W
Broome, Australia 60 C3 18 0S 122 15 E
Brora, U.K. 12 C5 58 0N 3 52W
Brora →, U.K. 12 C5 58 0N 3 51W
Brosna →, Ireland 13 C4 53 14N 7 58W
Brothers, U.S.A. 82 E3 43 49N 120 36W
Brough, U.K. 10 C5 54 32N 2 18W
Brough Hd., U.K. 12 B5 59 8N 3 20W
Broughton Island = Qikiqtarjuaq, Canada 69 B13 67 33N 63 0W
Brown, L., Australia 61 F2 31 5S 118 15 E
Brown, Pt., Australia 63 E1 32 32S 133 50 E
Brown City, U.S.A. 78 C2 43 13N 82 59W
Brown Willy, U.K. 11 G3 50 35N 4 37W
Brownfield, U.S.A. 81 J3 33 11N 102 17W
Browning, U.S.A. 82 B7 48 34N 113 1W
Brownsville, Oreg., U.S.A. 82 D2 44 24N 122 59W
Brownsville, Pa., U.S.A. 78 F5 40 1N 79 53W
Brownsville, Tenn., U.S.A. 81 H10 35 36N 89 16W
Brownsville, Tex., U.S.A. 81 N6 25 54N 97 30W
Brownville, U.S.A. 79 C9 44 0N 75 59W
Brownwood, U.S.A. 81 K5 31 43N 98 59W
Browse I., Australia 60 B3 14 7S 123 33 E
Bruas, Malaysia 39 K3 4 30N 100 47 E
Bruay-la-Buissière, France 18 A5 50 29N 2 33 E
Bruce, Mt., Australia 60 D2 22 37S 118 8 E
Bruce Pen., Canada 78 B3 45 0N 81 30W
Bruce Rock, Australia 61 F2 31 52S 118 8 E
Bruck an der Leitha, Austria 17 D9 48 1N 16 47 E
Bruck an der Mur, Austria 16 E8 47 24N 15 16 E
Brue →, U.K. 11 F5 51 13N 2 59W
Bruges = Brugge, Belgium 15 C3 51 13N 3 13 E
Brugge, Belgium 15 C3 51 13N 3 13 E
Bruin, U.S.A. 78 E5 41 3N 79 43W
Brûlé, Canada 72 C5 53 15N 117 58W
Brumado, Brazil 93 F10 14 14S 41 40W
Brumunddal, Norway 9 F14 60 53N 10 56 E
Bruneau, U.S.A. 82 E6 42 53N 115 48W
Bruneau →, U.S.A. 82 E6 42 56N 115 57W
Brunei = Bandar Seri Begawan, Brunei 36 C4 4 52N 115 0 E
Brunei ■, Asia 36 D4 4 50N 115 0 E
Brunner, L., N.Z. 59 K3 42 37S 171 27 E
Brunssum, Neths. 15 D5 50 57N 5 59 E
Brunswick = Braunschweig, Germany 16 B6 52 15N 10 31 E
Brunswick, Ga., U.S.A. 77 K5 31 10N 81 30W
Brunswick, Maine, U.S.A. 77 D11 43 55N 69 58W
Brunswick, Md., U.S.A. 76 F7 39 19N 77 38W
Brunswick, Mo., U.S.A. 80 F8 39 26N 93 8W
Brunswick, Ohio, U.S.A. 78 E3 41 14N 81 51W
Brunswick, Pen. de, Chile 96 G2 53 30S 71 30W
Brunswick B., Australia 60 C3 15 15S 124 50 E
Brunswick Junction, Australia 61 F2 33 15S 115 50 E
Bruny I., Australia 62 G4 43 20S 147 15 E
Brus Laguna, Honduras 88 C3 15 47N 84 35W
Brush, U.S.A. 80 E3 40 15N 103 37W
Brushton, U.S.A. 79 B10 44 50N 74 31W
Brusque, Brazil 95 B6 27 5S 49 0W
Brussel, Belgium 15 D4 50 51N 4 21 E
Brussels = Brussel, Belgium 15 D4 50 51N 4 21 E
Brussels, Canada 78 C3 43 44N 81 15W
Bruthen, Australia 63 F4 37 42S 147 50 E
Bruxelles = Brussel, Belgium 15 D4 50 51N 4 21 E
Bryan, Ohio, U.S.A. 76 E3 41 28N 84 33W
Bryan, Tex., U.S.A. 81 K6 30 40N 96 22W
Bryan, Mt., Australia 63 E2 33 30S 139 0 E
Bryansk, Russia 24 D4 53 13N 34 25 E
Bryce Canyon Nat. Park, U.S.A. 83 H7 37 30N 112 10W
Bryne, Norway 9 G11 58 44N 5 38 E
Bryson City, U.S.A. 77 H4 35 26N 83 27W
Bsharri, Lebanon 47 A5 34 15N 36 0 E
Bū Baqarah, U.A.E. 45 E8 25 35N 56 25 E
Bu Craa, W. Sahara 50 C3 26 45N 12 50W
Bū Ḥasā, U.A.E. 45 F7 23 30N 53 20 E
Bua Yai, Thailand 38 E4 15 33N 102 26 E
Buapinang, Indonesia 37 E6 4 40S 121 30 E
Bubanza, Burundi 54 C2 3 6S 29 23 E
Būbiyān, Kuwait 46 B4 29 45N 48 15 E
Bucaramanga, Colombia 92 B4 7 0N 73 0W
Bucasia, Australia 62 C4 21 2S 149 10 E
Buccaneer Arch., Australia 60 C3 16 7S 123 20 E
Buchach, Ukraine 17 D13 49 5N 25 25 E
Buchan, U.K. 12 D6 57 32N 2 21W
Buchan Ness, U.K. 12 D7 57 29N 1 46W
Buchanan, Canada 73 C8 51 40N 102 45W
Buchanan, Liberia 50 G3 5 57N 10 2W
Buchanan, L., Queens., Australia 62 C4 21 35S 145 52 E
Buchanan, L., W. Austral., Australia 61 E3 25 33S 123 2 E
Buchanan, L., U.S.A. 81 K5 30 45N 98 25W
Buchanan Cr. →, Australia 62 B2 19 13S 136 33 E
Buchans, Canada 71 C8 48 50N 56 52W
Bucharest = Bucureşti, Romania 17 F14 44 27N 26 10 E
Buchon, Pt., U.S.A. 84 K6 35 15N 120 54W
Buck Hill Falls, U.S.A. 79 E9 41 11N 75 16W
Buckeye, U.S.A. 83 K7 33 22N 112 35W
Buckeye Lake, U.S.A. 78 G2 39 55N 82 29W
Buckhannon, U.S.A. 76 F5 39 0N 80 8W
Buckhaven, U.K. 12 E5 56 11N 3 3W
Buckhorn L., Canada 78 B6 44 29N 78 23W
Buckie, U.K. 12 D6 57 41N 2 58W
Buckingham, Canada 70 C4 45 37N 75 24W
Buckingham, U.K. 11 F7 51 59N 0 57W
Buckingham B., Australia 62 A2 12 10S 135 40 E
Buckinghamshire □, U.K. 11 F7 51 53N 0 55W
Buckle Hd., Australia 60 B4 14 26S 127 52 E
Buckleboo, Australia 63 E2 32 54S 136 12 E
Buckley, U.K. 10 D4 53 10N 3 5W
Buckley →, Australia 62 C2 20 10S 138 49 E
Bucklin, U.S.A. 81 G5 37 33N 99 38W
Bucks L., U.S.A. 84 F5 39 54N 121 12W
Bucktouche, Canada 71 C7 46 30N 64 45W
Bucureşti, Romania 17 F14 44 27N 26 10 E
Bucyrus, U.S.A. 76 E4 40 48N 82 59W
Budalin, Burma 41 H19 22 20N 95 10 E
Budapest, Hungary 17 E10 47 29N 19 5 E
Budaun, India 43 E8 28 5N 79 10 E
Budd Coast, Antarctica 5 C8 68 0S 112 0 E
Bude, U.K. 11 G3 50 49N 4 34W
Budennovsk, Russia 25 F7 44 50N 44 10 E
Budge Budge = Baj Baj, India 43 H13 22 30N 88 5 E
Budgewoi, Australia 63 E5 33 13S 151 34 E
Budjala, Dem. Rep. of the Congo 52 D3 2 50N 19 40 E
Buellton, U.S.A. 85 L6 34 37N 120 12W
Buena Esperanza, Argentina 94 C2 34 45S 65 15W
Buena Park, U.S.A. 85 M9 33 52N 117 59W
Buena Vista, Colo., U.S.A. 83 G10 38 51N 106 8W
Buena Vista, Va., U.S.A. 76 G6 37 44N 79 21W
Buena Vista Lake Bed, U.S.A. 85 K7 35 12N 119 18W
Buenaventura, Colombia 92 C3 3 53N 77 4W
Buenaventura, Mexico 86 B3 29 50N 107 30W
Buenos Aires, Argentina 94 C4 34 30S 58 20W
Buenos Aires, Costa Rica 88 E3 9 10N 83 20W
Buenos Aires □, Argentina 94 D4 36 30S 60 0W
Buenos Aires, L., Chile 96 F2 46 35S 72 30W
Buffalo, Mo., U.S.A. 81 G8 37 39N 93 6W
Buffalo, N.Y., U.S.A. 78 D6 42 53N 78 53W
Buffalo, Okla., U.S.A. 81 G5 36 50N 99 38W
Buffalo, S. Dak., U.S.A. 80 C3 45 35N 103 33W
Buffalo, Wyo., U.S.A. 82 D10 44 21N 106 42W
Buffalo →, Canada 72 A5 60 5N 115 5W
Buffalo →, S. Africa 57 D5 28 43S 30 37 E
Buffalo Head Hills, Canada 72 B5 57 25N 115 55W
Buffalo L., Alta., Canada 72 C6 52 27N 112 54W
Buffalo L., N.W.T., Canada 72 A5 60 12N 115 25W
Buffalo Narrows, Canada 73 B7 55 51N 108 29W
Buffels →, S. Africa 56 D2 29 36S 17 3 E
Buford, U.S.A. 77 H4 34 10N 84 0W
Bug = Buh →, Ukraine 25 E5 46 59N 31 58 E
Bug →, Poland 17 B11 52 31N 21 5 E
Buga, Colombia 92 C3 4 0N 76 15W
Bugala I., Uganda 54 C3 0 40S 32 20 E
Buganda, Uganda 54 C3 0 0 31 30 E
Buganga, Uganda 54 C3 0 3S 32 0 E
Bugel, Tanjung, Indonesia 37 G14 6 26S 111 3 E
Bugibba, Malta 23 D1 35 57N 14 25 E
Bugsuk, Phil. 36 C5 8 15N 117 15 E
Bugulma, Russia 24 D9 54 33N 52 48 E
Bugun Shara, Mongolia 32 B5 49 0N 104 0 E
Buguruslan, Russia 24 D9 53 39N 52 26 E
Buh →, Ukraine 25 E5 46 59N 31 58 E
Buhera, Zimbabwe 57 B5 19 18S 31 29 E
Buhl, U.S.A. 82 E6 42 36N 114 46W
Builth Wells, U.K. 11 E4 52 9N 3 25W
Buir Nur, Mongolia 33 B6 47 50N 117 42 E
Bujumbura, Burundi 54 C2 3 16S 29 18 E
Bukachacha, Russia 27 D12 52 55N 116 50 E
Bukama, Dem. Rep. of the Congo 55 D2 9 10S 25 50 E
Bukavu, Dem. Rep. of the Congo 54 C2 2 20S 28 52 E
Bukene, Tanzania 54 C3 4 15S 32 48 E
Bukhara = Bukhoro, Uzbekistan 26 F7 39 48N 64 25 E
Bukhoro, Uzbekistan 26 F7 39 48N 64 25 E
Bukima, Tanzania 54 C3 1 50S 33 25 E
Bukit Mertajam, Malaysia 39 K3 5 22N 100 28 E
Bukittinggi, Indonesia 36 E2 0 20S 100 20 E
Bukoba, Tanzania 54 C3 1 20S 31 49 E
Bukuya, Uganda 54 B3 0 40N 31 52 E
Būl, Kuh-e, Iran 45 D7 30 48N 52 45 E
Bula, Indonesia 37 E8 3 6S 130 30 E
Bulahdelah, Australia 63 E5 32 23S 152 13 E
Bulan, Phil. 37 B6 12 40N 123 52 E
Bulandshahr, India 42 E7 28 28N 77 51 E
Bulawayo, Zimbabwe 55 G2 20 7S 28 32 E
Buldan, Turkey 21 E13 38 2N 28 50 E
Bulgar, Russia 24 D8 54 57N 49 4 E
Bulgaria ■, Europe 21 C11 42 35N 25 30 E
Buli, Teluk, Indonesia 37 D7 0 48N 128 25 E
Buliluyan, C., Phil. 36 C5 8 20N 117 15 E
Bulkley →, Canada 72 B3 55 15N 127 40W
Bull Shoals L., U.S.A. 81 G8 36 22N 92 35W
Bullhead City, U.S.A. 85 K12 35 8N 114 32W
Büllingen, Belgium 15 D6 50 25N 6 16 E
Bullock Creek, Australia 62 B3 17 43S 144 31 E
Bulloo →, Australia 63 D3 28 43S 142 30 E
Bulloo L., Australia 63 D3 28 43S 142 25 E
Bulls, N.Z. 59 J5 40 10S 175 24 E
Bulnes, Chile 94 D1 36 42S 72 19W
Bulsar = Valsad, India 40 J8 20 40N 72 58 E
Bultfontein, S. Africa 56 D4 28 18S 26 10 E
Bulukumba, Indonesia 37 F6 5 33S 120 11 E
Bulun, Russia 27 B13 70 37N 127 30 E
Bumba, Dem. Rep. of the Congo 52 D4 2 13N 22 30 E
Bumbiri I., Tanzania 54 C3 1 40S 31 55 E
Bumhpa Bum, Burma 41 F20 26 51N 97 14 E
Bumi →, Zimbabwe 55 F2 17 0S 28 20 E
Buna, Kenya 54 B4 2 58N 39 30 E
Bunazi, Tanzania 54 C3 1 3S 31 23 E
Bunbury, Australia 61 F2 33 20S 115 35 E
Bunclody, Ireland 13 D5 52 39N 6 40W
Buncrana, Ireland 13 A4 55 8N 7 27W
Bundaberg, Australia 63 C5 24 54S 152 22 E
Bundey →, Australia 62 C2 21 46S 135 37 E
Bundi, India 42 G6 25 30N 75 35 E
Bundoran, Ireland 13 B3 54 28N 8 16W
Bung Kan, Thailand 38 C4 18 23N 103 37 E
Bungay, U.K. 11 E9 52 27N 1 28 E
Bungil Cr. →, Australia 63 D4 27 5S 149 5 E
Bungo-Suidō, Japan 31 H6 33 0N 132 15 E
Bungoma, Kenya 54 B3 0 34N 34 34 E
Bungotakada, Japan 31 H5 33 35N 131 25 E
Bungu, Tanzania 54 D4 7 35S 39 0 E
Bunia, Dem. Rep. of the Congo 54 B3 1 35N 30 20 E
Bunji, Pakistan 43 B6 35 45N 74 40 E
Bunkie, U.S.A. 81 K8 30 57N 92 11W
Bunnell, U.S.A. 77 L5 29 28N 81 16W
Buntok, Indonesia 36 E4 1 40S 114 58 E
Bunyu, Indonesia 36 D5 3 35N 117 50 E
Buol, Indonesia 37 D6 1 15N 121 32 E
Buon Brieng, Vietnam 38 F7 13 9N 108 12 E
Buon Ma Thuot, Vietnam 38 F7 12 40N 108 3 E
Buong Long, Cambodia 38 F6 13 44N 106 59 E
Buorkhaya, Mys, Russia 27 B14 71 50N 132 40 E
Buqayq, Si. Arabia 45 E6 26 0N 49 45 E
Bur Acaba, Somali Rep. 46 G3 3 12N 44 20 E
Būr Safāga, Egypt 44 E2 26 43N 33 57 E
Būr Sa'īd, Egypt 51 B12 31 16N 32 18 E
Būr Sūdān, Sudan 51 E13 19 32N 37 9 E
Bura, Kenya 54 C4 1 4S 39 58 E
Burakin, Australia 61 F2 30 31S 117 10 E
Burao, Somali Rep. 46 F4 9 32N 45 32 E
Burāq, Syria 47 B5 33 11N 36 29 E
Buraydah, Si. Arabia 44 E4 26 20N 43 59 E
Burbank, U.S.A. 85 L8 34 11N 118 19W
Burda, India 42 G6 25 50N 77 35 E
Burdekin →, Australia 62 B4 19 38S 147 25 E
Burdur, Turkey 25 G5 37 45N 30 17 E
Burdwan = Barddhaman, India 43 H12 23 14N 87 39 E
Bure, Ethiopia 46 E2 10 40N 37 4 E
Bure →, U.K. 10 E9 52 38N 1 43 E
Bureya →, Russia 27 E13 49 27N 129 30 E
Burford, Canada 78 C4 43 7N 80 27W
Burgas, Bulgaria 21 C12 42 33N 27 29 E
Burgeo, Canada 71 C8 47 37N 57 38W
Burgersdorp, S. Africa 56 E4 31 0S 26 20 E
Burges, Mt., Australia 61 F3 30 50S 121 5 E
Burgos, Spain 19 A4 42 21N 3 41W
Burgsvik, Sweden 9 H18 57 3N 18 19 E
Burgundy = Bourgogne, France 18 C6 47 0N 4 50 E
Burhaniye, Turkey 21 E12 39 30N 26 58 E
Burhanpur, India 40 J10 21 18N 76 14 E
Burhi Gandak →, India 43 G12 25 20N 86 37 E
Burhner →, India 43 H9 22 43N 80 31 E
Burias I., Phil. 37 B6 12 55N 123 5 E
Burica, Pta., Costa Rica 88 E3 8 3N 82 51W
Burien, U.S.A. 84 C4 47 28N 122 21W
Burigi, L., Tanzania 54 C3 2 2S 31 22 E
Burin, Canada 71 C8 47 1N 55 14W
Buriram, Thailand 38 E4 15 0N 103 0 E
Burj Sāfitā, Syria 44 C3 34 48N 36 7 E
Burkburnett, U.S.A. 81 H5 34 6N 98 34W
Burke →, Australia 62 C2 23 12S 139 33 E
Burke Chan., Canada 72 C3 52 10N 127 30W
Burketown, Australia 62 B2 17 45S 139 33 E
Burkina Faso ■, Africa 50 F5 12 0N 1 0W
Burk's Falls, Canada 70 C4 45 37N 79 24W
Burleigh Falls, Canada 78 B6 44 33N 78 12W
Burley, U.S.A. 82 E7 42 32N 113 48W
Burlingame, U.S.A. 84 H4 37 35N 122 21W
Burlington, Canada 78 C5 43 18N 79 45W
Burlington, Colo., U.S.A. 80 F3 39 18N 102 16W
Burlington, Iowa, U.S.A. 80 E9 40 49N 91 14W
Burlington, Kans., U.S.A. 80 F7 38 12N 95 45W
Burlington, N.C., U.S.A. 77 G6 36 6N 79 26W
Burlington, N.J., U.S.A. 79 F10 40 4N 74 51W
Burlington, Vt., U.S.A. 79 B11 44 29N 73 12W
Burlington, Wash., U.S.A. 84 B4 48 28N 122 20W
Burlington, Wis., U.S.A. 76 D1 42 41N 88 17W
Burlyu-Tyube, Kazakstan 26 E8 46 30N 79 10 E
Burma ■, Asia 41 J20 21 0N 96 30 E
Burnaby I., Canada 72 C2 52 25N 131 19W
Burnet, U.S.A. 81 K5 30 45N 98 14W
Burney, U.S.A. 82 F3 40 53N 121 40W
Burnham, U.S.A. 78 F7 40 38N 77 34W
Burnham-on-Sea, U.K. 11 F5 51 14N 3 0W
Burnie, Australia 62 G4 41 4S 145 56 E
Burnley, U.K. 10 D5 53 47N 2 14W
Burns, U.S.A. 82 E4 43 35N 119 3W
Burns Lake, Canada 72 C3 54 20N 125 45W
Burnside →, Canada 68 B9 66 51N 108 4W
Burnside, L., Australia 61 E3 25 22S 123 0 E
Burnsville, U.S.A. 80 C8 44 47N 93 17W
Burnt L., Canada 71 B7 53 35N 64 4W
Burnt River, Canada 78 B6 44 41N 78 42W
Burntwood →, Canada 73 B9 56 8N 96 34W
Burntwood L., Canada 73 B8 55 22N 100 26W
Burqān, Kuwait 44 D5 29 0N 47 57 E
Burra, Australia 63 E2 33 40S 138 55 E
Burray, U.K. 12 C6 58 51N 2 54W
Burren Junction, Australia 63 E4 30 7S 148 59 E
Burrinjuck Res., Australia 63 F4 35 0S 148 36 E
Burro, Serranías del, Mexico 86 B4 29 0N 102 0W
Burrow Hd., U.K. 12 G4 54 41N 4 24W
Burruyacú, Argentina 94 B3 26 30S 64 40W
Burry Port, U.K. 11 F3 51 41N 4 15W
Bursa, Turkey 21 D13 40 15N 29 5 E
Burstall, Canada 73 C7 50 39N 109 54W
Burton, Ohio, U.S.A. 78 E3 41 28N 81 8W
Burton, S.C., U.S.A. 77 J5 32 25N 80 45W
Burton, L., Canada 70 B4 54 45N 78 20W
Burton upon Trent, U.K. 10 E6 52 48N 1 38W
Buru, Indonesia 37 E7 3 30S 126 30 E
Burûn, Râs, Egypt 47 D2 31 14N 33 7 E
Burundi ■, Africa 54 C3 3 15S 30 0 E
Bururi, Burundi 54 C2 3 57S 29 37 E
Burutu, Nigeria 50 G7 5 20N 5 29 E
Burwell, U.S.A. 80 E5 41 47N 99 8W
Burwick, U.K. 12 C5 58 45N 2 58W
Bury, U.K. 10 D5 53 35N 2 17W
Bury St. Edmunds, U.K. 11 E8 52 15N 0 43 E
Buryatia □, Russia 27 D11 53 0N 110 0 E
Busan = Pusan, S. Korea 35 G15 35 5N 129 0 E
Busango Swamp, Zambia 55 E2 14 15S 25 45 E
Buşayrah, Syria 44 C4 35 9N 40 26 E
Būshehr, Iran 45 D6 28 55N 50 55 E
Būshehr □, Iran 45 D6 28 20N 51 45 E
Bushell, Canada 73 B7 59 31N 108 45W
Bushenyi, Uganda 54 C3 0 35S 30 10 E
Bushire = Būshehr, Iran 45 D6 28 55N 50 55 E
Businga, Dem. Rep. of the Congo 52 D4 3 16N 20 59 E
Buşra ash Shām, Syria 47 C5 32 30N 36 25 E
Busselton, Australia 61 F2 33 42S 115 15 E
Bussum, Neths. 15 B5 52 16N 5 10 E
Busto Arsizio, Italy 18 D8 45 37N 8 51 E
Busu Djanoa, Dem. Rep. of the Congo 52 D4 1 43N 21 23 E
Busuanga I., Phil. 37 B5 12 10N 120 0 E
Buta, Dem. Rep. of the Congo 54 B1 2 50N 24 53 E
Butare, Rwanda 54 C2 2 31S 29 52 E
Butaritari, Kiribati 64 G9 3 30N 174 0 E
Bute, U.K. 12 F3 55 48N 5 2W
Bute Inlet, Canada 72 C4 50 40N 124 53W
Butemba, Uganda 54 B3 1 9N 31 37 E
Butembo, Dem. Rep. of the Congo 54 B2 0 9N 29 18 E
Butha Qi, China 33 B7 48 0N 122 32 E
Butiaba, Uganda 54 B3 1 50N 31 20 E
Butler, Mo., U.S.A. 80 F7 38 16N 94 20W
Butler, Pa., U.S.A. 78 F5 40 52N 79 54W
Buton, Indonesia 37 E6 5 0S 122 45 E
Butte, Mont., U.S.A. 82 C7 46 0N 112 32W
Butte, Nebr., U.S.A. 80 D5 42 58N 98 51W

Canyonville

Butte Creek →, U.S.A. **84 F5** 39 12N 121 56W
Butterworth = Gcuwa, S. Africa **57 E4** 32 20S 28 11 E
Butterworth, Malaysia **39 K3** 5 24N 100 23 E
Buttevant, Ireland **13 D3** 52 14N 8 40W
Buttfield, Mt., Australia **61 D4** 24 45S 128 9 E
Button B., Canada **73 B10** 58 45N 94 23W
Buttonwillow, U.S.A. **85 K7** 35 24N 119 28W
Butty Hd., Australia **61 F3** 33 54S 121 39 E
Butuan, Phil. **37 C7** 8 57N 125 33 E
Butung = Buton, Indonesia **37 E6** 5 0S 122 45 E
Buturlinovka, Russia **25 D7** 50 50N 40 35 E
Buur Hakaba = Bur Acaba, Somali Rep. **46 G3** 3 12N 44 20 E
Buxa Duar, India **43 F13** 27 45N 89 35 E
Buxar, India **43 G10** 25 34N 83 58 E
Buxtehude, Germany **16 B5** 53 28N 9 39 E
Buxton, U.K. **10 D6** 53 16N 1 54W
Buy, Russia **24 C7** 58 28N 41 28 E
Büyük Menderes →, Turkey **21 F12** 37 28N 27 11 E
Büyükçekmece, Turkey **21 D13** 41 2N 28 35 E
Buzău, Romania **17 F14** 45 10N 26 50 E
Buzău →, Romania **17 F14** 45 26N 27 44 E
Buzen, Japan **31 H5** 33 35N 131 5 E
Buzi →, Mozam. **55 F3** 19 50S 34 43 E
Buzuluk, Russia **24 D9** 52 48N 52 12 E
Buzzards B., U.S.A. **79 E14** 41 45N 70 37W
Buzzards Bay, U.S.A. **79 E14** 41 44N 70 37W
Bwana Mkubwe, Dem. Rep. of the Congo **55 E2** 13 8S 28 38 E
Byarezina →, Belarus **17 B16** 52 33N 30 14 E
Byaroza, Belarus **17 B13** 52 31N 24 51 E
Bydgoszcz, Poland **17 B9** 53 10N 18 0 E
Byelarus = Belarus ■, Europe **17 B14** 53 30N 27 0 E
Byelorussia = Belarus ■, Europe **17 B14** 53 30N 27 0 E
Byers, U.S.A. **80 F2** 39 43N 104 14W
Byesville, U.S.A. **78 G3** 39 58N 81 32W
Byford, Australia **61 F2** 32 15S 116 0 E
Bykhaw, Belarus **17 B16** 53 31N 30 14 E
Bykhov = Bykhaw, Belarus **17 B16** 53 31N 30 14 E
Bylas, U.S.A. **83 K8** 33 8N 110 7W
Bylot, Canada **73 B10** 58 25N 94 8W
Bylot I., Canada **69 A12** 73 13N 78 34W
Byrd, C., Antarctica **5 C17** 69 38S 76 7W
Byrock, Australia **63 E4** 30 40S 146 27 E
Byron Bay, Australia **63 D5** 28 43S 153 37 E
Byrranga, Gory, Russia **27 B11** 75 0N 100 0 E
Byrranga Mts. = Byrranga, Gory, Russia **27 B11** 75 0N 100 0 E
Byske, Sweden **8 D19** 64 57N 21 11 E
Byske älv →, Sweden **8 D19** 64 57N 21 13 E
Bytom, Poland **17 C10** 50 25N 18 54 E
Bytów, Poland **17 A9** 54 10N 17 30 E
Byumba, Rwanda **54 C3** 1 35S 30 4 E

C

Ca →, Vietnam **38 C5** 18 45N 105 45 E
Ca Mau, Vietnam **39 H5** 9 7N 105 8 E
Ca Mau, Mui, Vietnam **39 H5** 8 38N 104 44 E
Ca Na, Vietnam **39 G7** 11 20N 108 54 E
Caacupé, Paraguay **94 B4** 25 23S 57 5W
Caála, Angola **53 G3** 12 46S 15 30 E
Caamano Sd., Canada **72 C3** 52 55N 129 25W
Caazapá, Paraguay **94 B4** 26 8S 56 19W
Caazapá □, Paraguay **95 B4** 26 10S 56 0W
Caballeria, C. de, Spain **22 A11** 40 5N 4 5 E
Cabanatuan, Phil. **37 A6** 15 30N 120 58 E
Cabano, Canada **71 C6** 47 40N 68 56W
Cabazon, U.S.A. **85 M10** 33 55N 116 47W
Cabedelo, Brazil **93 E12** 7 0S 34 50W
Cabildo, Chile **94 C1** 32 30S 71 5W
Cabimas, Venezuela **92 A4** 10 23N 71 25W
Cabinda, Angola **52 F2** 5 33S 12 11 E
Cabinda □, Angola **52 F2** 5 0S 12 30 E
Cabinet Mts., U.S.A. **82 C6** 48 0N 115 30W
Cabo Blanco, Argentina **96 F3** 47 15S 65 47W
Cabo Frio, Brazil **95 A7** 22 51S 42 3W
Cabo Pantoja, Peru **92 D3** 1 0S 75 10W
Cabonga, Réservoir, Canada **70 C4** 47 20N 76 40W
Cabool, U.S.A. **81 G8** 37 7N 92 6W
Caboolture, Australia **63 D5** 27 5S 152 58 E
Cabora Bassa Dam = Cahora Bassa, Représa de, Mozam. **55 F3** 15 20S 32 50 E
Caborca, Mexico **86 A2** 30 40N 112 10W
Cabot, Mt., U.S.A. **79 B13** 44 30N 71 25W
Cabot Hd., Canada **78 A3** 45 14N 81 17W
Cabot Str., Canada **71 C8** 47 15N 59 40W
Cabra, Spain **19 D3** 37 30N 4 28W
Cabrera, Spain **22 B9** 39 8N 2 57 E
Cabri, Canada **73 C7** 50 35N 108 25W
Cabriel →, Spain **19 C5** 39 14N 1 3W
Caçador, Brazil **95 B5** 26 47S 51 0W
Čačak, Serbia, Yug. **21 C9** 43 54N 20 20 E
Caçapava do Sul, Brazil **95 C5** 30 30S 53 30W
Cáceres, Brazil **92 G7** 16 5S 57 40W
Cáceres, Spain **19 C2** 39 26N 6 23W
Cache Bay, Canada **70 C4** 46 22N 80 0W
Cache Cr. →, U.S.A. **84 G5** 38 42N 121 42W
Cache Creek, Canada **72 C4** 50 48N 121 19W
Cachi, Argentina **94 B2** 25 5S 66 10W
Cachimbo, Serra do, Brazil **93 E7** 9 30S 55 30W
Cachinal de la Sierra, Chile **94 A2** 24 58S 69 32W
Cachoeira, Brazil **93 F11** 12 30S 39 0W
Cachoeira do Sul, Brazil **95 C5** 30 3S 52 53W
Cachoeiro de Itapemirim, Brazil **95 A7** 20 51S 41 7W
Cacoal, Brazil **92 F6** 11 32S 61 18W
Cacólo, Angola **52 G3** 10 9S 19 21 E
Caconda, Angola **53 G3** 13 48S 15 8 E
Caddo, U.S.A. **81 H6** 34 7N 96 16W
Cader Idris, U.K. **11 E4** 52 42N 3 53W
Cadereyta, Mexico **86 B5** 25 36N 100 0W
Cadibarrawirracanna, L., Australia **63 D2** 28 52S 135 27 E
Cadillac, U.S.A. **76 C3** 44 15N 85 24W
Cadiz, Phil. **37 B6** 10 57N 123 15 E
Cádiz, Spain **19 D2** 36 30N 6 20W
Cadiz, Calif., U.S.A. **85 L11** 34 30N 115 28W
Cadiz, Ohio, U.S.A. **78 F4** 40 22N 81 0W
Cádiz, G. de, Spain **19 D2** 36 40N 7 0W
Cadiz L., U.S.A. **83 J6** 34 18N 115 24W
Cadney Park, Australia **63 D1** 27 55S 134 3 E
Cadomin, Canada **72 C5** 53 2N 117 20W
Cadotte Lake, Canada **72 B5** 56 26N 116 23W
Cadoux, Australia **61 F2** 30 46S 117 7 E
Caen, France **18 B3** 49 10N 0 22W
Caernarfon, U.K. **10 D3** 53 8N 4 16W
Caernarfon B., U.K. **10 D3** 53 4N 4 40W
Caernarvon = Caernarfon, U.K. **10 D3** 53 8N 4 16W
Caerphilly, U.K. **11 F4** 51 35N 3 13W
Caerphilly □, U.K. **11 F4** 51 37N 3 12W
Caesarea, Israel **47 C3** 32 30N 34 53 E
Caetité, Brazil **93 F10** 13 50S 42 32W
Cafayate, Argentina **94 B2** 26 2S 66 0W
Cafu, Angola **56 B2** 16 30S 15 8 E
Cagayan de Oro, Phil. **37 C6** 8 30N 124 40 E
Cagayan Is., Phil. **37 C5** 9 40N 121 16 E
Cágliari, Italy **20 E3** 39 13N 9 7 E
Cágliari, G. di, Italy **20 E3** 39 8N 9 11 E
Caguán →, Colombia **92 D4** 0 8S 74 18W
Caguas, Puerto Rico **89 C6** 18 14N 66 2W
Caha Mts., Ireland **13 E2** 51 45N 9 40W
Cahama, Angola **56 B1** 16 17S 14 19 E
Caher, Ireland **13 D4** 52 22N 7 56W
Cahersiveen, Ireland **13 E1** 51 56N 10 14W
Cahora Bassa, L. de, Mozam. **55 F3** 15 35S 32 0 E
Cahora Bassa, Représa de, Mozam. **55 F3** 15 20S 32 50 E
Cahore Pt., Ireland **13 D5** 52 33N 6 12W
Cahors, France **18 D4** 44 27N 1 27 E
Cahul, Moldova **17 F15** 45 50N 28 15 E
Cai Bau, Dao, Vietnam **38 B6** 21 10N 107 27 E
Cai Nuoc, Vietnam **39 H5** 8 56N 105 1 E
Caia, Mozam. **55 F4** 17 51S 35 24 E
Caianda, Angola **55 E1** 11 2S 23 31 E
Caibarién, Cuba **88 B4** 22 30N 79 30W
Caicara, Venezuela **92 B5** 7 38N 66 10W
Caicó, Brazil **93 E11** 6 20S 37 0W
Caicos Is., Turks & Caicos **89 B5** 21 40N 71 40W
Caicos Passage, W. Indies **89 B5** 22 45N 72 45W
Caird Coast, Antarctica **5 D1** 75 0S 25 0W
Cairn Gorm, U.K. **12 D5** 57 7N 3 39W
Cairngorm Mts., U.K. **12 D5** 57 6N 3 42W
Cairnryan, U.K. **12 G3** 54 59N 5 1W
Cairns, Australia **62 B4** 16 57S 145 45 E
Cairns L., Canada **73 C10** 51 42N 94 30W
Cairo = El Qâhira, Egypt **51 B12** 30 1N 31 14 E
Cairo, Ga., U.S.A. **77 K3** 30 52N 84 13W
Cairo, Ill., U.S.A. **81 G10** 37 0N 89 11W
Cairo, N.Y., U.S.A. **79 D11** 42 18N 74 0W
Caithness, Ord of, U.K. **12 C5** 58 8N 3 36W
Cajamarca, Peru **92 E3** 7 5S 78 28W
Cajàzeiras, Brazil **93 E11** 6 52S 38 30W
Cala d'Or, Spain **22 B10** 39 23N 3 14 E
Cala en Porter, Spain **22 B11** 39 52N 4 8 E
Cala Figuera, C. de, Spain **22 B9** 39 27N 2 31 E
Cala Forcat, Spain **22 B10** 40 0N 3 47 E
Cala Major, Spain **22 B9** 39 33N 2 37 E
Cala Mezquida = Sa Mesquida, Spain **22 B11** 39 55N 4 16 E
Cala Millor, Spain **22 B10** 39 35N 3 22 E
Cala Ratjada, Spain **22 B10** 39 43N 3 27 E
Cala Santa Galdana, Spain **22 B10** 39 56N 3 58 E
Calabar, Nigeria **50 H7** 4 57N 8 20 E
Calabogie, Canada **79 A8** 45 18N 76 43W
Calabozo, Venezuela **92 B5** 9 0N 67 28W
Calábria □, Italy **20 E7** 39 0N 16 30 E
Calafate, Argentina **96 G2** 50 19S 72 15W
Calahorra, Spain **19 A5** 42 18N 1 59W
Calais, France **18 A4** 50 57N 1 56 E
Calais, U.S.A. **77 C12** 45 11N 67 17W
Calalaste, Cord. de, Argentina **94 B2** 25 0S 67 0W
Calama, Brazil **92 E6** 8 0S 62 50W
Calama, Chile **94 A2** 22 30S 68 55W
Calamar, Colombia **92 A4** 10 15N 74 55W
Calamian Group, Phil. **37 B5** 11 50N 119 55 E
Calamocha, Spain **19 B5** 40 50N 1 17W
Calang, Indonesia **36 D1** 4 37N 95 37 E
Calapan, Phil. **37 B6** 13 25N 121 7 E
Călăraşi, Romania **17 F14** 44 12N 27 20 E
Calatayud, Spain **19 B5** 41 20N 1 40W
Calauag, Phil. **37 B6** 13 55N 122 15 E
Calavite, C., Phil. **37 B6** 13 26N 120 20 E
Calbayog, Phil. **37 B6** 12 4N 124 38 E
Calca, Peru **92 F4** 13 22S 72 0W
Calcasieu L., U.S.A. **81 L8** 29 55N 93 18W
Calcutta = Kolkata, India **43 H13** 22 36N 88 24 E
Calcutta, U.S.A. **78 F4** 40 40N 80 34W
Caldas da Rainha, Portugal **19 C1** 39 24N 9 8W
Calder →, U.K. **10 D6** 53 44N 1 22W
Caldera, Chile **94 B1** 27 5S 70 55W
Caldwell, Idaho, U.S.A. **82 E5** 43 40N 116 41W
Caldwell, Kans., U.S.A. **81 G6** 37 2N 97 37W
Caldwell, Tex., U.S.A. **81 K6** 30 32N 96 42W
Caledon, S. Africa **56 E2** 34 14S 19 26 E
Caledon →, S. Africa **56 E4** 30 31S 26 5 E
Caledon B., Australia **62 A2** 12 45S 137 0 E
Caledonia, Canada **78 C5** 43 7N 79 58W
Caledonia, U.S.A. **78 D7** 42 58N 77 51W
Calemba, Angola **56 B2** 16 0S 15 44 E
Calen, Australia **62 C4** 20 56S 148 48 E
Caletones, Chile **94 C1** 34 6S 70 27W
Calexico, U.S.A. **85 N11** 32 40N 115 30W
Calf of Man, U.K. **10 C3** 54 3N 4 48W
Calgary, Canada **72 C6** 51 0N 114 10W
Calheta, Madeira **22 D2** 32 44N 17 11W
Calhoun, U.S.A. **77 H3** 34 30N 84 57W
Cali, Colombia **92 C3** 3 25N 76 35W
Calicut, India **40 P9** 11 15N 75 43 E
Caliente, U.S.A. **83 H6** 37 37N 114 31W
California, Mo., U.S.A. **80 F8** 38 38N 92 34W
California, Pa., U.S.A. **78 F5** 40 4N 79 54W
California □, U.S.A. **84 H7** 37 30N 119 30W
California, Baja, Mexico **86 A1** 32 10N 115 12W
California, Baja, T.N. = Baja California □, Mexico **86 B2** 30 0N 115 0W
California, Baja, T.S. = Baja California Sur □, Mexico **86 B2** 25 50N 111 50W
California, G. de, Mexico **86 B2** 27 0N 111 0W
California City, U.S.A. **85 K9** 35 10N 117 55W
California Hot Springs, U.S.A. **85 K8** 35 51N 118 41W
Calingasta, Argentina **94 C2** 31 15S 69 30W
Calipatria, U.S.A. **85 M11** 33 8N 115 31W
Calistoga, U.S.A. **84 G4** 38 35N 122 35W
Calitzdorp, S. Africa **56 E3** 33 33S 21 42 E
Callabonna, L., Australia **63 D3** 29 40S 140 5 E
Callan, Ireland **13 D4** 52 32N 7 24W
Callander, U.K. **12 E4** 56 15N 4 13W
Callao, Peru **92 F3** 12 0S 77 0W
Calles, Mexico **87 C5** 23 2N 98 42W
Callicoon, U.S.A. **79 E9** 41 46N 75 3W
Calling Lake, Canada **72 B6** 55 15N 113 12W
Calliope, Australia **62 C5** 24 0S 151 16 E
Calne, U.K. **11 F6** 51 26N 2 0W
Calola, Angola **56 B2** 16 25S 17 48 E
Caloundra, Australia **63 D5** 26 45S 153 10 E
Calpella, U.S.A. **84 F3** 39 14N 123 12W
Calpine, U.S.A. **84 F6** 39 40N 120 27W
Calstock, Canada **70 C3** 49 47N 84 9W
Caltagirone, Italy **20 F6** 37 14N 14 31 E
Caltanissetta, Italy **20 F6** 37 29N 14 4 E
Calulo, Angola **52 G2** 10 1S 14 56 E
Calvert →, Australia **62 B2** 16 17S 137 44 E
Calvert I., Canada **72 C3** 51 30N 128 0W
Calvert Ra., Australia **60 D3** 24 0S 122 30 E
Calvi, France **18 E8** 42 34N 8 45 E
Calvià, Spain **19 C7** 39 34N 2 31 E
Calvillo, Mexico **86 C4** 21 51N 102 43W
Calvinia, S. Africa **56 E2** 31 28S 19 45 E
Calwa, U.S.A. **84 J7** 36 42N 119 46W
Cam →, U.K. **11 E8** 52 21N 0 16 E
Cam Lam, Vietnam **39 G7** 11 54N 109 10 E
Cam Pha, Vietnam **38 B6** 21 7N 107 18 E
Cam Ranh, Vietnam **39 G7** 11 54N 109 12 E
Cam Xuyen, Vietnam **38 C6** 18 15N 106 0 E
Camabatela, Angola **52 F3** 8 20S 15 26 E
Camacha, Madeira **22 D3** 32 41N 16 49W
Camacho, Mexico **86 C4** 24 25N 102 18W
Camacupa, Angola **53 G3** 11 58S 17 22 E
Camagüey, Cuba **88 B4** 21 20N 78 0W
Camaná, Peru **92 G4** 16 30S 72 50W
Camanche Reservoir, U.S.A. **84 G6** 38 14N 121 1W
Camaquã, Brazil **95 C5** 30 51S 51 49W
Camaquã →, Brazil **95 C5** 31 17S 51 47W
Câmara de Lobos, Madeira **22 D3** 32 39N 16 59W
Camargo, Mexico **87 B5** 26 19N 98 50W
Camargue, France **18 E6** 43 34N 4 34 E
Camarillo, U.S.A. **85 L7** 34 13N 119 2W
Camarón, C., Honduras **88 C2** 16 0N 85 5W
Camarones, Argentina **96 E3** 44 50S 65 40W
Camas, U.S.A. **84 E4** 45 35N 122 24W
Camas Valley, U.S.A. **82 E2** 43 2N 123 40W
Camballin, Australia **60 C3** 17 59S 124 12 E
Cambará, Brazil **95 A5** 23 2S 50 5W
Cambay = Khambhat, India **42 H5** 22 23N 72 33 E
Cambay, G. of = Khambhat, G. of, India **40 J8** 20 45N 72 30 E
Cambodia ■, Asia **38 F5** 12 15N 105 0 E
Camborne, U.K. **11 G2** 50 12N 5 19W
Cambrai, France **18 A5** 50 11N 3 14 E
Cambria, U.S.A. **84 K5** 35 34N 121 5W
Cambrian Mts., U.K. **11 E4** 52 3N 3 57W
Cambridge, Canada **78 C4** 43 23N 80 15W
Cambridge, Jamaica **88 C4** 18 18N 77 54W
Cambridge, N.Z. **59 G5** 37 54S 175 29 E
Cambridge, U.K. **11 E8** 52 12N 0 8 E
Cambridge, Mass., U.S.A. **79 D13** 42 22N 71 6W
Cambridge, Minn., U.S.A. **80 C8** 45 34N 93 13W
Cambridge, N.Y., U.S.A. **79 C11** 43 2N 73 22W
Cambridge, Nebr., U.S.A. **80 E4** 40 17N 100 10W
Cambridge, Ohio, U.S.A. **78 F3** 40 2N 81 35W
Cambridge Bay = Ikaluktutiak, Canada **68 B9** 69 10N 105 0W
Cambridge G., Australia **60 B4** 14 55S 128 15 E
Cambridge Springs, U.S.A. **78 E4** 41 48N 80 4W
Cambridgeshire □, U.K. **11 E7** 52 25N 0 7W
Cambuci, Brazil **95 A7** 21 35S 41 55W
Cambundi-Catembo, Angola **52 G3** 10 10S 17 35 E
Camden, Ala., U.S.A. **77 K2** 31 59N 87 17W
Camden, Ark., U.S.A. **81 J8** 33 35N 92 50W
Camden, Maine, U.S.A. **77 C11** 44 13N 69 4W
Camden, N.J., U.S.A. **79 G9** 39 56N 75 7W
Camden, N.Y., U.S.A. **79 C9** 43 20N 75 45W
Camden, S.C., U.S.A. **77 H5** 34 16N 80 36W
Camden Sd., Australia **60 C3** 15 27S 124 25 E
Camdenton, U.S.A. **81 F8** 38 1N 92 45W
Cameron, Ariz., U.S.A. **83 J8** 35 53N 111 25W
Cameron, La., U.S.A. **81 L8** 29 48N 93 20W
Cameron, Mo., U.S.A. **80 F7** 39 44N 94 14W
Cameron, Tex., U.S.A. **81 K6** 30 51N 96 59W
Cameron Highlands, Malaysia **39 K3** 4 27N 101 22 E
Cameron Hills, Canada **72 B5** 59 48N 118 0W
Cameroon ■, Africa **52 C2** 6 0N 12 30 E
Cameroun, Mt., Cameroon **52 D1** 4 13N 9 10 E
Cametá, Brazil **93 D9** 2 12S 49 30W
Camiguin I., Phil. **37 C6** 18 56N 121 55 E
Camilla, U.S.A. **77 K3** 31 14N 84 12W
Caminha, Portugal **19 B1** 41 50N 8 50W
Camino, U.S.A. **84 G6** 38 44N 120 41W
Camira Creek, Australia **63 D5** 29 15S 152 58 E
Cammal, U.S.A. **78 E7** 41 24N 77 28W
Camocim, Brazil **93 D10** 2 55S 40 50W
Camooweal, Australia **62 B2** 19 56S 138 7 E
Camopi, Fr. Guiana **93 C8** 3 12N 52 17W
Camp Borden, Canada **78 B5** 44 18N 79 56W
Camp Hill, U.S.A. **78 F8** 40 14N 76 55W
Camp Nelson, U.S.A. **85 J8** 36 8N 118 39W
Camp Pendleton, U.S.A. **85 M9** 33 16N 117 23W
Camp Verde, U.S.A. **83 J8** 34 34N 111 51W
Camp Wood, U.S.A. **81 L5** 29 40N 100 1W
Campana, Argentina **94 C4** 34 10S 58 55W
Campana, I., Chile **96 F1** 48 20S 75 20W
Campanário, Madeira **22 D2** 32 39N 17 2W
Campània □, Italy **20 D6** 41 0N 14 30 E
Campbell, S. Africa **56 D3** 28 48S 23 44 E
Campbell, Calif., U.S.A. **84 H5** 37 17N 121 57W
Campbell, Ohio, U.S.A. **78 E4** 41 5N 80 37W
Campbell I., Pac. Oc. **64 N8** 52 30S 169 0 E
Campbell L., Canada **73 A7** 63 14N 106 55W
Campbell River, Canada **72 C3** 50 5N 125 20W
Campbell Town, Australia **62 G4** 41 52S 147 30 E
Campbellford, Canada **78 B7** 44 18N 77 48W
Campbellpur, Pakistan **42 C5** 33 46N 72 26 E
Campbellsville, U.S.A. **76 G3** 37 21N 85 20W
Campbellton, Canada **71 C6** 47 57N 66 43W
Campbelltown, Australia **63 E5** 34 4S 150 49 E
Campbeltown, U.K. **12 F3** 55 26N 5 36W
Campeche, Mexico **87 D6** 19 50N 90 32W
Campeche □, Mexico **87 D6** 19 50N 90 32W
Campeche, Golfo de, Mexico **87 D6** 19 30N 93 0W
Camperdown, Australia **63 F3** 38 14S 143 9 E
Camperville, Canada **73 C8** 51 59N 100 9W
Câmpina, Romania **17 F13** 45 10N 25 45 E
Campina Grande, Brazil **93 E11** 7 20S 35 47W
Campinas, Brazil **95 A6** 22 50S 47 0W
Campo Grande, Brazil **93 H8** 20 25S 54 40W
Campo Maíor, Brazil **93 D10** 4 50S 42 12W
Campo Mourão, Brazil **95 A5** 24 3S 52 22W
Campobasso, Italy **20 D6** 41 34N 14 39 E
Campos, Brazil **95 A7** 21 50S 41 20W
Campos Belos, Brazil **93 F9** 13 10S 47 3W
Campos del Port, Spain **22 B10** 39 26N 3 1 E
Campos Novos, Brazil **95 B5** 27 21S 51 50W
Camptonville, U.S.A. **84 F5** 39 27N 121 3W
Camptown, U.S.A. **79 E8** 41 44N 76 14W
Câmpulung, Romania **17 F13** 45 17N 25 3 E
Camrose, Canada **72 C6** 53 0N 112 50W
Camsell Portage, Canada **73 B7** 59 37N 109 15W
Çan, Turkey **21 D12** 40 2N 27 3 E
Can Clavo, Spain **22 C7** 38 57N 1 27 E
Can Creu, Spain **22 C7** 38 58N 1 28 E
Can Gio, Vietnam **39 G6** 10 25N 106 58 E
Can Tho, Vietnam **39 G5** 10 2N 105 46 E
Canaan, U.S.A. **79 D11** 42 2N 73 20W
Canada ■, N. Amer. **68 C10** 60 0N 100 0W
Cañada de Gómez, Argentina **94 C3** 32 40S 61 30W
Canadian, U.S.A. **81 H4** 35 55N 100 23W
Canadian →, U.S.A. **81 H7** 35 28N 95 3W
Canajoharie, U.S.A. **79 D10** 42 54N 74 35W
Çanakkale, Turkey **21 D12** 40 8N 26 24 E
Çanakkale Boğazı, Turkey **21 D12** 40 17N 26 32 E
Canal Flats, Canada **72 C5** 50 10N 115 48W
Canalejas, Argentina **94 D2** 35 15S 66 34W
Canals, Argentina **94 C3** 33 35S 62 53W
Canandaigua, U.S.A. **78 D7** 42 54N 77 17W
Canandaigua L., U.S.A. **78 D7** 42 47N 77 19W
Cananea, Mexico **86 A2** 31 0N 110 20W
Canarias, Is., Atl. Oc. **22 F4** 28 30N 16 0W
Canarreos, Arch. de los, Cuba **88 B3** 21 35N 81 40W
Canary Is. = Canarias, Is., Atl. Oc. **22 F4** 28 30N 16 0W
Canaseraga, U.S.A. **78 D7** 42 27N 77 45W
Canatlán, Mexico **86 C4** 24 31N 104 47W
Canaveral, C., U.S.A. **77 L5** 28 27N 80 32W
Canavieiras, Brazil **93 G11** 15 39S 39 0W
Canberra, Australia **63 F4** 35 15S 149 8 E
Canby, Calif., U.S.A. **82 F3** 41 27N 120 52W
Canby, Minn., U.S.A. **80 C6** 44 43N 96 16W
Canby, Oreg., U.S.A. **84 E4** 45 16N 122 42W
Cancún, Mexico **87 C7** 21 8N 86 44W
Candelaria, Argentina **95 B4** 27 29S 55 44W
Candelaria, Canary Is. **22 F3** 28 22N 16 22W
Candelo, Australia **63 F4** 36 47S 149 43 E
Candia = Iráklion, Greece **23 D7** 35 20N 25 12 E
Candle L., Canada **73 C7** 53 50N 105 18W
Candlemas I., Antarctica **5 B1** 57 3S 26 40W
Cando, U.S.A. **80 A5** 48 32N 99 12W
Canea = Khaniá, Greece **23 D6** 35 30N 24 4 E
Canelones, Uruguay **95 C4** 34 32S 56 17W
Cañete, Chile **94 D1** 37 50S 73 30W
Cañete, Peru **92 F3** 13 8S 76 30W
Cangas de Narcea, Spain **19 A2** 43 10N 6 32W
Canguaretama, Brazil **93 E11** 6 20S 35 5W
Canguçu, Brazil **95 C5** 31 22S 52 43W
Canguçu, Serra do, Brazil **95 C5** 31 20S 52 40W
Cangzhou, China **34 E9** 38 19N 116 52 E
Caniapiscau →, Canada **71 A6** 56 40N 69 30W
Caniapiscau, Rés. de, Canada **71 B6** 54 10N 69 55W
Canicattì, Italy **20 F5** 37 21N 13 51 E
Canim Lake, Canada **72 C4** 51 47N 120 54W
Canindeyu □, Paraguay **95 A5** 24 10S 55 0W
Canisteo, U.S.A. **78 D7** 42 16N 77 36W
Canisteo →, U.S.A. **78 D7** 42 7N 77 8W
Cañitas, Mexico **86 C4** 23 36N 102 43W
Çankırı, Turkey **25 F5** 40 40N 33 37 E
Cankuzo, Burundi **54 C3** 3 10S 30 31 E
Canmore, Canada **72 C5** 51 7N 115 18W
Cann River, Australia **63 F4** 37 35S 149 7 E
Canna, U.K. **12 D2** 57 3N 6 33W
Cannanore, India **40 P9** 11 53N 75 27 E
Cannes, France **18 E7** 43 32N 7 1 E
Canning Town = Port Canning, India **43 H13** 22 23N 88 40 E
Cannington, Canada **78 B5** 44 20N 79 2W
Cannock, U.K. **11 E5** 52 41N 2 1W
Cannon Ball →, U.S.A. **80 B4** 46 20N 100 38W
Cannondale Mt., Australia **62 D4** 25 13S 148 57 E
Cannonsville Reservoir, U.S.A. **79 D9** 42 4N 75 22W
Cannonvale, Australia **62 C4** 20 17S 148 43 E
Canoas, Brazil **95 B5** 29 56S 51 11W
Canoe L., Canada **73 B7** 55 10N 108 15W
Canon City, U.S.A. **80 F2** 38 27N 105 14W
Canora, Canada **73 C8** 51 40N 102 30W
Canowindra, Australia **63 E4** 33 35S 148 38 E
Canso, Canada **71 C7** 45 20N 61 0W
Cantabria □, Spain **19 A4** 43 10N 4 0W
Cantabrian Mts. = Cantábrica, Cordillera, Spain **19 A3** 43 0N 5 10W
Cantábrica, Cordillera, Spain **19 A3** 43 0N 5 10W
Cantal, Plomb du, France **18 D5** 45 3N 2 45 E
Canterbury, Australia **62 D3** 25 23S 141 53 E
Canterbury, U.K. **11 F9** 51 16N 1 6 E
Canterbury Bight, N.Z. **59 L3** 44 16S 171 55 E
Canterbury Plains, N.Z. **59 K3** 43 55S 171 22 E
Cantil, U.S.A. **85 K9** 35 18N 117 58W
Canton = Guangzhou, China **33 D6** 23 5N 113 10 E
Canton, Ga., U.S.A. **77 H3** 34 14N 84 29W
Canton, Ill., U.S.A. **80 E9** 40 33N 90 2W
Canton, Miss., U.S.A. **81 J9** 32 37N 90 2W
Canton, Mo., U.S.A. **80 E9** 40 8N 91 32W
Canton, N.Y., U.S.A. **79 B9** 44 36N 75 10W
Canton, Ohio, U.S.A. **78 F3** 40 48N 81 23W
Canton, Pa., U.S.A. **78 E8** 41 39N 76 51W
Canton, S. Dak., U.S.A. **80 D6** 43 18N 96 35W
Canton L., U.S.A. **81 G5** 36 6N 98 35W
Canudos, Brazil **92 E7** 7 13S 58 5W
Canumã →, Brazil **92 D7** 3 55S 59 10W
Canutama, Brazil **92 E6** 6 30S 64 20W
Canutillo, U.S.A. **83 L10** 31 55N 106 36W
Canvey, U.K. **11 F8** 51 31N 0 37 E
Canyon, U.S.A. **81 H4** 34 59N 101 55W
Canyonlands Nat. Park, U.S.A. **83 G9** 38 15N 110 0W
Canyonville, U.S.A. **82 E2** 42 56N 123 17W

Cao Bang

Cao Bang, Vietnam 38 A6 22 40N 106 15 E
Cao He →, China 35 D13 40 10N 124 32 E
Cao Lanh, Vietnam 39 G5 10 27N 105 38 E
Cao Xian, China 34 G8 34 50N 115 35 E
Cap-aux-Meules, Canada 71 C7 47 23N 61 52W
Cap-Chat, Canada 71 C6 49 6N 66 40W
Cap-de-la-Madeleine, Canada 70 C5 46 22N 72 31W
Cap-Haïtien, Haiti 89 C5 19 40N 72 20W
Capac, U.S.A. 78 C2 43 1N 82 56W
Capanaparo →, Venezuela 92 B5 7 1N 67 7W
Cape →, Australia 62 C4 20 59S 146 51 E
Cape Barren I., Australia 62 G4 40 25S 148 15 E
Cape Breton Highlands Nat. Park, Canada 71 C7 46 50N 60 40W
Cape Breton I., Canada 71 C7 46 0N 60 30W
Cape Charles, U.S.A. 76 G8 37 16N 76 1W
Cape Coast, Ghana 50 G5 5 5N 1 15W
Cape Coral, U.S.A. 77 M5 26 33N 81 57W
Cape Dorset, Canada 69 B12 64 14N 76 32W
Cape Fear →, U.S.A. 77 H6 33 53N 78 1W
Cape Girardeau, U.S.A. 81 G10 37 19N 89 32W
Cape May, U.S.A. 76 F8 38 56N 74 56W
Cape May Point, U.S.A. 76 F8 38 56N 74 58W
Cape Province, S. Africa 53 L3 32 0S 23 0 E
Cape Tormentine, Canada 71 C7 46 8N 63 47W
Cape Town, S. Africa 56 E2 33 55S 18 22 E
Cape Verde Is. ■, Atl. Oc. 49 E1 16 0N 24 0W
Cape Vincent, U.S.A. 79 B8 44 8N 76 20W
Cape York Peninsula, Australia 62 A3 12 0S 142 30 E
Capela, Brazil 93 F11 10 30S 37 0W
Capella, Australia 62 C4 23 2S 148 1 E
Capim →, Brazil 93 D9 1 40S 47 47W
Capitan, U.S.A. 83 K11 33 35N 105 35W
Capitol Reef Nat. Park, U.S.A. 83 G8 38 15N 111 10W
Capitola, U.S.A. 84 J5 36 59N 121 57W
Capoche →, Mozam. 55 F3 15 35S 33 0 E
Capraia, Italy 18 E8 43 2N 9 50 E
Capreol, Canada 70 C3 46 43N 80 56W
Capri, Italy 20 D6 40 33N 14 14 E
Capricorn Group, Australia 62 C5 23 30S 151 55 E
Capricorn Ra., Australia 60 D2 23 20S 116 50 E
Caprivi Strip, Namibia 56 B3 18 0S 23 0 E
Captain's Flat, Australia 63 F4 35 35S 149 27 E
Caquetá →, Colombia 92 D5 1 15S 69 15W
Caracal, Romania 17 F13 44 8N 24 22 E
Caracas, Venezuela 92 A5 10 30N 66 55W
Caracol, Mato Grosso do Sul, Brazil 94 A4 22 18S 57 1W
Caracol, Piauí, Brazil 93 E10 9 15S 43 22W
Carajas, Brazil 93 E8 6 5S 50 23W
Carajás, Serra dos, Brazil 93 E8 6 0S 51 30W
Carangola, Brazil 95 A7 20 44S 42 5W
Caransebeş, Romania 17 F12 45 28N 22 18 E
Caraquet, Canada 71 C6 47 48N 64 57W
Caras, Peru 92 E3 9 3S 77 47W
Caratasca, L., Honduras 88 C3 15 20N 83 40W
Caratinga, Brazil 93 G10 19 50S 42 10W
Caraúbas, Brazil 93 E11 5 43S 37 33W
Caravaca = Caravaca de la Cruz, Spain 19 C5 38 8N 1 52W
Caravaca de la Cruz, Spain 19 C5 38 8N 1 52W
Caravelas, Brazil 93 G11 17 45S 39 15W
Caraveli, Peru 92 G4 15 45S 73 25W
Caràzinho, Brazil 95 B5 28 16S 52 46W
Carballo, Spain 19 A1 43 13N 8 41W
Carberry, Canada 73 D9 49 50N 99 25W
Carbó, Mexico 86 B2 29 42N 110 58W
Carbonara, C., Italy 20 E3 39 6N 9 31 E
Carbondale, Colo., U.S.A. 82 G10 39 24N 107 13W
Carbondale, Ill., U.S.A. 81 G10 37 44N 89 13W
Carbondale, Pa., U.S.A. 79 E9 41 35N 75 30W
Carbonear, Canada 71 C9 47 42N 53 13W
Carbónia, Italy 20 E3 39 10N 8 30 E
Carcajou, Canada 72 B5 57 47N 117 6W
Carcarana →, Argentina 94 C3 32 27S 60 48W
Carcasse, C., Haiti 89 C5 18 30N 74 28W
Carcassonne, France 18 E5 43 13N 2 20 E
Carcross, Canada 72 A2 60 13N 134 45W
Cardamon Hills, India 40 Q10 9 30N 77 15 E
Cárdenas, Cuba 88 B3 23 0N 81 30W
Cárdenas, San Luis Potosí, Mexico 87 C5 22 0N 99 41W
Cárdenas, Tabasco, Mexico 87 D6 17 59N 93 21W
Cardiff, U.K. 11 F4 51 29N 3 10W
Cardiff □, U.K. 11 F4 51 31N 3 12W
Cardiff-by-the-Sea, U.S.A. 85 M9 33 1N 117 17W
Cardigan, U.K. 11 E3 52 5N 4 40W
Cardigan B., U.K. 11 E3 52 30N 4 30W
Cardinal, Canada 79 B9 44 47N 75 23W
Cardona, Uruguay 94 C4 33 53S 57 18W
Cardoso, Ilha do, Brazil 95 B5 25 8S 47 58W
Cardston, Canada 72 D6 49 15N 113 20W
Cardwell, Australia 62 B4 18 14S 146 2 E
Careen L., Canada 73 B7 57 0N 108 11W
Carei, Romania 17 E12 47 40N 22 29 E
Careme = Ciremai, Indonesia 37 G13 6 55S 108 27 E
Carey, U.S.A. 82 E7 43 19N 113 57W
Carey, L., Australia 61 E3 29 0S 122 15 E
Carey L., Canada 73 A8 62 12N 102 55W
Carhué, Argentina 94 D3 37 10S 62 50W
Caria, Turkey 21 F13 37 20N 28 10 E
Cariacica, Brazil 93 H10 20 16S 40 25W
Caribbean Sea, W. Indies 89 D5 15 0N 75 0W
Cariboo Mts., Canada 72 C4 53 0N 121 0W
Caribou, U.S.A. 77 B12 46 52N 68 1W
Caribou →, Man., Canada 73 B10 59 20N 94 44W
Caribou →, N.W.T., Canada 72 A3 61 27N 125 45W
Caribou I., Canada 70 C2 47 22N 85 49W
Caribou Is., Canada 72 A6 61 55N 113 15W
Caribou L., Man., Canada 73 B9 59 21N 96 10W
Caribou L., Ont., Canada 70 B2 50 25N 89 5W
Caribou Mts., Canada 72 B5 59 12N 115 40W
Carichic, Mexico 86 B3 27 56N 107 3W
Carillo, Mexico 86 B4 26 50N 103 55W
Carinda, Australia 63 E4 30 28S 147 41 E
Carinhanha, Brazil 93 F10 14 15S 44 46W
Carinhanha →, Brazil 93 F10 14 20S 43 47W
Carinthia = Kärnten □, Austria 16 E8 46 52N 13 30 E
Caripito, Venezuela 92 A6 10 8N 63 6W
Carleton, Mt., Canada 71 C6 47 23N 66 53W
Carleton Place, Canada 79 A8 45 8N 76 9W
Carletonville, S. Africa 56 D4 26 23S 27 22 E
Carlin, U.S.A. 82 F5 40 43N 116 7W
Carlingford L., U.K. 13 B5 54 3N 6 9W
Carlinville, U.S.A. 80 F10 39 17N 89 53W
Carlisle, U.K. 10 C5 54 54N 2 56W
Carlisle, U.S.A. 78 F7 40 12N 77 12W
Carlos Casares, Argentina 94 D3 35 32S 61 20W
Carlos Tejedor, Argentina 94 D3 35 25S 62 25W
Carlow, Ireland 13 D5 52 50N 6 56W
Carlow □, Ireland 13 D5 52 43N 6 50W
Carlsbad, Calif., U.S.A. 85 M9 33 10N 117 21W
Carlsbad, N. Mex., U.S.A. 81 J2 32 25N 104 14W
Carlsbad Caverns Nat. Park, U.S.A. 81 J2 32 10N 104 35W
Carluke, U.K. 12 F5 55 45N 3 50W
Carlyle, Canada 73 D8 49 40N 102 20W
Carmacks, Canada 68 B6 62 5N 136 16W
Carman, Canada 73 D9 49 30N 98 0W
Carmarthen, U.K. 11 F3 51 52N 4 19W
Carmarthen B., U.K. 11 F3 51 40N 4 30W
Carmarthenshire □, U.K. 11 F3 51 55N 4 13W
Carmaux, France 18 D5 44 3N 2 10 E
Carmel, U.S.A. 79 E11 41 26N 73 41W
Carmel-by-the-Sea, U.S.A. 84 J5 36 33N 121 55W
Carmel Valley, U.S.A. 84 J5 36 29N 121 43W
Carmelo, Uruguay 94 C4 34 0S 58 20W
Carmen, Colombia 92 B3 9 43N 75 8W
Carmen, Paraguay 95 B4 27 13S 56 12W
Carmen →, Mexico 86 A3 30 42N 106 29W
Carmen, I., Mexico 86 B2 26 0N 111 20W
Carmen de Patagones, Argentina 96 E4 40 50S 63 0W
Carmensa, Argentina 94 D2 35 15S 67 40W
Carmi, Canada 72 D5 49 36N 119 8W
Carmi, U.S.A. 76 F1 38 5N 88 10W
Carmichael, U.S.A. 84 G5 38 38N 121 19W
Carmila, Australia 62 C4 21 55S 149 24 E
Carmona, Costa Rica 88 E2 10 0N 85 15W
Carmona, Spain 19 D3 37 28N 5 42W
Carn Ban, U.K. 12 D4 57 7N 4 15W
Carn Eige, U.K. 12 D3 57 17N 5 8W
Carnac, France 18 C2 47 35N 3 6W
Carnamah, Australia 61 E2 29 41S 115 53 E
Carnarvon, Australia 61 D1 24 51S 113 42 E
Carnarvon, S. Africa 56 E3 30 56S 22 8 E
Carnarvon Ra., Queens., Australia 62 D4 25 15S 148 30 E
Carnarvon Ra., W. Austral., Australia 61 E3 25 20S 120 45 E
Carnation, U.S.A. 84 C5 47 39N 121 55W
Carndonagh, Ireland 13 A4 55 16N 7 15W
Carnduff, Canada 73 D8 49 10N 101 50W
Carnegie, U.S.A. 78 F4 40 24N 80 5W
Carnegie, L., Australia 61 E3 26 5S 122 30 E
Carnic Alps = Karnische Alpen, Europe 16 E7 46 36N 13 0 E
Carniche Alpi = Karnische Alpen, Europe 16 E7 46 36N 13 0 E
Carnot, C.A.R. 52 D3 4 59N 15 56 E
Carnot, C., Australia 63 E2 34 57S 135 38 E
Carnot B., Australia 60 C3 17 20S 122 15 E
Carnoustie, U.K. 12 E6 56 30N 2 42W
Carnsore Pt., Ireland 13 D5 52 10N 6 22W
Caro, U.S.A. 76 D4 43 29N 83 24W
Carol City, U.S.A. 77 N5 25 56N 80 16W
Carolina, Brazil 93 E9 7 10S 47 30W
Carolina, Puerto Rico 89 C6 18 23N 65 58W
Carolina, S. Africa 57 D5 26 5S 30 6 E
Caroline I., Kiribati 65 H12 9 58S 150 13W
Caroline Is., Micronesia 28 J17 8 0N 150 0 E
Caroni →, Venezuela 92 B6 8 21N 62 43W
Caronie = Nébrodi, Monti, Italy 20 F6 37 54N 14 35 E
Caroona, Australia 63 E5 31 24S 150 26 E
Carpathians, Europe 17 D11 49 30N 21 0 E
Carpaţii Meridionali, Romania 17 F13 45 30N 25 0 E
Carpentaria, G. of, Australia 62 A2 14 0S 139 0 E
Carpentras, France 18 D6 44 3N 5 2 E
Carpi, Italy 20 B4 44 47N 10 53 E
Carpinteria, U.S.A. 85 L7 34 24N 119 31W
Carr Boyd Ra., Australia 60 C4 16 15S 128 35 E
Carrabelle, U.S.A. 77 L3 29 51N 84 40W
Carranza, Presa V., Mexico 86 B4 27 20N 100 50W
Carrara, Italy 18 D9 44 5N 10 6 E
Carrauntoohill, Ireland 13 D2 52 0N 9 45W
Carrick-on-Shannon, Ireland 13 C3 53 57N 8 5W
Carrick-on-Suir, Ireland 13 D4 52 21N 7 24W
Carrickfergus, U.K. 13 B6 54 43N 5 49W
Carrickmacross, Ireland 13 C5 53 59N 6 43W
Carrieton, Australia 63 E2 32 25S 138 31 E
Carrington, U.S.A. 80 B5 47 27N 99 8W
Carrizal Bajo, Chile 94 B1 28 5S 71 20W
Carrizalillo, Chile 94 B1 29 5S 71 30W
Carrizo Cr. →, U.S.A. 81 G3 36 55N 103 55W
Carrizo Springs, U.S.A. 81 L5 28 31N 99 52W
Carrizozo, U.S.A. 83 K11 33 38N 105 53W
Carroll, U.S.A. 80 D7 42 4N 94 52W
Carrollton, Ga., U.S.A. 77 J3 33 35N 85 5W
Carrollton, Ill., U.S.A. 80 F9 39 18N 90 24W
Carrollton, Ky., U.S.A. 76 F3 38 41N 85 11W
Carrollton, Mo., U.S.A. 80 F8 39 22N 93 30W
Carrollton, Ohio, U.S.A. 78 F3 40 34N 81 5W
Carron →, U.K. 12 D4 57 53N 4 22W
Carron, L., U.K. 12 D3 57 22N 5 35W
Carrot →, Canada 73 C8 53 50N 101 17W
Carrot River, Canada 73 C8 53 17N 103 35W
Carruthers, Canada 73 C7 52 52N 109 16W
Carson, Calif., U.S.A. 85 M8 33 48N 118 17W
Carson, N. Dak., U.S.A. 80 B4 46 25N 101 34W
Carson →, U.S.A. 84 F8 39 45N 118 40W
Carson City, U.S.A. 84 F7 39 10N 119 46W
Carson Sink, U.S.A. 82 G4 39 50N 118 25W
Cartagena, Colombia 92 A3 10 25N 75 33W
Cartagena, Spain 19 D5 37 38N 0 59W
Cartago, Colombia 92 C3 4 45N 75 55W
Cartago, Costa Rica 88 E3 9 50N 83 55W
Cartersville, U.S.A. 77 H3 34 10N 84 48W
Carterton, N.Z. 59 J5 41 2S 175 31 E
Carthage, Tunisia 51 A8 36 50N 10 21 E
Carthage, Ill., U.S.A. 80 E9 40 25N 91 8W
Carthage, Mo., U.S.A. 81 G7 37 11N 94 19W
Carthage, N.Y., U.S.A. 76 D8 43 59N 75 37W
Carthage, Tex., U.S.A. 81 J7 32 9N 94 20W
Cartier I., Australia 60 B3 12 31S 123 29 E
Cartwright, Canada 71 B8 53 41N 56 58W
Caruaru, Brazil 93 E11 8 15S 35 55W
Carúpano, Venezuela 92 A6 10 39N 63 15W
Caruthersville, U.S.A. 81 G10 36 11N 89 39W
Carvoeiro, Brazil 92 D6 1 30S 61 59W
Carvoeiro, C., Portugal 19 C1 39 21N 9 24W
Cary, U.S.A. 77 H6 35 47N 78 46W
Casa Grande, U.S.A. 83 K8 32 53N 111 45W
Casablanca, Chile 94 C1 33 20S 71 25W
Casablanca, Morocco 50 B4 33 36N 7 36W
Cascade, Idaho, U.S.A. 82 D5 44 31N 116 2W
Cascade, Mont., U.S.A. 82 C8 47 16N 111 42W
Cascade Locks, U.S.A. 84 E5 45 40N 121 54W
Cascade Ra., U.S.A. 84 D5 47 0N 121 30W
Cascade Reservoir, U.S.A. 82 D5 44 32N 116 3W
Cascais, Portugal 19 C1 38 41N 9 25W
Cascavel, Brazil 95 A5 24 57S 53 28W
Cáscina, Italy 20 C4 43 41N 10 33 E
Casco B., U.S.A. 77 D10 43 45N 70 0W
Caserta, Italy 20 D6 41 4N 14 20 E
Caseyr, Raas = Asir, Ras, Somali Rep. 46 E5 11 55N 51 10 E
Cashel, Ireland 13 D4 52 30N 7 53W
Casiguran, Phil. 37 A6 16 22N 122 7 E
Casilda, Argentina 94 C3 33 10S 61 10W
Casino, Australia 63 D5 28 52S 153 3 E
Casiquiare →, Venezuela 92 C5 2 1N 67 7W
Casma, Peru 92 E3 9 30S 78 20W
Casmalia, U.S.A. 85 L6 34 50N 120 32W
Caspe, Spain 19 B5 41 14N 0 1W
Casper, U.S.A. 82 E10 42 51N 106 19W
Caspian Depression, Eurasia 25 E8 47 0N 48 0 E
Caspian Sea, Eurasia 25 F9 43 0N 50 0 E
Cass Lake, U.S.A. 80 B7 47 23N 94 37W
Cassadaga, U.S.A. 78 D5 42 20N 79 19W
Casselman, Canada 79 A9 45 19N 75 5W
Casselton, U.S.A. 80 B6 46 54N 97 13W
Cassiar, Canada 72 B3 59 16N 129 40W
Cassiar Mts., Canada 72 B2 59 30N 130 30W
Cassino, Italy 20 D5 41 30N 13 49 E
Cassville, U.S.A. 81 G8 36 41N 93 52W
Castaic, U.S.A. 85 L8 34 30N 118 38W
Castalia, U.S.A. 78 E2 41 24N 82 49W
Castanhal, Brazil 93 D9 1 18S 47 55W
Castellammare di Stábia, Italy 20 D6 40 42N 14 29 E
Castelli, Argentina 94 D4 36 7S 57 47W
Castelló de la Plana, Spain 19 C5 39 58N 0 3W
Castelo, Brazil 95 A7 20 33S 41 14W
Castelo Branco, Portugal 19 C2 39 50N 7 31W
Castelsarrasin, France 18 E4 44 2N 1 7 E
Castelvetrano, Italy 20 F5 37 41N 12 47 E
Casterton, Australia 63 F3 37 30S 141 30 E
Castile, U.S.A. 78 D6 42 38N 78 3W
Castilla-La Mancha □, Spain 19 C4 39 30N 3 30W
Castilla y Leon □, Spain 19 B3 42 0N 5 0W
Castillos, Uruguay 95 C5 34 12S 53 52W
Castle Dale, U.S.A. 82 G8 39 13N 111 1W
Castle Douglas, U.K. 12 G5 54 56N 3 56W
Castle Rock, Colo., U.S.A. 80 F2 39 22N 104 51W
Castle Rock, Wash., U.S.A. 84 D4 46 17N 122 54W
Castlebar, Ireland 13 C2 53 52N 9 18W
Castleblaney, Ireland 13 B5 54 7N 6 44W
Castlederg, U.K. 13 B4 54 42N 7 35W
Castleford, U.K. 10 D6 53 43N 1 21W
Castlegar, Canada 72 D5 49 20N 117 40W
Castlemaine, Australia 63 F3 37 2S 144 12 E
Castlepollard, Ireland 13 C4 53 41N 7 19W
Castlerea, Ireland 13 C3 53 46N 8 29W
Castlereagh →, Australia 63 E4 30 12S 147 32 E
Castlereagh B., Australia 62 A2 12 10S 135 10 E
Castleton, U.S.A. 79 C11 43 37N 73 11W
Castletown, U.K. 10 C3 54 5N 4 38W
Castletown Bearhaven, Ireland 13 E2 51 39N 9 55W
Castor, Canada 72 C6 52 15N 111 50W
Castor →, Canada 70 B4 53 24N 78 58W
Castorland, U.S.A. 79 C9 43 53N 75 31W
Castres, France 18 E5 43 37N 2 13 E
Castricum, Neths. 15 B4 52 33N 4 40 E
Castries, St. Lucia 89 D7 14 2N 60 58W
Castro, Brazil 95 A6 24 45S 50 0W
Castro, Chile 96 E2 42 30S 73 50W
Castro Alves, Brazil 93 F11 12 46S 39 33W
Castroville, U.S.A. 84 J5 36 46N 121 45W
Castuera, Spain 19 C3 38 43N 5 37W
Cat Ba, Dao, Vietnam 38 B6 20 50N 107 0 E
Cat I., Bahamas 89 B4 24 30N 75 30W
Cat L., Canada 70 B1 51 40N 91 50W
Cat Lake, Canada 70 B1 51 40N 91 50W
Catacamas, Honduras 88 D2 14 54N 85 56W
Cataguases, Brazil 95 A7 21 23S 42 39W
Catalão, Brazil 93 G9 18 10S 47 57W
Çatalca, Turkey 21 D13 41 8N 28 27 E
Catalina, Canada 71 C9 48 31N 53 4W
Catalina, Chile 94 B2 25 13S 69 43W
Catalina, U.S.A. 83 K8 32 30N 110 50W
Catalonia = Cataluña □, Spain 19 B6 41 40N 1 15 E
Cataluña □, Spain 19 B6 41 40N 1 15 E
Catamarca, Argentina 94 B2 28 30S 65 50W
Catamarca □, Argentina 94 B2 27 0S 65 50W
Catanduanes □, Phil. 37 B6 13 50N 124 20 E
Catanduva, Brazil 95 A6 21 5S 48 58W
Catánia, Italy 20 F6 37 30N 15 6 E
Catanzaro, Italy 20 E7 38 54N 16 35 E
Catarman, Phil. 37 B6 12 28N 124 35 E
Cateel, Phil. 37 C7 7 47N 126 24 E
Catembe, Mozam. 57 D5 26 0S 32 33 E
Caterham, U.K. 11 F7 51 15N 0 4W
Cathcart, S. Africa 56 E4 32 18S 27 10 E
Cathlamet, U.S.A. 84 D3 46 12N 123 23W
Catlettsburg, U.S.A. 76 F4 38 25N 82 36W
Catoche, C., Mexico 87 C7 21 40N 87 8W
Catriló, Argentina 94 D3 36 26S 63 24W
Catrimani, Brazil 92 C6 0 27N 61 41W
Catrimani →, Brazil 92 C6 0 28N 61 44W
Catskill, U.S.A. 79 D11 42 14N 73 52W
Catskill Mts., U.S.A. 79 D10 42 10N 74 25W
Catt, Mt., Australia 62 A1 13 49S 134 23 E
Cattaraugus, U.S.A. 78 D6 42 22N 78 52W
Catuala, Angola 56 B2 16 25S 19 2 E
Catuane, Mozam. 57 D5 26 48S 32 18 E
Catur, Mozam. 55 E4 13 45S 35 30 E
Catwick Is., Vietnam 39 G7 10 0N 109 0 E
Cauca →, Colombia 92 B4 8 54N 74 28W
Caucaia, Brazil 93 D11 3 40S 38 35W
Caucasus Mountains, Eurasia 25 F7 42 50N 44 0 E
Caungula, Angola 52 F3 8 26S 18 38 E
Cauquenes, Chile 94 D1 36 0S 72 22W
Caura →, Venezuela 92 B6 7 38N 64 53W
Cauresi →, Mozam. 55 F3 17 8S 33 0 E
Causapscal, Canada 71 C6 48 19N 67 12W
Cauvery →, India 40 P11 11 9N 78 52 E
Caux, Pays de, France 18 B4 49 38N 0 35 E
Cavalier, U.S.A. 80 A6 48 48N 97 37W
Cavan, Ireland 13 B4 54 0N 7 22W
Cavan □, Ireland 13 C4 54 1N 7 16W
Cave Creek, U.S.A. 83 K7 33 50N 111 57W
Cavenagh Ra., Australia 61 E4 26 12S 127 55 E
Cavendish, Australia 63 F3 37 31S 142 2 E
Caviana, I., Brazil 93 C8 0 10N 50 10W
Cavite, Phil. 37 B6 14 29N 120 55 E
Cawndilla L., Australia 63 E3 32 30S 142 15 E
Cawnpore = Kanpur, India 43 F9 26 28N 80 20 E
Caxias, Brazil 93 D10 4 55S 43 20W
Caxias do Sul, Brazil 95 B5 29 10S 51 10W
Cay Sal Bank, Bahamas 88 B4 23 45N 80 0W
Cayambe, Ecuador 92 C3 0 3N 78 8W
Cayenne, Fr. Guiana 93 B8 5 5N 52 18W
Cayman Brac, Cayman Is. 88 C4 19 43N 79 49W
Cayman Is. ■, W. Indies 88 C3 19 40N 80 30W
Cayo Romano, Cuba 88 B4 22 0N 78 0W
Cayuga, Canada 78 D5 42 59N 79 50W
Cayuga, U.S.A. 79 D8 42 54N 76 44W
Cayuga L., U.S.A. 79 D8 42 41N 76 41W
Cazenovia, U.S.A. 79 D9 42 56N 75 51W
Cazombo, Angola 53 G4 11 54S 22 56 E
Ceanannus Mor, Ireland 13 C5 53 44N 6 53W
Ceará = Fortaleza, Brazil 93 D11 3 45S 38 35W
Ceará □, Brazil 93 E11 5 0S 40 0W
Ceará Mirim, Brazil 93 E11 5 38S 35 25W
Cebaco, I. de, Panama 88 E3 7 33N 81 9W
Cebollar, Argentina 94 B2 29 10S 66 35W
Cebu, Phil. 37 B6 10 18N 123 54 E
Cecil Plains, Australia 63 D5 27 30S 151 11 E
Cedar →, U.S.A. 80 E9 41 17N 91 21W
Cedar City, U.S.A. 83 H7 37 41N 113 4W
Cedar Creek Reservoir, U.S.A. 81 J6 32 11N 96 4W
Cedar Falls, Iowa, U.S.A. 80 D8 42 32N 92 27W
Cedar Falls, Wash., U.S.A. 84 C5 47 25N 121 45W
Cedar Key, U.S.A. 77 L4 29 8N 83 2W
Cedar L., Canada 73 C9 53 10N 100 0W
Cedar Rapids, U.S.A. 80 E9 41 59N 91 40W
Cedartown, U.S.A. 77 H3 34 1N 85 15W
Cedarvale, Canada 72 B3 55 1N 128 22W
Cedarville, S. Africa 57 E4 30 23S 29 3 E
Cedral, Mexico 86 C4 23 50N 100 42W
Cedro, Brazil 93 E11 6 34S 39 3W
Cedros, I. de, Mexico 86 B1 28 10N 115 20W
Ceduna, Australia 63 E1 32 7S 133 46 E
Ceerigaabo = Erigavo, Somali Rep. 46 E4 10 35N 47 20 E
Cefalù, Italy 20 E6 38 2N 14 1 E
Cegléd, Hungary 17 E10 47 11N 19 47 E
Celaya, Mexico 86 C4 20 31N 100 37W
Celebes Sea, Indonesia 37 D6 3 0N 123 0 E
Celina, U.S.A. 76 E3 40 33N 84 35W
Celje, Slovenia 16 E8 46 16N 15 18 E
Celle, Germany 16 B6 52 37N 10 4 E
Cenderwasih, Teluk, Indonesia 37 E9 3 0S 135 20 E
Center, N. Dak., U.S.A. 80 B4 47 7N 101 18W
Center, Tex., U.S.A. 81 K7 31 48N 94 11W
Centerburg, U.S.A. 78 F2 40 18N 82 42W
Centerville, Calif., U.S.A. 84 J7 36 44N 119 30W
Centerville, Iowa, U.S.A. 80 E8 40 44N 92 52W
Centerville, Pa., U.S.A. 78 F5 40 3N 79 59W
Centerville, Tenn., U.S.A. 77 H2 35 47N 87 28W
Centerville, Tex., U.S.A. 81 K7 31 16N 95 59W
Central □, Kenya 54 C4 0 30S 37 30 E
Central □, Malawi 55 E3 13 30S 33 30 E
Central □, Zambia 55 E2 14 25S 28 50 E
Central, Cordillera, Colombia 92 C4 5 0N 75 0W
Central, Cordillera, Costa Rica 88 D3 10 10N 84 5W
Central, Cordillera, Dom. Rep. 89 C5 19 15N 71 0W
Central African Rep. ■, Africa 52 C4 7 0N 20 0 E
Central America, America 66 H11 12 0N 85 0W
Central Butte, Canada 73 C7 50 48N 106 31W
Central City, Colo., U.S.A. 82 G11 39 48N 105 31W
Central City, Ky., U.S.A. 76 G2 37 18N 87 7W
Central City, Nebr., U.S.A. 80 E6 41 7N 98 0W
Central I., Kenya 54 B4 3 30N 36 0 E
Central Makran Range, Pakistan 40 F4 26 30N 64 15 E
Central Patricia, Canada 70 B1 51 30N 90 9W
Central Point, U.S.A. 82 E2 42 23N 122 55W
Central Russian Uplands, Europe 6 E13 54 0N 36 0 E
Central Siberian Plateau, Russia 28 C14 65 0N 105 0 E
Central Square, U.S.A. 79 C8 43 17N 76 9W
Centralia, Ill., U.S.A. 80 F10 38 32N 89 8W
Centralia, Mo., U.S.A. 80 F8 39 13N 92 8W
Centralia, Wash., U.S.A. 84 D4 46 43N 122 58W
Cephalonia = Kefallinía, Greece 21 E9 38 15N 20 30 E
Cepu, Indonesia 37 G14 7 9S 111 35 E
Ceram = Seram, Indonesia 37 E7 3 10S 129 0 E
Ceram Sea = Seram Sea, Indonesia 37 E7 2 30S 128 30 E
Ceredigion □, U.K. 11 E3 52 16N 4 15W
Ceres, Argentina 94 B3 29 55S 61 55W
Ceres, S. Africa 56 E2 33 21S 19 18 E
Ceres, U.S.A. 84 H6 37 35N 120 57W
Cerignola, Italy 20 D6 41 17N 15 53 E
Cerigo = Kíthira, Greece 21 F10 36 8N 23 0 E
Çerkezköy, Turkey 21 D12 41 17N 28 0 E
Cerralvo, I., Mexico 86 C3 24 20N 109 45W
Cerritos, Mexico 86 C4 22 27N 100 20W
Cerro Chato, Uruguay 95 C4 33 6S 55 8W
Cerventes, Australia 61 F2 30 31S 115 3 E
Cervera, Spain 19 B6 41 40N 1 16 E
Cesena, Italy 20 B5 44 8N 12 15 E
Cēsis, Latvia 9 H21 57 18N 25 15 E
České Budějovice, Czech Rep. 16 D8 48 55N 14 25 E
Českomoravská Vrchovina, Czech Rep. 16 D8 49 30N 15 40 E
Çeşme, Turkey 21 E12 38 20N 26 23 E
Cessnock, Australia 63 E5 32 50S 151 21 E
Cetinje, Montenegro, Yug. 21 C8 42 23N 18 59 E
Cetraro, Italy 20 E6 39 31N 15 55 E
Ceuta, N. Afr. 19 E3 35 52N 5 18W
Cévennes, France 18 D5 44 10N 3 50 E
Ceyhan, Turkey 44 B2 37 4N 35 47 E
Ceylon = Sri Lanka ■, Asia 40 R12 7 30N 80 50 E

Chicopee

Cha-am, Thailand 38 F2 12 48N 99 58 E
Cha Pa, Vietnam 38 A4 22 20N 103 47 E
Chacabuco, Argentina 94 C3 34 40S 60 27W
Chachapoyas, Peru 92 E3 6 15S 77 50W
Chachoengsao, Thailand 38 F3 13 42N 101 5 E
Chachran, Pakistan 40 E7 28 55N 70 30 E
Chachro, Pakistan 42 G4 25 5N 70 15 E
Chaco □, Argentina 94 B3 26 30S 61 0W
Chaco □, Paraguay 94 B4 26 0S 60 0W
Chaco →, U.S.A. 83 H9 36 46N 108 39W
Chaco Austral, S. Amer. 96 B4 27 0S 61 30W
Chaco Boreal, S. Amer. 92 H6 22 0S 60 0W
Chaco Central, S. Amer. 96 A4 24 0S 61 0W
Chacon, C., U.S.A. 72 C2 54 42N 132 0W
Chad ■, Africa 51 F8 15 0N 17 15 E
Chad, L. = Tchad, L., Chad 51 F8 13 30N 14 30 E
Chadan, Russia 27 D10 51 17N 91 35 E
Chadileuvú →, Argentina 94 D2 37 46S 66 0W
Chadiza, Zambia 55 E3 14 45S 32 27 E
Chadron, U.S.A. 80 D3 42 50N 103 0W
Chadyr-Lunga = Ciadâr-Lunga, Moldova 17 E15 46 3N 28 51 E
Chae Hom, Thailand 38 C2 18 43N 99 35 E
Chaem →, Thailand 38 C2 18 11N 98 38 E
Chaeryŏng, N. Korea 35 E13 38 24N 125 36 E
Chagai Hills = Chāh Gay Hills, Afghan. 40 E3 29 30N 64 0 E
Chagda, Russia 27 D14 58 45N 130 38 E
Chaghcharān, Afghan. 40 B4 34 31N 65 15 E
Chagos Arch., Ind. Oc. 29 K11 6 0S 72 0 E
Chagrin Falls, U.S.A. 78 E3 41 26N 81 24W
Chāh Ākhvor, Iran 45 C8 32 41N 59 40 E
Chāh Bahar, Iran 45 E9 25 20N 60 40 E
Chāh-e Kavīr, Iran 45 C8 34 29N 56 52 E
Chāh Gay Hills, Afghan. 40 E3 29 30N 64 0 E
Chahar Burjak, Afghan. 40 D3 30 15N 62 0 E
Chahār Mahāll va Bakhtīarī □, Iran 45 C6 32 0N 49 0 E
Chaibasa, India 41 H14 22 42N 85 49 E
Chainat, Thailand 38 E3 15 11N 100 8 E
Chaiya, Thailand 39 H2 9 23N 99 14 E
Chaj Doab, Pakistan 42 C5 32 15N 73 0 E
Chajari, Argentina 94 C4 30 42S 58 0W
Chak Amru, Pakistan 42 C6 32 22N 75 11 E
Chakar →, Pakistan 42 E3 29 29N 68 2 E
Chakari, Zimbabwe 57 B4 18 5S 29 51 E
Chake Chake, Tanzania 54 D4 5 15S 39 45 E
Chakhānsūr, Afghan. 40 D3 31 10N 62 0 E
Chakonipau, L., Canada 71 A6 56 18N 68 30W
Chakradharpur, India 43 H11 22 45N 85 40 E
Chakrata, India 42 D7 30 42N 77 51 E
Chakwal, Pakistan 42 C5 32 56N 72 53 E
Chala, Peru 92 G4 15 48S 74 20W
Chalchihuites, Mexico 86 C4 23 29N 103 53W
Chalcis = Khalkís, Greece 21 E10 38 27N 23 42 E
Chaleur B., Canada 71 C6 47 55N 65 30W
Chalfant, U.S.A. 84 H8 37 32N 118 21W
Chalhuanca, Peru 92 F4 14 15S 73 15W
Chalisgaon, India 40 J9 20 30N 75 10 E
Chalk River, Canada 70 C4 46 1N 77 27W
Chalky Inlet, N.Z. 59 M1 46 3S 166 31 E
Challapata, Bolivia 92 G5 18 53S 66 50W
Challis, U.S.A. 82 D6 44 30N 114 14W
Chalmette, U.S.A. 81 L10 29 56N 89 58W
Chalon-sur-Saône, France 18 C6 46 48N 4 50 E
Châlons-en-Champagne, France 18 B6 48 58N 4 20 E
Chalyaphum, Thailand 38 E4 15 48N 102 2 E
Cham, Cu Lao, Vietnam 38 E7 15 57N 108 30 E
Chama, U.S.A. 83 H10 36 54N 106 35W
Chamaicó, Argentina 94 D3 35 3S 64 58W
Chaman, Pakistan 40 D5 30 58N 66 25 E
Chamba, India 42 C7 32 35N 76 10 E
Chamba, Tanzania 55 E4 11 37S 37 0 E
Chambal →, India 43 F8 26 29N 79 15 E
Chamberlain, U.S.A. 80 D5 43 49N 99 20W
Chamberlain →, Australia 60 C4 15 30S 127 54 E
Chamberlain L., U.S.A. 77 B11 46 14N 69 19W
Chambers, U.S.A. 83 J9 35 11N 109 26W
Chambersburg, U.S.A. 76 F7 39 56N 77 40W
Chambéry, France 18 D6 45 34N 5 55 E
Chambeshi →, Zambia 52 G6 11 53S 29 48 E
Chambly, Canada 79 A11 45 27N 73 17W
Chambord, Canada 71 C5 48 25N 72 6W
Chamchamal, Iraq 44 C5 35 32N 44 50 E
Chamela, Mexico 86 D3 19 32N 105 5W
Chamical, Argentina 94 C2 30 22S 66 27W
Chamkar Luong, Cambodia 39 G4 11 0N 103 45 E
Chamoli, India 43 D8 30 24N 79 21 E
Chamonix-Mont Blanc, France 18 D7 45 55N 6 51 E
Chamouchouane →, Canada 70 C5 48 37N 72 20W
Champa, India 43 H10 22 2N 82 43 E
Champagne, Canada 72 A1 60 49N 136 30W
Champagne, France 18 B6 48 40N 4 20 E
Champaign, U.S.A. 76 E1 40 7N 88 15W
Champassak, Laos 38 E5 14 53N 105 52 E
Champawat, India 43 E9 29 20N 80 6 E
Champdoré, L., Canada 71 A6 55 55N 65 49W
Champion, U.S.A. 78 E4 41 19N 80 51W
Champlain, U.S.A. 79 B11 44 59N 73 27W
Champlain, L., U.S.A. 79 B11 44 40N 73 20W
Champotón, Mexico 87 D6 19 20N 90 50W
Champua, India 43 H11 22 5N 85 40 E
Chana, Thailand 39 J3 6 55N 100 44 E
Chañaral, Chile 94 B1 26 23S 70 40W
Chanārān, Iran 45 B8 36 39N 59 6 E
Chanasma, India 42 H5 23 44N 72 5 E
Chanco, Chile 94 D1 35 44S 72 32W
Chand, India 43 J8 21 57N 79 7 E
Chandan, India 43 G12 24 38N 86 40 E
Chandan Chauki, India 43 E9 28 33N 80 47 E
Chandannagar, India 43 H13 22 52N 88 24 E
Chandausi, India 43 E8 28 27N 78 49 E
Chandeleur Is., U.S.A. 81 L10 29 55N 88 57W
Chandeleur Sd., U.S.A. 81 L10 29 55N 89 0W
Chandigarh, India 42 D7 30 43N 76 47 E
Chandil, India 43 H12 22 58N 86 3 E
Chandler, Australia 63 D1 27 0S 133 19 E
Chandler, Canada 71 C7 48 18N 64 46W
Chandler, Ariz., U.S.A. 83 K8 33 18N 111 50W
Chandler, Okla., U.S.A. 81 H6 35 42N 96 53W
Chandod, India 42 J5 21 59N 73 28 E
Chandpur, Bangla. 41 H17 23 8N 90 45 E
Chandrapur, India 40 K11 19 57N 79 25 E
Chānf, Iran 45 E9 26 38N 60 29 E
Chang, Pakistan 42 F3 26 59N 68 30 E
Chang, Ko, Thailand 39 F4 12 0N 102 23 E
Ch'ang Chiang = Chang Jiang →, China 33 C7 31 48N 121 10 E
Chang Jiang →, China 33 C7 31 48N 121 10 E
Changa, India 43 C7 33 53N 77 35 E
Changanacheri, India 40 Q10 9 25N 76 31 E
Changane →, Mozam. 57 C5 24 30S 33 30 E
Changbai, China 35 D15 41 25N 128 5 E
Changbai Shan, China 35 C15 42 20N 129 0 E
Changchiak'ou = Zhangjiakou, China 34 D8 40 48N 114 55 E
Ch'angchou = Changzhou, China 33 C6 31 47N 119 58 E
Changchun, China 35 C13 43 57N 125 17 E
Changchunling, China 35 B13 45 18N 125 27 E
Changde, China 33 D6 29 4N 111 35 E
Changdo-ri, N. Korea 35 E14 38 30N 127 40 E
Changhai = Shanghai, China 33 C7 31 15N 121 26 E
Changhua, Taiwan 33 D7 24 2N 120 30 E
Changhŭng, S. Korea 35 G14 34 41N 126 52 E
Changhŭngni, N. Korea 35 D15 40 24N 128 19 E
Changjiang, China 38 C7 19 20N 108 55 E
Changjin, N. Korea 35 D14 40 23N 127 15 E
Changjin-chŏsuji, N. Korea 35 D14 40 30N 127 15 E
Changli, China 35 E10 39 40N 119 13 E
Changling, China 35 B12 44 20N 123 58 E
Changlun, Malaysia 39 J3 6 25N 100 26 E
Changping, China 34 D9 40 14N 116 12 E
Changsha, China 33 D6 28 12N 113 0 E
Changwu, China 34 G4 35 10N 107 45 E
Changyi, China 35 F10 36 40N 119 30 E
Changyŏn, N. Korea 35 E13 38 15N 125 6 E
Changyuan, China 34 G8 35 15N 114 42 E
Changzhi, China 34 F7 36 10N 113 6 E
Changzhou, China 33 C6 31 47N 119 58 E
Chanhanga, Angola 56 B1 16 0S 14 8 E
Channapatna, India 40 N10 12 40N 77 15 E
Channel Is., U.K. 11 H5 49 19N 2 24W
Channel Is., U.S.A. 85 M7 33 40N 119 15W
Channel Islands Nat. Park, U.S.A. 85 M8 33 30N 119 0W
Channel-Port aux Basques, Canada 71 C8 47 30N 59 9W
Channel Tunnel, Europe 11 F9 51 0N 1 30 E
Channing, U.S.A. 81 H3 35 41N 102 20W
Chantada, Spain 19 A2 42 36N 7 46W
Chanthaburi, Thailand 38 F4 12 38N 102 12 E
Chantrey Inlet, Canada 68 B10 67 48N 96 20W
Chanute, U.S.A. 81 G7 37 41N 95 27W
Chao Phraya →, Thailand 38 F3 13 32N 100 36 E
Chao Phraya Lowlands, Thailand 38 E3 15 30N 100 0 E
Chaocheng, China 34 F8 36 4N 115 37 E
Chaoyang, China 35 D11 41 35N 120 22 E
Chaozhou, China 33 D6 23 42N 116 32 E
Chapais, Canada 70 C5 49 47N 74 51W
Chapala, Mozam. 55 F4 15 50S 37 35 E
Chapala, L. de, Mexico 86 C4 20 10N 103 20W
Chapayev, Kazakstan 25 D9 50 25N 51 10 E
Chapayevsk, Russia 24 D8 53 0N 49 40 E
Chapecó, Brazil 95 B5 27 14S 52 41W
Chapel Hill, U.S.A. 77 H6 35 55N 79 4W
Chapleau, Canada 70 C3 47 50N 83 24W
Chaplin, Canada 73 C7 50 28N 106 40W
Chaplin L., Canada 73 C7 50 22N 106 36W
Chappell, U.S.A. 80 E3 41 6N 102 28W
Chapra = Chhapra, India 43 G11 25 48N 84 44 E
Chara, Russia 27 D12 56 54N 118 20 E
Charadai, Argentina 94 B4 27 35S 59 55W
Charagua, Bolivia 92 G6 19 45S 63 10W
Charambirá, Punta, Colombia 92 C3 4 16N 77 32W
Charaña, Bolivia 92 G5 17 30S 69 25W
Charanwala, India 42 F5 27 51N 72 10 E
Charata, Argentina 94 B3 27 13S 61 14W
Charcas, Mexico 86 C4 23 10N 101 20W
Chard, U.K. 11 G5 50 52N 2 58W
Chardon, U.S.A. 78 E3 41 35N 81 12W
Chardzhou = Chärjew, Turkmenistan 26 F7 39 6N 63 34 E
Charente →, France 18 D3 45 57N 1 5W
Chari →, Chad 51 F8 12 58N 14 31 E
Chārīkār, Afghan. 40 B6 35 0N 69 10 E
Chariton →, U.S.A. 80 F8 39 19N 92 58W
Chärjew, Turkmenistan 26 F7 39 6N 63 34 E
Charkhari, India 43 G8 25 24N 79 45 E
Charkhi Dadri, India 42 E7 28 37N 76 17 E
Charleroi, Belgium 15 D4 50 24N 4 27 E
Charleroi, U.S.A. 78 F5 40 9N 79 57W
Charles, C., U.S.A. 76 G8 37 7N 75 58W
Charles City, U.S.A. 80 D8 43 4N 92 41W
Charles L., Canada 73 B6 59 50N 110 33W
Charles Town, U.S.A. 76 F7 39 17N 77 52W
Charleston, Ill., U.S.A. 76 F1 39 30N 88 10W
Charleston, Miss., U.S.A. 81 H9 34 1N 90 4W
Charleston, Mo., U.S.A. 81 G10 36 55N 89 21W
Charleston, S.C., U.S.A. 77 J6 32 46N 79 56W
Charleston, W. Va., U.S.A. 76 F5 38 21N 81 38W
Charleston L., Canada 79 B9 44 32N 76 0W
Charleston Peak, U.S.A. 85 J11 36 16N 115 42W
Charlestown, Ireland 13 C3 53 58N 8 48W
Charlestown, S. Africa 57 D4 27 26S 29 53 E
Charlestown, Ind., U.S.A. 76 F3 38 27N 85 40W
Charlestown, N.H., U.S.A. 79 C12 43 14N 72 25W
Charleville = Rath Luirc, Ireland 13 D3 52 21N 8 40W
Charleville, Australia 63 D4 26 24S 146 15 E
Charleville-Mézières, France 18 B6 49 44N 4 40 E
Charlevoix, U.S.A. 76 C3 45 19N 85 16W
Charlotte, Mich., U.S.A. 76 D3 42 34N 84 50W
Charlotte, N.C., U.S.A. 77 H5 35 13N 80 51W
Charlotte, Vt., U.S.A. 79 B11 44 19N 73 14W
Charlotte Amalie, U.S. Virgin Is. 89 C7 18 21N 64 56W
Charlotte Harbor, U.S.A. 77 M4 26 50N 82 10W
Charlotte L., Canada 72 C3 52 12N 125 19W
Charlottesville, U.S.A. 76 F6 38 2N 78 30W
Charlottetown, Nfld., Canada 71 B8 52 46N 56 7W
Charlottetown, P.E.I., Canada 71 C7 46 14N 63 8W
Charlton, Australia 63 F3 36 16S 143 24 E
Charlton, U.S.A. 80 E8 40 59N 93 20W
Charlton I., Canada 70 B4 52 0N 79 20W
Charny, Canada 71 C5 46 43N 71 15W
Charolles, France 18 C6 46 27N 4 16 E
Charre, Mozam. 55 F4 17 13S 35 10 E
Charsadda, Pakistan 42 B4 34 7N 71 45 E
Charters Towers, Australia 62 C4 20 5S 146 13 E
Chartres, France 18 B4 48 29N 1 30 E
Chascomús, Argentina 94 D4 35 30S 58 0W
Chasefu, Zambia 55 E3 11 55S 33 8 E
Chashma Barrage, Pakistan 42 C4 32 27N 71 20 E
Chāt, Iran 45 B7 37 59N 55 16 E
Châteaubriant, France 18 C3 47 43N 1 23W
Chateaugay, U.S.A. 79 B10 44 56N 74 5W
Châteauguay, L., Canada 71 A5 56 26N 70 3W
Châteaulin, France 18 B1 48 11N 4 8W
Châteauroux, France 18 C4 46 50N 1 40 E
Châtellerault, France 18 C4 46 50N 0 30 E
Chatham = Miramichi, Canada 71 C6 47 2N 65 28W
Chatham, Canada 78 D2 42 24N 82 11W
Chatham, U.K. 11 F8 51 22N 0 32 E
Chatham, U.S.A. 79 D11 42 21N 73 36W
Chatham Is., Pac. Oc. 64 M10 44 0S 176 40W
Chatmohar, Bangla. 43 G13 24 15N 89 15 E
Chatra, India 43 G11 24 12N 84 56 E
Chatrapur, India 41 K14 19 22N 85 2 E
Chats, L. des, Canada 79 A8 45 30N 76 20W
Chatsu, India 42 F6 26 36N 75 57 E
Chatsworth, Canada 78 B4 44 27N 80 54W
Chatsworth, Zimbabwe 55 F3 19 38S 31 13 E
Chattahoochee, U.S.A. 77 K3 30 42N 84 51W
Chattahoochee →, U.S.A. 77 K3 30 54N 84 57W
Chattanooga, U.S.A. 77 H3 35 3N 85 19W
Chatteris, U.K. 11 E8 52 28N 0 2 E
Chaturat, Thailand 38 E3 15 40N 101 51 E
Chau Doc, Vietnam 39 G5 10 42N 105 7 E
Chaukan Pass, Burma 41 F20 27 0N 97 15 E
Chaumont, France 18 B6 48 7N 5 8 E
Chaumont, U.S.A. 79 B8 44 4N 76 8W
Chautauqua L., U.S.A. 78 D5 42 10N 79 24W
Chauvin, Canada 73 C6 52 45N 110 10W
Chaves, Brazil 93 D9 0 15S 49 55W
Chaves, Portugal 19 B2 41 45N 7 32W
Chawang, Thailand 39 H2 8 25N 99 30 E
Chaykovskiy, Russia 24 C9 56 47N 54 9 E
Chazy, U.S.A. 79 B11 44 53N 73 26W
Cheb, Czech Rep. 16 C7 50 9N 12 28 E
Cheboksary, Russia 24 C8 56 8N 47 12 E
Cheboygan, U.S.A. 76 C3 45 39N 84 29W
Chech, Erg, Africa 50 D5 25 0N 2 15W
Chechenia □, Russia 25 F8 43 30N 45 29 E
Checheno-Ingush Republic = Chechenia □, Russia 25 F8 43 30N 45 29 E
Chechnya = Chechenia □, Russia 25 F8 43 30N 45 29 E
Chech'ŏn, S. Korea 35 F15 37 8N 128 12 E
Checotah, U.S.A. 81 H7 35 28N 95 31W
Chedabucto B., Canada 71 C7 45 25N 61 8W
Cheduba I., Burma 41 K18 18 45N 93 40 E
Cheepie, Australia 63 D4 26 33S 145 1 E
Chegdomyn, Russia 27 D14 51 7N 133 1 E
Chegga, Mauritania 50 C4 25 27N 5 40W
Chegutu, Zimbabwe 55 F3 18 10S 30 14 E
Chehalis, U.S.A. 84 D4 46 40N 122 58W
Chehalis →, U.S.A. 84 D3 46 57N 123 50W
Cheju do, S. Korea 35 H14 33 29N 126 34 E
Chekiang = Zhejiang □, China 33 D7 29 0N 120 0 E
Chela, Sa. da, Angola 56 B1 16 20S 13 20 E
Chelan, U.S.A. 82 C4 47 51N 120 1W
Chelan, L., U.S.A. 82 B3 48 11N 120 30W
Cheleken, Turkmenistan 25 G9 39 34N 53 16 E
Cheleken Yarymadasy, Turkmenistan 45 B7 39 30N 53 15 E
Chelforó, Argentina 96 D3 39 0S 66 33W
Chelkar = Shalqar, Kazakstan 26 E6 47 48N 59 39 E
Chelkar Tengiz, Solonchak, Kazakstan 26 E7 48 5N 63 7 E
Chełm, Poland 17 C12 51 8N 23 30 E
Chełmno, Poland 17 B10 53 20N 18 30 E
Chelmsford, U.K. 11 F8 51 44N 0 29 E
Chelsea, U.S.A. 79 C12 43 59N 72 27W
Cheltenham, U.K. 11 F5 51 54N 2 4W
Chelyabinsk, Russia 26 D7 55 10N 61 24 E
Chelyuskin, C., Russia 28 B14 77 30N 103 0 E
Chemainus, Canada 84 B3 48 55N 123 42W
Chemba, Mozam. 53 H6 17 9S 34 53 E
Chemnitz, Germany 16 C7 50 51N 12 54 E
Chemult, U.S.A. 82 E3 43 14N 121 47W
Chen, Gora, Russia 27 C15 65 16N 141 50 E
Chenab →, Pakistan 42 D4 30 23N 71 2 E
Chenango Forks, U.S.A. 79 D9 42 15N 75 51W
Cheney, U.S.A. 82 C5 47 30N 117 35W
Cheng Xian, China 34 H3 33 43N 105 42 E
Chengcheng, China 34 G5 35 8N 109 56 E
Chengchou = Zhengzhou, China 34 G7 34 45N 113 34 E
Chengde, China 35 D9 40 59N 117 58 E
Chengdu, China 32 C5 30 38N 104 2 E
Chenggu, China 34 H4 33 10N 107 21 E
Chengjiang, China 32 D5 24 39N 103 0 E
Ch'engmai, China 38 C7 19 50N 109 58 E
Ch'engtu = Chengdu, China 32 C5 30 38N 104 2 E
Chengwu, China 34 G8 34 58N 115 50 E
Chengyang, China 35 F11 36 18N 120 21 E
Chenjiagang, China 35 G10 34 23N 119 47 E
Chenkán, Mexico 87 D6 19 8N 90 58W
Chennai, India 40 N12 13 8N 80 19 E
Cheo Reo, Vietnam 36 B3 13 25N 108 28 E
Cheom Ksan, Cambodia 38 E5 14 13N 104 56 E
Chepén, Peru 92 E3 7 15S 79 23W
Chepes, Argentina 94 C2 31 20S 66 35W
Chepo, Panama 88 E4 9 10N 79 6W
Chepstow, U.K. 11 F5 51 38N 2 41W
Cheptulil, Mt., Kenya 54 B4 1 25N 35 35 E
Chequamegon B., U.S.A. 80 B9 46 40N 90 30W
Cher →, France 18 C4 47 21N 0 29 E
Cheraw, U.S.A. 77 H6 34 42N 79 53W
Cherbourg, France 18 B3 49 39N 1 40W
Cherdyn, Russia 24 B10 60 24N 56 29 E
Cheremkhovo, Russia 27 D11 53 8N 103 1 E
Cherepanovo, Russia 26 D9 54 15N 83 30 E
Cherepovets, Russia 24 C6 59 5N 37 55 E
Chergui, Chott ech, Algeria 50 B6 34 21N 0 25 E
Cherikov = Cherykaw, Belarus 17 B16 53 32N 31 20 E
Cherkasy, Ukraine 25 E5 49 27N 32 4 E
Cherkessk, Russia 25 F7 44 15N 42 5 E
Cherlak, Russia 26 D8 54 15N 74 55 E
Chernaya, Russia 27 B9 70 30N 89 10 E
Chernigov = Chernihiv, Ukraine 24 D5 51 28N 31 20 E
Chernihiv, Ukraine 24 D5 51 28N 31 20 E
Chernivtsi, Ukraine 17 D13 48 15N 25 52 E
Chernobyl = Chornobyl, Ukraine 17 C16 51 20N 30 15 E
Chernogorsk, Russia 27 D10 53 49N 91 18 E
Chernovtsy = Chernivtsi, Ukraine 17 D13 48 15N 25 52 E
Chernyakhovsk, Russia 9 J19 54 36N 21 48 E
Chernysheyskiy, Russia 27 C12 63 0N 112 30 E
Cherokee, Iowa, U.S.A. 80 D7 42 45N 95 33W
Cherokee, Okla., U.S.A. 81 G5 36 45N 98 21W
Cherokee Village, U.S.A. 81 G9 36 17N 91 30W
Cherokees, Grand Lake O' The, U.S.A. 81 G7 36 28N 95 2W
Cherrapunji, India 41 G17 25 17N 91 47 E
Cherry Valley, Calif., U.S.A. 85 M10 33 59N 116 57W
Cherry Valley, N.Y., U.S.A. 79 D10 42 48N 74 45W
Cherskiy, Russia 27 C17 68 45N 161 18 E
Cherskogo Khrebet, Russia 27 C15 65 0N 143 0 E
Cherven, Belarus 17 B15 53 45N 28 28 E
Chervonohrad, Ukraine 17 C13 50 25N 24 10 E
Cherwell →, U.K. 11 F6 51 44N 1 14W
Cherykaw, Belarus 17 B16 53 32N 31 20 E
Chesapeake, U.S.A. 76 G7 36 50N 76 17W
Chesapeake B., U.S.A. 76 G7 38 0N 76 10W
Cheshire □, U.K. 10 D5 53 14N 2 30W
Cheshskaya Guba, Russia 24 A8 67 20N 47 0 E
Cheshunt, U.K. 11 F7 51 43N 0 1W
Chesil Beach, U.K. 11 G5 50 37N 2 33W
Chesley, Canada 78 B3 44 17N 81 5W
Chester, U.K. 10 D5 53 12N 2 53W
Chester, Calif., U.S.A. 82 F3 40 19N 121 14W
Chester, Ill., U.S.A. 81 G10 37 55N 89 49W
Chester, Mont., U.S.A. 82 B8 48 31N 110 58W
Chester, Pa., U.S.A. 76 F8 39 51N 75 22W
Chester, S.C., U.S.A. 77 H5 34 43N 81 12W
Chester, Vt., U.S.A. 79 C12 43 16N 72 36W
Chester, W. Va., U.S.A. 78 F4 40 37N 80 34W
Chester-le-Street, U.K. 10 C6 54 51N 1 34W
Chesterfield, U.K. 10 D6 53 15N 1 25W
Chesterfield, Is., N. Cal. 64 J7 19 52S 158 15 E
Chesterfield Inlet, Canada 68 B10 63 30N 90 45W
Chesterton Ra., Australia 63 D4 25 30S 147 27 E
Chestertown, U.S.A. 79 C11 43 40N 73 48W
Chesterville, Canada 79 A9 45 6N 75 14W
Chestnut Ridge, U.S.A. 78 F5 40 20N 79 10W
Chesuncook L., U.S.A. 77 C11 46 0N 69 21W
Chéticamp, Canada 71 C7 46 37N 60 59W
Chetumal, Mexico 87 D7 18 30N 88 20W
Chetumal, B. de, Mexico 87 D7 18 40N 88 10W
Chetwynd, Canada 72 B4 55 45N 121 36W
Cheviot, The, U.K. 10 B5 55 29N 2 9W
Cheviot Hills, U.K. 10 B5 55 20N 2 30W
Cheviot Ra., Australia 62 D3 25 20S 143 45 E
Chew Bahir, Ethiopia 46 G2 4 40N 36 50 E
Chewelah, U.S.A. 82 B5 48 17N 117 43W
Cheyenne, Okla., U.S.A. 81 H5 35 37N 99 40W
Cheyenne, Wyo., U.S.A. 80 E2 41 8N 104 49W
Cheyenne →, U.S.A. 80 C4 44 41N 101 18W
Cheyenne Wells, U.S.A. 80 F3 38 49N 102 21W
Cheyne B., Australia 61 F2 34 35S 118 50 E
Chhabra, India 42 G7 24 40N 76 54 E
Chhaktala, India 42 H6 22 6N 74 11 E
Chhapra, India 43 G11 25 48N 84 44 E
Chhata, India 42 F7 27 42N 77 30 E
Chhatarpur, Jharkhand, India 43 G11 24 23N 84 11 E
Chhatarpur, Mad. P., India 43 G8 24 55N 79 35 E
Chhattisgarh □, India 43 J10 22 0N 82 0 E
Chhep, Cambodia 38 F5 13 45N 105 24 E
Chhindwara, Mad. P., India 43 H8 23 3N 79 29 E
Chhindwara, Mad. P., India 43 H8 22 2N 78 59 E
Chhlong, Cambodia 39 F5 12 15N 105 58 E
Chhota Tawa →, India 42 H7 22 14N 76 36 E
Chhoti Kali Sindh →, India 42 G6 24 2N 75 31 E
Chhuikhadan, India 43 J9 21 32N 80 59 E
Chhuk, Cambodia 39 G5 10 46N 104 28 E
Chi →, Thailand 38 E5 15 11N 104 43 E
Chiai, Taiwan 33 D7 23 29N 120 25 E
Chiamboni, Somali Rep. 52 E8 1 39S 41 35 E
Chiamussu = Jiamusi, China 33 B8 46 40N 130 26 E
Chiang Dao, Thailand 38 C2 19 22N 98 58 E
Chiang Kham, Thailand 38 C3 19 32N 100 18 E
Chiang Khan, Thailand 38 D3 17 52N 101 36 E
Chiang Khong, Thailand 38 B3 20 17N 100 24 E
Chiang Mai, Thailand 38 C2 18 47N 98 59 E
Chiang Rai, Thailand 38 C2 19 52N 99 50 E
Chiang Saen, Thailand 38 B3 20 16N 100 5 E
Chiapa →, Mexico 87 D6 16 42N 93 0W
Chiapa de Corzo, Mexico 87 D6 16 42N 93 0W
Chiapas □, Mexico 87 D6 17 0N 92 45W
Chiautla, Mexico 87 D5 18 18N 98 34W
Chiávari, Italy 18 D8 44 19N 9 19 E
Chiavenna, Italy 18 C8 46 19N 9 24 E
Chiba, Japan 31 G10 35 30N 140 7 E
Chiba □, Japan 31 G10 35 30N 140 20 E
Chibabava, Mozam. 57 C5 20 17S 33 35 E
Chibemba, Cunene, Angola 53 H2 15 48S 14 8 E
Chibemba, Huila, Angola 56 B2 16 20S 15 20 E
Chibi, Zimbabwe 57 C5 20 18S 30 25 E
Chibia, Angola 53 H2 15 10S 13 42 E
Chibougamau, Canada 70 C5 49 56N 74 24W
Chibougamau, L., Canada 70 C5 49 50N 74 20W
Chibuto, Mozam. 57 C5 24 40S 33 33 E
Chic-Chocs, Mts., Canada 71 C6 48 55N 66 0W
Chicacole = Srikakulam, India 41 K13 18 14N 83 58 E
Chicago, U.S.A. 76 E2 41 53N 87 38W
Chicago Heights, U.S.A. 76 E2 41 30N 87 38W
Chichagof I., U.S.A. 68 C6 57 30N 135 30W
Chichén-Itzá, Mexico 87 C7 20 40N 88 36W
Chicheng, China 34 D8 40 55N 115 55 E
Chichester, U.K. 11 G7 50 50N 0 47W
Chichester Ra., Australia 60 D2 22 12S 119 15 E
Chichibu, Japan 31 F9 35 59N 139 10 E
Ch'ich'ihaerh = Qiqihar, China 27 E13 47 26N 124 0 E
Chicholi, India 42 H8 22 1N 77 40 E
Chickasha, U.S.A. 81 H6 35 3N 97 58W
Chiclana de la Frontera, Spain 19 D2 36 26N 6 9W
Chiclayo, Peru 92 E3 6 42S 79 50W
Chico, U.S.A. 84 F5 39 44N 121 50W
Chico →, Chubut, Argentina 96 E3 44 0S 67 0W
Chico →, Santa Cruz, Argentina 96 G3 50 0S 68 30W
Chicomo, Mozam. 57 C5 24 31S 34 6 E
Chicontepec, Mexico 87 C5 20 58N 98 10W
Chicopee, U.S.A. 79 D12 42 9N 72 37W

Coral Sea

Clinch →, U.S.A. 77 H3 35 53N 84 29W
Clingmans Dome, U.S.A. 77 H4 35 34N 83 30W
Clint, U.S.A. 83 L10 31 35N 106 14W
Clinton, B.C., Canada 72 C4 51 6N 121 35W
Clinton, Ont., Canada 78 C3 43 37N 81 32W
Clinton, N.Z. 59 M2 46 12S 169 23 E
Clinton, Ark., U.S.A. 81 H8 35 36N 92 28W
Clinton, Conn., U.S.A. 79 E12 41 17N 72 32W
Clinton, Ill., U.S.A. 80 E10 40 9N 88 57W
Clinton, Ind., U.S.A. 76 F2 39 40N 87 24W
Clinton, Iowa, U.S.A. 80 E9 41 51N 90 12W
Clinton, Mass., U.S.A. 79 D13 42 25N 71 41W
Clinton, Miss., U.S.A. 81 J9 32 20N 90 20W
Clinton, Mo., U.S.A. 80 F8 38 22N 93 46W
Clinton, N.C., U.S.A. 77 H6 35 0N 78 22W
Clinton, Okla., U.S.A. 81 H5 35 31N 98 58W
Clinton, S.C., U.S.A. 77 H5 34 29N 81 53W
Clinton, Tenn., U.S.A. 77 G3 36 6N 84 8W
Clinton, Wash., U.S.A. 84 C4 47 59N 122 21W
Clinton C., Australia 62 C5 22 30S 150 45 E
Clinton Colden L., Canada 68 B9 63 58N 107 27W
Clintonville, U.S.A. 80 C10 44 37N 88 46W
Clipperton, I., Pac. Oc. 65 F17 10 18N 109 13W
Clisham, U.K. 12 D2 57 57N 6 49W
Clitheroe, U.K. 10 D5 53 53N 2 22W
Clo-oose, Canada 84 B2 48 39N 124 49W
Cloates, Pt., Australia 60 D1 22 43S 113 40 E
Clocolan, S. Africa 57 D4 28 55S 27 34 E
Clodomira, Argentina 94 B3 27 35S 64 14W
Clogher Hd., Ireland 13 C5 53 48N 6 14W
Clonakilty, Ireland 13 E3 51 37N 8 53W
Clonakilty B., Ireland 13 E3 51 35N 8 51W
Cloncurry, Australia 62 C3 20 40S 140 28 E
Cloncurry →, Australia 62 B3 18 37S 140 40 E
Clondalkin, Ireland 13 C5 53 19N 6 25W
Clones, Ireland 13 B4 54 11N 7 15W
Clonmel, Ireland 13 D4 52 21N 7 42W
Cloquet, U.S.A. 80 B8 46 43N 92 28W
Clorinda, Argentina 94 B4 25 16S 57 45W
Cloud Bay, Canada 70 C2 48 5N 89 26W
Cloud Peak, U.S.A. 82 D10 44 23N 107 11W
Cloudcroft, U.S.A. 83 K11 32 58N 105 45W
Cloverdale, U.S.A. 84 G4 38 48N 123 1W
Clovis, Calif., U.S.A. 84 J7 36 49N 119 42W
Clovis, N. Mex., U.S.A. 81 H3 34 24N 103 12W
Cloyne, Canada 78 B7 44 49N 77 11W
Cluj-Napoca, Romania 17 E12 46 47N 23 38 E
Clunes, Australia 63 F3 37 20S 143 45 E
Clutha →, N.Z. 59 M2 46 20S 169 49 E
Clwyd →, U.K. 10 D4 53 19N 3 31W
Clyde, Canada 72 C6 54 9N 113 39W
Clyde, N.Z. 59 L2 45 12S 169 20 E
Clyde, U.S.A. 78 C8 43 5N 76 52W
Clyde →, U.K. 12 F4 55 55N 4 30W
Clyde, Firth of, U.K. 12 F3 55 22N 5 1W
Clyde River, Canada 69 A13 70 30N 68 30W
Clydebank, U.K. 12 F4 55 54N 4 23W
Clymer, N.Y., U.S.A. 78 D5 42 1N 79 37W
Clymer, Pa., U.S.A. 78 D5 40 40N 79 1W
Coachella, U.S.A. 85 M10 33 41N 116 10W
Coachella Canal, U.S.A. 85 N12 32 43N 114 57W
Coahoma, U.S.A. 81 J4 32 18N 101 18W
Coahuayana →, Mexico 86 D4 18 41N 103 45W
Coahuila □, Mexico 86 B4 27 0N 103 0W
Coal →, Canada 72 B3 59 39N 126 57W
Coalane, Mozam. 55 F4 17 48S 37 2 E
Coalcomán, Mexico 86 D4 18 40N 103 10W
Coaldale, Canada 72 D6 49 45N 112 35W
Coalgate, U.S.A. 81 H6 34 32N 96 13W
Coalinga, U.S.A. 84 J6 36 9N 120 21W
Coalisland, U.K. 13 B5 54 33N 6 42W
Coalville, U.K. 10 E6 52 44N 1 23W
Coalville, U.S.A. 82 F8 40 55N 111 24W
Coari, Brazil 92 D6 4 8S 63 7W
Coast □, Kenya 54 C4 2 40S 39 45 E
Coast Mts., Canada 72 C3 55 0N 129 20W
Coast Ranges, U.S.A. 84 G4 39 0N 123 0W
Coatbridge, U.K. 12 F4 55 52N 4 6W
Coatepec, Mexico 87 D5 19 27N 96 58W
Coatepeque, Guatemala 88 D1 14 46N 91 55W
Coatesville, U.S.A. 76 F8 39 59N 75 50W
Coaticook, Canada 79 A13 45 10N 71 46W
Coats I., Canada 69 B11 62 30N 83 0W
Coats Land, Antarctica 5 D1 77 0S 25 0W
Coatzacoalcos, Mexico 87 D6 18 7N 94 25W
Cobalt, Canada 70 C4 47 25N 79 42W
Cobán, Guatemala 88 C1 15 30N 90 21W
Cobar, Australia 63 E4 31 27S 145 48 E
Cóbh, Ireland 13 E3 51 51N 8 17W
Cobija, Bolivia 92 F5 11 0S 68 50W
Cobleskill, U.S.A. 79 D10 42 41N 74 29W
Coboconk, Canada 78 B6 44 39N 78 48W
Cobourg, Canada 78 C6 43 58N 78 10W
Cobourg Pen., Australia 60 B5 11 20S 132 15 E
Cobram, Australia 63 F4 35 54S 145 40 E
Cóbué, Mozam. 55 E3 12 0S 34 58 E
Coburg, Germany 16 C6 50 15N 10 58 E
Cocanada = Kakinada, India 41 L13 16 57N 82 11 E
Cochabamba, Bolivia 92 G5 17 26S 66 10W
Cochemane, Mozam. 55 F3 17 0S 32 54 E
Cochin, India 40 Q10 9 58N 76 20 E
Cochin China = Nam-Phan, Vietnam 39 G6 10 30N 106 0 E
Cochran, U.S.A. 77 J4 32 23N 83 21W
Cochrane, Alta., Canada 72 C6 51 11N 114 30W
Cochrane, Ont., Canada 70 C3 49 0N 81 0W
Cochrane, Chile 96 F2 47 15S 72 33W
Cochrane →, Canada 73 B8 59 0N 103 40W
Cochrane, L., Chile 96 F2 47 10S 72 0W
Cochranton, U.S.A. 78 E4 41 31N 80 3W
Cockburn, Australia 63 E3 32 5S 141 0 E
Cockburn, Canal, Chile 96 G2 54 30S 72 0W
Cockburn I., Canada 70 C3 45 55N 83 22W
Cockburn Ra., Australia 60 C4 15 46S 128 0 E
Cockermouth, U.K. 10 C4 54 40N 3 22W
Cocklebiddy, Australia 61 F4 32 0S 126 3 E
Coco →, Cent. Amer. 88 D3 15 0N 83 8W
Coco, I. del, Pac. Oc. 65 G19 5 25N 87 55W
Cocoa, U.S.A. 77 L5 28 21N 80 44W
Cocobeach, Gabon 52 D1 0 59N 9 34 E
Cocos Is., Ind. Oc. 64 J1 12 10S 96 55 E
Cod, C., U.S.A. 76 D10 42 5N 70 10W
Codajás, Brazil 92 D6 3 55S 62 0W
Codó, Brazil 93 D10 4 30S 43 55W
Cody, U.S.A. 82 D9 44 32N 109 3W
Coe Hill, Canada 78 B7 44 52N 77 50W
Coelemu, Chile 94 D1 36 30S 72 48W
Coen, Australia 62 A3 13 52S 143 12 E
Cœur d'Alene, U.S.A. 82 C5 47 45N 116 51W
Cœur d'Alene L., U.S.A. 82 C5 47 32N 116 48W
Coevorden, Neths. 15 B6 52 40N 6 44 E
Cofete, Canary Is. 22 F5 28 6N 14 23W
Coffeyville, U.S.A. 81 G7 37 2N 95 37W
Coffin B., Australia 63 E2 34 38S 135 28 E
Coffin Bay, Australia 63 E2 34 37S 135 29 E
Coffin Bay Peninsula, Australia 63 E2 34 32S 135 15 E
Coffs Harbour, Australia 63 E5 30 16S 153 5 E
Cognac, France 18 D3 45 41N 0 20W
Cohocton, U.S.A. 78 D7 42 30N 77 30W
Cohocton →, U.S.A. 78 D7 42 9N 77 6W
Cohoes, U.S.A. 79 D11 42 46N 73 42W
Cohuna, Australia 63 F3 35 45S 144 15 E
Coiba, I., Panama 88 E3 7 30N 81 40W
Coig →, Argentina 96 G3 51 0S 69 10W
Coigeach, Rubha, U.K. 12 C3 58 6N 5 26W
Coihaique, Chile 96 F2 45 30S 71 45W
Coimbatore, India 40 P10 11 2N 76 59 E
Coimbra, Brazil 92 G7 19 55S 57 48W
Coimbra, Portugal 19 B1 40 15N 8 27W
Coín, Spain 19 D3 36 40N 4 48W
Coipasa, Salar de, Bolivia 92 G5 19 26S 68 9W
Cojimies, Ecuador 92 C3 0 20N 80 0W
Cojutepequé, El Salv. 88 D2 13 41N 88 54W
Cokeville, U.S.A. 82 E8 42 5N 110 57W
Colac, Australia 63 F3 38 21S 143 35 E
Colatina, Brazil 93 G10 19 32S 40 37W
Colbeck, C., Antarctica 5 D13 77 6S 157 48W
Colborne, Canada 78 C7 44 0N 77 53W
Colby, U.S.A. 80 F4 39 24N 101 3W
Colchester, U.K. 11 F8 51 54N 0 55 E
Cold L., Canada 73 C7 54 33N 110 5W
Coldstream, Canada 72 C5 50 13N 119 11W
Coldstream, U.K. 12 F6 55 39N 2 15W
Coldwater, Canada 78 B5 44 42N 79 40W
Coldwater, Kans., U.S.A. 81 G5 37 16N 99 20W
Coldwater, Mich., U.S.A. 76 E3 41 57N 85 0W
Colebrook, U.S.A. 79 B13 44 54N 71 30W
Coleman, U.S.A. 81 K5 31 50N 99 26W
Coleman →, Australia 62 B3 15 6S 141 38 E
Colenso, S. Africa 57 D4 28 44S 29 50 E
Coleraine, Australia 63 F3 37 36S 141 40 E
Coleraine, U.K. 13 A5 55 8N 6 41W
Coleridge, L., N.Z. 59 K3 43 17S 171 30 E
Colesberg, S. Africa 56 E4 30 45S 25 5 E
Coleville, U.S.A. 84 G7 38 34N 119 30W
Colfax, Calif., U.S.A. 84 F6 39 6N 120 57W
Colfax, La., U.S.A. 81 K8 31 31N 92 42W
Colfax, Wash., U.S.A. 82 C5 46 53N 117 22W
Colhué Huapi, L., Argentina 96 F3 45 30S 69 0W
Coligny, S. Africa 57 D4 26 17S 26 15 E
Colima, Mexico 86 D4 19 14N 103 43W
Colima □, Mexico 86 D4 19 10N 103 40W
Colima, Nevado de, Mexico 86 D4 19 35N 103 45W
Colina, Chile 94 C1 33 13S 70 45W
Colinas, Brazil 93 E10 6 0S 44 10W
Coll, U.K. 12 E2 56 39N 6 34W
Collaguasi, Chile 94 A2 21 5S 68 45W
Collarenebri, Australia 63 D4 29 33S 148 34 E
Colleen Bawn, Zimbabwe 55 G2 21 0S 29 12 E
College Park, U.S.A. 77 J3 33 40N 84 27W
College Station, U.S.A. 81 K6 30 37N 96 21W
Collie, Australia 61 F2 33 22S 116 8 E
Collier B., Australia 60 C3 16 10S 124 15 E
Collier Ra., Australia 61 D2 24 45S 119 10 E
Collina, Passo di, Italy 20 B4 44 2N 10 56 E
Collingwood, Canada 78 B4 44 29N 80 13W
Collingwood, N.Z. 59 J4 40 41S 172 40 E
Collins, Canada 70 B2 50 17N 89 27W
Collinsville, Australia 62 C4 20 30S 147 56 E
Collipulli, Chile 94 D1 37 55S 72 30W
Collooney, Ireland 13 B3 54 11N 8 29W
Colmar, France 18 B7 48 5N 7 20 E
Colo →, Australia 63 E5 33 25S 150 52 E
Cologne = Köln, Germany 16 C4 50 56N 6 57 E
Colom, I. d'en, Spain 22 B11 39 58N 4 16 E
Coloma, U.S.A. 84 G6 38 48N 120 53W
Colomb-Béchar = Béchar, Algeria 50 B5 31 38N 2 18W
Colombia ■, S. Amer. 92 C4 3 45N 73 0W
Colombian Basin, S. Amer. 66 H12 14 0N 76 0W
Colombo, Sri Lanka 40 R11 6 56N 79 58 E
Colón, Buenos Aires, Argentina 94 C3 33 53S 61 7W
Colón, Entre Ríos, Argentina 94 C4 32 12S 58 10W
Colón, Cuba 88 B3 22 42N 80 54W
Colón, Panama 88 E4 9 20N 79 54W
Colònia de Sant Jordi, Spain 22 B9 39 19N 2 59 E
Colonia del Sacramento, Uruguay 94 C4 34 25S 57 50W
Colonia Dora, Argentina 94 B3 28 34S 62 59W
Colonial Beach, U.S.A. 76 F7 38 15N 76 58W
Colonie, U.S.A. 79 D11 42 43N 73 50W
Colonsay, Canada 73 C7 51 59N 105 52W
Colonsay, U.K. 12 E2 56 5N 6 12W
Colorado □, U.S.A. 83 G10 39 30N 105 30W
Colorado →, Argentina 96 D4 39 50S 62 8W
Colorado →, N. Amer. 83 L6 31 45N 114 40W
Colorado →, U.S.A. 81 L7 28 36N 95 59W
Colorado City, U.S.A. 81 J4 32 24N 100 52W
Colorado Plateau, U.S.A. 83 H8 37 0N 111 0W
Colorado River Aqueduct, U.S.A. 85 L12 34 17N 114 10W
Colorado Springs, U.S.A. 80 F2 38 50N 104 49W
Colotlán, Mexico 86 C4 22 6N 103 16W
Colstrip, U.S.A. 82 D10 45 53N 106 38W
Colton, U.S.A. 79 B10 44 33N 74 56W
Columbia, Ky., U.S.A. 76 G3 37 6N 85 18W
Columbia, La., U.S.A. 81 J8 32 6N 92 5W
Columbia, Miss., U.S.A. 81 K10 31 15N 89 50W
Columbia, Mo., U.S.A. 80 F8 38 57N 92 20W
Columbia, Pa., U.S.A. 79 F8 40 2N 76 30W
Columbia, S.C., U.S.A. 77 J5 34 0N 81 2W
Columbia, Tenn., U.S.A. 77 H2 35 37N 87 2W
Columbia →, N. Amer. 84 D2 46 15N 124 5W
Columbia, C., Canada 4 A4 83 0N 70 0W
Columbia, District of □, U.S.A. 76 F7 38 55N 77 0W
Columbia, Mt., Canada 72 C5 52 8N 117 20W
Columbia Basin, U.S.A. 82 C4 46 45N 119 5W
Columbia Falls, U.S.A. 82 B6 48 23N 114 11W
Columbia Mts., Canada 72 C5 52 0N 119 0W
Columbia Plateau, U.S.A. 82 D5 44 0N 117 30W
Columbiana, U.S.A. 78 F4 40 53N 80 42W
Columbretes, Is., Spain 19 C6 39 50N 0 50 E
Columbus, Ga., U.S.A. 77 J3 32 28N 84 59W
Columbus, Ind., U.S.A. 76 F3 39 13N 85 55W
Columbus, Kans., U.S.A. 81 G7 37 10N 94 50W
Columbus, Miss., U.S.A. 77 J1 33 30N 88 25W
Columbus, Mont., U.S.A. 82 D9 45 38N 109 15W
Columbus, N. Mex., U.S.A. 83 L10 31 50N 107 38W
Columbus, Nebr., U.S.A. 80 E6 41 26N 97 22W
Columbus, Ohio, U.S.A. 76 F4 39 58N 83 0W
Columbus, Tex., U.S.A. 81 L6 29 42N 96 33W
Colusa, U.S.A. 84 F4 39 13N 122 1W
Colville, U.S.A. 82 B5 48 33N 117 54W
Colville →, U.S.A. 68 A4 70 25N 150 30W
Colville, C., N.Z. 59 G5 36 29S 175 21 E
Colwood, Canada 84 B3 48 26N 123 29W
Colwyn Bay, U.K. 10 D4 53 18N 3 44W
Comácchio, Italy 20 B5 44 42N 12 11 E
Comalcalco, Mexico 87 D6 18 16N 93 13W
Comallo, Argentina 96 E2 41 0S 70 5W
Comanche, U.S.A. 81 K5 31 54N 98 36W
Comayagua, Honduras 88 D2 14 25N 87 37W
Combahee →, U.S.A. 77 J5 32 30N 80 31W
Combarbalá, Chile 94 C1 31 11S 71 2W
Comber, Canada 78 D2 42 14N 82 33W
Comber, U.K. 13 B6 54 33N 5 45W
Combermere, Canada 78 A7 45 22N 77 37W
Comblain-au-Pont, Belgium 15 D5 50 29N 5 35 E
Comeragh Mts., Ireland 13 D4 52 18N 7 34W
Comet, Australia 62 C4 23 36S 148 38 E
Comilla, Bangla. 41 H17 23 28N 91 10 E
Comino, Malta 23 C1 36 1N 14 20 E
Comino, C., Italy 20 D3 40 32N 9 49 E
Comitán, Mexico 87 D6 16 18N 92 9W
Commerce, Ga., U.S.A. 77 H4 34 12N 83 28W
Commerce, Tex., U.S.A. 81 J7 33 15N 95 54W
Committee B., Canada 69 B11 68 30N 86 30W
Commonwealth B., Antarctica 5 C10 67 0S 144 0 E
Commoron Cr. →, Australia 63 D5 28 22S 150 8 E
Communism Pk. = Kommunizma, Pik, Tajikistan 26 F8 39 0N 72 2 E
Como, Italy 18 D8 45 47N 9 5 E
Como, Lago di, Italy 18 D8 46 0N 9 11 E
Comodoro Rivadavia, Argentina 96 F3 45 50S 67 40W
Comorin, C., India 40 Q10 8 3N 77 40 E
Comoro Is. = Comoros ■, Ind. Oc. 49 H8 12 10S 44 15 E
Comoros ■, Ind. Oc. 49 H8 12 10S 44 15 E
Comox, Canada 72 D4 49 42N 124 55W
Compiègne, France 18 B5 49 24N 2 50 E
Compostela, Mexico 86 C4 21 15N 104 53W
Comprida, I., Brazil 95 A6 24 50S 47 42W
Compton, Canada 79 A13 45 14N 71 49W
Compton, U.S.A. 85 M8 33 54N 118 13W
Comrat, Moldova 17 E15 46 18N 28 40 E
Con Cuong, Vietnam 38 C5 19 2N 104 54 E
Con Son, Vietnam 39 H6 8 41N 106 37 E
Conakry, Guinea 50 G3 9 29N 13 49W
Conara, Australia 62 G4 41 50S 147 26 E
Concarneau, France 18 C2 47 52N 3 56W
Conceição, Mozam. 55 F4 18 47S 36 7 E
Conceição da Barra, Brazil 93 G11 18 35S 39 45W
Conceição do Araguaia, Brazil 93 E9 8 0S 49 2W
Concepción, Argentina 94 B2 27 20S 65 35W
Concepción, Bolivia 92 G6 16 15S 62 8W
Concepción, Chile 94 D1 36 50S 73 0W
Concepción, Mexico 87 D6 18 15N 90 5W
Concepción, Paraguay 94 A4 23 22S 57 26W
Concepción □, Chile 94 D1 37 0S 72 30W
Concepción →, Mexico 86 A2 30 32N 113 2W
Concepción, Est. de, Chile 96 G2 50 30S 74 55W
Concepción, L., Bolivia 92 G6 17 20S 61 20W
Concepción, Punta, Mexico 86 B2 26 55N 111 59W
Concepción del Oro, Mexico 86 C4 24 40N 101 30W
Concepción del Uruguay, Argentina 94 C4 32 35S 58 20W
Conception, Pt., U.S.A. 85 L6 34 27N 120 28W
Conception B., Canada 71 C9 47 45N 53 0W
Conception B., Namibia 56 C1 23 55S 14 22 E
Conception I., Bahamas 89 B4 23 52N 75 9W
Concession, Zimbabwe 55 F3 17 27S 30 56 E
Conchas Dam, U.S.A. 81 H2 35 22N 104 11W
Concho, U.S.A. 83 J9 34 28N 109 36W
Concho →, U.S.A. 81 K5 31 34N 99 43W
Conchos →, Chihuahua, Mexico 86 B4 29 32N 105 0W
Conchos →, Tamaulipas, Mexico 87 B5 25 9N 98 35W
Concord, Calif., U.S.A. 84 H4 37 59N 122 2W
Concord, N.C., U.S.A. 77 H5 35 25N 80 35W
Concord, N.H., U.S.A. 79 C13 43 12N 71 32W
Concordia, Argentina 94 C4 31 20S 58 2W
Concórdia, Brazil 92 D5 4 36S 66 36W
Concordia, Mexico 86 C3 23 18N 106 2W
Concordia, U.S.A. 80 F6 39 34N 97 40W
Concrete, U.S.A. 82 B3 48 32N 121 45W
Condamine, Australia 63 D5 26 56S 150 9 E
Conde, U.S.A. 80 C5 45 9N 98 6W
Condeúba, Brazil 93 F10 14 52S 42 0W
Condobolin, Australia 63 E4 33 4S 147 6 E
Condon, U.S.A. 82 D3 45 14N 120 11W
Conegliano, Italy 20 B5 45 53N 12 18 E
Conejera, I. = Conills, I. des, Spain 22 B9 39 11N 2 58 E
Conejos, Mexico 86 B4 26 14N 103 53W
Confuso →, Paraguay 94 B4 25 9S 57 34W
Congleton, U.K. 10 D5 53 10N 2 13W
Congo (Kinshasa) = Congo, Dem. Rep. of the ■, Africa 52 E4 3 0S 23 0 E
Congo ■, Africa 52 E3 1 0S 16 0 E
Congo →, Africa 52 F2 6 4S 12 24 E
Congo, Dem. Rep. of the ■, Africa 52 E4 3 0S 23 0 E
Congo Basin, Africa 52 E4 0 10S 24 30 E
Congonhas, Brazil 95 A7 20 30S 43 52W
Congress, U.S.A. 83 J7 34 9N 112 51W
Conills, I. des, Spain 22 B9 39 11N 2 58 E
Coniston, Canada 70 C3 46 29N 80 51W
Conjeeveram = Kanchipuram, India 40 N11 12 52N 79 45 E
Conklin, Canada 73 B6 55 38N 111 5W
Conklin, U.S.A. 79 D9 42 2N 75 49W
Conn, L., Ireland 13 B2 54 3N 9 15W
Connacht □, Ireland 13 C2 53 43N 9 12W
Conneaut, U.S.A. 78 E4 41 57N 80 34W
Connecticut □, U.S.A. 79 E12 41 30N 72 45W
Connecticut →, U.S.A. 79 E12 41 16N 72 20W
Connell, U.S.A. 82 C4 46 40N 118 52W
Connellsville, U.S.A. 78 F5 40 1N 79 35W
Connemara, Ireland 13 C2 53 29N 9 45W
Connemaugh →, U.S.A. 78 F5 40 28N 79 19W
Connersville, U.S.A. 76 F3 39 39N 85 8W
Connors Ra., Australia 62 C4 21 40S 149 10 E
Conquest, Canada 73 C7 51 32N 107 14W
Conrad, U.S.A. 82 B8 48 10N 111 57W
Conran, C., Australia 63 F4 37 49S 148 44 E
Conroe, U.S.A. 81 K7 30 19N 95 27W
Consecon, Canada 78 C7 44 0N 77 31W
Conselheiro Lafaiete, Brazil 95 A7 20 40S 43 48W
Consett, U.K. 10 C6 54 51N 1 50W
Consort, Canada 73 C6 52 1N 110 46W
Constance = Konstanz, Germany 16 E5 47 40N 9 10 E
Constance, L. = Bodensee, Europe 18 C8 47 35N 9 25 E
Constanța, Romania 17 F15 44 14N 28 38 E
Constantia, U.S.A. 79 C8 43 15N 76 1W
Constantine, Algeria 50 A7 36 25N 6 42 E
Constitución, Chile 94 D1 35 20S 72 30W
Constitución, Uruguay 94 C4 31 0S 57 50W
Consul, Canada 73 D7 49 20N 109 30W
Contact, U.S.A. 82 F6 41 46N 114 45W
Contai, India 43 J12 21 54N 87 46 E
Contamana, Peru 92 E4 7 19S 74 55W
Contas →, Brazil 93 F11 14 17S 39 1W
Contoocook, U.S.A. 79 C13 43 13N 71 45W
Contra Costa, Mozam. 57 D5 25 9S 33 30 E
Contwoyto L., Canada 68 B8 65 42N 110 50W
Conway = Conwy, U.K. 10 D4 53 17N 3 50W
Conway = Conwy →, U.K. 10 D4 53 17N 3 50W
Conway, Ark., U.S.A. 81 H8 35 5N 92 26W
Conway, N.H., U.S.A. 79 C13 43 59N 71 7W
Conway, S.C., U.S.A. 77 J6 33 51N 79 3W
Conway, L., Australia 63 D2 28 17S 135 35 E
Conwy, U.K. 10 D4 53 17N 3 50W
Conwy □, U.K. 10 D4 53 10N 3 44W
Conwy →, U.K. 10 D4 53 17N 3 50W
Coober Pedy, Australia 63 D1 29 1S 134 43 E
Cooch Behar = Koch Bihar, India 41 F16 26 22N 89 29 E
Cooinda, Australia 60 B5 13 15S 130 5 E
Cook, Australia 61 F5 30 37S 130 25 E
Cook, U.S.A. 80 B8 47 49N 92 39W
Cook, B., Chile 96 H3 55 10S 70 0W
Cook, C., Canada 72 C3 50 8N 127 55W
Cook, Mt. = Aoraki Mount Cook, N.Z. 59 K3 43 36S 170 9 E
Cook Inlet, U.S.A. 68 C4 60 0N 152 0W
Cook Is., Pac. Oc. 65 J12 17 0S 160 0W
Cook Strait, N.Z. 59 J5 41 15S 174 29 E
Cookeville, U.S.A. 77 G3 36 10N 85 30W
Cookhouse, S. Africa 56 E4 32 44S 25 47 E
Cookshire, Canada 79 A13 45 25N 71 38W
Cookstown, U.K. 13 B5 54 39N 6 45W
Cooksville, Canada 78 C5 43 36N 79 35W
Cooktown, Australia 62 B4 15 30S 145 16 E
Coolabah, Australia 63 E4 31 1S 146 43 E
Cooladdi, Australia 63 D4 26 37S 145 23 E
Coolah, Australia 63 E4 31 48S 149 41 E
Coolamon, Australia 63 E4 34 46S 147 8 E
Coolgardie, Australia 61 F3 30 55S 121 8 E
Coolidge, U.S.A. 83 K8 32 59N 111 31W
Coolidge Dam, U.S.A. 83 K8 33 0N 110 20W
Cooma, Australia 63 F4 36 12S 149 8 E
Coon Rapids, U.S.A. 80 C8 45 9N 93 19W
Coonabarabran, Australia 63 E4 31 14S 149 18 E
Coonamble, Australia 63 E4 30 56S 148 27 E
Coonana, Australia 61 F3 31 0S 123 0 E
Coondapoor, India 40 N9 13 42N 74 40 E
Cooninie, L., Australia 63 D2 26 4S 139 59 E
Cooper, U.S.A. 81 J7 33 23N 95 42W
Cooper Cr. →, Australia 63 D2 28 29S 137 46 E
Cooperstown, N. Dak., U.S.A. 80 B5 47 27N 98 8W
Cooperstown, N.Y., U.S.A. 79 D10 42 42N 74 56W
Coorabie, Australia 61 F5 31 54S 132 18 E
Coorong, The, Australia 63 F2 35 50S 139 20 E
Coorow, Australia 61 E2 29 53S 116 2 E
Cooroy, Australia 63 D5 26 22S 152 54 E
Coos Bay, U.S.A. 82 E1 43 22N 124 13W
Coosa →, U.S.A. 77 J2 32 30N 86 16W
Cootamundra, Australia 63 E4 34 36S 148 1 E
Cootehill, Ireland 13 B4 54 4N 7 5W
Copahue Paso, Argentina 94 D1 37 49S 71 8W
Copainalá, Mexico 87 D6 17 8N 93 11W
Copake Falls, U.S.A. 79 D11 42 7N 73 31W
Copán, Honduras 88 D2 14 50N 89 9W
Cope, U.S.A. 80 F3 39 40N 102 51W
Copenhagen = København, Denmark 9 J15 55 41N 12 34 E
Copenhagen, U.S.A. 79 C9 43 54N 75 41W
Copiapó, Chile 94 B1 27 30S 70 20W
Copiapó →, Chile 94 B1 27 19S 70 56W
Coplay, U.S.A. 79 F9 40 44N 75 29W
Copp L., Canada 72 A6 60 14N 114 40W
Coppename →, Surinam 93 B7 5 48N 55 55W
Copper Harbor, U.S.A. 76 B2 47 28N 87 53W
Copper Queen, Zimbabwe 55 F2 17 29S 29 18 E
Copperas Cove, U.S.A. 81 K6 31 8N 97 54W
Copperbelt □, Zambia 55 E2 13 15S 27 30 E
Coppermine = Kugluktuk, Canada 68 B8 67 50N 115 5W
Coppermine →, Canada 68 B8 67 49N 116 4W
Copperopolis, U.S.A. 84 H6 37 58N 120 38W
Coquet →, U.K. 10 B6 55 20N 1 32W
Coquille, U.S.A. 82 E1 43 11N 124 11W
Coquimbo, Chile 94 C1 30 0S 71 20W
Coquimbo □, Chile 94 C1 31 0S 71 0W
Corabia, Romania 17 G13 43 48N 24 30 E
Coracora, Peru 92 G4 15 5S 73 45W
Coraki, Australia 63 D5 28 59S 153 17 E
Coral, U.S.A. 78 F5 40 29N 79 10W
Coral Gables, U.S.A. 77 N5 25 45N 80 16W
Coral Harbour = Salliq, Canada 69 B11 64 8N 83 10W
Coral Sea, Pac. Oc. 64 J7 15 0S 150 0 E

D

Denau

Dacca = Dhaka, *Bangla.* .... **43 H14** 23 43N 90 26 E
Dacca = Dhaka □, *Bangla.* .. **43 G14** 24 25N 90 25 E
Dachau, *Germany* .... **16 D6** 48 15N 11 26 E
Dadanawa, *Guyana* **92 C7** 2 50N 59 30W
Dade City, *U.S.A.* **77 L4** 28 22N 82 11W
Dadhar, *Pakistan* .... **42 E2** 29 28N 67 39 E
Dadra & Nagar Haveli □, *India* .... **40 J8** 20 5N 73 0 E
Dadri = Charkhi Dadri, *India* . **42 E7** 28 37N 76 17 E
Dadu, *Pakistan* .... **42 F2** 26 45N 67 45 E
Daet, *Phil.* .... **37 B6** 14 2N 122 55 E
Dagana, *Senegal* **50 E2** 16 30N 15 35W
Dagestan □, *Russia* .... **25 F8** 42 30N 47 0 E
Daggett, *U.S.A.* **85 L10** 34 52N 116 52W
Daghestan Republic = Dagestan □, *Russia* ...... **25 F8** 42 30N 47 0 E
Dağlıq Qarabağ = Nagorno-Karabakh, *Azerbaijan* .... **25 F8** 39 55N 46 45 E
Dagö = Hiiumaa, *Estonia* ... **9 G20** 58 50N 22 45 E
Dagu, *China* .... **35 E9** 38 59N 117 40 E
Dagupan, *Phil.* .... **37 A6** 16 3N 120 20 E
Daguragu, *Australia* .... **60 C5** 17 33S 130 30 E
Dahlak Kebir, *Eritrea* .... **46 D3** 15 50N 40 10 E
Dahlonega, *U.S.A.* **77 H4** 34 32N 83 59W
Dahod, *India* .... **42 H6** 22 50N 74 15 E
Dahomey = Benin ■, *Africa* . **50 G6** 10 0N 2 0 E
Dahūk, *Iraq* .... **44 B3** 36 50N 43 1 E
Dai Hao, *Vietnam* .... **38 C6** 18 1N 106 25 E
Dai-Sen, *Japan* .... **31 G6** 35 22N 133 32 E
Dai Xian, *China* .... **34 E7** 39 4N 112 58 E
Daicheng, *China* .... **34 E9** 38 42N 116 38 E
Daingean, *Ireland* **13 C4** 53 18N 7 17W
Daintree, *Australia* .... **62 B4** 16 20S 145 20 E
Daiō-Misaki, *Japan* .... **31 G8** 34 15N 136 45 E
Daisetsu-Zan, *Japan* .... **30 C11** 43 30N 142 57 E
Dajarra, *Australia* .... **62 C2** 21 42S 139 30 E
Dak Dam, *Cambodia* .... **38 F6** 12 20N 107 21 E
Dak Nhe, *Vietnam* .... **38 E6** 15 28N 107 48 E
Dak Pek, *Vietnam* .... **38 E6** 15 4N 107 44 E
Dak Song, *Vietnam* .... **39 F6** 12 19N 107 35 E
Dak Sui, *Vietnam* .... **38 E6** 14 55N 107 43 E
Dakar, *Senegal* **50 F2** 14 34N 17 29W
Dakhla, *W. Sahara* **50 D2** 23 50N 15 53W
Dakhla, El Wâhât el-, *Egypt* . **51 C11** 25 30N 28 50 E
Dakor, *India* .... **42 H5** 22 45N 73 11 E
Dakota City, *U.S.A.* **80 D6** 42 25N 96 25W
Đakovica, *Kosovo, Yug.* .... **21 C9** 42 22N 20 26 E
Dalachi, *China* .... **34 F3** 36 48N 105 0 E
Dalai Nur, *China* .... **34 C9** 43 20N 116 45 E
Dālakī, *Iran* .... **45 D6** 29 26N 51 17 E
Dalälven →, *Sweden* .... **9 F17** 60 12N 16 43 E
Dalaman →, *Turkey* .... **21 F13** 36 41N 28 43 E
Dalandzadgad, *Mongolia* ... **34 C3** 43 27N 104 30 E
Dalap-Uliga-Darrit, *Marshall Is.* .... **64 G9** 7 7N 171 24 E
Dalarna, *Sweden* .... **9 F16** 61 0N 14 0 E
Dālbandīn, *Pakistan* .... **40 E4** 29 0N 64 23 E
Dalbeattie, *U.K.* **12 G5** 54 56N 3 50W
Dalbeg, *Australia* .... **62 C4** 20 16S 147 18 E
Dalby, *Australia* .... **63 D5** 27 10S 151 17 E
Dale City, *U.S.A.* **76 F7** 38 38N 77 18W
Dale Hollow L., *U.S.A.* **77 G3** 36 32N 85 27W
Dalgān, *Iran* .... **45 E8** 27 31N 59 19 E
Dalhart, *U.S.A.* **81 G3** 36 4N 102 31W
Dalhousie, *Canada* **71 C6** 48 5N 66 26W
Dalhousie, *India* .... **42 C6** 32 38N 75 58 E
Dali, *Shaanxi, China* .... **34 G5** 34 48N 109 58 E
Dali, *Yunnan, China* .... **32 D5** 25 40N 100 10 E
Dalian, *China* .... **35 E11** 38 50N 121 40 E
Daliang Shan, *China* .... **32 D5** 28 0N 102 45 E
Daling He →, *China* .... **35 D11** 40 55N 121 40 E
Dāliyat el Karmel, *Israel* .... **47 C4** 32 43N 35 2 E
Dalkeith, *U.K.* **12 F5** 55 54N 3 4W
Dallas, *Oreg., U.S.A.* **82 D2** 44 55N 123 19W
Dallas, *Tex., U.S.A.* **81 J6** 32 47N 96 49W
Dalmā, *U.A.E.* .... **45 E7** 24 30N 52 20 E
Dalmacija, *Croatia* .... **20 C7** 43 20N 17 0 E
Dalmas, L., *Canada* **71 B5** 53 30N 71 50W
Dalmatia = Dalmacija, *Croatia* **20 C7** 43 20N 17 0 E
Dalmau, *India* .... **43 F9** 26 4N 81 2 E
Dalmellington, *U.K.* **12 F4** 55 19N 4 23W
Dalnegorsk, *Russia* .... **27 E14** 44 32N 135 33 E
Dalnerechensk, *Russia* ..... **27 E14** 45 50N 133 40 E
Daloa, *Ivory C.* **50 G4** 7 0N 6 30W
Dalry, *U.K.* **12 F4** 55 42N 4 43W
Dalrymple, L., *Australia* .... **62 C4** 20 40S 147 0 E
Dalsland, *Sweden* .... **9 G14** 58 50N 12 15 E
Daltenganj, *India* .... **43 H11** 24 0N 84 4 E
Dalton, *Ga., U.S.A.* **77 H3** 34 46N 84 58W
Dalton, *Mass., U.S.A.* **79 D11** 42 28N 73 11W
Dalton, *Nebr., U.S.A.* **80 E3** 41 25N 102 58W
Dalton-in-Furness, *U.K.* **10 C4** 54 10N 3 11W
Dalvík, *Iceland* **8 D4** 65 58N 18 32W
Dalwallinu, *Australia* .... **61 F2** 30 17S 116 40 E
Daly →, *Australia* .... **60 B5** 13 35S 130 19 E
Daly City, *U.S.A.* **84 H4** 37 42N 122 28W
Daly L., *Canada* **73 B7** 56 32N 105 39W
Daly River, *Australia* .... **60 B5** 13 46S 130 42 E
Daly Waters, *Australia* ..... **62 B1** 16 15S 133 24 E
Dam Doi, *Vietnam* .... **39 H5** 8 50N 105 12 E
Dam Ha, *Vietnam* .... **38 B6** 21 21N 107 36 E
Daman, *India* .... **40 J8** 20 25N 72 57 E
Dāmaneh, *Iran* .... **45 C6** 33 1N 50 29 E
Damanhûr, *Egypt* .... **51 B12** 31 0N 30 30 E
Damant L., *Canada* **73 A7** 61 45N 105 5W
Damanzhuang, *China* .... **34 E9** 38 5N 116 35 E
Damar, *Indonesia* .... **37 F7** 7 7S 128 40 E
Damaraland, *Namibia* ...... **56 C2** 20 0S 15 0 E
Damascus = Dimashq, *Syria* **47 B5** 33 30N 36 18 E
Damāvand, *Iran* .... **45 C7** 35 47N 52 0 E
Damāvand, Qolleh-ye, *Iran* . **45 C7** 35 56N 52 10 E
Damba, *Angola* .... **52 F3** 6 44S 15 20 E
Dâmboviţa →, *Romania* .... **17 F14** 44 12N 26 26 E
Dame Marie, *Haiti* **89 C5** 18 36N 74 26W
Dāmghān, *Iran* .... **45 B7** 36 10N 54 17 E
Damiel, *Spain* **19 C4** 39 4N 3 37W
Damietta = Dumyât, *Egypt* .. **51 B12** 31 24N 31 48 E
Daming, *China* .... **34 F8** 36 15N 115 6 E
Damir Qābū, *Syria* .... **44 B4** 36 58N 41 51 E
Dammam = Ad Dammām, *Si. Arabia* .... **45 E6** 26 20N 50 5 E
Damodar →, *India* .... **43 H12** 23 17N 87 35 E
Damoh, *India* .... **43 H8** 23 50N 79 28 E
Dampier, *Australia* .... **60 D2** 20 41S 116 42 E
Dampier, Selat, *Indonesia* .. **37 E8** 0 40S 131 0 E
Dampier Arch., *Australia* ... **60 D2** 20 38S 116 32 E
Damrei, Chuor Phnum, *Cambodia* .... **39 G4** 11 30N 103 0 E
Dan Xian, *China* .... **38 C7** 19 31N 109 33 E
Dana, *Indonesia* .... **37 F6** 11 0S 122 52 E
Dana, L., *Canada* **70 B4** 50 53N 77 20W
Dana, Mt., *U.S.A.* **84 H7** 37 54N 119 12W
Danakil Desert, *Ethiopia* .... **46 E3** 12 45N 41 0 E
Danané, *Ivory C.* **50 G4** 7 16N 8 9W
Danau Poso, *Indonesia* ..... **37 E6** 1 52S 120 35 E
Danbury, *U.S.A.* **79 E11** 41 24N 73 28W
Danby L., *U.S.A.* **83 J6** 34 13N 115 5W
Dand, *Afghan.* .... **42 D1** 31 28N 65 32 E
Dandeldhura, *Nepal* .... **43 E9** 29 20N 80 35 E
Dandeli, *India* .... **40 M9** 15 5N 74 30 E
Dandenong, *Australia* ...... **63 F4** 38 0S 145 15 E
Dandong, *China* .... **35 D13** 40 10N 124 20 E
Danfeng, *China* .... **34 H6** 33 45N 110 25 E
Danger Is. = Pukapuka, *Cook Is.* **65 J11** 10 53S 165 49W
Danger Pt., *S. Africa* .... **56 E2** 34 40S 19 17 E
Dangla Shan = Tanggula Shan, *China* .... **32 C4** 32 40N 92 10 E
Dangrek, Phnom, *Thailand* .. **38 E5** 14 15N 105 0 E
Dangriga, *Belize* **87 D7** 17 0N 88 13W
Dangshan, *China* .... **34 G9** 34 27N 116 22 E
Daniel, *U.S.A.* **82 E8** 42 52N 110 4W
Daniel's Harbour, *Canada* .. **71 B8** 50 13N 57 35W
Danielskuil, *S. Africa* .... **56 D3** 28 11S 23 33 E
Danielson, *U.S.A.* **79 E13** 41 48N 71 53W
Danilov, *Russia* .... **24 C7** 58 16N 40 13 E
Daning, *China* .... **34 F6** 36 28N 110 45 E
Danissa, *Kenya* .... **54 B5** 3 15N 40 58 E
Dank, *Oman* .... **45 F8** 23 33N 56 16 E
Dankhar Gompa, *India* ..... **40 C11** 32 10N 78 10 E
Danlí, *Honduras* **88 D2** 14 4N 86 35W
Dannemora, *U.S.A.* **79 B11** 44 43N 73 44W
Dannevirke, *N.Z.* .... **59 J6** 40 12S 176 8 E
Dannhauser, *S. Africa* ...... **57 D5** 28 0S 30 3 E
Dansville, *U.S.A.* **78 D7** 42 34N 77 42W
Danta, *India* .... **42 G5** 24 11N 72 46 E
Dantan, *India* .... **43 J12** 21 57N 87 20 E
Danube = Dunărea →, *Europe* **17 F15** 45 20N 29 40 E
Danvers, *U.S.A.* **79 D14** 42 34N 70 56W
Danville, *Ill., U.S.A.* **76 E2** 40 8N 87 37W
Danville, *Ky., U.S.A.* **76 G3** 37 39N 84 46W
Danville, *Pa., U.S.A.* **79 F8** 40 58N 76 37W
Danville, *Va., U.S.A.* **77 G6** 36 36N 79 23W
Danville, *Vt., U.S.A.* **79 B12** 44 25N 72 9W
Danzig = Gdańsk, *Poland* ... **17 A10** 54 22N 18 40 E
Dapaong, *Togo* .... **50 F6** 10 55N 0 16 E
Daqing Shan, *China* .... **34 D6** 40 40N 111 0 E
Dar Banda, *Africa* .... **48 F6** 8 0N 23 0 E
Dar el Beida = Casablanca, *Morocco* **50 B4** 33 36N 7 36W
Dar es Salaam, *Tanzania* ... **54 D4** 6 50S 39 12 E
Dar Mazār, *Iran* .... **45 D8** 29 14N 57 20 E
Dar'ā, *Syria* .... **47 C5** 32 36N 36 7 E
Dar'ā □, *Syria* .... **47 C5** 32 55N 36 10 E
Dārāb, *Iran* .... **45 D7** 28 50N 54 30 E
Daraban, *Pakistan* .... **42 D4** 31 44N 70 20 E
Daraina, *Madag.* .... **57 A8** 13 12S 49 40 E
Daraj, *Libya* .... **51 B8** 30 10N 10 28 E
Dārān, *Iran* .... **45 C6** 32 59N 50 24 E
Dārayyā, *Syria* .... **47 B5** 33 28N 36 15 E
Darband, *Pakistan* .... **42 B5** 34 20N 72 50 E
Darband, Kūh-e, *Iran* ...... **45 D8** 31 34N 57 8 E
Darbhanga, *India* .... **43 F11** 26 15N 85 55 E
D'Arcy, *Canada* **72 C4** 50 27N 122 35W
Dardanelle, *Ark., U.S.A.* ... **81 H8** 35 13N 93 9W
Dardanelle, *Calif., U.S.A.* .. **84 G7** 38 20N 119 50W
Dardanelles = Çanakkale Boğazı, *Turkey* .... **21 D12** 40 17N 26 32 E
Dārestān, *Iran* .... **45 D8** 29 9N 58 42 E
Dârfûr, *Sudan* .... **51 F10** 13 40N 24 0 E
Dargai, *Pakistan* .... **42 B4** 34 25N 71 55 E
Dargan Ata, *Turkmenistan* .. **26 E7** 40 29N 62 10 E
Dargaville, *N.Z.* .... **59 F4** 35 57S 173 52 E
Darhan, *Mongolia* .... **32 B5** 49 37N 106 21 E
Darhan Muminggan Lianheqi, *China* .... **34 D6** 41 40N 110 28 E
Darıca, *Turkey* .... **21 D13** 40 45N 29 23 E
Darién, G. del, *Colombia* ... **92 B3** 9 0N 77 0W
Dariganga = Ovoot, *Mongolia* **34 B7** 45 21N 113 45 E
Darjeeling = Darjiling, *India* . **43 F13** 27 3N 88 18 E
Darjiling, *India* .... **43 F13** 27 3N 88 18 E
Darkan, *Australia* .... **61 F2** 33 20S 116 43 E
Darkhana, *Pakistan* .... **42 D5** 30 39N 72 11 E
Darkhazīneh, *Iran* .... **45 D6** 31 54N 48 39 E
Darkot Pass, *Pakistan* ..... **43 A5** 36 45N 73 26 E
Darling →, *Australia* ...... **63 E3** 34 4S 141 54 E
Darling Downs, *Australia* ... **63 D5** 27 30S 150 30 E
Darling Ra., *Australia* ...... **61 F2** 32 30S 116 0 E
Darlington, *U.K.* **10 C6** 54 32N 1 33W
Darlington, *U.S.A.* **77 H6** 34 18N 79 52W
Darlington □, *U.K.* **10 C6** 54 32N 1 33W
Darlington, L., *S. Africa* .... **56 E4** 33 10S 25 9 E
Darlot, L., *Australia* .... **61 E3** 27 48S 121 35 E
Darłowo, *Poland* .... **16 A9** 54 25N 16 25 E
Darmstadt, *Germany* .... **16 D5** 49 51N 8 39 E
Darnah, *Libya* .... **51 B10** 32 45N 22 45 E
Darnall, *S. Africa* .... **57 D5** 29 23S 31 18 E
Darnley, C., *Antarctica* ..... **5 C6** 68 0S 69 0 E
Darnley B., *Canada* **68 B7** 69 30N 123 30W
Darr →, *Australia* .... **62 C3** 23 39S 143 50 E
Darra Pezu, *Pakistan* ...... **42 C4** 32 19N 70 44 E
Darrequeira, *Argentina* **94 D3** 37 42S 63 10W
Darrington, *U.S.A.* **82 B3** 48 15N 121 36W
Dart →, *U.K.* **11 G4** 50 24N 3 39W
Dart, C., *Antarctica* **5 D14** 73 6S 126 20W
Dartford, *U.K.* .... **11 F8** 51 26N 0 13 E
Dartmoor, *U.K.* **11 G4** 50 38N 3 57W
Dartmouth, *Canada* **71 D7** 44 40N 63 30W
Dartmouth, *U.K.* **11 G4** 50 21N 3 36W
Dartmouth, L., *Australia* .... **63 D4** 26 4S 145 18 E
Dartuch, C. = Artrutx, C. de, *Spain* .... **22 B10** 39 55N 3 49 E
Darvaza, *Turkmenistan* ..... **26 E6** 40 11N 58 24 E
Darvel, Teluk = Lahad Datu, Teluk, *Malaysia* .... **37 D5** 4 50N 118 20 E
Darwen, *U.K.* **10 D5** 53 42N 2 29W
Darwendale, *Zimbabwe* .... **57 B5** 17 41S 30 33 E
Darwha, *India* .... **40 J10** 20 15N 77 45 E
Darwin, *Australia* .... **60 B5** 12 25S 130 51 E
Darwin, *U.S.A.* **85 J9** 36 15N 117 35W
Darya Khan, *Pakistan* ...... **42 D4** 31 48N 71 6 E
Daryoi Amu = Amudarya →, *Uzbekistan* .... **26 E6** 43 58N 59 34 E
Dās, *U.A.E.* .... **45 E7** 25 20N 53 30 E
Dashen, Ras, *Ethiopia* ...... **46 E2** 13 8N 38 26 E
Dashetai, *China* .... **34 D5** 41 0N 109 5 E
Dashhowuz, *Turkmenistan* .. **26 E6** 41 49N 59 58 E
Dashköpri, *Turkmenistan* ... **45 B9** 36 16N 62 8 E
Dasht, *Iran* .... **45 B8** 37 17N 56 7 E
Dasht →, *Pakistan* .... **40 G2** 25 10N 61 40 E
Daska, *Pakistan* .... **42 C6** 32 20N 74 20 E
Dasuya, *India* .... **42 D6** 31 49N 75 38 E
Datça, *Turkey* .... **21 F12** 36 46N 27 40 E
Datia, *India* .... **43 G8** 25 39N 78 27 E
Datong, *China* .... **34 D7** 40 6N 113 18 E
Dattakhel, *Pakistan* .... **42 C3** 32 54N 69 46 E
Datu, Tanjung, *Indonesia* ... **36 D3** 2 5N 109 39 E
Datu Piang, *Phil.* .... **37 C6** 7 2N 124 30 E
Datuk, Tanjong = Datu, Tanjung, *Indonesia* ...... **36 D3** 2 5N 109 39 E
Daud Khel, *Pakistan* .... **42 C4** 32 53N 71 34 E
Daudnagar, *India* .... **43 G11** 25 2N 84 24 E
Daugava →, *Latvia* .... **9 H21** 57 4N 24 3 E
Daugavpils, *Latvia* .... **9 J22** 55 53N 26 32 E
Daulpur, *India* .... **42 F7** 26 45N 77 59 E
Dauphin, *Canada* **73 C8** 51 9N 100 5W
Dauphin, *U.S.A.* **78 F8** 40 22N 76 56W
Dauphin L., *Canada* **73 C9** 51 20N 99 45W
Dauphiné, *France* .... **18 D6** 45 15N 5 25 E
Dausa, *India* .... **42 F7** 26 52N 76 20 E
Davangere, *India* .... **40 M9** 14 25N 75 55 E
Davao, *Phil.* .... **37 C7** 7 0N 125 40 E
Davao G., *Phil.* .... **37 C7** 6 30N 125 48 E
Dāvar Panāh, *Iran* .... **45 E9** 27 25N 62 15 E
Davenport, *Calif., U.S.A.* ... **84 H4** 37 1N 122 12W
Davenport, *Iowa, U.S.A.* **80 E9** 41 32N 90 35W
Davenport, *Wash., U.S.A.* ... **82 C4** 47 39N 118 9W
Davenport Ra., *Australia* ... **62 C1** 20 28S 134 0 E
Daventry, *U.K.* **11 E6** 52 16N 1 10W
David, *Panama* **88 E3** 8 30N 82 30W
David City, *U.S.A.* **80 E6** 41 15N 97 8W
David Gorodok = Davyd Haradok, *Belarus* .... **17 B14** 52 4N 27 8 E
Davidson, *Canada* **73 C7** 51 16N 105 59W
Davis, *U.S.A.* **84 G5** 38 33N 121 44W
Davis Dam, *U.S.A.* **85 K12** 35 11N 114 34W
Davis Inlet, *Canada* **71 A7** 55 50N 60 59W
Davis Mts., *U.S.A.* **81 K2** 30 50N 103 55W
Davis Sea, *Antarctica* ...... **5 C7** 66 0S 92 0 E
Davis Str., *N. Amer.* **69 B14** 65 0N 58 0W
Davos, *Switz.* .... **18 C8** 46 48N 9 49 E
Davy L., *Canada* **73 B7** 58 53N 108 18W
Davyd Haradok, *Belarus* .... **17 B14** 52 4N 27 8 E
Dawei, *Burma* .... **38 E2** 14 2N 98 12 E
Dawes Ra., *Australia* ...... **62 C5** 24 40S 150 40 E
Dawlish, *U.K.* **11 G4** 50 35N 3 28W
Dawna Ra., *Burma* .... **38 D2** 16 30N 98 30 E
Dawros Hd., *Ireland* **13 B3** 54 50N 8 33W
Dawson, *Canada* **68 B6** 64 10N 139 30W
Dawson, *U.S.A.* **77 K3** 31 46N 84 27W
Dawson, I., *Chile* **96 G2** 53 50S 70 50W
Dawson B., *Canada* **73 C8** 52 53N 100 49W
Dawson Creek, *Canada* **72 B4** 55 45N 120 15W
Dawson Inlet, *Canada* **73 A10** 61 50N 93 25W
Dawson Ra., *Australia* ..... **62 C4** 24 30S 149 48 E
Dax, *France* **18 E3** 43 44N 1 3W
Daxian, *China* .... **32 C5** 31 15N 107 23 E
Daxindian, *China* .... **35 F11** 37 30N 120 50 E
Daxinggou, *China* .... **35 C15** 43 25N 129 40 E
Daxue Shan, *China* .... **32 C5** 30 30N 101 30 E
Daylesford, *Australia* ...... **63 F3** 37 21S 144 9 E
Dayr az Zawr, *Syria* .... **44 C4** 35 20N 40 5 E
Daysland, *Canada* **72 C6** 52 50N 112 20W
Dayton, *Nev., U.S.A.* **84 F7** 39 14N 119 36W
Dayton, *Ohio, U.S.A.* **76 F3** 39 45N 84 12W
Dayton, *Pa., U.S.A.* **78 F5** 40 53N 79 15W
Dayton, *Tenn., U.S.A.* **77 H3** 35 30N 85 1W
Dayton, *Wash., U.S.A.* **82 C4** 46 19N 117 59W
Dayton, *Wyo., U.S.A.* **82 D10** 44 53N 107 16W
Daytona Beach, *U.S.A.* **77 L5** 29 13N 81 1W
Dayville, *U.S.A.* **82 D4** 44 28N 119 32W
De Aar, *S. Africa* .... **56 E3** 30 39S 24 0 E
De Funiak Springs, *U.S.A.* .. **77 K2** 30 43N 86 7W
De Grey →, *Australia* ..... **60 D2** 20 12S 119 13 E
De Haan, *Belgium* .... **15 C3** 51 16N 3 2 E
De Kalb, *U.S.A.* **80 E10** 41 56N 88 46W
De Land, *U.S.A.* **77 L5** 29 2N 81 18W
De Leon, *U.S.A.* **81 J5** 32 7N 98 32W
De Panne, *Belgium* .... **15 C2** 51 6N 2 34 E
De Pere, *U.S.A.* **76 C1** 44 27N 88 4W
De Queen, *U.S.A.* **81 H7** 34 2N 94 21W
De Quincy, *U.S.A.* **81 K8** 30 27N 93 26W
De Ridder, *U.S.A.* **81 K8** 30 51N 93 17W
De Smet, *U.S.A.* **80 C6** 44 23N 97 33W
De Soto, *U.S.A.* **80 F9** 38 8N 90 34W
De Tour Village, *U.S.A.* **76 C4** 46 0N 83 56W
De Witt, *U.S.A.* **81 H9** 34 18N 91 20W
Dead Sea, *Asia* .... **47 D4** 31 30N 35 30 E
Deadwood, *U.S.A.* **80 C3** 44 23N 103 44W
Deadwood L., *Canada* **72 B3** 59 10N 128 30W
Deal, *U.K.* .... **11 F9** 51 13N 1 25 E
Deal I., *Australia* .... **62 F4** 39 30S 147 20 E
Dealesville, *S. Africa* ...... **56 D4** 28 41S 25 44 E
Dean →, *Canada* **72 C3** 52 49N 126 58W
Dean, Forest of, *U.K.* **11 F5** 51 45N 2 33W
Dean Chan., *Canada* **72 C3** 52 30N 127 15W
Deán Funes, *Argentina* **94 C3** 30 20S 64 20W
Dease →, *Canada* **72 B3** 59 56N 128 32W
Dease L., *Canada* **72 B2** 58 40N 130 5W
Dease Lake, *Canada* **72 B2** 58 25N 130 6W
Death Valley, *U.S.A.* **85 J10** 36 15N 116 50W
Death Valley Junction, *U.S.A.* **85 J10** 36 20N 116 25W
Death Valley Nat. Park, *U.S.A.* **85 J10** 36 45N 117 15W
Debar, *Macedonia* .... **21 D9** 41 31N 20 30 E
Debden, *Canada* **73 C7** 53 30N 106 50W
Dębica, *Poland* .... **17 C11** 50 2N 21 25 E
Debolt, *Canada* **72 B5** 55 12N 118 1W
Deborah East, L., *Australia* .. **61 F2** 30 45S 119 0 E
Deborah West, L., *Australia* . **61 F2** 30 45S 118 50 E
Debre Markos, *Ethiopia* .... **46 E2** 10 20N 37 40 E
Debre Tabor, *Ethiopia* ..... **46 E2** 11 50N 38 26 E
Debre Zeyit, *Ethiopia* ...... **46 F2** 11 48N 38 30 E
Debrecen, *Hungary* .... **17 E11** 47 33N 21 42 E
Decatur, *Ala., U.S.A.* **77 H2** 34 36N 86 59W
Decatur, *Ga., U.S.A.* **77 J3** 33 47N 84 18W
Decatur, *Ill., U.S.A.* **80 F10** 39 51N 88 57W
Decatur, *Ind., U.S.A.* **76 E3** 40 50N 84 56W
Decatur, *Tex., U.S.A.* **81 J6** 33 14N 97 35W
Deccan, *India* .... **40 L11** 18 0N 79 0 E
Deception Bay, *Australia* ... **63 D5** 27 10S 153 5 E
Deception L., *Canada* **73 B8** 56 33N 104 13W
Dechhu, *India* .... **42 F5** 26 46N 72 20 E
Děčín, *Czech Rep.* .... **16 C8** 50 47N 14 12 E
Deckerville, *U.S.A.* **78 C2** 43 32N 82 44W
Decorah, *U.S.A.* **80 D9** 43 18N 91 48W
Dedéagach = Alexandroúpolis, *Greece* . **21 D11** 40 50N 25 54 E
Dedham, *U.S.A.* **79 D13** 42 15N 71 10W
Dedza, *Malawi* .... **55 E3** 14 20S 34 20 E
Dee →, *Aberds., U.K.* **12 D6** 57 9N 2 5W
Dee →, *Dumf. & Gall., U.K.* . **12 G4** 54 51N 4 3W
Dee →, *Wales, U.K.* **10 D4** 53 22N 3 17W
Deep B., *Canada* **72 A5** 61 15N 116 35W
Deepwater, *Australia* ...... **63 D5** 29 25S 151 51 E
Deer →, *Canada* **73 B10** 58 23N 94 13W
Deer L., *Canada* **73 C10** 52 40N 94 20W
Deer Lake, *Nfld., Canada* ... **71 C8** 49 11N 57 27W
Deer Lake, *Ont., Canada* **73 C10** 52 36N 94 20W
Deer Lodge, *U.S.A.* **82 C7** 46 24N 112 44W
Deer Park, *U.S.A.* **82 C5** 47 57N 117 28W
Deer River, *U.S.A.* **80 B8** 47 20N 93 48W
Deeragun, *Australia* .... **62 B4** 19 16S 146 33 E
Deerdepoort, *S. Africa* ..... **56 C4** 24 37S 26 27 E
Deferiet, *U.S.A.* **79 B9** 44 2N 75 41W
Defiance, *U.S.A.* **76 E3** 41 17N 84 22W
Degana, *India* .... **42 F6** 26 50N 74 20 E
Dégelis, *Canada* **71 C6** 47 30N 68 35W
Deggendorf, *Germany* ...... **16 D7** 48 50N 12 57 E
Degh →, *Pakistan* .... **42 D5** 31 3N 73 21 E
Deh Bid, *Iran* .... **45 D7** 30 39N 53 11 E
Deh-e Shīr, *Iran* .... **45 D7** 31 29N 53 45 E
Dehaj, *Iran* .... **45 D7** 30 42N 54 53 E
Dehak, *Iran* .... **45 E9** 27 11N 62 37 E
Dehdez, *Iran* .... **45 D6** 31 43N 50 17 E
Dehej, *India* .... **42 J5** 21 44N 72 40 E
Dehestān, *Iran* .... **45 D7** 28 30N 55 35 E
Dehgolān, *Iran* .... **44 C5** 35 17N 47 25 E
Dehibat, *Tunisia* .... **51 B8** 32 0N 10 47 E
Dehlorān, *Iran* .... **44 C5** 32 41N 47 16 E
Dehnow-e Kūhestān, *Iran* ... **45 E8** 27 58N 58 32 E
Dehra Dun, *India* .... **42 D8** 30 20N 78 4 E
Dehri, *India* .... **43 G11** 24 50N 84 15 E
Dehui, *China* .... **35 B13** 44 30N 125 40 E
Deinze, *Belgium* .... **15 D3** 50 59N 3 32 E
Dej, *Romania* .... **17 E12** 47 10N 23 52 E
Deka →, *Zimbabwe* .... **56 B4** 18 4S 26 42 E
Dekese, *Dem. Rep. of the Congo* .... **52 E4** 3 24S 21 24 E
Del Mar, *U.S.A.* **85 N9** 32 58N 117 16W
Del Norte, *U.S.A.* **83 H10** 37 41N 106 21W
Del Rio, *U.S.A.* **81 L4** 29 22N 100 54W
Delambre I., *Australia* ...... **60 D2** 20 26S 117 5 E
Delano, *U.S.A.* **85 K7** 35 46N 119 15W
Delano Peak, *U.S.A.* **83 G7** 38 22N 112 22W
Delareyville, *S. Africa* ..... **56 D4** 26 41S 25 26 E
Delaronde L., *Canada* **73 C7** 54 3N 107 3W
Delavan, *U.S.A.* **80 D10** 42 38N 88 39W
Delaware, *U.S.A.* **76 E4** 40 18N 83 4W
Delaware □, *U.S.A.* **76 F8** 39 0N 75 20W
Delaware →, *U.S.A.* **79 G9** 39 15N 75 20W
Delaware B., *U.S.A.* **76 F8** 39 0N 75 10W
Delay →, *Canada* **71 A5** 56 56N 71 28W
Delegate, *Australia* .... **63 F4** 37 4S 148 56 E
Delevan, *U.S.A.* **78 D6** 42 29N 78 29W
Delft, *Neths.* .... **15 B4** 52 1N 4 22 E
Delfzijl, *Neths.* .... **15 A6** 53 20N 6 55 E
Delgado, C., *Mozam.* ...... **55 E5** 10 45S 40 40 E
Delgerhet, *Mongolia* .... **34 B6** 45 50N 110 30 E
Delgo, *Sudan* .... **51 D12** 20 6N 30 40 E
Delhi, *Canada* **78 D4** 42 51N 80 30W
Delhi, *India* .... **42 E7** 28 38N 77 17 E
Delhi, *La., U.S.A.* **81 J9** 32 28N 91 30W
Delhi, *N.Y., U.S.A.* **79 D10** 42 17N 74 55W
Delia, *Canada* **72 C6** 51 38N 112 23W
Delice, *Turkey* .... **25 G5** 39 54N 34 2 E
Delicias, *Mexico* **86 B3** 28 10N 105 30W
Delījān, *Iran* .... **45 C6** 33 59N 50 40 E
Déline, *Canada* **68 B7** 65 10N 123 30W
Delisle, *Canada* **73 C7** 51 55N 107 8W
Dell City, *U.S.A.* **83 L11** 31 56N 105 12W
Dell Rapids, *U.S.A.* **80 D6** 43 50N 96 43W
Delmar, *U.S.A.* **79 D11** 42 37N 73 47W
Delmenhorst, *Germany* ..... **16 B5** 53 3N 8 37 E
Delonga, Ostrova, *Russia* .. **27 B15** 76 40N 149 20 E
Deloraine, *Australia* .... **62 G4** 41 30S 146 40 E
Deloraine, *Canada* **73 D8** 49 15N 100 29W
Delphi, *U.S.A.* **76 E2** 40 36N 86 41W
Delphos, *U.S.A.* **76 E3** 40 51N 84 21W
Delportshoop, *S. Africa* .... **56 D3** 28 22S 24 20 E
Delray Beach, *U.S.A.* **77 M5** 26 28N 80 4W
Delta, *Colo., U.S.A.* **83 G9** 38 44N 108 4W
Delta, *Utah, U.S.A.* **82 G7** 39 21N 112 35W
Delta Junction, *U.S.A.* **68 B5** 64 2N 145 44W
Deltona, *U.S.A.* **77 L5** 28 54N 81 16W
Delungra, *Australia* .... **63 D5** 29 39S 150 51 E
Delvada, *India* .... **42 J4** 20 46N 71 2 E
Delvinë, *Albania* .... **21 E9** 39 59N 20 6 E
Demak, *Indonesia* .... **37 G14** 6 53S 110 38 E
Demanda, Sierra de la, *Spain* **19 A4** 42 15N 3 0W
Demavand = Damāvand, *Iran* **45 C7** 35 47N 52 0 E
Dembia, *Dem. Rep. of the Congo* .... **54 B2** 3 33N 25 48 E
Dembidolo, *Ethiopia* .... **46 F1** 8 34N 34 50 E
Demchok, *India* .... **43 C8** 32 42N 79 29 E
Demer →, *Belgium* .... **15 D4** 50 57N 4 42 E
Deming, *N. Mex., U.S.A.* ... **83 K10** 32 16N 107 46W
Deming, *Wash., U.S.A.* **84 B4** 48 50N 122 13W
Demini →, *Brazil* **92 D6** 0 46S 62 56W
Demirci, *Turkey* .... **21 E13** 39 2N 28 38 E
Demirköy, *Turkey* .... **21 D12** 41 49N 27 45 E
Demopolis, *U.S.A.* **77 J2** 32 31N 87 50W
Dempo, *Indonesia* .... **36 E2** 4 2S 103 15 E
Den Burg, *Neths.* .... **15 A4** 53 3N 4 47 E
Den Chai, *Thailand* .... **38 D3** 17 59N 100 4 E
Den Haag = 's-Gravenhage, *Neths.* .... **15 B4** 52 7N 4 17 E
Den Helder, *Neths.* .... **15 B4** 52 57N 4 45 E
Den Oever, *Neths.* .... **15 B5** 52 56N 5 2 E
Denair, *U.S.A.* **84 H6** 37 32N 120 48W
Denau, *Uzbekistan* .... **26 F7** 38 16N 67 54 E

Denbigh

Denbigh, *Canada* 78 A7 45 8N 77 15W
Denbigh, *U.K.* 10 D4 53 12N 3 25W
Denbighshire □, *U.K.* 10 D4 53 8N 3 22W
Dendang, *Indonesia* 36 E3 3 7S 107 56 E
Dendermonde, *Belgium* 15 C4 51 2N 4 5 E
Dengfeng, *China* 34 G7 34 25N 113 2 E
Dengkou, *China* 34 D4 40 18N 106 55 E
Denham, *Australia* 61 E1 25 56S 113 31 E
Denham Ra., *Australia* 62 C4 21 55S 147 46 E
Denham Sd., *Australia* 61 E1 25 45S 113 15 E
Denholm, *Canada* 73 C7 52 39N 108 1W
Denia, *Spain* 19 C6 38 49N 0 8 E
Denial B., *Australia* 63 E1 32 14S 133 32 E
Deniliquin, *Australia* 63 F3 35 30S 144 58 E
Denison, *Iowa, U.S.A.* 80 E7 42 1N 95 21W
Denison, *Tex., U.S.A.* 81 J6 33 45N 96 33W
Denison Plains, *Australia* 60 C4 18 35S 128 0 E
Denizli, *Turkey* 25 G4 37 42N 29 2 E
Denman Glacier, *Antarctica* 5 C7 66 45S 99 25 E
Denmark, *Australia* 61 F2 34 59S 117 25 E
Denmark ■, *Europe* 9 J13 55 45N 10 0 E
Denmark Str., *Atl. Oc.* 4 C6 66 0N 30 0W
Dennison, *U.S.A.* 78 F3 40 24N 81 19W
Denny, *U.K.* 12 E5 56 1N 3 55W
Denpasar, *Indonesia* 36 F5 8 39S 115 13 E
Denton, *Mont., U.S.A.* 82 C9 47 19N 109 57W
Denton, *Tex., U.S.A.* 81 J6 33 13N 97 8W
D'Entrecasteaux, Pt., *Australia* 61 F2 34 50S 115 57 E
Denver, *Colo., U.S.A.* 80 F2 39 44N 104 59W
Denver, *Pa., U.S.A.* 79 F8 40 14N 76 8W
Denver City, *U.S.A.* 81 J3 32 58N 102 50W
Deoband, *India* 42 E7 29 42N 77 43 E
Deogarh, *India* 42 G5 25 32N 73 54 E
Deoghar, *India* 43 G12 24 30N 86 42 E
Deolali, *India* 40 K8 19 58N 73 50 E
Deoli = Devli, *India* 42 G6 25 50N 75 20 E
Deora, *India* 42 F4 26 22N 70 55 E
Deori, *India* 43 H8 23 24N 79 1 E
Deoria, *India* 43 F10 26 31N 83 48 E
Deosai Mts., *Pakistan* 43 B6 35 40N 75 0 E
Deosri, *India* 43 F14 26 46N 90 29 E
Depalpur, *India* 42 H6 22 51N 75 33 E
Deping, *China* 35 F9 37 25N 116 58 E
Deposit, *U.S.A.* 79 D9 42 4N 75 25W
Depuch I., *Australia* 60 D2 20 37S 117 44 E
Deputatskiy, *Russia* 27 C14 69 18N 139 54 E
Dera Ghazi Khan, *Pakistan* 42 D4 30 5N 70 43 E
Dera Ismail Khan, *Pakistan* 42 D4 31 50N 70 50 E
Derabugti, *Pakistan* 42 E3 29 2N 69 9 E
Derawar Fort, *Pakistan* 42 E4 28 46N 71 20 E
Derbent, *Russia* 25 F8 42 5N 48 15 E
Derby, *Australia* 60 C3 17 18S 123 38 E
Derby, *U.K.* 10 E6 52 56N 1 28W
Derby, *Conn., U.S.A.* 79 E11 41 19N 73 5W
Derby, *Kans., U.S.A.* 81 G6 37 33N 97 16W
Derby, *N.Y., U.S.A.* 78 D6 42 41N 78 58W
Derby City □, *U.K.* 10 E6 52 56N 1 28W
Derby Line, *U.S.A.* 79 B12 45 0N 72 6W
Derbyshire □, *U.K.* 10 D6 53 11N 1 38W
Derg →, *U.K.* 13 B4 54 44N 7 26W
Derg, L., *Ireland* 13 D3 53 0N 8 20W
Dergaon, *India* 41 F19 26 45N 94 0 E
Dermott, *U.S.A.* 81 J9 33 32N 91 26W
Derry = Londonderry, *U.K.* 13 B4 55 0N 7 20W
Derry = Londonderry □, *U.K.* 13 B4 55 0N 7 20W
Derry, *N.H., U.S.A.* 79 D13 42 53N 71 19W
Derry, *Pa., U.S.A.* 78 F5 40 20N 79 18W
Derryveagh Mts., *Ireland* 13 B3 54 56N 8 11W
Derwent →, *Cumb., U.K.* 10 C4 54 39N 3 33W
Derwent →, *Derby, U.K.* 10 E6 52 57N 1 28W
Derwent →, *N. Yorks., U.K.* 10 D7 53 45N 0 58W
Derwent Water, *U.K.* 10 C4 54 35N 3 9W
Des Moines, *Iowa, U.S.A.* 80 E8 41 35N 93 37W
Des Moines, *N. Mex., U.S.A.* 81 G3 36 46N 103 50W
Des Moines →, *U.S.A.* 80 E9 40 23N 91 25W
Desaguadero →, *Argentina* 94 C2 34 30S 66 46W
Desaguadero →, *Bolivia* 92 G5 16 35S 69 5W
Descanso, Pta., *Mexico* 85 N9 32 21N 117 3W
Deschaillons, *Canada* 71 C5 46 32N 72 7W
Deschambault L., *Canada* 73 C8 54 50N 103 30W
Deschutes →, *U.S.A.* 82 D3 45 38N 120 55W
Dese, *Ethiopia* 46 E2 11 5N 39 40 E
Deseado →, *Argentina* 96 F3 47 45S 65 54W
Desert Center, *U.S.A.* 85 M11 33 43N 115 24W
Desert Hot Springs, *U.S.A.* 85 M10 33 58N 116 30W
Deshnok, *India* 42 F5 27 48N 73 21 E
Desna →, *Ukraine* 17 C16 50 33N 30 32 E
Desolación, I., *Chile* 96 G2 53 0S 74 0W
Despeñaperros, Paso, *Spain* 19 C4 38 24N 3 30W
Dessau, *Germany* 16 C7 51 51N 12 14 E
Dessye = Dese, *Ethiopia* 46 E2 11 5N 39 40 E
D'Estrees B., *Australia* 63 F2 35 55S 137 45 E
Desuri, *India* 42 G5 25 18N 73 35 E
Det Udom, *Thailand* 38 E5 14 54N 105 5 E
Dete, *Zimbabwe* 56 B4 18 38S 26 50 E
Detmold, *Germany* 16 C5 51 56N 8 52 E
Detour, Pt., *U.S.A.* 76 C2 45 40N 86 40W
Detroit, *U.S.A.* 78 D1 42 20N 83 3W
Detroit Lakes, *U.S.A.* 80 B7 46 49N 95 51W
Deurne, *Neths.* 15 C5 51 27N 5 49 E
Deutsche Bucht, *Germany* 16 A5 54 15N 8 0 E
Deva, *Romania* 17 F12 45 53N 22 55 E
Devakottai, *India* 40 Q11 9 55N 78 45 E
Devaprayag, *India* 43 D8 30 13N 78 35 E
Deventer, *Neths.* 15 B6 52 15N 6 10 E
Deveron →, *U.K.* 12 D6 57 41N 2 32W
Devgadh Bariya, *India* 42 H5 22 40N 73 55 E
Devikot, *India* 42 F4 26 42N 71 12 E
Devils Den, *U.S.A.* 84 K7 35 46N 119 58W
Devils Lake, *U.S.A.* 80 A5 48 7N 98 52W
Devils Paw, *Canada* 72 B2 58 47N 134 0W
Devils Tower Junction, *U.S.A.* 80 C2 44 31N 104 57W
Devine, *U.S.A.* 81 L5 29 8N 98 54W
Devizes, *U.K.* 11 F6 51 22N 1 58W
Devli, *India* 42 G6 25 50N 75 20 E
Devon, *Canada* 72 C6 53 24N 113 44W
Devon □, *U.K.* 11 G4 50 50N 3 40W
Devon I., *Canada* 4 B3 75 10N 85 0W
Devonport, *Australia* 62 G4 41 10S 146 22 E
Devonport, *N.Z.* 59 G5 36 49S 174 49 E
Dewas, *India* 42 H7 22 59N 76 3 E
Dewetsdorp, *S. Africa* 56 D4 29 33S 26 39 E
Dexter, *Maine, U.S.A.* 77 C11 45 1N 69 18W
Dexter, *Mo., U.S.A.* 81 G10 36 48N 89 57W
Dexter, *N. Mex., U.S.A.* 81 J2 33 12N 104 22W
Dey-Dey, L., *Australia* 61 E5 29 12S 131 4 E
Deyhūk, *Iran* 45 C8 33 15N 57 30 E
Deyyer, *Iran* 45 E6 27 55N 51 55 E
Dezadeash L., *Canada* 72 A1 60 28N 136 58W
Dezfūl, *Iran* 45 C6 32 20N 48 30 E
Dezhneva, Mys, *Russia* 27 C19 66 5N 169 40W
Dezhou, *China* 34 F9 37 26N 116 18 E
Dhadhar →, *India* 43 G11 24 56N 85 24 E
Dháfni, *Greece* 23 D7 35 13N 25 3 E
Dhahiriya = Az̧ Z̧āhirīyah, *West Bank* 47 D3 31 25N 34 58 E
Dhahran = Az̧ Z̧ahrān, *Si. Arabia* 45 E6 26 10N 50 7 E
Dhak, *Pakistan* 42 C5 32 25N 72 33 E
Dhaka, *Bangla.* 43 H14 23 43N 90 26 E
Dhaka □, *Bangla.* 43 G14 24 25N 90 25 E
Dhali, *Cyprus* 23 D12 35 1N 33 25 E
Dhampur, *India* 43 E8 29 19N 78 33 E
Dhamtari, *India* 41 J12 20 42N 81 35 E
Dhanbad, *India* 43 H12 23 50N 86 30 E
Dhangarhi, *Nepal* 41 E12 28 55N 80 40 E
Dhankuta, *Nepal* 43 F12 26 55N 87 40 E
Dhar, *India* 42 H6 22 35N 75 26 E
Dharampur, *India* 42 H6 22 13N 75 18 E
Dharamsala = Dharmsala, *India* 42 C7 32 16N 76 23 E
Dhariwal, *India* 42 D6 31 57N 75 19 E
Dharla →, *Bangla.* 43 G13 25 46N 89 42 E
Dharmapuri, *India* 40 N11 12 10N 78 10 E
Dharmjaygarh, *India* 43 H10 22 28N 83 13 E
Dharmsala, *India* 42 C7 32 16N 76 23 E
Dharni, *India* 42 J7 21 33N 76 53 E
Dharwad, *India* 40 M9 15 30N 75 4 E
Dhasan →, *India* 43 G8 25 48N 79 24 E
Dhaulagiri, *Nepal* 43 E10 28 39N 83 28 E
Dhebar, L., *India* 42 G6 24 10N 74 0 E
Dheftera, *Cyprus* 23 D12 35 5N 33 16 E
Dhenkanal, *India* 41 J14 20 45N 85 35 E
Dherinia, *Cyprus* 23 D12 35 3N 33 57 E
Dhiarrizos →, *Cyprus* 23 E11 34 41N 32 34 E
Dhībān, *Jordan* 47 D4 31 30N 35 46 E
Dhikti Óros, *Greece* 23 D7 35 8N 25 30 E
Dhilwan, *India* 42 D6 31 31N 75 21 E
Dhimarkhera, *India* 43 H9 23 28N 80 22 E
Dhírfis Óros, *Greece* 21 E10 38 40N 23 54 E
Dhodhekánisos, *Greece* 21 F12 36 35N 27 0 E
Dholka, *India* 42 H5 22 44N 72 29 E
Dhoraji, *India* 42 J4 21 45N 70 37 E
Dhráhstis, Ákra, *Greece* 23 A3 39 48N 19 40 E
Dhrangadhra, *India* 42 H4 22 59N 71 31 E
Dhrápanon, Ákra, *Greece* 23 D6 35 28N 24 14 E
Dhrol, *India* 42 H4 22 33N 70 25 E
Dhuburi, *India* 41 F16 26 2N 89 59 E
Dhule, *India* 40 J9 20 58N 74 50 E
Di Linh, *Vietnam* 39 G7 11 35N 108 4 E
Di Linh, Cao Nguyen, *Vietnam* 39 G7 11 30N 108 0 E
Dia, *Greece* 23 D7 35 28N 25 14 E
Diablo, Mt., *U.S.A.* 84 H5 37 53N 121 56W
Diablo Range, *U.S.A.* 84 J5 37 20N 121 25W
Diafarabé, *Mali* 50 F5 14 9N 4 57W
Diamante, *Argentina* 94 C3 32 5S 60 40W
Diamante →, *Argentina* 94 C2 34 30S 66 46W
Diamantina, *Brazil* 93 G10 18 17S 43 40W
Diamantina →, *Australia* 63 D2 26 45S 139 10 E
Diamantino, *Brazil* 93 F7 14 30S 56 30W
Diamond Bar, *U.S.A.* 85 L9 34 1N 117 48W
Diamond Harbour, *India* 43 H13 22 11N 88 14 E
Diamond Is., *Australia* 62 B5 17 25S 151 5 E
Diamond Mts., *U.S.A.* 82 G6 39 50N 115 30W
Diamond Springs, *U.S.A.* 84 G6 38 42N 120 49W
Dībā, *Oman* 45 E8 25 45N 56 16 E
Dibai, *India* 42 E8 28 13N 78 15 E
Dibaya-Lubue, *Dem. Rep. of the Congo* 52 E3 4 12S 19 54 E
Dibete, *Botswana* 56 C4 23 45S 26 32 E
Dibrugarh, *India* 41 F19 27 29N 94 55 E
Dickens, *U.S.A.* 81 J4 33 37N 100 50W
Dickinson, *U.S.A.* 80 B3 46 53N 102 47W
Dickson = Dikson, *Russia* 26 B9 73 40N 80 5 E
Dickson, *U.S.A.* 77 G2 36 5N 87 23W
Dickson City, *U.S.A.* 79 E9 41 29N 75 40W
Didiéni, *Mali* 50 F4 13 53N 8 6W
Didsbury, *Canada* 72 C6 51 35N 114 10W
Didwana, *India* 42 F6 27 23N 74 36 E
Diefenbaker, L., *Canada* 73 C7 51 0N 106 55W
Diego de Almagro, *Chile* 94 B1 26 22S 70 3W
Diego Garcia, *Ind. Oc.* 3 E13 7 50S 72 50 E
Diekirch, *Lux.* 15 E6 49 52N 6 10 E
Dien Ban, *Vietnam* 38 E7 15 53N 108 16 E
Dien Bien, *Vietnam* 38 B4 21 20N 103 0 E
Dien Khanh, *Vietnam* 39 F7 12 15N 109 6 E
Dieppe, *France* 18 B4 49 54N 1 4 E
Dierks, *U.S.A.* 81 H8 34 7N 94 1W
Diest, *Belgium* 15 D5 50 58N 5 4 E
Dif, *Somali Rep.* 46 G3 0 59N 0 56 E
Differdange, *Lux.* 15 E5 49 31N 5 54 E
Dig, *India* 42 F7 27 28N 77 20 E
Digba, *Dem. Rep. of the Congo* 54 B2 4 25N 25 48 E
Digby, *Canada* 71 D6 44 38N 65 50W
Diggi, *India* 42 F6 26 22N 75 26 E
Dighinala, *Bangla.* 41 H18 23 15N 92 5 E
Dighton, *U.S.A.* 80 F4 38 29N 100 28W
Digne-les-Bains, *France* 18 D7 44 5N 6 12 E
Digos, *Phil.* 37 C7 6 45N 125 20 E
Digranes, *Iceland* 8 C6 66 4N 14 44W
Digul →, *Indonesia* 37 F9 7 7S 138 42 E
Dihang = Brahmaputra →, *India* 41 F19 27 48N 95 30 E
Dijlah, Nahr →, *Asia* 44 D5 31 0N 47 25 E
Dijon, *France* 18 C6 47 20N 5 3 E
Dikhil, *Djibouti* 46 E3 11 8N 42 20 E
Dikkil = Dikhil, *Djibouti* 46 E3 11 8N 42 20 E
Diksmuide, *Belgium* 15 C2 51 2N 2 52 E
Dikson, *Russia* 26 B9 73 40N 80 5 E
Dila, *Ethiopia* 46 F2 6 21N 38 22 E
Dili, *E. Timor* 37 F7 8 39S 125 34 E
Dilley, *U.S.A.* 81 L5 28 40N 99 10W
Dillingham, *U.S.A.* 68 C4 59 3N 158 28W
Dillon, *Canada* 73 B7 55 56N 108 35W
Dillon, *Mont., U.S.A.* 82 D7 45 13N 112 38W
Dillon, *S.C., U.S.A.* 77 H6 34 25N 79 22W
Dillon →, *Canada* 73 B7 55 56N 108 56W
Dillsburg, *U.S.A.* 78 F7 40 7N 77 2W
Dilolo, *Dem. Rep. of the Congo* 52 G4 10 28S 22 18 E
Dimas, *Mexico* 86 C3 23 43N 106 47W
Dimashq, *Syria* 47 B5 33 30N 36 18 E
Dimashq □, *Syria* 47 B5 33 30N 36 30 E
Dimbaza, *S. Africa* 57 E4 32 50S 27 14 E
Dimboola, *Australia* 63 F3 36 28S 142 7 E
Dîmboviţa = Dâmboviţa →, *Romania* 17 F14 44 12N 26 26 E
Dimbulah, *Australia* 62 B4 17 8S 145 4 E
Dimitrovgrad, *Bulgaria* 21 C11 42 5N 25 35 E
Dimitrovgrad, *Russia* 24 D8 54 14N 49 39 E
Dimitrovo = Pernik, *Bulgaria* 21 C10 42 35N 23 2 E
Dimmitt, *U.S.A.* 81 H3 34 33N 102 19W
Dimona, *Israel* 47 D4 31 2N 35 1 E
Dinagat, *Phil.* 37 B7 10 10N 125 40 E
Dinajpur, *Bangla.* 41 G16 25 33N 88 43 E
Dinan, *France* 18 B2 48 28N 2 2W
Dīnān Āb, *Iran* 45 C8 32 4N 56 49 E
Dinant, *Belgium* 15 D4 50 16N 4 55 E
Dinapur, *India* 43 G11 25 38N 85 5 E
Dīnār, Kūh-e, *Iran* 45 D6 30 42N 51 46 E
Dinara Planina, *Croatia* 20 C7 44 0N 16 30 E
Dinard, *France* 18 B2 48 38N 2 6W
Dinaric Alps = Dinara Planina, *Croatia* 20 C7 44 0N 16 30 E
Dindigul, *India* 40 P11 10 25N 78 0 E
Dindori, *India* 43 H9 22 57N 81 5 E
Ding Xian = Dingzhou, *China* 34 E8 38 30N 114 59 E
Dinga, *Pakistan* 42 G2 25 26N 67 10 E
Dingbian, *China* 34 F4 37 35N 107 32 E
Dingle, *Ireland* 13 D1 52 9N 10 17W
Dingle B., *Ireland* 13 D1 52 3N 10 20W
Dingmans Ferry, *U.S.A.* 79 E10 41 13N 74 55W
Dingo, *Australia* 62 C4 23 38S 149 19 E
Dingtao, *China* 34 G8 35 5N 115 35 E
Dingwall, *U.K.* 12 D4 57 36N 4 26W
Dingxi, *China* 34 G3 35 30N 104 33 E
Dingxiang, *China* 34 E7 38 30N 112 58 E
Dingzhou, *China* 34 E8 38 30N 114 59 E
Dinh, Mui, *Vietnam* 39 G7 11 22N 109 1 E
Dinh Lap, *Vietnam* 38 B6 21 33N 107 6 E
Dinokwe, *Botswana* 56 C4 23 29S 26 37 E
Dinorwic, *Canada* 73 D10 49 41N 92 30W
Dinosaur Nat. Monument, *U.S.A.* 82 F9 40 30N 108 45W
Dinosaur Prov. Park, *Canada* 72 C6 50 47N 111 30W
Dinuba, *U.S.A.* 84 J7 36 32N 119 23W
Dipalpur, *Pakistan* 42 D5 30 40N 73 39 E
Diplo, *Pakistan* 42 G3 24 35N 69 35 E
Dipolog, *Phil.* 37 C6 8 36N 123 20 E
Dir, *Pakistan* 40 B7 35 8N 71 59 E
Dire Dawa, *Ethiopia* 46 F3 9 35N 41 45 E
Diriamba, *Nic.* 88 D2 11 51N 86 19W
Dirk Hartog I., *Australia* 61 E1 25 50S 113 5 E
Dirranbandi, *Australia* 63 D4 28 33S 148 17 E
Disa, *India* 42 G5 24 18N 72 10 E
Disappointment, C., *U.S.A.* 82 C2 46 18N 124 5W
Disappointment, L., *Australia* 60 D3 23 20S 122 40 E
Disaster B., *Australia* 63 F4 37 15S 149 58 E
Discovery B., *Australia* 63 F3 38 10S 140 40 E
Disko = Qeqertarsuaq, *Greenland* 69 B5 69 45N 53 30W
Disko Bugt, *Greenland* 4 C5 69 10N 52 0W
Diss, *U.K.* 11 E9 52 23N 1 7 E
Disteghil Sar, *Pakistan* 43 A6 36 20N 75 12 E
Distrito Federal □, *Brazil* 93 G9 15 45S 47 45W
Distrito Federal □, *Mexico* 87 D5 19 15N 99 10W
Diu, *India* 42 J4 20 45N 70 58 E
Divāndarreh, *Iran* 44 C5 35 55N 47 2 E
Divide, *U.S.A.* 82 D7 45 45N 112 45W
Dividing Ra., *Australia* 61 E2 27 45S 116 0 E
Divinópolis, *Brazil* 93 H10 20 10S 44 54W
Divnoye, *Russia* 25 E7 45 55N 43 21 E
Divo, *Ivory C.* 50 G4 5 48N 5 15W
Diwāl Kol, *Afghan.* 42 B2 34 23N 67 52 E
Dixie Mt., *U.S.A.* 84 F6 39 55N 120 16W
Dixon, *Calif., U.S.A.* 84 G5 38 27N 121 49W
Dixon, *Ill., U.S.A.* 80 E10 41 50N 89 29W
Dixon Entrance, *U.S.A.* 68 C6 54 30N 132 0W
Dixville, *Canada* 79 A13 45 4N 71 46W
Diyālā →, *Iraq* 44 C5 33 14N 44 31 E
Diyarbakır, *Turkey* 25 G7 37 55N 40 18 E
Diyodar, *India* 42 G4 24 8N 71 50 E
Djakarta = Jakarta, *Indonesia* 36 F3 6 9S 106 49 E
Djamba, *Angola* 56 B1 16 45S 13 58 E
Djambala, *Congo* 52 E2 2 32S 14 30 E
Djanet, *Algeria* 50 D7 24 35N 9 32 E
Djawa = Jawa, *Indonesia* 36 F3 7 0S 110 0 E
Djelfa, *Algeria* 50 B6 34 40N 3 15 E
Djema, *C.A.R.* 54 A2 6 3N 25 15 E
Djerba, I. de, *Tunisia* 51 B8 33 50N 10 48 E
Djerid, Chott, *Tunisia* 50 B7 33 42N 8 30 E
Djibouti, *Djibouti* 46 E3 11 30N 43 5 E
Djibouti ■, *Africa* 46 E3 12 0N 43 0 E
Djolu, *Dem. Rep. of the Congo* 52 D4 0 35N 22 5 E
Djoum, *Cameroon* 52 D2 2 41N 12 35 E
Djourab, Erg du, *Chad* 51 E9 16 40N 18 50 E
Djugu, *Dem. Rep. of the Congo* 54 B3 1 55N 30 35 E
Djúpivogur, *Iceland* 8 D6 64 39N 14 17W
Dmitriya Lapteva, Proliv, *Russia* 27 B15 73 0N 140 0 E
Dnepr → = Dnipro →, *Ukraine* 25 E5 46 30N 32 18 E
Dneprodzerzhinsk = Dniprodzerzhynsk, *Ukraine* 25 E5 48 32N 34 37 E
Dnepropetrovsk = Dnipropetrovsk, *Ukraine* 25 E6 48 30N 35 0 E
Dnestr → = Dnister →, *Europe* 17 E16 46 18N 30 17 E
Dnestrovski = Belgorod, *Russia* 25 D6 50 35N 36 35 E
Dnieper = Dnipro →, *Ukraine* 25 E5 46 30N 32 18 E
Dniester = Dnister →, *Europe* 17 E16 46 18N 30 17 E
Dnipro →, *Ukraine* 25 E5 46 30N 32 18 E
Dniprodzerzhynsk, *Ukraine* 25 E5 48 32N 34 37 E
Dnipropetrovsk, *Ukraine* 25 E6 48 30N 35 0 E
Dnister →, *Europe* 17 E16 46 18N 30 17 E
Dnistrovskyy Lyman, *Ukraine* 17 E16 46 15N 30 17 E
Dno, *Russia* 24 C4 57 50N 29 58 E
Dnyapro = Dnipro →, *Ukraine* 25 E5 46 30N 32 18 E
Doaktown, *Canada* 71 C6 46 33N 66 8W
Doan Hung, *Vietnam* 38 B5 21 30N 105 10 E
Doany, *Madag.* 57 A8 14 21S 49 30 E
Doba, *Chad* 51 G9 8 40N 16 50 E
Dobandi, *Pakistan* 42 D2 31 13N 66 50 E
Dobbyn, *Australia* 62 B3 19 44S 140 2 E
Dobele, *Latvia* 9 H20 56 37N 23 16 E
Doberai, Jazirah, *Indonesia* 37 E8 1 25S 133 0 E
Doblas, *Argentina* 94 D3 37 5S 64 0W
Dobo, *Indonesia* 37 F8 5 45S 134 15 E
Doboj, *Bos.-H.* 21 B8 44 46N 18 4 E
Dobrich, *Bulgaria* 21 C12 43 37N 27 49 E
Dobruja, *Europe* 17 F15 44 30N 28 15 E
Dobrush, *Belarus* 17 B16 52 25N 31 22 E
Doc, Mui, *Vietnam* 38 D6 17 58N 106 30 E
Docker River, *Australia* 61 D4 24 52S 129 5 E
Doctor Arroyo, *Mexico* 86 C4 23 40N 100 11W
Doda, *India* 43 C6 33 10N 75 34 E
Doda, L., *Canada* 70 C4 49 25N 75 13W
Dodecanese = Dhodhekánisos, *Greece* 21 F12 36 35N 27 0 E
Dodge City, *U.S.A.* 81 G5 37 45N 100 1W
Dodge L., *Canada* 73 B7 59 50N 105 36W
Dodgeville, *U.S.A.* 80 D9 42 58N 90 8W
Dodoma, *Tanzania* 54 D4 6 8S 35 45 E
Dodoma □, *Tanzania* 54 D4 6 0S 36 0 E
Dodsland, *Canada* 73 C7 51 50N 108 45W
Dodson, *U.S.A.* 82 B9 48 24N 108 15W
Doesburg, *Neths.* 15 B6 52 1N 6 9 E
Doetinchem, *Neths.* 15 C6 51 59N 6 18 E
Dog Creek, *Canada* 72 C4 51 35N 122 14W
Dog L., *Man., Canada* 73 C9 51 2N 98 31W
Dog L., *Ont., Canada* 70 C2 48 48N 89 30W
Dogran, *Pakistan* 42 D5 31 48N 73 35 E
Doğubayazıt, *Turkey* 44 B5 39 31N 44 5 E
Doha = Ad Dawḩah, *Qatar* 46 B5 25 15N 51 35 E
Dohazari, *Bangla.* 41 H18 22 10N 92 5 E
Dohrighat, *India* 43 F10 26 16N 83 31 E
Doi, *Indonesia* 37 D7 2 14N 127 49 E
Doi Luang, *Thailand* 38 C3 18 30N 101 0 E
Doi Saket, *Thailand* 38 C2 18 52N 99 9 E
Dois Irmãos, Sa., *Brazil* 93 E10 9 0S 42 30W
Dokkum, *Neths.* 15 A5 53 20N 5 59 E
Dokri, *Pakistan* 42 F3 27 25N 68 7 E
Dolak, Pulau, *Indonesia* 37 F9 8 0S 138 30 E
Dolbeau, *Canada* 71 C5 48 53N 72 18W
Dole, *France* 18 C6 47 7N 5 31 E
Dolgellau, *U.K.* 10 E4 52 45N 3 53W
Dolgelley = Dolgellau, *U.K.* 10 E4 52 45N 3 53W
Dollard, *Neths.* 15 A7 53 20N 7 10 E
Dolo, *Ethiopia* 46 G3 4 11N 42 3 E
Dolomites = Dolomiti, *Italy* 20 A4 46 23N 11 51 E
Dolomiti, *Italy* 20 A4 46 23N 11 51 E
Dolores, *Argentina* 94 D4 36 20S 57 40W
Dolores, *Uruguay* 94 C4 33 34S 58 15W
Dolores, *U.S.A.* 83 H9 37 28N 108 30W
Dolores →, *U.S.A.* 83 G9 38 49N 109 17W
Dolphin, C., *Falk. Is.* 96 G5 51 10S 59 0W
Dolphin and Union Str., *Canada* 68 B8 69 5N 114 45W
Dom Pedrito, *Brazil* 95 C5 31 0S 54 40W
Domariaganj →, *India* 43 F10 26 17N 83 44 E
Domasi, *Malawi* 55 F4 15 15S 35 22 E
Dombarovskiy, *Russia* 26 D6 50 46N 59 32 E
Dombås, *Norway* 9 E13 62 4N 9 8 E
Domel I. = Letsôk-aw Kyun, *Burma* 39 G2 11 30N 98 25 E
Domeyko, *Chile* 94 B1 29 0S 71 0W
Domeyko, Cordillera, *Chile* 94 A2 24 30S 69 0W
Dominador, *Chile* 94 A2 24 21S 69 20W
Dominica ■, *W. Indies* 89 C7 15 20N 61 20W
Dominica Passage, *W. Indies* 89 C7 15 10N 61 20W
Dominican Rep. ■, *W. Indies* 89 C5 19 0N 70 30W
Domodóssola, *Italy* 18 C8 46 7N 8 17 E
Domville, Mt., *Australia* 63 D5 28 1S 151 15 E
Don →, *Russia* 25 E6 47 4N 39 18 E
Don →, *Aberds., U.K.* 12 D6 57 11N 2 5W
Don →, *S. Yorks., U.K.* 10 D7 53 41N 0 52W
Don, C., *Australia* 60 B5 11 18S 131 46 E
Don Benito, *Spain* 19 C3 38 53N 5 51W
Dona Ana = Nhamaabué, *Mozam.* 55 F4 17 25S 35 5 E
Donaghadee, *U.K.* 13 B6 54 39N 5 33W
Donald, *Australia* 63 F3 36 23S 143 0 E
Donaldsonville, *U.S.A.* 81 K9 30 6N 90 59W
Donalsonville, *U.S.A.* 77 K3 31 3N 84 53W
Donau = Dunărea →, *Europe* 17 F15 45 20N 29 40 E
Donau →, *Austria* 15 D3 48 10N 17 0 E
Donauwörth, *Germany* 16 D6 48 43N 10 47 E
Doncaster, *U.K.* 10 D6 53 32N 1 6W
Dondo, *Mozam.* 55 F3 19 33S 34 46 E
Dondo, Teluk, *Indonesia* 37 D6 0 50N 120 30 E
Dondra Head, *Sri Lanka* 40 S12 5 55N 80 40 E
Donegal, *Ireland* 13 B3 54 39N 8 5W
Donegal □, *Ireland* 13 B4 54 53N 8 0W
Donegal B., *Ireland* 13 B3 54 31N 8 49W
Donets →, *Russia* 25 E7 47 33N 40 55 E
Donetsk, *Ukraine* 25 E6 48 0N 37 45 E
Dong Ba Thin, *Vietnam* 39 F7 12 8N 109 13 E
Dong Dang, *Vietnam* 38 B6 21 54N 106 42 E
Dong Giam, *Vietnam* 38 C5 19 25N 105 31 E
Dong Ha, *Vietnam* 38 D6 16 55N 107 8 E
Dong Hene, *Laos* 38 D5 16 40N 105 18 E
Dong Hoi, *Vietnam* 38 D6 17 29N 106 36 E
Dong Khe, *Vietnam* 38 A6 22 26N 106 27 E
Dong Ujimqin Qi, *China* 34 B9 45 32N 116 55 E
Dong Van, *Vietnam* 38 A5 23 16N 105 22 E
Dong Xoai, *Vietnam* 39 G6 11 32N 106 55 E
Dongara, *Australia* 61 E1 29 14S 114 57 E
Dongbei, *China* 35 D13 45 0N 125 0 E
Dongchuan, *China* 32 D5 26 8N 103 1 E
Dongfang, *China* 38 C7 18 50N 108 33 E
Dongfeng, *China* 35 C13 42 40N 125 34 E
Donggala, *Indonesia* 37 E5 0 30S 119 40 E
Donggou, *China* 35 E13 39 52N 124 10 E
Dongguang, *China* 34 F9 37 50N 116 30 E
Dongjingcheng, *China* 35 B15 44 5N 129 10 E
Dongning, *China* 35 B16 44 2N 131 5 E
Dongola, *Sudan* 51 E12 19 9N 30 22 E
Dongping, *China* 34 G9 35 55N 116 20 E
Dongsheng, *China* 34 E6 39 50N 110 0 E
Dongtai, *China* 35 H11 32 51N 120 21 E
Dongting Hu, *China* 33 D6 29 18N 112 45 E
Donington, C., *Australia* 63 E2 34 45S 136 0 E
Doniphan, *U.S.A.* 81 G9 36 37N 90 50W
Dønna, *Norway* 8 C15 66 6N 12 30 E
Donna, *U.S.A.* 81 M5 26 9N 98 4W
Donnaconna, *Canada* 71 C5 46 41N 71 41W
Donnelly's Crossing, *N.Z.* 59 F4 35 42S 173 38 E

East Providence

Donnybrook, *Australia* ..... **61 F2** 33 34S 115 48 E
Donnybrook, *S. Africa* ..... **57 D4** 29 59S 29 48 E
Donora, *U.S.A.* **78 F5** 40 11N 79 52W
Donostia = Donostia-San Sebastián, *Spain* **19 A5** 43 17N 1 58W
Donostia-San Sebastián, *Spain* **19 A5** 43 17N 1 58W
Doon →, *U.K.* **12 F4** 55 27N 4 39W
Dora, L., *Australia* .... **60 D3** 22 0S 123 0 E
Dora Báltea →, *Italy* .... **18 D8** 45 11N 8 3 E
Doran L., *Canada* **73 A7** 61 13N 108 6W
Dorchester, *U.K.* **11 G5** 50 42N 2 27W
Dorchester, C., *Canada* **69 B12** 65 27N 77 27W
Dordabis, *Namibia* .... **56 C2** 22 52S 17 38 E
Dordogne →, *France* **18 D3** 45 2N 0 36W
Dordrecht, *Neths.* .... **15 C4** 51 48N 4 39 E
Dordrecht, *S. Africa* .... **56 E4** 31 20S 27 3 E
Doré L., *Canada* **73 C7** 54 46N 107 17W
Doré Lake, *Canada* **73 C7** 54 38N 107 36W
Dori, *Burkina Faso* **50 F5** 14 3N 0 2W
Doring →, *S. Africa* .... **56 E2** 31 54S 18 39 E
Doringbos, *S. Africa* .... **56 E2** 31 59S 19 16 E
Dorion, *Canada* **79 A10** 45 23N 74 3W
Dornbirn, *Austria* .... **16 E5** 47 25N 9 45 E
Dornie, *U.K.* **12 D3** 57 17N 5 31W
Dornoch, *U.K.* **12 D4** 57 53N 4 2W
Dornoch Firth, *U.K.* **12 D4** 57 51N 4 4W
Dornogovĭ □, *Mongolia* .... **34 C6** 44 0N 110 0 E
Dorohoi, *Romania* .... **17 E14** 47 56N 26 23 E
Döröö Nuur, *Mongolia* ..... **32 B4** 48 0N 93 0 E
Dorr, *Iran* .... **45 C6** 33 17N 50 38 E
Dorre I., *Australia* .... **61 E1** 25 13S 113 12 E
Dorrigo, *Australia* .... **63 E5** 30 20S 152 44 E
Dorris, *U.S.A.* **82 F3** 41 58N 121 55W
Dorset, *Canada* **78 A6** 45 14N 78 54W
Dorset, *U.S.A.* **78 E4** 41 40N 80 40W
Dorset □, *U.K.* **11 G5** 50 45N 2 26W
Dortmund, *Germany* .... **16 C4** 51 30N 7 28 E
Doruma, *Dem. Rep. of the Congo* .... **54 B2** 4 42N 27 33 E
Dorūneh, *Iran* .... **45 C8** 35 10N 57 18 E
Dos Bahías, C., *Argentina* ... **96 E3** 44 58S 65 32W
Dos Hermanas, *Spain* **19 D3** 37 16N 5 55W
Dos Palos, *U.S.A.* **84 J6** 36 59N 120 37W
Dosso, *Niger* .... **50 F6** 13 0N 3 13 E
Dothan, *U.S.A.* **77 K3** 31 13N 85 24W
Doty, *U.S.A.* **84 D3** 46 38N 123 17W
Douai, *France* .... **18 A5** 50 21N 3 4 E
Douala, *Cameroon* .... **52 D1** 4 0N 9 45 E
Douarnenez, *France* **18 B1** 48 6N 4 21W
Double Island Pt., *Australia* . **63 D5** 25 56S 153 11 E
Double Mountain Fork →, *U.S.A.* **81 J4** 33 16N 100 0W
Doubs →, *France* .... **18 C6** 46 53N 5 1 E
Doubtful Sd., *N.Z.* .... **59 L1** 45 20S 166 49 E
Doubtless B., *N.Z.* .... **59 F4** 34 55S 173 26 E
Douglas, *S. Africa* .... **56 D3** 29 4S 23 46 E
Douglas, *U.K.* **10 C3** 54 10N 4 28W
Douglas, *Ariz., U.S.A.* **83 L9** 31 21N 109 33W
Douglas, *Ga., U.S.A.* **77 K4** 31 31N 82 51W
Douglas, *Wyo., U.S.A.* **80 D2** 42 45N 105 24W
Douglas Chan., *Canada* **72 C3** 53 40N 129 20W
Douglas Pt., *Canada* **78 B3** 44 19N 81 37W
Douglasville, *U.S.A.* **77 J3** 33 45N 84 45W
Dounreay, *U.K.* **12 C5** 58 35N 3 44W
Dourada, Serra, *Brazil* **93 F9** 13 10S 48 45W
Dourados, *Brazil* **95 A5** 22 9S 54 50W
Dourados →, *Brazil* **95 A5** 21 58S 54 18W
Dourados, Serra dos, *Brazil* . **95 A5** 23 30S 53 30W
Douro →, *Europe* **19 B1** 41 8N 8 40W
Dove →, *U.K.* **10 E6** 52 51N 1 36W
Dove Creek, *U.S.A.* **83 H9** 37 46N 108 54W
Dover, *Australia* .... **62 G4** 43 18S 147 2 E
Dover, *U.K.* .... **11 F9** 51 7N 1 19 E
Dover, *Del., U.S.A.* **76 F8** 39 10N 75 32W
Dover, *N.H., U.S.A.* **79 C14** 43 12N 70 56W
Dover, *N.J., U.S.A.* **79 F10** 40 53N 74 34W
Dover, *Ohio, U.S.A.* **78 F3** 40 32N 81 29W
Dover, Pt., *Australia* .... **61 F4** 32 32S 125 32 E
Dover, Str. of, *Europe* ...... **11 G9** 51 0N 1 30 E
Dover-Foxcroft, *U.S.A.* **77 C11** 45 11N 69 13W
Dover Plains, *U.S.A.* **79 E11** 41 43N 73 35W
Dovey = Dyfi →, *U.K.* **11 E3** 52 32N 4 3W
Dovrefjell, *Norway* .... **9 E13** 62 15N 9 33 E
Dow Rūd, *Iran* .... **45 C6** 33 28N 49 4 E
Dowa, *Malawi* .... **55 E3** 13 38S 33 58 E
Dowagiac, *U.S.A.* **76 E2** 41 59N 86 6W
Dowerin, *Australia* .... **61 F2** 31 12S 117 2 E
Dowgha'i, *Iran* .... **45 B8** 36 54N 58 32 E
Dowlatābād, *Iran* .... **45 D8** 28 20N 56 40 E
Down □, *U.K.* **13 B5** 54 23N 6 2W
Downey, *Calif., U.S.A.* **85 M8** 33 56N 118 7W
Downey, *Idaho, U.S.A.* **82 E7** 42 26N 112 7W
Downham Market, *U.K.* ..... **11 E8** 52 37N 0 23 E
Downieville, *U.S.A.* **84 F6** 39 34N 120 50W
Downpatrick, *U.K.* **13 B6** 54 20N 5 43W
Downpatrick Hd., *Ireland* ... **13 B2** 54 20N 9 21W
Downsville, *U.S.A.* **79 D10** 42 5N 74 50W
Downton, Mt., *Canada* **72 C4** 52 42N 124 52W
Dowsārī, *Iran* .... **45 D8** 28 25N 57 59 E
Doyle, *U.S.A.* **84 E6** 40 2N 120 6W
Doylestown, *U.S.A.* **79 F9** 40 21N 75 10W
Dozois, Rés., *Canada* **70 C4** 47 30N 77 5W
Dra Khel, *Pakistan* .... **42 F2** 27 58N 66 45 E
Drachten, *Neths.* .... **15 A6** 53 7N 6 5 E
Drăgăşani, *Romania* .... **17 F13** 44 39N 24 17 E
Dragichyn, *Belarus* .... **17 B13** 52 15N 25 8 E
Dragoman, Prokhod, *Bulgaria* **21 C10** 42 58N 22 53 E
Draguignan, *France* .... **18 E7** 43 32N 6 27 E
Drain, *U.S.A.* **82 E2** 43 40N 123 19W
Drake, *U.S.A.* **80 B4** 47 55N 100 23W
Drake Passage, *S. Ocean* ... **5 B17** 58 0S 68 0W
Drakensberg, *S. Africa* ..... **57 D4** 31 0S 28 0 E
Dráma, *Greece* .... **21 D11** 41 9N 24 10 E
Drammen, *Norway* .... **9 G14** 59 42N 10 12 E
Drangajökull, *Iceland* **8 C2** 66 9N 22 15W
Dras, *India* .... **43 B6** 34 25N 75 48 E
Drau = Drava →, *Croatia* ... **21 B8** 45 33N 18 55 E
Drava →, *Croatia* .... **21 B8** 45 33N 18 55 E
Drayton Valley, *Canada* **72 C6** 53 12N 114 58W
Drenthe □, *Neths.* .... **15 B6** 52 52N 6 40 E
Drepanum, C., *Cyprus* ..... **23 E11** 34 54N 32 19 E
Dresden, *Canada* **78 D2** 42 35N 82 11W
Dresden, *Germany* .... **16 C7** 51 3N 13 44 E
Dreux, *France* .... **18 B4** 48 44N 1 23 E
Driffield, *U.K.* **10 C7** 54 0N 0 26W
Driftwood, *U.S.A.* **78 E6** 41 20N 78 8W
Driggs, *U.S.A.* **82 E8** 43 44N 111 6W
Drin →, *Albania* .... **21 C8** 42 1N 19 38 E
Drina →, *Bos.-H.* .... **21 B8** 44 53N 19 21 E
Drøbak, *Norway* .... **9 G14** 59 39N 10 39 E
Drobeta-Turnu Severin, *Romania* .... **17 F12** 44 39N 22 41 E
Drochia, *Moldova* .... **17 D14** 48 2N 27 48 E
Drogheda, *Ireland* **13 C5** 53 43N 6 22W
Drogichin = Dragichyn, *Belarus* .... **17 B13** 52 15N 25 8 E
Drogobych = Drohobych, *Ukraine* .... **17 D12** 49 20N 23 30 E
Drohobych, *Ukraine* .... **17 D12** 49 20N 23 30 E
Droichead Atha = Drogheda, *Ireland* **13 C5** 53 43N 6 22W
Droichead Nua, *Ireland* **13 C5** 53 11N 6 48W
Droitwich, *U.K.* **11 E5** 52 16N 2 8W
Dromedary, C., *Australia* ... **63 F5** 36 17S 150 10 E
Dromore, *U.K.* **13 B4** 54 31N 7 28W
Dromore West, *Ireland* **13 B3** 54 15N 8 52W
Dronfield, *U.K.* **10 D6** 53 19N 1 27W
Dronten, *Neths.* .... **15 B5** 52 32N 5 43 E
Drumbo, *Canada* **78 C4** 43 16N 80 35W
Drumheller, *Canada* **72 C6** 51 25N 112 40W
Drummond, *U.S.A.* **82 C7** 46 40N 113 9W
Drummond I., *U.S.A.* **76 C4** 46 1N 83 39W
Drummond Pt., *Australia* ... **63 E2** 34 9S 135 16 E
Drummond Ra., *Australia* ... **62 C4** 23 45S 147 10 E
Drummondville, *Canada* **70 C5** 45 55N 72 25W
Drumright, *U.S.A.* **81 H6** 35 59N 96 36W
Druskininkai, *Lithuania* ..... **9 J20** 54 3N 23 58 E
Drut →, *Belarus* .... **17 B16** 53 8N 30 5 E
Druzhina, *Russia* .... **27 C15** 68 14N 145 18 E
Dry Tortugas, *U.S.A.* **88 B3** 24 38N 82 55W
Dryden, *Canada* **73 D10** 49 47N 92 50W
Dryden, *U.S.A.* **79 D8** 42 30N 76 18W
Drygalski I., *Antarctica* ..... **5 C7** 66 0S 92 0 E
Drysdale →, *Australia* ..... **60 B4** 13 59S 126 51 E
Drysdale I., *Australia* ...... **62 A2** 11 41S 136 0 E
Du Bois, *U.S.A.* **78 E6** 41 8N 78 46W
Du Gué →, *Canada* **70 A5** 57 21N 70 45W
Du Quoin, *U.S.A.* **80 G10** 38 1N 89 14W
Duanesburg, *U.S.A.* **79 D10** 42 45N 74 11W
Duaringa, *Australia* .... **62 C4** 23 42S 149 42 E
Ḑubā, *Si. Arabia* .... **44 E2** 27 10N 35 40 E
Dubai = Dubayy, *U.A.E.* ..... **46 B6** 25 18N 55 20 E
Dubăsari, *Moldova* .... **17 E15** 47 15N 29 10 E
Dubăsari Vdkhr., *Moldova* ... **17 E15** 47 30N 29 0 E
Dubawnt →, *Canada* **73 A8** 63 8N 101 28W
Dubawnt, L., *Canada* **73 A8** 63 4N 101 42W
Dubayy, *U.A.E.* .... **46 B6** 25 18N 55 20 E
Dubbo, *Australia* .... **63 E4** 32 11S 148 35 E
Dubele, *Dem. Rep. of the Congo* .... **54 B2** 2 56N 29 35 E
Dublin, *Ireland* **13 C5** 53 21N 6 15W
Dublin, *Ga., U.S.A.* **77 J4** 32 32N 82 54W
Dublin, *Tex., U.S.A.* **81 J5** 32 5N 98 21W
Dublin □, *Ireland* **13 C5** 53 24N 6 20W
Dubno, *Ukraine* .... **17 C13** 50 25N 25 45 E
Dubois, *U.S.A.* **82 D7** 44 10N 112 14W
Dubossary = Dubăsari, *Moldova* .... **17 E15** 47 15N 29 10 E
Dubossary Vdkhr. = Dubăsari Vdkhr., *Moldova* .... **17 E15** 47 30N 29 0 E
Dubovka, *Russia* .... **25 E7** 49 5N 44 50 E
Dubrajpur, *India* .... **43 H12** 23 48N 87 25 E
Dubréka, *Guinea* **50 G3** 9 46N 13 31W
Dubrovitsa = Dubrovytsya, *Ukraine* .... **17 C14** 51 31N 26 35 E
Dubrovnik, *Croatia* .... **21 C8** 42 39N 18 6 E
Dubrovytsya, *Ukraine* ...... **17 C14** 51 31N 26 35 E
Dubuque, *U.S.A.* **80 D9** 42 30N 90 41W
Duchesne, *U.S.A.* **82 F8** 40 10N 110 24W
Duchess, *Australia* .... **62 C2** 21 20S 139 50 E
Ducie I., *Pac. Oc.* **65 K15** 24 40S 124 48W
Duck →, *U.S.A.* **77 G2** 36 2N 87 52W
Duck Cr. →, *Australia* ...... **60 D2** 22 37S 116 53 E
Duck Lake, *Canada* **73 C7** 52 50N 106 16W
Duck Mountain Prov. Park, *Canada* **73 C8** 51 45N 101 0W
Duckwall, Mt., *U.S.A.* **84 H6** 37 58N 120 7W
Dudhi, *India* .... **41 G13** 24 15N 83 10 E
Dudinka, *Russia* .... **27 C9** 69 30N 86 13 E
Dudley, *U.K.* **11 E5** 52 31N 2 5W
Dudwa, *India* .... **43 E9** 28 30N 80 41 E
Duero = Douro →, *Europe* .. **19 B1** 41 8N 8 40W
Dufftown, *U.K.* **12 D5** 57 27N 3 8W
Dūghī Kalā, *Afghan.* .... **40 C3** 32 20N 62 50 E
Dugi Otok, *Croatia* .... **16 G8** 44 0N 15 3 E
Duifken Pt., *Australia* ...... **62 A3** 12 33S 141 38 E
Duisburg, *Germany* .... **16 C4** 51 26N 6 45 E
Duiwelskloof, *S. Africa* ..... **57 C5** 23 42S 30 10 E
Dūkdamīn, *Iran* .... **45 C8** 35 59N 57 43 E
Dukelský Průsmyk, *Slovak Rep.* .... **17 D11** 49 25N 21 42 E
Dukhān, *Qatar* .... **45 E6** 25 25N 50 50 E
Duki, *Pakistan* .... **40 D6** 30 14N 68 25 E
Duku, *Nigeria* .... **51 F8** 10 43N 10 43 E
Dulce, *U.S.A.* **83 H10** 36 56N 107 0W
Dulce →, *Argentina* **94 C3** 30 32S 62 33W
Dulce, G., *Costa Rica* **88 E3** 8 40N 83 20W
Dulf, *Iraq* .... **44 C5** 35 7N 45 51 E
Dulit, Banjaran, *Malaysia* ... **36 D4** 3 15N 114 30 E
Duliu, *China* .... **34 E9** 39 2N 116 55 E
Dullewala, *Pakistan* .... **42 D4** 31 50N 71 25 E
Dullstroom, *S. Africa* ...... **57 D5** 25 27S 30 7 E
Dulq Maghār, *Syria* .... **44 B3** 36 22N 38 39 E
Duluth, *U.S.A.* **80 B8** 46 47N 92 6W
Dum Dum, *India* .... **43 H13** 22 39N 88 33 E
Dum Duma, *India* .... **41 F19** 27 40N 95 40 E
Dūmā, *Syria* .... **47 B5** 33 34N 36 24 E
Dumaguete, *Phil.* .... **37 C6** 9 17N 123 15 E
Dumai, *Indonesia* .... **36 D2** 1 35N 101 28 E
Dumaran, *Phil.* .... **37 B5** 10 33N 119 50 E
Dumas, *Ark., U.S.A.* **81 J9** 33 53N 91 29W
Dumas, *Tex., U.S.A.* **81 H4** 35 52N 101 58W
Ḑumayr, *Syria* .... **47 B5** 33 39N 36 42 E
Dumbarton, *U.K.* **12 F4** 55 57N 4 33W
Dumbleyung, *Australia* ..... **61 F2** 33 17S 117 42 E
Dumfries, *U.K.* **12 F5** 55 4N 3 37W
Dumfries & Galloway □, *U.K.* **12 F5** 55 9N 3 58W
Dumka, *India* .... **43 G12** 24 12N 87 15 E
Dumoine →, *Canada* **70 C4** 46 13N 77 51W
Dumoine, L., *Canada* **70 C4** 46 55N 77 55W
Dumraon, *India* .... **43 G11** 25 33N 84 8 E
Dumyât, *Egypt* .... **51 B12** 31 24N 31 48 E
Dún Dealgan = Dundalk, *Ireland* **13 B5** 54 1N 6 24W
Dun Laoghaire, *Ireland* **13 C5** 53 17N 6 8W
Duna = Dunărea →, *Europe* . **17 F15** 45 20N 29 40 E
Dunagiri, *India* .... **43 D8** 30 31N 79 52 E
Dunaj = Dunărea →, *Europe* **17 F15** 45 20N 29 40 E
Dunakeszi, *Hungary* .... **17 E10** 47 37N 19 8 E
Dunărea →, *Europe* .... **17 F15** 45 20N 29 40 E
Dunaújváros, *Hungary* ..... **17 E10** 46 58N 18 57 E
Dunav = Dunărea →, *Europe* **17 F15** 45 20N 29 40 E
Dunay, *Russia* .... **30 C6** 42 52N 132 22 E
Dunback, *N.Z.* .... **59 L3** 45 23S 170 36 E
Dunbar, *U.K.* **12 E6** 56 0N 2 31W
Dunblane, *U.K.* **12 E5** 56 11N 3 58W
Duncan, *Canada* **72 D4** 48 45N 123 40W
Duncan, *Ariz., U.S.A.* **83 K9** 32 43N 109 6W
Duncan, *Okla., U.S.A.* **81 H6** 34 30N 97 57W
Duncan, L., *Canada* **70 B4** 53 29N 77 58W
Duncan L., *Canada* **72 A6** 62 51N 113 58W
Duncan Town, *Bahamas* ... **88 B4** 22 15N 75 45W
Duncannon, *U.S.A.* **78 F7** 40 23N 77 2W
Duncansby Head, *U.K.* **12 C5** 58 38N 3 1W
Duncansville, *U.S.A.* **78 F6** 40 25N 78 26W
Dundalk, *Canada* **78 B4** 44 10N 80 24W
Dundalk, *Ireland* **13 B5** 54 1N 6 24W
Dundalk, *U.S.A.* **76 F7** 39 16N 76 32W
Dundalk Bay, *Ireland* **13 C5** 53 55N 6 15W
Dundas, *Canada* **78 C5** 43 17N 79 59W
Dundas, L., *Australia* ...... **61 F3** 32 35S 121 50 E
Dundas I., *Canada* **72 C2** 54 30N 130 50W
Dundas Str., *Australia* ..... **60 B5** 11 15S 131 35 E
Dundee, *S. Africa* .... **57 D5** 28 11S 30 15 E
Dundee, *U.K.* **12 E6** 56 28N 2 59W
Dundee, *U.S.A.* **78 D8** 42 32N 76 59W
Dundee City □, *U.K.* **12 E6** 56 30N 2 58W
Dundgovĭ □, *Mongolia* ..... **34 B4** 45 10N 106 0 E
Dundrum, *U.K.* **13 B6** 54 16N 5 52W
Dundrum B., *U.K.* **13 B6** 54 13N 5 47W
Dunedin, *N.Z.* .... **59 L3** 45 50S 170 33 E
Dunedin, *U.S.A.* **77 L4** 28 1N 82 47W
Dunfermline, *U.K.* **12 E5** 56 5N 3 27W
Dungannon, *Canada* **78 C3** 43 51N 81 36W
Dungannon, *U.K.* **13 B5** 54 31N 6 46W
Dungarpur, *India* .... **42 H5** 23 52N 73 45 E
Dungarvan, *Ireland* **13 D4** 52 5N 7 37W
Dungarvan Harbour, *Ireland* **13 D4** 52 4N 7 35W
Dungeness, *U.K.* .... **11 G8** 50 54N 0 59 E
Dungo, L. do, *Angola* ...... **56 B2** 17 15S 19 0 E
Dungog, *Australia* .... **63 E5** 32 22S 151 46 E
Dungu, *Dem. Rep. of the Congo* .... **54 B2** 3 40N 28 32 E
Dungun, *Malaysia* .... **39 K4** 4 45N 103 25 E
Dunhua, *China* .... **35 C15** 43 20N 128 14 E
Dunhuang, *China* .... **32 B4** 40 8N 94 36 E
Dunk I., *Australia* .... **62 B4** 17 59S 146 29 E
Dunkeld, *Australia* .... **63 E4** 33 25S 149 29 E
Dunkeld, *U.K.* **12 E5** 56 34N 3 35W
Dunkerque, *France* .... **18 A5** 51 2N 2 20 E
Dunkery Beacon, *U.K.* **11 F4** 51 9N 3 36W
Dunkirk = Dunkerque, *France* **18 A5** 51 2N 2 20 E
Dunkirk, *U.S.A.* **78 D5** 42 29N 79 20W
Dúnleary = Dun Laoghaire, *Ireland* **13 C5** 53 17N 6 8W
Dunleer, *Ireland* **13 C5** 53 50N 6 24W
Dunmanus B., *Ireland* **13 E2** 51 31N 9 50W
Dunmanway, *Ireland* **13 E2** 51 43N 9 6W
Dunmara, *Australia* .... **62 B1** 16 42S 133 25 E
Dunmore, *U.S.A.* **79 E9** 41 25N 75 38W
Dunmore Hd., *Ireland* **13 D1** 52 10N 10 35W
Dunmore Town, *Bahamas* .. **88 A4** 25 30N 76 39W
Dunn, *U.S.A.* **77 H6** 35 19N 78 37W
Dunnellon, *U.S.A.* **77 L4** 29 3N 82 28W
Dunnet Hd., *U.K.* **12 C5** 58 40N 3 21W
Dunning, *U.S.A.* **80 E4** 41 50N 100 6W
Dunnville, *Canada* **78 D5** 42 54N 79 36W
Dunolly, *Australia* .... **63 F3** 36 51S 143 44 E
Dunoon, *U.K.* **12 F4** 55 57N 4 56W
Dunphy, *U.S.A.* **82 F5** 40 42N 116 31W
Duns, *U.K.* **12 F6** 55 47N 2 20W
Dunseith, *U.S.A.* **80 A4** 48 50N 100 3W
Dunsmuir, *U.S.A.* **82 F2** 41 13N 122 16W
Dunstable, *U.K.* **11 F7** 51 53N 0 32W
Dunstan Mts., *N.Z.* .... **59 L2** 44 53S 169 35 E
Dunster, *Canada* **72 C5** 53 8N 119 50W
Dunvegan L., *Canada* **73 A7** 60 8N 107 10W
Duolun, *China* .... **34 C9** 42 12N 116 28 E
Duong Dong, *Vietnam* ...... **39 G4** 10 13N 103 58 E
Dupree, *U.S.A.* **80 C4** 45 4N 101 35W
Dupuyer, *U.S.A.* **82 B7** 48 13N 112 30W
Duque de Caxias, *Brazil* **95 A7** 22 45S 43 19W
Durack →, *Australia* .... **60 C4** 15 33S 127 52 E
Durack Ra., *Australia* ...... **60 C4** 16 50S 127 40 E
Durance →, *France* .... **18 E6** 43 55N 4 45 E
Durand, *U.S.A.* **80 C9** 44 38N 91 58W
Durango, *Mexico* **86 C4** 24 3N 104 39W
Durango, *U.S.A.* **83 H10** 37 16N 107 53W
Durango □, *Mexico* **86 C4** 25 0N 105 0W
Durant, *Miss., U.S.A.* **81 J10** 33 4N 89 51W
Durant, *Okla., U.S.A.* **81 J6** 33 59N 96 25W
Durazno, *Uruguay* **94 C4** 33 25S 56 31W
Durazzo = Durrës, *Albania* .. **21 D8** 41 19N 19 28 E
Durban, *S. Africa* .... **57 D5** 29 49S 31 1 E
Durbuy, *Belgium* .... **15 D5** 50 21N 5 28 E
Düren, *Germany* .... **16 C4** 50 48N 6 29 E
Durg, *India* .... **41 J12** 21 15N 81 22 E
Durgapur, *India* .... **43 H12** 23 30N 87 20 E
Durham, *Canada* **78 B4** 44 10N 80 49W
Durham, *U.K.* **10 C6** 54 47N 1 34W
Durham, *Calif., U.S.A.* **84 F5** 39 39N 121 48W
Durham, *N.C., U.S.A.* **77 H6** 35 59N 78 54W
Durham, *N.H., U.S.A.* **79 C14** 43 8N 70 56W
Durham □, *U.K.* **10 C6** 54 42N 1 45W
Ḑurmā, *Si. Arabia* .... **44 E5** 24 37N 46 8 E
Durmitor, *Montenegro, Yug.* **21 C8** 43 10N 19 0 E
Durness, *U.K.* **12 C4** 58 34N 4 45W
Durrës, *Albania* .... **21 D8** 41 19N 19 28 E
Durrow, *Ireland* **13 D4** 52 51N 7 24W
Dursey I., *Ireland* **13 E1** 51 36N 10 12W
Dursunbey, *Turkey* .... **21 E13** 39 35N 28 37 E
Duru, *Dem. Rep. of the Congo* **54 B2** 4 14N 28 50 E
Durūz, Jabal ad, *Jordan* .... **47 C5** 32 35N 36 40 E
D'Urville, Tanjung, *Indonesia* **37 E9** 1 28S 137 54 E
D'Urville I., *N.Z.* .... **59 J4** 40 50S 173 55 E
Duryea, *U.S.A.* **79 E9** 41 20N 75 45W
Dushak, *Turkmenistan* ..... **26 F7** 37 13N 60 1 E
Dushanbe, *Tajikistan* ...... **26 F7** 38 33N 68 48 E
Dushore, *U.S.A.* **79 E8** 41 31N 76 24W
Dusky Sd., *N.Z.* .... **59 L1** 45 47S 166 30 E
Dussejour, C., *Australia* .... **60 B4** 14 45S 128 13 E
Düsseldorf, *Germany* ...... **16 C4** 51 14N 6 47 E
Dutch Harbor, *U.S.A.* **68 C3** 53 53N 166 32W
Dutlwe, *Botswana* .... **56 C3** 23 58S 23 46 E
Dutton, *Canada* **78 D3** 42 39N 81 30W
Dutton →, *Australia* ...... **62 C3** 20 44S 143 10 E
Duwayhin, Khawr, *U.A.E.* ... **45 E6** 24 20N 51 25 E
Duyun, *China* .... **32 D5** 26 18N 107 29 E
Duzdab = Zāhedān, *Iran* .... **45 D9** 29 30N 60 50 E
Dvina, Severnaya →, *Russia* **24 B7** 64 32N 40 30 E
Dvinsk = Daugavpils, *Latvia* . **9 J22** 55 53N 26 32 E
Dvinskaya Guba, *Russia* .... **24 B6** 65 0N 39 0 E
Dwarka, *India* .... **42 H3** 22 18N 69 8 E
Dwellingup, *Australia* ...... **61 F2** 32 43S 116 4 E
Dwight, *Canada* **78 A5** 45 20N 79 1W
Dwight, *U.S.A.* **76 E1** 41 5N 88 26W
Dyatlovo = Dzyatlava, *Belarus* **17 B13** 53 28N 25 28 E
Dyce, *U.K.* **12 D6** 57 13N 2 12W
Dyer, C., *Canada* **69 B13** 66 40N 61 0W
Dyer Bay, *Canada* **78 A3** 45 10N 81 20W
Dyer Plateau, *Antarctica* **5 D17** 70 45S 65 30W
Dyersburg, *U.S.A.* **81 G10** 36 3N 89 23W
Dyfi →, *U.K.* **11 E3** 52 32N 4 3W
Dymer, *Ukraine* .... **17 C16** 50 47N 30 18 E
Dysart, *Australia* .... **62 C4** 22 32S 148 23 E
Dzamin Üüd = Borhoyn Tal, *Mongolia* .... **34 C6** 43 50N 111 58 E
Dzerzhinsk, *Russia* .... **24 C7** 56 14N 43 30 E
Dzhalinda, *Russia* .... **27 D13** 53 26N 124 0 E
Dzhambul = Taraz, *Kazakstan* **26 E8** 42 54N 71 22 E
Dzhankoy, *Ukraine* .... **25 E5** 45 40N 34 20 E
Dzhezkazgan = Zhezqazghan, *Kazakstan* .... **26 E7** 47 44N 67 40 E
Dzhizak = Jizzakh, *Uzbekistan* **26 E7** 40 6N 67 50 E
Dzhugdzur, Khrebet, *Russia* . **27 D14** 57 30N 138 0 E
Dzhungarskiye Vorota = Dzungarian Gates, *Asia* ... **32 B3** 45 0N 82 0 E
Działdowo, *Poland* .... **17 B11** 53 15N 20 15 E
Dzibilchaltun, *Mexico* **87 C7** 21 5N 89 36W
Dzierżoniów, *Poland* .... **17 C9** 50 45N 16 39 E
Dzilam de Bravo, *Mexico* ... **87 C7** 21 24N 88 53W
Dzungaria = Junggar Pendi, *China* .... **32 B3** 44 30N 86 0 E
Dzungarian Gates, *Asia* ..... **32 B3** 45 0N 82 0 E
Dzuumod, *Mongolia* .... **32 B5** 47 45N 106 58 E
Dzyarzhynsk, *Belarus* ...... **17 B14** 53 40N 27 1 E
Dzyatlava, *Belarus* .... **17 B13** 53 28N 25 28 E

E

Eabamet L., *Canada* **70 B2** 51 30N 87 46W
Eads, *U.S.A.* **80 F3** 38 29N 102 47W
Eagar, *U.S.A.* **83 J9** 34 6N 109 17W
Eagle, *Alaska, U.S.A.* **68 B5** 64 47N 141 12W
Eagle, *Colo., U.S.A.* **82 G10** 39 39N 106 50W
Eagle →, *Canada* **71 B8** 53 36N 57 26W
Eagle Butte, *U.S.A.* **80 C4** 45 0N 101 10W
Eagle Grove, *U.S.A.* **80 D8** 42 40N 93 54W
Eagle L., *Canada* **73 D10** 49 42N 93 13W
Eagle L., *Calif., U.S.A.* **82 F3** 40 39N 120 45W
Eagle L., *Maine, U.S.A.* **77 B11** 46 20N 69 22W
Eagle Lake, *Canada* **78 A6** 45 8N 78 29W
Eagle Lake, *Maine, U.S.A.* ... **77 B11** 47 3N 68 36W
Eagle Lake, *Tex., U.S.A.* **81 L6** 29 35N 96 20W
Eagle Mountain, *U.S.A.* **85 M11** 33 49N 115 27W
Eagle Nest, *U.S.A.* **83 H11** 36 33N 105 16W
Eagle Pass, *U.S.A.* **81 L4** 28 43N 100 30W
Eagle Pk., *U.S.A.* **84 G7** 38 10N 119 25W
Eagle Pt., *Australia* .... **60 C3** 16 11S 124 23 E
Eagle River, *Mich., U.S.A.* .. **76 B1** 47 24N 88 18W
Eagle River, *Wis., U.S.A.* ... **80 C10** 45 55N 89 15W
Eaglehawk, *Australia* ...... **63 F3** 36 44S 144 15 E
Eagles Mere, *U.S.A.* **79 E8** 41 25N 76 33W
Ealing □, *U.K.* **11 F7** 51 31N 0 20W
Ear Falls, *Canada* **73 C10** 50 38N 93 13W
Earle, *U.S.A.* **81 H9** 35 16N 90 28W
Earlimart, *U.S.A.* **85 K7** 35 53N 119 16W
Earn →, *U.K.* **12 E5** 56 21N 3 18W
Earn, L., *U.K.* **12 E4** 56 23N 4 13W
Earnslaw, Mt., *N.Z.* .... **59 L2** 44 32S 168 27 E
Earth, *U.S.A.* **81 H3** 34 14N 102 24W
Easley, *U.S.A.* **77 H4** 34 50N 82 36W
East Anglia, *U.K.* .... **10 E9** 52 30N 1 0 E
East Angus, *Canada* **71 C5** 45 30N 71 40W
East Aurora, *U.S.A.* **78 D6** 42 46N 78 37W
East Ayrshire □, *U.K.* **12 F4** 55 26N 4 11W
East Bengal, *Bangla.* .... **41 H17** 24 0N 90 0 E
East Beskids = Vychodné Beskydy, *Europe* .... **17 D11** 49 20N 22 0 E
East Brady, *U.S.A.* **78 F5** 40 59N 79 36W
East C., *N.Z.* .... **59 G7** 37 42S 178 35 E
East Chicago, *U.S.A.* **76 E2** 41 38N 87 27W
East China Sea, *Asia* .... **33 D7** 30 0N 126 0 E
East Coulee, *Canada* **72 C6** 51 23N 112 27W
East Dereham, *U.K.* .... **11 E8** 52 41N 0 57 E
East Dunbartonshire □, *U.K.* **12 F4** 55 57N 4 13W
East Falkland, *Falk. Is.* **96 G5** 51 30S 58 30W
East Grand Forks, *U.S.A.* ... **80 B6** 47 56N 97 1W
East Greenwich, *U.S.A.* **79 E13** 41 40N 71 27W
East Grinstead, *U.K.* **11 F8** 51 7N 0 0W
East Hartford, *U.S.A.* **79 E12** 41 46N 72 39W
East Helena, *U.S.A.* **82 C8** 46 35N 111 56W
East Indies, *Asia* .... **28 K15** 0 0 120 0 E
East Kilbride, *U.K.* **12 F4** 55 47N 4 11W
East Lansing, *U.S.A.* **76 D3** 42 44N 84 29W
East Liverpool, *U.S.A.* **78 F4** 40 37N 80 35W
East London, *S. Africa* ..... **57 E4** 33 0S 27 55 E
East Lothian □, *U.K.* **12 F6** 55 58N 2 44W
East Main = Eastmain, *Canada* **70 B4** 52 10N 78 30W
East Northport, *U.S.A.* **79 F11** 40 53N 73 20W
East Orange, *U.S.A.* **79 F10** 40 46N 74 13W
East Pacific Ridge, *Pac. Oc.* **65 J17** 15 0S 110 0W
East Palestine, *U.S.A.* **78 F4** 40 50N 80 33W
East Pine, *Canada* **72 B4** 55 48N 120 12W
East Point, *U.S.A.* **77 J3** 33 41N 84 27W
East Providence, *U.S.A.* **79 E13** 41 49N 71 23W

East Pt.

East Pt., *Canada* **71 C7** 46 27N 61 58W
East Renfrewshire □, *U.K.* **12 F4** 55 46N 4 21W
East Retford = Retford, *U.K.* **10 D7** 53 19N 0 56W
East Riding of Yorkshire □, *U.K.* **10 D7** 53 55N 0 30W
East Rochester, *U.S.A.* **78 C7** 43 7N 77 29W
East St. Louis, *U.S.A.* **80 F9** 38 37N 90 9W
East Schelde = Oosterschelde →, *Neths.* **15 C4** 51 33N 4 0 E
East Sea = Japan, Sea of, *Asia* **30 E7** 40 0N 135 0 E
East Siberian Sea, *Russia* **27 B17** 73 0N 160 0 E
East Stroudsburg, *U.S.A.* **79 E9** 41 1N 75 11W
East Sussex □, *U.K.* **11 G8** 50 56N 0 19 E
East Tawas, *U.S.A.* **76 C4** 44 17N 83 29W
East Timor ■, *Asia* **37 F7** 8 50S 126 0 E
East Toorale, *Australia* **63 E4** 30 27S 145 28 E
East Walker →, *U.S.A.* **84 G7** 38 52N 119 10W
East Windsor, *U.S.A.* **79 F10** 40 17N 74 34W
Eastbourne, *N.Z.* **59 J5** 41 19S 174 55 E
Eastbourne, *U.K.* **11 G8** 50 46N 0 18 E
Eastend, *Canada* **73 D7** 49 32N 108 50W
Easter I. = Pascua, I. de, *Chile* **65 K17** 27 7S 109 23W
Eastern □, *Kenya* **54 C4** 0 0 38 30 E
Eastern Cape □, *S. Africa* **56 E4** 32 0S 26 0 E
Eastern Cr. →, *Australia* **62 C3** 20 40S 141 35 E
Eastern Ghats, *India* **40 N11** 14 0N 78 50 E
Eastern Group = Lau Group, *Fiji* **59 C9** 17 0S 178 30W
Eastern Group, *Australia* **61 F3** 33 30S 124 30 E
Eastern Transvaal = Mpumalanga □, *S. Africa* **57 B5** 26 0S 30 0 E
Easterville, *Canada* **73 C9** 53 8N 99 49W
Easthampton, *U.S.A.* **79 D12** 42 16N 72 40W
Eastlake, *U.S.A.* **78 E3** 41 40N 81 26W
Eastland, *U.S.A.* **81 J5** 32 24N 98 49W
Eastleigh, *U.K.* **11 G6** 50 58N 1 21W
Eastmain, *Canada* **70 B4** 52 10N 78 30W
Eastmain →, *Canada* **70 B4** 52 27N 78 26W
Eastman, *Canada* **79 A12** 45 18N 72 19W
Eastman, *U.S.A.* **77 J4** 32 12N 83 11W
Easton, *Md., U.S.A.* **76 F7** 38 47N 76 5W
Easton, *Pa., U.S.A.* **79 F9** 40 41N 75 13W
Easton, *Wash., U.S.A.* **84 C5** 47 14N 121 11W
Eastpointe, *U.S.A.* **78 D2** 42 27N 82 56W
Eastport, *U.S.A.* **77 C12** 44 56N 67 0W
Eastsound, *U.S.A.* **84 B4** 48 42N 122 55W
Eaton, *U.S.A.* **80 E2** 40 32N 104 42W
Eatonia, *Canada* **73 C7** 51 13N 109 25W
Eatonton, *U.S.A.* **77 J4** 33 20N 83 23W
Eatontown, *U.S.A.* **79 F10** 40 19N 74 4W
Eatonville, *U.S.A.* **84 D4** 46 52N 122 16W
Eau Claire, *U.S.A.* **80 C9** 44 49N 91 30W
Eau Claire, L. à l', *Canada* **70 A5** 56 10N 74 25W
Ebbw Vale, *U.K.* **11 F4** 51 46N 3 12W
Ebeltoft, *Denmark* **9 H14** 56 12N 10 41 E
Ebensburg, *U.S.A.* **78 F6** 40 29N 78 44W
Eberswalde-Finow, *Germany* **16 B7** 52 50N 13 49 E
Ebetsu, *Japan* **30 C10** 43 7N 141 34 E
Ebolowa, *Cameroon* **52 D2** 2 55N 11 10 E
Ebro →, *Spain* **19 B6** 40 43N 0 54 E
Eceabat, *Turkey* **21 D12** 40 11N 26 21 E
Ech Chéliff, *Algeria* **50 A6** 36 10N 1 20 E
Echigo-Sammyaku, *Japan* **31 F9** 36 50N 139 50 E
Echizen-Misaki, *Japan* **31 G7** 35 59N 135 57 E
Echo Bay, *N.W.T., Canada* **68 B8** 66 5N 117 55W
Echo Bay, *Ont., Canada* **70 C3** 46 29N 84 4W
Echoing →, *Canada* **70 B1** 55 51N 92 5W
Echternach, *Lux.* **15 E6** 49 49N 6 25 E
Echuca, *Australia* **63 F3** 36 10S 144 45 E
Ecija, *Spain* **19 D3** 37 30N 5 10W
Eclipse Is., *Australia* **60 B4** 13 54S 126 19 E
Eclipse Sd., *Canada* **69 A11** 72 38N 79 0W
Ecuador ■, *S. Amer.* **92 D3** 2 0S 78 0W
Ed Damazin, *Sudan* **51 F12** 11 46N 34 21 E
Ed Debba, *Sudan* **51 E12** 18 0N 30 51 E
Ed Dueim, *Sudan* **51 F12** 14 0N 32 10 E
Edam, *Canada* **73 C7** 53 11N 108 46W
Edam, *Neths.* **15 B5** 52 31N 5 3 E
Eday, *U.K.* **12 B6** 59 11N 2 47W
Eddrachillis B., *U.K.* **12 C3** 58 17N 5 14W
Eddystone Pt., *Australia* **62 G4** 40 59S 148 20 E
Ede, *Neths.* **15 B5** 52 4N 5 40 E
Edehon L., *Canada* **73 A9** 60 25N 97 15W
Eden, *Australia* **63 F4** 37 3S 149 55 E
Eden, *N.C., U.S.A.* **77 G6** 36 29N 79 53W
Eden, *N.Y., U.S.A.* **78 D6** 42 39N 78 55W
Eden, *Tex., U.S.A.* **81 K5** 31 13N 99 51W
Eden →, *U.K.* **10 C4** 54 57N 3 1W
Edenburg, *S. Africa* **56 D4** 29 43S 25 58 E
Edendale, *S. Africa* **57 D5** 29 39S 30 18 E
Edenderry, *Ireland* **13 C4** 53 21N 7 4W
Edenton, *U.S.A.* **77 G7** 36 4N 76 39W
Edenville, *S. Africa* **57 D4** 27 37S 27 34 E
Eder →, *Germany* **16 C5** 51 12N 9 28 E
Edgar, *U.S.A.* **80 E6** 40 22N 97 58W
Edgartown, *U.S.A.* **79 E14** 41 23N 70 31W
Edge Hill, *U.K.* **11 E6** 52 8N 1 26W
Edgefield, *U.S.A.* **77 J5** 33 47N 81 56W
Edgeley, *U.S.A.* **80 B5** 46 22N 98 43W
Edgemont, *U.S.A.* **80 D3** 43 18N 103 50W
Edgeøya, *Svalbard* **4 B9** 77 45N 22 30 E
Édhessa, *Greece* **21 D10** 40 48N 22 5 E
Edievale, *N.Z.* **59 L2** 45 49S 169 22 E
Edina, *U.S.A.* **80 E8** 40 10N 92 11W
Edinboro, *U.S.A.* **78 E4** 41 52N 80 8W
Edinburg, *U.S.A.* **81 M5** 26 18N 98 10W
Edinburgh, *U.K.* **12 F5** 55 57N 3 13W
Edinburgh, City of □, *U.K.* **12 F5** 55 57N 3 17W
Edineţ, *Moldova* **17 D14** 48 9N 27 18 E
Edirne, *Turkey* **21 D12** 41 40N 26 34 E
Edison, *U.S.A.* **84 B4** 48 33N 122 27W
Edithburgh, *Australia* **63 F2** 35 5S 137 43 E
Edmeston, *U.S.A.* **79 D9** 42 42N 75 15W
Edmond, *U.S.A.* **81 H6** 35 39N 97 29W
Edmonds, *U.S.A.* **84 C4** 47 49N 122 23W
Edmonton, *Australia* **62 B4** 17 2S 145 46 E
Edmonton, *Canada* **72 C6** 53 30N 113 30W
Edmund L., *Canada* **70 B1** 54 45N 93 17W
Edmundston, *Canada* **71 C6** 47 23N 68 20W
Edna, *U.S.A.* **81 L6** 28 59N 96 39W
Edremit, *Turkey* **21 E12** 39 34N 27 0 E
Edremit Körfezi, *Turkey* **21 E12** 39 30N 26 45 E
Edson, *Canada* **72 C5** 53 35N 116 28W
Eduardo Castex, *Argentina* **94 D3** 35 50S 64 18W
Edward →, *Australia* **63 F3** 35 5S 143 30 E
Edward, L., *Africa* **54 C2** 0 25S 29 40 E
Edward River, *Australia* **62 A3** 14 59S 141 26 E
Edward VII Land, *Antarctica* **5 E13** 80 0S 150 0W
Edwards, *Calif., U.S.A.* **85 L9** 34 55N 117 51W
Edwards, *N.Y., U.S.A.* **79 B9** 44 20N 75 15W
Edwards Air Force Base, *U.S.A.* **85 L9** 34 50N 117 40W
Edwards Plateau, *U.S.A.* **81 K4** 30 45N 101 20W
Edwardsville, *U.S.A.* **79 E9** 41 15N 75 56W
Edzo, *Canada* **72 A5** 62 49N 116 4W
Eeklo, *Belgium* **15 C3** 51 11N 3 33 E
Effingham, *U.S.A.* **76 F1** 39 7N 88 33W
Égadi, Ísole, *Italy* **20 F5** 37 55N 12 16 E
Egan Range, *U.S.A.* **82 G6** 39 35N 114 55W
Eganville, *Canada* **78 A7** 45 32N 77 5W
Eger = Cheb, *Czech Rep.* **16 C7** 50 9N 12 28 E
Eger, *Hungary* **17 E11** 47 53N 20 27 E
Egersund, *Norway* **9 G12** 58 26N 6 1 E
Egg L., *Canada* **73 B7** 55 5N 105 30W
Éghezée, *Belgium* **15 D4** 50 35N 4 55 E
Egmont, *Canada* **72 D4** 49 45N 123 56W
Egmont, C., *N.Z.* **59 H4** 39 16S 173 45 E
Egmont, Mt. = Taranaki, Mt., *N.Z.* **59 H5** 39 17S 174 5 E
Egra, *India* **43 J12** 21 54N 87 32 E
Eğridir, *Turkey* **25 G5** 37 52N 30 51 E
Eğridir Gölü, *Turkey* **25 G5** 37 53N 30 50 E
Egvekinot, *Russia* **27 C19** 66 19N 179 50W
Egypt ■, *Africa* **51 C12** 28 0N 31 0 E
Ehime □, *Japan* **31 H6** 33 30N 132 40 E
Ehrenberg, *U.S.A.* **85 M12** 33 36N 114 31W
Eibar, *Spain* **19 A4** 43 11N 2 28W
Eidsvold, *Australia* **63 D5** 25 25S 151 12 E
Eidsvoll, *Norway* **9 F14** 60 19N 11 14 E
Eifel, *Germany* **16 C4** 50 15N 6 50 E
Eiffel Flats, *Zimbabwe* **55 F3** 18 20S 30 0 E
Eigg, *U.K.* **12 E2** 56 54N 6 10W
Eighty Mile Beach, *Australia* **60 C3** 19 30S 120 40 E
Eil, *Somali Rep.* **46 F4** 8 0N 49 50 E
Eil, L., *U.K.* **12 E3** 56 51N 5 16W
Eildon, L., *Australia* **63 F4** 37 10S 146 0 E
Einasleigh, *Australia* **62 B3** 18 32S 144 5 E
Einasleigh →, *Australia* **62 B3** 17 30S 142 17 E
Eindhoven, *Neths.* **15 C5** 51 26N 5 28 E
Eire = Ireland ■, *Europe* **13 C4** 53 50N 7 52W
Eiríksjökull, *Iceland* **8 D3** 64 46N 20 24W
Eirunepé, *Brazil* **92 E5** 6 35S 69 53W
Eiseb →, *Namibia* **56 C2** 20 33S 20 59 E
Eisenach, *Germany* **16 C6** 50 58N 10 19 E
Eisenerz, *Austria* **16 E8** 47 32N 14 54 E
Eivissa, *Spain* **22 C7** 38 54N 1 26 E
Ejeda, *Madag.* **57 C7** 24 20S 44 31 E
Ejutla, *Mexico* **87 D5** 16 34N 96 44W
Ekalaka, *U.S.A.* **80 C2** 45 53N 104 33W
Eketahuna, *N.Z.* **59 J5** 40 38S 175 43 E
Ekibastuz, *Kazakstan* **26 D8** 51 50N 75 10 E
Ekoli, *Dem. Rep. of the Congo* **54 C1** 0 23S 24 13 E
Eksjö, *Sweden* **9 H16** 57 40N 14 58 E
Ekuma →, *Namibia* **56 B2** 18 40S 16 2 E
Ekwan →, *Canada* **70 B3** 53 12N 82 15W
Ekwan Pt., *Canada* **70 B3** 53 16N 82 7W
El Aaiún, *W. Sahara* **50 C3** 27 9N 13 12W
El Abanico, *Chile* **94 D1** 37 20S 71 31W
El 'Agrûd, *Egypt* **47 E3** 30 14N 34 24 E
El Alamein, *Egypt* **51 B11** 30 48N 28 58 E
El 'Aqaba, W. →, *Egypt* **47 E2** 30 7N 33 54 E
El Arīḥā, *West Bank* **47 D4** 31 52N 35 27 E
El 'Arîsh, *Egypt* **47 D2** 31 8N 33 50 E
El 'Arîsh, W. →, *Egypt* **47 D2** 31 8N 33 47 E
El Asnam = Ech Chéliff, *Algeria* **50 A6** 36 10N 1 20 E
El Bayadh, *Algeria* **50 B6** 33 40N 1 1 E
El Bluff, *Nic.* **88 D3** 11 59N 83 40W
El Brûk, W. →, *Egypt* **47 E2** 30 15N 33 50 E
El Cajon, *U.S.A.* **85 N10** 32 48N 116 58W
El Campo, *U.S.A.* **81 L6** 29 12N 96 16W
El Centro, *U.S.A.* **85 N11** 32 48N 115 34W
El Cerro, *Bolivia* **92 G6** 17 30S 61 40W
El Compadre, *Mexico* **85 N10** 32 20N 116 14W
El Cuy, *Argentina* **96 D3** 39 55S 68 25W
El Cuyo, *Mexico* **87 C7** 21 30N 87 40W
El Daheir, *Egypt* **47 D3** 31 13N 34 10 E
El Dátil, *Mexico* **86 B2** 30 7N 112 15W
El Dere, *Somali Rep.* **46 G4** 3 50N 47 8 E
El Descanso, *Mexico* **85 N10** 32 12N 116 58W
El Desemboque, *Mexico* **86 A2** 30 30N 112 57W
El Diviso, *Colombia* **92 C3** 1 22N 78 14W
El Djouf, *Mauritania* **50 D4** 20 0N 9 0W
El Dorado, *Ark., U.S.A.* **81 J8** 33 12N 92 40W
El Dorado, *Kans., U.S.A.* **81 G6** 37 49N 96 52W
El Dorado, *Venezuela* **92 B6** 6 55N 61 37W
El Escorial, *Spain* **19 B3** 40 35N 4 7W
El Faiyûm, *Egypt* **51 C12** 29 19N 30 50 E
El Fâsher, *Sudan* **51 F11** 13 33N 25 26 E
El Ferrol = Ferrol, *Spain* **19 A1** 43 29N 8 15W
El Fuerte, *Mexico* **86 B3** 26 30N 108 40W
El Gal, *Somali Rep.* **46 E5** 10 58N 50 20 E
El Geneina = Al Junaynah, *Sudan* **51 F10** 13 27N 22 45 E
El Gîza, *Egypt* **51 C12** 30 0N 31 10 E
El Goléa, *Algeria* **50 B6** 30 30N 2 50 E
El Iskandarîya, *Egypt* **51 B11** 31 13N 29 58 E
El Istiwa'iya, *Sudan* **51 G11** 5 0N 28 0 E
El Jadida, *Morocco* **50 B4** 33 11N 8 17W
El Jardal, *Honduras* **88 D2** 14 54N 88 50W
El Kabrît, G., *Egypt* **47 F2** 29 42N 33 16 E
El Khârga, *Egypt* **51 C12** 25 30N 30 33 E
El Khartûm, *Sudan* **51 E12** 15 31N 32 35 E
El Kuntilla, *Egypt* **47 E3** 30 1N 34 45 E
El Maestrazgo, *Spain* **19 B5** 40 30N 0 25W
El Mahalla el Kubra, *Egypt* **51 B12** 31 0N 31 0 E
El Mansûra, *Egypt* **51 B12** 31 0N 31 19 E
El Medano, *Canary Is.* **22 F3** 28 3N 16 32W
El Milagro, *Argentina* **94 C2** 30 59S 65 59W
El Minyâ, *Egypt* **51 C12** 28 7N 30 33 E
El Monte, *U.S.A.* **85 L8** 34 4N 118 1W
El Obeid, *Sudan* **51 F12** 13 8N 30 10 E
El Odaiya, *Sudan* **51 F11** 12 8N 28 12 E
El Oro, *Mexico* **87 D4** 19 48N 100 8W
El Oued, *Algeria* **50 B7** 33 20N 6 58 E
El Palmito, Presa, *Mexico* **86 B3** 25 40N 105 30W
El Paso, *U.S.A.* **83 L10** 31 45N 106 29W
El Paso Robles, *U.S.A.* **84 K6** 35 38N 120 41W
El Portal, *U.S.A.* **84 H7** 37 41N 119 47W
El Porvenir, *Mexico* **86 A3** 31 15N 105 51W
El Prat de Llobregat, *Spain* **19 B7** 41 18N 2 3 E
El Progreso, *Honduras* **88 C2** 15 26N 87 51W
El Pueblito, *Mexico* **86 B3** 29 3N 105 4W
El Pueblo, *Canary Is.* **22 F2** 28 36N 17 47W
El Puerto de Santa María, *Spain* **19 D2** 36 36N 6 13W
El Qâhira, *Egypt* **51 B12** 30 1N 31 14 E
El Qantara, *Egypt* **47 E1** 30 51N 32 20 E
El Quseima, *Egypt* **47 E3** 30 40N 34 15 E
El Real, *Panama* **92 B3** 8 0N 77 40W
El Reno, *U.S.A.* **81 H6** 35 32N 97 57W
El Rio, *U.S.A.* **85 L7** 34 14N 119 10W
El Roque, Pta., *Canary Is.* **22 F4** 28 10N 15 25W
El Rosarito, *Mexico* **86 B2** 28 38N 114 4W
El Saheira, W. →, *Egypt* **47 E2** 30 5N 33 25 E
El Salto, *Mexico* **86 C3** 23 47N 105 22W
El Salvador ■, *Cent. Amer.* **88 D2** 13 50N 89 0W
El Sauce, *Nic.* **88 D2** 13 0N 86 40W
El Sueco, *Mexico* **86 B3** 29 54N 106 24W
El Suweis, *Egypt* **51 C12** 29 58N 32 31 E
El Tamarâni, W. →, *Egypt* **47 E3** 30 7N 34 43 E
El Thamad, *Egypt* **47 F3** 29 40N 34 28 E
El Tigre, *Venezuela* **92 B6** 8 44N 64 15W
El Tîh, Gebal, *Egypt* **47 F2** 29 40N 33 50 E
El Tîna, Khalîg, *Egypt* **47 D1** 31 10N 32 40 E
El Tofo, *Chile* **94 B1** 29 22S 71 18W
El Tránsito, *Chile* **94 B1** 28 52S 70 17W
El Tûr, *Egypt* **44 D2** 28 14N 33 36 E
El Turbio, *Argentina* **96 G2** 51 45S 72 5W
El Uqsur, *Egypt* **51 C12** 25 41N 32 38 E
El Venado, *Mexico* **86 C4** 22 56N 101 10W
El Vergel, *Mexico* **86 B3** 26 28N 106 22W
El Vigia, *Venezuela* **92 B4** 8 38N 71 39W
El Wabeira, *Egypt* **47 F2** 29 34N 33 6 E
El Wak, *Kenya* **54 B5** 2 49N 40 56 E
El Wuz, *Sudan* **51 E12** 15 5N 30 7 E
Elat, *Israel* **47 F3** 29 30N 34 56 E
Elâzığ, *Turkey* **25 G6** 38 37N 39 14 E
Elba, *Italy* **20 C4** 42 46N 10 17 E
Elba, *U.S.A.* **77 K2** 31 25N 86 4W
Elbasan, *Albania* **21 D9** 41 9N 20 9 E
Elbe, *U.S.A.* **84 D4** 46 45N 122 10W
Elbe →, *Europe* **16 B5** 53 50N 9 0 E
Elbert, Mt., *U.S.A.* **83 G10** 39 7N 106 27W
Elberton, *U.S.A.* **77 H4** 34 7N 82 52W
Elbeuf, *France* **18 B4** 49 17N 1 2 E
Elbidtan, *Turkey* **44 B3** 38 13N 37 12 E
Elbing = Elbląg, *Poland* **17 A10** 54 10N 19 25 E
Elbląg, *Poland* **17 A10** 54 10N 19 25 E
Elbow, *Canada* **73 C7** 51 7N 106 35W
Elbrus, *Asia* **25 F7** 43 21N 42 30 E
Elburz Mts. = Alborz, Reshteh-ye Kūhhā-ye, *Iran* **45 C7** 36 0N 52 0 E
Elche, *Spain* **19 C5** 38 15N 0 42W
Elcho I., *Australia* **62 A2** 11 55S 135 45 E
Elda, *Spain* **19 C5** 38 29N 0 47W
Elde →, *Germany* **16 B6** 53 7N 11 15 E
Eldon, *Mo., U.S.A.* **80 F8** 38 21N 92 35W
Eldon, *Wash., U.S.A.* **84 C3** 47 33N 123 3W
Eldora, *U.S.A.* **80 D8** 42 22N 93 5W
Eldorado, *Argentina* **95 B5** 26 28S 54 43W
Eldorado, *Canada* **78 B7** 44 35N 77 31W
Eldorado, *Mexico* **86 C3** 24 20N 107 22W
Eldorado, *Ill., U.S.A.* **76 G1** 37 49N 88 26W
Eldorado, *Tex., U.S.A.* **81 K4** 30 52N 100 36W
Eldorado Springs, *U.S.A.* **81 G8** 37 52N 94 1W
Eldoret, *Kenya* **54 B4** 0 30N 35 17 E
Eldred, *U.S.A.* **78 E6** 41 58N 78 23W
Elea, C., *Cyprus* **23 D13** 35 19N 34 4 E
Eleanora, Pk., *Australia* **61 F3** 32 57S 121 9 E
Elefantes →, *Mozam.* **57 C5** 24 10S 32 40 E
Elektrostal, *Russia* **24 C6** 55 41N 38 32 E
Elephant Butte Reservoir, *U.S.A.* **83 K10** 33 9N 107 11W
Elephant I., *Antarctica* **5 C18** 61 0S 55 0W
Eleuthera, *Bahamas* **88 B4** 25 0N 76 20W
Elgin, *Canada* **79 B8** 44 36N 76 13W
Elgin, *U.K.* **12 D5** 57 39N 3 19W
Elgin, *Ill., U.S.A.* **76 D1** 42 2N 88 17W
Elgin, *N. Dak., U.S.A.* **80 B4** 46 24N 101 51W
Elgin, *Oreg., U.S.A.* **82 D5** 45 34N 117 55W
Elgin, *Tex., U.S.A.* **81 K6** 30 21N 97 22W
Elgon, Mt., *Africa* **54 B3** 1 10N 34 30 E
Eliase, *Indonesia* **37 F8** 8 21S 130 48 E
Elim, *Namibia* **56 B2** 17 48S 15 31 E
Elim, *S. Africa* **56 E2** 34 35S 19 45 E
Elista, *Russia* **25 E7** 46 16N 44 14 E
Elizabeth, *Australia* **63 E2** 34 42S 138 41 E
Elizabeth, *N.J., U.S.A.* **79 F10** 40 39N 74 13W
Elizabeth, *N.J., U.S.A.* **79 F10** 40 40N 74 13W
Elizabeth City, *U.S.A.* **77 G7** 36 18N 76 14W
Elizabethton, *U.S.A.* **77 G4** 36 21N 82 13W
Elizabethtown, *Ky., U.S.A.* **76 G3** 37 42N 85 52W
Elizabethtown, *N.Y., U.S.A.* **79 B11** 44 13N 73 36W
Elizabethtown, *Pa., U.S.A.* **79 F8** 40 9N 76 36W
Ełk, *Poland* **17 B12** 53 50N 22 21 E
Elk →, *Canada* **72 C5** 49 11N 115 14W
Elk →, *U.S.A.* **77 H2** 34 46N 87 16W
Elk City, *U.S.A.* **81 H5** 35 25N 99 25W
Elk Creek, *U.S.A.* **84 F4** 39 36N 122 32W
Elk Grove, *U.S.A.* **84 G5** 38 25N 121 22W
Elk Island Nat. Park, *Canada* **72 C6** 53 35N 112 59W
Elk Lake, *Canada* **70 C3** 47 40N 80 25W
Elk Point, *Canada* **73 C6** 53 54N 110 55W
Elk River, *Idaho, U.S.A.* **82 C5** 46 47N 116 11W
Elk River, *Minn., U.S.A.* **80 C8** 45 18N 93 35W
Elkedra →, *Australia* **62 C2** 21 8S 136 22 E
Elkhart, *Ind., U.S.A.* **76 E3** 41 41N 85 58W
Elkhart, *Kans., U.S.A.* **81 G4** 37 0N 101 54W
Elkhorn, *Canada* **73 D8** 49 59N 101 14W
Elkhorn →, *U.S.A.* **80 E6** 41 8N 96 19W
Elkhovo, *Bulgaria* **21 C12** 42 10N 26 35 E
Elkin, *U.S.A.* **77 G5** 36 15N 80 51W
Elkins, *U.S.A.* **76 F6** 38 55N 79 51W
Elkland, *U.S.A.* **78 E7** 41 59N 77 19W
Elko, *Canada* **72 D5** 49 20N 115 10W
Elko, *U.S.A.* **82 F6** 40 50N 115 46W
Elkton, *U.S.A.* **78 C1** 43 49N 83 11W
Ell, L., *Australia* **61 E4** 29 13S 127 46 E
Ellef Ringnes I., *Canada* **4 B2** 78 30N 102 2W
Ellen, Mt., *U.S.A.* **79 B12** 44 9N 72 56W
Ellenburg, *U.S.A.* **79 B11** 44 54N 73 48W
Ellendale, *U.S.A.* **80 B5** 46 0N 98 32W
Ellensburg, *U.S.A.* **82 C3** 46 59N 120 34W
Ellenville, *U.S.A.* **79 E10** 41 43N 74 24W
Ellery, Mt., *Australia* **63 F4** 37 28S 148 47 E
Ellesmere, L., *N.Z.* **59 M4** 43 47S 172 28 E
Ellesmere I., *Canada* **4 B4** 79 30N 80 0W
Ellesmere Port, *U.K.* **10 D5** 53 17N 2 54W
Ellice Is. = Tuvalu ■, *Pac. Oc.* **64 H9** 8 0S 178 0 E
Ellicottville, *U.S.A.* **78 D6** 42 17N 78 40W
Elliot, *Australia* **62 B1** 17 33S 133 32 E
Elliot, *S. Africa* **57 E4** 31 22S 27 48 E
Elliot Lake, *Canada* **70 C3** 46 25N 82 35W
Elliotdale = Xhora, *S. Africa* **57 E4** 31 55S 28 38 E
Ellis, *U.S.A.* **80 F5** 38 56N 99 34W
Elliston, *Australia* **63 E1** 33 39S 134 53 E
Ellisville, *U.S.A.* **81 K10** 31 36N 89 12W
Ellon, *U.K.* **12 D6** 57 22N 2 4W
Ellore = Eluru, *India* **41 L12** 16 48N 81 8 E
Ellsworth, *Kans., U.S.A.* **80 F5** 38 44N 98 14W
Ellsworth, *Maine, U.S.A.* **77 C11** 44 33N 68 25W
Ellsworth Land, *Antarctica* **5 D16** 76 0S 89 0W
Ellsworth Mts., *Antarctica* **5 D16** 78 30S 85 0W
Ellwood City, *U.S.A.* **78 F4** 40 52N 80 17W
Elma, *Canada* **73 D9** 49 52N 95 55W
Elma, *U.S.A.* **84 D3** 47 0N 123 25W
Elmalı, *Turkey* **25 G4** 36 44N 29 56 E
Elmhurst, *U.S.A.* **76 E2** 41 53N 87 56W
Elmira, *Canada* **78 C4** 43 36N 80 33W
Elmira, *U.S.A.* **78 D8** 42 6N 76 48W
Elmira Heights, *U.S.A.* **78 D8** 42 8N 76 50W
Elmore, *Australia* **63 F3** 36 30S 144 37 E
Elmore, *U.S.A.* **85 M11** 33 7N 115 49W
Elmshorn, *Germany* **16 B5** 53 43N 9 40 E
Elmvale, *Canada* **78 B5** 44 35N 79 52W
Elora, *Canada* **78 C4** 43 41N 80 26W
Eloúnda, *Greece* **23 D7** 35 16N 25 42 E
Eloy, *U.S.A.* **83 K8** 32 45N 111 33W
Elrose, *Canada* **73 C7** 51 12N 108 0W
Elsie, *U.S.A.* **84 E3** 45 52N 123 36W
Elsinore = Helsingør, *Denmark* **9 H15** 56 2N 12 35 E
Eltham, *N.Z.* **59 H5** 39 26S 174 19 E
Eluru, *India* **41 L12** 16 48N 81 8 E
Elvas, *Portugal* **19 C2** 38 50N 7 10W
Elverum, *Norway* **9 F14** 60 53N 11 34 E
Elvire →, *Australia* **60 C4** 17 51S 128 11 E
Elvire, Mt., *Australia* **61 E2** 29 22S 119 36 E
Elwell, L., *U.S.A.* **82 B8** 48 22N 111 17W
Elwood, *Ind., U.S.A.* **76 E3** 40 17N 85 50W
Elwood, *Nebr., U.S.A.* **80 E5** 40 36N 99 52W
Elx = Elche, *Spain* **19 C5** 38 15N 0 42W
Ely, *U.K.* **11 E8** 52 24N 0 16 E
Ely, *Minn., U.S.A.* **80 B9** 47 55N 91 51W
Ely, *Nev., U.S.A.* **82 G6** 39 15N 114 54W
Elyria, *U.S.A.* **78 E2** 41 22N 82 7W
Emāmrūd, *Iran* **45 B7** 36 30N 55 0 E
Emba, *Kazakstan* **26 E6** 48 50N 58 8 E
Emba →, *Kazakstan* **25 E9** 46 55N 53 28 E
Embarcación, *Argentina* **94 A3** 23 10S 64 0W
Embarras Portage, *Canada* **73 B6** 58 27N 111 28W
Embetsu, *Japan* **30 B10** 44 44N 141 47 E
Embi = Emba, *Kazakstan* **26 E6** 48 50N 58 8 E
Embi = Emba →, *Kazakstan* **25 E9** 46 55N 53 28 E
Embóna, *Greece* **23 C9** 36 13N 27 51 E
Embrun, *France* **18 D7** 44 34N 6 30 E
Embu, *Kenya* **54 C4** 0 32S 37 38 E
Emden, *Germany* **16 B4** 53 21N 7 12 E
Emerald, *Australia* **62 C4** 23 32S 148 10 E
Emerson, *Canada* **73 D9** 49 0N 97 10W
Emet, *Turkey* **21 E13** 39 20N 29 15 E
Emi Koussi, *Chad* **51 E9** 19 45N 18 55 E
Eminabad, *Pakistan* **42 C6** 32 2N 74 8 E
Emine, Nos, *Bulgaria* **21 C12** 42 40N 27 56 E
Emissi, Tarso, *Chad* **51 D9** 21 27N 18 36 E
Emlenton, *U.S.A.* **78 E5** 41 11N 79 43W
Emmaus, *S. Africa* **56 D4** 29 2S 25 15 E
Emmaus, *U.S.A.* **79 F9** 40 32N 75 30W
Emmeloord, *Neths.* **15 B5** 52 44N 5 46 E
Emmen, *Neths.* **15 B6** 52 48N 6 57 E
Emmet, *Australia* **62 C3** 24 45S 144 30 E
Emmetsburg, *U.S.A.* **80 D7** 43 7N 94 41W
Emmett, *Idaho, U.S.A.* **82 E5** 43 52N 116 30W
Emmett, *Mich., U.S.A.* **78 D2** 42 59N 82 46W
Emmonak, *U.S.A.* **68 B3** 62 46N 164 30W
Emo, *Canada* **73 D10** 48 38N 93 50W
Empalme, *Mexico* **86 B2** 28 1N 110 49W
Empangeni, *S. Africa* **57 D5** 28 50S 31 52 E
Empedrado, *Argentina* **94 B4** 28 0S 58 46W
Emperor Seamount Chain, *Pac. Oc.* **64 D9** 40 0N 170 0 E
Emporia, *Kans., U.S.A.* **80 F6** 38 25N 96 11W
Emporia, *Va., U.S.A.* **77 G7** 36 42N 77 32W
Emporium, *U.S.A.* **78 E6** 41 31N 78 14W
Empress, *Canada* **73 C7** 50 57N 110 0W
Empty Quarter = Rub' al Khālī, *Si. Arabia* **46 D4** 19 0N 48 0 E
Ems →, *Germany* **16 B4** 53 20N 7 12 E
Emsdale, *Canada* **78 A5** 45 32N 79 19W
Emu, *China* **35 C15** 43 40N 128 6 E
Emu Park, *Australia* **62 C5** 23 13S 150 50 E
'En 'Avrona, *Israel* **47 F4** 29 43N 35 0 E
En Nahud, *Sudan* **51 F11** 12 45N 28 25 E
Ena, *Japan* **31 G8** 35 25N 137 25 E
Enana, *Namibia* **56 B2** 17 30S 16 23 E
Enard B., *U.K.* **12 C3** 58 5N 5 20W
Enare = Inarijärvi, *Finland* **8 B22** 69 0N 28 0 E
Enarotali, *Indonesia* **37 E9** 3 55S 136 21 E
Encampment, *U.S.A.* **82 F10** 41 12N 106 47W
Encantadas, Serra, *Brazil* **95 C5** 30 40S 53 0W
Encarnación, *Paraguay* **95 B4** 27 15S 55 50W
Encarnación de Diaz, *Mexico* **86 C4** 21 30N 102 13W
Encinitas, *U.S.A.* **85 M9** 33 3N 117 17W
Encino, *U.S.A.* **83 J11** 34 39N 105 28W
Encounter B., *Australia* **63 F2** 35 45S 138 45 E
Endako, *Canada* **72 C3** 54 6N 125 2W
Ende, *Indonesia* **37 F6** 8 45S 121 40 E
Endeavour Str., *Australia* **62 A3** 10 45S 142 0 E
Enderbury I., *Kiribati* **64 H10** 3 8S 171 5W
Enderby, *Canada* **72 C5** 50 35N 119 10W
Enderby I., *Australia* **60 D2** 20 35S 116 30 E
Enderby Land, *Antarctica* **5 C5** 66 0S 53 0 E
Enderlin, *U.S.A.* **80 B6** 46 38N 97 36W
Endicott, *U.S.A.* **79 D8** 42 6N 76 4W
Endwell, *U.S.A.* **79 D8** 42 6N 76 2W
Endyalgout I., *Australia* **60 B5** 11 40S 132 35 E
Eneabba, *Australia* **61 E2** 29 49S 115 16 E
Enewetak Atoll, *Marshall Is.* **64 F8** 11 30N 162 15 E
Enez, *Turkey* **21 D12** 40 45N 26 5 E

Farmington

Enfield, Canada 71 D7 44 56N 63 32W
Enfield, Conn., U.S.A. 79 E12 41 58N 72 36W
Enfield, N.H., U.S.A. 79 C12 43 39N 72 9W
Engadin, Switz. 18 C9 46 45N 10 10 E
Engaño, C., Dom. Rep. 89 C6 18 30N 68 20W
Engaño, C., Phil. 37 A6 18 35N 122 23 E
Engaru, Japan 30 B11 44 3N 143 31 E
Engcobo, S. Africa 57 E4 31 37S 28 0 E
Engels, Russia 25 D8 51 28N 46 6 E
Engemann L., Canada 73 B7 58 0N 106 55W
Enggano, Indonesia 36 F2 5 20S 102 40 E
England, U.S.A. 81 H9 34 33N 91 58W
England □, U.K. 10 D7 53 0N 2 0W
Englee, Canada 71 B8 50 45N 56 5W
Englehart, Canada 70 C4 47 49N 79 52W
Englewood, U.S.A. 80 F2 39 39N 104 59W
English →, Canada 73 C10 50 35N 93 30W
English Bazar = Ingraj Bazar, India 43 G13 24 58N 88 10 E
English Channel, Europe 11 G6 50 0N 2 0W
English River, Canada 70 C1 49 14N 91 0W
Enid, U.S.A. 81 G6 36 24N 97 53W
Enkhuizen, Neths. 15 B5 52 42N 5 17 E
Enna, Italy 20 F6 37 34N 14 16 E
Ennadai, Canada 73 A8 61 8N 100 53W
Ennadai L., Canada 73 A8 61 0N 101 0W
Ennedi, Chad 51 E10 17 15N 22 0 E
Enngonia, Australia 63 D4 29 21S 145 50 E
Ennis, Ireland 13 D3 52 51N 8 59W
Ennis, Mont., U.S.A. 82 D8 45 21N 111 44W
Ennis, Tex., U.S.A. 81 J6 32 20N 96 38W
Enniscorthy, Ireland 13 D5 52 30N 6 34W
Enniskillen, U.K. 13 B4 54 21N 7 39W
Ennistimon, Ireland 13 D2 52 57N 9 17W
Enns →, Austria 16 D8 48 14N 14 32 E
Enontekiö, Finland 8 B20 68 23N 23 37 E
Enosburg Falls, U.S.A. 79 B12 44 55N 72 48W
Enriquillo, L., Dom. Rep. 89 C5 18 20N 72 5W
Enschede, Neths. 15 B6 52 13N 6 53 E
Ensenada, Argentina 94 C4 34 55S 57 55W
Ensenada, Mexico 86 A1 31 50N 116 50W
Ensenada de los Muertos, Mexico 86 C2 23 59N 109 50W
Ensiola, Pta. de n', Spain 22 B9 39 7N 2 55 E
Entebbe, Uganda 54 B3 0 4N 32 28 E
Enterprise, Canada 72 A5 60 47N 115 45W
Enterprise, Ala., U.S.A. 77 K3 31 19N 85 51W
Enterprise, Oreg., U.S.A. 82 D5 45 25N 117 17W
Entre Ríos, Bolivia 94 A3 21 30S 64 25W
Entre Ríos □, Argentina 94 C4 30 30S 58 30W
Entroncamento, Portugal 19 C1 39 28N 8 28W
Enugu, Nigeria 50 G7 6 30N 7 30 E
Enumclaw, U.S.A. 84 C5 47 12N 121 59W
Eólie, Ís., Italy 20 E6 38 30N 14 57 E
Epe, Neths. 15 B5 52 21N 5 59 E
Epernay, France 18 B5 49 3N 3 56 E
Ephesus, Turkey 21 F12 37 55N 27 22 E
Ephraim, U.S.A. 82 G8 39 22N 111 35W
Ephrata, Pa., U.S.A. 79 F8 40 11N 76 11W
Ephrata, Wash., U.S.A. 82 C4 47 19N 119 33W
Épinal, France 18 B7 48 10N 6 27 E
Episkopi, Cyprus 23 E11 34 40N 32 54 E
Episkopi, Greece 23 D6 35 20N 24 20 E
Episkopi Bay, Cyprus 23 E11 34 35N 32 50 E
Epsom, U.K. 11 F7 51 19N 0 16W
Epukiro, Namibia 56 C2 21 40S 19 9 E
Equatorial Guinea ■, Africa 52 D1 2 0N 8 0 E
Er Rachidia, Morocco 50 B5 31 58N 4 20W
Er Rahad, Sudan 51 F12 12 45N 30 32 E
Er Rif, Morocco 50 A5 35 1N 4 1W
Erāwadī Myit = Irrawaddy →, Burma 41 M19 15 50N 95 6 E
Erāwadī Myitwanya = Irrawaddy, Mouths of the, Burma 41 M19 15 30N 95 0 E
Erbil = Arbīl, Iraq 44 B5 36 15N 44 5 E
Erçek, Turkey 44 B4 38 39N 43 36 E
Erciyaş Dağı, Turkey 25 G6 38 30N 35 30 E
Érd, Hungary 17 E10 47 22N 18 56 E
Erdao Jiang →, China 35 C14 43 0N 127 0 E
Erdek, Turkey 21 D12 40 23N 27 47 E
Erdene = Ulaan-Uul, Mongolia 34 B6 44 13N 111 10 E
Erdenetsogt, Mongolia 34 C4 42 55N 106 5 E
Erebus, Mt., Antarctica 5 D11 77 35S 167 0 E
Erechim, Brazil 95 B5 27 35S 52 15W
Ereğli, Konya, Turkey 25 G5 37 31N 34 4 E
Ereğli, Zonguldak, Turkey 25 F5 41 15N 31 24 E
Erenhot, China 34 C7 43 48N 112 2 E
Eresma →, Spain 19 B3 41 26N 4 45W
Erfenisdam, S. Africa 56 D4 28 30S 26 50 E
Erfurt, Germany 16 C6 50 58N 11 2 E
Erg Iguidi, Africa 50 C4 27 0N 7 0 E
Ergani, Turkey 44 B3 38 17N 39 49 E
Ergel, Mongolia 34 C5 43 8N 109 5 E
Ergeni Vozvyshennost, Russia 25 E7 47 0N 44 0 E
Ērgli, Latvia 9 H21 56 54N 25 38 E
Eriboll, L., U.K. 12 C4 58 30N 4 42W
Érice, Italy 20 E5 38 2N 12 35 E
Erie, U.S.A. 78 D4 42 8N 80 5W
Erie, L., N. Amer. 78 D4 42 15N 81 0W
Erie Canal, U.S.A. 78 C7 43 5N 78 43W
Erieau, Canada 78 D3 42 16N 81 57W
Erigavo, Somali Rep. 46 E4 10 35N 47 20 E
Erikoúsa, Greece 23 A3 39 53N 19 34 E
Eriksdale, Canada 73 C9 50 52N 98 7W
Erímanthos, Greece 21 F9 37 57N 21 50 E
Erimo-misaki, Japan 30 D11 41 50N 143 15 E
Erinpura, India 42 G5 25 9N 73 3 E
Eriskay, U.K. 12 D1 57 4N 7 18W
Eritrea ■, Africa 46 D2 14 0N 38 30 E
Erlangen, Germany 16 D6 49 36N 11 0 E
Erldunda, Australia 62 D1 25 14S 133 12 E
Ermelo, Neths. 15 B5 52 18N 5 35 E
Ermelo, S. Africa 57 D4 26 31S 29 59 E
Ermenek, Turkey 44 B2 36 38N 33 0 E
Ermones, Greece 23 A3 39 37N 19 46 E
Ermoúpolis = Síros, Greece 21 F11 37 28N 24 57 E
Erne →, Ireland 13 B3 54 30N 8 16W
Erne, Lower L., U.K. 13 B4 54 28N 7 47W
Erne, Upper L., U.K. 13 B4 54 14N 7 32W
Ernest Giles Ra., Australia 61 E3 27 0S 123 45 E
Erode, India 40 P10 11 24N 77 45 E
Eromanga, Australia 63 D3 26 40S 143 11 E
Erongo, Namibia 56 C2 21 39S 15 58 E
Erramala Hills, India 40 M11 15 30N 78 15 E
Errigal, Ireland 13 A3 55 2N 8 6W
Erris Hd., Ireland 13 B1 54 19N 10 0W
Erskine, U.S.A. 80 B7 47 40N 96 0W
Ertis = Irtysh →, Russia 26 C7 61 4N 68 52 E
Erwin, U.S.A. 77 G4 36 9N 82 25W
Erzgebirge, Germany 16 C7 50 27N 12 55 E
Erzin, Russia 27 D10 50 15N 95 10 E
Erzincan, Turkey 25 G6 39 46N 39 30 E
Erzurum, Turkey 25 G7 39 57N 41 15 E
Es Caló, Spain 22 C8 38 40N 1 30 E
Es Canar, Spain 22 B8 39 2N 1 36 E
Es Mercadal, Spain 22 B11 39 59N 4 5 E
Es Migjorn Gran, Spain 22 B11 39 57N 4 3 E
Es Sahrâ' Esh Sharqîya, Egypt 51 C12 27 30N 32 30 E
Es Sînâ', Egypt 47 F3 29 0N 34 0 E
Es Vedrà, Spain 22 C7 38 52N 1 12 E
Esambo, Dem. Rep. of the Congo 54 C1 3 48S 23 30 E
Esan-Misaki, Japan 30 D10 41 40N 141 10 E
Esashi, Hokkaidō, Japan 30 B11 44 56N 142 35 E
Esashi, Hokkaidō, Japan 30 D10 41 52N 140 7 E
Esbjerg, Denmark 9 J13 55 29N 8 29 E
Escalante, U.S.A. 83 H8 37 47N 111 36W
Escalante →, U.S.A. 83 H8 37 24N 110 57W
Escalón, Mexico 86 B4 26 46N 104 20W
Escambia →, U.S.A. 77 K2 30 32N 87 11W
Escanaba, U.S.A. 76 C2 45 45N 87 4W
Esch-sur-Alzette, Lux. 18 B6 49 32N 6 0 E
Escondido, U.S.A. 85 M9 33 7N 117 5W
Escuinapa, Mexico 86 C3 22 50N 105 50W
Escuintla, Guatemala 88 D1 14 20N 90 48W
Esenguly, Turkmenistan 26 F6 37 37N 53 59 E
Eşfahān, Iran 45 C6 32 39N 51 43 E
Eşfahān □, Iran 45 C6 32 50N 51 50 E
Esfarāyen, Iran 45 B8 37 4N 57 30 E
Esfideh, Iran 45 C8 33 39N 59 46 E
Esh Sham = Dimashq, Syria 47 B5 33 30N 36 18 E
Esha Ness, U.K. 12 A7 60 29N 1 38W
Esher, U.K. 11 F7 51 21N 0 20W
Eshowe, S. Africa 57 D5 28 50S 31 30 E
Esigodini, Zimbabwe 57 C4 20 18S 28 56 E
Esil = Ishim →, Russia 26 D8 57 45N 71 10 E
Esira, Madag. 57 C8 24 20S 46 42 E
Esk →, Cumb., U.K. 12 G5 54 58N 3 2W
Esk →, N. Yorks., U.K. 10 C7 54 30N 0 37W
Eskān, Iran 45 E9 26 48N 63 9 E
Esker, Canada 71 B6 53 53N 66 25W
Eskifjörður, Iceland 8 D7 65 3N 13 55W
Eskilstuna, Sweden 9 G17 59 22N 16 32 E
Eskimo Pt., Canada 68 B10 61 10N 94 15W
Eskişehir, Turkey 25 G5 39 50N 30 30 E
Esla →, Spain 19 B2 41 29N 6 3W
Eslāmābād-e Gharb, Iran 44 C5 34 10N 46 30 E
Eslāmshahr, Iran 45 C6 35 40N 51 10 E
Eşme, Turkey 21 E13 38 23N 28 58 E
Esmeraldas, Ecuador 92 C3 1 0N 79 40W
Esnagi L., Canada 70 C3 48 36N 84 33W
Espanola, Canada 70 C3 46 15N 81 46W
Espanola, U.S.A. 83 H10 35 59N 106 5W
Esparta, Costa Rica 88 E3 9 59N 84 40W
Esperance, Australia 61 F3 33 45S 121 55 E
Esperance B., Australia 61 F3 33 48S 121 55 E
Esperanza, Argentina 94 C3 31 29S 61 3W
Espichel, C., Portugal 19 C1 38 22N 9 16W
Espigão, Serra do, Brazil 95 B5 26 35S 50 30W
Espinazo, Sierra del = Espinhaço, Serra do, Brazil 93 G10 17 30S 43 30W
Espinhaço, Serra do, Brazil 93 G10 17 30S 43 30W
Espinilho, Serra do, Brazil 95 B5 28 30S 55 0W
Espírito Santo □, Brazil 93 H10 20 0S 40 45W
Espíritu Santo, Vanuatu 64 J8 15 15S 166 50 E
Espíritu Santo, B. del, Mexico 87 D7 19 15N 87 0W
Espíritu Santo, I., Mexico 86 C2 24 30N 110 23W
Espita, Mexico 87 C7 21 1N 88 19W
Espoo, Finland 9 F21 60 12N 24 40 E
Espungabera, Mozam. 57 C5 20 29S 32 45 E
Esquel, Argentina 96 E2 42 55S 71 20W
Esquimalt, Canada 72 D4 48 26N 123 25W
Esquina, Argentina 94 C4 30 0S 59 30W
Essaouira, Morocco 50 B4 31 32N 9 42W
Essebie, Dem. Rep. of the Congo 54 B3 2 58N 30 40 E
Essen, Belgium 15 C4 51 28N 4 28 E
Essen, Germany 16 C4 51 28N 7 2 E
Essendon, Mt., Australia 61 E3 25 0S 120 29 E
Essequibo →, Guyana 92 B7 6 50N 58 30W
Essex, Canada 78 D2 42 10N 82 49W
Essex, Calif., U.S.A. 85 L11 34 44N 115 15W
Essex, N.Y., U.S.A. 79 B11 44 19N 73 21W
Essex □, U.K. 11 F8 51 54N 0 27 E
Essex Junction, U.S.A. 79 B11 44 29N 73 7W
Esslingen, Germany 16 D5 48 44N 9 18 E
Estados, I. de Los, Argentina 96 G4 54 40S 64 30W
Eşţahbānāt, Iran 45 D7 29 8N 54 4 E
Estância, Brazil 93 F11 11 16S 37 26W
Estancia, U.S.A. 83 J10 34 46N 106 4W
Estārm, Iran 45 D8 28 21N 58 21 E
Estcourt, S. Africa 57 D4 29 0S 29 53 E
Estelí, Nic. 88 D2 13 9N 86 22W
Estellencs, Spain 22 B9 39 39N 2 29 E
Esterhazy, Canada 73 C8 50 37N 102 5W
Estevan, Canada 73 D8 49 10N 102 59W
Estevan Group, Canada 72 C3 53 3N 129 38W
Estherville, U.S.A. 80 D7 43 24N 94 50W
Eston, Canada 73 C7 51 8N 108 40W
Estonia ■, Europe 9 G21 58 30N 25 30 E
Estreito, Brazil 93 E9 6 32S 47 25W
Estrela, Serra da, Portugal 19 B2 40 10N 7 45W
Estremoz, Portugal 19 C2 38 51N 7 39W
Estrondo, Serra do, Brazil 93 E9 7 20S 48 0W
Esztergom, Hungary 17 E10 47 47N 18 44 E
Etah, India 43 F8 27 35N 78 40 E
Étampes, France 18 B5 48 26N 2 10 E
Etanga, Namibia 56 B1 17 55S 13 0 E
Etawah, India 43 F8 26 48N 79 6 E
Etawney L., Canada 73 B9 57 50N 96 50W
Ethel, U.S.A. 84 D4 46 32N 122 46W
Ethelbert, Canada 73 C8 51 32N 100 25W
Ethiopia ■, Africa 46 F3 8 0N 40 0 E
Ethiopian Highlands, Ethiopia 28 J7 10 0N 37 0 E
Etive, L., U.K. 12 E3 56 29N 5 10W
Etna, Italy 20 F6 37 50N 14 55 E
Etoile, Dem. Rep. of the Congo 55 E2 11 33S 27 30 E
Etosha Pan, Namibia 56 B2 18 40S 16 30 E
Etowah, U.S.A. 77 H3 35 20N 84 32W
Ettelbruck, Lux. 15 E6 49 51N 6 5 E
Ettrick Water →, U.K. 12 F6 55 31N 2 55W
Etuku, Dem. Rep. of the Congo 54 C2 3 42S 25 45 E
Etzatlán, Mexico 86 C4 20 48N 104 5W
Etzná, Mexico 87 D6 19 35N 90 15W
Euboea = Évvoia, Greece 21 E11 38 30N 24 0 E
Eucla, Australia 61 F4 31 41S 128 52 E
Euclid, U.S.A. 78 E3 41 34N 81 32W
Eucumbene, L., Australia 63 F4 36 2S 148 40 E
Eudora, U.S.A. 81 J9 33 7N 91 16W
Eufaula, Ala., U.S.A. 77 K3 31 54N 85 9W
Eufaula, Okla., U.S.A. 81 H7 35 17N 95 35W
Eufaula L., U.S.A. 81 H7 35 18N 95 21W
Eugene, U.S.A. 82 E2 44 5N 123 4W
Eugowra, Australia 63 E4 33 22S 148 24 E
Eulo, Australia 63 D4 28 10S 145 3 E
Eunice, La., U.S.A. 81 K8 30 30N 92 25W
Eunice, N. Mex., U.S.A. 81 J3 32 26N 103 10W
Eupen, Belgium 15 D6 50 37N 6 3 E
Euphrates = Furāt, Nahr al →, Asia 44 D5 31 0N 47 25 E
Eureka, Canada 4 B3 80 0N 85 56W
Eureka, Calif., U.S.A. 82 F1 40 47N 124 9W
Eureka, Kans., U.S.A. 81 G6 37 49N 96 17W
Eureka, Mont., U.S.A. 82 B6 48 53N 115 3W
Eureka, Nev., U.S.A. 82 G5 39 31N 115 58W
Eureka, S. Dak., U.S.A. 80 C5 45 46N 99 38W
Eureka, Mt., Australia 61 E3 26 35S 121 35 E
Euroa, Australia 63 F4 36 44S 145 35 E
Europa, Île, Ind. Oc. 53 J8 22 20S 40 22 E
Europa, Picos de, Spain 19 A3 43 10N 4 49W
Europa, Pta. de, Gib. 19 D3 36 3N 5 21W
Europe 6 E10 50 0N 20 0 E
Europoort, Neths. 15 C4 51 57N 4 10 E
Eustis, U.S.A. 77 L5 28 51N 81 41W
Eutsuk L., Canada 72 C3 53 20N 126 45W
Evale, Angola 56 B2 16 33S 15 44 E
Evans, U.S.A. 80 E2 40 23N 104 41W
Evans, L., Canada 70 B4 50 50N 77 0W
Evans City, U.S.A. 78 F4 40 46N 80 4W
Evans Head, Australia 63 D5 29 7S 153 27 E
Evans Mills, U.S.A. 79 B9 44 6N 75 48W
Evansburg, Canada 72 C5 53 36N 114 59W
Evanston, Ill., U.S.A. 76 E2 42 3N 87 41W
Evanston, Wyo., U.S.A. 82 F8 41 16N 110 58W
Evansville, U.S.A. 76 G2 37 58N 87 35W
Evaz, Iran 45 E7 27 46N 53 59 E
Eveleth, U.S.A. 80 B8 47 28N 92 32W
Evensk, Russia 27 C16 62 12N 159 30 E
Everard, L., Australia 63 E2 31 30S 135 0 E
Everard Ranges, Australia 61 E5 27 5S 132 28 E
Everest, Mt., Nepal 43 E12 28 5N 86 58 E
Everett, Pa., U.S.A. 78 F6 40 1N 78 23W
Everett, Wash., U.S.A. 84 C4 47 59N 122 12W
Everglades, The, U.S.A. 77 N5 25 50N 81 0W
Everglades City, U.S.A. 77 N5 25 52N 81 23W
Everglades Nat. Park, U.S.A. 77 N5 25 30N 81 0W
Evergreen, Ala., U.S.A. 77 K2 31 26N 86 57W
Evergreen, Mont., U.S.A. 82 B6 48 9N 114 13W
Evesham, U.K. 11 E6 52 6N 1 56W
Evje, Norway 9 G12 58 36N 7 51 E
Évora, Portugal 19 C2 38 33N 7 57W
Evowghlī, Iran 44 B5 38 43N 45 13 E
Évreux, France 18 B4 49 3N 1 8 E
Évros →, Greece 21 D12 41 40N 26 34 E
Évry, France 18 B5 48 38N 2 27 E
Évvoia, Greece 21 E11 38 30N 24 0 E
Ewe, L., U.K. 12 D3 57 49N 5 38W
Ewing, U.S.A. 80 D5 42 16N 98 21W
Ewo, Congo 52 E2 0 48S 14 45 E
Exaltación, Bolivia 92 F5 13 10S 65 20W
Excelsior Springs, U.S.A. 80 F7 39 20N 94 13W
Exe →, U.K. 11 G4 50 41N 3 29W
Exeter, Canada 78 C3 43 21N 81 29W
Exeter, U.K. 11 G4 50 43N 3 31W
Exeter, Calif., U.S.A. 84 J7 36 18N 119 9W
Exeter, N.H., U.S.A. 79 D14 42 59N 70 57W
Exmoor, U.K. 11 F4 51 12N 3 45W
Exmouth, Australia 60 D1 21 54S 114 10 E
Exmouth, U.K. 11 G4 50 37N 3 25W
Exmouth G., Australia 60 D1 22 15S 114 15 E
Expedition Ra., Australia 62 C4 24 30S 149 12 E
Extremadura □, Spain 19 C2 39 30N 6 5W
Exuma Sound, Bahamas 88 B4 24 30N 76 20W
Eyasi, L., Tanzania 54 C4 3 30S 35 0 E
Eye Pen., U.K. 12 C2 58 13N 6 10W
Eyemouth, U.K. 12 F6 55 52N 2 5W
Eyjafjörður, Iceland 8 C4 66 15N 18 30W
Eyre (North), L., Australia 63 D2 28 30S 137 20 E
Eyre (South), L., Australia 63 D2 29 18S 137 25 E
Eyre Mts., N.Z. 59 L2 45 25S 168 25 E
Eyre Pen., Australia 63 E2 33 30S 136 17 E
Eysturoy, Færoe Is. 8 E9 62 13N 6 54W
Eyvānkī, Iran 45 C6 35 24N 51 56 E
Ezine, Turkey 21 E12 39 48N 26 20 E
Ezouza →, Cyprus 23 E11 34 44N 32 27 E

F

F.Y.R.O.M. = Macedonia ■, Europe 21 D9 41 53N 21 40 E
Fabala, Guinea 50 G4 9 44N 9 5W
Fabens, U.S.A. 83 L10 31 30N 106 10W
Fabriano, Italy 20 C5 43 20N 12 54 E
Fachi, Niger 51 E8 18 6N 11 34 E
Fada, Chad 51 E10 17 13N 21 34 E
Fada-n-Gourma, Burkina Faso 50 F6 12 10N 0 30 E
Faddeyevskiy, Ostrov, Russia 27 B15 76 0N 144 0 E
Fadghāmī, Syria 44 C4 35 53N 40 52 E
Faenza, Italy 20 B4 44 17N 11 53 E
Færoe Is. = Føroyar, Atl. Oc. 8 F9 62 0N 7 0W
Făgăras, Romania 17 F13 45 48N 24 58 E
Fagersta, Sweden 9 F16 60 1N 15 46 E
Fagnano, L., Argentina 96 G3 54 30S 68 0W
Fahliān, Iran 45 D6 30 11N 51 28 E
Fahraj, Kermān, Iran 45 D8 29 0N 59 0 E
Fahraj, Yazd, Iran 45 D7 31 46N 54 36 E
Faial, Madeira 22 D3 32 47N 16 53W
Fair Haven, U.S.A. 76 D9 43 36N 73 16W
Fair Hd., U.K. 13 A5 55 14N 6 9W
Fair Oaks, U.S.A. 84 G5 38 39N 121 16W
Fairbanks, U.S.A. 68 B5 64 51N 147 43W
Fairbury, U.S.A. 80 E6 40 8N 97 11W
Fairfax, U.S.A. 79 B11 44 40N 73 1W
Fairfield, Ala., U.S.A. 77 J2 33 29N 86 55W
Fairfield, Calif., U.S.A. 84 G4 38 15N 122 3W
Fairfield, Conn., U.S.A. 79 E11 41 9N 73 16W
Fairfield, Idaho, U.S.A. 82 E6 43 21N 114 44W
Fairfield, Ill., U.S.A. 76 F1 38 23N 88 22W
Fairfield, Iowa, U.S.A. 80 E9 40 56N 91 57W
Fairfield, Tex., U.S.A. 81 K7 31 44N 96 10W
Fairford, Canada 73 C9 51 37N 98 38W
Fairhope, U.S.A. 77 K2 30 31N 87 54W
Fairlie, N.Z. 59 L3 44 5S 170 49 E
Fairmead, U.S.A. 84 H6 37 5N 120 10W
Fairmont, Minn., U.S.A. 80 D7 43 39N 94 28W
Fairmont, W. Va., U.S.A. 76 F5 39 29N 80 9W
Fairmount, Calif., U.S.A. 85 L8 34 45N 118 26W
Fairmount, N.Y., U.S.A. 79 C8 43 5N 76 12W
Fairplay, U.S.A. 83 G11 39 15N 106 2W
Fairport, U.S.A. 78 C7 43 6N 77 27W
Fairport Harbor, U.S.A. 78 E3 41 45N 81 17W
Fairview, Canada 72 B5 56 5N 118 25W
Fairview, Mont., U.S.A. 80 B2 47 51N 104 3W
Fairview, Okla., U.S.A. 81 G5 36 16N 98 29W
Fairweather, Mt., U.S.A. 72 B1 58 55N 137 32W
Faisalabad, Pakistan 42 D5 31 30N 73 5 E
Faith, U.S.A. 80 C3 45 2N 102 2W
Faizabad, India 43 F10 26 45N 82 10 E
Fajardo, Puerto Rico 89 C6 18 20N 65 39W
Fajr, W. →, Si. Arabia 44 D3 29 10N 38 10 E
Fakenham, U.K. 10 E8 52 51N 0 51 E
Fakfak, Indonesia 37 E8 2 55S 132 18 E
Faku, China 35 C12 42 32N 123 21 E
Falaise, France 18 B3 48 54N 0 12W
Falaise, Mui, Vietnam 38 C5 19 6N 105 45 E
Falam, Burma 41 H18 23 0N 93 45 E
Falcó, C. des, Spain 22 C7 38 50N 1 23 E
Falcón, Presa, Mexico 87 B5 26 35N 99 10W
Falcon Lake, Canada 73 D9 49 42N 95 15W
Falcon Reservoir, U.S.A. 81 M5 26 34N 99 10W
Falconara Marittima, Italy 20 C5 43 37N 13 24 E
Falcone, C. del, Italy 20 D3 40 58N 8 12 E
Falconer, U.S.A. 78 D5 42 7N 79 13W
Faleshty = Făleşti, Moldova 17 E14 47 32N 27 44 E
Făleşti, Moldova 17 E14 47 32N 27 44 E
Falfurrias, U.S.A. 81 M5 27 14N 98 9W
Falher, Canada 72 B5 55 44N 117 15W
Faliraki, Greece 23 C10 36 22N 28 12 E
Falkenberg, Sweden 9 H15 56 54N 12 30 E
Falkirk, U.K. 12 F5 56 0N 3 47W
Falkirk □, U.K. 12 F5 55 58N 3 49W
Falkland, U.K. 12 E5 56 16N 3 12W
Falkland Is. □, Atl. Oc. 96 G5 51 30S 59 0W
Falkland Sd., Falk. Is. 96 G5 52 0S 60 0W
Falköping, Sweden 9 G15 58 12N 13 33 E
Fall River, U.S.A. 79 E13 41 43N 71 10W
Fallbrook, U.S.A. 85 M9 33 23N 117 15W
Fallon, U.S.A. 82 G4 39 28N 118 47W
Falls City, U.S.A. 80 E7 40 3N 95 36W
Falls Creek, U.S.A. 78 E6 41 9N 78 48W
Falmouth, Jamaica 88 C4 18 30N 77 40W
Falmouth, U.K. 11 G2 50 9N 5 5W
Falmouth, U.S.A. 79 E14 41 33N 70 37W
Falsa, Pta., Mexico 86 B1 27 51N 115 3W
False B., S. Africa 56 E2 34 15S 18 40 E
Falso, C., Honduras 88 C3 15 12N 83 21W
Falster, Denmark 9 J14 54 45N 11 55 E
Falsterbo, Sweden 9 J15 55 23N 12 50 E
Fălticeni, Romania 17 E14 47 21N 26 20 E
Falun, Sweden 9 F16 60 37N 15 37 E
Famagusta, Cyprus 23 D12 35 8N 33 55 E
Famagusta Bay, Cyprus 23 D13 35 15N 34 0 E
Famalé, Niger 50 F6 14 33N 1 5 E
Famatina, Sierra de, Argentina 94 B2 27 30S 68 0W
Family L., Canada 73 C9 51 54N 95 27W
Famoso, U.S.A. 85 K7 35 37N 119 12W
Fan Xian, China 34 G8 35 55N 115 38 E
Fanad Hd., Ireland 13 A4 55 17N 7 38W
Fandriana, Madag. 57 C8 20 14S 47 21 E
Fang, Thailand 38 C2 19 55N 99 13 E
Fangcheng, China 34 H7 33 18N 112 59 E
Fangshan, China 34 E6 38 3N 111 25 E
Fangzi, China 35 F10 36 33N 119 10 E
Fanjakana, Madag. 57 C8 21 10S 46 53 E
Fanjiatun, China 35 C13 43 40N 125 15 E
Fannich, L., U.K. 12 D4 57 38N 4 59W
Fannūj, Iran 45 E8 26 35N 59 38 E
Fanø, Denmark 9 J13 55 25N 8 25 E
Fano, Italy 20 C5 43 50N 13 1 E
Fanshi, China 34 E7 39 12N 113 20 E
Fao = Al Fāw, Iraq 45 D6 30 0N 48 30 E
Faqirwali, Pakistan 42 E5 29 27N 73 0 E
Faradje, Dem. Rep. of the Congo 54 B2 3 50N 29 45 E
Farafangana, Madag. 57 C8 22 49S 47 50 E
Farāh, Afghan. 40 C3 32 20N 62 7 E
Farāh □, Afghan. 40 C3 32 25N 62 10 E
Farahalana, Madag. 57 A9 14 26S 50 10 E
Faranah, Guinea 50 F3 10 3N 10 45W
Farasān, Jazā'ir, Si. Arabia 46 D3 16 45N 41 55 E
Farasan Is. = Farasān, Jazā'ir, Si. Arabia 46 D3 16 45N 41 55 E
Faratsiho, Madag. 57 B8 19 24S 46 57 E
Fareham, U.K. 11 G6 50 51N 1 11W
Farewell, C., N.Z. 59 J4 40 29S 172 43 E
Farewell C. = Nunap Isua, Greenland 69 C15 59 48N 43 55W
Farghona, Uzbekistan 26 E8 40 23N 71 19 E
Fargo, U.S.A. 80 B6 46 53N 96 48W
Fār'iah, W. al →, West Bank 47 C4 32 12N 35 27 E
Faribault, U.S.A. 80 C8 44 18N 93 16W
Faridabad, India 42 E6 28 26N 77 19 E
Faridkot, India 42 D6 30 44N 74 45 E
Faridpur, Bangla. 43 H13 23 15N 89 55 E
Faridpur, India 43 E8 28 13N 79 33 E
Farimān, Iran 45 C8 35 40N 59 49 E
Farina, Australia 63 E2 30 3S 138 15 E
Fariones, Pta., Canary Is. 22 E6 29 13N 13 28W
Farmerville, U.S.A. 81 J8 32 47N 92 24W
Farmingdale, U.S.A. 79 F10 40 12N 74 10W
Farmington, Canada 72 B4 55 54N 120 30W
Farmington, Calif., U.S.A. 84 H6 37 55N 120 59W
Farmington, Maine, U.S.A. 77 C10 44 40N 70 9W

Farmington

Farmington, Mo., U.S.A. 81 G9 37 47N 90 25W
Farmington, N.H., U.S.A. 79 C13 43 24N 71 4W
Farmington, N. Mex., U.S.A. 83 H9 36 44N 108 12W
Farmington, Utah, U.S.A. 82 F8 41 0N 111 12W
Farmington →, U.S.A. 79 E12 41 51N 72 38W
Farmville, U.S.A. 76 G6 37 18N 78 24W
Farne Is., U.K. 10 B6 55 38N 1 37W
Farnham, Canada 79 A12 45 17N 72 59W
Farnham, Mt., Canada 72 C5 50 29N 116 30W
Faro, Brazil 93 D7 2 10S 56 39W
Faro, Canada 68 B6 62 11N 133 22W
Faro, Portugal 19 D2 37 2N 7 55W
Fårö, Sweden 9 H18 57 55N 19 5 E
Farquhar, C., Australia 61 D1 23 50S 113 36 E
Farrars Cr. →, Australia 62 D3 25 35S 140 43 E
Farrāshband, Iran 45 D7 28 57N 52 5 E
Farrell, U.S.A. 78 E4 41 13N 80 30W
Farrokhī, Iran 45 C8 33 50N 59 31 E
Farruch, C. = Ferrutx, C., Spain 22 B10 39 47N 3 21 E
Fārs □, Iran 45 D7 29 30N 55 0 E
Fársala, Greece 21 E10 39 17N 22 23 E
Farson, U.S.A. 82 E9 42 6N 109 27W
Farsund, Norway 9 G12 58 5N 6 55 E
Fartak, Râs, Si. Arabia 44 D2 28 5N 34 34 E
Fartak, Ra's, Yemen 46 D5 15 38N 52 15 E
Fartura, Serra da, Brazil 95 B5 26 21S 52 52W
Fārūj, Iran 45 B8 37 14N 58 14 E
Farvel, Kap = Nunap Isua, Greenland 69 C15 59 48N 43 55W
Farwell, U.S.A. 81 H3 34 23N 103 2W
Fāryāb □, Afghan. 40 B4 36 0N 65 0 E
Fasā, Iran 45 D7 29 0N 53 39 E
Fasano, Italy 20 D7 40 50N 17 22 E
Fastiv, Ukraine 17 C15 50 7N 29 57 E
Fastov = Fastiv, Ukraine 17 C15 50 7N 29 57 E
Fatagar, Tanjung, Indonesia 37 E8 2 46S 131 57 E
Fatehabad, Haryana, India 42 E6 29 31N 75 27 E
Fatehabad, Ut. P., India 42 F8 27 1N 78 19 E
Fatehgarh, India 43 F8 27 25N 79 35 E
Fatehpur, Bihar, India 43 G11 24 38N 85 14 E
Fatehpur, Raj., India 42 F6 28 0N 74 40 E
Fatehpur, Ut. P., India 43 G9 25 56N 81 13 E
Fatehpur, Ut. P., India 43 F9 27 10N 81 13 E
Fatehpur Sikri, India 42 F6 27 6N 77 40 E
Fatima, Canada 71 C7 47 24N 61 53W
Faulkton, U.S.A. 80 C5 45 2N 99 8W
Faure I., Australia 61 E1 25 52S 113 50 E
Fauresmith, S. Africa 56 D4 29 44S 25 17 E
Fauske, Norway 8 C16 67 17N 15 25 E
Favara, Italy 20 F5 37 19N 13 39 E
Favàritx, C. de, Spain 22 B11 40 0N 4 15 E
Favignana, Italy 20 F5 37 56N 12 20 E
Fawcett, Pt., Australia 60 B5 11 46S 130 2 E
Fawn →, Canada 70 A2 55 20N 87 35W
Fawnskin, U.S.A. 85 L10 34 16N 116 56W
Faxaflói, Iceland 8 D2 64 29N 23 0W
Faya-Largeau, Chad 51 E9 17 58N 19 6 E
Fayd, Si. Arabia 44 E4 27 1N 42 52 E
Fayette, Ala., U.S.A. 77 J2 33 41N 87 50W
Fayette, Mo., U.S.A. 80 F8 39 9N 92 41W
Fayetteville, Ark., U.S.A. 81 G7 36 4N 94 10W
Fayetteville, N.C., U.S.A. 77 H6 35 3N 78 53W
Fayetteville, Tenn., U.S.A. 77 H2 35 9N 86 34W
Fazilka, India 42 D6 30 27N 74 2 E
Fazilpur, Pakistan 42 E4 29 18N 70 29 E
Fdérik, Mauritania 50 D3 22 40N 12 45W
Feale →, Ireland 13 D2 52 27N 9 37W
Fear, C., U.S.A. 77 J7 33 50N 77 58W
Feather →, U.S.A. 82 G3 38 47N 121 36W
Feather Falls, U.S.A. 84 F5 39 36N 121 16W
Featherston, N.Z. 59 J5 41 6S 175 20 E
Featherstone, Zimbabwe 55 F3 18 42S 30 55 E
Fécamp, France 18 B4 49 45N 0 22 E
Fedala = Mohammedia, Morocco 50 B4 33 44N 7 21W
Federación, Argentina 94 C4 31 0S 57 55W
Féderal, Argentina 96 C5 30 57S 58 48W
Federal Way, U.S.A. 84 C4 47 18N 122 19W
Fedeshkūh, Iran 45 D7 28 49N 53 50 E
Fehmarn, Germany 16 A6 54 27N 11 7 E
Fehmarn Bælt, Europe 9 J14 54 35N 11 20 E
Fehmarn Belt = Fehmarn Bælt, Europe 9 J14 54 35N 11 20 E
Fei Xian, China 35 G9 35 18N 117 59 E
Feijó, Brazil 92 E4 8 9S 70 21W
Feilding, N.Z. 59 J5 40 13S 175 35 E
Feira de Santana, Brazil 93 F11 12 15S 38 57W
Feixiang, China 34 F8 36 30N 114 45 E
Felanitx, Spain 22 B10 39 28N 3 9 E
Feldkirch, Austria 16 E5 47 15N 9 37 E
Felipe Carrillo Puerto, Mexico 87 D7 19 38N 88 3W
Felixburg, Zimbabwe 57 B5 19 29S 30 51 E
Felixstowe, U.K. 11 F9 51 58N 1 23 E
Felton, U.S.A. 84 H4 37 3N 122 4W
Femer Bælt = Fehmarn Bælt, Europe 9 J14 54 35N 11 20 E
Femunden, Norway 9 E14 62 10N 11 53 E
Fen He →, China 34 G6 35 36N 110 42 E
Fenelon Falls, Canada 78 B6 44 32N 78 45W
Feng Xian, Jiangsu, China 34 G9 34 43N 116 35 E
Feng Xian, Shaanxi, China 34 H4 33 54N 106 40 E
Fengcheng, China 35 D13 40 28N 124 5 E
Fengfeng, China 34 F8 36 28N 114 8 E
Fengning, China 34 D9 41 10N 116 33 E
Fengqiu, China 34 G8 35 2N 114 25 E
Fengrun, China 35 E10 39 48N 118 8 E
Fengtai, China 34 E9 39 50N 116 18 E
Fengxiang, China 34 G4 34 29N 107 25 E
Fengyang, China 35 H9 32 51N 117 29 E
Fengzhen, China 34 D7 40 25N 113 2 E
Fenoarivo, Fianarantsoa, Madag. 57 C8 21 43S 46 24 E
Fenoarivo, Fianarantsoa, Madag. 57 C8 20 52S 46 53 E
Fenoarivo Afovoany, Madag. 57 B8 18 26S 46 34 E
Fenoarivo Atsinanana, Madag. 57 B8 17 22S 49 25 E
Fens, The, U.K. 10 E7 52 38N 0 2W
Fenton, U.S.A. 76 D4 42 48N 83 42W
Fenxi, China 34 F6 36 40N 111 31 E
Fenyang, China 34 F6 37 18N 111 48 E
Feodosiya, Ukraine 25 E6 45 2N 35 16 E
Ferdows, Iran 45 C8 33 58N 58 2 E
Ferfer, Somali Rep. 46 F4 5 4N 45 9 E
Fergana = Farghona, Uzbekistan 26 E8 40 23N 71 19 E
Fergus, Canada 78 C4 43 43N 80 24W
Fergus Falls, U.S.A. 80 B6 46 17N 96 4W
Ferkéssédougou, Ivory C. 50 G4 9 35N 5 6W
Ferland, Canada 70 B2 50 19N 88 27W
Fermanagh □, U.K. 13 B4 54 21N 7 40W
Fermo, Italy 20 C5 43 9N 13 43 E
Fermont, Canada 71 B6 52 47N 67 5W
Fermoy, Ireland 13 D3 52 9N 8 16W
Fernández, Argentina 94 B3 27 55S 63 50W
Fernandina Beach, U.S.A. 77 K5 30 40N 81 27W
Fernando de Noronha, Brazil 93 D12 4 0S 33 10W
Fernando Póo = Bioko, Eq. Guin. 52 D1 3 30N 8 40 E
Ferndale, U.S.A. 84 B4 48 51N 122 36W
Fernie, Canada 72 D5 49 30N 115 5W
Fernlees, Australia 62 C4 23 51S 148 7 E
Fernley, U.S.A. 82 G4 39 36N 119 15W
Ferozepore = Firozpur, India 42 D6 30 55N 74 40 E
Ferrara, Italy 20 B4 44 50N 11 35 E
Ferreñafe, Peru 92 E3 6 42S 79 50W
Ferrerias, Spain 22 B11 39 59N 4 1 E
Ferret, C., France 18 D3 44 38N 1 15W
Ferriday, U.S.A. 81 K9 31 38N 91 33W
Ferrol, Spain 19 A1 43 29N 8 15W
Ferron, U.S.A. 83 G8 39 5N 111 8W
Ferrutx, C., Spain 22 B10 39 47N 3 21 E
Ferryland, Canada 71 C9 47 2N 52 53W
Fertile, U.S.A. 80 B6 47 32N 96 17W
Fès, Morocco 50 B5 34 0N 5 0W
Fessenden, U.S.A. 80 B5 47 39N 99 38W
Festus, U.S.A. 80 F9 38 13N 90 24W
Feteşti, Romania 17 F14 44 22N 27 51 E
Fethiye, Turkey 25 G4 36 36N 29 6 E
Fetlar, U.K. 12 A8 60 36N 0 52W
Feuilles →, Canada 69 C12 58 47N 70 4W
Fez = Fès, Morocco 50 B5 34 0N 5 0W
Fezzan, Libya 51 C8 27 0N 13 0 E
Fiambalá, Argentina 94 B2 27 45S 67 37W
Fianarantsoa, Madag. 57 C8 21 26S 47 5 E
Fianarantsoa □, Madag. 57 B8 19 30S 47 0 E
Ficksburg, S. Africa 57 D4 28 51S 27 53 E
Field →, Australia 62 C2 23 48S 138 0 E
Field I., Australia 60 B5 12 5S 132 23 E
Fier, Albania 21 D8 40 43N 19 33 E
Fife □, U.K. 12 E5 56 16N 3 1W
Fife Ness, U.K. 12 E6 56 17N 2 35W
Fifth Cataract, Sudan 51 E12 18 22N 33 50 E
Figeac, France 18 D5 44 37N 2 2 E
Figtree, Zimbabwe 55 G2 20 22S 28 20 E
Figueira da Foz, Portugal 19 B1 40 7N 8 54W
Figueres, Spain 19 A7 42 18N 2 58 E
Figuig, Morocco 50 B5 32 5N 1 11W
Fihaonana, Madag. 57 B8 18 36S 47 12 E
Fiherenana, Madag. 57 B8 18 29S 48 24 E
Fiherenana →, Madag. 57 C7 23 19S 43 37 E
Fiji ■, Pac. Oc. 59 C8 17 20S 179 0 E
Filabusi, Zimbabwe 57 C4 20 34S 29 20 E
Filey, U.K. 10 C7 54 12N 0 18W
Filey B., U.K. 10 C7 54 12N 0 15W
Filfla, Malta 23 D1 35 47N 14 24 E
Filiatrá, Greece 21 F9 37 9N 21 35 E
Filingué, Niger 50 F6 14 21N 3 22 E
Filipstad, Sweden 9 G16 59 43N 14 9 E
Fillmore, Calif., U.S.A. 85 L8 34 24N 118 55W
Fillmore, Utah, U.S.A. 83 G7 38 58N 112 20W
Finch, Canada 79 A9 45 11N 75 7W
Findhorn →, U.K. 12 D5 57 38N 3 38W
Findlay, U.S.A. 76 E4 41 2N 83 39W
Finger L., Canada 70 B1 53 33N 93 30W
Finger Lakes, U.S.A. 79 D8 42 40N 76 30W
Fingoè, Mozam. 55 E3 14 55S 31 50 E
Finisterre, C. = Fisterra, C., Spain 19 A1 42 50N 9 19W
Finke, Australia 62 D1 25 34S 134 35 E
Finland ■, Europe 8 E22 63 0N 27 0 E
Finland, G. of, Europe 9 G21 60 0N 26 0 E
Finlay →, Canada 72 B3 57 0N 125 10W
Finley, Australia 63 F4 35 38S 145 35 E
Finley, U.S.A. 80 B6 47 31N 97 50W
Finn →, Ireland 13 B4 54 51N 7 28W
Finnigan, Mt., Australia 62 B4 15 49S 145 17 E
Finniss, C., Australia 63 E1 33 8S 134 51 E
Finnmark, Norway 8 B20 69 37N 23 57 E
Finnsnes, Norway 8 B18 69 14N 18 0 E
Finspång, Sweden 9 G16 58 43N 15 47 E
Fiora →, Italy 20 C4 42 20N 11 34 E
Fiq, Syria 47 C4 32 46N 35 41 E
Firat = Furāt, Nahr al →, Asia 44 D5 31 0N 47 25 E
Firebag →, Canada 73 B6 57 45N 111 21W
Firebaugh, U.S.A. 84 J6 36 52N 120 27W
Firedrake L., Canada 73 A8 61 25N 104 30W
Firenze, Italy 20 C4 43 46N 11 15 E
Firk →, Iraq 44 D5 30 59N 44 34 E
Firozabad, India 43 F8 27 10N 78 25 E
Firozpur, India 42 D6 30 55N 74 40 E
Firozpur-Jhirka, India 42 F7 27 48N 76 57 E
Fīrūzābād, Iran 45 D7 28 52N 52 35 E
Fīrūzkūh, Iran 45 C7 35 50N 52 50 E
Firvale, Canada 72 C3 52 27N 126 13W
Fish →, Namibia 56 D2 28 7S 17 10 E
Fish →, S. Africa 56 E3 31 30S 20 16 E
Fish River Canyon, Namibia 56 D2 27 40S 17 35 E
Fisher, Australia 61 F5 30 30S 131 0 E
Fisher B., Canada 73 C9 51 35N 97 13W
Fishers I., U.S.A. 79 E13 41 15N 72 0W
Fishguard, U.K. 11 E3 52 0N 4 58W
Fishing L., Canada 73 C9 52 10N 95 24W
Fishkill, U.S.A. 79 E11 41 32N 73 53W
Fisterra, C., Spain 19 A1 42 50N 9 19W
Fitchburg, U.S.A. 79 D13 42 35N 71 48W
Fitz Roy, Argentina 96 F3 47 0S 67 0W
Fitzgerald, Canada 72 B6 59 51N 111 36W
Fitzgerald, U.S.A. 77 K4 31 43N 83 15W
Fitzmaurice →, Australia 60 B5 14 45S 130 5 E
Fitzroy →, Queens., Australia 62 C5 23 32S 150 52 E
Fitzroy →, W. Austral., Australia 60 C3 17 31S 123 35 E
Fitzroy, Mte., Argentina 96 F2 49 17S 73 5W
Fitzroy Crossing, Australia 60 C4 18 9S 125 38 E
Fitzwilliam I., Canada 78 A3 45 30N 81 45W
Fiume = Rijeka, Croatia 16 F8 45 20N 14 21 E
Five Points, U.S.A. 84 J6 36 26N 120 6W
Fizi, Dem. Rep. of the Congo 54 C2 4 17S 28 55 E
Flagstaff, U.S.A. 83 J8 35 12N 111 39W
Flagstaff L., U.S.A. 77 C10 45 12N 70 18W
Flaherty I., Canada 70 A4 56 15N 79 15W
Flåm, Norway 9 F12 60 50N 7 7 E
Flambeau →, U.S.A. 80 C9 45 18N 91 14W
Flamborough Hd., U.K. 10 C7 54 7N 0 5W
Flaming Gorge Reservoir, U.S.A. 82 F9 41 10N 109 25W
Flamingo, Teluk, Indonesia 37 F9 5 30S 138 0 E
Flanders = Flandre, Europe 18 A5 50 50N 2 30 E
Flandre, Europe 18 A5 50 50N 2 30 E
Flandre Occidentale = West-Vlaanderen □, Belgium 15 D2 51 0N 3 0 E
Flandre Orientale = Oost-Vlaanderen □, Belgium 15 C3 51 5N 3 50 E
Flandreau, U.S.A. 80 C6 44 3N 96 36W
Flanigan, U.S.A. 84 E7 40 10N 119 53W
Flannan Is., U.K. 12 C1 58 9N 7 52W
Flåsjön, Sweden 8 D16 64 5N 15 40 E
Flat →, Canada 72 A3 61 33N 125 18W
Flathead L., U.S.A. 82 C7 47 51N 114 8W
Flattery, C., Australia 62 A4 14 58S 145 21 E
Flattery, C., U.S.A. 84 B2 48 23N 124 29W
Flatwoods, U.S.A. 76 F4 38 31N 82 43W
Fleetwood, U.K. 10 D4 53 55N 3 1W
Fleetwood, U.S.A. 79 F9 40 27N 75 49W
Flekkefjord, Norway 9 G12 58 18N 6 39 E
Flemington, U.S.A. 78 E7 41 7N 77 28W
Flensburg, Germany 16 A5 54 47N 9 27 E
Flers, France 18 B3 48 47N 0 33W
Flesherton, Canada 78 B4 44 16N 80 33W
Flesko, Tanjung, Indonesia 37 D6 0 29N 124 30 E
Fleurieu Pen., Australia 63 F2 35 40S 138 5 E
Flevoland □, Neths. 15 B5 52 30N 5 30 E
Flin Flon, Canada 73 C8 54 46N 101 53W
Flinders →, Australia 62 B3 17 36S 140 36 E
Flinders B., Australia 61 F2 34 19S 115 19 E
Flinders Group, Australia 62 A3 14 11S 144 15 E
Flinders I., S. Austral., Australia 63 E1 33 44S 134 41 E
Flinders I., Tas., Australia 62 G4 40 0S 148 0 E
Flinders Ranges, Australia 63 E2 31 30S 138 30 E
Flinders Reefs, Australia 62 B4 17 37S 148 31 E
Flint, U.K. 10 D4 53 15N 3 8W
Flint, U.S.A. 76 D4 43 1N 83 41W
Flint →, U.S.A. 77 K3 30 57N 84 34W
Flint I., Kiribati 65 J12 11 26S 151 48W
Flintshire □, U.K. 10 D4 53 17N 3 17W
Flodden, U.K. 10 B5 55 37N 2 8W
Floodwood, U.S.A. 80 B8 46 55N 92 55W
Flora, U.S.A. 76 F1 38 40N 88 29W
Florala, U.S.A. 77 K2 31 0N 86 20W
Florence = Firenze, Italy 20 C4 43 46N 11 15 E
Florence, Ala., U.S.A. 77 H2 34 48N 87 41W
Florence, Ariz., U.S.A. 83 K8 33 2N 111 23W
Florence, Colo., U.S.A. 80 F2 38 23N 105 8W
Florence, Oreg., U.S.A. 82 E1 43 58N 124 7W
Florence, S.C., U.S.A. 77 H6 34 12N 79 46W
Florence, L., Australia 63 D2 28 53S 138 9 E
Florencia, Colombia 92 C3 1 36N 75 36W
Florennes, Belgium 15 D4 50 15N 4 35 E
Florenville, Belgium 15 E5 49 40N 5 19 E
Flores, Guatemala 88 C2 16 59N 89 50W
Flores, Indonesia 37 F6 8 35S 121 0 E
Flores I., Canada 72 D3 49 20N 126 10W
Flores Sea, Indonesia 37 F6 6 30S 120 0 E
Floreşti, Moldova 17 E15 47 53N 28 17 E
Floresville, U.S.A. 81 L5 29 8N 98 10W
Floriano, Brazil 93 E10 6 50S 43 0W
Florianópolis, Brazil 95 B6 27 30S 48 30W
Florida, Cuba 88 B4 21 32N 78 14W
Florida, Uruguay 95 C4 34 7S 56 10W
Florida □, U.S.A. 77 L5 28 0N 82 0W
Florida, Straits of, U.S.A. 88 B4 25 0N 80 0W
Florida B., U.S.A. 88 B3 25 0N 80 45W
Florida Keys, U.S.A. 77 N5 24 40N 81 0W
Flórina, Greece 21 D9 40 48N 21 26 E
Florø, Norway 9 F11 61 35N 5 1 E
Flower Station, Canada 79 A8 45 10N 76 41W
Flowerpot I., Canada 78 A3 45 18N 81 38W
Floydada, U.S.A. 81 J4 33 59N 101 20W
Fluk, Indonesia 37 E7 1 42S 127 44 E
Flushing = Vlissingen, Neths. 15 C3 51 26N 3 34 E
Flying Fish, C., Antarctica 5 D15 72 6S 102 29W
Foam Lake, Canada 73 C8 51 40N 103 32W
Foça, Turkey 21 E12 38 39N 26 46 E
Focşani, Romania 17 F14 45 41N 27 15 E
Fóggia, Italy 20 D6 41 27N 15 34 E
Fogo, Canada 71 C9 49 43N 54 17W
Fogo I., Canada 71 C9 49 40N 54 5W
Föhr, Germany 16 A5 54 43N 8 30 E
Foix, France 18 E4 42 58N 1 38 E
Folda, Nord-Trøndelag, Norway 8 D14 64 32N 10 30 E
Folda, Nordland, Norway 8 C16 67 38N 14 50 E
Foley, Botswana 56 C4 21 34S 27 21 E
Foley, U.S.A. 77 K2 30 24N 87 41W
Foleyet, Canada 70 C3 48 15N 82 25W
Folgefonni, Norway 9 F12 60 3N 6 23 E
Foligno, Italy 20 C5 42 57N 12 42 E
Folkestone, U.K. 11 F9 51 5N 1 12 E
Folkston, U.S.A. 77 K5 30 50N 82 0W
Follansbee, U.S.A. 78 F4 40 19N 80 35W
Folsom L., U.S.A. 84 G5 38 42N 121 9W
Fond-du-Lac, Canada 73 B7 59 19N 107 12W
Fond du Lac, U.S.A. 80 D10 43 47N 88 27W
Fond-du-Lac →, Canada 73 B7 59 17N 106 0W
Fonda, U.S.A. 79 D10 42 57N 74 22W
Fondi, Italy 20 D5 41 21N 13 25 E
Fongafale, Tuvalu 64 H9 8 31S 179 13 E
Fonsagrada = A Fonsagrada, Spain 19 A2 43 8N 7 4W
Fonseca, G. de, Cent. Amer. 88 D2 13 10N 87 40W
Fontainebleau, France 18 B5 48 24N 2 40 E
Fontana, U.S.A. 85 L9 34 6N 117 26W
Fontas →, Canada 72 B4 58 14N 121 48W
Fonte Boa, Brazil 92 D5 2 33S 66 0W
Fontenay-le-Comte, France 18 C3 46 28N 0 48W
Fontenelle Reservoir, U.S.A. 82 E8 42 1N 110 3W
Fontur, Iceland 8 C6 66 23N 14 32W
Foochow = Fuzhou, China 33 D6 26 5N 119 16 E
Foping, China 34 H5 33 41N 108 0 E
Forbes, Australia 63 E4 33 22S 148 5 E
Forbesganj, India 43 F12 26 17N 87 18 E
Ford City, Calif., U.S.A. 85 K7 35 9N 119 27W
Ford City, Pa., U.S.A. 78 F5 40 46N 79 32W
Førde, Norway 9 F11 61 27N 5 53 E
Ford's Bridge, Australia 63 D4 29 41S 145 29 E
Fordyce, U.S.A. 81 J8 33 49N 92 25W
Forel, Mt., Greenland 4 C6 66 52N 36 55W
Foremost, Canada 72 D6 49 26N 111 34W
Forest, Canada 78 C3 43 6N 82 0W
Forest, U.S.A. 81 J10 32 22N 89 29W
Forest City, Iowa, U.S.A. 80 D8 43 16N 93 39W
Forest City, N.C., U.S.A. 77 H5 35 20N 81 52W
Forest City, Pa., U.S.A. 79 E9 41 39N 75 28W
Forest Grove, U.S.A. 84 E3 45 31N 123 7W
Forestburg, Canada 72 C6 52 35N 112 1W
Foresthill, U.S.A. 84 F6 39 1N 120 49W
Forestier Pen., Australia 62 G4 43 0S 148 0 E
Forestville, Canada 71 C6 48 48N 69 2W
Forestville, Calif., U.S.A. 84 G4 38 28N 122 54W
Forestville, N.Y., U.S.A. 78 D5 42 28N 79 10W
Forfar, U.K. 12 E6 56 39N 2 53W
Forks, U.S.A. 84 C2 47 57N 124 23W
Forksville, U.S.A. 79 E8 41 29N 76 35W
Forlì, Italy 20 B5 44 13N 12 3 E
Forman, U.S.A. 80 B6 46 7N 97 38W
Formby Pt., U.K. 10 D4 53 33N 3 6W
Formentera, Spain 22 C7 38 43N 1 27 E
Formentor, C. de, Spain 22 B10 39 58N 3 13 E
Former Yugoslav Republic of Macedonia = Macedonia ■, Europe 21 D9 41 53N 21 40 E
Fórmia, Italy 20 D5 41 15N 13 37 E
Formosa = Taiwan ■, Asia 33 D7 23 30N 121 0 E
Formosa, Argentina 94 B4 26 15S 58 10W
Formosa, Brazil 93 G9 15 32S 47 20W
Formosa □, Argentina 94 B4 25 0S 60 0W
Formosa, Serra, Brazil 93 F8 12 0S 55 0W
Formosa Bay, Kenya 54 C5 2 40S 40 20 E
Fornells, Spain 22 A11 40 3N 4 7 E
Føroyar, Atl. Oc. 8 F9 62 0N 7 0W
Forres, U.K. 12 D5 57 37N 3 37W
Forrest, Australia 61 F4 30 51S 128 6 E
Forrest, Mt., Australia 61 D4 24 48S 127 45 E
Forrest City, U.S.A. 81 H9 35 1N 90 47W
Forsayth, Australia 62 B3 18 33S 143 34 E
Forssa, Finland 9 F20 60 49N 23 38 E
Forst, Germany 16 C8 51 45N 14 37 E
Forsyth, U.S.A. 82 C10 46 16N 106 41W
Fort Abbas, Pakistan 42 E5 29 12N 72 52 E
Fort Albany, Canada 70 B3 52 15N 81 35W
Fort Ann, U.S.A. 79 C11 43 25N 73 30W
Fort Assiniboine, Canada 72 C6 54 20N 114 45W
Fort Augustus, U.K. 12 D4 57 9N 4 42W
Fort Beaufort, S. Africa 56 E4 32 46S 26 40 E
Fort Benton, U.S.A. 82 C8 47 49N 110 40W
Fort Bragg, U.S.A. 82 G2 39 26N 123 48W
Fort Bridger, U.S.A. 82 F8 41 19N 110 23W
Fort Chipewyan, Canada 73 B6 58 42N 111 8W
Fort Collins, U.S.A. 80 E2 40 35N 105 5W
Fort-Coulonge, Canada 70 C4 45 50N 76 45W
Fort Covington, U.S.A. 79 B10 44 59N 74 29W
Fort Davis, U.S.A. 81 K3 30 35N 103 54W
Fort-de-France, Martinique 89 D7 14 36N 61 2W
Fort Defiance, U.S.A. 83 J9 35 45N 109 5W
Fort Dodge, U.S.A. 80 D7 42 30N 94 11W
Fort Edward, U.S.A. 79 C11 43 16N 73 35W
Fort Erie, Canada 78 D6 42 54N 78 56W
Fort Fairfield, U.S.A. 77 B12 46 46N 67 50W
Fort Frances, Canada 73 D10 48 36N 93 24W
Fort Garland, U.S.A. 83 H11 37 26N 105 26W
Fort George = Chisasibi, Canada 70 B4 53 50N 79 0W
Fort Good-Hope, Canada 68 B7 66 14N 128 40W
Fort Hancock, U.S.A. 83 L11 31 18N 105 51W
Fort Hertz = Putao, Burma 41 F20 27 28N 97 30 E
Fort Hope, Canada 70 B2 51 30N 88 0W
Fort Irwin, U.S.A. 85 K10 35 16N 116 34W
Fort Kent, U.S.A. 77 B11 47 15N 68 36W
Fort Klamath, U.S.A. 82 E3 42 42N 122 0W
Fort Laramie, U.S.A. 80 D2 42 13N 104 31W
Fort Lauderdale, U.S.A. 77 M5 26 7N 80 8W
Fort Liard, Canada 72 A4 60 14N 123 30W
Fort Liberté, Haiti 89 C5 19 42N 71 51W
Fort Lupton, U.S.A. 80 E2 40 5N 104 49W
Fort Mackay, Canada 72 B6 57 12N 111 41W
Fort Macleod, Canada 72 D6 49 45N 113 30W
Fort McMurray, Canada 72 B6 56 44N 111 7W
Fort McPherson, Canada 68 B6 67 30N 134 55W
Fort Madison, U.S.A. 80 E9 40 38N 91 27W
Fort Meade, U.S.A. 77 M5 27 45N 81 48W
Fort Morgan, U.S.A. 80 E3 40 15N 103 48W
Fort Myers, U.S.A. 77 M5 26 39N 81 52W
Fort Nelson, Canada 72 B4 58 50N 122 44W
Fort Nelson →, Canada 72 B4 59 32N 124 0W
Fort Norman = Tulita, Canada 68 B7 64 57N 125 30W
Fort Payne, U.S.A. 77 H3 34 26N 85 43W
Fort Peck, U.S.A. 82 B10 48 1N 106 27W
Fort Peck Dam, U.S.A. 82 C10 48 0N 106 26W
Fort Peck L., U.S.A. 82 C10 48 0N 106 26W
Fort Pierce, U.S.A. 77 M5 27 27N 80 20W
Fort Pierre, U.S.A. 80 C4 44 21N 100 22W
Fort Plain, U.S.A. 79 D10 42 56N 74 37W
Fort Portal, Uganda 54 B3 0 40N 30 20 E
Fort Providence, Canada 72 A5 61 3N 117 40W
Fort Qu'Appelle, Canada 73 C8 50 45N 103 50W
Fort Resolution, Canada 72 A6 61 10N 113 40W
Fort Rixon, Zimbabwe 55 G2 20 2S 29 17 E
Fort Ross, U.S.A. 84 G3 38 32N 123 13W
Fort Rupert = Waskaganish, Canada 70 B4 51 30N 78 40W
Fort St. James, Canada 72 C4 54 30N 124 10W
Fort St. John, Canada 72 B4 56 15N 120 50W
Fort Saskatchewan, Canada 72 C6 53 40N 113 15W
Fort Scott, U.S.A. 81 G7 37 50N 94 42W
Fort Severn, Canada 70 A2 56 0N 87 40W
Fort Shevchenko, Kazakstan 25 F9 44 35N 50 23 E
Fort Simpson, Canada 72 A4 61 45N 121 15W
Fort Smith, Canada 72 B6 60 0N 111 51W
Fort Smith, U.S.A. 81 H7 35 23N 94 25W
Fort Stockton, U.S.A. 81 K3 30 53N 102 53W
Fort Sumner, U.S.A. 81 H2 34 28N 104 15W
Fort Thompson, U.S.A. 80 C5 44 3N 99 26W
Fort Valley, U.S.A. 77 J4 32 33N 83 53W
Fort Vermilion, Canada 72 B5 58 24N 116 0W
Fort Walton Beach, U.S.A. 77 K2 30 25N 86 36W
Fort Wayne, U.S.A. 76 E3 41 4N 85 9W
Fort William, U.K. 12 E3 56 49N 5 7W
Fort Worth, U.S.A. 81 J6 32 45N 97 18W
Fort Yates, U.S.A. 80 B4 46 5N 100 38W

G

Garden Grove

Garden Grove, *U.S.A.* **85 M9** 33 47N 117 55W
Gardēz, *Afghan.* . . . . **42 C3** 33 37N 69 9 E
Gardiner, *Maine, U.S.A.* **77 C11** 44 14N 69 47W
Gardiner, *Mont., U.S.A.* **82 D8** 45 2N 110 22W
Gardiners I., *U.S.A.* **79 E12** 41 6N 72 6W
Gardner, *U.S.A.* **79 D13** 42 34N 71 59W
Gardner Canal, *Canada* **72 C3** 53 27N 128 8W
Gardnerville, *U.S.A.* **84 G7** 38 56N 119 45W
Gardo, *Somali Rep.* . . . . **46 F4** 9 30N 49 6 E
Garey, *U.S.A.* **85 L6** 34 53N 120 19W
Garfield, *U.S.A.* **82 C5** 47 1N 117 9W
Garforth, *U.K.* **10 D6** 53 47N 1 24W
Gargano, Mte., *Italy* . . . . **20 D6** 41 43N 15 43 E
Garibaldi Prov. Park, *Canada* **72 D4** 49 50N 122 40W
Gariep, L., *S. Africa* . . . . **56 E4** 30 40S 25 40 E
Garies, *S. Africa* . . . . **56 E2** 30 32S 17 59 E
Garigliano →, *Italy* . . . . **20 D5** 41 13N 13 45 E
Garissa, *Kenya* . . . . **54 C4** 0 25S 39 40 E
Garland, *Tex., U.S.A.* **81 J6** 32 55N 96 38W
Garland, *Utah, U.S.A.* **82 F7** 41 47N 112 10W
Garm, *Tajikistan* . . . . **26 F8** 39 0N 70 20 E
Garmāb, *Iran* . . . . **45 C8** 35 25N 56 45 E
Garmisch-Partenkirchen, *Germany* . . . . **16 E6** 47 30N 11 6 E
Garmo, Qullai = Kommunizma, Pik, *Tajikistan* . . . . **26 F8** 39 0N 72 2 E
Garmsār, *Iran* . . . . **45 C7** 35 20N 52 25 E
Garner, *U.S.A.* **80 D8** 43 6N 93 36W
Garnett, *U.S.A.* **80 F7** 38 17N 95 14W
Garo Hills, *India* . . . . **43 G14** 25 30N 90 30 E
Garoe, *Somali Rep.* . . . . **46 F4** 8 25N 48 33 E
Garonne →, *France* **18 D3** 45 2N 0 36W
Garoowe = Garoe, *Somali Rep.* . . . . **46 F4** 8 25N 48 33 E
Garot, *India* . . . . **42 G6** 24 19N 75 41 E
Garoua, *Cameroon* . . . . **51 G8** 9 19N 13 21 E
Garrauli, *India* . . . . **43 G8** 25 5N 79 22 E
Garrison, *Mont., U.S.A.* **82 C7** 46 31N 112 49W
Garrison, *N. Dak., U.S.A.* . . . **80 B4** 47 40N 101 25W
Garrison Res. = Sakakawea, L., *U.S.A.* **80 B4** 47 30N 101 25W
Garron Pt., *U.K.* **13 A6** 55 3N 5 59W
Garry →, *U.K.* **12 E5** 56 44N 3 47W
Garry, L., *Canada* **68 B9** 65 58N 100 18W
Garsen, *Kenya* . . . . **54 C5** 2 20S 40 5 E
Garson L., *Canada* **73 B6** 56 19N 110 2W
Garu, *India* . . . . **43 H11** 23 40N 84 14 E
Garub, *Namibia* . . . . **56 D2** 26 37S 16 0 E
Garut, *Indonesia* . . . . **37 G12** 7 14S 107 53 E
Garvie Mts., *N.Z.* . . . . **59 L2** 45 30S 168 50 E
Garwa = Garoua, *Cameroon* . . **51 G8** 9 19N 13 21 E
Garwa, *India* . . . . **43 G10** 24 11N 83 47 E
Gary, *U.S.A.* **76 E2** 41 36N 87 20W
Garzê, *China* . . . . **32 C5** 31 38N 100 1 E
Garzón, *Colombia* **92 C3** 2 10N 75 40W
Gas-San, *Japan* . . . . **30 E10** 38 32N 140 1 E
Gasan Kuli = Esenguly, *Turkmenistan* . . . . **26 F6** 37 37N 53 59 E
Gascogne, *France* . . . . **18 E4** 43 45N 0 20 E
Gascogne, G. de, *Europe* . . . **18 D2** 44 0N 2 0W
Gascony = Gascogne, *France* **18 E4** 43 45N 0 20 E
Gascoyne →, *Australia* . . . . **61 D1** 24 52S 113 37 E
Gascoyne Junction, *Australia* **61 E2** 25 2S 115 17 E
Gashaka, *Nigeria* . . . . **51 G8** 7 20N 11 29 E
Gasherbrum, *Pakistan* . . . . . **43 B7** 35 40N 76 40 E
Gashua, *Nigeria* . . . . **51 F8** 12 54N 11 0 E
Gaspé, *Canada* **71 C7** 48 52N 64 30W
Gaspé, C. de, *Canada* **71 C7** 48 48N 64 7W
Gaspé, Pén. de, *Canada* **71 C6** 48 45N 65 40W
Gaspésie, Parc de Conservation de la, *Canada* **71 C6** 48 55N 65 50W
Gasteiz = Vitoria-Gasteiz, *Spain* **19 A4** 42 50N 2 41W
Gastonia, *U.S.A.* **77 H5** 35 16N 81 11W
Gastre, *Argentina* **96 E3** 42 20S 69 15W
Gata, C., *Cyprus* . . . . **23 E12** 34 34N 33 2 E
Gata, C. de, *Spain* **19 D4** 36 41N 2 13W
Gata, Sierra de, *Spain* **19 B2** 40 20N 6 45W
Gataga →, *Canada* **72 B3** 58 35N 126 59W
Gatehouse of Fleet, *U.K.* . . . **12 G4** 54 53N 4 12W
Gates, *U.S.A.* **78 C7** 43 9N 77 42W
Gateshead, *U.K.* **10 C6** 54 57N 1 35W
Gatesville, *U.S.A.* **81 K6** 31 26N 97 45W
Gaths, *Zimbabwe* . . . . **55 G3** 20 2S 30 32 E
Gatico, *Chile* **94 A1** 22 29S 70 20W
Gatineau, *Canada* **79 A9** 45 29N 75 38W
Gatineau →, *Canada* **70 C4** 45 27N 75 42W
Gatineau, Parc Nat. de la, *Canada* **70 C4** 45 40N 76 0W
Gatton, *Australia* . . . . **63 D5** 27 32S 152 17 E
Gatun, L., *Panama* **88 E4** 9 7N 79 56W
Gatyana, *S. Africa* . . . . **57 E4** 32 16S 28 31 E
Gau, *Fiji* . . . . **59 D8** 18 2S 179 18 E
Gauer L., *Canada* **73 B9** 57 0N 97 50W
Gauhati = Guwahati, *India* . . **41 F17** 26 10N 91 45 E
Gauja →, *Latvia* . . . . **9 H21** 57 10N 24 16 E
Gaula →, *Norway* . . . . **8 E14** 63 21N 10 14 E
Gauri Phanta, *India* . . . . **43 E9** 28 41N 80 36 E
Gausta, *Norway* . . . . **9 G13** 59 48N 8 40 E
Gauteng □, *S. Africa* . . . . **57 D4** 26 0S 28 0 E
Gāv Koshī, *Iran* . . . . **45 D8** 28 38N 57 12 E
Gāvakān, *Iran* . . . . **45 D7** 29 37N 53 10 E
Gavāter, *Iran* . . . . **45 E9** 25 10N 61 31 E
Gāvbandī, *Iran* . . . . **45 E7** 27 12N 53 4 E
Gavdhopoúla, *Greece* . . . . . . **23 E6** 34 56N 24 0 E
Gávdhos, *Greece* . . . . **23 E6** 34 50N 24 5 E
Gaviota, *U.S.A.* **85 L6** 34 29N 120 13W
Gāvkhūnī, Bāṭlāq-e, *Iran* . . . **45 C7** 32 6N 52 52 E
Gävle, *Sweden* . . . . **9 F17** 60 40N 17 9 E
Gawachab, *Namibia* . . . . **56 D2** 27 4S 17 55 E
Gawilgarh Hills, *India* . . . . . **40 J10** 21 15N 76 45 E
Gawler, *Australia* . . . . **63 E2** 34 30S 138 42 E
Gaxun Nur, *China* . . . . **32 B5** 42 22N 100 30 E
Gay, *Russia* . . . . **24 D10** 51 27N 58 27 E
Gaya, *India* . . . . **43 G11** 24 47N 85 4 E
Gaya, *Niger* . . . . **50 F6** 11 52N 3 28 E
Gaylord, *U.S.A.* **76 C3** 45 2N 84 41W
Gayndah, *Australia* . . . . **63 D5** 25 35S 151 32 E
Gaysin = Haysyn, *Ukraine* . . **17 D15** 48 57N 29 25 E
Gayvoron = Hayvoron, *Ukraine* . . . . **17 D15** 48 22N 29 52 E
Gaza, *Gaza Strip* . . . . **47 D3** 31 30N 34 28 E
Gaza □, *Mozam.* . . . . **57 C5** 23 10S 32 45 E
Gaza Strip □, *Asia* . . . . **47 D3** 31 29N 34 25 E
Gazanjyk, *Turkmenistan* . . . . **45 B7** 39 16N 55 32 E
Gāzbor, *Iran* . . . . **45 D8** 28 5N 58 51 E
Gazi, *Dem. Rep. of the Congo* **54 B1** 1 3N 24 30 E
Gaziantep, *Turkey* . . . . **25 G6** 37 6N 37 23 E
Gcoverega, *Botswana* . . . . . . **56 B3** 19 8S 24 18 E
Gcuwa, *S. Africa* . . . . **57 E4** 32 20S 28 11 E
Gdańsk, *Poland* . . . . **17 A10** 54 22N 18 40 E
Gdańska, Zatoka, *Poland* . . . **17 A10** 54 30N 19 20 E
Gdov, *Russia* . . . . **9 G22** 58 48N 27 55 E
Gdynia, *Poland* . . . . **17 A10** 54 35N 18 33 E
Gebe, *Indonesia* . . . . **37 D7** 0 5N 129 25 E
Gebze, *Turkey* . . . . **21 D13** 40 47N 29 25 E
Gedaref, *Sudan* . . . . **51 F13** 14 2N 35 28 E
Gediz →, *Turkey* . . . . **21 E12** 38 35N 26 48 E
Gedser, *Denmark* . . . . **9 J14** 54 35N 11 55 E
Geegully Cr. →, *Australia* . . **60 C3** 18 32S 123 41 E
Geel, *Belgium* . . . . **15 C4** 51 10N 4 59 E
Geelong, *Australia* . . . . **63 F3** 38 10S 144 22 E
Geelvink B. = Cenderwasih, Teluk, *Indonesia* . . . . **37 E9** 3 0S 135 20 E
Geelvink Chan., *Australia* . . . **61 E1** 28 30S 114 0 E
Geesthacht, *Germany* . . . . . . **16 B6** 53 26N 10 22 E
Geidam, *Nigeria* . . . . **51 F8** 12 57N 11 57 E
Geikie →, *Canada* **73 B8** 57 45N 103 52W
Geistown, *U.S.A.* **78 F6** 40 18N 78 52W
Geita, *Tanzania* . . . . **54 C3** 2 48S 32 12 E
Gejiu, *China* . . . . **32 D5** 23 20N 103 10 E
Gel, Meydān-e, *Iran* . . . . **45 D7** 29 4N 54 50 E
Gela, *Italy* . . . . **20 F6** 37 4N 14 15 E
Gelderland □, *Neths.* . . . . . . **15 B6** 52 5N 6 10 E
Geldrop, *Neths.* . . . . **15 C5** 51 25N 5 32 E
Geleen, *Neths.* . . . . **15 D5** 50 57N 5 49 E
Gelibolu, *Turkey* . . . . **21 D12** 40 28N 26 43 E
Gelsenkirchen, *Germany* . . . **16 C4** 51 32N 7 6 E
Gemas, *Malaysia* . . . . **39 L4** 2 37N 102 36 E
Gembloux, *Belgium* . . . . **15 D4** 50 34N 4 43 E
Gemena, *Dem. Rep. of the Congo* . . . . **52 D3** 3 13N 19 48 E
Gemerek, *Turkey* . . . . **44 B3** 39 15N 36 10 E
Gemlik, *Turkey* . . . . **21 D13** 40 26N 29 9 E
Genale →, *Ethiopia* . . . . **46 F2** 6 2N 39 1 E
General Acha, *Argentina* . . . **94 D3** 37 20S 64 38W
General Alvear, *Buenos Aires, Argentina* **94 D4** 36 0S 60 0W
General Alvear, *Mendoza, Argentina* **94 D2** 35 0S 67 40W
General Artigas, *Paraguay* . . **94 B4** 26 52S 56 16W
General Belgrano, *Argentina* **94 D4** 36 35S 58 47W
General Cabrera, *Argentina* . **94 C3** 32 53S 63 52W
General Cepeda, *Mexico* . . . **86 B4** 25 23N 101 27W
General Guido, *Argentina* . . **94 D4** 36 40S 57 50W
General Juan Madariaga, *Argentina* **94 D4** 37 0S 57 0W
General La Madrid, *Argentina* **94 D3** 37 17S 61 20W
General MacArthur, *Phil.* . . . **37 B7** 11 18N 125 28 E
General Martin Miguel de Güemes, *Argentina* **94 A3** 24 50S 65 0W
General Paz, *Argentina* **94 B4** 27 45S 57 36W
General Pico, *Argentina* **94 D3** 35 45S 63 50W
General Pinedo, *Argentina* . . **94 B3** 27 15S 61 20W
General Pinto, *Argentina* . . . **94 C3** 34 45S 61 50W
General Roca, *Argentina* . . . **96 D3** 39 2S 67 35W
General Santos, *Phil.* . . . . . . **37 C7** 6 5N 125 14 E
General Trevino, *Mexico* **87 B5** 26 14N 99 29W
General Trías, *Mexico* **86 B3** 28 21N 106 22W
General Viamonte, *Argentina* **94 D3** 35 1S 61 3W
General Villegas, *Argentina* . **94 D3** 35 5S 63 0W
Genesee, *Idaho, U.S.A.* **82 C5** 46 33N 116 56W
Genesee, *Pa., U.S.A.* **78 E7** 41 59N 77 54W
Genesee →, *U.S.A.* **78 C7** 43 16N 77 36W
Geneseo, *Ill., U.S.A.* **80 E9** 41 27N 90 9W
Geneseo, *N.Y., U.S.A.* **78 D7** 42 48N 77 49W
Geneva = Genève, *Switz.* . . . **18 C7** 46 12N 6 9 E
Geneva, *Ala., U.S.A.* **77 K3** 31 2N 85 52W
Geneva, *N.Y., U.S.A.* **78 D8** 42 52N 76 59W
Geneva, *Nebr., U.S.A.* **80 E6** 40 32N 97 36W
Geneva, *Ohio, U.S.A.* **78 E4** 41 48N 80 57W
Geneva, L. = Léman, L., *Europe* . . . . **18 C7** 46 26N 6 30 E
Geneva, L., *U.S.A.* **76 D1** 42 38N 88 30W
Genève, *Switz.* . . . . **18 C7** 46 12N 6 9 E
Genil →, *Spain* **19 D3** 37 42N 5 19W
Genk, *Belgium* . . . . **15 D5** 50 58N 5 32 E
Gennargentu, Mti. del, *Italy* . **20 D3** 40 1N 9 19 E
Genoa = Génova, *Italy* . . . . . **18 D8** 44 25N 8 57 E
Genoa, *Australia* . . . . **63 F4** 37 29S 149 35 E
Genoa, *N.Y., U.S.A.* **79 D8** 42 40N 76 32W
Genoa, *Nebr., U.S.A.* **80 E6** 41 27N 97 44W
Genoa, *Nev., U.S.A.* **84 F7** 39 2N 119 50W
Génova, *Italy* . . . . **18 D8** 44 25N 8 57 E
Génova, G. di, *Italy* . . . . **20 C3** 44 0N 9 0 E
Genriyetty, Ostrov, *Russia* . . **27 B16** 77 6N 156 30 E
Gent, *Belgium* . . . . **15 C3** 51 2N 3 42 E
Genteng, *Indonesia* . . . . **37 G12** 7 22S 106 24 E
Genyem, *Indonesia* . . . . **37 E10** 2 46S 140 12 E
Geographe B., *Australia* . . . . **61 F2** 33 30S 115 15 E
Geographe Chan., *Australia* . **61 D1** 24 30S 113 0 E
Georga, Zemlya, *Russia* . . . . **26 A5** 80 30N 49 0 E
George, *S. Africa* . . . . **56 E3** 33 58S 22 29 E
George →, *Canada* **71 A6** 58 49N 66 10W
George, L., *N.S.W., Australia* **63 F4** 35 10S 149 25 E
George, L., *S. Austral., Australia* . . . . **63 F3** 37 25S 140 0 E
George, L., *W. Austral., Australia* . . . . **60 D3** 22 45S 123 40 E
George, L., *Uganda* . . . . **54 B3** 0 5N 30 10 E
George, L., *Fla., U.S.A.* **77 L5** 29 17N 81 36W
George, L., *N.Y., U.S.A.* **79 C11** 43 37N 73 33W
George Gill Ra., *Australia* . . . **60 D5** 24 22S 131 45 E
George River = Kangiqsualujjuaq, *Canada* **69 C13** 58 30N 65 59W
George Sound, *N.Z.* . . . . **59 L1** 44 52S 167 25 E
George Town, *Australia* . . . . **62 G4** 41 6S 146 49 E
George Town, *Bahamas* **88 B4** 23 33N 75 47W
George Town, *Cayman Is.* . . . **88 C3** 19 20N 81 24W
George Town, *Malaysia* . . . . . **39 K3** 5 25N 100 20 E
George V Land, *Antarctica* . . . **5 C10** 69 0S 148 0 E
George VI Sound, *Antarctica* . **5 D17** 71 0S 68 0W
George West, *U.S.A.* **81 L5** 28 20N 98 7W
Georgetown, *Australia* . . . . . **62 B3** 18 17S 143 33 E
Georgetown, *Ont., Canada* . . . **78 C5** 43 40N 79 56W
Georgetown, *P.E.I., Canada* . . **71 C7** 46 13N 62 24W
Georgetown, *Gambia* **50 F3** 13 30N 14 47W
Georgetown, *Guyana* **92 B7** 6 50N 58 12W
Georgetown, *Calif., U.S.A.* . . . **84 G6** 38 54N 120 50W
Georgetown, *Colo., U.S.A.* . . . **82 G11** 39 42N 105 42W
Georgetown, *Ky., U.S.A.* **76 F3** 38 13N 84 33W
Georgetown, *N.Y., U.S.A.* **79 D9** 42 46N 75 44W
Georgetown, *Ohio, U.S.A.* **76 F4** 38 52N 83 54W
Georgetown, *S.C., U.S.A.* **77 J6** 33 23N 79 17W
Georgetown, *Tex., U.S.A.* **81 K6** 30 38N 97 41W
Georgia □, *U.S.A.* **77 K5** 32 50N 83 15W
Georgia ■, *Asia* . . . . **25 F7** 42 0N 43 0 E
Georgia, Str. of, *Canada* **72 D4** 49 25N 124 0W
Georgian B., *Canada* **78 A4** 45 15N 81 0W
Georgina →, *Australia* . . . . . **62 C2** 23 30S 139 47 E
Georgina I., *Canada* **78 B5** 44 22N 79 17W
Georgiu-Dezh = Liski, *Russia* **25 D6** 51 3N 39 30 E
Georgiyevsk, *Russia* . . . . **25 F7** 44 12N 43 28 E
Gera, *Germany* . . . . **16 C7** 50 53N 12 4 E
Geraardsbergen, *Belgium* . . . **15 D3** 50 45N 3 53 E
Geral, Serra, *Brazil* **95 B6** 26 25S 50 0W
Geral de Goiás, Serra, *Brazil* **93 F9** 12 0S 46 0W
Geraldine, *U.S.A.* **82 C8** 47 36N 110 16W
Geraldton, *Australia* . . . . **61 E1** 28 48S 114 32 E
Geraldton, *Canada* **70 C2** 49 44N 86 59W
Gereshk, *Afghan.* . . . . **40 D4** 31 47N 64 35 E
Gerik, *Malaysia* . . . . **39 K3** 5 50N 101 15 E
Gering, *U.S.A.* **80 E3** 41 50N 103 40W
Gerlach, *U.S.A.* **82 F4** 40 39N 119 21W
Germansen Landing, *Canada* **72 B4** 55 43N 124 40W
Germantown, *U.S.A.* **81 M10** 35 5N 89 49W
Germany ■, *Europe* . . . . **16 C6** 51 0N 10 0 E
Germī, *Iran* . . . . **45 B6** 39 1N 48 3 E
Germiston, *S. Africa* . . . . **57 D4** 26 15S 28 10 E
Gernika-Lumo, *Spain* **19 A4** 43 19N 2 40W
Gero, *Japan* . . . . **31 G8** 35 48N 137 14 E
Gerona = Girona, *Spain* . . . . **19 B7** 41 58N 2 46 E
Gerrard, *Canada* **72 C5** 50 30N 117 17W
Geser, *Indonesia* . . . . **37 E8** 3 50S 130 54 E
Getafe, *Spain* **19 B4** 40 18N 3 44W
Gettysburg, *Pa., U.S.A.* **76 F7** 39 50N 77 14W
Gettysburg, *S. Dak., U.S.A.* . . **80 C5** 45 1N 99 57W
Getxo, *Spain* **19 A4** 43 21N 2 59W
Getz Ice Shelf, *Antarctica* . . . **5 D14** 75 0S 130 0W
Geyser, *U.S.A.* **82 C8** 47 16N 110 30W
Geyserville, *U.S.A.* **84 G4** 38 42N 122 54W
Ghaggar →, *India* . . . . **42 E6** 29 30N 74 53 E
Ghaghara →, *India* . . . . **43 G11** 25 45N 84 40 E
Ghaghat →, *Bangla.* . . . . **43 G13** 25 19N 89 38 E
Ghagra, *India* . . . . **43 H11** 23 17N 84 33 E
Ghagra →, *India* . . . . **43 F9** 27 29N 81 9 E
Ghana ■, *W. Afr.* **50 G5** 8 0N 1 0W
Ghansor, *India* . . . . **43 H9** 22 39N 80 1 E
Ghanzi, *Botswana* . . . . **56 C3** 21 50S 21 34 E
Ghardaïa, *Algeria* . . . . **50 B6** 32 20N 3 37 E
Gharyān, *Libya* . . . . **51 B8** 32 10N 13 0 E
Ghat, *Libya* . . . . **51 D8** 24 59N 10 11 E
Ghatal, *India* . . . . **43 H12** 22 40N 87 46 E
Ghatampur, *India* . . . . **43 F9** 26 8N 80 13 E
Ghatsila, *India* . . . . **43 H12** 22 36N 86 29 E
Ghaṭṭī, *Si. Arabia* . . . . **44 D3** 31 16N 37 31 E
Ghawdex = Gozo, *Malta* . . . . **23 C1** 36 3N 14 15 E
Ghazal, Bahr el →, *Chad* . . . . **51 F9** 13 0N 15 47 E
Ghazâl, Bahr el →, *Sudan* . . . **51 G12** 9 31N 30 25 E
Ghaziabad, *India* . . . . **42 E7** 28 42N 77 26 E
Ghazipur, *India* . . . . **43 G10** 25 38N 83 35 E
Ghaznī, *Afghan.* . . . . **42 C3** 33 30N 68 28 E
Ghaznī □, *Afghan.* . . . . **40 C6** 32 10N 68 20 E
Ghent = Gent, *Belgium* . . . . . **15 C3** 51 2N 3 42 E
Gheorghe Gheorghiu-Dej = Oneşti, *Romania* . . . . **17 E14** 46 17N 26 47 E
Ghīnah, Wādī al →, *Si. Arabia* **44 D3** 30 27N 38 14 E
Ghizao, *Afghan.* . . . . **42 C1** 33 20N 65 44 E
Ghizar →, *Pakistan* . . . . **43 A5** 36 15N 73 43 E
Ghotaru, *India* . . . . **42 F4** 27 20N 70 1 E
Ghotki, *Pakistan* . . . . **42 E3** 28 5N 69 21 E
Ghowr □, *Afghan.* . . . . **40 C4** 34 0N 64 20 E
Ghudaf, W. al →, *Iraq* . . . . . . **44 C4** 32 56N 43 30 E
Ghudāmis, *Libya* . . . . **49 C4** 30 11N 9 29 E
Ghughri, *India* . . . . **43 H9** 22 39N 80 41 E
Ghugus, *India* . . . . **40 K11** 19 58N 79 12 E
Ghulam Mohammad Barrage, *Pakistan* . . . . **42 G3** 25 30N 68 20 E
Ghūrīān, *Afghan.* . . . . **40 B2** 34 17N 61 25 E
Gia Dinh, *Vietnam* . . . . **39 G6** 10 49N 106 42 E
Gia Lai = Plei Ku, *Vietnam* . . . **38 F7** 13 57N 108 0 E
Gia Nghia, *Vietnam* . . . . **39 G6** 11 58N 107 42 E
Gia Ngoc, *Vietnam* . . . . **38 E7** 14 50N 108 58 E
Gia Vuc, *Vietnam* . . . . **38 E7** 14 42N 108 34 E
Giant Forest, *U.S.A.* **84 J8** 36 36N 118 43W
Giants Causeway, *U.K.* **13 A5** 55 16N 6 29W
Giarabub = Al Jaghbūb, *Libya* **51 C10** 29 42N 24 38 E
Giarre, *Italy* . . . . **20 F6** 37 43N 15 11 E
Gibara, *Cuba* **88 B4** 21 9N 76 11W
Gibb River, *Australia* . . . . . . **60 C4** 16 26S 126 26 E
Gibbon, *U.S.A.* **80 E5** 40 45N 98 51W
Gibeon, *Namibia* . . . . **56 D2** 25 9S 17 43 E
Gibraltar ■, *Europe* **19 D3** 36 7N 5 22W
Gibraltar, Str. of, *Medit. S.* . . **19 E3** 35 55N 5 40W
Gibson Desert, *Australia* . . . **60 D4** 24 0S 126 0 E
Gibsons, *Canada* **72 D4** 49 24N 123 32W
Gibsonville, *U.S.A.* **84 F6** 39 46N 120 54W
Giddings, *U.S.A.* **81 K6** 30 11N 96 56W
Giebnegáisi = Kebnekaise, *Sweden* . . . . **8 C18** 67 53N 18 33 E
Giessen, *Germany* . . . . **16 C5** 50 34N 8 41 E
Gīfān, *Iran* . . . . **45 B8** 37 54N 57 28 E
Gift Lake, *Canada* **72 B5** 55 53N 115 49W
Gifu, *Japan* . . . . **31 G8** 35 30N 136 45 E
Gifu □, *Japan* . . . . **31 G8** 35 40N 137 0 E
Giganta, Sa. de la, *Mexico* . . . **86 B2** 25 30N 111 30W
Gigha, *U.K.* **12 F3** 55 42N 5 44W
Giglio, *Italy* . . . . **20 C4** 42 20N 10 52 E
Gijón, *Spain* **19 A3** 43 32N 5 42W
Gil I., *Canada* **72 C3** 53 12N 129 15W
Gila →, *U.S.A.* **83 K6** 32 43N 114 33W
Gila Bend, *U.S.A.* **83 K7** 32 57N 112 43W
Gila Bend Mts., *U.S.A.* **83 K7** 33 10N 113 0W
Gīlān □, *Iran* . . . . **45 B6** 37 0N 50 0 E
Gilbert →, *Australia* . . . . **62 B3** 16 35S 141 15 E
Gilbert Is., *Kiribati* . . . . **64 G9** 1 0N 172 0 E
Gilbert River, *Australia* . . . . . **62 B3** 18 9S 142 52 E
Gilead, *U.S.A.* **79 B14** 44 24N 70 59W
Gilford I., *Canada* **72 C3** 50 40N 126 30W
Gilgandra, *Australia* . . . . **63 E4** 31 43S 148 39 E
Gilgil, *Kenya* . . . . **54 C4** 0 30S 36 20 E
Gilgit, *India* . . . . **43 B6** 35 50N 74 15 E
Gilgit →, *Pakistan* . . . . **43 B6** 35 44N 74 37 E
Gillam, *Canada* **73 B10** 56 20N 94 40W
Gillen, L., *Australia* . . . . **61 E3** 26 11S 124 38 E
Gilles, L., *Australia* . . . . **63 E2** 32 50S 136 45 E
Gillette, *U.S.A.* **80 C2** 44 18N 105 30W
Gilliat, *Australia* . . . . **62 C3** 20 40S 141 28 E
Gillingham, *U.K.* . . . . **11 F8** 51 23N 0 33 E
Gilmer, *U.S.A.* **81 J7** 32 44N 94 57W
Gilmore, L., *Australia* . . . . . . **61 F3** 32 29S 121 37 E
Gilroy, *U.S.A.* **84 H5** 37 1N 121 34W
Gimli, *Canada* **73 C9** 50 40N 97 0W
Gin Gin, *Australia* . . . . **63 D5** 25 0S 151 58 E
Gingin, *Australia* . . . . **61 F2** 31 22S 115 54 E
Gingindlovu, *S. Africa* . . . . . . **57 D5** 29 2S 31 30 E
Ginir, *Ethiopia* . . . . **46 F3** 7 6N 40 40 E
Gióna, Óros, *Greece* . . . . **21 E10** 38 38N 22 14 E
Gir Hills, *India* . . . . **42 J4** 21 0N 71 0 E
Girab, *India* . . . . **42 F4** 26 2N 70 38 E
Girâfi, W. →, *Egypt* . . . . **47 F3** 29 58N 34 39 E
Girard, *Kans., U.S.A.* **81 G7** 37 31N 94 51W
Girard, *Ohio, U.S.A.* **78 E4** 41 9N 80 42W
Girard, *Pa., U.S.A.* **78 E4** 42 0N 80 19W
Girdle Ness, *U.K.* **12 D6** 57 9N 2 3W
Giresun, *Turkey* . . . . **25 F6** 40 55N 38 30 E
Girga, *Egypt* . . . . **51 C12** 26 17N 31 55 E
Giri →, *India* . . . . **42 D7** 30 28N 77 41 E
Giridih, *India* . . . . **43 G12** 24 10N 86 21 E
Girne = Kyrenia, *Cyprus* . . . . **23 D12** 35 20N 33 20 E
Girona, *Spain* . . . . **19 B7** 41 58N 2 46 E
Gironde →, *France* **18 D3** 45 32N 1 7W
Giru, *Australia* . . . . **62 B4** 19 30S 147 5 E
Girvan, *U.K.* **12 F4** 55 14N 4 51W
Gisborne, *N.Z.* . . . . **59 H7** 38 39S 178 5 E
Gisenyi, *Rwanda* . . . . **54 C2** 1 41S 29 15 E
Gislaved, *Sweden* . . . . **9 H15** 57 19N 13 32 E
Gitega, *Burundi* . . . . **54 C2** 3 26S 29 56 E
Giuba →, *Somali Rep.* . . . . . . **46 G3** 1 30N 42 35 E
Giurgiu, *Romania* . . . . **17 G13** 43 52N 25 57 E
Giza = El Gîza, *Egypt* . . . . . . **51 C12** 30 0N 31 10 E
Gizhiga, *Russia* . . . . **27 C17** 62 3N 160 30 E
Gizhiginskaya Guba, *Russia* . **27 C16** 61 0N 158 0 E
Giżycko, *Poland* . . . . **17 A11** 54 2N 21 48 E
Gjirokastër, *Albania* . . . . **21 D9** 40 7N 20 10 E
Gjoa Haven, *Canada* **68 B10** 68 20N 96 8W
Gjøvik, *Norway* . . . . **9 F14** 60 47N 10 43 E
Glace Bay, *Canada* **71 C8** 46 11N 59 58W
Glacier Bay Nat. Park and Preserve, *U.S.A.* **72 B1** 58 45N 136 30W
Glacier Nat. Park, *Canada* . . . **72 C5** 51 15N 117 30W
Glacier Nat. Park, *U.S.A.* **82 B7** 48 30N 113 18W
Glacier Peak, *U.S.A.* **82 B3** 48 7N 121 7W
Gladewater, *U.S.A.* **81 J7** 32 33N 94 56W
Gladstone, *Queens., Australia* **62 C5** 23 52S 151 16 E
Gladstone, *S. Austral., Australia* . . . . **63 E2** 33 15S 138 22 E
Gladstone, *Canada* **73 C9** 50 13N 98 57W
Gladstone, *U.S.A.* **76 C2** 45 51N 87 1W
Gladwin, *U.S.A.* **76 D3** 43 59N 84 29W
Glåma = Glomma →, *Norway* **9 G14** 59 12N 10 57 E
Gláma, *Iceland* **8 D2** 65 48N 23 0W
Glamis, *U.S.A.* **85 N11** 32 55N 115 5W
Glasco, *Kans., U.S.A.* **80 F6** 39 22N 97 50W
Glasco, *N.Y., U.S.A.* **79 D11** 42 3N 73 57W
Glasgow, *U.K.* **12 F4** 55 51N 4 15W
Glasgow, *Ky., U.S.A.* **76 G3** 37 0N 85 55W
Glasgow, *Mont., U.S.A.* **82 B10** 48 12N 106 38W
Glasgow, City of □, *U.K.* **12 F4** 55 51N 4 12W
Glaslyn, *Canada* **73 C7** 53 22N 108 21W
Glastonbury, *U.K.* **11 F5** 51 9N 2 43W
Glastonbury, *U.S.A.* **79 E12** 41 43N 72 37W
Glazov, *Russia* . . . . **24 C9** 58 9N 52 40 E
Gleichen, *Canada* **72 C6** 50 52N 113 3W
Gleiwitz = Gliwice, *Poland* . . **17 C10** 50 22N 18 41 E
Glen, *U.S.A.* **79 B13** 44 7N 71 11W
Glen Affric, *U.K.* **12 D3** 57 17N 5 1W
Glen Canyon, *U.S.A.* **83 H8** 37 30N 110 40W
Glen Canyon Dam, *U.S.A.* . . . **83 H8** 36 57N 111 29W
Glen Canyon Nat. Recr. Area, *U.S.A.* **83 H8** 37 15N 111 0W
Glen Coe, *U.K.* **12 E3** 56 40N 5 0W
Glen Cove, *U.S.A.* **79 F11** 40 52N 73 38W
Glen Garry, *U.K.* **12 D3** 57 3N 5 7W
Glen Innes, *Australia* . . . . . . **63 D5** 29 44S 151 44 E
Glen Lyon, *U.S.A.* **79 E8** 41 10N 76 5W
Glen Mor, *U.K.* **12 D4** 57 9N 4 37W
Glen Moriston, *U.K.* **12 D4** 57 11N 4 52W
Glen Robertson, *Canada* **79 A10** 45 22N 74 30W
Glen Spean, *U.K.* **12 E4** 56 53N 4 40W
Glen Ullin, *U.S.A.* **80 B4** 46 49N 101 50W
Glencoe, *Canada* **78 D3** 42 45N 81 43W
Glencoe, *S. Africa* . . . . **57 D5** 28 11S 30 11 E
Glencoe, *U.S.A.* **80 C7** 44 46N 94 9W
Glendale, *Ariz., U.S.A.* **83 K7** 33 32N 112 11W
Glendale, *Calif., U.S.A.* **85 L8** 34 9N 118 15W
Glendale, *Zimbabwe* . . . . **55 F3** 17 22S 31 5 E
Glendive, *U.S.A.* **80 B2** 47 7N 104 43W
Glendo, *U.S.A.* **80 D2** 42 30N 105 2W
Glenelg →, *Australia* . . . . . . **63 F3** 38 4S 140 59 E
Glenfield, *U.S.A.* **79 C9** 43 43N 75 24W
Glengarriff, *Ireland* **13 E2** 51 45N 9 34W
Glenmont, *U.S.A.* **78 F2** 40 31N 82 6W
Glenmorgan, *Australia* . . . . . **63 D4** 27 14S 149 42 E
Glenn, *U.S.A.* **84 F4** 39 31N 122 1W
Glennallen, *U.S.A.* **68 B5** 62 7N 145 33W
Glennamaddy, *Ireland* **13 C3** 53 37N 8 33W
Glenns Ferry, *U.S.A.* **82 E6** 42 57N 115 18W
Glenore, *Australia* . . . . **62 B3** 17 50S 141 12 E
Glenreagh, *Australia* . . . . **63 E5** 30 2S 153 1 E
Glenrock, *U.S.A.* **82 E11** 42 52N 105 52W
Glenrothes, *U.K.* **12 E5** 56 12N 3 10W
Glens Falls, *U.S.A.* **79 C11** 43 19N 73 39W
Glenside, *U.S.A.* **79 F9** 40 6N 75 9W
Glenties, *Ireland* **13 B3** 54 49N 8 16W
Glenville, *U.S.A.* **76 F5** 38 56N 80 50W
Glenwood, *Canada* **71 C9** 49 0N 54 58W
Glenwood, *Ark., U.S.A.* **81 H8** 34 20N 93 33W
Glenwood, *Iowa, U.S.A.* **80 E7** 41 3N 95 45W
Glenwood, *Minn., U.S.A.* **80 C7** 45 39N 95 23W
Glenwood, *Wash., U.S.A.* **84 D5** 46 1N 121 17W
Glenwood Springs, *U.S.A.* . . . **82 G10** 39 33N 107 19W
Glettinganes, *Iceland* **8 D7** 65 30N 13 37W
Gliwice, *Poland* . . . . **17 C10** 50 22N 18 41 E
Globe, *U.S.A.* **83 K8** 33 24N 110 47W
Głogów, *Poland* . . . . **16 C9** 51 37N 16 5 E
Glomma →, *Norway* . . . . **9 G14** 59 12N 10 57 E
Glorieuses, Is., *Ind. Oc.* . . . . **57 A8** 11 30S 47 20 E
Glossop, *U.K.* **10 D6** 53 27N 1 56W

Green River

Gloucester, *Australia* ...... **63 E5** 32 0S 151 59 E
Gloucester, *U.K.* **11 F5** 51 53N 2 15W
Gloucester, *U.S.A.* **79 D14** 42 37N 70 40W
Gloucester I., *Australia* ..... **62 C4** 20 0S 148 30 E
Gloucester Point, *U.S.A.* **76 G7** 37 15N 76 29W
Gloucestershire □, *U.K.* **11 F5** 51 46N 2 15W
Gloversville, *U.S.A.* **79 C10** 43 3N 74 21W
Glovertown, *Canada* **71 C9** 48 40N 54 3W
Glusk, *Belarus* .... **17 B15** 52 53N 28 41 E
Gmünd, *Austria* .... **16 D8** 48 45N 15 0 E
Gmunden, *Austria* .... **16 E7** 47 55N 13 48 E
Gniezno, *Poland* .... **17 B9** 52 30N 17 35 E
Gnowangerup, *Australia* ... **61 F2** 33 58S 117 59 E
Go Cong, *Vietnam* .... **39 G6** 10 22N 106 40 E
Gō-no-ura, *Japan* .... **31 H4** 33 44N 129 40 E
Goa, *India* .... **40 M8** 15 33N 73 59 E
Goa □, *India* .... **40 M8** 15 33N 73 59 E
Goalen Hd., *Australia* ...... **63 F5** 36 33S 150 4 E
Goalpara, *India* .... **41 F17** 26 10N 90 40 E
Goaltor, *India* .... **43 H12** 22 43N 87 10 E
Goalundo Ghat, *Bangla.* .... **43 H13** 23 50N 89 47 E
Goat Fell, *U.K.* **12 F3** 55 38N 5 11W
Goba, *Ethiopia* .... **46 F2** 7 1N 39 59 E
Goba, *Mozam.* .... **57 D5** 26 15S 32 13 E
Gobabis, *Namibia* .... **56 C2** 22 30S 19 0 E
Gobi, *Asia* .... **34 C6** 44 0N 110 0 E
Gobō, *Japan* .... **31 H7** 33 53N 135 10 E
Gochas, *Namibia* .... **56 C2** 24 59S 18 55 E
Godavari →, *India* .... **41 L13** 16 25N 82 18 E
Godavari Pt., *India* .... **41 L13** 17 0N 82 20 E
Godbout, *Canada* **71 C6** 49 20N 67 38W
Godda, *India* .... **43 G12** 24 50N 87 13 E
Goderich, *Canada* **78 C3** 43 45N 81 41W
Godfrey Ra., *Australia* ..... **61 D2** 24 0S 117 0 E
Godhavn = Qeqertarsuaq, *Greenland* **4 C5** 69 15N 53 38W
Godhra, *India* .... **42 H5** 22 49N 73 40 E
Godoy Cruz, *Argentina* **94 C2** 32 56S 68 52W
Gods →, *Canada* **70 A1** 56 22N 92 51W
Gods L., *Canada* **70 B1** 54 40N 94 15W
Gods River, *Canada* **73 C10** 54 50N 94 5W
Godthåb = Nuuk, *Greenland* **69 B14** 64 10N 51 35W
Godwin Austen = K2, *Pakistan* **43 B7** 35 58N 76 32 E
Goeie Hoop, Kaap die = Good Hope, C. of, *S. Africa* ..... **56 E2** 34 24S 18 30 E
Goéland, L. au, *Canada* **70 C4** 49 50N 76 48W
Goeree, *Neths.* .... **15 C3** 51 50N 4 0 E
Goes, *Neths.* .... **15 C3** 51 30N 3 55 E
Goffstown, *U.S.A.* **79 C13** 43 1N 71 36W
Gogama, *Canada* **70 C3** 47 35N 81 43W
Gogebic, L., *U.S.A.* **80 B10** 46 30N 89 35W
Gogra = Ghaghara →, *India* . **43 G11** 25 45N 84 40 E
Gogriâl, *Sudan* .... **51 G11** 8 30N 28 8 E
Gohana, *India* .... **42 E7** 29 8N 76 42 E
Goharganj, *India* .... **42 H7** 23 1N 77 41 E
Goi →, *India* .... **42 H6** 22 4N 74 46 E
Goiânia, *Brazil* **93 G9** 16 43S 49 20W
Goiás, *Brazil* **93 G8** 15 55S 50 10W
Goiás □, *Brazil* **93 F9** 12 10S 48 0W
Goio-Erê, *Brazil* **95 A5** 24 12S 53 1W
Gojō, *Japan* .... **31 G7** 34 21N 135 42 E
Gojra, *Pakistan* .... **42 D5** 31 10N 72 40 E
Gökçeada, *Turkey* .... **21 D11** 40 10N 25 50 E
Gökova Körfezi, *Turkey* .... **21 F12** 36 55N 27 50 E
Gokteik, *Burma* .... **41 H20** 22 26N 97 0 E
Gokurt, *Pakistan* .... **42 E2** 29 40N 67 26 E
Gokwe, *Zimbabwe* .... **57 B4** 18 7S 28 58 E
Gola, *India* .... **43 E9** 28 3N 80 32 E
Golakganj, *India* .... **43 F13** 26 8N 89 52 E
Golan Heights = Hagolan, *Syria* .... **47 C4** 33 0N 35 45 E
Golāshkerd, *Iran* .... **45 E8** 27 59N 57 16 E
Golchikha, *Russia* .... **4 B12** 71 45N 83 30 E
Golconda, *U.S.A.* **82 F5** 40 58N 117 30W
Gold, *U.S.A.* **78 E7** 41 52N 77 50W
Gold Beach, *U.S.A.* **82 E1** 42 25N 124 25W
Gold Coast, *W. Afr.* **50 H5** 4 0N 1 40W
Gold Hill, *U.S.A.* **82 E2** 42 26N 123 3W
Gold River, *Canada* **72 D3** 49 46N 126 3W
Golden, *Canada* **72 C5** 51 20N 116 59W
Golden B., *N.Z.* .... **59 J4** 40 40S 172 50 E
Golden Gate, *U.S.A.* **82 H2** 37 54N 122 30W
Golden Hinde, *Canada* **72 D3** 49 40N 125 44W
Golden Lake, *Canada* **78 A7** 45 34N 77 21W
Golden Vale, *Ireland* **13 D3** 52 33N 8 17W
Goldendale, *U.S.A.* **82 D3** 45 49N 120 50W
Goldfield, *U.S.A.* **83 H5** 37 42N 117 14W
Goldsand L., *Canada* **73 B8** 57 2N 101 8W
Goldsboro, *U.S.A.* **77 H7** 35 23N 77 59W
Goldsmith, *U.S.A.* **81 K3** 31 59N 102 37W
Goldsworthy, *Australia* ..... **60 D2** 20 21S 119 30 E
Goldthwaite, *U.S.A.* **81 K5** 31 27N 98 34W
Goleniów, *Poland* .... **16 B8** 53 35N 14 50 E
Golestānak, *Iran* .... **45 D7** 30 36N 54 14 E
Goleta, *U.S.A.* **85 L7** 34 27N 119 50W
Golfito, *Costa Rica* **88 E3** 8 41N 83 5W
Golfo Aranci, *Italy* .... **20 D3** 40 59N 9 38 E
Goliad, *U.S.A.* **81 L6** 28 40N 97 23W
Golpāyegān, *Iran* .... **45 C6** 33 27N 50 18 E
Golra, *Pakistan* .... **42 C5** 33 37N 72 56 E
Golspie, *U.K.* **12 D5** 57 58N 3 59W
Goma, *Dem. Rep. of the Congo* .... **54 C2** 1 37S 29 10 E
Gomal Pass, *Pakistan* ...... **42 D3** 31 56N 69 20 E
Gomati →, *India* .... **43 G10** 25 32N 83 11 E
Gombari, *Dem. Rep. of the Congo* .... **54 B2** 2 45N 29 3 E
Gombe, *Nigeria* .... **51 F8** 10 19N 11 2 E
Gombe →, *Tanzania* .... **54 C3** 4 38S 31 40 E
Gomel = Homyel, *Belarus* .. **17 B16** 52 28N 31 0 E
Gomera, *Canary Is.* **22 F2** 28 7N 17 14W
Gómez Palacio, *Mexico* **86 B4** 25 40N 104 0W
Gomīshān, *Iran* .... **45 B7** 37 4N 54 6 E
Gomogomo, *Indonesia* ..... **37 F8** 6 39S 134 43 E
Gomoh, *India* .... **41 H15** 23 52N 86 10 E
Gompa = Ganta, *Liberia* **50 G4** 7 15N 8 59W
Gonābād, *Iran* .... **45 C8** 34 15N 58 45 E
Gonaïves, *Haiti* **89 C5** 19 20N 72 42W
Gonâve, G. de la, *Haiti* **89 C5** 19 29N 72 42W
Gonâve, I. de la, *Haiti* **89 C5** 18 45N 73 0W
Gonbad-e Kāvūs, *Iran* ...... **45 B7** 37 20N 55 25 E
Gonda, *India* .... **43 F9** 27 9N 81 58 E
Gondal, *India* .... **42 J4** 21 58N 70 52 E
Gonder, *Ethiopia* .... **46 E2** 12 39N 37 30 E
Gondia, *India* .... **40 J12** 21 23N 80 10 E
Gondola, *Mozam.* .... **55 F3** 19 10S 33 37 E
Gönen, *Turkey* .... **21 D12** 40 6N 27 39 E
Gonghe, *China* .... **32 C5** 36 18N 100 32 E
Gongolgon, *Australia* ...... **63 E4** 30 21S 146 54 E
Gongzhuling, *China* .... **35 C13** 43 30N 124 40 E
Gonzales, *Calif., U.S.A.* **84 J5** 36 30N 121 26W
Gonzales, *Tex., U.S.A.* **81 L6** 29 30N 97 27W
González Chaves, *Argentina* **94 D3** 38 2S 60 5W
Good Hope, C. of, *S. Africa* . **56 E2** 34 24S 18 30 E
Gooderham, *Canada* **78 B6** 44 54N 78 21W
Goodhouse, *S. Africa* ...... **56 D2** 28 57S 18 13 E
Gooding, *U.S.A.* **82 E6** 42 56N 114 43W
Goodland, *U.S.A.* **80 F4** 39 21N 101 43W
Goodlow, *Canada* **72 B4** 56 20N 120 8W
Goodooga, *Australia* ...... **63 D4** 29 3S 147 28 E
Goodsprings, *U.S.A.* **85 K11** 35 49N 115 27W
Goole, *U.K.* **10 D7** 53 42N 0 53W
Goolgowi, *Australia* .... **63 E4** 33 58S 145 41 E
Goomalling, *Australia* ..... **61 F2** 31 15S 116 49 E
Goomeri, *Australia* .... **63 D5** 26 12S 152 6 E
Goonda, *Mozam.* .... **55 F3** 19 48S 33 57 E
Goondiwindi, *Australia* ..... **63 D5** 28 30S 150 21 E
Goongarrie, L., *Australia* ... **61 F3** 30 3S 121 9 E
Goonyella, *Australia* .... **62 C4** 21 47S 147 58 E
Goose →, *Canada* **71 B7** 53 20N 60 35W
Goose Creek, *U.S.A.* **77 J5** 32 59N 80 2W
Goose L., *U.S.A.* **82 F3** 41 56N 120 26W
Gop, *India* .... **40 H6** 22 5N 69 50 E
Gopalganj, *India* .... **43 F11** 26 28N 84 30 E
Göppingen, *Germany* ...... **16 D5** 48 42N 9 39 E
Gorakhpur, *India* .... **43 F10** 26 47N 83 23 E
Goražde, *Bos.-H.* .... **21 C8** 43 38N 18 58 E
Gorda, *U.S.A.* **84 K5** 35 53N 121 26W
Gorda, Pta., *Canary Is.* **22 F2** 28 45N 18 0W
Gorda, Pta., *Nic.* **88 D3** 14 20N 83 10W
Gordan B., *Australia* ...... **60 B5** 11 35S 130 10 E
Gordon, *U.S.A.* **80 D3** 42 48N 102 12W
Gordon →, *Australia* ...... **62 G4** 42 27S 145 30 E
Gordon L., *Alta., Canada* ... **73 B6** 56 30N 110 25W
Gordon L., *N.W.T., Canada* . **72 A6** 63 5N 113 11W
Gordonvale, *Australia* ..... **62 B4** 17 5S 145 50 E
Gore, *Ethiopia* .... **46 F2** 8 12N 35 32 E
Gore, *N.Z.* .... **59 M2** 46 5S 168 58 E
Gore Bay, *Canada* **70 C3** 45 57N 82 28W
Gorey, *Ireland* **13 D5** 52 41N 6 18W
Gorg, *Iran* .... **45 D8** 29 29N 59 43 E
Gorgān, *Iran* .... **45 B7** 36 50N 54 29 E
Gorgona, I., *Colombia* **92 C3** 3 0N 78 10W
Gorham, *U.S.A.* **79 B13** 44 23N 71 10W
Goriganga →, *India* .... **43 E9** 29 45N 80 23 E
Gorinchem, *Neths.* .... **15 C4** 51 50N 4 59 E
Goris, *Armenia* .... **25 G8** 39 31N 46 22 E
Gorízia, *Italy* .... **20 B5** 45 56N 13 37 E
Gorki = Nizhniy Novgorod, *Russia* .... **24 C7** 56 20N 44 0 E
Gorkiy = Nizhniy Novgorod, *Russia* .... **24 C7** 56 20N 44 0 E
Gorkovskoye Vdkhr., *Russia* **24 C7** 57 2N 43 4 E
Görlitz, *Germany* .... **16 C8** 51 9N 14 58 E
Gorlovka = Horlivka, *Ukraine* **25 E6** 48 19N 38 5 E
Gorman, *U.S.A.* **85 L8** 34 47N 118 51W
Gorna Dzhumayo = Blagoevgrad, *Bulgaria* ... **21 C10** 42 2N 23 5 E
Gorna Oryakhovitsa, *Bulgaria* **21 C11** 43 7N 25 40 E
Gorno-Altay □, *Russia* ..... **26 D9** 51 0N 86 0 E
Gorno-Altaysk, *Russia* ..... **26 D9** 51 50N 86 5 E
Gornyatski, *Russia* .... **24 A11** 67 32N 64 3 E
Gornyy, *Russia* .... **30 B6** 44 57N 133 59 E
Gorodenka = Horodenka, *Ukraine* .... **17 D13** 48 41N 25 29 E
Gorodok = Horodok, *Ukraine* **17 D12** 49 46N 23 32 E
Gorokhov = Horokhiv, *Ukraine* .... **17 C13** 50 30N 24 45 E
Goromonzi, *Zimbabwe* ..... **55 F3** 17 52S 31 22 E
Gorong, Kepulauan, *Indonesia* .... **37 E8** 3 59S 131 25 E
Gorongose →, *Mozam.* .... **57 C5** 20 30S 34 40 E
Gorongoza, *Mozam.* .... **55 F3** 18 44S 34 2 E
Gorongoza, Sa. da, *Mozam.* . **55 F3** 18 27S 34 2 E
Gorontalo, *Indonesia* ...... **37 D6** 0 35N 123 5 E
Gort, *Ireland* **13 C3** 53 3N 8 49W
Gortis, *Greece* .... **23 D6** 35 4N 24 58 E
Gorzów Wielkopolski, *Poland* **16 B8** 52 43N 15 15 E
Gosford, *Australia* .... **63 E5** 33 23S 151 18 E
Goshen, *Calif., U.S.A.* **84 J7** 36 21N 119 25W
Goshen, *Ind., U.S.A.* **76 E3** 41 35N 85 50W
Goshen, *N.Y., U.S.A.* **79 E10** 41 24N 74 20W
Goshogawara, *Japan* ...... **30 D10** 40 48N 140 27 E
Goslar, *Germany* .... **16 C6** 51 54N 10 25 E
Gospič, *Croatia* .... **16 F8** 44 35N 15 23 E
Gosport, *U.K.* **11 G6** 50 48N 1 9W
Gosse →, *Australia* .... **62 B1** 19 32S 134 37 E
Göta älv →, *Sweden* .... **9 H14** 57 42N 11 54 E
Göta kanal, *Sweden* .... **9 G16** 58 30N 15 58 E
Götaland, *Sweden* .... **9 G15** 57 30N 14 30 E
Göteborg, *Sweden* .... **9 H14** 57 43N 11 59 E
Gotha, *Germany* .... **16 C6** 50 56N 10 42 E
Gothenburg = Göteborg, *Sweden* .... **9 H14** 57 43N 11 59 E
Gothenburg, *U.S.A.* **80 E4** 40 56N 100 10W
Gotland, *Sweden* .... **9 H18** 57 30N 18 33 E
Gotō-Rettō, *Japan* .... **31 H4** 32 55N 129 5 E
Gotska Sandön, *Sweden* .... **9 G18** 58 24N 19 15 E
Gōtsu, *Japan* .... **31 G6** 35 0N 132 14 E
Gott Pk., *Canada* **72 C4** 50 18N 122 16W
Göttingen, *Germany* .... **16 C5** 51 31N 9 55 E
Gottwaldov = Zlin, *Czech Rep.* **17 D9** 49 14N 17 40 E
Goubangzi, *China* .... **35 D11** 41 20N 121 52 E
Gouda, *Neths.* .... **15 B4** 52 1N 4 42 E
Goúdhoura, Ákra, *Greece* ... **23 E8** 34 59N 26 6 E
Gough I., *Atl. Oc.* **2 G9** 40 10S 9 45W
Gouin, Rés., *Canada* **70 C5** 48 35N 74 40W
Goulburn, *Australia* .... **63 E4** 34 44S 149 44 E
Goulburn Is., *Australia* ..... **62 A1** 11 40S 133 20 E
Goulimine, *Morocco* **50 C3** 28 56N 10 0W
Gourits →, *S. Africa* ...... **56 E3** 34 21S 21 52 E
Goúrnais, *Greece* .... **23 D7** 35 19N 25 16 E
Gouverneur, *U.S.A.* **79 B9** 44 20N 75 28W
Gouviá, *Greece* .... **23 A3** 39 39N 19 50 E
Governador Valadares, *Brazil* **93 G10** 18 15S 41 57W
Governor's Harbour, *Bahamas* **88 A4** 25 10N 76 14W
Govindgarh, *India* .... **43 G9** 24 23N 81 18 E
Gowan Ra., *Australia* ...... **62 D4** 25 0S 145 0 E
Gowanda, *U.S.A.* **78 D6** 42 28N 78 56W
Gower, *U.K.* **11 F3** 51 35N 4 10W
Gowna, L., *Ireland* **13 C4** 53 51N 7 34W
Goya, *Argentina* **94 B4** 29 10S 59 10W
Goyder Lagoon, *Australia* .. **63 D2** 27 3S 138 58 E
Goyllarisquisga, *Peru* **92 F3** 10 31S 76 24W
Goz Beïda, *Chad* .... **51 F10** 12 10N 21 20 E
Gozo, *Malta* .... **23 C1** 36 3N 14 15 E
Graaff-Reinet, *S. Africa* .... **56 E3** 32 13S 24 32 E
Gračac, *Croatia* .... **16 F8** 44 18N 15 57 E
Gracias a Dios, C., *Honduras* **88 D3** 15 0N 83 10W
Graciosa, I., *Canary Is.* **22 E6** 29 15N 13 32W
Grado, *Spain* **19 A2** 43 23N 6 4W
Grady, *U.S.A.* **81 H3** 34 49N 103 19W
Grafham Water, *U.K.* **11 E7** 52 19N 0 18W
Grafton, *Australia* .... **63 D5** 29 38S 152 58 E
Grafton, *N. Dak., U.S.A.* **80 A6** 48 25N 97 25W
Grafton, *W. Va., U.S.A.* **76 F5** 39 21N 80 2W
Graham, *Canada* **70 C1** 49 20N 90 30W
Graham, *U.S.A.* **81 J5** 33 6N 98 35W
Graham, Mt., *U.S.A.* **83 K9** 32 42N 109 52W
Graham Bell, Ostrov = Greem-Bell, Ostrov, *Russia* **26 A7** 81 0N 62 0 E
Graham I., *Canada* **72 C2** 53 40N 132 30W
Graham Land, *Antarctica* ... **5 C17** 65 0S 64 0W
Grahamstown, *S. Africa* ... **56 E4** 33 19S 26 31 E
Grahamsville, *U.S.A.* **79 E10** 41 51N 74 33W
Grain Coast, *W. Afr.* **50 H3** 4 20N 10 0W
Grajaú, *Brazil* **93 E9** 5 50S 46 4W
Grajaú →, *Brazil* **93 D10** 3 41S 44 48W
Grampian, *U.S.A.* **78 F6** 40 58N 78 37W
Grampian Highlands = Grampian Mts., *U.K.* **12 E5** 56 50N 4 0W
Grampian Mts., *U.K.* **12 E5** 56 50N 4 0W
Grampians, The, *Australia* .. **63 F3** 37 0S 142 20 E
Gran Canaria, *Canary Is.* ... **22 G4** 27 55N 15 35W
Gran Chaco, *S. Amer.* **94 B3** 25 0S 61 0W
Gran Paradiso, *Italy* .... **18 D7** 45 33N 7 17 E
Gran Sasso d'Itália, *Italy* ... **20 C5** 42 27N 13 42 E
Granada, *Nic.* **88 D2** 11 58N 86 0W
Granada, *Spain* **19 D4** 37 10N 3 35W
Granada, *U.S.A.* **81 F3** 38 4N 102 19W
Granadilla de Abona, *Canary Is.* **22 F3** 28 7N 16 33W
Granard, *Ireland* **13 C4** 53 47N 7 30W
Granbury, *U.S.A.* **81 J6** 32 27N 97 47W
Granby, *Canada* **79 A12** 45 25N 72 45W
Granby, *U.S.A.* **82 F11** 40 5N 105 56W
Grand →, *Canada* **78 D5** 42 51N 79 34W
Grand →, *Mo., U.S.A.* **80 F8** 39 23N 93 7W
Grand →, *S. Dak., U.S.A.* .. **80 C4** 45 40N 100 45W
Grand Bahama, *Bahamas* ... **88 A4** 26 40N 78 30W
Grand Bank, *Canada* **71 C8** 47 6N 55 48W
Grand Bassam, *Ivory C.* **50 G5** 5 10N 3 49W
Grand-Bourg, *Guadeloupe* .. **89 C7** 15 53N 61 19W
Grand Canal = Yun Ho →, *China* .... **35 E9** 39 10N 117 10 E
Grand Canyon, *U.S.A.* **83 H7** 36 3N 112 9W
Grand Canyon Nat. Park, *U.S.A.* **83 H7** 36 15N 112 30W
Grand Cayman, *Cayman Is.* . **88 C3** 19 20N 81 20W
Grand Centre, *Canada* **73 C6** 54 25N 110 13W
Grand Coulee, *U.S.A.* **82 C4** 47 57N 119 0W
Grand Coulee Dam, *U.S.A.* .. **82 C4** 47 57N 118 59W
Grand Falls, *Canada* **71 C6** 47 3N 67 44W
Grand Falls-Windsor, *Canada* **71 C8** 48 56N 55 40W
Grand Forks, *Canada* **72 D5** 49 0N 118 30W
Grand Forks, *U.S.A.* **80 B6** 47 55N 97 3W
Grand Gorge, *U.S.A.* **79 D10** 42 21N 74 29W
Grand Haven, *U.S.A.* **76 D2** 43 4N 86 13W
Grand I., *Mich., U.S.A.* **76 B2** 46 31N 86 40W
Grand I., *N.Y., U.S.A.* **78 D6** 43 0N 78 58W
Grand Island, *U.S.A.* **80 E5** 40 55N 98 21W
Grand Isle, *La., U.S.A.* **81 L9** 29 14N 90 0W
Grand Isle, *Vt., U.S.A.* **79 B11** 44 43N 73 18W
Grand Junction, *U.S.A.* **83 G9** 39 4N 108 33W
Grand L., *N.B., Canada* **71 C6** 45 57N 66 7W
Grand L., *Nfld., Canada* **71 C8** 49 0N 57 30W
Grand L., *Nfld., Canada* **71 B7** 53 40N 60 30W
Grand L., *U.S.A.* **81 L8** 29 55N 92 47W
Grand Lake, *U.S.A.* **82 F11** 40 15N 105 49W
Grand Manan I., *Canada* ... **71 D6** 44 45N 66 52W
Grand Marais, *Canada* **80 B9** 47 45N 90 25W
Grand Marais, *U.S.A.* **76 B3** 46 40N 85 59W
Grand-Mère, *Canada* **70 C5** 46 36N 72 40W
Grand Portage, *U.S.A.* **80 B10** 47 58N 89 41W
Grand Prairie, *U.S.A.* **81 J6** 32 47N 97 0W
Grand Rapids, *Canada* **73 C9** 53 12N 99 19W
Grand Rapids, *Mich., U.S.A.* **76 D2** 42 58N 85 40W
Grand Rapids, *Minn., U.S.A.* **80 B8** 47 14N 93 31W
Grand St-Bernard, Col du, *Europe* .... **18 D7** 45 50N 7 10 E
Grand Teton, *U.S.A.* **82 E8** 43 54N 111 50W
Grand Teton Nat. Park, *U.S.A.* **82 D8** 43 50N 110 50W
Grand Union Canal, *U.K.* ... **11 E7** 52 7N 0 53W
Grand View, *Canada* **73 C8** 51 10N 100 42W
Grande →, *Jujuy, Argentina* **94 A2** 24 20S 65 2W
Grande →, *Mendoza, Argentina* **94 D2** 36 52S 69 45W
Grande →, *Bolivia* **92 G6** 15 51S 64 39W
Grande →, *Bahia, Brazil* ... **93 F10** 11 30S 44 30W
Grande →, *Minas Gerais, Brazil* **93 H8** 20 6S 51 4W
Grande, B., *Argentina* **96 G3** 50 30S 68 20W
Grande, Rio →, *U.S.A.* **81 N6** 25 58N 97 9W
Grande Baleine, R. de la →, *Canada* **70 A4** 55 16N 77 47W
Grande Cache, *Canada* **72 C5** 53 53N 119 8W
Grande-Entrée, *Canada* **71 C7** 47 30N 61 40W
Grande Prairie, *Canada* **72 B5** 55 10N 118 50W
Grande-Rivière, *Canada* **71 C7** 48 26N 64 30W
Grande-Vallée, *Canada* **71 C6** 49 14N 65 8W
Grandfalls, *U.S.A.* **81 K3** 31 20N 102 51W
Grandview, *U.S.A.* **82 C4** 46 15N 119 54W
Graneros, *Chile* **94 C1** 34 5S 70 45W
Grangemouth, *U.K.* **12 E5** 56 1N 3 42W
Granger, *U.S.A.* **82 F9** 41 35N 109 58W
Grangeville, *U.S.A.* **82 D5** 45 56N 116 7W
Granisle, *Canada* **72 C3** 54 53N 126 13W
Granite City, *U.S.A.* **80 F9** 38 42N 90 9W
Granite Falls, *U.S.A.* **80 C7** 44 49N 95 33W
Granite L., *Canada* **71 C8** 48 8N 57 5W
Granite Mt., *U.S.A.* **85 M10** 33 5N 116 28W
Granite Pk., *U.S.A.* **82 D9** 45 10N 109 48W
Graniteville, *U.S.A.* **79 B12** 44 8N 72 29W
Granity, *N.Z.* .... **59 J3** 41 39S 171 51 E
Granja, *Brazil* **93 D10** 3 7S 40 50W
Granollers, *Spain* .... **19 B7** 41 39N 2 18 E
Grant, *U.S.A.* **80 E4** 40 53N 101 42W
Grant, Mt., *U.S.A.* **82 G4** 38 34N 118 48W
Grant City, *U.S.A.* **80 E7** 40 29N 94 25W
Grant I., *Australia* .... **60 B5** 11 10S 132 52 E
Grant Range, *U.S.A.* **83 G6** 38 30N 115 25W
Grantham, *U.K.* **10 E7** 52 55N 0 38W
Grantown-on-Spey, *U.K.* ... **12 D5** 57 20N 3 36W
Grants, *U.S.A.* **83 J10** 35 9N 107 52W
Grants Pass, *U.S.A.* **82 E2** 42 26N 123 19W
Grantsville, *U.S.A.* **82 F7** 40 36N 112 28W
Granville, *France* **18 B3** 48 50N 1 35W
Granville, *N. Dak., U.S.A.* ... **80 A4** 48 16N 100 47W
Granville, *N.Y., U.S.A.* **79 C11** 43 24N 73 16W
Granville, *Ohio, U.S.A.* **78 F2** 40 4N 82 31W
Granville L., *Canada* **73 B8** 56 18N 100 30W
Graskop, *S. Africa* .... **57 C5** 24 56S 30 49 E
Grass →, *Canada* **73 B9** 56 3N 96 33W
Grass Range, *U.S.A.* **82 C9** 47 0N 109 0W
Grass River Prov. Park, *Canada* **73 C8** 54 40N 100 50W
Grass Valley, *Calif., U.S.A.* .. **84 F6** 39 13N 121 4W
Grass Valley, *Oreg., U.S.A.* . **82 D3** 45 22N 120 47W
Grasse, *France* .... **18 E7** 43 38N 6 56 E
Grassflat, *U.S.A.* **78 F6** 41 0N 78 6W
Grasslands Nat. Park, *Canada* **73 D7** 49 11N 107 38W
Grassy, *Australia* .... **62 G3** 40 3S 144 5 E
Graulhet, *France* .... **18 E4** 43 45N 1 59 E
Gravelbourg, *Canada* **73 D7** 49 50N 106 35W
's-Gravenhage, *Neths.* ..... **15 B4** 52 7N 4 17 E
Gravenhurst, *Canada* **78 B5** 44 52N 79 20W
Gravesend, *Australia* ...... **63 D5** 29 35S 150 20 E
Gravesend, *U.K.* .... **11 F8** 51 26N 0 22 E
Gravois, Pointe-à-, *Haiti* ... **89 C5** 18 15N 73 56W
Grayling, *U.S.A.* **76 C3** 44 40N 84 43W
Grays Harbor, *U.S.A.* **82 C1** 46 59N 124 1W
Grays L., *U.S.A.* **82 E8** 43 4N 111 26W
Grays River, *U.S.A.* **84 D3** 46 21N 123 37W
Graz, *Austria* .... **16 E8** 47 4N 15 27 E
Greasy L., *Canada* **72 A4** 62 55N 122 12W
Great Abaco I., *Bahamas* ... **88 A4** 26 25N 77 10W
Great Artesian Basin, *Australia* .... **62 C3** 23 0S 144 0 E
Great Australian Bight, *Australia* .... **61 F5** 33 30S 130 0 E
Great Bahama Bank, *Bahamas* **88 B4** 23 15N 78 0W
Great Barrier I., *N.Z.* ...... **59 G5** 36 11S 175 25 E
Great Barrier Reef, *Australia* **62 B4** 18 0S 146 50 E
Great Barrington, *U.S.A.* ... **79 D11** 42 12N 73 22W
Great Basin, *U.S.A.* **82 G5** 40 0N 117 0W
Great Basin Nat. Park, *U.S.A.* **82 G6** 38 55N 114 14W
Great Bear →, *Canada* **68 B7** 65 0N 124 0W
Great Bear L., *Canada* **68 B7** 65 30N 120 0W
Great Belt = Store Bælt, *Denmark* .... **9 J14** 55 20N 11 0 E
Great Bend, *Kans., U.S.A.* .. **80 F5** 38 22N 98 46W
Great Bend, *Pa., U.S.A.* **79 E9** 41 58N 75 45W
Great Blasket I., *Ireland* ... **13 D1** 52 6N 10 32W
Great Britain, *Europe* **6 E5** 54 0N 2 15W
Great Codroy, *Canada* **71 C8** 47 51N 59 16W
Great Dividing Ra., *Australia* **62 C4** 23 0S 146 0 E
Great Driffield = Driffield, *U.K.* **10 C7** 54 0N 0 26W
Great Exuma I., *Bahamas* ... **88 B4** 23 30N 75 50W
Great Falls, *U.S.A.* **82 C8** 47 30N 111 17W
Great Fish = Groot Vis →, *S. Africa* .... **56 E4** 33 28S 27 5 E
Great Guana Cay, *Bahamas* . **88 B4** 24 0N 76 20W
Great Inagua I., *Bahamas* .. **89 B5** 21 0N 73 20W
Great Indian Desert = Thar Desert, *India* .... **42 F5** 28 0N 72 0 E
Great Karoo, *S. Africa* ..... **56 E3** 31 55S 21 0 E
Great Lake, *Australia* ...... **62 G4** 41 50S 146 40 E
Great Lakes, *N. Amer.* **66 E11** 46 0N 84 0W
Great Malvern, *U.K.* **11 E5** 52 7N 2 18W
Great Miami →, *U.S.A.* **76 F3** 39 20N 84 40W
Great Ormes Head, *U.K.* **10 D4** 53 20N 3 52W
Great Ouse →, *U.K.* .... **10 E8** 52 48N 0 21 E
Great Palm I., *Australia* .... **62 B4** 18 45S 146 40 E
Great Plains, *N. Amer.* **74 A6** 47 0N 105 0W
Great Ruaha →, *Tanzania* .. **54 D4** 7 56S 37 52 E
Great Sacandaga Res., *U.S.A.* **79 C10** 43 6N 74 16W
Great Saint Bernard Pass = Grand St-Bernard, Col du, *Europe* .... **18 D7** 45 50N 7 10 E
Great Salt L., *U.S.A.* **82 F7** 41 15N 112 40W
Great Salt Lake Desert, *U.S.A.* **82 F7** 40 50N 113 30W
Great Salt Plains L., *U.S.A.* .. **81 G5** 36 45N 98 8W
Great Sandy Desert, *Australia* **60 D3** 21 0S 124 0 E
Great Sangi = Sangihe, Pulau, *Indonesia* .... **37 D7** 3 35N 125 30 E
Great Skellig, *Ireland* **13 E1** 51 47N 10 33W
Great Slave L., *Canada* **72 A5** 61 23N 115 38W
Great Smoky Mts. Nat. Park, *U.S.A.* **77 H4** 35 40N 83 40W
Great Snow Mt., *Canada* ... **72 B4** 57 26N 124 0W
Great Stour = Stour →, *U.K.* **11 F9** 51 18N 1 22 E
Great Victoria Desert, *Australia* .... **61 E4** 29 30S 126 30 E
Great Wall, *China* .... **34 E5** 38 30N 109 30 E
Great Whernside, *U.K.* **10 C6** 54 10N 1 58W
Great Yarmouth, *U.K.* ...... **11 E9** 52 37N 1 44 E
Greater Antilles, *W. Indies* .. **89 C5** 17 40N 74 0W
Greater London □, *U.K.* **11 F7** 51 31N 0 6W
Greater Manchester □, *U.K.* . **10 D5** 53 30N 2 15W
Greater Sunda Is., *Indonesia* **36 F4** 7 0S 112 0 E
Greco, C., *Cyprus* .... **23 E13** 34 57N 34 5 E
Gredos, Sierra de, *Spain* ... **19 B3** 40 20N 5 0W
Greece, *U.S.A.* **78 C7** 43 13N 77 41W
Greece ■, *Europe* .... **21 E9** 40 0N 23 0 E
Greeley, *Colo., U.S.A.* **80 E2** 40 25N 104 42W
Greeley, *Nebr., U.S.A.* **80 E5** 41 33N 98 32W
Greem-Bell, Ostrov, *Russia* . **26 A7** 81 0N 62 0 E
Green, *U.S.A.* **82 E2** 43 9N 123 22W
Green →, *Ky., U.S.A.* **76 G2** 37 54N 87 30W
Green →, *Utah, U.S.A.* **83 G9** 38 11N 109 53W
Green B., *U.S.A.* **76 C2** 45 0N 87 30W
Green Bay, *U.S.A.* **76 C2** 44 31N 88 0W
Green C., *Australia* .... **63 F5** 37 13S 150 1 E
Green Cove Springs, *U.S.A.* . **77 L5** 29 59N 81 42W
Green Lake, *Canada* **73 C7** 54 17N 107 47W
Green Mts., *U.S.A.* **79 C12** 43 45N 72 45W
Green River, *Utah, U.S.A.* .. **83 G8** 38 59N 110 10W
Green River, *Wyo., U.S.A.* .. **82 F9** 41 32N 109 28W

Hefei

Hailuoto, *Finland* **8 D21** 65 3N 24 45 E
Hainan □, *China* **33 E5** 19 0N 109 30 E
Hainaut □, *Belgium* **15 D4** 50 30N 4 0 E
Haines, *Alaska, U.S.A.* **72 B1** 59 14N 135 26W
Haines, *Oreg., U.S.A.* **82 D5** 44 55N 117 56W
Haines City, *U.S.A.* **77 L5** 28 7N 81 38W
Haines Junction, *Canada* **72 A1** 60 45N 137 30W
Haiphong, *Vietnam* **32 D5** 20 47N 106 41 E
Haiti ■, *W. Indies* **89 C5** 19 0N 72 30W
Haiya, *Sudan* **51 E13** 18 20N 36 21 E
Haiyang, *China* **35 F11** 36 47N 121 9 E
Haiyuan, *China* **34 F3** 36 35N 105 52 E
Haizhou, *China* **35 G10** 34 37N 119 7 E
Haizhou Wan, *China* **35 G10** 34 50N 119 20 E
Hajdúböszörmény, *Hungary* **17 E11** 47 40N 21 30 E
Hajipur, *India* **43 G11** 25 45N 85 13 E
Hājjī Muḩsin, *Iraq* **44 C5** 32 35N 45 29 E
Ḩājjīābād, *Iran* **45 D7** 28 19N 55 55 E
Ḩājjīābād-e Zarrīn, *Iran* **45 C7** 33 9N 54 51 E
Hajnówka, *Poland* **17 B12** 52 47N 23 35 E
Hakansson, Mts., *Dem. Rep. of the Congo* **55 D2** 8 40S 25 45 E
Hakkâri, *Turkey* **44 B4** 37 34N 43 44 E
Hakken-Zan, *Japan* **31 G7** 34 10N 135 54 E
Hakodate, *Japan* **30 D10** 41 45N 140 44 E
Hakos, *Namibia* **56 C2** 23 13S 16 21 E
Haku-San, *Japan* **31 F8** 36 9N 136 46 E
Hakui, *Japan* **31 F8** 36 53N 136 47 E
Hala, *Pakistan* **40 G6** 25 43N 68 20 E
Ḩalab, *Syria* **44 B3** 36 10N 37 15 E
Ḩalabjah, *Iraq* **44 C5** 35 10N 45 58 E
Halaib, *Sudan* **51 D13** 22 12N 36 30 E
Hālat 'Ammār, *Si. Arabia* **44 D3** 29 10N 36 4 E
Halbā, *Lebanon* **47 A5** 34 34N 36 6 E
Halberstadt, *Germany* **16 C6** 51 54N 11 3 E
Halcombe, *N.Z.* **59 J5** 40 8S 175 30 E
Halcon, *Phil.* **37 B6** 13 0N 121 30 E
Halden, *Norway* **9 G14** 59 9N 11 23 E
Haldia, *India* **41 H16** 22 5N 88 3 E
Haldwani, *India* **43 E8** 29 31N 79 30 E
Hale →, *Australia* **62 C2** 24 56S 135 53 E
Halesowen, *U.K.* **11 E5** 52 27N 2 3W
Haleyville, *U.S.A.* **77 H2** 34 14N 87 37W
Halfmoon Bay, *N.Z.* **59 M2** 46 50S 168 5 E
Halfway →, *Canada* **72 B4** 56 12N 121 32W
Halia, *India* **43 G10** 24 50N 82 19 E
Haliburton, *Canada* **78 A6** 45 3N 78 30W
Halifax, *Australia* **62 B4** 18 32S 146 22 E
Halifax, *Canada* **71 D7** 44 38N 63 35W
Halifax, *U.K.* **10 D6** 53 43N 1 52W
Halifax, *U.S.A.* **78 F8** 40 25N 76 55W
Halifax B., *Australia* **62 B4** 18 50S 147 0 E
Halifax I., *Namibia* **56 D2** 26 38S 15 4 E
Halīl →, *Iran* **45 E8** 27 40N 58 30 E
Halkirk, *U.K.* **12 C5** 58 30N 3 29W
Hall Beach = Sanirajak, *Canada* **69 B11** 68 46N 81 12W
Hall Pen., *Canada* **69 B13** 63 30N 66 0W
Hall Pt., *Australia* **60 C3** 15 40S 124 23 E
Halland, *Sweden* **9 H15** 57 8N 12 47 E
Halle, *Belgium* **15 D4** 50 44N 4 13 E
Halle, *Germany* **16 C6** 51 30N 11 56 E
Hällefors, *Sweden* **9 G16** 59 47N 14 31 E
Hallett, *Australia* **63 E2** 33 25S 138 55 E
Hallettsville, *U.S.A.* **81 L6** 29 27N 96 57W
Hallim, *S. Korea* **35 H14** 33 24N 126 15 E
Hallingdalselvi →, *Norway* **9 F13** 60 23N 9 35 E
Hallock, *U.S.A.* **80 A6** 48 47N 96 57W
Halls Creek, *Australia* **60 C4** 18 16S 127 38 E
Hallsberg, *Sweden* **9 G16** 59 5N 15 7 E
Hallstead, *U.S.A.* **79 E9** 41 58N 75 45W
Halmahera, *Indonesia* **37 D7** 0 40N 128 0 E
Halmstad, *Sweden* **9 H15** 56 41N 12 52 E
Hälsingborg = Helsingborg, *Sweden* **9 H15** 56 3N 12 42 E
Hälsingland, *Sweden* **9 F16** 61 40N 16 5 E
Halstead, *U.K.* **11 F8** 51 57N 0 40 E
Halti, *Finland* **8 B19** 69 17N 21 18 E
Halton □, *U.K.* **10 D5** 53 22N 2 45W
Haltwhistle, *U.K.* **10 C5** 54 58N 2 26W
Ḩālūl, *Qatar* **45 E7** 25 40N 52 40 E
Halvad, *India* **42 H4** 23 1N 71 11 E
Ḩalvān, *Iran* **45 C8** 33 57N 56 15 E
Ham Tan, *Vietnam* **39 G6** 10 40N 107 45 E
Ham Yen, *Vietnam* **38 A5** 22 4N 105 3 E
Hamab, *Namibia* **56 D2** 28 7S 19 16 E
Hamada, *Japan* **31 G6** 34 56N 132 4 E
Hamadān, *Iran* **45 C6** 34 52N 48 32 E
Hamadān □, *Iran* **45 C6** 35 0N 49 0 E
Hamāh, *Syria* **44 C3** 35 5N 36 40 E
Hamamatsu, *Japan* **31 G8** 34 45N 137 45 E
Hamar, *Norway* **9 F14** 60 48N 11 7 E
Hamâta, Gebel, *Egypt* **44 E2** 24 17N 35 0 E
Hambantota, *Sri Lanka* **40 R12** 6 10N 81 10 E
Hamber Prov. Park, *Canada* **72 C5** 52 20N 118 0W
Hamburg, *Germany* **16 B5** 53 33N 9 59 E
Hamburg, *Ark., U.S.A.* **81 J9** 33 14N 91 48W
Hamburg, *N.Y., U.S.A.* **78 D6** 42 43N 78 50W
Hamburg, *Pa., U.S.A.* **79 F9** 40 33N 75 59W
Ḩamḑ, W. al →, *Si. Arabia* **44 E3** 24 55N 36 20 E
Hamden, *U.S.A.* **79 E12** 41 23N 72 54W
Häme, *Finland* **9 F20** 61 38N 25 10 E
Hämeenlinna, *Finland* **9 F21** 61 0N 24 28 E
Hamelin Pool, *Australia* **61 E1** 26 22S 114 20 E
Hameln, *Germany* **16 B5** 52 6N 9 21 E
Hamerkaz □, *Israel* **47 C3** 32 15N 34 55 E
Hamersley Ra., *Australia* **60 D2** 22 0S 117 45 E
Hamhung, *N. Korea* **35 E14** 39 54N 127 30 E
Hami, *China* **32 B4** 42 55N 93 25 E
Hamilton, *Australia* **63 F3** 37 45S 142 2 E
Hamilton, *Canada* **78 C5** 43 15N 79 50W
Hamilton, *N.Z.* **59 G5** 37 47S 175 19 E
Hamilton, *U.K.* **12 F4** 55 46N 4 2W
Hamilton, *Ala., U.S.A.* **77 H1** 34 9N 87 59W
Hamilton, *Mont., U.S.A.* **82 C6** 46 15N 114 10W
Hamilton, *N.Y., U.S.A.* **79 D9** 42 50N 75 33W
Hamilton, *Ohio, U.S.A.* **76 F3** 39 24N 84 34W
Hamilton, *Tex., U.S.A.* **81 K5** 31 42N 98 7W
Hamilton →, *Australia* **62 C2** 23 30S 139 47 E
Hamilton City, *U.S.A.* **84 F4** 39 45N 122 1W
Hamilton Inlet, *Canada* **71 B8** 54 0N 57 30W
Hamilton Mt., *U.S.A.* **79 C10** 43 25N 74 22W
Hamina, *Finland* **9 F22** 60 34N 27 12 E
Hamirpur, *H.P., India* **42 D7** 31 41N 76 31 E
Hamirpur, *Ut. P., India* **43 G9** 25 57N 80 9 E
Hamlet, *U.S.A.* **77 H6** 34 53N 79 42W
Hamley Bridge, *Australia* **63 E2** 34 17S 138 35 E
Hamlin = Hameln, *Germany* **16 B5** 52 6N 9 21 E
Hamlin, *N.Y., U.S.A.* **78 C7** 43 17N 77 55W
Hamlin, *Tex., U.S.A.* **81 J4** 32 53N 100 8W
Hamm, *Germany* **16 C4** 51 40N 7 50 E
Ḩammār, Hawr al, *Iraq* **44 D5** 30 50N 47 10 E
Hammerfest, *Norway* **8 A20** 70 39N 23 41 E
Hammond, *Ind., U.S.A.* **76 E2** 41 38N 87 30W
Hammond, *La., U.S.A.* **81 K9** 30 30N 90 28W
Hammond, *N.Y., U.S.A.* **79 B9** 44 27N 75 42W
Hammondsport, *U.S.A.* **78 D7** 42 25N 77 13W
Hammonton, *U.S.A.* **76 F8** 39 39N 74 48W
Hampden, *N.Z.* **59 L3** 45 18S 170 50 E
Hampshire □, *U.K.* **11 F6** 51 7N 1 23W
Hampshire Downs, *U.K.* **11 F6** 51 15N 1 10W
Hampton, *N.B., Canada* **71 C6** 45 32N 65 51W
Hampton, *Ont., Canada* **78 C6** 43 58N 78 45W
Hampton, *Ark., U.S.A.* **81 J8** 33 32N 92 28W
Hampton, *Iowa, U.S.A.* **80 D8** 42 45N 93 13W
Hampton, *N.H., U.S.A.* **79 D14** 42 57N 70 50W
Hampton, *S.C., U.S.A.* **77 J5** 32 52N 81 7W
Hampton, *Va., U.S.A.* **76 G7** 37 2N 76 21W
Hampton Bays, *U.S.A.* **79 F12** 40 53N 72 30W
Hampton Tableland, *Australia* **61 F4** 32 0S 127 0 E
Hamyang, *S. Korea* **35 G14** 35 32N 127 42 E
Han Pijesak, *Bos.-H.* **21 B8** 44 5N 18 57 E
Hanak, *Si. Arabia* **44 E3** 25 32N 37 0 E
Hanamaki, *Japan* **30 E10** 39 23N 141 7 E
Hanang, *Tanzania* **54 C4** 4 30S 35 25 E
Hanau, *Germany* **16 C5** 50 7N 8 56 E
Hanbogd = Ihbulag, *Mongolia* **34 C4** 43 11N 107 10 E
Hancheng, *China* **34 G6** 35 31N 110 25 E
Hancock, *Mich., U.S.A.* **80 B10** 47 8N 88 35W
Hancock, *N.Y., U.S.A.* **79 E9** 41 57N 75 17W
Handa, *Japan* **31 G8** 34 53N 136 55 E
Handan, *China* **34 F8** 36 35N 114 28 E
Handeni, *Tanzania* **54 D4** 5 25S 38 2 E
Handwara, *India* **43 B6** 34 21N 74 20 E
Hanegev, *Israel* **47 E4** 30 50N 35 0 E
Hanford, *U.S.A.* **84 J7** 36 20N 119 39W
Hang Chat, *Thailand* **38 C2** 18 20N 99 21 E
Hang Dong, *Thailand* **38 C2** 18 41N 98 55 E
Hangang →, *S. Korea* **35 F14** 37 50N 126 30 E
Hangayn Nuruu, *Mongolia* **32 B4** 47 30N 99 0 E
Hangchou = Hangzhou, *China* **33 C7** 30 18N 120 11 E
Hanggin Houqi, *China* **34 D4** 40 58N 107 4 E
Hanggin Qi, *China* **34 E5** 39 52N 108 50 E
Hangu, *China* **35 E9** 39 18N 117 53 E
Hangzhou, *China* **33 C7** 30 18N 120 11 E
Hangzhou Wan, *China* **33 C7** 30 15N 120 45 E
Hanhongor, *Mongolia* **34 C3** 43 55N 104 28 E
Ḩanīdh, *Si. Arabia* **45 E6** 26 35N 48 38 E
Ḩanīsh, *Yemen* **46 E3** 13 45N 42 46 E
Hankinson, *U.S.A.* **80 B6** 46 4N 96 54W
Hanko, *Finland* **9 G20** 59 50N 22 57 E
Hanksville, *U.S.A.* **83 G8** 38 22N 110 43W
Hanle, *India* **43 C8** 32 42N 79 4 E
Hanmer Springs, *N.Z.* **59 K4** 42 32S 172 50 E
Hann →, *Australia* **60 C4** 17 26S 126 17 E
Hann, Mt., *Australia* **60 C4** 15 45S 126 0 E
Hanna, *Canada* **72 C6** 51 40N 111 54W
Hanna, *U.S.A.* **82 F10** 41 52N 106 34W
Hannah B., *Canada* **70 B4** 51 40N 80 0W
Hannibal, *Mo., U.S.A.* **80 F9** 39 42N 91 22W
Hannibal, *N.Y., U.S.A.* **79 C8** 43 19N 76 35W
Hannover, *Germany* **16 B5** 52 22N 9 46 E
Hanoi, *Vietnam* **32 D5** 21 5N 105 55 E
Hanover = Hannover, *Germany* **16 B5** 52 22N 9 46 E
Hanover, *Canada* **78 B3** 44 9N 81 2W
Hanover, *S. Africa* **56 E3** 31 4S 24 29 E
Hanover, *N.H., U.S.A.* **79 C12** 43 42N 72 17W
Hanover, *Ohio, U.S.A.* **78 F2** 40 4N 82 16W
Hanover, *Pa., U.S.A.* **76 F7** 39 48N 76 59W
Hanover, I., *Chile* **96 G2** 51 0S 74 50W
Hansdiha, *India* **43 G12** 24 36N 87 5 E
Hansi, *India* **42 E6** 29 10N 75 57 E
Hanson, L., *Australia* **63 E2** 31 0S 136 15 E
Hantsavichy, *Belarus* **17 B14** 52 49N 26 30 E
Hanumangarh, *India* **42 E6** 29 35N 74 19 E
Hanzhong, *China* **34 H4** 33 10N 107 1 E
Hanzhuang, *China* **35 G9** 34 33N 117 23 E
Haora, *India* **43 H13** 22 37N 88 20 E
Haparanda, *Sweden* **8 D21** 65 52N 24 8 E
Happy, *U.S.A.* **81 H4** 34 45N 101 52W
Happy Camp, *U.S.A.* **82 F2** 41 48N 123 23W
Happy Valley-Goose Bay, *Canada* **71 B7** 53 15N 60 20W
Hapsu, *N. Korea* **35 D15** 41 13N 128 51 E
Hapur, *India* **42 E7** 28 45N 77 45 E
Ḩaql, *Si. Arabia* **47 F3** 29 10N 34 58 E
Har, *Indonesia* **37 F8** 5 16S 133 14 E
Har-Ayrag, *Mongolia* **34 B5** 45 47N 109 16 E
Har Hu, *China* **32 C4** 38 20N 97 38 E
Har Us Nuur, *Mongolia* **32 B4** 48 0N 92 0 E
Har Yehuda, *Israel* **47 D3** 31 35N 34 57 E
Ḩaraḑ, *Si. Arabia* **46 C4** 24 22N 49 0 E
Haranomachi, *Japan* **30 F10** 37 38N 140 58 E
Harare, *Zimbabwe* **55 F3** 17 43S 31 2 E
Harbin, *China* **35 B14** 45 48N 126 40 E
Harbor Beach, *U.S.A.* **78 C2** 43 51N 82 39W
Harbour Breton, *Canada* **71 C8** 47 29N 55 50W
Harbour Deep, *Canada* **71 B8** 50 25N 56 32W
Harda, *India* **42 H7** 22 27N 77 5 E
Hardangerfjorden, *Norway* **9 F12** 60 5N 6 0 E
Hardangervidda, *Norway* **9 F12** 60 7N 7 20 E
Hardap Dam, *Namibia* **56 C2** 24 32S 17 50 E
Hardenberg, *Neths.* **15 B6** 52 34N 6 37 E
Harderwijk, *Neths.* **15 B5** 52 21N 5 38 E
Hardey →, *Australia* **60 D2** 22 45S 116 8 E
Hardin, *U.S.A.* **82 D10** 45 44N 107 37W
Harding, *S. Africa* **57 E4** 30 35S 29 55 E
Harding Ra., *Australia* **60 C3** 16 17S 124 55 E
Hardisty, *Canada* **72 C6** 52 40N 111 18W
Hardoi, *India* **43 F9** 27 26N 80 6 E
Hardwar = Haridwar, *India* **42 E8** 29 58N 78 9 E
Hardwick, *U.S.A.* **79 B12** 44 30N 72 22W
Hardy, Pen., *Chile* **96 H3** 55 30S 68 20W
Hare B., *Canada* **71 B8** 51 15N 55 45W
Hareid, *Norway* **9 E12** 62 22N 6 1 E
Harer, *Ethiopia* **46 F3** 9 20N 42 8 E
Hargeisa, *Somali Rep.* **46 F3** 9 30N 44 2 E
Hari →, *Indonesia* **36 E2** 1 16S 104 5 E
Haria, *Canary Is.* **22 E6** 29 8N 13 32W
Haridwar, *India* **42 E8** 29 58N 78 9 E
Harim, Jabal al, *Oman* **45 E8** 25 58N 56 14 E
Haringhata →, *Bangla.* **41 J16** 22 0N 89 58 E
Harīrūd →, *Asia* **40 A2** 37 24N 60 38 E
Härjedalen, *Sweden* **9 E15** 62 22N 13 5 E
Harlan, *Iowa, U.S.A.* **80 E7** 41 39N 95 19W
Harlan, *Ky., U.S.A.* **77 G4** 36 51N 83 19W
Harlech, *U.K.* **10 E3** 52 52N 4 6W
Harlem, *U.S.A.* **82 B9** 48 32N 108 47W
Harlingen, *Neths.* **15 A5** 53 11N 5 25 E
Harlingen, *U.S.A.* **81 M6** 26 12N 97 42W
Harlow, *U.K.* **11 F8** 51 46N 0 8 E
Harlowton, *U.S.A.* **82 C9** 46 26N 109 50W
Harnai, *Pakistan* **42 D2** 30 6N 67 56 E
Harney Basin, *U.S.A.* **82 E4** 43 30N 119 0W
Harney L., *U.S.A.* **82 E4** 43 14N 119 8W
Harney Peak, *U.S.A.* **80 D3** 43 52N 103 32W
Härnösand, *Sweden* **9 E17** 62 38N 17 55 E
Haroldswick, *U.K.* **12 A8** 60 48N 0 50W
Harp L., *Canada* **71 A7** 55 5N 61 50W
Harper, *Liberia* **50 H4** 4 25N 7 43W
Harrai, *India* **43 H8** 22 37N 79 13 E
Harrand, *Pakistan* **42 E4** 29 28N 70 3 E
Harricana →, *Canada* **70 B4** 50 56N 79 32W
Harriman, *U.S.A.* **77 H3** 35 56N 84 33W
Harrington Harbour, *Canada* **71 B8** 50 31N 59 30W
Harris, *U.K.* **12 D2** 57 50N 6 55W
Harris, Sd. of, *U.K.* **12 D1** 57 44N 7 6W
Harris L., *Australia* **63 E2** 31 10S 135 10 E
Harris Pt., *Canada* **78 C2** 43 6N 82 9W
Harrisburg, *Ill., U.S.A.* **81 G10** 37 44N 88 32W
Harrisburg, *Nebr., U.S.A.* **80 E3** 41 33N 103 44W
Harrisburg, *Pa., U.S.A.* **78 F8** 40 16N 76 53W
Harrismith, *S. Africa* **57 D4** 28 15S 29 8 E
Harrison, *Ark., U.S.A.* **81 G8** 36 14N 93 7W
Harrison, *Maine, U.S.A.* **79 B14** 44 7N 70 39W
Harrison, *Nebr., U.S.A.* **80 D3** 42 41N 103 53W
Harrison, C., *Canada* **71 B8** 54 55N 57 55W
Harrison L., *Canada* **72 D4** 49 33N 121 50W
Harrisonburg, *U.S.A.* **76 F6** 38 27N 78 52W
Harrisonville, *U.S.A.* **80 F7** 38 39N 94 21W
Harriston, *Canada* **78 C4** 43 57N 80 53W
Harrisville, *Mich., U.S.A.* **78 B1** 44 39N 83 17W
Harrisville, *N.Y., U.S.A.* **79 B9** 44 9N 75 19W
Harrisville, *Pa., U.S.A.* **78 E5** 41 8N 80 0W
Harrodsburg, *U.S.A.* **76 G3** 37 46N 84 51W
Harrogate, *U.K.* **10 C6** 54 0N 1 33W
Harrow □, *U.K.* **11 F7** 51 35N 0 21W
Harrowsmith, *Canada* **79 B8** 44 24N 76 40W
Harry S. Truman Reservoir, *U.S.A.* **80 F7** 38 16N 93 24W
Harsīn, *Iran* **44 C5** 34 18N 47 33 E
Harstad, *Norway* **8 B17** 68 48N 16 30 E
Harsud, *India* **42 H7** 22 6N 76 44 E
Hart, *U.S.A.* **76 D2** 43 42N 86 22W
Hart, L., *Australia* **63 E2** 31 10S 136 25 E
Hartbees →, *S. Africa* **56 D3** 28 45S 20 32 E
Hartford, *Conn., U.S.A.* **79 E12** 41 46N 72 41W
Hartford, *Ky., U.S.A.* **76 G2** 37 27N 86 55W
Hartford, *S. Dak., U.S.A.* **80 D6** 43 38N 96 57W
Hartford, *Wis., U.S.A.* **80 D10** 43 19N 88 22W
Hartford City, *U.S.A.* **76 E3** 40 27N 85 22W
Hartland, *Canada* **71 C6** 46 20N 67 32W
Hartland Pt., *U.K.* **11 F3** 51 1N 4 32W
Hartlepool, *U.K.* **10 C6** 54 42N 1 13W
Hartlepool □, *U.K.* **10 C6** 54 42N 1 17W
Hartley Bay, *Canada* **72 C3** 53 25N 129 15W
Hartmannberge, *Namibia* **56 B1** 17 0S 13 0 E
Hartney, *Canada* **73 D8** 49 30N 100 35W
Harts →, *S. Africa* **56 D3** 28 24S 24 17 E
Hartselle, *U.S.A.* **77 H2** 34 27N 86 56W
Hartshorne, *U.S.A.* **81 H7** 34 51N 95 34W
Hartstown, *U.S.A.* **78 E4** 41 33N 80 23W
Hartsville, *U.S.A.* **77 H5** 34 23N 80 4W
Hartswater, *S. Africa* **56 D3** 27 34S 24 43 E
Hartwell, *U.S.A.* **77 H4** 34 21N 82 56W
Harunabad, *Pakistan* **42 E5** 29 35N 73 8 E
Harvand, *Iran* **45 D7** 28 25N 55 43 E
Harvey, *Australia* **61 F2** 33 5S 115 54 E
Harvey, *Ill., U.S.A.* **76 E2** 41 36N 87 50W
Harvey, *N. Dak., U.S.A.* **80 B5** 47 47N 99 56W
Harwich, *U.K.* **11 F9** 51 56N 1 17 E
Haryana □, *India* **42 E7** 29 0N 76 10 E
Haryn →, *Belarus* **17 B14** 52 7N 27 17 E
Harz, *Germany* **16 C6** 51 38N 10 44 E
Hasa □, *Si. Arabia* **45 E6** 25 50N 49 0 E
Ḩasanābād, *Iran* **45 C7** 32 8N 52 44 E
Hasdo →, *India* **43 J10** 21 44N 82 44 E
Hashimoto, *Japan* **31 G7** 34 19N 135 37 E
Hashtjerd, *Iran* **45 C6** 35 52N 50 40 E
Haskell, *U.S.A.* **81 J5** 33 10N 99 44W
Haslemere, *U.K.* **11 F7** 51 5N 0 43W
Hasselt, *Belgium* **15 D5** 50 56N 5 21 E
Hassi Messaoud, *Algeria* **50 B7** 31 51N 6 1 E
Hässleholm, *Sweden* **9 H15** 56 10N 13 46 E
Hastings, *N.Z.* **59 H6** 39 39S 176 52 E
Hastings, *U.K.* **11 G8** 50 51N 0 35 E
Hastings, *Mich., U.S.A.* **76 D3** 42 39N 85 17W
Hastings, *Minn., U.S.A.* **80 C8** 44 44N 92 51W
Hastings, *Nebr., U.S.A.* **80 E5** 40 35N 98 23W
Hastings Ra., *Australia* **63 E5** 31 15S 152 14 E
Hat Yai, *Thailand* **39 J3** 7 1N 100 27 E
Hatanbulag = Ergel, *Mongolia* **34 C5** 43 8N 109 5 E
Hatay = Antalya, *Turkey* **25 G5** 36 52N 30 45 E
Hatch, *U.S.A.* **83 K10** 32 40N 107 9W
Hatchet L., *Canada* **73 B8** 58 36N 103 40W
Hateruma-Shima, *Japan* **31 M1** 24 3N 123 47 E
Hatfield P.O., *Australia* **63 E3** 33 54S 143 49 E
Hatgal, *Mongolia* **32 A5** 50 26N 100 9 E
Hathras, *India* **42 F8** 27 36N 78 6 E
Hatia, *Bangla.* **41 H17** 22 30N 91 5 E
Hato Mayor, *Dom. Rep.* **89 C6** 18 46N 69 15W
Hatta, *India* **43 G8** 24 7N 79 36 E
Hattah, *Australia* **63 E3** 34 48S 142 17 E
Hatteras, C., *U.S.A.* **77 H8** 35 14N 75 32W
Hattiesburg, *U.S.A.* **81 K10** 31 20N 89 17W
Hatvan, *Hungary* **17 E10** 47 40N 19 45 E
Hau Bon = Cheo Reo, *Vietnam* **36 B3** 13 25N 108 28 E
Hau Duc, *Vietnam* **38 E7** 15 20N 108 13 E
Haugesund, *Norway* **9 G11** 59 23N 5 13 E
Haukipudas, *Finland* **8 D21** 65 12N 25 20 E
Haultain →, *Canada* **73 B7** 55 51N 106 46W
Hauraki G., *N.Z.* **59 G5** 36 35S 175 5 E
Haut Atlas, *Morocco* **50 B4** 32 30N 5 0W
Haut-Zaïre = Orientale □, *Dem. Rep. of the Congo* **54 B2** 2 20N 26 0 E
Hautes Fagnes = Hohe Venn, *Belgium* **15 D6** 50 30N 6 5 E
Hauts Plateaux, *Algeria* **48 C4** 35 0N 1 0 E
Havana = La Habana, *Cuba* **88 B3** 23 8N 82 22W
Havana, *U.S.A.* **80 E9** 40 18N 90 4W
Havant, *U.K.* **11 G7** 50 51N 0 58W
Havasu, L., *U.S.A.* **85 L12** 34 18N 114 28W
Havel →, *Germany* **16 B7** 52 50N 12 3 E
Havelian, *Pakistan* **42 B5** 34 2N 73 10 E
Havelock, *Canada* **78 B7** 44 26N 77 53W
Havelock, *N.Z.* **59 J4** 41 17S 173 48 E
Havelock, *U.S.A.* **77 H7** 34 53N 76 54W
Haverfordwest, *U.K.* **11 F3** 51 48N 4 58W
Haverhill, *U.S.A.* **79 D13** 42 47N 71 5W
Haverstraw, *U.S.A.* **79 E11** 41 12N 73 58W
Havirga, *Mongolia* **34 B7** 45 41N 113 5 E
Havířov, *Czech Rep.* **17 D10** 49 46N 18 20 E
Havlíčkův Brod, *Czech Rep.* **16 D8** 49 36N 15 33 E
Havre, *U.S.A.* **82 B9** 48 33N 109 41W
Havre-Aubert, *Canada* **71 C7** 47 12N 61 56W
Havre-St.-Pierre, *Canada* **71 B7** 50 18N 63 33W
Haw →, *U.S.A.* **77 H6** 35 36N 79 3W
Hawaii □, *U.S.A.* **74 H16** 19 30N 156 30W
Hawaii I., *Pac. Oc.* **74 J17** 20 0N 155 0W
Hawaiian Is., *Pac. Oc.* **74 H17** 20 30N 156 0W
Hawaiian Ridge, *Pac. Oc.* **65 E11** 24 0N 165 0W
Hawarden, *U.S.A.* **80 D6** 43 0N 96 29W
Hawea, L., *N.Z.* **59 L2** 44 28S 169 19 E
Hawera, *N.Z.* **59 H5** 39 35S 174 19 E
Hawick, *U.K.* **12 F6** 55 26N 2 47W
Hawk Junction, *Canada* **70 C3** 48 5N 84 38W
Hawke B., *N.Z.* **59 H6** 39 25S 177 20 E
Hawker, *Australia* **63 E2** 31 59S 138 22 E
Hawkesbury, *Canada* **70 C5** 45 37N 74 37W
Hawkesbury I., *Canada* **72 C3** 53 37N 129 3W
Hawkesbury Pt., *Australia* **62 A1** 11 55S 134 5 E
Hawkinsville, *U.S.A.* **77 J4** 32 17N 83 28W
Hawley, *Minn., U.S.A.* **80 B6** 46 53N 96 19W
Hawley, *Pa., U.S.A.* **79 E9** 41 28N 75 11W
Ḩawrān, W. →, *Iraq* **44 C4** 33 58N 42 34 E
Hawsh Mūssá, *Lebanon* **47 B4** 33 45N 35 55 E
Hawthorne, *U.S.A.* **82 G4** 38 32N 118 38W
Hay, *Australia* **63 E3** 34 30S 144 51 E
Hay →, *Australia* **62 C2** 24 50S 138 0 E
Hay →, *Canada* **72 A5** 60 50N 116 26W
Hay, C., *Australia* **60 B4** 14 5S 129 29 E
Hay I., *Canada* **78 B4** 44 53N 80 58W
Hay L., *Canada* **72 B5** 58 50N 118 50W
Hay-on-Wye, *U.K.* **11 E4** 52 5N 3 8W
Hay River, *Canada* **72 A5** 60 51N 115 44W
Hay Springs, *U.S.A.* **80 D3** 42 41N 102 41W
Haya = Tehoru, *Indonesia* **37 E7** 3 23S 129 30 E
Hayachine-San, *Japan* **30 E10** 39 34N 141 29 E
Hayden, *U.S.A.* **82 F10** 40 30N 107 16W
Haydon, *Australia* **62 B3** 18 0S 141 30 E
Hayes, *U.S.A.* **80 C4** 44 23N 101 1W
Hayes →, *Canada* **70 A1** 57 3N 92 12W
Hayes Creek, *Australia* **60 B5** 13 43S 131 22 E
Hayle, *U.K.* **11 G2** 50 11N 5 26W
Hayling I., *U.K.* **11 G7** 50 48N 0 59W
Hayrabolu, *Turkey* **21 D12** 41 12N 27 5 E
Hays, *Canada* **72 C6** 50 6N 111 48W
Hays, *U.S.A.* **80 F5** 38 53N 99 20W
Haysyn, *Ukraine* **17 D15** 48 57N 29 25 E
Hayvoron, *Ukraine* **17 D15** 48 22N 29 52 E
Hayward, *Calif., U.S.A.* **84 H4** 37 40N 122 5W
Hayward, *Wis., U.S.A.* **80 B9** 46 1N 91 29W
Haywards Heath, *U.K.* **11 G7** 51 0N 0 5W
Haẕafon □, *Israel* **47 C4** 32 40N 35 20 E
Hazārān, Kūh-e, *Iran* **45 D8** 29 35N 57 20 E
Hazard, *U.S.A.* **76 G4** 37 15N 83 12W
Hazaribag, *India* **43 H11** 23 58N 85 26 E
Hazaribag Road, *India* **43 G11** 24 12N 85 57 E
Hazelton, *Canada* **72 B3** 55 20N 127 42W
Hazelton, *U.S.A.* **80 B4** 46 29N 100 17W
Hazen, *U.S.A.* **80 B4** 47 18N 101 38W
Hazlehurst, *Ga., U.S.A.* **77 K4** 31 52N 82 36W
Hazlehurst, *Miss., U.S.A.* **81 K9** 31 52N 90 24W
Hazlet, *U.S.A.* **79 F10** 40 25N 74 12W
Hazleton, *U.S.A.* **79 F9** 40 57N 75 59W
Hazlett, L., *Australia* **60 D4** 21 30S 128 48 E
Hazro, *Turkey* **44 B4** 38 15N 40 47 E
Head of Bight, *Australia* **61 F5** 31 30S 131 25 E
Headlands, *Zimbabwe* **55 F3** 18 15S 32 2 E
Healdsburg, *U.S.A.* **84 G4** 38 37N 122 52W
Healdton, *U.S.A.* **81 H6** 34 14N 97 29W
Healesville, *Australia* **63 F4** 37 35S 145 30 E
Heany Junction, *Zimbabwe* **57 C4** 20 6S 28 54 E
Heard I., *Ind. Oc.* **3 G13** 53 0S 74 0 E
Hearne, *U.S.A.* **81 K6** 30 53N 96 36W
Hearst, *Canada* **70 C3** 49 40N 83 41W
Heart →, *U.S.A.* **80 B4** 46 46N 100 50W
Heart's Content, *Canada* **71 C9** 47 54N 53 27W
Heath Pt., *Canada* **71 C7** 49 8N 61 40W
Heavener, *U.S.A.* **81 H7** 34 53N 94 36W
Hebbronville, *U.S.A.* **81 M5** 27 18N 98 41W
Hebei □, *China* **34 E9** 39 0N 116 0 E
Hebel, *Australia* **63 D4** 28 58S 147 47 E
Heber, *U.S.A.* **85 N11** 32 44N 115 32W
Heber City, *U.S.A.* **82 F8** 40 31N 111 25W
Heber Springs, *U.S.A.* **81 H9** 35 30N 92 2W
Hebert, *Canada* **73 C7** 50 30N 107 10W
Hebgen L., *U.S.A.* **82 D8** 44 52N 111 20W
Hebi, *China* **34 G8** 35 57N 114 7 E
Hebrides, *U.K.* **6 D4** 57 30N 7 0W
Hebrides, Sea of the, *U.K.* **12 D2** 57 5N 7 0W
Hebron = Al Khalīl, *West Bank* **47 D4** 31 32N 35 6 E
Hebron, *Canada* **69 C13** 58 5N 62 30W
Hebron, *N. Dak., U.S.A.* **80 B3** 46 54N 102 3W
Hebron, *Nebr., U.S.A.* **80 E6** 40 10N 97 35W
Hecate Str., *Canada* **72 C2** 53 10N 130 30W
Heceta I., *U.S.A.* **72 B2** 55 46N 133 40W
Hechi, *China* **32 D5** 24 40N 108 2 E
Hechuan, *China* **32 C5** 30 2N 106 12 E
Hecla, *U.S.A.* **80 C5** 45 53N 98 9W
Hecla I., *Canada* **73 C9** 51 10N 96 43W
Hede, *Sweden* **9 E15** 62 23N 13 30 E
Hedemora, *Sweden* **9 F16** 60 18N 15 58 E
Heerde, *Neths.* **15 B6** 52 24N 6 2 E
Heerenveen, *Neths.* **15 B5** 52 57N 5 55 E
Heerhugowaard, *Neths.* **15 B4** 52 40N 4 51 E
Heerlen, *Neths.* **18 A6** 50 55N 5 58 E
Ḫefa, *Israel* **47 C4** 32 46N 35 0 E
Ḫefa □, *Israel* **47 C4** 32 40N 35 0 E
Hefei, *China* **33 C6** 31 52N 117 18 E

Hegang

Hegang, *China* 33 B8 47 20N 130 19 E
Heichengzhen, *China* 34 F4 36 24N 106 3 E
Heidelberg, *Germany* 16 D5 49 24N 8 42 E
Heidelberg, *S. Africa* 56 E3 34 6S 20 59 E
Heilbron, *S. Africa* 57 D4 27 16S 27 59 E
Heilbronn, *Germany* 16 D5 49 9N 9 13 E
Heilongjiang □, *China* 33 B7 48 0N 126 0 E
Heilunkiang = Heilongjiang □, *China* 33 B7 48 0N 126 0 E
Heimaey, *Iceland* 8 E3 63 26N 20 17W
Heinola, *Finland* 9 F22 61 13N 26 2 E
Heinze Kyun, *Burma* 38 E1 14 25N 97 45 E
Heishan, *China* 35 D12 41 40N 122 5 E
Heishui, *China* 35 C10 42 8N 119 30 E
Hejaz = Ḥijāz □, *Si. Arabia* 46 C2 24 0N 40 0 E
Hejian, *China* 34 E9 38 25N 116 5 E
Hejin, *China* 34 G6 35 35N 110 42 E
Hekımhan, *Turkey* 44 B3 38 50N 37 55 E
Hekla, *Iceland* 8 E4 63 56N 19 35W
Hekou, *China* 32 D5 22 30N 103 59 E
Helan Shan, *China* 34 E3 38 30N 105 55 E
Helen Atoll, *Pac. Oc.* 37 D8 2 40N 132 0 E
Helena, *Ark., U.S.A.* 81 H9 34 32N 90 36W
Helena, *Mont., U.S.A.* 82 C7 46 36N 112 2W
Helendale, *U.S.A.* 85 L9 34 44N 117 19W
Helensburgh, *U.K.* 12 E4 56 1N 4 43W
Helensville, *N.Z.* 59 G5 36 41S 174 29 E
Helenvale, *Australia* 62 B4 15 43S 145 14 E
Helgeland, *Norway* 8 C15 66 7N 13 29 E
Helgoland, *Germany* 16 A4 54 10N 7 53 E
Heligoland = Helgoland, *Germany* 16 A4 54 10N 7 53 E
Heligoland B. = Deutsche Bucht, *Germany* 16 A5 54 15N 8 0 E
Hella, *Iceland* 8 E3 63 50N 20 24W
Hellertown, *U.S.A.* 79 F9 40 35N 75 21W
Hellespont = Çanakkale Boğazı, *Turkey* 21 D12 40 17N 26 32 E
Hellevoetsluis, *Neths.* 15 C4 51 50N 4 8 E
Hellín, *Spain* 19 C5 38 31N 1 40W
Helmand □, *Afghan.* 40 D4 31 20N 64 0 E
Helmand →, *Afghan.* 40 D2 31 12N 61 34 E
Helmeringhausen, *Namibia* 56 D2 25 54S 16 57 E
Helmond, *Neths.* 15 C5 51 29N 5 41 E
Helmsdale, *U.K.* 12 C5 58 7N 3 39W
Helmsdale →, *U.K.* 12 C5 58 7N 3 40W
Helong, *China* 35 C15 42 40N 129 0 E
Helper, *U.S.A.* 82 G8 39 41N 110 51W
Helsingborg, *Sweden* 9 H15 56 3N 12 42 E
Helsingfors = Helsinki, *Finland* 9 F21 60 15N 25 3 E
Helsingør, *Denmark* 9 H15 56 2N 12 35 E
Helsinki, *Finland* 9 F21 60 15N 25 3 E
Helston, *U.K.* 11 G2 50 6N 5 17W
Helvellyn, *U.K.* 10 C4 54 32N 3 1W
Helwân, *Egypt* 51 C12 29 50N 31 20 E
Hemel Hempstead, *U.K.* 11 F7 51 44N 0 28W
Hemet, *U.S.A.* 85 M10 33 45N 116 58W
Hemingford, *U.S.A.* 80 D3 42 19N 103 4W
Hemmingford, *Canada* 79 A11 45 3N 73 35W
Hempstead, *U.S.A.* 81 K6 30 6N 96 5W
Hemse, *Sweden* 9 H18 57 15N 18 22 E
Henan □, *China* 34 H8 34 0N 114 0 E
Henares →, *Spain* 19 B4 40 24N 3 30W
Henashi-Misaki, *Japan* 30 D9 40 37N 139 51 E
Henderson, *Argentina* 94 D3 36 18S 61 43W
Henderson, *Ky., U.S.A.* 76 G2 37 50N 87 35W
Henderson, *N.C., U.S.A.* 77 G6 36 20N 78 25W
Henderson, *Nev., U.S.A.* 85 J12 36 2N 114 59W
Henderson, *Tenn., U.S.A.* 77 H1 35 26N 88 38W
Henderson, *Tex., U.S.A.* 81 J7 32 9N 94 48W
Hendersonville, *N.C., U.S.A.* 77 H4 35 19N 82 28W
Hendersonville, *Tenn., U.S.A.* 77 G2 36 18N 86 37W
Hendījān, *Iran* 45 D6 30 14N 49 43 E
Hendorābī, *Iran* 45 E7 26 40N 53 37 E
Hengcheng, *China* 34 E4 38 18N 106 28 E
Hengdaohezi, *China* 35 B15 44 52N 129 0 E
Hengelo, *Neths.* 15 B6 52 16N 6 48 E
Hengshan, *China* 34 F5 37 58N 109 5 E
Hengshui, *China* 34 F8 37 41N 115 40 E
Hengyang, *China* 33 D6 26 59N 112 22 E
Henlopen, C., *U.S.A.* 76 F8 38 48N 75 6W
Hennenman, *S. Africa* 56 D4 27 59S 27 1 E
Hennessey, *U.S.A.* 81 G6 36 6N 97 54W
Henrietta, *U.S.A.* 81 J5 33 49N 98 12W
Henrietta, Ostrov = Genriyetty, Ostrov, *Russia* 27 B16 77 6N 156 30 E
Henrietta Maria, C., *Canada* 70 A3 55 9N 82 20W
Henry, *U.S.A.* 80 E10 41 7N 89 22W
Henryetta, *U.S.A.* 81 H7 35 27N 95 59W
Henryville, *Canada* 79 A11 45 8N 73 11W
Hensall, *Canada* 78 C3 43 26N 81 30W
Hentiesbaai, *Namibia* 56 C1 22 8S 14 18 E
Hentiyn Nuruu, *Mongolia* 33 B5 48 30N 108 30 E
Henty, *Australia* 63 F4 35 30S 147 0 E
Henzada, *Burma* 41 L19 17 38N 95 26 E
Heppner, *U.S.A.* 82 D4 45 21N 119 33W
Hepworth, *Canada* 78 B3 44 37N 81 9W
Hequ, *China* 34 E6 39 20N 111 15 E
Héraðsflói, *Iceland* 8 D6 65 42N 14 12W
Héraðsvötn →, *Iceland* 8 D4 65 45N 19 25W
Herald Cays, *Australia* 62 B4 16 58S 149 9 E
Herāt, *Afghan.* 40 B3 34 20N 62 7 E
Herāt □, *Afghan.* 40 B3 35 0N 62 0 E
Herbert →, *Australia* 62 B4 18 31S 146 17 E
Herberton, *Australia* 62 B4 17 20S 145 25 E
Herbertsdale, *S. Africa* 56 E3 34 1S 21 46 E
Herceg-Novi, *Montenegro, Yug.* 21 C8 42 30N 18 33 E
Herchmer, *Canada* 73 B10 57 22N 94 10W
Herðubreið, *Iceland* 8 D5 65 11N 16 21W
Hereford, *U.K.* 11 E5 52 4N 2 43W
Hereford, *U.S.A.* 81 H3 34 49N 102 24W
Herefordshire □, *U.K.* 11 E5 52 8N 2 40W
Herentals, *Belgium* 15 C4 51 12N 4 51 E
Herford, *Germany* 16 B5 52 7N 8 39 E
Herington, *U.S.A.* 80 F6 38 40N 96 57W
Herkimer, *U.S.A.* 79 D10 43 0N 74 59W
Herlong, *U.S.A.* 84 E6 40 8N 120 8W
Herm, *U.K.* 11 H5 49 30N 2 28W
Hermann, *U.S.A.* 80 F9 38 42N 91 27W
Hermannsburg, *Australia* 60 D5 23 57S 132 45 E
Hermanus, *S. Africa* 56 E2 34 27S 19 12 E
Hermidale, *Australia* 63 E4 31 30S 146 42 E
Hermiston, *U.S.A.* 82 D4 45 51N 119 17W
Hermite, I., *Chile* 96 H3 55 50S 68 0W
Hermon, *U.S.A.* 79 B9 44 28N 75 14W
Hermon, Mt. = Shaykh, J. ash, *Lebanon* 47 B4 33 25N 35 50 E
Hermosillo, *Mexico* 86 B2 29 10N 111 0W
Hernád →, *Hungary* 17 D11 47 56N 21 8 E
Hernandarias, *Paraguay* 95 B5 25 20S 54 40W
Hernandez, *U.S.A.* 84 J6 36 24N 120 46W
Hernando, *Argentina* 94 C3 32 28S 63 40W
Hernando, *U.S.A.* 81 H10 34 50N 90 0W
Herndon, *U.S.A.* 78 F8 40 43N 76 51W
Herne, *Germany* 15 C7 51 32N 7 14 E
Herne Bay, *U.K.* 11 F9 51 21N 1 8 E
Herning, *Denmark* 9 H13 56 8N 8 58 E
Heroica = Caborca, *Mexico* 86 A2 30 40N 112 10W
Heroica Nogales = Nogales, *Mexico* 86 A2 31 20N 110 56W
Heron Bay, *Canada* 70 C2 48 40N 86 25W
Herradura, Pta. de la, *Canary Is.* 22 F5 28 26N 14 8W
Herreid, *U.S.A.* 80 C4 45 50N 100 4W
Herrin, *U.S.A.* 81 G10 37 48N 89 2W
Herriot, *Canada* 73 B8 56 22N 101 16W
Hershey, *U.S.A.* 79 F8 40 17N 76 39W
Hersonissos, *Greece* 23 D7 35 18N 25 22 E
Herstal, *Belgium* 15 D5 50 40N 5 38 E
Hertford, *U.K.* 11 F7 51 48N 0 4W
Hertfordshire □, *U.K.* 11 F7 51 51N 0 5W
's-Hertogenbosch, *Neths.* 15 C5 51 42N 5 17 E
Hertzogville, *S. Africa* 56 D4 28 9S 25 30 E
Hervey B., *Australia* 62 C5 25 0S 152 52 E
Herzliyya, *Israel* 47 C3 32 10N 34 50 E
Ḥeṣār, *Fārs, Iran* 45 D6 29 52N 50 16 E
Ḥeṣār, *Markazī, Iran* 45 C6 35 50N 49 12 E
Heshui, *China* 34 G5 35 48N 108 0 E
Heshun, *China* 34 F7 37 22N 113 32 E
Hesperia, *U.S.A.* 85 L9 34 25N 117 18W
Hesse = Hessen □, *Germany* 16 C5 50 30N 9 0 E
Hessen □, *Germany* 16 C5 50 30N 9 0 E
Hetch Hetchy Aqueduct, *U.S.A.* 84 H5 37 29N 122 19W
Hettinger, *U.S.A.* 80 C3 46 0N 102 42W
Heuvelton, *U.S.A.* 79 B9 44 37N 75 25W
Hewitt, *U.S.A.* 81 K6 31 27N 97 11W
Hexham, *U.K.* 10 C5 54 58N 2 4W
Hexigten Qi, *China* 35 C9 43 18N 117 30 E
Heydarābād, *Iran* 45 D7 30 33N 55 38 E
Heysham, *U.K.* 10 C5 54 3N 2 53W
Heywood, *Australia* 63 F3 38 8S 141 37 E
Heze, *China* 34 G8 35 14N 115 20 E
Hi Vista, *U.S.A.* 85 L9 34 45N 117 46W
Hialeah, *U.S.A.* 77 N5 25 50N 80 17W
Hiawatha, *U.S.A.* 80 F7 39 51N 95 32W
Hibbing, *U.S.A.* 80 B8 47 25N 92 56W
Hibbs B., *Australia* 62 G4 42 35S 145 15 E
Hibernia Reef, *Australia* 60 B3 12 0S 123 23 E
Hickman, *U.S.A.* 81 G10 36 34N 89 11W
Hickory, *U.S.A.* 77 H5 35 44N 81 21W
Hicks, Pt., *Australia* 63 F4 37 49S 149 17 E
Hicks L., *Canada* 73 A9 61 25N 100 0W
Hicksville, *U.S.A.* 79 F11 40 46N 73 32W
Hida-Gawa →, *Japan* 31 G8 35 26N 137 3 E
Hida-Sammyaku, *Japan* 31 F8 36 30N 137 40 E
Hidaka-Sammyaku, *Japan* 30 C11 42 35N 142 45 E
Hidalgo, *Mexico* 87 C5 24 15N 99 26W
Hidalgo □, *Mexico* 87 C5 20 30N 99 10W
Hidalgo, Presa M., *Mexico* 86 B3 26 30N 108 35W
Hidalgo del Parral, *Mexico* 86 B3 26 58N 105 40W
Hierro, *Canary Is.* 22 G1 27 44N 18 0W
Higashiajima-San, *Japan* 30 F10 37 40N 140 10 E
Higashiōsaka, *Japan* 31 G7 34 40N 135 37 E
Higgins, *U.S.A.* 81 G4 36 7N 100 2W
Higgins Corner, *U.S.A.* 84 F5 39 2N 121 5W
High Atlas = Haut Atlas, *Morocco* 50 B4 32 30N 5 0W
High Bridge, *U.S.A.* 79 F10 40 40N 74 54W
High Level, *Canada* 72 B5 58 31N 117 8W
High Point, *U.S.A.* 77 H6 35 57N 80 0W
High Prairie, *Canada* 72 B5 55 30N 116 30W
High River, *Canada* 72 C6 50 30N 113 50W
High Tatra = Tatry, *Slovak Rep.* 17 D11 49 20N 20 0 E
High Veld, *Africa* 48 J6 27 0S 27 0 E
High Wycombe, *U.K.* 11 F7 51 37N 0 45W
Highland □, *U.K.* 12 D4 57 17N 4 21W
Highland Park, *U.S.A.* 76 D2 42 11N 87 48W
Highmore, *U.S.A.* 80 C5 44 31N 99 27W
Highrock L., *Man., Canada* 73 B8 55 45N 100 30W
Highrock L., *Sask., Canada* 73 B7 57 5N 105 32W
Higüey, *Dom. Rep.* 89 C6 18 37N 68 42W
Hiiumaa, *Estonia* 9 G20 58 50N 22 45 E
Ḥijāz □, *Si. Arabia* 46 C2 24 0N 40 0 E
Hijo = Tagum, *Phil.* 37 C7 7 33N 125 53 E
Hikari, *Japan* 31 H5 33 58N 131 58 E
Hiko, *U.S.A.* 84 H11 37 32N 115 14W
Hikone, *Japan* 31 G8 35 15N 136 10 E
Hikurangi, *Gisborne, N.Z.* 59 H6 37 55S 178 4 E
Hikurangi, *Northland, N.Z.* 59 F5 35 36S 174 17 E
Hildesheim, *Germany* 16 B5 52 9N 9 56 E
Hill →, *Australia* 61 F2 30 23S 115 3 E
Hill City, *Idaho, U.S.A.* 82 E6 43 18N 115 3W
Hill City, *Kans., U.S.A.* 80 F5 39 22N 99 51W
Hill City, *S. Dak., U.S.A.* 80 D3 43 56N 103 35W
Hill Island L., *Canada* 73 A7 60 30N 109 50W
Hillcrest Center, *U.S.A.* 85 K8 35 23N 118 57W
Hillegom, *Neths.* 15 B4 52 18N 4 35 E
Hillerød, *Denmark* 9 J15 55 56N 12 19 E
Hillsboro, *Kans., U.S.A.* 80 F6 38 21N 97 12W
Hillsboro, *N. Dak., U.S.A.* 80 B6 47 26N 97 3W
Hillsboro, *N.H., U.S.A.* 79 C13 43 7N 71 54W
Hillsboro, *Ohio, U.S.A.* 76 F4 39 12N 83 37W
Hillsboro, *Oreg., U.S.A.* 84 E4 45 31N 122 59W
Hillsboro, *Tex., U.S.A.* 81 J6 32 1N 97 8W
Hillsborough, *Grenada* 89 D7 12 28N 61 28W
Hillsdale, *Mich., U.S.A.* 76 E3 41 56N 84 38W
Hillsdale, *N.Y., U.S.A.* 79 D11 42 11N 73 30W
Hillsport, *Canada* 70 C2 49 27N 85 34W
Hillston, *Australia* 63 E4 33 30S 145 31 E
Hilo, *U.S.A.* 74 J17 19 44N 155 5W
Hilton, *U.S.A.* 78 C7 43 17N 77 48W
Hilton Head Island, *U.S.A.* 77 J5 32 13N 80 45W
Hilversum, *Neths.* 15 B5 52 14N 5 10 E
Himachal Pradesh □, *India* 42 D7 31 30N 77 0 E
Himalaya, *Asia* 43 E11 29 0N 84 0 E
Himatnagar, *India* 40 H8 23 37N 72 57 E
Himeji, *Japan* 31 G7 34 50N 134 40 E
Himi, *Japan* 31 F8 36 50N 136 55 E
Ḥimş, *Syria* 47 A5 34 40N 36 45 E
Ḥimş □, *Syria* 47 A6 34 30N 37 0 E
Hinche, *Haiti* 89 C5 19 9N 72 1W
Hinchinbrook I., *Australia* 62 B4 18 20S 146 15 E
Hinckley, *U.K.* 11 E6 52 33N 1 22W
Hinckley, *U.S.A.* 80 B8 46 1N 92 56W
Hindaun, *India* 42 F7 26 44N 77 5 E
Hindmarsh, L., *Australia* 63 F3 36 5S 141 55 E
Hindu Bagh, *Pakistan* 42 D2 30 56N 67 50 E
Hindu Kush, *Asia* 40 B7 36 0N 71 0 E
Hindubagh, *Pakistan* 40 D5 30 56N 67 57 E
Hindupur, *India* 40 N10 13 49N 77 32 E
Hines Creek, *Canada* 72 B5 56 20N 118 40W
Hinesville, *U.S.A.* 77 K5 31 51N 81 36W
Hinganghat, *India* 40 J11 20 30N 78 52 E
Hingham, *U.S.A.* 82 B8 48 33N 110 25W
Hingir, *India* 43 J10 21 57N 83 41 E
Hingoli, *India* 40 K10 19 41N 77 15 E
Hinna = Imi, *Ethiopia* 46 F3 6 28N 42 10 E
Hinnøya, *Norway* 8 B16 68 35N 15 50 E
Hinojosa del Duque, *Spain* 19 C3 38 30N 5 9W
Hinsdale, *U.S.A.* 79 D12 42 47N 72 29W
Hinton, *Canada* 72 C5 53 26N 117 34W
Hinton, *U.S.A.* 76 G5 37 40N 80 54W
Hirado, *Japan* 31 H4 33 22N 129 33 E
Hirakud Dam, *India* 41 J13 21 32N 83 45 E
Hiran →, *India* 43 H8 23 6N 79 21 E
Hirapur, *India* 43 G8 24 22N 79 13 E
Hiratsuka, *Japan* 31 G9 35 19N 139 21 E
Hiroo, *Japan* 30 C11 42 17N 143 19 E
Hirosaki, *Japan* 30 D10 40 34N 140 28 E
Hiroshima, *Japan* 31 G6 34 24N 132 30 E
Hiroshima □, *Japan* 31 G6 34 50N 133 0 E
Hisar, *India* 42 E6 29 12N 75 45 E
Hisb →, *Iraq* 44 D5 31 45N 44 17 E
Ḥismá, *Si. Arabia* 44 D3 28 30N 36 0 E
Hispaniola, *W. Indies* 89 C5 19 0N 71 0W
Hīt, *Iraq* 44 C4 33 38N 42 49 E
Hita, *Japan* 31 H5 33 20N 130 58 E
Hitachi, *Japan* 31 F10 36 36N 140 39 E
Hitchin, *U.K.* 11 F7 51 58N 0 16W
Hitoyoshi, *Japan* 31 H5 32 13N 130 45 E
Hitra, *Norway* 8 E13 63 30N 8 45 E
Hixon, *Canada* 72 C4 53 25N 122 35W
Ḥiyyon, N. →, *Israel* 47 E4 30 25N 35 10 E
Hjalmar L., *Canada* 73 A7 61 33N 109 25W
Hjälmaren, *Sweden* 9 G16 59 18N 15 40 E
Hjørring, *Denmark* 9 H13 57 29N 9 59 E
Hkakabo Razi, *Burma* 41 E20 28 25N 97 23 E
Hlobane, *S. Africa* 57 D5 27 42S 31 0 E
Hluhluwe, *S. Africa* 57 D5 28 1S 32 15 E
Hlyboka, *Ukraine* 17 D13 48 5N 25 56 E
Ho Chi Minh City = Thanh Pho Ho Chi Minh, *Vietnam* 39 G6 10 58N 106 40 E
Ho Thuong, *Vietnam* 38 C5 19 32N 105 48 E
Hoa Binh, *Vietnam* 38 B5 20 50N 105 20 E
Hoa Da, *Vietnam* 39 G7 11 16N 108 40 E
Hoa Hiep, *Vietnam* 39 G5 11 34N 105 51 E
Hoai Nhon, *Vietnam* 38 E7 14 28N 109 1 E
Hoang Lien Son, *Vietnam* 38 A4 22 0N 104 0 E
Hoanib →, *Namibia* 56 B2 19 27S 12 46 E
Hoare B., *Canada* 69 B13 65 17N 62 30W
Hoarusib →, *Namibia* 56 B2 19 3S 12 36 E
Hobart, *Australia* 62 G4 42 50S 147 21 E
Hobart, *U.S.A.* 81 H5 35 1N 99 6W
Hobbs, *U.S.A.* 81 J3 32 42N 103 8W
Hobbs Coast, *Antarctica* 5 D14 74 50S 131 0W
Hobe Sound, *U.S.A.* 77 M5 27 4N 80 8W
Hoboken, *U.S.A.* 79 F10 40 45N 74 4W
Hobro, *Denmark* 9 H13 56 39N 9 46 E
Hoburgen, *Sweden* 9 H18 56 55N 18 7 E
Hochfeld, *Namibia* 56 C2 21 28S 17 58 E
Hodaka-Dake, *Japan* 31 F8 36 17N 137 39 E
Hodeida = Al Ḥudaydah, *Yemen* 46 E3 14 50N 43 0 E
Hodgeville, *Canada* 73 C7 50 7N 106 58W
Hodgson, *Canada* 73 C9 51 13N 97 36W
Hódmezővásárhely, *Hungary* 17 E11 46 28N 20 22 E
Hodna, Chott el, *Algeria* 50 A6 35 26N 4 43 E
Hodonín, *Czech Rep.* 17 D9 48 50N 17 10 E
Hoeamdong, *N. Korea* 35 C16 42 30N 130 16 E
Hoek van Holland, *Neths.* 15 C4 52 0N 4 7 E
Hoengsŏng, *S. Korea* 35 F14 37 29N 127 59 E
Hoeryong, *N. Korea* 35 C15 42 30N 129 45 E
Hoeyang, *N. Korea* 35 E14 38 43N 127 36 E
Hof, *Germany* 16 C6 50 19N 11 55 E
Hofmeyr, *S. Africa* 56 E4 31 39S 25 50 E
Höfn, *Iceland* 8 D6 64 15N 15 13W
Hofors, *Sweden* 9 F17 60 31N 16 15 E
Hofsjökull, *Iceland* 8 D4 64 49N 18 48W
Hōfu, *Japan* 31 G5 34 3N 131 34 E
Hogan Group, *Australia* 63 F4 39 13S 147 1 E
Hogarth, Mt., *Australia* 62 C2 21 48S 136 58 E
Hoggar = Ahaggar, *Algeria* 50 D7 23 0N 6 30 E
Hogsty Reef, *Bahamas* 89 B5 21 41N 73 48W
Hoh →, *U.S.A.* 84 C2 47 45N 124 29W
Hohe Venn, *Belgium* 15 D6 50 30N 6 5 E
Hohenwald, *U.S.A.* 77 H2 35 33N 87 33W
Hoher Rhön = Rhön, *Germany* 16 C5 50 24N 9 58 E
Hohhot, *China* 34 D6 40 52N 111 40 E
Hóhlakas, *Greece* 23 D9 35 57N 27 53 E
Hoi An, *Vietnam* 38 E7 15 30N 108 19 E
Hoi Xuan, *Vietnam* 38 B5 20 25N 105 9 E
Hoisington, *U.S.A.* 80 F5 38 31N 98 47W
Hōjō, *Japan* 31 H6 33 58N 132 46 E
Hokianga Harbour, *N.Z.* 59 F4 35 31S 173 22 E
Hokitika, *N.Z.* 59 K3 42 42S 171 0 E
Hokkaidō □, *Japan* 30 C11 43 30N 143 0 E
Holbrook, *Australia* 63 F4 35 42S 147 18 E
Holbrook, *U.S.A.* 83 J8 34 54N 110 10W
Holden, *U.S.A.* 82 G7 39 6N 112 16W
Holdenville, *U.S.A.* 81 H6 35 5N 96 24W
Holdrege, *U.S.A.* 80 E5 40 26N 99 23W
Holguín, *Cuba* 88 B4 20 50N 76 20W
Hollams Bird I., *Namibia* 56 C1 24 40S 14 30 E
Holland, *Mich., U.S.A.* 76 D2 42 47N 86 7W
Holland, *N.Y., U.S.A.* 78 D6 42 38N 78 32W
Hollandale, *U.S.A.* 81 J9 33 10N 90 51W
Hollandia = Jayapura, *Indonesia* 37 E10 2 28S 140 38 E
Holley, *U.S.A.* 78 C6 43 14N 78 2W
Hollidaysburg, *U.S.A.* 78 F6 40 26N 78 24W
Hollis, *U.S.A.* 81 H5 34 41N 99 55W
Hollister, *Calif., U.S.A.* 84 J5 36 51N 121 24W
Hollister, *Idaho, U.S.A.* 82 E6 42 21N 114 35W
Holly Hill, *U.S.A.* 77 L5 29 16N 81 3W
Holly Springs, *U.S.A.* 81 H10 34 46N 89 27W
Hollywood, *U.S.A.* 77 N5 26 1N 80 9W
Holman, *Canada* 68 A8 70 44N 117 44W
Hólmavík, *Iceland* 8 D3 65 42N 21 40W
Holmen, *U.S.A.* 80 D9 43 58N 91 15W
Holmes Reefs, *Australia* 62 B4 16 27S 148 0 E
Holmsund, *Sweden* 8 E19 63 41N 20 20 E
Holroyd →, *Australia* 62 A3 14 10S 141 36 E
Holstebro, *Denmark* 9 H13 56 22N 8 37 E
Holsworthy, *U.K.* 11 G3 50 48N 4 22W
Holton, *Canada* 71 B8 54 31N 57 12W
Holton, *U.S.A.* 80 F7 39 28N 95 44W
Holtville, *U.S.A.* 85 N11 32 49N 115 23W
Holwerd, *Neths.* 15 A5 53 22N 5 54 E
Holy I., *Angl., U.K.* 10 D3 53 17N 4 37W
Holy I., *Northumb., U.K.* 10 B6 55 40N 1 47W
Holyhead, *U.K.* 10 D3 53 18N 4 38W
Holyoke, *Colo., U.S.A.* 80 E3 40 35N 102 18W
Holyoke, *Mass., U.S.A.* 79 D12 42 12N 72 37W
Holyrood, *Canada* 71 C9 47 27N 53 8W
Homa Bay, *Kenya* 54 C3 0 36S 34 30 E
Homalin, *Burma* 41 G19 24 55N 95 0 E
Homand, *Iran* 45 C8 32 28N 59 37 E
Homathko →, *Canada* 72 C4 51 0N 124 56W
Hombori, *Mali* 50 E5 15 20N 1 38W
Home B., *Canada* 69 B13 68 40N 67 10W
Home Hill, *Australia* 62 B4 19 43S 147 25 E
Homedale, *U.S.A.* 82 E5 43 37N 116 56W
Homer, *Alaska, U.S.A.* 68 C4 59 39N 151 33W
Homer, *La., U.S.A.* 81 J8 32 48N 93 4W
Homer City, *U.S.A.* 78 F5 40 32N 79 10W
Homestead, *Australia* 62 C4 20 20S 145 40 E
Homestead, *U.S.A.* 77 N5 25 28N 80 29W
Homewood, *U.S.A.* 84 F6 39 4N 120 8W
Homoine, *Mozam.* 57 C6 23 55S 35 8 E
Homs = Ḥimş, *Syria* 47 A5 34 40N 36 45 E
Homyel, *Belarus* 17 B16 52 28N 31 0 E
Hon Chong, *Vietnam* 39 G5 10 25N 104 30 E
Hon Me, *Vietnam* 38 C5 19 23N 105 56 E
Honan = Henan □, *China* 34 H8 34 0N 114 0 E
Honbetsu, *Japan* 30 C11 43 7N 143 37 E
Honcut, *U.S.A.* 84 F5 39 20N 121 32W
Hondeklipbaai, *S. Africa* 56 E2 30 19S 17 17 E
Hondo, *Japan* 31 H5 32 27N 130 12 E
Hondo, *U.S.A.* 81 L5 29 21N 99 9W
Hondo →, *Belize* 87 D7 18 25N 88 21W
Honduras ■, *Cent. Amer.* 88 D2 14 40N 86 30W
Honduras, G. de, *Caribbean* 88 C2 16 50N 87 0W
Hønefoss, *Norway* 9 F14 60 10N 10 18 E
Honesdale, *U.S.A.* 79 E9 41 34N 75 16W
Honey L., *U.S.A.* 84 E6 40 15N 120 19W
Honfleur, *France* 18 B4 49 25N 0 13 E
Hong →, *Vietnam* 32 D5 22 0N 104 0 E
Hong Gai, *Vietnam* 38 B6 20 57N 107 5 E
Hong He →, *China* 34 H8 32 25N 115 35 E
Hong Kong □, *China* 33 D6 22 11N 114 14 E
Hongch'ŏn, *S. Korea* 35 F14 37 44N 127 53 E
Hongjiang, *China* 33 D5 27 7N 109 59 E
Hongliu He →, *China* 34 F5 38 0N 109 50 E
Hongor, *Mongolia* 34 B7 45 45N 112 50 E
Hongsa, *Laos* 38 C3 19 43N 101 20 E
Hongshui He →, *China* 33 D5 23 48N 109 30 E
Hongsŏng, *S. Korea* 35 F14 36 37N 126 38 E
Hongtong, *China* 34 F6 36 16N 111 40 E
Honguedo, Détroit d', *Canada* 71 C7 49 15N 64 0W
Hongwon, *N. Korea* 35 E14 40 0N 127 56 E
Hongze Hu, *China* 35 H10 33 15N 118 35 E
Honiara, *Solomon Is.* 64 H7 9 27S 159 57 E
Honiton, *U.K.* 11 G4 50 47N 3 11W
Honjō, *Japan* 30 E10 39 23N 140 3 E
Honningsvåg, *Norway* 8 A21 70 59N 25 59 E
Honolulu, *U.S.A.* 74 H16 21 19N 157 52W
Honshū, *Japan* 33 C8 36 0N 138 0 E
Hood, Mt., *U.S.A.* 82 D3 45 23N 121 42W
Hood, Pt., *Australia* 61 F2 34 23S 119 34 E
Hood River, *U.S.A.* 82 D3 45 43N 121 31W
Hoodsport, *U.S.A.* 84 C3 47 24N 123 9W
Hoogeveen, *Neths.* 15 B6 52 44N 6 28 E
Hoogezand-Sappemeer, *Neths.* 15 A6 53 9N 6 45 E
Hooghly = Hugli →, *India* 43 J13 21 56N 88 4 E
Hooghly-Chinsura = Chunchura, *India* 43 H13 22 53N 88 27 E
Hook Hd., *Ireland* 13 D5 52 7N 6 56W
Hook I., *Australia* 62 C4 20 4S 149 0 E
Hook of Holland = Hoek van Holland, *Neths.* 15 C4 52 0N 4 7 E
Hooker, *U.S.A.* 81 G4 36 52N 101 13W
Hooker Creek, *Australia* 60 C5 18 23S 130 38 E
Hoonah, *U.S.A.* 72 B1 58 7N 135 27W
Hooper Bay, *U.S.A.* 68 B3 61 32N 166 6W
Hoopeston, *U.S.A.* 76 E2 40 28N 87 40W
Hoopstad, *S. Africa* 56 D4 27 50S 25 55 E
Hoorn, *Neths.* 15 B5 52 38N 5 4 E
Hoover, *U.S.A.* 77 J2 33 20N 86 11W
Hoover Dam, *U.S.A.* 85 K12 36 1N 114 44W
Hooversville, *U.S.A.* 78 F6 40 9N 78 55W
Hop Bottom, *U.S.A.* 79 E9 41 42N 75 46W
Hope, *Canada* 72 D4 49 25N 121 25W
Hope, *Ariz., U.S.A.* 85 M13 33 43N 113 42W
Hope, *Ark., U.S.A.* 81 J8 33 40N 93 36W
Hope, L., *S. Austral., Australia* 63 D2 28 24S 139 18 E
Hope, L., *W. Austral., Australia* 61 F3 32 35S 120 15 E
Hope I., *Canada* 78 B4 44 55N 80 11W
Hope Town, *Bahamas* 88 A4 26 35N 76 57W
Hopedale, *Canada* 71 A7 55 28N 60 13W
Hopedale, *U.S.A.* 79 D13 42 8N 71 33W
Hopefield, *S. Africa* 56 E2 33 3S 18 22 E
Hopei = Hebei □, *China* 34 E9 39 0N 116 0 E
Hopelchén, *Mexico* 87 D7 19 46N 89 50W
Hopetoun, *Vic., Australia* 63 F3 35 42S 142 22 E
Hopetoun, *W. Austral., Australia* 61 F3 33 57S 120 7 E
Hopetown, *S. Africa* 56 D3 29 34S 24 3 E
Hopevale, *Australia* 62 B4 15 16S 145 20 E
Hopewell, *U.S.A.* 76 G7 37 18N 77 17W
Hopkins, L., *Australia* 60 D4 24 15S 128 35 E
Hopkinsville, *U.S.A.* 77 G2 36 52N 87 29W
Hopland, *U.S.A.* 84 G3 38 58N 123 7W
Hoquiam, *U.S.A.* 84 C3 46 59N 123 53W
Horden Hills, *Australia* 60 D5 20 15S 130 0 E

Ilfracombe

Horinger, *China* **34 D6** 40 28N 111 48 E
Horlick Mts., *Antarctica* **5 E15** 84 0S 102 0W
Horlivka, *Ukraine* **25 E6** 48 19N 38 5 E
Hormak, *Iran* **45 D9** 29 58N 60 51 E
Hormoz, *Iran* **45 E7** 27 35N 55 0 E
Hormoz, Jaz.-ye, *Iran* **45 E8** 27 8N 56 28 E
Hormozgān □, *Iran* **45 E8** 27 30N 56 0 E
Hormuz, Kūh-e, *Iran* **45 E7** 27 27N 55 10 E
Hormuz, Str. of, *The Gulf* **45 E8** 26 30N 56 30 E
Horn, *Austria* **16 D8** 48 39N 15 40 E
Horn, *Iceland* **8 C2** 66 28N 22 28W
Horn ➝, *Canada* **72 A5** 61 30N 118 1W
Horn, Cape = Hornos, C. de, *Chile* **96 H3** 55 50S 67 30W
Horn Head, *Ireland* **13 A3** 55 14N 8 0W
Horn I., *Australia* **62 A3** 10 37S 142 17 E
Horn Mts., *Canada* **72 A5** 62 15N 119 15W
Hornavan, *Sweden* **8 C17** 66 15N 17 30 E
Hornbeck, *U.S.A.* **81 K8** 31 20N 93 24W
Hornbrook, *U.S.A.* **82 F2** 41 55N 122 33W
Horncastle, *U.K.* **10 D7** 53 13N 0 7W
Hornell, *U.S.A.* **78 D7** 42 20N 77 40W
Hornell L., *Canada* **72 A5** 62 20N 119 25W
Hornepayne, *Canada* **70 C3** 49 14N 84 48W
Hornings Mills, *Canada* **78 B4** 44 9N 80 12W
Hornitos, *U.S.A.* **84 H6** 37 30N 120 14W
Hornos, C. de, *Chile* **96 H3** 55 50S 67 30W
Hornsea, *U.K.* **10 D7** 53 55N 0 11W
Horobetsu, *Japan* **30 C10** 42 24N 141 6 E
Horodenka, *Ukraine* **17 D13** 48 41N 25 29 E
Horodok, *Khmelnytskyy, Ukraine* **17 D14** 49 10N 26 34 E
Horodok, *Lviv, Ukraine* **17 D12** 49 46N 23 32 E
Horokhiv, *Ukraine* **17 C13** 50 30N 24 45 E
Horqin Youyi Qianqi, *China* **35 A12** 46 5N 122 3 E
Horqueta, *Paraguay* **94 A4** 23 15S 56 55W
Horse Creek, *U.S.A.* **80 E3** 41 57N 105 10W
Horse Is., *Canada* **71 B8** 50 15N 55 50W
Horsefly L., *Canada* **72 C4** 52 25N 121 0W
Horseheads, *U.S.A.* **78 D8** 42 10N 76 49W
Horsens, *Denmark* **9 J13** 55 52N 9 51 E
Horsham, *Australia* **63 F3** 36 44S 142 13 E
Horsham, *U.K.* **11 F7** 51 4N 0 20W
Horten, *Norway* **9 G14** 59 25N 10 32 E
Horton, *U.S.A.* **80 F7** 39 40N 95 32W
Horton ➝, *Canada* **68 B7** 69 56N 126 52W
Horwood L., *Canada* **70 C3** 48 5N 82 20W
Hose, Gunung-Gunung, *Malaysia* **36 D4** 2 5N 114 6 E
Hoseynābād, *Khuzestān, Iran* **45 C6** 32 45N 48 20 E
Hoseynābād, *Kordestān, Iran* **44 C5** 35 33N 47 8 E
Hoshangabad, *India* **42 H7** 22 45N 77 45 E
Hoshiarpur, *India* **42 D6** 31 30N 75 58 E
Hospet, *India* **40 M10** 15 15N 76 20 E
Hoste, I., *Chile* **96 H3** 55 0S 69 0W
Hot, *Thailand* **38 C2** 18 8N 98 29 E
Hot Creek Range, *U.S.A.* **82 G6** 38 40N 116 20W
Hot Springs, *Ark., U.S.A.* **81 H8** 34 31N 93 3W
Hot Springs, *S. Dak., U.S.A.* **80 D3** 43 26N 103 29W
Hotagen, *Sweden* **8 E16** 63 50N 14 30 E
Hotan, *China* **32 C2** 37 25N 79 55 E
Hotazel, *S. Africa* **56 D3** 27 17S 22 58 E
Hotchkiss, *U.S.A.* **83 G10** 38 48N 107 43W
Hotham, C., *Australia* **60 B5** 12 2S 131 18 E
Hoting, *Sweden* **8 D17** 64 8N 16 15 E
Hotte, Massif de la, *Haiti* **89 C5** 18 30N 73 45W
Hottentotsbaai, *Namibia* **56 D1** 26 8S 14 59 E
Houei Sai, *Laos* **38 B3** 20 18N 100 26 E
Houffalize, *Belgium* **15 D5** 50 8N 5 48 E
Houghton, *Mich., U.S.A.* **80 B10** 47 7N 88 34W
Houghton, *N.Y., U.S.A.* **78 D6** 42 25N 78 10W
Houghton L., *U.S.A.* **76 C3** 44 21N 84 44W
Houhora Heads, *N.Z.* **59 F4** 34 49S 173 9 E
Houlton, *U.S.A.* **77 B12** 46 8N 67 51W
Houma, *U.S.A.* **81 L9** 29 36N 90 43W
Housatonic ➝, *U.S.A.* **79 E11** 41 10N 73 7W
Houston, *Canada* **72 C3** 54 25N 126 39W
Houston, *Mo., U.S.A.* **81 G9** 37 22N 91 58W
Houston, *Tex., U.S.A.* **81 L7** 29 46N 95 22W
Hout ➝, *S. Africa* **57 C4** 23 4S 29 36 E
Houtkraal, *S. Africa* **56 E3** 30 23S 24 5 E
Houtman Abrolhos, *Australia* **61 E1** 28 43S 113 48 E
Hovd, *Mongolia* **32 B4** 48 2N 91 37 E
Hove, *U.K.* **11 G7** 50 50N 0 10W
Hoveyzeh, *Iran* **45 D6** 31 27N 48 4 E
Hövsgöl, *Mongolia* **34 C5** 43 37N 109 39 E
Hövsgöl Nuur, *Mongolia* **32 A5** 51 0N 100 30 E
Howard, *Australia* **63 D5** 25 16S 152 32 E
Howard, *Pa., U.S.A.* **78 F7** 41 1N 77 40W
Howard, *S. Dak., U.S.A.* **80 C6** 44 1N 97 32W
Howe, *U.S.A.* **82 E7** 43 48N 113 0W
Howe, C., *Australia* **63 F5** 37 30S 150 0 E
Howe I., *Canada* **79 B8** 44 16N 76 17W
Howell, *U.S.A.* **76 D4** 42 36N 83 56W
Howick, *Canada* **79 A11** 45 11N 73 51W
Howick, *S. Africa* **57 D5** 29 28S 30 14 E
Howick Group, *Australia* **62 A4** 14 20S 145 30 E
Howitt, L., *Australia* **63 D2** 27 40S 138 40 E
Howland I., *Pac. Oc.* **64 G10** 0 48N 176 38W
Howrah = Haora, *India* **43 H13** 22 37N 88 20 E
Howth Hd., *Ireland* **13 C5** 53 22N 6 3W
Höxter, *Germany* **16 C5** 51 46N 9 22 E
Hoy, *U.K.* **12 C5** 58 50N 3 15W
Høyanger, *Norway* **9 F12** 61 13N 6 4 E
Hoyerswerda, *Germany* **16 C8** 51 26N 14 14 E
Hoylake, *U.K.* **10 D4** 53 24N 3 10W
Hpa-an = Pa-an, *Burma* **41 L20** 16 51N 97 40 E
Hpungan Pass, *Burma* **41 F20** 27 30N 96 55 E
Hradec Králové, *Czech Rep.* **16 C8** 50 15N 15 50 E
Hrodna, *Belarus* **17 B12** 53 42N 23 52 E
Hrodzyanka, *Belarus* **17 B15** 53 31N 28 42 E
Hron ➝, *Slovak Rep.* **17 E10** 47 49N 18 45 E
Hrvatska = Croatia ■, *Europe* **16 F9** 45 20N 16 0 E
Hrymayliv, *Ukraine* **17 D14** 49 20N 26 5 E
Hsenwi, *Burma* **41 H20** 23 22N 97 55 E
Hsiamen = Xiamen, *China* **33 D6** 24 25N 118 4 E
Hsian = Xi'an, *China* **34 G5** 34 15N 109 0 E
Hsinchu, *Taiwan* **33 D7** 24 48N 120 58 E
Hsinhailien = Lianyungang, *China* **35 G10** 34 40N 119 11 E
Hsüchou = Xuzhou, *China* **35 G9** 34 18N 117 10 E
Hu Xian, *China* **34 G5** 34 8N 108 42 E
Hua Hin, *Thailand* **38 F2** 12 34N 99 58 E
Hua Xian, *Henan, China* **34 G8** 35 30N 114 30 E
Hua Xian, *Shaanxi, China* **34 G5** 34 30N 109 48 E
Huab ➝, *Namibia* **56 B2** 20 52S 13 25 E
Huachinera, *Mexico* **86 A3** 30 9N 108 55W
Huacho, *Peru* **92 F3** 11 10S 77 35W
Huade, *China* **34 D7** 41 55N 113 59 E
Huadian, *China* **35 C14** 43 0N 126 40 E
Huai He ➝, *China* **33 C6** 33 0N 118 30 E
Huai Yot, *Thailand* **39 J2** 7 45N 99 37 E
Huai'an, *Hebei, China* **34 D8** 40 30N 114 20 E
Huai'an, *Jiangsu, China* **35 H10** 33 30N 119 10 E
Huaibei, *China* **34 G9** 34 0N 116 48 E
Huaide = Gongzhuling, *China* **35 C13** 43 30N 124 40 E
Huaidezhen, *China* **35 C13** 43 48N 124 50 E
Huainan, *China* **33 C6** 32 38N 116 58 E
Huairen, *China* **34 E7** 39 48N 113 20 E
Huairou, *China* **34 D9** 40 20N 116 35 E
Huaiyang, *China* **34 H8** 33 40N 114 52 E
Huaiyin, *China* **35 H10** 33 30N 119 2 E
Huaiyuan, *China* **35 H9** 32 55N 117 10 E
Huajianzi, *China* **35 D13** 41 23N 125 20 E
Huajuapan de Leon, *Mexico* **87 D5** 17 50N 97 48W
Hualapai Peak, *U.S.A.* **83 J7** 35 5N 113 54W
Huallaga ➝, *Peru* **92 E3** 5 15S 75 30W
Huambo, *Angola* **53 G3** 12 42S 15 54 E
Huan Jiang ➝, *China* **34 G5** 34 28N 109 0 E
Huan Xian, *China* **34 F4** 36 33N 107 7 E
Huancabamba, *Peru* **92 E3** 5 10S 79 15W
Huancane, *Peru* **92 G5** 15 10S 69 44W
Huancavelica, *Peru* **92 F3** 12 50S 75 5W
Huancayo, *Peru* **92 F3** 12 5S 75 12W
Huanchaca, *Bolivia* **92 H5** 20 15S 66 40W
Huang Hai = Yellow Sea, *China* **35 G12** 35 0N 123 0 E
Huang He ➝, *China* **35 F10** 37 55N 118 50 E
Huang Xian, *China* **35 F11** 37 38N 120 30 E
Huangling, *China* **34 G5** 35 34N 109 15 E
Huanglong, *China* **34 G5** 35 30N 109 59 E
Huangshan, *China* **33 D6** 29 42N 118 25 E
Huangshi, *China* **33 C6** 30 10N 115 3 E
Huangsongdian, *China* **35 C14** 43 45N 127 25 E
Huantai, *China* **35 F9** 36 58N 117 56 E
Huánuco, *Peru* **92 E3** 9 55S 76 15W
Huaraz, *Peru* **92 E3** 9 30S 77 32W
Huarmey, *Peru* **92 F3** 10 5S 78 5W
Huascarán, *Peru* **92 E3** 9 8S 77 36W
Huasco, *Chile* **94 B1** 28 30S 71 15W
Huasco ➝, *Chile* **94 B1** 28 27S 71 13W
Huasna, *U.S.A.* **85 K6** 35 6N 120 24W
Huatabampo, *Mexico* **86 B3** 26 50N 109 50W
Huauchinango, *Mexico* **87 C5** 20 11N 98 3W
Huautla de Jiménez, *Mexico* **87 D5** 18 8N 96 51W
Huay Namota, *Mexico* **86 C4** 21 56N 104 30W
Huayin, *China* **34 G6** 34 35N 110 5 E
Hubbard, *Ohio, U.S.A.* **78 E4** 41 9N 80 34W
Hubbard, *Tex., U.S.A.* **81 K6** 31 51N 96 48W
Hubbart Pt., *Canada* **73 B10** 59 21N 94 41W
Hubei □, *China* **33 C6** 31 0N 112 0 E
Huch'ang, *N. Korea* **35 D14** 41 25N 127 2 E
Hucknall, *U.K.* **10 D6** 53 3N 1 13W
Huddersfield, *U.K.* **10 D6** 53 39N 1 47W
Hudiksvall, *Sweden* **9 F17** 61 43N 17 10 E
Hudson, *Canada* **70 B1** 50 6N 92 9W
Hudson, *Mass., U.S.A.* **79 D13** 42 23N 71 34W
Hudson, *N.Y., U.S.A.* **79 D11** 42 15N 73 46W
Hudson, *Wis., U.S.A.* **80 C8** 44 58N 92 45W
Hudson, *Wyo., U.S.A.* **82 E9** 42 54N 108 35W
Hudson ➝, *U.S.A.* **79 F10** 40 42N 74 2W
Hudson Bay, *Nunavut, Canada* **69 C11** 60 0N 86 0W
Hudson Bay, *Sask., Canada* **73 C8** 52 51N 102 23W
Hudson Falls, *U.S.A.* **79 C11** 43 18N 73 35W
Hudson Mts., *Antarctica* **5 D16** 74 32S 99 20W
Hudson Str., *Canada* **69 B13** 62 0N 70 0W
Hudson's Hope, *Canada* **72 B4** 56 0N 121 54W
Hue, *Vietnam* **38 D6** 16 30N 107 35 E
Huehuetenango, *Guatemala* **88 C1** 15 20N 91 28W
Huejúcar, *Mexico* **86 C4** 22 21N 103 13W
Huelva, *Spain* **19 D2** 37 18N 6 57W
Huentelauquén, *Chile* **94 C1** 31 38S 71 33W
Huerta, Sa. de la, *Argentina* **94 C2** 31 10S 67 30W
Huesca, *Spain* **19 A5** 42 8N 0 25W
Huetamo, *Mexico* **86 D4** 18 36N 100 54W
Hugh ➝, *Australia* **62 D1** 25 1S 134 1 E
Hughenden, *Australia* **62 C3** 20 52S 144 10 E
Hughes, *Australia* **61 F4** 30 42S 129 31 E
Hughesville, *U.S.A.* **79 E8** 41 14N 76 44W
Hugli ➝, *India* **43 J13** 21 56N 88 4 E
Hugo, *Colo., U.S.A.* **80 F3** 39 8N 103 28W
Hugo, *Okla., U.S.A.* **81 H7** 34 1N 95 31W
Hugoton, *U.S.A.* **81 G4** 37 11N 101 21W
Hui Xian = Huixian, *China* **34 G7** 35 27N 113 12 E
Hui Xian, *China* **34 H4** 33 50N 106 4 E
Hui'anbu, *China* **34 F4** 37 28N 106 38 E
Huichapán, *Mexico* **87 C5** 20 24N 99 40W
Huifa He ➝, *China* **35 C14** 43 0N 127 50 E
Huila, Nevado del, *Colombia* **92 C3** 3 0N 76 0W
Huimin, *China* **35 F9** 37 27N 117 28 E
Huinan, *China* **35 C14** 42 40N 126 2 E
Huinca Renancó, *Argentina* **94 C3** 34 51S 64 22W
Huining, *China* **34 G3** 35 38N 105 0 E
Huinong, *China* **34 E4** 39 5N 106 35 E
Huisache, *Mexico* **86 C4** 22 55N 100 25W
Huiting, *China* **34 G9** 34 5N 116 5 E
Huixian, *China* **34 G7** 35 27N 113 12 E
Huixtla, *Mexico* **87 D6** 15 9N 92 28W
Huize, *China* **32 D5** 26 24N 103 15 E
Hukawng Valley, *Burma* **41 F20** 26 30N 96 30 E
Hŭksan-chedo, *S. Korea* **35 G13** 34 40N 125 30 E
Hukuntsi, *Botswana* **56 C3** 23 58S 21 45 E
Ḥulayfā', *Si. Arabia* **44 E4** 25 58N 40 45 E
Huld = Ulaanjirem, *Mongolia* **34 B3** 45 5N 105 30 E
Hulin He ➝, *China* **35 B12** 45 0N 122 10 E
Hull = Kingston upon Hull, *U.K.* **10 D7** 53 45N 0 21W
Hull, *Canada* **79 A9** 45 25N 75 44W
Hull ➝, *U.K.* **10 D7** 53 44N 0 20W
Hulst, *Neths.* **15 C4** 51 17N 4 2 E
Hulun Nur, *China* **33 B6** 49 0N 117 30 E
Humahuaca, *Argentina* **94 A2** 23 10S 65 25W
Humaitá, *Brazil* **92 E6** 7 35S 63 1W
Humaitá, *Paraguay* **94 B4** 27 2S 58 31W
Humansdorp, *S. Africa* **56 E3** 34 2S 24 46 E
Humbe, *Angola* **56 B1** 16 40S 14 55 E
Humber ➝, *U.K.* **10 D7** 53 42N 0 27W
Humboldt, *Canada* **73 C7** 52 15N 105 9W
Humboldt, *Iowa, U.S.A.* **80 D7** 42 44N 94 13W
Humboldt, *Tenn., U.S.A.* **81 H10** 35 50N 88 55W
Humboldt ➝, *U.S.A.* **82 F4** 39 59N 118 36W
Humboldt Gletscher, *Greenland* **4 B4** 79 30N 62 0W
Hume, *U.S.A.* **84 J8** 36 48N 118 54W
Hume, L., *Australia* **63 F4** 36 0S 147 5 E
Humenné, *Slovak Rep.* **17 D11** 48 55N 21 50 E
Humphreys, Mt., *U.S.A.* **84 H8** 37 17N 118 40W
Humphreys Peak, *U.S.A.* **83 J8** 35 21N 111 41W
Humptulips, *U.S.A.* **84 C3** 47 14N 123 57W
Hūn, *Libya* **51 C9** 29 2N 16 0 E
Hun Jiang ➝, *China* **35 D13** 40 50N 125 38 E
Húnaflói, *Iceland* **8 D3** 65 50N 20 50W
Hunan □, *China* **33 D6** 27 30N 112 0 E
Hunchun, *China* **35 C16** 42 52N 130 28 E
Hundewali, *Pakistan* **42 D5** 31 55N 72 38 E
Hundred Mile House, *Canada* **72 C4** 51 38N 121 18W
Hunedoara, *Romania* **17 F12** 45 40N 22 50 E
Hung Yen, *Vietnam* **38 B6** 20 39N 106 4 E
Hungary ■, *Europe* **17 E10** 47 20N 19 20 E
Hungary, Plain of, *Europe* **6 F10** 47 0N 20 0 E
Hungerford, *Australia* **63 D3** 28 58S 144 24 E
Hŭngnam, *N. Korea* **35 E14** 39 49N 127 45 E
Hunsberge, *Namibia* **56 D2** 27 45S 17 12 E
Hunsrück, *Germany* **16 D4** 49 56N 7 27 E
Hunstanton, *U.K.* **10 E8** 52 56N 0 29 E
Hunter, *U.S.A.* **79 D10** 42 13N 74 13W
Hunter I., *Australia* **62 G3** 40 30S 144 45 E
Hunter I., *Canada* **72 C3** 51 55N 128 0W
Hunter Ra., *Australia* **63 E5** 32 45S 150 15 E
Hunters Road, *Zimbabwe* **55 F2** 19 9S 29 49 E
Hunterville, *N.Z.* **59 H5** 39 56S 175 35 E
Huntingburg, *U.S.A.* **76 F2** 38 18N 86 57W
Huntingdon, *Canada* **70 C5** 45 6N 74 10W
Huntingdon, *U.K.* **11 E7** 52 20N 0 11W
Huntingdon, *U.S.A.* **78 F6** 40 30N 78 1W
Huntington, *Ind., U.S.A.* **76 E3** 40 53N 85 30W
Huntington, *Oreg., U.S.A.* **82 D5** 44 21N 117 16W
Huntington, *Utah, U.S.A.* **82 G8** 39 20N 110 58W
Huntington, *W. Va., U.S.A.* **76 F4** 38 25N 82 27W
Huntington Beach, *U.S.A.* **85 M9** 33 40N 118 5W
Huntington Station, *U.S.A.* **79 F11** 40 52N 73 26W
Huntly, *N.Z.* **59 G5** 37 34S 175 11 E
Huntly, *U.K.* **12 D6** 57 27N 2 47W
Huntsville, *Canada* **78 A5** 45 20N 79 14W
Huntsville, *Ala., U.S.A.* **77 H2** 34 44N 86 35W
Huntsville, *Tex., U.S.A.* **81 K7** 30 43N 95 33W
Hunyani ➝, *Zimbabwe* **55 F3** 15 57S 30 39 E
Hunyuan, *China* **34 E7** 39 42N 113 42 E
Hunza ➝, *India* **43 B6** 35 54N 74 20 E
Huo Xian = Huozhou, *China* **34 F6** 36 36N 111 42 E
Huong Hoa, *Vietnam* **38 D6** 16 37N 106 45 E
Huong Khe, *Vietnam* **38 C5** 18 13N 105 41 E
Huonville, *Australia* **62 G4** 43 0S 147 5 E
Huozhou, *China* **34 F6** 36 36N 111 42 E
Hupeh = Hubei □, *China* **33 C6** 31 0N 112 0 E
Ḥūr, *Iran* **45 D8** 30 50N 57 7 E
Hurd, C., *Canada* **78 A3** 45 13N 81 44W
Hure Qi, *China* **35 C11** 42 45N 121 45 E
Hurghada, *Egypt* **51 C12** 27 15N 33 50 E
Hurley, *N. Mex., U.S.A.* **83 K9** 32 42N 108 8W
Hurley, *Wis., U.S.A.* **80 B9** 46 27N 90 11W
Huron, *Calif., U.S.A.* **84 J6** 36 12N 120 6W
Huron, *Ohio, U.S.A.* **78 E2** 41 24N 82 33W
Huron, *S. Dak., U.S.A.* **80 C5** 44 22N 98 13W
Huron, L., *U.S.A.* **78 B2** 44 30N 82 40W
Hurricane, *U.S.A.* **83 H7** 37 11N 113 17W
Hurunui ➝, *N.Z.* **59 K4** 42 54S 173 18 E
Húsavík, *Iceland* **8 C5** 66 3N 17 21W
Huşi, *Romania* **17 E15** 46 41N 28 7 E
Huskvarna, *Sweden* **9 H16** 57 47N 14 15 E
Hustadvika, *Norway* **8 E12** 63 0N 7 0 E
Hustontown, *U.S.A.* **78 F6** 40 3N 78 2W
Hutchinson, *Kans., U.S.A.* **81 F6** 38 5N 97 56W
Hutchinson, *Minn., U.S.A.* **80 C7** 44 54N 94 22W
Hutte Sauvage, L. de la, *Canada* **71 A7** 56 15N 64 45W
Hutton, Mt., *Australia* **63 D4** 25 51S 148 20 E
Huy, *Belgium* **15 D5** 50 31N 5 15 E
Huzhou, *China* **33 C7** 30 51N 120 8 E
Hvammstangi, *Iceland* **8 D3** 65 24N 20 57W
Hvar, *Croatia* **20 C7** 43 11N 16 28 E
Hvítá ➝, *Iceland* **8 D3** 64 30N 21 58W
Hwachŏn-chŏsuji, *S. Korea* **35 E14** 38 5N 127 50 E
Hwang Ho = Huang He ➝, *China* **35 F10** 37 55N 118 50 E
Hwange, *Zimbabwe* **55 F2** 18 18S 26 30 E
Hwange Nat. Park, *Zimbabwe* **56 B4** 19 0S 26 30 E
Hyannis, *Mass., U.S.A.* **76 E10** 41 39N 70 17W
Hyannis, *Nebr., U.S.A.* **80 E4** 42 0N 101 46W
Hyargas Nuur, *Mongolia* **32 B4** 49 0N 93 0 E
Hydaburg, *U.S.A.* **72 B2** 55 15N 132 50W
Hyde Park, *U.S.A.* **79 E11** 41 47N 73 56W
Hyden, *Australia* **61 F2** 32 24S 118 53 E
Hyder, *U.S.A.* **72 B2** 55 55N 130 5W
Hyderabad, *India* **40 L11** 17 22N 78 29 E
Hyderabad, *Pakistan* **42 G3** 25 23N 68 24 E
Hyères, *France* **18 E7** 43 8N 6 9 E
Hyères, Îs. d', *France* **18 E7** 43 0N 6 20 E
Hyesan, *N. Korea* **35 D15** 41 20N 128 10 E
Hyland ➝, *Canada* **72 B3** 59 52N 128 12W
Hymia, *India* **43 C8** 33 40N 78 2 E
Hyndman Peak, *U.S.A.* **82 E6** 43 45N 114 8W
Hyōgo □, *Japan* **31 G7** 35 15N 134 50 E
Hyrum, *U.S.A.* **82 F8** 41 38N 111 51W
Hysham, *U.S.A.* **82 C10** 46 18N 107 14W
Hythe, *U.K.* **11 F9** 51 4N 1 5 E
Hyūga, *Japan* **31 H5** 32 25N 131 35 E
Hyvinge = Hyvinkää, *Finland* **9 F21** 60 38N 24 50 E
Hyvinkää, *Finland* **9 F21** 60 38N 24 50 E

I

I-n-Gall, *Niger* **50 E7** 16 51N 7 1 E
Iaco ➝, *Brazil* **92 E5** 9 3S 68 34W
Iakora, *Madag.* **57 C8** 23 6S 46 40 E
Ialomiţa ➝, *Romania* **17 F14** 44 42N 27 51 E
Iaşi, *Romania* **17 E14** 47 10N 27 40 E
Ib ➝, *India* **43 J10** 21 34N 83 48 E
Iba, *Phil.* **37 A6** 15 22N 120 0 E
Ibadan, *Nigeria* **50 G6** 7 22N 3 58 E
Ibagué, *Colombia* **92 C3** 4 20N 75 20W
Ibar ➝, *Serbia, Yug.* **21 C9** 43 43N 20 45 E
Ibaraki □, *Japan* **31 F10** 36 10N 140 10 E
Ibarra, *Ecuador* **92 C3** 0 21N 78 7W
Ibembo, *Dem. Rep. of the Congo* **54 B1** 2 35N 23 35 E
Ibera, L., *Argentina* **94 B4** 28 30S 57 9W
Iberian Peninsula, *Europe* **6 H5** 40 0N 5 0W
Iberville, *Canada* **79 A11** 45 19N 73 17W
Iberville, Lac d', *Canada* **70 A5** 55 55N 73 15W
Ibiá, *Brazil* **93 G9** 19 30S 46 30W
Ibiapaba, Sa. da, *Brazil* **93 D10** 4 0S 41 30W
Ibicui ➝, *Brazil* **95 B4** 29 25S 56 47W
Ibicuy, *Argentina* **94 C4** 33 55S 59 10W
Ibiza = Eivissa, *Spain* **22 C7** 38 54N 1 26 E
Ibo, *Mozam.* **55 E5** 12 22S 40 40 E
Ibonma, *Indonesia* **37 E8** 3 29S 133 31 E
Ibotirama, *Brazil* **93 F10** 12 13S 43 12W
Ibrāhīm ➝, *Lebanon* **47 A4** 34 4N 35 38 E
'Ibrī, *Oman* **45 F8** 23 14N 56 30 E
Ibu, *Indonesia* **37 D7** 1 35N 127 33 E
Ibusuki, *Japan* **31 J5** 31 12N 130 40 E
Ica, *Peru* **92 F3** 14 0S 75 48W
Iça ➝, *Brazil* **92 D5** 2 55S 67 58W
Içana, *Brazil* **92 C5** 0 21N 67 19W
Içana ➝, *Brazil* **92 C5** 0 26N 67 19W
İçel = Mersin, *Turkey* **25 G5** 36 51N 34 36 E
Iceland ■, *Europe* **8 D4** 64 45N 19 0W
Ich'ang = Yichang, *China* **33 C6** 30 40N 111 20 E
Ichchapuram, *India* **41 K14** 19 10N 84 40 E
Ichhawar, *India* **42 H7** 23 1N 77 1 E
Ichihara, *Japan* **31 G10** 35 28N 140 5 E
Ichikawa, *Japan* **31 G9** 35 44N 139 55 E
Ichilo ➝, *Bolivia* **92 G6** 15 57S 64 50W
Ichinohe, *Japan* **30 D10** 40 13N 141 17 E
Ichinomiya, *Japan* **31 G8** 35 18N 136 48 E
Ichinoseki, *Japan* **30 E10** 38 55N 141 8 E
Icod, *Canary Is.* **22 F3** 28 22N 16 43W
Ida Grove, *U.S.A.* **80 D7** 42 21N 95 28W
Idabel, *U.S.A.* **81 J7** 33 54N 94 50W
Idaho □, *U.S.A.* **82 D7** 45 0N 115 0W
Idaho City, *U.S.A.* **82 E6** 43 50N 115 50W
Idaho Falls, *U.S.A.* **82 E7** 43 30N 112 2W
Idar-Oberstein, *Germany* **16 D4** 49 43N 7 16 E
Idfû, *Egypt* **51 D12** 24 55N 32 49 E
Ídhi Óros, *Greece* **23 D6** 35 15N 24 45 E
Ídhra, *Greece* **21 F10** 37 20N 23 28 E
Idi, *Indonesia* **36 C1** 5 2N 97 37 E
Idiofa, *Dem. Rep. of the Congo* **52 E3** 4 55S 19 42 E
Idlib, *Syria* **44 C3** 35 55N 36 36 E
Idria, *U.S.A.* **84 J6** 36 25N 120 41W
Idutywa, *S. Africa* **57 E4** 32 8S 28 18 E
Ieper, *Belgium* **15 D2** 50 51N 2 53 E
Ierápetra, *Greece* **23 E7** 35 1N 25 44 E
Iesi, *Italy* **20 C5** 43 31N 13 14 E
Ifakara, *Tanzania* **52 F7** 8 8S 36 41 E
'Ifāl, W. al ➝, *Si. Arabia* **44 D2** 28 7N 35 3 E
Ifanadiana, *Madag.* **57 C8** 21 19S 47 39 E
Ife, *Nigeria* **50 G6** 7 30N 4 31 E
Iffley, *Australia* **62 B3** 18 53S 141 12 E
Iforas, Adrar des, *Africa* **50 E6** 19 40N 1 40 E
Ifould, L., *Australia* **61 F5** 30 52S 132 6 E
Iganga, *Uganda* **54 B3** 0 37N 33 28 E
Igarapava, *Brazil* **93 H9** 20 3S 47 47W
Igarka, *Russia* **26 C9** 67 30N 86 33 E
Igatimi, *Paraguay* **95 A4** 24 5S 55 40W
Iggesund, *Sweden* **9 F17** 61 39N 17 10 E
Iglésias, *Italy* **20 E3** 39 19N 8 32 E
Igloolik, *Canada* **69 B11** 69 20N 81 49W
Igluligaarjuk, *Canada* **69 B10** 63 21N 90 42W
Iglulik = Igloolik, *Canada* **69 B11** 69 20N 81 49W
Ignace, *Canada* **70 C1** 49 30N 91 40W
İğneada Burnu, *Turkey* **21 D13** 41 53N 28 2 E
Igoumenitsa, *Greece* **21 E9** 39 32N 20 18 E
Iguaçu ➝, *Brazil* **95 B5** 25 36S 54 36W
Iguaçu, Cat. del, *Brazil* **95 B5** 25 41S 54 26W
Iguaçu Falls = Iguaçu, Cat. del, *Brazil* **95 B5** 25 41S 54 26W
Iguala, *Mexico* **87 D5** 18 20N 99 40W
Igualada, *Spain* **19 B6** 41 37N 1 37 E
Iguassu = Iguaçu ➝, *Brazil* **95 B5** 25 36S 54 36W
Iguatu, *Brazil* **93 E11** 6 20S 39 18W
Iharana, *Madag.* **57 A9** 13 25S 50 0 E
Ihbulag, *Mongolia* **34 C4** 43 11N 107 10 E
Iheya-Shima, *Japan* **31 L3** 27 4N 127 58 E
Ihosy, *Madag.* **57 C8** 22 24S 46 8 E
Ihotry, Farihy, *Madag.* **57 C7** 21 56S 43 41 E
Ii, *Finland* **8 D21** 65 19N 25 22 E
Ii-Shima, *Japan* **31 L3** 26 43N 127 47 E
Iida, *Japan* **31 G8** 35 35N 137 50 E
Iijoki ➝, *Finland* **8 D21** 65 20N 25 20 E
Iisalmi, *Finland* **8 E22** 63 32N 27 10 E
Iiyama, *Japan* **31 F9** 36 51N 138 22 E
Iizuka, *Japan* **31 H5** 33 38N 130 42 E
Ijebu-Ode, *Nigeria* **50 G6** 6 47N 3 58 E
IJmuiden, *Neths.* **15 B4** 52 28N 4 35 E
IJssel ➝, *Neths.* **15 B5** 52 35N 5 50 E
IJsselmeer, *Neths.* **15 B5** 52 45N 5 20 E
Ijuí, *Brazil* **95 B5** 28 23S 53 55W
Ijuí ➝, *Brazil* **95 B4** 27 58S 55 20W
Ikalamavony, *Madag.* **57 C8** 21 9S 46 35 E
Ikaluktutiak, *Canada* **68 B9** 69 10N 105 0W
Ikare, *Nigeria* **50 G7** 7 32N 5 40 E
Ikaria, *Greece* **21 F12** 37 35N 26 10 E
Ikeda, *Japan* **31 G6** 34 1N 133 48 E
Ikela, *Dem. Rep. of the Congo* **52 E4** 1 6S 23 6 E
Iki, *Japan* **31 H4** 33 45N 129 42 E
Ikimba L., *Tanzania* **54 C3** 1 30S 31 20 E
Ikongo, *Madag.* **57 C8** 21 52S 47 27 E
Ikopa ➝, *Madag.* **57 B8** 16 45S 46 40 E
Ikungu, *Tanzania* **54 C3** 1 33S 33 42 E
Ilagan, *Phil.* **37 A6** 17 7N 121 53 E
Ilaka, *Madag.* **57 B8** 19 33S 48 52 E
Īlām, *Iran* **44 C5** 33 36N 46 36 E
Ilam, *Nepal* **43 F12** 26 58N 87 58 E
Īlām □, *Iran* **44 C5** 33 0N 47 0 E
Ilanskiy, *Russia* **27 D10** 56 14N 96 3 E
Iława, *Poland* **17 B10** 53 36N 19 34 E
Ile ➝, *Kazakstan* **26 E8** 45 53N 77 10 E
Ile-à-la-Crosse, *Canada* **73 B7** 55 27N 107 53W
Ile-à-la-Crosse, Lac, *Canada* **73 B7** 55 40N 107 45W
Île-de-France □, *France* **18 B5** 49 0N 2 20 E
Ilebo, *Dem. Rep. of the Congo* **52 E4** 4 17S 20 55 E
Ilek, *Russia* **26 D6** 51 32N 53 21 E
Ilek ➝, *Russia* **24 D9** 51 30N 53 22 E
Ilesha, *Nigeria* **50 G6** 7 37N 4 40 E
Ilford, *Canada* **73 B9** 56 4N 95 35W
Ilfracombe, *Australia* **62 C3** 23 30S 144 30 E

Ilfracombe

Ilfracombe, U.K. 11 F3 51 12N 4 8W
Ilhéus, Brazil 93 F11 14 49S 39 2W
Ili = Ile →, Kazakstan 26 E8 45 53N 77 10 E
Iliamna L., U.S.A. 68 C4 59 30N 155 0W
Iligan, Phil. 37 C6 8 12N 124 13 E
Ilion, U.S.A. 79 D9 43 1N 75 2W
Ilkeston, U.K. 10 E6 52 58N 1 19W
Ilkley, U.K. 10 D6 53 56N 1 48W
Illampu = Ancohuma, Nevada, Bolivia 92 G5 16 0S 68 50W
Illana B., Phil. 37 C6 7 35N 123 45 E
Illapel, Chile 94 C1 32 0S 71 10W
Iller →, Germany 16 D6 48 23N 9 58 E
Illetas, Spain 22 B9 39 32N 2 35 E
Illimani, Nevado, Bolivia 92 G5 16 30S 67 50W
Illinois □, U.S.A. 80 E10 40 15N 89 30W
Illinois →, U.S.A. 80 F9 38 58N 90 28W
Illium = Troy, Turkey 21 E12 39 57N 26 12 E
Illizi, Algeria 50 C7 26 31N 8 32 E
Ilmajoki, Finland 9 E20 62 44N 22 34 E
Ilmen, Ozero, Russia 24 C5 58 15N 31 10 E
Ilo, Peru 92 G4 17 40S 71 20W
Iloilo, Phil. 37 B6 10 45N 122 33 E
Ilorin, Nigeria 50 G6 8 30N 4 35 E
Ilwaco, U.S.A. 84 D2 46 19N 124 3W
Ilwaki, Indonesia 37 F7 7 55S 126 30 E
Imabari, Japan 31 G6 34 4N 133 0 E
Imaloto →, Madag. 57 C8 23 27S 45 13 E
Imandra, Ozero, Russia 24 A5 67 30N 33 0 E
Imanombo, Madag. 57 C8 24 26S 45 49 E
Imari, Japan 31 H4 33 15N 129 52 E
Imatra, Finland 24 B4 61 12N 28 48 E
Imbil, Australia 63 D5 26 22S 152 32 E
imeni 26 Bakinskikh Komissarov = Neftçala, Azerbaijan 25 G8 39 19N 49 12 E
imeni 26 Bakinskikh Komissarov, Turkmenistan 45 B7 39 22N 54 10 E
Imeri, Serra, Brazil 92 C5 0 50N 65 25W
Imerimandroso, Madag. 57 B8 17 26S 48 35 E
Imi, Ethiopia 46 F3 6 28N 42 10 E
Imlay, U.S.A. 82 F4 40 40N 118 9W
Imlay City, U.S.A. 78 D1 43 2N 83 5W
Immingham, U.K. 10 D7 53 37N 0 13W
Immokalee, U.S.A. 77 M5 26 25N 81 25W
Ímola, Italy 20 B4 44 20N 11 42 E
Imperatriz, Brazil 93 E9 5 30S 47 29W
Impéria, Italy 18 E8 43 53N 8 3 E
Imperial, Canada 73 C7 51 21N 105 28W
Imperial, Calif., U.S.A. 85 N11 32 51N 115 34W
Imperial, Nebr., U.S.A. 80 E4 40 31N 101 39W
Imperial Beach, U.S.A. 85 N9 32 35N 117 8W
Imperial Dam, U.S.A. 85 N12 32 55N 114 25W
Imperial Reservoir, U.S.A. 85 N12 32 53N 114 28W
Imperial Valley, U.S.A. 85 N11 33 0N 115 30W
Imperieuse Reef, Australia 60 C2 17 36S 118 50 E
Impfondo, Congo 52 D3 1 40N 18 0 E
Imphal, India 41 G18 24 48N 93 56 E
İmroz = Gökçeada, Turkey 21 D11 40 10N 25 50 E
Imuris, Mexico 86 A2 30 47N 110 52W
Imuruan B., Phil. 37 B5 10 40N 119 10 E
In Salah, Algeria 50 C6 27 10N 2 32 E
Ina, Japan 31 G8 35 50N 137 55 E
Inangahua, N.Z. 59 J3 41 52S 171 59 E
Inanwatan, Indonesia 37 E8 2 8S 132 10 E
Iñapari, Peru 92 F5 11 0S 69 40W
Inari, Finland 8 B22 68 54N 27 5 E
Inarijärvi, Finland 8 B22 69 0N 28 0 E
Inawashiro-Ko, Japan 30 F10 37 29N 140 6 E
Inca, Spain 22 B9 39 43N 2 54 E
Inca de Oro, Chile 94 B2 26 45S 69 54W
Incaguasi, Chile 94 B1 29 12S 71 5W
İnce Burun, Turkey 25 F5 42 7N 34 56 E
Incesu, Turkey 44 B2 38 38N 35 11 E
Inch'ŏn, S. Korea 35 F14 37 27N 126 40 E
İncirliova, Turkey 21 F12 37 50N 27 41 E
Incline Village, U.S.A. 82 G4 39 10N 119 58W
Incomáti →, Mozam. 57 D5 25 46S 32 43 E
Indalsälven →, Sweden 9 E17 62 36N 17 30 E
Indaw, Burma 41 G20 24 15N 96 5 E
Independence, Calif., U.S.A. 84 J8 36 48N 118 12W
Independence, Iowa, U.S.A. 80 D9 42 28N 91 54W
Independence, Kans., U.S.A. 81 G7 37 14N 95 42W
Independence, Ky., U.S.A. 76 F3 38 57N 84 33W
Independence, Mo., U.S.A. 80 F7 39 6N 94 25W
Independence Fjord, Greenland 4 A6 82 10N 29 0W
Independence Mts., U.S.A. 82 F5 41 20N 116 0W
Index, U.S.A. 84 C5 47 50N 121 33W
India ■, Asia 40 K11 20 0N 78 0 E
Indian →, U.S.A. 77 M5 27 59N 80 34W
Indian Cabins, Canada 72 B5 59 52N 117 40W
Indian Harbour, Canada 71 B8 54 27N 57 13W
Indian Head, Canada 73 C8 50 30N 103 41W
Indian Lake, U.S.A. 79 C10 43 47N 74 16W
Indian Ocean 28 K11 5 0S 75 0 E
Indian Springs, U.S.A. 85 J11 36 35N 115 40W
Indiana, U.S.A. 78 F5 40 37N 79 9W
Indiana □, U.S.A. 76 F3 40 0N 86 0W
Indianapolis, U.S.A. 76 F2 39 46N 86 9W
Indianola, Iowa, U.S.A. 80 E8 41 22N 93 34W
Indianola, Miss., U.S.A. 81 J9 33 27N 90 39W
Indiga, Russia 24 A8 67 38N 49 9 E
Indigirka →, Russia 27 B15 70 48N 148 54 E
Indio, U.S.A. 85 M10 33 43N 116 13W
Indo-China, Asia 28 H14 15 0N 102 0 E
Indonesia ■, Asia 36 F5 5 0S 115 0 E
Indore, India 42 H6 22 42N 75 53 E
Indramayu, Indonesia 37 G13 6 20S 108 19 E
Indravati →, India 41 K12 19 20N 80 20 E
Indre →, France 18 C4 47 16N 0 11 E
Indulkana, Australia 63 D1 26 58S 133 5 E
Indus →, Pakistan 42 G2 24 20N 67 47 E
Indus, Mouths of the, Pakistan 42 H3 24 0N 68 0 E
İnebolu, Turkey 25 F5 41 55N 33 40 E
Infiernillo, Presa del, Mexico 86 D4 18 9N 102 0W
Ingenio, Canary Is. 22 G4 27 55N 15 26W
Ingenio Santa Ana, Argentina 94 B2 27 25S 65 40W
Ingersoll, Canada 78 C4 43 4N 80 55W
Ingham, Australia 62 B4 18 43S 146 10 E
Ingleborough, U.K. 10 C5 54 10N 2 22W
Inglewood, Queens., Australia 63 D5 28 25S 151 2 E
Inglewood, Vic., Australia 63 F3 36 29S 143 53 E
Inglewood, N.Z. 59 H5 39 9S 174 14 E
Inglewood, U.S.A. 85 M8 33 58N 118 21W
Ingólfshöfði, Iceland 8 E5 63 48N 16 39W
Ingolstadt, Germany 16 D6 48 46N 11 26 E
Ingomar, U.S.A. 82 C10 46 35N 107 23W
Ingonish, Canada 71 C7 46 42N 60 18W
Ingraj Bazar, India 43 G13 24 58N 88 10 E
Ingrid Christensen Coast, Antarctica 5 C6 69 30S 76 0 E
Ingulec = Inhulec, Ukraine 25 E5 47 42N 33 14 E
Ingushetia □, Russia 25 F8 43 20N 44 50 E
Ingwavuma, S. Africa 57 D5 27 9S 31 59 E
Inhaca, Mozam. 57 D5 26 1S 32 57 E
Inhafenga, Mozam. 57 C5 20 36S 33 53 E
Inhambane, Mozam. 57 C6 23 54S 35 30 E
Inhambane □, Mozam. 57 C5 22 30S 34 20 E
Inhaminga, Mozam. 55 F4 18 26S 35 0 E
Inharrime, Mozam. 57 C6 24 30S 35 0 E
Inharrime →, Mozam. 57 C6 24 30S 35 0 E
Inhulec, Ukraine 25 E5 47 42N 33 14 E
Ining = Yining, China 26 E9 43 58N 81 10 E
Inírida →, Colombia 92 C5 3 55N 67 52W
Inishbofin, Ireland 13 C1 53 37N 10 13W
Inisheer, Ireland 13 C2 53 3N 9 32W
Inishfree B., Ireland 13 A3 55 4N 8 23W
Inishkea North, Ireland 13 B1 54 9N 10 11W
Inishkea South, Ireland 13 B1 54 7N 10 12W
Inishmaan, Ireland 13 C2 53 5N 9 35W
Inishmore, Ireland 13 C2 53 8N 9 45W
Inishowen Pen., Ireland 13 A4 55 14N 7 15W
Inishshark, Ireland 13 C1 53 37N 10 16W
Inishturk, Ireland 13 C1 53 42N 10 7W
Inishvickillane, Ireland 13 D1 52 3N 10 37W
Injune, Australia 63 D4 25 53S 148 32 E
Inklin →, Canada 72 B2 58 50N 133 10W
Inland Sea = Setonaikai, Japan 31 G6 34 20N 133 30 E
Inle L., Burma 41 J20 20 30N 96 58 E
Inlet, U.S.A. 79 C10 43 45N 74 48W
Inn →, Austria 16 D7 48 35N 13 28 E
Innamincka, Australia 63 D3 27 44S 140 46 E
Inner Hebrides, U.K. 12 E2 57 0N 6 30W
Inner Mongolia = Nei Monggol Zizhiqu □, China 34 D7 42 0N 112 0 E
Inner Sound, U.K. 12 D3 57 30N 5 55W
Innerkip, Canada 78 C4 43 13N 80 42W
Innetalling I., Canada 70 A4 56 0N 79 0W
Innisfail, Australia 62 B4 17 33S 146 5 E
Innisfail, Canada 72 C6 52 0N 113 57W
In'noshima, Japan 31 G6 34 19N 133 10 E
Innsbruck, Austria 16 E6 47 16N 11 23 E
Inny →, Ireland 13 C4 53 30N 7 50W
Inongo, Dem. Rep. of the Congo 52 E3 1 55S 18 30 E
Inoucdjouac = Inukjuak, Canada 69 C12 58 25N 78 15W
Inowrocław, Poland 17 B10 52 50N 18 12 E
Inpundong, N. Korea 35 D14 41 25N 126 34 E
Inscription, C., Australia 61 E1 25 29S 112 59 E
Insein, Burma 41 L20 16 50N 96 5 E
Inta, Russia 24 A11 66 5N 60 8 E
Intendente Alvear, Argentina 94 D3 35 12S 63 32W
Interlaken, Switz. 18 C7 46 41N 7 50 E
Interlaken, U.S.A. 79 D8 42 37N 76 44W
International Falls, U.S.A. 80 A8 48 36N 93 25W
Intiyaco, Argentina 94 B3 28 43S 60 5W
Inukjuak, Canada 69 C12 58 25N 78 15W
Inútil, B., Chile 96 G2 53 30S 70 15W
Inuvik, Canada 68 B6 68 16N 133 40W
Inveraray, U.K. 12 E3 56 14N 5 5W
Inverbervie, U.K. 12 E6 56 51N 2 17W
Invercargill, N.Z. 59 M2 46 24S 168 24 E
Inverclyde □, U.K. 12 F4 55 55N 4 49W
Inverell, Australia 63 D5 29 45S 151 8 E
Invergordon, U.K. 12 D4 57 41N 4 10W
Inverloch, Australia 63 F4 38 38S 145 45 E
Invermere, Canada 72 C5 50 30N 116 2W
Inverness, Canada 71 C7 46 15N 61 19W
Inverness, U.K. 12 D4 57 29N 4 13W
Inverness, U.S.A. 77 L4 28 50N 82 20W
Inverurie, U.K. 12 D6 57 17N 2 23W
Investigator Group, Australia 63 E1 34 45S 134 20 E
Investigator Str., Australia 63 F2 35 30S 137 0 E
Inya, Russia 26 D9 50 28N 86 37 E
Inyanga, Zimbabwe 55 F3 18 12S 32 40 E
Inyangani, Zimbabwe 55 F3 18 5S 32 50 E
Inyantue, Zimbabwe 56 B4 18 33S 26 39 E
Inyo Mts., U.S.A. 84 J9 36 40N 118 0W
Inyokern, U.S.A. 85 K9 35 39N 117 49W
Inza, Russia 24 D8 53 55N 46 25 E
Iō-Jima, Japan 31 J5 30 48N 130 18 E
Ioánnina, Greece 21 E9 39 42N 20 47 E
Iola, U.S.A. 81 G7 37 55N 95 24W
Iona, U.K. 12 E2 56 20N 6 25W
Ione, U.S.A. 84 G6 38 21N 120 56W
Ionia, U.S.A. 76 D3 42 59N 85 4W
Ionian Is. = Iónioi Nisoi, Greece 21 E9 38 40N 20 0 E
Ionian Sea, Medit. S. 21 E7 37 30N 17 30 E
Iónioi Nisoi, Greece 21 E9 38 40N 20 0 E
Íos, Greece 21 F11 36 41N 25 20 E
Iowa □, U.S.A. 80 D8 42 18N 93 30W
Iowa →, U.S.A. 80 E9 41 10N 91 1W
Iowa City, U.S.A. 80 E9 41 40N 91 32W
Iowa Falls, U.S.A. 80 D8 42 31N 93 16W
Iowa Park, U.S.A. 81 J5 33 57N 98 40W
Ipala, Tanzania 54 C3 4 30S 32 52 E
Ipameri, Brazil 93 G9 17 44S 48 9W
Ipatinga, Brazil 93 G10 19 32S 42 30W
Ipiales, Colombia 92 C3 0 50N 77 37W
Ipin = Yibin, China 32 D5 28 45N 104 32 E
Ipixuna, Brazil 92 E4 7 0S 71 40W
Ipoh, Malaysia 39 K3 4 35N 101 5 E
Ippy, C.A.R. 52 C4 6 5N 21 7 E
Ipsala, Turkey 21 D12 40 55N 26 23 E
Ipswich, Australia 63 D5 27 35S 152 40 E
Ipswich, U.K. 11 E9 52 4N 1 10 E
Ipswich, Mass., U.S.A. 79 D14 42 41N 70 50W
Ipswich, S. Dak., U.S.A. 80 C5 45 27N 99 2W
Ipu, Brazil 93 D10 4 23S 40 44W
Iqaluit, Canada 69 B13 63 44N 68 31W
Iquique, Chile 92 H4 20 19S 70 5W
Iquitos, Peru 92 D4 3 45S 73 10W
Irabu-Jima, Japan 31 M2 24 50N 125 10 E
Iracoubo, Fr. Guiana 93 B8 5 30N 53 10W
Īrafshān, Iran 45 E9 26 42N 61 56 E
Iráklion, Greece 23 D7 35 20N 25 12 E
Iráklion □, Greece 23 D7 35 10N 25 10 E
Irala, Paraguay 95 B5 25 55S 54 35W
Iran ■, Asia 45 C7 33 0N 53 0 E
Iran, Gunung-Gunung, Malaysia 36 D4 2 20N 114 50 E
Iran, Plateau of, Asia 28 F9 32 0N 55 0 E
Iran Ra. = Iran, Gunung-Gunung, Malaysia 36 D4 2 20N 114 50 E
Īrānshahr, Iran 45 E9 27 15N 60 40 E
Irapuato, Mexico 86 C4 20 40N 101 30W
Iraq ■, Asia 44 C5 33 0N 44 0 E
Irati, Brazil 95 B5 25 25S 50 38W
Irbid, Jordan 47 C4 32 35N 35 48 E
Irbid □, Jordan 47 C5 32 15N 36 35 E
Ireland ■, Europe 13 C4 53 50N 7 52W
Irhyangdong, N. Korea 35 D15 41 15N 129 30 E
Iri, S. Korea 35 G14 35 59N 127 0 E
Irian Jaya □, Indonesia 37 E9 4 0S 137 0 E
Iringa, Tanzania 54 D4 7 48S 35 43 E
Iringa □, Tanzania 54 D4 7 48S 35 43 E
Iriomote-Jima, Japan 31 M1 24 19N 123 48 E
Iriona, Honduras 88 C2 15 57N 85 11W
Iriri →, Brazil 93 D8 3 52S 52 37W
Irish Republic ■, Europe 13 C3 53 0N 8 0W
Irish Sea, U.K. 10 D3 53 38N 4 48W
Irkutsk, Russia 27 D11 52 18N 104 20 E
Irma, Canada 73 C6 52 55N 111 14W
Irō-Zaki, Japan 31 G9 34 36N 138 51 E
Iron Baron, Australia 63 E2 32 58S 137 11 E
Iron Gate = Portile de Fier, Europe 17 F12 44 44N 22 30 E
Iron Knob, Australia 63 E2 32 46S 137 8 E
Iron Mountain, U.S.A. 76 C1 45 49N 88 4W
Iron River, U.S.A. 80 B10 46 6N 88 39W
Irondequoit, U.S.A. 78 C7 43 13N 77 35W
Ironton, Mo., U.S.A. 81 G9 37 36N 90 38W
Ironton, Ohio, U.S.A. 76 F4 38 32N 82 41W
Ironwood, U.S.A. 80 B9 46 27N 90 9W
Iroquois, Canada 79 B9 44 51N 75 19W
Iroquois Falls, Canada 70 C3 48 46N 80 41W
Irpin, Ukraine 17 C16 50 30N 30 15 E
Irrara Cr. →, Australia 63 D4 29 35S 145 31 E
Irrawaddy □, Burma 41 L19 17 0N 95 0 E
Irrawaddy →, Burma 41 M19 15 50N 95 6 E
Irrawaddy, Mouths of the, Burma 41 M19 15 30N 95 0 E
Irricana, Canada 72 C6 51 19N 113 37W
Irtysh →, Russia 26 C7 61 4N 68 52 E
Irumu, Dem. Rep. of the Congo 54 B2 1 32N 29 53 E
Irún, Spain 19 A5 43 20N 1 52W
Irunea = Pamplona, Spain 19 A5 42 48N 1 38W
Irvine, U.K. 12 F4 55 37N 4 41W
Irvine, Calif., U.S.A. 85 M9 33 41N 117 46W
Irvine, Ky., U.S.A. 76 G4 37 42N 83 58W
Irvinestown, U.K. 13 B4 54 28N 7 39W
Irving, U.S.A. 81 J6 32 49N 96 56W
Irvona, U.S.A. 78 F6 40 46N 78 33W
Irwin →, Australia 61 E1 29 15S 114 54 E
Irymple, Australia 63 E3 34 14S 142 8 E
Isa Khel, Pakistan 42 C4 32 41N 71 17 E
Isaac →, Australia 62 C4 22 55S 149 20 E
Isabel, U.S.A. 80 C4 45 24N 101 26W
Isabela, Phil. 37 C6 6 40N 121 59 E
Isabela, I., Mexico 86 C3 21 51N 105 55W
Isabelia, Cord., Nic. 88 D2 13 30N 85 25W
Isabella Ra., Australia 60 D3 21 0S 121 4 E
Ísafjarðardjúp, Iceland 8 C2 66 10N 23 0W
Ísafjörður, Iceland 8 C2 66 5N 23 9W
Isagarh, India 42 G7 24 48N 77 51 E
Isahaya, Japan 31 H5 32 52N 130 2 E
Isaka, Tanzania 54 C3 3 56S 32 59 E
Isan →, India 43 F9 26 51N 80 7 E
Isana = Içana →, Brazil 92 C5 0 26N 67 19W
Isar →, Germany 16 D7 48 48N 12 57 E
Íschia, Italy 20 D5 40 44N 13 57 E
Isdell →, Australia 60 C3 16 27S 124 51 E
Ise, Japan 31 G8 34 25N 136 45 E
Ise-Wan, Japan 31 G8 34 43N 136 43 E
Iseramagazi, Tanzania 54 C3 4 37S 32 10 E
Isère →, France 18 D6 44 59N 4 51 E
Isérnia, Italy 20 D6 41 36N 14 14 E
Isfahan = Eşfahān, Iran 45 C6 32 39N 51 43 E
Ishigaki-Shima, Japan 31 M2 24 20N 124 10 E
Ishikari-Gawa →, Japan 30 C10 43 15N 141 23 E
Ishikari-Sammyaku, Japan 30 C11 43 30N 143 0 E
Ishikari-Wan, Japan 30 C10 43 25N 141 1 E
Ishikawa □, Japan 31 F8 36 30N 136 30 E
Ishim, Russia 26 D7 56 10N 69 30 E
Ishim →, Russia 26 D8 57 45N 71 10 E
Ishinomaki, Japan 30 E10 38 32N 141 20 E
Ishioka, Japan 31 F10 36 11N 140 16 E
Ishkuman, Pakistan 43 A5 36 30N 73 50 E
Ishpeming, U.S.A. 76 B2 46 29N 87 40W
Isil Kul, Russia 26 D8 54 55N 71 16 E
Isiolo, Kenya 54 B4 0 24N 37 33 E
Isiro, Dem. Rep. of the Congo 54 B2 2 53N 27 40 E
Isisford, Australia 62 C3 24 15S 144 21 E
İskenderun, Turkey 25 G6 36 32N 36 10 E
İskenderun Körfezi, Turkey 25 G6 36 40N 35 50 E
Iskŭr →, Bulgaria 21 C11 43 45N 24 25 E
Iskut →, Canada 72 B2 56 45N 131 49W
Isla →, U.K. 12 E5 56 32N 3 20W
Isla Vista, U.S.A. 85 L7 34 25N 119 53W
Islam Headworks, Pakistan 42 E5 29 49N 72 33 E
Islamabad, Pakistan 42 C5 33 40N 73 10 E
Islamgarh, Pakistan 42 F4 27 51N 70 48 E
Islamkot, Pakistan 42 G4 24 42N 70 13 E
Islampur, India 43 G11 25 9N 85 12 E
Island L., Canada 73 C10 53 47N 94 25W
Island Lagoon, Australia 63 E2 31 30S 136 40 E
Island Pond, U.S.A. 79 B13 44 49N 71 53W
Islands, B. of, Canada 71 C8 49 11N 58 15W
Islands, B. of, N.Z. 59 F5 35 15S 174 6 E
Islay, U.K. 12 F2 55 46N 6 10W
Isle →, France 18 D3 44 55N 0 15W
Isle aux Morts, Canada 71 C8 47 35N 59 0W
Isle of Wight □, U.K. 11 G6 50 41N 1 17W
Isle Royale Nat. Park, U.S.A. 80 B10 48 0N 88 55W
Isleton, U.S.A. 84 G5 38 10N 121 37W
Ismail = Izmayil, Ukraine 17 F15 45 22N 28 46 E
Ismâ'ilîya, Egypt 51 B12 30 37N 32 18 E
Isoanala, Madag. 57 C8 23 50S 45 44 E
Isogstalo, India 43 B8 34 15N 78 46 E
Isparta, Turkey 25 G5 37 47N 30 30 E
Íspica, Italy 20 F6 36 47N 14 55 E
Israel ■, Asia 47 D3 32 0N 34 50 E
Issoire, France 18 D5 45 32N 3 15 E
Issyk-Kul, Ozero = Ysyk-Köl, Kyrgyzstan 26 E8 42 25N 77 15 E
İstanbul, Turkey 21 D13 41 0N 29 0 E
İstanbul Boğazı, Turkey 21 D13 41 10N 29 10 E
Istiaía, Greece 21 E10 38 57N 23 9 E
Istokpoga, L., U.S.A. 77 M5 27 23N 81 17W
Istra, Croatia 16 F7 45 10N 14 0 E
Istres, France 18 E6 43 31N 4 59 E
Istria = Istra, Croatia 16 F7 45 10N 14 0 E
Itá, Paraguay 94 B4 25 29S 57 21W
Itaberaba, Brazil 93 F10 12 32S 40 18W
Itabira, Brazil 93 G10 19 37S 43 13W
Itabirito, Brazil 95 A7 20 15S 43 48W
Itabuna, Brazil 93 F11 14 48S 39 16W
Itacaunas →, Brazil 93 E9 5 21S 49 8W
Itacoatiara, Brazil 92 D7 3 8S 58 25W
Itaipú, Représa de, Brazil 95 B5 25 30S 54 30W
Itaituba, Brazil 93 D7 4 10S 55 50W
Itajaí, Brazil 95 B6 27 50S 48 39W
Itajubá, Brazil 95 A6 22 24S 45 30W
Itaka, Tanzania 55 D3 8 50S 32 49 E
Italy ■, Europe 20 C5 42 0N 13 0 E
Itamaraju, Brazil 93 G11 17 5S 39 31W
Itampolo, Madag. 57 C7 24 41S 43 57 E
Itandrano, Madag. 57 C8 21 47S 45 17 E
Itapecuru-Mirim, Brazil 93 D10 3 24S 44 20W
Itaperuna, Brazil 95 A7 21 10S 41 54W
Itapetininga, Brazil 95 A6 23 36S 48 7W
Itapeva, Brazil 95 A6 23 59S 48 59W
Itapicuru →, Bahia, Brazil 93 F11 11 47S 37 32W
Itapicuru →, Maranhão, Brazil 93 D10 2 52S 44 12W
Itapipoca, Brazil 93 D11 3 30S 39 35W
Itapuá □, Paraguay 95 B4 26 40S 55 40W
Itaquari, Brazil 95 A7 20 20S 40 25W
Itaquí, Brazil 94 B4 29 8S 56 30W
Itararé, Brazil 95 A6 24 6S 49 23W
Itarsi, India 42 H7 22 36N 77 51 E
Itatí, Argentina 94 B4 27 16S 58 15W
Itchen →, U.K. 11 G6 50 55N 1 22W
Itezhi Tezhi, L., Zambia 55 F2 15 30S 25 30 E
Ithaca = Itháki, Greece 21 E9 38 25N 20 40 E
Ithaca, U.S.A. 79 D8 42 27N 76 30W
Itháki, Greece 21 E9 38 25N 20 40 E
Itiquira →, Brazil 93 G7 17 18S 56 44W
Itō, Japan 31 G9 34 58N 139 5 E
Itoigawa, Japan 31 F8 37 2N 137 51 E
Itonamas →, Bolivia 92 F6 12 28S 64 24W
Ittoqqortoormiit, Greenland 4 B6 70 20N 23 0W
Itu, Brazil 95 A6 23 17S 47 15W
Itu Aba I., S. China Sea 36 B4 10 23N 114 21 E
Ituiutaba, Brazil 93 G9 19 0S 49 25W
Itumbiara, Brazil 93 G9 18 20S 49 10W
Ituna, Canada 73 C8 51 10N 103 24W
Itunge Port, Tanzania 55 D3 9 40S 33 55 E
Iturbe, Argentina 94 A2 23 0S 65 25W
Ituri →, Dem. Rep. of the Congo 54 B2 1 40N 27 1 E
Iturup, Ostrov, Russia 27 E15 45 0N 148 0 E
Ituxi →, Brazil 92 E6 7 18S 64 51W
Ituyuro →, Argentina 94 A3 22 40S 63 50W
Itzehoe, Germany 16 B5 53 55N 9 31 E
Ivahona, Madag. 57 C8 23 27S 46 10 E
Ivaí →, Brazil 95 A5 23 18S 53 42W
Ivalo, Finland 8 B22 68 38N 27 35 E
Ivalojoki →, Finland 8 B22 68 40N 27 40 E
Ivanava, Belarus 17 B13 52 7N 25 29 E
Ivanhoe, Australia 63 E3 32 56S 144 20 E
Ivanhoe, Calif., U.S.A. 84 J7 36 23N 119 13W
Ivanhoe, Minn., U.S.A. 80 C6 44 28N 96 15W
Ivano-Frankivsk, Ukraine 17 D13 48 40N 24 40 E
Ivano-Frankovsk = Ivano-Frankivsk, Ukraine 17 D13 48 40N 24 40 E
Ivanovo = Ivanava, Belarus 17 B13 52 7N 25 29 E
Ivanovo, Russia 24 C7 57 5N 41 0 E
Ivato, Madag. 57 C8 20 37S 47 10 E
Ivatsevichy, Belarus 17 B13 52 43N 25 21 E
Ivdel, Russia 24 B11 60 42N 60 24 E
Ivinheima →, Brazil 95 A5 23 14S 53 42W
Ivinhema, Brazil 95 A5 22 10S 53 37W
Ivohibe, Madag. 57 C8 22 31S 46 57 E
Ivory Coast, W. Afr. 50 H4 4 20N 5 0W
Ivory Coast ■, Africa 50 G4 7 30N 5 0W
Ivrea, Italy 18 D7 45 28N 7 52 E
Ivujivik, Canada 69 B12 62 24N 77 55W
Ivybridge, U.K. 11 G4 50 23N 3 56W
Iwaizumi, Japan 30 E10 39 50N 141 45 E
Iwaki, Japan 31 F10 37 3N 140 55 E
Iwakuni, Japan 31 G6 34 15N 132 8 E
Iwamizawa, Japan 30 C10 43 12N 141 46 E
Iwanai, Japan 30 C10 42 58N 140 30 E
Iwata, Japan 31 G8 34 42N 137 51 E
Iwate □, Japan 30 E10 39 30N 141 30 E
Iwate-San, Japan 30 E10 39 51N 141 0 E
Iwo, Nigeria 50 G6 7 39N 4 9 E
Ixiamas, Bolivia 92 F5 13 50S 68 5W
Ixopo, S. Africa 57 E5 30 11S 30 5 E
Ixtepec, Mexico 87 D5 16 32N 95 10W
Ixtlán del Río, Mexico 86 C4 21 5N 104 21W
Iyo, Japan 31 H6 33 45N 132 45 E
Izabal, L. de, Guatemala 88 C2 15 30N 89 10W
Izamal, Mexico 87 C7 20 56N 89 1W
Izena-Shima, Japan 31 L3 26 56N 127 56 E
Izhevsk, Russia 24 C9 56 51N 53 14 E
Izhma →, Russia 24 A9 65 19N 52 54 E
Izmayil, Ukraine 17 F15 45 22N 28 46 E
İzmir, Turkey 21 E12 38 25N 27 8 E
İzmit = Kocaeli, Turkey 25 F4 40 45N 29 50 E
İznik Gölü, Turkey 21 D13 40 27N 29 30 E
Izra, Syria 47 C5 32 51N 36 15 E
Izu-Shotō, Japan 31 G10 34 30N 140 0 E
Izúcar de Matamoros, Mexico 87 D5 18 36N 98 28W
Izumi-Sano, Japan 31 G7 34 23N 135 18 E
Izumo, Japan 31 G6 35 20N 132 46 E
Izyaslav, Ukraine 17 C14 50 5N 26 50 E

J

Jabalpur, India 43 H8 23 9N 79 58 E
Jabbūl, Syria 44 B3 36 4N 37 30 E
Jabiru, Australia 60 B5 12 40S 132 53 E
Jablah, Syria 44 C3 35 20N 36 0 E
Jablonec nad Nisou, Czech Rep. 16 C8 50 43N 15 10 E

Joplin

Jaboatão, *Brazil* **93 E11** 8 7S 35 1W
Jaboticabal, *Brazil* **95 A6** 21 15S 48 17W
Jaca, *Spain* **19 A5** 42 35N 0 33W
Jacarei, *Brazil* **95 A6** 23 20S 46 0W
Jacarèzinho, *Brazil* **95 A6** 23 5S 49 58W
Jackman, *U.S.A.* **77 C10** 45 35N 70 17W
Jacksboro, *U.S.A.* **81 J5** 33 14N 98 15W
Jackson, *Ala., U.S.A.* **77 K2** 31 31N 87 53W
Jackson, *Calif., U.S.A.* **84 G6** 38 21N 120 46W
Jackson, *Ky., U.S.A.* **76 G4** 37 33N 83 23W
Jackson, *Mich., U.S.A.* **76 D3** 42 15N 84 24W
Jackson, *Minn., U.S.A.* **80 D7** 43 37N 95 1W
Jackson, *Miss., U.S.A.* **81 J9** 32 18N 90 12W
Jackson, *Mo., U.S.A.* **81 G10** 37 23N 89 40W
Jackson, *N.H., U.S.A.* **79 B13** 44 10N 71 11W
Jackson, *Ohio, U.S.A.* **76 F4** 39 3N 82 39W
Jackson, *Tenn., U.S.A.* **77 H1** 35 37N 88 49W
Jackson, *Wyo., U.S.A.* **82 E8** 43 29N 110 46W
Jackson B., *N.Z.* **59 K2** 43 58S 168 42 E
Jackson L., *U.S.A.* **82 E8** 43 52N 110 36W
Jacksons, *N.Z.* **59 K3** 42 46S 171 32 E
Jackson's Arm, *Canada* **71 C8** 49 52N 56 47W
Jacksonville, *Ala., U.S.A.* **77 J3** 33 49N 85 46W
Jacksonville, *Ark., U.S.A.* **81 H8** 34 52N 92 7W
Jacksonville, *Calif., U.S.A.* **84 H6** 37 52N 120 24W
Jacksonville, *Fla., U.S.A.* **77 K5** 30 20N 81 39W
Jacksonville, *Ill., U.S.A.* **80 F9** 39 44N 90 14W
Jacksonville, *N.C., U.S.A.* **77 H7** 34 45N 77 26W
Jacksonville, *Tex., U.S.A.* **81 K7** 31 58N 95 17W
Jacksonville Beach, *U.S.A.* **77 K5** 30 17N 81 24W
Jacmel, *Haiti* **89 C5** 18 14N 72 32W
Jacob Lake, *U.S.A.* **83 H7** 36 43N 112 13W
Jacobabad, *Pakistan* **42 E3** 28 20N 68 29 E
Jacobina, *Brazil* **93 F10** 11 11S 40 30W
Jacques Cartier, Dét. de, *Canada* **71 C7** 50 0N 63 30W
Jacques Cartier, Mt., *Canada* **71 C6** 48 57N 66 0W
Jacques Cartier, Parc Prov., *Canada* **71 C5** 47 15N 71 33W
Jacuí →, *Brazil* **95 C5** 30 2S 51 15W
Jacumba, *U.S.A.* **85 N10** 32 37N 116 11W
Jacundá →, *Brazil* **93 D8** 1 57S 50 26W
Jadotville = Likasi, *Dem. Rep. of the Congo* **55 E2** 10 55S 26 48 E
Jaén, *Peru* **92 E3** 5 25S 78 40W
Jaén, *Spain* **19 D4** 37 44N 3 43W
Jafarabad, *India* **42 J4** 20 52N 71 22 E
Jaffa = Tel Aviv-Yafo, *Israel* **47 C3** 32 4N 34 48 E
Jaffa, C., *Australia* **63 F2** 36 58S 139 40 E
Jaffna, *Sri Lanka* **40 Q12** 9 45N 80 2 E
Jaffrey, *U.S.A.* **79 D12** 42 49N 72 2W
Jagadhri, *India* **42 D7** 30 10N 77 20 E
Jagadishpur, *India* **43 G11** 25 30N 84 21 E
Jagdalpur, *India* **41 K13** 19 3N 82 0 E
Jagersfontein, *S. Africa* **56 D4** 29 44S 25 27 E
Jaghīn →, *Iran* **45 E8** 27 17N 57 13 E
Jagodina, *Serbia, Yug.* **21 C9** 44 5N 21 15 E
Jagraon, *India* **40 D9** 30 50N 75 25 E
Jagtial, *India* **40 K11** 18 50N 79 0 E
Jaguariaíva, *Brazil* **95 A6** 24 10S 49 50W
Jaguaribe →, *Brazil* **93 D11** 4 25S 37 45W
Jagüey Grande, *Cuba* **88 B3** 22 35N 81 7W
Jahanabad, *India* **43 G11** 25 13N 84 59 E
Jahazpur, *India* **42 G6** 25 37N 75 17 E
Jahrom, *Iran* **45 D7** 28 30N 53 31 E
Jaijon, *India* **42 D7** 31 21N 76 9 E
Jailolo, *Indonesia* **37 D7** 1 5N 127 30 E
Jailolo, Selat, *Indonesia* **37 D7** 0 5N 129 5 E
Jaipur, *India* **42 F6** 27 0N 75 50 E
Jais, *India* **43 F9** 26 15N 81 32 E
Jaisalmer, *India* **42 F4** 26 55N 70 54 E
Jaisinghnagar, *India* **43 H8** 23 38N 78 34 E
Jaitaran, *India* **42 F5** 26 12N 73 56 E
Jaithari, *India* **43 H8** 23 14N 78 37 E
Jājarm, *Iran* **45 B8** 36 58N 56 27 E
Jakam →, *India* **42 H6** 23 54N 74 13 E
Jakarta, *Indonesia* **36 F3** 6 9S 106 49 E
Jakhal, *India* **42 E6** 29 48N 75 50 E
Jakhau, *India* **42 H3** 23 13N 68 43 E
Jakobstad = Pietarsaari, *Finland* **8 E20** 63 40N 22 43 E
Jal, *U.S.A.* **81 J3** 32 7N 103 12W
Jalālābād, *Afghan.* **42 B4** 34 30N 70 29 E
Jalalabad, *India* **43 F8** 27 41N 79 42 E
Jalalpur Jattan, *Pakistan* **42 C6** 32 38N 74 11 E
Jalama, *U.S.A.* **85 L6** 34 29N 120 29W
Jalapa, *Guatemala* **88 D2** 14 39N 89 59W
Jalapa Enríquez, *Mexico* **87 D5** 19 32N 96 55W
Jalasjärvi, *Finland* **9 E20** 62 29N 22 47 E
Jalaun, *India* **43 F8** 26 8N 79 25 E
Jaldhaka →, *Bangla.* **43 F13** 26 16N 89 16 E
Jalesar, *India* **42 F8** 27 29N 78 19 E
Jaleswar, *Nepal* **43 F11** 26 38N 85 48 E
Jalgaon, *India* **40 J9** 21 0N 75 42 E
Jalībah, *Iraq* **44 D5** 30 35N 46 32 E
Jalisco □, *Mexico* **86 D4** 20 0N 104 0W
Jalkot, *Pakistan* **43 B5** 35 14N 73 24 E
Jalna, *India* **40 K9** 19 48N 75 38 E
Jalón →, *Spain* **19 B5** 41 47N 1 4W
Jalor, *India* **42 G5** 25 21N 72 37 E
Jalpa, *Mexico* **86 C4** 21 38N 102 58W
Jalpaiguri, *India* **41 F16** 26 32N 88 46 E
Jaluit I., *Marshall Is.* **64 G8** 6 0N 169 30 E
Jalūlā, *Iraq* **44 C5** 34 16N 45 10 E
Jamaica ■, *W. Indies* **88 C4** 18 10N 77 30W
Jamalpur, *Bangla.* **41 G16** 24 52N 89 56 E
Jamalpur, *India* **43 G12** 25 18N 86 28 E
Jamalpurganj, *India* **43 H13** 23 2N 87 59 E
Jamanxim →, *Brazil* **93 D7** 4 43S 56 18W
Jambi, *Indonesia* **36 E2** 1 38S 103 30 E
Jambi □, *Indonesia* **36 E2** 1 30S 102 30 E
Jambusar, *India* **42 H5** 22 3N 72 51 E
James →, *S. Dak., U.S.A.* **80 D6** 42 52N 97 18W
James →, *Va., U.S.A.* **76 G7** 36 56N 76 27W
James B., *Canada* **70 B3** 54 0N 80 0W
James Ranges, *Australia* **60 D5** 24 10S 132 30 E
James Ross I., *Antarctica* **5 C18** 63 58S 57 50W
Jamesabad, *Pakistan* **42 G3** 25 17N 69 15 E
Jamestown, *Australia* **63 E2** 33 10S 138 32 E
Jamestown, *S. Africa* **56 E4** 31 6S 26 45 E
Jamestown, *N. Dak., U.S.A.* **80 B5** 46 54N 98 42W
Jamestown, *N.Y., U.S.A.* **78 D5** 42 6N 79 14W
Jamestown, *Pa., U.S.A.* **78 E4** 41 29N 80 27W
Jamīlābād, *Iran* **45 C6** 34 24N 48 28 E
Jamiltepec, *Mexico* **87 D5** 16 17N 97 49W
Jamira →, *India* **43 J13** 21 35N 88 28 E
Jamkhandi, *India* **40 L9** 16 30N 75 15 E
Jammu, *India* **42 C6** 32 43N 74 54 E
Jammu & Kashmir □, *India* **43 B7** 34 25N 77 0 E
Jamnagar, *India* **42 H4** 22 30N 70 6 E
Jamni →, *India* **43 G8** 25 13N 78 35 E
Jampur, *Pakistan* **42 E4** 29 39N 70 40 E
Jamrud, *Pakistan* **42 C4** 33 59N 71 24 E
Jämsä, *Finland* **9 F21** 61 53N 25 10 E
Jamshedpur, *India* **43 H12** 22 44N 86 12 E
Jamtara, *India* **43 H12** 23 59N 86 49 E
Jämtland, *Sweden* **8 E15** 63 31N 14 0 E
Jan L., *Canada* **73 C8** 54 56N 102 55W
Jan Mayen, *Arctic* **4 B7** 71 0N 9 0W
Janakkala, *Finland* **9 F21** 60 54N 24 36 E
Janaúba, *Brazil* **93 G10** 15 48S 43 19W
Jand, *Pakistan* **42 C5** 33 30N 72 6 E
Jandaq, *Iran* **45 C7** 34 3N 54 22 E
Jandia, *Canary Is.* **22 F5** 28 6N 14 21W
Jandia, Pta. de, *Canary Is.* **22 F5** 28 3N 14 31W
Jandola, *Pakistan* **42 C4** 32 20N 70 9 E
Jandowae, *Australia* **63 D5** 26 45S 151 7 E
Janesville, *U.S.A.* **80 D10** 42 41N 89 1W
Jangamo, *Mozam.* **57 C6** 24 6S 35 21 E
Janghai, *India* **43 G10** 25 33N 82 19 E
Janin, *West Bank* **47 C4** 32 28N 35 18 E
Janjgir, *India* **43 J10** 22 1N 82 34 E
Janjina, *Madag.* **57 C8** 20 30S 45 50 E
Janos, *Mexico* **86 A3** 30 45N 108 10W
Januária, *Brazil* **93 G10** 15 25S 44 25W
Janubio, *Canary Is.* **22 F6** 28 56N 13 50W
Jaora, *India* **42 H6** 23 40N 75 10 E
Japan ■, *Asia* **31 G8** 36 0N 136 0 E
Japan, Sea of, *Asia* **30 E7** 40 0N 135 0 E
Japan Trench, *Pac. Oc.* **28 F18** 32 0N 142 0 E
Japen = Yapen, *Indonesia* **37 E9** 1 50S 136 0 E
Japla, *India* **43 G11** 24 33N 84 1 E
Japurá →, *Brazil* **92 D5** 3 8S 65 46W
Jaquarão, *Brazil* **95 C5** 32 34S 53 23W
Jaqué, *Panama* **88 E4** 7 27N 78 8W
Jarābulus, *Syria* **44 B3** 36 49N 38 1 E
Jarama →, *Spain* **19 B4** 40 24N 3 32W
Jaranwala, *Pakistan* **42 D5** 31 15N 73 26 E
Jarash, *Jordan* **47 C4** 32 17N 35 54 E
Jardim, *Brazil* **94 A4** 21 28S 56 2W
Jardines de la Reina, Arch. de los, *Cuba* **88 B4** 20 50N 78 50W
Jargalang, *China* **35 C12** 43 5N 122 55 E
Jargalant = Hovd, *Mongolia* **32 B4** 48 2N 91 37 E
Jari →, *Brazil* **93 D8** 1 9S 51 54W
Jarīr, W. al →, *Si. Arabia* **44 E4** 25 38N 42 30 E
Jarosław, *Poland* **17 C12** 50 2N 22 42 E
Jarrahdale, *Australia* **61 F2** 32 24S 116 5 E
Jarrahi →, *Iran* **45 D6** 30 49N 48 48 E
Jarres, Plaine des, *Laos* **38 C4** 19 27N 103 10 E
Jartai, *China* **34 E3** 39 45N 105 48 E
Jarud Qi, *China* **35 B11** 44 28N 120 50 E
Järvenpää, *Finland* **9 F21** 60 29N 25 5 E
Jarvis, *Canada* **78 D4** 42 53N 80 6W
Jarvis I., *Pac. Oc.* **65 H12** 0 15S 160 5W
Jarwa, *India* **43 F10** 27 38N 82 30 E
Jasdan, *India* **42 H4** 22 2N 71 12 E
Jashpurnagar, *India* **43 H11** 22 54N 84 9 E
Jasidih, *India* **43 G12** 24 31N 86 39 E
Jāsimīyah, *Iraq* **44 C5** 33 45N 44 41 E
Jasin, *Malaysia* **39 L4** 2 20N 102 26 E
Jāsk, *Iran* **45 E8** 25 38N 57 45 E
Jasło, *Poland* **17 D11** 49 45N 21 30 E
Jaso, *India* **43 G9** 24 30N 80 29 E
Jasper, *Alta., Canada* **72 C5** 52 55N 118 5W
Jasper, *Ont., Canada* **79 B9** 44 52N 75 57W
Jasper, *Ala., U.S.A.* **77 J2** 33 50N 87 17W
Jasper, *Fla., U.S.A.* **77 K4** 30 31N 82 57W
Jasper, *Ind., U.S.A.* **76 F2** 38 24N 86 56W
Jasper, *Tex., U.S.A.* **81 K8** 30 56N 94 1W
Jasper Nat. Park, *Canada* **72 C5** 52 50N 118 8W
Jasrasar, *India* **42 F5** 27 43N 73 49 E
Jászberény, *Hungary* **17 E10** 47 30N 19 55 E
Jataí, *Brazil* **93 G8** 17 58S 51 48W
Jati, *Pakistan* **42 G3** 24 20N 68 19 E
Jatibarang, *Indonesia* **37 G13** 6 28S 108 18 E
Jatinegara, *Indonesia* **37 G12** 6 13S 106 52 E
Játiva = Xátiva, *Spain* **19 C5** 38 59N 0 32W
Jaú, *Brazil* **95 A6** 22 10S 48 30W
Jauja, *Peru* **92 F3** 11 45S 75 15W
Jaunpur, *India* **43 G10** 25 46N 82 44 E
Java = Jawa, *Indonesia* **36 F3** 7 0S 110 0 E
Java Barat □, *Indonesia* **37 G12** 7 0S 107 0 E
Java Sea, *Indonesia* **36 E3** 4 35S 107 15 E
Java Tengah □, *Indonesia* **37 G14** 7 0S 110 0 E
Java Timur □, *Indonesia* **37 G15** 8 0S 113 0 E
Java Trench, *Ind. Oc.* **36 F3** 9 0S 105 0 E
Javhlant = Ulyasutay, *Mongolia* **32 B4** 47 56N 97 28 E
Jawa, *Indonesia* **36 F3** 7 0S 110 0 E
Jawad, *India* **42 G6** 24 36N 74 51 E
Jay Peak, *U.S.A.* **79 B12** 44 55N 72 32W
Jaya, Puncak, *Indonesia* **37 E9** 3 57S 137 17 E
Jayanti, *India* **41 F16** 26 45N 89 40 E
Jayapura, *Indonesia* **37 E10** 2 28S 140 38 E
Jayawijaya, Pegunungan, *Indonesia* **37 E9** 5 0S 139 0 E
Jaynagar, *India* **41 F15** 26 43N 86 9 E
Jayrūd, *Syria* **44 C3** 33 49N 36 44 E
Jayton, *U.S.A.* **81 J4** 33 15N 100 34W
Jāz Mūrīān, Hāmūn-e, *Iran* **45 E8** 27 20N 58 55 E
Jazīreh-ye Shīf, *Iran* **45 D6** 29 4N 50 54 E
Jazminal, *Mexico* **86 C4** 24 56N 101 25W
Jazzīn, *Lebanon* **47 B4** 33 31N 35 35 E
Jean, *U.S.A.* **85 K11** 35 47N 115 20W
Jean Marie River, *Canada* **72 A4** 61 32N 120 38W
Jean Rabel, *Haiti* **89 C5** 19 50N 73 5W
Jeanerette, *U.S.A.* **81 L9** 29 55N 91 40W
Jeanette, Ostrov = Zhannetty, Ostrov, *Russia* **27 B16** 76 43N 158 0 E
Jeannette, *U.S.A.* **78 F5** 40 20N 79 36W
Jebāl Bārez, Kūh-e, *Iran* **45 D8** 28 30N 58 20 E
Jebel, Bahr el →, *Sudan* **51 G12** 9 30N 30 25 E
Jedburgh, *U.K.* **12 F6** 55 29N 2 33W
Jedda = Jiddah, *Si. Arabia* **46 C2** 21 29N 39 10 E
Jeddore L., *Canada* **71 C8** 48 3N 55 55W
Jędrzejów, *Poland* **17 C11** 50 35N 20 15 E
Jefferson, *Iowa, U.S.A.* **80 D7** 42 1N 94 23W
Jefferson, *Ohio, U.S.A.* **78 E4** 41 44N 80 46W
Jefferson, *Tex., U.S.A.* **81 J7** 32 46N 94 21W
Jefferson, Mt., *Nev., U.S.A.* **82 G5** 38 51N 117 0W
Jefferson, Mt., *Oreg., U.S.A.* **82 D3** 44 41N 121 48W
Jefferson City, *Mo., U.S.A.* **80 F8** 38 34N 92 10W
Jefferson City, *Tenn., U.S.A.* **77 G4** 36 7N 83 30W
Jeffersontown, *U.S.A.* **76 F3** 38 12N 85 35W
Jeffersonville, *U.S.A.* **76 F3** 38 17N 85 44W
Jeffrey City, *U.S.A.* **82 E10** 42 30N 107 49W
Jega, *Nigeria* **50 F6** 12 15N 4 23 E
Jēkabpils, *Latvia* **9 H21** 56 29N 25 57 E
Jekyll I., *U.S.A.* **77 K5** 31 4N 81 25W
Jelenia Góra, *Poland* **16 C8** 50 50N 15 45 E
Jelgava, *Latvia* **9 H20** 56 41N 23 49 E
Jemaja, *Indonesia* **39 L5** 3 5N 105 45 E
Jemaluang, *Malaysia* **39 L4** 2 16N 103 52 E
Jember, *Indonesia* **37 H15** 8 11S 113 41 E
Jembongan, *Malaysia* **36 C5** 6 45N 117 20 E
Jena, *Germany* **16 C6** 50 54N 11 35 E
Jena, *U.S.A.* **81 K8** 31 41N 92 8W
Jenkins, *U.S.A.* **76 G4** 37 10N 82 38W
Jenner, *U.S.A.* **84 G3** 38 27N 123 7W
Jennings, *U.S.A.* **81 K8** 30 13N 92 40W
Jepara, *Indonesia* **37 G14** 7 40S 109 14 E
Jeparit, *Australia* **63 F3** 36 8S 142 1 E
Jequié, *Brazil* **93 F10** 13 51S 40 5W
Jequitinhonha, *Brazil* **93 G10** 16 30S 41 0W
Jequitinhonha →, *Brazil* **93 G11** 15 51S 38 53W
Jerantut, *Malaysia* **39 L4** 3 56N 102 22 E
Jérémie, *Haiti* **89 C5** 18 40N 74 10W
Jerez, Punta, *Mexico* **87 C5** 22 58N 97 40W
Jerez de García Salinas, *Mexico* **86 C4** 22 39N 103 0W
Jerez de la Frontera, *Spain* **19 D2** 36 41N 6 7W
Jerez de los Caballeros, *Spain* **19 C2** 38 20N 6 45W
Jericho = El Arīḥā, *West Bank* **47 D4** 31 52N 35 27 E
Jericho, *Australia* **62 C4** 23 38S 146 6 E
Jerid, Chott el = Djerid, Chott, *Tunisia* **50 B7** 33 42N 8 30 E
Jerilderie, *Australia* **63 F4** 35 20S 145 41 E
Jermyn, *U.S.A.* **79 E9** 41 31N 75 31W
Jerome, *U.S.A.* **82 E6** 42 44N 114 31W
Jerramungup, *Australia* **61 F2** 33 55S 118 55 E
Jersey, *U.K.* **11 H5** 49 11N 2 7W
Jersey City, *U.S.A.* **79 F10** 40 44N 74 4W
Jersey Shore, *U.S.A.* **78 E7** 41 12N 77 15W
Jerseyville, *U.S.A.* **80 F9** 39 7N 90 20W
Jerusalem, *Israel* **47 D4** 31 47N 35 10 E
Jervis B., *Australia* **63 F5** 35 8S 150 46 E
Jervis Inlet, *Canada* **72 C4** 50 0N 123 57W
Jesi = Iesi, *Italy* **20 C5** 43 31N 13 14 E
Jesselton = Kota Kinabalu, *Malaysia* **36 C5** 6 0N 116 4 E
Jessore, *Bangla.* **41 H16** 23 10N 89 10 E
Jesup, *U.S.A.* **77 K5** 31 36N 81 53W
Jesús Carranza, *Mexico* **87 D5** 17 28N 95 1W
Jesús María, *Argentina* **94 C3** 30 59S 64 5W
Jetmore, *U.S.A.* **81 F5** 38 4N 99 54W
Jetpur, *India* **42 J4** 21 45N 70 10 E
Jevnaker, *Norway* **9 F14** 60 15N 10 26 E
Jewett, *U.S.A.* **78 F3** 40 22N 81 2W
Jewett City, *U.S.A.* **79 E13** 41 36N 72 0W
Jeyḥūnābād, *Iran* **45 C6** 34 58N 48 59 E
Jeypore, *India* **41 K13** 18 50N 82 38 E
Jha Jha, *India* **43 G12** 24 46N 86 22 E
Jhaarkand = Jharkhand □, *India* **43 H11** 24 0N 85 50 E
Jhabua, *India* **42 H6** 22 46N 74 36 E
Jhajjar, *India* **42 E7** 28 37N 76 42 E
Jhal, *Pakistan* **42 E2** 28 17N 67 27 E
Jhal Jhao, *Pakistan* **40 F4** 26 20N 65 35 E
Jhalawar, *India* **42 G7** 24 40N 76 10 E
Jhalida, *India* **43 H11** 23 22N 85 58 E
Jhalrapatan, *India* **42 G7** 24 33N 76 10 E
Jhang Maghiana, *Pakistan* **42 D5** 31 15N 72 22 E
Jhansi, *India* **43 G8** 25 30N 78 36 E
Jhargram, *India* **43 H12** 22 27N 86 59 E
Jharia, *India* **43 H12** 23 45N 86 26 E
Jharkhand □, *India* **43 H11** 24 0N 85 50 E
Jharsuguda, *India* **41 J14** 21 56N 84 5 E
Jhelum, *Pakistan* **42 C5** 33 0N 73 45 E
Jhelum →, *Pakistan* **42 D5** 31 20N 72 10 E
Jhilmilli, *India* **43 H10** 23 24N 82 51 E
Jhudo, *Pakistan* **42 G3** 24 58N 69 18 E
Jhunjhunu, *India* **42 E6** 28 10N 75 30 E
Ji-Paraná, *Brazil* **92 F6** 10 52S 62 57W
Ji Xian, *Hebei, China* **34 F8** 37 35N 115 30 E
Ji Xian, *Henan, China* **34 G8** 35 22N 114 5 E
Ji Xian, *Shanxi, China* **34 F6** 36 7N 110 40 E
Jia Xian, *Henan, China* **34 H7** 33 59N 113 12 E
Jia Xian, *Shaanxi, China* **34 E6** 38 12N 110 28 E
Jiamusi, *China* **33 B8** 46 40N 130 26 E
Ji'an, *Jiangxi, China* **33 D6** 27 6N 114 59 E
Ji'an, *Jilin, China* **35 D14** 41 5N 126 10 E
Jianchang, *China* **35 D11** 40 55N 120 35 E
Jianchangying, *China* **35 D10** 40 10N 118 50 E
Jiangcheng, *China* **32 D5** 22 36N 101 52 E
Jiangmen, *China* **33 D6** 22 32N 113 0 E
Jiangsu □, *China* **35 H11** 33 0N 120 0 E
Jiangxi □, *China* **33 D6** 27 30N 116 0 E
Jiao Xian = Jiaozhou, *China* **35 F11** 36 18N 120 1 E
Jiaohe, *Hebei, China* **34 E9** 38 2N 116 20 E
Jiaohe, *Jilin, China* **35 C14** 43 40N 127 22 E
Jiaozhou, *China* **35 F11** 36 18N 120 1 E
Jiaozhou Wan, *China* **35 F11** 36 5N 120 10 E
Jiaozuo, *China* **34 G7** 35 16N 113 12 E
Jiawang, *China* **35 G9** 34 28N 117 26 E
Jiaxiang, *China* **34 G9** 35 25N 116 20 E
Jiaxing, *China* **33 C7** 30 49N 120 45 E
Jiayi = Chiai, *Taiwan* **33 D7** 23 29N 120 25 E
Jibuti = Djibouti ■, *Africa* **46 E3** 12 0N 43 0 E
Jicarón, I., *Panama* **88 E3** 7 10N 81 50W
Jiddah, *Si. Arabia* **46 C2** 21 29N 39 10 E
Jido, *India* **41 E19** 29 2N 94 58 E
Jieshou, *China* **34 H8** 33 18N 115 22 E
Jiexiu, *China* **34 F6** 37 2N 111 55 E
Jiggalong, *Australia* **60 D3** 23 21S 120 47 E
Jigni, *India* **43 G8** 25 45N 79 25 E
Jihlava, *Czech Rep.* **16 D8** 49 28N 15 35 E
Jihlava →, *Czech Rep.* **17 D9** 48 55N 16 36 E
Jijiga, *Ethiopia* **46 F3** 9 20N 42 50 E
Jilin, *China* **35 C14** 43 44N 126 30 E
Jilin □, *China* **35 C14** 44 0N 127 0 E
Jilong = Chilung, *Taiwan* **33 D7** 25 3N 121 45 E
Jim Thorpe, *U.S.A.* **79 F9** 40 52N 75 44W
Jima, *Ethiopia* **46 F2** 7 40N 36 47 E
Jiménez, *Mexico* **86 B4** 27 10N 104 54W
Jimo, *China* **35 F11** 36 23N 120 30 E
Jin Xian = Jinzhou, *China* **34 E8** 38 2N 115 2 E
Jin Xian, *China* **35 E11** 38 55N 121 42 E
Jinan, *China* **34 F9** 36 38N 117 1 E
Jinchang, *China* **32 C5** 38 30N 102 10 E
Jincheng, *China* **34 G7** 35 29N 112 50 E
Jind, *India* **42 E7** 29 19N 76 22 E
Jindabyne, *Australia* **63 F4** 36 25S 148 35 E
Jindřichův Hradec, *Czech Rep.* **16 D8** 49 10N 15 2 E
Jing He →, *China* **34 G5** 34 27N 109 4 E
Jingbian, *China* **34 F5** 37 20N 108 30 E
Jingchuan, *China* **34 G4** 35 20N 107 20 E
Jingdezhen, *China* **33 D6** 29 20N 117 11 E
Jinggu, *China* **32 D5** 23 35N 100 41 E
Jinghai, *China* **34 E9** 38 55N 116 55 E
Jingle, *China* **34 E6** 38 20N 111 55 E
Jingning, *China* **34 G3** 35 30N 105 43 E
Jingpo Hu, *China* **35 C15** 43 55N 128 55 E
Jingtai, *China* **34 F3** 37 10N 104 6 E
Jingxing, *China* **34 E8** 38 2N 114 8 E
Jingyang, *China* **34 G5** 34 30N 108 50 E
Jingyu, *China* **35 C14** 42 25N 126 45 E
Jingyuan, *China* **34 F3** 36 30N 104 40 E
Jingziguan, *China* **34 H6** 33 15N 111 0 E
Jinhua, *China* **33 D6** 29 8N 119 38 E
Jining, *Nei Monggol Zizhiqu, China* **34 D7** 41 5N 113 0 E
Jining, *Shandong, China* **34 G9** 35 22N 116 34 E
Jinja, *Uganda* **54 B3** 0 25N 33 12 E
Jinjang, *Malaysia* **39 L3** 3 13N 101 39 E
Jinji, *China* **34 F4** 37 58N 106 8 E
Jinnah Barrage, *Pakistan* **40 C7** 32 58N 71 33 E
Jinotega, *Nic.* **88 D2** 13 6N 85 59W
Jinotepe, *Nic.* **88 D2** 11 50N 86 10W
Jinsha Jiang →, *China* **32 D5** 28 50N 104 36 E
Jinxi, *China* **35 D11** 40 52N 120 50 E
Jinxiang, *China* **34 G9** 35 5N 116 22 E
Jinzhou, *Hebei, China* **34 E8** 38 2N 115 2 E
Jinzhou, *Liaoning, China* **35 D11** 41 5N 121 3 E
Jiparaná →, *Brazil* **92 E6** 8 3S 62 52W
Jipijapa, *Ecuador* **92 D2** 1 0S 80 40W
Jiquilpan, *Mexico* **86 D4** 19 57N 102 42W
Jishan, *China* **34 G6** 35 34N 110 58 E
Jisr ash Shughūr, *Syria* **44 C3** 35 49N 36 18 E
Jitarning, *Australia* **61 F2** 32 48S 117 57 E
Jitra, *Malaysia* **39 J3** 6 16N 100 25 E
Jiu →, *Romania* **17 F12** 43 47N 23 48 E
Jiudengkou, *China* **34 E4** 39 56N 106 40 E
Jiujiang, *China* **33 D6** 29 42N 115 58 E
Jiutai, *China* **35 B13** 44 10N 125 50 E
Jiuxincheng, *China* **34 E8** 39 17N 115 59 E
Jixi, *China* **35 B16** 45 20N 130 50 E
Jiyang, *China* **35 F9** 37 0N 117 12 E
Jiyuan, *China* **34 G7** 35 7N 112 57 E
Jīzān, *Si. Arabia* **46 D3** 17 0N 42 20 E
Jize, *China* **34 F8** 36 54N 114 56 E
Jīzl, Wādī al, *Si. Arabia* **44 E3** 25 39N 38 25 E
Jizō-Zaki, *Japan* **31 G6** 35 34N 133 20 E
Jizzakh, *Uzbekistan* **26 E7** 40 6N 67 50 E
Joaçaba, *Brazil* **95 B5** 27 5S 51 31W
João Pessoa, *Brazil* **93 E12** 7 10S 34 52W
Joaquín V. González, *Argentina* **94 B3** 25 10S 64 0W
Jobat, *India* **42 H6** 22 25N 74 34 E
Jodhpur, *India* **42 F5** 26 23N 73 8 E
Jodiya, *India* **42 H4** 22 42N 70 18 E
Joensuu, *Finland* **24 B4** 62 37N 29 49 E
Jōetsu, *Japan* **31 F9** 37 12N 138 10 E
Jofane, *Mozam.* **57 C5** 21 15S 34 18 E
Jogbani, *India* **43 F12** 26 25N 87 15 E
Jõgeva, *Estonia* **9 G22** 58 45N 26 24 E
Jogjakarta = Yogyakarta, *Indonesia* **36 F4** 7 49S 110 22 E
Johannesburg, *S. Africa* **57 D4** 26 10S 28 2 E
Johannesburg, *U.S.A.* **85 K9** 35 22N 117 38W
Johilla →, *India* **43 H9** 23 37N 81 14 E
John Day, *U.S.A.* **82 D4** 44 25N 118 57W
John Day →, *U.S.A.* **82 D3** 45 44N 120 39W
John D'Or Prairie, *Canada* **72 B5** 58 30N 115 8W
John H. Kerr Reservoir, *U.S.A.* **77 G6** 36 36N 78 18W
John o' Groats, *U.K.* **12 C5** 58 38N 3 4W
Johnnie, *U.S.A.* **85 J10** 36 25N 116 5W
John's Ra., *Australia* **62 C1** 21 55S 133 23 E
Johnson, *Kans., U.S.A.* **81 G4** 37 34N 101 45W
Johnson, *Vt., U.S.A.* **79 B12** 44 38N 72 41W
Johnson City, *N.Y., U.S.A.* **79 D9** 42 7N 75 58W
Johnson City, *Tenn., U.S.A.* **77 G4** 36 19N 82 21W
Johnson City, *Tex., U.S.A.* **81 K5** 30 17N 98 25W
Johnsonburg, *U.S.A.* **78 E6** 41 29N 78 41W
Johnsondale, *U.S.A.* **85 K8** 35 58N 118 32W
Johnson's Crossing, *Canada* **72 A2** 60 29N 133 18W
Johnston, L., *Australia* **61 F3** 32 25S 120 30 E
Johnston Falls = Mambilima Falls, *Zambia* **55 E2** 10 31S 28 45 E
Johnston I., *Pac. Oc.* **65 F11** 17 10N 169 8W
Johnstone Str., *Canada* **72 C3** 50 28N 126 0W
Johnstown, *N.Y., U.S.A.* **79 C10** 43 0N 74 22W
Johnstown, *Ohio, U.S.A.* **78 F2** 40 9N 82 41W
Johnstown, *Pa., U.S.A.* **78 F6** 40 20N 78 55W
Johor Baharu, *Malaysia* **39 M4** 1 28N 103 46 E
Jõhvi, *Estonia* **9 G22** 59 22N 27 27 E
Joinville, *Brazil* **95 B6** 26 15S 48 55W
Joinville I., *Antarctica* **5 C18** 65 0S 55 30W
Jojutla, *Mexico* **87 D5** 18 37N 99 11W
Jokkmokk, *Sweden* **8 C18** 66 35N 19 50 E
Jökulsá á Bru →, *Iceland* **8 D6** 65 40N 14 16W
Jökulsá á Fjöllum →, *Iceland* **8 C5** 66 10N 16 30W
Jolfā, *Āzarbājān-e Sharqī, Iran* **44 B5** 38 57N 45 38 E
Jolfā, *Eşfahan, Iran* **45 C6** 32 58N 51 37 E
Joliet, *U.S.A.* **76 E1** 41 32N 88 5W
Joliette, *Canada* **70 C5** 46 3N 73 24W
Jolo, *Phil.* **37 C6** 6 0N 121 0 E
Jolon, *U.S.A.* **84 K5** 35 58N 121 9W
Jombang, *Indonesia* **37 G15** 7 33S 112 14 E
Jonava, *Lithuania* **9 J21** 55 8N 24 12 E
Jones Sound, *Canada* **4 B3** 76 0N 85 0W
Jonesboro, *Ark., U.S.A.* **81 H9** 35 50N 90 42W
Jonesboro, *La., U.S.A.* **81 J8** 32 15N 92 43W
Joniškis, *Lithuania* **9 H20** 56 13N 23 35 E
Jönköping, *Sweden* **9 H16** 57 45N 14 8 E
Jonquière, *Canada* **71 C5** 48 27N 71 14W
Joplin, *U.S.A.* **81 G7** 37 6N 94 31W

Keene

Kandy, Sri Lanka 40 R12 7 18N 80 43 E
Kane, U.S.A. 78 E6 41 40N 78 49W
Kane Basin, Greenland 4 B4 79 1N 70 0W
Kaneohe, U.S.A. 74 H16 21 25N 157 48W
Kang, Botswana 56 C3 23 41S 22 50 E
Kangān, Fārs, Iran 45 E7 27 50N 52 3 E
Kangān, Hormozgān, Iran 45 E8 25 48N 57 28 E
Kangar, Malaysia 39 J3 6 27N 100 12 E
Kangaroo I., Australia 63 F2 35 45S 137 0 E
Kangaroo Mts., Australia 62 C3 23 29S 141 51 E
Kangasala, Finland 9 F21 61 28N 24 4 E
Kangāvar, Iran 45 C6 34 40N 48 0 E
Kangdong, N. Korea 35 E14 39 9N 126 5 E
Kangean, Kepulauan, Indonesia 36 F5 6 55S 115 23 E
Kangean Is. = Kangean, Kepulauan, Indonesia 36 F5 6 55S 115 23 E
Kanggye, N. Korea 35 D14 41 0N 126 35 E
Kanggyŏng, S. Korea 35 F14 36 10N 127 0 E
Kanghwa, S. Korea 35 F14 37 45N 126 30 E
Kangikajik, Greenland 4 B6 70 7N 22 0W
Kangiqsliniq = Rankin Inlet, Canada 68 B10 62 30N 93 0W
Kangiqsualujjuaq, Canada 69 C13 58 30N 65 59W
Kangiqsujuaq, Canada 69 B12 61 30N 72 0W
Kangiqtugaapik = Clyde River, Canada 69 A13 70 30N 68 30W
Kangirsuk, Canada 69 B13 60 0N 70 0W
Kangnŭng, S. Korea 35 F15 37 45N 128 54 E
Kangping, China 35 C12 42 43N 123 18 E
Kangra, India 42 C7 32 6N 76 16 E
Kangto, India 41 F18 27 50N 92 35 E
Kanhar →, India 43 G10 24 28N 83 8 E
Kaniama, Dem. Rep. of the Congo 54 D1 7 30S 24 12 E
Kaniapiskau = Caniapiscau →, Canada 71 A6 56 40N 69 30W
Kaniapiskau, Res. = Caniapiscau, Rés. de, Canada 71 B6 54 10N 69 55W
Kanin, Poluostrov, Russia 24 A8 68 0N 45 0 E
Kanin Nos, Mys, Russia 24 A7 68 39N 43 32 E
Kanin Pen. = Kanin, Poluostrov, Russia 24 A8 68 0N 45 0 E
Kaniva, Australia 63 F3 36 22S 141 18 E
Kanjut Sar, Pakistan 43 A6 36 7N 75 25 E
Kankaanpää, Finland 9 F20 61 44N 22 50 E
Kankakee, U.S.A. 76 E2 41 7N 87 52W
Kankakee →, U.S.A. 76 E1 41 23N 88 15W
Kankan, Guinea 50 F4 10 23N 9 15W
Kankendy = Xankändi, Azerbaijan 25 G8 39 52N 46 49 E
Kanker, India 41 J12 20 10N 81 40 E
Kankroli, India 42 G5 25 4N 73 53 E
Kannapolis, U.S.A. 77 H5 35 30N 80 37W
Kannauj, India 43 F8 27 3N 79 56 E
Kannod, India 40 H10 22 45N 76 40 E
Kano, Nigeria 50 F7 12 2N 8 30 E
Kan'onji, Japan 31 G6 34 7N 133 39 E
Kanowit, Malaysia 36 D4 2 14N 112 20 E
Kanoya, Japan 31 J5 31 25N 130 50 E
Kanpetlet, Burma 41 J18 21 10N 93 59 E
Kanpur, India 43 F9 26 28N 80 20 E
Kansas □, U.S.A. 80 F6 38 30N 99 0W
Kansas →, U.S.A. 80 F7 39 7N 94 37W
Kansas City, Kans., U.S.A. 80 F7 39 7N 94 38W
Kansas City, Mo., U.S.A. 80 F7 39 6N 94 35W
Kansenia, Dem. Rep. of the Congo 55 E2 10 20S 26 0 E
Kansk, Russia 27 D10 56 20N 95 37 E
Kansŏng, S. Korea 35 E15 38 24N 128 30 E
Kansu = Gansu □, China 34 G3 36 0N 104 0 E
Kantaphor, India 42 H7 22 35N 76 34 E
Kantharalak, Thailand 38 E5 14 39N 104 39 E
Kantli →, India 42 E6 28 20N 75 30 E
Kantō □, Japan 31 F9 36 15N 139 30 E
Kantō-Sanchi, Japan 31 G9 35 59N 138 50 E
Kanturk, Ireland 13 D3 52 11N 8 54W
Kanuma, Japan 31 F9 36 34N 139 42 E
Kanus, Namibia 56 D2 27 50S 18 39 E
Kanye, Botswana 56 C4 24 55S 25 28 E
Kanzenze, Dem. Rep. of the Congo 55 E2 10 30S 25 12 E
Kanzi, Ras, Tanzania 54 D4 7 1S 39 33 E
Kaohsiung, Taiwan 33 D7 22 35N 120 16 E
Kaokoveld, Namibia 56 B1 19 15S 14 30 E
Kaolack, Senegal 50 F2 14 5N 16 8W
Kaoshan, China 35 B13 44 38N 124 50 E
Kapaa, U.S.A. 74 G15 22 5N 159 19W
Kapadvanj, India 42 H5 23 5N 73 0 E
Kapan, Armenia 25 G8 39 18N 46 27 E
Kapanga, Dem. Rep. of the Congo 52 F4 8 30S 22 40 E
Kapchagai = Qapshaghay, Kazakstan 26 E8 43 51N 77 14 E
Kapela = Velika Kapela, Croatia 16 F8 45 10N 15 5 E
Kapema, Dem. Rep. of the Congo 55 E2 10 45S 28 22 E
Kapfenberg, Austria 16 E8 47 26N 15 18 E
Kapiri Mposhi, Zambia 55 E2 13 59S 28 43 E
Kāpīsā □, Afghan. 40 B6 35 0N 69 20 E
Kapiskau →, Canada 70 B3 52 47N 81 55W
Kapit, Malaysia 36 D4 2 0N 112 55 E
Kapiti I., N.Z. 59 J5 40 50S 174 56 E
Kaplan, U.S.A. 81 K8 30 0N 92 17W
Kapoe, Thailand 39 H2 9 34N 98 32 E
Kapoeta, Sudan 51 H12 4 50N 33 35 E
Kaposvár, Hungary 17 E9 46 25N 17 47 E
Kapowsin, U.S.A. 84 D4 46 59N 122 13W
Kapps, Namibia 56 C2 22 32S 17 18 E
Kapsan, N. Korea 35 D15 41 4N 128 19 E
Kapsukas = Marijampolė, Lithuania 9 J20 54 33N 23 19 E
Kapuas →, Indonesia 36 E3 0 25S 109 20 E
Kapuas Hulu, Pegunungan, Malaysia 36 D4 1 30N 113 30 E
Kapuas Hulu Ra. = Kapuas Hulu, Pegunungan, Malaysia 36 D4 1 30N 113 30 E
Kapulo, Dem. Rep. of the Congo 55 D2 8 18S 29 15 E
Kapunda, Australia 63 E2 34 20S 138 56 E
Kapuni, N.Z. 59 H5 39 29S 174 8 E
Kapurthala, India 42 D6 31 23N 75 25 E
Kapuskasing, Canada 70 C3 49 25N 82 30W
Kapuskasing →, Canada 70 C3 49 49N 82 0W
Kaputar, Australia 63 E5 30 15S 150 10 E
Kaputir, Kenya 54 B4 2 5N 35 28 E
Kara, Russia 26 C7 69 10N 65 0 E
Kara Bogaz Gol, Zaliv = Garabogazköl Aylagy, Turkmenistan 25 F9 41 0N 53 30 E
Kara Kalpak Republic = Qoraqalpoghistan □, Uzbekistan 26 E6 43 0N 58 0 E
Kara Kum, Turkmenistan 26 F6 39 30N 60 0 E
Kara Sea, Russia 26 B7 75 0N 70 0 E
Karabiğa, Turkey 21 D12 40 23N 27 17 E
Karabük, Turkey 25 F5 41 12N 32 37 E
Karaburun, Turkey 21 E12 38 41N 26 28 E
Karabutak = Qarabutaq, Kazakstan 26 E7 49 59N 60 14 E
Karacabey, Turkey 21 D13 40 12N 28 21 E
Karacasu, Turkey 21 F13 37 43N 28 35 E
Karachey-Cherkessia □, Russia 25 F7 43 40N 41 30 E
Karachi, Pakistan 42 G2 24 53N 67 0 E
Karad, India 40 L9 17 15N 74 10 E
Karaganda = Qaraghandy, Kazakstan 26 E8 49 50N 73 10 E
Karagayly, Kazakstan 26 E8 49 26N 76 0 E
Karaginskiy, Ostrov, Russia 27 D17 58 45N 164 0 E
Karagiye, Vpadina, Kazakstan 25 F9 43 27N 51 45 E
Karagiye Depression = Karagiye, Vpadina, Kazakstan 25 F9 43 27N 51 45 E
Karagola Road, India 43 G12 25 29N 87 23 E
Karaikal, India 40 P11 10 59N 79 50 E
Karaikkudi, India 40 P11 10 5N 78 45 E
Karaj, Iran 45 C6 35 48N 51 0 E
Karak, Malaysia 39 L4 3 25N 102 2 E
Karakalpakstan = Qoraqalpoghistan □, Uzbekistan 26 E6 43 0N 58 0 E
Karakelong, Indonesia 37 D7 4 35N 126 50 E
Karakitang, Indonesia 37 D7 3 14N 125 28 E
Karaklis = Vanadzor, Armenia 25 F7 40 48N 44 30 E
Karakol, Kyrgyzstan 26 E8 42 30N 78 20 E
Karakoram Pass, Asia 43 B7 35 33N 77 50 E
Karakoram Ra., Pakistan 43 B7 35 30N 77 0 E
Karakuwisa, Namibia 56 B2 18 56S 19 40 E
Karalon, Russia 27 D12 57 5N 115 50 E
Karama, Jordan 47 D4 31 57N 35 35 E
Karaman, Turkey 25 G5 37 14N 33 13 E
Karamay, China 32 B3 45 30N 84 58 E
Karambu, Indonesia 36 E5 3 53S 116 6 E
Karamea Bight, N.Z. 59 J3 41 22S 171 40 E
Karamnasa →, India 43 G10 25 31N 83 52 E
Karand, Iran 44 C5 34 16N 46 15 E
Karanganyar, Indonesia 37 G13 7 38S 109 37 E
Karanjia, India 43 J11 21 47N 85 58 E
Karasburg, Namibia 56 D2 28 0S 18 44 E
Karasino, Russia 26 C9 66 50N 86 50 E
Karasjok, Norway 8 B21 69 27N 25 30 E
Karasuk, Russia 26 D8 53 44N 78 2 E
Karasuyama, Japan 31 F10 36 39N 140 9 E
Karatau, Khrebet = Qaratau, Kazakstan 26 E7 43 30N 69 30 E
Karatsu, Japan 31 H4 33 26N 129 58 E
Karaul, Russia 26 B9 70 6N 82 15 E
Karauli, India 42 F7 26 30N 77 4 E
Karavostasi, Cyprus 23 D11 35 8N 32 50 E
Karawang, Indonesia 37 G12 6 30S 107 15 E
Karawanken, Europe 16 E8 46 30N 14 40 E
Karayazı, Turkey 25 G7 39 41N 42 9 E
Karazhal, Kazakstan 26 E8 48 2N 70 49 E
Karbalā', Iraq 44 C5 32 36N 44 3 E
Karcag, Hungary 17 E11 47 19N 20 57 E
Karcha →, Pakistan 43 B7 34 45N 76 10 E
Karchana, India 43 G9 25 17N 81 56 E
Kardhítsa, Greece 21 E9 39 23N 21 54 E
Kärdla, Estonia 9 G20 58 50N 22 40 E
Kareeberge, S. Africa 56 E3 30 59S 21 50 E
Kareha →, India 43 G12 25 44N 86 21 E
Kareima, Sudan 51 E12 18 30N 31 49 E
Karelia □, Russia 24 A5 65 30N 32 30 E
Karelian Republic = Karelia □, Russia 24 A5 65 30N 32 30 E
Karera, India 42 G8 25 32N 78 9 E
Kārevāndar, Iran 45 E9 27 53N 60 44 E
Kargasok, Russia 26 D9 59 3N 80 53 E
Kargat, Russia 26 D9 55 10N 80 15 E
Kargil, India 43 B7 34 32N 76 12 E
Kargopol, Russia 24 B6 61 30N 38 58 E
Karhal, India 43 F8 27 1N 78 57 E
Karīān, Iran 45 E8 26 57N 57 14 E
Karianga, Madag. 57 C8 22 25S 47 22 E
Kariba, Zimbabwe 55 F2 16 28S 28 50 E
Kariba, L., Zimbabwe 55 F2 16 40S 28 25 E
Kariba Dam, Zimbabwe 55 F2 16 30S 28 35 E
Kariba Gorge, Zambia 55 F2 16 30S 28 50 E
Karibib, Namibia 56 C2 22 0S 15 56 E
Karimata, Kepulauan, Indonesia 36 E3 1 25S 109 0 E
Karimata, Selat, Indonesia 36 E3 2 0S 108 40 E
Karimata Is. = Karimata, Kepulauan, Indonesia 36 E3 1 25S 109 0 E
Karimnagar, India 40 K11 18 26N 79 10 E
Karimunjawa, Kepulauan, Indonesia 36 F4 5 50S 110 30 E
Karin, Somali Rep. 46 E4 10 50N 45 52 E
Karīt, Iran 45 C8 33 29N 56 55 E
Kariya, Japan 31 G8 34 58N 137 1 E
Kariyangwe, Zimbabwe 57 B4 18 0S 27 38 E
Karkaralinsk = Qarqaraly, Kazakstan 26 E8 49 26N 75 30 E
Karkheh →, Iran 44 D5 31 2N 47 29 E
Karkinitska Zatoka, Ukraine 25 E5 45 56N 33 0 E
Karkinitskiy Zaliv = Karkinitska Zatoka, Ukraine 25 E5 45 56N 33 0 E
Karl-Marx-Stadt = Chemnitz, Germany 16 C7 50 51N 12 54 E
Karlovac, Croatia 16 F8 45 31N 15 36 E
Karlovo, Bulgaria 21 C11 42 38N 24 47 E
Karlovy Vary, Czech Rep. 16 C7 50 13N 12 51 E
Karlsbad = Karlovy Vary, Czech Rep. 16 C7 50 13N 12 51 E
Karlsborg, Sweden 9 G16 58 33N 14 33 E
Karlshamn, Sweden 9 H16 56 10N 14 51 E
Karlskoga, Sweden 9 G16 59 28N 14 33 E
Karlskrona, Sweden 9 H16 56 10N 15 35 E
Karlsruhe, Germany 16 D5 49 0N 8 23 E
Karlstad, Sweden 9 G15 59 23N 13 30 E
Karlstad, U.S.A. 80 A6 48 35N 96 31W
Karmi'el, Israel 47 C4 32 55N 35 18 E
Karnak, Egypt 51 C12 25 43N 32 39 E
Karnal, India 42 E7 29 42N 77 2 E
Karnali →, Nepal 43 E9 28 45N 81 16 E
Karnaphuli Res., Bangla. 41 H18 22 40N 92 20 E
Karnaprayag, India 43 D8 30 16N 79 15 E
Karnataka □, India 40 N10 13 15N 77 0 E
Karnes City, U.S.A. 81 L6 28 53N 97 54W
Karnische Alpen, Europe 16 E7 46 36N 13 0 E
Kärnten □, Austria 16 E8 46 52N 13 30 E
Karoi, Zimbabwe 55 F2 16 48S 29 45 E
Karonga, Malawi 55 D3 9 57S 33 55 E
Karoonda, Australia 63 F2 35 1S 139 59 E
Karor, Pakistan 42 D4 31 15N 70 59 E
Karora, Sudan 51 E13 17 44N 38 15 E
Karpasia, Cyprus 23 D13 35 32N 34 15 E
Kárpathos, Greece 21 G12 35 37N 27 10 E
Karpinsk, Russia 24 C11 59 45N 60 1 E
Karpogory, Russia 24 B7 64 0N 44 27 E
Karpuz Burnu = Apostolos Andreas, C., Cyprus 23 D13 35 42N 34 35 E
Karratha, Australia 60 D2 20 53S 116 40 E
Kars, Turkey 25 F7 40 40N 43 5 E
Karsakpay, Kazakstan 26 E7 47 55N 66 40 E
Karshi = Qarshi, Uzbekistan 26 F7 38 53N 65 48 E
Karsiyang, India 43 F13 26 56N 88 18 E
Karsog, India 42 D7 31 23N 77 12 E
Kartaly, Russia 26 D7 53 3N 60 40 E
Kartapur, India 42 D6 31 27N 75 32 E
Karthaus, U.S.A. 78 E6 41 8N 78 9W
Karufa, Indonesia 37 E8 3 50S 133 20 E
Karumba, Australia 62 B3 17 31S 140 50 E
Karumo, Tanzania 54 C3 2 25S 32 50 E
Karumwa, Tanzania 54 C3 3 12S 32 38 E
Kārūn →, Iran 45 D6 30 26N 48 10 E
Karungu, Kenya 54 C3 0 50S 34 10 E
Karviná, Czech Rep. 17 D10 49 53N 18 31 E
Karwan →, India 42 F8 27 26N 78 4 E
Karwar, India 40 M9 14 55N 74 13 E
Karwi, India 43 G9 25 12N 80 57 E
Kasache, Malawi 55 E3 13 25S 34 20 E
Kasai →, Dem. Rep. of the Congo 52 E3 3 30S 16 10 E
Kasai-Oriental □, Dem. Rep. of the Congo 54 D1 5 0S 24 30 E
Kasaji, Dem. Rep. of the Congo 55 E1 10 25S 23 27 E
Kasama, Zambia 55 E3 10 16S 31 9 E
Kasan-dong, N. Korea 35 D14 41 18N 126 55 E
Kasane, Namibia 56 B3 17 34S 24 50 E
Kasanga, Tanzania 55 D3 8 30S 31 10 E
Kasaragod, India 40 N9 12 30N 74 58 E
Kasba L., Canada 73 A8 60 20N 102 10W
Kāseh Garān, Iran 44 C5 34 5N 46 2 E
Kasempa, Zambia 55 E2 13 30S 25 44 E
Kasenga, Dem. Rep. of the Congo 55 E2 10 20S 28 45 E
Kasese, Uganda 54 B3 0 13N 30 3 E
Kasewa, Zambia 55 E2 14 28S 28 53 E
Kasganj, India 43 F8 27 48N 78 42 E
Kashabowie, Canada 70 C1 48 40N 90 26W
Kashaf, Iran 45 C9 35 58N 61 7 E
Kāshān, Iran 45 C6 34 5N 51 30 E
Kashechewan, Canada 70 B3 52 18N 81 37W
Kashgar = Kashi, China 32 C2 39 30N 76 2 E
Kashi, China 32 C2 39 30N 76 2 E
Kashimbo, Dem. Rep. of the Congo 55 E2 11 12S 26 19 E
Kashipur, India 43 E8 29 15N 79 0 E
Kashiwazaki, Japan 31 F9 37 22N 138 33 E
Kashk-e Kohneh, Afghan. 40 B3 34 55N 62 30 E
Kashkū'iyeh, Iran 45 D7 30 31N 55 40 E
Kāshmar, Iran 45 C8 35 16N 58 26 E
Kashmir, Asia 43 C7 34 0N 76 0 E
Kashmor, Pakistan 42 E3 28 28N 69 32 E
Kashun Noerh = Gaxun Nur, China 32 B5 42 22N 100 30 E
Kasiari, India 43 H12 22 8N 87 14 E
Kasimov, Russia 24 D7 54 55N 41 20 E
Kasinge, Dem. Rep. of the Congo 54 D2 6 15S 26 58 E
Kasiruta, Indonesia 37 E7 0 25S 127 12 E
Kaskaskia →, U.S.A. 80 G10 37 58N 89 57W
Kaskattama →, Canada 73 B10 57 3N 90 4W
Kaskinen, Finland 9 E19 62 22N 21 15 E
Kaslo, Canada 72 D5 49 55N 116 55W
Kasmere L., Canada 73 B8 59 34N 101 10W
Kasongo, Dem. Rep. of the Congo 54 C2 4 30S 26 33 E
Kasongo Lunda, Dem. Rep. of the Congo 52 F3 6 35S 16 49 E
Kásos, Greece 21 G12 35 20N 26 55 E
Kassalâ, Sudan 51 E13 15 30N 36 0 E
Kassel, Germany 16 C5 51 18N 9 26 E
Kassiópi, Greece 23 A3 39 48N 19 53 E
Kasson, U.S.A. 80 C8 44 2N 92 45W
Kastamonu, Turkey 25 F5 41 25N 33 43 E
Kastélli, Greece 23 D5 35 29N 23 38 E
Kastéllion, Greece 23 D7 35 12N 25 20 E
Kasterlee, Belgium 15 C4 51 15N 4 59 E
Kastoria, Greece 21 D9 40 30N 21 19 E
Kasulu, Tanzania 54 C3 4 37S 30 5 E
Kasumi, Japan 31 G7 35 38N 134 38 E
Kasungu, Malawi 55 E3 13 0S 33 29 E
Kasur, Pakistan 42 D6 31 5N 74 25 E
Kataba, Zambia 55 F2 16 5S 25 10 E
Katahdin, Mt., U.S.A. 77 C11 45 54N 68 56W
Katako Kombe, Dem. Rep. of the Congo 54 C1 3 25S 24 20 E
Katale, Tanzania 54 C3 4 52S 31 7 E
Katanda, Katanga, Dem. Rep. of the Congo 54 D1 7 52S 24 13 E
Katanda, Nord-Kivu, Dem. Rep. of the Congo 54 C2 0 55S 29 21 E
Katanga □, Dem. Rep. of the Congo 54 D2 8 0S 25 0 E
Katangi, India 40 J11 21 56N 79 50 E
Katanning, Australia 61 F2 33 40S 117 33 E
Katavi Swamp, Tanzania 54 D3 6 50S 31 10 E
Katerini, Greece 21 D10 40 18N 22 37 E
Katghora, India 43 H10 22 30N 82 33 E
Katha, Burma 41 G20 24 10N 96 30 E
Katherîna, Gebel, Egypt 44 D2 28 30N 33 57 E
Katherine, Australia 60 B5 14 27S 132 20 E
Katherine Gorge, Australia 60 B5 14 18S 132 28 E
Kathi, India 42 J6 21 47N 74 3 E
Kathiawar, India 42 H4 22 20N 71 0 E
Kathikas, Cyprus 23 E11 34 55N 32 25 E
Kathmandu = Katmandu, Nepal 43 F11 27 45N 85 20 E
Kathua, India 42 C6 32 23N 75 34 E
Katihar, India 43 G12 25 34N 87 36 E
Katima Mulilo, Zambia 56 B3 17 28S 24 13 E
Katimbira, Malawi 55 E3 12 40S 34 0 E
Katingan = Mendawai →, Indonesia 36 E4 3 30S 113 0 E
Katiola, Ivory C. 50 G4 8 10N 5 10W
Katmandu, Nepal 43 F11 27 45N 85 20 E
Katni, India 43 H9 23 51N 80 24 E
Káto Arkhánai, Greece 23 D7 35 15N 25 10 E
Káto Khorió, Greece 23 D7 35 3N 25 47 E
Káto Pyrgos, Cyprus 23 D11 35 11N 32 41 E
Katompe, Dem. Rep. of the Congo 54 D2 6 2S 26 23 E
Katonga →, Uganda 54 B3 0 34N 31 50 E
Katoomba, Australia 63 E5 33 41S 150 19 E
Katowice, Poland 17 C10 50 17N 19 5 E
Katrine, L., U.K. 12 E4 56 15N 4 30W
Katrineholm, Sweden 9 G17 59 9N 16 12 E
Katsepe, Madag. 57 B8 15 45S 46 15 E
Katsina, Nigeria 50 F7 13 0N 7 32 E
Katsumoto, Japan 31 H4 33 51N 129 42 E
Katsuura, Japan 31 G10 35 10N 140 20 E
Katsuyama, Japan 31 F8 36 3N 136 30 E
Kattaviá, Greece 23 D9 35 57N 27 46 E
Kattegat, Denmark 9 H14 56 40N 11 20 E
Katumba, Dem. Rep. of the Congo 54 D2 7 40S 25 17 E
Katungu, Kenya 54 C5 2 55S 40 3 E
Katwa, India 43 H13 23 30N 88 5 E
Katwijk, Neths. 15 B4 52 12N 4 24 E
Kauai, U.S.A. 74 H15 22 3N 159 30W
Kauai Channel, U.S.A. 74 H15 21 45N 158 50W
Kaufman, U.S.A. 81 J6 32 35N 96 19W
Kauhajoki, Finland 9 E20 62 25N 22 10 E
Kaukauna, U.S.A. 76 C1 44 17N 88 17W
Kaukauveld, Namibia 56 C3 20 0S 20 15 E
Kaunakakai, U.S.A. 74 H16 21 6N 157 1W
Kaunas, Lithuania 9 J20 54 54N 23 54 E
Kaunia, Bangla. 43 G13 25 46N 89 26 E
Kautokeino, Norway 8 B20 69 0N 23 4 E
Kauwapur, India 43 F10 27 31N 82 18 E
Kavacha, Russia 27 C17 60 16N 169 51 E
Kavalerovo, Russia 30 B7 44 15N 135 4 E
Kavali, India 40 M12 14 55N 80 1 E
Kavála, Greece 21 D11 40 57N 24 28 E
Kavār, Iran 45 D7 29 11N 52 44 E
Kavi, India 42 H5 22 12N 72 38 E
Kavimba, Botswana 56 B3 18 2S 24 38 E
Kavīr, Dasht-e, Iran 45 C7 34 30N 55 0 E
Kavos, Greece 23 B4 39 23N 20 3 E
Kaw, Fr. Guiana 93 C8 4 30N 52 15W
Kawagama L., Canada 78 A6 45 18N 78 45W
Kawagoe, Japan 31 G9 35 55N 139 29 E
Kawaguchi, Japan 31 G9 35 52N 139 45 E
Kawambwa, Zambia 55 D2 9 48S 29 3 E
Kawanoe, Japan 31 G6 34 1N 133 34 E
Kawardha, India 43 J9 22 0N 81 17 E
Kawasaki, Japan 31 G9 35 35N 139 42 E
Kawasi, Indonesia 37 E7 1 38S 127 28 E
Kawerau, N.Z. 59 H6 38 7S 176 42 E
Kawhia Harbour, N.Z. 59 H5 38 5S 174 51 E
Kawio, Kepulauan, Indonesia 37 D7 4 30N 125 30 E
Kawnro, Burma 41 H21 22 48N 99 8 E
Kawthaung, Burma 39 H2 10 5N 98 36 E
Kawthoolei = Kayin □, Burma 41 L20 18 0N 97 30 E
Kawthule = Kayin □, Burma 41 L20 18 0N 97 30 E
Kaya, Burkina Faso 50 F5 13 4N 1 10W
Kayah □, Burma 41 K20 19 15N 97 15 E
Kayan →, Indonesia 36 D5 2 55N 117 35 E
Kaycee, U.S.A. 82 E10 43 43N 106 38W
Kayeli, Indonesia 37 E7 3 20S 127 10 E
Kayenta, U.S.A. 83 H8 36 44N 110 15W
Kayes, Mali 50 F3 14 25N 11 30W
Kayin □, Burma 41 L20 18 0N 97 30 E
Kayoa, Indonesia 37 D7 0 1N 127 28 E
Kayomba, Zambia 55 E1 13 11S 24 2 E
Kayseri, Turkey 25 G6 38 45N 35 30 E
Kaysville, U.S.A. 82 F8 41 2N 111 56W
Kazachye, Russia 27 B14 70 52N 135 58 E
Kazakstan ■, Asia 26 E7 50 0N 70 0 E
Kazan, Russia 24 C8 55 50N 49 10 E
Kazan →, Canada 73 A9 64 3N 95 35W
Kazan-Rettō, Pac. Oc. 64 E6 25 0N 141 0 E
Kazanlŭk, Bulgaria 21 C11 42 38N 25 20 E
Kazatin = Kozyatyn, Ukraine 17 D15 49 45N 28 50 E
Kāzerūn, Iran 45 D6 29 38N 51 40 E
Kazi Magomed = Qazimämmäd, Azerbaijan 45 A6 40 3N 49 0 E
Kazuno, Japan 30 D10 40 10N 140 45 E
Kazym →, Russia 26 C7 63 54N 65 50 E
Kéa, Greece 21 F11 37 35N 24 22 E
Keady, U.K. 13 B5 54 15N 6 42W
Kearney, U.S.A. 80 E5 40 42N 99 5W
Kearny, U.S.A. 83 K8 33 3N 110 55W
Kearsarge, Mt., U.S.A. 79 C13 43 22N 71 50W
Keban, Turkey 25 G6 38 50N 38 50 E
Keban Barajı, Turkey 25 G6 38 41N 38 33 E
Kebnekaise, Sweden 8 C18 67 53N 18 33 E
Kebri Dehar, Ethiopia 46 F3 6 45N 44 17 E
Kebumen, Indonesia 37 G13 7 42S 109 40 E
Kechika →, Canada 72 B3 59 41N 127 12W
Kecskemét, Hungary 17 E10 46 57N 19 42 E
Kėdainiai, Lithuania 9 J21 55 15N 24 2 E
Kedarnath, India 43 D8 30 44N 79 4 E
Kedgwick, Canada 71 C6 47 40N 67 20W
Kédhros Óros, Greece 23 D6 35 11N 24 37 E
Kediri, Indonesia 36 F4 7 51S 112 1 E
Keeler, U.S.A. 84 J9 36 29N 117 52W
Keeley L., Canada 73 C7 54 54N 108 8W
Keeling Is. = Cocos Is., Ind. Oc. 64 J1 12 10S 96 55 E
Keelung = Chilung, Taiwan 33 D7 25 3N 121 45 E
Keene, Canada 78 B6 44 15N 78 10W
Keene, Calif., U.S.A. 85 K8 35 13N 118 33W
Keene, N.H., U.S.A. 79 D12 42 56N 72 17W

Keene

Keene, N.Y., U.S.A. 79 B11 44 16N 73 46W
Keeper Hill, Ireland 13 D3 52 45N 8 16W
Keer-Weer, C., Australia 62 A3 14 0S 141 32 E
Keeseville, U.S.A. 79 B11 44 29N 73 30W
Keetmanshoop, Namibia 56 D2 26 35S 18 8 E
Keewatin, Canada 73 D10 49 46N 94 34W
Keewatin →, Canada 73 B8 56 29N 100 46W
Kefallinía, Greece 21 E9 38 15N 20 30 E
Kefamenanu, Indonesia 37 F6 9 28S 124 29 E
Kefar Sava, Israel 47 C3 32 11N 34 54 E
Keffi, Nigeria 50 G7 8 55N 7 43 E
Keflavík, Iceland 8 D2 64 2N 22 35W
Keg River, Canada 72 B5 57 54N 117 55W
Kegaska, Canada 71 B7 50 9N 61 18W
Keighley, U.K. 10 D6 53 52N 1 54W
Keila, Estonia 9 G21 59 18N 24 25 E
Keimoes, S. Africa 56 D3 28 41S 20 59 E
Keitele, Finland 8 E22 63 10N 26 20 E
Keith, Australia 63 F3 36 6S 140 20 E
Keith, U.K. 12 D6 57 32N 2 57W
Keizer, U.S.A. 82 D2 44 57N 123 1W
Kejimkujik Nat. Park, Canada 71 D6 44 25N 65 25W
Kejserr Franz Joseph Fd., Greenland 4 B6 73 30N 24 30W
Kekri, India 42 G6 26 0N 75 10 E
Kelan, China 34 E6 38 43N 111 31 E
Kelang, Malaysia 39 L3 3 2N 101 26 E
Kelantan →, Malaysia 39 J4 6 13N 102 14 E
Kelkit →, Turkey 25 F6 40 45N 36 32 E
Kellerberrin, Australia 61 F2 31 36S 117 38 E
Kellett, C., Canada 4 B1 72 0N 126 0W
Kelleys I., U.S.A. 78 E2 41 36N 82 42W
Kellogg, U.S.A. 82 C5 47 32N 116 7W
Kells = Ceanannus Mor, Ireland 13 C5 53 44N 6 53W
Kelokedhara, Cyprus 23 E11 34 48N 32 39 E
Kelowna, Canada 72 D5 49 50N 119 25W
Kelseyville, U.S.A. 84 G4 38 59N 122 50W
Kelso, N.Z. 59 L2 45 54S 169 15 E
Kelso, U.K. 12 F6 55 36N 2 26W
Kelso, U.S.A. 84 D4 46 9N 122 54W
Keluang, Malaysia 39 L4 2 3N 103 18 E
Kelvington, Canada 73 C8 52 10N 103 30W
Kem, Russia 24 B5 65 0N 34 38 E
Kem →, Russia 24 B5 64 57N 34 41 E
Kema, Indonesia 37 D7 1 22N 125 8 E
Kemah, Turkey 44 B3 39 32N 39 5 E
Kemaman, Malaysia 36 D2 4 12N 103 18 E
Kemano, Canada 72 C3 53 35N 128 0W
Kemasik, Malaysia 39 K4 4 25N 103 27 E
Kemerovo, Russia 26 D9 55 20N 86 5 E
Kemi, Finland 8 D21 65 44N 24 34 E
Kemi älv = Kemijoki →, Finland 8 D21 65 47N 24 32 E
Kemijärvi, Finland 8 C22 66 43N 27 22 E
Kemijoki →, Finland 8 D21 65 47N 24 32 E
Kemmerer, U.S.A. 82 F8 41 48N 110 32W
Kemmuna = Comino, Malta 23 C1 36 1N 14 20 E
Kemp, L., U.S.A. 81 J5 33 46N 99 9W
Kemp Land, Antarctica 5 C5 69 0S 55 0 E
Kempsey, Australia 63 E5 31 1S 152 50 E
Kempt, L., Canada 70 C5 47 25N 74 22W
Kempten, Germany 16 E6 47 45N 10 17 E
Kempton, Australia 62 G4 42 31S 147 12 E
Kemptville, Canada 79 B9 45 0N 75 38W
Ken →, India 43 G9 25 13N 80 27 E
Kenai, U.S.A. 68 B4 60 33N 151 16W
Kendai, India 43 H10 22 45N 82 37 E
Kendal, Indonesia 37 G14 6 56S 110 14 E
Kendal, U.K. 10 C5 54 20N 2 44W
Kendall, Australia 63 E5 31 35S 152 44 E
Kendall, U.S.A. 77 N5 25 41N 80 19W
Kendall →, Australia 62 A3 14 4S 141 35 E
Kendallville, U.S.A. 76 E3 41 27N 85 16W
Kendari, Indonesia 37 E6 3 50S 122 30 E
Kendawangan, Indonesia 36 E4 2 32S 110 17 E
Kendrapara, India 41 J15 20 35N 86 30 E
Kendrew, S. Africa 56 E3 32 32S 24 30 E
Kene Thao, Laos 38 D3 17 44N 101 10 E
Kenedy, U.S.A. 81 L6 28 49N 97 51W
Kenema, S. Leone 50 G3 7 50N 11 14W
Keng Kok, Laos 38 D5 16 26N 105 12 E
Keng Tawng, Burma 41 J21 20 45N 98 18 E
Keng Tung, Burma 41 J21 21 0N 99 30 E
Kengeja, Tanzania 54 D4 5 26S 39 45 E
Kenhardt, S. Africa 56 D3 29 19S 21 12 E
Kenitra, Morocco 50 B4 34 15N 6 40W
Kenli, China 35 F10 37 30N 118 20 E
Kenmare, Ireland 13 E2 51 53N 9 36W
Kenmare, U.S.A. 80 A3 48 41N 102 5W
Kenmare River, Ireland 13 E2 51 48N 9 51W
Kennebago Lake, U.S.A. 79 A14 45 4N 70 40W
Kennebec, U.S.A. 80 D5 43 54N 99 52W
Kennebec →, U.S.A. 77 D11 43 45N 69 46W
Kennebunk, U.S.A. 79 C14 43 23N 70 33W
Kennedy, Zimbabwe 56 B4 18 52S 27 10 E
Kennedy Ra., Australia 61 D2 24 45S 115 10 E
Kennedy Taungdeik, Burma 41 H18 23 15N 93 45 E
Kenner, U.S.A. 81 L9 29 59N 90 15W
Kennet →, U.K. 11 F7 51 27N 0 57W
Kenneth Ra., Australia 61 D2 23 50S 117 8 E
Kennett, U.S.A. 81 G9 36 14N 90 3W
Kennewick, U.S.A. 82 C4 46 12N 119 7W
Kenogami →, Canada 70 B3 51 6N 84 28W
Kenora, Canada 73 D10 49 47N 94 29W
Kenosha, U.S.A. 76 D2 42 35N 87 49W
Kensington, Canada 71 C7 46 28N 63 34W
Kent, Ohio, U.S.A. 78 E3 41 9N 81 22W
Kent, Tex., U.S.A. 81 K2 31 4N 104 13W
Kent, Wash., U.S.A. 84 C4 47 23N 122 14W
Kent □, U.K. 11 F8 51 12N 0 40 E
Kent Group, Australia 62 F4 39 30S 147 20 E
Kent Pen., Canada 68 B9 68 30N 107 0W
Kentaū, Kazakstan 26 E7 43 32N 68 36 E
Kentland, U.S.A. 76 E2 40 46N 87 27W
Kenton, U.S.A. 76 E4 40 39N 83 37W
Kentucky □, U.S.A. 76 G3 37 0N 84 0W
Kentucky →, U.S.A. 76 F3 38 41N 85 11W
Kentucky L., U.S.A. 77 G2 37 1N 88 16W
Kentville, Canada 71 C7 45 6N 64 29W
Kentwood, U.S.A. 81 K9 30 56N 90 31W
Kenya ■, Africa 54 B4 1 0N 38 0 E
Kenya, Mt., Kenya 54 C4 0 10S 37 18 E
Keo Neua, Deo, Vietnam 38 C5 18 23N 105 10 E
Keokuk, U.S.A. 80 E9 40 24N 91 24W
Keonjhargarh, India 43 J11 21 28N 85 35 E
Kep, Cambodia 39 G5 10 29N 104 19 E
Kep, Vietnam 38 B6 21 24N 106 16 E
Kepi, Indonesia 37 F9 6 32S 139 19 E
Kerala □, India 40 P10 11 0N 76 15 E
Kerama-Rettō, Japan 31 L3 26 5N 127 15 E
Keran, Pakistan 43 B5 34 35N 73 59 E
Kerang, Australia 63 F3 35 40S 143 55 E
Keraudren, C., Australia 60 C2 19 58S 119 45 E
Kerava, Finland 9 F21 60 25N 25 5 E
Kerch, Ukraine 25 E6 45 20N 36 20 E
Kerguelen, Ind. Oc. 3 G13 49 15S 69 10 E
Kericho, Kenya 54 C4 0 22S 35 15 E
Kerinci, Indonesia 36 E2 1 40S 101 15 E
Kerki, Turkmenistan 26 F7 37 50N 65 12 E
Kérkira, Greece 23 A3 39 38N 19 50 E
Kerkrade, Neths. 15 D6 50 53N 6 4 E
Kermadec Is., Pac. Oc. 64 L10 30 0S 178 15W
Kermadec Trench, Pac. Oc. 64 L10 30 30S 176 0W
Kermān, Iran 45 D8 30 15N 57 1 E
Kerman, U.S.A. 84 J6 36 43N 120 4W
Kermān □, Iran 45 D8 30 0N 57 0 E
Kermān, Biābān-e, Iran 45 D8 28 45N 59 45 E
Kermānshāh = Bākhtarān, Iran 44 C5 34 23N 47 0 E
Kermit, U.S.A. 81 K3 31 52N 103 6W
Kern →, U.S.A. 85 K7 35 16N 119 18W
Kernow = Cornwall □, U.K. 11 G3 50 26N 4 40W
Kernville, U.S.A. 85 K8 35 45N 118 26W
Keroh, Malaysia 39 K3 5 43N 101 1 E
Kerrera, U.K. 12 E3 56 24N 5 33W
Kerrobert, Canada 73 C7 51 56N 109 8W
Kerrville, U.S.A. 81 K5 30 3N 99 8W
Kerry □, Ireland 13 D2 52 7N 9 35W
Kerry Hd., Ireland 13 D2 52 25N 9 56W
Kerulen →, Asia 33 B6 48 48N 117 0 E
Kerzaz, Algeria 50 C5 29 29N 1 37W
Kesagami →, Canada 70 B4 51 40N 79 45W
Kesagami L., Canada 70 B3 50 23N 80 15W
Keşan, Turkey 21 D12 40 49N 26 38 E
Kesennuma, Japan 30 E10 38 54N 141 35 E
Keshit, Iran 45 D8 29 43N 58 17 E
Kestell, S. Africa 57 D4 28 17S 28 42 E
Kestenga, Russia 24 A5 65 50N 31 45 E
Keswick, U.K. 10 C4 54 36N 3 8W
Ket →, Russia 26 D9 58 55N 81 32 E
Ketapang, Indonesia 36 E4 1 55S 110 0 E
Ketchikan, U.S.A. 72 B2 55 21N 131 39W
Ketchum, U.S.A. 82 E6 43 41N 114 22W
Ketef, Khalîg Umm el, Egypt 44 F2 23 40N 35 35 E
Keti Bandar, Pakistan 42 G2 24 8N 67 27 E
Ketri, India 42 E6 28 1N 75 50 E
Kętrzyn, Poland 17 A11 54 7N 21 22 E
Kettering, U.K. 11 E7 52 24N 0 43W
Kettering, U.S.A. 76 F3 39 41N 84 10W
Kettle →, Canada 73 B11 56 40N 89 34W
Kettle Falls, U.S.A. 82 B4 48 37N 118 3W
Kettle Pt., Canada 78 C2 43 13N 82 1W
Kettleman City, U.S.A. 84 J7 36 1N 119 58W
Keuka L., U.S.A. 78 D7 42 30N 77 9W
Keuruu, Finland 9 E21 62 16N 24 41 E
Kewanee, U.S.A. 80 E10 41 14N 89 56W
Kewaunee, U.S.A. 76 C2 44 27N 87 31W
Keweenaw B., U.S.A. 76 B1 47 0N 88 15W
Keweenaw Pen., U.S.A. 76 B2 47 30N 88 0W
Keweenaw Pt., U.S.A. 76 B2 47 25N 87 43W
Key Largo, U.S.A. 77 N5 25 5N 80 27W
Key West, U.S.A. 75 F10 24 33N 81 48W
Keynsham, U.K. 11 F5 51 24N 2 29W
Keyser, U.S.A. 76 F6 39 26N 78 59W
Kezhma, Russia 27 D11 58 59N 101 9 E
Kezi, Zimbabwe 57 C4 20 58S 28 32 E
Khabarovsk, Russia 27 E14 48 30N 135 5 E
Khabr, Iran 45 D8 28 51N 56 22 E
Khābūr →, Syria 44 C4 35 17N 40 35 E
Khachmas = Xaçmaz, Azerbaijan 25 F8 41 31N 48 42 E
Khachrod, India 42 H6 23 25N 75 20 E
Khadro, Pakistan 42 F3 26 11N 68 50 E
Khadzhilyangar, China 43 B8 35 45N 79 20 E
Khaga, India 43 G9 25 47N 81 7 E
Khagaria, India 43 G12 25 30N 86 32 E
Khaipur, Pakistan 42 E5 29 34N 72 17 E
Khair, India 42 F7 27 57N 77 46 E
Khairabad, India 43 F9 27 33N 80 47 E
Khairagarh, India 43 J9 21 27N 81 2 E
Khairpur, Pakistan 42 F3 27 32N 68 49 E
Khairpur Nathan Shah, Pakistan 42 F2 27 6N 67 44 E
Khairwara, India 42 H5 23 58N 73 38 E
Khaisor →, Pakistan 42 D3 31 17N 68 59 E
Khajuri Kach, Pakistan 42 C3 32 4N 69 51 E
Khakassia □, Russia 26 D9 53 0N 90 0 E
Khakhea, Botswana 56 C3 24 48S 23 22 E
Khalafābād, Iran 45 D6 30 54N 49 24 E
Khalilabad, India 43 F10 26 48N 83 5 E
Khalīlī, Iran 45 E7 27 38N 53 17 E
Khalkhāl, Iran 45 B6 37 37N 48 32 E
Khalkís, Greece 21 E10 38 27N 23 42 E
Khalmer-Sede = Tazovskiy, Russia 26 C8 67 30N 78 44 E
Khalmer Yu, Russia 26 C7 67 58N 65 1 E
Khalturin, Russia 24 C8 58 40N 48 50 E
Khalūf, Oman 46 C6 20 30N 58 13 E
Kham Keut, Laos 38 C5 18 15N 104 43 E
Khamaria, India 43 H9 23 5N 80 48 E
Khambhaliya, India 42 H3 22 14N 69 41 E
Khambhat, India 42 H5 22 23N 72 33 E
Khambhat, G. of, India 40 J8 20 45N 72 30 E
Khamīr, Iran 45 E7 26 57N 55 36 E
Khamir, Yemen 46 D3 16 2N 44 0 E
Khamsa, Egypt 47 E1 30 27N 32 23 E
Khan →, Namibia 56 C2 22 37S 14 56 E
Khān Abū Shāmat, Syria 47 B5 33 39N 36 53 E
Khān Azād, Iraq 44 C5 33 7N 44 22 E
Khān Mujiddah, Iraq 44 C4 32 21N 43 48 E
Khān Shaykhūn, Syria 44 C3 35 26N 36 38 E
Khān Yūnis, Gaza Strip 47 D3 31 21N 34 18 E
Khanai, Pakistan 42 D2 30 30N 67 8 E
Khānaqīn, Iraq 44 C5 34 23N 45 25 E
Khānbāghī, Iran 45 B7 36 10N 55 25 E
Khandwa, India 40 J10 21 49N 76 22 E
Khandyga, Russia 27 C14 62 42N 135 35 E
Khāneh, Iran 44 B5 36 41N 45 8 E
Khanewal, Pakistan 42 D4 30 20N 71 55 E
Khangah Dogran, Pakistan 42 D5 31 50N 73 37 E
Khanh Duong, Vietnam 38 F7 12 44N 108 44 E
Khaniá, Greece 23 D6 35 30N 24 4 E
Khaniá □, Greece 23 D6 35 30N 24 0 E
Khaniadhana, India 42 G8 25 1N 78 8 E
Khanion, Kólpos, Greece 23 D5 35 33N 23 55 E
Khanka, L., Asia 27 E14 45 0N 132 24 E
Khankendy = Xankändi, Azerbaijan 25 G8 39 52N 46 49 E
Khanna, India 42 D7 30 42N 76 16 E
Khanozai, Pakistan 42 D2 30 37N 67 19 E
Khanpur, Pakistan 42 E4 28 42N 70 35 E
Khanty-Mansiysk, Russia 26 C7 61 0N 69 0 E
Khapalu, Pakistan 43 B7 35 10N 76 20 E
Khapcheranga, Russia 27 E12 49 42N 112 24 E
Kharaghoda, India 42 H4 23 11N 71 46 E
Kharagpur, India 43 H12 22 20N 87 25 E
Khárakas, Greece 23 D7 35 1N 25 7 E
Kharan Kalat, Pakistan 40 E4 28 34N 65 21 E
Kharānaq, Iran 45 C7 32 20N 54 45 E
Kharda, India 40 K9 18 40N 75 34 E
Khardung La, India 43 B7 34 20N 77 43 E
Khârga, El Wâhât-el, Egypt 51 C12 25 10N 30 35 E
Khargon, India 40 J9 21 45N 75 40 E
Khari →, India 42 G6 25 54N 74 31 E
Kharian, Pakistan 42 C5 32 49N 73 52 E
Khārk, Jazīreh-ye, Iran 45 D6 29 15N 50 28 E
Kharkiv, Ukraine 25 E6 49 58N 36 20 E
Kharkov = Kharkiv, Ukraine 25 E6 49 58N 36 20 E
Kharovsk, Russia 24 C7 59 56N 40 13 E
Kharsawangarh, India 43 H11 22 48N 85 50 E
Kharta, Turkey 21 D13 40 55N 29 7 E
Khartoum = El Khartûm, Sudan 51 E12 15 31N 32 35 E
Khasan, Russia 30 C5 42 25N 130 40 E
Khāsh, Iran 40 E2 28 15N 61 15 E
Khashm el Girba, Sudan 51 F13 14 59N 35 58 E
Khaskovo, Bulgaria 21 D11 41 56N 25 30 E
Khatanga, Russia 27 B11 72 0N 102 20 E
Khatanga →, Russia 27 B11 72 55N 106 0 E
Khatauli, India 42 E7 29 17N 77 43 E
Khatra, India 43 H12 22 59N 86 51 E
Khātūnābād, Iran 45 D7 30 1N 55 25 E
Khatyrka, Russia 27 C18 62 3N 175 15 E
Khavda, India 42 H3 23 51N 69 43 E
Khaybar, Ḥarrat, Si. Arabia 44 E4 25 45N 40 0 E
Khayelitsha, S. Africa 53 L3 34 5S 18 42 E
Khāzimiyah, Iraq 44 C4 34 46N 43 37 E
Khe Bo, Vietnam 38 C5 19 8N 104 41 E
Khe Long, Vietnam 38 B5 21 29N 104 46 E
Khed Brahma, India 40 G8 24 7N 73 5 E
Khekra, India 42 E7 28 52N 77 20 E
Khemarak Phouminville, Cambodia 39 G4 11 37N 102 59 E
Khemisset, Morocco 50 B4 33 50N 6 1W
Khemmarat, Thailand 38 D5 16 10N 105 15 E
Khenāmān, Iran 45 D8 30 27N 56 29 E
Khenchela, Algeria 50 A7 35 28N 7 11 E
Khersān →, Iran 45 D6 31 33N 50 22 E
Kherson, Ukraine 25 E5 46 35N 32 35 E
Khersónisos Akrotíri, Greece 23 D6 35 30N 24 10 E
Kheta →, Russia 27 B11 71 54N 102 6 E
Khewari, Pakistan 42 F3 26 36N 68 52 E
Khilchipur, India 42 G7 24 2N 76 34 E
Khilok, Russia 27 D12 51 30N 110 45 E
Khios, Greece 21 E12 38 27N 26 9 E
Khirsadoh, India 43 H8 22 11N 78 47 E
Khiuma = Hiiumaa, Estonia 9 G20 58 50N 22 45 E
Khiva, Uzbekistan 26 E7 41 30N 60 18 E
Khīyāv, Iran 44 B5 38 30N 47 45 E
Khlong Khlung, Thailand 38 D2 16 12N 99 43 E
Khmelnik, Ukraine 17 D14 49 33N 27 58 E
Khmelnitskiy = Khmelnytskyy, Ukraine 17 D14 49 23N 27 0 E
Khmelnytskyy, Ukraine 17 D14 49 23N 27 0 E
Khmer Rep. = Cambodia ■, Asia 38 F5 12 15N 105 0 E
Khoai, Hon, Vietnam 39 H5 8 26N 104 50 E
Khodoriv, Ukraine 17 D13 49 24N 24 19 E
Khodzent = Khŭjand, Tajikistan 26 E7 40 17N 69 37 E
Khojak Pass, Afghan. 42 D2 30 51N 66 34 E
Khok Kloi, Thailand 39 H2 8 17N 98 19 E
Khok Pho, Thailand 39 J3 6 43N 101 6 E
Kholm, Russia 24 C5 57 10N 31 15 E
Kholmsk, Russia 27 E15 47 40N 142 5 E
Khomas Hochland, Namibia 56 C2 22 40S 16 0 E
Khomeyn, Iran 45 C6 33 40N 50 7 E
Khomeyni Shahr, Iran 45 C6 32 41N 51 31 E
Khomodino, Botswana 56 C3 22 46S 23 52 E
Khon Kaen, Thailand 38 D4 16 30N 102 47 E
Khong →, Cambodia 38 F5 13 32N 105 58 E
Khong Sedone, Laos 38 E5 15 34N 105 49 E
Khonuu, Russia 27 C15 66 30N 143 12 E
Khoper →, Russia 25 D6 49 30N 42 20 E
Khóra Sfakion, Greece 23 D6 35 15N 24 9 E
Khorāsān □, Iran 45 C8 34 0N 58 0 E
Khorat = Nakhon Ratchasima, Thailand 38 E4 14 59N 102 12 E
Khorat, Cao Nguyen, Thailand 38 E4 15 30N 102 50 E
Khorixas, Namibia 56 C1 20 16S 14 59 E
Khorramābād, Khorāsān, Iran 45 C8 35 6N 57 57 E
Khorramābād, Lorestān, Iran 45 C6 33 30N 48 25 E
Khorrāmshahr, Iran 45 D6 30 29N 48 15 E
Khorugh, Tajikistan 26 F8 37 30N 71 36 E
Khosravī, Iran 45 D6 30 48N 51 28 E
Khosrowābād, Khuzestān, Iran 45 D6 30 10N 48 25 E
Khosrowābād, Kordestān, Iran 44 C5 35 31N 47 38 E
Khost, Pakistan 42 D2 30 13N 67 35 E
Khosūyeh, Iran 45 D7 28 32N 54 26 E
Khotyn, Ukraine 17 D14 48 31N 26 27 E
Khouribga, Morocco 50 B4 32 58N 6 57W
Khowst, Afghan. 42 C3 33 22N 69 58 E
Khowst □, Afghan. 40 C6 33 20N 70 0 E
Khoyniki, Belarus 17 C15 51 54N 29 55 E
Khrysokhou B., Cyprus 23 D11 35 6N 32 25 E
Khu Khan, Thailand 38 E5 14 42N 104 12 E
Khudzhand = Khŭjand, Tajikistan 26 E7 40 17N 69 37 E
Khuff, Si. Arabia 44 E5 24 55N 44 53 E
Khūgiānī, Afghan. 42 D2 31 34N 66 32 E
Khuis, Botswana 56 D3 26 40S 21 49 E
Khuiyala, India 42 F4 27 9N 70 25 E
Khŭjand, Tajikistan 26 E7 40 17N 69 37 E
Khujner, India 42 H7 23 47N 76 36 E
Khulna, Bangla. 41 H16 22 45N 89 34 E
Khulna □, Bangla. 41 H16 22 25N 89 35 E
Khumago, Botswana 56 C3 20 26S 24 32 E
Khūnsorkh, Iran 45 E8 27 9N 56 7 E
Khunti, India 43 H11 23 5N 85 17 E
Khūr, Iran 45 C8 32 55N 58 18 E
Khurai, India 42 G8 24 3N 78 23 E
Khurayş, Si. Arabia 45 E6 25 6N 48 2 E
Khurīyā Murīyā, Jazā'ir, Oman 46 D6 17 30N 55 58 E
Khurja, India 42 E7 28 15N 77 58 E
Khūrmāl, Iraq 44 C5 35 18N 46 2 E
Khurr, Wādī al, Iraq 44 C4 32 3N 43 52 E
Khūsf, Iran 45 C8 32 46N 58 53 E
Khush, Afghan. 40 C3 32 55N 62 10 E
Khushab, Pakistan 42 C5 32 20N 72 20 E
Khust, Ukraine 17 D12 48 10N 23 18 E
Khuzdar, Pakistan 42 F2 27 52N 66 30 E
Khūzestān □, Iran 45 D6 31 0N 49 0 E
Khvāf, Iran 45 C9 34 33N 60 8 E
Khvājeh, Iran 44 B5 38 9N 46 35 E
Khvānsār, Iran 45 D7 29 56N 54 8 E
Khvor, Iran 45 C7 33 45N 55 0 E
Khvorgū, Iran 45 E8 27 34N 56 27 E
Khvormūj, Iran 45 D6 28 40N 51 30 E
Khvoy, Iran 44 B5 38 35N 45 0 E
Khyber Pass, Afghan. 42 B4 34 10N 71 8 E
Kiabukwa, Dem. Rep. of the Congo 55 D1 8 40S 24 48 E
Kiama, Australia 63 E5 34 40S 150 50 E
Kiamba, Phil. 37 C6 6 2N 124 46 E
Kiambi, Dem. Rep. of the Congo 54 D2 7 15S 28 0 E
Kiambu, Kenya 54 C4 1 8S 36 50 E
Kiangara, Madag. 57 B8 17 58S 47 2 E
Kiangsi = Jiangxi □, China 33 D6 27 30N 116 0 E
Kiangsu = Jiangsu □, China 35 H11 33 0N 120 0 E
Kibanga Port, Uganda 54 B3 0 10N 32 58 E
Kibara, Tanzania 54 C3 2 8S 33 30 E
Kibare, Mts., Dem. Rep. of the Congo 54 D2 8 25S 27 10 E
Kibombo, Dem. Rep. of the Congo 54 C2 3 57S 25 53 E
Kibondo, Tanzania 54 C3 3 35S 30 45 E
Kibre Mengist, Ethiopia 46 F2 5 54N 38 59 E
Kibumbu, Burundi 54 C2 3 32S 29 45 E
Kibungo, Rwanda 54 C3 2 10S 30 32 E
Kibuye, Burundi 54 C2 3 39S 29 59 E
Kibuye, Rwanda 54 C2 2 3S 29 21 E
Kibwesa, Tanzania 54 D2 6 30S 29 58 E
Kibwezi, Kenya 54 C4 2 27S 37 57 E
Kichha, India 43 E8 28 53N 79 30 E
Kichha →, India 43 E8 28 41N 79 18 E
Kichmengskiy Gorodok, Russia 24 B8 59 59N 45 48 E
Kicking Horse Pass, Canada 72 C5 51 28N 116 16W
Kidal, Mali 50 E6 18 26N 1 22 E
Kidderminster, U.K. 11 E5 52 24N 2 15W
Kidete, Tanzania 54 D4 6 25S 37 17 E
Kidnappers, C., N.Z. 59 H6 39 38S 177 5 E
Kidsgrove, U.K. 10 D5 53 5N 2 14W
Kidston, Australia 62 B3 18 52S 144 8 E
Kidugallo, Tanzania 54 D4 6 49S 38 15 E
Kiel, Germany 16 A6 54 19N 10 8 E
Kiel Canal = Nord-Ostsee-Kanal, Germany 16 A5 54 12N 9 32 E
Kielce, Poland 17 C11 50 52N 20 42 E
Kielder Water, U.K. 10 B5 55 11N 2 31W
Kieler Bucht, Germany 16 A6 54 35N 10 25 E
Kien Binh, Vietnam 39 H5 9 55N 105 19 E
Kien Tan, Vietnam 39 G5 10 7N 105 17 E
Kienge, Dem. Rep. of the Congo 55 E2 10 30S 27 30 E
Kiev = Kyyiv, Ukraine 17 C16 50 30N 30 28 E
Kiffa, Mauritania 50 E3 16 37N 11 24W
Kifri, Iraq 44 C5 34 45N 45 0 E
Kigali, Rwanda 54 C3 1 59S 30 4 E
Kigarama, Tanzania 54 C3 1 1S 31 50 E
Kigoma □, Tanzania 54 D3 5 0S 30 0 E
Kigoma-Ujiji, Tanzania 54 C2 4 55S 29 36 E
Kigomasha, Ras, Tanzania 54 C4 4 58S 38 58 E
Kığzı, Turkey 44 B4 38 18N 43 25 E
Kihei, U.S.A. 74 H16 20 47N 156 28W
Kihnu, Estonia 9 G21 58 9N 24 1 E
Kii-Sanchi, Japan 31 G8 34 20N 136 0 E
Kii-Suidō, Japan 31 H7 33 40N 134 45 E
Kikaiga-Shima, Japan 31 K4 28 19N 129 59 E
Kikinda, Serbia, Yug. 21 B9 45 50N 20 30 E
Kikládhes, Greece 21 F11 37 0N 24 30 E
Kikwit, Dem. Rep. of the Congo 52 E3 5 0S 18 45 E
Kilar, India 42 C7 33 6N 76 25 E
Kilauea Crater, U.S.A. 74 J17 19 25N 155 17W
Kilbrannan Sd., U.K. 12 F3 55 37N 5 26W
Kilchu, N. Korea 35 D15 40 57N 129 25 E
Kilcoy, Australia 63 D5 26 59S 152 30 E
Kildare, Ireland 13 C5 53 9N 6 55W
Kildare □, Ireland 13 C5 53 10N 6 50W
Kilfinnane, Ireland 13 D3 52 21N 8 28W
Kilgore, U.S.A. 81 J7 32 23N 94 53W
Kilifi, Kenya 54 C4 3 40S 39 48 E
Kilimanjaro, Tanzania 54 C4 3 7S 37 20 E
Kilimanjaro □, Tanzania 54 C4 4 0S 38 0 E
Kilindini, Kenya 54 C4 4 4S 39 40 E
Kilis, Turkey 44 B3 36 42N 37 6 E
Kiliya, Ukraine 17 F15 45 28N 29 16 E
Kilkee, Ireland 13 D2 52 41N 9 39W
Kilkeel, U.K. 13 B5 54 4N 6 0W
Kilkenny, Ireland 13 D4 52 39N 7 15W
Kilkenny □, Ireland 13 D4 52 35N 7 15W
Kilkieran B., Ireland 13 C2 53 20N 9 41W
Kilkís, Greece 21 D10 40 58N 22 57 E
Killala, Ireland 13 B2 54 13N 9 12W
Killala B., Ireland 13 B2 54 16N 9 8W
Killaloe, Ireland 13 D3 52 48N 8 28W
Killaloe Station, Canada 78 A7 45 33N 77 25W
Killarney, Australia 63 D5 28 20S 152 18 E
Killarney, Canada 73 D9 49 10N 99 40W
Killarney, Ireland 13 D2 52 4N 9 30W
Killary Harbour, Ireland 13 C2 53 38N 9 52W
Killdeer, U.S.A. 80 B3 47 26N 102 48W
Killeen, U.S.A. 81 K6 31 7N 97 44W
Killin, U.K. 12 E4 56 28N 4 19W
Killini, Greece 21 F10 37 54N 22 25 E
Killorglin, Ireland 13 D2 52 6N 9 47W
Killybegs, Ireland 13 B3 54 38N 8 26W
Kilmarnock, U.K. 12 F4 55 37N 4 29W
Kilmore, Australia 63 F3 37 25S 144 53 E

Kondratyevo

Kilondo, Tanzania 55 D3 9 45S 34 20 E
Kilosa, Tanzania 54 D4 6 48S 37 0 E
Kilrush, Ireland 13 D2 52 38N 9 29W
Kilwa Kisiwani, Tanzania 55 D4 8 58S 39 32 E
Kilwa Kivinje, Tanzania 55 D4 8 45S 39 25 E
Kilwa Masoko, Tanzania 55 D4 8 55S 39 30 E
Kilwinning, U.K. 12 F4 55 39N 4 43W
Kim, U.S.A. 81 G3 37 15N 103 21W
Kimaam, Indonesia 37 F9 7 58S 138 53 E
Kimamba, Tanzania 54 D4 6 45S 37 10 E
Kimba, Australia 63 E2 33 8S 136 23 E
Kimball, Nebr., U.S.A. 80 E3 41 14N 103 40W
Kimball, S. Dak., U.S.A. 80 D5 43 45N 98 57W
Kimberley, Australia 60 C4 16 20S 127 0 E
Kimberley, Canada 72 D5 49 40N 115 59W
Kimberley, S. Africa 56 D3 28 43S 24 46 E
Kimberly, U.S.A. 82 E6 42 32N 114 22W
Kimch'aek, N. Korea 35 D15 40 40N 129 10 E
Kimch'ŏn, S. Korea 35 F15 36 11N 128 4 E
Kimje, S. Korea 35 G14 35 48N 126 45 E
Kimmirut, Canada 69 B13 62 50N 69 50W
Kimpese, Dem. Rep. of the Congo 52 F2 5 35S 14 26 E
Kimry, Russia 24 C6 56 55N 37 15 E
Kinabalu, Gunong, Malaysia 36 C5 6 3N 116 14 E
Kinaskan L., Canada 72 B2 57 38N 130 8W
Kinbasket L., Canada 72 C5 52 0N 118 10W
Kincardine, Canada 78 B3 44 10N 81 40W
Kincolith, Canada 72 B3 55 0N 129 57W
Kinda, Dem. Rep. of the Congo 55 D2 9 18S 25 4 E
Kinde, U.S.A. 78 C2 43 56N 83 0W
Kinder Scout, U.K. 10 D6 53 24N 1 52W
Kindersley, Canada 73 C7 51 30N 109 10W
Kindia, Guinea 50 F3 10 0N 12 52W
Kindu, Dem. Rep. of the Congo 54 C2 2 55S 25 50 E
Kineshma, Russia 24 C7 57 30N 42 5 E
Kinesi, Tanzania 54 C3 1 25S 33 50 E
King, L., Australia 61 F2 33 10S 119 35 E
King, Mt., Australia 62 D4 25 10S 147 30 E
King City, U.S.A. 84 J5 36 13N 121 8W
King Cr. →, Australia 62 C2 24 35S 139 30 E
King Edward →, Australia 60 B4 14 14S 126 35 E
King Frederick VI Land = Kong Frederik VI Kyst, Greenland 4 C5 63 0N 43 0W
King George B., Falk. Is. 96 G4 51 30S 60 30W
King George I., Antarctica 5 C18 60 0S 60 0W
King George Is., Canada 69 C11 57 20N 80 30W
King I. = Kadan Kyun, Burma 38 F2 12 30N 98 20 E
King I., Australia 62 F3 39 50S 144 0 E
King I., Canada 72 C3 52 10N 127 40W
King Leopold Ranges, Australia 60 C4 17 30S 125 45 E
King of Prussia, U.S.A. 79 F9 40 5N 75 23W
King Sd., Australia 60 C3 16 50S 123 20 E
King William I., Canada 68 B10 69 10N 97 25W
King William's Town, S. Africa 56 E4 32 51S 27 22 E
Kingaok = Bathurst Inlet, Canada 68 B9 66 50N 108 1W
Kingaroy, Australia 63 D5 26 32S 151 51 E
Kingfisher, U.S.A. 81 H6 35 52N 97 56W
Kingirbān, Iraq 44 C5 34 40N 44 54 E
Kingisepp = Kuressaare, Estonia 9 G20 58 15N 22 30 E
Kingman, Ariz., U.S.A. 85 K12 35 12N 114 4W
Kingman, Kans., U.S.A. 81 G5 37 39N 98 7W
Kingoonya, Australia 63 E2 30 55S 135 19 E
Kingri, Pakistan 42 D3 30 27N 69 49 E
Kings →, U.S.A. 84 J7 36 3N 119 50W
Kings Canyon Nat. Park, U.S.A. 84 J8 36 50N 118 40W
King's Lynn, U.K. 10 E8 52 45N 0 24 E
Kings Mountain, U.S.A. 77 H5 35 15N 81 20W
Kings Park, U.S.A. 79 F11 40 53N 73 16W
King's Peak, U.S.A. 82 F8 40 46N 110 27W
Kingsbridge, U.K. 11 G4 50 17N 3 47W
Kingsburg, U.S.A. 84 J7 36 31N 119 33W
Kingscote, Australia 63 F2 35 40S 137 38 E
Kingscourt, Ireland 13 C5 53 55N 6 48W
Kingsford, U.S.A. 76 C1 45 48N 88 4W
Kingsland, U.S.A. 77 K5 30 48N 81 41W
Kingsley, U.S.A. 80 D7 42 35N 95 58W
Kingsport, U.S.A. 77 G4 36 33N 82 33W
Kingston, Canada 79 B8 44 14N 76 30W
Kingston, Jamaica 88 C4 18 0N 76 50W
Kingston, N.Z. 59 L2 45 20S 168 43 E
Kingston, N.H., U.S.A. 79 D13 42 56N 71 3W
Kingston, N.Y., U.S.A. 79 E11 41 56N 73 59W
Kingston, Pa., U.S.A. 79 E9 41 16N 75 54W
Kingston, R.I., U.S.A. 79 E13 41 29N 71 30W
Kingston Pk., U.S.A. 85 K11 35 45N 115 54W
Kingston South East, Australia 63 F2 36 51S 139 55 E
Kingston upon Hull, U.K. 10 D7 53 45N 0 21W
Kingston upon Hull □, U.K. 10 D7 53 45N 0 21W
Kingston-upon-Thames □, U.K. 11 F7 51 24N 0 17W
Kingstown, St. Vincent 89 D7 13 10N 61 10W
Kingstree, U.S.A. 77 J6 33 40N 79 50W
Kingsville, Canada 78 D2 42 2N 82 45W
Kingsville, U.S.A. 81 M6 27 31N 97 52W
Kingussie, U.K. 12 D4 57 6N 4 2W
Kingwood, U.S.A. 81 K7 29 54N 95 18W
Kınık, Turkey 21 E12 39 6N 27 24 E
Kinistino, Canada 73 C7 52 57N 105 2W
Kinkala, Congo 52 E2 4 18S 14 49 E
Kinki □, Japan 31 H8 33 45N 136 0 E
Kinleith, N.Z. 59 H5 38 20S 175 56 E
Kinmount, Canada 78 B6 44 48N 78 45W
Kinna, Sweden 9 H15 57 32N 12 42 E
Kinnairds Hd., U.K. 12 D6 57 43N 2 1W
Kinnarodden, Norway 6 A11 71 8N 27 40 E
Kinngait = Cape Dorset, Canada 69 B12 64 14N 76 32W
Kino, Mexico 86 B2 28 45N 111 59W
Kinoje →, Canada 70 B3 52 8N 81 25W
Kinomoto, Japan 31 G8 35 30N 136 13 E
Kinoni, Uganda 54 C3 0 41S 30 28 E
Kinoosao, Canada 73 B8 57 5N 102 1W
Kinross, U.K. 12 E5 56 13N 3 25W
Kinsale, Ireland 13 E3 51 42N 8 31W
Kinsale, Old Hd. of, Ireland 13 E3 51 37N 8 33W
Kinsha = Chang Jiang →, China 33 C7 31 48N 121 10 E
Kinshasa, Dem. Rep. of the Congo 52 E3 4 20S 15 15 E
Kinsley, U.S.A. 81 G5 37 55N 99 25W
Kinsman, U.S.A. 78 E4 41 26N 80 35W
Kinston, U.S.A. 77 H7 35 16N 77 35W
Kintore Ra., Australia 60 D4 23 15S 128 47 E
Kintyre, U.K. 12 F3 55 30N 5 35W
Kintyre, Mull of, U.K. 12 F3 55 17N 5 47W
Kinushseo →, Canada 70 A3 55 15N 83 45W
Kinuso, Canada 72 B5 55 20N 115 25W
Kinyangiri, Tanzania 54 C3 4 25S 34 37 E
Kinzua, U.S.A. 78 E6 41 52N 78 58W
Kinzua Dam, U.S.A. 78 E6 41 53N 79 0W
Kiosk, Canada 70 C4 46 6N 78 53W
Kiowa, Kans., U.S.A. 81 G5 37 1N 98 29W
Kiowa, Okla., U.S.A. 81 H7 34 43N 95 54W
Kipahigan L., Canada 73 B8 55 20N 101 55W
Kipanga, Tanzania 54 D4 6 15S 35 20 E
Kiparissia, Greece 21 F9 37 15N 21 40 E
Kiparissiakós Kólpos, Greece 21 F9 37 25N 21 25 E
Kipawa, L., Canada 70 C4 46 50N 79 0W
Kipembawe, Tanzania 54 D3 7 38S 33 27 E
Kipengere Ra., Tanzania 55 D3 9 12S 34 15 E
Kipili, Tanzania 54 D3 7 28S 30 32 E
Kipini, Kenya 54 C5 2 30S 40 32 E
Kipling, Canada 73 C8 50 6N 102 38W
Kippure, Ireland 13 C5 53 11N 6 21W
Kipushi, Dem. Rep. of the Congo 55 E2 11 48S 27 12 E
Kiranomena, Madag. 57 B8 18 17S 46 2 E
Kirensk, Russia 27 D11 57 50N 107 55 E
Kirghizia = Kyrgyzstan ■, Asia 26 E8 42 0N 75 0 E
Kirghizstan = Kyrgyzstan ■, Asia 26 E8 42 0N 75 0 E
Kirgiziya Steppe, Eurasia 25 E10 50 0N 55 0 E
Kiribati ■, Pac. Oc. 64 H10 5 0S 180 0 E
Kırıkkale, Turkey 25 G5 39 51N 33 32 E
Kirillov, Russia 24 C6 59 49N 38 24 E
Kirin = Jilin, China 35 C14 43 44N 126 30 E
Kirinyaga = Kenya, Mt., Kenya 54 C4 0 10S 37 18 E
Kiritimati, Kiribati 65 G12 1 58N 157 27W
Kirkby, U.K. 10 D5 53 30N 2 54W
Kirkby Lonsdale, U.K. 10 C5 54 12N 2 36W
Kirkcaldy, U.K. 12 E5 56 7N 3 9W
Kirkcudbright, U.K. 12 G4 54 50N 4 2W
Kirkee, India 40 K8 18 34N 73 56 E
Kirkenes, Norway 8 B23 69 40N 30 5 E
Kirkfield, Canada 78 B6 44 34N 78 59W
Kirkjubæjarklaustur, Iceland 8 E4 63 47N 18 4W
Kirkkonummi, Finland 9 F21 60 8N 24 26 E
Kirkland Lake, Canada 70 C3 48 9N 80 2W
Kırklareli, Turkey 21 D12 41 44N 27 15 E
Kirksville, U.S.A. 80 E8 40 12N 92 35W
Kirkūk, Iraq 44 C5 35 30N 44 21 E
Kirkwall, U.K. 12 C6 58 59N 2 58W
Kirkwood, S. Africa 56 E4 33 22S 25 15 E
Kirov, Russia 24 C8 58 35N 49 40 E
Kirovabad = Gäncä, Azerbaijan 25 F8 40 45N 46 20 E
Kirovakan = Vanadzor, Armenia 25 F7 40 48N 44 30 E
Kirovograd = Kirovohrad, Ukraine 25 E5 48 35N 32 20 E
Kirovohrad, Ukraine 25 E5 48 35N 32 20 E
Kirovsk = Babadayhan, Turkmenistan 26 F7 37 42N 60 23 E
Kirovsk, Russia 24 A5 67 32N 33 41 E
Kirovskiy, Kamchatka, Russia 27 D16 54 27N 155 42 E
Kirovskiy, Primorsk, Russia 30 B6 45 7N 133 30 E
Kirriemuir, U.K. 12 E5 56 41N 3 1W
Kirsanov, Russia 24 D7 52 35N 42 40 E
Kırşehir, Turkey 25 G5 39 14N 34 5 E
Kirthar Range, Pakistan 42 F2 27 0N 67 0 E
Kirtland, U.S.A. 83 H9 36 44N 108 21W
Kiruna, Sweden 8 C19 67 52N 20 15 E
Kirundu, Dem. Rep. of the Congo 54 C2 0 50S 25 35 E
Kiryū, Japan 31 F9 36 24N 139 20 E
Kisaga, Tanzania 54 C3 4 30S 34 23 E
Kisalaya, Nic. 88 D3 14 40N 84 3W
Kisámou, Kólpos, Greece 23 D5 35 30N 23 38 E
Kisanga, Dem. Rep. of the Congo 54 B2 2 30N 26 35 E
Kisangani, Dem. Rep. of the Congo 54 B2 0 35N 25 15 E
Kisar, Indonesia 37 F7 8 5S 127 10 E
Kisarawe, Tanzania 54 D4 6 53S 39 0 E
Kisarazu, Japan 31 G9 35 23N 139 55 E
Kishanganga →, Pakistan 43 B5 34 18N 73 28 E
Kishanganj, India 43 F13 26 3N 88 14 E
Kishangarh, Raj., India 42 F6 26 34N 74 52 E
Kishangarh, Raj., India 42 F4 27 50N 70 30 E
Kishinev = Chişinău, Moldova 17 E15 47 2N 28 50 E
Kishiwada, Japan 31 G7 34 28N 135 22 E
Kishtwar, India 43 C6 33 20N 75 48 E
Kisii, Kenya 54 C3 0 40S 34 45 E
Kisiju, Tanzania 54 D4 7 23S 39 19 E
Kisizi, Uganda 54 C2 1 0S 29 58 E
Kiskőrös, Hungary 17 E10 46 37N 19 20 E
Kiskunfélegyháza, Hungary 17 E10 46 42N 19 53 E
Kiskunhalas, Hungary 17 E10 46 28N 19 37 E
Kislovodsk, Russia 25 F7 43 50N 42 45 E
Kismayu = Chisimaio, Somali Rep. 49 G8 0 22S 42 32 E
Kiso-Gawa →, Japan 31 G8 35 20N 136 45 E
Kiso-Sammyaku, Japan 31 G8 35 45N 137 45 E
Kisofukushima, Japan 31 G8 35 52N 137 43 E
Kisoro, Uganda 54 C2 1 17S 29 48 E
Kissidougou, Guinea 50 G3 9 5N 10 5W
Kissimmee, U.S.A. 77 L5 28 18N 81 24W
Kissimmee →, U.S.A. 77 M5 27 9N 80 52W
Kississing L., Canada 73 B8 55 10N 101 20W
Kissónerga, Cyprus 23 E11 34 49N 32 24 E
Kisumu, Kenya 54 C3 0 3S 34 45 E
Kiswani, Tanzania 54 C4 4 5S 37 57 E
Kiswere, Tanzania 55 D4 9 27S 39 30 E
Kit Carson, U.S.A. 80 F3 38 46N 102 48W
Kita, Mali 50 F4 13 5N 9 25W
Kitaibaraki, Japan 31 F10 36 50N 140 45 E
Kitakami, Japan 30 E10 39 20N 141 10 E
Kitakami-Gawa →, Japan 30 E10 38 25N 141 19 E
Kitakami-Sammyaku, Japan 30 E10 39 30N 141 30 E
Kitakata, Japan 30 F9 37 39N 139 52 E
Kitakyūshū, Japan 31 H5 33 50N 130 50 E
Kitale, Kenya 54 B4 1 0N 35 0 E
Kitami, Japan 30 C11 43 48N 143 54 E
Kitami-Sammyaku, Japan 30 B11 44 22N 142 43 E
Kitangiri, L., Tanzania 54 C3 4 5S 34 20 E
Kitaya, Tanzania 55 E5 10 38S 40 8 E
Kitchener, Canada 78 C4 43 27N 80 29W
Kitega = Gitega, Burundi 54 C2 3 26S 29 56 E
Kitengo, Dem. Rep. of the Congo 54 D1 7 26S 24 8 E
Kitgum, Uganda 54 B3 3 17N 32 52 E
Kíthira, Greece 21 F10 36 8N 23 0 E
Kíthnos, Greece 21 F11 37 26N 24 27 E
Kiti, Cyprus 23 E12 34 50N 33 34 E
Kiti, C., Cyprus 23 E12 34 48N 33 36 E
Kitimat, Canada 72 C3 54 3N 128 38W
Kitinen →, Finland 8 C22 67 14N 27 27 E
Kitsuki, Japan 31 H5 33 25N 131 37 E
Kittakittaooloo, L., Australia 63 D2 28 3S 138 14 E
Kittanning, U.S.A. 78 F5 40 49N 79 31W
Kittatinny Mts., U.S.A. 79 F10 41 0N 75 0W
Kittery, U.S.A. 77 D10 43 5N 70 45W
Kittilä, Finland 8 C21 67 40N 24 51 E
Kitui, Kenya 54 C4 1 17S 38 0 E
Kitwanga, Canada 72 B3 55 6N 128 4W
Kitwe, Zambia 55 E2 12 54S 28 13 E
Kivarli, India 42 G5 24 33N 72 46 E
Kivertsi, Ukraine 17 C13 50 50N 25 28 E
Kividhes, Cyprus 23 E11 34 46N 32 51 E
Kivu, L., Dem. Rep. of the Congo 54 C2 1 48S 29 0 E
Kiyev = Kyyiv, Ukraine 17 C16 50 30N 30 28 E
Kiyevskoye Vdkhr. = Kyyivske Vdskh., Ukraine 17 C16 51 0N 30 25 E
Kizel, Russia 24 C10 59 3N 57 40 E
Kiziguru, Rwanda 54 C3 1 46S 30 23 E
Kızıl Irmak →, Turkey 25 F6 41 44N 35 58 E
Kizil Jilga, China 43 B8 35 26N 78 50 E
Kızıltepe, Turkey 44 B4 37 12N 40 35 E
Kizimkazi, Tanzania 54 D4 6 28S 39 30 E
Kizlyar, Russia 25 F8 43 51N 46 40 E
Kizyl-Arvat = Gyzylarbat, Turkmenistan 26 F6 39 4N 56 23 E
Kjölur, Iceland 8 D4 64 50N 19 25W
Kladno, Czech Rep. 16 C8 50 10N 14 7 E
Klaeng, Thailand 38 F3 12 47N 101 39 E
Klagenfurt, Austria 16 E8 46 38N 14 20 E
Klaipėda, Lithuania 9 J19 55 43N 21 10 E
Klaksvík, Færoe Is. 8 E9 62 14N 6 35W
Klamath →, U.S.A. 82 F1 41 33N 124 5W
Klamath Falls, U.S.A. 82 E3 42 13N 121 46W
Klamath Mts., U.S.A. 82 F2 41 20N 123 0W
Klamono, Indonesia 37 E8 1 8S 131 30 E
Klappan →, Canada 72 B3 58 0N 129 43W
Klarälven →, Sweden 9 G15 59 23N 13 32 E
Klatovy, Czech Rep. 16 D7 49 23N 13 18 E
Klawer, S. Africa 56 E2 31 44S 18 36 E
Klazienaveen, Neths. 15 B6 52 44N 7 0 E
Kleena Kleene, Canada 72 C4 52 0N 124 59W
Klein-Karas, Namibia 56 D2 27 33S 18 7 E
Klerksdorp, S. Africa 56 D4 26 53S 26 38 E
Kletsk = Klyetsk, Belarus 17 B14 53 5N 26 45 E
Kletskiy, Russia 25 E7 49 16N 43 11 E
Klickitat, U.S.A. 82 D3 45 49N 121 9W
Klickitat →, U.S.A. 84 E5 45 42N 121 17W
Klidhes, Cyprus 23 D13 35 42N 34 36 E
Klinaklini →, Canada 72 C3 51 21N 125 40W
Klip →, S. Africa 57 D4 27 3S 29 3 E
Klipdale, S. Africa 56 E2 34 19S 19 57 E
Klipplaat, S. Africa 56 E3 33 1S 24 22 E
Kłodzko, Poland 17 C9 50 28N 16 38 E
Klouto, Togo 50 G6 6 57N 0 44 E
Kluane L., Canada 68 B6 61 15N 138 40W
Kluane Nat. Park, Canada 72 A1 60 45N 139 30W
Kluczbork, Poland 17 C10 50 58N 18 12 E
Klukwan, U.S.A. 72 B1 59 24N 135 54W
Klyetsk, Belarus 17 B14 53 5N 26 45 E
Klyuchevskaya, Gora, Russia 27 D17 55 50N 160 30 E
Knaresborough, U.K. 10 C6 54 1N 1 28W
Knee L., Man., Canada 70 A1 55 3N 94 45W
Knee L., Sask., Canada 73 B7 55 51N 107 0W
Knight Inlet, Canada 72 C3 50 45N 125 40W
Knighton, U.K. 11 E4 52 21N 3 3W
Knights Ferry, U.S.A. 84 H6 37 50N 120 40W
Knights Landing, U.S.A. 84 G5 38 48N 121 43W
Knob, C., Australia 61 F2 34 32S 119 16 E
Knock, Ireland 13 C3 53 48N 8 55W
Knockmealdown Mts., Ireland 13 D4 52 14N 7 56W
Knokke-Heist, Belgium 15 C3 51 21N 3 17 E
Knossós, Greece 23 D7 35 16N 25 10 E
Knowlton, Canada 79 A12 45 13N 72 31W
Knox, U.S.A. 76 E2 41 18N 86 37W
Knox Coast, Antarctica 5 C8 66 30S 108 0 E
Knoxville, Iowa, U.S.A. 80 E8 41 19N 93 6W
Knoxville, Pa., U.S.A. 78 E7 41 57N 77 27W
Knoxville, Tenn., U.S.A. 77 H4 35 58N 83 55W
Knysna, S. Africa 56 E3 34 2S 23 2 E
Ko Kha, Thailand 38 C2 18 11N 99 24 E
Koartac = Quaqtaq, Canada 69 B13 60 55N 69 40W
Koba, Indonesia 37 F8 6 37S 134 37 E
Kobarid, Slovenia 16 E7 46 15N 13 30 E
Kobayashi, Japan 31 J5 31 56N 130 59 E
Kobdo = Hovd, Mongolia 32 B4 48 2N 91 37 E
Kōbe, Japan 31 G7 34 45N 135 10 E
København, Denmark 9 J15 55 41N 12 34 E
Kōbi-Sho, Japan 31 M1 25 56N 123 41 E
Koblenz, Germany 16 C4 50 21N 7 36 E
Kobryn, Belarus 17 B13 52 15N 24 22 E
Kocaeli, Turkey 25 F4 40 45N 29 50 E
Kočani, Macedonia 21 D10 41 55N 22 25 E
Koch Bihar, India 41 F16 26 22N 89 29 E
Kochang, S. Korea 35 G14 35 41N 127 55 E
Kochas, India 43 G10 25 15N 83 56 E
Kochi = Cochin, India 40 Q10 9 58N 76 20 E
Kōchi, Japan 31 H6 33 30N 133 35 E
Kōchi □, Japan 31 H6 33 40N 133 30 E
Kochiu = Gejiu, China 32 D5 23 20N 103 10 E
Kodarma, India 43 G11 24 28N 85 36 E
Kodiak, U.S.A. 68 C4 57 47N 152 24W
Kodiak I., U.S.A. 68 C4 57 30N 152 45W
Kodinar, India 42 J4 20 46N 70 46 E
Koedoesberge, S. Africa 56 E3 32 40S 20 11 E
Koes, Namibia 56 D2 26 0S 19 15 E
Koffiefontein, S. Africa 56 D4 29 30S 25 0 E
Kofiau, Indonesia 37 E7 1 11S 129 50 E
Koforidua, Ghana 50 G5 6 3N 0 17W
Kōfu, Japan 31 G9 35 40N 138 30 E
Koga, Japan 31 F9 36 11N 139 43 E
Kogaluk →, Canada 71 A7 56 12N 61 44W
Køge, Denmark 9 J15 55 27N 12 11 E
Koh-i-Khurd, Afghan. 42 C1 33 30N 65 59 E
Koh-i-Maran, Pakistan 42 E2 29 18N 66 50 E
Kohat, Pakistan 42 C4 33 40N 71 29 E
Kohima, India 41 G19 25 35N 94 10 E
Kohkīlūyeh va Būyer Aḥmadi □, Iran 45 D6 31 30N 50 30 E
Kohler Ra., Antarctica 5 D15 77 0S 110 0W
Kohlu, Pakistan 42 E3 29 54N 69 15 E
Kohtla-Järve, Estonia 9 G22 59 20N 27 20 E
Koillismaa, Finland 8 D23 65 44N 28 36 E
Koin-dong, N. Korea 35 D14 40 28N 126 18 E
Kojō, N. Korea 35 E14 38 58N 127 58 E
Kojonup, Australia 61 F2 33 48S 117 10 E
Kojūr, Iran 45 B6 36 23N 51 43 E
Kokand = Qūqon, Uzbekistan 26 E8 40 30N 70 57 E
Kokas, Indonesia 37 E8 2 42S 132 26 E
Kokchetav = Kökshetaū, Kazakstan 26 D7 53 20N 69 25 E
Kokemäenjoki →, Finland 9 F19 61 32N 21 44 E
Kokkola, Finland 8 E20 63 50N 23 8 E
Koko Kyunzu, Burma 41 M18 14 10N 93 25 E
Kokomo, U.S.A. 76 E2 40 29N 86 8W
Koksan, N. Korea 35 E14 38 46N 126 40 E
Kökshetaū, Kazakstan 26 D7 53 20N 69 25 E
Koksoak →, Canada 69 C13 58 30N 68 10W
Kokstad, S. Africa 57 E4 30 32S 29 29 E
Kokubu, Japan 31 J5 31 44N 130 46 E
Kola, Indonesia 37 F8 5 35S 134 30 E
Kola, Russia 24 A5 68 45N 33 8 E
Kola Pen. = Kolskiy Poluostrov, Russia 24 A6 67 30N 38 0 E
Kolachi →, Pakistan 42 F2 27 8N 67 2 E
Kolahoi, India 43 B6 34 12N 75 22 E
Kolaka, Indonesia 37 E6 4 3S 121 46 E
Kolar, India 40 N11 13 12N 78 15 E
Kolar Gold Fields, India 40 N11 12 58N 78 16 E
Kolaras, India 42 G6 25 14N 77 36 E
Kolari, Finland 8 C20 67 20N 23 48 E
Kolayat, India 40 F8 27 50N 72 50 E
Kolchugino = Leninsk-Kuznetskiy, Russia 26 D9 54 44N 86 10 E
Kolding, Denmark 9 J13 55 30N 9 29 E
Kolepom = Dolak, Pulau, Indonesia 37 F9 8 0S 138 30 E
Kolguyev, Ostrov, Russia 24 A8 69 20N 48 30 E
Kolhapur, India 40 L9 16 43N 74 15 E
Kolín, Czech Rep. 16 C8 50 2N 15 9 E
Kolkas rags, Latvia 9 H20 57 46N 22 37 E
Kolkata, India 43 H13 22 36N 88 24 E
Kollam = Quilon, India 40 Q10 8 50N 76 38 E
Kollum, Neths. 15 A6 53 17N 6 10 E
Kolmanskop, Namibia 56 D2 26 45S 15 14 E
Köln, Germany 16 C4 50 56N 6 57 E
Koło, Poland 17 B10 52 14N 18 40 E
Kołobrzeg, Poland 16 A8 54 10N 15 35 E
Kolomna, Russia 24 C6 55 8N 38 45 E
Kolomyya, Ukraine 17 D13 48 31N 25 2 E
Kolonodale, Indonesia 37 E6 2 0S 121 19 E
Kolosib, India 41 G18 24 15N 92 45 E
Kolpashevo, Russia 26 D9 58 20N 83 5 E
Kolpino, Russia 24 C5 59 44N 30 39 E
Kolskiy Poluostrov, Russia 24 A6 67 30N 38 0 E
Kolskiy Zaliv, Russia 24 A5 69 23N 34 0 E
Kolwezi, Dem. Rep. of the Congo 55 E2 10 40S 25 25 E
Kolyma →, Russia 27 C17 69 30N 161 0 E
Kolymskoye Nagorye, Russia 27 C16 63 0N 157 0 E
Kôm Ombo, Egypt 51 D12 24 25N 32 52 E
Komandorskiye Is. = Komandorskiye Ostrova, Russia 27 D17 55 0N 167 0 E
Komandorskiye Ostrova, Russia 27 D17 55 0N 167 0 E
Komárno, Slovak Rep. 17 E10 47 49N 18 5 E
Komatipoort, S. Africa 57 D5 25 25S 31 55 E
Komatou Yialou, Cyprus 23 D13 35 25N 34 8 E
Komatsu, Japan 31 F8 36 25N 136 30 E
Komatsushima, Japan 31 H7 34 0N 134 35 E
Komi □, Russia 24 B10 64 0N 55 0 E
Kommunarsk = Alchevsk, Ukraine 25 E6 48 30N 38 45 E
Kommunizma, Pik, Tajikistan 26 F8 39 0N 72 2 E
Komodo, Indonesia 37 F5 8 37S 119 20 E
Komoran, Pulau, Indonesia 37 F9 8 18S 138 45 E
Komoro, Japan 31 F9 36 19N 138 26 E
Komotini, Greece 21 D11 41 9N 25 26 E
Kompasberg, S. Africa 56 E3 31 45S 24 32 E
Kompong Bang, Cambodia 39 F5 12 24N 104 40 E
Kompong Cham, Cambodia 39 F5 12 0N 105 30 E
Kompong Chhnang = Kampang Chhnang, Cambodia 39 F5 12 20N 104 35 E
Kompong Chikreng, Cambodia 38 F5 13 5N 104 18 E
Kompong Kleang, Cambodia 38 F5 13 6N 104 8 E
Kompong Luong, Cambodia 39 G5 11 49N 104 48 E
Kompong Pranak, Cambodia 38 F5 13 35N 104 55 E
Kompong Som = Kampong Saom, Cambodia 39 G4 10 38N 103 30 E
Kompong Som, Chhung = Kampong Saom, Chaak, Cambodia 39 G4 10 50N 103 32 E
Kompong Speu, Cambodia 39 G5 11 26N 104 32 E
Kompong Sralao, Cambodia 38 E5 14 5N 105 46 E
Kompong Thom, Cambodia 38 F5 12 35N 104 51 E
Kompong Trabeck, Cambodia 38 F5 13 6N 105 14 E
Kompong Trabeck, Cambodia 39 G5 11 9N 105 28 E
Kompong Trach, Cambodia 39 G5 11 25N 105 48 E
Kompong Tralach, Cambodia 39 G5 11 54N 104 47 E
Komrat = Comrat, Moldova 17 E15 46 18N 28 40 E
Komsberg, S. Africa 56 E3 32 40S 20 45 E
Komsomolets, Ostrov, Russia 27 A10 80 30N 95 0 E
Komsomolsk, Russia 27 D14 50 30N 137 0 E
Kon Tum, Vietnam 38 E7 14 24N 108 0 E
Kon Tum, Plateau du, Vietnam 38 E7 14 30N 108 30 E
Konarhā □, Afghan. 40 B7 34 30N 71 3 E
Konāri, Iran 45 D6 28 13N 51 36 E
Konch, India 43 G8 26 0N 79 10 E
Konde, Tanzania 54 C4 4 57S 39 45 E
Kondinin, Australia 61 F2 32 34S 118 8 E
Kondoa, Tanzania 54 C4 4 55S 35 50 E
Kondókali, Greece 23 A3 39 38N 19 51 E
Kondopaga, Russia 24 B5 62 12N 34 17 E
Kondratyevo, Russia 27 D10 57 22N 98 15 E

Köneürgench

Lanak La

Kut, Ko, *Thailand* .... **39 G4** 11 40N 102 35 E
Kütahya, *Turkey* .... **25 G5** 39 30N 30 2 E
Kutaisi, *Georgia* .... **25 F7** 42 19N 42 40 E
Kutaraja = Banda Aceh, *Indonesia* .... **36 C1** 5 35N 95 20 E
Kutch, Gulf of = Kachchh, Gulf of, *India* .... **42 H3** 22 50N 69 15 E
Kutch, Rann of = Kachchh, Rann of, *India* .... **42 H4** 24 0N 70 0 E
Kutiyana, *India* .... **42 J4** 21 36N 70 2 E
Kutno, *Poland* .... **17 B10** 52 15N 19 23 E
Kutse, *Botswana* .... **56 C3** 21 7S 22 16 E
Kutu, *Dem. Rep. of the Congo* **52 E3** 2 40S 18 11 E
Kutum, *Sudan* .... **51 F10** 14 10N 24 40 E
Kuujjuaq, *Canada* **69 C13** 58 6N 68 15W
Kuujjuarapik, *Canada* **70 A4** 55 20N 77 35W
Kuŭp-tong, *N. Korea* .... **35 D14** 40 45N 126 1 E
Kuusamo, *Finland* .... **8 D23** 65 57N 29 8 E
Kuusankoski, *Finland* .... **9 F22** 60 55N 26 38 E
Kuwait = Al Kuwayt, *Kuwait* . **46 B4** 29 30N 48 0 E
Kuwait ■, *Asia* .... **46 B4** 29 30N 47 30 E
Kuwana, *Japan* .... **31 G8** 35 5N 136 43 E
Kuwana →, *India* .... **43 F10** 26 25N 83 15 E
Kuybyshev = Samara, *Russia* **24 D9** 53 8N 50 6 E
Kuybyshev, *Russia* .... **26 D8** 55 27N 78 19 E
Kuybyshevskoye Vdkhr., *Russia* .... **24 C8** 55 2N 49 30 E
Kuye He →, *China* .... **34 E6** 38 23N 110 46 E
Kūyeh, *Iran* .... **44 B5** 38 45N 47 57 E
Kūysanjaq, *Iraq* .... **44 B5** 36 5N 44 38 E
Kuyto, Ozero, *Russia* .... **24 B5** 65 6N 31 20 E
Kuyumba, *Russia* .... **27 C10** 60 58N 96 59 E
Kuzey Anadolu Dağları, *Turkey* .... **25 F6** 41 30N 35 0 E
Kuznetsk, *Russia* .... **24 D8** 53 12N 46 40 E
Kuzomen, *Russia* .... **24 A6** 66 22N 36 50 E
Kvænangen, *Norway* .... **8 A19** 70 5N 21 15 E
Kvaløy, *Norway* .... **8 B18** 69 40N 18 30 E
Kvarner, *Croatia* .... **16 F8** 44 50N 14 10 E
Kvarnerič, *Croatia* .... **16 F8** 44 43N 14 37 E
Kwa-Nobuhle, *S. Africa* .... **53 L5** 33 50S 25 22 E
Kwabhaca, *S. Africa* .... **57 E4** 30 51S 29 0 E
Kwakhanai, *Botswana* .... **56 C3** 21 39S 21 16 E
Kwakoegron, *Surinam* **93 B7** 5 12N 55 25W
Kwale, *Kenya* .... **54 C4** 4 15S 39 31 E
KwaMashu, *S. Africa* .... **57 D5** 29 45S 30 58 E
Kwando →, *Africa* .... **56 B3** 18 27S 23 32 E
Kwangdaeri, *N. Korea* .... **35 D14** 40 31N 127 32 E
Kwangju, *S. Korea* .... **35 G14** 35 9N 126 54 E
Kwango →, *Dem. Rep. of the Congo* .... **52 E3** 3 14S 17 22 E
Kwangsi-Chuang = Guangxi Zhuangzu Zizhiqu □, *China* **33 D5** 24 0N 109 0 E
Kwangtung = Guangdong □, *China* .... **33 D6** 23 0N 113 0 E
Kwataboahegan →, *Canada* **70 B3** 51 9N 80 50W
Kwatisore, *Indonesia* .... **37 E8** 3 18S 134 50 E
KwaZulu Natal □, *S. Africa* . **57 D5** 29 0S 30 0 E
Kweichow = Guizhou □, *China* .... **32 D5** 27 0N 107 0 E
Kwekwe, *Zimbabwe* .... **55 F2** 18 58S 29 48 E
Kwidzyn, *Poland* .... **17 B10** 53 44N 18 55 E
Kwinana New Town, *Australia* .... **61 F2** 32 15S 115 47 E
Kwoka, *Indonesia* .... **37 E8** 0 31S 132 27 E
Kyabra Cr. →, *Australia* .... **63 D3** 25 36S 142 55 E
Kyabram, *Australia* .... **63 F4** 36 19S 145 4 E
Kyaikto, *Burma* .... **38 D1** 17 20N 97 3 E
Kyakhta, *Russia* .... **27 D11** 50 30N 106 25 E
Kyancutta, *Australia* .... **63 E2** 33 8S 135 33 E
Kyaukpadaung, *Burma* .... **41 J19** 20 52N 95 8 E
Kyaukpyu, *Burma* .... **41 K18** 19 28N 93 30 E
Kyaukse, *Burma* .... **41 J20** 21 36N 96 10 E
Kyburz, *U.S.A.* **84 G6** 38 47N 120 18W
Kyelang, *India* .... **42 C7** 32 35N 77 2 E
Kyenjojo, *Uganda* .... **54 B3** 0 40N 30 37 E
Kyle, *Canada* **73 C7** 50 50N 108 2W
Kyle Dam, *Zimbabwe* .... **55 G3** 20 15S 31 0 E
Kyle of Lochalsh, *U.K.* **12 D3** 57 17N 5 44W
Kymijoki →, *Finland* .... **9 F22** 60 30N 26 55 E
Kyneton, *Australia* .... **63 F3** 37 10S 144 29 E
Kynuna, *Australia* .... **62 C3** 21 37S 141 55 E
Kyō-ga-Saki, *Japan* .... **31 G7** 35 45N 135 15 E
Kyoga, L., *Uganda* .... **54 B3** 1 35N 33 0 E
Kyogle, *Australia* .... **63 D5** 28 40S 153 0 E
Kyongju, *S. Korea* .... **35 G15** 35 51N 129 14 E
Kyongpyaw, *Burma* .... **41 L19** 17 12N 95 10 E
Kyŏngsŏng, *N. Korea* .... **35 D15** 41 35N 129 36 E
Kyōto, *Japan* .... **31 G7** 35 0N 135 45 E
Kyōto □, *Japan* .... **31 G7** 35 15N 135 45 E
Kyparissovouno, *Cyprus* .... **23 D12** 35 19N 33 10 E
Kyperounda, *Cyprus* .... **23 E11** 34 56N 32 58 E
Kyrenia, *Cyprus* .... **23 D12** 35 20N 33 20 E
Kyrgyzstan ■, *Asia* .... **26 E8** 42 0N 75 0 E
Kyrönjoki →, *Finland* .... **8 E19** 63 14N 21 45 E
Kystatyam, *Russia* .... **27 C13** 67 20N 123 10 E
Kythréa, *Cyprus* .... **23 D12** 35 15N 33 29 E
Kyunhla, *Burma* .... **41 H19** 23 25N 95 15 E
Kyuquot Sound, *Canada* **72 D3** 50 2N 127 22W
Kyūshū, *Japan* .... **31 H5** 33 0N 131 0 E
Kyūshū □, *Japan* .... **31 H5** 33 0N 131 0 E
Kyūshū-Sanchi, *Japan* .... **31 H5** 32 35N 131 17 E
Kyustendil, *Bulgaria* .... **21 C10** 42 16N 22 41 E
Kyusyur, *Russia* .... **27 B13** 70 19N 127 30 E
Kyyiv, *Ukraine* .... **17 C16** 50 30N 30 28 E
Kyyivske Vdskh., *Ukraine* .... **17 C16** 51 0N 30 25 E
Kyzyl, *Russia* .... **27 D10** 51 50N 94 30 E
Kyzyl Kum, *Uzbekistan* .... **26 E7** 42 30N 65 0 E
Kyzyl-Kyya, *Kyrgyzstan* .... **26 E8** 40 16N 72 8 E
Kzyl-Orda = Qyzylorda, *Kazakstan* .... **26 E7** 44 48N 65 28 E

L

La Alcarria, *Spain* **19 B4** 40 31N 2 45W
La Asunción, *Venezuela* **92 A6** 11 2N 63 53W
La Baie, *Canada* **71 C5** 48 19N 70 53W
La Banda, *Argentina* **94 B3** 27 45S 64 10W
La Barca, *Mexico* **86 C4** 20 20N 102 40W
La Barge, *U.S.A.* **82 E8** 42 16N 110 12W
La Belle, *U.S.A.* **77 M5** 26 46N 81 26W
La Biche →, *Canada* **72 B4** 59 57N 123 50W
La Biche, L., *Canada* **72 C6** 54 50N 112 5W
La Bomba, *Mexico* **86 A1** 31 53N 115 2W
La Calera, *Chile* **94 C1** 32 50S 71 10W
La Canal = Sa Canal, *Spain* . **22 C7** 38 51N 1 23 E
La Carlota, *Argentina* **94 C3** 33 30S 63 20W
La Ceiba, *Honduras* **88 C2** 15 40N 86 50W
La Chaux-de-Fonds, *Switz.* . . **18 C7** 47 7N 6 50 E
La Chorrera, *Panama* **88 E4** 8 53N 79 47W
La Cocha, *Argentina* **94 B2** 27 50S 65 40W
La Concepción, *Panama* **88 E3** 8 31N 82 37W
La Concordia, *Mexico* **87 D6** 16 8N 92 38W
La Coruña = A Coruña, *Spain* **19 A1** 43 20N 8 25W
La Crescent, *U.S.A.* **80 D9** 43 50N 91 18W
La Crete, *Canada* **72 B5** 58 11N 116 24W
La Crosse, *Kans., U.S.A.* **80 F5** 38 32N 99 18W
La Crosse, *Wis., U.S.A.* **80 D9** 43 48N 91 15W
La Cruz, *Costa Rica* **88 D2** 11 4N 85 39W
La Cruz, *Mexico* **86 C3** 23 55N 106 54W
La Désirade, *Guadeloupe* **89 C7** 16 18N 61 3W
La Escondida, *Mexico* **86 C5** 24 6N 99 55W
La Esmeralda, *Paraguay* **94 A3** 22 16S 62 33W
La Esperanza, *Cuba* **88 B3** 22 46N 83 44W
La Esperanza, *Honduras* **88 D2** 14 15N 88 10W
La Estrada = A Estrada, *Spain* **19 A1** 42 43N 8 27W
La Fayette, *U.S.A.* **77 H3** 34 42N 85 17W
La Fé, *Cuba* **88 B3** 22 2N 84 15W
La Follette, *U.S.A.* **77 G3** 36 23N 84 7W
La Grande, *U.S.A.* **82 D4** 45 20N 118 5W
La Grande →, *Canada* **70 B5** 53 50N 79 0W
La Grande Deux, Rés., *Canada* **70 B4** 53 40N 76 55W
La Grande Quatre, Rés., *Canada* **70 B5** 54 0N 73 15W
La Grande Trois, Rés., *Canada* **70 B4** 53 40N 75 10W
La Grange, *Calif., U.S.A.* **84 H6** 37 42N 120 27W
La Grange, *Ga., U.S.A.* **77 J3** 33 2N 85 2W
La Grange, *Ky., U.S.A.* **76 F3** 38 25N 85 23W
La Grange, *Tex., U.S.A.* **81 L6** 29 54N 96 52W
La Guaira, *Venezuela* **92 A5** 10 36N 66 56W
La Habana, *Cuba* **88 B3** 23 8N 82 22W
La Independencia, *Mexico* .. **87 D6** 16 31N 91 47W
La Isabela, *Dom. Rep.* **89 C5** 19 58N 71 2W
La Junta, *U.S.A.* **81 F3** 37 59N 103 33W
La Laguna, *Canary Is.* **22 F3** 28 28N 16 18W
La Libertad, *Guatemala* **88 C1** 16 47N 90 7W
La Libertad, *Mexico* **86 B2** 29 55N 112 41W
La Ligua, *Chile* **94 C1** 32 30S 71 16W
La Línea de la Concepción, *Spain* **19 D3** 36 15N 5 23W
La Loche, *Canada* **73 B7** 56 29N 109 26W
La Louvière, *Belgium* .... **15 D4** 50 27N 4 10 E
La Malbaie, *Canada* **71 C5** 47 40N 70 10W
La Mancha, *Spain* **19 C4** 39 10N 2 54W
La Martre, L., *Canada* **72 A5** 63 15N 117 55W
La Mesa, *U.S.A.* **85 N9** 32 46N 117 3W
La Misión, *Mexico* **86 A1** 32 5N 116 50W
La Moure, *U.S.A.* **80 B5** 46 21N 98 18W
La Negra, *Chile* **94 A1** 23 46S 70 18W
La Oliva, *Canary Is.* **22 F6** 28 36N 13 57W
La Orotava, *Canary Is.* **22 F3** 28 22N 16 31W
La Oroya, *Peru* **92 F3** 11 32S 75 54W
La Palma, *Canary Is.* **22 F2** 28 40N 17 50W
La Palma, *Panama* **88 E4** 8 15N 78 0W
La Palma del Condado, *Spain* **19 D2** 37 21N 6 38W
La Paloma, *Chile* **94 C1** 30 35S 71 0W
La Pampa □, *Argentina* **94 D2** 36 50S 66 0W
La Paragua, *Venezuela* **92 B6** 6 50N 63 20W
La Paz, *Entre Ríos, Argentina* **94 C4** 30 50S 59 45W
La Paz, *San Luis, Argentina* . **94 C2** 33 30S 67 20W
La Paz, *Bolivia* **92 G5** 16 20S 68 10W
La Paz, *Honduras* **88 D2** 14 20N 87 47W
La Paz, *Mexico* **86 C2** 24 10N 110 20W
La Paz Centro, *Nic.* **88 D2** 12 20N 86 41W
La Pedrera, *Colombia* **92 D5** 1 18S 69 43W
La Pérade, *Canada* **71 C5** 46 35N 72 12W
La Perouse Str., *Asia* .... **30 B11** 45 40N 142 0 E
La Pesca, *Mexico* **87 C5** 23 46N 97 47W
La Piedad, *Mexico* **86 C4** 20 20N 102 1W
La Pine, *U.S.A.* **82 E3** 43 40N 121 30W
La Plata, *Argentina* **94 D4** 35 0S 57 55W
La Pocatière, *Canada* **71 C5** 47 22N 70 2W
La Porte, *Ind., U.S.A.* **76 E2** 41 36N 86 43W
La Porte, *Tex., U.S.A.* **81 L7** 29 39N 95 1W
La Purísima, *Mexico* **86 B2** 26 10N 112 4W
La Push, *U.S.A.* **84 C2** 47 55N 124 38W
La Quiaca, *Argentina* **94 A2** 22 5S 65 35W
La Restinga, *Canary Is.* **22 G2** 27 38N 17 59W
La Rioja, *Argentina* **94 B2** 29 20S 67 0W
La Rioja □, *Argentina* **94 B2** 29 30S 67 0W
La Rioja □, *Spain* **19 A4** 42 20N 2 20W
La Robla, *Spain* **19 A3** 42 50N 5 41W
La Roche-en-Ardenne, *Belgium* .... **15 D5** 50 11N 5 35 E
La Roche-sur-Yon, *France* **18 C3** 46 40N 1 25W
La Rochelle, *France* **18 C3** 46 10N 1 9W
La Roda, *Spain* **19 C4** 39 13N 2 15W
La Romana, *Dom. Rep.* **89 C6** 18 27N 68 57W
La Ronge, *Canada* **73 B7** 55 5N 105 20W
La Rumorosa, *Mexico* **85 N10** 32 33N 116 4W
La Sabina = Sa Savina, *Spain* **22 C7** 38 44N 1 25 E
La Salle, *U.S.A.* **80 E10** 41 20N 89 6W
La Santa, *Canary Is.* **22 E6** 29 5N 13 40W
La Sarre, *Canada* **70 C4** 48 45N 79 15W
La Scie, *Canada* **71 C8** 49 57N 55 36W
La Selva Beach, *U.S.A.* **84 J5** 36 56N 121 51W
La Serena, *Chile* **94 B1** 29 55S 71 10W
La Seu d'Urgell, *Spain* .... **19 A6** 42 22N 1 23 E
La Seyne-sur-Mer, *France* .... **18 E6** 43 7N 5 52 E
La Soufrière, *St. Vincent* **89 D7** 13 20N 61 11W
La Spézia, *Italy* .... **18 D8** 44 7N 9 50 E
La Tagua, *Colombia* **92 C4** 0 3N 74 40W
La Tortuga, *Venezuela* **89 D6** 11 0N 65 22W
La Tuque, *Canada* **70 C5** 47 30N 72 50W
La Unión, *Chile* **96 E2** 40 10S 73 0W
La Unión, *El Salv.* **88 D2** 13 20N 87 50W
La Unión, *Mexico* **86 D4** 17 58N 101 49W
La Urbana, *Venezuela* **92 B5** 7 8N 66 56W
La Vall d'Uixó, *Spain* **19 C5** 39 49N 0 15W
La Vega, *Dom. Rep.* **89 C5** 19 20N 70 30W
La Vela de Coro, *Venezuela* .. **92 A5** 11 27N 69 34W
La Venta, *Mexico* **87 D6** 18 8N 94 3W
La Ventura, *Mexico* **86 C4** 24 38N 100 54W
Laas Caanood = Las Anod, *Somali Rep.* .... **46 F4** 8 26N 47 19 E
Labasa, *Fiji* .... **59 C8** 16 30S 179 27 E
Labe = Elbe →, *Europe* .... **16 B5** 53 50N 9 0 E
Labé, *Guinea* **50 F3** 11 24N 12 16W
Laberge, L., *Canada* **72 A1** 61 11N 135 12W
Labinsk, *Russia* .... **25 F7** 44 40N 40 48 E
Labis, *Malaysia* .... **39 L4** 2 22N 103 2 E
Laboulaye, *Argentina* **94 C3** 34 10S 63 30W
Labrador, *Canada* **71 B7** 53 20N 61 0W
Labrador City, *Canada* **71 B6** 52 57N 66 55W
Labrador Sea, *Atl. Oc.* **69 C14** 57 0N 54 0W
Lábrea, *Brazil* **92 E6** 7 15S 64 51W
Labuan, *Malaysia* .... **36 C5** 5 20N 115 14 E
Labuan, Pulau, *Malaysia* .... **36 C5** 5 21N 115 13 E
Labuha, *Indonesia* .... **37 E7** 0 30S 127 30 E
Labuhan, *Indonesia* .... **37 G11** 6 22S 105 50 E
Labuhanbajo, *Indonesia* .... **37 F6** 8 28S 119 54 E
Labuk, Telok, *Malaysia* .... **36 C5** 6 10N 117 50 E
Labyrinth, L., *Australia* .... **63 E2** 30 40S 135 11 E
Labytnangi, *Russia* .... **26 C7** 66 39N 66 21 E
Lac Bouchette, *Canada* **71 C5** 48 16N 72 11W
Lac Édouard, *Canada* **70 C5** 47 40N 72 16W
Lac La Biche, *Canada* **72 C6** 54 45N 111 58W
Lac la Martre = Wha Ti, *Canada* **68 B8** 63 8N 117 16W
Lac La Ronge Prov. Park, *Canada* **73 B7** 55 9N 104 41W
Lac-Mégantic, *Canada* **71 C5** 45 35N 70 53W
Lac Thien, *Vietnam* .... **38 F7** 12 25N 108 11 E
Lacanau, *France* **18 D3** 44 58N 1 5W
Lacantúm →, *Mexico* **87 D6** 16 36N 90 40W
Laccadive Is. = Lakshadweep Is., *India* .... **29 H11** 10 0N 72 30 E
Lacepede B., *Australia* .... **63 F2** 36 40S 139 40 E
Lacepede Is., *Australia* .... **60 C3** 16 55S 122 0 E
Lacerdónia, *Mozam.* .... **55 F4** 18 3S 35 35 E
Lacey, *U.S.A.* **84 C4** 47 7N 122 49W
Lachhmangarh, *India* .... **42 F6** 27 50N 75 4 E
Lachi, *Pakistan* .... **42 C4** 33 25N 71 20 E
Lachine, *Canada* **79 A11** 45 30N 73 40W
Lachlan →, *Australia* .... **63 E3** 34 22S 143 55 E
Lachute, *Canada* **70 C5** 45 39N 74 21W
Lackawanna, *U.S.A.* **78 D6** 42 50N 78 50W
Lackawaxen, *U.S.A.* **79 E10** 41 29N 74 59W
Lacolle, *Canada* **79 A11** 45 5N 73 22W
Lacombe, *Canada* **72 C6** 52 30N 113 44W
Lacona, *U.S.A.* **79 C8** 43 39N 76 10W
Laconia, *U.S.A.* **79 C13** 43 32N 71 28W
Ladakh Ra., *India* .... **43 C8** 34 0N 78 0 E
Ladismith, *S. Africa* .... **56 E3** 33 28S 21 15 E
Lādīz, *Iran* .... **45 D9** 28 55N 61 15 E
Ladnun, *India* .... **42 F6** 27 38N 74 25 E
Ladoga, L. = Ladozhskoye Ozero, *Russia* .... **24 B5** 61 15N 30 30 E
Ladozhskoye Ozero, *Russia* . **24 B5** 61 15N 30 30 E
Lady Elliott I., *Australia* .... **62 C5** 24 7S 152 42 E
Lady Grey, *S. Africa* .... **56 E4** 30 43S 27 13 E
Ladybrand, *S. Africa* .... **56 D4** 29 9S 27 29 E
Ladysmith, *Canada* **72 D4** 49 0N 123 49W
Ladysmith, *S. Africa* .... **57 D4** 28 32S 29 46 E
Ladysmith, *U.S.A.* **80 C9** 45 28N 91 12W
Lae, *Papua N. G.* .... **64 H6** 6 40S 147 2 E
Laem Ngop, *Thailand* .... **39 F4** 12 10N 102 26 E
Laem Pho, *Thailand* .... **39 J3** 6 55N 101 19 E
Læsø, *Denmark* .... **9 H14** 57 15N 11 5 E
Lafayette, *Colo., U.S.A.* **80 F2** 39 58N 105 12W
Lafayette, *Ind., U.S.A.* **76 E2** 40 25N 86 54W
Lafayette, *La., U.S.A.* **81 K9** 30 14N 92 1W
Lafayette, *Tenn., U.S.A.* **77 G2** 36 31N 86 2W
Laferte →, *Canada* **72 A5** 61 53N 117 44W
Lafia, *Nigeria* .... **50 G7** 8 30N 8 34 E
Lafleche, *Canada* **73 D7** 49 45N 106 40W
Lagan →, *U.K.* **13 B6** 54 36N 5 55W
Lagarfljót →, *Iceland* **8 D6** 65 40N 14 18W
Lågen →, *Oppland, Norway* .. **9 F14** 61 8N 10 25 E
Lågen →, *Vestfold, Norway* . **9 G14** 59 3N 10 3 E
Laghouat, *Algeria* .... **50 B6** 33 50N 2 59 E
Lagoa Vermelha, *Brazil* **95 B5** 28 13S 51 32W
Lagonoy G., *Phil.* .... **37 B6** 13 35N 123 50 E
Lagos, *Nigeria* .... **50 G6** 6 25N 3 27 E
Lagos, *Portugal* **19 D1** 37 5N 8 41W
Lagos de Moreno, *Mexico* **86 C4** 21 21N 101 55W
Lagrange, *Australia* .... **60 C3** 18 45S 121 43 E
Lagrange B., *Australia* .... **60 C3** 18 38S 121 42 E
Laguna, *Brazil* **95 B6** 28 30S 48 50W
Laguna, *U.S.A.* **83 J10** 35 2N 107 25W
Laguna Beach, *U.S.A.* **85 M9** 33 33N 117 47W
Laguna Limpia, *Argentina* **94 B4** 26 32S 59 45W
Lagunas, *Chile* **94 A2** 21 0S 69 45W
Lagunas, *Peru* **92 E3** 5 10S 75 35W
Lahad Datu, *Malaysia* .... **37 C5** 5 0N 118 20 E
Lahad Datu, Teluk, *Malaysia* **37 D5** 4 50N 118 20 E
Lahan Sai, *Thailand* .... **38 E4** 14 25N 102 52 E
Lahanam, *Laos* .... **38 D5** 16 16N 105 16 E
Lahar, *India* .... **43 F8** 26 12N 78 57 E
Laharpur, *India* .... **43 F9** 27 43N 80 56 E
Lahat, *Indonesia* .... **36 E2** 3 45S 103 30 E
Lahewa, *Indonesia* .... **36 D1** 1 22N 97 12 E
Lāhījān, *Iran* .... **45 B6** 37 10N 50 6 E
Lahn →, *Germany* .... **16 C4** 50 19N 7 37 E
Laholm, *Sweden* .... **9 H15** 56 30N 13 2 E
Lahore, *Pakistan* .... **42 D6** 31 32N 74 22 E
Lahri, *Pakistan* .... **42 E3** 29 11N 68 13 E
Lahti, *Finland* .... **9 F21** 60 58N 25 40 E
Lahtis = Lahti, *Finland* .... **9 F21** 60 58N 25 40 E
Laï, *Chad* .... **51 G9** 9 25N 16 18 E
Lai Chau, *Vietnam* .... **38 A4** 22 5N 103 3 E
Laila = Layla, *Si. Arabia* .... **46 C4** 22 10N 46 40 E
Laingsburg, *S. Africa* .... **56 E3** 33 9S 20 52 E
Lainio älv →, *Sweden* .... **8 C20** 67 35N 22 40 E
Lairg, *U.K.* **12 C4** 58 2N 4 24W
Laishui, *China* .... **34 E8** 39 23N 115 45 E
Laiwu, *China* .... **35 F9** 36 15N 117 40 E
Laixi, *China* .... **35 F11** 36 50N 120 31 E
Laiyang, *China* .... **35 F11** 36 59N 120 45 E
Laiyuan, *China* .... **34 E8** 39 20N 114 40 E
Laizhou, *China* .... **35 F10** 37 8N 119 57 E
Laizhou Wan, *China* .... **35 F10** 37 30N 119 30 E
Laja →, *Mexico* **86 C4** 20 55N 100 46W
Lajes, *Brazil* **95 B5** 27 48S 50 20W
Lak Sao, *Laos* .... **38 C5** 18 11N 104 59 E
Lakaband, *Pakistan* .... **42 D3** 31 2N 69 15 E
Lake Alpine, *U.S.A.* **84 G7** 38 29N 120 0W
Lake Andes, *U.S.A.* **80 D5** 43 9N 98 32W
Lake Arthur, *U.S.A.* **81 K8** 30 5N 92 41W
Lake Cargelligo, *Australia* .. **63 E4** 33 15S 146 22 E
Lake Charles, *U.S.A.* **81 K8** 30 14N 93 13W
Lake City, *Colo., U.S.A.* **83 G10** 38 2N 107 19W
Lake City, *Fla., U.S.A.* **77 K4** 30 11N 82 38W
Lake City, *Mich., U.S.A.* **76 C3** 44 20N 85 13W
Lake City, *Minn., U.S.A.* **80 C8** 44 27N 92 16W
Lake City, *Pa., U.S.A.* **78 D4** 42 1N 80 21W
Lake City, *S.C., U.S.A.* **77 J6** 33 52N 79 45W
Lake Cowichan, *Canada* **72 D4** 48 49N 124 3W
Lake District, *U.K.* .... **10 C4** 54 35N 3 20 E
Lake Elsinore, *U.S.A.* **85 M9** 33 38N 117 20W
Lake George, *U.S.A.* **79 C11** 43 26N 73 43W
Lake Grace, *Australia* .... **61 F2** 33 7S 118 28 E
Lake Harbour = Kimmirut, *Canada* **69 B13** 62 50N 69 50W
Lake Havasu City, *U.S.A.* **85 L12** 34 27N 114 22W
Lake Hughes, *U.S.A.* **85 L8** 34 41N 118 26W
Lake Isabella, *U.S.A.* **85 K8** 35 38N 118 28W
Lake Jackson, *U.S.A.* **81 L7** 29 3N 95 27W
Lake Junction, *U.S.A.* **82 D8** 44 35N 110 28W
Lake King, *Australia* .... **61 F2** 33 5S 119 45 E
Lake Lenore, *Canada* **73 C8** 52 24N 104 59W
Lake Louise, *Canada* **72 C5** 51 30N 116 10W
Lake Mead Nat. Recr. Area, *U.S.A.* **85 K12** 36 15N 114 30W
Lake Mills, *U.S.A.* **80 D8** 43 25N 93 32W
Lake Placid, *U.S.A.* **79 B11** 44 17N 73 59W
Lake Pleasant, *U.S.A.* **79 C10** 43 28N 74 25W
Lake Providence, *U.S.A.* **81 J9** 32 48N 91 10W
Lake St. Peter, *Canada* **78 A6** 45 18N 78 2W
Lake Superior Prov. Park, *Canada* **70 C3** 47 45N 84 45W
Lake Village, *U.S.A.* **81 J9** 33 20N 91 17W
Lake Wales, *U.S.A.* **77 M5** 27 54N 81 35W
Lake Worth, *U.S.A.* **77 M5** 26 37N 80 3W
Lakeba, *Fiji* **59 D9** 18 13S 178 47W
Lakefield, *Canada* **78 B6** 44 25N 78 16W
Lakehurst, *U.S.A.* **79 F10** 40 1N 74 19W
Lakeland, *Australia* .... **62 B3** 15 49S 144 57 E
Lakeland, *U.S.A.* **77 M5** 28 3N 81 57W
Lakemba = Lakeba, *Fiji* **59 D9** 18 13S 178 47W
Lakeport, *Calif., U.S.A.* **84 F4** 39 3N 122 55W
Lakeport, *Mich., U.S.A.* **78 C2** 43 7N 82 30W
Lakes Entrance, *Australia* .... **63 F4** 37 50S 148 0 E
Lakeside, *Ariz., U.S.A.* **83 J9** 34 9N 109 58W
Lakeside, *Calif., U.S.A.* **85 N10** 32 52N 116 55W
Lakeside, *Nebr., U.S.A.* **80 D3** 42 3N 102 26W
Lakeside, *Ohio, U.S.A.* **78 E2** 41 32N 82 46W
Lakeview, *U.S.A.* **82 E3** 42 11N 120 21W
Lakeville, *U.S.A.* **80 C8** 44 39N 93 14W
Lakewood, *Colo., U.S.A.* **80 F2** 39 44N 105 5W
Lakewood, *N.J., U.S.A.* **79 F10** 40 6N 74 13W
Lakewood, *N.Y., U.S.A.* **78 D5** 42 6N 79 19W
Lakewood, *Ohio, U.S.A.* **78 E3** 41 29N 81 48W
Lakewood, *Wash., U.S.A.* **84 C4** 47 11N 122 32W
Lakha, *India* .... **42 F4** 26 9N 70 54 E
Lakhaniá, *Greece* .... **23 D9** 35 58N 27 54 E
Lakhimpur, *India* .... **43 F9** 27 57N 80 46 E
Lakhnadon, *India* .... **43 H8** 22 36N 79 36 E
Lakhonpheng, *Laos* .... **38 E5** 15 54N 105 34 E
Lakhpat, *India* .... **42 H3** 23 48N 68 47 E
Lakin, *U.S.A.* **81 G4** 37 57N 101 15W
Lakitusaki →, *Canada* **70 B3** 54 21N 82 25W
Lakki, *Pakistan* .... **42 C4** 32 36N 70 55 E
Lákkoi, *Greece* .... **23 D5** 35 24N 23 57 E
Lakonikós Kólpos, *Greece* .. **21 F10** 36 40N 22 40 E
Lakor, *Indonesia* .... **37 F7** 8 15S 128 17 E
Lakota, *Ivory C.* **50 G4** 5 50N 5 30W
Lakota, *U.S.A.* **80 A5** 48 2N 98 21W
Laksar, *India* .... **42 E8** 29 46N 78 3 E
Laksefjorden, *Norway* .... **8 A22** 70 45N 26 50 E
Lakselv, *Norway* .... **8 A21** 70 2N 25 0 E
Lakshadweep Is., *India* .... **29 H11** 10 0N 72 30 E
Lakshmanpur, *India* .... **43 H10** 22 58N 83 3 E
Lakshmikantapur, *India* .... **43 H13** 22 5N 88 20 E
Lala Ghat, *India* .... **41 G18** 24 30N 92 40 E
Lala Musa, *Pakistan* .... **42 C5** 32 40N 73 57 E
Lalago, *Tanzania* .... **54 C3** 3 28S 33 58 E
Lalapanzi, *Zimbabwe* .... **55 F3** 19 20S 30 15 E
L'Albufera, *Spain* **19 C5** 39 20N 0 27W
Lalganj, *India* .... **43 G11** 25 52N 85 13 E
Lalgola, *India* .... **43 G13** 24 25N 88 15 E
Lālī, *Iran* .... **45 C6** 32 21N 49 6 E
Lalibela, *Ethiopia* .... **46 E2** 12 2N 39 2 E
Lalin, *China* .... **35 B14** 45 12N 127 0 E
Lalín, *Spain* **19 A1** 42 40N 8 5W
Lalin He →, *China* .... **35 B13** 45 32N 125 40 E
Lalitapur, *Nepal* .... **43 F11** 27 40N 85 20 E
Lalitpur, *India* .... **43 G8** 24 42N 78 28 E
Lalkua, *India* .... **43 E8** 29 5N 79 31 E
Lalsot, *India* .... **42 F7** 26 34N 76 20 E
Lam, *Vietnam* .... **38 B6** 21 21N 106 31 E
Lam Pao Res., *Thailand* .... **38 D4** 16 50N 103 15 E
Lamaing, *Burma* .... **41 M20** 15 25N 97 53 E
Lamar, *Colo., U.S.A.* **80 F3** 38 5N 102 37W
Lamar, *Mo., U.S.A.* **81 G7** 37 30N 94 16W
Lamas, *Peru* **92 E3** 6 28S 76 31W
Lambaréné, *Gabon* .... **52 E2** 0 41S 10 12 E
Lambasa = Labasa, *Fiji* .... **59 C8** 16 30S 179 27 E
Lambay I., *Ireland* **13 C5** 53 29N 6 1W
Lambert Glacier, *Antarctica* . **5 D6** 71 0S 70 0 E
Lambert's Bay, *S. Africa* .... **56 E2** 32 5S 18 17 E
Lambeth, *Canada* **78 D3** 42 54N 81 18W
Lambomakondro, *Madag.* .... **57 C7** 22 41S 44 44 E
Lame Deer, *U.S.A.* **82 D10** 45 37N 106 40W
Lamego, *Portugal* **19 B2** 41 5N 7 52W
Lamèque, *Canada* **71 C7** 47 45N 64 38W
Lameroo, *Australia* .... **63 F3** 35 19S 140 33 E
Lamesa, *U.S.A.* **81 J4** 32 44N 101 58W
Lamía, *Greece* .... **21 E10** 38 55N 22 26 E
Lammermuir Hills, *U.K.* **12 F6** 55 50N 2 40W
Lamoille →, *U.S.A.* **79 B11** 44 38N 73 13W
Lamon B., *Phil.* .... **37 B6** 14 30N 122 20 E
Lamont, *Canada* **72 C6** 53 46N 112 50W
Lamont, *Calif., U.S.A.* **85 K8** 35 15N 118 55W
Lamont, *Wyo., U.S.A.* **82 E10** 42 13N 107 29W
Lampa, *Peru* **92 G4** 15 22S 70 22W
Lampang, *Thailand* .... **38 C2** 18 16N 99 32 E
Lampasas, *U.S.A.* **81 K5** 31 4N 98 11W
Lampazos de Naranjo, *Mexico* **86 B4** 27 2N 100 32W
Lampedusa, *Medit. S.* .... **20 G5** 35 36N 12 40 E
Lampeter, *U.K.* **11 E3** 52 7N 4 4W
Lampione, *Medit. S.* .... **20 G5** 35 33N 12 20 E
Lampman, *Canada* **73 D8** 49 25N 102 50W
Lampung □, *Indonesia* .... **36 F2** 5 30S 104 30 E
Lamta, *India* .... **43 H9** 22 8N 80 7 E
Lamu, *Kenya* .... **54 C5** 2 16S 40 55 E
Lamy, *U.S.A.* **83 J11** 35 29N 105 53W
Lan Xian, *China* .... **34 E6** 38 15N 111 35 E
Lanak La, *China* .... **43 B8** 34 27N 79 32 E

Lokitaung

Lewiston, *Maine, U.S.A.* **77 C11** 44 6N 70 13W
Lewiston, *N.Y., U.S.A.* **78 C5** 43 11N 79 3W
Lewistown, *Mont., U.S.A.* ... **82 C9** 47 4N 109 26W
Lewistown, *Pa., U.S.A.* **78 F7** 40 36N 77 34W
Lexington, *Ill., U.S.A.* **80 E10** 40 39N 88 47W
Lexington, *Ky., U.S.A.* **76 F3** 38 3N 84 30W
Lexington, *Mich., U.S.A.* **78 C2** 43 16N 82 32W
Lexington, *Mo., U.S.A.* **80 F8** 39 11N 93 52W
Lexington, *N.C., U.S.A.* **77 H5** 35 49N 80 15W
Lexington, *N.Y., U.S.A.* **79 D10** 42 15N 74 22W
Lexington, *Nebr., U.S.A.* **80 E5** 40 47N 99 45W
Lexington, *Ohio, U.S.A.* **78 F2** 40 41N 82 35W
Lexington, *Tenn., U.S.A.* ... **77 H1** 35 39N 88 24W
Lexington, *Va., U.S.A.* **76 G6** 37 47N 79 27W
Lexington Park, *U.S.A.* **76 F7** 38 16N 76 27W
Leyburn, *U.K.* **10 C6** 54 19N 1 48W
Leyland, *U.K.* **10 D5** 53 42N 2 43W
Leyte □, *Phil.* .... **37 B6** 11 0N 125 0 E
Lezhë, *Albania* .... **21 D8** 41 47N 19 39 E
Lhasa, *China* .... **32 D4** 29 25N 90 58 E
Lhazê, *China* .... **32 D3** 29 5N 87 38 E
Lhokkruet, *Indonesia* ...... **36 D1** 4 55N 95 24 E
Lhokseumawe, *Indonesia* ... **36 C1** 5 10N 97 10 E
L'Hospitalet de Llobregat, *Spain* .... **19 B7** 41 21N 2 6 E
Li, *Thailand* .... **38 D2** 17 48N 98 57 E
Li Xian, *Gansu, China* ...... **34 G3** 34 10N 105 5 E
Li Xian, *Hebei, China* ...... **34 E8** 38 30N 115 35 E
Lianga, *Phil.* .... **37 C7** 8 38N 126 6 E
Liangcheng, *Nei Monggol Zizhiqu, China* **34 D7** 40 28N 112 25 E
Liangcheng, *Shandong, China* **35 G10** 35 32N 119 37 E
Liangdang, *China* .... **34 H4** 33 56N 106 18 E
Liangpran, *Indonesia* ...... **36 D4** 1 4N 114 23 E
Lianshanguan, *China* ...... **35 D12** 40 53N 123 43 E
Lianshui, *China* .... **35 H10** 33 42N 119 20 E
Lianyungang, *China* .... **35 G10** 34 40N 119 11 E
Liao He →, *China* .... **35 D11** 41 0N 121 50 E
Liaocheng, *China* .... **34 F8** 36 28N 115 58 E
Liaodong Bandao, *China* ... **35 E12** 40 0N 122 30 E
Liaodong Wan, *China* ...... **35 D11** 40 20N 121 10 E
Liaoning □, *China* .... **35 D12** 41 40N 122 30 E
Liaoyang, *China* .... **35 D12** 41 15N 122 58 E
Liaoyuan, *China* .... **35 C13** 42 58N 125 2 E
Liaozhong, *China* .... **35 D12** 41 23N 122 50 E
Liard →, *Canada* **72 A4** 61 51N 121 18W
Liard River, *Canada* **72 B3** 59 25N 126 5W
Liari, *Pakistan* .... **42 G2** 25 37N 66 30 E
Libau = Liepāja, *Latvia* .... **9 H19** 56 30N 21 0 E
Libby, *U.S.A.* **82 B6** 48 23N 115 33W
Libenge, *Dem. Rep. of the Congo* .... **52 D3** 3 40N 18 55 E
Liberal, *U.S.A.* **81 G4** 37 3N 100 55W
Liberec, *Czech Rep.* .... **16 C8** 50 47N 15 7 E
Liberia, *Costa Rica* **88 D2** 10 40N 85 30W
Liberia ■, *W. Afr.* **50 G4** 6 30N 9 30W
Liberty, *Mo., U.S.A.* **80 F7** 39 15N 94 25W
Liberty, *N.Y., U.S.A.* **79 E10** 41 48N 74 45W
Liberty, *Pa., U.S.A.* **78 E7** 41 34N 77 6W
Liberty, *Tex., U.S.A.* **81 K7** 30 3N 94 48W
Lībīya, Sahrâ', *Africa* ...... **51 C10** 25 0N 25 0 E
Libobo, Tanjung, *Indonesia* . **37 E7** 0 54S 128 28 E
Libode, *S. Africa* .... **57 E4** 31 33S 29 2 E
Libourne, *France* **18 D3** 44 55N 0 14W
Libramont, *Belgium* .... **15 E5** 49 55N 5 23 E
Libreville, *Gabon* .... **52 D1** 0 25N 9 26 E
Libya ■, *N. Afr.* .... **51 C9** 27 0N 17 0 E
Libyan Desert = Lībīya, Sahrâ', *Africa* .... **51 C10** 25 0N 25 0 E
Licantén, *Chile* **94 D1** 35 55S 72 0W
Licata, *Italy* .... **20 F5** 37 6N 13 56 E
Licheng, *China* .... **34 F7** 36 28N 113 20 E
Lichfield, *U.K.* **11 E6** 52 41N 1 49W
Lichinga, *Mozam.* .... **55 E4** 13 13S 35 11 E
Lichtenburg, *S. Africa* ..... **56 D4** 26 8S 26 8 E
Licking →, *U.S.A.* **76 F3** 39 6N 84 30W
Licungo →, *Mozam.* .... **55 F4** 17 40S 37 15 E
Lida, *Belarus* .... **9 K21** 53 53N 25 15 E
Lidköping, *Sweden* .... **9 G15** 58 31N 13 7 E
Liebig, Mt., *Australia* ...... **60 D5** 23 18S 131 22 E
Liechtenstein ■, *Europe* ... **18 C8** 47 8N 9 35 E
Liège, *Belgium* .... **15 D5** 50 38N 5 35 E
Liège □, *Belgium* .... **15 D5** 50 32N 5 35 E
Liegnitz = Legnica, *Poland* . **16 C9** 51 12N 16 10 E
Lienart, *Dem. Rep. of the Congo* .... **54 B2** 3 3N 25 31 E
Lienyünchiangshih = Lianyungang, *China* ..... **35 G10** 34 40N 119 11 E
Lienz, *Austria* .... **16 E7** 46 50N 12 46 E
Liepāja, *Latvia* .... **9 H19** 56 30N 21 0 E
Lier, *Belgium* .... **15 C4** 51 7N 4 34 E
Lièvre →, *Canada* **70 C4** 45 31N 75 26W
Liffey →, *Ireland* **13 C5** 53 21N 6 13W
Lifford, *Ireland* **13 B4** 54 51N 7 29W
Lifudzin, *Russia* .... **30 B7** 44 21N 134 58 E
Lightning Ridge, *Australia* .. **63 D4** 29 22S 148 0 E
Ligonha →, *Mozam.* .... **55 F4** 16 54S 39 9 E
Ligonier, *U.S.A.* **78 F5** 40 15N 79 14W
Liguria □, *Italy* .... **18 D8** 44 30N 8 50 E
Ligurian Sea, *Medit. S.* .... **20 C3** 43 20N 9 0 E
Lihou Reefs and Cays, *Australia* .... **62 B5** 17 25S 151 40 E
Lihue, *U.S.A.* **74 H15** 21 59N 159 23W
Lijiang, *China* .... **32 D5** 26 55N 100 20 E
Likasi, *Dem. Rep. of the Congo* .... **55 E2** 10 55S 26 48 E
Likoma I., *Malawi* .... **55 E3** 12 3S 34 45 E
Likumburu, *Tanzania* ...... **55 D4** 9 43S 35 8 E
Lille, *France* .... **18 A5** 50 38N 3 3 E
Lille Bælt, *Denmark* .... **9 J13** 55 20N 9 45 E
Lillehammer, *Norway* ...... **9 F14** 61 8N 10 30 E
Lillesand, *Norway* .... **9 G13** 58 15N 8 23 E
Lillian Pt., *Australia* .... **61 E4** 27 40S 126 6 E
Lillooet, *Canada* **72 C4** 50 44N 121 57W
Lillooet →, *Canada* **72 D4** 49 15N 121 57W
Lilongwe, *Malawi* .... **55 E3** 14 0S 33 48 E
Liloy, *Phil.* .... **37 C6** 8 4N 122 39 E
Lim →, *Bos.-H.* .... **21 C8** 43 45N 19 15 E
Lima, *Indonesia* .... **37 E7** 3 39S 127 58 E
Lima, *Peru* **92 F3** 12 0S 77 0W
Lima, *Mont., U.S.A.* **82 D7** 44 38N 112 36W
Lima, *Ohio, U.S.A.* **76 E3** 40 44N 84 6W
Lima →, *Portugal* **19 B1** 41 41N 8 50W
Liman, *Indonesia* .... **37 G14** 7 48S 111 45 E
Limassol, *Cyprus* .... **23 E12** 34 42N 33 1 E
Limavady, *U.K.* **13 A5** 55 3N 6 56W
Limay →, *Argentina* **96 D3** 39 0S 68 0W
Limay Mahuida, *Argentina* .. **94 D2** 37 10S 66 45W
Limbang, *Brunei* .... **36 D5** 4 42N 115 6 E
Limbaži, *Latvia* .... **9 H21** 57 31N 24 42 E
Limbdi, *India* .... **42 H4** 22 34N 71 51 E
Limbe, *Cameroon* .... **52 D1** 4 1N 9 10 E
Limburg, *Germany* .... **16 C5** 50 22N 8 4 E
Limburg □, *Belgium* .... **15 C5** 51 2N 5 25 E
Limburg □, *Neths.* .... **15 C5** 51 20N 5 55 E
Limeira, *Brazil* **95 A6** 22 35S 47 28W
Limerick, *Ireland* **13 D3** 52 40N 8 37W
Limerick, *U.S.A.* **79 C14** 43 41N 70 48W
Limerick □, *Ireland* **13 D3** 52 30N 8 50W
Limestone, *U.S.A.* **78 D6** 42 2N 78 38W
Limestone →, *Canada* **73 B10** 56 31N 94 7W
Limfjorden, *Denmark* ...... **9 H13** 56 55N 9 0 E
Limia = Lima →, *Portugal* .. **19 B1** 41 41N 8 50W
Limingen, *Norway* .... **8 D15** 64 48N 13 35 E
Limmen Bight, *Australia* .... **62 A2** 14 40S 135 35 E
Limmen Bight →, *Australia* . **62 B2** 15 7S 135 44 E
Límnos, *Greece* .... **21 E11** 39 50N 25 5 E
Limoges, *Canada* **79 A9** 45 20N 75 16W
Limoges, *France* .... **18 D4** 45 50N 1 15 E
Limón, *Costa Rica* **88 E3** 10 0N 83 2W
Limon, *U.S.A.* **80 F3** 39 16N 103 41W
Limousin, *France* .... **18 D4** 45 30N 1 30 E
Limoux, *France* .... **18 E5** 43 4N 2 12 E
Limpopo →, *Africa* .... **57 D5** 25 5S 33 30 E
Limuru, *Kenya* .... **54 C4** 1 2S 36 35 E
Lin Xian, *China* .... **34 F6** 37 57N 110 58 E
Linares, *Chile* **94 D1** 35 50S 71 40W
Linares, *Mexico* **87 C5** 24 50N 99 40W
Linares, *Spain* **19 C4** 38 10N 3 40W
Lincheng, *China* .... **34 F8** 37 25N 114 30 E
Lincoln, *Argentina* **94 C3** 34 55S 61 30W
Lincoln, *N.Z.* .... **59 K4** 43 38S 172 30 E
Lincoln, *U.K.* **10 D7** 53 14N 0 32W
Lincoln, *Calif., U.S.A.* **84 G5** 38 54N 121 17W
Lincoln, *Ill., U.S.A.* **80 E10** 40 9N 89 22W
Lincoln, *Kans., U.S.A.* **80 F5** 39 3N 98 9W
Lincoln, *Maine, U.S.A.* **77 C11** 45 22N 68 30W
Lincoln, *N.H., U.S.A.* **79 B13** 44 3N 71 40W
Lincoln, *N. Mex., U.S.A.* ... **83 K11** 33 30N 105 23W
Lincoln, *Nebr., U.S.A.* **80 E6** 40 49N 96 41W
Lincoln City, *U.S.A.* **82 D1** 44 57N 124 1W
Lincoln Hav = Lincoln Sea, *Arctic* **4 A5** 84 0N 55 0W
Lincoln Sea, *Arctic* **4 A5** 84 0N 55 0W
Lincolnshire □, *U.K.* **10 D7** 53 14N 0 32W
Lincolnshire Wolds, *U.K.* ... **10 D7** 53 26N 0 13W
Lincolnton, *U.S.A.* **77 H5** 35 29N 81 16W
Lind, *U.S.A.* **82 C4** 46 58N 118 37W
Linda, *U.S.A.* **84 F5** 39 8N 121 34W
Linden, *Guyana* **92 B7** 6 0N 58 10W
Linden, *Ala., U.S.A.* **77 J2** 32 18N 87 48W
Linden, *Calif., U.S.A.* **84 G5** 38 1N 121 5W
Linden, *Tex., U.S.A.* **81 J7** 33 1N 94 22W
Lindenhurst, *U.S.A.* **79 F11** 40 41N 73 23W
Lindesnes, *Norway* .... **9 H12** 57 58N 7 3 E
Líndhos, *Greece* .... **23 C10** 36 6N 28 4 E
Líndhos, Ákra, *Greece* .... **23 C10** 36 4N 28 10 E
Lindi, *Tanzania* .... **55 D4** 9 58S 39 38 E
Lindi □, *Tanzania* .... **55 D4** 9 40S 38 30 E
Lindi →, *Dem. Rep. of the Congo* .... **54 B2** 0 33N 25 5 E
Lindsay, *Canada* **78 B6** 44 22N 78 43W
Lindsay, *Calif., U.S.A.* **84 J7** 36 12N 119 5W
Lindsay, *Okla., U.S.A.* **81 H6** 34 50N 97 38W
Lindsborg, *U.S.A.* **80 F6** 38 35N 97 40W
Linesville, *U.S.A.* **78 E4** 41 39N 80 26W
Linfen, *China* .... **34 F6** 36 3N 111 30 E
Ling Xian, *China* .... **34 F9** 37 22N 116 30 E
Lingao, *China* .... **38 C7** 19 56N 109 42 E
Lingayen, *Phil.* .... **37 A6** 16 1N 120 14 E
Lingayen G., *Phil.* .... **37 A6** 16 10N 120 15 E
Lingbi, *China* .... **35 H9** 33 33N 117 33 E
Lingchuan, *China* .... **34 G7** 35 45N 113 12 E
Lingen, *Germany* .... **16 B4** 52 31N 7 19 E
Lingga, *Indonesia* .... **36 E2** 0 12S 104 37 E
Lingga, Kepulauan, *Indonesia* **36 E2** 0 10S 104 30 E
Lingga Arch. = Lingga, Kepulauan, *Indonesia* .... **36 E2** 0 10S 104 30 E
Lingle, *U.S.A.* **80 D2** 42 8N 104 21W
Lingqiu, *China* .... **34 E8** 39 28N 114 22 E
Lingshi, *China* .... **34 F6** 36 48N 111 48 E
Lingshou, *China* .... **34 E8** 38 20N 114 20 E
Lingshui, *China* .... **38 C8** 18 27N 110 0 E
Lingtai, *China* .... **34 G4** 35 0N 107 40 E
Linguère, *Senegal* **50 E2** 15 25N 15 5W
Lingwu, *China* .... **34 E4** 38 6N 106 20 E
Lingyuan, *China* .... **35 D10** 41 10N 119 15 E
Linhai, *China* .... **33 D7** 28 50N 121 8 E
Linhares, *Brazil* **93 G10** 19 25S 40 4W
Linhe, *China* .... **34 D4** 40 48N 107 20 E
Linjiang, *China* .... **35 D14** 41 50N 127 0 E
Linköping, *Sweden* .... **9 G16** 58 28N 15 36 E
Linkou, *China* .... **35 B16** 45 15N 130 18 E
Linnhe, L., *U.K.* **12 E3** 56 36N 5 25W
Linosa, *Medit. S.* .... **20 G5** 35 51N 12 50 E
Linqi, *China* .... **34 G7** 35 45N 113 52 E
Linqing, *China* .... **34 F8** 36 50N 115 42 E
Linqu, *China* .... **35 F10** 36 25N 118 30 E
Linru, *China* .... **34 G7** 34 11N 112 52 E
Lins, *Brazil* **95 A6** 21 40S 49 44W
Linta →, *Madag.* .... **57 D7** 25 2S 44 5 E
Linton, *Ind., U.S.A.* **76 F2** 39 2N 87 10W
Linton, *N. Dak., U.S.A.* **80 B4** 46 16N 100 14W
Lintong, *China* .... **34 G5** 34 20N 109 10 E
Linwood, *Canada* **78 C4** 43 35N 80 43W
Linxi, *China* .... **35 C10** 43 36N 118 2 E
Linxia, *China* .... **32 C5** 35 36N 103 10 E
Linyanti →, *Africa* .... **56 B4** 17 50S 25 5 E
Linyi, *China* .... **35 G10** 35 5N 118 21 E
Linz, *Austria* .... **16 D8** 48 18N 14 18 E
Linzhenzhen, *China* ...... **34 F5** 36 30N 109 59 E
Linzi, *China* .... **35 F10** 36 50N 118 20 E
Lion, G. du, *France* .... **18 E6** 43 10N 4 0 E
Lionárisso, *Cyprus* .... **23 D13** 35 28N 34 8 E
Lions, G. of = Lion, G. du, *France* .... **18 E6** 43 10N 4 0 E
Lion's Den, *Zimbabwe* ..... **55 F3** 17 15S 30 5 E
Lion's Head, *Canada* **78 B3** 44 58N 81 15W
Lipa, *Phil.* .... **37 B6** 13 57N 121 10 E
Lipali, *Mozam.* .... **55 F4** 15 50S 35 50 E
Lípari, *Italy* .... **20 E6** 38 26N 14 58 E
Lípari, Is. = Eólie, Ís., *Italy* ... **20 E6** 38 30N 14 57 E
Lipcani, *Moldova* .... **17 D14** 48 14N 26 48 E
Lipetsk, *Russia* .... **24 D6** 52 37N 39 35 E
Lipkany = Lipcani, *Moldova* . **17 D14** 48 14N 26 48 E
Lipovcy Manzovka, *Russia* .. **30 B6** 44 12N 132 26 E
Lipovets, *Ukraine* .... **17 D15** 49 12N 29 1 E
Lippe →, *Germany* .... **16 C4** 51 39N 6 36 E
Lipscomb, *U.S.A.* **81 G4** 36 14N 100 16W
Liptrap C., *Australia* .... **63 F4** 38 50S 145 55 E
Lira, *Uganda* .... **54 B3** 2 17N 32 57 E
Liria = Llíria, *Spain* **19 C5** 39 37N 0 35W
Lisala, *Dem. Rep. of the Congo* .... **52 D4** 2 12N 21 38 E
Lisboa, *Portugal* **19 C1** 38 42N 9 10W
Lisbon = Lisboa, *Portugal* .. **19 C1** 38 42N 9 10W
Lisbon, *N. Dak., U.S.A.* **80 B6** 46 27N 97 41W
Lisbon, *N.H., U.S.A.* **79 B13** 44 13N 71 55W
Lisbon, *Ohio, U.S.A.* **78 F4** 40 46N 80 46W
Lisbon Falls, *U.S.A.* **77 D10** 44 0N 70 4W
Lisburn, *U.K.* **13 B5** 54 31N 6 3W
Liscannor B., *Ireland* **13 D2** 52 55N 9 24W
Lishi, *China* .... **34 F6** 37 31N 111 8 E
Lishu, *China* .... **35 C13** 43 20N 124 18 E
Lisianski I., *Pac. Oc.* **64 E10** 26 2N 174 0W
Lisichansk = Lysychansk, *Ukraine* .... **25 E6** 48 55N 38 30 E
Lisieux, *France* .... **18 B4** 49 10N 0 12 E
Liski, *Russia* .... **25 D6** 51 3N 39 30 E
Lismore, *Australia* .... **63 D5** 28 44S 153 21 E
Lismore, *Ireland* **13 D4** 52 8N 7 55W
Lista, *Norway* .... **9 G12** 58 7N 6 39 E
Lister, Mt., *Antarctica* ..... **5 D11** 78 0S 162 0 E
Liston, *Australia* .... **63 D5** 28 39S 152 6 E
Listowel, *Canada* **78 C4** 43 44N 80 58W
Listowel, *Ireland* **13 D2** 52 27N 9 29W
Litani →, *Lebanon* .... **47 B4** 33 20N 35 15 E
Litchfield, *Calif., U.S.A.* **84 E6** 40 24N 120 23W
Litchfield, *Conn., U.S.A.* ... **79 E11** 41 45N 73 11W
Litchfield, *Ill., U.S.A.* **80 F10** 39 11N 89 39W
Litchfield, *Minn., U.S.A.* ... **80 C7** 45 8N 94 32W
Lithgow, *Australia* .... **63 E5** 33 25S 150 8 E
Líthinon, Ákra, *Greece* .... **23 E6** 34 55N 24 44 E
Lithuania ■, *Europe* .... **9 J20** 55 30N 24 0 E
Lititz, *U.S.A.* **79 F8** 40 9N 76 18W
Litoměřice, *Czech Rep.* .... **16 C8** 50 33N 14 10 E
Little Abaco I., *Bahamas* **88 A4** 26 50N 77 30W
Little Barrier I., *N.Z.* .... **59 G5** 36 12S 175 8 E
Little Belt Mts., *U.S.A.* **82 C8** 46 40N 110 45W
Little Blue →, *U.S.A.* **80 F6** 39 42N 96 41W
Little Buffalo →, *Canada* ... **72 A6** 61 0N 113 46W
Little Cayman, *Cayman Is.* .. **88 C3** 19 41N 80 3W
Little Churchill →, *Canada* . **73 B9** 57 30N 95 22W
Little Colorado →, *U.S.A.* .. **83 H8** 36 12N 111 48W
Little Current, *Canada* **70 C3** 45 55N 82 0W
Little Current →, *Canada* ... **70 B3** 50 57N 84 36W
Little Falls, *Minn., U.S.A.* .. **80 C7** 45 59N 94 22W
Little Falls, *N.Y., U.S.A.* ... **79 C10** 43 3N 74 51W
Little Fork →, *U.S.A.* **80 A8** 48 31N 93 35W
Little Grand Rapids, *Canada* **73 C9** 52 0N 95 29W
Little Humboldt →, *U.S.A.* .. **82 F5** 41 1N 117 43W
Little Inagua I., *Bahamas* ... **89 B5** 21 40N 73 50W
Little Karoo, *S. Africa* ...... **56 E3** 33 45S 21 0 E
Little Lake, *U.S.A.* **85 K9** 35 56N 117 55W
Little Laut Is. = Laut Kecil, Kepulauan, *Indonesia* .... **36 E5** 4 45S 115 40 E
Little Mecatina = Petit-Mécatina →, *Canada* **71 B8** 50 40N 59 30W
Little Minch, *U.K.* **12 D2** 57 35N 6 45W
Little Missouri →, *U.S.A.* ... **80 B3** 47 36N 102 25W
Little Ouse →, *U.K.* .... **11 E9** 52 22N 1 12 E
Little Rann, *India* .... **42 H4** 23 25N 71 25 E
Little Red →, *U.S.A.* **81 H9** 35 11N 91 27W
Little River, *N.Z.* .... **59 K4** 43 45S 172 49 E
Little Rock, *U.S.A.* **81 H8** 34 45N 92 17W
Little Ruaha →, *Tanzania* .. **54 D4** 7 57S 37 53 E
Little Sable Pt., *U.S.A.* **76 D2** 43 38N 86 33W
Little Sioux →, *U.S.A.* **80 E6** 41 48N 96 4W
Little Smoky →, *Canada* ... **72 C5** 54 44N 117 11W
Little Snake →, *U.S.A.* **82 F9** 40 27N 108 26W
Little Valley, *U.S.A.* **78 D6** 42 15N 78 48W
Little Wabash →, *U.S.A.* ... **76 G1** 37 55N 88 5W
Little White →, *U.S.A.* **80 D4** 43 40N 100 40W
Littlefield, *U.S.A.* **81 J3** 33 55N 102 20W
Littlehampton, *U.K.* **11 G7** 50 49N 0 32W
Littleton, *U.S.A.* **79 B13** 44 18N 71 46W
Liu He →, *China* .... **35 D11** 40 55N 121 35 E
Liuba, *China* .... **34 H4** 33 38N 106 55 E
Liugou, *China* .... **35 D10** 40 57N 118 15 E
Liuhe, *China* .... **35 C13** 42 17N 125 43 E
Liukang Tenggaja = Sabalana, Kepulauan, *Indonesia* .... **37 F5** 6 45S 118 50 E
Liuli, *Tanzania* .... **55 E3** 11 3S 34 38 E
Liuwa Plain, *Zambia* .... **53 G4** 14 20S 22 30 E
Liuzhou, *China* .... **33 D5** 24 22N 109 22 E
Liuzhuang, *China* .... **35 H11** 33 12N 120 18 E
Livadhia, *Cyprus* .... **23 E12** 34 57N 33 38 E
Live Oak, *Calif., U.S.A.* **84 F5** 39 17N 121 40W
Live Oak, *Fla., U.S.A.* **77 K4** 30 18N 82 59W
Liveras, *Cyprus* .... **23 D11** 35 23N 32 57 E
Livermore, *U.S.A.* **84 H5** 37 41N 121 47W
Livermore, Mt., *U.S.A.* **81 K2** 30 38N 104 11W
Livermore Falls, *U.S.A.* **77 C11** 44 29N 70 11W
Liverpool, *Canada* **71 D7** 44 5N 64 41W
Liverpool, *U.K.* **10 D4** 53 25N 3 0W
Liverpool, *U.S.A.* **79 C8** 43 6N 76 13W
Liverpool Bay, *U.K.* **10 D4** 53 30N 3 20W
Liverpool Plains, *Australia* .. **63 E5** 31 15S 150 15 E
Liverpool Ra., *Australia* .... **63 E5** 31 50S 150 30 E
Livingston, *Guatemala* **88 C2** 15 50N 88 50W
Livingston, *U.K.* **12 F5** 55 54N 3 30W
Livingston, *Ala., U.S.A.* **77 J1** 32 35N 88 11W
Livingston, *Calif., U.S.A.* ... **84 H6** 37 23N 120 43W
Livingston, *Mont., U.S.A.* ... **82 D8** 45 40N 110 34W
Livingston, *S.C., U.S.A.* **77 J5** 33 32N 80 53W
Livingston, *Tenn., U.S.A.* ... **77 G3** 36 23N 85 19W
Livingston, *Tex., U.S.A.* **81 K7** 30 43N 94 56W
Livingston, L., *U.S.A.* **81 K7** 30 50N 95 10W
Livingston Manor, *U.S.A.* ... **79 E10** 41 54N 74 50W
Livingstone, *Zambia* .... **55 F2** 17 46S 25 52 E
Livingstone Mts., *Tanzania* . **55 D3** 9 40S 34 20 E
Livingstonia, *Malawi* .... **55 E3** 10 38S 34 5 E
Livny, *Russia* .... **24 D6** 52 30N 37 30 E
Livonia, *Mich., U.S.A.* **76 D4** 42 23N 83 23W
Livonia, *N.Y., U.S.A.* **78 D7** 42 49N 77 40W
Livorno, *Italy* .... **20 C4** 43 33N 10 19 E
Livramento, *Brazil* **95 C4** 30 55S 55 30W
Liwale, *Tanzania* .... **55 D4** 9 48S 37 58 E
Lizard I., *Australia* .... **62 A4** 14 42S 145 30 E
Lizard Pt., *U.K.* **11 H2** 49 57N 5 13W
Ljubljana, *Slovenia* .... **16 E8** 46 4N 14 33 E
Ljungan →, *Sweden* .... **9 E17** 62 18N 17 23 E
Ljungby, *Sweden* .... **9 H15** 56 49N 13 55 E
Ljusdal, *Sweden* .... **9 F17** 61 46N 16 3 E
Ljusnan →, *Sweden* .... **9 F17** 61 12N 17 8 E
Ljusne, *Sweden* .... **9 F17** 61 13N 17 7 E
Llancanelo, Salina, *Argentina* **94 D2** 35 40S 69 8W
Llandeilo, *U.K.* **11 F4** 51 53N 3 59W
Llandovery, *U.K.* **11 F4** 51 59N 3 48W
Llandrindod Wells, *U.K.* **11 E4** 52 14N 3 22W
Llandudno, *U.K.* **10 D4** 53 19N 3 50W
Llanelli, *U.K.* **11 F3** 51 41N 4 10W
Llanes, *Spain* **19 A3** 43 25N 4 50W
Llangollen, *U.K.* **10 E4** 52 58N 3 11W
Llanidloes, *U.K.* **11 E4** 52 27N 3 31W
Llano, *U.S.A.* **81 K5** 30 45N 98 41W
Llano →, *U.S.A.* **81 K5** 30 39N 98 26W
Llano Estacado, *U.S.A.* **81 J3** 33 30N 103 0W
Llanos, *S. Amer.* **92 C4** 5 0N 71 35W
Llanquihue, L., *Chile* **96 E1** 41 10S 72 50W
Llanwrtyd Wells, *U.K.* **11 E4** 52 7N 3 38W
Llebeig, C. des, *Spain* ..... **22 B9** 39 33N 2 18 E
Lleida, *Spain* .... **19 B6** 41 37N 0 39 E
Llentrisca, C., *Spain* ...... **22 C7** 38 52N 1 15 E
Llera, *Mexico* **87 C5** 23 19N 99 1W
Lleyn Peninsula, *U.K.* **10 E3** 52 51N 4 36W
Llico, *Chile* **94 C1** 34 46S 72 5W
Llíria, *Spain* **19 C5** 39 37N 0 35W
Llobregat →, *Spain* .... **19 B7** 41 19N 2 9 E
Lloret de Mar, *Spain* ...... **19 B7** 41 41N 2 53 E
Lloyd B., *Australia* .... **62 A3** 12 45S 143 27 E
Lloyd L., *Canada* **73 B7** 57 22N 108 57W
Lloydminster, *Canada* **73 C7** 53 17N 110 0W
Llucmajor, *Spain* .... **22 B9** 39 29N 2 53 E
Llullaillaco, Volcán, *S. Amer.* **94 A2** 24 43S 68 30W
Lo →, *Vietnam* .... **38 B5** 21 18N 105 25 E
Loa, *U.S.A.* **83 G8** 38 24N 111 39W
Loa →, *Chile* **94 A1** 21 26S 70 41W
Loaita I., *S. China Sea* ..... **36 B4** 10 41N 114 25 E
Loange →, *Dem. Rep. of the Congo* .... **52 E4** 4 17S 20 2 E
Lobatse, *Botswana* .... **56 D4** 25 12S 25 40 E
Lobería, *Argentina* **94 D4** 38 10S 58 40W
Lobito, *Angola* .... **53 G2** 12 18S 13 35 E
Lobos, *Argentina* **94 D4** 35 10S 59 0W
Lobos, I., *Mexico* **86 B2** 27 15N 110 30W
Lobos, I. de, *Canary Is.* **22 F6** 28 45N 13 50W
Loc Binh, *Vietnam* .... **38 B6** 21 46N 106 54 E
Loc Ninh, *Vietnam* .... **39 G6** 11 50N 106 34 E
Locarno, *Switz.* .... **18 C8** 46 10N 8 47 E
Loch Baghasdail = Lochboisdale, *U.K.* **12 D1** 57 9N 7 20W
Loch Garman = Wexford, *Ireland* **13 D5** 52 20N 6 28W
Loch Nam Madadh = Lochmaddy, *U.K.* **12 D1** 57 36N 7 10W
Lochaber, *U.K.* **12 E3** 56 59N 5 1W
Locharbriggs, *U.K.* **12 F5** 55 7N 3 35W
Lochboisdale, *U.K.* **12 D1** 57 9N 7 20W
Loche, L. La, *Canada* **73 B7** 56 30N 109 30W
Lochem, *Neths.* .... **15 B6** 52 9N 6 26 E
Loches, *France* .... **18 C4** 47 7N 1 0 E
Lochgilphead, *U.K.* **12 E3** 56 2N 5 26W
Lochinver, *U.K.* **12 C3** 58 9N 5 14W
Lochmaddy, *U.K.* **12 D1** 57 36N 7 10W
Lochnagar, *Australia* ...... **62 C4** 23 33S 145 38 E
Lochnagar, *U.K.* **12 E5** 56 57N 3 15W
Lochy, L., *U.K.* **12 E4** 57 0N 4 53W
Lock, *Australia* .... **63 E2** 33 34S 135 46 E
Lock Haven, *U.S.A.* **78 E7** 41 8N 77 28W
Lockeford, *U.S.A.* **84 G5** 38 10N 121 9W
Lockeport, *Canada* **71 D6** 43 47N 65 4W
Lockerbie, *U.K.* **12 F5** 55 7N 3 21W
Lockhart, *U.S.A.* **81 L6** 29 53N 97 40W
Lockhart, L., *Australia* ..... **61 F2** 33 15S 119 3 E
Lockhart River, *Australia* .. **62 A3** 12 58S 143 30 E
Lockney, *U.S.A.* **81 H4** 34 7N 101 27W
Lockport, *U.S.A.* **78 C6** 43 10N 78 42W
Lod, *Israel* .... **47 D3** 31 57N 34 54 E
Lodeinoye Pole, *Russia* .... **24 B5** 60 44N 33 33 E
Lodge Bay, *Canada* **71 B8** 52 14N 55 51W
Lodge Grass, *U.S.A.* **82 D10** 45 19N 107 22W
Lodgepole Cr. →, *U.S.A.* .. **80 E2** 41 20N 104 30W
Lodhran, *Pakistan* .... **42 E4** 29 32N 71 30 E
Lodi, *Italy* .... **18 D8** 45 19N 9 30 E
Lodi, *Calif., U.S.A.* **84 G5** 38 8N 121 16W
Lodi, *Ohio, U.S.A.* **78 E3** 41 2N 82 0W
Lodja, *Dem. Rep. of the Congo* .... **54 C1** 3 30S 23 23 E
Lodwar, *Kenya* .... **54 B4** 3 10N 35 40 E
Łódź, *Poland* .... **17 C10** 51 45N 19 27 E
Loei, *Thailand* .... **38 D3** 17 29N 101 35 E
Loengo, *Dem. Rep. of the Congo* .... **54 C2** 4 48S 26 30 E
Loeriesfontein, *S. Africa* ... **56 E2** 31 0S 19 26 E
Lofoten, *Norway* .... **8 B15** 68 30N 14 0 E
Logan, *Iowa, U.S.A.* **80 E7** 41 39N 95 47W
Logan, *Ohio, U.S.A.* **76 F4** 39 32N 82 25W
Logan, *Utah, U.S.A.* **82 F8** 41 44N 111 50W
Logan, *W. Va., U.S.A.* **76 G5** 37 51N 81 59W
Logan, Mt., *Canada* **68 B5** 60 31N 140 22W
Logandale, *U.S.A.* **85 J12** 36 36N 114 29W
Logansport, *Ind., U.S.A.* ... **76 E2** 40 45N 86 22W
Logansport, *La., U.S.A.* **81 K8** 31 58N 94 0W
Logone →, *Chad* .... **51 F9** 12 6N 15 2 E
Logroño, *Spain* **19 A4** 42 28N 2 27W
Lohardaga, *India* .... **43 H11** 23 27N 84 45 E
Loharia, *India* .... **42 H6** 23 45N 74 14 E
Loharu, *India* .... **42 E6** 28 27N 75 49 E
Lohja, *Finland* .... **9 F21** 60 12N 24 5 E
Lohri Wah →, *Pakistan* .... **42 F2** 27 27N 67 37 E
Loi-kaw, *Burma* .... **41 K20** 19 40N 97 17 E
Loimaa, *Finland* .... **9 F20** 60 50N 23 5 E
Loir →, *France* **18 C3** 47 33N 0 32W
Loire →, *France* **18 C2** 47 16N 2 10W
Loja, *Ecuador* **92 D3** 3 59S 79 16W
Loja, *Spain* **19 D3** 37 10N 4 10W
Loji = Kawasi, *Indonesia* ... **37 E7** 1 38S 127 28 E
Lokandu, *Dem. Rep. of the Congo* .... **54 C2** 2 30S 25 45 E
Lokeren, *Belgium* .... **15 C3** 51 6N 3 59 E
Lokgwabe, *Botswana* ...... **56 C3** 24 10S 21 50 E
Lokichokio, *Kenya* .... **54 B3** 4 19N 34 13 E
Lokitaung, *Kenya* .... **54 B4** 4 12N 35 48 E

Lokkan tekojärvi

Lokkan tekojärvi, Finland 8 C22 67 55N 27 35 E
Lokoja, Nigeria 50 G7 7 47N 6 45 E
Lola, Mt., U.S.A. 84 F6 39 26N 120 22W
Loliondo, Tanzania 54 C4 2 2S 35 39 E
Lolland, Denmark 9 J14 54 45N 11 30 E
Lolo, U.S.A. 82 C6 46 45N 114 5W
Lom, Bulgaria 21 C10 43 48N 23 12 E
Lom Kao, Thailand 38 D3 16 53N 101 14 E
Lom Sak, Thailand 38 D3 16 47N 101 15 E
Loma, U.S.A. 82 C8 47 56N 110 30W
Loma Linda, U.S.A. 85 L9 34 3N 117 16W
Lomami →, Dem. Rep. of the Congo 54 B1 0 46N 24 16 E
Lomas de Zamóra, Argentina 94 C4 34 45S 58 25W
Lombadina, Australia 60 C3 16 31S 122 54 E
Lombárdia □, Italy 18 D8 45 40N 9 30 E
Lombardy = Lombárdia □, Italy 18 D8 45 40N 9 30 E
Lomblen, Indonesia 37 F6 8 30S 123 32 E
Lombok, Indonesia 36 F5 8 45S 116 30 E
Lomé, Togo 50 G6 6 9N 1 20 E
Lomela, Dem. Rep. of the Congo 52 E4 2 19S 23 15 E
Lomela →, Dem. Rep. of the Congo 52 E4 0 15S 20 40 E
Lommel, Belgium 15 C5 51 14N 5 19 E
Lomond, Canada 72 C6 50 24N 112 36W
Lomond, L., U.K. 12 E4 56 8N 4 38W
Lomphat, Cambodia 38 F6 13 30N 106 59 E
Lompobatang, Indonesia 37 F5 5 24S 119 56 E
Lompoc, U.S.A. 85 L6 34 38N 120 28W
Łomża, Poland 17 B12 53 10N 22 2 E
Loncoche, Chile 96 D2 39 20S 72 50W
Londa, India 40 M9 15 30N 74 30 E
Londiani, Kenya 54 C4 0 10S 35 33 E
London, Canada 78 D3 42 59N 81 15W
London, U.K. 11 F7 51 30N 0 3W
London, Ky., U.S.A. 76 G3 37 8N 84 5W
London, Ohio, U.S.A. 76 F4 39 53N 83 27W
London, Greater □, U.K. 11 F7 51 36N 0 5W
Londonderry, U.K. 13 B4 55 0N 7 20W
Londonderry □, U.K. 13 B4 55 0N 7 20W
Londonderry, C., Australia 60 B4 13 45S 126 55 E
Londonderry, I., Chile 96 H2 55 0S 71 0W
Londres, Argentina 96 B3 27 43S 67 7W
Londrina, Brazil 95 A5 23 18S 51 10W
Lone Pine, U.S.A. 84 J8 36 36N 118 4W
Lonely Mine, Zimbabwe 57 B4 19 30S 28 49 E
Long B., U.S.A. 77 J6 33 35N 78 45W
Long Beach, Calif., U.S.A. 85 M8 33 47N 118 11W
Long Beach, N.Y., U.S.A. 79 F11 40 35N 73 39W
Long Beach, Wash., U.S.A. 84 D2 46 21N 124 3W
Long Branch, U.S.A. 79 F11 40 18N 74 0W
Long Creek, U.S.A. 82 D4 44 43N 119 6W
Long Eaton, U.K. 10 E6 52 53N 1 15W
Long I., Australia 62 C4 20 22S 148 51 E
Long I., Bahamas 89 B4 23 20N 75 10W
Long I., Canada 70 B4 54 50N 79 20W
Long I., Ireland 13 E2 51 30N 9 34W
Long I., U.S.A. 79 F11 40 45N 73 30W
Long Island Sd., U.S.A. 79 E12 41 10N 73 0W
Long L., Canada 70 C2 49 30N 86 50W
Long Lake, U.S.A. 79 C10 43 58N 74 25W
Long Point B., Canada 78 D4 42 40N 80 10W
Long Prairie →, U.S.A. 80 C7 46 20N 94 36W
Long Pt., Canada 78 D4 42 35N 80 2W
Long Range Mts., Canada 71 C8 49 30N 57 30W
Long Reef, Australia 60 B4 14 1S 125 48 E
Long Spruce, Canada 73 B10 56 24N 94 21W
Long Str. = Longa, Proliv, Russia 4 C16 70 0N 175 0 E
Long Thanh, Vietnam 39 G6 10 47N 106 57 E
Long Xian, China 34 G4 34 55N 106 55 E
Long Xuyen, Vietnam 39 G5 10 19N 105 28 E
Longa, Proliv, Russia 4 C16 70 0N 175 0 E
Longbenton, U.K. 10 B6 55 1N 1 31W
Longboat Key, U.S.A. 77 M4 27 23N 82 39W
Longde, China 34 G4 35 30N 106 20 E
Longford, Australia 62 G4 41 32S 147 3 E
Longford, Ireland 13 C4 53 43N 7 49W
Longford □, Ireland 13 C4 53 42N 7 45W
Longhua, China 35 D9 41 18N 117 45 E
Longido, Tanzania 54 C4 2 43S 36 42 E
Longiram, Indonesia 36 E5 0 5S 115 45 E
Longkou, China 35 F11 37 40N 120 18 E
Longlac, Canada 70 C2 49 45N 86 25W
Longmeadow, U.S.A. 79 D12 42 3N 72 34W
Longmont, U.S.A. 80 E2 40 10N 105 6W
Longnawan, Indonesia 36 D4 1 51N 114 55 E
Longreach, Australia 62 C3 23 28S 144 14 E
Longueuil, Canada 79 A11 45 32N 73 28W
Longview, Tex., U.S.A. 81 J7 32 30N 94 44W
Longview, Wash., U.S.A. 84 D4 46 8N 122 57W
Longxi, China 34 G3 34 53N 104 40 E
Lonoke, U.S.A. 81 H9 34 47N 91 54W
Lonquimay, Chile 96 D2 38 26S 71 14W
Lons-le-Saunier, France 18 C6 46 40N 5 31 E
Looe, U.K. 11 G3 50 22N 4 28W
Lookout, C., Canada 70 A3 55 18N 83 56W
Lookout, C., U.S.A. 77 H7 34 35N 76 32W
Loolmalasin, Tanzania 54 C4 3 0S 35 53 E
Loon →, Alta., Canada 72 B5 57 8N 115 3W
Loon →, Man., Canada 73 B8 55 53N 101 59W
Loon Lake, Canada 73 C7 54 2N 109 10W
Loongana, Australia 61 F4 30 52S 127 5 E
Loop Hd., Ireland 13 D2 52 34N 9 56W
Lop Buri, Thailand 38 E3 14 48N 100 37 E
Lop Nor = Lop Nur, China 32 B4 40 20N 90 10 E
Lop Nur, China 32 B4 40 20N 90 10 E
Lopatina, Gora, Russia 27 D15 50 47N 143 10 E
Lopez, U.S.A. 79 E8 41 27N 76 20W
Lopez, C., Gabon 52 E1 0 47S 8 40 E
Lopphavet, Norway 8 A19 70 27N 21 15 E
Lora →, Afghan. 40 D4 31 35N 66 32 E
Lora, Hāmūn-i-, Pakistan 40 E4 29 38N 64 58 E
Lora Cr. →, Australia 63 D2 28 10S 135 22 E
Lora del Río, Spain 19 D3 37 39N 5 33W
Lorain, U.S.A. 78 E2 41 28N 82 11W
Loralai, Pakistan 42 D3 30 20N 68 41 E
Lorca, Spain 19 D5 37 41N 1 42W
Lord Howe I., Pac. Oc. 64 L7 31 33S 159 6 E
Lord Howe Ridge, Pac. Oc. 64 L8 30 0S 162 30 E
Lordsburg, U.S.A. 83 K9 32 21N 108 43W
Lorestan □, Iran 45 C6 33 30N 48 40 E
Loreto, Brazil 93 E9 7 5S 45 10W
Loreto, Mexico 86 B2 26 1N 111 21W
Lorient, France 18 C2 47 45N 3 23W
Lormi, India 43 H9 22 17N 81 41 E
Lorn, U.K. 12 E3 56 26N 5 10W
Lorn, Firth of, U.K. 12 E3 56 20N 5 40W
Lorne, Australia 63 F3 38 33S 143 59 E
Lorovouno, Cyprus 23 D11 35 8N 32 36 E
Lorraine □, France 18 B7 48 53N 6 0 E
Los Alamos, Calif., U.S.A. 85 L6 34 44N 120 17W
Los Alamos, N. Mex., U.S.A. 83 J10 35 53N 106 19W
Los Altos, U.S.A. 84 H4 37 23N 122 7W
Los Andes, Chile 94 C1 32 50S 70 40W
Los Angeles, Chile 94 D1 37 28S 72 23W
Los Angeles, U.S.A. 85 M8 34 4N 118 15W
Los Angeles, Bahia de, Mexico 86 B2 28 56N 113 34W
Los Angeles Aqueduct, U.S.A. 85 K9 35 22N 118 5W
Los Banos, U.S.A. 84 H6 37 4N 120 51W
Los Blancos, Argentina 94 A3 23 40S 62 30W
Los Chiles, Costa Rica 88 D3 11 2N 84 43W
Los Cristianos, Canary Is. 22 F3 28 3N 16 42W
Los Gatos, U.S.A. 84 H5 37 14N 121 59W
Los Hermanos Is., Venezuela 89 D7 11 45N 64 25W
Los Islotes, Canary Is. 22 E6 29 4N 13 44W
Los Llanos de Aridane, Canary Is. 22 F2 28 38N 17 54W
Los Loros, Chile 94 B1 27 50S 70 6W
Los Lunas, U.S.A. 83 J10 34 48N 106 44W
Los Mochis, Mexico 86 B3 25 45N 108 57W
Los Olivos, U.S.A. 85 L6 34 40N 120 7W
Los Palacios, Cuba 88 B3 22 35N 83 15W
Los Reyes, Mexico 86 D4 19 34N 102 30W
Los Roques Is., Venezuela 89 D6 11 50N 66 45W
Los Teques, Venezuela 92 A5 10 21N 67 2W
Los Testigos, Is., Venezuela 92 A6 11 23N 63 6W
Los Vilos, Chile 94 C1 32 10S 71 30W
Lošinj, Croatia 16 F8 44 30N 14 30 E
Loskop Dam, S. Africa 57 D4 25 23S 29 20 E
Lossiemouth, U.K. 12 D5 57 42N 3 17W
Lostwithiel, U.K. 11 G3 50 24N 4 41W
Lot →, France 18 D4 44 18N 0 20 E
Lota, Chile 94 D1 37 5S 73 10W
Loṭfābād, Iran 45 B8 37 32N 59 20 E
Lothair, S. Africa 57 D5 26 22S 30 27 E
Loubomo, Congo 52 E2 4 9S 12 47 E
Loudonville, U.S.A. 78 F2 40 38N 82 14W
Louga, Senegal 50 E2 15 45N 16 5W
Loughborough, U.K. 10 E6 52 47N 1 11W
Loughrea, Ireland 13 C3 53 12N 8 33W
Loughros More B., Ireland 13 B3 54 48N 8 32W
Louis Trichardt, S. Africa 57 C4 23 1S 29 43 E
Louis XIV, Pte., Canada 70 B4 54 37N 79 45W
Louisa, U.S.A. 76 F4 38 7N 82 36W
Louisbourg, Canada 71 C8 45 55N 60 0W
Louise I., Canada 72 C2 52 55N 131 50W
Louiseville, Canada 70 C5 46 20N 72 56W
Louisiade Arch., Papua N. G. 64 J7 11 10S 153 0 E
Louisiana, U.S.A. 80 F9 39 27N 91 3W
Louisiana □, U.S.A. 81 K9 30 50N 92 0W
Louisville, Ky., U.S.A. 76 F3 38 15N 85 46W
Louisville, Miss., U.S.A. 81 J10 33 7N 89 3W
Louisville, Ohio, U.S.A. 78 F3 40 50N 81 16W
Loulé, Portugal 19 D1 37 9N 8 0W
Loup City, U.S.A. 80 E5 41 17N 98 58W
Loups Marins, Lacs des, Canada 70 A5 56 30N 73 45W
Lourdes, France 18 E3 43 6N 0 3W
Louth, Australia 63 E4 30 30S 145 8 E
Louth, Ireland 13 C5 53 58N 6 32W
Louth, U.K. 10 D7 53 22N 0 1W
Louth □, Ireland 13 C5 53 56N 6 34W
Louvain = Leuven, Belgium 15 D4 50 52N 4 42 E
Louwsburg, S. Africa 57 D5 27 37S 31 7 E
Lovech, Bulgaria 21 C11 43 8N 24 42 E
Loveland, U.S.A. 80 E2 40 24N 105 5W
Lovell, U.S.A. 82 D9 44 50N 108 24W
Lovelock, U.S.A. 82 F4 40 11N 118 28W
Loviisa, Finland 9 F22 60 28N 26 12 E
Loving, U.S.A. 81 J2 32 17N 104 6W
Lovington, U.S.A. 81 J3 32 57N 103 21W
Lovisa = Loviisa, Finland 9 F22 60 28N 26 12 E
Low, L., Canada 70 B4 52 29N 76 17W
Low Pt., Australia 61 F4 32 25S 127 25 E
Low Tatra = Nízké Tatry, Slovak Rep. 17 D10 48 55N 19 30 E
Lowa, Dem. Rep. of the Congo 54 C2 1 25S 25 47 E
Lowa →, Dem. Rep. of the Congo 54 C2 1 24S 25 51 E
Lowell, U.S.A. 79 D13 42 38N 71 19W
Lowellville, U.S.A. 78 E4 41 2N 80 32W
Löwen →, Namibia 56 D2 26 51S 18 17 E
Lower Alkali L., U.S.A. 82 F3 41 16N 120 2W
Lower Arrow L., Canada 72 D5 49 40N 118 5W
Lower California = Baja California, Mexico 86 A1 31 10N 115 12W
Lower Hutt, N.Z. 59 J5 41 10S 174 55 E
Lower Lake, U.S.A. 84 G4 38 55N 122 37W
Lower Manitou L., Canada 73 D10 49 15N 93 0W
Lower Post, Canada 72 B3 59 58N 128 30W
Lower Red L., U.S.A. 80 B7 47 58N 95 0W
Lower Saxony = Niedersachsen □, Germany 16 B5 52 50N 9 0 E
Lower Tunguska = Tunguska, Nizhnyaya →, Russia 27 C9 65 48N 88 4 E
Lowestoft, U.K. 11 E9 52 29N 1 45 E
Lowgar □, Afghan. 40 B6 34 0N 69 0 E
Łowicz, Poland 17 B10 52 6N 19 55 E
Lowville, U.S.A. 79 C9 43 47N 75 29W
Loxton, Australia 63 E3 34 28S 140 31 E
Loxton, S. Africa 56 E3 31 30S 22 22 E
Loyalton, U.S.A. 84 F6 39 41N 120 14W
Loyalty Is. = Loyauté, Îs., N. Cal. 64 K8 20 50S 166 30 E
Loyang = Luoyang, China 34 G7 34 40N 112 26 E
Loyauté, Îs., N. Cal. 64 K8 20 50S 166 30 E
Loyev = Loyew, Belarus 17 C16 51 56N 30 46 E
Loyew, Belarus 17 C16 51 56N 30 46 E
Loyoro, Uganda 54 B3 3 22N 34 14 E
Luachimo, Angola 52 F4 7 23S 20 48 E
Luajan →, India 43 G11 24 44N 85 1 E
Lualaba →, Dem. Rep. of the Congo 54 B2 0 26N 25 20 E
Luampa, Zambia 55 F1 15 4S 24 20 E
Luan Chau, Vietnam 38 B4 21 38N 103 24 E
Luan He →, China 35 E10 39 20N 119 5 E
Luan Xian, China 35 E10 39 40N 118 40 E
Luancheng, China 34 F8 37 53N 114 40 E
Luanda, Angola 52 F2 8 50S 13 15 E
Luang, Thale, Thailand 39 J3 7 30N 100 15 E
Luang Prabang, Laos 38 C4 19 52N 102 10 E
Luangwa, Zambia 55 F3 15 35S 30 16 E
Luangwa →, Zambia 55 E3 14 25S 30 25 E
Luangwa Valley, Zambia 55 E3 13 30S 31 30 E
Luanne, China 35 D9 40 55N 117 40 E
Luanping, China 35 D9 40 53N 117 23 E
Luanshya, Zambia 55 E2 13 3S 28 28 E
Luapula □, Zambia 55 E2 11 0S 29 0 E
Luapula →, Africa 55 D2 9 26S 28 33 E
Luarca, Spain 19 A2 43 32N 6 32W
Luashi, Dem. Rep. of the Congo 55 E1 10 50S 23 36 E
Luau, Angola 52 G4 10 40S 22 10 E
Lubana, Ozero = Lubānas Ezers, Latvia 9 H22 56 45N 27 0 E
Lubānas Ezers, Latvia 9 H22 56 45N 27 0 E
Lubang Is., Phil. 37 B6 13 50N 120 12 E
Lubao, Dem. Rep. of the Congo 54 D2 5 17S 25 42 E
Lubbock, U.S.A. 81 J4 33 35N 101 51W
Lübeck, Germany 16 B6 53 52N 10 40 E
Lubefu, Dem. Rep. of the Congo 54 C1 4 47S 24 27 E
Lubefu →, Dem. Rep. of the Congo 54 C1 4 10S 23 0 E
Lubero = Luofu, Dem. Rep. of the Congo 54 C2 0 10S 29 15 E
Lubicon L., Canada 72 B5 56 23N 115 56W
Lubilash →, Dem. Rep. of the Congo 52 F4 6 2S 23 45 E
Lubin, Poland 16 C9 51 24N 16 11 E
Lublin, Poland 17 C12 51 12N 22 38 E
Lubnān, Jabal, Lebanon 47 B4 33 45N 35 40 E
Lubny, Ukraine 26 D4 50 3N 32 58 E
Lubongola, Dem. Rep. of the Congo 54 C2 2 35S 27 50 E
Lubudi, Dem. Rep. of the Congo 52 F5 9 57S 25 58 E
Lubudi →, Dem. Rep. of the Congo 55 D2 9 0S 25 35 E
Lubuklinggau, Indonesia 36 E2 3 15S 102 55 E
Lubuksikaping, Indonesia 36 D2 0 10N 100 15 E
Lubumbashi, Dem. Rep. of the Congo 55 E2 11 40S 27 28 E
Lubunda, Dem. Rep. of the Congo 54 D2 5 12S 26 41 E
Lubungu, Zambia 55 E2 14 35S 26 24 E
Lubutu, Dem. Rep. of the Congo 54 C2 0 45S 26 30 E
Luc An Chau, Vietnam 38 A5 22 6N 104 43 E
Lucan, Canada 78 C3 43 11N 81 24W
Lucania, Mt., Canada 68 B5 61 1N 140 29W
Lucas Channel, Canada 78 A3 45 21N 81 45W
Lucca, Italy 20 C4 43 50N 10 29 E
Luce Bay, U.K. 12 G4 54 45N 4 48W
Lucea, Jamaica 88 C4 18 25N 78 10W
Lucedale, U.S.A. 77 K1 30 56N 88 35W
Lucena, Phil. 37 B6 13 56N 121 37 E
Lucena, Spain 19 D3 37 27N 4 31W
Lučenec, Slovak Rep. 17 D10 48 18N 19 42 E
Lucerne = Luzern, Switz. 18 C8 47 3N 8 18 E
Lucerne, U.S.A. 84 F4 39 6N 122 48W
Lucerne Valley, U.S.A. 85 L10 34 27N 116 57W
Lucero, Mexico 86 A3 30 49N 106 30W
Lucheng, China 34 F7 36 20N 113 11 E
Lucheringo →, Mozam. 55 E4 11 43S 36 17 E
Lucia, U.S.A. 84 J5 36 2N 121 33W
Lucinda, Australia 62 B4 18 32S 146 20 E
Luckenwalde, Germany 16 B7 52 5N 13 10 E
Luckhoff, S. Africa 56 D3 29 44S 24 43 E
Lucknow, Canada 78 C3 43 57N 81 31W
Lucknow, India 43 F9 26 50N 81 0 E
Lüda = Dalian, China 35 E11 38 50N 121 40 E
Lüderitz, Namibia 56 D2 26 41S 15 8 E
Lüderitzbaai, Namibia 56 D2 26 36S 15 8 E
Ludhiana, India 42 D6 30 57N 75 56 E
Ludington, U.S.A. 76 D2 43 57N 86 27W
Ludlow, U.K. 11 E5 52 22N 2 42W
Ludlow, Calif., U.S.A. 85 L10 34 43N 116 10W
Ludlow, Pa., U.S.A. 78 E6 41 43N 78 56W
Ludlow, Vt., U.S.A. 79 C12 43 24N 72 42W
Ludvika, Sweden 9 F16 60 8N 15 14 E
Ludwigsburg, Germany 16 D5 48 53N 9 11 E
Ludwigshafen, Germany 16 D5 49 29N 8 26 E
Lueki, Dem. Rep. of the Congo 54 C2 3 20S 25 48 E
Luena, Dem. Rep. of the Congo 55 D2 9 28S 25 43 E
Luena, Zambia 55 E3 10 40S 30 25 E
Lüeyang, China 34 H4 33 22N 106 10 E
Lufira →, Dem. Rep. of the Congo 55 D2 9 30S 27 0 E
Lufkin, U.S.A. 81 K7 31 21N 94 44W
Lufupa, Dem. Rep. of the Congo 55 E1 10 37S 24 56 E
Luga, Russia 24 C4 58 40N 29 55 E
Lugano, Switz. 18 C8 46 1N 8 57 E
Lugansk = Luhansk, Ukraine 25 E6 48 38N 39 15 E
Lugard's Falls, Kenya 54 C4 3 6S 38 41 E
Lugela, Mozam. 55 F4 16 25S 36 43 E
Lugenda →, Mozam. 55 E4 11 25S 38 33 E
Lugh Ganana, Somali Rep. 46 G3 3 48N 42 34 E
Lugnaquilla, Ireland 13 D5 52 58N 6 28W
Lugo, Italy 20 B4 44 25N 11 54 E
Lugo, Spain 19 A2 43 2N 7 35W
Lugoj, Romania 17 F11 45 42N 21 57 E
Lugovoy = Qulan, Kazakstan 26 E8 42 55N 72 43 E
Luhansk, Ukraine 25 E6 48 38N 39 15 E
Luiana, Angola 56 B3 17 25S 22 59 E
Luimneach = Limerick, Ireland 13 D3 52 40N 8 37W
Luing, U.K. 12 E3 56 14N 5 39W
Luís Correia, Brazil 93 D10 3 0S 41 35W
Luitpold Coast, Antarctica 5 D1 78 30S 32 0W
Luiza, Dem. Rep. of the Congo 52 F4 7 40S 22 30 E
Luizi, Dem. Rep. of the Congo 54 D2 6 0S 27 25 E
Luján, Argentina 94 C4 34 45S 59 5W
Lukanga Swamp, Zambia 55 E2 14 30S 27 40 E
Lukenie →, Dem. Rep. of the Congo 52 E3 3 0S 18 50 E
Lukhisaral, India 43 G12 25 11N 86 5 E
Lukolela, Dem. Rep. of the Congo 54 D1 5 23S 24 32 E
Lukosi, Zimbabwe 55 F2 18 30S 26 30 E
Łuków, Poland 17 C12 51 55N 22 23 E
Lule älv →, Sweden 8 D19 65 35N 22 10 E
Luleå, Sweden 8 D20 65 35N 22 10 E
Lüleburgaz, Turkey 21 D12 41 23N 27 22 E
Luling, U.S.A. 81 L6 29 41N 97 39W
Lulong, China 35 E10 39 53N 118 51 E
Lulonga →, Dem. Rep. of the Congo 52 D3 1 0N 18 10 E
Lulua →, Dem. Rep. of the Congo 52 E4 4 30S 20 30 E
Lumajang, Indonesia 37 H15 8 8S 113 13 E
Lumbala N'guimbo, Angola 53 G4 14 18S 21 18 E
Lumberton, U.S.A. 77 H6 34 37N 79 0W
Lumbwa, Kenya 54 C4 0 12S 35 28 E
Lumsden, Canada 73 C8 50 39N 104 52W
Lumsden, N.Z. 59 L2 45 44S 168 27 E
Lumut, Malaysia 39 K3 4 13N 100 37 E
Lumut, Tanjung, Indonesia 36 E3 3 50S 105 58 E
Luna, India 42 H3 23 43N 69 16 E
Lunavada, India 42 H5 23 8N 73 37 E
Lund, Sweden 9 J15 55 44N 13 12 E
Lundazi, Zambia 55 E3 12 20S 33 7 E
Lundi →, Zimbabwe 55 G3 21 43S 32 34 E
Lundu, Malaysia 36 D3 1 40N 109 50 E
Lundy, U.K. 11 F3 51 10N 4 41W
Lune →, U.K. 10 C5 54 0N 2 51W
Lüneburg, Germany 16 B6 53 15N 10 24 E
Lüneburg Heath = Lüneburger Heide, Germany 16 B6 53 10N 10 12 E
Lüneburger Heide, Germany 16 B6 53 10N 10 12 E
Lunenburg, Canada 71 D7 44 22N 64 18W
Lunéville, France 18 B7 48 36N 6 30 E
Lunga →, Zambia 55 E2 14 34S 26 25 E
Lunglei, India 41 H18 22 55N 92 45 E
Luni, India 42 G5 26 0N 73 6 E
Luni →, India 42 G4 24 41N 71 14 E
Luninets = Luninyets, Belarus 17 B14 52 15N 26 50 E
Luning, U.S.A. 82 G4 38 30N 118 11W
Luninyets, Belarus 17 B14 52 15N 26 50 E
Lunkaransar, India 42 E5 28 29N 73 44 E
Lunsemfwa →, Zambia 55 E3 14 54S 30 12 E
Lunsemfwa Falls, Zambia 55 E2 14 30S 29 6 E
Luo He →, China 34 G6 34 35N 110 20 E
Luochuan, China 34 G5 35 45N 109 26 E
Luofu, Dem. Rep. of the Congo 54 C2 0 10S 29 15 E
Luohe, China 34 H8 33 32N 114 2 E
Luonan, China 34 G6 34 5N 110 10 E
Luoning, China 34 G6 34 35N 111 40 E
Luoyang, China 34 G7 34 40N 112 26 E
Luozigou, China 35 C16 43 42N 130 18 E
Lupanshui, China 32 D5 26 38N 104 48 E
Lupilichi, Mozam. 55 E4 11 47S 35 13 E
Luque, Paraguay 94 B4 25 19S 57 25W
Luray, U.S.A. 76 F6 38 40N 78 28W
Lurgan, U.K. 13 B5 54 28N 6 19W
Lusaka, Zambia 55 F2 15 28S 28 16 E
Lusambo, Dem. Rep. of the Congo 54 C1 4 58S 23 28 E
Lusangaye, Dem. Rep. of the Congo 54 C2 4 54S 26 0 E
Luseland, Canada 73 C7 52 5N 109 24W
Lushan, China 34 H7 33 45N 112 55 E
Lushi, China 34 G6 34 3N 111 3 E
Lushnjë, Albania 21 D8 40 55N 19 41 E
Lushoto, Tanzania 54 C4 4 47S 38 20 E
Lüshun, China 35 E11 38 45N 121 15 E
Lusk, U.S.A. 80 D2 42 46N 104 27W
Lūt, Dasht-e, Iran 45 D8 31 30N 58 0 E
Luta = Dalian, China 35 E11 38 50N 121 40 E
Lutherstadt Wittenberg, Germany 16 C7 51 53N 12 39 E
Luton, U.K. 11 F7 51 53N 0 24W
Luton □, U.K. 11 F7 51 53N 0 24W
Lutselke, Canada 73 A6 62 24N 110 44W
Lutsk, Ukraine 17 C13 50 50N 25 15 E
Lützow Holmbukta, Antarctica 5 C4 69 10S 37 30 E
Lutzputs, S. Africa 56 D3 28 3S 20 40 E
Luuq = Lugh Ganana, Somali Rep. 46 G3 3 48N 42 34 E
Luverne, Ala., U.S.A. 77 K2 31 43N 86 16W
Luverne, Minn., U.S.A. 80 D6 43 39N 96 13W
Luvua, Dem. Rep. of the Congo 55 D2 8 48S 25 17 E
Luvua →, Dem. Rep. of the Congo 54 D2 6 50S 27 30 E
Luvuvhu →, S. Africa 57 C5 22 25S 31 18 E
Luwegu →, Tanzania 55 D4 8 31S 37 23 E
Luwuk, Indonesia 37 E6 0 56S 122 47 E
Luxembourg, Lux. 18 B7 49 37N 6 9 E
Luxembourg □, Belgium 15 E5 49 58N 5 30 E
Luxembourg ■, Europe 18 B7 49 45N 6 0 E
Luxi, China 32 D4 24 27N 98 36 E
Luxor = El Uqsur, Egypt 51 C12 25 41N 32 38 E
Luyi, China 34 H8 33 50N 115 35 E
Luza, Russia 24 B8 60 39N 47 10 E
Luzern, Switz. 18 C8 47 3N 8 18 E
Luzhou, China 32 D5 28 52N 105 20 E
Luziânia, Brazil 93 G9 16 20S 48 0W
Luzon, Phil. 37 A6 16 0N 121 0 E
Lviv, Ukraine 17 D13 49 50N 24 0 E
Lvov = Lviv, Ukraine 17 D13 49 50N 24 0 E
Lyakhavichy, Belarus 17 B14 53 2N 26 32 E
Lyakhovskiye, Ostrova, Russia 27 B15 73 40N 141 0 E
Lyal I., Canada 78 B3 44 57N 81 24W
Lyallpur = Faisalabad, Pakistan 42 D5 31 30N 73 5 E
Lybster, U.K. 12 C5 58 18N 3 15W
Lycksele, Sweden 8 D18 64 38N 18 40 E
Lydda = Lod, Israel 47 D3 31 57N 34 54 E
Lydenburg, S. Africa 57 D5 25 10S 30 29 E
Lydia, Turkey 21 E13 38 48N 28 19 E
Lyell, N.Z. 59 J4 41 48S 172 4 E
Lyell I., Canada 72 C2 52 40N 131 35W
Lyepyel, Belarus 24 D4 54 50N 28 40 E
Lykens, U.S.A. 79 F8 40 34N 76 42W
Lyman, U.S.A. 82 F8 41 20N 110 18W
Lyme B., U.K. 11 G4 50 42N 2 53W
Lyme Regis, U.K. 11 G5 50 43N 2 57W
Lymington, U.K. 11 G6 50 45N 1 32W
Łyna →, Poland 9 J19 54 37N 21 14 E
Lynchburg, U.S.A. 76 G6 37 25N 79 9W
Lynd →, Australia 62 B3 16 28S 143 18 E
Lynd Ra., Australia 63 D4 25 30S 149 20 E
Lynden, Canada 78 C4 43 14N 80 9W

Mal B., *Ireland* **13 D2** 52 50N 9 30W
Mala, Pta., *Panama* **88 E3** 7 28N 80 2W
Malabar Coast, *India* **40 P9** 11 0N 75 0 E
Malabo = Rey Malabo, *Eq. Guin.* **52 D1** 3 45N 8 50 E
Malacca, Str. of, *Indonesia* **39 L3** 3 0N 101 0 E
Malad City, *U.S.A.* **82 E7** 42 12N 112 15W
Maladzyechna, *Belarus* **17 A14** 54 20N 26 50 E
Málaga, *Spain* **19 D3** 36 43N 4 23W
Malagarasi, *Tanzania* **54 D3** 5 5S 30 50 E
Malagarasi →, *Tanzania* **54 D2** 5 12S 29 47 E
Malagasy Rep. = Madagascar ■, *Africa* **57 C8** 20 0S 47 0 E
Malahide, *Ireland* **13 C5** 53 26N 6 9W
Malaimbandy, *Madag.* **57 C8** 20 20S 45 36 E
Malakâl, *Sudan* **51 G12** 9 33N 31 40 E
Malakand, *Pakistan* **42 B4** 34 40N 71 55 E
Malakwal, *Pakistan* **42 C5** 32 34N 73 13 E
Malamala, *Indonesia* **37 E6** 3 21S 120 55 E
Malanda, *Australia* **62 B4** 17 22S 145 35 E
Malang, *Indonesia* **36 F4** 7 59S 112 45 E
Malangen, *Norway* **8 B18** 69 24N 18 37 E
Malanje, *Angola* **52 F3** 9 36S 16 17 E
Mälaren, *Sweden* **9 G17** 59 30N 17 10 E
Malargüe, *Argentina* **94 D2** 35 32S 69 30W
Malartic, *Canada* **70 C4** 48 9N 78 9W
Malaryta, *Belarus* **17 C13** 51 50N 24 3 E
Malatya, *Turkey* **25 G6** 38 25N 38 20 E
Malawi ■, *Africa* **55 E3** 11 55S 34 0 E
Malawi, L. = Nyasa, L., *Africa* **55 E3** 12 30S 34 30 E
Malay Pen., *Asia* **39 J3** 7 25N 100 0 E
Malaya Vishera, *Russia* **24 C5** 58 55N 32 25 E
Malaybalay, *Phil.* **37 C7** 8 5N 125 7 E
Malāyer, *Iran* **45 C6** 34 19N 48 51 E
Malaysia ■, *Asia* **39 K4** 5 0N 110 0 E
Malazgirt, *Turkey* **25 G7** 39 10N 42 33 E
Malbon, *Australia* **62 C3** 21 5S 140 17 E
Malbooma, *Australia* **63 E1** 30 41S 134 11 E
Malbork, *Poland* **17 B10** 54 3N 19 1 E
Malcolm, *Australia* **61 E3** 28 51S 121 25 E
Malcolm, Pt., *Australia* **61 F3** 33 48S 123 45 E
Maldah, *India* **43 G13** 25 2N 88 9 E
Maldegem, *Belgium* **15 C3** 51 14N 3 26 E
Malden, *Mass., U.S.A.* **79 D13** 42 26N 71 4W
Malden, *Mo., U.S.A.* **81 G10** 36 34N 89 57W
Malden I., *Kiribati* **65 H12** 4 3S 155 1W
Maldives ■, *Ind. Oc.* **29 J11** 5 0N 73 0 E
Maldonado, *Uruguay* **95 C5** 34 59S 55 0W
Maldonado, Punta, *Mexico* **87 D5** 16 19N 98 35W
Malé, *Maldives* **29 J11** 4 0N 73 28 E
Malé Karpaty, *Slovak Rep.* **17 D9** 48 30N 17 20 E
Maléa, Ákra, *Greece* **21 F10** 36 28N 23 7 E
Malegaon, *India* **40 J9** 20 30N 74 38 E
Malei, *Mozam.* **55 F4** 17 12S 36 58 E
Malek Kandī, *Iran* **44 B5** 37 9N 46 6 E
Malela, *Dem. Rep. of the Congo* **54 C2** 4 22S 26 8 E
Malema, *Mozam.* **55 E4** 14 57S 37 20 E
Maléme, *Greece* **23 D5** 35 31N 23 49 E
Maleny, *Australia* **63 D5** 26 45S 152 52 E
Malerkotla, *India* **42 D6** 30 32N 75 58 E
Máles, *Greece* **23 D7** 35 6N 25 35 E
Malgomaj, *Sweden* **8 D17** 64 40N 16 30 E
Malha, *Sudan* **51 E11** 15 8N 25 10 E
Malhargarh, *India* **42 G6** 24 17N 74 59 E
Malheur →, *U.S.A.* **82 D5** 44 4N 116 59W
Malheur L., *U.S.A.* **82 E4** 43 20N 118 48W
Mali ■, *Africa* **50 E5** 17 0N 3 0W
Mali →, *Burma* **41 G20** 25 40N 97 40 E
Mali Kyun, *Burma* **38 F2** 13 0N 98 20 E
Malibu, *U.S.A.* **85 L8** 34 2N 118 41W
Maliku, *Indonesia* **37 E6** 0 39S 123 16 E
Malili, *Indonesia* **37 E6** 2 42S 121 6 E
Malimba, Mts., *Dem. Rep. of the Congo* **54 D2** 7 30S 29 30 E
Malin Hd., *Ireland* **13 A4** 55 23N 7 23W
Malin Pen., *Ireland* **13 A4** 55 20N 7 17W
Malindi, *Kenya* **54 C5** 3 12S 40 5 E
Malines = Mechelen, *Belgium* **15 C4** 51 2N 4 29 E
Malino, *Indonesia* **37 D6** 1 0N 121 0 E
Malinyi, *Tanzania* **55 D4** 8 56S 36 0 E
Malita, *Phil.* **37 C7** 6 19N 125 39 E
Maliwun, *Burma* **36 B1** 10 17N 98 40 E
Maliya, *India* **42 H4** 23 5N 70 46 E
Malkara, *Turkey* **21 D12** 40 53N 26 53 E
Mallacoota Inlet, *Australia* **63 F4** 37 34S 149 40 E
Mallaig, *U.K.* **12 D3** 57 0N 5 50W
Mallawan, *India* **43 F9** 27 4N 80 12 E
Mallawi, *Egypt* **51 C12** 27 44N 30 44 E
Mállia, *Greece* **23 D7** 35 17N 25 32 E
Mallión, Kólpos, *Greece* **23 D7** 35 19N 25 27 E
Mallorca, *Spain* **22 B10** 39 30N 3 0 E
Mallorytown, *Canada* **79 B9** 44 29N 75 53W
Mallow, *Ireland* **13 D3** 52 8N 8 39W
Malmberget, *Sweden* **8 C19** 67 11N 20 40 E
Malmédy, *Belgium* **15 D6** 50 25N 6 2 E
Malmesbury, *S. Africa* **56 E2** 33 28S 18 41 E
Malmö, *Sweden* **9 J15** 55 36N 12 59 E
Malolos, *Phil.* **37 B6** 14 50N 120 49 E
Malombe L., *Malawi* **55 E4** 14 40S 35 15 E
Malone, *U.S.A.* **79 B10** 44 51N 74 18W
Måløy, *Norway* **9 F11** 61 57N 5 6 E
Malpaso, *Canary Is.* **22 G1** 27 43N 18 3W
Malpelo, I. de, *Colombia* **92 C2** 4 3N 81 35W
Malpur, *India* **42 H5** 23 21N 73 27 E
Malpura, *India* **42 F6** 26 17N 75 23 E
Malta, *Idaho, U.S.A.* **82 E7** 42 18N 113 22W
Malta, *Mont., U.S.A.* **82 B10** 48 21N 107 52W
Malta ■, *Europe* **23 D2** 35 55N 14 26 E
Maltahöhe, *Namibia* **56 C2** 24 55S 17 0 E
Malton, *Canada* **78 C5** 43 42N 79 38W
Malton, *U.K.* **10 C7** 54 8N 0 49W
Maluku, *Indonesia* **37 E7** 1 0S 127 0 E
Maluku □, *Indonesia* **37 E7** 3 0S 128 0 E
Maluku Sea = Molucca Sea, *Indonesia* **37 E6** 0 0 125 0 E
Malvan, *India* **40 L8** 16 2N 73 30 E
Malvern, *U.S.A.* **81 H8** 34 22N 92 49W
Malvern Hills, *U.K.* **11 E5** 52 0N 2 19W
Malvinas, Is. = Falkland Is. □, *Atl. Oc.* **96 G5** 51 30S 59 0W
Malya, *Tanzania* **54 C3** 3 5S 33 38 E
Malyn, *Ukraine* **17 C15** 50 46N 29 3 E
Malyy Lyakhovskiy, Ostrov, *Russia* **27 B15** 74 7N 140 36 E
Mama, *Russia* **27 D12** 58 18N 112 54 E
Mamanguape, *Brazil* **93 E11** 6 50S 35 4W
Mamarr Mitlā, *Egypt* **47 E1** 30 2N 32 54 E
Mamasa, *Indonesia* **37 E5** 2 55S 119 20 E
Mambasa, *Dem. Rep. of the Congo* **54 B2** 1 22N 29 3 E
Mamberamo →, *Indonesia* **37 E9** 2 0S 137 50 E
Mambilima Falls, *Zambia* **55 E2** 10 31S 28 45 E
Mambirima, *Dem. Rep. of the Congo* **55 E2** 11 25S 27 33 E
Mambo, *Tanzania* **54 C4** 4 52S 38 22 E
Mambrui, *Kenya* **54 C5** 3 5S 40 5 E
Mamburao, *Phil.* **37 B6** 13 13N 120 39 E
Mameigwess L., *Canada* **70 B2** 52 35N 87 50W
Mammoth, *U.S.A.* **83 K8** 32 43N 110 39W
Mammoth Cave Nat. Park, *U.S.A.* **76 G3** 37 8N 86 13W
Mamoré →, *Bolivia* **92 F5** 10 23S 65 53W
Mamou, *Guinea* **50 F3** 10 15N 12 0W
Mamoudzou, *Mayotte* **49 H8** 12 48S 45 14 E
Mampikony, *Madag.* **57 B8** 16 6S 47 38 E
Mamuju, *Indonesia* **37 E5** 2 41S 118 50 E
Mamuno, *Botswana* **56 C3** 22 16S 20 1 E
Man, *Ivory C.* **50 G4** 7 30N 7 40W
Man, I. of, *U.K.* **10 C3** 54 15N 4 30W
Man-Bazar, *India* **43 H12** 23 4N 86 39 E
Man Na, *Burma* **41 H20** 23 27N 97 19 E
Mana →, *Fr. Guiana* **93 B8** 5 45N 53 55W
Manaar, G. of = Mannar, G. of, *Asia* **40 Q11** 8 30N 79 0 E
Manacapuru, *Brazil* **92 D6** 3 16S 60 37W
Manacor, *Spain* **22 B10** 39 34N 3 13 E
Manado, *Indonesia* **37 D6** 1 29N 124 51 E
Managua, *Nic.* **88 D2** 12 6N 86 20W
Managua, L. de, *Nic.* **88 D2** 12 20N 86 30W
Manakara, *Madag.* **57 C8** 22 8S 48 1 E
Manali, *India* **42 C7** 32 16N 77 10 E
Manama = Al Manāmah, *Bahrain* **46 B5** 26 10N 50 30 E
Manambao →, *Madag.* **57 B7** 17 35S 44 0 E
Manambato, *Madag.* **57 A8** 13 43S 49 7 E
Manambolo →, *Madag.* **57 B7** 19 18S 44 22 E
Manambolosy, *Madag.* **57 B8** 16 2S 49 40 E
Mananara, *Madag.* **57 B8** 16 10S 49 46 E
Mananara →, *Madag.* **57 C8** 23 21S 47 42 E
Mananjary, *Madag.* **57 C8** 21 13S 48 20 E
Manantenina, *Madag.* **57 C8** 24 17S 47 19 E
Manaos = Manaus, *Brazil* **92 D7** 3 0S 60 0W
Manapire →, *Venezuela* **92 B5** 7 42N 66 7W
Manapouri, *N.Z.* **59 L1** 45 34S 167 39 E
Manapouri, L., *N.Z.* **59 L1** 45 32S 167 32 E
Manār, Jabal, *Yemen* **46 E3** 14 2N 44 17 E
Manaravolo, *Madag.* **57 C8** 23 59S 45 39 E
Manas, *China* **32 B3** 44 17N 85 56 E
Manas →, *India* **41 F17** 26 12N 90 40 E
Manaslu, *Nepal* **43 E11** 28 33N 84 33 E
Manasquan, *U.S.A.* **79 F10** 40 8N 74 3W
Manassa, *U.S.A.* **83 H11** 37 11N 105 56W
Manaung, *Burma* **41 K18** 18 45N 93 40 E
Manaus, *Brazil* **92 D7** 3 0S 60 0W
Manawan L., *Canada* **73 B8** 55 24N 103 14W
Manbij, *Syria* **44 B3** 36 31N 37 57 E
Manchegorsk, *Russia* **26 C4** 67 54N 32 58 E
Manchester, *U.K.* **10 D5** 53 29N 2 12W
Manchester, *Calif., U.S.A.* **84 G3** 38 58N 123 41W
Manchester, *Conn., U.S.A.* **79 E12** 41 47N 72 31W
Manchester, *Ga., U.S.A.* **77 J3** 32 51N 84 37W
Manchester, *Iowa, U.S.A.* **80 D9** 42 29N 91 27W
Manchester, *Ky., U.S.A.* **76 G4** 37 9N 83 46W
Manchester, *N.H., U.S.A.* **79 D13** 42 59N 71 28W
Manchester, *N.Y., U.S.A.* **78 D7** 42 56N 77 16W
Manchester, *Pa., U.S.A.* **79 F8** 40 4N 76 43W
Manchester, *Tenn., U.S.A.* **77 H2** 35 29N 86 5W
Manchester, *Vt., U.S.A.* **79 C11** 43 10N 73 5W
Manchester L., *Canada* **73 A7** 61 28N 107 29W
Manchhar L., *Pakistan* **42 F2** 26 25N 67 39 E
Manchuria = Dongbei, *China* **35 D13** 45 0N 125 0 E
Manchurian Plain, *China* **28 E16** 47 0N 124 0 E
Mand →, *India* **43 J10** 21 42N 83 15 E
Mand →, *Iran* **45 D7** 28 20N 52 30 E
Manda, *Ludewe, Tanzania* **55 E3** 10 30S 34 40 E
Manda, *Mbeya, Tanzania* **54 D3** 7 58S 32 29 E
Manda, *Mbeya, Tanzania* **55 D3** 8 30S 32 49 E
Mandabé, *Madag.* **57 C7** 21 0S 44 55 E
Mandaguari, *Brazil* **95 A5** 23 32S 51 42W
Mandah = Töhöm, *Mongolia* **34 B5** 44 27N 108 2 E
Mandal, *Norway* **9 G12** 58 2N 7 25 E
Mandala, Puncak, *Indonesia* **37 E10** 4 44S 140 20 E
Mandalay, *Burma* **41 J20** 22 0N 96 4 E
Mandale = Mandalay, *Burma* **41 J20** 22 0N 96 4 E
Mandalgarh, *India* **42 G6** 25 12N 75 6 E
Mandalgovi, *Mongolia* **34 B4** 45 45N 106 10 E
Mandalī, *Iraq* **44 C5** 33 43N 45 28 E
Mandan, *U.S.A.* **80 B4** 46 50N 100 54W
Mandar, Teluk, *Indonesia* **37 E5** 3 35S 119 15 E
Mandaue, *Phil.* **37 B6** 10 20N 123 56 E
Mandera, *Kenya* **54 B5** 3 55N 41 53 E
Mandi, *India* **42 D7** 31 39N 76 58 E
Mandi Dabwali, *India* **42 E6** 29 58N 74 42 E
Mandimba, *Mozam.* **55 E4** 14 20S 35 40 E
Mandioli, *Indonesia* **37 E7** 0 40S 127 20 E
Mandla, *India* **43 H9** 22 39N 80 30 E
Mandorah, *Australia* **60 B5** 12 32S 130 42 E
Mandoto, *Madag.* **57 B8** 19 34S 46 17 E
Mandra, *Pakistan* **42 C5** 33 23N 73 12 E
Mandrare →, *Madag.* **57 D8** 25 10S 46 30 E
Mandritsara, *Madag.* **57 B8** 15 50S 48 49 E
Mandronarivo, *Madag.* **57 C8** 21 7S 45 38 E
Mandsaur, *India* **42 G6** 24 3N 75 8 E
Mandurah, *Australia* **61 F2** 32 36S 115 48 E
Mandvi, *India* **42 H3** 22 51N 69 22 E
Mandya, *India* **40 N10** 12 30N 77 0 E
Mandzai, *Pakistan* **42 D2** 30 55N 67 6 E
Maneh, *Iran* **45 B8** 37 39N 57 7 E
Manera, *Madag.* **57 C7** 22 55S 44 20 E
Maneroo Cr. →, *Australia* **62 C3** 23 21S 143 53 E
Manfalût, *Egypt* **51 C12** 27 20N 30 52 E
Manfredónia, *Italy* **20 D6** 41 38N 15 55 E
Mangabeiras, Chapada das, *Brazil* **93 F9** 10 0S 46 30W
Mangalia, *Romania* **17 G15** 43 50N 28 35 E
Mangalore, *India* **40 N9** 12 55N 74 47 E
Mangan, *India* **43 F13** 27 31N 88 32 E
Mangaung, *S. Africa* **53 K5** 29 10S 26 25 E
Mangawan, *India* **43 G9** 24 41N 81 33 E
Mangaweka, *N.Z.* **59 H5** 39 48S 175 47 E
Manggar, *Indonesia* **36 E3** 2 50S 108 10 E
Manggawitu, *Indonesia* **37 E8** 4 8S 133 32 E
Mangindrano, *Madag.* **57 A8** 14 17S 48 58 E
Mangkalihat, Tanjung, *Indonesia* **37 D5** 1 2N 118 59 E
Mangla, *Pakistan* **42 C5** 33 7N 73 39 E
Mangla Dam, *Pakistan* **43 C5** 33 9N 73 44 E
Manglaur, *India* **42 E7** 29 44N 77 49 E
Mangnai, *China* **32 C4** 37 52N 91 43 E
Mango, *Togo* **50 F6** 10 20N 0 30 E
Mangoche, *Malawi* **55 E4** 14 25S 35 16 E
Mangoky →, *Madag.* **57 C7** 21 29S 43 41 E
Mangole, *Indonesia* **37 E6** 1 50S 125 55 E
Mangombe, *Dem. Rep. of the Congo* **54 C2** 1 20S 26 48 E
Mangonui, *N.Z.* **59 F4** 35 1S 173 32 E
Mangoro →, *Madag.* **57 B8** 20 0S 48 45 E
Mangrol, *Mad. P., India* **42 J4** 21 7N 70 7 E
Mangrol, *Raj., India* **42 G6** 25 20N 76 31 E
Mangueira, L. da, *Brazil* **95 C5** 33 0S 52 50W
Mangum, *U.S.A.* **81 H5** 34 53N 99 30W
Mangyshlak Poluostrov, *Kazakstan* **26 E6** 44 30N 52 30 E
Manhattan, *U.S.A.* **80 F6** 39 11N 96 35W
Manhiça, *Mozam.* **57 D5** 25 23S 32 49 E
Mania →, *Madag.* **57 B8** 19 42S 45 22 E
Manica, *Mozam.* **57 B5** 18 58S 32 59 E
Manica □, *Mozam.* **57 B5** 19 10S 33 45 E
Manicaland □, *Zimbabwe* **55 F3** 19 0S 32 30 E
Manicoré, *Brazil* **92 E6** 5 48S 61 16W
Manicouagan →, *Canada* **71 C6** 49 30N 68 30W
Manicouagan, Rés., *Canada* **71 B6** 51 5N 68 40W
Maniema □, *Dem. Rep. of the Congo* **54 C2** 3 0S 26 0 E
Manīfah, *Si. Arabia* **45 E6** 27 44N 49 0 E
Manifold, C., *Australia* **62 C5** 22 41S 150 50 E
Manigotagan, *Canada* **73 C9** 51 6N 96 18W
Manigotagan →, *Canada* **73 C9** 51 7N 96 20W
Manihari, *India* **43 G12** 25 21N 87 38 E
Manihiki, *Cook Is.* **65 J11** 10 24S 161 1W
Manika, Plateau de la, *Dem. Rep. of the Congo* **55 E2** 10 0S 25 5 E
Manikpur, *India* **43 G9** 25 4N 81 7 E
Manila, *Phil.* **37 B6** 14 40N 121 3 E
Manila, *U.S.A.* **82 F9** 40 59N 109 43W
Manila B., *Phil.* **37 B6** 14 40N 120 35 E
Manilla, *Australia* **63 E5** 30 45S 150 43 E
Maningrida, *Australia* **62 A1** 12 3S 134 13 E
Manipur □, *India* **41 G19** 25 0N 94 0 E
Manipur →, *Burma* **41 H19** 23 45N 94 20 E
Manisa, *Turkey* **21 E12** 38 38N 27 30 E
Manistee, *U.S.A.* **76 C2** 44 15N 86 19W
Manistee →, *U.S.A.* **76 C2** 44 15N 86 21W
Manistique, *U.S.A.* **76 C2** 45 57N 86 15W
Manito L., *Canada* **73 C7** 52 43N 109 43W
Manitoba □, *Canada* **73 B9** 55 30N 97 0W
Manitoba, L., *Canada* **73 C9** 51 0N 98 45W
Manitou, *Canada* **73 D9** 49 15N 98 32W
Manitou, L., *Canada* **71 B6** 50 55N 65 17W
Manitou Is., *U.S.A.* **76 C3** 45 8N 86 0W
Manitou Springs, *U.S.A.* **80 F2** 38 52N 104 55W
Manitoulin I., *Canada* **70 C3** 45 40N 82 30W
Manitouwadge, *Canada* **70 C2** 49 8N 85 48W
Manitowoc, *U.S.A.* **76 C2** 44 5N 87 40W
Manizales, *Colombia* **92 B3** 5 5N 75 32W
Manja, *Madag.* **57 C7** 21 26S 44 20 E
Manjacaze, *Mozam.* **57 C5** 24 45S 34 0 E
Manjakandriana, *Madag.* **57 B8** 18 55S 47 47 E
Manjhand, *Pakistan* **42 G3** 25 50N 68 10 E
Manjil, *Iran* **45 B6** 36 46N 49 30 E
Manjimup, *Australia* **61 F2** 34 15S 116 6 E
Manjra →, *India* **40 K10** 18 49N 77 52 E
Mankato, *Kans., U.S.A.* **80 F5** 39 47N 98 13W
Mankato, *Minn., U.S.A.* **80 C8** 44 10N 94 0W
Mankayane, *Swaziland* **57 D5** 26 40S 31 4 E
Mankera, *Pakistan* **42 D4** 31 23N 71 26 E
Mankota, *Canada* **73 D7** 49 25N 107 5W
Manlay = Üydzin, *Mongolia* **34 B4** 44 9N 107 0 E
Manmad, *India* **40 J9** 20 18N 74 28 E
Mann Ranges, *Australia* **61 E5** 26 6S 130 5 E
Manna, *Indonesia* **36 E2** 4 25S 102 55 E
Mannahill, *Australia* **63 E3** 32 25S 140 0 E
Mannar, *Sri Lanka* **40 Q11** 9 1N 79 54 E
Mannar, G. of, *Asia* **40 Q11** 8 30N 79 0 E
Mannar I., *Sri Lanka* **40 Q11** 9 5N 79 45 E
Mannheim, *Germany* **16 D5** 49 29N 8 29 E
Manning, *Canada* **72 B5** 56 53N 117 39W
Manning, *Oreg., U.S.A.* **84 E3** 45 45N 123 13W
Manning, *S.C., U.S.A.* **77 J5** 33 42N 80 13W
Manning Prov. Park, *Canada* **72 D4** 49 5N 120 45W
Mannum, *Australia* **63 E2** 34 50S 139 20 E
Manoharpur, *India* **43 H11** 22 23N 85 12 E
Manokwari, *Indonesia* **37 E8** 0 54S 134 0 E
Manombo, *Madag.* **57 C7** 22 57S 43 28 E
Manono, *Dem. Rep. of the Congo* **54 D2** 7 15S 27 25 E
Manosque, *France* **18 E6** 43 49N 5 47 E
Manotick, *Canada* **79 A9** 45 13N 75 41W
Manouane →, *Canada* **71 C5** 49 30N 71 10W
Manouane, L., *Canada* **71 B5** 50 45N 70 45W
Manp'o, *N. Korea* **35 D14** 41 6N 126 24 E
Manpojin = Manp'o, *N. Korea* **35 D14** 41 6N 126 24 E
Manpur, *Chhattisgarh, India* **43 H10** 23 17N 83 35 E
Manpur, *Mad. P., India* **42 H6** 22 26N 75 37 E
Manresa, *Spain* **19 B6** 41 48N 1 50 E
Mansa, *Gujarat, India* **42 H5** 23 27N 72 45 E
Mansa, *Punjab, India* **42 E6** 30 0N 75 27 E
Mansa, *Zambia* **55 E2** 11 13S 28 55 E
Mansehra, *Pakistan* **42 B5** 34 20N 73 15 E
Mansel I., *Canada* **69 B11** 62 0N 80 0W
Mansfield, *Australia* **63 F4** 37 4S 146 6 E
Mansfield, *U.K.* **10 D6** 53 9N 1 11W
Mansfield, *La., U.S.A.* **81 J8** 32 2N 93 43W
Mansfield, *Mass., U.S.A.* **79 D13** 42 2N 71 13W
Mansfield, *Ohio, U.S.A.* **78 F2** 40 45N 82 31W
Mansfield, *Pa., U.S.A.* **78 E7** 41 48N 77 5W
Mansfield, Mt., *U.S.A.* **79 B12** 44 33N 72 49W
Manson Creek, *Canada* **72 B4** 55 37N 124 32W
Manta, *Ecuador* **92 D2** 1 0S 80 40W
Mantalingajan, Mt., *Phil.* **36 C5** 8 55N 117 45 E
Mantare, *Tanzania* **54 C3** 2 42S 33 13 E
Manteca, *U.S.A.* **84 H5** 37 48N 121 13W
Manteo, *U.S.A.* **77 H8** 35 55N 75 40W
Mantes-la-Jolie, *France* **18 B4** 48 58N 1 41 E
Manthani, *India* **40 K11** 18 40N 79 35 E
Manti, *U.S.A.* **82 G8** 39 16N 111 38W
Mantiqueira, Serra da, *Brazil* **95 A7** 22 0S 44 0W
Manton, *U.S.A.* **76 C3** 44 25N 85 24W
Mántova, *Italy* **20 B4** 45 9N 10 48 E
Mänttä, *Finland* **9 E21** 62 0N 24 40 E
Mantua = Mántova, *Italy* **20 B4** 45 9N 10 48 E
Manu, *Peru* **92 F4** 12 10S 70 51W
Manu →, *Peru* **92 F4** 12 16S 70 55W
Manu'a Is., *Amer. Samoa* **59 B14** 14 13S 169 35W
Manuel Alves →, *Brazil* **93 F9** 11 19S 48 28W
Manui, *Indonesia* **37 E6** 3 35S 123 5 E
Manukau, *N.Z.* **59 G5** 40 43S 175 13 E
Manuripi →, *Bolivia* **92 F5** 11 6S 67 36W
Many, *U.S.A.* **81 K8** 31 34N 93 29W
Manyara, L., *Tanzania* **54 C4** 3 40S 35 50 E
Manych-Gudilo, Ozero, *Russia* **25 E7** 46 24N 42 38 E
Manyonga →, *Tanzania* **54 C3** 4 10S 34 15 E
Manyoni, *Tanzania* **54 D3** 5 45S 34 55 E
Manzai, *Pakistan* **42 C4** 32 12N 70 15 E
Manzanares, *Spain* **19 C4** 39 2N 3 22W
Manzanillo, *Cuba* **88 B4** 20 20N 77 31W
Manzanillo, *Mexico* **86 D4** 19 0N 104 20W
Manzanillo, Pta., *Panama* **88 E4** 9 30N 79 40W
Manzano Mts., *U.S.A.* **83 J10** 34 40N 106 20W
Manzarīyeh, *Iran* **45 C6** 34 53N 50 50 E
Manzhouli, *China* **33 B6** 49 35N 117 25 E
Manzini, *Swaziland* **57 D5** 26 30S 31 25 E
Mao, *Chad* **51 F9** 14 4N 15 19 E
Maó, *Spain* **22 B11** 39 53N 4 16 E
Maoke, Pegunungan, *Indonesia* **37 E9** 3 40S 137 30 E
Maolin, *China* **35 C12** 43 58N 123 30 E
Maoming, *China* **33 D6** 21 50N 110 54 E
Maoxing, *China* **35 B13** 45 28N 124 40 E
Mapam Yumco, *China* **32 C3** 30 45N 81 28 E
Mapastepec, *Mexico* **87 D6** 15 26N 92 54W
Mapia, Kepulauan, *Indonesia* **37 D8** 0 50N 134 20 E
Mapimí, *Mexico* **86 B4** 25 50N 103 50W
Mapimí, Bolsón de, *Mexico* **86 B4** 27 30N 104 15W
Mapinga, *Tanzania* **54 D4** 6 40S 39 12 E
Mapinhane, *Mozam.* **57 C6** 22 20S 35 0 E
Maple Creek, *Canada* **73 D7** 49 55N 109 29W
Maple Valley, *U.S.A.* **84 C4** 47 25N 122 3W
Mapleton, *U.S.A.* **82 D2** 44 2N 123 52W
Mapuera →, *Brazil* **92 D7** 1 5S 57 2W
Mapulanguene, *Mozam.* **57 C5** 24 29S 32 6 E
Maputo, *Mozam.* **57 D5** 25 58S 32 32 E
Maputo □, *Mozam.* **57 D5** 26 0S 32 25 E
Maputo, B. de, *Mozam.* **57 D5** 25 50S 32 45 E
Maqiaohe, *China* **35 B16** 44 40N 130 30 E
Maqnā, *Si. Arabia* **44 D2** 28 25N 34 50 E
Maquela do Zombo, *Angola* **52 F3** 6 0S 15 15 E
Maquinchao, *Argentina* **96 E3** 41 15S 68 50W
Maquoketa, *U.S.A.* **80 D9** 42 4N 90 40W
Mar, Serra do, *Brazil* **95 B6** 25 30S 49 0W
Mar Chiquita, L., *Argentina* **94 C3** 30 40S 62 50W
Mar del Plata, *Argentina* **94 D4** 38 0S 57 30W
Mar Menor, *Spain* **19 D5** 37 40N 0 45W
Mara, *Tanzania* **54 C3** 1 30S 34 32 E
Mara □, *Tanzania* **54 C3** 1 45S 34 20 E
Maraã, *Brazil* **92 D5** 1 52S 65 25W
Marabá, *Brazil* **93 E9** 5 20S 49 5W
Maracá, I. de, *Brazil* **93 C8** 2 10N 50 30W
Maracaibo, *Venezuela* **92 A4** 10 40N 71 37W
Maracaibo, L. de, *Venezuela* **92 B4** 9 40N 71 30W
Maracaju, *Brazil* **95 A4** 21 38S 55 9W
Maracay, *Venezuela* **92 A5** 10 15N 67 28W
Maradi, *Niger* **50 F7** 13 29N 7 20 E
Marāgheh, *Iran* **44 B5** 37 30N 46 12 E
Marāh, *Si. Arabia* **44 E5** 25 0N 45 35 E
Marajó, I. de, *Brazil* **93 D9** 1 0S 49 30W
Marākand, *Iran* **44 B5** 38 51N 45 16 E
Maralal, *Kenya* **54 B4** 1 0N 36 38 E
Maralinga, *Australia* **61 F5** 30 13S 131 32 E
Maran, *Malaysia* **39 L4** 3 35N 102 45 E
Marana, *U.S.A.* **83 K8** 32 27N 111 13W
Maranboy, *Australia* **60 B5** 14 40S 132 39 E
Marand, *Iran* **44 B5** 38 30N 45 45 E
Marang, *Malaysia* **39 K4** 5 12N 103 13 E
Maranguape, *Brazil* **93 D11** 3 55S 38 50W
Maranhão = São Luís, *Brazil* **93 D10** 2 39S 44 15W
Maranhão □, *Brazil* **93 E9** 5 0S 46 0W
Maranoa →, *Australia* **63 D4** 27 50S 148 37 E
Marañón →, *Peru* **92 D4** 4 30S 73 35W
Marão, *Mozam.* **57 C5** 24 18S 34 2 E
Maraş = Kahramanmaraş, *Turkey* **25 G6** 37 37N 36 53 E
Marathasa, *Cyprus* **23 E11** 34 59N 32 51 E
Marathon, *Australia* **62 C3** 20 51S 143 32 E
Marathon, *Canada* **70 C2** 48 44N 86 23W
Marathon, *N.Y., U.S.A.* **79 D8** 42 27N 76 2W
Marathon, *Tex., U.S.A.* **81 K3** 30 12N 103 15W
Marathóvouno, *Cyprus* **23 D12** 35 13N 33 37 E
Maratua, *Indonesia* **37 D5** 2 10N 118 35 E
Maravatío, *Mexico* **86 D4** 19 51N 100 25W
Marāwih, *U.A.E.* **45 E7** 24 18N 53 18 E
Marbella, *Spain* **19 D3** 36 30N 4 57W
Marble Bar, *Australia* **60 D2** 21 9S 119 44 E
Marble Falls, *U.S.A.* **81 K5** 30 35N 98 16W
Marblehead, *U.S.A.* **79 D14** 42 30N 70 51W
Marburg, *Germany* **16 C5** 50 47N 8 46 E
March, *U.K.* **11 E8** 52 33N 0 5 E
Marche, *France* **18 C4** 46 5N 1 20 E
Marche-en-Famenne, *Belgium* **15 D5** 50 14N 5 19 E
Marchena, *Spain* **19 D3** 37 18N 5 23W
Marco, *U.S.A.* **77 N5** 25 58N 81 44W
Marcos Juárez, *Argentina* **94 C3** 32 42S 62 5W
Marcus I. = Minami-Tori-Shima, *Pac. Oc.* **64 E7** 24 20N 153 58 E
Marcus Necker Ridge, *Pac. Oc.* **64 F9** 20 0N 175 0 E
Marcy, Mt., *U.S.A.* **79 B11** 44 7N 73 56W
Mardan, *Pakistan* **42 B5** 34 20N 72 0 E
Mardin, *Turkey* **25 G7** 37 20N 40 43 E
Maree, L., *U.K.* **12 D3** 57 40N 5 26W
Mareeba, *Australia* **62 B4** 16 59S 145 28 E
Mareetsane, *S. Africa* **56 D4** 26 9S 25 25 E
Marek = Stanke Dimitrov, *Bulgaria* **21 C10** 42 17N 23 9 E
Marengo, *U.S.A.* **80 E8** 41 48N 92 4W
Marenyi, *Kenya* **54 C4** 4 22S 39 8 E
Marerano, *Madag.* **57 C7** 21 23S 44 52 E
Marfa, *U.S.A.* **81 K2** 30 19N 104 1W
Marfa Pt., *Malta* **23 D1** 35 59N 14 19 E
Margaret →, *Australia* **60 C4** 18 9S 125 41 E
Margaret Bay, *Canada* **72 C3** 51 20N 127 35W

Mbala

Margaret L., Canada 72 B5 58 56N 115 25W
Margaret River, Australia 61 F2 33 57S 115 4 E
Margarita, I. de, Venezuela 92 A6 11 0N 64 0W
Margaritovo, Russia 30 C7 43 25N 134 45 E
Margate, S. Africa 57 E5 30 50S 30 20 E
Margate, U.K. 11 F9 51 23N 1 23 E
Mārgow, Dasht-e, Afghan. 40 D3 30 40N 62 30 E
Marguerite, Canada 72 C4 52 30N 122 25W
Mari El □, Russia 24 C8 56 30N 48 0 E
Mari Indus, Pakistan 42 C4 32 57N 71 34 E
Mari Republic = Mari El □, Russia 24 C8 56 30N 48 0 E
María Elena, Chile 94 A2 22 18S 69 40W
María Grande, Argentina 94 C4 31 45S 59 55W
Maria I., N. Terr., Australia 62 A2 14 52S 135 45 E
Maria I., Tas., Australia 62 G4 42 35S 148 0 E
Maria van Diemen, C., N.Z. 59 F4 34 29S 172 40 E
Mariakani, Kenya 54 C4 3 50S 39 27 E
Marian, Australia 62 C4 21 9S 148 57 E
Marian L., Canada 72 A5 63 0N 116 15W
Mariana Trench, Pac. Oc. 28 H18 13 0N 145 0 E
Marianao, Cuba 88 B3 23 8N 82 24W
Marianna, Ark., U.S.A. 81 H9 34 46N 90 46W
Marianna, Fla., U.S.A. 77 K3 30 46N 85 14W
Marias →, U.S.A. 82 C8 47 56N 110 30W
Mariato, Punta, Panama 88 E3 7 12N 80 52W
Maribor, Slovenia 16 E8 46 36N 15 40 E
Marico →, Africa 56 C4 23 35S 26 57 E
Maricopa, Ariz., U.S.A. 83 K7 33 4N 112 3W
Maricopa, Calif., U.S.A. 85 K7 35 4N 119 24W
Marié →, Brazil 92 D5 0 27S 66 26W
Marie Byrd Land, Antarctica 5 D14 79 30S 125 0W
Marie-Galante, Guadeloupe 89 C7 15 56N 61 16W
Mariecourt = Kangiqsujuaq, Canada 69 B12 61 30N 72 0W
Mariembourg, Belgium 15 D4 50 6N 4 31 E
Mariental, Namibia 56 C2 24 36S 18 0 E
Marienville, U.S.A. 78 E5 41 28N 79 8W
Mariestad, Sweden 9 G15 58 43N 13 50 E
Marietta, Ga., U.S.A. 77 J3 33 57N 84 33W
Marietta, Ohio, U.S.A. 76 F5 39 25N 81 27W
Marieville, Canada 79 A11 45 26N 73 10W
Mariinsk, Russia 26 D9 56 10N 87 20 E
Marijampolė, Lithuania 9 J20 54 33N 23 19 E
Marília, Brazil 95 A6 22 13S 50 0W
Marín, Spain 19 A1 42 23N 8 42W
Marina, U.S.A. 84 J5 36 41N 121 48W
Marinduque, Phil. 37 B6 13 25N 122 0 E
Marine City, U.S.A. 78 D2 42 43N 82 30W
Marinette, U.S.A. 76 C2 45 6N 87 38W
Maringá, Brazil 95 A5 23 26S 52 2W
Marion, Ala., U.S.A. 77 J2 32 38N 87 19W
Marion, Ill., U.S.A. 81 G10 37 44N 88 56W
Marion, Ind., U.S.A. 76 E3 40 32N 85 40W
Marion, Iowa, U.S.A. 80 D9 42 2N 91 36W
Marion, Kans., U.S.A. 80 F6 38 21N 97 1W
Marion, N.C., U.S.A. 77 H5 35 41N 82 1W
Marion, Ohio, U.S.A. 76 E4 40 35N 83 8W
Marion, S.C., U.S.A. 77 H6 34 11N 79 24W
Marion, Va., U.S.A. 77 G5 36 50N 81 31W
Marion, L., U.S.A. 77 J5 33 28N 80 10W
Mariposa, U.S.A. 84 H7 37 29N 119 58W
Mariscal Estigarribia, Paraguay 94 A3 22 3S 60 40W
Maritime Alps = Maritimes, Alpes, Europe 18 D7 44 10N 7 10 E
Maritimes, Alpes, Europe 18 D7 44 10N 7 10 E
Maritsa = Évros →, Greece 21 D12 41 40N 26 34 E
Maritsá, Greece 23 C10 36 22N 28 8 E
Mariupol, Ukraine 25 E6 47 5N 37 31 E
Marīvān, Iran 44 C5 35 30N 46 25 E
Marj 'Uyūn, Lebanon 47 B4 33 20N 35 35 E
Marka = Merca, Somali Rep. 46 G3 1 48N 44 50 E
Markazī □, Iran 45 C6 35 0N 49 30 E
Markdale, Canada 78 B4 44 19N 80 39W
Marked Tree, U.S.A. 81 H9 35 32N 90 25W
Market Drayton, U.K. 10 E5 52 54N 2 29W
Market Harborough, U.K. 11 E7 52 29N 0 55W
Market Rasen, U.K. 10 D7 53 24N 0 20W
Markham, Canada 78 C5 43 52N 79 16W
Markham, Mt., Antarctica 5 E11 83 0S 164 0 E
Markleeville, U.S.A. 84 G7 38 42N 119 47W
Markovo, Russia 27 C17 64 40N 170 24 E
Marks, Russia 24 D8 51 45N 46 50 E
Marksville, U.S.A. 81 K8 31 8N 92 4W
Marla, Australia 63 D1 27 19S 133 33 E
Marlbank, Canada 78 B7 44 26N 77 6W
Marlboro, Mass., U.S.A. 79 D13 42 19N 71 33W
Marlboro, N.Y., U.S.A. 79 E11 41 36N 73 59W
Marlborough, Australia 62 C4 22 46S 149 52 E
Marlborough, U.K. 11 F6 51 25N 1 43W
Marlborough Downs, U.K. 11 F6 51 27N 1 53W
Marlin, U.S.A. 81 K6 31 18N 96 54W
Marlow, U.S.A. 81 H6 34 39N 97 58W
Marmagao, India 40 M8 15 25N 73 56 E
Marmara, Turkey 21 D12 40 35N 27 34 E
Marmara, Sea of = Marmara Denizi, Turkey 21 D13 40 45N 28 15 E
Marmara Denizi, Turkey 21 D13 40 45N 28 15 E
Marmaris, Turkey 21 F13 36 50N 28 14 E
Marmion, Mt., Australia 61 E2 29 16S 119 50 E
Marmion L., Canada 70 C1 48 55N 91 20W
Marmolada, Mte., Italy 20 A4 46 26N 11 51 E
Marmora, Canada 78 B7 44 28N 77 41W
Marne →, France 18 B5 48 48N 2 24 E
Maroala, Madag. 57 B8 15 23S 47 59 E
Maroantsetra, Madag. 57 B8 15 26S 49 44 E
Maroelaboom, Namibia 56 B2 19 15S 18 53 E
Marofandilia, Madag. 57 C7 20 7S 44 34 E
Marolambo, Madag. 57 C8 20 2S 48 7 E
Maromandia, Madag. 57 A8 14 13S 48 5 E
Marondera, Zimbabwe 55 F3 18 5S 31 42 E
Maroni →, Fr. Guiana 93 B8 5 30N 54 0W
Maroochydore, Australia 63 D5 26 29S 153 5 E
Maroona, Australia 63 F3 37 27S 142 54 E
Marosakoa, Madag. 57 B8 15 26S 46 38 E
Maroseranana, Madag. 57 B8 18 32S 48 51 E
Marotandrano, Madag. 57 B8 16 10S 48 50 E
Marotaolano, Madag. 57 A8 12 47S 49 15 E
Maroua, Cameroon 51 F8 10 40N 14 20 E
Marovato, Madag. 57 B8 15 48S 48 5 E
Marovoay, Madag. 57 B8 16 6S 46 39 E
Marquard, S. Africa 56 D4 28 40S 27 28 E
Marquesas Is. = Marquises, Is., Pac. Oc. 65 H14 9 30S 140 0W
Marquette, U.S.A. 76 B2 46 33N 87 24W
Marquises, Is., Pac. Oc. 65 H14 9 30S 140 0W
Marra, Djebel, Sudan 51 F10 13 10N 24 22 E
Marracuene, Mozam. 57 D5 25 45S 32 35 E
Marrakech, Morocco 50 B4 31 9N 8 0W
Marrawah, Australia 62 G3 40 55S 144 42 E
Marree, Australia 63 D2 29 39S 138 1 E
Marrero, U.S.A. 81 L9 29 54N 90 6W
Marrimane, Mozam. 57 C5 22 58S 33 34 E
Marromeu, Mozam. 57 B6 18 15S 36 25 E
Marrowie Cr. →, Australia 63 E4 33 23S 145 40 E
Marrubane, Mozam. 55 F4 18 0S 37 0 E
Marrupa, Mozam. 55 E4 13 8S 37 30 E
Mars Hill, U.S.A. 77 B12 46 31N 67 52W
Marsá Matrûh, Egypt 51 B11 31 19N 27 9 E
Marsabit, Kenya 54 B4 2 18N 38 0 E
Marsala, Italy 20 F5 37 48N 12 26 E
Marsalforn, Malta 23 C1 36 4N 14 16 E
Marsden, Australia 63 E4 33 47S 147 32 E
Marseille, France 18 E6 43 18N 5 23 E
Marseilles = Marseille, France 18 E6 43 18N 5 23 E
Marsh I., U.S.A. 81 L9 29 34N 91 53W
Marshall, Ark., U.S.A. 81 H8 35 55N 92 38W
Marshall, Mich., U.S.A. 76 D3 42 16N 84 58W
Marshall, Minn., U.S.A. 80 C7 44 25N 95 45W
Marshall, Mo., U.S.A. 80 F8 39 7N 93 12W
Marshall, Tex., U.S.A. 81 J7 32 33N 94 23W
Marshall →, Australia 62 C2 22 59S 136 59 E
Marshall Is. ■, Pac. Oc. 64 G9 9 0N 171 0 E
Marshalltown, U.S.A. 80 D8 42 3N 92 55W
Marshbrook, Zimbabwe 57 B5 18 33S 31 9 E
Marshfield, Mo., U.S.A. 81 G8 37 15N 92 54W
Marshfield, Vt., U.S.A. 79 B12 44 20N 72 20W
Marshfield, Wis., U.S.A. 80 C9 44 40N 90 10W
Marshūn, Iran 45 B6 36 19N 49 23 E
Märsta, Sweden 9 G17 59 37N 17 52 E
Mart, U.S.A. 81 K6 31 33N 96 50W
Martaban, Burma 41 L20 16 30N 97 35 E
Martaban, G. of, Burma 41 L20 16 5N 96 30 E
Martapura, Kalimantan, Indonesia 36 E4 3 22S 114 47 E
Martapura, Sumatera, Indonesia 36 E2 4 19S 104 22 E
Martelange, Belgium 15 E5 49 49N 5 43 E
Martha's Vineyard, U.S.A. 79 E14 41 25N 70 38W
Martigny, Switz. 18 C7 46 6N 7 3 E
Martigues, France 18 E6 43 24N 5 4 E
Martin, Slovak Rep. 17 D10 49 6N 18 58 E
Martin, S. Dak., U.S.A. 80 D4 43 11N 101 44W
Martin, Tenn., U.S.A. 81 G10 36 21N 88 51W
Martin, L., U.S.A. 77 J3 32 41N 85 55W
Martina Franca, Italy 20 D7 40 42N 17 20 E
Martinborough, N.Z. 59 J5 41 14S 175 29 E
Martinez, Calif., U.S.A. 84 G4 38 1N 122 8W
Martinez, Ga., U.S.A. 77 J4 33 31N 82 4W
Martinique ■, W. Indies 89 D7 14 40N 61 0W
Martinique Passage, W. Indies 89 C7 15 15N 61 0W
Martinópolis, Brazil 95 A5 22 11S 51 12W
Martins Ferry, U.S.A. 78 F4 40 6N 80 44W
Martinsburg, Pa., U.S.A. 78 F6 40 19N 78 20W
Martinsburg, W. Va., U.S.A. 76 F7 39 27N 77 58W
Martinsville, Ind., U.S.A. 76 F2 39 26N 86 25W
Martinsville, Va., U.S.A. 77 G6 36 41N 79 52W
Marton, N.Z. 59 J5 40 4S 175 23 E
Martos, Spain 19 D4 37 44N 3 58W
Marudi, Malaysia 36 D4 4 11N 114 19 E
Maruf, Afghan. 40 D5 31 30N 67 6 E
Marugame, Japan 31 G6 34 15N 133 40 E
Marunga, Angola 56 B3 17 28S 20 2 E
Marungu, Mts., Dem. Rep. of the Congo 54 D3 7 30S 30 0 E
Marv Dasht, Iran 45 D7 29 50N 52 40 E
Marvast, Iran 45 D7 30 30N 54 15 E
Marvel Loch, Australia 61 F2 31 28S 119 29 E
Marwar, India 42 G5 25 43N 73 45 E
Mary, Turkmenistan 26 F7 37 40N 61 50 E
Maryborough = Port Laoise, Ireland 13 C4 53 2N 7 18W
Maryborough, Queens., Australia 63 D5 25 31S 152 37 E
Maryborough, Vic., Australia 63 F3 37 0S 143 44 E
Maryfield, Canada 73 D8 49 50N 101 35W
Maryland □, U.S.A. 76 F7 39 0N 76 30W
Maryland Junction, Zimbabwe 55 F3 17 45S 30 31 E
Maryport, U.K. 10 C4 54 44N 3 28W
Mary's Harbour, Canada 71 B8 52 18N 55 51W
Marystown, Canada 71 C8 47 10N 55 10W
Marysville, Canada 72 D5 49 35N 116 0W
Marysville, Calif., U.S.A. 84 F5 39 9N 121 35W
Marysville, Kans., U.S.A. 80 F6 39 51N 96 39W
Marysville, Mich., U.S.A. 78 D2 42 54N 82 29W
Marysville, Ohio, U.S.A. 76 E4 40 14N 83 22W
Marysville, Wash., U.S.A. 84 B4 48 3N 122 11W
Maryville, Mo., U.S.A. 80 E7 40 21N 94 52W
Maryville, Tenn., U.S.A. 77 H4 35 46N 83 58W
Marzūq, Libya 51 C8 25 53N 13 57 E
Masahunga, Tanzania 54 C3 2 6S 33 18 E
Masai Steppe, Tanzania 54 C4 4 30S 36 30 E
Masaka, Uganda 54 C3 0 21S 31 45 E
Masalembo, Kepulauan, Indonesia 36 F4 5 35S 114 30 E
Masalima, Kepulauan, Indonesia 36 F5 5 4S 117 5 E
Masamba, Indonesia 37 E6 2 30S 120 15 E
Masan, S. Korea 35 G15 35 11N 128 32 E
Masandam, Ra's, Oman 46 B6 26 30N 56 30 E
Masasi, Tanzania 55 E4 10 45S 38 52 E
Masaya, Nic. 88 D2 12 0N 86 7W
Masbate, Phil. 37 B6 12 21N 123 36 E
Mascara, Algeria 50 A6 35 26N 0 6 E
Mascota, Mexico 86 C4 20 30N 104 50W
Masela, Indonesia 37 F7 8 9S 129 51 E
Maseru, Lesotho 56 D4 29 18S 27 30 E
Mashaba, Zimbabwe 55 G3 20 2S 30 29 E
Mashābih, Si. Arabia 44 E3 25 35N 36 30 E
Masherbrum, Pakistan 43 B7 35 38N 76 18 E
Mashhad, Iran 45 B8 36 20N 59 35 E
Mashīz, Iran 45 D8 29 56N 56 37 E
Māshkel, Hāmūn-i-, Pakistan 40 E3 28 20N 62 56 E
Mashki Chāh, Pakistan 40 E3 29 5N 62 30 E
Mashonaland, Zimbabwe 53 H6 16 30S 31 0 E
Mashonaland Central □, Zimbabwe 57 B5 17 30S 31 0 E
Mashonaland East □, Zimbabwe 57 B5 18 0S 32 0 E
Mashonaland West □, Zimbabwe 57 B4 17 30S 29 30 E
Mashrakh, India 43 F11 26 7N 84 48 E
Masindi, Uganda 54 B3 1 40N 31 43 E
Masindi Port, Uganda 54 B3 1 43N 32 2 E
Maşīrah, Oman 46 C6 21 0N 58 50 E
Maşīrah, Khalīj, Oman 46 C6 20 10N 58 10 E
Masisi, Dem. Rep. of the Congo 54 C2 1 23S 28 49 E
Masjed Soleyman, Iran 45 D6 31 55N 49 18 E
Mask, L., Ireland 13 C2 53 36N 9 22W
Maskin, Oman 45 F8 23 30N 56 50 E
Masoala, Tanjon' i, Madag. 57 B9 15 59S 50 13 E
Masoarivo, Madag. 57 B7 19 3S 44 19 E
Masohi = Amahai, Indonesia 37 E7 3 20S 128 55 E
Masomeloka, Madag. 57 C8 20 17S 48 37 E
Mason, Nev., U.S.A. 84 G7 38 56N 119 8W
Mason, Tex., U.S.A. 81 K5 30 45N 99 14W
Mason City, U.S.A. 80 D8 43 9N 93 12W
Maspalomas, Canary Is. 22 G4 27 46N 15 35W
Maspalomas, Pta., Canary Is. 22 G4 27 43N 15 36W
Masqat, Oman 46 C6 23 37N 58 36 E
Massa, Italy 18 D9 44 1N 10 9 E
Massachusetts □, U.S.A. 79 D13 42 30N 72 0W
Massachusetts B., U.S.A. 79 D14 42 20N 70 50W
Massakory, Chad 51 F9 13 0N 15 49 E
Massanella, Spain 22 B9 39 48N 2 51 E
Massangena, Mozam. 57 C5 21 34S 33 0 E
Massango, Angola 52 F3 8 2S 16 21 E
Massawa = Mitsiwa, Eritrea 46 D2 15 35N 39 25 E
Massena, U.S.A. 79 B10 44 56N 74 54W
Massénya, Chad 51 F9 11 21N 16 9 E
Masset, Canada 72 C2 54 2N 132 10W
Massif Central, France 18 D5 44 55N 3 0 E
Massillon, U.S.A. 78 F3 40 48N 81 32W
Massinga, Mozam. 57 C6 23 15S 35 22 E
Massingir, Mozam. 57 C5 23 51S 32 4 E
Masson, Canada 79 A9 45 32N 75 25W
Masson I., Antarctica 5 C7 66 10S 93 20 E
Mastanli = Momchilgrad, Bulgaria 21 D11 41 33N 25 23 E
Masterton, N.Z. 59 J5 40 56S 175 39 E
Mastic, U.S.A. 79 F12 40 47N 72 54W
Mastuj, Pakistan 43 A5 36 20N 72 36 E
Mastung, Pakistan 40 E5 29 50N 66 56 E
Masty, Belarus 17 B13 53 27N 24 38 E
Masuda, Japan 31 G5 34 40N 131 51 E
Masvingo, Zimbabwe 55 G3 20 8S 30 49 E
Masvingo □, Zimbabwe 55 G3 21 0S 31 30 E
Maşyāf, Syria 44 C3 35 4N 36 20 E
Matabeleland, Zimbabwe 53 H5 18 0S 27 0 E
Matabeleland North □, Zimbabwe 55 F2 19 0S 28 0 E
Matabeleland South □, Zimbabwe 55 G2 21 0S 29 0 E
Matachewan, Canada 70 C3 47 56N 80 39W
Matadi, Dem. Rep. of the Congo 52 F2 5 52S 13 31 E
Matagalpa, Nic. 88 D2 13 0N 85 58W
Matagami, Canada 70 C4 49 45N 77 34W
Matagami, L., Canada 70 C4 49 50N 77 40W
Matagorda B., U.S.A. 81 L6 28 40N 96 0W
Matagorda I., U.S.A. 81 L6 28 15N 96 30W
Matak, Indonesia 39 L6 3 18N 106 16 E
Mátala, Greece 23 E6 34 59N 24 45 E
Matam, Senegal 50 E3 15 34N 13 17W
Matamoros, Campeche, Mexico 87 D6 18 50N 90 50W
Matamoros, Coahuila, Mexico 86 B4 25 33N 103 15W
Matamoros, Tamaulipas, Mexico 87 B5 25 50N 97 30W
Ma'ţan as Sarra, Libya 51 D10 21 45N 22 0 E
Matandu →, Tanzania 55 D3 8 45S 34 19 E
Matane, Canada 71 C6 48 50N 67 33W
Matanomadh, India 42 H3 23 33N 68 57 E
Matanzas, Cuba 88 B3 23 0N 81 40W
Matapa, Botswana 56 C3 23 11S 24 39 E
Matapan, C. = Tainaron, Ákra, Greece 21 F10 36 22N 22 27 E
Matapédia, Canada 71 C6 48 0N 66 59W
Matara, Sri Lanka 40 S12 5 58N 80 30 E
Mataram, Indonesia 36 F5 8 35S 116 7 E
Matarani, Peru 92 G4 17 0S 72 10W
Mataranka, Australia 60 B5 14 55S 133 4 E
Matarma, Râs, Egypt 47 E1 30 27N 32 44 E
Mataró, Spain 19 B7 41 32N 2 29 E
Matatiele, S. Africa 57 E4 30 20S 28 49 E
Mataura, N.Z. 59 M2 46 11S 168 51 E
Matehuala, Mexico 86 C4 23 40N 100 40W
Mateke Hills, Zimbabwe 55 G3 21 48S 31 0 E
Matera, Italy 20 D7 40 40N 16 36 E
Matetsi, Zimbabwe 55 F2 18 12S 26 0 E
Mathis, U.S.A. 81 L6 28 6N 97 50W
Mathráki, Greece 23 A3 39 48N 19 31 E
Mathura, India 42 F7 27 30N 77 40 E
Mati, Phil. 37 C7 6 55N 126 15 E
Matiali, India 43 F13 26 56N 88 49 E
Matías Romero, Mexico 87 D5 16 53N 95 2W
Matibane, Mozam. 55 E5 14 49S 40 45 E
Matima, Botswana 56 C3 20 15S 24 26 E
Matiri Ra., N.Z. 59 J4 41 38S 172 20 E
Matjiesfontein, S. Africa 56 E3 33 14S 20 35 E
Matla →, India 43 J13 21 40N 88 40 E
Matlamanyane, Botswana 56 B4 19 33S 25 57 E
Matli, Pakistan 42 G3 25 2N 68 39 E
Matlock, U.K. 10 D6 53 9N 1 33W
Mato Grosso □, Brazil 93 F8 14 0S 55 0W
Mato Grosso, Planalto do, Brazil 93 G8 15 0S 55 0W
Mato Grosso do Sul □, Brazil 93 G8 18 0S 55 0W
Matochkin Shar, Russia 26 B6 73 10N 56 40 E
Matopo Hills, Zimbabwe 55 G2 20 36S 28 20 E
Matopos, Zimbabwe 55 G2 20 20S 28 29 E
Matosinhos, Portugal 19 B1 41 11N 8 42W
Matroosberg, S. Africa 56 E2 33 23S 19 40 E
Maţruḩ, Oman 46 C6 23 37N 58 30 E
Matsue, Japan 31 G6 35 25N 133 10 E
Matsumae, Japan 30 D10 41 26N 140 7 E
Matsumoto, Japan 31 F9 36 15N 138 0 E
Matsusaka, Japan 31 G8 34 34N 136 32 E
Matsuura, Japan 31 H4 33 20N 129 49 E
Matsuyama, Japan 31 H6 33 45N 132 45 E
Mattagami →, Canada 70 B3 50 43N 81 29W
Mattancheri, India 40 Q10 9 50N 76 15 E
Mattawa, Canada 70 C4 46 20N 78 45W
Matterhorn, Switz. 18 D7 45 58N 7 39 E
Matthew Town, Bahamas 89 B5 20 57N 73 40W
Matthew's Ridge, Guyana 92 B6 7 37N 60 10W
Mattice, Canada 70 C3 49 40N 83 20W
Mattituck, U.S.A. 79 F12 40 59N 72 32W
Mattō, Japan 31 F8 36 31N 136 34 E
Mattoon, U.S.A. 76 F1 39 29N 88 23W
Matuba, Mozam. 57 C5 24 28S 32 49 E
Matucana, Peru 92 F3 11 55S 76 25W
Matūn = Khowst, Afghan. 42 C3 33 22N 69 58 E
Maturín, Venezuela 92 B6 9 45N 63 11W
Mau, Mad. P., India 43 F8 26 17N 78 41 E
Mau, Ut. P., India 43 G10 25 56N 83 33 E
Mau, Ut. P., India 43 G9 25 17N 81 23 E
Mau Escarpment, Kenya 54 C4 0 40S 36 0 E
Mau Ranipur, India 43 G8 25 16N 79 8 E
Maubeuge, France 18 A6 50 17N 3 57 E
Maud, Pt., Australia 60 D1 23 6S 113 45 E
Maude, Australia 63 E3 34 29S 144 18 E
Maudin Sun, Burma 41 M19 16 0N 94 30 E
Maués, Brazil 92 D7 3 20S 57 45W
Mauganj, India 41 G12 24 50N 81 55 E
Maughold Hd., U.K. 10 C3 54 18N 4 18W
Maui, U.S.A. 74 H16 20 48N 156 20W
Maulamyaing = Moulmein, Burma 41 L20 16 30N 97 40 E
Maule □, Chile 94 D1 36 5S 72 30W
Maumee, U.S.A. 76 E4 41 34N 83 39W
Maumee →, U.S.A. 76 E4 41 42N 83 28W
Maumere, Indonesia 37 F6 8 38S 122 13 E
Maun, Botswana 56 C3 20 0S 23 26 E
Mauna Kea, U.S.A. 74 J17 19 50N 155 28W
Mauna Loa, U.S.A. 74 J17 19 30N 155 35W
Maungmagan Kyunzu, Burma 38 E1 14 0N 97 48 E
Maupin, U.S.A. 82 D3 45 11N 121 5W
Maurepas, L., U.S.A. 81 K9 30 15N 90 30W
Maurice, L., Australia 61 E5 29 30S 131 0 E
Mauricie, Parc Nat. de la, Canada 70 C5 46 45N 73 0W
Mauritania ■, Africa 50 E3 20 50N 10 0W
Mauritius ■, Ind. Oc. 49 J9 20 0S 57 0 E
Mauston, U.S.A. 80 D9 43 48N 90 5W
Mavli, India 42 G5 24 45N 73 55 E
Mavuradonha Mts., Zimbabwe 55 F3 16 30S 31 30 E
Mawa, Dem. Rep. of the Congo 54 B2 2 45N 26 40 E
Mawai, India 43 H9 22 30N 81 4 E
Mawana, India 42 E7 29 6N 77 58 E
Mawand, Pakistan 42 E3 29 33N 68 38 E
Mawk Mai, Burma 41 J20 20 14N 97 37 E
Mawlaik, Burma 41 H19 23 40N 94 26 E
Mawlamyine = Moulmein, Burma 41 L20 16 30N 97 40 E
Mawqaq, Si. Arabia 44 E4 27 25N 41 8 E
Mawson Coast, Antarctica 5 C6 68 30S 63 0 E
Max, U.S.A. 80 B4 47 49N 101 18W
Maxcanú, Mexico 87 C6 20 40N 92 0W
Maxesibeni, S. Africa 57 E4 30 49S 29 23 E
Maxhamish L., Canada 72 B4 59 50N 123 17W
Maxixe, Mozam. 57 C6 23 54S 35 17 E
Maxville, Canada 79 A10 45 17N 74 51W
Maxwell, U.S.A. 84 F4 39 17N 122 11W
Maxwelton, Australia 62 C3 20 43S 142 41 E
May, C., U.S.A. 76 F8 38 56N 74 58W
May Pen, Jamaica 88 C4 17 58N 77 15W
Maya →, Russia 27 D14 60 28N 134 28 E
Maya Mts., Belize 87 D7 16 30N 89 0W
Mayaguana, Bahamas 89 B5 22 30N 72 44W
Mayagüez, Puerto Rico 89 C6 18 12N 67 9W
Mayāmey, Iran 45 B7 36 24N 55 42 E
Mayanup, Australia 61 F2 33 57S 116 27 E
Mayapan, Mexico 87 C7 20 30N 89 25W
Mayarí, Cuba 89 B4 20 40N 75 41W
Maybell, U.S.A. 82 F9 40 31N 108 5W
Maybole, U.K. 12 F4 55 21N 4 42W
Maydān, Iraq 44 C5 34 55N 45 37 E
Maydena, Australia 62 G4 42 45S 146 30 E
Mayenne →, France 18 C3 47 30N 0 32W
Mayer, U.S.A. 83 J7 34 24N 112 14W
Mayerthorpe, Canada 72 C5 53 57N 115 8W
Mayfield, Ky., U.S.A. 77 G1 36 44N 88 38W
Mayfield, N.Y., U.S.A. 79 C10 43 6N 74 16W
Mayhill, U.S.A. 83 K11 32 53N 105 29W
Maykop, Russia 25 F7 44 35N 40 10 E
Maymyo, Burma 38 A1 22 2N 96 28 E
Maynard, Mass., U.S.A. 79 D13 42 26N 71 27W
Maynard, Wash., U.S.A. 84 C4 47 59N 122 55W
Maynard Hills, Australia 61 E2 28 28S 119 49 E
Mayne →, Australia 62 C3 23 40S 141 55 E
Maynooth, Ireland 13 C5 53 23N 6 34W
Mayo, Canada 68 B6 63 38N 135 57W
Mayo □, Ireland 13 C2 53 53N 9 3W
Mayon Volcano, Phil. 37 B6 13 15N 123 41 E
Mayor I., N.Z. 59 G6 37 16S 176 17 E
Mayotte, Ind. Oc. 53 G9 12 50S 45 10 E
Maysville, U.S.A. 76 F4 38 39N 83 46W
Mayu, Indonesia 37 D7 1 30N 126 30 E
Mayville, N. Dak., U.S.A. 80 B6 47 30N 97 20W
Mayville, N.Y., U.S.A. 78 D5 42 15N 79 30W
Mayya, Russia 27 C14 61 44N 130 18 E
Mazabuka, Zambia 55 F2 15 52S 27 44 E
Mazagán = El Jadida, Morocco 50 B4 33 11N 8 17W
Mazagão, Brazil 93 D8 0 7S 51 16W
Mazán, Peru 92 D4 3 30S 73 0W
Māzandarān □, Iran 45 B7 36 30N 52 0 E
Mazapil, Mexico 86 C4 24 38N 101 34W
Mazara del Vallo, Italy 20 F5 37 39N 12 35 E
Mazarrón, Spain 19 D5 37 38N 1 19W
Mazaruni →, Guyana 92 B7 6 25N 58 35W
Mazatán, Mexico 86 B2 29 0N 110 8W
Mazatenango, Guatemala 88 D1 14 35N 91 30W
Mazatlán, Mexico 86 C3 23 13N 106 25W
Mažeikiai, Lithuania 9 H20 56 20N 22 20 E
Māzhān, Iran 45 C8 32 30N 59 0 E
Mazīnān, Iran 45 B8 36 19N 56 56 E
Mazoe, Mozam. 55 F3 16 42S 33 7 E
Mazoe →, Mozam. 55 F3 16 20S 33 30 E
Mazowe, Zimbabwe 55 F3 17 28S 30 58 E
Mazurian Lakes = Mazurski, Pojezierze, Poland 17 B11 53 50N 21 0 E
Mazurski, Pojezierze, Poland 17 B11 53 50N 21 0 E
Mazyr, Belarus 17 B15 51 59N 29 15 E
Mbabane, Swaziland 57 D5 26 18S 31 6 E
Mbaïki, C.A.R. 52 D3 3 53N 18 1 E
Mbala, Zambia 55 D3 8 46S 31 24 E

143

Najmah

Mtama, *Tanzania* .... **55 E4** 10 17S 39 21 E
Mtamvuna →, *S. Africa* .... **57 E5** 31 6S 30 12 E
Mtilikwe →, *Zimbabwe* .... **55 G3** 21 9S 31 30 E
Mtubatuba, *S. Africa* .... **57 D5** 28 30S 32 8 E
Mtwalume, *S. Africa* .... **57 E5** 30 30S 30 38 E
Mtwara-Mikindani, *Tanzania* **55 E5** 10 20S 40 20 E
Mu Gia, Deo, *Vietnam* ..... **38 D5** 17 40N 105 47 E
Mu Us Shamo, *China* ...... **34 E5** 39 0N 109 0 E
Muang Chiang Rai = Chiang Rai, *Thailand* .... **38 C2** 19 52N 99 50 E
Muang Khong, *Laos* .... **38 E5** 14 7N 105 51 E
Muang Lamphun, *Thailand* . **38 C2** 18 40N 99 2 E
Muang Pak Beng, *Laos* ..... **38 C3** 19 54N 101 8 E
Muar, *Malaysia* .... **39 L4** 2 3N 102 34 E
Muarabungo, *Indonesia* .... **36 E2** 1 28S 102 52 E
Muaraenim, *Indonesia* ..... **36 E2** 3 40S 103 50 E
Muarajuloi, *Indonesia* ...... **36 E4** 0 12S 114 3 E
Muarakaman, *Indonesia* .... **36 E5** 0 2S 116 45 E
Muaratebo, *Indonesia* ...... **36 E2** 1 30S 102 26 E
Muaratembesi, *Indonesia* ... **36 E2** 1 42S 103 8 E
Muaratewe, *Indonesia* ..... **36 E4** 0 58S 114 52 E
Mubarakpur, *India* .... **43 F10** 26 6N 83 18 E
Mubarraz = Al Mubarraz, *Si. Arabia* .... **45 E6** 25 30N 49 40 E
Mubende, *Uganda* .... **54 B3** 0 33N 31 22 E
Mubi, *Nigeria* .... **51 F8** 10 18N 13 16 E
Mubur, Pulau, *Indonesia* ... **39 L6** 3 20N 106 12 E
Mucajaí →, *Brazil* **92 C6** 2 25N 60 52W
Muchachos, Roque de los, *Canary Is.* **22 F2** 28 44N 17 52W
Muchinga Mts., *Zambia* .... **55 E3** 11 30S 31 30 E
Muck, *U.K.* **12 E2** 56 50N 6 15W
Muckadilla, *Australia* ...... **63 D4** 26 35S 148 23 E
Mucuri, *Brazil* **93 G11** 18 0S 39 36W
Mucusso, *Angola* .... **56 B3** 18 1S 21 25 E
Muda, *Canary Is.* **22 F6** 28 34N 13 57W
Mudanjiang, *China* .... **35 B15** 44 38N 129 30 E
Mudanya, *Turkey* .... **21 D13** 40 25N 28 50 E
Muddy Cr. →, *U.S.A.* **83 H8** 38 24N 110 42W
Mudgee, *Australia* .... **63 E4** 32 32S 149 31 E
Mudjatik →, *Canada* **73 B7** 56 1N 107 36W
Muecate, *Mozam.* .... **55 E4** 14 55S 39 40 E
Mueda, *Mozam.* .... **55 E4** 11 36S 39 28 E
Mueller Ra., *Australia* ..... **60 C4** 18 18S 126 46 E
Muende, *Mozam.* .... **55 E3** 14 28S 33 0 E
Muerto, Mar, *Mexico* **87 D6** 16 10N 94 10W
Mufulira, *Zambia* .... **55 E2** 12 32S 28 15 E
Mufumbiro Range, *Africa* ... **54 C2** 1 25S 29 30 E
Mughal Sarai, *India* .... **43 G10** 25 18N 83 7 E
Mughayrā', *Si. Arabia* ...... **44 D3** 29 17N 37 41 E
Mugi, *Japan* .... **31 H7** 33 40N 134 25 E
Mugila, Mts., *Dem. Rep. of the Congo* .... **54 D2** 7 0S 28 50 E
Muğla, *Turkey* .... **21 F13** 37 15N 28 22 E
Mugu, *Nepal* .... **43 E10** 29 45N 82 30 E
Muhammad, Râs, *Egypt* .... **44 E2** 27 44N 34 16 E
Muhammad Qol, *Sudan* .... **51 D13** 20 53N 37 9 E
Muhammadabad, *India* .... **43 F10** 26 4N 83 25 E
Muhesi →, *Tanzania* ...... **54 D4** 7 0S 35 20 E
Mühlhausen, *Germany* ..... **16 C6** 51 12N 10 27 E
Mühlig Hofmann fjell, *Antarctica* .... **5 D3** 72 30S 5 0 E
Muhos, *Finland* .... **8 D22** 64 47N 25 59 E
Muhu, *Estonia* .... **9 G20** 58 36N 23 11 E
Muhutwe, *Tanzania* .... **54 C3** 1 35S 31 45 E
Muine Bheag, *Ireland* **13 D5** 52 42N 6 58W
Muir, L., *Australia* .... **61 F2** 34 30S 116 40 E
Mujnak = Muynak, *Uzbekistan* **26 E6** 43 44N 59 10 E
Mukacheve, *Ukraine* .... **17 D12** 48 27N 22 45 E
Mukachevo = Mukacheve, *Ukraine* .... **17 D12** 48 27N 22 45 E
Mukah, *Malaysia* .... **36 D4** 2 55N 112 5 E
Mukandwara, *India* .... **42 G6** 24 49N 75 59 E
Mukdahan, *Thailand* .... **38 D5** 16 32N 104 43 E
Mukden = Shenyang, *China* . **35 D12** 41 48N 123 27 E
Mukerian, *India* .... **42 D6** 31 57N 75 37 E
Mukhtuya = Lensk, *Russia* .. **27 C12** 60 48N 114 55 E
Mukinbudin, *Australia* ..... **61 F2** 30 55S 118 5 E
Mukishi, *Dem. Rep. of the Congo* .... **55 D1** 8 30S 24 44 E
Mukomuko, *Indonesia* ..... **36 E2** 2 30S 101 10 E
Mukomwenze, *Dem. Rep. of the Congo* .... **54 D2** 6 49S 27 15 E
Muktsar, *India* .... **42 D6** 30 30N 74 30 E
Mukur = Moqor, *Afghan.* ... **42 C2** 32 50N 67 42 E
Mukutawa →, *Canada* **73 C9** 53 10N 97 24W
Mukwela, *Zambia* .... **55 F2** 17 0S 26 40 E
Mula, *Spain* **19 C5** 38 3N 1 33W
Mula →, *Pakistan* .... **42 F2** 27 57N 67 36 E
Mulange, *Dem. Rep. of the Congo* .... **54 C2** 3 40S 27 10 E
Mulanje, *Malawi* .... **55 F4** 16 2S 35 33 E
Mulchén, *Chile* **94 D1** 37 45S 72 20W
Mulde →, *Germany* .... **16 C7** 51 53N 12 15 E
Mule Creek Junction, *U.S.A.* **80 D2** 43 19N 104 8W
Muleba, *Tanzania* .... **54 C3** 1 50S 31 37 E
Mulejé, *Mexico* **86 B2** 26 53N 112 1W
Muleshoe, *U.S.A.* **81 H3** 34 13N 102 43W
Mulgrave, *Canada* **71 C7** 45 38N 61 31W
Mulhacén, *Spain* **19 D4** 37 4N 3 20W
Mülheim, *Germany* .... **33 C6** 51 25N 6 54 E
Mulhouse, *France* .... **18 C7** 47 40N 7 20 E
Muling, *China* .... **35 B16** 44 35N 130 10 E
Mull, *U.K.* **12 E3** 56 25N 5 56W
Mull, Sound of, *U.K.* **12 E3** 56 30N 5 50W
Mullaittivu, *Sri Lanka* ...... **40 Q12** 9 15N 80 49 E
Mullen, *U.S.A.* **80 D4** 42 3N 101 1W
Mullens, *U.S.A.* **76 G5** 37 35N 81 23W
Muller, Pegunungan, *Indonesia* .... **36 D4** 0 30N 113 30 E
Mullet Pen., *Ireland* **13 B1** 54 13N 10 2W
Mullewa, *Australia* .... **61 E2** 28 29S 115 30 E
Mulligan →, *Australia* ..... **62 D2** 25 0S 139 0 E
Mullingar, *Ireland* **13 C4** 53 31N 7 21W
Mullins, *U.S.A.* **77 H6** 34 12N 79 15W
Mullumbimby, *Australia* .... **63 D5** 28 30S 153 30 E
Mulobezi, *Zambia* .... **55 F2** 16 45S 25 7 E
Mulroy B., *Ireland* **13 A4** 55 15N 7 46W
Multan, *Pakistan* .... **42 D4** 30 15N 71 36 E
Mulumbe, Mts., *Dem. Rep. of the Congo* .... **55 D2** 8 40S 27 30 E
Mulungushi Dam, *Zambia* .. **55 E2** 14 48S 28 48 E
Mulvane, *U.S.A.* **81 G6** 37 29N 97 15W
Mumbai, *India* .... **40 K8** 18 55N 72 50 E
Mumbwa, *Zambia* .... **55 F2** 15 0S 27 0 E
Mun →, *Thailand* .... **38 E5** 15 19N 105 30 E
Muna, *Indonesia* .... **37 F6** 5 0S 122 30 E
Munabao, *India* .... **42 G4** 25 45N 70 17 E
Munamagi, *Estonia* .... **9 H22** 57 43N 27 4 E
München, *Germany* .... **16 D6** 48 8N 11 34 E
Munchen-Gladbach = Mönchengladbach, *Germany* .... **16 C4** 51 11N 6 27 E
Muncho Lake, *Canada* **72 B3** 59 0N 125 50W
Munch'ŏn, *N. Korea* .... **35 E14** 39 14N 127 19 E
Muncie, *U.S.A.* **76 E3** 40 12N 85 23W
Muncoonie, L., *Australia* ... **62 D2** 25 12S 138 40 E
Mundabbera, *Australia* ..... **63 D5** 25 36S 151 18 E
Munday, *U.S.A.* **81 J5** 33 27N 99 38W
Münden, *Germany* .... **16 C5** 51 25N 9 38 E
Mundiwindi, *Australia* ..... **60 D3** 23 47S 120 9 E
Mundo Novo, *Brazil* **93 F10** 11 50S 40 29W
Mundra, *India* .... **42 H3** 22 54N 69 48 E
Mundrabilla, *Australia* ..... **61 F4** 31 52S 127 51 E
Mungallala, *Australia* ...... **63 D4** 26 28S 147 34 E
Mungallala Cr. →, *Australia* . **63 D4** 28 53S 147 5 E
Mungana, *Australia* .... **62 B3** 17 8S 144 27 E
Mungaoli, *India* .... **42 G8** 24 24N 78 7 E
Mungari, *Mozam.* .... **55 F3** 17 12S 33 30 E
Mungbere, *Dem. Rep. of the Congo* .... **54 B2** 2 36N 28 28 E
Mungeli, *India* .... **43 H9** 22 4N 81 41 E
Munger, *India* .... **43 G12** 25 23N 86 30 E
Munich = München, *Germany* **16 D6** 48 8N 11 34 E
Munising, *U.S.A.* **76 B2** 46 25N 86 40W
Munku-Sardyk, *Russia* ..... **27 D11** 51 45N 100 20 E
Muñoz Gamero, Pen., *Chile* . **96 G2** 52 30S 73 5W
Munroe L., *Canada* **73 B9** 59 13N 98 35W
Munsan, *S. Korea* .... **35 F14** 37 51N 126 48 E
Münster, *Germany* .... **16 C4** 51 58N 7 37 E
Munster □, *Ireland* **13 D3** 52 18N 8 44W
Muntadgin, *Australia* ...... **61 F2** 31 45S 118 33 E
Muntok, *Indonesia* .... **36 E3** 2 5S 105 10 E
Munyama, *Zambia* .... **55 F2** 16 5S 28 31 E
Muong Beng, *Laos* .... **38 B3** 20 23N 101 46 E
Muong Boum, *Vietnam* .... **38 A4** 22 24N 102 49 E
Muong Et, *Laos* .... **38 B5** 20 49N 104 1 E
Muong Hai, *Laos* .... **38 B3** 21 3N 101 49 E
Muong Hiem, *Laos* .... **38 B4** 20 5N 103 22 E
Muong Houn, *Laos* .... **38 B3** 20 8N 101 23 E
Muong Hung, *Vietnam* ..... **38 B4** 20 56N 103 53 E
Muong Kau, *Laos* .... **38 E5** 15 6N 105 47 E
Muong Khao, *Laos* .... **38 C4** 19 38N 103 32 E
Muong Khoua, *Laos* .... **38 B4** 21 5N 102 31 E
Muong Liep, *Laos* .... **38 C3** 18 29N 101 40 E
Muong May, *Laos* .... **38 E6** 14 49N 106 56 E
Muong Ngeun, *Laos* .... **38 B3** 20 36N 101 3 E
Muong Ngoi, *Laos* .... **38 B4** 20 43N 102 41 E
Muong Nhie, *Vietnam* ..... **38 A4** 22 12N 102 28 E
Muong Nong, *Laos* .... **38 D6** 16 22N 106 30 E
Muong Ou Tay, *Laos* .... **38 A3** 22 7N 101 48 E
Muong Oua, *Laos* .... **38 C3** 18 18N 101 20 E
Muong Peun, *Laos* .... **38 B4** 20 13N 103 52 E
Muong Phalane, *Laos* ...... **38 D5** 16 39N 105 34 E
Muong Phieng, *Laos* .... **38 C3** 19 6N 101 32 E
Muong Phine, *Laos* .... **38 D6** 16 32N 106 2 E
Muong Sai, *Laos* .... **38 B3** 20 42N 101 59 E
Muong Saiapoun, *Laos* ..... **38 C3** 18 24N 101 31 E
Muong Sen, *Vietnam* ...... **38 C5** 19 24N 104 8 E
Muong Sing, *Laos* .... **38 B3** 21 11N 101 9 E
Muong Son, *Laos* .... **38 B4** 20 27N 103 19 E
Muong Soui, *Laos* .... **38 C4** 19 33N 102 52 E
Muong Va, *Laos* .... **38 B4** 21 53N 102 19 E
Muong Xia, *Vietnam* .... **38 B5** 20 19N 104 50 E
Muonio, *Finland* .... **8 C20** 67 57N 23 40 E
Muonionjoki →, *Finland* ... **8 C20** 67 11N 23 34 E
Muping, *China* .... **35 F11** 37 22N 121 36 E
Muqdisho, *Somali Rep.* ..... **46 G4** 2 2N 45 25 E
Mur →, *Austria* .... **17 E9** 46 18N 16 52 E
Murakami, *Japan* .... **30 E9** 38 14N 139 29 E
Murallón, Cerro, *Chile* **96 F2** 49 48S 73 30W
Muranda, *Rwanda* .... **54 C2** 1 52S 29 20 E
Murang'a, *Kenya* .... **54 C4** 0 45S 37 9 E
Murashi, *Russia* .... **24 C8** 59 30N 49 0 E
Murat →, *Turkey* .... **25 G7** 38 46N 40 0 E
Muratlı, *Turkey* .... **21 D12** 41 10N 27 29 E
Murayama, *Japan* .... **30 E10** 38 30N 140 25 E
Murchison →, *Australia* ... **61 E1** 27 45S 114 0 E
Murchison, Mt., *Antarctica* . **5 D11** 73 0S 168 0 E
Murchison Falls, *Uganda* ... **54 B3** 2 15N 31 30 E
Murchison Ra., *Australia* ... **62 C1** 20 0S 134 10 E
Murchison Rapids, *Malawi* .. **55 F3** 15 55S 34 35 E
Murcia, *Spain* **19 D5** 38 5N 1 10W
Murcia □, *Spain* **19 D5** 37 50N 1 30W
Murdo, *U.S.A.* **80 D4** 43 53N 100 43W
Murdoch Pt., *Australia* ..... **62 A3** 14 37S 144 55 E
Mureş →, *Romania* .... **17 E11** 46 15N 20 13 E
Mureşul = Mureş →, *Romania* .... **17 E11** 46 15N 20 13 E
Murewa, *Zimbabwe* .... **57 B5** 17 39S 31 47 E
Murfreesboro, *N.C., U.S.A.* . **77 G7** 36 27N 77 6W
Murfreesboro, *Tenn., U.S.A.* **77 H2** 35 51N 86 24W
Murgab = Murghob, *Tajikistan* .... **26 F8** 38 10N 74 2 E
Murgab →, *Turkmenistan* .. **45 B9** 38 18N 61 12 E
Murgenella, *Australia* ...... **60 B5** 11 34S 132 56 E
Murgha Kibzai, *Pakistan* ... **42 D3** 30 44N 69 25 E
Murghob, *Tajikistan* .... **26 F8** 38 10N 74 2 E
Murgon, *Australia* .... **63 D5** 26 15S 151 54 E
Muri, *India* .... **43 H11** 23 22N 85 52 E
Muria, *Indonesia* .... **37 G14** 6 36S 110 53 E
Muriaé, *Brazil* **95 A7** 21 8S 42 23W
Muriel Mine, *Zimbabwe* .... **55 F3** 17 14S 30 40 E
Müritz, *Germany* .... **16 B7** 53 25N 12 42 E
Murka, *Kenya* .... **54 C4** 3 27S 38 0 E
Murliganj, *India* .... **43 G12** 25 54N 86 59 E
Murmansk, *Russia* .... **24 A5** 68 57N 33 10 E
Muro, *Spain* .... **22 B10** 39 44N 3 3 E
Murom, *Russia* .... **24 C7** 55 35N 42 3 E
Muroran, *Japan* .... **30 C10** 42 25N 141 0 E
Muroto, *Japan* .... **31 H7** 33 18N 134 9 E
Muroto-Misaki, *Japan* ..... **31 H7** 33 15N 134 10 E
Murphy, *U.S.A.* **82 E5** 43 13N 116 33W
Murphys, *U.S.A.* **84 G6** 38 8N 120 28W
Murray, *Ky., U.S.A.* **77 G1** 36 37N 88 19W
Murray, *Utah, U.S.A.* **82 F8** 40 40N 111 53W
Murray →, *Australia* ...... **63 F2** 35 20S 139 22 E
Murray, L., *U.S.A.* **77 H5** 34 3N 81 13W
Murray Bridge, *Australia* ... **63 F2** 35 6S 139 14 E
Murray Harbour, *Canada* ... **71 C7** 46 0N 62 28W
Murraysburg, *S. Africa* .... **56 E3** 31 58S 23 47 E
Murree, *Pakistan* .... **42 C5** 33 56N 73 28 E
Murrieta, *U.S.A.* **85 M9** 33 33N 117 13W
Murrumbidgee →, *Australia* **63 E3** 34 43S 143 12 E
Murrumburrah, *Australia* ... **63 E4** 34 32S 148 22 E
Murrurundi, *Australia* ...... **63 E5** 31 42S 150 51 E
Murshidabad, *India* .... **43 G13** 24 11N 88 19 E
Murtle L., *Canada* **72 C5** 52 8N 119 38W
Murtoa, *Australia* .... **63 F3** 36 35S 142 28 E
Murungu, *Tanzania* .... **54 C3** 4 12S 31 10 E
Mururoa, *Pac. Oc.* **65 K14** 21 52S 138 55W
Murwara, *India* .... **43 H9** 23 46N 80 28 E
Murwillumbah, *Australia* ... **63 D5** 28 18S 153 27 E
Mürzzuschlag, *Austria* ..... **16 E8** 47 36N 15 41 E
Muş, *Turkey* .... **25 G7** 38 45N 41 30 E
Mûsa, Gebel, *Egypt* .... **44 D2** 28 33N 33 59 E
Musa Khel, *Pakistan* .... **42 D3** 30 59N 69 52 E
Mūsá Qal'eh, *Afghan.* ...... **40 C4** 32 20N 64 50 E
Musafirkhana, *India* .... **43 F9** 26 22N 81 48 E
Musala, *Bulgaria* .... **21 C10** 42 13N 23 37 E
Musala, *Indonesia* .... **36 D1** 1 41N 98 28 E
Musan, *N. Korea* .... **35 C15** 42 12N 129 12 E
Musangu, *Dem. Rep. of the Congo* .... **55 E1** 10 28S 23 55 E
Musasa, *Tanzania* .... **54 C3** 3 25S 31 30 E
Musay'id, *Qatar* .... **45 E6** 25 0N 51 33 E
Muscat = Masqat, *Oman* .... **46 C6** 23 37N 58 36 E
Muscat & Oman = Oman ■, *Asia* .... **46 C6** 23 0N 58 0 E
Muscatine, *U.S.A.* **80 E9** 41 25N 91 3W
Musgrave Harbour, *Canada* . **71 C9** 49 27N 53 58W
Musgrave Ranges, *Australia* **61 E5** 26 0S 132 0 E
Mushie, *Dem. Rep. of the Congo* .... **52 E3** 2 56S 16 55 E
Musi →, *Indonesia* .... **36 E2** 2 20S 104 56 E
Muskeg →, *Canada* **72 A4** 60 20N 123 20W
Muskegon, *U.S.A.* **76 D2** 43 14N 86 16W
Muskegon →, *U.S.A.* **76 D2** 43 14N 86 21W
Muskegon Heights, *U.S.A.* .. **76 D2** 43 12N 86 16W
Muskogee, *U.S.A.* **81 H7** 35 45N 95 22W
Muskoka, L., *Canada* **78 B5** 45 0N 79 25W
Muskwa →, *Canada* **72 B4** 58 47N 122 48W
Muslīmiyah, *Syria* .... **44 B3** 36 19N 37 12 E
Musofu, *Zambia* .... **55 E2** 13 30S 29 0 E
Musoma, *Tanzania* .... **54 C3** 1 30S 33 48 E
Musquaro, L., *Canada* **71 B7** 50 38N 61 5W
Musquodoboit Harbour, *Canada* **71 D7** 44 50N 63 9W
Musselburgh, *U.K.* **12 F5** 55 57N 3 2W
Musselshell →, *U.S.A.* **82 C10** 47 21N 107 57W
Mussoorie, *India* .... **42 D8** 30 27N 78 6 E
Mussuco, *Angola* .... **56 B2** 17 2S 19 3 E
Mustafakemalpaşa, *Turkey* . **21 D13** 40 2N 28 24 E
Mustang, *Nepal* .... **43 E10** 29 10N 83 55 E
Musters, L., *Argentina* **96 F3** 45 20S 69 25W
Musudan, *N. Korea* .... **35 D15** 40 50N 129 43 E
Muswellbrook, *Australia* ... **63 E5** 32 16S 150 56 E
Mût, *Egypt* .... **51 C11** 25 28N 28 58 E
Mut, *Turkey* .... **44 B2** 36 40N 33 28 E
Mutanda, *Mozam.* .... **57 C5** 21 0S 33 34 E
Mutanda, *Zambia* .... **55 E2** 12 24S 26 13 E
Mutare, *Zimbabwe* .... **55 F3** 18 58S 32 38 E
Muting, *Indonesia* .... **37 F10** 7 23S 140 20 E
Mutoko, *Zimbabwe* .... **57 B5** 17 24S 32 13 E
Mutoray, *Russia* .... **27 C11** 60 56N 101 0 E
Mutshatsha, *Dem. Rep. of the Congo* .... **55 E1** 10 35S 24 20 E
Mutsu, *Japan* .... **30 D10** 41 5N 140 55 E
Mutsu-Wan, *Japan* .... **30 D10** 41 5N 140 55 E
Muttaburra, *Australia* ...... **62 C3** 22 38S 144 29 E
Mutton I., *Ireland* **13 D2** 52 49N 9 32W
Mutuáli, *Mozam.* .... **55 E4** 14 55S 37 0 E
Muweilih, *Egypt* .... **47 E3** 30 42N 34 19 E
Muy Muy, *Nic.* **88 D2** 12 39N 85 36W
Muyinga, *Burundi* .... **54 C3** 3 14S 30 33 E
Muynak, *Uzbekistan* .... **26 E6** 43 44N 59 10 E
Muzaffarabad, *Pakistan* .... **43 B5** 34 25N 73 30 E
Muzaffargarh, *Pakistan* ..... **42 D4** 30 5N 71 14 E
Muzaffarnagar, *India* .... **42 E7** 29 26N 77 40 E
Muzaffarpur, *India* .... **43 F11** 26 7N 85 23 E
Muzafirpur, *Pakistan* ...... **42 D3** 30 58N 69 9 E
Muzhi, *Russia* .... **24 A11** 65 25N 64 40 E
Mvuma, *Zimbabwe* .... **55 F3** 19 16S 30 30 E
Mvurwi, *Zimbabwe* .... **55 F3** 17 0S 30 57 E
Mwadui, *Tanzania* .... **54 C3** 3 26S 33 32 E
Mwambo, *Tanzania* .... **55 E5** 10 30S 40 22 E
Mwandi, *Zambia* .... **55 F1** 17 30S 24 51 E
Mwanza, *Dem. Rep. of the Congo* .... **54 D2** 7 55S 26 43 E
Mwanza, *Tanzania* .... **54 C3** 2 30S 32 58 E
Mwanza, *Zambia* .... **55 F1** 16 58S 24 28 E
Mwanza □, *Tanzania* ...... **54 C3** 2 0S 33 0 E
Mwaya, *Tanzania* .... **55 D3** 9 32S 33 55 E
Mweelrea, *Ireland* **13 C2** 53 39N 9 49W
Mweka, *Dem. Rep. of the Congo* .... **52 E4** 4 50S 21 34 E
Mwenezi, *Zimbabwe* .... **55 G3** 21 15S 30 48 E
Mwenezi →, *Mozam.* ...... **55 G3** 22 40S 31 50 E
Mwenga, *Dem. Rep. of the Congo* .... **54 C2** 3 1S 28 28 E
Mweru, L., *Zambia* .... **55 D2** 9 0S 28 40 E
Mweza Range, *Zimbabwe* .. **55 G3** 21 0S 30 0 E
Mwilambwe, *Dem. Rep. of the Congo* .... **54 D2** 8 7S 25 5 E
Mwimbi, *Tanzania* .... **55 D3** 8 38S 31 39 E
Mwinilunga, *Zambia* .... **55 E1** 11 43S 24 25 E
My Tho, *Vietnam* .... **39 G6** 10 29N 106 23 E
Myajlar, *India* .... **42 F4** 26 15N 70 20 E
Myanaung, *Burma* .... **41 K19** 18 18N 95 22 E
Myanmar = Burma ■, *Asia* .. **41 J20** 21 0N 96 30 E
Myaungmya, *Burma* .... **41 L19** 16 30N 94 40 E
Mycenæ, *Greece* .... **21 F10** 37 39N 22 52 E
Myeik Kyunzu, *Burma* ..... **39 G1** 11 30N 97 30 E
Myers Chuck, *U.S.A.* **72 B2** 55 44N 132 11W
Myerstown, *U.S.A.* **79 F8** 40 22N 76 19W
Myingyan, *Burma* .... **41 J19** 21 30N 95 20 E
Myitkyina, *Burma* .... **41 G20** 25 24N 97 26 E
Mykines, *Færoe Is.* **8 E9** 62 7N 7 35W
Mykolayiv, *Ukraine* .... **25 E5** 46 58N 32 0 E
Mymensingh, *Bangla.* ...... **41 G17** 24 45N 90 24 E
Mynydd Du, *U.K.* **11 F4** 51 52N 3 50W
Mýrdalsjökull, *Iceland* **8 E4** 63 40N 19 6W
Myrtle Beach, *U.S.A.* **77 J6** 33 42N 78 53W
Myrtle Creek, *U.S.A.* **82 E2** 43 1N 123 17W
Myrtle Point, *U.S.A.* **82 E1** 43 4N 124 8W
Myrtou, *Cyprus* .... **23 D12** 35 18N 33 4 E
Mysia, *Turkey* .... **21 E12** 39 50N 27 0 E
Mysore = Karnataka □, *India* **40 N10** 13 15N 77 0 E
Mysore, *India* .... **40 N10** 12 17N 76 41 E
Mystic, *U.S.A.* **79 E13** 41 21N 71 58W
Myszków, *Poland* .... **17 C10** 50 45N 19 22 E
Mytishchi, *Russia* .... **24 C6** 55 50N 37 50 E
Mývatn, *Iceland* **8 D5** 65 36N 17 0W
Mzimba, *Malawi* .... **55 E3** 11 55S 33 39 E
Mzimkulu →, *S. Africa* ..... **57 E5** 30 44S 30 28 E
Mzimvubu →, *S. Africa* .... **57 E4** 31 38S 29 33 E
Mzuzu, *Malawi* .... **55 E3** 11 30S 33 55 E

N

Na Hearadh = Harris, *U.K.* .. **12 D2** 57 50N 6 55W
Na Noi, *Thailand* .... **38 C3** 18 19N 100 43 E
Na Phao, *Laos* .... **38 D5** 17 35N 105 44 E
Na Sam, *Vietnam* .... **38 A6** 22 3N 106 37 E
Na San, *Vietnam* .... **38 B5** 21 12N 104 2 E
Naab →, *Germany* .... **16 D6** 49 1N 12 2 E
Naantali, *Finland* .... **9 F19** 60 29N 22 2 E
Naas, *Ireland* **13 C5** 53 12N 6 40W
Nababeep, *S. Africa* .... **56 D2** 29 36S 17 46 E
Nabadwip = Navadwip, *India* **43 H13** 23 34N 88 20 E
Nabari, *Japan* .... **31 G8** 34 37N 136 5 E
Nabawa, *Australia* .... **61 E1** 28 30S 114 48 E
Nabberu, L., *Australia* ..... **61 E3** 25 50S 120 30 E
Naberezhnyye Chelny, *Russia* **24 C9** 55 42N 52 19 E
Nabeul, *Tunisia* .... **51 A8** 36 30N 10 44 E
Nabha, *India* .... **42 D7** 30 26N 76 14 E
Nabīd, *Iran* .... **45 D8** 29 40N 57 38 E
Nabire, *Indonesia* .... **37 E9** 3 15S 135 26 E
Nabisar, *Pakistan* .... **42 G3** 25 8N 69 40 E
Nabisipi →, *Canada* **71 B7** 50 14N 62 13W
Nabiswera, *Uganda* .... **54 B3** 1 27N 32 15 E
Nablus = Nābulus, *West Bank* **47 C4** 32 14N 35 15 E
Naboomspruit, *S. Africa* .... **57 C4** 24 32S 28 40 E
Nābulus, *West Bank* .... **47 C4** 32 14N 35 15 E
Nacala, *Mozam.* .... **55 E5** 14 31S 40 34 E
Nacala-Velha, *Mozam.* ..... **55 E5** 14 32S 40 34 E
Nacaome, *Honduras* **88 D2** 13 31N 87 30W
Nacaroa, *Mozam.* .... **55 E4** 14 22S 39 56 E
Naches, *U.S.A.* **82 C3** 46 44N 120 42W
Naches →, *U.S.A.* **84 D6** 46 38N 120 31W
Nachicapau, L., *Canada* **71 A6** 56 40N 68 5W
Nachingwea, *Tanzania* ..... **55 E4** 10 23S 38 49 E
Nachna, *India* .... **42 F4** 27 34N 71 41 E
Nacimiento L., *U.S.A.* **84 K6** 35 46N 120 53W
Naco, *Mexico* **86 A3** 31 20N 109 56W
Nacogdoches, *U.S.A.* **81 K7** 31 36N 94 39W
Nácori Chico, *Mexico* **86 B3** 29 39N 109 1W
Nacozari, *Mexico* **86 A3** 30 24N 109 39W
Nadi, *Fiji* .... **59 C7** 17 42S 177 20 E
Nadiad, *India* .... **42 H5** 22 41N 72 56 E
Nador, *Morocco* **50 B5** 35 14N 2 58W
Nadur, *Malta* .... **23 C1** 36 2N 14 18 E
Nadūshan, *Iran* .... **45 C7** 32 2N 53 35 E
Nadvirna, *Ukraine* .... **17 D13** 48 37N 24 30 E
Nadvoitsy, *Russia* .... **24 B5** 63 52N 34 14 E
Nadvornaya = Nadvirna, *Ukraine* .... **17 D13** 48 37N 24 30 E
Nadym, *Russia* .... **26 C8** 65 35N 72 42 E
Nadym →, *Russia* .... **26 C8** 66 12N 72 0 E
Nærbø, *Norway* .... **9 G11** 58 40N 5 39 E
Næstved, *Denmark* .... **9 J14** 55 13N 11 44 E
Naft-e Safīd, *Iran* .... **45 D6** 31 40N 49 17 E
Naftshahr, *Iran* .... **44 C5** 34 0N 45 30 E
Nafud Desert = An Nafūd, *Si. Arabia* .... **44 D4** 28 15N 41 0 E
Naga, *Phil.* .... **37 B6** 13 38N 123 15 E
Nagahama, *Japan* .... **31 G8** 35 23N 136 16 E
Nagai, *Japan* .... **30 E10** 38 6N 140 2 E
Nagaland □, *India* .... **41 G19** 26 0N 94 30 E
Nagano, *Japan* .... **31 F9** 36 40N 138 10 E
Nagano □, *Japan* .... **31 F9** 36 15N 138 0 E
Nagaoka, *Japan* .... **31 F9** 37 27N 138 51 E
Nagappattinam, *India* ...... **40 P11** 10 46N 79 51 E
Nagar →, *Bangla.* .... **43 G13** 24 27N 89 12 E
Nagar Parkar, *Pakistan* .... **42 G4** 24 28N 70 46 E
Nagasaki, *Japan* .... **31 H4** 32 47N 129 50 E
Nagasaki □, *Japan* .... **31 H4** 32 50N 129 40 E
Nagato, *Japan* .... **31 G5** 34 19N 131 5 E
Nagaur, *India* .... **42 F5** 27 15N 73 45 E
Nagda, *India* .... **42 H6** 23 27N 75 25 E
Nagercoil, *India* .... **40 Q10** 8 12N 77 26 E
Nagina, *India* .... **43 E8** 29 30N 78 30 E
Nagīneh, *Iran* .... **45 C8** 34 20N 57 15 E
Nagir, *Pakistan* .... **43 A6** 36 12N 74 42 E
Nagod, *India* .... **43 G9** 24 34N 80 36 E
Nagoorin, *Australia* .... **62 C5** 24 17S 151 15 E
Nagorno-Karabakh, *Azerbaijan* .... **25 F8** 39 55N 46 45 E
Nagornyy, *Russia* .... **27 D13** 55 58N 124 57 E
Nagoya, *Japan* .... **31 G8** 35 10N 136 50 E
Nagpur, *India* .... **40 J11** 21 8N 79 10 E
Nagua, *Dom. Rep.* **89 C6** 19 23N 69 50W
Nagykanizsa, *Hungary* ..... **17 E9** 46 28N 17 0 E
Nagykőrös, *Hungary* .... **17 E10** 47 5N 19 48 E
Naha, *Japan* .... **31 L3** 26 13N 127 42 E
Nahan, *India* .... **42 D7** 30 33N 77 18 E
Nahanni Butte, *Canada* **72 A4** 61 2N 123 31W
Nahanni Nat. Park, *Canada* . **72 A4** 61 15N 125 0W
Nahargarh, *Mad. P., India* .. **42 G6** 24 10N 75 14 E
Nahargarh, *Raj., India* ..... **42 G7** 24 55N 76 50 E
Nahariyya, *Israel* .... **44 C2** 33 1N 35 5 E
Nahāvand, *Iran* .... **45 C6** 34 10N 48 22 E
Naicá, *Mexico* **86 B3** 27 53N 105 31W
Naicam, *Canada* **73 C8** 52 30N 104 30W
Naikoon Prov. Park, *Canada* . **72 C2** 53 55N 131 55W
Naimisharanya, *India* ...... **43 F9** 27 21N 80 30 E
Nain, *Canada* **71 A7** 56 34N 61 40W
Nā'īn, *Iran* .... **45 C7** 32 54N 53 0 E
Naini Tal, *India* .... **43 E8** 29 30N 79 30 E
Nainpur, *India* .... **40 H12** 22 30N 80 10 E
Nainwa, *India* .... **42 G6** 25 46N 75 51 E
Nairn, *U.K.* **12 D5** 57 35N 3 53W
Nairobi, *Kenya* .... **54 C4** 1 17S 36 48 E
Naissaar, *Estonia* .... **9 G21** 59 34N 24 29 E
Naivasha, *Kenya* .... **54 C4** 0 40S 36 30 E
Naivasha, L., *Kenya* .... **54 C4** 0 48S 36 20 E
Najafābād, *Iran* .... **45 C6** 32 40N 51 15 E
Najd, *Si. Arabia* .... **46 B3** 26 30N 42 0 E
Najibabad, *India* .... **42 E8** 29 40N 78 20 E
Najin, *N. Korea* .... **35 C16** 42 12N 130 15 E
Najmah, *Si. Arabia* .... **45 E6** 26 42N 50 6 E

Norderstedt

Neuquén □, Argentina 94 D2 38 0S 69 50W
Neuruppin, Germany 16 B7 52 55N 12 48 E
Neuse →, U.S.A. 77 H7 35 6N 76 29W
Neusiedler See, Austria 17 E9 47 50N 16 47 E
Neustrelitz, Germany 16 B7 53 21N 13 4 E
Neva →, Russia 24 C5 59 50N 30 30 E
Nevada, Iowa, U.S.A. 80 D8 42 1N 93 27W
Nevada, Mo., U.S.A. 81 G7 37 51N 94 22W
Nevada □, U.S.A. 82 G5 39 0N 117 0W
Nevada City, U.S.A. 84 F6 39 16N 121 1W
Nevado, Cerro, Argentina 94 D2 35 30S 68 32W
Nevel, Russia 24 C4 56 0N 29 55 E
Nevers, France 18 C5 47 0N 3 9 E
Nevertire, Australia 63 E4 31 50S 147 44 E
Neville, Canada 73 D7 49 58N 107 39W
Nevinnomyssk, Russia 25 F7 44 40N 42 0 E
Nevis, St. Kitts & Nevis 89 C7 17 0N 62 30W
Nevşehir, Turkey 44 B2 38 33N 34 40 E
Nevyansk, Russia 24 C11 57 30N 60 13 E
New →, U.S.A. 76 F5 38 10N 81 12W
New Aiyansh, Canada 72 B3 55 12N 129 4W
New Albany, Ind., U.S.A. 76 F3 38 18N 85 49W
New Albany, Miss., U.S.A. 81 H10 34 29N 89 0W
New Albany, Pa., U.S.A. 79 E8 41 36N 76 27W
New Amsterdam, Guyana 92 B7 6 15N 57 36W
New Angledool, Australia 63 D4 29 5S 147 55 E
New Baltimore, U.S.A. 78 D2 42 41N 82 44W
New Bedford, U.S.A. 79 E14 41 38N 70 56W
New Berlin, N.Y., U.S.A. 79 D9 42 37N 75 20W
New Berlin, Pa., U.S.A. 78 F8 40 50N 76 57W
New Bern, U.S.A. 77 H7 35 7N 77 3W
New Bethlehem, U.S.A. 78 F5 41 0N 79 20W
New Bloomfield, U.S.A. 78 F7 40 25N 77 11W
New Boston, U.S.A. 81 J7 33 28N 94 25W
New Braunfels, U.S.A. 81 L5 29 42N 98 8W
New Brighton, N.Z. 59 K4 43 29S 172 43 E
New Brighton, U.S.A. 78 F4 40 42N 80 19W
New Britain, Papua N. G. 64 H7 5 50S 150 20 E
New Britain, U.S.A. 79 E12 41 40N 72 47W
New Brunswick, U.S.A. 79 F10 40 30N 74 27W
New Brunswick □, Canada 71 C6 46 50N 66 30W
New Caledonia ■, Pac. Oc. 64 K8 21 0S 165 0 E
New Castile = Castilla-La Mancha □, Spain 19 C4 39 30N 3 30W
New Castle, Ind., U.S.A. 76 F3 39 55N 85 22W
New Castle, Pa., U.S.A. 78 F4 41 0N 80 21W
New City, U.S.A. 79 E11 41 9N 73 59W
New Concord, U.S.A. 78 G3 39 59N 81 54W
New Cumberland, U.S.A. 78 F4 40 30N 80 36W
New Cuyama, U.S.A. 85 L7 34 57N 119 38W
New Delhi, India 42 E7 28 37N 77 13 E
New Denver, Canada 72 D5 50 0N 117 25W
New Don Pedro Reservoir, U.S.A. 84 H6 37 43N 120 24W
New England, U.S.A. 80 B3 46 32N 102 52W
New England Ra., Australia 63 E5 30 20S 151 45 E
New Forest, U.K. 11 G6 50 53N 1 34W
New Galloway, U.K. 12 F4 55 5N 4 9W
New Glasgow, Canada 71 C7 45 35N 62 36W
New Guinea, Oceania 28 K17 4 0S 136 0 E
New Hamburg, Canada 78 C4 43 23N 80 42W
New Hampshire □, U.S.A. 79 C13 44 0N 71 30W
New Hampton, U.S.A. 80 D8 43 3N 92 19W
New Hanover, S. Africa 57 D5 29 22S 30 31 E
New Hartford, U.S.A. 79 C9 43 4N 75 18W
New Haven, Conn., U.S.A. 79 E12 41 18N 72 55W
New Haven, Mich., U.S.A. 78 D2 42 44N 82 48W
New Hazelton, Canada 72 B3 55 20N 127 30W
New Hebrides = Vanuatu ■, Pac. Oc. 64 J8 15 0S 168 0 E
New Holland, U.S.A. 79 F8 40 6N 76 5W
New Iberia, U.S.A. 81 K9 30 1N 91 49W
New Ireland, Papua N. G. 64 H7 3 20S 151 50 E
New Jersey □, U.S.A. 76 E8 40 0N 74 30W
New Kensington, U.S.A. 78 F5 40 34N 79 46W
New Lexington, U.S.A. 76 F4 39 43N 82 13W
New Liskeard, Canada 70 C4 47 31N 79 41W
New London, Conn., U.S.A. 79 E12 41 22N 72 6W
New London, Ohio, U.S.A. 78 E2 41 5N 82 24W
New London, Wis., U.S.A. 80 C10 44 23N 88 45W
New Madrid, U.S.A. 81 G10 36 36N 89 32W
New Martinsville, U.S.A. 76 F5 39 39N 80 52W
New Meadows, U.S.A. 82 D5 44 58N 116 18W
New Melones L., U.S.A. 84 H6 37 57N 120 31W
New Mexico □, U.S.A. 83 J10 34 30N 106 0W
New Milford, Conn., U.S.A. 79 E11 41 35N 73 25W
New Milford, Pa., U.S.A. 79 E9 41 52N 75 44W
New Norcia, Australia 61 F2 30 57S 116 13 E
New Norfolk, Australia 62 G4 42 46S 147 2 E
New Orleans, U.S.A. 81 L9 29 58N 90 4W
New Philadelphia, U.S.A. 78 F3 40 30N 81 27W
New Plymouth, N.Z. 59 H5 39 4S 174 5 E
New Plymouth, U.S.A. 82 E5 43 58N 116 49W
New Port Richey, U.S.A. 77 L4 28 16N 82 43W
New Providence, Bahamas 88 A4 25 25N 78 35W
New Quay, U.K. 11 E3 52 13N 4 21W
New Radnor, U.K. 11 E4 52 15N 3 9W
New Richmond, Canada 71 C6 48 15N 65 45W
New Richmond, U.S.A. 80 C8 45 7N 92 32W
New Roads, U.S.A. 81 K9 30 42N 91 26W
New Rochelle, U.S.A. 79 F11 40 55N 73 47W
New Rockford, U.S.A. 80 B5 47 41N 99 8W
New Romney, U.K. 11 G8 50 59N 0 57 E
New Ross, Ireland 13 D5 52 23N 6 57W
New Salem, U.S.A. 80 B4 46 51N 101 25W
New Scone, U.K. 12 E5 56 25N 3 24W
New Siberian I. = Novaya Sibir, Ostrov, Russia 27 B16 75 10N 150 0 E
New Siberian Is. = Novosibirskiye Ostrova, Russia 27 B15 75 0N 142 0 E
New Smyrna Beach, U.S.A. 77 L5 29 1N 80 56W
New South Wales □, Australia 63 E4 33 0S 146 0 E
New Town, U.S.A. 80 B3 47 59N 102 30W
New Tredegar, U.K. 11 F4 51 44N 3 16W
New Ulm, U.S.A. 80 C7 44 19N 94 28W
New Waterford, Canada 71 C7 46 13N 60 4W
New Westminster, Canada 84 A4 49 13N 122 55W
New York, U.S.A. 79 F11 40 45N 74 0W
New York □, U.S.A. 79 D9 43 0N 75 0W
New York Mts., U.S.A. 83 J6 35 0N 115 20W
New Zealand ■, Oceania 59 J6 40 0S 176 0 E
Newaj →, India 42 G7 24 24N 76 49 E
Newala, Tanzania 55 E4 10 58S 39 18 E
Newark, Del., U.S.A. 76 F8 39 41N 75 46W
Newark, N.J., U.S.A. 79 F10 40 44N 74 10W
Newark, N.Y., U.S.A. 78 C7 43 3N 77 6W
Newark, Ohio, U.S.A. 78 F2 40 3N 82 24W
Newark-on-Trent, U.K. 10 D7 53 5N 0 48W
Newark Valley, U.S.A. 79 D8 42 14N 76 11W
Newberg, U.S.A. 82 D2 45 18N 122 58W
Newberry, Mich., U.S.A. 76 B3 46 21N 85 30W
Newberry, S.C., U.S.A. 77 H5 34 17N 81 37W
Newberry Springs, U.S.A. 85 L10 34 50N 116 41W
Newboro L., Canada 79 B8 44 38N 76 20W
Newbridge = Droichead Nua, Ireland 13 C5 53 11N 6 48W
Newburgh, Canada 78 B8 44 19N 76 52W
Newburgh, U.S.A. 79 E10 41 30N 74 1W
Newbury, U.K. 11 F6 51 24N 1 20W
Newbury, N.H., U.S.A. 79 B12 43 19N 72 3W
Newbury, Vt., U.S.A. 79 B12 44 5N 72 4W
Newburyport, U.S.A. 77 D10 42 49N 70 53W
Newcastle, Australia 63 E5 33 0S 151 46 E
Newcastle, N.B., Canada 71 C6 47 1N 65 38W
Newcastle, Ont., Canada 70 D4 43 55N 78 35W
Newcastle, S. Africa 57 D4 27 45S 29 58 E
Newcastle, U.K. 13 B6 54 13N 5 54W
Newcastle, Calif., U.S.A. 84 G5 38 53N 121 8W
Newcastle, Wyo., U.S.A. 80 D2 43 50N 104 11W
Newcastle Emlyn, U.K. 11 E3 52 2N 4 28W
Newcastle Ra., Australia 60 C5 15 45S 130 15 E
Newcastle-under-Lyme, U.K. 10 D5 53 1N 2 14W
Newcastle-upon-Tyne, U.K. 10 C6 54 58N 1 36W
Newcastle Waters, Australia 62 B1 17 30S 133 28 E
Newcastle West, Ireland 13 D2 52 27N 9 3W
Newcomb, U.S.A. 79 C10 43 58N 74 10W
Newcomerstown, U.S.A. 78 F3 40 16N 81 36W
Newdegate, Australia 61 F2 33 6S 119 0 E
Newell, Australia 62 B4 16 20S 145 16 E
Newell, U.S.A. 80 C3 44 43N 103 25W
Newfane, U.S.A. 78 C6 43 17N 78 43W
Newfield, U.S.A. 79 D8 42 18N 76 33W
Newfound L., U.S.A. 79 C13 43 40N 71 47W
Newfoundland, Canada 66 E14 49 0N 55 0W
Newfoundland, U.S.A. 79 E9 41 18N 75 19W
Newfoundland □, Canada 71 B8 53 0N 58 0W
Newhall, U.S.A. 85 L8 34 23N 118 32W
Newhaven, U.K. 11 G8 50 47N 0 3 E
Newkirk, U.S.A. 81 G6 36 53N 97 3W
Newlyn, U.K. 11 G2 50 6N 5 34W
Newman, Australia 60 D2 23 18S 119 45 E
Newman, U.S.A. 84 H5 37 19N 121 1W
Newmarket, Canada 78 B5 44 3N 79 28W
Newmarket, Ireland 13 D2 52 13N 9 0W
Newmarket, U.K. 11 E8 52 15N 0 25 E
Newmarket, U.S.A. 79 C14 43 4N 70 56W
Newnan, U.S.A. 77 J3 33 23N 84 48W
Newport, Ireland 13 C2 53 53N 9 33W
Newport, I. of W., U.K. 11 G6 50 42N 1 17W
Newport, Newp., U.K. 11 F5 51 35N 3 0W
Newport, Ark., U.S.A. 81 H9 35 37N 91 16W
Newport, Ky., U.S.A. 76 F3 39 5N 84 30W
Newport, N.H., U.S.A. 79 C12 43 22N 72 10W
Newport, N.Y., U.S.A. 79 C9 43 11N 75 1W
Newport, Oreg., U.S.A. 82 D1 44 39N 124 3W
Newport, Pa., U.S.A. 78 F7 40 29N 77 8W
Newport, R.I., U.S.A. 79 E13 41 29N 71 19W
Newport, Tenn., U.S.A. 77 H4 35 58N 83 11W
Newport, Vt., U.S.A. 79 B12 44 56N 72 13W
Newport, Wash., U.S.A. 82 B5 48 11N 117 3W
Newport □, U.K. 11 F4 51 33N 3 1W
Newport Beach, U.S.A. 85 M9 33 37N 117 56W
Newport News, U.S.A. 76 G7 36 59N 76 25W
Newport Pagnell, U.K. 11 E7 52 5N 0 43W
Newquay, U.K. 11 G2 50 25N 5 6W
Newry, U.K. 13 B5 54 11N 6 21W
Newton, Ill., U.S.A. 80 F10 38 59N 88 10W
Newton, Iowa, U.S.A. 80 E8 41 42N 93 3W
Newton, Kans., U.S.A. 81 F6 38 3N 97 21W
Newton, Mass., U.S.A. 79 D13 42 21N 71 12W
Newton, Miss., U.S.A. 81 J10 32 19N 89 10W
Newton, N.C., U.S.A. 77 H5 35 40N 81 13W
Newton, N.J., U.S.A. 79 E10 41 3N 74 45W
Newton, Tex., U.S.A. 81 K8 30 51N 93 46W
Newton Abbot, U.K. 11 G4 50 32N 3 37W
Newton Aycliffe, U.K. 10 C6 54 37N 1 34W
Newton Falls, U.S.A. 78 E4 41 11N 80 59W
Newton Stewart, U.K. 12 G4 54 57N 4 30W
Newtonmore, U.K. 12 D4 57 4N 4 8W
Newtown, U.K. 11 E4 52 31N 3 19W
Newtownabbey, U.K. 13 B6 54 40N 5 56W
Newtownards, U.K. 13 B6 54 36N 5 42W
Newtownbarry = Bunclody, Ireland 13 D5 52 39N 6 40W
Newtownstewart, U.K. 13 B4 54 43N 7 23W
Newville, U.S.A. 78 F7 40 10N 77 24W
Neya, Russia 24 C7 58 21N 43 49 E
Neyriz, Iran 45 D7 29 15N 54 19 E
Neyshābūr, Iran 45 B8 36 10N 58 50 E
Nezhin = Nizhyn, Ukraine 25 D5 51 5N 31 55 E
Nezperce, U.S.A. 82 C5 46 14N 116 14W
Ngabang, Indonesia 36 D3 0 23N 109 55 E
Ngabordamlu, Tanjung, Indonesia 37 F8 6 56S 134 11 E
N'Gage, Angola 52 F3 7 46S 15 16 E
Ngami Depression, Botswana 56 C3 20 30S 22 46 E
Ngamo, Zimbabwe 55 F2 19 3S 27 32 E
Nganglong Kangri, China 41 C12 33 0N 81 0 E
Ngao, Thailand 38 C2 18 46N 99 59 E
Ngaoundéré, Cameroon 52 C2 7 15N 13 35 E
Ngapara, N.Z. 59 L3 44 57S 170 46 E
Ngara, Tanzania 54 C3 2 29S 30 40 E
Ngawi, Indonesia 37 G14 7 24S 111 26 E
Nghia Lo, Vietnam 38 B5 21 33N 104 28 E
Ngoma, Malawi 55 E3 13 8S 33 45 E
Ngomahura, Zimbabwe 55 G3 20 26S 30 43 E
Ngomba, Tanzania 55 D3 8 20S 32 53 E
Ngoring Hu, China 32 C4 34 55N 97 5 E
Ngorongoro, Tanzania 54 C4 3 11S 35 32 E
Ngozi, Burundi 54 C2 2 54S 29 50 E
Ngudu, Tanzania 54 C3 2 58S 33 25 E
Nguigmi, Niger 51 F8 14 20N 13 20 E
Nguiu, Australia 60 B5 11 46S 130 38 E
Ngukurr, Australia 62 A1 14 44S 134 44 E
Ngunga, Tanzania 54 C3 3 37S 33 37 E
Nguru, Nigeria 51 F8 12 56N 10 29 E
Nguru Mts., Tanzania 54 D4 6 0S 37 30 E
Ngusi, Malawi 55 E3 14 0S 34 50 E
Nguyen Binh, Vietnam 38 A5 22 39N 105 56 E
Nha Trang, Vietnam 39 F7 12 16N 109 10 E
Nhacoongo, Mozam. 57 C6 24 18S 35 14 E
Nhamaabué, Mozam. 55 F4 17 25S 35 5 E
Nhamundá →, Brazil 93 D7 2 12S 56 41W
Nhangulaze, L., Mozam. 57 C5 24 0S 34 30 E
Nhill, Australia 63 F3 36 18S 141 40 E
Nho Quan, Vietnam 38 B5 20 18N 105 45 E
Nhulunbuy, Australia 62 A2 12 10S 137 20 E
Nia-nia, Dem. Rep. of the Congo 54 B2 1 30N 27 40 E
Niagara Falls, Canada 78 C5 43 7N 79 5W
Niagara Falls, U.S.A. 78 C6 43 5N 79 4W
Niagara-on-the-Lake, Canada 78 C5 43 15N 79 4W
Niah, Malaysia 36 D4 3 58N 113 46 E
Niamey, Niger 50 F6 13 27N 2 6 E
Niangara, Dem. Rep. of the Congo 54 B2 3 42N 27 50 E
Niantic, U.S.A. 79 E12 41 20N 72 11W
Nias, Indonesia 36 D1 1 0N 97 30 E
Niassa □, Mozam. 55 E4 13 30S 36 0 E
Nibāk, Si. Arabia 45 E7 24 25N 50 50 E
Nicaragua ■, Cent. Amer. 88 D2 11 40N 85 30W
Nicaragua, L. de, Nic. 88 D2 12 0N 85 30W
Nicastro, Italy 20 E7 38 59N 16 19 E
Nice, France 18 E7 43 42N 7 14 E
Niceville, U.S.A. 77 K2 30 31N 86 30W
Nichicun, L., Canada 71 B5 53 5N 71 0W
Nichinan, Japan 31 J5 31 38N 131 23 E
Nicholás, Canal, W. Indies 88 B3 23 30N 80 5W
Nicholasville, U.S.A. 76 G3 37 53N 84 34W
Nichols, U.S.A. 79 D8 42 1N 76 22W
Nicholson, Australia 60 C4 18 2S 128 54 E
Nicholson, U.S.A. 79 E9 41 37N 75 47W
Nicholson →, Australia 62 B2 17 31S 139 36 E
Nicholson L., Canada 73 A8 62 40N 102 40W
Nicholson Ra., Australia 61 E2 27 15S 116 45 E
Nicholville, U.S.A. 79 B10 44 41N 74 39W
Nicobar Is., Ind. Oc. 29 J13 8 0N 93 30 E
Nicola, Canada 72 C4 50 12N 120 40W
Nicolls Town, Bahamas 88 A4 25 8N 78 0W
Nicosia, Cyprus 23 D12 35 10N 33 25 E
Nicoya, Costa Rica 88 D2 10 9N 85 27W
Nicoya, G. de, Costa Rica 88 E3 10 0N 85 0W
Nicoya, Pen. de, Costa Rica 88 E2 9 45N 85 40W
Nidd →, U.K. 10 D6 53 59N 1 23W
Niedersachsen □, Germany 16 B5 52 50N 9 0 E
Niekerkshoop, S. Africa 56 D3 29 19S 22 51 E
Niemba, Dem. Rep. of the Congo 54 D2 5 58S 28 24 E
Niemen = Neman →, Lithuania 9 J19 55 25N 21 10 E
Nienburg, Germany 16 B5 52 39N 9 13 E
Nieu Bethesda, S. Africa 56 E3 31 51S 24 34 E
Nieuw Amsterdam, Surinam 93 B7 5 53N 55 5W
Nieuw Nickerie, Surinam 93 B7 6 0N 56 59W
Nieuwoudtville, S. Africa 56 E2 31 23S 19 7 E
Nieuwpoort, Belgium 15 C2 51 8N 2 45 E
Nieves, Pico de las, Canary Is. 22 G4 27 57N 15 35W
Niğde, Turkey 25 G5 37 58N 34 40 E
Nigel, S. Africa 57 D4 26 27S 28 25 E
Niger ■, W. Afr. 50 E7 17 30N 10 0 E
Niger →, W. Afr. 50 G7 5 33N 6 33 E
Nigeria ■, W. Afr. 50 G7 8 30N 8 0 E
Nighasin, India 43 E9 28 14N 80 52 E
Nightcaps, N.Z. 59 L2 45 57S 168 2 E
Nii-Jima, Japan 31 G9 34 20N 139 15 E
Niigata, Japan 30 F9 37 58N 139 0 E
Niigata □, Japan 31 F9 37 15N 138 45 E
Niihama, Japan 31 H6 33 55N 133 16 E
Niihau, U.S.A. 74 H14 21 54N 160 9W
Niimi, Japan 31 G6 34 59N 133 28 E
Niitsu, Japan 30 F9 37 48N 139 7 E
Nijil, Jordan 47 E4 30 32N 35 33 E
Nijkerk, Neths. 15 B5 52 13N 5 30 E
Nijmegen, Neths. 15 C5 51 50N 5 52 E
Nijverdal, Neths. 15 B6 52 22N 6 28 E
Nīk Pey, Iran 45 B6 36 50N 48 10 E
Nikiniki, Indonesia 37 F6 9 49S 124 30 E
Nikkō, Japan 31 F9 36 45N 139 35 E
Nikolayev = Mykolayiv, Ukraine 25 E5 46 58N 32 0 E
Nikolayevsk, Russia 25 E8 50 0N 45 35 E
Nikolayevsk-na-Amur, Russia 27 D15 53 8N 140 44 E
Nikolskoye, Russia 27 D17 55 12N 166 0 E
Nikopol, Ukraine 25 E5 47 35N 34 25 E
Nīkshahr, Iran 45 E9 26 15N 60 10 E
Nikšić, Montenegro, Yug. 21 C8 42 50N 18 57 E
Nîl, Nahr en →, Africa 51 B12 30 10N 31 6 E
Nîl el Abyad →, Sudan 51 E12 15 38N 32 31 E
Nîl el Azraq →, Sudan 51 E12 15 38N 32 31 E
Nila, Indonesia 37 F7 6 44S 129 31 E
Niland, U.S.A. 85 M11 33 14N 115 31W
Nile = Nîl, Nahr en →, Africa 51 B12 30 10N 31 6 E
Niles, Mich., U.S.A. 76 E2 41 50N 86 15W
Niles, Ohio, U.S.A. 78 E4 41 11N 80 46W
Nim Ka Thana, India 42 F6 27 44N 75 48 E
Nimach, India 42 G6 24 30N 74 56 E
Nimbahera, India 42 G6 24 37N 74 45 E
Nîmes, France 18 E6 43 50N 4 23 E
Nimfaíon, Ákra = Pinnes, Ákra, Greece 21 D11 40 5N 24 20 E
Nimmitabel, Australia 63 F4 36 29S 149 15 E
Nīnawá, Iraq 44 B4 36 25N 43 10 E
Nindigully, Australia 63 D4 28 21S 148 50 E
Nineveh = Nīnawá, Iraq 44 B4 36 25N 43 10 E
Ning Xian, China 34 G4 35 30N 107 58 E
Ning'an, China 35 B15 44 22N 129 20 E
Ningbo, China 33 D7 29 51N 121 28 E
Ningcheng, China 35 D10 41 32N 119 53 E
Ningjin, China 34 F8 37 35N 114 57 E
Ningjing Shan, China 32 D4 30 0N 98 20 E
Ningling, China 34 G8 34 25N 115 22 E
Ningpo = Ningbo, China 33 D7 29 51N 121 28 E
Ningqiang, China 34 H4 32 47N 106 15 E
Ningshan, China 34 H5 33 21N 108 21 E
Ningsia Hui A.R. = Ningxia Huizu Zizhiqu □, China 34 F4 38 0N 106 0 E
Ningwu, China 34 E7 39 0N 112 18 E
Ningxia Huizu Zizhiqu □, China 34 F4 38 0N 106 0 E
Ningyang, China 34 G9 35 47N 116 45 E
Ninh Binh, Vietnam 38 B5 20 15N 105 55 E
Ninh Giang, Vietnam 38 B6 20 44N 106 24 E
Ninh Hoa, Vietnam 38 F7 12 30N 109 7 E
Ninh Ma, Vietnam 38 F7 12 48N 109 21 E
Ninove, Belgium 15 D4 50 51N 4 2 E
Nioaque, Brazil 95 A4 21 5S 55 50W
Niobrara, U.S.A. 80 D6 42 45N 98 2W
Niobrara →, U.S.A. 80 D6 42 46N 98 3W
Nioro du Sahel, Mali 50 E4 15 15N 9 30W
Niort, France 18 C3 46 19N 0 29W
Nipawin, Canada 73 C8 53 20N 104 0W
Nipigon, Canada 70 C2 49 0N 88 17W
Nipigon, L., Canada 70 C2 49 50N 88 30W
Nipishish L., Canada 71 B7 54 12N 60 45W
Nipissing, L., Canada 70 C4 46 20N 80 0W
Nipomo, U.S.A. 85 K6 35 3N 120 29W
Nipton, U.S.A. 85 K11 35 28N 115 16W
Niquelândia, Brazil 93 F9 14 33S 48 23W
Nīr, Iran 44 B5 38 2N 47 59 E
Nirasaki, Japan 31 G9 35 42N 138 27 E
Nirmal, India 40 K11 19 3N 78 20 E
Nirmali, India 43 F12 26 20N 86 35 E
Niš, Serbia, Yug. 21 C9 43 19N 21 58 E
Nişāb, Si. Arabia 44 D5 29 11N 44 43 E
Nişāb, Yemen 46 E4 14 25N 46 29 E
Nishinomiya, Japan 31 G7 34 45N 135 20 E
Nishino'omote, Japan 31 J5 30 43N 130 59 E
Nishiwaki, Japan 31 G7 34 59N 134 58 E
Niskibi →, Canada 70 A2 56 29N 88 9W
Nisqually →, U.S.A. 84 C4 47 6N 122 42W
Nissáki, Greece 23 A3 39 43N 19 52 E
Nissum Bredning, Denmark 9 H13 56 40N 8 20 E
Nistru = Dnister →, Europe 17 E16 46 18N 30 17 E
Nisutlin →, Canada 72 A2 60 14N 132 34W
Nitchequon, Canada 71 B5 53 10N 70 58W
Niterói, Brazil 95 A7 22 52S 43 0W
Nith →, Canada 78 C4 43 12N 80 23W
Nith →, U.K. 12 F5 55 14N 3 33W
Nitra, Slovak Rep. 17 D10 48 19N 18 4 E
Nitra →, Slovak Rep. 17 E10 47 46N 18 10 E
Niuafo'ou, Tonga 59 B11 15 30S 175 58W
Niue, Cook Is. 65 J11 19 2S 169 54W
Niut, Indonesia 36 D4 0 55N 110 6 E
Niuzhuang, China 35 D12 40 58N 122 28 E
Nivala, Finland 8 E21 63 56N 24 57 E
Nivelles, Belgium 15 D4 50 35N 4 20 E
Nivernais, France 18 C5 47 15N 3 30 E
Niwas, India 43 H9 23 3N 80 26 E
Nixon, U.S.A. 81 L6 29 16N 97 46W
Nizamabad, India 40 K11 18 45N 78 7 E
Nizamghat, India 41 E19 28 20N 95 45 E
Nizhne Kolymsk, Russia 27 C17 68 34N 160 55 E
Nizhnekamsk, Russia 24 C9 55 38N 51 49 E
Nizhneudinsk, Russia 27 D10 54 54N 99 3 E
Nizhnevartovsk, Russia 26 C8 60 56N 76 38 E
Nizhniy Novgorod, Russia 24 C7 56 20N 44 0 E
Nizhniy Tagil, Russia 24 C10 57 55N 59 57 E
Nizhyn, Ukraine 25 D5 51 5N 31 55 E
Nizip, Turkey 44 B3 37 5N 37 50 E
Nízké Tatry, Slovak Rep. 17 D10 48 55N 19 30 E
Njakwa, Malawi 55 E3 11 1S 33 56 E
Njanji, Zambia 55 E3 14 25S 31 46 E
Njinjo, Tanzania 55 D4 8 48S 38 54 E
Njombe, Tanzania 55 D3 9 20S 34 50 E
Njombe →, Tanzania 54 D4 6 56S 35 6 E
Nkana, Zambia 55 E2 12 50S 28 8 E
Nkandla, S. Africa 57 D5 28 37S 31 5 E
Nkayi, Zimbabwe 55 F2 19 41S 29 20 E
Nkhotakota, Malawi 55 E3 12 56S 34 15 E
Nkongsamba, Cameroon 52 D1 4 55N 9 55 E
Nkurenkuru, Namibia 56 B2 17 42S 18 32 E
Nmai →, Burma 41 G20 25 30N 97 25 E
Noakhali = Maijdi, Bangla. 41 H17 22 48N 91 10 E
Nobel, Canada 78 A4 45 25N 80 6W
Nobeoka, Japan 31 H5 32 36N 131 41 E
Noblesville, U.S.A. 76 E3 40 3N 86 1W
Nocera Inferiore, Italy 20 D6 40 44N 14 38 E
Nocona, U.S.A. 81 J6 33 47N 97 44W
Noda, Japan 31 G9 35 56N 139 52 E
Nogales, Mexico 86 A2 31 20N 110 56W
Nogales, U.S.A. 83 L8 31 20N 110 56W
Nōgata, Japan 31 H5 33 48N 130 44 E
Noggerup, Australia 61 F2 33 32S 116 5 E
Noginsk, Russia 27 C10 64 30N 90 50 E
Nogoa →, Australia 62 C4 23 40S 147 55 E
Nogoyá, Argentina 94 C4 32 24S 59 48W
Nohar, India 42 E6 29 11N 74 49 E
Nohta, India 43 H8 23 40N 79 34 E
Noires, Mts., France 18 B2 48 11N 3 40W
Noirmoutier, Î. de, France 18 C2 46 58N 2 10W
Nojane, Botswana 56 C3 23 15S 20 14 E
Nojima-Zaki, Japan 31 G9 34 54N 139 53 E
Nok Kundi, Pakistan 40 E3 28 50N 62 45 E
Nokaneng, Botswana 56 B3 19 40S 22 17 E
Nokia, Finland 9 F20 61 30N 23 30 E
Nokomis, Canada 73 C8 51 35N 105 0W
Nokomis L., Canada 73 B8 57 0N 103 0W
Nola, C.A.R. 52 D3 3 35N 16 4 E
Noma Omuramba →, Namibia 56 B3 18 52S 20 53 E
Nombre de Dios, Panama 88 E4 9 34N 79 28W
Nome, U.S.A. 68 B3 64 30N 165 25W
Nomo-Zaki, Japan 31 H4 32 35N 129 44 E
Nonacho L., Canada 73 A7 61 42N 109 40W
Nonda, Australia 62 C3 20 40S 142 28 E
Nong Chang, Thailand 38 E2 15 23N 99 51 E
Nong Het, Laos 38 C4 19 29N 103 59 E
Nong Khai, Thailand 38 D4 17 50N 102 46 E
Nong'an, China 35 B13 44 25N 125 5 E
Nongoma, S. Africa 57 D5 27 58S 31 35 E
Nonoava, Mexico 86 B3 27 28N 106 44W
Nonoava →, Mexico 86 B3 27 29N 106 45W
Nonthaburi, Thailand 38 F3 13 51N 100 34 E
Noonamah, Australia 60 B5 12 40S 131 4 E
Noord Brabant □, Neths. 15 C5 51 40N 5 0 E
Noord Holland □, Neths. 15 B4 52 30N 4 45 E
Noordbeveland, Neths. 15 C3 51 35N 3 50 E
Noordoostpolder, Neths. 15 B5 52 45N 5 45 E
Noordwijk, Neths. 15 B4 52 14N 4 26 E
Nootka I., Canada 72 D3 49 32N 126 42W
Nopiming Prov. Park, Canada 73 C9 50 30N 95 37W
Noralee, Canada 72 C3 53 59N 126 26W
Noranda = Rouyn-Noranda, Canada 70 C4 48 20N 79 0W
Norco, U.S.A. 85 M9 33 56N 117 33W
Nord-Kivu □, Dem. Rep. of the Congo 54 C2 1 0S 29 0 E
Nord-Ostsee-Kanal, Germany 16 A5 54 12N 9 32 E
Nordaustlandet, Svalbard 4 B9 79 14N 23 0 E
Nordegg, Canada 72 C5 52 29N 116 5W
Norderney, Germany 16 B4 53 42N 7 9 E
Norderstedt, Germany 16 B5 53 42N 10 1 E

O

Orford Ness

Oakhurst, U.S.A. **84 H7** 37 19N 119 40W
Oakland, U.S.A. **84 H4** 37 49N 122 16W
Oakley, Idaho, U.S.A. **82 E7** 42 15N 113 53W
Oakley, Kans., U.S.A. **80 F4** 39 8N 100 51W
Oakover →, Australia **60 D3** 21 0S 120 40 E
Oakridge, U.S.A. **82 E2** 43 45N 122 28W
Oakville, Canada **78 C5** 43 27N 79 41W
Oakville, U.S.A. **84 D3** 46 51N 123 14W
Oamaru, N.Z. **59 L3** 45 5S 170 59 E
Oasis, Calif., U.S.A. **85 M10** 33 28N 116 6W
Oasis, Nev., U.S.A. **84 H9** 37 29N 117 55W
Oates Land, Antarctica **5 C11** 69 0S 160 0 E
Oatlands, Australia **62 G4** 42 17S 147 21 E
Oatman, U.S.A. **85 K12** 35 1N 114 19W
Oaxaca, Mexico **87 D5** 17 2N 96 40W
Oaxaca □, Mexico **87 D5** 17 0N 97 0W
Ob →, Russia **26 C7** 66 45N 69 30 E
Oba, Canada **70 C3** 49 4N 84 7W
Obama, Japan **31 G7** 35 30N 135 45 E
Oban, U.K. **12 E3** 56 25N 5 29W
Obbia, Somali Rep. **46 F4** 5 25N 48 30 E
Obera, Argentina **95 B4** 27 21S 55 2W
Oberhausen, Germany **16 C4** 51 28N 6 51 E
Oberlin, Kans., U.S.A. **80 F4** 39 49N 100 32W
Oberlin, La., U.S.A. **81 K8** 30 37N 92 46W
Oberlin, Ohio, U.S.A. **78 E2** 41 18N 82 13W
Oberon, Australia **63 E4** 33 45S 149 52 E
Obi, Indonesia **37 E7** 1 23S 127 45 E
Óbidos, Brazil **93 D7** 1 50S 55 30W
Obihiro, Japan **30 C11** 42 56N 143 12 E
Obilatu, Indonesia **37 E7** 1 25S 127 20 E
Obluchye, Russia **27 E14** 49 1N 131 4 E
Obo, C.A.R. **54 A2** 5 20N 26 32 E
Oboa, Mt., Uganda **54 B3** 1 45N 34 45 E
Oboyan, Russia **26 D4** 51 15N 36 21 E
Obozerskaya = Obozerskiy, Russia **24 B7** 63 34N 40 21 E
Obozerskiy, Russia **24 B7** 63 34N 40 21 E
Observatory Inlet, Canada **72 B3** 55 10N 129 54W
Obshchi Syrt, Russia **6 E16** 52 0N 53 0 E
Obskaya Guba, Russia **26 C8** 69 0N 73 0 E
Obuasi, Ghana **50 G5** 6 17N 1 40W
Ocala, U.S.A. **77 L4** 29 11N 82 8W
Ocampo, Chihuahua, Mexico **86 B3** 28 9N 108 24W
Ocampo, Tamaulipas, Mexico **87 C5** 22 50N 99 20W
Ocaña, Spain **19 C4** 39 55N 3 30W
Ocanomowoc, U.S.A. **80 D10** 43 7N 88 30W
Occidental, Cordillera, Colombia **92 C3** 5 0N 76 0W
Occidental, Grand Erg, Algeria **50 B6** 30 20N 1 0 E
Ocean City, Md., U.S.A. **76 F8** 38 20N 75 5W
Ocean City, N.J., U.S.A. **76 F8** 39 17N 74 35W
Ocean City, Wash., U.S.A. **84 C2** 47 4N 124 10W
Ocean Falls, Canada **72 C3** 52 18N 127 48W
Ocean I. = Banaba, Kiribati **64 H8** 0 45S 169 50 E
Ocean Park, U.S.A. **84 D2** 46 30N 124 3W
Oceano, U.S.A. **85 K6** 35 6N 120 37W
Oceanport, U.S.A. **79 F10** 40 19N 74 3W
Oceanside, U.S.A. **85 M9** 33 12N 117 23W
Ochil Hills, U.K. **12 E5** 56 14N 3 40W
Ocilla, U.S.A. **77 K4** 31 36N 83 15W
Ocmulgee →, U.S.A. **77 K4** 31 58N 82 33W
Ocniţa, Moldova **17 D14** 48 25N 27 30 E
Oconee →, U.S.A. **77 K4** 31 58N 82 33W
Oconto, U.S.A. **76 C2** 44 53N 87 52W
Oconto Falls, U.S.A. **76 C1** 44 52N 88 9W
Ocosingo, Mexico **87 D6** 17 10N 92 15W
Ocotal, Nic. **88 D2** 13 41N 86 31W
Ocotlán, Mexico **86 C4** 20 21N 102 42W
Ocotlán de Morelos, Mexico **87 D5** 16 48N 96 40W
Ōda, Japan **31 G6** 35 11N 132 30 E
Ódáðahraun, Iceland **8 D5** 65 5N 17 0W
Odate, Japan **30 D10** 40 16N 140 34 E
Odawara, Japan **31 G9** 35 20N 139 6 E
Odda, Norway **9 F12** 60 3N 6 35 E
Odei →, Canada **73 B9** 56 6N 96 54W
Ödemiş, Turkey **21 E13** 38 15N 28 0 E
Odendaalsrus, S. Africa **56 D4** 27 48S 26 45 E
Odense, Denmark **9 J14** 55 22N 10 23 E
Oder →, Europe **16 B8** 53 33N 14 38 E
Odesa, Ukraine **25 E5** 46 30N 30 45 E
Odessa = Odesa, Ukraine **25 E5** 46 30N 30 45 E
Odessa, Canada **79 B8** 44 17N 76 43W
Odessa, Tex., U.S.A. **81 K3** 31 52N 102 23W
Odessa, Wash., U.S.A. **82 C4** 47 20N 118 41W
Odiakwe, Botswana **56 C4** 20 12S 25 17 E
Odienné, Ivory C. **50 G4** 9 30N 7 34W
Odintsovo, Russia **24 C6** 55 39N 37 15 E
O'Donnell, U.S.A. **81 J4** 32 58N 101 50W
Odorheiu Secuiesc, Romania **17 E13** 46 21N 25 21 E
Odra = Oder →, Europe **16 B8** 53 33N 14 38 E
Odzi, Zimbabwe **57 B5** 19 0S 32 20 E
Odzi →, Zimbabwe **57 B5** 19 45S 32 23 E
Oeiras, Brazil **93 E10** 7 0S 42 8W
Oelrichs, U.S.A. **80 D3** 43 11N 103 14W
Oelwein, U.S.A. **80 D9** 42 41N 91 55W
Oenpelli, Australia **60 B5** 12 20S 133 4 E
Ofanto →, Italy **20 D7** 41 22N 16 13 E
Offa, Nigeria **50 G6** 8 13N 4 42 E
Offaly □, Ireland **13 C4** 53 15N 7 30W
Offenbach, Germany **16 C5** 50 6N 8 44 E
Offenburg, Germany **16 D4** 48 28N 7 56 E
Ofotfjorden, Norway **8 B17** 68 27N 17 0 E
Ōfunato, Japan **30 E10** 39 4N 141 43 E
Oga, Japan **30 E9** 39 55N 139 50 E
Oga-Hantō, Japan **30 E9** 39 58N 139 47 E
Ogaden, Ethiopia **46 F3** 7 30N 45 30 E
Ōgaki, Japan **31 G8** 35 21N 136 37 E
Ogallala, U.S.A. **80 E4** 41 8N 101 43W
Ogasawara Gunto, Pac. Oc. **28 G18** 27 0N 142 0 E
Ogbomosho, Nigeria **50 G6** 8 1N 4 11 E
Ogden, U.S.A. **82 F7** 41 13N 111 58W
Ogdensburg, U.S.A. **79 B9** 44 42N 75 30W
Ogeechee →, U.S.A. **77 K5** 31 50N 81 3W
Ogilby, U.S.A. **85 N12** 32 49N 114 50W
Oglio →, Italy **20 B4** 45 2N 10 39 E
Ogmore, Australia **62 C4** 22 37S 149 35 E
Ogoki, Canada **70 B2** 51 38N 85 58W
Ogoki →, Canada **70 B2** 51 38N 85 57W
Ogoki L., Canada **70 B2** 50 50N 87 10W
Ogoki Res., Canada **70 B2** 50 45N 88 15W
Ogooué →, Gabon **52 E1** 1 0S 9 0 E
Ogowe = Ogooué →, Gabon **52 E1** 1 0S 9 0 E
Ogre, Latvia **9 H21** 56 49N 24 36 E
Ogurchinskiy, Ostrov, Turkmenistan **45 B7** 38 55N 53 2 E
Ohai, N.Z. **59 L2** 45 55S 168 0 E
Ohakune, N.Z. **59 H5** 39 24S 175 24 E
Ohata, Japan **30 D10** 41 24N 141 10 E
Ohau, L., N.Z. **59 L2** 44 15S 169 53 E
Ohio □, U.S.A. **78 F2** 40 15N 82 45W
Ohio →, U.S.A. **76 G1** 36 59N 89 8W
Ohře →, Czech Rep. **16 C8** 50 30N 14 10 E
Ohrid, Macedonia **21 D9** 41 8N 20 52 E
Ohridsko Jezero, Macedonia **21 D9** 41 8N 20 52 E
Ohrigstad, S. Africa **57 C5** 24 39S 30 36 E
Oiapoque, Brazil **93** 3 50N 51 50W
Oikou, China **35 E9** 38 35N 117 42 E
Oil City, U.S.A. **78 E5** 41 26N 79 42W
Oil Springs, Canada **78 D2** 42 47N 82 7W
Oildale, U.S.A. **85 K7** 35 25N 119 1W
Oise →, France **18 B5** 49 0N 2 4 E
Ōita, Japan **31 H5** 33 14N 131 36 E
Ōita □, Japan **31 H5** 33 15N 131 30 E
Oiticica, Brazil **93 E10** 5 3S 41 5W
Ojacaliente, Mexico **86 C4** 22 34N 102 15W
Ojai, U.S.A. **85 L7** 34 27N 119 15W
Ojinaga, Mexico **86 B4** 29 34N 104 25W
Ojiya, Japan **31 F9** 37 18N 138 48 E
Ojos del Salado, Cerro, Argentina **94 B2** 27 0S 68 40W
Oka →, Russia **24 C7** 56 20N 43 59 E
Okaba, Indonesia **37 F9** 8 6S 139 42 E
Okahandja, Namibia **56 C2** 22 0S 16 59 E
Okanagan L., Canada **72 D5** 50 0N 119 30W
Okanogan, U.S.A. **82 B4** 48 22N 119 35W
Okanogan →, U.S.A. **82 B4** 48 6N 119 44W
Okaputa, Namibia **56 C2** 20 5S 17 0 E
Okara, Pakistan **42 D5** 30 50N 73 31 E
Okaukuejo, Namibia **56 B2** 19 10S 16 0 E
Okavango Delta, Botswana **56 B3** 18 45S 22 45 E
Okavango Swamp = Okavango Delta, Botswana **56 B3** 18 45S 22 45 E
Okaya, Japan **31 F9** 36 5N 138 10 E
Okayama, Japan **31 G6** 34 40N 133 54 E
Okayama □, Japan **31 G6** 35 0N 133 50 E
Okazaki, Japan **31 G8** 34 57N 137 10 E
Okeechobee, U.S.A. **77 M5** 27 15N 80 50W
Okeechobee, L., U.S.A. **77 M5** 27 0N 80 50W
Okefenokee Swamp, U.S.A. **77 K4** 30 40N 82 20W
Okehampton, U.K. **11 G4** 50 44N 4 0W
Okha, India **42 H3** 22 27N 69 4 E
Okha, Russia **27 D15** 53 40N 143 0 E
Okhotsk, Russia **27 D15** 59 20N 143 10 E
Okhotsk, Sea of, Asia **27 D15** 55 0N 145 0 E
Okhotskiy Perevoz, Russia **27 C14** 61 52N 135 35 E
Okhtyrka, Ukraine **25 D5** 50 25N 35 0 E
Oki-Shotō, Japan **31 F6** 36 5N 133 15 E
Okiep, S. Africa **56 D2** 29 39S 17 53 E
Okinawa □, Japan **31 L4** 26 40N 128 0 E
Okinawa-Guntō, Japan **31 L4** 26 40N 128 0 E
Okinawa-Jima, Japan **31 L4** 26 32N 128 0 E
Okino-erabu-Shima, Japan **31 L4** 27 21N 128 33 E
Oklahoma □, U.S.A. **81 H6** 35 20N 97 30W
Oklahoma City, U.S.A. **81 H6** 35 30N 97 30W
Okmulgee, U.S.A. **81 H7** 35 37N 95 58W
Oknitsa = Ocniţa, Moldova **17 D14** 48 25N 27 30 E
Okolo, Uganda **54 B3** 2 37N 31 8 E
Okolona, U.S.A. **81 J10** 34 0N 88 45W
Okombahe, Namibia **56 C2** 21 23S 15 22 E
Okotoks, Canada **72 C6** 50 43N 113 58W
Oksibil, Indonesia **37 E10** 4 59S 140 35 E
Oksovskiy, Russia **24 B6** 62 33N 39 57 E
Oktabrsk = Oktyabrsk, Kazakstan **25 E10** 49 28N 57 25 E
Oktyabrsk, Kazakstan **25 E10** 49 28N 57 25 E
Oktyabrskiy = Aktsyabrski, Belarus **17 B15** 52 38N 28 53 E
Oktyabrskiy, Russia **24 D9** 54 28N 53 28 E
Oktyabrskoy Revolyutsii, Ostrov, Russia **27 B10** 79 30N 97 0 E
Okuru, N.Z. **59 K2** 43 55S 168 55 E
Okushiri-Tō, Japan **30 C9** 42 15N 139 30 E
Okwa →, Botswana **56 C3** 22 30S 23 0 E
Ola, U.S.A. **81 H8** 35 2N 93 13W
Ólafsfjörður, Iceland **8 C4** 66 4N 18 39W
Ólafsvík, Iceland **8 D2** 64 53N 23 43W
Olancha, U.S.A. **85 J8** 36 17N 118 1W
Olancha Pk., U.S.A. **85 J8** 36 15N 118 7W
Olanchito, Honduras **88 C2** 15 30N 86 30W
Öland, Sweden **9 H17** 56 45N 16 38 E
Olary, Australia **63 E3** 32 18S 140 19 E
Olascoaga, Argentina **94 D3** 35 15S 60 39W
Olathe, U.S.A. **80 F7** 38 53N 94 49W
Olavarría, Argentina **94 D3** 36 55S 60 20W
Oława, Poland **17 C9** 50 57N 17 20 E
Ólbia, Italy **20 D3** 40 55N 9 31 E
Olcott, U.S.A. **78 C6** 43 20N 78 42W
Old Bahama Chan. = Bahama, Canal Viejo de, W. Indies **88 B4** 22 10N 77 30W
Old Baldy Pk. = San Antonio, Mt., U.S.A. **85 L9** 34 17N 117 38W
Old Castile = Castilla y Leon □, Spain **19 B3** 42 0N 5 0W
Old Crow, Canada **68 B6** 67 30N 139 55W
Old Dale, U.S.A. **85 L11** 34 8N 115 47W
Old Forge, N.Y., U.S.A. **79 C10** 43 43N 74 58W
Old Forge, Pa., U.S.A. **79 E9** 41 22N 75 45W
Old Perlican, Canada **71 C9** 48 5N 53 1W
Old Shinyanga, Tanzania **54 C3** 3 33S 33 27 E
Old Speck Mt., U.S.A. **79 B14** 44 34N 70 57W
Old Town, U.S.A. **77 C11** 44 56N 68 39W
Old Washington, U.S.A. **78 F3** 40 2N 81 27W
Old Wives L., Canada **73 C7** 50 5N 106 0W
Oldbury, U.K. **11 F5** 51 38N 2 33W
Oldcastle, Ireland **13 C4** 53 46N 7 10W
Oldeani, Tanzania **54 C4** 3 22S 35 35 E
Oldenburg, Germany **16 B5** 53 9N 8 13 E
Oldenzaal, Neths. **15 B6** 52 19N 6 53 E
Oldham, U.K. **10 D5** 53 33N 2 7W
Oldman →, Canada **72 D6** 49 57N 111 42W
Oldmeldrum, U.K. **12 D6** 57 20N 2 19W
Olds, Canada **72 C6** 51 50N 114 10W
Oldziyt, Mongolia **34 B5** 44 40N 109 1 E
Olean, U.S.A. **78 D6** 42 5N 78 26W
Olekma →, Russia **27 C13** 60 22N 120 42 E
Olekminsk, Russia **27 C13** 60 25N 120 30 E
Oleksandriya, Ukraine **17 C14** 50 37N 26 19 E
Olema, U.S.A. **84 G4** 38 3N 122 47W
Olenegorsk, Russia **24 A5** 68 9N 33 18 E
Olenek, Russia **27 C12** 68 28N 112 18 E
Olenek →, Russia **27 B13** 73 0N 120 10 E
Oléron, Î. d', France **18 D3** 45 55N 1 15W
Oleśnica, Poland **17 C9** 51 13N 17 22 E
Olevsk, Ukraine **17 C14** 51 12N 27 39 E
Olga, Russia **27 E14** 43 50N 135 14 E
Olga, L., Canada **70 C4** 49 47N 77 15W
Olga, Mt., Australia **61 E5** 25 20S 130 50 E
Olhão, Portugal **19 D2** 37 3N 7 48W
Olifants →, Africa **57 C5** 23 57S 31 58 E
Olifants →, Namibia **56 C2** 25 30S 19 30 E
Olifantshoek, S. Africa **56 D3** 27 57S 22 42 E
Ólimbos, Óros, Greece **21 D10** 40 6N 22 23 E
Olímpia, Brazil **95 A6** 20 44S 48 54W
Olinda, Brazil **93 E12** 8 1S 34 51W
Oliva, Argentina **94 C3** 32 0S 63 38W
Olivehurst, U.S.A. **84 F5** 39 6N 121 34W
Olivenza, Spain **19 C2** 38 41N 7 9W
Oliver, Canada **72 D5** 49 13N 119 37W
Oliver L., Canada **73 B8** 56 56N 103 22W
Ollagüe, Chile **94 A2** 21 15S 68 10W
Olney, Ill., U.S.A. **76 F1** 38 44N 88 5W
Olney, Tex., U.S.A. **81 J5** 33 22N 98 45W
Olomane →, Canada **71 B7** 50 14N 60 37W
Olomouc, Czech Rep. **17 D9** 49 38N 17 12 E
Olonets, Russia **24 B5** 61 0N 32 54 E
Olongapo, Phil. **37 B6** 14 50N 120 18 E
Olot, Spain **19 A7** 42 11N 2 30 E
Olovyannaya, Russia **27 D12** 50 58N 115 35 E
Oloy →, Russia **27 C16** 66 29N 159 29 E
Olsztyn, Poland **17 B11** 53 48N 20 29 E
Olt →, Romania **17 G13** 43 43N 24 51 E
Olteniţa, Romania **17 F14** 44 7N 26 42 E
Olton, U.S.A. **81 H3** 34 11N 102 8W
Olymbos, Cyprus **23 D12** 35 21N 33 45 E
Olympia, Greece **21 F9** 37 39N 21 39 E
Olympia, U.S.A. **84 D4** 47 3N 122 53W
Olympic Dam, Australia **63 E2** 30 30S 136 55 E
Olympic Mts., U.S.A. **84 C3** 47 55N 123 45W
Olympic Nat. Park, U.S.A. **84 C3** 47 48N 123 30W
Olympus, Cyprus **23 E11** 34 56N 32 52 E
Olympus, Mt. = Ólimbos, Óros, Greece **21 D10** 40 6N 22 23 E
Olympus, Mt. = Uludağ, Turkey **21 D13** 40 4N 29 13 E
Olympus, Mt., U.S.A. **84 C3** 47 48N 123 43W
Olyphant, U.S.A. **79 E9** 41 27N 75 36W
Om →, Russia **26 D8** 54 59N 73 22 E
Om Koi, Thailand **38 D2** 17 48N 98 22 E
Ōma, Japan **30 D10** 41 45N 141 5 E
Ōmachi, Japan **31 F8** 36 30N 137 50 E
Omae-Zaki, Japan **31 G9** 34 36N 138 14 E
Ōmagari, Japan **30 E10** 39 27N 140 29 E
Omagh, U.K. **13 B4** 54 36N 7 19W
Omagh □, U.K. **13 B4** 54 35N 7 15W
Omaha, U.S.A. **80 E7** 41 17N 95 58W
Omak, U.S.A. **82 B4** 48 25N 119 31W
Omalos, Greece **23 D5** 35 19N 23 55 E
Oman ■, Asia **46 C6** 23 0N 58 0 E
Oman, G. of, Asia **45 E8** 24 30N 58 30 E
Omaruru, Namibia **56 C2** 21 26S 16 0 E
Omaruru →, Namibia **56 C1** 22 7S 14 15 E
Omate, Peru **92 G4** 16 45S 71 0W
Ombai, Selat, Indonesia **37 F6** 8 30S 124 50 E
Omboué, Gabon **52 E1** 1 35S 9 15 E
Ombrone →, Italy **20 C4** 42 42N 11 5 E
Omdurmân, Sudan **51 E12** 15 40N 32 28 E
Omemee, Canada **78 B6** 44 18N 78 33W
Omeonga, Dem. Rep. of the Congo **54 C1** 3 40S 24 22 E
Ometepe, I. de, Nic. **88 D2** 11 32N 85 35W
Ometepec, Mexico **87 D5** 16 39N 98 23W
Ominato, Japan **30 D10** 41 17N 141 10 E
Omineca →, Canada **72 B4** 56 3N 124 16W
Omitara, Namibia **56 C2** 22 16S 18 2 E
Ōmiya, Japan **31 G9** 35 54N 139 38 E
Ommen, Neths. **15 B6** 52 31N 6 26 E
Ömnögovĭ □, Mongolia **34 C3** 43 15N 104 0 E
Omo →, Ethiopia **46 F2** 6 25N 36 10 E
Omodhos, Cyprus **23 E11** 34 51N 32 48 E
Omolon →, Russia **27 C16** 68 42N 158 36 E
Omono-Gawa →, Japan **30 E10** 39 46N 140 3 E
Omsk, Russia **26 D8** 55 0N 73 12 E
Omsukchan, Russia **27 C16** 62 32N 155 48 E
Ōmu, Japan **30 B11** 44 34N 142 58 E
Omul, Vf., Romania **17 F13** 45 27N 25 29 E
Ōmura, Japan **31 H4** 32 56N 129 57 E
Omuramba Omatako →, Namibia **56 B2** 17 45S 20 25 E
Omuramba Ovambo →, Namibia **56 B2** 18 45S 16 59 E
Ōmuta, Japan **31 H5** 33 5N 130 26 E
Onaga, U.S.A. **80 F6** 39 29N 96 10W
Onalaska, U.S.A. **80 D9** 43 53N 91 14W
Onancock, U.S.A. **76 G8** 37 43N 75 45W
Onang, Indonesia **37 E5** 3 2S 118 49 E
Onaping L., Canada **70 C3** 47 3N 81 30W
Onavas, Mexico **86 B3** 28 28N 109 30W
Onawa, U.S.A. **80 D6** 42 2N 96 6W
Oncócua, Angola **56 B1** 16 30S 13 25 E
Onda, Spain **19 C5** 39 55N 0 17W
Ondaejin, N. Korea **35 D15** 41 34N 129 40 E
Ondangwa, Namibia **56 B2** 17 57S 16 4 E
Ondjiva, Angola **56 B2** 16 48S 15 50 E
Öndörshil, Mongolia **34 B5** 45 13N 108 5 E
Öndverðarnes, Iceland **8 D1** 64 52N 24 0W
One Tree, Australia **63 E3** 34 11S 144 43 E
Onega, Russia **24 B6** 64 0N 38 10 E
Onega →, Russia **24 B6** 63 58N 38 2 E
Onega, G. of = Onezhskaya Guba, Russia **24 B6** 64 24N 36 38 E
Onega, L. = Onezhskoye Ozero, Russia **24 B6** 61 44N 35 22 E
Oneida, U.S.A. **79 C9** 43 6N 75 39W
Oneida L., U.S.A. **79 C9** 43 12N 75 54W
O'Neill, U.S.A. **80 D5** 42 27N 98 39W
Onekotan, Ostrov, Russia **27 E16** 49 25N 154 45 E
Onema, Dem. Rep. of the Congo **54 C1** 4 35S 24 30 E
Oneonta, U.S.A. **79 D9** 42 27N 75 4W
Oneşti, Romania **17 E14** 46 17N 26 47 E
Onezhskaya Guba, Russia **24 B6** 64 24N 36 38 E
Onezhskoye Ozero, Russia **24 B6** 61 44N 35 22 E
Ongarue, N.Z. **59 H5** 38 42S 175 19 E
Ongers →, S. Africa **56 E3** 31 4S 23 13 E
Ongerup, Australia **61 F2** 33 58S 118 28 E
Ongjin, N. Korea **35 F13** 37 56N 125 21 E
Ongkharak, Thailand **38 E3** 14 8N 101 1 E
Ongniud Qi, China **35 C10** 43 0N 118 38 E
Ongoka, Dem. Rep. of the Congo **54 C2** 1 20S 26 0 E
Ongole, India **40 M12** 15 33N 80 2 E
Ongon = Havirga, Mongolia **34 B7** 45 41N 113 5 E
Onida, U.S.A. **80 C4** 44 42N 100 4W
Onilahy →, Madag. **57 C7** 23 34S 43 45 E
Onitsha, Nigeria **50 G7** 6 6N 6 42 E
Onoda, Japan **31 G5** 33 59N 131 11 E
Onpyŏng-ni, S. Korea **35 H14** 33 25N 126 55 E
Onslow, Australia **60 D2** 21 40S 115 12 E
Onslow B., U.S.A. **77 H7** 34 20N 77 15W
Ontake-San, Japan **31 G8** 35 53N 137 29 E
Ontario, Calif., U.S.A. **85 L9** 34 4N 117 39W
Ontario, Oreg., U.S.A. **82 D5** 44 2N 116 58W
Ontario □, Canada **70 B2** 48 0N 83 0W
Ontario, L., N. Amer. **78 C7** 43 20N 78 0W
Ontonagon, U.S.A. **80 B10** 46 52N 89 19W
Onyx, U.S.A. **85 K8** 35 41N 118 14W
Oodnadatta, Australia **63 D2** 27 33S 135 30 E
Ooldea, Australia **61 F5** 30 27S 131 50 E
Oombulgurri, Australia **60 C4** 15 15S 127 45 E
Oorindi, Australia **62 C3** 20 40S 141 1 E
Oost-Vlaanderen □, Belgium **15 C3** 51 5N 3 50 E
Oostende, Belgium **15 C2** 51 15N 2 54 E
Oosterhout, Neths. **15 C4** 51 39N 4 47 E
Oosterschelde →, Neths. **15 C4** 51 33N 4 0 E
Oosterwolde, Neths. **15 B6** 53 0N 6 17 E
Ootacamund = Udagamandalam, India **40 P10** 11 30N 76 44 E
Ootsa L., Canada **72 C3** 53 50N 126 2W
Opala, Dem. Rep. of the Congo **54 C1** 0 40S 24 20 E
Opanake, Sri Lanka **40 R12** 6 35N 80 40 E
Opasatika, Canada **70 C3** 49 30N 82 50W
Opasquia Prov. Park, Canada **70 B1** 53 33N 93 5W
Opava, Czech Rep. **17 D9** 49 57N 17 58 E
Opelika, U.S.A. **77 J3** 32 39N 85 23W
Opelousas, U.S.A. **81 K8** 30 32N 92 5W
Opémisca, L., Canada **70 C5** 49 56N 74 52W
Opheim, U.S.A. **82 B10** 48 51N 106 24W
Ophthalmia Ra., Australia **60 D2** 23 15S 119 30 E
Opinaca →, Canada **70 B4** 52 15N 78 2W
Opinaca, Rés., Canada **70 B4** 52 39N 76 20W
Opinnagau →, Canada **70 B3** 54 12N 82 25W
Opiscoteo, L., Canada **71 B6** 53 10N 68 10W
Opole, Poland **17 C9** 50 42N 17 58 E
Oponono L., Namibia **56 B2** 18 8S 15 45 E
Oporto = Porto, Portugal **19 B1** 41 8N 8 40W
Opotiki, N.Z. **59 H6** 38 1S 177 19 E
Opp, U.S.A. **77 K2** 31 17N 86 16W
Oppdal, Norway **9 E13** 62 35N 9 41 E
Opportunity, U.S.A. **82 C5** 47 39N 117 15W
Opua, N.Z. **59 F5** 35 19S 174 9 E
Opunake, N.Z. **59 H4** 39 26S 173 52 E
Opuwo, Namibia **56 B1** 18 3S 13 45 E
Ora, Cyprus **23 E12** 34 51N 33 12 E
Oracle, U.S.A. **83 K8** 32 37N 110 46W
Oradea, Romania **17 E11** 47 2N 21 58 E
Öræfajökull, Iceland **8 D5** 64 2N 16 39W
Orai, India **43 G8** 25 58N 79 30 E
Oral = Zhayyq →, Kazakstan **25 E9** 47 0N 51 48 E
Oral, Kazakstan **25 D9** 51 20N 51 20 E
Oran, Algeria **50 A5** 35 45N 0 39W
Orange, Australia **63 E4** 33 15S 149 7 E
Orange, France **18 D6** 44 8N 4 47 E
Orange, Calif., U.S.A. **85 M9** 33 47N 117 51W
Orange, Mass., U.S.A. **79 D12** 42 35N 72 19W
Orange, Tex., U.S.A. **81 K8** 30 6N 93 44W
Orange, Va., U.S.A. **76 F6** 38 15N 78 7W
Orange →, S. Africa **56 D2** 28 41S 16 28 E
Orange, C., Brazil **93 C8** 4 20N 51 30W
Orange Cove, U.S.A. **84 J7** 36 38N 119 19W
Orange Free State = Free State □, S. Africa **56 D4** 28 30S 27 0 E
Orange Grove, U.S.A. **81 M6** 27 58N 97 56W
Orange Walk, Belize **87 D7** 18 6N 88 33W
Orangeburg, U.S.A. **77 J5** 33 30N 80 52W
Orangeville, Canada **78 C4** 43 55N 80 5W
Oranienburg, Germany **16 B7** 52 45N 13 14 E
Oranje = Orange →, S. Africa **56 D2** 28 41S 16 28 E
Oranje Vrystaat = Free State □, S. Africa **56 D4** 28 30S 27 0 E
Oranjemund, Namibia **56 D2** 28 38S 16 29 E
Oranjerivier, S. Africa **56 D3** 29 40S 24 12 E
Oranjestad, Aruba **89 D5** 12 32N 70 2W
Orapa, Botswana **53 J5** 21 15S 25 30 E
Oras, Phil. **37 B7** 12 9N 125 28 E
Oraşul Stalin = Braşov, Romania **17 F13** 45 38N 25 35 E
Orbetello, Italy **20 C4** 42 27N 11 13 E
Orbisonia, U.S.A. **78 F7** 40 15N 77 54W
Orbost, Australia **63 F4** 37 40S 148 29 E
Orcas I., U.S.A. **84 B4** 48 42N 122 56W
Orchard City, U.S.A. **83 G10** 38 50N 107 58W
Orchila, I., Venezuela **89 D6** 11 48N 66 10W
Orcutt, U.S.A. **85 L6** 34 52N 120 27W
Ord, U.S.A. **80 E5** 41 36N 98 56W
Ord →, Australia **60 C4** 15 33S 128 15 E
Ord, Mt., Australia **60 C4** 17 20S 125 34 E
Orderville, U.S.A. **83 H7** 37 17N 112 38W
Ordos = Mu Us Shamo, China **34 E5** 39 0N 109 0 E
Ordu, Turkey **25 F6** 40 55N 37 53 E
Ordway, U.S.A. **80 F3** 38 13N 103 46W
Ordzhonikidze = Vladikavkaz, Russia **25 F7** 43 0N 44 35 E
Ore, Dem. Rep. of the Congo **54 B2** 3 17N 29 30 E
Ore Mts. = Erzgebirge, Germany **16 C7** 50 27N 12 55 E
Örebro, Sweden **9 G16** 59 20N 15 18 E
Oregon, U.S.A. **80 D10** 42 1N 89 20W
Oregon □, U.S.A. **82 E3** 44 0N 121 0W
Oregon City, U.S.A. **84 E4** 45 21N 122 36W
Orekhovo-Zuyevo, Russia **24 C6** 55 50N 38 55 E
Orel, Russia **24 D6** 52 57N 36 3 E
Orem, U.S.A. **82 F8** 40 19N 111 42W
Ören, Turkey **21 F12** 37 3N 27 57 E
Orenburg, Russia **24 D10** 51 45N 55 6 E
Orense = Ourense, Spain **19 A2** 42 19N 7 55W
Orepuki, N.Z. **59 M1** 46 19S 167 46 E
Orestiás, Greece **21 D12** 41 30N 26 33 E
Orestos Pereyra, Mexico **86 B3** 26 31N 105 40W
Orford Ness, U.K. **11 E9** 52 5N 1 35 E

Peera Peera Poolanna L.

Palma de Mallorca, Spain **22 B9** 39 35N 2 39 E
Palma Soriano, Cuba **88 B4** 20 15N 76 0W
Palmares, Brazil **93 E11** 8 41S 35 28W
Palmas, Brazil **95 B5** 26 29S 52 0W
Palmas, C., Liberia **50 H4** 4 27N 7 46W
Pálmas, G. di, Italy **20 E3** 39 0N 8 30 E
Palmdale, U.S.A. **85 L8** 34 35N 118 7W
Palmeira das Missões, Brazil **95 B5** 27 55S 53 17W
Palmeira dos Índios, Brazil **93 E11** 9 25S 36 37W
Palmer, U.S.A. **68 B5** 61 36N 149 7W
Palmer →, Australia **62 B3** 16 0S 142 26 E
Palmer Arch., Antarctica **5 C17** 64 15S 65 0W
Palmer Lake, U.S.A. **80 F2** 39 7N 104 55W
Palmer Land, Antarctica **5 D18** 73 0S 63 0W
Palmerston, Canada **78 C4** 43 50N 80 51W
Palmerston, N.Z. **59 L3** 45 29S 170 43 E
Palmerston North, N.Z. **59 J5** 40 21S 175 39 E
Palmerton, U.S.A. **79 F9** 40 48N 75 37W
Palmetto, U.S.A. **77 M4** 27 31N 82 34W
Palmi, Italy **20 E6** 38 21N 15 51 E
Palmira, Argentina **94 C2** 32 59S 68 34W
Palmira, Colombia **92 C3** 3 32N 76 16W
Palmyra = Tudmur, Syria **44 C3** 34 36N 38 15 E
Palmyra, Mo., U.S.A. **80 F9** 39 48N 91 32W
Palmyra, N.J., U.S.A. **79 F9** 40 1N 75 1W
Palmyra, N.Y., U.S.A. **78 C7** 43 5N 77 18W
Palmyra, Pa., U.S.A. **79 F8** 40 18N 76 36W
Palmyra Is., Pac. Oc. **65 G11** 5 52N 162 5W
Palo Alto, U.S.A. **84 H4** 37 27N 122 10W
Palo Verde, U.S.A. **85 M12** 33 26N 114 44W
Palopo, Indonesia **37 E6** 3 0S 120 16 E
Palos, C. de, Spain **19 D5** 37 38N 0 40W
Palos Verdes, U.S.A. **85 M8** 33 48N 118 23W
Palos Verdes, Pt., U.S.A. **85 M8** 33 43N 118 26W
Palu, Indonesia **37 E5** 1 0S 119 52 E
Palu, Turkey **25 G7** 38 45N 40 0 E
Palwal, India **42 E7** 28 8N 77 19 E
Pamanukan, Indonesia **37 G12** 6 16S 107 49 E
Pamiers, France **18 E4** 43 7N 1 39 E
Pamir, Tajikistan **26 F8** 37 40N 73 0 E
Pamlico →, U.S.A. **77 H7** 35 20N 76 28W
Pamlico Sd., U.S.A. **77 H8** 35 20N 76 0W
Pampa, U.S.A. **81 H4** 35 32N 100 58W
Pampa de las Salinas, Argentina **94 C2** 32 1S 66 58W
Pampanua, Indonesia **37 E6** 4 16S 120 8 E
Pampas, Argentina **94 D3** 35 0S 63 0W
Pampas, Peru **92 F4** 12 20S 74 50W
Pamplona, Colombia **92 B4** 7 23N 72 39W
Pamplona, Spain **19 A5** 42 48N 1 38W
Pampoenpoort, S. Africa **56 E3** 31 3S 22 40 E
Pana, U.S.A. **80 F10** 39 23N 89 5W
Panaca, U.S.A. **83 H6** 37 47N 114 23W
Panaitan, Indonesia **37 G11** 6 36S 105 12 E
Panaji, India **40 M8** 15 25N 73 50 E
Panamá, Panama **88 E4** 9 0N 79 25W
Panama ■, Cent. Amer. **88 E4** 8 48N 79 55W
Panamá, G. de, Panama **88 E4** 8 4N 79 20W
Panama Canal, Panama **88 E4** 9 10N 79 37W
Panama City, U.S.A. **77 K3** 30 10N 85 40W
Panamint Range, U.S.A. **85 J9** 36 20N 117 20W
Panamint Springs, U.S.A. **85 J9** 36 20N 117 28W
Panão, Peru **92 E3** 9 55S 75 55W
Panare, Thailand **39 J3** 6 51N 101 30 E
Panay, Phil. **37 B6** 11 10N 122 30 E
Panay, G., Phil. **37 B6** 11 0N 122 30 E
Pančevo, Serbia, Yug. **21 B9** 44 52N 20 41 E
Panda, Mozam. **57 C5** 24 2S 34 45 E
Pandan, Phil. **37 B6** 11 45N 122 10 E
Pandegelang, Indonesia **37 G12** 6 25S 106 5 E
Pandhana, India **42 J7** 21 42N 76 13 E
Pandharpur, India **40 L9** 17 41N 75 20 E
Pando, Uruguay **95 C4** 34 44S 56 0W
Pando, L. = Hope, L., Australia **63 D2** 28 24S 139 18 E
Pandokrátor, Greece **23 A3** 39 45N 19 50 E
Pandora, Costa Rica **88 E3** 9 43N 83 3W
Panevėžys, Lithuania **9 J21** 55 42N 24 25 E
Panfilov, Kazakstan **26 E8** 44 10N 80 0 E
Pang-Long, Burma **41 H21** 23 11N 98 45 E
Pang-Yang, Burma **41 H21** 22 7N 98 48 E
Panga, Dem. Rep. of the Congo **54 B2** 1 52N 26 18 E
Pangalanes, Canal des = Ampangalana, Lakandranon', Madag. **57 C8** 22 48S 47 50 E
Pangani, Tanzania **54 D4** 5 25S 38 58 E
Pangani →, Tanzania **54 D4** 5 26S 38 58 E
Pangfou = Bengbu, China **35 H9** 32 58N 117 20 E
Pangil, Dem. Rep. of the Congo **54 C2** 3 10S 26 35 E
Pangkah, Tanjung, Indonesia **37 G15** 6 51S 112 33 E
Pangkajene, Indonesia **37 E5** 4 46S 119 34 E
Pangkalanbrandan, Indonesia **36 D1** 4 1N 98 20 E
Pangkalanbuun, Indonesia **36 E4** 2 41S 111 37 E
Pangkalpinang, Indonesia **36 E3** 2 0S 106 0 E
Pangnirtung, Canada **69 B13** 66 8N 65 54W
Pangong Tso, India **42 B8** 34 40N 78 40 E
Panguitch, U.S.A. **83 H7** 37 50N 112 26W
Pangutaran Group, Phil. **37 C6** 6 18N 120 34 E
Panhandle, U.S.A. **81 H4** 35 21N 101 23W
Pani Mines, India **42 H5** 22 29N 73 50 E
Pania-Mutombo, Dem. Rep. of the Congo **54 D1** 5 11S 23 51 E
Panikota I., India **42 J4** 20 46N 71 21 E
Panipat, India **42 E7** 29 25N 77 2 E
Panjal Range = Pir Panjal Range, India **42 C7** 32 30N 76 50 E
Panjang, Hon, Vietnam **39 H4** 9 20N 103 28 E
Panjgur, Pakistan **40 F4** 27 0N 64 5 E
Panjim = Panaji, India **40 M8** 15 25N 73 50 E
Panjin, China **35 D12** 41 3N 122 2 E
Panjinad Barrage, Pakistan **40 E7** 29 22N 71 15 E
Panjnad →, Pakistan **42 E4** 28 57N 70 30 E
Panjwai, Afghan. **42 D1** 31 26N 65 27 E
Panmunjŏm, N. Korea **35 F14** 37 59N 126 38 E
Panna, India **43 G9** 24 40N 80 15 E
Panna Hills, India **43 G9** 24 40N 81 15 E
Pannawonica, Australia **60 D2** 21 39S 116 19 E
Pannirtuuq = Pangnirtung, Canada **69 B13** 66 8N 65 54W
Pano Akil, Pakistan **42 F3** 27 51N 69 7 E
Pano Lefkara, Cyprus **23 E12** 34 53N 33 20 E
Pano Panayia, Cyprus **23 E11** 34 55N 32 38 E
Panorama, Brazil **95 A5** 21 21S 51 51W
Pánormon, Greece **23 D6** 35 25N 24 41 E
Pansemal, India **42 J6** 21 39N 74 42 E
Panshan = Panjin, China **35 D12** 41 3N 122 2 E
Panshi, China **35 C14** 42 58N 126 5 E
Pantanal, Brazil **92 H7** 17 30S 57 40W
Pantar, Indonesia **37 F6** 8 28S 124 10 E
Pante Macassar, E. Timor **37 F6** 9 30S 123 58 E
Pante Makasar = Pante Macassar, E. Timor **37 F6** 9 30S 123 58 E
Pantelleria, Italy **20 F4** 36 50N 11 57 E
Pánuco, Mexico **87 C5** 22 0N 98 15W
Paola, Malta **23 D2** 35 52N 14 30 E
Paola, U.S.A. **80 F7** 38 35N 94 53W
Paonia, U.S.A. **83 G10** 38 52N 107 36W
Paoting = Baoding, China **34 E8** 38 50N 115 28 E
Paot'ou = Baotou, China **34 D6** 40 32N 110 2 E
Paoua, C.A.R. **52 C3** 7 9N 16 20 E
Pápa, Hungary **17 E9** 47 22N 17 30 E
Papa Stour, U.K. **12 A7** 60 20N 1 42W
Papa Westray, U.K. **12 B6** 59 20N 2 55W
Papagayo →, Mexico **87 D5** 16 36N 99 43W
Papagayo, G. de, Costa Rica **88 D2** 10 30N 85 50W
Papakura, N.Z. **59 G5** 37 4S 174 59 E
Papantla, Mexico **87 C5** 20 30N 97 30W
Papar, Malaysia **36 C5** 5 45N 116 0 E
Papeete, Tahiti **65 J13** 17 32S 149 34W
Paphos, Cyprus **23 E11** 34 46N 32 25 E
Papien Chiang = Da →, Vietnam **38 B5** 21 15N 105 20 E
Papigochic →, Mexico **86 B3** 29 9N 109 40W
Paposo, Chile **94 B1** 25 0S 70 30W
Papoutsa, Cyprus **23 E12** 34 54N 33 4 E
Papua New Guinea ■, Oceania **64 H6** 8 0S 145 0 E
Papudo, Chile **94 C1** 32 29S 71 27W
Papun, Burma **41 K20** 18 2N 97 30 E
Papunya, Australia **60 D5** 23 15S 131 54 E
Pará = Belém, Brazil **93 D9** 1 20S 48 30W
Pará □, Brazil **93 D8** 3 20S 52 0W
Paraburdoo, Australia **60 D2** 23 14S 117 32 E
Paracatu, Brazil **93 G9** 17 10S 46 50W
Paracel Is., S. China Sea **36 A4** 15 50N 112 0 E
Parachilna, Australia **63 E2** 31 10S 138 21 E
Parachinar, Pakistan **42 C4** 33 55N 70 5 E
Paradhísi, Greece **23 C10** 36 18N 28 7 E
Paradip, India **41 J15** 20 15N 86 35 E
Paradise, Calif., U.S.A. **84 F5** 39 46N 121 37W
Paradise, Nev., U.S.A. **85 J11** 36 9N 115 10W
Paradise →, Canada **71 B8** 53 27N 57 19W
Paradise Hill, Canada **73 C7** 53 32N 109 28W
Paradise River, Canada **71 B8** 53 27N 57 17W
Paradise Valley, U.S.A. **82 F5** 41 30N 117 32W
Parado, Indonesia **37 F5** 8 42S 118 30 E
Paragould, U.S.A. **81 G9** 36 3N 90 29W
Paragua →, Venezuela **92 B6** 6 55N 62 55W
Paraguaçu →, Brazil **93 F11** 12 45S 38 54W
Paraguaçu Paulista, Brazil **95 A5** 22 22S 50 35W
Paraguaná, Pen. de, Venezuela **92 A5** 12 0N 70 0W
Paraguari, Paraguay **94 B4** 25 36S 57 0W
Paraguarí □, Paraguay **94 B4** 26 0S 57 10W
Paraguay ■, S. Amer. **94 A4** 23 0S 57 0W
Paraguay →, Paraguay **94 B4** 27 18S 58 38W
Paraíba = João Pessoa, Brazil **93 E12** 7 10S 34 52W
Paraíba □, Brazil **93 E11** 7 0S 36 0W
Paraíba do Sul →, Brazil **95 A7** 21 37S 41 3W
Parainen, Finland **9 F20** 60 18N 22 18 E
Paraíso, Mexico **87 D6** 18 24N 93 14W
Parak, Iran **45 E7** 27 38N 52 25 E
Parakou, Benin **50 G6** 9 25N 2 40 E
Paralimni, Cyprus **23 D12** 35 2N 33 58 E
Paramaribo, Surinam **93 B7** 5 50N 55 10W
Paramushir, Ostrov, Russia **27 D16** 50 24N 156 0 E
Paran →, Israel **47 E4** 30 20N 35 10 E
Paraná, Argentina **94 C3** 31 45S 60 30W
Paranã, Brazil **93 F9** 12 30S 47 48W
Paraná □, Brazil **95 A5** 24 30S 51 0W
Paraná →, Argentina **94 C4** 33 43S 59 15W
Paranaguá, Brazil **95 B6** 25 30S 48 30W
Paranaíba, Brazil **93 G8** 19 40S 51 11W
Paranaíba →, Brazil **93 H8** 20 6S 51 4W
Paranapanema →, Brazil **95 A5** 22 40S 53 9W
Paranapiacaba, Serra do, Brazil **95 A6** 24 31S 48 35W
Paranavaí, Brazil **95 A5** 23 4S 52 56W
Parang, Maguindanao, Phil. **37 C6** 7 23N 124 16 E
Parang, Sulu, Phil. **37 C6** 5 55N 120 54 E
Parângul Mare, Vf., Romania **17 F12** 45 20N 23 37 E
Paraparaumu, N.Z. **59 J5** 40 57S 175 3 E
Parbati →, Mad. P., India **42 G7** 25 50N 76 30 E
Parbati →, Raj., India **42 F7** 26 54N 77 53 E
Parbhani, India **40 K10** 19 8N 76 52 E
Parchim, Germany **16 B6** 53 26N 11 52 E
Pardes Hanna-Karkur, Israel **47 C3** 32 28N 34 57 E
Pardo →, Bahia, Brazil **93 G11** 15 40S 39 0W
Pardo →, Mato Grosso, Brazil **95 A5** 21 46S 52 9W
Pardubice, Czech Rep. **16 C8** 50 3N 15 45 E
Pare, Indonesia **37 G15** 7 43S 112 12 E
Pare Mts., Tanzania **54 C4** 4 0S 37 45 E
Parecis, Serra dos, Brazil **92 F7** 13 0S 60 0W
Paren, Russia **27 C17** 62 30N 163 15 E
Parent, Canada **70 C5** 47 55N 74 35W
Parent, L., Canada **70 C4** 48 31N 77 1W
Parepare, Indonesia **37 E5** 4 0S 119 40 E
Párga, Greece **21 E9** 39 15N 20 29 E
Pargo, Pta. do, Madeira **22 D2** 32 49N 17 17W
Pariaguán, Venezuela **92 B6** 8 51N 64 34W
Paricutín, Cerro, Mexico **86 D4** 19 28N 102 15W
Parigi, Indonesia **37 E6** 0 50S 120 5 E
Parika, Guyana **92 B7** 6 50N 58 20W
Parima, Serra, Brazil **92 C6** 2 30N 64 0W
Parinari, Peru **92 D4** 4 35S 74 25W
Pariñas, Pta., S. Amer. **90 D2** 4 30S 82 0W
Parintins, Brazil **93 D7** 2 40S 56 50W
Pariparit Kyun, Burma **41 M18** 14 55N 93 45 E
Paris, Canada **78 C4** 43 12N 80 25W
Paris, France **18 B5** 48 50N 2 20 E
Paris, Idaho, U.S.A. **82 E8** 42 14N 111 24W
Paris, Ky., U.S.A. **76 F3** 38 13N 84 15W
Paris, Tenn., U.S.A. **77 G1** 36 18N 88 19W
Paris, Tex., U.S.A. **81 J7** 33 40N 95 33W
Parish, U.S.A. **79 C8** 43 25N 76 8W
Parishville, U.S.A. **79 B10** 44 38N 74 49W
Park, U.S.A. **84 B4** 48 45N 122 18W
Park City, U.S.A. **81 G6** 37 48N 97 20W
Park Falls, U.S.A. **80 C9** 45 56N 90 27W
Park Head, Canada **78 B3** 44 36N 81 9W
Park Hills, U.S.A. **81 G9** 37 53N 90 28W
Park Range, U.S.A. **82 G10** 40 0N 106 30W
Park Rapids, U.S.A. **80 B7** 46 55N 95 4W
Park River, U.S.A. **80 A6** 48 24N 97 45W
Park Rynie, S. Africa **57 E5** 30 25S 30 45 E
Parkā Bandar, Iran **45 E8** 25 55N 59 35 E
Parkano, Finland **9 E20** 62 1N 23 0 E
Parker, Ariz., U.S.A. **85 L12** 34 9N 114 17W
Parker, Pa., U.S.A. **78 E5** 41 5N 79 41W
Parker Dam, U.S.A. **85 L12** 34 18N 114 8W
Parkersburg, U.S.A. **76 F5** 39 16N 81 34W
Parkes, Australia **63 E4** 33 9S 148 11 E
Parkfield, U.S.A. **84 K6** 35 54N 120 26W
Parkhill, Canada **78 C3** 43 15N 81 38W
Parkland, U.S.A. **84 C4** 47 9N 122 26W
Parkston, U.S.A. **80 D6** 43 24N 97 59W
Parksville, Canada **72 D4** 49 20N 124 21W
Parla, Spain **19 B4** 40 14N 3 46W
Parma, Italy **18 D9** 44 48N 10 20 E
Parma, Idaho, U.S.A. **82 E5** 43 47N 116 57W
Parma, Ohio, U.S.A. **78 E3** 41 23N 81 43W
Parnaguá, Brazil **93 F10** 10 10S 44 38W
Parnaíba, Brazil **93 D10** 2 54S 41 47W
Parnaíba →, Brazil **93 D10** 3 0S 41 50W
Parnassós, Greece **21 E10** 38 35N 22 30 E
Pärnu, Estonia **9 G21** 58 28N 24 33 E
Paroo →, Australia **63 E3** 31 28S 143 32 E
Páros, Greece **21 F11** 37 5N 25 12 E
Parowan, U.S.A. **83 H7** 37 51N 112 50W
Parral, Chile **94 D1** 36 10S 71 52W
Parras, Mexico **86 B4** 25 30N 102 20W
Parrett →, U.K. **11 F4** 51 12N 3 1W
Parris I., U.S.A. **77 J5** 32 20N 80 41W
Parrsboro, Canada **71 C7** 45 30N 64 25W
Parry I., Canada **78 A4** 45 18N 80 10W
Parry Is., Canada **4 B2** 77 0N 110 0W
Parry Sound, Canada **78 A5** 45 20N 80 0W
Parsnip →, Canada **72 B4** 55 10N 123 2W
Parsons, U.S.A. **81 G7** 37 20N 95 16W
Parsons Ra., Australia **62 A2** 13 30S 135 15 E
Partinico, Italy **20 E5** 38 3N 13 7 E
Partridge I., Canada **70 A2** 55 59N 87 37W
Paru →, Brazil **93 D8** 1 33S 52 38W
Parvān □, Afghan. **40 B6** 35 0N 69 0 E
Parvatipuram, India **41 K13** 18 50N 83 25 E
Parvatsar, India **42 F6** 26 52N 74 49 E
Parys, S. Africa **56 D4** 26 52S 27 29 E
Pas, Pta. des, Spain **22 C7** 38 46N 1 26 E
Pasadena, Canada **71 C8** 49 1N 57 36W
Pasadena, Calif., U.S.A. **85 L8** 34 9N 118 9W
Pasadena, Tex., U.S.A. **81 L7** 29 43N 95 13W
Pasaje →, Argentina **94 B3** 25 39S 63 56W
Pascagoula, U.S.A. **81 K10** 30 21N 88 33W
Pascagoula →, U.S.A. **81 K10** 30 23N 88 37W
Paşcani, Romania **17 E14** 47 14N 26 45 E
Pasco, U.S.A. **82 C4** 46 14N 119 6W
Pasco, Cerro de, Peru **92 F3** 10 45S 76 10W
Pasco I., Australia **60 D2** 20 57S 115 20 E
Pascoag, U.S.A. **79 E13** 41 57N 71 42W
Pascua, I. de, Chile **65 K17** 27 7S 109 23W
Pasfield L., Canada **73 B7** 58 24N 105 20W
Pashmakli = Smolyan, Bulgaria **21 D11** 41 36N 24 38 E
Pasir Mas, Malaysia **39 J4** 6 2N 102 8 E
Pasir Putih, Malaysia **39 K4** 5 50N 102 24 E
Pasirian, Indonesia **37 H15** 8 13S 113 8 E
Pasirkuning, Indonesia **36 E2** 0 30S 104 33 E
Pasküh, Iran **45 E9** 27 34N 61 39 E
Pasley, C., Australia **61 F3** 33 52S 123 35 E
Pašman, Croatia **16 G8** 43 58N 15 20 E
Pasni, Pakistan **40 G3** 25 15N 63 27 E
Paso Cantinela, Mexico **85 N11** 32 33N 115 47W
Paso de Indios, Argentina **96 E3** 43 55S 69 0W
Paso de los Libres, Argentina **94 B4** 29 44S 57 10W
Paso de los Toros, Uruguay **94 C4** 32 45S 56 30W
Paso Robles, U.S.A. **83 J3** 35 38N 120 41W
Paspébiac, Canada **71 C6** 48 3N 65 17W
Pasrur, Pakistan **42 C6** 32 16N 74 43 E
Passage West, Ireland **13 E3** 51 52N 8 21W
Passaic, U.S.A. **79 F10** 40 51N 74 7W
Passau, Germany **16 D7** 48 34N 13 28 E
Passero, C., Italy **20 F6** 36 41N 15 10 E
Passo Fundo, Brazil **95 B5** 28 10S 52 20W
Passos, Brazil **93 H9** 20 45S 46 37W
Pastavy, Belarus **9 J22** 55 4N 26 50 E
Pastaza →, Peru **92 D3** 4 50S 76 52W
Pasto, Colombia **92 C3** 1 13N 77 17W
Pasuruan, Indonesia **37 G15** 7 40S 112 44 E
Patagonia, Argentina **96 F3** 45 0S 69 0W
Patagonia, U.S.A. **83 L8** 31 33N 110 45W
Patambar, Iran **45 D9** 29 45N 60 17 E
Patan = Lalitapur, Nepal **43 F11** 27 40N 85 20 E
Patan, Gujarat, India **40 H8** 23 54N 72 14 E
Patan, Maharashtra, India **42 H5** 23 54N 72 14 E
Patani, Indonesia **37 D7** 0 20N 128 50 E
Pataudi, India **42 E7** 28 18N 76 48 E
Patchewollock, Australia **63 F3** 35 22S 142 12 E
Patchogue, U.S.A. **79 F11** 40 46N 73 1W
Patea, N.Z. **59 H5** 39 45S 174 30 E
Patensie, S. Africa **56 E3** 33 46S 24 49 E
Paternò, Italy **20 F6** 37 34N 14 54 E
Pateros, U.S.A. **82 B4** 48 3N 119 54W
Paterson, U.S.A. **79 F10** 40 55N 74 11W
Paterson Ra., Australia **60 D3** 21 45S 122 10 E
Pathankot, India **42 C6** 32 18N 75 45 E
Pathein = Bassein, Burma **41 L19** 16 45N 94 30 E
Pathfinder Reservoir, U.S.A. **82 E10** 42 28N 106 51W
Pathiu, Thailand **39 G2** 10 42N 99 19 E
Pathum Thani, Thailand **38 E3** 14 1N 100 32 E
Pati, Indonesia **37 G14** 6 45S 111 1 E
Patía →, Colombia **92 C3** 2 13N 78 40W
Patiala, Punjab, India **42 D7** 30 23N 76 26 E
Patiala, Ut. P., India **43 F8** 27 43N 79 1 E
Patkai Bum, India **41 F19** 27 0N 95 30 E
Pátmos, Greece **21 F12** 37 21N 26 36 E
Patna, India **43 G11** 25 35N 85 12 E
Pato Branco, Brazil **95 B5** 26 13S 52 40W
Patonga, Uganda **54 B3** 2 45N 33 15 E
Patos, Brazil **93 E11** 6 55S 37 16W
Patos, L. dos, Brazil **95 C5** 31 20S 51 0W
Patos, Río de los →, Argentina **94 C2** 31 18S 69 25W
Patos de Minas, Brazil **93 G9** 18 35S 46 32W
Patquía, Argentina **94 C2** 30 2S 66 55W
Pátrai, Greece **21 E9** 38 14N 21 47 E
Pátraikós Kólpos, Greece **21 E9** 38 17N 21 30 E
Patras = Pátrai, Greece **21 E9** 38 14N 21 47 E
Patrocínio, Brazil **93 G9** 18 57S 47 0W
Patta, Kenya **54 C5** 2 10S 41 0 E
Pattani, Thailand **39 J3** 6 48N 101 15 E
Pattaya, Thailand **36 B2** 12 52N 100 55 E
Patten, U.S.A. **77 C11** 46 0N 68 38W
Patterson, Calif., U.S.A. **84 H5** 37 28N 121 8W
Patterson, La., U.S.A. **81 L9** 29 42N 91 18W
Patterson, Mt., U.S.A. **84 G7** 38 29N 119 20W
Patti, Punjab, India **42 D6** 31 17N 74 54 E
Patti, Ut. P., India **43 G10** 25 55N 82 12 E
Pattoki, Pakistan **42 D5** 31 5N 73 52 E
Patton, U.S.A. **78 F6** 40 38N 78 39W
Patuakhali, Bangla. **41 H17** 22 20N 90 25 E
Patuanak, Canada **73 B7** 55 55N 107 43W
Patuca →, Honduras **88 C3** 15 50N 84 18W
Patuca, Punta, Honduras **88 C3** 15 49N 84 14W
Pátzcuaro, Mexico **86 D4** 19 30N 101 40W
Pau, France **18 E3** 43 19N 0 25W
Pauk, Burma **41 J19** 21 27N 94 30 E
Paul I., Canada **71 A7** 56 30N 61 20W
Paul Smiths, U.S.A. **79 B10** 44 26N 74 15W
Paulatuk, Canada **68 B7** 69 25N 124 0W
Paulis = Isiro, Dem. Rep. of the Congo **54 B2** 2 53N 27 40 E
Paulistana, Brazil **93 E10** 8 9S 41 9W
Paulo Afonso, Brazil **93 E11** 9 21S 38 15W
Paulpietersburg, S. Africa **57 D5** 27 23S 30 50 E
Pauls Valley, U.S.A. **81 H6** 34 44N 97 13W
Pauma Valley, U.S.A. **85 M10** 33 16N 116 58W
Pauri, India **43 D8** 30 9N 78 47 E
Pāveh, Iran **44 C5** 35 3N 46 22 E
Pavia, Italy **18 D8** 45 7N 9 8 E
Pavilion, U.S.A. **78 D6** 42 52N 78 1W
Pāvilosta, Latvia **9 H19** 56 53N 21 14 E
Pavlodar, Kazakstan **26 D8** 52 33N 77 0 E
Pavlograd = Pavlohrad, Ukraine **25 E6** 48 30N 35 52 E
Pavlohrad, Ukraine **25 E6** 48 30N 35 52 E
Pavlovo, Russia **24 C7** 55 58N 43 5 E
Pavlovsk, Russia **25 D7** 50 26N 40 5 E
Pavlovskaya, Russia **25 E6** 46 17N 39 47 E
Pawayan, India **43 E9** 28 4N 80 6 E
Pawhuska, U.S.A. **81 G6** 36 40N 96 20W
Pawling, U.S.A. **79 E11** 41 34N 73 36W
Pawnee, U.S.A. **81 G6** 36 20N 96 48W
Pawnee City, U.S.A. **80 E6** 40 7N 96 9W
Pawtucket, U.S.A. **79 E13** 41 53N 71 23W
Paximádhia, Greece **23 E6** 35 0N 24 35 E
Paxoí, Greece **21 E9** 39 14N 20 12 E
Paxton, Ill., U.S.A. **76 E1** 40 27N 88 6W
Paxton, Nebr., U.S.A. **80 E4** 41 7N 101 21W
Payakumbuh, Indonesia **36 E2** 0 20S 100 35 E
Payette, U.S.A. **82 D5** 44 5N 116 56W
Payne Bay = Kangirsuk, Canada **69 B13** 60 0N 70 0W
Payne L., Canada **69 C12** 59 30N 74 30W
Paynes Find, Australia **61 E2** 29 15S 117 42 E
Paynesville, U.S.A. **80 C7** 45 23N 94 43W
Paysandú, Uruguay **94 C4** 32 19S 58 8W
Payson, U.S.A. **83 J8** 34 14N 111 20W
Paz →, Guatemala **88 D1** 13 44N 90 10W
Paz, B. de la, Mexico **86 C2** 24 15N 110 25W
Pāzanān, Iran **45 D6** 30 35N 49 59 E
Pazardzhik, Bulgaria **21 C11** 42 12N 24 20 E
Pe Ell, U.S.A. **84 D3** 46 34N 123 18W
Peabody, U.S.A. **79 D14** 42 31N 70 56W
Peace →, Canada **72 B6** 59 0N 111 25W
Peace Point, Canada **72 B6** 59 7N 112 27W
Peace River, Canada **72 B5** 56 15N 117 18W
Peach Springs, U.S.A. **83 J7** 35 32N 113 25W
Peachland, Canada **72 D5** 49 47N 119 45W
Peachtree City, U.S.A. **77 J3** 33 25N 84 35W
Peak, The = Kinder Scout, U.K. **10 D6** 53 24N 1 52W
Peak District, U.K. **10 D6** 53 10N 1 50W
Peak Hill, N.S.W., Australia **63 E4** 32 47S 148 11 E
Peak Hill, W. Austral., Australia **61 E2** 25 35S 118 43 E
Peak Ra., Australia **62 C4** 22 50S 148 20 E
Peake Cr. →, Australia **63 D2** 28 2S 136 7 E
Peale, Mt., U.S.A. **83 G9** 38 26N 109 14W
Pearblossom, U.S.A. **85 L9** 34 30N 117 55W
Pearl →, U.S.A. **81 K10** 30 11N 89 32W
Pearl City, U.S.A. **74 H16** 21 24N 157 59W
Pearl Harbor, U.S.A. **74 H16** 21 21N 157 57W
Pearl River, U.S.A. **79 E10** 41 4N 74 2W
Pearsall, U.S.A. **81 L5** 28 54N 99 6W
Peary Land, Greenland **4 A6** 82 40N 33 0W
Pease →, U.S.A. **81 H5** 34 12N 99 2W
Peawanuck, Canada **69 C11** 55 15N 85 12W
Pebane, Mozam. **55 F4** 17 10S 38 8 E
Pebas, Peru **92 D4** 3 10S 71 46W
Pebble Beach, U.S.A. **84 J5** 36 34N 121 57W
Peć, Kosovo, Yug. **21 C9** 42 40N 20 17 E
Pechenga, Russia **24 A5** 69 29N 31 4 E
Pechenizhyn, Ukraine **17 D13** 48 30N 24 48 E
Pechiguera, Pta., Canary Is. **22 F6** 28 51N 13 53W
Pechora, Russia **24 A10** 65 10N 57 11 E
Pechora →, Russia **24 A9** 68 13N 54 15 E
Pechorskaya Guba, Russia **24 A9** 68 40N 54 0 E
Pečory, Russia **9 H22** 57 48N 27 40 E
Pecos, U.S.A. **81 K3** 31 26N 103 30W
Pecos →, U.S.A. **81 L3** 29 42N 101 22W
Pécs, Hungary **17 E10** 46 5N 18 15 E
Pedder, L., Australia **62 G4** 42 55S 146 10 E
Peddie, S. Africa **57 E4** 33 14S 27 7 E
Pedernales, Dom. Rep. **89 C5** 18 2N 71 44W
Pedieos →, Cyprus **23 D12** 35 10N 33 54 E
Pedirka, Australia **63 D2** 26 40S 135 14 E
Pedra Azul, Brazil **93 G10** 16 2S 41 17W
Pedreiras, Brazil **93 D10** 4 32S 44 40W
Pedro Afonso, Brazil **93 E9** 9 0S 48 10W
Pedro Cays, Jamaica **88 C4** 17 5N 77 48W
Pedro de Valdivia, Chile **94 A2** 22 55S 69 38W
Pedro Juan Caballero, Paraguay **95 A4** 22 30S 55 40W
Pee Dee →, U.S.A. **77 J6** 33 22N 79 16W
Peebinga, Australia **63 E3** 34 52S 140 57 E
Peebles, U.K. **12 F5** 55 40N 3 11W
Peekskill, U.S.A. **79 E11** 41 17N 73 55W
Peel, U.K. **10 C3** 54 13N 4 40W
Peel →, Australia **63 E5** 30 50S 150 29 E
Peel →, Canada **68 B6** 67 0N 135 0W
Peel Sound, Canada **68 A10** 73 0N 96 0W
Peera Peera Poolanna L., Australia **63 D2** 26 30S 138 0 E

153

Randalstown

Purcell, *U.S.A.* **81 H6** 35 1N 97 22W
Purcell Mts., *Canada* **72 D5** 49 55N 116 15W
Puri, *India* **41 K14** 19 50N 85 58 E
Purmerend, *Neths.* **15 B4** 52 32N 4 58 E
Purnia, *India* **43 G12** 25 45N 87 31 E
Pursat = Pouthisat, *Cambodia* **38 F4** 12 34N 103 50 E
Purukcahu, *Indonesia* **36 E4** 0 35S 114 35 E
Puruliya, *India* **43 H12** 23 17N 86 24 E
Purus →, *Brazil* **92 D6** 3 42S 61 28W
Purvis, *U.S.A.* **81 K10** 31 9N 89 25W
Purwa, *India* **43 F9** 26 28N 80 47 E
Purwakarta, *Indonesia* **37 G12** 6 35S 107 29 E
Purwodadi, *Indonesia* **37 G14** 7 7S 110 55 E
Purwokerto, *Indonesia* **37 G13** 7 25S 109 14 E
Puryŏng, *N. Korea* **35 C15** 42 5N 129 43 E
Pusa, *India* **43 G11** 25 59N 85 41 E
Pusan, *S. Korea* **35 G15** 35 5N 129 0 E
Pushkino, *Russia* **25 D8** 51 16N 47 0 E
Putahow L., *Canada* **73 B8** 59 54N 100 40W
Putao, *Burma* **41 F20** 27 28N 97 30 E
Putaruru, *N.Z.* **59 H5** 38 2S 175 50 E
Putignano, *Italy* **20 D7** 40 51N 17 7 E
Puting, Tanjung, *Indonesia* **36 E4** 3 31S 111 46 E
Putnam, *U.S.A.* **79 E13** 41 55N 71 55W
Putorana, Gory, *Russia* **27 C10** 69 0N 95 0 E
Puttalam, *Sri Lanka* **40 Q11** 8 1N 79 55 E
Puttgarden, *Germany* **16 A6** 54 30N 11 10 E
Putumayo →, *S. Amer.* **92 D5** 3 7S 67 58W
Putussibau, *Indonesia* **36 D4** 0 50N 112 56 E
Puvirnituq, *Canada* **69 B12** 60 2N 77 10W
Puy-de-Dôme, *France* **18 D5** 45 46N 2 57 E
Puyallup, *U.S.A.* **84 C4** 47 12N 122 18W
Puyang, *China* **34 G8** 35 40N 115 1 E
Pūzeh Rīg, *Iran* **45 E8** 27 20N 58 40 E
Pwani □, *Tanzania* **54 D4** 7 0S 39 0 E
Pweto, *Dem. Rep. of the Congo* **55 D2** 8 25S 28 51 E
Pwllheli, *U.K.* **10 E3** 52 53N 4 25W
Pya-ozero, *Russia* **24 A5** 66 5N 30 58 E
Pyapon, *Burma* **41 L19** 16 20N 95 40 E
Pyasina →, *Russia* **27 B9** 73 30N 87 0 E
Pyatigorsk, *Russia* **25 F7** 44 2N 43 6 E
Pyè = Prome, *Burma* **41 K19** 18 49N 95 13 E
Pyetrikaw, *Belarus* **17 B15** 52 11N 28 29 E
Pyhäjoki, *Finland* **8 D21** 64 28N 24 14 E
Pyinmana, *Burma* **41 K20** 19 45N 96 12 E
Pyla, C., *Cyprus* **23 E12** 34 56N 33 51 E
Pymatuning Reservoir, *U.S.A.* **78 E4** 41 30N 80 28W
Pyŏktong, *N. Korea* **35 D13** 40 50N 125 50 E
Pyŏnggang, *N. Korea* **35 E14** 38 24N 127 17 E
P'yŏngt'aek, *S. Korea* **35 F14** 37 1N 127 4 E
P'yŏngyang, *N. Korea* **35 E13** 39 0N 125 30 E
Pyote, *U.S.A.* **81 K3** 31 32N 103 8W
Pyramid L., *U.S.A.* **82 G4** 40 1N 119 35W
Pyramid Pk., *U.S.A.* **85 J10** 36 25N 116 37W
Pyrénées, *Europe* **18 E4** 42 45N 0 18 E
Pyu, *Burma* **41 K20** 18 30N 96 28 E

Q

Qaanaaq, *Greenland* **4 B4** 77 40N 69 0W
Qachasnek, *S. Africa* **57 E4** 30 6S 28 42 E
Qa'el Jafr, *Jordan* **47 E5** 30 20N 36 25 E
Qa'emābād, *Iran* **45 D9** 31 44N 60 2 E
Qā'emshahr, *Iran* **45 B7** 36 30N 52 53 E
Qagan Nur, *China* **34 C8** 43 30N 114 55 E
Qahar Youyi Zhongqi, *China* **34 D7** 41 12N 112 40 E
Qahremānshahr = Bākhtarān, *Iran* **44 C5** 34 23N 47 0 E
Qaidam Pendi, *China* **32 C4** 37 0N 95 0 E
Qajarīyeh, *Iran* **45 D6** 31 1N 48 22 E
Qala, Ras il, *Malta* **23 C1** 36 2N 14 20 E
Qala-i-Jadid = Spīn Būldak, *Afghan.* **42 D2** 31 1N 66 25 E
Qala Point = Qala, Ras il, *Malta* **23 C1** 36 2N 14 20 E
Qala Viala, *Pakistan* **42 D2** 30 49N 67 17 E
Qala Yangi, *Afghan.* **42 B2** 34 20N 66 30 E
Qal'at al Akhḑar, *Si. Arabia* **44 E3** 28 0N 37 10 E
Qal'at Dīzah, *Iraq* **44 B5** 36 11N 45 7 E
Qal'at Şāliḩ, *Iraq* **44 D5** 31 31N 47 16 E
Qal'at Sukkar, *Iraq* **44 D5** 31 51N 46 5 E
Qamani'tuaq = Baker Lake, *Canada* **68 B10** 64 20N 96 3W
Qamdo, *China* **32 C4** 31 15N 97 6 E
Qamruddin Karez, *Pakistan* **42 D3** 31 45N 68 20 E
Qandahār, *Afghan.* **40 D4** 31 32N 65 43 E
Qandahār □, *Afghan.* **40 D4** 31 0N 65 0 E
Qapān, *Iran* **45 B7** 37 40N 55 47 E
Qapshaghay, *Kazakstan* **26 E8** 43 51N 77 14 E
Qaqortoq, *Greenland* **69 B6** 60 43N 46 0W
Qara Qash →, *China* **43 B8** 35 0N 78 30 E
Qarabutaq, *Kazakstan* **26 E7** 49 59N 60 14 E
Qaraghandy, *Kazakstan* **26 E8** 49 50N 73 10 E
Qārah, *Si. Arabia* **44 D4** 29 55N 40 3 E
Qarataū, *Kazakstan* **26 E8** 43 10N 70 28 E
Qarataū, *Kazakstan* **26 E7** 43 30N 69 30 E
Qardho = Gardo, *Somali Rep.* **46 F4** 9 30N 49 6 E
Qareh →, *Iran* **44 B5** 39 25N 47 22 E
Qareh Tekān, *Iran* **45 B6** 36 38N 49 29 E
Qarqan He →, *China* **32 C3** 39 30N 88 30 E
Qarqaraly, *Kazakstan* **26 E8** 49 26N 75 30 E
Qarshi, *Uzbekistan* **26 F7** 38 53N 65 48 E
Qartabā, *Lebanon* **47 A4** 34 4N 35 50 E
Qaryat al Gharab, *Iraq* **44 D5** 31 27N 44 48 E
Qaryat al 'Ulyā, *Si. Arabia* **44 E5** 27 33N 47 42 E
Qasr 'Amra, *Jordan* **44 D3** 31 48N 36 35 E
Qaşr-e Qand, *Iran* **45 E9** 26 15N 60 45 E
Qasr Farâfra, *Egypt* **51 C11** 27 0N 28 1 E
Qatanā, *Syria* **47 B5** 33 26N 36 4 E
Qatar ■, *Asia* **46 B5** 25 30N 51 15 E
Qatlīsh, *Iran* **45 B8** 37 50N 57 19 E
Qattâra, Munkhafed el, *Egypt* **51 C11** 29 30N 27 30 E
Qattâra Depression = Qattâra, Munkhafed el, *Egypt* **51 C11** 29 30N 27 30 E
Qawām al Ḩamzah, *Iraq* **44 D5** 31 43N 44 58 E
Qāyen, *Iran* **45 C8** 33 40N 59 10 E
Qazaqstan = Kazakstan ■, *Asia* **26 E7** 50 0N 70 0 E
Qazimämmäd, *Azerbaijan* **45 A6** 40 3N 49 0 E
Qazvin, *Iran* **45 B6** 36 15N 50 0 E
Qena, *Egypt* **51 C12** 26 10N 32 43 E
Qeqertarsuaq, *Greenland* **4 C5** 69 15N 53 38W
Qeqertarsuaq, *Greenland* **69 B5** 69 45N 53 30W
Qeshlāq, *Iran* **44 C5** 34 55N 46 28 E
Qeshm, *Iran* **45 E8** 26 55N 56 10 E
Qeys, *Iran* **45 E7** 26 32N 53 58 E
Qezel Owzen →, *Iran* **45 B6** 36 45N 49 22 E
Qezi'ot, *Israel* **47 E3** 30 52N 34 26 E
Qi Xian, *China* **34 G8** 34 40N 114 48 E
Qian Gorlos, *China* **35 B13** 45 5N 124 42 E
Qian Xian, *China* **34 G5** 34 31N 108 15 E
Qianyang, *China* **34 G4** 34 40N 107 8 E
Qikiqtarjuaq, *Canada* **69 B13** 67 33N 63 0W
Qila Safed, *Pakistan* **40 E2** 29 0N 61 30 E
Qila Saifullāh, *Pakistan* **42 D3** 30 45N 68 17 E
Qilian Shan, *China* **32 C4** 38 30N 96 0 E
Qin He →, *China* **34 G7** 35 1N 113 22 E
Qin Ling = Qinling Shandi, *China* **34 H5** 33 50N 108 10 E
Qin'an, *China* **34 G3** 34 48N 105 40 E
Qing Xian, *China* **34 E9** 38 35N 116 45 E
Qingcheng, *China* **35 F9** 37 15N 117 40 E
Qingdao, *China* **35 F11** 36 5N 120 20 E
Qingfeng, *China* **34 G8** 35 52N 115 8 E
Qinghai □, *China* **32 C4** 36 0N 98 0 E
Qinghai Hu, *China* **32 C5** 36 40N 100 10 E
Qinghecheng, *China* **35 D13** 41 28N 124 15 E
Qinghemen, *China* **35 D11** 41 48N 121 25 E
Qingjian, *China* **34 F6** 37 8N 110 8 E
Qingjiang = Huaiyin, *China* **35 H10** 33 30N 119 2 E
Qingshui, *China* **34 G4** 34 48N 106 8 E
Qingshuihe, *China* **34 E6** 39 55N 111 35 E
Qingtongxia Shuiku, *China* **34 F3** 37 50N 105 58 E
Qingxu, *China* **34 F7** 37 34N 112 22 E
Qingyang, *China* **34 F4** 36 2N 107 55 E
Qingyuan, *China* **35 C13** 42 10N 124 55 E
Qingyun, *China* **35 F9** 37 45N 117 20 E
Qinhuangdao, *China* **35 E10** 39 56N 119 30 E
Qinling Shandi, *China* **34 H5** 33 50N 108 10 E
Qinshui, *China* **34 G7** 35 40N 112 8 E
Qinyang = Jiyuan, *China* **34 G7** 35 7N 112 57 E
Qinyuan, *China* **34 F7** 36 29N 112 20 E
Qinzhou, *China* **32 D5** 21 58N 108 38 E
Qionghai, *China* **38 C8** 19 15N 110 26 E
Qiongzhou Haixia, *China* **38 B8** 20 10N 110 15 E
Qiqihar, *China* **27 E13** 47 26N 124 0 E
Qiraîya, W. →, *Egypt* **47 E3** 30 27N 34 0 E
Qiryat Ata, *Israel* **47 C4** 32 47N 35 6 E
Qiryat Gat, *Israel* **47 D3** 31 32N 34 46 E
Qiryat Mal'akhi, *Israel* **47 D3** 31 44N 34 44 E
Qiryat Shemona, *Israel* **47 B4** 33 13N 35 35 E
Qiryat Yam, *Israel* **47 C4** 32 51N 35 4 E
Qishan, *China* **34 G4** 34 25N 107 38 E
Qitai, *China* **32 B3** 44 2N 89 35 E
Qixia, *China* **35 F11** 37 17N 120 52 E
Qızılağac Körfäzi, *Azerbaijan* **45 B6** 39 9N 49 0 E
Qojūr, *Iran* **44 B5** 36 12N 47 55 E
Qom, *Iran* **45 C6** 34 40N 51 0 E
Qomolangma Feng = Everest, Mt., *Nepal* **43 E12** 28 5N 86 58 E
Qomsheh, *Iran* **45 D6** 32 0N 51 55 E
Qoraqalpoghistan □, *Uzbekistan* **26 E6** 43 0N 58 0 E
Qostanay, *Kazakstan* **26 D7** 53 10N 63 35 E
Quabbin Reservoir, *U.S.A.* **79 D12** 42 20N 72 20W
Quairading, *Australia* **61 F2** 32 0S 117 21 E
Quakertown, *U.S.A.* **79 F9** 40 26N 75 21W
Qualicum Beach, *Canada* **72 D4** 49 22N 124 26W
Quambatook, *Australia* **63 F3** 35 49S 143 34 E
Quambone, *Australia* **63 E4** 30 57S 147 53 E
Quamby, *Australia* **62 C3** 20 22S 140 17 E
Quan Long = Ca Mau, *Vietnam* **39 H5** 9 7N 105 8 E
Quanah, *U.S.A.* **81 H5** 34 18N 99 44W
Quang Ngai, *Vietnam* **38 E7** 15 13N 108 58 E
Quang Tri, *Vietnam* **38 D6** 16 45N 107 13 E
Quang Yen, *Vietnam* **38 B6** 20 56N 106 52 E
Quantock Hills, *U.K.* **11 F4** 51 8N 3 10W
Quanzhou, *China* **33 D6** 24 55N 118 34 E
Qu'Appelle, *Canada* **73 C8** 50 33N 103 53W
Quaqtaq, *Canada* **69 B13** 60 55N 69 40W
Quaraí, *Brazil* **94 C4** 30 15S 56 20W
Quartu Sant'Elena, *Italy* **20 E3** 39 15N 9 10 E
Quartzsite, *U.S.A.* **85 M12** 33 40N 114 13W
Quatsino Sd., *Canada* **72 C3** 50 25N 127 58W
Quba, *Azerbaijan* **25 F8** 41 21N 48 32 E
Qūchān, *Iran* **45 B8** 37 10N 58 27 E
Queanbeyan, *Australia* **63 F4** 35 17S 149 14 E
Québec, *Canada* **71 C5** 46 52N 71 13W
Québec □, *Canada* **71 C6** 48 0N 74 0W
Queen Alexandra Ra., *Antarctica* **5 E11** 85 0S 170 0 E
Queen Charlotte City, *Canada* **72 C2** 53 15N 132 2W
Queen Charlotte Is., *Canada* **72 C2** 53 20N 132 10W
Queen Charlotte Sd., *Canada* **72 C3** 51 0N 128 0W
Queen Charlotte Strait, *Canada* **72 C3** 50 45N 127 10W
Queen Elizabeth Is., *Canada* **66 B10** 76 0N 95 0W
Queen Mary Land, *Antarctica* **5 D7** 70 0S 95 0 E
Queen Maud G., *Canada* **68 B9** 68 15N 102 30W
Queen Maud Land, *Antarctica* **5 D3** 72 30S 12 0 E
Queen Maud Mts., *Antarctica* **5 E13** 86 0S 160 0W
Queens Chan., *Australia* **60 C4** 15 0S 129 30 E
Queenscliff, *Australia* **63 F3** 38 16S 144 39 E
Queensland □, *Australia* **62 C3** 22 0S 142 0 E
Queenstown, *Australia* **62 G4** 42 4S 145 35 E
Queenstown, *N.Z.* **59 L2** 45 1S 168 40 E
Queenstown, *S. Africa* **56 E4** 31 52S 26 52 E
Queets, *U.S.A.* **84 C2** 47 32N 124 20W
Queguay Grande →, *Uruguay* **94 C4** 32 9S 58 9W
Queimadas, *Brazil* **93 F11** 11 0S 39 38W
Quelimane, *Mozam.* **55 F4** 17 53S 36 58 E
Quellón, *Chile* **96 E2** 43 7S 73 37W
Quelpart = Cheju do, *S. Korea* **35 H14** 33 29N 126 34 E
Quemado, *N. Mex., U.S.A.* **83 J9** 34 20N 108 30W
Quemado, *Tex., U.S.A.* **81 L4** 28 58N 100 35W
Quemú-Quemú, *Argentina* **94 D3** 36 3S 63 36W
Quequén, *Argentina* **94 D4** 38 30S 58 30W
Querétaro, *Mexico* **86 C4** 20 36N 100 23W
Querétaro □, *Mexico* **86 C5** 20 30N 100 0W
Queshan, *China* **34 H8** 32 55N 114 2 E
Quesnel, *Canada* **72 C4** 53 0N 122 30W
Quesnel →, *Canada* **72 C4** 52 58N 122 29W
Quesnel L., *Canada* **72 C4** 52 30N 121 20W
Questa, *U.S.A.* **83 H11** 36 42N 105 36W
Quetico Prov. Park, *Canada* **70 C1** 48 30N 91 45W
Quetta, *Pakistan* **42 D2** 30 15N 66 55 E
Quezaltenango, *Guatemala* **88 D1** 14 50N 91 30W
Quezon City, *Phil.* **37 B6** 14 38N 121 0 E
Qufār, *Si. Arabia* **44 E4** 27 26N 41 37 E
Qui Nhon, *Vietnam* **38 F7** 13 40N 109 13 E
Quibaxe, *Angola* **52 F2** 8 24S 14 27 E
Quibdo, *Colombia* **92 B3** 5 42N 76 40W
Quiberon, *France* **18 C2** 47 29N 3 9W
Quiet L., *Canada* **72 A2** 61 5N 133 5W
Quiindy, *Paraguay* **94 B4** 25 58S 57 14W
Quila, *Mexico* **86 C3** 24 23N 107 13W
Quilán, C., *Chile* **96 E2** 43 15S 74 30W
Quilcene, *U.S.A.* **84 C4** 47 49N 122 53W
Quilimarí, *Chile* **94 C1** 32 5S 71 30W
Quilino, *Argentina* **94 C3** 30 14S 64 29W
Quill Lakes, *Canada* **73 C8** 51 55N 104 13W
Quillabamba, *Peru* **92 F4** 12 50S 72 50W
Quillagua, *Chile* **94 A2** 21 40S 69 40W
Quillaicillo, *Chile* **94 C1** 31 17S 71 40W
Quillota, *Chile* **94 C1** 32 54S 71 16W
Quilmes, *Argentina* **94 C4** 34 43S 58 15W
Quilon, *India* **40 Q10** 8 50N 76 38 E
Quilpie, *Australia* **63 D3** 26 35S 144 11 E
Quilpué, *Chile* **94 C1** 33 5S 71 33W
Quilua, *Mozam.* **55 F4** 16 17S 39 54 E
Quimilí, *Argentina* **94 B3** 27 40S 62 30W
Quimper, *France* **18 B1** 48 0N 4 9W
Quimperlé, *France* **18 C2** 47 53N 3 33W
Quinault →, *U.S.A.* **84 C2** 47 21N 124 18W
Quincy, *Calif., U.S.A.* **84 F6** 39 56N 120 57W
Quincy, *Fla., U.S.A.* **77 K3** 30 35N 84 34W
Quincy, *Ill., U.S.A.* **80 F9** 39 56N 91 23W
Quincy, *Mass., U.S.A.* **79 D14** 42 15N 71 0W
Quincy, *Wash., U.S.A.* **82 C4** 47 22N 119 56W
Quines, *Argentina* **94 C2** 32 13S 65 48W
Quinga, *Mozam.* **55 F5** 15 49S 40 15 E
Quinns Rocks, *Australia* **61 F2** 31 40S 115 42 E
Quintana Roo □, *Mexico* **87 D7** 19 0N 88 0W
Quintanar de la Orden, *Spain* **19 C4** 39 36N 3 5W
Quintero, *Chile* **94 C1** 32 45S 71 30W
Quirihue, *Chile* **94 D1** 36 15S 72 35W
Quirindi, *Australia* **63 E5** 31 28S 150 40 E
Quirinópolis, *Brazil* **93 G8** 18 32S 50 30W
Quissanga, *Mozam.* **55 E5** 12 24S 40 28 E
Quissico, *Mozam.* **57 C5** 24 42S 34 44 E
Quitilipi, *Argentina* **94 B3** 26 50S 60 13W
Quitman, *U.S.A.* **77 K4** 30 47N 83 34W
Quito, *Ecuador* **92 D3** 0 15S 78 35W
Quixadá, *Brazil* **93 D11** 4 55S 39 0W
Quixaxe, *Mozam.* **55 F5** 15 17S 40 4 E
Qulan, *Kazakstan* **26 E8** 42 55N 72 43 E
Qul'ân, Jazâ'ir, *Egypt* **44 E2** 24 22N 35 31 E
Qumbu, *S. Africa* **57 E4** 31 10S 28 48 E
Quneitra, *Syria* **47 B4** 33 7N 35 48 E
Qŭnghirot, *Uzbekistan* **26 E6** 43 6N 58 54 E
Quoin I., *Australia* **60 B4** 14 54S 129 32 E
Quoin Pt., *S. Africa* **56 E2** 34 46S 19 37 E
Quorn, *Australia* **63 E2** 32 25S 138 5 E
Qŭqon, *Uzbekistan* **26 E8** 40 30N 70 57 E
Qurnat as Sawdā', *Lebanon* **47 A5** 34 18N 36 6 E
Quşaybā', *Si. Arabia* **44 E4** 26 53N 43 35 E
Qusaybah, *Iraq* **44 C4** 34 24N 40 59 E
Quseir, *Egypt* **44 E2** 26 7N 34 16 E
Qūshchī, *Iran* **44 B5** 37 59N 45 3 E
Quthing, *Lesotho* **57 E4** 30 25S 27 36 E
Qūţīābād, *Iran* **45 C6** 35 47N 48 30 E
Quwo, *China* **34 G6** 35 38N 111 25 E
Quyang, *China* **34 E8** 38 35N 114 40 E
Quynh Nhai, *Vietnam* **38 B4** 21 49N 103 33 E
Quyon, *Canada* **79 A8** 45 31N 76 14W
Quzhou, *China* **33 D6** 28 57N 118 54 E
Quzi, *China* **34 F4** 36 20N 107 20 E
Qyzylorda, *Kazakstan* **26 E7** 44 48N 65 28 E

R

Ra, Ko, *Thailand* **39 H2** 9 13N 98 16 E
Raahe, *Finland* **8 D21** 64 40N 24 28 E
Raalte, *Neths.* **15 B6** 52 23N 6 16 E
Raasay, *U.K.* **12 D2** 57 25N 6 4W
Raasay, Sd. of, *U.K.* **12 D2** 57 30N 6 8W
Raba, *Indonesia* **37 F5** 8 36S 118 55 E
Rába →, *Hungary* **17 E9** 47 38N 17 38 E
Rabai, *Kenya* **54 C4** 3 50S 39 31 E
Rabat = Victoria, *Malta* **23 C1** 36 3N 14 14 E
Rabat, *Malta* **23 D1** 35 53N 14 24 E
Rabat, *Morocco* **50 B4** 34 2N 6 48W
Rabaul, *Papua N. G.* **64 H7** 4 24S 152 18 E
Rābigh, *Si. Arabia* **46 C2** 22 50N 39 5 E
Râbniţa, *Moldova* **17 E15** 47 45N 29 0 E
Rābor, *Iran* **45 D8** 29 17N 56 55 E
Race, C., *Canada* **71 C9** 46 40N 53 5W
Rach Gia, *Vietnam* **39 G5** 10 5N 105 5 E
Rachid, *Mauritania* **50 E3** 18 45N 11 35W
Racibórz, *Poland* **17 C10** 50 7N 18 18 E
Racine, *U.S.A.* **76 D2** 42 41N 87 51W
Rackerby, *U.S.A.* **84 F5** 39 26N 121 22W
Radama, Nosy, *Madag.* **57 A8** 14 0S 47 47 E
Radama, Saikanosy, *Madag.* **57 A8** 14 16S 47 53 E
Rădăuţi, *Romania* **17 E13** 47 50N 25 59 E
Radcliff, *U.S.A.* **76 G3** 37 51N 85 57W
Radekhiv, *Ukraine* **17 C13** 50 25N 24 32 E
Radekhov = Radekhiv, *Ukraine* **17 C13** 50 25N 24 32 E
Radford, *U.S.A.* **76 G5** 37 8N 80 34W
Radhanpur, *India* **42 H4** 23 50N 71 38 E
Radhwa, Jabal, *Si. Arabia* **44 E3** 24 34N 38 18 E
Radisson, *Qué., Canada* **70 B4** 53 47N 77 37W
Radisson, *Sask., Canada* **73 C7** 52 30N 107 20W
Radium Hot Springs, *Canada* **72 C5** 50 35N 116 2W
Radnor Forest, *U.K.* **11 E4** 52 17N 3 10W
Radom, *Poland* **17 C11** 51 23N 21 12 E
Radomsko, *Poland* **17 C10** 51 5N 19 28 E
Radomyshl, *Ukraine* **17 C15** 50 30N 29 12 E
Radstock, C., *Australia* **63 E1** 33 12S 134 20 E
Radviliškis, *Lithuania* **9 J20** 55 49N 23 33 E
Radville, *Canada* **73 D8** 49 30N 104 15W
Rae, *Canada* **72 A5** 62 50N 116 3W
Rae Bareli, *India* **43 F9** 26 18N 81 20 E
Rae Isthmus, *Canada* **69 B11** 66 40N 87 30W
Raeren, *Belgium* **15 D6** 50 41N 6 7 E
Raeside, L., *Australia* **61 E3** 29 20S 122 0 E
Raetihi, *N.Z.* **59 H5** 39 25S 175 17 E
Rafaela, *Argentina* **94 C3** 31 10S 61 30W
Rafah, *Gaza Strip* **47 D3** 31 18N 34 14 E
Rafai, *C.A.R.* **54 B1** 4 59N 23 58 E
Rafḩā, *Si. Arabia* **44 D4** 29 35N 43 35 E
Rafsanjān, *Iran* **45 D8** 30 30N 56 5 E
Raft Pt., *Australia* **60 C3** 16 4S 124 26 E
Râgâ, *Sudan* **51 G11** 8 28N 25 41 E
Ragachow, *Belarus* **17 B16** 53 8N 30 5 E
Ragama, *Sri Lanka* **40 R11** 7 0N 79 50 E
Ragged, Mt., *Australia* **61 F3** 33 27S 123 25 E
Raghunathpalli, *India* **43 H11** 22 14N 84 48 E
Raghunathpur, *India* **43 H12** 23 33N 86 40 E
Raglan, *N.Z.* **59 G5** 37 55S 174 55 E
Ragusa, *Italy* **20 F6** 36 55N 14 44 E
Raha, *Indonesia* **37 E6** 4 55S 123 0 E
Rahaeng = Tak, *Thailand* **38 D2** 16 52N 99 8 E
Rahatgarh, *India* **43 H8** 23 47N 78 22 E
Rahimyar Khan, *Pakistan* **42 E4** 28 30N 70 25 E
Rāhjerd, *Iran* **45 C6** 34 22N 50 22 E
Rahon, *India* **42 D7** 31 3N 76 7 E
Raichur, *India* **40 L10** 16 10N 77 20 E
Raiganj, *India* **43 G13** 25 37N 88 10 E
Raigarh, *India* **41 J13** 21 56N 83 25 E
Raijua, *Indonesia* **37 F6** 10 37S 121 36 E
Raikot, *India* **42 D6** 30 41N 75 42 E
Railton, *Australia* **62 G4** 41 25S 146 28 E
Rainbow Lake, *Canada* **72 B5** 58 30N 119 23W
Rainier, *U.S.A.* **84 D4** 46 53N 122 41W
Rainier, Mt., *U.S.A.* **84 D5** 46 52N 121 46W
Rainy L., *Canada* **73 D10** 48 42N 93 10W
Rainy River, *Canada* **73 D10** 48 43N 94 29W
Raippaluoto, *Finland* **8 E19** 63 13N 21 14 E
Raipur, *India* **41 J12** 21 17N 81 45 E
Raisen, *India* **42 H8** 23 20N 77 48 E
Raisio, *Finland* **9 F20** 60 28N 22 11 E
Raj Nandgaon, *India* **41 J12** 21 5N 81 5 E
Raj Nilgiri, *India* **43 J12** 21 28N 86 46 E
Raja, Ujung, *Indonesia* **36 D1** 3 40N 96 25 E
Raja Ampat, Kepulauan, *Indonesia* **37 E7** 0 30S 130 0 E
Rajahmundry, *India* **41 L12** 17 1N 81 48 E
Rajang →, *Malaysia* **36 D4** 2 30N 112 0 E
Rajanpur, *Pakistan* **42 E4** 29 6N 70 19 E
Rajapalaiyam, *India* **40 Q10** 9 25N 77 35 E
Rajasthan □, *India* **42 F5** 26 45N 73 30 E
Rajasthan Canal, *India* **42 F5** 28 0N 72 0 E
Rajauri, *India* **43 C6** 33 25N 74 21 E
Rajgarh, *Mad. P., India* **42 G7** 24 2N 76 45 E
Rajgarh, *Raj., India* **42 F7** 27 14N 76 38 E
Rajgarh, *Raj., India* **42 E6** 28 40N 75 25 E
Rajgir, *India* **43 G11** 25 2N 85 25 E
Rajkot, *India* **42 H4** 22 15N 70 56 E
Rajmahal Hills, *India* **43 G12** 24 30N 87 30 E
Rajpipla, *India* **40 J8** 21 50N 73 30 E
Rajpur, *India* **42 H6** 22 18N 74 21 E
Rajpura, *India* **42 D7** 30 25N 76 32 E
Rajshahi, *Bangla.* **41 G16** 24 22N 88 39 E
Rajshahi □, *Bangla.* **43 G13** 25 0N 89 0 E
Rajula, *India* **42 J4** 21 3N 71 26 E
Rakaia, *N.Z.* **59 K4** 43 45S 172 1 E
Rakaia →, *N.Z.* **59 K4** 43 36S 172 15 E
Rakan, Ra's, *Qatar* **45 E6** 26 10N 51 20 E
Rakaposhi, *Pakistan* **43 A6** 36 10N 74 25 E
Rakata, Pulau, *Indonesia* **36 F3** 6 10S 105 20 E
Rakhiv, *Ukraine* **17 D13** 48 3N 24 12 E
Rakhni, *Pakistan* **42 D3** 30 4N 69 56 E
Rakhni →, *Pakistan* **42 E3** 29 31N 69 36 E
Rakitnoye, *Russia* **30 B7** 45 36N 134 17 E
Rakops, *Botswana* **56 C3** 21 1S 24 28 E
Rakvere, *Estonia* **9 G22** 59 20N 26 25 E
Raleigh, *U.S.A.* **77 H6** 35 47N 78 39W
Ralls, *U.S.A.* **81 J4** 33 41N 101 24W
Ralston, *U.S.A.* **78 E8** 41 30N 76 57W
Ram →, *Canada* **72 A4** 62 1N 123 41W
Rām Allāh, *West Bank* **47 D4** 31 55N 35 10 E
Rama, *Nic.* **88 D3** 12 9N 84 15W
Ramakona, *India* **43 J8** 21 43N 78 50 E
Raman, *Thailand* **39 J3** 6 29N 101 18 E
Ramanathapuram, *India* **40 Q11** 9 25N 78 55 E
Ramanetaka, B. de, *Madag.* **57 A8** 14 13S 47 52 E
Ramanujganj, *India* **43 H10** 23 48N 83 42 E
Ramat Gan, *Israel* **47 C3** 32 4N 34 48 E
Ramatlhabama, *S. Africa* **56 D4** 25 37S 25 33 E
Ramban, *India* **43 C6** 33 14N 75 12 E
Rambipuji, *Indonesia* **37 H15** 8 12S 113 37 E
Rame Hd., *Australia* **63 F4** 37 47S 149 30 E
Ramechhap, *Nepal* **43 F12** 27 25N 86 10 E
Ramganga →, *India* **43 F8** 27 5N 79 58 E
Ramgarh, *Jharkhand, India* **43 H11** 23 40N 85 35 E
Ramgarh, *Raj., India* **42 F6** 27 16N 75 14 E
Ramgarh, *Raj., India* **42 F4** 27 30N 70 36 E
Rāmhormoz, *Iran* **45 D6** 31 15N 49 35 E
Ramīān, *Iran* **45 B7** 37 3N 55 16 E
Ramingining, *Australia* **62 A2** 12 19S 135 3 E
Ramla, *Israel* **47 D3** 31 55N 34 52 E
Ramnad = Ramanathapuram, *India* **40 Q11** 9 25N 78 55 E
Ramnagar, *Jammu & Kashmir, India* **43 C6** 32 47N 75 18 E
Ramnagar, *Uttaranchal, India* **43 E8** 29 24N 79 7 E
Râmnicu Sărat, *Romania* **17 F14** 45 26N 27 3 E
Râmnicu Vâlcea, *Romania* **17 F13** 45 9N 24 21 E
Ramona, *U.S.A.* **85 M10** 33 2N 116 52W
Ramore, *Canada* **70 C3** 48 30N 80 25W
Ramotswa, *Botswana* **56 C4** 24 50S 25 52 E
Rampur, *H.P., India* **42 D7** 31 26N 77 43 E
Rampur, *Mad. P., India* **42 H5** 23 25N 73 53 E
Rampur, *Ut. P., India* **43 E8** 28 50N 79 5 E
Rampur Hat, *India* **43 G12** 24 10N 87 50 E
Rampura, *India* **42 G6** 24 30N 75 27 E
Ramrama Tola, *India* **43 J8** 21 52N 79 55 E
Ramree I., *Burma* **41 K19** 19 0N 93 40 E
Rāmsar, *Iran* **45 B6** 36 53N 50 41 E
Ramsey, *U.K.* **10 C3** 54 20N 4 22W
Ramsey, *U.S.A.* **79 E10** 41 4N 74 9W
Ramsey L., *Canada* **70 C3** 47 13N 82 15W
Ramsgate, *U.K.* **11 F9** 51 20N 1 25 E
Ramtek, *India* **40 J11** 21 20N 79 15 E
Rana Pratap Sagar Dam, *India* **42 G6** 24 58N 75 38 E
Ranaghat, *India* **43 H13** 23 15N 88 35 E
Ranahu, *Pakistan* **42 G3** 25 55N 69 45 E
Ranau, *Malaysia* **36 C5** 6 2N 116 40 E
Rancagua, *Chile* **94 C1** 34 10S 70 50W
Rancheria →, *Canada* **72 A3** 60 13N 129 7W
Ranchester, *U.S.A.* **82 D10** 44 54N 107 10W
Ranchi, *India* **43 H11** 23 19N 85 27 E
Rancho Cucamonga, *U.S.A.* **85 L9** 34 10N 117 30W
Randalstown, *U.K.* **13 B5** 54 45N 6 19W

Russell

Russell, N.Y., U.S.A. 79 B9 44 27N 75 9W
Russell, Pa., U.S.A. 78 E5 41 56N 79 8W
Russell L., Man., Canada 73 B8 56 15N 101 30W
Russell L., N.W.T., Canada 72 A5 63 5N 115 44W
Russellkonda, India 41 K14 19 57N 84 42 E
Russellville, Ala., U.S.A. 77 H2 34 30N 87 44W
Russellville, Ark., U.S.A. 81 H8 35 17N 93 8W
Russellville, Ky., U.S.A. 77 G2 36 51N 86 53W
Russia ■, Eurasia 27 C11 62 0N 105 0 E
Russian →, U.S.A. 84 G3 38 27N 123 8W
Russkoye Ustie, Russia 4 B15 71 0N 149 0 E
Rustam, Pakistan 42 B5 34 25N 72 13 E
Rustam Shahr, Pakistan 42 F2 26 58N 66 6 E
Rustavi, Georgia 25 F8 41 30N 45 0 E
Rustenburg, S. Africa 56 D4 25 41S 27 14 E
Ruston, U.S.A. 81 J8 32 32N 92 38W
Rutana, Burundi 54 C3 3 55S 30 0 E
Ruteng, Indonesia 37 F6 8 35S 120 30 E
Ruth, U.S.A. 78 C2 43 42N 82 45W
Rutherford, U.S.A. 84 G4 38 26N 122 24W
Rutland, U.S.A. 79 C12 43 37N 72 58W
Rutland □, U.K. 11 E7 52 38N 0 40W
Rutland Water, U.K. 11 E7 52 39N 0 38W
Rutledge →, Canada 73 A6 61 4N 112 0W
Rutledge L., Canada 73 A6 61 33N 110 47W
Rutshuru, Dem. Rep. of the Congo 54 C2 1 13S 29 25 E
Ruvu, Tanzania 54 D4 6 49S 38 43 E
Ruvu →, Tanzania 54 D4 6 23S 38 52 E
Ruvuma □, Tanzania 55 E4 10 20S 36 0 E
Ruvuma →, Tanzania 55 E5 10 29S 40 28 E
Ruwais, U.A.E. 45 E7 24 5N 52 50 E
Ruwenzori, Africa 54 B2 0 30N 29 55 E
Ruya →, Zimbabwe 57 B5 16 27S 32 5 E
Ruyigi, Burundi 54 C3 3 29S 30 15 E
Ružomberok, Slovak Rep. 17 D10 49 3N 19 17 E
Rwanda ■, Africa 54 C3 2 0S 30 0 E
Ryan, L., U.K. 12 G3 55 0N 5 2W
Ryazan, Russia 24 D6 54 40N 39 40 E
Ryazhsk, Russia 24 D7 53 45N 40 3 E
Rybache = Rybachye, Kazakstan 26 E9 46 40N 81 20 E
Rybachiy Poluostrov, Russia 24 A5 69 43N 32 0 E
Rybachye, Kazakstan 26 E9 46 40N 81 20 E
Rybinsk, Russia 24 C6 58 5N 38 50 E
Rybinskoye Vdkhr., Russia 24 C6 58 30N 38 25 E
Rybnitsa = Râbniţa, Moldova 17 E15 47 45N 29 0 E
Rycroft, Canada 72 B5 55 45N 118 40W
Ryde, U.K. 11 G6 50 43N 1 9W
Ryderwood, U.S.A. 84 D3 46 23N 123 3W
Rye, U.K. 11 G8 50 57N 0 45 E
Rye →, U.K. 10 C7 54 11N 0 44W
Rye Bay, U.K. 11 G8 50 52N 0 49 E
Rye Patch Reservoir, U.S.A. 82 F4 40 28N 118 19W
Ryegate, U.S.A. 82 C9 46 18N 109 15W
Ryley, Canada 72 C6 53 17N 112 26W
Rylstone, Australia 63 E4 32 46S 149 58 E
Ryōtsu, Japan 30 E9 38 5N 138 26 E
Rypin, Poland 17 B10 53 3N 19 25 E
Ryūgasaki, Japan 31 G10 35 54N 140 11 E
Ryūkyū Is. = Ryūkyū-rettō, Japan 31 M3 26 0N 126 0 E
Ryūkyū-rettō, Japan 31 M3 26 0N 126 0 E
Rzeszów, Poland 17 C11 50 5N 21 58 E
Rzhev, Russia 24 C5 56 20N 34 20 E

S

Sa, Thailand 38 C3 18 34N 100 45 E
Sa Canal, Spain 22 C7 38 51N 1 23 E
Sa Conillera, Spain 22 C7 38 59N 1 13 E
Sa Dec, Vietnam 39 G5 10 20N 105 46 E
Sa Dragonera, Spain 22 B9 39 35N 2 19 E
Sa Mesquida, Spain 22 B11 39 55N 4 16 E
Sa Savina, Spain 22 C7 38 44N 1 25 E
Sa'ādatābād, Fārs, Iran 45 D7 30 10N 53 5 E
Sa'ādatābād, Hormozgān, Iran 45 D7 28 3N 55 53 E
Sa'ādatābād, Kermān, Iran 45 D7 29 40N 55 51 E
Saale →, Germany 16 C6 51 56N 11 54 E
Saalfeld, Germany 16 C6 50 38N 11 21 E
Saar →, Europe 18 B7 49 41N 6 32 E
Saarbrücken, Germany 16 D4 49 14N 6 59 E
Saaremaa, Estonia 9 G20 58 30N 22 30 E
Saarijärvi, Finland 9 E21 62 43N 25 16 E
Saariselkä, Finland 8 B23 68 16N 28 15 E
Sab 'Âbar, Syria 44 C3 33 46N 37 41 E
Saba, W. Indies 89 C7 17 42N 63 26W
Šabac, Serbia, Yug. 21 B8 44 48N 19 42 E
Sabadell, Spain 19 B7 41 28N 2 7 E
Sabah □, Malaysia 36 C5 6 0N 117 0 E
Sabak Bernam, Malaysia 39 L3 3 46N 100 58 E
Sabalān, Kūhhā-ye, Iran 44 B5 38 15N 47 45 E
Sabalana, Kepulauan, Indonesia 37 F5 6 45S 118 50 E
Sábana de la Mar, Dom. Rep. 89 C6 19 7N 69 24W
Sábanalarga, Colombia 92 A4 10 38N 74 55W
Sabang, Indonesia 36 C1 5 50N 95 15 E
Sabará, Brazil 93 G10 19 55S 43 46W
Sabarmati →, India 42 H5 22 18N 72 22 E
Sabattis, U.S.A. 79 B10 44 6N 74 40W
Saberania, Indonesia 37 E9 2 5S 138 18 E
Sabhah, Libya 51 C8 27 9N 14 29 E
Sabi →, India 42 E7 28 29N 76 44 E
Sabie, S. Africa 57 D5 25 10S 30 48 E
Sabinal, Mexico 86 A3 30 58N 107 25W
Sabinal, U.S.A. 81 L5 29 19N 99 28W
Sabinas, Mexico 86 B4 27 50N 101 10W
Sabinas →, Mexico 86 B4 27 37N 100 42W
Sabinas Hidalgo, Mexico 86 B4 26 33N 100 10W
Sabine →, U.S.A. 81 L8 29 59N 93 47W
Sabine L., U.S.A. 81 L8 29 53N 93 51W
Sabine Pass, U.S.A. 81 L8 29 44N 93 54W
Sabinsville, U.S.A. 78 E7 41 52N 77 31W
Sablayan, Phil. 37 B6 12 50N 120 50 E
Sable, Canada 71 A6 55 30N 68 21W
Sable, C., Canada 71 D6 43 29N 65 38W
Sable, C., U.S.A. 75 E10 25 9N 81 8W
Sable I., Canada 71 D8 44 0N 60 0W
Sabrina Coast, Antarctica 5 C9 68 0S 120 0 E
Sabulubbek, Indonesia 36 E1 1 36S 98 40 E
Sabzevār, Iran 45 B8 36 15N 57 40 E
Sabzvārān, Iran 45 D8 28 45N 57 50 E
Sac City, U.S.A. 80 D7 42 25N 95 0W
Săcele, Romania 17 F13 45 37N 25 41 E
Sachigo →, Canada 70 A2 55 6N 88 58W
Sachigo, L., Canada 70 B1 53 50N 92 12W
Sachsen □, Germany 16 C7 50 55N 13 10 E
Sachsen-Anhalt □, Germany 16 C7 52 0N 12 0 E
Sackets Harbor, U.S.A. 79 C8 43 57N 76 7W
Sackville, Canada 71 C7 45 54N 64 22W
Saco, Maine, U.S.A. 77 D10 43 30N 70 27W
Saco, Mont., U.S.A. 82 B10 48 28N 107 21W
Sacramento, U.S.A. 84 G5 38 35N 121 29W
Sacramento →, U.S.A. 84 G5 38 3N 121 56W
Sacramento Mts., U.S.A. 83 K11 32 30N 105 30W
Sacramento Valley, U.S.A. 84 G5 39 30N 122 0W
Sada-Misaki, Japan 31 H6 33 20N 132 1 E
Sadabad, India 42 F8 27 27N 78 3 E
Sadani, Tanzania 54 D4 5 58S 38 35 E
Sadao, Thailand 39 J3 6 38N 100 26 E
Sadd el Aali, Egypt 51 D12 23 54N 32 54 E
Saddle Mt., U.S.A. 84 E3 45 58N 123 41W
Sadimi, Dem. Rep. of the Congo 55 D1 9 25S 23 32 E
Sado, Japan 30 F9 38 0N 138 25 E
Sadon, Burma 41 G20 25 28N 97 55 E
Sadra, India 42 H5 23 21N 72 43 E
Sadri, India 42 G5 25 11N 73 26 E
Sæby, Denmark 9 H14 57 21N 10 30 E
Saegertown, U.S.A. 78 E4 41 43N 80 9W
Şafājah, Si. Arabia 44 E3 26 25N 39 0 E
Säffle, Sweden 9 G15 59 8N 12 55 E
Safford, U.S.A. 83 K9 32 50N 109 43W
Saffron Walden, U.K. 11 E8 52 1N 0 16 E
Safi, Morocco 50 B4 32 18N 9 20W
Şafīābād, Iran 45 B8 36 45N 57 58 E
Safīd Dasht, Iran 45 C6 33 27N 48 11 E
Safīd Kūh, Afghan. 40 B3 34 45N 63 0 E
Safīd Rūd →, Iran 45 B6 37 23N 50 11 E
Safipur, India 43 F9 26 44N 80 21 E
Safwān, Iraq 44 D5 30 7N 47 43 E
Sag Harbor, U.S.A. 79 F12 41 0N 72 18W
Saga, Japan 31 H5 33 15N 130 16 E
Saga □, Japan 31 H5 33 15N 130 20 E
Sagae, Japan 30 E10 38 22N 140 17 E
Sagamore, U.S.A. 78 F5 40 46N 79 14W
Sagar, Karnataka, India 40 M9 14 14N 75 6 E
Sagar, Mad. P., India 43 H8 23 50N 78 44 E
Sagara, L., Tanzania 54 D3 5 20S 31 0 E
Saginaw, U.S.A. 76 D4 43 26N 83 56W
Saginaw →, U.S.A. 76 D4 43 39N 83 51W
Saginaw B., U.S.A. 76 D4 43 50N 83 40W
Saglouc = Salluit, Canada 69 B12 62 14N 75 38W
Sagŏ-ri, S. Korea 35 G14 35 25N 126 49 E
Sagua la Grande, Cuba 88 B3 22 50N 80 10W
Saguache, U.S.A. 83 G10 38 5N 106 8W
Saguaro Nat. Park, U.S.A. 83 K8 32 12N 110 38W
Saguenay →, Canada 71 C5 48 22N 71 0W
Sagunt, Spain 19 C5 39 42N 0 18W
Sagunto = Sagunt, Spain 19 C5 39 42N 0 18W
Sagwara, India 42 H6 23 41N 74 1 E
Sahagún, Spain 19 A3 42 18N 5 2W
Şaḩam al Jawlān, Syria 47 C4 32 45N 35 55 E
Sahamandrevo, Madag. 57 C8 23 15S 45 35 E
Sahand, Kūh-e, Iran 44 B5 37 44N 46 27 E
Sahara, Africa 50 D6 23 0N 5 0 E
Saharan Atlas = Saharien, Atlas, Algeria 50 B6 33 30N 1 0 E
Saharanpur, India 42 E7 29 58N 77 33 E
Saharien, Atlas, Algeria 50 B6 33 30N 1 0 E
Saharsa, India 43 G12 25 53N 86 36 E
Sahasinaka, Madag. 57 C8 21 49S 47 49 E
Sahaswan, India 43 E8 28 5N 78 45 E
Sahel, Africa 50 E5 16 0N 5 0 E
Sahibganj, India 43 G12 25 12N 87 40 E
Şāḩilīyah, Iraq 44 C4 33 43N 42 42 E
Sahiwal, Pakistan 42 D5 30 45N 73 8 E
Şaḩneh, Iran 44 C5 34 29N 47 41 E
Sahuaripa, Mexico 86 B3 29 0N 109 13W
Sahuarita, U.S.A. 83 L8 31 57N 110 58W
Sahuayo, Mexico 86 C4 20 4N 102 43W
Sai →, India 43 G10 25 39N 82 47 E
Sai Buri, Thailand 39 J3 6 43N 101 45 E
Sa'id Bundas, Sudan 51 G10 8 24N 24 48 E
Sa'īdābād, Kermān, Iran 45 D7 29 30N 55 45 E
Sa'īdābād, Semnān, Iran 45 B7 36 8N 54 11 E
Sa'īdīyeh, Iran 45 B6 36 20N 48 55 E
Saidpur, Bangla. 41 G16 25 48N 89 0 E
Saidpur, India 43 G10 25 33N 83 11 E
Saidu, Pakistan 43 B5 34 43N 72 24 E
Saigon = Thanh Pho Ho Chi Minh, Vietnam 39 G6 10 58N 106 40 E
Saijō, Japan 31 H6 33 55N 133 11 E
Saikanosy Masoala, Madag. 57 B9 15 45S 50 10 E
Saikhoa Ghat, India 41 F19 27 50N 95 40 E
Saiki, Japan 31 H5 32 58N 131 51 E
Sailana, India 42 H6 23 28N 74 55 E
Sailolof, Indonesia 37 E8 1 15S 130 46 E
Saimaa, Finland 9 F23 61 15N 28 15 E
Şa'in Dezh, Iran 44 B5 36 40N 46 25 E
St. Abb's Head, U.K. 12 F6 55 55N 2 8W
St. Alban's, Canada 71 C8 47 51N 55 50W
St. Albans, U.K. 11 F7 51 45N 0 19W
St. Albans, Vt., U.S.A. 79 B11 44 49N 73 5W
St. Albans, W. Va., U.S.A. 76 F5 38 23N 81 50W
St. Alban's Head, U.K. 11 G5 50 34N 2 4W
St. Albert, Canada 72 C6 53 37N 113 32W
St. Andrew's, Canada 71 C8 47 45N 59 15W
St. Andrews, U.K. 12 E6 56 20N 2 47W
St-Anicet, Canada 79 A10 45 8N 74 22W
St. Ann B., Canada 71 C7 46 22N 60 25W
St. Ann's Bay, Jamaica 88 C4 18 26N 77 15W
St. Anthony, Canada 71 B8 51 22N 55 35W
St. Anthony, U.S.A. 82 E8 43 58N 111 41W
St. Antoine, Canada 71 C7 46 22N 64 45W
St. Arnaud, Australia 63 F3 36 40S 143 16 E
St-Augustin →, Canada 71 B8 51 16N 58 40W
St-Augustin-Saguenay, Canada 71 B8 51 13N 58 38W
St. Augustine, U.S.A. 77 L5 29 54N 81 19W
St. Austell, U.K. 11 G3 50 20N 4 47W
St. Barbe, Canada 71 B8 51 12N 56 46W
St-Barthélemy, W. Indies 89 C7 17 50N 62 50W
St. Bees Hd., U.K. 10 C4 54 31N 3 38W
St. Bride's, Canada 71 C9 46 56N 54 10W
St. Brides B., U.K. 11 F2 51 49N 5 9W
St-Brieuc, France 18 B2 48 30N 2 46W
St. Catharines, Canada 78 C5 43 10N 79 15W
St. Catherines I., U.S.A. 77 K5 31 40N 81 10W
St. Catherine's Pt., U.K. 11 G6 50 34N 1 18W
St-Chamond, France 18 D6 45 28N 4 31 E
St. Charles, Ill., U.S.A. 76 E1 41 54N 88 19W
St. Charles, Mo., U.S.A. 80 F9 38 47N 90 29W
St. Charles, Va., U.S.A. 76 F7 36 48N 83 4W
St. Christopher-Nevis = St. Kitts & Nevis ■, W. Indies 89 C7 17 20N 62 40W
St. Clair, Mich., U.S.A. 78 D2 42 50N 82 30W
St. Clair, Pa., U.S.A. 79 F8 40 43N 76 12W
St. Clair →, U.S.A. 78 D2 42 38N 82 31W
St. Clair, L., Canada 70 D3 42 30N 82 45W
St. Clair, L., U.S.A. 78 D2 42 27N 82 39W
St. Clairsville, U.S.A. 78 F4 40 5N 80 54W
St. Claude, Canada 73 D9 49 40N 98 20W
St-Clet, Canada 79 A10 45 21N 74 13W
St. Cloud, Fla., U.S.A. 77 L5 28 15N 81 17W
St. Cloud, Minn., U.S.A. 80 C7 45 34N 94 10W
St. Cricq, C., Australia 61 E1 25 17S 113 6 E
St. Croix, U.S. Virgin Is. 89 C7 17 45N 64 45W
St. Croix →, U.S.A. 80 C8 44 45N 92 48W
St. Croix Falls, U.S.A. 80 C8 45 24N 92 38W
St. David's, Canada 71 C8 48 12N 58 52W
St. David's, U.K. 11 F2 51 53N 5 16W
St. David's Head, U.K. 11 F2 51 54N 5 19W
St-Denis, France 18 B5 48 56N 2 22 E
St-Dizier, France 18 B6 48 38N 4 56 E
St. Elias, Mt., U.S.A. 68 B5 60 18N 140 56W
St. Elias Mts., Canada 72 A1 60 33N 139 28W
St. Elias Mts., U.S.A. 68 C6 60 0N 138 0W
St-Étienne, France 18 D6 45 27N 4 22 E
St. Eugène, Canada 79 A10 45 30N 74 28W
St. Eustatius, W. Indies 89 C7 17 20N 63 0W
St-Félicien, Canada 70 C5 48 40N 72 25W
St-Flour, France 18 D5 45 2N 3 6 E
St. Francis, U.S.A. 80 F4 39 47N 101 48W
St. Francis →, U.S.A. 81 H9 34 38N 90 36W
St. Francis, C., S. Africa 56 E3 34 14S 24 49 E
St. Francisville, U.S.A. 81 K9 30 47N 91 23W
St-François, L., Canada 79 A10 45 10N 74 22W
St-Gabriel, Canada 70 C5 46 17N 73 24W
St. Gallen = Sankt Gallen, Switz. 18 C8 47 26N 9 22 E
St-Gaudens, France 18 E4 43 6N 0 44 E
St. George, Australia 63 D4 28 1S 148 30 E
St. George, Canada 71 C6 45 11N 66 50W
St. George, S.C., U.S.A. 77 J5 33 11N 80 35W
St. George, Utah, U.S.A. 83 H7 37 6N 113 35W
St. George, C., Canada 71 C8 48 30N 59 16W
St. George, C., U.S.A. 77 L3 29 40N 85 5W
St. George Ra., Australia 60 C4 18 40S 125 0 E
St. George's, Canada 71 C8 48 26N 58 31W
St-Georges, Canada 71 C5 46 8N 70 40W
St. George's, Grenada 89 D7 12 5N 61 43W
St. George's B., Canada 71 C8 48 24N 58 53W
St. Georges Basin, N.S.W., Australia 63 F5 35 7S 150 36 E
St. Georges Basin, W. Austral., Australia 60 C4 15 23S 125 2 E
St. George's Channel, Europe 13 E6 52 0N 6 0W
St. Georges Hd., Australia 63 F5 35 12S 150 42 E
St. Gotthard P. = San Gottardo, P. del, Switz. 18 C8 46 33N 8 33 E
St. Helena, Atl. Oc. 48 H3 15 58S 5 42W
St. Helena, U.S.A. 82 G2 38 30N 122 28W
St. Helena, Mt., U.S.A. 84 G4 38 40N 122 36W
St. Helena B., S. Africa 56 E2 32 40S 18 10 E
St. Helens, Australia 62 G4 41 20S 148 15 E
St. Helens, U.K. 10 D5 53 27N 2 44W
St. Helens, U.S.A. 84 E4 45 52N 122 48W
St. Helens, Mt., U.S.A. 84 D4 46 12N 122 12W
St. Helier, U.K. 11 H5 49 10N 2 7W
St-Hubert, Belgium 15 D5 50 2N 5 23 E
St-Hyacinthe, Canada 70 C5 45 40N 72 58W
St. Ignace, U.S.A. 76 C3 45 52N 84 44W
St. Ignace I., Canada 70 C2 48 45N 88 0W
St. Ignatius, U.S.A. 82 C6 47 19N 114 6W
St. Ives, U.K. 11 G2 50 12N 5 30W
St. James, U.S.A. 80 D7 43 59N 94 38W
St-Jean →, Canada 71 B7 50 17N 64 20W
St-Jean, L., Canada 71 C5 48 40N 72 0W
St-Jean-Port-Joli, Canada 71 C5 47 15N 70 13W
St-Jean-sur-Richelieu, Canada 79 A11 45 20N 73 20W
St-Jérôme, Canada 70 C5 45 47N 74 0W
St. John, Canada 71 C6 45 20N 66 8W
St. John, U.S.A. 81 G5 38 0N 98 46W
St. John →, U.S.A. 77 C12 45 12N 66 5W
St. John, C., Canada 71 C8 50 0N 55 32W
St. John's, Antigua 89 C7 17 6N 61 51W
St. John's, Canada 71 C9 47 35N 52 40W
St. Johns, Ariz., U.S.A. 83 J9 34 30N 109 22W
St. Johns, Mich., U.S.A. 76 D3 43 0N 84 33W
St. Johns →, U.S.A. 77 K5 30 24N 81 24W
St. John's Pt., Ireland 13 B3 54 34N 8 27W
St. Johnsbury, U.S.A. 79 B12 44 25N 72 1W
St. Johnsville, U.S.A. 79 D10 43 0N 74 43W
St. Joseph, La., U.S.A. 81 K9 31 55N 91 14W
St. Joseph, Mo., U.S.A. 80 F7 39 46N 94 50W
St. Joseph →, U.S.A. 76 D2 42 7N 86 29W
St. Joseph, I., Canada 70 C3 46 12N 83 58W
St. Joseph, L., Canada 70 B1 51 10N 90 35W
St-Jovite, Canada 70 C5 46 8N 74 38W
St. Kitts & Nevis ■, W. Indies 89 C7 17 20N 62 40W
St. Laurent, Canada 73 C9 50 25N 97 58W
St. Lawrence, Australia 62 C4 22 16S 149 31 E
St. Lawrence, Canada 71 C8 46 54N 55 23W
St. Lawrence →, Canada 71 C6 49 30N 66 0W
St. Lawrence, Gulf of, Canada 71 C7 48 25N 62 0W
St. Lawrence I., U.S.A. 68 B3 63 30N 170 30W
St. Leonard, Canada 71 C6 47 12N 67 58W
St. Lewis →, Canada 71 B8 52 26N 56 11W
St-Lô, France 18 B3 49 7N 1 5W
St. Louis, Senegal 50 E2 16 8N 16 27W
St. Louis, U.S.A. 80 F9 38 37N 90 12W
St. Louis →, U.S.A. 80 B8 47 15N 92 45W
St. Lucia ■, W. Indies 89 D7 14 0N 60 50W
St. Lucia, L., S. Africa 57 D5 28 5S 32 30 E
St. Lucia Channel, W. Indies 89 D7 14 15N 61 0W
St. Maarten, W. Indies 89 C7 18 0N 63 5W
St. Magnus B., U.K. 12 A7 60 25N 1 35W
St-Malo, France 18 B2 48 39N 2 1W
St-Marc, Haiti 89 C5 19 10N 72 41W
St. Maries, U.S.A. 82 C5 47 19N 116 35W
St-Martin, W. Indies 89 C7 18 0N 63 0W
St. Martin, L., Canada 73 C9 51 40N 98 30W
St. Mary Pk., Australia 63 E2 31 32S 138 34 E
St. Marys, Australia 62 G4 41 35S 148 11 E
St. Marys, Canada 78 C3 43 20N 81 10W
St. Mary's, Corn., U.K. 11 H1 49 55N 6 18W
St. Mary's, Orkney, U.K. 12 C6 58 54N 2 54W
St. Marys, Ga., U.S.A. 77 K5 30 44N 81 33W
St. Marys, Pa., U.S.A. 78 E6 41 26N 78 34W
St. Mary's, C., Canada 71 C9 46 50N 54 12W
St. Mary's B., Canada 71 C9 46 50N 53 50W
St. Marys Bay, Canada 71 D6 44 25N 66 10W
St-Mathieu, Pte., France 18 B1 48 20N 4 45W
St. Matthew I., U.S.A. 68 B2 60 24N 172 42W
St. Matthews, I. = Zadetkyi Kyun, Burma 39 G1 10 0N 98 25 E
St-Maurice →, Canada 70 C5 46 21N 72 31W
St-Nazaire, France 18 C2 47 17N 2 12W
St. Neots, U.K. 11 E7 52 14N 0 15W
St-Niklaas, Belgium 15 C4 51 10N 4 8 E
St-Omer, France 18 A5 50 45N 2 15 E
St-Pamphile, Canada 71 C6 46 58N 69 48W
St. Pascal, Canada 71 C6 47 32N 69 48W
St. Paul, Canada 72 C6 54 0N 111 17W
St. Paul, Minn., U.S.A. 80 C8 44 57N 93 6W
St. Paul, Nebr., U.S.A. 80 E5 41 13N 98 27W
St-Paul →, Canada 71 B8 51 27N 57 42W
St. Paul, I., Ind. Oc. 3 F13 38 55S 77 34 E
St. Paul I., Canada 71 C7 47 12N 60 9W
St. Peter, U.S.A. 80 C8 44 20N 93 57W
St. Peter Port, U.K. 11 H5 49 26N 2 33W
St. Peters, N.S., Canada 71 C7 45 40N 60 53W
St. Peters, P.E.I., Canada 71 C7 46 25N 62 35W
St. Petersburg = Sankt-Peterburg, Russia 24 C5 59 55N 30 20 E
St. Petersburg, U.S.A. 77 M4 27 46N 82 39W
St-Pie, Canada 79 A12 45 30N 72 54W
St-Pierre, St- P. & M. 71 C8 46 46N 56 12W
St-Pierre, L., Canada 70 C5 46 12N 72 52W
St-Pierre et Miquelon □, St- P. & M. 71 C8 46 55N 56 10W
St. Quentin, Canada 71 C6 47 30N 67 23W
St-Quentin, France 18 B5 49 50N 3 16 E
St. Regis, U.S.A. 82 C6 47 18N 115 6W
St. Sebastien, Tanjon' i, Madag. 57 A8 12 26S 48 44 E
St-Siméon, Canada 71 C6 47 51N 69 54W
St. Simons I., U.S.A. 77 K5 31 12N 81 15W
St. Simons Island, U.S.A. 77 K5 31 9N 81 22W
St. Stephen, Canada 71 C6 45 16N 67 17W
St. Thomas, Canada 78 D3 42 45N 81 10W
St. Thomas I., U.S. Virgin Is. 89 C7 18 20N 64 55W
St-Tite, Canada 70 C5 46 45N 72 34W
St-Tropez, France 18 E7 43 17N 6 38 E
St. Troud = St. Truiden, Belgium 15 D5 50 48N 5 10 E
St. Truiden, Belgium 15 D5 50 48N 5 10 E
St. Vincent, G., Australia 63 F2 35 0S 138 0 E
St. Vincent & the Grenadines ■, W. Indies 89 D7 13 0N 61 10W
St. Vincent Passage, W. Indies 89 D7 13 30N 61 0W
St-Vith, Belgium 15 D6 50 17N 6 9 E
St. Walburg, Canada 73 C7 53 39N 109 12W
Ste-Agathe-des-Monts, Canada 70 C5 46 3N 74 17W
Ste-Anne, L., Canada 71 B6 50 0N 67 42W
Ste-Anne-des-Monts, Canada 71 C6 49 8N 66 30W
Ste. Genevieve, U.S.A. 80 G9 37 59N 90 2W
Ste-Marguerite →, Canada 71 B6 50 9N 66 36W
Ste-Marie, Martinique 89 D7 14 48N 61 1W
Ste-Marie de la Madeleine, Canada 71 C5 46 26N 71 0W
Ste-Rose, Guadeloupe 89 C7 16 20N 61 45W
Ste. Rose du Lac, Canada 73 C9 51 4N 99 30W
Saintes, France 18 D3 45 45N 0 37W
Saintes, I. des, Guadeloupe 89 C7 15 50N 61 35W
Saintfield, U.K. 13 B6 54 28N 5 49W
Saintonge, France 18 D3 45 40N 0 50W
Saipan, Pac. Oc. 64 F6 15 12N 145 45 E
Sairang, India 41 H18 23 50N 92 45 E
Sairecábur, Cerro, Bolivia 94 A2 22 43S 67 54W
Saitama □, Japan 31 F9 36 25N 139 30 E
Saiyid, Pakistan 42 C5 33 7N 73 2 E
Sajama, Bolivia 92 G5 18 7S 69 0W
Sajószentpéter, Hungary 17 D11 48 12N 20 44 E
Sajum, India 43 C8 33 20N 79 0 E
Sak →, S. Africa 56 E3 30 52S 20 25 E
Sakai, Japan 31 G7 34 30N 135 30 E
Sakaide, Japan 31 G6 34 19N 133 50 E
Sakaiminato, Japan 31 G6 35 38N 133 11 E
Sakākah, Si. Arabia 44 D4 30 0N 40 8 E
Sakakawea, L., U.S.A. 80 B4 47 30N 101 25W
Sakami →, Canada 70 B4 53 40N 76 40W
Sakami, L., Canada 70 B4 53 15N 77 0W
Sakania, Dem. Rep. of the Congo 55 E2 12 43S 28 30 E
Sakaraha, Madag. 57 C7 22 55S 44 32 E
Sakarya, Turkey 25 F5 40 48N 30 25 E
Sakashima-Guntō, Japan 31 M2 24 46N 124 0 E
Sakata, Japan 30 E9 38 55N 139 50 E
Sakchu, N. Korea 35 D13 40 23N 125 2 E
Sakeny →, Madag. 57 C8 20 0S 45 25 E
Sakha □, Russia 27 C13 66 0N 130 0 E
Sakhalin, Russia 27 D15 51 0N 143 0 E
Sakhalinskiy Zaliv, Russia 27 D15 54 0N 141 0 E
Šakiai, Lithuania 9 J20 54 59N 23 2 E
Sakon Nakhon, Thailand 38 D5 17 10N 104 9 E
Sakrand, Pakistan 42 F3 26 10N 68 15 E
Sakri, India 43 F12 26 13N 86 5 E
Sakrivier, S. Africa 56 E3 30 54S 20 28 E
Sakti, India 43 H10 22 2N 82 58 E
Sakuma, Japan 31 G8 35 3N 137 49 E
Sakurai, Japan 31 G7 34 30N 135 51 E
Sala, Sweden 9 G17 59 58N 16 35 E
Sala Consilina, Italy 20 D6 40 23N 15 36 E
Sala-y-Gómez, Pac. Oc. 65 K17 26 28S 105 28W
Salaberry-de-Valleyfield, Canada 79 A10 45 15N 74 8W
Saladas, Argentina 94 B4 28 15S 58 40W
Saladillo, Argentina 94 D4 35 40S 59 55W
Salado →, Buenos Aires, Argentina 94 D4 35 44S 57 22W
Salado →, La Pampa, Argentina 96 D3 37 30S 67 0W
Salado →, Santa Fe, Argentina 94 C3 31 40S 60 41W
Salado →, Mexico 81 M5 26 52N 99 19W
Salaga, Ghana 50 G5 8 31N 0 31W
Sālah, Syria 47 C5 32 40N 36 45 E

Söderhamn

Sierra Blanca, U.S.A. 83 L11 31 11N 105 22W
Sierra Blanca Peak, U.S.A. 83 K11 33 23N 105 49W
Sierra City, U.S.A. 84 F6 39 34N 120 38W
Sierra Colorada, Argentina 96 E3 40 35S 67 50W
Sierra Gorda, Chile 94 A2 22 50S 69 15W
Sierra Leone ■, W. Afr. 50 G3 9 0N 12 0W
Sierra Madre, Mexico 87 D6 16 0N 93 0W
Sierra Mojada, Mexico 86 B4 27 19N 103 42W
Sierra Nevada, Spain 19 D4 37 3N 3 15W
Sierra Nevada, U.S.A. 84 H8 39 0N 120 30W
Sierra Vista, U.S.A. 83 L8 31 33N 110 18W
Sierraville, U.S.A. 84 F6 39 36N 120 22W
Sífnos, Greece 21 F11 37 0N 24 45 E
Sifton, Canada 73 C8 51 21N 100 8W
Sifton Pass, Canada 72 B3 57 52N 126 15W
Sighetu-Marmaţiei, Romania 17 E12 47 57N 23 52 E
Sighişoara, Romania 17 E13 46 12N 24 50 E
Sigli, Indonesia 36 C1 5 25N 96 0 E
Siglufjörður, Iceland 8 C4 66 12N 18 55W
Signal, U.S.A. 85 L13 34 30N 113 38W
Signal Pk., U.S.A. 85 M12 33 20N 114 2W
Sigsig, Ecuador 92 D3 3 0S 78 50W
Sigüenza, Spain 19 B4 41 3N 2 40W
Siguiri, Guinea 50 F4 11 31N 9 10W
Sigulda, Latvia 9 H21 57 10N 24 55 E
Sihanoukville = Kampong Saom, Cambodia 39 G4 10 38N 103 30 E
Sihora, India 43 H9 23 29N 80 6 E
Siikajoki →, Finland 8 D21 64 50N 24 43 E
Siilinjärvi, Finland 8 E22 63 4N 27 39 E
Sijarira Ra. = Chizarira, Zimbabwe 55 F2 17 36S 27 45 E
Sika, India 42 H3 22 26N 69 47 E
Sikao, Thailand 39 J2 7 34N 99 21 E
Sikar, India 42 F6 27 33N 75 10 E
Sikasso, Mali 50 F4 11 18N 5 35W
Sikeston, U.S.A. 81 G10 36 53N 89 35W
Sikhote Alin, Khrebet, Russia 27 E14 45 0N 136 0 E
Sikhote Alin Ra. = Sikhote Alin, Khrebet, Russia 27 E14 45 0N 136 0 E
Síkinos, Greece 21 F11 36 40N 25 8 E
Sikkani Chief →, Canada 72 B4 57 47N 122 15W
Sikkim □, India 41 F16 27 50N 88 30 E
Sikotu-Ko, Japan 30 C10 42 45N 141 25 E
Sil →, Spain 19 A2 42 27N 7 43W
Silacayoapan, Mexico 87 D5 17 30N 98 9W
Silawad, India 42 J6 21 54N 74 54 E
Silchar, India 41 G18 24 49N 92 48 E
Siler City, U.S.A. 77 H6 35 44N 79 28W
Silesia = Śląsk, Poland 16 C9 51 0N 16 30 E
Silgarhi Doti, Nepal 43 E9 29 15N 81 0 E
Silghat, India 41 F18 26 35N 93 0 E
Silifke, Turkey 25 G5 36 22N 33 58 E
Siliguri = Shiliguri, India 41 F16 26 45N 88 25 E
Siling Co, China 32 C3 31 50N 89 20 E
Silistra, Bulgaria 21 B12 44 6N 27 19 E
Silivri, Turkey 21 D13 41 4N 28 14 E
Siljan, Sweden 9 F16 60 55N 14 45 E
Silkeborg, Denmark 9 H13 56 10N 9 32 E
Silkwood, Australia 62 B4 17 45S 146 2 E
Sillajhuay, Cordillera, Chile 92 G5 19 46S 68 40W
Sillamäe, Estonia 9 G22 59 24N 27 45 E
Silloth, U.K. 10 C4 54 52N 3 23W
Siloam Springs, U.S.A. 81 G7 36 11N 94 32W
Silsbee, U.S.A. 81 K7 30 21N 94 11W
Šilutė, Lithuania 9 J19 55 21N 21 33 E
Silva Porto = Kuito, Angola 53 G3 12 22S 16 55 E
Silvani, India 43 H8 23 18N 78 25 E
Silver City, U.S.A. 83 K9 32 46N 108 17W
Silver Cr. →, U.S.A. 82 E4 43 16N 119 13W
Silver Creek, U.S.A. 78 D5 42 33N 79 10W
Silver L., U.S.A. 84 G6 38 39N 120 6W
Silver Lake, Calif., U.S.A. 85 K10 35 21N 116 7W
Silver Lake, Oreg., U.S.A. 82 E3 43 8N 121 3W
Silverton, Colo., U.S.A. 83 H10 37 49N 107 40W
Silverton, Tex., U.S.A. 81 H4 34 28N 101 19W
Silvies →, U.S.A. 82 E4 43 34N 119 2W
Simaltala, India 43 G12 24 43N 86 33 E
Simanggang = Bandar Sri Aman, Malaysia 36 D4 1 15N 111 32 E
Simard, L., Canada 70 C4 47 40N 78 40W
Simav, Turkey 21 E13 39 4N 28 58 E
Simba, Tanzania 54 C4 2 10S 37 36 E
Simbirsk, Russia 24 D8 54 20N 48 25 E
Simbo, Tanzania 54 C2 4 51S 29 41 E
Simcoe, Canada 78 D4 42 50N 80 20W
Simcoe, L., Canada 78 B5 44 25N 79 20W
Simdega, India 43 H11 22 37N 84 31 E
Simeria, Romania 17 F12 45 51N 23 1 E
Simeulue, Indonesia 36 D1 2 45N 95 45 E
Simferopol, Ukraine 25 F5 44 55N 34 3 E
Sími, Greece 21 F12 36 35N 27 50 E
Simi Valley, U.S.A. 85 L8 34 16N 118 47W
Simikot, Nepal 43 E9 30 0N 81 50 E
Simla, India 42 D7 31 2N 77 9 E
Simmie, Canada 73 D7 49 56N 108 6W
Simmler, U.S.A. 85 K7 35 21N 119 59W
Simojoki →, Finland 8 D21 65 35N 25 1 E
Simojovel, Mexico 87 D6 17 12N 92 38W
Simonette →, Canada 72 B5 55 9N 118 15W
Simonstown, S. Africa 56 E2 34 14S 18 26 E
Simplonpass, Switz. 18 C8 46 15N 8 3 E
Simpson Desert, Australia 62 D2 25 0S 137 0 E
Simpson Pen., Canada 69 B11 68 34N 88 45W
Simpungdong, N. Korea 35 D15 40 56N 129 29 E
Simrishamn, Sweden 9 J16 55 33N 14 22 E
Simsbury, U.S.A. 79 E12 41 53N 72 48W
Simushir, Ostrov, Russia 27 E16 46 50N 152 30 E
Sin Cowe I., S. China Sea 36 C4 9 53N 114 19 E
Sinabang, Indonesia 36 D1 2 30N 96 24 E
Sinadogo, Somali Rep. 46 F4 5 50N 47 0 E
Sinai = Es Sînâ', Egypt 47 F3 29 0N 34 0 E
Sinai, Mt. = Mûsa, Gebel, Egypt 44 D2 28 33N 33 59 E
Sinai Peninsula, Egypt 47 F3 29 30N 34 0 E
Sinaloa □, Mexico 86 C3 25 0N 107 30W
Sinaloa de Leyva, Mexico 86 B3 25 50N 108 20W
Sinarádhes, Greece 23 A3 39 34N 19 51 E
Sincelejo, Colombia 92 B3 9 18N 75 24W
Sinch'ang, N. Korea 35 D15 40 7N 128 28 E
Sinchang-ni, N. Korea 35 E14 39 24N 126 8 E
Sinclair, U.S.A. 82 F10 41 47N 107 7W
Sinclair Mills, Canada 72 C4 54 5N 121 40W
Sinclair's B., U.K. 12 C5 58 31N 3 5W
Sinclairville, U.S.A. 78 D5 42 16N 79 16W
Sincorá, Serra do, Brazil 93 F10 13 30S 41 0W
Sind, Pakistan 42 G3 26 0N 68 30 E
Sind □, Pakistan 42 G3 26 0N 69 0 E
Sind →, Jammu & Kashmir, India 43 B6 34 18N 74 45 E
Sind →, Mad. P., India 43 F8 26 26N 79 13 E
Sind Sagar Doab, Pakistan 42 D4 32 0N 71 30 E
Sindangan, Phil. 37 C6 8 10N 123 5 E
Sindangbarang, Indonesia 37 G12 7 27S 107 1 E
Sinde, Zambia 55 F2 17 28S 25 51 E
Sindh = Sind □, Pakistan 42 G3 26 0N 69 0 E
Sindri, India 43 H12 23 45N 86 42 E
Sines, Portugal 19 D1 37 56N 8 51W
Sines, C. de, Portugal 19 D1 37 58N 8 53W
Sineu, Spain 22 B10 39 38N 3 1 E
Sing Buri, Thailand 38 E3 14 53N 100 25 E
Singa, Sudan 51 F12 13 10N 33 57 E
Singapore ■, Asia 39 M4 1 17N 103 51 E
Singapore, Straits of, Asia 39 M5 1 15N 104 0 E
Singaraja, Indonesia 36 F5 8 7S 115 6 E
Singida, Tanzania 54 C3 4 49S 34 48 E
Singida □, Tanzania 54 D3 6 0S 34 30 E
Singitikós Kólpos, Greece 21 D11 40 6N 24 0 E
Singkaling Hkamti, Burma 41 G19 26 0N 95 39 E
Singkang, Indonesia 37 E6 4 8S 120 1 E
Singkawang, Indonesia 36 D3 1 0N 108 57 E
Singkep, Indonesia 36 E2 0 30S 104 25 E
Singleton, Australia 63 E5 32 33S 151 0 E
Singleton, Mt., N. Terr., Australia 60 D5 22 0S 130 46 E
Singleton, Mt., W. Austral., Australia 61 E2 29 27S 117 15 E
Singoli, India 42 G6 25 0N 75 22 E
Singora = Songkhla, Thailand 39 J3 7 13N 100 37 E
Singosan, N. Korea 35 E14 38 52N 127 25 E
Sinhung, N. Korea 35 D14 40 11N 127 34 E
Sinjai, Indonesia 37 F6 5 7S 120 20 E
Sinjār, Iraq 44 B4 36 19N 41 52 E
Sinkat, Sudan 51 E13 18 55N 36 49 E
Sinkiang Uighur = Xinjiang Uygur Zizhiqu □, China 32 C3 42 0N 86 0 E
Sinmak, N. Korea 35 E14 38 25N 126 14 E
Sinnamary, Fr. Guiana 93 B8 5 25N 53 0W
Sinni →, Italy 20 D7 40 8N 16 41 E
Sinop, Turkey 25 F6 42 1N 35 11 E
Sinor, India 42 J5 21 55N 73 20 E
Sinp'o, N. Korea 35 E15 40 0N 128 13 E
Sinsk, Russia 27 C13 61 8N 126 48 E
Sintang, Indonesia 36 D4 0 5N 111 35 E
Sinton, U.S.A. 81 L6 28 2N 97 31W
Sintra, Portugal 19 C1 38 47N 9 25W
Sinŭiju, N. Korea 35 D13 40 5N 124 24 E
Siocon, Phil. 37 C6 7 40N 122 10 E
Siófok, Hungary 17 E10 46 54N 18 3 E
Sion, Switz. 18 C7 46 14N 7 20 E
Sion Mills, U.K. 13 B4 54 48N 7 29W
Sioux City, U.S.A. 80 D6 42 30N 96 24W
Sioux Falls, U.S.A. 80 D6 43 33N 96 44W
Sioux Lookout, Canada 70 B1 50 10N 91 50W
Sioux Narrows, Canada 73 D10 49 25N 94 10W
Siping, China 35 C13 43 8N 124 21 E
Sipiwesk L., Canada 73 B9 55 5N 97 35W
Sipra →, India 42 H6 23 55N 75 28 E
Sipura, Indonesia 36 E1 2 18S 99 40 E
Siquia →, Nic. 88 D3 12 10N 84 20W
Siquijor, Phil. 37 C6 9 12N 123 35 E
Siquirres, Costa Rica 88 D3 10 6N 83 30W
Şīr Banī Yās, U.A.E. 45 E7 24 19N 52 37 E
Sir Edward Pellew Group, Australia 62 B2 15 40S 137 10 E
Sir Graham Moore Is., Australia 60 B4 13 53S 126 34 E
Sir James MacBrien, Mt., Canada 68 B7 62 8N 127 40W
Sira →, Norway 9 G12 58 23N 6 34 E
Siracusa, Italy 20 F6 37 4N 15 17 E
Sirajganj, Bangla. 43 G13 24 25N 89 47 E
Sirathu, India 43 G9 25 39N 81 19 E
Sīrdān, Iran 45 B6 36 39N 49 12 E
Sirdaryo = Syrdarya →, Kazakstan 26 E7 46 3N 61 0 E
Siren, U.S.A. 80 C8 45 47N 92 24W
Sirer, Spain 22 C7 38 56N 1 22 E
Siret →, Romania 17 F14 45 24N 28 1 E
Sirghāyā, Syria 47 B5 33 51N 36 8 E
Sirmaur, India 43 G9 24 51N 81 23 E
Sirohi, India 42 G5 24 52N 72 53 E
Sironj, India 42 G7 24 5N 77 39 E
Síros, Greece 21 F11 37 28N 24 57 E
Sirretta Pk., U.S.A. 85 K8 35 56N 118 19W
Sirri, Iran 45 E7 25 55N 54 32 E
Sirsa, India 42 E6 29 33N 75 4 E
Sirsa →, India 43 F8 26 51N 79 4 E
Sisak, Croatia 16 F9 45 30N 16 21 E
Sisaket, Thailand 38 E5 15 8N 104 23 E
Sishen, S. Africa 56 D3 27 47S 22 59 E
Sishui, Henan, China 34 G7 34 48N 113 15 E
Sishui, Shandong, China 35 G9 35 42N 117 18 E
Sisipuk L., Canada 73 B8 55 45N 101 50W
Sisophon, Cambodia 38 F4 13 38N 102 59 E
Sisseton, U.S.A. 80 C6 45 40N 97 3W
Sīstān, Asia 45 D9 30 50N 61 0 E
Sīstān, Daryācheh-ye, Iran 45 D9 31 0N 61 0 E
Sīstān va Balūchestān □, Iran 45 E9 27 0N 62 0 E
Sisters, U.S.A. 82 D3 44 18N 121 33W
Siswa Bazar, India 43 F10 27 9N 83 46 E
Sitamarhi, India 43 F11 26 37N 85 30 E
Sitampiky, Madag. 57 B8 16 41S 46 6 E
Sitapur, India 43 F9 27 38N 80 45 E
Siteki, Swaziland 57 D5 26 32S 31 58 E
Sitges, Spain 19 B6 41 17N 1 47 E
Sitía, Greece 23 D8 35 13N 26 6 E
Sitka, U.S.A. 72 B1 57 3N 135 20W
Sitoti, Botswana 56 C3 23 15S 23 40 E
Sittang Myit →, Burma 41 L20 17 20N 96 45 E
Sittard, Neths. 15 C5 51 0N 5 52 E
Sittingbourne, U.K. 11 F8 51 21N 0 45 E
Sittoung = Sittang Myit →, Burma 41 L20 17 20N 96 45 E
Sittwe, Burma 41 J18 20 18N 92 45 E
Situbondo, Indonesia 37 G16 7 42S 114 0 E
Siuna, Nic. 88 D3 13 37N 84 45W
Siuri, India 43 H12 23 50N 87 34 E
Sīvand, Iran 45 D7 30 5N 52 55 E
Sivas, Turkey 25 G6 39 43N 36 58 E
Siverek, Turkey 44 B3 37 50N 39 19 E
Sivomaskinskiy, Russia 24 A11 66 40N 62 35 E
Sivrihisar, Turkey 25 G5 39 30N 31 35 E
Sîwa, Egypt 51 C11 29 11N 25 31 E
Siwa, El Wâhât es, Egypt 48 D6 29 10N 25 30 E
Siwa Oasis = Siwa, El Wâhât es, Egypt 48 D6 29 10N 25 30 E
Siwalik Range, Nepal 43 F10 28 0N 83 0 E
Siwan, India 43 F11 26 13N 84 21 E
Siwana, India 42 G5 25 38N 72 25 E
Sixmilebridge, Ireland 13 D3 52 44N 8 46W
Sixth Cataract, Sudan 51 E12 16 20N 32 42 E
Siziwang Qi, China 34 D6 41 25N 111 40 E
Sjælland, Denmark 9 J14 55 30N 11 30 E
Sjumen = Shumen, Bulgaria 21 C12 43 18N 26 55 E
Skadarsko Jezero, Montenegro, Yug. 21 C8 42 10N 19 20 E
Skaftafell, Iceland 8 D5 64 1N 17 0W
Skagafjörður, Iceland 8 D4 65 54N 19 35W
Skagastølstindane, Norway 9 F12 61 28N 7 52 E
Skagaströnd, Iceland 8 D3 65 50N 20 19W
Skagen, Denmark 9 H14 57 43N 10 35 E
Skagerrak, Denmark 9 H13 57 30N 9 0 E
Skagit →, U.S.A. 84 B4 48 23N 122 22W
Skagway, U.S.A. 68 C6 59 28N 135 19W
Skala-Podilska, Ukraine 17 D14 48 50N 26 15 E
Skala Podolskaya = Skala-Podilska, Ukraine 17 D14 48 50N 26 15 E
Skalat, Ukraine 17 D13 49 23N 25 55 E
Skåne, Sweden 9 J15 55 59N 13 30 E
Skaneateles, U.S.A. 79 D8 42 57N 76 26W
Skaneateles L., U.S.A. 79 D8 42 51N 76 22W
Skara, Sweden 9 G15 58 25N 13 30 E
Skardu, Pakistan 43 B6 35 20N 75 44 E
Skarżysko-Kamienna, Poland 17 C11 51 7N 20 52 E
Skeena →, Canada 72 C2 54 9N 130 5W
Skeena Mts., Canada 72 B3 56 40N 128 30W
Skegness, U.K. 10 D8 53 9N 0 20 E
Skeldon, Guyana 92 B7 5 55N 57 20W
Skellefte älv →, Sweden 8 D19 64 45N 21 10 E
Skellefteå, Sweden 8 D19 64 45N 20 50 E
Skelleftehamn, Sweden 8 D19 64 40N 21 9 E
Skerries, The, U.K. 10 D3 53 25N 4 36W
Ski, Norway 9 G14 59 43N 10 52 E
Skíathos, Greece 21 E10 39 12N 23 30 E
Skibbereen, Ireland 13 E2 51 33N 9 16W
Skiddaw, U.K. 10 C4 54 39N 3 9W
Skidegate, Canada 72 C2 53 15N 132 1W
Skien, Norway 9 G13 59 12N 9 35 E
Skierniewice, Poland 17 C11 51 58N 20 10 E
Skikda, Algeria 50 A7 36 50N 6 58 E
Skilloura, Cyprus 23 D12 35 14N 33 10 E
Skipton, U.K. 10 D5 53 58N 2 3W
Skirmish Pt., Australia 62 A1 11 59S 134 17 E
Skíros, Greece 21 E11 38 55N 24 34 E
Skive, Denmark 9 H13 56 33N 9 2 E
Skjálfandafljót →, Iceland 8 D5 65 59N 17 25W
Skjálfandi, Iceland 8 C5 66 5N 17 30W
Skoghall, Sweden 9 G15 59 20N 13 30 E
Skole, Ukraine 17 D12 49 3N 23 30 E
Skópelos, Greece 21 E10 39 9N 23 47 E
Skopí, Greece 23 D8 35 11N 26 2 E
Skopje, Macedonia 21 C9 42 1N 21 26 E
Skövde, Sweden 9 G15 58 24N 13 50 E
Skovorodino, Russia 27 D13 54 0N 124 0 E
Skowhegan, U.S.A. 77 C11 44 46N 69 43W
Skull, Ireland 13 E2 51 32N 9 34W
Skunk →, U.S.A. 80 E9 40 42N 91 7W
Skuodas, Lithuania 9 H19 56 16N 21 33 E
Skvyra, Ukraine 17 D15 49 44N 29 40 E
Skye, U.K. 12 D2 57 15N 6 10W
Skykomish, U.S.A. 82 C3 47 42N 121 22W
Skyros = Skíros, Greece 21 E11 38 55N 24 34 E
Slættaratindur, Færoe Is. 8 E9 62 18N 7 1W
Slagelse, Denmark 9 J14 55 23N 11 19 E
Slamet, Indonesia 37 G13 7 16S 109 8 E
Slaney →, Ireland 13 D5 52 26N 6 33W
Slangberge, S. Africa 56 E3 31 32S 20 48 E
Śląsk, Poland 16 C9 51 0N 16 30 E
Slate Is., Canada 70 C2 48 40N 87 0W
Slatina, Romania 17 F13 44 28N 24 22 E
Slatington, U.S.A. 79 F9 40 45N 75 37W
Slaton, U.S.A. 81 J4 33 26N 101 39W
Slave →, Canada 72 A6 61 18N 113 39W
Slave Coast, W. Afr. 50 G6 6 0N 2 30 E
Slave Lake, Canada 72 B6 55 17N 114 43W
Slave Pt., Canada 72 A5 61 11N 115 56W
Slavgorod, Russia 26 D8 53 1N 78 37 E
Slavonski Brod, Croatia 21 B8 45 11N 18 1 E
Slavuta, Ukraine 17 C14 50 15N 27 2 E
Slavyanka, Russia 30 C5 42 53N 131 21 E
Slavyansk = Slovyansk, Ukraine 25 E6 48 55N 37 36 E
Slawharad, Belarus 17 B16 53 27N 31 0 E
Sleaford, U.K. 10 D7 53 0N 0 24W
Sleaford B., Australia 63 E2 34 55S 135 45 E
Sleat, Sd. of, U.K. 12 D3 57 5N 5 47W
Sleeper Is., Canada 69 C11 58 30N 81 0W
Sleepy Eye, U.S.A. 80 C7 44 18N 94 43W
Slemon L., Canada 72 A5 63 13N 116 4W
Slide Mt., U.S.A. 79 E10 42 0N 74 25W
Slidell, U.S.A. 81 K10 30 17N 89 47W
Sliema, Malta 23 D2 35 55N 14 30 E
Slieve Aughty, Ireland 13 C3 53 4N 8 30W
Slieve Bloom, Ireland 13 C4 53 4N 7 40W
Slieve Donard, U.K. 13 B6 54 11N 5 55W
Slieve Gamph, Ireland 13 B3 54 6N 9 0W
Slieve Gullion, U.K. 13 B5 54 7N 6 26W
Slieve Mish, Ireland 13 D2 52 12N 9 50W
Slievenamon, Ireland 13 D4 52 25N 7 34W
Sligeach = Sligo, Ireland 13 B3 54 16N 8 28W
Sligo, Ireland 13 B3 54 16N 8 28W
Sligo, U.S.A. 78 E5 41 6N 79 29W
Sligo □, Ireland 13 B3 54 8N 8 42W
Sligo B., Ireland 13 B3 54 18N 8 40W
Slippery Rock, U.S.A. 78 E4 41 3N 80 3W
Slite, Sweden 9 H18 57 42N 18 48 E
Sliven, Bulgaria 21 C12 42 42N 26 19 E
Sloan, U.S.A. 85 K11 35 57N 115 13W
Sloansville, U.S.A. 79 D10 42 45N 74 22W
Slobodskoy, Russia 24 C9 58 40N 50 6 E
Slobozia, Romania 17 F14 44 34N 27 23 E
Slocan, Canada 72 D5 49 48N 117 28W
Slonim, Belarus 17 B13 53 4N 25 19 E
Slough, U.K. 11 F7 51 30N 0 36W
Slough □, U.K. 11 F7 51 30N 0 36W
Sloughhouse, U.S.A. 84 G5 38 26N 121 12W
Slovak Rep. ■, Europe 17 D10 48 30N 20 0 E
Slovakia = Slovak Rep. ■, Europe 17 D10 48 30N 20 0 E
Slovakian Ore Mts. = Slovenské Rudohorie, Slovak Rep. 17 D10 48 45N 20 0 E
Slovenia ■, Europe 16 F8 45 58N 14 30 E
Slovenija = Slovenia ■, Europe 16 F8 45 58N 14 30 E
Slovenské Rudohorie, Slovak Rep. 17 D10 48 45N 20 0 E
Slovyansk, Ukraine 25 E6 48 55N 37 36 E
Sluch →, Ukraine 17 C14 51 37N 26 38 E
Sluis, Neths. 15 C3 51 18N 3 23 E
Słupsk, Poland 17 A9 54 30N 17 3 E
Slurry, S. Africa 56 D4 25 49S 25 42 E
Slutsk, Belarus 17 B14 53 2N 27 31 E
Slyne Hd., Ireland 13 C1 53 25N 10 10W
Slyudyanka, Russia 27 D11 51 40N 103 40 E
Småland, Sweden 9 H16 57 15N 15 25 E
Smalltree L., Canada 73 A8 61 0N 105 0W
Smallwood Res., Canada 71 B7 54 0N 64 0W
Smarhon, Belarus 17 A14 54 20N 26 24 E
Smartt Syndicate Dam, S. Africa 56 E3 30 45S 23 10 E
Smartville, U.S.A. 84 F5 39 13N 121 18W
Smeaton, Canada 73 C8 53 30N 104 49W
Smederevo, Serbia, Yug. 21 B9 44 40N 20 57 E
Smerwick Harbour, Ireland 13 D1 52 12N 10 23W
Smethport, U.S.A. 78 E6 41 49N 78 27W
Smidovich, Russia 27 E14 48 36N 133 49 E
Smith, Canada 72 B6 55 10N 114 0W
Smith Center, U.S.A. 80 F5 39 47N 98 47W
Smith Sund, Greenland 4 B4 78 30N 74 0W
Smithburne →, Australia 62 B3 17 3S 140 57 E
Smithers, Canada 72 C3 54 45N 127 10W
Smithfield, S. Africa 57 E4 30 9S 26 30 E
Smithfield, N.C., U.S.A. 77 H6 35 31N 78 21W
Smithfield, Utah, U.S.A. 82 F8 41 50N 111 50W
Smiths Falls, Canada 79 B9 44 55N 76 0W
Smithton, Australia 62 G4 40 53S 145 6 E
Smithville, Canada 78 C5 43 6N 79 33W
Smithville, U.S.A. 81 K6 30 1N 97 10W
Smoky →, Canada 72 B5 56 10N 117 21W
Smoky Bay, Australia 63 E1 32 22S 134 13 E
Smoky Hill →, U.S.A. 80 F6 39 4N 96 48W
Smoky Hills, U.S.A. 80 F5 39 15N 99 30W
Smoky Lake, Canada 72 C6 54 10N 112 30W
Smøla, Norway 8 E13 63 23N 8 3 E
Smolensk, Russia 24 D5 54 45N 32 5 E
Smolikas, Óros, Greece 21 D9 40 9N 20 58 E
Smolyan, Bulgaria 21 D11 41 36N 24 38 E
Smooth Rock Falls, Canada 70 C3 49 17N 81 37W
Smoothstone L., Canada 73 C7 54 40N 106 50W
Smorgon = Smarhon, Belarus 17 A14 54 20N 26 24 E
Smyrna = İzmir, Turkey 21 E12 38 25N 27 8 E
Smyrna, U.S.A. 76 F8 39 18N 75 36W
Snæfell, Iceland 8 D6 64 48N 15 34W
Snaefell, U.K. 10 C3 54 16N 4 27W
Snæfellsjökull, Iceland 8 D2 64 49N 23 46W
Snake →, U.S.A. 82 C4 46 12N 119 2W
Snake I., Australia 63 F4 38 47S 146 33 E
Snake Range, U.S.A. 82 G6 39 0N 114 20W
Snake River Plain, U.S.A. 82 E7 42 50N 114 0W
Snåsavatnet, Norway 8 D14 64 12N 12 0 E
Sneek, Neths. 15 A5 53 2N 5 40 E
Sneeuberge, S. Africa 56 E3 31 46S 24 20 E
Snelling, U.S.A. 84 H6 37 31N 120 26W
Sněžka, Europe 16 C8 50 41N 15 50 E
Snizort, L., U.K. 12 D2 57 33N 6 28W
Snøhetta, Norway 9 E13 62 19N 9 16 E
Snohomish, U.S.A. 84 C4 47 55N 122 6W
Snoul, Cambodia 39 F6 12 4N 106 26 E
Snow Hill, U.S.A. 76 F8 38 11N 75 24W
Snow Lake, Canada 73 C8 54 52N 100 3W
Snow Mt., Calif., U.S.A. 84 F4 39 23N 122 45W
Snow Mt., Maine, U.S.A. 79 A14 45 18N 70 48W
Snow Shoe, U.S.A. 78 E7 41 2N 77 57W
Snowbird L., Canada 73 A8 60 45N 103 0W
Snowdon, U.K. 10 D3 53 4N 4 5W
Snowdrift →, Canada 73 A6 62 24N 110 44W
Snowflake, U.S.A. 83 J8 34 30N 110 5W
Snowshoe Pk., U.S.A. 82 B6 48 13N 115 41W
Snowtown, Australia 63 E2 33 46S 138 14 E
Snowville, U.S.A. 82 F7 41 58N 112 43W
Snowy →, Australia 63 F4 37 46S 148 30 E
Snowy Mt., U.S.A. 79 C10 43 42N 74 23W
Snowy Mts., Australia 63 F4 36 30S 148 20 E
Snug Corner, Bahamas 89 B5 22 33N 73 52W
Snyatyn, Ukraine 17 D13 48 27N 25 38 E
Snyder, Okla., U.S.A. 81 H5 34 40N 98 57W
Snyder, Tex., U.S.A. 81 J4 32 44N 100 55W
Soahanina, Madag. 57 B7 18 42S 44 13 E
Soalala, Madag. 57 B8 16 6S 45 20 E
Soaloka, Madag. 57 B8 18 32S 45 15 E
Soamanonga, Madag. 57 C7 23 52S 44 47 E
Soan →, Pakistan 42 C4 33 1N 71 44 E
Soanierana-Ivongo, Madag. 57 B8 16 55S 49 35 E
Soanindraniny, Madag. 57 B8 19 54S 47 14 E
Soavina, Madag. 57 C8 20 23S 46 56 E
Soavinandriana, Madag. 57 B8 19 9S 46 45 E
Sobat, Nahr →, Sudan 51 G12 9 22N 31 33 E
Sobhapur, India 42 H8 22 47N 78 17 E
Sobradinho, Reprêsa de, Brazil 93 E10 9 30S 42 0 E
Sobral, Brazil 93 D10 3 50S 40 20W
Soc Giang, Vietnam 38 A6 22 54N 106 1 E
Soc Trang, Vietnam 39 H5 9 37N 105 50 E
Socastee, U.S.A. 77 J6 33 41N 79 1W
Soch'e = Shache, China 32 C2 38 20N 77 10 E
Sochi, Russia 25 F6 43 35N 39 40 E
Société, Is. de la, Pac. Oc. 65 J12 17 0S 151 0W
Society Is. = Société, Is. de la, Pac. Oc. 65 J12 17 0S 151 0W
Socompa, Portezuelo de, Chile 94 A2 24 27S 68 18W
Socorro, N. Mex., U.S.A. 83 J10 34 4N 106 54W
Socorro, Tex., U.S.A. 83 L10 31 39N 106 18W
Socorro, I., Mexico 86 D2 18 45N 110 58W
Socotra, Yemen 46 E5 12 30N 54 0 E
Soda L., U.S.A. 83 J5 35 10N 116 4W
Soda Plains, India 43 B8 35 30N 79 0 E
Soda Springs, U.S.A. 82 E8 42 39N 111 36W
Sodankylä, Finland 8 C22 67 29N 26 40 E
Soddy-Daisy, U.S.A. 77 H3 35 17N 85 10W
Söderhamn, Sweden 9 F17 61 18N 17 10 E

Söderköping

Söderköping, Sweden 9 G17 58 31N 16 20 E
Södermanland, Sweden 9 G17 58 56N 16 55 E
Södertälje, Sweden 9 G17 59 12N 17 39 E
Sodiri, Sudan 51 F11 14 27N 29 0 E
Sodus, U.S.A. 78 C7 43 14N 77 4W
Soekmekaar, S. Africa 57 C4 23 30S 29 55 E
Soest, Neths. 15 B5 52 9N 5 19 E
Sofala □, Mozam. 57 B5 19 30S 34 30 E
Sofia = Sofiya, Bulgaria 21 C10 42 45N 23 20 E
Sofia →, Madag. 57 B8 15 27S 47 23 E
Sofiya, Bulgaria 21 C10 42 45N 23 20 E
Sōfu-Gan, Japan 31 K10 29 49N 140 21 E
Sogamoso, Colombia 92 B4 5 43N 72 56W
Sogār, Iran 45 E8 25 53N 58 6 E
Sogndalsfjøra, Norway 9 F12 61 14N 7 5 E
Søgne, Norway 9 G12 58 5N 7 48 E
Sognefjorden, Norway 9 F11 61 10N 5 50 E
Sŏgwipo, S. Korea 35 H14 33 13N 126 34 E
Soh, Iran 45 C6 33 26N 51 27 E
Sohâg, Egypt 51 C12 26 33N 31 43 E
Sohagpur, India 42 H8 22 42N 78 12 E
Sŏhori, N. Korea 35 D15 40 7N 128 23 E
Soignies, Belgium 15 D4 50 35N 4 5 E
Soissons, France 18 B5 49 25N 3 19 E
Sōja, Japan 31 G6 34 40N 133 45 E
Sojat, India 42 G5 25 55N 73 45 E
Sokal, Ukraine 17 C13 50 31N 24 15 E
Söke, Turkey 21 F12 37 48N 27 28 E
Sokelo, Dem. Rep. of the Congo 55 D1 9 55S 24 36 E
Sokhumi, Georgia 25 F7 43 0N 41 0 E
Sokodé, Togo 50 G6 9 0N 1 11 E
Sokol, Russia 24 C7 59 30N 40 5 E
Sokółka, Poland 17 B12 53 25N 23 30 E
Sokołów Podlaski, Poland 17 B12 52 25N 22 15 E
Sokoto, Nigeria 50 F7 13 2N 5 16 E
Sol Iletsk, Russia 24 D10 51 10N 55 0 E
Solai, Kenya 54 B4 0 2N 36 12 E
Solan, India 42 D7 30 55N 77 7 E
Solano, Phil. 37 A6 16 31N 121 15 E
Solapur, India 40 L9 17 43N 75 56 E
Soldotna, U.S.A. 68 B4 60 29N 151 3W
Soléa, Cyprus 23 D12 35 5N 33 4 E
Soledad, Colombia 92 A4 10 55N 74 46W
Soledad, U.S.A. 84 J5 36 26N 121 20W
Soledad, Venezuela 92 B6 8 10N 63 34W
Solent, The, U.K. 11 G6 50 45N 1 25W
Solfonn, Norway 9 F12 60 2N 6 57 E
Solhan, Turkey 44 B4 38 57N 41 3 E
Soligalich, Russia 24 C7 59 5N 42 10 E
Soligorsk = Salihorsk, Belarus 17 B14 52 51N 27 27 E
Solihull, U.K. 11 E6 52 26N 1 47W
Solikamsk, Russia 24 C10 59 38N 56 50 E
Solila, Madag. 57 C8 21 25S 46 37 E
Solimões = Amazonas →, S. Amer. 93 D9 0 5S 50 0W
Solingen, Germany 16 C4 51 10N 7 5 E
Sollefteå, Sweden 8 E17 63 12N 17 20 E
Sóller, Spain 22 B9 39 46N 2 43 E
Solo →, Indonesia 37 G15 6 47S 112 22 E
Sologne, France 18 C4 47 40N 1 45 E
Solok, Indonesia 36 E2 0 45S 100 40 E
Sololá, Guatemala 88 D1 14 49N 91 10W
Solomon, N. Fork →, U.S.A. 80 F5 39 29N 98 26W
Solomon, S. Fork →, U.S.A. 80 F5 39 25N 99 12W
Solomon Is. ■, Pac. Oc. 64 H7 6 0S 155 0 E
Solon, China 33 B7 46 32N 121 10 E
Solon Springs, U.S.A. 80 B9 46 22N 91 49W
Solor, Indonesia 37 F6 8 27S 123 0 E
Solothurn, Switz. 18 C7 47 13N 7 32 E
Šolta, Croatia 20 C7 43 24N 16 15 E
Solţānābād, Khorāsān, Iran 45 C8 34 13N 59 58 E
Solţānābād, Khorāsān, Iran 45 B8 36 29N 58 5 E
Solunska Glava, Macedonia 21 D9 41 44N 21 31 E
Solvang, U.S.A. 85 L6 34 36N 120 8W
Solvay, U.S.A. 79 C8 43 3N 76 13W
Sölvesborg, Sweden 9 H16 56 5N 14 35 E
Solvychegodsk, Russia 24 B8 61 21N 46 56 E
Solway Firth, U.K. 10 C4 54 49N 3 35W
Solwezi, Zambia 55 E2 12 11S 26 21 E
Sōma, Japan 30 F10 37 40N 140 50 E
Soma, Turkey 21 E12 39 10N 27 35 E
Somabhula, Zimbabwe 57 B4 19 42S 29 40 E
Somali Pen., Africa 48 F8 7 0N 46 0 E
Somali Rep. ■, Africa 46 F4 7 0N 47 0 E
Somalia = Somali Rep. ■, Africa 46 F4 7 0N 47 0 E
Sombor, Serbia, Yug. 21 B8 45 46N 19 9 E
Sombra, Canada 78 D2 42 43N 82 29W
Sombrerete, Mexico 86 C4 23 40N 103 40W
Sombrero, Anguilla 89 C7 18 37N 63 30W
Somdari, India 42 G5 25 47N 72 38 E
Somers, U.S.A. 82 B6 48 5N 114 13W
Somerset, Ky., U.S.A. 76 G3 37 5N 84 36W
Somerset, Mass., U.S.A. 79 E13 41 47N 71 8W
Somerset, Pa., U.S.A. 78 F5 40 1N 79 5W
Somerset □, U.K. 11 F5 51 9N 3 0W
Somerset East, S. Africa 56 E4 32 42S 25 35 E
Somerset I., Canada 68 A10 73 30N 93 0W
Somerset West, S. Africa 56 E2 34 8S 18 50 E
Somersworth, U.S.A. 79 C14 43 16N 70 52W
Somerton, U.S.A. 83 K6 32 36N 114 43W
Somerville, U.S.A. 79 F10 40 35N 74 38W
Someş →, Romania 17 D12 47 49N 22 43 E
Somme →, France 18 A4 50 11N 1 38 E
Somnath, India 42 J4 20 53N 70 22 E
Somosierra, Puerto de, Spain 19 B4 41 4N 3 35W
Somoto, Nic. 88 D2 13 28N 86 37W
Somport, Puerto de, Spain 18 E3 42 48N 0 31W
Son →, India 43 G11 25 42N 84 52 E
Son Ha, Vietnam 38 E7 15 3N 108 34 E
Son Hoa, Vietnam 38 F7 13 2N 108 58 E
Son La, Vietnam 38 B4 21 20N 103 50 E
Son Serra, Spain 22 B10 39 43N 3 13 E
Son Tay, Vietnam 38 B5 21 8N 105 30 E
Soná, Panama 88 E3 8 0N 81 20W
Sonamarg, India 43 B6 34 18N 75 21 E
Sonamukhi, India 43 H12 23 18N 87 27 E
Sonar →, India 43 G8 24 24N 79 56 E
Sŏnch'ŏn, N. Korea 35 E13 39 48N 124 55 E
Sondags →, S. Africa 56 E4 33 44S 25 51 E
Sondar, India 43 C6 33 28N 75 56 E
Sønderborg, Denmark 9 J13 54 55N 9 49 E
Sóndrio, Italy 18 C8 46 10N 9 52 E
Sone, Mozam. 55 F3 17 23S 34 55 E
Sonepur, India 41 J13 20 55N 83 50 E
Song, Thailand 38 C3 18 28N 100 11 E
Song Cau, Vietnam 38 F7 13 27N 109 18 E
Song Xian, China 34 G7 34 12N 112 8 E
Songch'ŏn, N. Korea 35 E14 39 12N 126 15 E
Songea, Tanzania 55 E4 10 40S 35 40 E
Songhua Hu, China 35 C14 43 35N 126 50 E
Songhua Jiang →, China 33 B8 47 45N 132 30 E
Songjin, N. Korea 35 D15 40 40N 129 10 E
Songjŏng-ni, S. Korea 35 G14 35 8N 126 47 E
Songkhla, Thailand 39 J3 7 13N 100 37 E
Songnim, N. Korea 35 E13 38 45N 125 39 E
Songo, Mozam. 53 H6 15 34S 32 38 E
Songo, Sudan 51 G10 9 47N 24 21 E
Songpan, China 32 C5 32 40N 103 30 E
Songwe, Dem. Rep. of the Congo 54 C2 3 20S 26 16 E
Songwe →, Africa 55 D3 9 44S 33 58 E
Sonhat, India 43 H10 23 29N 82 31 E
Sonid Youqi, China 34 C7 42 45N 112 48 E
Sonipat, India 42 E7 29 0N 77 5 E
Sonkach, India 42 H7 22 59N 76 21 E
Sonmiani, Pakistan 42 G2 25 25N 66 40 E
Sonmiani B., Pakistan 42 G2 25 15N 66 30 E
Sono →, Brazil 93 E9 9 58S 48 11W
Sonoma, U.S.A. 84 G4 38 18N 122 28W
Sonora, Calif., U.S.A. 84 H6 37 59N 120 23W
Sonora, Tex., U.S.A. 81 K4 30 34N 100 39W
Sonora □, Mexico 86 B2 29 0N 111 0W
Sonora →, Mexico 86 B2 28 50N 111 33W
Sonoran Desert, U.S.A. 85 L12 33 40N 114 15W
Sonoyta, Mexico 86 A2 31 51N 112 50W
Sŏnsan, S. Korea 35 F15 36 14N 128 17 E
Sonsonate, El Salv. 88 D2 13 43N 89 44W
Soochow = Suzhou, China 33 C7 31 19N 120 38 E
Sooke, Canada 84 B3 48 13N 123 43W
Sop Hao, Laos 38 B5 20 33N 104 27 E
Sop Prap, Thailand 38 D2 17 53N 99 20 E
Sopi, Indonesia 37 D7 2 34N 128 28 E
Sopot, Poland 17 A10 54 27N 18 31 E
Sopron, Hungary 17 E9 47 45N 16 32 E
Sopur, India 43 B6 34 18N 74 27 E
Sør-Rondane, Antarctica 5 D4 72 0S 25 0 E
Sorah, Pakistan 42 F3 27 13N 68 56 E
Soraon, India 43 G9 25 37N 81 51 E
Sorel, Canada 70 C5 46 0N 73 10W
Sórgono, Italy 20 D3 40 1N 9 6 E
Soria, Spain 19 B4 41 43N 2 32W
Soriano, Uruguay 94 C4 33 24S 58 19W
Sorkh, Kuh-e, Iran 45 C8 35 40N 58 30 E
Soroca, Moldova 17 D15 48 8N 28 12 E
Sorocaba, Brazil 95 A6 23 31S 47 27W
Sorochinsk, Russia 24 D9 52 26N 53 10 E
Soroki = Soroca, Moldova 17 D15 48 8N 28 12 E
Sorong, Indonesia 37 E8 0 55S 131 15 E
Soroní, Greece 23 C10 36 21N 28 1 E
Soroti, Uganda 54 B3 1 43N 33 35 E
Sørøya, Norway 8 A20 70 40N 22 30 E
Sørøysundet, Norway 8 A20 70 25N 23 0 E
Sorrell, Australia 62 G4 42 47S 147 34 E
Sorsele, Sweden 8 D17 65 31N 17 30 E
Sorsogon, Phil. 37 B6 13 0N 124 0 E
Sortavala, Russia 24 B5 61 42N 30 41 E
Sortland, Norway 8 B16 68 42N 15 25 E
Sŏsan, S. Korea 35 F14 36 47N 126 27 E
Soscumica, L., Canada 70 B4 50 15N 77 27W
Sosnogorsk, Russia 24 B9 63 37N 53 51 E
Sosnowiec, Poland 17 C10 50 20N 19 10 E
Sossus Vlei, Namibia 56 C2 24 40S 15 23 E
Sŏsura, N. Korea 35 C16 42 16N 130 36 E
Sot →, India 43 F8 27 27N 79 37 E
Sotkamo, Finland 8 D23 64 8N 28 23 E
Soto la Marina →, Mexico 87 C5 23 40N 97 40W
Sotuta, Mexico 87 C7 20 29N 89 43W
Souanké, Congo 52 D2 2 10N 14 3 E
Souderton, U.S.A. 79 F9 40 19N 75 19W
Soúdha, Greece 23 D6 35 29N 24 4 E
Soúdhas, Kólpos, Greece 23 D6 35 25N 24 10 E
Soufrière, St. Lucia 89 D7 13 51N 61 3W
Soukhouma, Laos 38 E5 14 38N 105 48 E
Sŏul, S. Korea 35 F14 37 31N 126 58 E
Sound, The, U.K. 11 G3 50 20N 4 10W
Sources, Mt. aux, Lesotho 57 D4 28 45S 28 50 E
Soure, Brazil 93 D9 0 35S 48 30W
Souris, Man., Canada 73 D8 49 40N 100 20W
Souris, P.E.I., Canada 71 C7 46 21N 62 15W
Souris →, Canada 80 A5 49 40N 99 34W
Sousa, Brazil 93 E11 6 45S 38 10W
Sousse, Tunisia 51 A8 35 50N 10 38 E
Sout →, S. Africa 56 E2 31 35S 18 24 E
South Africa ■, Africa 56 E3 32 0S 23 0 E
South America 90 E5 10 0S 60 0W
South Atlantic Ocean 90 H7 20 0S 10 0W
South Aulatsivik I., Canada 71 A7 56 45N 61 30W
South Australia □, Australia 63 E2 32 0S 139 0 E
South Ayrshire □, U.K. 12 F4 55 18N 4 41W
South Baldy, U.S.A. 83 J10 33 59N 107 11W
South Bass I., U.S.A. 78 E2 41 38N 82 53W
South Bend, Ind., U.S.A. 76 E2 41 41N 86 15W
South Bend, Wash., U.S.A. 84 D3 46 40N 123 48W
South Boston, U.S.A. 77 G6 36 42N 78 54W
South Branch, Canada 71 C8 47 55N 59 2W
South Brook, Canada 71 C8 49 26N 56 5W
South Carolina □, U.S.A. 77 J5 34 0N 81 0W
South Charleston, U.S.A. 76 F5 38 22N 81 44W
South China Sea, Asia 36 C4 10 0N 113 0 E
South Dakota □, U.S.A. 80 C5 44 15N 100 0W
South Deerfield, U.S.A. 79 D12 42 29N 72 37W
South Downs, U.K. 11 G7 50 52N 0 25W
South East C., Australia 62 G4 43 40S 146 50 E
South East Is., Australia 61 F3 34 17S 123 30 E
South Esk →, U.K. 12 E6 56 43N 2 31W
South Foreland, U.K. 11 F9 51 8N 1 24 E
South Fork American →, U.S.A. 84 G5 38 45N 121 5W
South Fork Feather →, U.S.A. 84 F5 39 17N 121 36W
South Fork Grand →, U.S.A. 80 C3 45 43N 102 17W
South Fork Republican →, U.S.A. 80 E4 40 3N 101 31W
South Georgia, Antarctica 96 G9 54 30S 37 0W
South Gloucestershire □, U.K. 11 F5 51 32N 2 28W
South Hadley, U.S.A. 79 D12 42 16N 72 35W
South Haven, U.S.A. 76 D2 42 24N 86 16W
South Henik, L., Canada 73 A9 61 30N 97 30W
South Honshu Ridge, Pac. Oc. 64 E6 23 0N 143 0 E
South Horr, Kenya 54 B4 2 12N 36 56 E
South I., Kenya 54 B4 2 35N 36 35 E
South I., N.Z. 59 L3 44 0S 170 0 E
South Indian Lake, Canada 73 B9 56 47N 98 56W
South Invercargill, N.Z. 59 M2 46 26S 168 23 E
South Knife →, Canada 73 B10 58 55N 94 37W
South Koel →, India 43 H11 22 32N 85 14 E
South Korea ■, Asia 35 G15 36 0N 128 0 E
South Lake Tahoe, U.S.A. 84 G6 38 57N 119 59W
South Lanarkshire □, U.K. 12 F5 55 37N 3 53W
South Loup →, U.S.A. 80 E5 41 4N 98 39W
South Magnetic Pole, Antarctica 5 C9 64 8S 138 8 E
South Milwaukee, U.S.A. 76 D2 42 55N 87 52W
South Molton, U.K. 11 F4 51 1N 3 51W
South Moose L., Canada 73 C8 53 46N 100 8W
South Nahanni →, Canada 72 A4 61 3N 123 21W
South Nation →, Canada 79 A9 45 34N 75 6W
South Natuna Is. = Natuna Selatan, Kepulauan, Indonesia 39 L7 2 45N 109 0 E
South Negril Pt., Jamaica 88 C4 18 14N 78 30W
South Orkney Is., Antarctica 5 C18 63 0S 45 0W
South Ossetia □, Georgia 25 F7 42 21N 44 2 E
South Pagai, I. = Pagai Selatan, Pulau, Indonesia 36 E2 3 0S 100 15 E
South Paris, U.S.A. 79 B14 44 14N 70 31W
South Pittsburg, U.S.A. 77 H3 35 1N 85 42W
South Platte →, U.S.A. 80 E4 41 7N 100 42W
South Pole, Antarctica 5 E 90 0S 0 0W
South Porcupine, Canada 70 C3 48 30N 81 12W
South Portland, U.S.A. 77 D10 43 38N 70 15W
South Pt., U.S.A. 78 B1 44 52N 83 19W
South River, Canada 70 C4 45 52N 79 23W
South River, U.S.A. 79 F10 40 27N 74 23W
South Ronaldsay, U.K. 12 C6 58 48N 2 58W
South Sandwich Is., Antarctica 5 B1 57 0S 27 0W
South Saskatchewan →, Canada 73 C7 53 15N 105 5W
South Seal →, Canada 73 B9 58 48N 98 8W
South Shetland Is., Antarctica 5 C18 62 0S 59 0W
South Shields, U.K. 10 C6 55 0N 1 25W
South Sioux City, U.S.A. 80 D6 42 28N 96 24W
South Taranaki Bight, N.Z. 59 H5 39 40S 174 5 E
South Thompson →, Canada 72 C4 50 40N 120 20W
South Twin I., Canada 70 B4 53 7N 79 52W
South Tyne →, U.K. 10 C5 54 59N 2 8W
South Uist, U.K. 12 D1 57 20N 7 15W
South West Africa = Namibia ■, Africa 56 C2 22 0S 18 9 E
South West C., Australia 62 G4 43 34S 146 3 E
South Williamsport, U.S.A. 78 E8 41 13N 77 0W
South Yorkshire □, U.K. 10 D6 53 27N 1 36W
Southampton, Canada 78 B3 44 30N 81 25W
Southampton, U.K. 11 G6 50 54N 1 23W
Southampton, U.S.A. 79 F12 40 53N 72 23W
Southampton □, U.K. 11 G6 50 54N 1 23W
Southampton I., Canada 69 B11 64 30N 84 0W
Southaven, U.S.A. 81 H9 34 59N 90 2W
Southbank, Canada 72 C3 54 2N 125 46W
Southbridge, N.Z. 59 K4 43 48S 172 16 E
Southbridge, U.S.A. 79 D12 42 5N 72 2W
Southend, Canada 73 B8 56 19N 103 22W
Southend-on-Sea, U.K. 11 F8 51 32N 0 44 E
Southend-on-Sea □, U.K. 11 F8 51 32N 0 44 E
Southern □, Malawi 55 F4 15 0S 35 0 E
Southern □, Zambia 55 F2 16 20S 26 20 E
Southern Alps, N.Z. 59 K3 43 41S 170 11 E
Southern Cross, Australia 61 F2 31 12S 119 15 E
Southern Indian L., Canada 73 B9 57 10N 98 30W
Southern Ocean, Antarctica 5 C6 62 0S 60 0 E
Southern Pines, U.S.A. 77 H6 35 11N 79 24W
Southern Uplands, U.K. 12 F5 55 28N 3 52W
Southington, U.S.A. 79 E12 41 36N 72 53W
Southland □, N.Z. 59 L1 45 30S 168 0 E
Southold, U.S.A. 79 E12 41 4N 72 26W
Southport, Australia 63 D5 27 58S 153 25 E
Southport, U.K. 10 D4 53 39N 3 0W
Southport, Fla., U.S.A. 77 K3 30 17N 85 38W
Southport, N.Y., U.S.A. 78 D8 42 3N 76 49W
Southwest C., N.Z. 59 M1 47 17S 167 28 E
Southwold, U.K. 11 E9 52 20N 1 41 E
Soutpansberg, S. Africa 57 C4 23 0S 29 30 E
Sovetsk, Kaliningd., Russia 9 J19 55 6N 21 50 E
Sovetsk, Kirov, Russia 24 C8 57 38N 48 53 E
Sovetskaya Gavan = Vanino, Russia 27 E15 48 50N 140 5 E
Soweto, S. Africa 57 D4 26 14S 27 54 E
Sōya-Kaikyō = La Perouse Str., Asia 30 B11 45 40N 142 0 E
Sōya-Misaki, Japan 30 B10 45 30N 141 55 E
Sozh →, Belarus 17 B16 51 57N 30 48 E
Spa, Belgium 15 D5 50 29N 5 53 E
Spain ■, Europe 19 B4 39 0N 4 0W
Spalding, Australia 63 E2 33 30S 138 37 E
Spalding, U.K. 10 E7 52 48N 0 9W
Spangler, U.S.A. 78 F6 40 39N 78 48W
Spanish, Canada 70 C3 46 12N 82 20W
Spanish Fork, U.S.A. 82 F8 40 7N 111 39W
Spanish Town, Jamaica 88 C4 18 0N 76 57W
Sparks, U.S.A. 84 F7 39 32N 119 45W
Sparta = Spárti, Greece 21 F10 37 5N 22 25 E
Sparta, Mich., U.S.A. 76 D3 43 10N 85 42W
Sparta, N.J., U.S.A. 79 E10 41 2N 74 38W
Sparta, Wis., U.S.A. 80 D9 43 56N 90 49W
Spartanburg, U.S.A. 77 H5 34 56N 81 57W
Spartansburg, U.S.A. 78 E5 41 49N 79 41W
Spárti, Greece 21 F10 37 5N 22 25 E
Spartivento, C., Calabria, Italy 20 F7 37 55N 16 4 E
Spartivento, C., Sard., Italy 20 E3 38 53N 8 50 E
Sparwood, Canada 72 D6 49 44N 114 53W
Spassk Dalniy, Russia 27 E14 44 40N 132 48 E
Spátha, Ákra, Greece 23 D5 35 42N 23 43 E
Spatsizi →, Canada 72 B3 57 42N 128 7W
Spatsizi Plateau Wilderness Park, Canada 72 B3 57 40N 128 0W
Spean →, U.K. 12 E4 56 55N 4 59W
Spearfish, U.S.A. 80 C3 44 30N 103 52W
Spearman, U.S.A. 81 G4 36 12N 101 12W
Speculator, U.S.A. 79 C10 43 30N 74 25W
Speightstown, Barbados 89 D8 13 15N 59 39W
Speke Gulf, Tanzania 54 C3 2 20S 32 50 E
Spencer, Idaho, U.S.A. 82 D7 44 22N 112 11W
Spencer, Iowa, U.S.A. 80 D7 43 9N 95 9W
Spencer, N.Y., U.S.A. 79 D8 42 13N 76 30W
Spencer, Nebr., U.S.A. 80 D5 42 53N 98 42W
Spencer, C., Australia 63 F2 35 20S 136 53 E
Spencer B., Namibia 56 D1 25 30S 14 47 E
Spencer G., Australia 63 E2 34 0S 137 20 E
Spencerville, Canada 79 B9 44 51N 75 33W
Spences Bridge, Canada 72 C4 50 25N 121 20W
Spennymoor, U.K. 10 C6 54 42N 1 36W
Spenser Mts., N.Z. 59 K4 42 15S 172 45 E
Sperrin Mts., U.K. 13 B5 54 50N 7 0W
Spey →, U.K. 12 D5 57 40N 3 6W
Speyer, Germany 16 D5 49 29N 8 25 E
Spezand, Pakistan 42 E2 29 59N 67 0 E
Spíli, Greece 23 D6 35 13N 24 31 E
Spīn Būldak, Afghan. 42 D2 31 1N 66 25 E
Spinalónga, Greece 23 D7 35 18N 25 44 E
Spirit Lake, U.S.A. 84 D4 46 15N 122 9W
Spirit River, Canada 72 B5 55 45N 118 50W
Spiritwood, Canada 73 C7 53 24N 107 33W
Spithead, U.K. 11 G6 50 45N 1 10W
Spitzbergen = Svalbard, Arctic 4 B8 78 0N 17 0 E
Spjelkavik, Norway 9 E12 62 28N 6 22 E
Split, Croatia 20 C7 43 31N 16 26 E
Split L., Canada 73 B9 56 8N 96 15W
Split Lake, Canada 73 B9 56 8N 96 15W
Spofford, U.S.A. 81 L4 29 10N 100 25W
Spokane, U.S.A. 82 C5 47 40N 117 24W
Spoleto, Italy 20 C5 42 44N 12 44 E
Spooner, U.S.A. 80 C9 45 50N 91 53W
Sporyy Navolok, Mys, Russia 26 B7 75 50N 68 40 E
Sprague, U.S.A. 82 C5 47 18N 117 59W
Spratly I., S. China Sea 36 C4 8 38N 111 55 E
Spratly Is., S. China Sea 36 C4 8 20N 112 0 E
Spray, U.S.A. 82 D4 44 50N 119 48W
Spree →, Germany 16 B7 52 32N 13 13 E
Sprengisandur, Iceland 8 D5 64 52N 18 7W
Spring City, U.S.A. 79 F9 40 11N 75 33W
Spring Creek, U.S.A. 82 F6 40 45N 115 38W
Spring Garden, U.S.A. 84 F6 39 52N 120 47W
Spring Hill, U.S.A. 77 L4 28 27N 82 41W
Spring Mts., U.S.A. 83 H6 36 0N 115 45W
Spring Valley, U.S.A. 85 N10 32 45N 117 5W
Springbok, S. Africa 56 D2 29 42S 17 54 E
Springboro, U.S.A. 78 E4 41 48N 80 22W
Springdale, Canada 71 C8 49 30N 56 6W
Springdale, U.S.A. 81 G7 36 11N 94 8W
Springer, U.S.A. 81 G2 36 22N 104 36W
Springerville, U.S.A. 83 J9 34 8N 109 17W
Springfield, Canada 78 D4 42 50N 80 56W
Springfield, N.Z. 59 K3 43 19S 171 56 E
Springfield, Colo., U.S.A. 81 G3 37 24N 102 37W
Springfield, Ill., U.S.A. 80 F10 39 48N 89 39W
Springfield, Mass., U.S.A. 79 D12 42 6N 72 35W
Springfield, Mo., U.S.A. 81 G8 37 13N 93 17W
Springfield, Ohio, U.S.A. 76 F4 39 55N 83 49W
Springfield, Oreg., U.S.A. 82 D2 44 3N 123 1W
Springfield, Tenn., U.S.A. 77 G2 36 31N 86 53W
Springfield, Vt., U.S.A. 79 C12 43 18N 72 29W
Springfontein, S. Africa 56 E4 30 15S 25 40 E
Springhill, Canada 71 C7 45 40N 64 4W
Springhill, U.S.A. 81 J8 33 0N 93 28W
Springhouse, Canada 72 C4 51 56N 122 7W
Springs, S. Africa 57 D4 26 13S 28 25 E
Springsure, Australia 62 C4 24 8S 148 6 E
Springvale, U.S.A. 79 C14 43 28N 70 48W
Springville, Calif., U.S.A. 84 J8 36 8N 118 49W
Springville, N.Y., U.S.A. 78 D6 42 31N 78 40W
Springville, Utah, U.S.A. 82 F8 40 10N 111 37W
Springwater, U.S.A. 78 D7 42 38N 77 35W
Spruce-Creek, U.S.A. 78 F6 40 36N 78 9W
Spruce Mt., U.S.A. 79 B12 44 12N 72 19W
Spur, U.S.A. 81 J4 33 28N 100 52W
Spurn Hd., U.K. 10 D8 53 35N 0 8 E
Spuzzum, Canada 72 D4 49 37N 121 23W
Squam L., U.S.A. 79 C13 43 45N 71 32W
Squamish, Canada 72 D4 49 45N 123 10W
Square Islands, Canada 71 B8 52 47N 55 47W
Squires, Mt., Australia 61 E4 26 14S 127 28 E
Srbija = Serbia □, Yugoslavia 21 C9 43 30N 21 0 E
Sre Ambel, Cambodia 39 G4 11 8N 103 46 E
Sre Khtum, Cambodia 39 F6 12 10N 106 52 E
Sre Umbell = Sre Ambel, Cambodia 39 G4 11 8N 103 46 E
Srebrenica, Bos.-H. 21 B8 44 6N 19 18 E
Sredinny Ra. = Sredinnyy Khrebet, Russia 27 D16 57 0N 160 0 E
Sredinnyy Khrebet, Russia 27 D16 57 0N 160 0 E
Srednekolymsk, Russia 27 C16 67 27N 153 40 E
Śrem, Poland 17 B9 52 6N 17 2 E
Sremska Mitrovica, Serbia, Yug. 21 B8 44 59N 19 38 E
Srepok →, Cambodia 38 F6 13 33N 106 16 E
Sretensk, Russia 27 D12 52 10N 117 40 E
Sri Lanka ■, Asia 40 R12 7 30N 80 50 E
Srikakulam, India 41 K13 18 14N 83 58 E
Srinagar, India 43 B6 34 5N 74 50 E
Staaten →, Australia 62 B3 16 24S 141 17 E
Stade, Germany 16 B5 53 35N 9 29 E
Stadskanaal, Neths. 15 A6 53 4N 6 55 E
Staffa, U.K. 12 E2 56 27N 6 21W
Stafford, U.K. 10 E5 52 49N 2 7W
Stafford, U.S.A. 81 G5 37 58N 98 36W
Stafford Springs, U.S.A. 79 E12 41 57N 72 18W
Staffordshire □, U.K. 10 E5 52 53N 2 10W
Staines, U.K. 11 F7 51 26N 0 29W
Stakhanov, Ukraine 25 E6 48 35N 38 40 E
Stalingrad = Volgograd, Russia 25 E7 48 40N 44 25 E
Staliniri = Tskhinvali, Georgia 25 F7 42 14N 44 1 E
Stalino = Donetsk, Ukraine 25 E6 48 0N 37 45 E
Stalinogorsk = Novomoskovsk, Russia 24 D6 54 5N 38 15 E
Stalis, Greece 23 D7 35 17N 25 25 E
Stalowa Wola, Poland 17 C12 50 34N 22 3 E
Stalybridge, U.K. 10 D5 53 28N 2 3W
Stamford, Australia 62 C3 21 15S 143 46 E
Stamford, U.K. 11 E7 52 39N 0 29W
Stamford, Conn., U.S.A. 79 E11 41 3N 73 32W
Stamford, N.Y., U.S.A. 79 D10 42 25N 74 38W
Stamford, Tex., U.S.A. 81 J5 32 57N 99 48W
Stampriet, Namibia 56 C2 24 20S 18 28 E
Stamps, U.S.A. 81 J8 33 22N 93 30W
Standerton, S. Africa 57 D4 26 55S 29 7 E
Standish, U.S.A. 76 D4 43 59N 83 57W
Stanford, S. Africa 56 E2 34 26S 19 29 E
Stanford, U.S.A. 82 C8 47 9N 110 13W
Stanger, S. Africa 57 D5 29 27S 31 14 E
Stanislaus →, U.S.A. 84 H5 37 40N 121 14W

Suphan Buri

Stanislav = Ivano-Frankivsk, Ukraine 17 D13 48 40N 24 40 E
Stanke Dimitrov, Bulgaria 21 C10 42 17N 23 9 E
Stanley, Australia 62 G4 40 46S 145 19 E
Stanley, Canada 73 B8 55 24N 104 22W
Stanley, Falk. Is. 96 G5 51 40S 59 51W
Stanley, U.K. 10 C6 54 53N 1 41W
Stanley, Idaho, U.S.A. 82 D6 44 13N 114 56W
Stanley, N. Dak., U.S.A. 80 A3 48 19N 102 23W
Stanley, N.Y., U.S.A. 78 D7 42 48N 77 6W
Stanovoy Khrebet, Russia 27 D13 55 0N 130 0 E
Stanovoy Ra. = Stanovoy Khrebet, Russia 27 D13 55 0N 130 0 E
Stansmore Ra., Australia 60 D4 21 23S 128 33 E
Stanthorpe, Australia 63 D5 28 36S 151 59 E
Stanton, U.S.A. 81 J4 32 8N 101 48W
Stanwood, U.S.A. 84 B4 48 15N 122 23W
Staples, U.S.A. 80 B7 46 21N 94 48W
Star City, Canada 73 C8 52 50N 104 20W
Star Lake, U.S.A. 79 B9 44 10N 75 2W
Stara Planina, Bulgaria 21 C10 43 15N 23 0 E
Stara Zagora, Bulgaria 21 C11 42 26N 25 39 E
Starachowice, Poland 17 C11 51 3N 21 2 E
Staraya Russa, Russia 24 C5 57 58N 31 23 E
Starbuck I., Kiribati 65 H12 5 37S 155 55W
Stargard Szczeciński, Poland 16 B8 53 20N 15 0 E
Staritsa, Russia 24 C5 56 33N 34 55 E
Starke, U.S.A. 77 L4 29 57N 82 7W
Starogard Gdański, Poland 17 B10 53 59N 18 30 E
Starokonstantinov = Starokonstyantyniv, Ukraine 17 D14 49 48N 27 10 E
Starokonstyantyniv, Ukraine 17 D14 49 48N 27 10 E
Start Pt., U.K. 11 G4 50 13N 3 39W
Staryy Chartoriysk, Ukraine 17 C13 51 15N 25 54 E
Staryy Oskol, Russia 25 D6 51 19N 37 55 E
State College, U.S.A. 78 F7 40 48N 77 52W
Stateline, U.S.A. 84 G7 38 57N 119 56W
Staten, I. = Estados, I. de Los, Argentina 96 G4 54 40S 64 30W
Staten I., U.S.A. 79 F10 40 35N 74 9W
Statesboro, U.S.A. 77 J5 32 27N 81 47W
Statesville, U.S.A. 77 H5 35 47N 80 53W
Stauffer, U.S.A. 85 L7 34 45N 119 3W
Staunton, Ill., U.S.A. 80 F10 39 1N 89 47W
Staunton, Va., U.S.A. 76 F6 38 9N 79 4W
Stavanger, Norway 9 G11 58 57N 5 40 E
Staveley, N.Z. 59 K3 43 40S 171 32 E
Stavelot, Belgium 15 D5 50 23N 5 55 E
Stavern, Norway 9 G14 59 0N 10 1 E
Stavoren, Neths. 15 B5 52 53N 5 22 E
Stavropol, Russia 25 E7 45 5N 42 0 E
Stavros, Cyprus 23 D11 35 1N 32 38 E
Stavrós, Greece 23 D6 35 12N 24 45 E
Stavrós, Ákra, Greece 23 D6 35 26N 24 58 E
Stawell, Australia 63 F3 37 5S 142 47 E
Stawell →, Australia 62 C3 20 20S 142 55 E
Stayner, Canada 78 B4 44 25N 80 5W
Stayton, U.S.A. 82 D2 44 48N 122 48W
Steamboat Springs, U.S.A. 82 F10 40 29N 106 50W
Steele, U.S.A. 80 B5 46 51N 99 55W
Steelton, U.S.A. 78 F8 40 14N 76 50W
Steen River, Canada 72 B5 59 40N 117 12W
Steenkool = Bintuni, Indonesia 37 E8 2 7S 133 32 E
Steens Mt., U.S.A. 82 E4 42 35N 118 40W
Steenwijk, Neths. 15 B6 52 47N 6 7 E
Steep Pt., Australia 61 E1 26 8S 113 8 E
Steep Rock, Canada 73 C9 51 30N 98 48W
Stefanie L. = Chew Bahir, Ethiopia 46 G2 4 40N 36 50 E
Stefansson Bay, Antarctica 5 C5 67 20S 59 8 E
Steiermark □, Austria 16 E8 47 26N 15 0 E
Steilacoom, U.S.A. 84 C4 47 10N 122 36W
Steilrandberge, Namibia 56 B1 17 45S 13 20 E
Steinbach, Canada 73 D9 49 32N 96 40W
Steinhausen, Namibia 56 C2 21 49S 18 20 E
Steinkjer, Norway 8 D14 64 1N 11 31 E
Steinkopf, S. Africa 56 D2 29 18S 17 43 E
Stellarton, Canada 71 C7 45 32N 62 30W
Stellenbosch, S. Africa 56 E2 33 58S 18 50 E
Stendal, Germany 16 B6 52 36N 11 53 E
Steornabhaigh = Stornoway, U.K. 12 C2 58 13N 6 23W
Stepanakert = Xankändi, Azerbaijan 25 G8 39 52N 46 49 E
Stephens Creek, Australia 63 E3 31 50S 141 30 E
Stephens I., Canada 72 C2 54 10N 130 45W
Stephens L., Canada 73 B9 56 32N 95 0W
Stephenville, Canada 71 C8 48 31N 58 35W
Stephenville, U.S.A. 81 J5 32 13N 98 12W
Stepnoi = Elista, Russia 25 E7 46 16N 44 14 E
Steppe, Asia 28 D9 50 0N 50 0 E
Sterkstroom, S. Africa 56 E4 31 32S 26 32 E
Sterling, Colo., U.S.A. 80 E3 40 37N 103 13W
Sterling, Ill., U.S.A. 80 E10 41 48N 89 42W
Sterling, Kans., U.S.A. 80 F5 38 13N 98 12W
Sterling City, U.S.A. 81 K4 31 51N 101 0W
Sterling Heights, U.S.A. 76 D4 42 35N 83 0W
Sterling Run, U.S.A. 78 E6 41 25N 78 12W
Sterlitamak, Russia 24 D10 53 40N 56 0 E
Stérnes, Greece 23 D6 35 30N 24 9 E
Stettin = Szczecin, Poland 16 B8 53 27N 14 27 E
Stettiner Haff, Germany 16 B8 53 47N 14 15 E
Stettler, Canada 72 C6 52 19N 112 40W
Steubenville, U.S.A. 78 F4 40 22N 80 37W
Stevenage, U.K. 11 F7 51 55N 0 13W
Stevens Point, U.S.A. 80 C10 44 31N 89 34W
Stevenson, U.S.A. 84 E5 45 42N 121 53W
Stevenson L., Canada 73 C9 53 55N 96 0W
Stevensville, U.S.A. 82 C6 46 30N 114 5W
Stewart, Canada 72 B3 55 56N 129 57W
Stewart, U.S.A. 84 F7 39 5N 119 46W
Stewart →, Canada 68 B6 63 19N 139 26W
Stewart, C., Australia 62 A1 11 57S 134 56 E
Stewart, I., Chile 96 G2 54 50S 71 15W
Stewart I., N.Z. 59 M1 46 58S 167 54 E
Stewarts Point, U.S.A. 84 G3 38 39N 123 24W
Stewartville, U.S.A. 80 D8 43 51N 92 29W
Stewiacke, Canada 71 C7 45 9N 63 22W
Steynsburg, S. Africa 56 E4 31 15S 25 49 E
Steyr, Austria 16 D8 48 3N 14 25 E
Steytlerville, S. Africa 56 E3 33 17S 24 19 E
Stigler, U.S.A. 81 H7 35 15N 95 8W
Stikine →, Canada 72 B2 56 40N 132 30W
Stilfontein, S. Africa 56 D4 26 51S 26 50 E
Stillwater, N.Z. 59 K3 42 27S 171 20 E
Stillwater, Minn., U.S.A. 80 C8 45 3N 92 49W
Stillwater, N.Y., U.S.A. 79 D11 42 55N 73 41W
Stillwater, Okla., U.S.A. 81 G6 36 7N 97 4W
Stillwater Range, U.S.A. 82 G4 39 50N 118 5W
Stillwater Reservoir, U.S.A. 79 C9 43 54N 75 3W
Stilwell, U.S.A. 81 H7 35 49N 94 38W
Štip, Macedonia 21 D10 41 42N 22 10 E
Stirling, Canada 78 B7 44 18N 77 33W
Stirling, U.K. 12 E5 56 8N 3 57W
Stirling □, U.K. 12 E4 56 12N 4 18W
Stirling Ra., Australia 61 F2 34 23S 118 0 E
Stittsville, Canada 79 A9 45 15N 75 55W
Stjernøya, Norway 8 A20 70 20N 22 40 E
Stjørdalshalsen, Norway 8 E14 63 29N 10 51 E
Stockerau, Austria 16 D9 48 24N 16 12 E
Stockholm, Sweden 9 G18 59 20N 18 3 E
Stockport, U.K. 10 D5 53 25N 2 9W
Stocksbridge, U.K. 10 D6 53 29N 1 35W
Stockton, Calif., U.S.A. 84 H5 37 58N 121 17W
Stockton, Kans., U.S.A. 80 F5 39 26N 99 16W
Stockton, Mo., U.S.A. 81 G8 37 42N 93 48W
Stockton-on-Tees, U.K. 10 C6 54 35N 1 19W
Stockton-on-Tees □, U.K. 10 C6 54 35N 1 19W
Stockton Plateau, U.S.A. 81 K3 30 30N 102 30W
Stoeng Treng, Cambodia 38 F5 13 31N 105 58 E
Stoer, Pt. of, U.K. 12 C3 58 16N 5 23W
Stoke-on-Trent, U.K. 10 D5 53 1N 2 11W
Stoke-on-Trent □, U.K. 10 D5 53 1N 2 11W
Stokes Pt., Australia 62 G3 40 10S 143 56 E
Stokes Ra., Australia 60 C5 15 50S 130 50 E
Stokksnes, Iceland 8 D6 64 14N 14 58W
Stokmarknes, Norway 8 B16 68 34N 14 54 E
Stolac, Bos.-H. 21 C7 43 5N 17 59 E
Stolbovoy, Ostrov, Russia 27 B14 74 44N 135 14 E
Stolbtsy = Stowbtsy, Belarus 17 B14 53 30N 26 43 E
Stolin, Belarus 17 C14 51 53N 26 50 E
Stomion, Greece 23 D5 35 21N 23 32 E
Stone, U.K. 10 E5 52 55N 2 9W
Stoneboro, U.S.A. 78 E4 41 20N 80 7W
Stonehaven, U.K. 12 E6 56 59N 2 12W
Stonehenge, Australia 62 C3 24 22S 143 17 E
Stonehenge, U.K. 11 F6 51 9N 1 45W
Stonewall, Canada 73 C9 50 10N 97 19W
Stony L., Man., Canada 73 B9 58 51N 98 40W
Stony L., Ont., Canada 78 B6 44 30N 78 5W
Stony Point, U.S.A. 79 E11 41 14N 73 59W
Stony Pt., U.S.A. 79 C8 43 50N 76 18W
Stony Rapids, Canada 73 B7 59 16N 105 50W
Stony Tunguska = Tunguska, Podkamennaya →, Russia 27 C10 61 50N 90 13 E
Stonyford, U.S.A. 84 F4 39 23N 122 33W
Stora Lulevatten, Sweden 8 C18 67 10N 19 30 E
Storavan, Sweden 8 D18 65 45N 18 10 E
Stord, Norway 9 G11 59 52N 5 23 E
Store Bælt, Denmark 9 J14 55 20N 11 0 E
Storm B., Australia 62 G4 43 10S 147 30 E
Storm Lake, U.S.A. 80 D7 42 39N 95 13W
Stormberge, S. Africa 56 E4 31 16S 26 17 E
Stormsrivier, S. Africa 56 E3 33 59S 23 52 E
Stornoway, U.K. 12 C2 58 13N 6 23W
Storozhinets = Storozhynets, Ukraine 17 D13 48 14N 25 45 E
Storozhynets, Ukraine 17 D13 48 14N 25 45 E
Storrs, U.S.A. 79 E12 41 49N 72 15W
Storsjön, Sweden 8 E16 63 9N 14 30 E
Storuman, Sweden 8 D17 65 5N 17 10 E
Storuman, sjö, Sweden 8 D17 65 13N 16 50 E
Stouffville, Canada 78 C5 43 58N 79 15W
Stoughton, Canada 73 D8 49 40N 103 0W
Stour →, Dorset, U.K. 11 G6 50 43N 1 47W
Stour →, Kent, U.K. 11 F9 51 18N 1 22 E
Stour →, Suffolk, U.K. 11 F9 51 57N 1 4 E
Stourbridge, U.K. 11 E5 52 28N 2 8W
Stout L., Canada 73 C10 52 0N 94 40W
Stove Pipe Wells Village, U.S.A. 85 J9 36 35N 117 11W
Stow, U.S.A. 78 E3 41 10N 81 27W
Stowbtsy, Belarus 17 B14 53 30N 26 43 E
Stowmarket, U.K. 11 E9 52 12N 1 0 E
Strabane, U.K. 13 B4 54 50N 7 27W
Strahan, Australia 62 G4 42 9S 145 20 E
Stralsund, Germany 16 A7 54 18N 13 4 E
Strand, S. Africa 56 E2 34 9S 18 48 E
Stranda, Møre og Romsdal, Norway 9 E12 62 19N 6 58 E
Stranda, Nord-Trøndelag, Norway 8 E14 63 33N 10 14 E
Strangford L., U.K. 13 B6 54 30N 5 37W
Stranraer, U.K. 12 G3 54 54N 5 1W
Strasbourg, Canada 73 C8 51 4N 104 55W
Strasbourg, France 18 B7 48 35N 7 42 E
Stratford, Canada 78 C4 43 23N 81 0W
Stratford, N.Z. 59 H5 39 20S 174 19 E
Stratford, Calif., U.S.A. 84 J7 36 11N 119 49W
Stratford, Conn., U.S.A. 79 E11 41 12N 73 8W
Stratford, Tex., U.S.A. 81 G3 36 20N 102 4W
Stratford-upon-Avon, U.K. 11 E6 52 12N 1 42W
Strath Spey, U.K. 12 D5 57 9N 3 49W
Strathalbyn, Australia 63 F2 35 13S 138 53 E
Strathaven, U.K. 12 F4 55 40N 4 5W
Strathcona Prov. Park, Canada 72 D3 49 38N 125 40W
Strathmore, Canada 72 C6 51 5N 113 18W
Strathmore, U.K. 12 E5 56 37N 3 7W
Strathmore, U.S.A. 84 J7 36 9N 119 4W
Strathnaver, Canada 72 C4 53 20N 122 33W
Strathpeffer, U.K. 12 D4 57 35N 4 32W
Strathroy, Canada 78 D3 42 58N 81 38W
Strathy Pt., U.K. 12 C4 58 36N 4 1W
Strattanville, U.S.A. 78 E5 41 12N 79 19W
Stratton, U.S.A. 79 A14 45 8N 70 26W
Stratton Mt., U.S.A. 79 C12 43 4N 72 55W
Straubing, Germany 16 D7 48 52N 12 34 E
Straumnes, Iceland 8 C2 66 26N 23 8W
Strawberry →, U.S.A. 82 F8 40 10N 110 24W
Streaky B., Australia 63 E1 32 48S 134 13 E
Streaky Bay, Australia 63 E1 32 51S 134 18 E
Streator, U.S.A. 80 E10 41 8N 88 50W
Streetsboro, U.S.A. 78 E3 41 14N 81 21W
Streetsville, Canada 78 C5 43 35N 79 42W
Strelka, Russia 27 D10 58 5N 93 3 E
Streng →, Cambodia 38 F4 13 12N 103 37 E
Streymoy, Færoe Is. 8 E9 62 8N 7 5W
Strezhevoy, Russia 26 C8 60 42N 77 34 E
Strimón →, Greece 21 D10 40 46N 23 51 E
Strimonikós Kólpos, Greece 21 D11 40 33N 24 0 E
Stroma, U.K. 12 C5 58 41N 3 7W
Strómboli, Italy 20 E6 38 47N 15 13 E
Stromeferry, U.K. 12 D3 57 21N 5 33W
Stromness, U.K. 12 C5 58 58N 3 17W
Stromsburg, U.S.A. 80 E6 41 7N 97 36W
Strömstad, Sweden 9 G14 58 56N 11 10 E
Strömsund, Sweden 8 E16 63 51N 15 33 E
Strongsville, U.S.A. 78 E3 41 19N 81 50W
Stronsay, U.K. 12 B6 59 7N 2 35W
Stroud, U.K. 11 F5 51 45N 2 13W
Stroud Road, Australia 63 E5 32 18S 151 57 E
Stroudsburg, U.S.A. 79 F9 40 59N 75 12W
Stroumbi, Cyprus 23 E11 34 53N 32 29 E
Struer, Denmark 9 H13 56 30N 8 35 E
Strumica, Macedonia 21 D10 41 28N 22 41 E
Struthers, Canada 70 C2 48 41N 85 51W
Struthers, U.S.A. 78 E4 41 4N 80 39W
Stryker, U.S.A. 82 B6 48 41N 114 46W
Stryy, Ukraine 17 D12 49 16N 23 48 E
Strzelecki Cr. →, Australia 63 D2 29 37S 139 59 E
Stuart, Fla., U.S.A. 77 M5 27 12N 80 15W
Stuart, Nebr., U.S.A. 80 D5 42 36N 99 8W
Stuart →, Canada 72 C4 54 0N 123 35W
Stuart Bluff Ra., Australia 60 D5 22 50S 131 52 E
Stuart L., Canada 72 C4 54 30N 124 30W
Stuart Ra., Australia 63 D1 29 10S 134 56 E
Stull L., Canada 70 B1 54 24N 92 34W
Stung Treng = Stoeng Treng, Cambodia 38 F5 13 31N 105 58 E
Stupart →, Canada 70 A1 56 0N 93 25W
Sturgeon B., Canada 73 C9 52 0N 97 50W
Sturgeon Bay, U.S.A. 76 C2 44 50N 87 23W
Sturgeon Falls, Canada 70 C4 46 25N 79 57W
Sturgeon L., Alta., Canada 72 B5 55 6N 117 32W
Sturgeon L., Ont., Canada 70 C1 50 0N 90 45W
Sturgeon L., Ont., Canada 78 B6 44 28N 78 43W
Sturgis, Canada 73 C8 51 56N 102 36W
Sturgis, Mich., U.S.A. 76 E3 41 48N 85 25W
Sturgis, S. Dak., U.S.A. 80 C3 44 25N 103 31W
Sturt Cr. →, Australia 60 C4 19 8S 127 50 E
Stutterheim, S. Africa 56 E4 32 33S 27 28 E
Stuttgart, Germany 16 D5 48 48N 9 11 E
Stuttgart, U.S.A. 81 H9 34 30N 91 33W
Stuyvesant, U.S.A. 79 D11 42 23N 73 45W
Stykkishólmur, Iceland 8 D2 65 2N 22 40W
Styria = Steiermark □, Austria 16 E8 47 26N 15 0 E
Su Xian = Suzhou, China 34 H9 33 41N 116 59 E
Suakin, Sudan 51 E13 19 8N 37 20 E
Suan, N. Korea 35 E14 38 42N 126 22 E
Suaqui, Mexico 86 B3 29 12N 109 41W
Suar, India 43 E8 29 2N 79 3 E
Subang, Indonesia 37 G12 6 34S 107 45 E
Subansiri →, India 41 F18 26 48N 93 50 E
Subarnarekha →, India 43 H12 22 34N 87 24 E
Subayhah, Si. Arabia 44 D3 30 2N 38 50 E
Subi, Indonesia 39 L7 2 58N 108 50 E
Subotica, Serbia, Yug. 21 A8 46 6N 19 39 E
Suceava, Romania 17 E14 47 38N 26 16 E
Suchan, Russia 30 C6 43 8N 133 9 E
Suchitoto, El Salv. 88 D2 13 56N 89 0W
Suchou = Suzhou, China 33 C7 31 19N 120 38 E
Süchow = Xuzhou, China 35 G9 34 18N 117 10 E
Suck →, Ireland 13 C3 53 17N 8 3W
Sucre, Bolivia 92 G5 19 0S 65 15W
Sucuriú →, Brazil 93 H8 20 47S 51 38W
Sud, Pte. du, Canada 71 C7 49 3N 62 14W
Sud-Kivu □, Dem. Rep. of the Congo 54 C2 3 0S 28 30 E
Sud-Ouest, Pte. du, Canada 71 C7 49 23N 63 36W
Sudan, U.S.A. 81 H3 34 4N 102 32W
Sudan ■, Africa 51 E11 15 0N 30 0 E
Sudbury, Canada 70 C3 46 30N 81 0W
Sudbury, U.K. 11 E8 52 2N 0 45 E
Sûdd, Sudan 51 G12 8 20N 30 0 E
Sudeten Mts. = Sudety, Europe 17 C9 50 20N 16 45 E
Sudety, Europe 17 C9 50 20N 16 45 E
Suðuroy, Færoe Is. 8 F9 61 32N 6 50W
Sudi, Tanzania 55 E4 10 11S 39 57 E
Sudirman, Pegunungan, Indonesia 37 E9 4 30S 137 0 E
Sueca, Spain 19 C5 39 12N 0 21W
Suemez I., U.S.A. 72 B2 55 15N 133 20W
Suez = El Suweis, Egypt 51 C12 29 58N 32 31 E
Suez, G. of = Suweis, Khalîg el, Egypt 51 C12 28 40N 33 0 E
Suez Canal = Suweis, Qanâ es, Egypt 51 B12 31 0N 32 20 E
Suffield, Canada 72 C6 50 12N 111 10W
Suffolk, U.S.A. 76 G7 36 44N 76 35W
Suffolk □, U.K. 11 E9 52 16N 1 0 E
Sugargrove, U.S.A. 78 E5 41 59N 79 21W
Sugarive →, India 43 F12 26 16N 86 24 E
Sugluk = Salluit, Canada 69 B12 62 14N 75 38W
Şuḥār, Oman 46 C6 24 20N 56 40 E
Sühbaatar □, Mongolia 34 B8 45 30N 114 0 E
Suhl, Germany 16 C6 50 36N 10 42 E
Sui, Pakistan 42 E3 28 37N 69 19 E
Sui Xian, China 34 G8 34 25N 115 2 E
Suide, China 34 F6 37 30N 110 12 E
Suifenhe, China 35 B16 44 25N 131 10 E
Suihua, China 33 B7 46 32N 126 55 E
Suining, China 35 H9 33 56N 117 58 E
Suiping, China 34 H7 33 10N 113 59 E
Suir →, Ireland 13 D4 52 16N 7 9W
Suisun City, U.S.A. 84 G4 38 15N 122 2W
Suiyang, China 35 B16 44 30N 130 56 E
Suizhong, China 35 D11 40 21N 120 20 E
Sujangarh, India 42 F6 27 42N 74 31 E
Sukabumi, Indonesia 37 G12 6 56S 106 50 E
Sukadana, Indonesia 36 E3 1 10S 110 0 E
Sukagawa, Japan 31 F10 37 17N 140 23 E
Sukaraja, Indonesia 36 E4 2 28S 110 25 E
Sukarnapura = Jayapura, Indonesia 37 E10 2 28S 140 38 E
Sukch'ŏn, N. Korea 35 E13 39 22N 125 35 E
Sukhona →, Russia 24 C6 61 15N 46 39 E
Sukhothai, Thailand 38 D2 17 1N 99 49 E
Sukhumi = Sokhumi, Georgia 25 F7 43 0N 41 0 E
Sukkur, Pakistan 42 F3 27 42N 68 54 E
Sukkur Barrage, Pakistan 42 F3 27 40N 68 50 E
Sukri →, India 42 G4 25 4N 71 43 E
Sukumo, Japan 31 H6 32 56N 132 44 E
Sukunka →, Canada 72 B4 55 45N 121 15W
Sula, Kepulauan, Indonesia 37 E7 1 45S 125 0 E
Sulaco →, Honduras 88 C2 15 2N 87 44W
Sulaiman Range, Pakistan 42 D3 30 30N 69 50 E
Sūlār, Iran 45 D6 31 53N 51 54 E
Sulawesi Sea = Celebes Sea, Indonesia 37 D6 3 0N 123 0 E
Sulawesi Selatan □, Indonesia 37 E6 2 30S 120 0 E
Sulawesi Utara □, Indonesia 37 D6 1 0N 122 30 E
Sulima, S. Leone 50 G3 6 58N 11 32W
Sulina, Romania 17 F15 45 10N 29 40 E
Sulitjelma, Norway 8 C17 67 9N 16 3 E
Sullana, Peru 92 D2 4 52S 80 39W
Sullivan, Ill., U.S.A. 80 F10 39 36N 88 37W
Sullivan, Ind., U.S.A. 76 F2 39 6N 87 24W
Sullivan, Mo., U.S.A. 80 F9 38 13N 91 10W
Sullivan Bay, Canada 72 C3 50 55N 126 50W
Sullivan I. = Lanbi Kyun, Burma 39 G2 10 50N 98 20 E
Sulphur, La., U.S.A. 81 K8 30 14N 93 23W
Sulphur, Okla., U.S.A. 81 H6 34 31N 96 58W
Sulphur Pt., Canada 72 A6 60 56N 114 48W
Sulphur Springs, U.S.A. 81 J7 33 8N 95 36W
Sultan, Canada 70 C3 47 36N 82 47W
Sultan, U.S.A. 84 C5 47 52N 121 49W
Sultanpur, Mad. P., India 42 H8 23 9N 77 56 E
Sultanpur, Punjab, India 42 D6 31 13N 75 11 E
Sultanpur, Ut. P., India 43 F10 26 18N 82 4 E
Sulu Arch., Phil. 37 C6 6 0N 121 0 E
Sulu Sea, E. Indies 37 C6 8 0N 120 0 E
Suluq, Libya 51 B10 31 44N 20 14 E
Sulzberger Ice Shelf, Antarctica 5 D10 78 0S 150 0 E
Sumalata, Indonesia 37 D6 1 0N 122 31 E
Sumampa, Argentina 94 B3 29 25S 63 29W
Sumatera □, Indonesia 36 D2 0 40N 100 20 E
Sumatera Barat □, Indonesia 36 E2 1 0S 101 0 E
Sumatera Utara □, Indonesia 36 D1 2 30N 98 0 E
Sumatra = Sumatera □, Indonesia 36 D2 0 40N 100 20 E
Sumba, Indonesia 37 F5 9 45S 119 35 E
Sumba, Selat, Indonesia 37 F5 9 0S 118 40 E
Sumbawa, Indonesia 36 F5 8 26S 117 30 E
Sumbawa Besar, Indonesia 36 F5 8 30S 117 26 E
Sumbawanga □, Tanzania 52 F6 8 0S 31 30 E
Sumbe, Angola 52 G2 11 10S 13 48 E
Sumburgh Hd., U.K. 12 B7 59 52N 1 17W
Sumdeo, India 43 D8 31 26N 78 44 E
Sumdo, China 43 B8 35 6N 78 41 E
Sumedang, Indonesia 37 G12 6 52S 107 55 E
Šumen = Shumen, Bulgaria 21 C12 43 18N 26 55 E
Sumenep, Indonesia 37 G15 7 1S 113 52 E
Sumgait = Sumqayıt, Azerbaijan 25 F8 40 34N 49 38 E
Summer L., U.S.A. 82 E3 42 50N 120 45W
Summerland, Canada 72 D5 49 32N 119 41W
Summerside, Canada 71 C7 46 24N 63 47W
Summersville, U.S.A. 76 F5 38 17N 80 51W
Summerville, Ga., U.S.A. 77 H3 34 29N 85 21W
Summerville, S.C., U.S.A. 77 J5 33 1N 80 11W
Summit Lake, Canada 72 C4 54 20N 122 40W
Summit Peak, U.S.A. 83 H10 37 21N 106 42W
Sumner, Iowa, U.S.A. 80 D8 42 51N 92 6W
Sumner, Wash., U.S.A. 84 C4 47 12N 122 14W
Sumoto, Japan 31 G7 34 21N 134 54 E
Šumperk, Czech Rep. 17 D9 49 59N 16 59 E
Sumqayıt, Azerbaijan 25 F8 40 34N 49 38 E
Sumter, U.S.A. 77 J5 33 55N 80 21W
Sumy, Ukraine 25 D5 50 57N 34 50 E
Sun City, S. Africa 56 D4 25 17S 27 3 E
Sun City, Ariz., U.S.A. 83 K7 33 36N 112 17W
Sun City, Calif., U.S.A. 85 M9 33 42N 117 11W
Sun City Center, U.S.A. 77 M4 27 43N 82 18W
Sun Lakes, U.S.A. 83 K8 33 10N 111 52W
Sun Valley, U.S.A. 82 E6 43 42N 114 21W
Sunagawa, Japan 30 C10 43 29N 141 55 E
Sunan, N. Korea 35 E13 39 15N 125 40 E
Sunart, L., U.K. 12 E3 56 42N 5 43W
Sunburst, U.S.A. 82 B8 48 53N 111 55W
Sunbury, Australia 63 F3 37 35S 144 44 E
Sunbury, U.S.A. 79 F8 40 52N 76 48W
Sunchales, Argentina 94 C3 30 58S 61 35W
Suncho Corral, Argentina 94 B3 27 55S 63 27W
Sunch'ŏn, S. Korea 35 G14 34 52N 127 31 E
Suncook, U.S.A. 79 C13 43 8N 71 27W
Sunda, Selat, Indonesia 36 F3 6 20S 105 30 E
Sunda Is., Indonesia 28 K14 5 0S 105 0 E
Sunda Str. = Sunda, Selat, Indonesia 36 F3 6 20S 105 30 E
Sundance, Canada 73 B10 56 32N 94 4W
Sundance, U.S.A. 80 C2 44 24N 104 23W
Sundar Nagar, India 42 D7 31 32N 76 53 E
Sundarbans, Asia 41 J16 22 0N 89 0 E
Sundargarh, India 41 H14 22 4N 84 5 E
Sundays = Sondags →, S. Africa 56 E4 33 44S 25 51 E
Sunderland, Canada 78 B5 44 16N 79 4W
Sunderland, U.K. 10 C6 54 55N 1 23W
Sundre, Canada 72 C6 51 49N 114 38W
Sundsvall, Sweden 9 E17 62 23N 17 17 E
Sung Hei, Vietnam 39 G6 10 20N 106 2 E
Sungai Kolok, Thailand 39 J3 6 2N 101 58 E
Sungai Lembing, Malaysia 39 L4 3 55N 103 3 E
Sungai Petani, Malaysia 39 K3 5 37N 100 30 E
Sungaigerong, Indonesia 36 E2 2 59S 104 52 E
Sungailiat, Indonesia 36 E3 1 51S 106 8 E
Sungaipenuh, Indonesia 36 E2 2 1S 101 20 E
Sungari = Songhua Jiang →, China 33 B8 47 45N 132 30 E
Sunghua Chiang = Songhua Jiang →, China 33 B8 47 45N 132 30 E
Sunland Park, U.S.A. 83 L10 31 50N 106 40W
Sunndalsøra, Norway 9 E13 62 40N 8 33 E
Sunnyside, U.S.A. 82 C3 46 20N 120 0W
Sunnyvale, U.S.A. 84 H4 37 23N 122 2W
Suntar, Russia 27 C12 62 15N 117 30 E
Suomenselkä, Finland 8 E21 62 52N 24 0 E
Suomussalmi, Finland 8 D23 64 54N 29 10 E
Suoyarvi, Russia 24 B5 62 3N 32 20 E
Supai, U.S.A. 83 H7 36 15N 112 41W
Supaul, India 43 F12 26 10N 86 40 E
Superior, Ariz., U.S.A. 83 K8 33 18N 111 6W
Superior, Mont., U.S.A. 82 C6 47 12N 114 53W
Superior, Nebr., U.S.A. 80 E5 40 1N 98 4W
Superior, Wis., U.S.A. 80 B8 46 44N 92 6W
Superior, L., N. Amer. 70 C2 47 0N 87 0W
Suphan Buri, Thailand 38 E3 14 14N 100 10 E

Suphan Dağı

Suphan Dağı, Turkey 44 B4 38 54N 42 48 E
Supiori, Indonesia 37 E9 1 0S 136 0 E
Supung Shuiku, China 35 D13 40 35N 124 50 E
Sūq Suwayq, Si. Arabia 44 E3 24 23N 38 27 E
Suqian, China 35 H10 33 54N 118 8 E
Sūr, Lebanon 47 B4 33 19N 35 16 E
Şūr, Oman 46 C6 22 34N 59 32 E
Sur, Pt., U.S.A. 84 J5 36 18N 121 54W
Sura →, Russia 24 C8 56 6N 46 0 E
Surab, Pakistan 42 E2 28 25N 66 15 E
Surabaja = Surabaya, Indonesia 36 F4 7 17S 112 45 E
Surabaya, Indonesia 36 F4 7 17S 112 45 E
Surakarta, Indonesia 36 F4 7 35S 110 48 E
Surat, Australia 63 D4 27 10S 149 6 E
Surat, India 40 J8 21 12N 72 55 E
Surat Thani, Thailand 39 H2 9 6N 99 20 E
Suratgarh, India 42 E5 29 18N 73 55 E
Surendranagar, India 42 H4 22 45N 71 40 E
Surf, U.S.A. 85 L6 34 41N 120 36W
Surgut, Russia 26 C8 61 14N 73 20 E
Suriapet, India 40 L11 17 10N 79 40 E
Surigao, Phil. 37 C7 9 47N 125 29 E
Surin, Thailand 38 E4 14 50N 103 34 E
Surin Nua, Ko, Thailand 39 H1 9 30N 97 55 E
Surinam ■, S. Amer. 93 C7 4 0N 56 0W
Suriname = Surinam ■, S. Amer. 93 C7 4 0N 56 0W
Suriname →, Surinam 93 B7 5 50N 55 15W
Sūrmaq, Iran 45 D7 31 3N 52 48 E
Surrey □, U.K. 11 F7 51 15N 0 31W
Sursand, India 43 F11 26 39N 85 43 E
Sursar →, India 43 F12 26 14N 87 3 E
Surt, Libya 51 B9 31 11N 16 39 E
Surt, Khalīj, Libya 51 B9 31 40N 18 30 E
Surtanahu, Pakistan 42 F4 26 22N 70 0 E
Surtsey, Iceland 8 E3 63 20N 20 30W
Suruga-Wan, Japan 31 G9 34 45N 138 30 E
Susaki, Japan 31 H6 33 22N 133 17 E
Sūsangerd, Iran 45 D6 31 35N 48 6 E
Susanville, U.S.A. 82 F3 40 25N 120 39W
Susner, India 42 H7 23 57N 76 5 E
Susquehanna, U.S.A. 79 E9 41 57N 75 36W
Susquehanna →, U.S.A. 79 G8 39 33N 76 5W
Susques, Argentina 94 A2 23 35S 66 25W
Sussex, Canada 71 C6 45 45N 65 37W
Sussex, U.S.A. 79 E10 41 13N 74 37W
Sussex, E. □, U.K. 11 G8 51 0N 0 20 E
Sussex, W. □, U.K. 11 G7 51 0N 0 30W
Sustut →, Canada 72 B3 56 20N 127 30W
Susuman, Russia 27 C15 62 47N 148 10 E
Susunu, Indonesia 37 E8 3 7S 133 39 E
Susurluk, Turkey 21 E13 39 54N 28 8 E
Sutherland, S. Africa 56 E3 32 24S 20 40 E
Sutherland, U.S.A. 80 E4 41 10N 101 8W
Sutherland Falls, N.Z. 59 L1 44 48S 167 46 E
Sutherlin, U.S.A. 82 E2 43 23N 123 19W
Suthri, India 42 H3 23 3N 68 55 E
Sutlej →, Pakistan 42 E4 29 23N 71 3 E
Sutter, U.S.A. 84 F5 39 10N 121 45W
Sutter Creek, U.S.A. 84 G6 38 24N 120 48W
Sutton, Canada 79 A12 45 6N 72 37W
Sutton, Nebr., U.S.A. 80 E6 40 36N 97 52W
Sutton, W. Va., U.S.A. 76 F5 38 40N 80 43W
Sutton →, Canada 70 A3 55 15N 83 45W
Sutton Coldfield, U.K. 11 E6 52 35N 1 49W
Sutton in Ashfield, U.K. 10 D6 53 8N 1 16W
Sutton L., Canada 70 B3 54 15N 84 42W
Suttor →, Australia 62 C4 21 36S 147 2 E
Suttsu, Japan 30 C10 42 48N 140 14 E
Suva, Fiji 59 D8 18 6S 178 30 E
Suva Planina, Serbia, Yug. 21 C10 43 10N 22 5 E
Suvorov Is. = Suwarrow Is., Cook Is. 65 J11 15 0S 163 0W
Suwałki, Poland 17 A12 54 8N 22 59 E
Suwannaphum, Thailand 38 E4 15 33N 103 47 E
Suwannee →, U.S.A. 77 L4 29 17N 83 10W
Suwanose-Jima, Japan 31 K4 29 38N 129 43 E
Suwarrow Is., Cook Is. 65 J11 15 0S 163 0W
Suwayq aş Şuqban, Iraq 44 D5 31 32N 46 7 E
Suweis, Khalîg el, Egypt 51 C12 28 40N 33 0 E
Suweis, Qanâ es, Egypt 51 B12 31 0N 32 20 E
Suwŏn, S. Korea 35 F14 37 17N 127 1 E
Suzdal, Russia 24 C7 56 29N 40 26 E
Suzhou, Anhui, China 34 H9 33 41N 116 59 E
Suzhou, Jiangsu, China 33 C7 31 19N 120 38 E
Suzu, Japan 31 F8 37 25N 137 17 E
Suzu-Misaki, Japan 31 F8 37 31N 137 21 E
Suzuka, Japan 31 G8 34 55N 136 36 E
Svalbard, Arctic 4 B8 78 0N 17 0 E
Svappavaara, Sweden 8 C19 67 40N 21 3 E
Svartisen, Norway 8 C15 66 40N 13 50 E
Svay Chek, Cambodia 38 F4 13 48N 102 58 E
Svay Rieng, Cambodia 39 G5 11 9N 105 45 E
Svealand □, Sweden 9 G16 60 20N 15 0 E
Sveg, Sweden 9 E16 62 2N 14 21 E
Svendborg, Denmark 9 J14 55 4N 10 35 E
Sverdlovsk = Yekaterinburg, Russia 26 D7 56 50N 60 30 E
Sverdrup Is., Canada 4 B3 79 0N 97 0W
Svetlaya, Russia 30 A9 46 33N 138 18 E
Svetlogorsk = Svyetlahorsk, Belarus 17 B15 52 38N 29 46 E
Svir →, Russia 24 B5 60 30N 32 48 E
Svishtov, Bulgaria 21 C11 43 36N 25 23 E
Svislach, Belarus 17 B13 53 3N 24 2 E
Svobodnyy, Russia 27 D13 51 20N 128 0 E
Svolvær, Norway 8 B16 68 15N 14 34 E
Svyetlahorsk, Belarus 17 B15 52 38N 29 46 E
Swabian Alps = Schwäbische Alb, Germany 16 D5 48 20N 9 30 E
Swainsboro, U.S.A. 77 J4 32 36N 82 20W
Swakop →, Namibia 56 C2 22 38S 14 36 E
Swakopmund, Namibia 56 C1 22 37S 14 30 E
Swale →, U.K. 10 C6 54 5N 1 20W
Swan →, Australia 61 F2 32 3S 115 45 E
Swan →, Canada 73 C8 52 30N 100 45W
Swan Hill, Australia 63 F3 35 20S 143 33 E
Swan Hills, Canada 72 C5 54 43N 115 24W
Swan Is. = Santanilla, Is., Honduras 88 C3 17 22N 83 57W
Swan L., Canada 73 C8 52 30N 100 40W
Swan Peak, U.S.A. 82 C7 47 43N 113 38W
Swan Ra., U.S.A. 82 C7 48 0N 113 45W
Swan River, Canada 73 C8 52 10N 101 16W
Swanage, U.K. 11 G6 50 36N 1 58W
Swansea, Australia 62 G4 42 8S 148 4 E
Swansea, Canada 78 C5 43 38N 79 28W
Swansea, U.K. 11 F4 51 37N 3 57W
Swansea □, U.K. 11 F3 51 38N 4 3W
Swar →, Pakistan 43 B5 34 40N 72 5 E
Swartberge, S. Africa 56 E3 33 20S 22 0 E
Swartmodder, S. Africa 56 D3 28 1S 20 32 E
Swartnossob →, Namibia 56 C2 23 8S 18 42 E
Swartruggens, S. Africa 56 D4 25 39S 26 42 E
Swastika, Canada 70 C3 48 7N 80 6W
Swatow = Shantou, China 33 D6 23 18N 116 40 E
Swaziland ■, Africa 57 D5 26 30S 31 30 E
Sweden ■, Europe 9 G16 57 0N 15 0 E
Sweet Home, U.S.A. 82 D2 44 24N 122 44W
Sweetgrass, U.S.A. 82 B8 48 59N 111 58W
Sweetwater, Nev., U.S.A. 84 G7 38 27N 119 9W
Sweetwater, Tenn., U.S.A. 77 H3 35 36N 84 28W
Sweetwater, Tex., U.S.A. 81 J4 32 28N 100 25W
Sweetwater →, U.S.A. 82 E10 42 31N 107 2W
Swellendam, S. Africa 56 E3 34 1S 20 26 E
Świdnica, Poland 17 C9 50 50N 16 30 E
Świdnik, Poland 17 C12 51 13N 22 39 E
Świebodzin, Poland 16 B8 52 15N 15 31 E
Świecie, Poland 17 B10 53 25N 18 30 E
Swift Current, Canada 73 C7 50 20N 107 45W
Swiftcurrent →, Canada 73 C7 50 38N 107 44W
Swilly, L., Ireland 13 A4 55 12N 7 33W
Swindon, U.K. 11 F6 51 34N 1 46W
Swindon □, U.K. 11 F6 51 34N 1 46W
Swinemünde = Świnoujście, Poland 16 B8 53 54N 14 16 E
Swinford, Ireland 13 C3 53 57N 8 58W
Świnoujście, Poland 16 B8 53 54N 14 16 E
Switzerland ■, Europe 18 C8 46 30N 8 0 E
Swords, Ireland 13 C5 53 28N 6 13W
Swoyerville, U.S.A. 79 E9 41 18N 75 53W
Sydenham →, Canada 78 D2 42 33N 82 25W
Sydney, Australia 63 E5 33 53S 151 10 E
Sydney, Canada 71 C7 46 7N 60 7W
Sydney L., Canada 73 C10 50 41N 94 25W
Sydney Mines, Canada 71 C7 46 18N 60 15W
Sydprøven = Alluitsup Paa, Greenland 4 C5 60 30N 45 35W
Sydra, G. of = Surt, Khalīj, Libya 51 B9 31 40N 18 30 E
Sykesville, U.S.A. 78 E6 41 3N 78 50W
Syktyvkar, Russia 24 B9 61 45N 50 40 E
Sylacauga, U.S.A. 77 J2 33 10N 86 15W
Sylarna, Sweden 8 E15 63 2N 12 13 E
Sylhet, Bangla. 41 G17 24 54N 91 52 E
Sylhet □, Bangla. 41 G17 24 50N 91 50 E
Sylt, Germany 16 A5 54 54N 8 22 E
Sylvan Beach, U.S.A. 79 C9 43 12N 75 44W
Sylvan Lake, Canada 72 C6 52 20N 114 3W
Sylvania, U.S.A. 77 J5 32 45N 81 38W
Sylvester, U.S.A. 77 K4 31 32N 83 50W
Sym, Russia 26 C9 60 20N 88 18 E
Symón, Mexico 86 C4 24 42N 102 35W
Synnott Ra., Australia 60 C4 16 30S 125 20 E
Syracuse, Kans., U.S.A. 81 G4 37 59N 101 45W
Syracuse, N.Y., U.S.A. 79 C8 43 3N 76 9W
Syracuse, Nebr., U.S.A. 80 E6 40 39N 96 11W
Syrdarya →, Kazakstan 26 E7 46 3N 61 0 E
Syria ■, Asia 44 C3 35 0N 38 0 E
Syrian Desert = Shām, Bādiyat ash, Asia 44 C3 32 0N 40 0 E
Syzran, Russia 24 D8 53 12N 48 30 E
Szczecin, Poland 16 B8 53 27N 14 27 E
Szczecinek, Poland 17 B9 53 43N 16 41 E
Szczeciński, Zalew = Stettiner Haff, Germany 16 B8 53 47N 14 15 E
Szczytno, Poland 17 B11 53 33N 21 0 E
Szechwan = Sichuan □, China 32 C5 30 30N 103 0 E
Szeged, Hungary 17 E11 46 16N 20 10 E
Székesfehérvár, Hungary 17 E10 47 15N 18 25 E
Szekszárd, Hungary 17 E10 46 22N 18 42 E
Szentes, Hungary 17 E11 46 39N 20 21 E
Szolnok, Hungary 17 E11 47 10N 20 15 E
Szombathely, Hungary 17 E9 47 14N 16 38 E

T

Ta Khli Khok, Thailand 38 E3 15 18N 100 20 E
Ta Lai, Vietnam 39 G6 11 24N 107 23 E
Tabacal, Argentina 94 A3 23 15S 64 15W
Tabaco, Phil. 37 B6 13 22N 123 44 E
Ţābah, Si. Arabia 44 E4 26 55N 42 38 E
Ţabas, Khorāsān, Iran 45 C9 32 48N 60 12 E
Ţabas, Khorāsān, Iran 45 C8 33 35N 56 55 E
Tabasará, Serranía de, Panama 88 E3 8 35N 81 40W
Tabasco □, Mexico 87 D6 17 45N 93 30W
Tabāsin, Iran 45 D8 31 12N 57 54 E
Tabatinga, Serra da, Brazil 93 F10 10 30S 44 0W
Taber, Canada 72 D6 49 47N 112 8W
Taberg, U.S.A. 79 C9 43 18N 75 37W
Tablas I., Phil. 37 B6 12 25N 122 2 E
Table B. = Tafelbaai, S. Africa 56 E2 33 35S 18 25 E
Table B., Canada 71 B8 53 40N 56 25W
Table Mt., S. Africa 56 E2 34 0S 18 22 E
Table Rock L., U.S.A. 81 G8 36 36N 93 19W
Tabletop, Mt., Australia 62 C4 23 24S 147 11 E
Tábor, Czech Rep. 16 D8 49 25N 14 39 E
Tabora, Tanzania 54 D3 5 2S 32 50 E
Tabora □, Tanzania 54 D3 5 0S 33 0 E
Tabou, Ivory C. 50 H4 4 30N 7 20W
Tabrīz, Iran 44 B5 38 7N 46 20 E
Tabuaeran, Kiribati 65 G12 3 51N 159 22W
Tabūk, Si. Arabia 44 D3 28 23N 36 36 E
Tacámbaro de Codallos, Mexico 86 D4 19 14N 101 28W
Tacheng, China 32 B3 46 40N 82 58 E
Tach'ing Shan = Daqing Shan, China 34 D6 40 40N 111 0 E
Tacloban, Phil. 37 B6 11 15N 124 58 E
Tacna, Peru 92 G4 18 0S 70 20W
Tacoma, U.S.A. 84 C4 47 14N 122 26W
Tacuarembó, Uruguay 95 C4 31 45S 56 0W
Tademaït, Plateau du, Algeria 50 C6 28 30N 2 30 E
Tadjoura, Djibouti 46 E3 11 50N 42 55 E
Tadmor, N.Z. 59 J4 41 27S 172 45 E
Tadoule, L., Canada 73 B9 58 36N 98 20W
Tadoussac, Canada 71 C6 48 11N 69 42W
Tadzhikistan = Tajikistan ■, Asia 26 F8 38 30N 70 0 E
Taechŏn-ni, S. Korea 35 F14 36 21N 126 36 E
Taegu, S. Korea 35 G15 35 50N 128 37 E
Taegwan, N. Korea 35 D13 40 13N 125 12 E
Taejŏn, S. Korea 35 F14 36 20N 127 28 E
Tafalla, Spain 19 A5 42 30N 1 41W
Tafelbaai, S. Africa 56 E2 33 35S 18 25 E
Tafermaar, Indonesia 37 F8 6 47S 134 10 E
Tafi Viejo, Argentina 94 B2 26 43S 65 17W
Tafīhān, Iran 45 D7 29 25N 52 39 E
Tafresh, Iran 45 C6 34 45N 49 57 E
Taft, Iran 45 D7 31 45N 54 14 E
Taft, Phil. 37 B7 11 57N 125 30 E
Taft, U.S.A. 85 K7 35 8N 119 28W
Taftān, Kūh-e, Iran 45 D9 28 40N 61 0 E
Taga Dzong, Bhutan 41 F16 27 5N 89 55 E
Taganrog, Russia 25 E6 47 12N 38 50 E
Tagbilaran, Phil. 37 C6 9 39N 123 51 E
Tagish, Canada 72 A2 60 19N 134 16W
Tagish L., Canada 72 A2 60 10N 134 20W
Tagliamento →, Italy 20 B5 45 38N 13 6 E
Tagomago, Spain 22 B8 39 2N 1 39 E
Taguatinga, Brazil 93 F10 12 16S 42 26W
Tagum, Phil. 37 C7 7 33N 125 53 E
Tagus = Tejo →, Europe 19 C1 38 40N 9 24W
Tahakopa, N.Z. 59 M2 46 30S 169 23 E
Tahan, Gunong, Malaysia 39 K4 4 34N 102 17 E
Tahat, Algeria 50 D7 23 18N 5 33 E
Tāherī, Iran 45 E7 27 43N 52 20 E
Tahiti, Pac. Oc. 65 J13 17 37S 149 27W
Tahlequah, U.S.A. 81 H7 35 55N 94 58W
Tahoe, L., U.S.A. 84 G6 39 6N 120 2W
Tahoe City, U.S.A. 84 F6 39 10N 120 9W
Tahoka, U.S.A. 81 J4 33 10N 101 48W
Taholah, U.S.A. 84 C2 47 21N 124 17W
Tahoua, Niger 50 F7 14 57N 5 16 E
Tahrūd, Iran 45 D8 29 26N 57 49 E
Tahsis, Canada 72 D3 49 55N 126 40W
Tahta, Egypt 51 C12 26 44N 31 32 E
Tahulandang, Indonesia 37 D7 2 27N 125 23 E
Tahuna, Indonesia 37 D7 3 38N 125 30 E
Tai Shan, China 35 F9 36 25N 117 20 E
Tai'an, China 35 F9 36 12N 117 8 E
Taibei = T'aipei, Taiwan 33 D7 25 2N 121 30 E
Taibique, Canary Is. 22 G2 27 42N 17 58W
Taibus Qi, China 34 D8 41 54N 115 22 E
T'aichung, Taiwan 33 D7 24 9N 120 37 E
Taieri →, N.Z. 59 M3 46 3S 170 12 E
Taigu, China 34 F7 37 28N 112 30 E
Taihang Shan, China 34 G7 36 0N 113 30 E
Taihape, N.Z. 59 H5 39 41S 175 48 E
Taihe, China 34 H8 33 20N 115 42 E
Taikang, China 34 G8 34 5N 114 50 E
Tailem Bend, Australia 63 F2 35 12S 139 29 E
Taimyr Peninsula = Taymyr, Poluostrov, Russia 27 B11 75 0N 100 0 E
Tain, U.K. 12 D4 57 49N 4 4W
T'ainan, Taiwan 33 D7 23 0N 120 10 E
Tainaron, Ákra, Greece 21 F10 36 22N 22 27 E
T'aipei, Taiwan 33 D7 25 2N 121 30 E
Taiping, Malaysia 39 K3 4 51N 100 44 E
Taipingzhen, China 34 H6 33 35N 111 42 E
Tairbeart = Tarbert, U.K. 12 D2 57 54N 6 49W
Taita Hills, Kenya 54 C4 3 25S 38 15 E
Taitao, Pen. de, Chile 96 F2 46 30S 75 0W
T'aitung, Taiwan 33 D7 22 43N 121 4 E
Taivalkoski, Finland 8 D23 65 33N 28 12 E
Taiwan ■, Asia 33 D7 23 30N 121 0 E
Taïyetos Óros, Greece 21 F10 37 0N 22 23 E
Taiyiba, Israel 47 C4 32 36N 35 27 E
Taiyuan, China 34 F7 37 52N 112 33 E
Taizhong = T'aichung, Taiwan 33 D7 24 9N 120 37 E
Ta'izz, Yemen 46 E3 13 35N 44 2 E
Tājābād, Iran 45 D7 30 2N 54 24 E
Tajikistan ■, Asia 26 F8 38 30N 70 0 E
Tajima, Japan 31 F9 37 12N 139 46 E
Tajo = Tejo →, Europe 19 C1 38 40N 9 24W
Tajrish, Iran 45 C6 35 48N 51 25 E
Tak, Thailand 38 D2 16 52N 99 8 E
Takāb, Iran 44 B5 36 24N 47 7 E
Takachiho, Japan 31 H5 32 42N 131 18 E
Takachu, Botswana 56 C3 22 37S 21 58 E
Takada, Japan 31 F9 37 7N 138 15 E
Takahagi, Japan 31 F10 36 43N 140 45 E
Takaka, N.Z. 59 J4 40 51S 172 50 E
Takamatsu, Japan 31 G7 34 20N 134 5 E
Takaoka, Japan 31 F8 36 47N 137 0 E
Takapuna, N.Z. 59 G5 36 47S 174 47 E
Takasaki, Japan 31 F9 36 20N 139 0 E
Takatsuki, Japan 31 G7 34 51N 135 37 E
Takaungu, Kenya 54 C4 3 38S 39 52 E
Takayama, Japan 31 F8 36 18N 137 11 E
Take-Shima, Japan 31 J5 30 49N 130 26 E
Takefu, Japan 31 G8 35 50N 136 10 E
Takengon, Indonesia 36 D1 4 45N 96 50 E
Takeo, Japan 31 H5 33 12N 130 1 E
Tākestān, Iran 45 C6 36 0N 49 40 E
Taketa, Japan 31 H5 32 58N 131 24 E
Takev, Cambodia 39 G5 10 59N 104 47 E
Takh, India 43 C7 33 6N 77 32 E
Takht-Sulaiman, Pakistan 42 D3 31 40N 69 58 E
Takikawa, Japan 30 C10 43 33N 141 54 E
Takla L., Canada 72 B3 55 15N 125 45W
Takla Landing, Canada 72 B3 55 30N 125 50W
Takla Makan = Taklamakan Shamo, China 32 C3 38 0N 83 0 E
Taklamakan Shamo, China 32 C3 38 0N 83 0 E
Taku →, Canada 72 B2 58 30N 133 50W
Tal Halāl, Iran 45 D7 28 54N 55 1 E
Tala, Uruguay 95 C4 34 21S 55 46W
Talagang, Pakistan 42 C5 32 55N 72 25 E
Talagante, Chile 94 C1 33 40S 70 50W
Talamanca, Cordillera de, Cent. Amer. 88 E3 9 20N 83 20W
Talara, Peru 92 D2 4 38S 81 18W
Talas, Kyrgyzstan 26 E8 42 30N 72 13 E
Talâta, Egypt 47 E1 30 36N 32 20 E
Talaud, Kepulauan, Indonesia 37 D7 4 30N 126 50 E
Talaud Is. = Talaud, Kepulauan, Indonesia 37 D7 4 30N 126 50 E
Talavera de la Reina, Spain 19 C3 39 55N 4 46W
Talayan, Phil. 37 C6 6 52N 124 24 E
Talbandh, India 43 H12 22 3N 86 20 E
Talbot, C., Australia 60 B4 13 48S 126 43 E
Talbragar →, Australia 63 E4 32 12S 148 37 E
Talca, Chile 94 D1 35 28S 71 40W
Talcahuano, Chile 94 D1 36 40S 73 10W
Talcher, India 41 J14 21 0N 85 18 E
Taldy Kurgan = Taldyqorghan, Kazakstan 26 E8 45 10N 78 45 E
Taldyqorghan, Kazakstan 26 E8 45 10N 78 45 E
Tālesh, Iran 45 B6 37 58N 48 58 E
Tālesh, Kūhhā-ye, Iran 45 B6 37 42N 48 55 E
Tali Post, Sudan 51 G12 5 55N 30 44 E
Taliabu, Indonesia 37 E6 1 50S 125 0 E
Talibon, Phil. 37 B6 10 9N 124 20 E
Talibong, Ko, Thailand 39 J2 7 15N 99 23 E
Talihina, U.S.A. 81 H7 34 45N 95 3W
Taliwang, Indonesia 36 F5 8 50S 116 55 E
Tall 'Afar, Iraq 44 B4 36 22N 42 27 E
Tall Kalakh, Syria 47 A5 34 41N 36 15 E
Talladega, U.S.A. 77 J2 33 26N 86 6W
Tallahassee, U.S.A. 77 K3 30 27N 84 17W
Tallangatta, Australia 63 F4 36 15S 147 19 E
Tallering Pk., Australia 61 E2 28 6S 115 37 E
Talli, Pakistan 42 E3 29 32N 68 8 E
Tallinn, Estonia 9 G21 59 22N 24 48 E
Tallmadge, U.S.A. 78 E3 41 6N 81 27W
Tallulah, U.S.A. 81 J9 32 25N 91 11W
Taloyoak, Canada 68 B10 69 32N 93 32W
Talpa de Allende, Mexico 86 C4 20 23N 104 51W
Talsi, Latvia 9 H20 57 10N 22 30 E
Taltal, Chile 94 B1 25 23S 70 33W
Taltson →, Canada 72 A6 61 24N 112 46W
Talurqjuak = Taloyoak, Canada 68 B10 69 32N 93 32W
Talwood, Australia 63 D4 28 29S 149 29 E
Talyawalka Cr. →, Australia 63 E3 32 28S 142 22 E
Tam Chau, Vietnam 39 G5 10 48N 105 12 E
Tam Ky, Vietnam 38 E7 15 34N 108 29 E
Tam Quan, Vietnam 38 E7 14 35N 109 3 E
Tama, U.S.A. 80 E8 41 58N 92 35W
Tamale, Ghana 50 G5 9 22N 0 50W
Tamano, Japan 31 G6 34 29N 133 59 E
Tamanrasset, Algeria 50 D7 22 50N 5 30 E
Tamaqua, U.S.A. 79 F9 40 48N 75 58W
Tamar →, U.K. 11 G3 50 27N 4 15W
Tamarinda, Spain 22 B10 39 55N 3 49 E
Tamashima, Japan 31 G6 34 32N 133 40 E
Tamaulipas □, Mexico 87 C5 24 0N 99 0W
Tamaulipas, Sierra de, Mexico 87 C5 23 30N 98 20W
Tamazula, Mexico 86 C3 24 55N 106 58W
Tamazunchale, Mexico 87 C5 21 16N 98 47W
Tambacounda, Senegal 50 F3 13 45N 13 40W
Tambelan, Kepulauan, Indonesia 36 D3 1 0N 107 30 E
Tambellup, Australia 61 F2 34 4S 117 37 E
Tambo, Australia 62 C4 24 54S 146 14 E
Tambo de Mora, Peru 92 F3 13 30S 76 8W
Tambohorano, Madag. 57 B7 17 30S 43 58 E
Tambora, Indonesia 36 F5 8 12S 118 5 E
Tambov, Russia 24 D7 52 45N 41 28 E
Tambuku, Indonesia 37 G15 7 8S 113 40 E
Tâmega →, Portugal 19 B1 41 5N 8 21W
Tamenglong, India 41 G18 25 0N 93 35 E
Tamiahua, L. de, Mexico 87 C5 21 30N 97 30W
Tamil Nadu □, India 40 P10 11 0N 77 0 E
Tamluk, India 43 H12 22 18N 87 58 E
Tammerfors = Tampere, Finland 9 F20 61 30N 23 50 E
Tammisaari, Finland 9 F20 60 0N 23 26 E
Tamo Abu, Pegunungan, Malaysia 36 D5 3 10N 115 5 E
Tampa, U.S.A. 77 M4 27 57N 82 27W
Tampa B., U.S.A. 77 M4 27 50N 82 30W
Tampere, Finland 9 F20 61 30N 23 50 E
Tampico, Mexico 87 C5 22 20N 97 50W
Tampin, Malaysia 39 L4 2 28N 102 13 E
Tamu, Burma 41 G19 24 13N 94 12 E
Tamworth, Australia 63 E5 31 7S 150 58 E
Tamworth, Canada 78 B8 44 29N 77 0W
Tamworth, U.K. 11 E6 52 39N 1 41W
Tamyang, S. Korea 35 G14 35 19N 126 59 E
Tan An, Vietnam 39 G6 10 32N 106 25 E
Tan-Tan, Morocco 50 C3 28 29N 11 1W
Tana →, Kenya 54 C5 2 32S 40 31 E
Tana →, Norway 8 A23 70 30N 28 14 E
Tana, L., Ethiopia 46 E2 13 5N 37 30 E
Tana River, Kenya 54 C4 2 0S 39 30 E
Tanabe, Japan 31 H7 33 44N 135 22 E
Tanafjorden, Norway 8 A23 70 45N 28 25 E
Tanaga, Pta., Canary Is. 22 G1 27 42N 18 10W
Tanahbala, Indonesia 36 E1 0 30S 98 30 E
Tanahgrogot, Indonesia 36 E5 1 55S 116 15 E
Tanahjampea, Indonesia 37 F6 7 10S 120 35 E
Tanahmasa, Indonesia 36 E1 0 12S 98 39 E
Tanahmerah, Indonesia 37 F10 6 5S 140 16 E
Tanakpur, India 43 E9 29 5N 80 7 E
Tanakura, Japan 31 F10 37 10N 140 20 E
Tanami, Australia 60 C4 19 59S 129 43 E
Tanami Desert, Australia 60 C5 18 50S 132 0 E
Tanana →, U.S.A. 68 B4 65 10N 151 58W
Tananarive = Antananarivo, Madag. 57 B8 18 55S 47 31 E
Tânaro →, Italy 18 D8 44 55N 8 40 E
Tancheng, China 35 G10 34 25N 118 20 E
Tanch'ŏn, N. Korea 35 D15 40 27N 128 54 E
Tanda, Ut. P., India 43 F10 26 33N 82 35 E
Tanda, Ut. P., India 43 E8 28 57N 78 56 E
Tandag, Phil. 37 C7 9 4N 126 9 E
Tandaia, Tanzania 55 D3 9 25S 34 15 E
Tandaué, Angola 56 B2 16 58S 18 5 E
Tandil, Argentina 94 D4 37 15S 59 6W
Tandil, Sa. del, Argentina 94 D4 37 30S 59 0W
Tandlianwala, Pakistan 42 D5 31 3N 73 9 E
Tando Adam, Pakistan 42 G3 25 45N 68 40 E
Tando Allahyar, Pakistan 42 G3 25 28N 68 43 E
Tando Bago, Pakistan 42 G3 24 47N 68 58 E
Tando Mohommed Khan, Pakistan 42 G3 25 8N 68 32 E
Tandou L., Australia 63 E3 32 40S 142 5 E
Tandragee, U.K. 13 B5 54 21N 6 24W
Tane-ga-Shima, Japan 31 J5 30 30N 131 0 E
Taneatua, N.Z. 59 H6 38 4S 177 1 E
Tanen Tong Dan = Dawna Ra., Burma 38 D2 16 30N 98 30 E
Tanezrouft, Algeria 50 D6 23 9N 0 11 E
Tang, Koh, Cambodia 39 G4 10 16N 103 7 E
Tang, Ra's-e, Iran 45 E8 25 21N 59 52 E
Tang Krasang, Cambodia 38 F5 12 34N 105 3 E

167

Tubruq

Toliara, Madag. 57 C7 23 21S 43 40 E
Toliara □, Madag. 57 C8 21 0S 45 0 E
Tolima, Colombia 92 C3 4 40N 75 19W
Tolitoli, Indonesia 37 D6 1 5N 120 50 E
Tollhouse, U.S.A. 84 H7 37 1N 119 24W
Tolo, Teluk, Indonesia 37 E6 2 20S 122 10 E
Toluca, Mexico 87 D5 19 20N 99 40W
Tom Burke, S. Africa 57 C4 23 5S 28 0 E
Tom Price, Australia 60 D2 22 40S 117 48 E
Tomah, U.S.A. 80 D9 43 59N 90 30W
Tomahawk, U.S.A. 80 C10 45 28N 89 44W
Tomakomai, Japan 30 C10 42 38N 141 36 E
Tomales, U.S.A. 84 G4 38 15N 122 53W
Tomales B., U.S.A. 84 G3 38 15N 123 58W
Tomar, Portugal 19 C1 39 36N 8 25W
Tomaszów Mazowiecki, Poland 17 C10 51 30N 20 2 E
Tomatlán, Mexico 86 D3 19 56N 105 15W
Tombador, Serra do, Brazil 92 F7 12 0S 58 0W
Tombigbee →, U.S.A. 77 K2 31 8N 87 57W
Tombouctou, Mali 50 E5 16 50N 3 0W
Tombstone, U.S.A. 83 L8 31 43N 110 4W
Tombua, Angola 56 B1 15 55S 11 55 E
Tomé, Chile 94 D1 36 36S 72 57W
Tomelloso, Spain 19 C4 39 10N 3 2W
Tomini, Indonesia 37 D6 0 30N 120 30 E
Tomini, Teluk, Indonesia 37 E6 0 10S 121 0 E
Tomintoul, U.K. 12 D5 57 15N 3 23W
Tomkinson Ranges, Australia 61 E4 26 11S 129 5 E
Tommot, Russia 27 D13 59 4N 126 20 E
Tomnop Ta Suos, Cambodia 39 G5 11 20N 104 15 E
Tomo →, Colombia 92 B5 5 20N 67 48W
Toms Place, U.S.A. 84 H8 37 34N 118 41W
Toms River, U.S.A. 79 G10 39 58N 74 12W
Tomsk, Russia 26 D9 56 30N 85 5 E
Tonalá, Mexico 87 D6 16 8N 93 41W
Tonantins, Brazil 92 D5 2 45S 67 45W
Tonasket, U.S.A. 82 B4 48 42N 119 26W
Tonawanda, U.S.A. 78 D6 43 1N 78 53W
Tonbridge, U.K. 11 F8 51 11N 0 17 E
Tondano, Indonesia 37 D6 1 35N 124 54 E
Tondoro, Namibia 56 B2 17 45S 18 50 E
Tone →, Australia 61 F2 34 25S 116 25 E
Tone-Gawa →, Japan 31 F9 35 44N 140 51 E
Tonekābon, Iran 45 B6 36 45N 51 12 E
Tong Xian, China 34 E9 39 55N 116 35 E
Tonga ■, Pac. Oc. 59 D11 19 50S 174 30W
Tonga Trench, Pac. Oc. 64 J10 18 0S 173 0W
Tongaat, S. Africa 57 D5 29 33S 31 9 E
Tongareva, Cook Is. 65 H12 9 0S 158 0W
Tongatapu Group, Tonga 59 E12 21 0S 175 0W
Tongchŏn-ni, N. Korea 35 E14 39 50N 127 25 E
Tongchuan, China 34 G5 35 6N 109 3 E
Tongeren, Belgium 15 D5 50 47N 5 28 E
Tongguan, China 34 G6 34 40N 110 25 E
Tonghua, China 35 D13 41 42N 125 58 E
Tongjosŏn Man, N. Korea 35 E15 39 30N 128 0 E
Tongking, G. of = Tonkin, G. of, Asia 32 E5 20 0N 108 0 E
Tongliao, China 35 C12 43 38N 122 18 E
Tongling, China 33 C6 30 55N 117 48 E
Tongnae, S. Korea 35 G15 35 12N 129 5 E
Tongobory, Madag. 57 C7 23 32S 44 20 E
Tongoy, Chile 94 C1 30 16S 71 31W
Tongres = Tongeren, Belgium 15 D5 50 47N 5 28 E
Tongsa Dzong, Bhutan 41 F17 27 31N 90 31 E
Tongue, U.K. 12 C4 58 29N 4 25W
Tongue →, U.S.A. 80 B2 46 25N 105 52W
Tongwei, China 34 G3 35 0N 105 5 E
Tongxin, China 34 F3 36 59N 105 58 E
Tongyang, N. Korea 35 E14 39 9N 126 53 E
Tongyu, China 35 B12 44 45N 123 4 E
Tonj, Sudan 51 G11 7 20N 28 44 E
Tonk, India 42 F6 26 6N 75 54 E
Tonkawa, U.S.A. 81 G6 36 41N 97 18W
Tonkin = Bac Phan, Vietnam 38 B5 22 0N 105 0 E
Tonkin, G. of, Asia 32 E5 20 0N 108 0 E
Tonle Sap, Cambodia 38 F4 13 0N 104 0 E
Tono, Japan 30 E10 39 19N 141 32 E
Tonopah, U.S.A. 83 G5 38 4N 117 14W
Tonosí, Panama 88 E3 7 20N 80 20W
Tons →, Haryana, India 42 D7 30 30N 77 39 E
Tons →, Ut. P., India 43 F10 26 1N 83 33 E
Tønsberg, Norway 9 G14 59 19N 10 25 E
Toobanna, Australia 62 B4 18 42S 146 9 E
Toodyay, Australia 61 F2 31 34S 116 28 E
Tooele, U.S.A. 82 F7 40 32N 112 18W
Toompine, Australia 63 D3 27 15S 144 19 E
Toora, Australia 63 F4 38 39S 146 23 E
Toora-Khem, Russia 27 D10 52 28N 96 17 E
Toowoomba, Australia 63 D5 27 32S 151 56 E
Top-ozero, Russia 24 A5 65 35N 32 0 E
Top Springs, Australia 60 C5 16 37S 131 51 E
Topaz, U.S.A. 84 G7 38 41N 119 30W
Topeka, U.S.A. 80 F7 39 3N 95 40W
Topley, Canada 72 C3 54 49N 126 18W
Topocalma, Pta., Chile 94 C1 34 10S 72 2W
Topock, U.S.A. 85 L12 34 46N 114 29W
Topol'čany, Slovak Rep. 17 D10 48 35N 18 12 E
Topolobampo, Mexico 86 B3 25 40N 109 4W
Toppenish, U.S.A. 82 C3 46 23N 120 19W
Toraka Vestale, Madag. 57 B7 16 20S 43 58 E
Torata, Peru 92 G4 17 23S 70 1W
Torbalı, Turkey 21 E12 38 10N 27 21 E
Torbat-e Heydārīyeh, Iran 45 C8 35 15N 59 12 E
Torbat-e Jām, Iran 45 C9 35 16N 60 35 E
Torbay, Canada 71 C9 47 40N 52 42W
Torbay □, U.K. 11 G4 50 26N 3 31W
Tordesillas, Spain 19 B3 41 30N 5 0W
Torfaen □, U.K. 11 F4 51 43N 3 3W
Torgau, Germany 16 C7 51 34N 13 0 E
Torhout, Belgium 15 C3 51 5N 3 7 E
Tori-Shima, Japan 31 J10 30 29N 140 19 E
Torin, Mexico 86 B2 27 33N 110 15W
Torino, Italy 18 D7 45 3N 7 40 E
Torit, Sudan 51 H12 4 27N 32 31 E
Torkamān, Iran 44 B5 37 35N 47 23 E
Tormes →, Spain 19 B2 41 18N 6 29W
Tornado Mt., Canada 72 D6 49 55N 114 40W
Torne älv →, Sweden 8 D21 65 50N 24 12 E
Torneå = Tornio, Finland 8 D21 65 50N 24 12 E
Torneträsk, Sweden 8 B18 68 24N 19 15 E
Tornio, Finland 8 D21 65 50N 24 12 E
Tornionjoki →, Finland 8 D21 65 50N 24 12 E
Tornquist, Argentina 94 D3 38 8S 62 15W
Toro, Spain 22 B11 39 59N 4 8 E
Toro, Cerro del, Chile 94 B2 29 10S 69 50W
Toro Pk., U.S.A. 85 M10 33 34N 116 24W
Toroníios Kólpos, Greece 21 D10 40 5N 23 30 E
Toronto, Canada 78 C5 43 39N 79 20W
Toronto, U.S.A. 78 F4 40 28N 80 36W
Toropets, Russia 24 C5 56 30N 31 40 E
Tororo, Uganda 54 B3 0 45N 34 12 E
Toros Dağları, Turkey 25 G5 37 0N 32 30 E
Torpa, India 43 H11 22 57N 85 6 E
Torquay, U.K. 11 G4 50 27N 3 32W
Torrance, U.S.A. 85 M8 33 50N 118 19W
Torre de Moncorvo, Portugal 19 B2 41 12N 7 8W
Torre del Greco, Italy 20 D6 40 47N 14 22 E
Torrejón de Ardoz, Spain 19 B4 40 27N 3 29W
Torrelavega, Spain 19 A3 43 20N 4 5W
Torremolinos, Spain 19 D3 36 38N 4 30W
Torrens, L., Australia 63 E2 31 0S 137 50 E
Torrens Cr. →, Australia 62 C4 22 23S 145 9 E
Torrens Creek, Australia 62 C4 20 48S 145 3 E
Torrent, Spain 19 C5 39 27N 0 28W
Torreón, Mexico 86 B4 25 33N 103 26W
Torres, Brazil 95 B5 29 21S 49 44W
Torres, Mexico 86 B2 28 46N 110 47W
Torres Strait, Australia 64 H6 9 50S 142 20 E
Torres Vedras, Portugal 19 C1 39 5N 9 15W
Torrevieja, Spain 19 D5 37 59N 0 42W
Torrey, U.S.A. 83 G8 38 18N 111 25W
Torridge →, U.K. 11 G3 51 0N 4 13W
Torridon, L., U.K. 12 D3 57 35N 5 50W
Torrington, Conn., U.S.A. 79 E11 41 48N 73 7W
Torrington, Wyo., U.S.A. 80 D2 42 4N 104 11W
Tórshavn, Færoe Is. 8 E9 62 5N 6 56W
Tortola, Br. Virgin Is. 89 C7 18 19N 64 45W
Tortosa, Spain 19 B6 40 49N 0 31 E
Tortosa, C., Spain 19 B6 40 41N 0 52 E
Tortue, I. de la, Haiti 89 B5 20 5N 72 57W
Torūd, Iran 45 C7 35 25N 55 5 E
Toruń, Poland 17 B10 53 2N 18 39 E
Tory I., Ireland 13 A3 55 16N 8 14W
Tosa, Japan 31 H6 33 24N 133 23 E
Tosa-Shimizu, Japan 31 H6 32 52N 132 58 E
Tosa-Wan, Japan 31 H6 33 15N 133 30 E
Toscana □, Italy 20 C4 43 25N 11 0 E
Toshkent, Uzbekistan 26 E7 41 20N 69 10 E
Tostado, Argentina 94 B3 29 15S 61 50W
Tostón, Pta. de, Canary Is. 22 F5 28 42N 14 2W
Tosu, Japan 31 H5 33 22N 130 31 E
Toteng, Botswana 56 C3 20 22S 22 58 E
Totma, Russia 24 C7 60 0N 42 40 E
Totnes, U.K. 11 G4 50 26N 3 42W
Totness, Surinam 93 B7 5 53N 56 19W
Totonicapán, Guatemala 88 D1 14 58N 91 12W
Totten Glacier, Antarctica 5 C8 66 45S 116 10 E
Tottenham, Australia 63 E4 32 14S 147 21 E
Tottenham, Canada 78 B5 44 1N 79 49W
Tottori, Japan 31 G7 35 30N 134 15 E
Tottori □, Japan 31 G7 35 30N 134 12 E
Toubkal, Djebel, Morocco 50 B4 31 0N 8 0W
Tougan, Burkina Faso 50 F5 13 11N 2 58W
Touggourt, Algeria 50 B7 33 6N 6 4 E
Toul, France 18 B6 48 40N 5 53 E
Toulon, France 18 E6 43 10N 5 55 E
Toulouse, France 18 E4 43 37N 1 27 E
Toummo, Niger 51 D8 22 45N 14 8 E
Toungoo, Burma 41 K20 19 0N 96 30 E
Touraine, France 18 C4 47 20N 0 30 E
Tourane = Da Nang, Vietnam 38 D7 16 4N 108 13 E
Tourcoing, France 18 A5 50 42N 3 10 E
Touriñán, C., Spain 19 A1 43 3N 9 18W
Tournai, Belgium 15 D3 50 35N 3 25 E
Tournon-sur-Rhône, France 18 D6 45 4N 4 50 E
Tours, France 18 C4 47 22N 0 40 E
Toussora, Mt., C.A.R. 52 C4 9 7N 23 14 E
Touws →, S. Africa 56 E3 33 45S 21 11 E
Touwsrivier, S. Africa 56 E3 33 20S 20 2 E
Towada, Japan 30 D10 40 37N 141 13 E
Towada-Ko, Japan 30 D10 40 28N 140 55 E
Towanda, U.S.A. 79 E8 41 46N 76 27W
Towang, India 41 F17 27 37N 91 50 E
Tower, U.S.A. 80 B8 47 48N 92 17W
Towerhill Cr. →, Australia 62 C3 22 28S 144 35 E
Towner, U.S.A. 80 A4 48 21N 100 25W
Townsend, U.S.A. 82 C8 46 19N 111 31W
Townshend I., Australia 62 C5 22 10S 150 31 E
Townsville, Australia 62 B4 19 15S 146 45 E
Towraghondī, Afghan. 40 B3 35 13N 62 16 E
Towson, U.S.A. 76 F7 39 24N 76 36W
Towuti, Danau, Indonesia 37 E6 2 45S 121 32 E
Toya-Ko, Japan 30 C10 42 35N 140 51 E
Toyama, Japan 31 F8 36 40N 137 15 E
Toyama □, Japan 31 F8 36 45N 137 30 E
Toyama-Wan, Japan 31 F8 37 0N 137 30 E
Toyohashi, Japan 31 G8 34 45N 137 25 E
Toyokawa, Japan 31 G8 34 48N 137 27 E
Toyonaka, Japan 31 G7 34 50N 135 28 E
Toyooka, Japan 31 G7 35 35N 134 48 E
Toyota, Japan 31 G8 35 3N 137 7 E
Tozeur, Tunisia 50 B7 33 56N 8 8 E
Trá Li = Tralee, Ireland 13 D2 52 16N 9 42W
Tra On, Vietnam 39 H5 9 58N 105 55 E
Trabzon, Turkey 25 F6 41 0N 39 45 E
Tracadie, Canada 71 C7 47 30N 64 55W
Tracy, Calif., U.S.A. 84 H5 37 44N 121 26W
Tracy, Minn., U.S.A. 80 C7 44 14N 95 37W
Trafalgar, C., Spain 19 D2 36 10N 6 2W
Trail, Canada 72 D5 49 5N 117 40W
Trainor L., Canada 72 A4 60 24N 120 17W
Trákhonas, Cyprus 23 D12 35 12N 33 21 E
Tralee, Ireland 13 D2 52 16N 9 42W
Tralee B., Ireland 13 D2 52 17N 9 55W
Tramore, Ireland 13 D4 52 10N 7 10W
Tramore B., Ireland 13 D4 52 9N 7 10W
Tran Ninh, Cao Nguyen, Laos 38 C4 19 30N 103 10 E
Tranås, Sweden 9 G16 58 3N 14 59 E
Trancas, Argentina 94 B2 26 11S 65 20W
Trang, Thailand 39 J2 7 33N 99 38 E
Trangahy, Madag. 57 B7 19 7S 44 31 E
Trangan, Indonesia 37 F8 6 40S 134 20 E
Trangie, Australia 63 E4 32 4S 148 0 E
Trani, Italy 20 D7 41 17N 16 25 E
Tranoroa, Madag. 57 C8 24 42S 45 4 E
Tranqueras, Uruguay 95 C4 31 13S 55 45W
Transantarctic Mts., Antarctica 5 E12 85 0S 170 0W
Transilvania, Romania 17 E12 46 30N 24 0 E
Transilvanian Alps = Carpaţii Meridionali, Romania 17 F13 45 30N 25 0 E
Transvaal, S. Africa 53 K5 25 0S 29 0 E
Transylvania = Transilvania, Romania 17 E12 46 30N 24 0 E
Trápani, Italy 20 E5 38 1N 12 29 E
Trapper Pk., U.S.A. 82 D6 45 54N 114 18W
Traralgon, Australia 63 F4 38 12S 146 34 E
Trasimeno, L., Italy 20 C5 43 8N 12 6 E
Trat, Thailand 39 F4 12 14N 102 33 E
Tratani →, Pakistan 42 E3 29 19N 68 20 E
Traun, Austria 16 D8 48 14N 14 15 E
Traveller's L., Australia 63 E3 33 20S 142 0 E
Travemünde, Germany 16 B6 53 57N 10 52 E
Travers, Mt., N.Z. 59 K4 42 1S 172 45 E
Traverse City, U.S.A. 76 C3 44 46N 85 38W
Travis, L., U.S.A. 81 K5 30 24N 97 55W
Travnik, Bos.-H. 21 B7 44 17N 17 39 E
Trébbia →, Italy 18 D8 45 4N 9 41 E
Třebíč, Czech Rep. 16 D8 49 14N 15 55 E
Trebinje, Bos.-H. 21 C8 42 44N 18 22 E
Trebonne, Australia 62 B4 18 37S 146 5 E
Tregaron, U.K. 11 E4 52 14N 3 56W
Tregrosse Is., Australia 62 B5 17 41S 150 43 E
Treherne, Canada 73 D9 49 38N 98 42W
Treinta y Tres, Uruguay 95 C5 33 16S 54 17W
Trelawney, Zimbabwe 57 B5 17 30S 30 30 E
Trelew, Argentina 96 E3 43 10S 65 20W
Trelleborg, Sweden 9 J15 55 20N 13 10 E
Tremadog Bay, U.K. 10 E3 52 51N 4 18W
Tremonton, U.S.A. 82 F7 41 43N 112 10W
Tremp, Spain 19 A6 42 10N 0 52 E
Trenche →, Canada 70 C5 47 46N 72 53W
Trenčín, Slovak Rep. 17 D10 48 52N 18 4 E
Trenggalek, Indonesia 37 H14 8 3S 111 43 E
Trenque Lauquen, Argentina 94 D3 36 5S 62 45W
Trent →, Canada 78 B7 44 6N 77 34W
Trent →, U.K. 10 D7 53 41N 0 42W
Trento, Italy 20 A4 46 4N 11 8 E
Trenton, Canada 78 B7 44 10N 77 34W
Trenton, Mo., U.S.A. 80 E8 40 5N 93 37W
Trenton, N.J., U.S.A. 79 F10 40 14N 74 46W
Trenton, Nebr., U.S.A. 80 E4 40 11N 101 1W
Trepassey, Canada 71 C9 46 43N 53 25W
Tres Arroyos, Argentina 94 D3 38 26S 60 20W
Três Corações, Brazil 95 A6 21 44S 45 15W
Três Lagoas, Brazil 93 H8 20 50S 51 43W
Tres Lomas, Argentina 94 D3 36 27S 62 51W
Tres Marías, Islas, Mexico 86 C3 21 25N 106 28W
Tres Montes, C., Chile 96 F1 46 50S 75 30W
Tres Pinos, U.S.A. 84 J5 36 48N 121 19W
Três Pontas, Brazil 95 A6 21 23S 45 29W
Tres Puentes, Chile 94 B1 27 50S 70 15W
Tres Puntas, C., Argentina 96 F3 47 0S 66 0W
Três Rios, Brazil 95 A7 22 6S 43 15W
Tres Valles, Mexico 87 D5 18 15N 96 8W
Tresco, U.K. 11 H1 49 57N 6 20W
Treviso, Italy 20 B5 45 40N 12 15 E
Triabunna, Australia 62 G4 42 30S 147 55 E
Triánda, Greece 23 C10 36 25N 28 10 E
Triangle, Zimbabwe 57 C5 21 2S 31 28 E
Tribal Areas □, Pakistan 42 C4 33 0N 70 0 E
Tribulation, C., Australia 62 B4 16 5S 145 29 E
Tribune, U.S.A. 80 F4 38 28N 101 45W
Trichinopoly = Tiruchchirappalli, India 40 P11 10 45N 78 45 E
Trichur, India 40 P10 10 30N 76 18 E
Trida, Australia 63 E4 33 1S 145 1 E
Trier, Germany 16 D4 49 45N 6 38 E
Trieste, Italy 20 B5 45 40N 13 46 E
Triglav, Slovenia 16 E7 46 21N 13 50 E
Trikkala, Greece 21 E9 39 34N 21 47 E
Trikomo, Cyprus 23 D12 35 17N 33 52 E
Trikora, Puncak, Indonesia 37 E9 4 15S 138 45 E
Trim, Ireland 13 C5 53 33N 6 48W
Trincomalee, Sri Lanka 40 Q12 8 38N 81 15 E
Trindade, Brazil 93 G9 16 40S 49 30W
Trindade, I., Atl. Oc. 2 F8 20 20S 29 50W
Trinidad, Bolivia 92 F6 14 46S 64 50W
Trinidad, Cuba 88 B4 21 48N 80 0W
Trinidad, Trin. & Tob. 89 D7 10 30N 61 15W
Trinidad, Uruguay 94 C4 33 30S 56 50W
Trinidad, U.S.A. 81 G2 37 10N 104 31W
Trinidad →, Mexico 87 D5 17 49N 95 9W
Trinidad & Tobago ■, W. Indies 89 D7 10 30N 61 20W
Trinity, Canada 71 C9 48 59N 53 55W
Trinity, U.S.A. 81 K7 30 57N 95 22W
Trinity →, Calif., U.S.A. 82 F2 41 11N 123 42W
Trinity →, Tex., U.S.A. 81 L7 29 45N 94 43W
Trinity B., Canada 71 C9 48 20N 53 10W
Trinity Is., U.S.A. 68 C4 56 33N 154 25W
Trinity Range, U.S.A. 82 F4 40 15N 118 45W
Trinkitat, Sudan 51 E13 18 45N 37 51 E
Trinway, U.S.A. 78 F2 40 9N 82 1W
Tripoli = Tarābulus, Lebanon 47 A4 34 31N 35 50 E
Tripoli = Tarābulus, Libya 51 B8 32 49N 13 7 E
Tripolis, Greece 21 F10 37 31N 22 25 E
Tripolitania, N. Afr. 51 B8 31 0N 13 0 E
Tripura □, India 41 H18 24 0N 92 0 E
Tripylos, Cyprus 23 E11 34 59N 32 41 E
Tristan da Cunha, Atl. Oc. 49 K2 37 6S 12 20W
Trisul, India 43 D8 30 19N 79 47 E
Trivandrum, India 40 Q10 8 41N 77 0 E
Trnava, Slovak Rep. 17 D9 48 23N 17 35 E
Trochu, Canada 72 C6 51 50N 113 13W
Trodely I., Canada 70 B4 52 15N 79 26W
Troglav, Croatia 20 C7 43 56N 16 36 E
Troilus, L., Canada 70 B5 50 50N 74 35W
Trois-Pistoles, Canada 71 C6 48 5N 69 10W
Trois-Rivières, Canada 70 C5 46 25N 72 34W
Troitsk, Russia 26 D7 54 10N 61 35 E
Troitsko Pechorsk, Russia 24 B10 62 40N 56 10 E
Trölladyngja, Iceland 8 D5 64 54N 17 16W
Trollhättan, Sweden 9 G15 58 17N 12 20 E
Trollheimen, Norway 8 E13 62 46N 9 1 E
Trombetas →, Brazil 93 D7 1 55S 55 35W
Tromsø, Norway 8 B18 69 40N 18 56 E
Trona, U.S.A. 85 K9 35 46N 117 23W
Tronador, Mte., Argentina 96 E2 41 10S 71 50W
Trøndelag, Norway 8 D14 64 17N 11 50 E
Trondheim, Norway 8 E14 63 36N 10 25 E
Trondheimsfjorden, Norway 8 E14 63 35N 10 30 E
Troodos, Cyprus 23 E11 34 55N 32 52 E
Troon, U.K. 12 F4 55 33N 4 39W
Tropic, U.S.A. 83 H7 37 37N 112 5W
Trostan, U.K. 13 A5 55 3N 6 10W
Trout →, Canada 72 A5 61 19N 119 51W
Trout L., N.W.T., Canada 72 A4 60 40N 121 14W
Trout L., Ont., Canada 73 C10 51 20N 93 15W
Trout Lake, Canada 72 B6 56 30N 114 32W
Trout Lake, U.S.A. 84 E5 46 0N 121 32W
Trout River, Canada 71 C8 49 29N 58 8W
Trout Run, U.S.A. 78 E7 41 23N 77 3W
Trouville-sur-Mer, France 18 B4 49 21N 0 5 E
Trowbridge, U.K. 11 F5 51 18N 2 12W
Troy, Turkey 21 E12 39 57N 26 12 E
Troy, Ala., U.S.A. 77 K3 31 48N 85 58W
Troy, Kans., U.S.A. 80 F7 39 47N 95 5W
Troy, Mo., U.S.A. 80 F9 38 59N 90 59W
Troy, Mont., U.S.A. 82 B6 48 28N 115 53W
Troy, N.Y., U.S.A. 79 D11 42 44N 73 41W
Troy, Ohio, U.S.A. 76 E3 40 2N 84 12W
Troy, Pa., U.S.A. 79 E8 41 47N 76 47W
Troyes, France 18 B6 48 19N 4 3 E
Truchas Peak, U.S.A. 81 H2 35 58N 105 39W
Trucial States = United Arab Emirates ■, Asia 46 C5 23 50N 54 0 E
Truckee, U.S.A. 84 F6 39 20N 120 11W
Trudovoye, Russia 30 C6 43 17N 132 5 E
Trujillo, Honduras 88 C2 16 0N 86 0W
Trujillo, Peru 92 E3 8 6S 79 0W
Trujillo, Spain 19 C3 39 28N 5 55W
Trujillo, U.S.A. 81 H2 35 32N 104 42W
Trujillo, Venezuela 92 B4 9 22N 70 38W
Truk, Micronesia 64 G7 7 25N 151 46 E
Trumann, U.S.A. 81 H9 35 41N 90 31W
Trumansburg, U.S.A. 79 D8 42 33N 76 40W
Trumbull, Mt., U.S.A. 83 H7 36 25N 113 8W
Trundle, Australia 63 E4 32 53S 147 35 E
Trung-Phan = Annam, Vietnam 38 E7 16 0N 108 0 E
Truro, Canada 71 C7 45 21N 63 14W
Truro, U.K. 11 G2 50 16N 5 4W
Truskavets, Ukraine 17 D12 49 17N 23 30 E
Trutch, Canada 72 B4 57 44N 122 57W
Truth or Consequences, U.S.A. 83 K10 33 8N 107 15W
Trutnov, Czech Rep. 16 C8 50 37N 15 54 E
Truxton, U.S.A. 79 D8 42 45N 76 2W
Tryonville, U.S.A. 78 E5 41 42N 79 48W
Tsandi, Namibia 56 B1 17 42S 14 50 E
Tsaratanana, Madag. 57 B8 16 47S 47 39 E
Tsaratanana, Mt. de, Madag. 57 A8 14 0S 49 0 E
Tsarevo = Michurin, Bulgaria 21 C12 42 9N 27 51 E
Tsau, Botswana 56 C3 20 8S 22 22 E
Tselinograd = Astana, Kazakstan 26 D8 51 10N 71 30 E
Tses, Namibia 56 D2 25 58S 18 8 E
Tsetserleg, Mongolia 32 B5 47 36N 101 32 E
Tshabong, Botswana 56 D3 26 2S 22 29 E
Tshane, Botswana 56 C3 24 5S 21 54 E
Tshela, Dem. Rep. of the Congo 52 E2 4 57S 13 4 E
Tshesebe, Botswana 57 C4 21 51S 27 32 E
Tshibeke, Dem. Rep. of the Congo 54 C2 2 40S 28 35 E
Tshibinda, Dem. Rep. of the Congo 54 C2 2 23S 28 43 E
Tshikapa, Dem. Rep. of the Congo 52 F4 6 28S 20 48 E
Tshilenge, Dem. Rep. of the Congo 54 D1 6 17S 23 48 E
Tshinsenda, Dem. Rep. of the Congo 55 E2 12 20S 28 0 E
Tshofa, Dem. Rep. of the Congo 54 D2 5 13S 25 16 E
Tshwane, Botswana 56 C3 22 24S 22 1 E
Tsigara, Botswana 56 C4 20 22S 25 54 E
Tsihombe, Madag. 57 D8 25 10S 45 41 E
Tsiigehtchic, Canada 68 B6 67 15N 134 0W
Tsimlyansk Res. = Tsimlyanskoye Vdkhr., Russia 25 E7 48 0N 43 0 E
Tsimlyanskoye Vdkhr., Russia 25 E7 48 0N 43 0 E
Tsinan = Jinan, China 34 F9 36 38N 117 1 E
Tsineng, S. Africa 56 D3 27 5S 23 5 E
Tsinghai = Qinghai □, China 32 C4 36 0N 98 0 E
Tsingtao = Qingdao, China 35 F11 36 5N 120 20 E
Tsinjoarivo, Madag. 57 B8 19 37S 47 40 E
Tsinjomitondraka, Madag. 57 B8 15 40S 47 8 E
Tsiroanomandidy, Madag. 57 B8 18 46S 46 2 E
Tsitondroina, Madag. 57 C8 21 19S 46 0 E
Tsivory, Madag. 57 C8 24 4S 46 5 E
Tskhinvali, Georgia 25 F7 42 14N 44 1 E
Tsna →, Russia 24 D7 54 55N 41 58 E
Tso Moriri, L., India 43 C8 32 50N 78 20 E
Tsobis, Namibia 56 B2 19 27S 17 30 E
Tsodilo Hill, Botswana 56 B3 18 49S 21 43 E
Tsogttsetsiy = Baruunsuu, Mongolia 34 C3 43 43N 105 35 E
Tsolo, S. Africa 57 E4 31 18S 28 37 E
Tsomo, S. Africa 57 E4 32 0S 27 42 E
Tsu, Japan 31 G8 34 45N 136 25 E
Tsu L., Canada 72 A6 60 40N 111 52W
Tsuchiura, Japan 31 F10 36 5N 140 15 E
Tsugaru-Kaikyō, Japan 30 D10 41 35N 141 0 E
Tsumeb, Namibia 56 B2 19 9S 17 44 E
Tsumis, Namibia 56 C2 23 39S 17 29 E
Tsuruga, Japan 31 G8 35 45N 136 2 E
Tsurugi-San, Japan 31 H7 33 51N 134 6 E
Tsuruoka, Japan 30 E9 38 44N 139 50 E
Tsushima, Gifu, Japan 31 G8 35 10N 136 43 E
Tsushima, Nagasaki, Japan 31 G4 34 20N 129 20 E
Tsuyama, Japan 31 G7 35 3N 134 0 E
Tsyelyakhany, Belarus 17 B13 52 30N 25 46 E
Tual, Indonesia 37 F8 5 38S 132 44 E
Tuam, Ireland 13 C3 53 31N 8 51W
Tuamotu Arch. = Tuamotu Is., Pac. Oc. 65 J13 17 0S 144 0W
Tuamotu Is., Pac. Oc. 65 J13 17 0S 144 0W
Tuamotu Ridge, Pac. Oc. 65 K14 20 0S 138 0W
Tuao, Phil. 37 A6 17 55N 121 22 E
Tuapse, Russia 25 F6 44 5N 39 10 E
Tuatapere, N.Z. 59 M1 46 8S 167 41 E
Tuba City, U.S.A. 83 H8 36 8N 111 14W
Tuban, Indonesia 37 G15 6 54S 112 3 E
Tubani, Botswana 56 C3 24 46S 24 18 E
Tubarão, Brazil 95 B6 28 30S 49 0W
Tūbās, West Bank 47 C4 32 20N 35 22 E
Tubas →, Namibia 56 C2 22 54S 14 35 E
Tübingen, Germany 16 D5 48 31N 9 4 E
Tubruq, Libya 51 B10 32 7N 23 55 E

Tubuai Is.

Tubuai Is., Pac. Oc. 65 K13 25 0S 150 0W
Tuc Trung, Vietnam 39 G6 11 1N 107 12 E
Tucacas, Venezuela 92 A5 10 48N 68 19W
Tuchodi →, Canada 72 B4 58 17N 123 42W
Tuckanarra, Australia 61 E2 27 7S 118 5 E
Tucson, U.S.A. 83 K8 32 13N 110 58W
Tucumán □, Argentina 94 B2 26 48S 66 2W
Tucumcari, U.S.A. 81 H3 35 10N 103 44W
Tucupita, Venezuela 92 B6 9 2N 62 3W
Tucuruí, Brazil 93 D9 3 42S 49 44W
Tucuruí, Represa de, Brazil 93 D9 4 0S 49 30W
Tudela, Spain 19 A5 42 4N 1 39W
Tudmur, Syria 44 C3 34 36N 38 15 E
Tudor, L., Canada 71 A6 55 50N 65 25W
Tugela →, S. Africa 57 D5 29 14S 31 30 E
Tuguegarao, Phil. 37 A6 17 35N 121 42 E
Tugur, Russia 27 D14 53 44N 136 45 E
Tui, Spain 19 A1 42 3N 8 39W
Tuineje, Canary Is. 22 F5 28 19N 14 3W
Tukangbesi, Kepulauan, Indonesia 37 F6 6 0S 124 0 E
Tukarak I., Canada 70 A4 56 15N 78 45W
Tukayyid, Iraq 44 D5 29 47N 45 36 E
Tuktoyaktuk, Canada 68 B6 69 27N 133 2W
Tukums, Latvia 9 H20 56 58N 23 10 E
Tukuyu, Tanzania 55 D3 9 17S 33 35 E
Tula, Hidalgo, Mexico 87 C5 20 5N 99 20W
Tula, Tamaulipas, Mexico 87 C5 23 0N 99 40W
Tula, Russia 24 D6 54 13N 37 38 E
Tulancingo, Mexico 87 C5 20 5N 99 22W
Tulare, U.S.A. 84 J7 36 13N 119 21W
Tulare Lake Bed, U.S.A. 84 K7 36 0N 119 48W
Tularosa, U.S.A. 83 K10 33 5N 106 1W
Tulbagh, S. Africa 56 E2 33 16S 19 6 E
Tulcán, Ecuador 92 C3 0 48N 77 43W
Tulcea, Romania 17 F15 45 13N 28 46 E
Tulchyn, Ukraine 17 D15 48 41N 28 49 E
Tüleh, Iran 45 C7 34 35N 52 33 E
Tulemalu L., Canada 73 A9 62 58N 99 25W
Tuli, Zimbabwe 55 G2 21 58S 29 13 E
Tulia, U.S.A. 81 H4 34 32N 101 46W
Tulita, Canada 68 B7 64 57N 125 30W
Ţūlkarm, West Bank 47 C4 32 19N 35 2 E
Tulla, Ireland 13 D3 52 53N 8 46W
Tullahoma, U.S.A. 77 H2 35 22N 86 13W
Tullamore, Australia 63 E4 32 39S 147 36 E
Tullamore, Ireland 13 C4 53 16N 7 31W
Tulle, France 18 D4 45 16N 1 46 E
Tullow, Ireland 13 D5 52 49N 6 45W
Tully, Australia 62 B4 17 56S 145 55 E
Tully, U.S.A. 79 D8 42 48N 76 7W
Tulsa, U.S.A. 81 G7 36 10N 95 55W
Tulsequah, Canada 72 B2 58 39N 133 35W
Tulua, Colombia 92 C3 4 6N 76 11W
Tulun, Russia 27 D11 54 32N 100 35 E
Tulungagung, Indonesia 37 H14 8 5S 111 54 E
Tuma →, Nic. 88 D3 13 6N 84 35W
Tumaco, Colombia 92 C3 1 50N 78 45W
Tumatumari, Guyana 92 B7 5 20N 58 55W
Tumba, Sweden 9 G17 59 12N 17 48 E
Tumba, L., Dem. Rep. of the Congo 52 E3 0 50S 18 0 E
Tumbarumba, Australia 63 F4 35 44S 148 0 E
Tumbaya, Argentina 94 A2 23 50S 65 26W
Tumbes, Peru 92 D2 3 37S 80 27W
Tumbwe, Dem. Rep. of the Congo 55 E2 11 25S 27 15 E
Tumby Bay, Australia 63 E2 34 21S 136 8 E
Tumd Youqi, China 34 D6 40 30N 110 30 E
Tumen, China 35 C15 43 0N 129 50 E
Tumen Jiang →, China 35 C16 42 20N 130 35 E
Tumeremo, Venezuela 92 B6 7 18N 61 30W
Tumkur, India 40 N10 13 18N 77 6 E
Tump, Pakistan 40 F3 26 7N 62 16 E
Tumpat, Malaysia 39 J4 6 11N 102 10 E
Tumu, Ghana 50 F5 10 56N 1 56W
Tumucumaque, Serra, Brazil 93 C8 2 0N 55 0W
Tumut, Australia 63 F4 35 16S 148 13 E
Tumwater, U.S.A. 84 C4 47 1N 122 54W
Tuna, India 42 H4 22 59N 70 5 E
Tunas de Zaza, Cuba 88 B4 21 39N 79 34W
Tunbridge Wells = Royal Tunbridge Wells, U.K. 11 F8 51 7N 0 16 E
Tuncurry, Australia 63 E5 32 17S 152 29 E
Tundla, India 42 F8 27 12N 78 17 E
Tunduru, Tanzania 55 E4 11 8S 37 25 E
Tundzha →, Bulgaria 21 C11 41 40N 26 35 E
Tungabhadra →, India 40 M11 15 57N 78 15 E
Tungla, Nic. 88 D3 13 24N 84 21W
Tungsten, Canada 72 A3 61 57N 128 16W
Tunguska, Nizhnyaya →, Russia 27 C9 65 48N 88 4 E
Tunguska, Podkamennaya →, Russia 27 C10 61 50N 90 13 E
Tunica, U.S.A. 81 H9 34 41N 90 23W
Tunis, Tunisia 50 A7 36 50N 10 11 E
Tunisia ■, Africa 50 B6 33 30N 9 10 E
Tunja, Colombia 92 B4 5 33N 73 25W
Tunkhannock, U.S.A. 79 E9 41 32N 75 57W
Tunliu, China 34 F7 36 13N 112 52 E
Tunnsjøen, Norway 8 D15 64 45N 13 25 E
Tunungayualok I., Canada 71 A7 56 0N 61 0W
Tununirusiq = Arctic Bay, Canada 69 A11 73 1N 85 7W
Tunuyán, Argentina 94 C2 33 35S 69 0W
Tunuyán →, Argentina 94 C2 33 33S 67 30W
Tuolumne, U.S.A. 84 H6 37 58N 120 15W
Tuolumne →, U.S.A. 84 H5 37 36N 121 13W
Tūp Āghāj, Iran 44 B5 36 3N 47 50 E
Tupã, Brazil 95 A5 21 57S 50 28W
Tupelo, U.S.A. 77 H1 34 16N 88 43W
Tupinambaranas, Brazil 92 D7 3 0S 58 0W
Tupiza, Bolivia 94 A2 21 30S 65 40W
Tupman, U.S.A. 85 K7 35 18N 119 21W
Tupper, Canada 72 B4 55 32N 120 1W
Tupper Lake, U.S.A. 79 B10 44 14N 74 28W
Tupungato, Cerro, S. Amer. 94 C2 33 15S 69 50W
Tuquan, China 35 B11 45 18N 121 38 E
Túquerres, Colombia 92 C3 1 5N 77 37W
Tura, Russia 27 C11 64 20N 100 17 E
Turabah, Si. Arabia 46 C3 28 20N 43 15 E
Tūrān, Iran 45 C8 35 39N 56 42 E
Turan, Russia 27 D10 51 55N 95 0 E
Ţurayf, Si. Arabia 44 D3 31 41N 38 39 E
Turda, Romania 17 E12 46 34N 23 47 E
Turek, Poland 17 B10 52 3N 18 30 E
Turen, Venezuela 92 B5 9 17N 69 6W
Turfan = Turpan, China 32 B3 43 58N 89 10 E
Turfan Depression = Turpan Hami, China 28 E12 42 40N 89 25 E
Turgeon →, Canada 70 C4 50 0N 78 56W
Türgovishte, Bulgaria 21 C12 43 17N 26 38 E
Turgutlu, Turkey 21 E12 38 30N 27 43 E
Turgwe →, Zimbabwe 57 C5 21 31S 32 15 E
Turia →, Spain 19 C5 39 27N 0 19W
Turiaçu, Brazil 93 D9 1 40S 45 19W
Turiaçu →, Brazil 93 D9 1 36S 45 19W
Turin = Torino, Italy 18 D7 45 3N 7 40 E
Turkana, L., Africa 54 B4 3 30N 36 5 E
Turkestan = Türkistan, Kazakstan 26 E7 43 17N 68 16 E
Turkey ■, Eurasia 25 G6 39 0N 36 0 E
Turkey Creek, Australia 60 C4 17 2S 128 12 E
Türkistan, Kazakstan 26 E7 43 17N 68 16 E
Türkmenbashi, Turkmenistan 25 G9 40 5N 53 5 E
Turkmenistan ■, Asia 26 F6 39 0N 59 0 E
Turks & Caicos Is. ■, W. Indies 89 B5 21 20N 71 20W
Turks Island Passage, W. Indies 89 B5 21 30N 71 30W
Turku, Finland 9 F20 60 30N 22 19 E
Turkwel →, Kenya 54 B4 3 6N 36 6 E
Turlock, U.S.A. 84 H6 37 30N 120 51W
Turnagain →, Canada 72 B3 59 12N 127 35W
Turnagain, C., N.Z. 59 J6 40 28S 176 38 E
Turneffe Is., Belize 87 D7 17 20N 87 50W
Turner, U.S.A. 82 B9 48 51N 108 24W
Turner Pt., Australia 62 A1 11 47S 133 32 E
Turner Valley, Canada 72 C6 50 40N 114 17W
Turners Falls, U.S.A. 79 D12 42 36N 72 33W
Turnhout, Belgium 15 C4 51 19N 4 57 E
Turnor L., Canada 73 B7 56 35N 108 35W
Tŭrnovo = Veliko Tŭrnovo, Bulgaria 21 C11 43 5N 25 41 E
Turnu Măgurele, Romania 17 G13 43 46N 24 56 E
Turnu Roşu, P., Romania 17 F13 45 33N 24 17 E
Turpan, China 32 B3 43 58N 89 10 E
Turpan Hami, China 28 E12 42 40N 89 25 E
Turriff, U.K. 12 D6 57 32N 2 27W
Tursāq, Iraq 44 C5 33 27N 45 47 E
Turtle Head I., Australia 62 A3 10 56S 142 37 E
Turtle L., Canada 73 C7 53 36N 108 38W
Turtle Lake, U.S.A. 80 B4 47 31N 100 53W
Turtleford, Canada 73 C7 53 23N 108 57W
Turukhansk, Russia 27 C9 65 21N 88 5 E
Tuscaloosa, U.S.A. 77 J2 33 12N 87 34W
Tuscany = Toscana □, Italy 20 C4 43 25N 11 0 E
Tuscarawas →, U.S.A. 78 F3 40 24N 81 25W
Tuscarora Mt., U.S.A. 78 F7 40 55N 77 55W
Tuscola, Ill., U.S.A. 76 F1 39 48N 88 17W
Tuscola, Tex., U.S.A. 81 J5 32 12N 99 48W
Tuscumbia, U.S.A. 77 H2 34 44N 87 42W
Tuskegee, U.S.A. 77 J3 32 25N 85 42W
Tustin, U.S.A. 85 M9 33 44N 117 49W
Tuticorin, India 40 Q11 8 50N 78 12 E
Tutóia, Brazil 93 D10 2 45S 42 20W
Tutong, Brunei 36 D4 4 47N 114 40 E
Tutrakan, Bulgaria 21 B12 44 2N 26 40 E
Tuttle Creek L., U.S.A. 80 F6 39 22N 96 40W
Tuttlingen, Germany 16 E5 47 58N 8 48 E
Tutuala, E. Timor 37 F7 8 25S 127 15 E
Tutuila, Amer. Samoa 59 B13 14 19S 170 50W
Tutume, Botswana 53 J5 20 30S 27 5 E
Tututepec, Mexico 87 D5 16 9N 97 38W
Tuva □, Russia 27 D10 51 30N 95 0 E
Tuvalu ■, Pac. Oc. 64 H9 8 0S 178 0 E
Tuxpan, Mexico 87 C5 20 58N 97 23W
Tuxtla Gutiérrez, Mexico 87 D6 16 50N 93 10W
Tuy = Tui, Spain 19 A1 42 3N 8 39W
Tuy An, Vietnam 38 F7 13 17N 109 16 E
Tuy Duc, Vietnam 39 F6 12 15N 107 27 E
Tuy Hoa, Vietnam 38 F7 13 5N 109 10 E
Tuy Phong, Vietnam 39 G7 11 14N 108 43 E
Tuya L., Canada 72 B2 59 7N 130 35W
Tuyen Hoa, Vietnam 38 D6 17 50N 106 10 E
Tuyen Quang, Vietnam 38 B5 21 50N 105 10 E
Tūysarkān, Iran 45 C6 34 33N 48 27 E
Tuz Gölü, Turkey 25 G5 38 42N 33 18 E
Ţūz Khurmātū, Iraq 44 C5 34 56N 44 38 E
Tuzla, Bos.-H. 21 B8 44 34N 18 41 E
Tver, Russia 24 C6 56 55N 35 55 E
Twain, U.S.A. 84 E5 40 1N 121 3W
Twain Harte, U.S.A. 84 G6 38 2N 120 14W
Tweed, Canada 78 B7 44 29N 77 19W
Tweed →, U.K. 12 F6 55 45N 2 0W
Tweed Heads, Australia 63 D5 28 10S 153 31 E
Tweedsmuir Prov. Park, Canada 72 C3 53 0N 126 20W
Twentynine Palms, U.S.A. 85 L10 34 8N 116 3W
Twillingate, Canada 71 C9 49 42N 54 45W
Twin Bridges, U.S.A. 82 D7 45 33N 112 20W
Twin Falls, Canada 71 B7 53 30N 64 32W
Twin Falls, U.S.A. 82 E6 42 34N 114 28W
Twin Valley, U.S.A. 80 B6 47 16N 96 16W
Twinsburg, U.S.A. 78 E3 41 18N 81 26W
Twitchell Reservoir, U.S.A. 85 L6 34 59N 120 19W
Two Harbors, U.S.A. 80 B9 47 2N 91 40W
Two Hills, Canada 72 C6 53 43N 111 52W
Two Rivers, U.S.A. 76 C2 44 9N 87 34W
Two Rocks, Australia 61 F2 31 30S 115 35 E
Twofold B., Australia 63 F4 37 8S 149 59 E
Tyachiv, Ukraine 17 D12 48 1N 23 35 E
Tychy, Poland 17 C10 50 9N 18 59 E
Tyler, Minn., U.S.A. 80 C6 44 18N 96 8W
Tyler, Tex., U.S.A. 81 J7 32 21N 95 18W
Tynda, Russia 27 D13 55 10N 124 43 E
Tyndall, U.S.A. 80 D6 43 0N 97 50W
Tyne →, U.K. 10 C6 54 59N 1 32W
Tyne & Wear □, U.K. 10 B6 55 6N 1 17W
Tynemouth, U.K. 10 B6 55 1N 1 26W
Tyre = Sūr, Lebanon 47 B4 33 19N 35 16 E
Tyrifjorden, Norway 9 F14 60 2N 10 8 E
Tyrol = Tirol □, Austria 16 E6 47 3N 10 43 E
Tyrone, U.S.A. 78 F6 40 40N 78 14W
Tyrone □, U.K. 13 B4 54 38N 7 11W
Tyrrell →, Australia 63 F3 35 26S 142 51 E
Tyrrell, L., Australia 63 F3 35 20S 142 50 E
Tyrrell L., Canada 73 A7 63 7N 105 27W
Tyrrhenian Sea, Medit. S. 20 E5 40 0N 12 30 E
Tysfjorden, Norway 8 B17 68 7N 16 25 E
Tyulgan, Russia 24 D10 52 22N 56 12 E
Tyumen, Russia 26 D7 57 11N 65 29 E
Tywi →, U.K. 11 F3 51 48N 4 21W
Tywyn, U.K. 11 E3 52 35N 4 5W
Tzaneen, S. Africa 57 C5 23 47S 30 9 E
Tzermiádhes, Greece 23 D7 35 12N 25 29 E
Tzukong = Zigong, China 32 D5 29 15N 104 48 E

U

U Taphao, Thailand 38 F3 12 35N 101 0 E
U.S.A. = United States of America ■, N. Amer. 74 C7 37 0N 96 0W
Uatumã →, Brazil 92 D7 2 26S 57 37W
Uaupés, Brazil 92 D5 0 8S 67 5W
Uaupés →, Brazil 92 C5 0 2N 67 16W
Uaxactún, Guatemala 88 C2 17 25N 89 29W
Ubá, Brazil 95 A7 21 8S 43 0W
Ubaitaba, Brazil 93 F11 14 18S 39 20W
Ubangi = Oubangi →, Dem. Rep. of the Congo 52 E3 0 30S 17 50 E
Ubauro, Pakistan 42 E3 28 15N 69 45 E
Ubayyiḑ, W. al →, Iraq 44 C4 32 34N 43 48 E
Ube, Japan 31 H5 33 56N 131 15 E
Úbeda, Spain 19 C4 38 3N 3 23W
Uberaba, Brazil 93 G9 19 50S 47 55W
Uberlândia, Brazil 93 G9 19 0S 48 20W
Ubolratna Res., Thailand 38 D4 16 45N 102 30 E
Ubombo, S. Africa 57 D5 27 31S 32 4 E
Ubon Ratchathani, Thailand 38 E5 15 15N 104 50 E
Ubondo, Dem. Rep. of the Congo 54 C2 0 55S 25 42 E
Ubort →, Belarus 17 B15 52 6N 28 30 E
Ubundu, Dem. Rep. of the Congo 54 C2 0 22S 25 30 E
Ucayali →, Peru 92 D4 4 30S 73 30W
Uchab, Namibia 56 B2 19 47S 17 42 E
Uchiura-Wan, Japan 30 C10 42 25N 140 40 E
Uchquduq, Uzbekistan 26 E7 41 50N 62 50 E
Uchur →, Russia 27 D14 58 48N 130 35 E
Ucluelet, Canada 72 D3 48 57N 125 32W
Uda →, Russia 27 D14 54 42N 135 14 E
Udagamandalam, India 40 P10 11 30N 76 44 E
Udainagar, India 42 H7 22 33N 76 13 E
Udaipur, India 42 G5 24 36N 73 44 E
Udaipur Garhi, Nepal 43 F12 27 0N 86 35 E
Udala, India 43 J12 21 35N 86 34 E
Uddevalla, Sweden 9 G14 58 21N 11 55 E
Uddjaur, Sweden 8 D17 65 56N 17 49 E
Uden, Neths. 15 C5 51 40N 5 37 E
Udgir, India 40 K10 18 25N 77 5 E
Udhampur, India 43 C6 33 0N 75 5 E
Údine, Italy 20 A5 46 3N 13 14 E
Udmurtia □, Russia 24 C9 57 30N 52 30 E
Udon Thani, Thailand 38 D4 17 29N 102 46 E
Udupi, India 40 N9 13 25N 74 42 E
Udzungwa Range, Tanzania 55 D4 9 30S 35 10 E
Ueda, Japan 31 F9 36 24N 138 16 E
Uedineniya, Os., Russia 4 B12 78 0N 85 0 E
Uele →, Dem. Rep. of the Congo 52 D4 3 45N 24 45 E
Uelen, Russia 27 C19 66 10N 170 0W
Uelzen, Germany 16 B6 52 57N 10 32 E
Ufa, Russia 24 D10 54 45N 55 55 E
Ufa →, Russia 24 D10 54 40N 56 0 E
Ugab →, Namibia 56 C1 20 55S 13 30 E
Ugalla →, Tanzania 54 D3 5 8S 30 42 E
Uganda ■, Africa 54 B3 2 0N 32 0 E
Ugie, S. Africa 57 E4 31 10S 28 13 E
Uglegorsk, Russia 27 E15 49 5N 142 2 E
Ugljan, Croatia 16 F8 44 12N 15 10 E
Uhlenhorst, Namibia 56 C2 23 45S 17 55 E
Uhrichsville, U.S.A. 78 F3 40 24N 81 21W
Uibhist a Deas = South Uist, U.K. 12 D1 57 20N 7 15W
Uibhist a Tuath = North Uist, U.K. 12 D1 57 40N 7 15W
Uig, U.K. 12 D2 57 35N 6 21W
Uíge, Angola 52 F2 7 30S 14 40 E
Uijŏngbu, S. Korea 35 F14 37 48N 127 0 E
Ŭiju, N. Korea 35 D13 40 15N 124 35 E
Uinta Mts., U.S.A. 82 F8 40 45N 110 30W
Uis, Namibia 56 B2 21 8S 14 49 E
Uitenhage, S. Africa 56 E4 33 40S 25 28 E
Uithuizen, Neths. 15 A6 53 24N 6 41 E
Ujh →, India 42 C6 32 10N 75 18 E
Ujhani, India 43 F8 28 0N 79 6 E
Uji-guntō, Japan 31 J4 31 15N 129 25 E
Ujjain, India 42 H6 23 9N 75 43 E
Ujung Pandang, Indonesia 37 F5 5 10S 119 20 E
Uka, Russia 27 D17 57 50N 162 0 E
Ukara I., Tanzania 54 C3 1 50S 33 0 E
Uke-Shima, Japan 31 K4 28 2N 129 14 E
Ukerewe I., Tanzania 54 C3 2 0S 33 0 E
Ukhrul, India 41 G19 25 10N 94 25 E
Ukhta, Russia 24 B9 63 34N 53 41 E
Ukiah, U.S.A. 84 F3 39 9N 123 13W
Ukki Fort, India 43 C7 33 28N 76 54 E
Ukmergė, Lithuania 9 J21 55 15N 24 45 E
Ukraine ■, Europe 25 E5 49 0N 32 0 E
Ukwi, Botswana 56 C3 23 29S 20 30 E
Ulaan-Uul, Mongolia 34 B6 44 13N 111 10 E
Ulaanbaatar, Mongolia 27 E11 47 55N 106 53 E
Ulaangom, Mongolia 32 A4 50 5N 92 10 E
Ulaanjirem, Mongolia 34 B3 45 5N 105 30 E
Ulamba, Dem. Rep. of the Congo 55 D1 9 3S 23 38 E
Ulan Bator = Ulaanbaatar, Mongolia 27 E11 47 55N 106 53 E
Ulan Ude, Russia 27 D11 51 45N 107 40 E
Ulaya, Morogoro, Tanzania 54 D4 7 3S 36 55 E
Ulaya, Tabora, Tanzania 54 C3 4 25S 33 30 E
Ulcinj, Montenegro, Yug. 21 D8 41 58N 19 10 E
Ulco, S. Africa 56 D3 28 21S 24 15 E
Ulefoss, Norway 9 G13 59 17N 9 16 E
Ulhasnagar, India 40 K8 19 15N 73 10 E
Uliastay = Ulyasutay, Mongolia 32 B4 47 56N 97 28 E
Ulladulla, Australia 63 F5 35 21S 150 29 E
Ullapool, U.K. 12 D3 57 54N 5 9W
Ullswater, U.K. 10 C5 54 34N 2 52W
Ullŭng-do, S. Korea 31 F5 37 30N 130 30 E
Ulm, Germany 16 D5 48 23N 9 58 E
Ulmarra, Australia 63 D5 29 37S 153 4 E
Ulonguè, Mozam. 55 E3 14 37S 34 19 E
Ulricehamn, Sweden 9 H15 57 46N 13 26 E
Ulsan, S. Korea 35 G15 35 20N 129 15 E
Ulsta, U.K. 12 A7 60 30N 1 9W
Ulster □, U.K. 13 B5 54 35N 6 30W
Ulubat Gölü, Turkey 21 D13 40 9N 28 35 E
Uludağ, Turkey 21 D13 40 4N 29 13 E
Uluguru Mts., Tanzania 54 D4 7 15S 37 40 E
Ulungur He →, China 32 B3 47 1N 87 24 E
Uluru = Ayers Rock, Australia 61 E5 25 23S 131 5 E
Ulutau, Kazakstan 26 E7 48 39N 67 1 E
Ulva, U.K. 12 E2 56 29N 6 13W
Ulverston, U.K. 10 C4 54 13N 3 5W
Ulverstone, Australia 62 G4 41 11S 146 11 E
Ulya, Russia 27 D15 59 10N 142 0 E
Ulyanovsk = Simbirsk, Russia 24 D8 54 20N 48 25 E
Ulyasutay, Mongolia 32 B4 47 56N 97 28 E
Ulysses, U.S.A. 81 G4 37 35N 101 22W
Umala, Bolivia 92 G5 17 25S 68 5W
Uman, Ukraine 17 D16 48 40N 30 12 E
Umaria, India 41 H12 23 35N 80 50 E
Umarkot, Pakistan 40 G6 25 15N 69 40 E
Umarpada, India 42 J5 21 27N 73 30 E
Umatilla, U.S.A. 82 D4 45 55N 119 21W
Umba, Russia 24 A5 66 42N 34 11 E
Umbagog L., U.S.A. 79 B13 44 46N 71 3W
Umbakumba, Australia 62 A2 13 47S 136 50 E
Umbrella Mts., N.Z. 59 L2 45 35S 169 5 E
Ume älv →, Sweden 8 E19 63 45N 20 20 E
Umeå, Sweden 8 E19 63 45N 20 20 E
Umera, Indonesia 37 E7 0 12S 129 37 E
Umfuli →, Zimbabwe 55 F2 17 30S 29 23 E
Umgusa, Zimbabwe 55 F2 19 29S 27 52 E
Umkomaas, S. Africa 57 E5 30 13S 30 48 E
Umlazi, S. Africa 53 L6 29 59S 30 54 E
Umm ad Daraj, J., Jordan 47 C4 32 18N 35 48 E
Umm al Qaywayn, U.A.E. 45 E7 25 30N 55 35 E
Umm al Qittayn, Jordan 47 C5 32 18N 36 40 E
Umm Bāb, Qatar 45 E6 25 12N 50 48 E
Umm el Fahm, Israel 47 C4 32 31N 35 9 E
Umm Keddada, Sudan 51 F11 13 33N 26 35 E
Umm Lajj, Si. Arabia 44 E3 25 0N 37 23 E
Umm Ruwaba, Sudan 51 F12 12 50N 31 20 E
Umnak I., U.S.A. 68 C3 53 15N 168 20W
Umniati →, Zimbabwe 55 F2 16 49S 28 45 E
Umpqua →, U.S.A. 82 E1 43 40N 124 12W
Umreth, India 42 H5 22 41N 73 4 E
Umtata, S. Africa 57 E4 31 36S 28 49 E
Umuarama, Brazil 95 A5 23 45S 53 20W
Umvukwe Ra., Zimbabwe 55 F3 16 45S 30 45 E
Umzimvubu, S. Africa 57 E4 31 38S 29 33 E
Umzingwane →, Zimbabwe 55 G2 22 12S 29 56 E
Umzinto, S. Africa 57 E5 30 15S 30 45 E
Una, India 42 J4 20 46N 71 8 E
Una →, Bos.-H. 16 F9 45 0N 16 20 E
Unadilla, U.S.A. 79 D9 42 20N 75 19W
Unalakleet, U.S.A. 68 B3 63 52N 160 47W
Unalaska, U.S.A. 68 C3 53 53N 166 32W
Unalaska I., U.S.A. 68 C3 53 35N 166 50W
'Unayzah, Si. Arabia 44 E4 26 6N 43 58 E
'Unāzah, J., Asia 44 C3 32 12N 39 18 E
Uncía, Bolivia 92 G5 18 25S 66 40W
Uncompahgre Peak, U.S.A. 83 G10 38 4N 107 28W
Uncompahgre Plateau, U.S.A. 83 G9 38 20N 108 15W
Underbool, Australia 63 F3 35 10S 141 51 E
Ungarie, Australia 63 E4 33 38S 146 56 E
Ungarra, Australia 63 E2 34 12S 136 2 E
Ungava, Pén. d', Canada 69 C12 60 0N 74 0W
Ungava B., Canada 69 C13 59 30N 67 30W
Ungeny = Ungheni, Moldova 17 E14 47 11N 27 51 E
Unggi, N. Korea 35 C16 42 16N 130 28 E
Ungheni, Moldova 17 E14 47 11N 27 51 E
União da Vitória, Brazil 95 B5 26 13S 51 5W
Unimak I., U.S.A. 68 C3 54 45N 164 0W
Union, Miss., U.S.A. 81 J10 32 34N 89 7W
Union, Mo., U.S.A. 80 F9 38 27N 91 0W
Union, S.C., U.S.A. 77 H5 34 43N 81 37W
Union City, Calif., U.S.A. 84 H4 37 36N 122 1W
Union City, N.J., U.S.A. 79 F10 40 45N 74 2W
Union City, Pa., U.S.A. 78 E5 41 54N 79 51W
Union City, Tenn., U.S.A. 81 G10 36 26N 89 3W
Union Gap, U.S.A. 82 C3 46 33N 120 28W
Union Springs, U.S.A. 77 J3 32 9N 85 43W
Uniondale, S. Africa 56 E3 33 39S 23 7 E
Uniontown, U.S.A. 76 F6 39 54N 79 44W
Unionville, U.S.A. 80 E8 40 29N 93 1W
United Arab Emirates ■, Asia 46 C5 23 50N 54 0 E
United Kingdom ■, Europe 7 E5 53 0N 2 0W
United States of America ■, N. Amer. 74 C7 37 0N 96 0W
Unity, Canada 73 C7 52 30N 109 5W
University Park, U.S.A. 83 K10 32 17N 106 45W
Unjha, India 42 H5 23 46N 72 24 E
Unnao, India 43 F9 26 35N 80 30 E
Unsengedsi →, Zimbabwe 55 F3 15 43S 31 14 E
Unst, U.K. 12 A8 60 44N 0 53W
Unuk →, Canada 72 B2 56 5N 131 3W
Uozu, Japan 31 F8 36 48N 137 24 E
Upata, Venezuela 92 B6 8 1N 62 24W
Upemba, L., Dem. Rep. of the Congo 55 D2 8 30S 26 20 E
Upernavik, Greenland 4 B5 72 49N 56 20W
Upington, S. Africa 56 D3 28 25S 21 15 E
Upleta, India 42 J4 21 46N 70 16 E
'Upolu, Samoa 59 A13 13 58S 172 0W
Upper Alkali L., U.S.A. 82 F3 41 47N 120 8W
Upper Arrow L., Canada 72 C5 50 30N 117 50W
Upper Foster L., Canada 73 B7 56 47N 105 20W
Upper Hutt, N.Z. 59 J5 41 8S 175 5 E
Upper Klamath L., U.S.A. 82 E3 42 25N 121 55W
Upper Lake, U.S.A. 84 F4 39 10N 122 54W
Upper Musquodoboit, Canada 71 C7 45 10N 62 58W
Upper Red L., U.S.A. 80 A7 48 8N 94 45W
Upper Sandusky, U.S.A. 76 E4 40 50N 83 17W
Upper Volta = Burkina Faso ■, Africa 50 F5 12 0N 1 0W
Uppland, Sweden 9 F17 59 59N 17 48 E
Uppsala, Sweden 9 G17 59 53N 17 38 E
Upshi, India 43 C7 33 48N 77 52 E
Upstart, C., Australia 62 B4 19 41S 147 45 E
Upton, U.S.A. 80 C2 44 6N 104 38W
Ur, Iraq 44 D5 30 55N 46 25 E
Urad Qianqi, China 34 D5 40 40N 108 30 E
Urakawa, Japan 30 C11 42 9N 142 47 E

Vero Beach

Ural = Zhayyq →, Kazakstan 25 E9 47 0N 51 48 E
Ural, Australia 63 E4 33 21S 146 12 E
Ural Mts. = Uralskie Gory, Eurasia 24 C10 60 0N 59 0 E
Uralla, Australia 63 E5 30 37S 151 29 E
Uralsk = Oral, Kazakstan 25 D9 51 20N 51 20 E
Uralskie Gory, Eurasia 24 C10 60 0N 59 0 E
Urambo, Tanzania 54 D3 5 4S 32 0 E
Urandangi, Australia 62 C2 21 32S 138 14 E
Uranium City, Canada 73 B7 59 34N 108 37W
Uraricoera →, Brazil 92 C6 3 2N 60 30W
Urawa, Japan 31 G9 35 50N 139 40 E
Uray, Russia 26 C7 60 5N 65 15 E
'Uray'irah, Si. Arabia 45 E6 25 57N 48 53 E
Urbana, Ill., U.S.A. 76 E1 40 7N 88 12W
Urbana, Ohio, U.S.A. 76 E4 40 7N 83 45W
Urbino, Italy 20 C5 43 43N 12 38 E
Urbión, Picos de, Spain 19 A4 42 1N 2 52W
Urcos, Peru 92 F4 13 40S 71 38W
Urdinarrain, Argentina 94 C4 32 37S 58 52W
Urdzhar, Kazakstan 26 E9 47 5N 81 38 E
Ure →, U.K. 10 C6 54 5N 1 20W
Ures, Mexico 86 B2 29 30N 110 30W
Urfa = Sanliurfa, Turkey 25 G6 37 12N 38 50 E
Urganch, Uzbekistan 26 E7 41 40N 60 41 E
Urgench = Urganch, Uzbekistan 26 E7 41 40N 60 41 E
Ürgüp, Turkey 44 B2 38 38N 34 56 E
Uri, India 43 B6 34 8N 74 2 E
Uribia, Colombia 92 A4 11 43N 72 16W
Uriondo, Bolivia 94 A3 21 41S 64 41W
Urique, Mexico 86 B3 27 13N 107 55W
Urique →, Mexico 86 B3 26 29N 107 58W
Urk, Neths. 15 B5 52 39N 5 36 E
Urla, Turkey 21 E12 38 20N 26 47 E
Urmia = Orūmīyeh, Iran 44 B5 37 40N 45 0 E
Urmia, L. = Orūmīyeh, Daryācheh-ye, Iran 44 B5 37 50N 45 30 E
Uroševac, Kosovo, Yug. 21 C9 42 23N 21 10 E
Uruaçu, Brazil 93 F9 14 30S 49 10W
Uruapan, Mexico 86 D4 19 30N 102 0W
Urubamba →, Peru 92 F4 10 43S 73 48W
Uruçara, Brazil 92 D7 2 32S 57 45W
Uruçuí, Brazil 93 E10 7 20S 44 28W
Uruguai →, Brazil 95 B5 26 0S 53 30W
Uruguaiana, Brazil 94 B4 29 50S 57 0W
Uruguay ■, S. Amer. 94 C4 32 30S 56 30W
Uruguay →, S. Amer. 94 C4 34 12S 58 18W
Urumchi = Ürümqi, China 26 E9 43 45N 87 45 E
Ürümqi, China 26 E9 43 45N 87 45 E
Urup, Ostrov, Russia 27 E16 46 0N 151 0 E
Usa →, Russia 24 A10 66 16N 59 49 E
Uşak, Turkey 25 G4 38 43N 29 28 E
Usakos, Namibia 56 C2 21 54S 15 31 E
Usedom, Germany 16 B8 53 55N 14 2 E
Useless Loop, Australia 61 E1 26 8S 113 23 E
Ush-Tobe, Kazakstan 26 E8 45 16N 78 0 E
Ushakova, Ostrov, Russia 4 A12 82 0N 80 0 E
Ushant = Ouessant, Î. d', France 18 B1 48 28N 5 6W
Ushashi, Tanzania 54 C3 1 59S 33 57 E
Ushibuka, Japan 31 H5 32 11N 130 1 E
Ushuaia, Argentina 96 G3 54 50S 68 23W
Ushumun, Russia 27 D13 52 47N 126 32 E
Usk, Canada 72 C3 54 38N 128 26W
Usk →, U.K. 11 F5 51 33N 2 58W
Uska, India 43 F10 27 12N 83 7 E
Usman, Russia 24 D6 52 5N 39 48 E
Usoke, Tanzania 54 D3 5 8S 32 24 E
Usolye Sibirskoye, Russia 27 D11 52 48N 103 40 E
Uspallata, P. de, Argentina 94 C2 32 37S 69 22W
Uspenskiy, Kazakstan 26 E8 48 41N 72 43 E
Ussuri →, Asia 30 A7 48 27N 135 0 E
Ussuriysk, Russia 27 E14 43 48N 131 59 E
Ussurka, Russia 30 B6 45 12N 133 31 E
Ust-Aldan = Batamay, Russia 27 C13 63 30N 129 15 E
Ust-Amginskoye = Khandyga, Russia 27 C14 62 42N 135 35 E
Ust-Bolsheretsk, Russia 27 D16 52 50N 156 15 E
Ust-Chaun, Russia 27 C18 68 47N 170 30 E
Ust-Ilimpeya = Yukta, Russia 27 C11 63 26N 105 42 E
Ust-Ilimsk, Russia 27 D11 58 3N 102 39 E
Ust-Ishim, Russia 26 D8 57 45N 71 10 E
Ust-Kamchatsk, Russia 27 D17 56 10N 162 28 E
Ust-Kamenogorsk = Öskemen, Kazakstan 26 E9 50 0N 82 36 E
Ust-Khayryuzovo, Russia 27 D16 57 15N 156 45 E
Ust-Kut, Russia 27 D11 56 50N 105 42 E
Ust-Kuyga, Russia 27 B14 70 1N 135 43 E
Ust-Maya, Russia 27 C14 60 30N 134 28 E
Ust-Mil, Russia 27 D14 59 40N 133 11 E
Ust-Nera, Russia 27 C15 64 35N 143 15 E
Ust-Nyukzha, Russia 27 D13 56 34N 121 37 E
Ust-Olenek, Russia 27 B12 73 0N 120 5 E
Ust-Omchug, Russia 27 C15 61 9N 149 38 E
Ust-Port, Russia 26 C9 69 40N 84 26 E
Ust-Tsilma, Russia 24 A9 65 28N 52 11 E
Ust Urt = Ustyurt Plateau, Asia 26 E6 44 0N 55 0 E
Ust-Usa, Russia 24 A10 66 2N 56 57 E
Ust-Vorkuta, Russia 24 A11 67 24N 64 0 E
Ústí nad Labem, Czech Rep. 16 C8 50 41N 14 3 E
Ústica, Italy 20 E5 38 42N 13 11 E
Ustinov = Izhevsk, Russia 24 C9 56 51N 53 14 E
Ustyurt Plateau, Asia 26 E6 44 0N 55 0 E
Usu, China 32 B3 44 27N 84 40 E
Usuki, Japan 31 H5 33 8N 131 49 E
Usulután, El Salv. 88 D2 13 25N 88 28W
Usumacinta →, Mexico 87 D6 17 0N 91 0W
Usumbura = Bujumbura, Burundi 54 C2 3 16S 29 18 E
Usure, Tanzania 54 C3 4 40S 34 22 E
Usutuo →, Mozam. 57 D5 26 48S 32 7 E
Uta, Indonesia 37 E9 4 33S 136 0 E
Utah □, U.S.A. 82 G8 39 20N 111 30W
Utah L., U.S.A. 82 F8 40 10N 111 58W
Utarni, India 42 F4 26 5N 71 58 E
Utatlan, Guatemala 88 C1 15 2N 91 11W
Ute Creek →, U.S.A. 81 H3 35 21N 103 50W
Utena, Lithuania 9 J21 55 27N 25 40 E
Utete, Tanzania 54 D4 8 0S 38 45 E
Uthai Thani, Thailand 38 E3 15 22N 100 3 E
Uthal, Pakistan 42 G2 25 44N 66 40 E
Utiariti, Brazil 92 F7 13 0S 58 10W
Utica, N.Y., U.S.A. 79 C9 43 6N 75 14W
Utica, Ohio, U.S.A. 78 F2 40 14N 82 27W
Utikuma L., Canada 72 B5 55 50N 115 30W
Utopia, Australia 62 C1 22 14S 134 33 E
Utraula, India 43 F10 27 19N 82 25 E
Utrecht, Neths. 15 B5 52 5N 5 8 E
Utrecht, S. Africa 57 D5 27 38S 30 20 E
Utrecht □, Neths. 15 B5 52 6N 5 7 E
Utrera, Spain 19 D3 37 12N 5 48W
Utsjoki, Finland 8 B22 69 51N 26 59 E
Utsunomiya, Japan 31 F9 36 30N 139 50 E
Uttar Pradesh □, India 43 F9 27 0N 80 0 E
Uttaradit, Thailand 38 D3 17 36N 100 5 E
Uttaranchal □, India 43 D8 30 0N 79 30 E
Uttoxeter, U.K. 10 E6 52 54N 1 52W
Uummannarsuaq = Nunap Isua, Greenland 69 C15 59 48N 43 55W
Uusikaarlepyy, Finland 8 E20 63 32N 22 31 E
Uusikaupunki, Finland 9 F19 60 47N 21 25 E
Uva, Russia 24 C9 56 59N 52 13 E
Uvalde, U.S.A. 81 L5 29 13N 99 47W
Uvat, Russia 26 D7 59 5N 68 50 E
Uvinza, Tanzania 54 D3 5 5S 30 24 E
Uvira, Dem. Rep. of the Congo 54 C2 3 22S 29 3 E
Uvs Nuur, Mongolia 32 A4 50 20N 92 30 E
'Uwairidh, Ḥarrat al, Si. Arabia 44 E3 26 50N 38 0 E
Uwajima, Japan 31 H6 33 10N 132 35 E
Uweinat, Jebel, Sudan 51 D10 21 54N 24 58 E
Uxbridge, Canada 78 B5 44 6N 79 7W
Uxin Qi, China 34 E5 38 50N 109 5 E
Uxmal, Mexico 87 C7 20 22N 89 46W
Üydzin, Mongolia 34 B4 44 9N 107 0 E
Uyo, Nigeria 50 G7 5 1N 7 53 E
Uyûn Mûsa, Egypt 47 F1 29 53N 32 40 E
Uyuni, Bolivia 92 H5 20 28S 66 47W
Uzbekistan ■, Asia 26 E7 41 30N 65 0 E
Uzen, Kazakstan 25 F9 43 29N 52 54 E
Uzen, Mal →, Kazakstan 25 E8 49 4N 49 44 E
Uzerche, France 18 D4 45 25N 1 34 E
Uzh →, Ukraine 17 C16 51 15N 30 12 E
Uzhgorod = Uzhhorod, Ukraine 17 D12 48 36N 22 18 E
Uzhhorod, Ukraine 17 D12 48 36N 22 18 E
Užice, Serbia, Yug. 21 C8 43 55N 19 50 E
Uzunköprü, Turkey 21 D12 41 16N 26 43 E

V

Vaal →, S. Africa 56 D3 29 4S 23 38 E
Vaal Dam, S. Africa 57 D4 27 0S 28 14 E
Vaalwater, S. Africa 57 C4 24 15S 28 8 E
Vaasa, Finland 8 E19 63 6N 21 38 E
Vác, Hungary 17 E10 47 49N 19 10 E
Vacaria, Brazil 95 B5 28 31S 50 52W
Vacaville, U.S.A. 84 G5 38 21N 121 59W
Vach = Vakh →, Russia 26 C8 60 45N 76 45 E
Vache, Î. à, Haiti 89 C5 18 2N 73 35W
Vadnagar, India 42 H5 23 47N 72 40 E
Vadodara, India 42 H5 22 20N 73 10 E
Vadsø, Norway 8 A23 70 3N 29 50 E
Vaduz, Liech. 18 C8 47 8N 9 31 E
Værøy, Norway 8 C15 67 40N 12 40 E
Vágar, Færoe Is. 8 E9 62 5N 7 15W
Vågsfjorden, Norway 8 B17 68 50N 16 50 E
Váh →, Slovak Rep. 17 D9 47 43N 18 7 E
Vahsel B., Antarctica 5 D1 75 0S 35 0W
Váï, Greece 23 D8 35 15N 26 18 E
Vaigach, Russia 26 B6 70 10N 59 0 E
Vail, U.S.A. 74 C5 39 40N 106 20W
Vaisali →, India 43 F8 26 28N 78 53 E
Vakh →, Russia 26 C8 60 45N 76 45 E
Val-d'Or, Canada 70 C4 48 7N 77 47W
Val Marie, Canada 73 D7 49 15N 107 45W
Valahia, Romania 17 F13 44 35N 25 0 E
Valandovo, Macedonia 21 D10 41 19N 22 34 E
Valcheta, Argentina 96 E3 40 40S 66 8W
Valdayskaya Vozvyshennost, Russia 24 C5 57 0N 33 30 E
Valdepeñas, Spain 19 C4 38 43N 3 25W
Valdés, Pen., Argentina 96 E4 42 30S 63 45W
Valdez, U.S.A. 68 B5 61 7N 146 16W
Valdivia, Chile 96 D2 39 50S 73 14W
Valdosta, U.S.A. 77 K4 30 50N 83 17W
Valdres, Norway 9 F13 61 5N 9 5 E
Vale, U.S.A. 82 E5 43 59N 117 15W
Vale of Glamorgan □, U.K. 11 F4 51 28N 3 25W
Valemount, Canada 72 C5 52 50N 119 15W
Valença, Brazil 93 F11 13 20S 39 5W
Valença do Piauí, Brazil 93 E10 6 20S 41 45W
Valence, France 18 D6 44 57N 4 54 E
Valencia, Spain 19 C5 39 27N 0 23W
Valencia, U.S.A. 83 J10 34 48N 106 43W
Valencia, Venezuela 92 A5 10 11N 68 0W
Valencia □, Spain 19 C5 39 20N 0 40W
Valencia, G. de, Spain 19 C6 39 30N 0 20 E
Valencia de Alcántara, Spain 19 C2 39 25N 7 14W
Valencia I., Ireland 13 E1 51 54N 10 22W
Valenciennes, France 18 A5 50 20N 3 34 E
Valentim, Sa. do, Brazil 93 E10 6 0S 43 30W
Valentin, Russia 30 C7 43 8N 134 17 E
Valentine, U.S.A. 81 K2 30 35N 104 30W
Valera, Venezuela 92 B4 9 19N 70 37W
Valga, Estonia 9 H22 57 47N 26 2 E
Valier, U.S.A. 82 B7 48 18N 112 16W
Valjevo, Serbia, Yug. 21 B8 44 18N 19 53 E
Valka, Latvia 9 H21 57 42N 25 57 E
Valkeakoski, Finland 9 F20 61 16N 24 2 E
Valkenswaard, Neths. 15 C5 51 21N 5 29 E
Vall de Uxó = La Vall d'Uixó, Spain 19 C5 39 49N 0 15W
Valladolid, Mexico 87 C7 20 40N 88 11W
Valladolid, Spain 19 B3 41 38N 4 43W
Valldemossa, Spain 22 B9 39 43N 2 37 E
Valle de la Pascua, Venezuela 92 B5 9 13N 66 0W
Valle de las Palmas, Mexico 85 N10 32 20N 116 43W
Valle de Santiago, Mexico 86 C4 20 25N 101 15W
Valle de Suchil, Mexico 86 C4 23 38N 103 55W
Valle de Zaragoza, Mexico 86 B3 27 28N 105 49W
Valle Fértil, Sierra del, Argentina 94 C2 30 20S 68 0W
Valle Hermoso, Mexico 87 B5 25 35N 97 40W
Valledupar, Colombia 92 A4 10 29N 73 15W
Vallehermoso, Canary Is. 22 F2 28 10N 17 15W
Vallejo, U.S.A. 84 G4 38 7N 122 14W
Vallenar, Chile 94 B1 28 30S 70 50W
Valletta, Malta 23 D2 35 54N 14 31 E
Valley Center, U.S.A. 85 M9 33 13N 117 2W
Valley City, U.S.A. 80 B6 46 55N 98 0W
Valley Falls, Oreg., U.S.A. 82 E3 42 29N 120 17W
Valley Falls, R.I., U.S.A. 79 E13 41 54N 71 24W
Valley Springs, U.S.A. 84 G6 38 12N 120 50W
Valley View, U.S.A. 79 F8 40 39N 76 33W
Valley Wells, U.S.A. 85 K11 35 27N 115 46W
Valleyview, Canada 72 B5 55 5N 117 17W
Vallimanca, Arroyo, Argentina 94 D4 35 40S 59 10W
Valls, Spain 19 B6 41 18N 1 15 E
Valmiera, Latvia 9 H21 57 37N 25 29 E
Valognes, France 18 B3 49 30N 1 28W
Valona = Vlorë, Albania 21 D8 40 32N 19 28 E
Valozhyn, Belarus 17 A14 54 3N 26 30 E
Valparaíso, Chile 94 C1 33 2S 71 40W
Valparaíso, Mexico 86 C4 22 50N 103 32W
Valparaiso, U.S.A. 76 E2 41 28N 87 4W
Valparaíso □, Chile 94 C1 33 2S 71 40W
Vals →, S. Africa 56 D4 27 23S 26 30 E
Vals, Tanjung, Indonesia 37 F9 8 26S 137 25 E
Valsad, India 40 J8 20 40N 72 58 E
Valverde, Canary Is. 22 G2 27 48N 17 55W
Valverde del Camino, Spain 19 D2 37 35N 6 47W
Vammala, Finland 9 F20 61 20N 22 54 E
Vámos, Greece 23 D6 35 24N 24 13 E
Van, Turkey 25 G7 38 30N 43 20 E
Van, L. = Van Gölü, Turkey 25 G7 38 30N 43 0 E
Van Alstyne, U.S.A. 81 J6 33 25N 96 35W
Van Blommestein Meer, Surinam 93 C7 4 45N 55 5W
Van Buren, Canada 71 C6 47 10N 67 55W
Van Buren, Ark., U.S.A. 81 H7 35 26N 94 21W
Van Buren, Maine, U.S.A. 77 B11 47 10N 67 58W
Van Buren, Mo., U.S.A. 81 G9 37 0N 91 1W
Van Canh, Vietnam 38 F7 13 37N 109 0 E
Van Diemen, C., N. Terr., Australia 60 B5 11 9S 130 24 E
Van Diemen, C., Queens., Australia 62 B2 16 30S 139 46 E
Van Diemen G., Australia 60 B5 11 45S 132 0 E
Van Gölü, Turkey 25 G7 38 30N 43 0 E
Van Horn, U.S.A. 81 K2 31 3N 104 50W
Van Ninh, Vietnam 38 F7 12 42N 109 14 E
Van Rees, Pegunungan, Indonesia 37 E9 2 35S 138 15 E
Van Wert, U.S.A. 76 E3 40 52N 84 35W
Van Yen, Vietnam 38 B5 21 4N 104 42 E
Vanadzor, Armenia 25 F7 40 48N 44 30 E
Vanavara, Russia 27 C11 60 22N 102 16 E
Vancouver, Canada 72 D4 49 15N 123 10W
Vancouver, U.S.A. 84 E4 45 38N 122 40W
Vancouver, C., Australia 61 G2 35 2S 118 11 E
Vancouver I., Canada 72 D3 49 50N 126 0W
Vandalia, Ill., U.S.A. 80 F10 38 58N 89 6W
Vandalia, Mo., U.S.A. 80 F9 39 19N 91 29W
Vandenburg, U.S.A. 85 L6 34 35N 120 33W
Vanderbijlpark, S. Africa 57 D4 26 42S 27 54 E
Vandergrift, U.S.A. 78 F5 40 36N 79 34W
Vanderhoof, Canada 72 C4 54 0N 124 0W
Vanderkloof Dam, S. Africa 56 E3 30 4S 24 40 E
Vanderlin I., Australia 62 B2 15 44S 137 2 E
Vänern, Sweden 9 G15 58 47N 13 30 E
Vänersborg, Sweden 9 G15 58 26N 12 19 E
Vang Vieng, Laos 38 C4 18 58N 102 32 E
Vanga, Kenya 54 C4 4 35S 39 12 E
Vangaindrano, Madag. 57 C8 23 21S 47 36 E
Vanguard, Canada 73 D7 49 55N 107 20W
Vanino, Russia 27 E15 48 50N 140 5 E
Vanna, Norway 8 A18 70 6N 19 50 E
Vännäs, Sweden 8 E18 63 58N 19 48 E
Vannes, France 18 C2 47 40N 2 47W
Vanrhynsdorp, S. Africa 56 E2 31 36S 18 44 E
Vansbro, Sweden 9 F16 60 32N 14 15 E
Vansittart B., Australia 60 B4 14 3S 126 17 E
Vantaa, Finland 9 F21 60 18N 24 58 E
Vanua Balavu, Fiji 59 C9 17 40S 178 57W
Vanua Levu, Fiji 59 C8 16 33S 179 15 E
Vanuatu ■, Pac. Oc. 64 J8 15 0S 168 0 E
Vanwyksvlei, S. Africa 56 E3 30 18S 21 49 E
Vanzylsrus, S. Africa 56 D3 26 52S 22 4 E
Vapnyarka, Ukraine 17 D15 48 32N 28 45 E
Varanasi, India 43 G10 25 22N 83 0 E
Varangerfjorden, Norway 8 A23 70 3N 29 25 E
Varangerhalvøya, Norway 8 A23 70 25N 29 30 E
Varaždin, Croatia 16 E9 46 20N 16 20 E
Varberg, Sweden 9 H15 57 6N 12 20 E
Vardak □, Afghan. 40 B6 34 0N 68 0 E
Vardar = Axiós →, Greece 21 D10 40 57N 22 35 E
Varde, Denmark 9 J13 55 38N 8 29 E
Vardø, Norway 8 A24 70 23N 31 5 E
Varella, Mui, Vietnam 38 F7 12 54N 109 26 E
Varėna, Lithuania 9 J21 54 12N 24 30 E
Varese, Italy 18 D8 45 48N 8 50 E
Varginha, Brazil 95 A6 21 33S 45 25W
Varillas, Chile 94 A1 24 0S 70 10W
Varkaus, Finland 9 E22 62 19N 27 50 E
Varna, Bulgaria 21 C12 43 13N 27 56 E
Värnamo, Sweden 9 H16 57 10N 14 3 E
Vars, Canada 79 A9 45 21N 75 21W
Varysburg, U.S.A. 78 D6 42 46N 78 19W
Varzaneh, Iran 45 C7 32 25N 52 40 E
Vasa Barris →, Brazil 93 F11 11 10S 37 10W
Vascongadas = País Vasco □, Spain 19 A4 42 50N 2 45W
Vasht = Khāsh, Iran 40 E2 28 15N 61 15 E
Vasilevichi, Belarus 17 B15 52 15N 29 50 E
Vasilkov = Vasylkiv, Ukraine 17 C16 50 7N 30 15 E
Vaslui, Romania 17 E14 46 38N 27 42 E
Vassar, Canada 73 D9 49 10N 95 55W
Vassar, U.S.A. 76 D4 43 22N 83 35W
Västerås, Sweden 9 G17 59 37N 16 38 E
Västerbotten, Sweden 8 D18 64 36N 20 4 E
Västerdalälven →, Sweden 9 F16 60 30N 14 7 E
Västervik, Sweden 9 H17 57 43N 16 33 E
Västmanland, Sweden 9 G16 59 45N 16 20 E
Vasto, Italy 20 C6 42 8N 14 40 E
Vasylkiv, Ukraine 17 C16 50 7N 30 15 E
Vatersay, U.K. 12 E1 56 55N 7 32W
Vatican City ■, Europe 20 D5 41 54N 12 27 E
Vatili, Cyprus 23 D12 35 6N 33 40 E
Vatnajökull, Iceland 8 D5 64 30N 16 48W
Vatoa, Fiji 59 D9 19 50S 178 13W
Vatólakkos, Greece 23 D5 35 27N 23 53 E
Vatoloha, Madag. 57 B8 17 52S 47 48 E
Vatomandry, Madag. 57 B8 19 20S 48 59 E
Vatra-Dornei, Romania 17 E13 47 22N 25 22 E
Vatrak →, India 42 H5 23 9N 73 2 E
Vättern, Sweden 9 G16 58 25N 14 30 E
Vaughn, Mont., U.S.A. 82 C8 47 33N 111 33W
Vaughn, N. Mex., U.S.A. 83 J11 34 36N 105 13W
Vaujours L., Canada 70 A5 55 27N 74 15W
Vaupés = Uaupés →, Brazil 92 C5 0 2N 67 16W
Vaupes □, Colombia 92 C4 1 0N 71 0W
Vauxhall, Canada 72 C6 50 5N 112 9W
Vav, India 42 G4 24 22N 71 31 E
Vavatenina, Madag. 57 B8 17 28S 49 12 E
Vava'u, Tonga 59 D12 18 36S 174 0W
Vawkavysk, Belarus 17 B13 53 9N 24 30 E
Växjö, Sweden 9 H16 56 52N 14 50 E
Vaygach, Ostrov, Russia 26 C6 70 0N 60 0 E
Váyia, Ákra, Greece 23 C10 36 15N 28 11 E
Vechte →, Neths. 15 B6 52 34N 6 6 E
Vedea →, Romania 17 G13 43 42N 25 41 E
Vedia, Argentina 94 C3 34 30S 61 31W
Veendam, Neths. 15 A6 53 5N 6 52 E
Veenendaal, Neths. 15 B5 52 2N 5 34 E
Vefsna →, Norway 8 D15 65 48N 13 10 E
Vega, Norway 8 D14 65 40N 11 55 E
Vega, U.S.A. 81 H3 35 15N 102 26W
Vegreville, Canada 72 C6 53 30N 112 5W
Vejer de la Frontera, Spain 19 D3 36 15N 5 59W
Vejle, Denmark 9 J13 55 43N 9 30 E
Velas, C., Costa Rica 88 D2 10 21N 85 52W
Velasco, Sierra de, Argentina 94 B2 29 20S 67 10W
Velddrif, S. Africa 56 E2 32 42S 18 11 E
Velebit Planina, Croatia 16 F8 44 50N 15 20 E
Veles, Macedonia 21 D9 41 46N 21 47 E
Vélez-Málaga, Spain 19 D3 36 48N 4 5W
Vélez Rubio, Spain 19 D4 37 41N 2 5W
Velhas →, Brazil 93 G10 17 13S 44 49W
Velika Kapela, Croatia 16 F8 45 10N 15 5 E
Velikaya →, Russia 24 C4 57 48N 28 10 E
Velikaya Kema, Russia 30 B8 45 30N 137 12 E
Veliki Ustyug, Russia 24 B8 60 47N 46 20 E
Velikiye Luki, Russia 24 C5 56 25N 30 32 E
Veliko Tŭrnovo, Bulgaria 21 C11 43 5N 25 41 E
Velikonda Range, India 40 M11 14 45N 79 10 E
Velletri, Italy 20 D5 41 41N 12 47 E
Vellore, India 40 N11 12 57N 79 10 E
Velsk, Russia 24 B7 61 10N 42 5 E
Velva, U.S.A. 80 A4 48 4N 100 56W
Venado Tuerto, Argentina 94 C3 33 50S 62 0W
Vendée □, France 18 C3 46 50N 1 35W
Vendôme, France 18 C4 47 47N 1 3 E
Venézia, Italy 20 B5 45 27N 12 21 E
Venézia, G. di, Italy 20 B5 45 15N 13 0 E
Venezuela ■, S. Amer. 92 B5 8 0N 66 0W
Venezuela, G. de, Venezuela 92 A4 11 30N 71 0W
Vengurla, India 40 M8 15 53N 73 45 E
Venice = Venézia, Italy 20 B5 45 27N 12 21 E
Venice, U.S.A. 77 M4 27 6N 82 27W
Venkatapuram, India 41 K12 18 20N 80 30 E
Venlo, Neths. 15 C6 51 22N 6 11 E
Vennesla, Norway 9 G12 58 15N 7 59 E
Venray, Neths. 15 C6 51 31N 6 0 E
Ventana, Punta de la, Mexico 86 C3 24 4N 109 48W
Ventana, Sa. de la, Argentina 94 D3 38 0S 62 30W
Ventersburg, S. Africa 56 D4 28 7S 27 9 E
Venterstad, S. Africa 56 E4 30 47S 25 48 E
Ventnor, U.K. 11 G6 50 36N 1 12W
Ventotène, Italy 20 D5 40 47N 13 25 E
Ventoux, Mt., France 18 D6 44 10N 5 17 E
Ventspils, Latvia 9 H19 57 25N 21 32 E
Ventuari →, Venezuela 92 C5 3 58N 67 2W
Ventucopa, U.S.A. 85 L7 34 50N 119 29W
Ventura, U.S.A. 85 L7 34 17N 119 18W
Venus B., Australia 63 F4 38 40S 145 42 E
Vera, Argentina 94 B3 29 30S 60 20W
Vera, Spain 19 D5 37 15N 1 51W
Veracruz, Mexico 87 D5 19 10N 96 10W
Veracruz □, Mexico 87 D5 19 0N 96 15W
Veraval, India 42 J4 20 53N 70 27 E
Verbánia, Italy 18 D8 45 56N 8 33 E
Vercelli, Italy 18 D8 45 19N 8 25 E
Verdalsøra, Norway 8 E14 63 48N 11 30 E
Verde →, Argentina 96 E3 41 56S 65 5W
Verde →, Goiás, Brazil 93 G8 18 1S 50 14W
Verde →, Mato Grosso do Sul, Brazil 93 H8 21 25S 52 20W
Verde →, Chihuahua, Mexico 86 B3 26 29N 107 58W
Verde →, Oaxaca, Mexico 87 D5 15 59N 97 50W
Verde →, Veracruz, Mexico 86 C4 21 10N 102 50W
Verde →, Paraguay 94 A4 23 9S 57 37W
Verde →, U.S.A. 74 D4 33 33N 111 40W
Verde, Cay, Bahamas 88 B4 23 0N 75 5W
Verden, Germany 16 B5 52 55N 9 14 E
Verdi, U.S.A. 84 F7 39 31N 119 59W
Verdun, France 18 B6 49 9N 5 24 E
Vereeniging, S. Africa 57 D4 26 38S 27 57 E
Verga, C., Guinea 50 F3 10 30N 14 10W
Vergara, Uruguay 95 C5 32 56S 53 57W
Vergemont Cr. →, Australia 62 C3 24 16S 143 16 E
Vergennes, U.S.A. 79 B11 44 10N 73 15W
Verín, Spain 19 B2 41 57N 7 27W
Verkhnevilyuysk, Russia 27 C13 63 27N 120 18 E
Verkhniy Baskunchak, Russia 25 E8 48 14N 46 44 E
Verkhoyansk, Russia 27 C14 67 35N 133 25 E
Verkhoyansk Ra. = Verkhoyanskiy Khrebet, Russia 27 C13 66 0N 129 0 E
Verkhoyanskiy Khrebet, Russia 27 C13 66 0N 129 0 E
Vermilion, Canada 73 C6 53 20N 110 50W
Vermilion, U.S.A. 78 E2 41 25N 82 22W
Vermilion →, Alta., Canada 73 C6 53 22N 110 51W
Vermilion →, Qué., Canada 70 C5 47 38N 72 56W
Vermilion, B., U.S.A. 81 L9 29 45N 91 55W
Vermilion Bay, Canada 73 D10 49 51N 93 34W
Vermilion L., U.S.A. 80 B8 47 53N 92 26W
Vermillion, U.S.A. 80 D6 42 47N 96 56W
Vermont □, U.S.A. 79 C12 44 0N 73 0W
Vernal, U.S.A. 82 F9 40 27N 109 32W
Vernalis, U.S.A. 84 H5 37 36N 121 17W
Verner, Canada 70 C3 46 25N 80 8W
Verneukpan, S. Africa 56 E3 30 0S 21 0 E
Vernon, Canada 72 C5 50 20N 119 15W
Vernon, U.S.A. 81 H5 34 9N 99 17W
Veronica, U.S.A. 84 E3 45 52N 123 11W
Vero Beach, U.S.A. 77 M5 27 38N 80 24W

Véroia

Véroia, Greece 21 D10 40 34N 22 12 E
Verona, Canada 79 B8 44 29N 76 42W
Verona, Italy 20 B4 45 27N 10 59 E
Verona, U.S.A. 80 D10 42 59N 89 32W
Versailles, France 18 B5 48 48N 2 8 E
Vert, C., Senegal 50 F2 14 45N 17 30W
Verulam, S. Africa 57 D5 29 38S 31 2 E
Verviers, Belgium 15 D5 50 37N 5 52 E
Veselovskoye Vdkhr., Russia 25 E7 46 58N 41 25 E
Vesoul, France 18 C7 47 40N 6 11 E
Vesterålen, Norway 8 B16 68 45N 15 0 E
Vestfjorden, Norway 8 C15 67 55N 14 0 E
Vestmannaeyjar, Iceland 8 E3 63 27N 20 15W
Vestspitsbergen, Svalbard 4 B8 78 40N 17 0 E
Vestvågøy, Norway 8 B15 68 18N 13 50 E
Vesuvio, Italy 20 D6 40 49N 14 26 E
Vesuvius, Mt. = Vesuvio, Italy 20 D6 40 49N 14 26 E
Veszprém, Hungary 17 E9 47 8N 17 57 E
Vetlanda, Sweden 9 H16 57 24N 15 3 E
Vetlugu →, Russia 24 C8 56 36N 46 4 E
Vettore, Mte., Italy 20 C5 42 49N 13 16 E
Veurne, Belgium 15 C2 51 5N 2 40 E
Veys, Iran 45 D6 31 30N 49 0 E
Vezhen, Bulgaria 21 C11 42 50N 24 20 E
Vi Thanh, Vietnam 39 H5 9 42N 105 26 E
Viacha, Bolivia 92 G5 16 39S 68 18W
Viamão, Brazil 95 C5 30 5S 51 0W
Viana, Brazil 93 D10 3 13S 44 55W
Viana do Alentejo, Portugal 19 C2 38 17N 7 59W
Viana do Castelo, Portugal 19 B1 41 42N 8 50W
Vianden, Lux. 15 E6 49 56N 6 12 E
Viangchan = Vientiane, Laos 38 D4 17 58N 102 36 E
Vianópolis, Brazil 93 G9 16 40S 48 35W
Viaréggio, Italy 20 C4 43 52N 10 14 E
Vibo Valéntia, Italy 20 E7 38 40N 16 6 E
Viborg, Denmark 9 H13 56 27N 9 23 E
Vic, Spain 19 B7 41 58N 2 19 E
Vicenza, Italy 20 B4 45 33N 11 33 E
Vich = Vic, Spain 19 B7 41 58N 2 19 E
Vichada →, Colombia 92 C5 4 55N 67 50W
Vichy, France 18 C5 46 9N 3 26 E
Vicksburg, Ariz., U.S.A. 85 M13 33 45N 113 45W
Vicksburg, Miss., U.S.A. 81 J9 32 21N 90 53W
Victor, India 42 J4 21 0N 71 30 E
Victor, U.S.A. 78 D7 42 58N 77 24W
Victor Harbor, Australia 63 F2 35 30S 138 37 E
Victoria = Labuan, Malaysia 36 C5 5 20N 115 14 E
Victoria, Argentina 94 C3 32 40S 60 10W
Victoria, Canada 72 D4 48 30N 123 25W
Victoria, Chile 96 D2 38 13S 72 20W
Victoria, Malta 23 C1 36 3N 14 14 E
Victoria, Kans., U.S.A. 80 F5 38 52N 99 9W
Victoria, Tex., U.S.A. 81 L6 28 48N 97 0W
Victoria □, Australia 63 F3 37 0S 144 0 E
Victoria →, Australia 60 C4 15 10S 129 40 E
Victoria, Grand L., Canada 70 C4 47 31N 77 30W
Victoria, L., Africa 54 C3 1 0S 33 0 E
Victoria, L., Australia 63 E3 33 57S 141 15 E
Victoria, Mt., Burma 41 J18 21 15N 93 55 E
Victoria Beach, Canada 73 C9 50 40N 96 35W
Victoria de Durango = Durango, Mexico 86 C4 24 3N 104 39W
Victoria de las Tunas, Cuba 88 B4 20 58N 76 59W
Victoria Falls, Zimbabwe 55 F2 17 58S 25 52 E
Victoria Harbour, Canada 78 B5 44 45N 79 45W
Victoria I., Canada 68 A8 71 0N 111 0W
Victoria L., Canada 71 C8 48 20N 57 27W
Victoria Ld., Antarctica 5 D11 75 0S 160 0 E
Victoria Nile →, Uganda 54 B3 2 14N 31 26 E
Victoria River, Australia 60 C5 16 25S 131 0 E
Victoria Str., Canada 68 B9 69 30N 100 0W
Victoria West, S. Africa 56 E3 31 25S 23 4 E
Victoriaville, Canada 71 C5 46 4N 71 56W
Victorica, Argentina 94 D2 36 20S 65 30W
Victorville, U.S.A. 85 L9 34 32N 117 18W
Vicuña, Chile 94 C1 30 0S 70 50W
Vicuña Mackenna, Argentina 94 C3 33 53S 64 25W
Vidal, U.S.A. 85 L12 34 7N 114 31W
Vidal Junction, U.S.A. 85 L12 34 11N 114 34W
Vidalia, U.S.A. 77 J4 32 13N 82 25W
Vídho, Greece 23 A3 39 38N 19 55 E
Vidin, Bulgaria 21 C10 43 59N 22 50 E
Vidisha, India 42 H7 23 28N 77 53 E
Vidzy, Belarus 9 J22 55 23N 26 37 E
Viedma, Argentina 96 E4 40 50S 63 0W
Viedma, L., Argentina 96 F2 49 30S 72 30W
Vielsalm, Belgium 15 D5 50 17N 5 54 E
Vieng Pou Kha, Laos 38 B3 20 41N 101 4 E
Vienna = Wien, Austria 16 D9 48 12N 16 22 E
Vienna, Ill., U.S.A. 81 G10 37 25N 88 54W
Vienna, Mo., U.S.A. 80 F9 38 11N 91 57W
Vienne, France 18 D6 45 31N 4 53 E
Vienne →, France 18 C4 47 13N 0 5 E
Vientiane, Laos 38 D4 17 58N 102 36 E
Vientos, Paso de los, Caribbean 89 C5 20 0N 74 0W
Vierzon, France 18 C5 47 13N 2 5 E
Vietnam ■, Asia 38 C6 19 0N 106 0 E
Vigan, Phil. 37 A6 17 35N 120 28 E
Vigévano, Italy 18 D8 45 19N 8 51 E
Vigia, Brazil 93 D9 0 50S 48 5W
Vigía Chico, Mexico 87 D7 19 46N 87 35W
Víglas, Ákra, Greece 23 D9 35 54N 27 51 E
Vigo, Spain 19 A1 42 12N 8 41W
Vihowa, Pakistan 42 D4 31 8N 70 30 E
Vihowa →, Pakistan 42 D4 31 8N 70 41 E
Vijayawada, India 41 L12 16 31N 80 39 E
Vijosë →, Albania 21 D8 40 37N 19 24 E
Vik, Iceland 8 E4 63 25N 19 1W
Vikeke = Viqueque, E. Timor 37 F7 8 52S 126 23 E
Viking, Canada 72 C6 53 7N 111 50W
Vikna, Norway 8 D14 64 55N 10 58 E
Vila da Maganja, Mozam. 55 F4 17 18S 37 30 E
Vila de João Belo = Xai-Xai, Mozam. 57 D5 25 6S 33 31 E
Vila do Bispo, Portugal 19 D1 37 5N 8 53W
Vila Franca de Xira, Portugal 19 C1 38 57N 8 59W
Vila Gamito, Mozam. 55 E3 14 12S 33 0 E
Vila Gomes da Costa, Mozam. 57 C5 24 20S 33 37 E
Vila Machado, Mozam. 55 F3 19 15S 34 14 E
Vila Mouzinho, Mozam. 55 E3 14 48S 34 25 E
Vila Nova de Gaia, Portugal 19 B1 41 8N 8 37W
Vila Real, Portugal 19 B2 41 17N 7 48W
Vila-real de los Infantes, Spain 19 C5 39 55N 0 3W
Vila Real de Santo António, Portugal 19 D2 37 10N 7 28W
Vila Vasco da Gama, Mozam. 55 E3 14 54S 32 14 E
Vila Velha, Brazil 95 A7 20 20S 40 17W
Vilagarcía de Arousa, Spain 19 A1 42 34N 8 46W
Vilaine →, France 18 C2 47 30N 2 27W
Vilanandro, Tanjona, Madag. 57 B7 16 11S 44 27 E
Vilanculos, Mozam. 57 C6 22 1S 35 17 E
Vilanova i la Geltrú, Spain 19 B6 41 13N 1 40 E
Vileyka, Belarus 17 A14 54 30N 26 53 E
Vilhelmina, Sweden 8 D17 64 35N 16 39 E
Vilhena, Brazil 92 F6 12 40S 60 5W
Viliga, Russia 27 C16 61 36N 156 56 E
Viliya →, Lithuania 9 J21 55 8N 24 16 E
Viljandi, Estonia 9 G21 58 28N 25 30 E
Vilkitskogo, Proliv, Russia 27 B11 78 0N 103 0 E
Vilkovo = Vylkove, Ukraine 17 F15 45 28N 29 32 E
Villa Abecia, Bolivia 94 A2 21 0S 68 18W
Villa Ahumada, Mexico 86 A3 30 38N 106 30W
Villa Ana, Argentina 94 B4 28 28S 59 40W
Villa Ángela, Argentina 94 B3 27 34S 60 45W
Villa Bella, Bolivia 92 F5 10 25S 65 22W
Villa Bens = Tarfaya, Morocco 50 C3 27 55N 12 55W
Villa Cañás, Argentina 94 C3 34 0S 61 35W
Villa Cisneros = Dakhla, W. Sahara 50 D2 23 50N 15 53W
Villa Colón, Argentina 94 C2 31 38S 68 20W
Villa Constitución, Argentina 94 C3 33 15S 60 20W
Villa de María, Argentina 94 B3 29 55S 63 43W
Villa Dolores, Argentina 94 C2 31 58S 65 15W
Villa Frontera, Mexico 86 B4 26 56N 101 27W
Villa Guillermina, Argentina 94 B4 28 15S 59 29W
Villa Hayes, Paraguay 94 B4 25 5S 57 20W
Villa Iris, Argentina 94 D3 38 12S 63 12W
Villa Juárez, Mexico 86 B4 27 37N 100 44W
Villa María, Argentina 94 C3 32 20S 63 10W
Villa Mazán, Argentina 94 B2 28 40S 66 30W
Villa Montes, Bolivia 94 A3 21 10S 63 30W
Villa Ocampo, Argentina 94 B4 28 30S 59 20W
Villa Ocampo, Mexico 86 B3 26 29N 105 30W
Villa Ojo de Agua, Argentina 94 B3 29 30S 63 44W
Villa San José, Argentina 94 C4 32 12S 58 15W
Villa San Martín, Argentina 94 B3 28 15S 64 9W
Villa Unión, Mexico 86 C3 23 12N 106 14W
Villacarlos, Spain 22 B11 39 53N 4 17 E
Villacarrillo, Spain 19 C4 38 7N 3 3W
Villach, Austria 16 E7 46 37N 13 51 E
Villafranca de los Caballeros, Spain 22 B10 39 34N 3 25 E
Villagrán, Mexico 87 C5 24 29N 99 29W
Villaguay, Argentina 94 C4 32 0S 59 0W
Villahermosa, Mexico 87 D6 17 59N 92 55W
Villajoyosa, Spain 19 C5 38 30N 0 12W
Villalba, Spain 19 A2 43 26N 7 40W
Villanueva, U.S.A. 81 H2 35 16N 105 22W
Villanueva de la Serena, Spain 19 C3 38 59N 5 50W
Villanueva y Geltrú = Vilanova i la Geltrú, Spain 19 B6 41 13N 1 40 E
Villarreal = Vila-real de los Infantes, Spain 19 C5 39 55N 0 3W
Villarrica, Chile 96 D2 39 15S 72 15W
Villarrica, Paraguay 94 B4 25 40S 56 30W
Villarrobledo, Spain 19 C4 39 18N 2 36W
Villavicencio, Argentina 94 C2 32 28S 69 0W
Villavicencio, Colombia 92 C4 4 9N 73 37W
Villaviciosa, Spain 19 A3 43 32N 5 27W
Villazón, Bolivia 94 A2 22 0S 65 35W
Ville-Marie, Canada 70 C4 47 20N 79 30W
Ville Platte, U.S.A. 81 K8 30 41N 92 17W
Villena, Spain 19 C5 38 39N 0 52W
Villeneuve-d'Ascq, France 18 A5 50 38N 3 9 E
Villeneuve-sur-Lot, France 18 D4 44 24N 0 42 E
Villiers, S. Africa 57 D4 27 2S 28 36 E
Villingen-Schwenningen, Germany 16 D5 48 3N 8 26 E
Vilna, Canada 72 C6 54 7N 111 55W
Vilnius, Lithuania 9 J21 54 38N 25 19 E
Vilvoorde, Belgium 15 D4 50 56N 4 26 E
Vilyuy →, Russia 27 C13 64 24N 126 26 E
Vilyuysk, Russia 27 C13 63 40N 121 35 E
Viña del Mar, Chile 94 C1 33 0S 71 30W
Vinarós, Spain 19 B6 40 30N 0 27 E
Vincennes, U.S.A. 76 F2 38 41N 87 32W
Vincent, U.S.A. 85 L8 34 33N 118 11W
Vinchina, Argentina 94 B2 28 45S 68 15W
Vindelälven →, Sweden 8 E18 63 55N 19 50 E
Vindeln, Sweden 8 D18 64 12N 19 43 E
Vindhya Ra., India 42 H7 22 50N 77 0 E
Vineland, U.S.A. 76 F8 39 29N 75 2W
Vinh, Vietnam 38 C5 18 45N 105 38 E
Vinh Linh, Vietnam 38 D6 17 4N 107 2 E
Vinh Long, Vietnam 39 G5 10 16N 105 57 E
Vinh Yen, Vietnam 38 B5 21 21N 105 35 E
Vinita, U.S.A. 81 G7 36 39N 95 9W
Vinkovci, Croatia 21 B8 45 19N 18 48 E
Vinnitsa = Vinnytsya, Ukraine 17 D15 49 15N 28 30 E
Vinnytsya, Ukraine 17 D15 49 15N 28 30 E
Vinton, Calif., U.S.A. 84 F6 39 48N 120 10W
Vinton, Iowa, U.S.A. 80 D8 42 10N 92 1W
Vinton, La., U.S.A. 81 K8 30 11N 93 35W
Viqueque, E. Timor 37 F7 8 52S 126 23 E
Virac, Phil. 37 B6 13 30N 124 20 E
Virachei, Cambodia 38 F6 13 59N 106 49 E
Virago Sd., Canada 72 C2 54 0N 132 30W
Viramgam, India 42 H5 23 5N 72 0 E
Viranşehir, Turkey 44 B3 37 13N 39 45 E
Virawah, Pakistan 42 G4 24 31N 70 46 E
Virden, Canada 73 D8 49 50N 100 56W
Vire, France 18 B3 48 50N 0 53W
Virgenes, C., Argentina 96 G3 52 19S 68 21W
Virgin →, U.S.A. 83 H6 36 28N 114 21W
Virgin Gorda, Br. Virgin Is. 89 C7 18 30N 64 26W
Virgin Is. (British) ■, W. Indies 89 C7 18 30N 64 30W
Virgin Is. (U.S.) ■, W. Indies 89 C7 18 20N 65 0W
Virginia, S. Africa 56 D4 28 8S 26 55 E
Virginia, U.S.A. 80 B8 47 31N 92 32W
Virginia □, U.S.A. 76 G7 37 30N 78 45W
Virginia Beach, U.S.A. 76 G8 36 51N 75 59W
Virginia City, Mont., U.S.A. 82 D8 45 18N 111 56W
Virginia City, Nev., U.S.A. 84 F7 39 19N 119 39W
Virginia Falls, Canada 72 A3 61 38N 125 42W
Virginiatown, Canada 70 C4 48 9N 79 36W
Viroqua, U.S.A. 80 D9 43 34N 90 53W
Virovitica, Croatia 20 B7 45 51N 17 21 E
Virpur, India 42 J4 21 51N 70 42 E
Virton, Belgium 15 E5 49 35N 5 32 E
Virudunagar, India 40 Q10 9 30N 77 58 E
Vis, Croatia 20 C7 43 4N 16 10 E
Visalia, U.S.A. 84 J7 36 20N 119 18W
Visayan Sea, Phil. 37 B6 11 30N 123 30 E
Visby, Sweden 9 H18 57 37N 18 18 E
Viscount Melville Sd., Canada 4 B2 74 10N 108 0W
Visé, Belgium 15 D5 50 44N 5 41 E
Višegrad, Bos.-H. 21 C8 43 47N 19 17 E
Viseu, Brazil 93 D9 1 10S 46 5W
Viseu, Portugal 19 B2 40 40N 7 55W
Vishakhapatnam, India 41 L13 17 45N 83 20 E
Visnagar, India 42 H5 23 45N 72 32 E
Viso, Mte., Italy 18 D7 44 38N 7 5 E
Visokoi I., Antarctica 5 B1 56 43S 27 15W
Vista, U.S.A. 85 M9 33 12N 117 14W
Vistula = Wisła →, Poland 17 A10 54 22N 18 55 E
Vitebsk = Vitsyebsk, Belarus 24 C5 55 10N 30 15 E
Viterbo, Italy 20 C5 42 25N 12 6 E
Viti Levu, Fiji 59 C7 17 30S 177 30 E
Vitigudino, Spain 19 B2 41 1N 6 26W
Vitim, Russia 27 D12 59 28N 112 35 E
Vitim →, Russia 27 D12 59 26N 112 34 E
Vitória, Brazil 93 H10 20 20S 40 22W
Vitória da Conquista, Brazil 93 F10 14 51S 40 51W
Vitória de São Antão, Brazil 93 E11 8 10S 35 20W
Vitoria-Gasteiz, Spain 19 A4 42 50N 2 41W
Vitsyebsk, Belarus 24 C5 55 10N 30 15 E
Vittória, Italy 20 F6 36 57N 14 32 E
Vittório Véneto, Italy 20 B5 45 59N 12 18 E
Viveiro, Spain 19 A2 43 39N 7 38W
Vivian, U.S.A. 81 J8 32 53N 93 59W
Vizcaíno, Desierto de, Mexico 86 B2 27 40N 113 50W
Vizcaíno, Sierra, Mexico 86 B2 27 30N 114 0W
Vize, Turkey 21 D12 41 34N 27 45 E
Vizianagaram, India 41 K13 18 6N 83 30 E
Vlaardingen, Neths. 15 C4 51 55N 4 21 E
Vladikavkaz, Russia 25 F7 43 0N 44 35 E
Vladimir, Russia 24 C7 56 15N 40 30 E
Vladimir Volynskiy = Volodymyr-Volynskyy, Ukraine 17 C13 50 50N 24 18 E
Vladivostok, Russia 27 E14 43 10N 131 53 E
Vlieland, Neths. 15 A4 53 16N 4 55 E
Vlissingen, Neths. 15 C3 51 26N 3 34 E
Vlorë, Albania 21 D8 40 32N 19 28 E
Vltava →, Czech Rep. 16 D8 50 21N 14 30 E
Vo Dat, Vietnam 39 G6 11 9N 107 31 E
Voe, U.K. 12 A7 60 21N 1 16W
Vogelkop = Doberai, Jazirah, Indonesia 37 E8 1 25S 133 0 E
Vogelsberg, Germany 16 C5 50 31N 9 12 E
Voghera, Italy 18 D8 44 59N 9 1 E
Vohibinany, Madag. 57 B8 18 49S 49 4 E
Vohilava, Madag. 57 C8 21 4S 48 0 E
Vohimarina = Iharana, Madag. 57 A9 13 25S 50 0 E
Vohimena, Tanjon' i, Madag. 57 D8 25 36S 45 8 E
Vohipeno, Madag. 57 C8 22 22S 47 51 E
Voi, Kenya 54 C4 3 25S 38 32 E
Voiron, France 18 D6 45 22N 5 35 E
Voisey B., Canada 71 A7 56 15N 61 50W
Vojmsjön, Sweden 8 D17 64 55N 16 40 E
Vojvodina □, Serbia, Yug. 21 B9 45 20N 20 0 E
Volborg, U.S.A. 80 C2 45 51N 105 41W
Volcano Is. = Kazan-Rettō, Pac. Oc. 64 E6 25 0N 141 0 E
Volda, Norway 9 E12 62 9N 6 5 E
Volga →, Russia 25 E8 46 0N 48 30 E
Volga Hts. = Privolzhskaya Vozvyshennost, Russia 25 D8 51 0N 46 0 E
Volgodonsk, Russia 25 E7 47 33N 42 5 E
Volgograd, Russia 25 E7 48 40N 44 25 E
Volgogradskoye Vdkhr., Russia 25 D8 50 0N 45 20 E
Volkhov →, Russia 24 B5 60 8N 32 20 E
Volkovysk = Vawkavysk, Belarus 17 B13 53 9N 24 30 E
Volksrust, S. Africa 57 D4 27 24S 29 53 E
Volochanka, Russia 27 B10 71 0N 94 28 E
Volodymyr-Volynskyy, Ukraine 17 C13 50 50N 24 18 E
Vologda, Russia 24 C6 59 10N 39 45 E
Vólos, Greece 21 E10 39 24N 22 59 E
Volovets, Ukraine 17 D12 48 43N 23 11 E
Volozhin = Valozhyn, Belarus 17 A14 54 3N 26 30 E
Volsk, Russia 24 D8 52 5N 47 22 E
Volta →, Ghana 48 F4 5 46N 0 41 E
Volta, L., Ghana 50 G6 7 30N 0 0W
Volta Redonda, Brazil 95 A7 22 31S 44 5W
Voltaire, C., Australia 60 B4 14 16S 125 35 E
Volterra, Italy 20 C4 43 24N 10 51 E
Volturno →, Italy 20 D5 41 1N 13 55 E
Volzhskiy, Russia 25 E7 48 56N 44 46 E
Vondrozo, Madag. 57 C8 22 49S 47 20 E
Vopnafjörður, Iceland 8 D6 65 45N 14 50W
Vóriai Sporádhes, Greece 21 E10 39 15N 23 30 E
Vorkuta, Russia 24 A11 67 48N 64 20 E
Vormsi, Estonia 9 G20 59 1N 23 13 E
Voronezh, Russia 25 D6 51 40N 39 10 E
Voroshilovgrad = Luhansk, Ukraine 25 E6 48 38N 39 15 E
Voroshilovsk = Alchevsk, Ukraine 25 E6 48 30N 38 45 E
Võrts Järv, Estonia 9 G22 58 16N 26 3 E
Võru, Estonia 9 H22 57 48N 26 54 E
Vosges, France 18 B7 48 20N 7 10 E
Voss, Norway 9 F12 60 38N 6 26 E
Vostok I., Kiribati 65 J12 10 5S 152 23W
Votkinsk, Russia 24 C9 57 0N 53 55 E
Votkinskoye Vdkhr., Russia 24 C10 57 22N 55 12 E
Votsuri-Shima, Japan 31 M1 25 45N 123 29 E
Vouga →, Portugal 19 B1 40 41N 8 40W
Voúxa, Ákra, Greece 23 D5 35 37N 23 32 E
Vozhe, Ozero, Russia 24 B6 60 45N 39 0 E
Voznesensk, Ukraine 25 E5 47 35N 31 21 E
Voznesenye, Russia 24 B6 61 0N 35 28 E
Vrangelya, Ostrov, Russia 27 B19 71 0N 180 0 E
Vranje, Serbia, Yug. 21 C9 42 34N 21 54 E
Vratsa, Bulgaria 21 C10 43 15N 23 30 E
Vrbas →, Bos.-H. 20 B7 45 8N 17 29 E
Vrede, S. Africa 57 D4 27 24S 29 6 E
Vredefort, S. Africa 56 D4 27 0S 27 22 E
Vredenburg, S. Africa 56 E2 32 56S 18 0 E
Vredendal, S. Africa 56 E2 31 41S 18 35 E
Vrindavan, India 42 F7 27 37N 77 40 E
Vrises, Greece 23 D6 35 23N 24 13 E
Vršac, Serbia, Yug. 21 B9 45 8N 21 20 E
Vryburg, S. Africa 56 D3 26 55S 24 45 E
Vryheid, S. Africa 57 D5 27 45S 30 47 E
Vu Liet, Vietnam 38 C5 18 43N 105 23 E
Vukovar, Croatia 21 B8 45 21N 18 59 E
Vulcan, Canada 72 C6 50 25N 113 15W
Vulcan, Romania 17 F12 45 23N 23 17 E
Vulcaneşti, Moldova 17 F15 45 41N 28 18 E
Vulcano, Italy 20 E6 38 24N 14 58 E
Vulkaneshty = Vulcaneşti, Moldova 17 F15 45 41N 28 18 E
Vunduzi →, Mozam. 55 F3 18 56S 34 1 E
Vung Tau, Vietnam 39 G6 10 21N 107 4 E
Vyatka = Kirov, Russia 24 C8 58 35N 49 40 E
Vyatka →, Russia 24 C9 55 37N 51 28 E
Vyatskiye Polyany, Russia 24 C9 56 14N 51 5 E
Vyazemskiy, Russia 27 E14 47 32N 134 45 E
Vyazma, Russia 24 C5 55 10N 34 15 E
Vyborg, Russia 24 B4 60 43N 28 47 E
Vychegda →, Russia 24 B8 61 18N 46 36 E
Vychodné Beskydy, Europe 17 D11 49 20N 22 0 E
Vyg-ozero, Russia 24 B5 63 47N 34 29 E
Vylkove, Ukraine 17 F15 45 28N 29 32 E
Vynohradiv, Ukraine 17 D12 48 9N 23 2 E
Vyrnwy, L., U.K. 10 E4 52 48N 3 31W
Vyshniy Volochek, Russia 24 C5 57 30N 34 30 E
Vyshzha = imeni 26 Bakinskikh Komissarov, Turkmenistan 45 B7 39 22N 54 10 E
Vyškov, Czech Rep. 17 D9 49 17N 17 0 E
Vytegra, Russia 24 B6 61 0N 36 27 E

W

W.A.C. Bennett Dam, Canada 72 B4 56 2N 122 6W
Waal →, Neths. 15 C5 51 37N 5 0 E
Waalwijk, Neths. 15 C5 51 42N 5 4 E
Wabana, Canada 71 C9 47 40N 53 0W
Wabasca →, Canada 72 B5 58 22N 115 20W
Wabasca-Desmarais, Canada 72 B6 55 57N 113 56W
Wabash, U.S.A. 76 E3 40 48N 85 49W
Wabash →, U.S.A. 76 G1 37 48N 88 2W
Wabigoon L., Canada 73 D10 49 44N 92 44W
Wabowden, Canada 73 C9 54 55N 98 38W
Wabuk Pt., Canada 70 A2 55 20N 85 5W
Wabush, Canada 71 B6 52 55N 66 52W
Waco, U.S.A. 81 K6 31 33N 97 9W
Waconichi, L., Canada 70 B5 50 8N 74 0W
Wad Hamid, Sudan 51 E12 16 30N 32 45 E
Wad Medanî, Sudan 51 F12 14 28N 33 30 E
Wad Thana, Pakistan 42 F2 27 22N 66 23 E
Wadai, Africa 48 E5 12 0N 19 0 E
Wadayama, Japan 31 G7 35 19N 134 52 E
Waddeneilanden, Neths. 15 A5 53 20N 5 10 E
Waddenzee, Neths. 15 A5 53 6N 5 10 E
Waddington, U.S.A. 79 B9 44 52N 75 12W
Waddington, Mt., Canada 72 C3 51 23N 125 15W
Waddy Pt., Australia 63 C5 24 58S 153 21 E
Wadebridge, U.K. 11 G3 50 31N 4 51W
Wadena, Canada 73 C8 51 57N 103 47W
Wadena, U.S.A. 80 B7 46 26N 95 8W
Wadeye, Australia 60 B4 14 28S 129 52 E
Wadhams, Canada 72 C3 51 30N 127 30W
Wādī as Sīr, Jordan 47 D4 31 56N 35 49 E
Wadi Halfa, Sudan 51 D12 21 53N 31 19 E
Wadsworth, Nev., U.S.A. 82 G4 39 38N 119 17W
Wadsworth, Ohio, U.S.A. 78 E3 41 2N 81 44W
Waegwan, S. Korea 35 G15 35 59N 128 23 E
Wafangdian, China 35 E11 39 38N 121 58 E
Wafrah, Si. Arabia 44 D5 28 33N 47 56 E
Wageningen, Neths. 15 C5 51 58N 5 40 E
Wager B., Canada 69 B11 65 26N 88 40W
Wagga Wagga, Australia 63 F4 35 7S 147 24 E
Waghete, Indonesia 37 E9 4 10S 135 50 E
Wagin, Australia 61 F2 33 17S 117 25 E
Wagner, U.S.A. 80 D5 43 5N 98 18W
Wagon Mound, U.S.A. 81 G2 36 1N 104 42W
Wagoner, U.S.A. 81 H7 35 58N 95 22W
Wah, Pakistan 42 C5 33 45N 72 40 E
Wahai, Indonesia 37 E7 2 48S 129 35 E
Wahiawa, U.S.A. 74 H15 21 30N 158 2W
Wâḥid, Egypt 47 E1 30 48N 32 21 E
Wahnai, Afghan. 42 C1 32 40N 65 50 E
Wahoo, U.S.A. 80 E6 41 13N 96 37W
Wahpeton, U.S.A. 80 B6 46 16N 96 36W
Waiau →, N.Z. 59 K4 42 47S 173 22 E
Waibeem, Indonesia 37 E8 0 30S 132 59 E
Waigeo, Indonesia 37 E8 0 20S 130 40 E
Waihi, N.Z. 59 G5 37 23S 175 52 E
Waihou →, N.Z. 59 G5 37 15S 175 40 E
Waika, Dem. Rep. of the Congo 54 C2 2 22S 25 42 E
Waikabubak, Indonesia 37 F5 9 45S 119 25 E
Waikari, N.Z. 59 K4 42 58S 172 41 E
Waikato →, N.Z. 59 G5 37 23S 174 43 E
Waikerie, Australia 63 E3 34 9S 140 0 E
Waikokopu, N.Z. 59 H6 39 3S 177 52 E
Waikouaiti, N.Z. 59 L3 45 36S 170 41 E
Wailuku, U.S.A. 74 H16 20 53N 156 30W
Waimakariri →, N.Z. 59 K4 43 24S 172 42 E
Waimate, N.Z. 59 L3 44 45S 171 3 E
Wainganga →, India 40 K11 18 50N 79 55 E
Waingapu, Indonesia 37 F6 9 35S 120 11 E
Waini →, Guyana 92 B7 8 20N 59 50W
Wainwright, Canada 73 C6 52 50N 110 50W
Waiouru, N.Z. 59 H5 39 28S 175 41 E
Waipara, N.Z. 59 K4 43 3S 172 46 E
Waipawa, N.Z. 59 H6 39 56S 176 38 E
Waipiro, N.Z. 59 H7 38 2S 178 22 E
Waipu, N.Z. 59 F5 35 59S 174 29 E
Waipukurau, N.Z. 59 J6 40 1S 176 33 E
Wairakei, N.Z. 59 H6 38 37S 176 6 E
Wairarapa, L., N.Z. 59 J5 41 14S 175 15 E
Wairoa, N.Z. 59 H6 39 3S 177 25 E
Waitaki →, N.Z. 59 L3 44 56S 171 7 E
Waitara, N.Z. 59 H5 38 59S 174 15 E
Waitsburg, U.S.A. 82 C5 46 16N 118 9W
Waiuku, N.Z. 59 G5 37 15S 174 45 E
Wajima, Japan 31 F8 37 30N 137 0 E

West Helena

Wajir, *Kenya* **54 B5** 1 42N 40 5 E
Wakasa, *Japan* **31 G7** 35 20N 134 24 E
Wakasa-Wan, *Japan* **31 G7** 35 40N 135 30 E
Wakatipu, L., *N.Z.* **59 L2** 45 5S 168 33 E
Wakaw, *Canada* **73 C7** 52 39N 105 44W
Wakayama, *Japan* **31 G7** 34 15N 135 15 E
Wakayama □, *Japan* **31 H7** 33 50N 135 30 E
Wake Forest, *U.S.A.* **77 H6** 35 59N 78 30W
Wake I., *Pac. Oc.* **64 F8** 19 18N 166 36 E
WaKeeney, *U.S.A.* **80 F5** 39 1N 99 53W
Wakefield, *N.Z.* **59 J4** 41 24S 173 5 E
Wakefield, *U.K.* **10 D6** 53 41N 1 29W
Wakefield, *Mass., U.S.A.* **79 D13** 42 30N 71 4W
Wakefield, *Mich., U.S.A.* **80 B10** 46 29N 89 56W
Wakkanai, *Japan* **30 B10** 45 28N 141 35 E
Wakkerstroom, *S. Africa* **57 D5** 27 24S 30 10 E
Wakool, *Australia* **63 F3** 35 28S 144 23 E
Wakool →, *Australia* **63 F3** 35 5S 143 33 E
Wakre, *Indonesia* **37 E8** 0 19S 131 5 E
Wakuach, L., *Canada* **71 A6** 55 34N 67 32W
Walamba, *Zambia* **55 E2** 13 30S 28 42 E
Wałbrzych, *Poland* **16 C9** 50 45N 16 18 E
Walbury Hill, *U.K.* **11 F6** 51 21N 1 28W
Walcha, *Australia* **63 E5** 30 55S 151 31 E
Walcheren, *Neths.* **15 C3** 51 30N 3 35 E
Walcott, *U.S.A.* **82 F10** 41 46N 106 51W
Wałcz, *Poland* **16 B9** 53 17N 16 27 E
Waldburg Ra., *Australia* **61 D2** 24 40S 117 35 E
Walden, *Colo., U.S.A.* **82 F10** 40 44N 106 17W
Walden, *N.Y., U.S.A.* **79 E10** 41 34N 74 11W
Waldport, *U.S.A.* **82 D1** 44 26N 124 4W
Waldron, *U.S.A.* **81 H7** 34 54N 94 5W
Walebing, *Australia* **61 F2** 30 41S 116 13 E
Wales □, *U.K.* **11 E3** 52 19N 4 43W
Walgett, *Australia* **63 E4** 30 0S 148 5 E
Walgreen Coast, *Antarctica* **5 D15** 75 15S 105 0W
Walker, *U.S.A.* **80 B7** 47 6N 94 35W
Walker, L., *Canada* **71 B6** 50 20N 67 11W
Walker L., *Canada* **73 C9** 54 42N 95 57W
Walker L., *U.S.A.* **82 G4** 38 42N 118 43W
Walkerston, *Australia* **62 C4** 21 11S 149 8 E
Walkerton, *Canada* **78 B3** 44 10N 81 10W
Wall, *U.S.A.* **80 D3** 44 0N 102 8W
Walla Walla, *U.S.A.* **82 C4** 46 4N 118 20W
Wallace, *Idaho, U.S.A.* **82 C6** 47 28N 115 56W
Wallace, *N.C., U.S.A.* **77 H7** 34 44N 77 59W
Wallaceburg, *Canada* **78 D2** 42 34N 82 23W
Wallachia = Valahia, *Romania* **17 F13** 44 35N 25 0 E
Wallal, *Australia* **63 D4** 26 32S 146 7 E
Wallam Cr. →, *Australia* **63 D4** 28 40S 147 20 E
Wallambin, L., *Australia* **61 F2** 30 57S 117 35 E
Wallangarra, *Australia* **63 D5** 28 56S 151 58 E
Wallaroo, *Australia* **63 E2** 33 56S 137 39 E
Wallenpaupack, L., *U.S.A.* **79 E9** 41 25N 75 15W
Wallingford, *U.S.A.* **79 E12** 41 27N 72 50W
Wallis & Futuna, Is., *Pac. Oc.* **64 J10** 13 18S 176 10W
Wallowa, *U.S.A.* **82 D5** 45 34N 117 32W
Wallowa Mts., *U.S.A.* **82 D5** 45 20N 117 30W
Walls, *U.K.* **12 A7** 60 14N 1 33W
Wallula, *U.S.A.* **82 C4** 46 5N 118 54W
Wallumbilla, *Australia* **63 D4** 26 33S 149 9 E
Walmsley, L., *Canada* **73 A7** 63 25N 108 36W
Walney, I. of, *U.K.* **10 C4** 54 6N 3 15W
Walnut Creek, *U.S.A.* **84 H4** 37 54N 122 4W
Walnut Ridge, *U.S.A.* **81 G9** 36 4N 90 57W
Walpole, *Australia* **61 F2** 34 58S 116 44 E
Walpole, *U.S.A.* **79 D13** 42 9N 71 15W
Walsall, *U.K.* **11 E6** 52 35N 1 58W
Walsenburg, *U.S.A.* **81 G2** 37 38N 104 47W
Walsh, *U.S.A.* **81 G3** 37 23N 102 17W
Walsh →, *Australia* **62 B3** 16 31S 143 42 E
Walterboro, *U.S.A.* **77 J5** 32 55N 80 40W
Walters, *U.S.A.* **81 H5** 34 22N 98 19W
Waltham, *U.S.A.* **79 D13** 42 23N 71 14W
Waltman, *U.S.A.* **82 E10** 43 4N 107 12W
Walton, *U.S.A.* **79 D9** 42 10N 75 8W
Walton-on-the-Naze, *U.K.* **11 F9** 51 51N 1 17 E
Walvis Bay, *Namibia* **56 C1** 23 0S 14 28 E
Walvisbaai = Walvis Bay, *Namibia* **56 C1** 23 0S 14 28 E
Wamba, *Dem. Rep. of the Congo* **54 B2** 2 10N 27 57 E
Wamba, *Kenya* **54 B4** 0 58N 37 19 E
Wamego, *U.S.A.* **80 F6** 39 12N 96 18W
Wamena, *Indonesia* **37 E9** 4 4S 138 57 E
Wamsutter, *U.S.A.* **82 F9** 41 40N 107 58W
Wamulan, *Indonesia* **37 E7** 3 27S 126 7 E
Wan Xian, *China* **34 E8** 38 47N 115 7 E
Wana, *Pakistan* **42 C3** 32 20N 69 32 E
Wanaaring, *Australia* **63 D3** 29 38S 144 9 E
Wanaka, *N.Z.* **59 L2** 44 42S 169 9 E
Wanaka, L., *N.Z.* **59 L2** 44 33S 169 7 E
Wanapitei L., *Canada* **70 C3** 46 45N 80 40W
Wandel Sea = McKinley Sea, *Arctic* **4 A7** 82 0N 0 0W
Wanderer, *Zimbabwe* **55 F3** 19 36S 30 1 E
Wandhari, *Pakistan* **42 F2** 27 42N 66 48 E
Wandoan, *Australia* **63 D4** 26 5S 149 55 E
Wanfu, *China* **35 D12** 40 8N 122 38 E
Wang →, *Thailand* **38 D2** 17 8N 99 2 E
Wang Noi, *Thailand* **38 E3** 14 13N 100 44 E
Wang Saphung, *Thailand* **38 D3** 17 18N 101 46 E
Wang Thong, *Thailand* **38 D3** 16 50N 100 26 E
Wanga, *Dem. Rep. of the Congo* **54 B2** 2 58N 29 12 E
Wangal, *Indonesia* **37 F8** 6 8S 134 9 E
Wanganella, *Australia* **63 F3** 35 6S 144 49 E
Wanganui, *N.Z.* **59 H5** 39 56S 175 3 E
Wangaratta, *Australia* **63 F4** 36 21S 146 19 E
Wangary, *Australia* **63 E2** 34 35S 135 29 E
Wangdu, *China* **34 E8** 38 40N 115 7 E
Wangerooge, *Germany* **16 B4** 53 47N 7 54 E
Wangi, *Kenya* **54 C5** 1 58S 40 58 E
Wangiwangi, *Indonesia* **37 F6** 5 22S 123 37 E
Wangqing, *China* **35 C15** 43 12N 129 42 E
Wankaner, *India* **42 H4** 22 35N 71 0 E
Wanless, *Canada* **73 C8** 54 11N 101 21W
Wanning, *China* **38 C8** 18 48N 110 22 E
Wanon Niwat, *Thailand* **38 D4** 17 38N 103 46 E
Wanquan, *China* **34 D8** 40 50N 114 40 E
Wanrong, *China* **34 G6** 35 25N 110 50 E
Wantage, *U.K.* **11 F6** 51 35N 1 25W
Wapakoneta, *U.S.A.* **76 E3** 40 34N 84 12W
Wapato, *U.S.A.* **82 C3** 46 27N 120 25W
Wapawekka L., *Canada* **73 C8** 54 55N 104 40W
Wapikopa L., *Canada* **70 B2** 52 56N 87 53W
Wapiti →, *Canada* **72 B5** 55 5N 118 18W
Wappingers Falls, *U.S.A.* **79 E11** 41 36N 73 55W
Wapsipinicon →, *U.S.A.* **80 E9** 41 44N 90 19W
Warangal, *India* **40 L11** 17 58N 79 35 E
Waraseoni, *India* **43 J9** 21 45N 80 2 E
Waratah, *Australia* **62 G4** 41 30S 145 30 E
Waratah B., *Australia* **63 F4** 38 54S 146 5 E
Warburton, *Vic., Australia* **63 F4** 37 47S 145 42 E
Warburton, *W. Austral., Australia* **61 E4** 26 8S 126 35 E
Warburton Ra., *Australia* **61 E4** 25 55S 126 28 E
Ward, *N.Z.* **59 J5** 41 49S 174 11 E
Ward →, *Australia* **63 D4** 26 28S 146 6 E
Ward Mt., *U.S.A.* **84 H8** 37 12N 118 54W
Warden, *S. Africa* **57 D4** 27 50S 29 0 E
Wardha, *India* **40 J11** 20 45N 78 39 E
Wardha →, *India* **40 K11** 19 57N 79 11 E
Ware, *Canada* **72 B3** 57 26N 125 41W
Ware, *U.S.A.* **79 D12** 42 16N 72 14W
Waregem, *Belgium* **15 D3** 50 53N 3 27 E
Wareham, *U.S.A.* **79 E14** 41 46N 70 43W
Waremme, *Belgium* **15 D5** 50 43N 5 15 E
Warialda, *Australia* **63 D5** 29 29S 150 33 E
Wariap, *Indonesia* **37 E8** 1 30S 134 5 E
Warin Chamrap, *Thailand* **38 E5** 15 12N 104 53 E
Warkopi, *Indonesia* **37 E8** 1 12S 134 9 E
Warm Springs, *U.S.A.* **83 G5** 38 10N 116 20W
Warman, *Canada* **73 C7** 52 19N 106 30W
Warmbad, *Namibia* **56 D2** 28 25S 18 42 E
Warmbad, *S. Africa* **57 C4** 24 51S 28 19 E
Warminster, *U.K.* **11 F5** 51 12N 2 10W
Warminster, *U.S.A.* **79 F9** 40 12N 75 6W
Warner Mts., *U.S.A.* **82 F3** 41 40N 120 15W
Warner Robins, *U.S.A.* **77 J4** 32 37N 83 36W
Waroona, *Australia* **61 F2** 32 50S 115 58 E
Warracknabeal, *Australia* **63 F3** 36 9S 142 26 E
Warragul, *Australia* **63 F4** 38 10S 145 58 E
Warrego →, *Australia* **63 E4** 30 24S 145 21 E
Warrego Ra., *Australia* **62 C4** 24 58S 146 0 E
Warren, *Australia* **63 E4** 31 42S 147 51 E
Warren, *Ark., U.S.A.* **81 J8** 33 37N 92 4W
Warren, *Mich., U.S.A.* **76 D4** 42 30N 83 0W
Warren, *Minn., U.S.A.* **80 A6** 48 12N 96 46W
Warren, *Ohio, U.S.A.* **78 E4** 41 14N 80 49W
Warren, *Pa., U.S.A.* **78 E5** 41 51N 79 9W
Warrenpoint, *U.K.* **13 B5** 54 6N 6 15W
Warrensburg, *Mo., U.S.A.* **80 F8** 38 46N 93 44W
Warrensburg, *N.Y., U.S.A.* **79 C11** 43 29N 73 46W
Warrenton, *S. Africa* **56 D3** 28 9S 24 47 E
Warrenton, *U.S.A.* **84 D3** 46 10N 123 56W
Warri, *Nigeria* **50 G7** 5 30N 5 41 E
Warrina, *Australia* **63 D2** 28 12S 135 50 E
Warrington, *U.K.* **10 D5** 53 24N 2 35W
Warrington, *U.S.A.* **77 K2** 30 23N 87 17W
Warrington □, *U.K.* **10 D5** 53 24N 2 35W
Warrnambool, *Australia* **63 F3** 38 25S 142 30 E
Warroad, *U.S.A.* **80 A7** 48 54N 95 19W
Warruwi, *Australia* **62 A1** 11 36S 133 20 E
Warsa, *Indonesia* **37 E9** 0 47S 135 55 E
Warsak Dam, *Pakistan* **42 B4** 34 11N 71 19 E
Warsaw = Warszawa, *Poland* **17 B11** 52 13N 21 0 E
Warsaw, *Ind., U.S.A.* **76 E3** 41 14N 85 51W
Warsaw, *N.Y., U.S.A.* **78 D6** 42 45N 78 8W
Warsaw, *Ohio, U.S.A.* **78 F3** 40 20N 82 0W
Warszawa, *Poland* **17 B11** 52 13N 21 0 E
Warta →, *Poland* **16 B8** 52 35N 14 39 E
Warthe = Warta →, *Poland* **16 B8** 52 35N 14 39 E
Waru, *Indonesia* **37 E8** 3 30S 130 36 E
Warwick, *Australia* **63 D5** 28 10S 152 1 E
Warwick, *U.K.* **11 E6** 52 18N 1 35W
Warwick, *N.Y., U.S.A.* **79 E10** 41 16N 74 22W
Warwick, *R.I., U.S.A.* **79 E13** 41 42N 71 28W
Warwickshire □, *U.K.* **11 E6** 52 14N 1 38W
Wasaga Beach, *Canada* **78 B4** 44 31N 80 1W
Wasagaming, *Canada* **73 C9** 50 39N 99 58W
Wasatch Ra., *U.S.A.* **82 F8** 40 30N 111 15W
Wasbank, *S. Africa* **57 D5** 28 15S 30 9 E
Wasco, *Calif., U.S.A.* **85 K7** 35 36N 119 20W
Wasco, *Oreg., U.S.A.* **82 D3** 45 36N 120 42W
Waseca, *U.S.A.* **80 C8** 44 5N 93 30W
Wasekamio L., *Canada* **73 B7** 56 45N 108 45W
Wash, The, *U.K.* **10 E8** 52 58N 0 20 E
Washago, *Canada* **78 B5** 44 45N 79 20W
Washburn, *N. Dak., U.S.A.* **80 B4** 47 17N 101 2W
Washburn, *Wis., U.S.A.* **80 B9** 46 40N 90 54W
Washim, *India* **40 J10** 20 3N 77 0 E
Washington, *U.K.* **10 C6** 54 55N 1 30W
Washington, *D.C., U.S.A.* **76 F7** 38 54N 77 2W
Washington, *Ga., U.S.A.* **77 J4** 33 44N 82 44W
Washington, *Ind., U.S.A.* **76 F2** 38 40N 87 10W
Washington, *Iowa, U.S.A.* **80 E9** 41 18N 91 42W
Washington, *Mo., U.S.A.* **80 F9** 38 33N 91 1W
Washington, *N.C., U.S.A.* **77 H7** 35 33N 77 3W
Washington, *N.J., U.S.A.* **79 F10** 40 46N 74 59W
Washington, *Pa., U.S.A.* **78 F4** 40 10N 80 15W
Washington, *Utah, U.S.A.* **83 H7** 37 8N 113 31W
Washington □, *U.S.A.* **82 C3** 47 30N 120 30W
Washington, Mt., *U.S.A.* **79 B13** 44 16N 71 18W
Washington Court House, *U.S.A.* **76 F4** 39 32N 83 26W
Washington I., *U.S.A.* **76 C2** 45 23N 86 54W
Washougal, *U.S.A.* **84 E4** 45 35N 122 21W
Wasian, *Indonesia* **37 E8** 1 47S 133 19 E
Wasilla, *U.S.A.* **68 B5** 61 35N 149 26W
Wasior, *Indonesia* **37 E8** 2 43S 134 30 E
Waskaganish, *Canada* **70 B4** 51 30N 78 40W
Waskaiowaka, L., *Canada* **73 B9** 56 33N 96 23W
Waskesiu Lake, *Canada* **73 C7** 53 55N 106 5W
Wasserkuppe, *Germany* **16 C5** 50 29N 9 55 E
Waswanipi, *Canada* **70 C4** 49 40N 76 29W
Waswanipi, L., *Canada* **70 C4** 49 35N 76 40W
Watampone, *Indonesia* **37 E6** 4 29S 120 25 E
Water Park Pt., *Australia* **62 C5** 22 56S 150 47 E
Water Valley, *U.S.A.* **81 H10** 34 10N 89 38W
Waterberge, *S. Africa* **57 C4** 24 10S 28 0 E
Waterbury, *Conn., U.S.A.* **79 E11** 41 33N 73 3W
Waterbury, *Vt., U.S.A.* **79 B12** 44 20N 72 46W
Waterbury L., *Canada* **73 B8** 58 10N 104 22W
Waterdown, *Canada* **78 C5** 43 20N 79 53W
Waterford, *Canada* **78 D4** 42 56N 80 17W
Waterford, *Ireland* **13 D4** 52 15N 7 8W
Waterford, *Calif., U.S.A.* **84 H6** 37 38N 120 46W
Waterford, *Pa., U.S.A.* **78 E5** 41 57N 79 59W
Waterford □, *Ireland* **13 D4** 52 10N 7 40W
Waterford Harbour, *Ireland* **13 D5** 52 8N 6 58W
Waterhen L., *Canada* **73 C9** 52 10N 99 40W
Waterloo, *Belgium* **15 D4** 50 43N 4 25 E
Waterloo, *Ont., Canada* **78 C4** 43 30N 80 32W
Waterloo, *Qué., Canada* **79 A12** 45 22N 72 32W
Waterloo, *Ill., U.S.A.* **80 F9** 38 20N 90 9W
Waterloo, *Iowa, U.S.A.* **80 D8** 42 30N 92 21W
Waterloo, *N.Y., U.S.A.* **78 D8** 42 54N 76 52W
Watersmeet, *U.S.A.* **80 B10** 46 16N 89 11W
Waterton Lakes Nat. Park, *U.S.A.* **82 B7** 48 45N 115 0W
Watertown, *Conn., U.S.A.* **79 E11** 41 36N 73 7W
Watertown, *N.Y., U.S.A.* **79 C9** 43 59N 75 55W
Watertown, *S. Dak., U.S.A.* **80 C6** 44 54N 97 7W
Watertown, *Wis., U.S.A.* **80 D10** 43 12N 88 43W
Waterval-Boven, *S. Africa* **57 D5** 25 40S 30 18 E
Waterville, *Canada* **79 A13** 45 16N 71 54W
Waterville, *Maine, U.S.A.* **77 C11** 44 33N 69 38W
Waterville, *N.Y., U.S.A.* **79 D9** 42 56N 75 23W
Waterville, *Pa., U.S.A.* **78 E7** 41 19N 77 21W
Waterville, *Wash., U.S.A.* **82 C3** 47 39N 120 4W
Watervliet, *U.S.A.* **79 D11** 42 44N 73 42W
Wates, *Indonesia* **37 G14** 7 51S 110 10 E
Watford, *Canada* **78 D3** 42 57N 81 53W
Watford, *U.K.* **11 F7** 51 40N 0 24W
Watford City, *U.S.A.* **80 B3** 47 48N 103 17W
Wathaman →, *Canada* **73 B8** 57 16N 102 59W
Wathaman L., *Canada* **73 B8** 56 58N 103 44W
Watheroo, *Australia* **61 F2** 30 15S 116 0 E
Wating, *China* **34 G4** 35 40N 106 38 E
Watkins Glen, *U.S.A.* **78 D8** 42 23N 76 52W
Watling I. = San Salvador I., *Bahamas* **89 B5** 24 0N 74 40W
Watonga, *U.S.A.* **81 H5** 35 51N 98 25W
Watrous, *Canada* **73 C7** 51 40N 105 25W
Watrous, *U.S.A.* **81 H2** 35 48N 104 59W
Watsa, *Dem. Rep. of the Congo* **54 B2** 3 4N 29 30 E
Watseka, *U.S.A.* **76 E2** 40 47N 87 44W
Watson, *Australia* **61 F5** 30 29S 131 31 E
Watson, *Canada* **73 C8** 52 10N 104 30W
Watson Lake, *Canada* **72 A3** 60 6N 128 49W
Watsontown, *U.S.A.* **78 E8** 41 5N 76 52W
Watsonville, *U.S.A.* **84 J5** 36 55N 121 45W
Wattiwarriganna Cr. →, *Australia* **63 D2** 28 57S 136 10 E
Watuata = Batuata, *Indonesia* **37 F6** 6 12S 122 42 E
Watubela, Kepulauan, *Indonesia* **37 E8** 4 28S 131 35 E
Watubela Is. = Watubela, Kepulauan, *Indonesia* **37 E8** 4 28S 131 35 E
Wau = Wâw, *Sudan* **51 G11** 7 45N 28 1 E
Waubamik, *Canada* **78 A4** 45 27N 80 1W
Waubay, *U.S.A.* **80 C6** 45 20N 97 18W
Wauchope, *N.S.W., Australia* **63 E5** 31 28S 152 45 E
Wauchope, *N. Terr., Australia* **62 C1** 20 36S 134 15 E
Wauchula, *U.S.A.* **77 M5** 27 33N 81 49W
Waukarlycarly, L., *Australia* **60 D3** 21 18S 121 56 E
Waukegan, *U.S.A.* **76 D2** 42 22N 87 50W
Waukesha, *U.S.A.* **76 D1** 43 1N 88 14W
Waukon, *U.S.A.* **80 D9** 43 16N 91 29W
Waupaca, *U.S.A.* **80 C10** 44 21N 89 5W
Waupun, *U.S.A.* **80 D10** 43 38N 88 44W
Waurika, *U.S.A.* **81 H6** 34 10N 98 0W
Wausau, *U.S.A.* **80 C10** 44 58N 89 38W
Wautoma, *U.S.A.* **80 C10** 44 4N 89 18W
Wauwatosa, *U.S.A.* **76 D2** 43 3N 88 0W
Waveney →, *U.K.* **11 E9** 52 35N 1 39 E
Waverley, *N.Z.* **59 H5** 39 46S 174 37 E
Waverly, *Iowa, U.S.A.* **80 D8** 42 44N 92 29W
Waverly, *N.Y., U.S.A.* **79 E8** 42 1N 76 32W
Wavre, *Belgium* **15 D4** 50 43N 4 38 E
Wâw, *Sudan* **51 G11** 7 45N 28 1 E
Wāw al Kabīr, *Libya* **51 C9** 25 20N 16 43 E
Wawa, *Canada* **70 C3** 47 59N 84 47W
Wawanesa, *Canada* **73 D9** 49 36N 99 40W
Wawona, *U.S.A.* **84 H7** 37 32N 119 39W
Waxahachie, *U.S.A.* **81 J6** 32 24N 96 51W
Way, L., *Australia* **61 E3** 26 45S 120 16 E
Waycross, *U.S.A.* **77 K4** 31 13N 82 21W
Wayland, *U.S.A.* **78 D7** 42 34N 77 35W
Wayne, *Nebr., U.S.A.* **80 D6** 42 14N 97 1W
Wayne, *W. Va., U.S.A.* **76 F4** 38 13N 82 27W
Waynesboro, *Ga., U.S.A.* **77 J4** 33 6N 82 1W
Waynesboro, *Miss., U.S.A.* **77 K1** 31 40N 88 39W
Waynesboro, *Pa., U.S.A.* **76 F7** 39 45N 77 35W
Waynesboro, *Va., U.S.A.* **76 F6** 38 4N 78 53W
Waynesburg, *U.S.A.* **76 F5** 39 54N 80 11W
Waynesville, *U.S.A.* **77 H4** 35 28N 82 58W
Waynoka, *U.S.A.* **81 G5** 36 35N 98 53W
Wazirabad, *Pakistan* **42 C6** 32 30N 74 8 E
We, *Indonesia* **36 C1** 5 51N 95 18 E
Weald, The, *U.K.* **11 F8** 51 4N 0 20 E
Wear →, *U.K.* **10 C6** 54 55N 1 23W
Weatherford, *Okla., U.S.A.* **81 H5** 35 32N 98 43W
Weatherford, *Tex., U.S.A.* **81 J6** 32 46N 97 48W
Weaverville, *U.S.A.* **82 F2** 40 44N 122 56W
Webb City, *U.S.A.* **81 G7** 37 9N 94 28W
Webequie, *Canada* **70 B2** 52 59N 87 21W
Webster, *Mass., U.S.A.* **79 D13** 42 3N 71 53W
Webster, *N.Y., U.S.A.* **78 C7** 43 13N 77 26W
Webster, *S. Dak., U.S.A.* **80 C6** 45 20N 97 31W
Webster City, *U.S.A.* **80 D8** 42 28N 93 49W
Webster Springs, *U.S.A.* **76 F5** 38 29N 80 25W
Weda, *Indonesia* **37 D7** 0 21N 127 50 E
Weda, Teluk, *Indonesia* **37 D7** 0 20N 128 0 E
Weddell I., *Falk. Is.* **96 G4** 51 50S 61 0W
Weddell Sea, *Antarctica* **5 D1** 72 30S 40 0W
Wedderburn, *Australia* **63 F3** 36 26S 143 33 E
Wedgeport, *Canada* **71 D6** 43 44N 65 59W
Wedza, *Zimbabwe* **55 F3** 18 40S 31 33 E
Wee Waa, *Australia* **63 E4** 30 11S 149 26 E
Weed, *U.S.A.* **82 F2** 41 25N 122 23W
Weed Heights, *U.S.A.* **84 G7** 38 59N 119 13W
Weedsport, *U.S.A.* **79 C8** 43 3N 76 35W
Weedville, *U.S.A.* **78 E6** 41 17N 78 30W
Weenen, *S. Africa* **57 D5** 28 48S 30 7 E
Weert, *Neths.* **15 C5** 51 15N 5 43 E
Wei He →, *Hebei, China* **34 F8** 36 10N 115 45 E
Wei He →, *Shaanxi, China* **34 G6** 34 38N 110 15 E
Weichang, *China* **35 D9** 41 58N 117 49 E
Weichuan, *China* **34 G7** 34 20N 113 59 E
Weiden, *Germany* **16 D7** 49 41N 12 10 E
Weifang, *China* **35 F10** 36 44N 119 7 E
Weihai, *China* **35 F12** 37 30N 122 6 E
Weimar, *Germany* **16 C6** 50 58N 11 19 E
Weinan, *China* **34 G5** 34 31N 109 29 E
Weipa, *Australia* **62 A3** 12 40S 141 50 E
Weir →, *Australia* **63 D4** 28 20S 149 50 E
Weir →, *Canada* **73 B10** 56 54N 93 21W
Weir River, *Canada* **73 B10** 56 49N 94 6W
Weirton, *U.S.A.* **78 F4** 40 24N 80 35W
Weiser, *U.S.A.* **82 D5** 44 10N 117 0W
Weishan, *China* **35 G9** 34 47N 117 5 E
Weiyuan, *China* **34 G3** 35 7N 104 10 E
Wejherowo, *Poland* **17 A10** 54 35N 18 12 E
Wekusko L., *Canada* **73 C9** 54 40N 99 50W
Welch, *U.S.A.* **76 G5** 37 26N 81 35W
Welkom, *S. Africa* **56 D4** 28 0S 26 46 E
Welland, *Canada* **78 D5** 43 0N 79 15W
Welland →, *U.K.* **11 E7** 52 51N 0 5W
Wellesley Is., *Australia* **62 B2** 16 42S 139 30 E
Wellingborough, *U.K.* **11 E7** 52 19N 0 41W
Wellington, *Australia* **63 E4** 32 35S 148 59 E
Wellington, *Canada* **78 C7** 43 57N 77 20W
Wellington, *N.Z.* **59 J5** 41 19S 174 46 E
Wellington, *S. Africa* **56 E2** 33 38S 19 1 E
Wellington, *Somst., U.K.* **11 G4** 50 58N 3 13W
Wellington, *Telford & Wrekin, U.K.* **11 E5** 52 42N 2 30W
Wellington, *Colo., U.S.A.* **80 E2** 40 42N 105 0W
Wellington, *Kans., U.S.A.* **81 G6** 37 16N 97 24W
Wellington, *Nev., U.S.A.* **84 G7** 38 45N 119 23W
Wellington, *Ohio, U.S.A.* **78 E2** 41 10N 82 13W
Wellington, *Tex., U.S.A.* **81 H4** 34 51N 100 13W
Wellington, I., *Chile* **96 F2** 49 30S 75 0W
Wellington, L., *Australia* **63 F4** 38 6S 147 20 E
Wells, *U.K.* **11 F5** 51 13N 2 39W
Wells, *Maine, U.S.A.* **79 C14** 43 20N 70 35W
Wells, *N.Y., U.S.A.* **79 C10** 43 24N 74 17W
Wells, *Nev., U.S.A.* **82 F6** 41 7N 114 58W
Wells, L., *Australia* **61 E3** 26 44S 123 15 E
Wells, Mt., *Australia* **60 C4** 17 25S 127 8 E
Wells Gray Prov. Park, *Canada* **72 C4** 52 30N 120 15W
Wells-next-the-Sea, *U.K.* **10 E8** 52 57N 0 51 E
Wells River, *U.S.A.* **79 B12** 44 9N 72 4W
Wellsboro, *U.S.A.* **78 E7** 41 45N 77 18W
Wellsburg, *U.S.A.* **78 F4** 40 16N 80 37W
Wellsville, *N.Y., U.S.A.* **78 D7** 42 7N 77 57W
Wellsville, *Ohio, U.S.A.* **78 F4** 40 36N 80 39W
Wellsville, *Utah, U.S.A.* **82 F8** 41 38N 111 56W
Wellton, *U.S.A.* **83 K6** 32 40N 114 8W
Wels, *Austria* **16 D8** 48 9N 14 1 E
Welshpool, *U.K.* **11 E4** 52 39N 3 8W
Welwyn Garden City, *U.K.* **11 F7** 51 48N 0 12W
Wem, *U.K.* **10 E5** 52 52N 2 44W
Wembere →, *Tanzania* **54 C3** 4 10S 34 15 E
Wemindji, *Canada* **70 B4** 53 0N 78 49W
Wen Xian, *China* **34 G7** 34 55N 113 5 E
Wenatchee, *U.S.A.* **82 C3** 47 25N 120 19W
Wenchang, *China* **38 C8** 19 38N 110 42 E
Wenchi, *Ghana* **50 G5** 7 46N 2 8W
Wenchow = Wenzhou, *China* **33 D7** 28 0N 120 38 E
Wenden, *U.S.A.* **85 M13** 33 49N 113 33W
Wendeng, *China* **35 F12** 37 15N 122 5 E
Wendesi, *Indonesia* **37 E8** 2 30S 134 17 E
Wendover, *U.S.A.* **82 F6** 40 44N 114 2W
Wenlock →, *Australia* **62 A3** 12 2S 141 55 E
Wenshan, *China* **32 D5** 23 20N 104 18 E
Wenshang, *China* **34 G9** 35 45N 116 30 E
Wenshui, *China* **34 F7** 37 26N 112 1 E
Wensleydale, *U.K.* **10 C6** 54 17N 2 0W
Wensu, *China* **32 B3** 41 15N 80 10 E
Wensum →, *U.K.* **10 E8** 52 40N 1 15 E
Wentworth, *Australia* **63 E3** 34 2S 141 54 E
Wentzel L., *Canada* **72 B6** 59 2N 114 28W
Wenut, *Indonesia* **37 E8** 3 11S 133 19 E
Wenxi, *China* **34 G6** 35 20N 111 10 E
Wenxian, *China* **34 H3** 32 43N 104 36 E
Wenzhou, *China* **33 D7** 28 0N 120 38 E
Weott, *U.S.A.* **82 F2** 40 20N 123 55W
Wepener, *S. Africa* **56 D4** 29 42S 27 3 E
Werda, *Botswana* **56 D3** 25 24S 23 15 E
Weri, *Indonesia* **37 E8** 3 10S 132 38 E
Werra →, *Germany* **16 C5** 51 24N 9 39 E
Werrimull, *Australia* **63 E3** 34 25S 141 38 E
Werris Creek, *Australia* **63 E5** 31 18S 150 38 E
Weser →, *Germany* **16 B5** 53 36N 8 28 E
Wesiri, *Indonesia* **37 F7** 7 30S 126 30 E
Weslemkoon L., *Canada* **78 A7** 45 2N 77 25W
Wesleyville, *Canada* **71 C9** 49 8N 53 36W
Wesleyville, *U.S.A.* **78 D4** 42 9N 80 0W
Wessel, C., *Australia* **62 A2** 10 59S 136 46 E
Wessel Is., *Australia* **62 A2** 11 10S 136 45 E
Wessington Springs, *U.S.A.* **80 C5** 44 5N 98 34W
West, *U.S.A.* **81 K6** 31 48N 97 6W
West →, *U.S.A.* **79 D12** 42 52N 72 33W
West Baines →, *Australia* **60 C4** 15 38S 129 59 E
West Bank □, *Asia* **47 C4** 32 6N 35 13 E
West Bend, *U.S.A.* **76 D1** 43 25N 88 11W
West Bengal □, *India* **43 H13** 23 0N 88 0 E
West Berkshire □, *U.K.* **11 F6** 51 25N 1 17W
West Beskids = Západné Beskydy, *Europe* **17 D10** 49 30N 19 0 E
West Branch, *U.S.A.* **76 C3** 44 17N 84 14W
West Branch Susquehanna →, *U.S.A.* **79 F8** 40 53N 76 48W
West Bromwich, *U.K.* **11 E6** 52 32N 1 59W
West Burra, *U.K.* **12 A7** 60 5N 1 21W
West Canada Cr. →, *U.S.A.* **79 C10** 43 1N 74 58W
West Cape Howe, *Australia* **61 G2** 35 8S 117 36 E
West Chazy, *U.S.A.* **79 B11** 44 49N 73 28W
West Chester, *U.S.A.* **79 G9** 39 58N 75 36W
West Columbia, *U.S.A.* **81 L7** 29 9N 95 39W
West Covina, *U.S.A.* **85 L9** 34 4N 117 54W
West Des Moines, *U.S.A.* **80 E8** 41 35N 93 43W
West Dunbartonshire □, *U.K.* **12 F4** 55 59N 4 30W
West End, *Bahamas* **88 A4** 26 41N 78 58W
West Falkland, *Falk. Is.* **96 G5** 51 40S 60 0W
West Fargo, *U.S.A.* **80 B6** 46 52N 96 54W
West Farmington, *U.S.A.* **78 E4** 41 23N 80 58W
West Fjord = Vestfjorden, *Norway* **8 C15** 67 55N 14 0 E
West Fork Trinity →, *U.S.A.* **81 J6** 32 48N 96 54W
West Frankfort, *U.S.A.* **80 G10** 37 54N 88 55W
West Hartford, *U.S.A.* **79 E12** 41 45N 72 44W
West Haven, *U.S.A.* **79 E12** 41 17N 72 57W
West Hazleton, *U.S.A.* **79 F9** 40 58N 76 0W
West Helena, *U.S.A.* **81 H9** 34 33N 90 38W

West Hurley

West Hurley, U.S.A. 79 E10 41 59N 74 7W
West Ice Shelf, Antarctica 5 C7 67 0S 85 0 E
West Indies, Cent. Amer. 89 D7 15 0N 65 0W
West Jordan, U.S.A. 82 F8 40 36N 111 56W
West Lorne, Canada 78 D3 42 36N 81 36W
West Lothian □, U.K. 12 F5 55 54N 3 36W
West Lunga →, Zambia 55 E1 13 6S 24 39 E
West Memphis, U.S.A. 81 H9 35 9N 90 11W
West Midlands □, U.K. 11 E6 52 26N 2 0W
West Mifflin, U.S.A. 78 F5 40 22N 79 52W
West Milton, U.S.A. 78 E8 41 1N 76 50W
West Monroe, U.S.A. 81 J8 32 31N 92 9W
West Newton, U.S.A. 78 F5 40 14N 79 46W
West Nicholson, Zimbabwe 55 G2 21 2S 29 20 E
West Palm Beach, U.S.A. 77 M5 26 43N 80 3W
West Plains, U.S.A. 81 G9 36 44N 91 51W
West Point, N.Y., U.S.A. 79 E11 41 24N 73 58W
West Point, Nebr., U.S.A. 80 E6 41 51N 96 43W
West Point, Va., U.S.A. 76 G7 37 32N 76 48W
West Pt. = Ouest, Pte. de l', Canada 71 C7 49 52N 64 40W
West Pt., Australia 63 F2 35 1S 135 56 E
West Road →, Canada 72 C4 53 18N 122 53W
West Rutland, U.S.A. 79 C11 43 38N 73 5W
West Schelde = Westerschelde →, Neths. 15 C3 51 25N 3 25 E
West Seneca, U.S.A. 78 D6 42 51N 78 48W
West Siberian Plain, Russia 28 C11 62 0N 75 0 E
West Sussex □, U.K. 11 G7 50 55N 0 30W
West-Terschelling, Neths. 15 A5 53 22N 5 13 E
West Valley City, U.S.A. 82 F8 40 42N 111 57W
West Virginia □, U.S.A. 76 F5 38 45N 80 30W
West-Vlaanderen □, Belgium 15 D2 51 0N 3 0 E
West Walker →, U.S.A. 84 G7 38 54N 119 9W
West Wyalong, Australia 63 E4 33 56S 147 10 E
West Yellowstone, U.S.A. 82 D8 44 40N 111 6W
West Yorkshire □, U.K. 10 D6 53 45N 1 40W
Westall Pt., Australia 63 E1 32 55S 134 4 E
Westbrook, U.S.A. 77 D10 43 41N 70 22W
Westbury, Australia 62 G4 41 30S 146 51 E
Westby, U.S.A. 80 A2 48 52N 104 3W
Westend, U.S.A. 85 K9 35 42N 117 24W
Westerland, Germany 9 J13 54 54N 8 17 E
Westerly, U.S.A. 79 E13 41 22N 71 50W
Western □, Kenya 54 B3 0 30N 34 30 E
Western □, Zambia 55 F1 15 0S 24 4 E
Western Australia □, Australia 61 E2 25 0S 118 0 E
Western Cape □, S. Africa 56 E3 34 0S 20 0 E
Western Dvina = Daugava →, Latvia 9 H21 57 4N 24 3 E
Western Ghats, India 40 N9 14 0N 75 0 E
Western Isles □, U.K. 12 D1 57 30N 7 10W
Western Sahara ■, Africa 50 D3 25 0N 13 0W
Western Samoa = Samoa ■, Pac. Oc. 59 B13 14 0S 172 0W
Westernport, U.S.A. 76 F6 39 29N 79 3W
Westerschelde →, Neths. 15 C3 51 25N 3 25 E
Westerwald, Germany 16 C4 50 38N 7 56 E
Westfield, Mass., U.S.A. 79 D12 42 7N 72 45W
Westfield, N.Y., U.S.A. 78 D5 42 20N 79 35W
Westfield, Pa., U.S.A. 78 E7 41 55N 77 32W
Westhill, U.K. 12 D6 57 9N 2 19W
Westhope, U.S.A. 80 A4 48 55N 101 1W
Westland Bight, N.Z. 59 K3 42 55S 170 5 E
Westlock, Canada 72 C6 54 9N 113 55W
Westmar, Australia 63 D4 27 55S 149 44 E
Westmeath □, Ireland 13 C4 53 33N 7 34W
Westminster, U.S.A. 76 F7 39 34N 76 59W
Westmont, U.S.A. 78 F6 40 19N 78 58W
Westmorland, U.S.A. 85 M11 33 2N 115 37W
Weston, Oreg., U.S.A. 82 D4 45 49N 118 26W
Weston, W. Va., U.S.A. 76 F5 39 2N 80 28W
Weston I., Canada 70 B4 52 33N 79 36W
Weston-super-Mare, U.K. 11 F5 51 21N 2 58W
Westover, U.S.A. 78 F6 40 45N 78 40W
Westport, Canada 79 B8 44 40N 76 25W
Westport, Ireland 13 C2 53 48N 9 31W
Westport, N.Z. 59 J3 41 46S 171 37 E
Westport, N.Y., U.S.A. 79 B11 44 11N 73 26W
Westport, Oreg., U.S.A. 84 D3 46 8N 123 23W
Westport, Wash., U.S.A. 84 D2 46 53N 124 6W
Westray, Canada 73 C8 53 36N 101 24W
Westray, U.K. 12 B5 59 18N 3 0W
Westree, Canada 70 C3 47 26N 81 34W
Westville, U.S.A. 84 F6 39 8N 120 42W
Westwood, U.S.A. 82 F3 40 18N 121 0W
Wetar, Indonesia 37 F7 7 48S 126 30 E
Wetaskiwin, Canada 72 C6 52 55N 113 24W
Wete, Tanzania 52 F7 5 4S 39 43 E
Wetherby, U.K. 10 D6 53 56N 1 23W
Wethersfield, U.S.A. 79 E12 41 42N 72 40W
Wetteren, Belgium 15 D3 51 0N 3 53 E
Wetzlar, Germany 16 C5 50 32N 8 31 E
Wewoka, U.S.A. 81 H6 35 9N 96 30W
Wexford, Ireland 13 D5 52 20N 6 28W
Wexford □, Ireland 13 D5 52 20N 6 25W
Wexford Harbour, Ireland 13 D5 52 20N 6 25W
Weyburn, Canada 73 D8 49 40N 103 50W
Weymouth, Canada 71 D6 44 30N 66 1W
Weymouth, U.K. 11 G5 50 37N 2 28W
Weymouth, U.S.A. 79 D14 42 13N 70 58W
Weymouth, C., Australia 62 A3 12 37S 143 27 E
Wha Ti, Canada 68 B8 63 8N 117 16W
Whakatane, N.Z. 59 G6 37 57S 177 1 E
Whale →, Canada 71 A6 58 15N 67 40W
Whale Cove, Canada 73 A10 62 11N 92 36W
Whales, B. of, Antarctica 5 D12 78 0S 165 0W
Whalsay, U.K. 12 A8 60 22N 0 59W
Whangamomona, N.Z. 59 H5 39 8S 174 44 E
Whangarei, N.Z. 59 F5 35 43S 174 21 E
Whangarei Harb., N.Z. 59 F5 35 45S 174 28 E
Wharfe →, U.K. 10 D6 53 51N 1 9W
Wharfedale, U.K. 10 C5 54 6N 2 1W
Wharton, N.J., U.S.A. 79 F10 40 54N 74 35W
Wharton, Pa., U.S.A. 78 E6 41 31N 78 1W
Wharton, Tex., U.S.A. 81 L6 29 19N 96 6W
Wheatland, Calif., U.S.A. 84 F5 39 1N 121 25W
Wheatland, Wyo., U.S.A. 80 D2 42 3N 104 58W
Wheatley, Canada 78 D2 42 6N 82 27W
Wheaton, Md., U.S.A. 76 F7 39 3N 77 3W
Wheaton, Minn., U.S.A. 80 C6 45 48N 96 30W
Wheelbarrow Pk., U.S.A. 84 H10 37 26N 116 5W
Wheeler, Oreg., U.S.A. 82 D2 45 41N 123 53W
Wheeler, Tex., U.S.A. 81 H4 35 27N 100 16W
Wheeler →, Canada 71 A6 57 2N 67 13W
Wheeler L., U.S.A. 77 H2 34 48N 87 23W
Wheeler Pk., N. Mex., U.S.A. 83 H11 36 34N 105 25W
Wheeler Pk., Nev., U.S.A. 83 G6 38 57N 114 15W
Wheeler Ridge, U.S.A. 85 L8 35 0N 118 57W
Wheeling, U.S.A. 78 F4 40 4N 80 43W
Whernside, U.K. 10 C5 54 14N 2 24W
Whiskey Jack L., Canada 73 B8 58 23N 101 55W
Whistleduck Cr. →, Australia 62 C2 20 15S 135 18 E
Whistler, Canada 72 C4 50 7N 122 58W
Whitby, Canada 78 C6 43 52N 78 56W
Whitby, U.K. 10 C7 54 29N 0 37W
White →, Ark., U.S.A. 81 J9 33 57N 91 5W
White →, Ind., U.S.A. 76 F2 38 25N 87 45W
White →, S. Dak., U.S.A. 80 D5 43 42N 99 27W
White →, Tex., U.S.A. 81 J4 33 14N 100 56W
White →, Utah, U.S.A. 82 F9 40 4N 109 41W
White →, Vt., U.S.A. 79 C12 43 37N 72 20W
White →, Wash., U.S.A. 84 C4 47 12N 122 15W
White, L., Australia 60 D4 21 9S 128 56 E
White B., Canada 71 C8 50 0N 56 35W
White Bird, U.S.A. 82 D5 45 46N 116 18W
White Butte, U.S.A. 80 B3 46 23N 103 18W
White City, U.S.A. 82 E2 42 26N 122 51W
White Cliffs, Australia 63 E3 30 50S 143 10 E
White Hall, U.S.A. 80 F9 39 26N 90 24W
White Haven, U.S.A. 79 E9 41 4N 75 47W
White Horse, Vale of, U.K. 11 F6 51 37N 1 30W
White I., N.Z. 59 G6 37 30S 177 13 E
White L., Canada 79 A8 45 18N 76 31W
White L., U.S.A. 81 L8 29 44N 92 30W
White Mountain Peak, U.S.A. 83 G4 37 38N 118 15W
White Mts., Calif., U.S.A. 84 H8 37 30N 118 15W
White Mts., N.H., U.S.A. 79 B13 44 15N 71 15W
White Nile = Nîl el Abyad →, Sudan 51 E12 15 38N 32 31 E
White Otter L., Canada 70 C1 49 5N 91 55W
White Pass, U.S.A. 84 D5 46 38N 121 24W
White Plains, U.S.A. 79 E11 41 2N 73 46W
White River, Canada 70 C2 48 35N 85 20W
White River, S. Africa 57 D5 25 20S 31 0 E
White River, U.S.A. 80 D4 43 34N 100 45W
White Rock, Canada 84 A4 49 2N 122 48W
White Russia = Belarus ■, Europe 17 B14 53 30N 27 0 E
White Sea = Beloye More, Russia 24 A6 66 30N 38 0 E
White Sulphur Springs, Mont., U.S.A. 82 C8 46 33N 110 54W
White Sulphur Springs, W. Va., U.S.A. 76 G5 37 48N 80 18W
White Swan, U.S.A. 84 D6 46 23N 120 44W
Whitecliffs, N.Z. 59 K3 43 26S 171 55 E
Whitecourt, Canada 72 C5 54 10N 115 45W
Whiteface Mt., U.S.A. 79 B11 44 22N 73 54W
Whitefield, U.S.A. 79 B13 44 23N 71 37W
Whitefish, U.S.A. 82 B6 48 25N 114 20W
Whitefish L., Canada 73 A7 62 41N 106 48W
Whitefish Point, U.S.A. 76 B3 46 45N 84 59W
Whitegull, L., Canada 71 A7 55 27N 64 17W
Whitehall, Mich., U.S.A. 76 D2 43 24N 86 21W
Whitehall, Mont., U.S.A. 82 D7 45 52N 112 6W
Whitehall, N.Y., U.S.A. 79 C11 43 33N 73 24W
Whitehall, Wis., U.S.A. 80 C9 44 22N 91 19W
Whitehaven, U.K. 10 C4 54 33N 3 35W
Whitehorse, Canada 72 A1 60 43N 135 3W
Whitemark, Australia 62 G4 40 7S 148 3 E
Whiteriver, U.S.A. 83 K9 33 50N 109 58W
Whitesand →, Canada 72 A5 60 9N 115 45W
Whitesands, S. Africa 56 E3 34 23S 20 50 E
Whitesboro, N.Y., U.S.A. 79 C9 43 7N 75 18W
Whitesboro, Tex., U.S.A. 81 J6 33 39N 96 54W
Whiteshell Prov. Park, Canada 73 D9 50 0N 95 40W
Whitesville, U.S.A. 78 D7 42 2N 77 46W
Whiteville, U.S.A. 77 H6 34 20N 78 42W
Whitewater, U.S.A. 76 D1 42 50N 88 44W
Whitewater Baldy, U.S.A. 83 K9 33 20N 108 39W
Whitewater L., Canada 70 B2 50 50N 89 10W
Whitewood, Australia 62 C3 21 28S 143 30 E
Whitewood, Canada 73 C8 50 20N 102 20W
Whithorn, U.K. 12 G4 54 44N 4 26W
Whitianga, N.Z. 59 G5 36 47S 175 41 E
Whitman, U.S.A. 79 D14 42 5N 70 56W
Whitney, Canada 78 A6 45 31N 78 14W
Whitney, Mt., U.S.A. 84 J8 36 35N 118 18W
Whitney Point, U.S.A. 79 D9 42 20N 75 58W
Whitstable, U.K. 11 F9 51 21N 1 3 E
Whitsunday I., Australia 62 C4 20 15S 149 4 E
Whittier, U.S.A. 85 M8 33 58N 118 3W
Whittlesea, Australia 63 F4 37 27S 145 9 E
Wholdaia L., Canada 73 A8 60 43N 104 20W
Whyalla, Australia 63 E2 33 2S 137 30 E
Wiarton, Canada 78 B3 44 40N 81 10W
Wiay, U.K. 12 D1 57 24N 7 13W
Wibaux, U.S.A. 80 B2 46 59N 104 11W
Wichian Buri, Thailand 38 E3 15 39N 101 7 E
Wichita, U.S.A. 81 G6 37 42N 97 20W
Wichita Falls, U.S.A. 81 J5 33 54N 98 30W
Wick, U.K. 12 C5 58 26N 3 5W
Wicked Pt., Canada 78 C7 43 52N 77 15W
Wickenburg, U.S.A. 83 K7 33 58N 112 44W
Wickepin, Australia 61 F2 32 50S 117 30 E
Wickham, Australia 60 D2 20 42S 117 11 E
Wickham, C., Australia 62 F3 39 35S 143 57 E
Wickliffe, U.S.A. 78 E3 41 36N 81 28W
Wicklow, Ireland 13 D5 52 59N 6 3W
Wicklow □, Ireland 13 D5 52 57N 6 25W
Wicklow Hd., Ireland 13 D6 52 58N 6 0W
Wicklow Mts., Ireland 13 C5 52 58N 6 26W
Widgeegoara Cr. →, Australia 63 D4 28 51S 146 34 E
Widgiemooltha, Australia 61 F3 31 30S 121 34 E
Widnes, U.K. 10 D5 53 23N 2 45W
Wieluń, Poland 17 C10 51 15N 18 34 E
Wien, Austria 16 D9 48 12N 16 22 E
Wiener Neustadt, Austria 16 E9 47 49N 16 16 E
Wiesbaden, Germany 16 C5 50 4N 8 14 E
Wigan, U.K. 10 D5 53 33N 2 38W
Wiggins, Colo., U.S.A. 80 E2 40 14N 104 4W
Wiggins, Miss., U.S.A. 81 K10 30 51N 89 8W
Wight, I. of □, U.K. 11 G6 50 40N 1 20W
Wigston, U.K. 11 E6 52 35N 1 6W
Wigton, U.K. 10 C4 54 50N 3 10W
Wigtown, U.K. 12 G4 54 53N 4 27W
Wigtown B., U.K. 12 G4 54 46N 4 15W
Wilber, U.S.A. 80 E6 40 29N 96 58W
Wilberforce, Canada 78 A6 45 2N 78 13W
Wilberforce, C., Australia 62 A2 11 54S 136 35 E
Wilburton, U.S.A. 81 H7 34 55N 95 19W
Wilcannia, Australia 63 E3 31 30S 143 26 E
Wilcox, U.S.A. 78 E6 41 35N 78 41W
Wildrose, U.S.A. 85 J9 36 14N 117 11W
Wildspitze, Austria 16 E6 46 53N 10 53 E
Wilge →, S. Africa 57 D4 27 3S 28 20 E
Wilhelm II Coast, Antarctica 5 C7 68 0S 90 0 E
Wilhelmshaven, Germany 16 B5 53 31N 8 7 E
Wilhelmstal, Namibia 56 C2 21 58S 16 21 E
Wilkes-Barre, U.S.A. 79 E9 41 15N 75 53W
Wilkie, Canada 73 C7 52 27N 108 42W
Wilkinsburg, U.S.A. 78 F5 40 26N 79 53W
Wilkinson Lakes, Australia 61 E5 29 40S 132 39 E
Willandra Creek →, Australia 63 E4 33 22S 145 52 E
Willapa B., U.S.A. 82 C2 46 40N 124 0W
Willapa Hills, U.S.A. 84 D3 46 35N 123 25W
Willard, N.Y., U.S.A. 78 D8 42 40N 76 50W
Willard, Ohio, U.S.A. 78 E2 41 3N 82 44W
Willcox, U.S.A. 83 K9 32 15N 109 50W
Willemstad, Neth. Ant. 89 D6 12 5N 69 0W
Willet, U.S.A. 79 D9 42 28N 75 55W
William →, Canada 73 B7 59 8N 109 19W
William 'Bill' Dannely Res., U.S.A. 77 J2 32 10N 87 10W
William Creek, Australia 63 D2 28 58S 136 22 E
Williams, Australia 61 F2 33 2S 116 52 E
Williams, Ariz., U.S.A. 83 J7 35 15N 112 11W
Williams, Calif., U.S.A. 84 F4 39 9N 122 9W
Williams Harbour, Canada 71 B8 52 33N 55 47W
Williams Lake, Canada 72 C4 52 10N 122 10W
Williamsburg, Ky., U.S.A. 77 G3 36 44N 84 10W
Williamsburg, Pa., U.S.A. 78 F6 40 28N 78 12W
Williamsburg, Va., U.S.A. 76 G7 37 17N 76 44W
Williamson, N.Y., U.S.A. 78 C7 43 14N 77 11W
Williamson, W. Va., U.S.A. 76 G4 37 41N 82 17W
Williamsport, U.S.A. 78 E7 41 15N 77 0W
Williamston, U.S.A. 77 H7 35 51N 77 4W
Williamstown, Australia 63 F3 37 51S 144 52 E
Williamstown, Ky., U.S.A. 76 F3 38 38N 84 34W
Williamstown, Mass., U.S.A. 79 D11 42 41N 73 12W
Williamstown, N.Y., U.S.A. 79 C9 43 26N 75 53W
Willimantic, U.S.A. 79 E12 41 43N 72 13W
Willingboro, U.S.A. 76 E8 40 3N 74 54W
Willis Group, Australia 62 B5 16 18S 150 0 E
Williston, S. Africa 56 E3 31 20S 20 53 E
Williston, Fla., U.S.A. 77 L4 29 23N 82 27W
Williston, N. Dak., U.S.A. 80 A3 48 9N 103 37W
Williston L., Canada 72 B4 56 0N 124 0W
Willits, U.S.A. 82 G2 39 25N 123 21W
Willmar, U.S.A. 80 C7 45 7N 95 3W
Willoughby, U.S.A. 78 E3 41 39N 81 24W
Willow Bunch, Canada 73 D7 49 20N 105 35W
Willow L., Canada 72 A5 62 10N 119 8W
Willow Wall, The, China 35 C12 42 10N 122 0 E
Willowick, U.S.A. 78 E3 41 38N 81 28W
Willowlake →, Canada 72 A4 62 42N 123 8W
Willowmore, S. Africa 56 E3 33 15S 23 30 E
Willows, U.S.A. 84 F4 39 31N 122 12W
Willowvale = Gatyana, S. Africa 57 E4 32 16S 28 31 E
Wills, L., Australia 60 D4 21 25S 128 51 E
Wills Cr. →, Australia 62 C3 22 43S 140 2 E
Willsboro, U.S.A. 79 B11 44 21N 73 24W
Willunga, Australia 63 F2 35 15S 138 30 E
Wilmette, U.S.A. 76 D2 42 5N 87 42W
Wilmington, Australia 63 E2 32 39S 138 7 E
Wilmington, Del., U.S.A. 76 F8 39 45N 75 33W
Wilmington, N.C., U.S.A. 77 H7 34 14N 77 55W
Wilmington, Ohio, U.S.A. 76 F4 39 27N 83 50W
Wilmington, Vt., U.S.A. 79 D12 42 52N 72 52W
Wilmslow, U.K. 10 D5 53 19N 2 13W
Wilpena Cr. →, Australia 63 E2 31 25S 139 29 E
Wilsall, U.S.A. 82 D8 45 59N 110 38W
Wilson, N.C., U.S.A. 77 H7 35 44N 77 55W
Wilson, N.Y., U.S.A. 78 C6 43 19N 78 50W
Wilson, Pa., U.S.A. 79 F9 40 41N 75 15W
Wilson →, Australia 60 C4 16 48S 128 16 E
Wilson Bluff, Australia 61 F4 31 41S 129 0 E
Wilson Inlet, Australia 61 G2 35 0S 117 22 E
Wilsons Promontory, Australia 63 F4 38 55S 146 25 E
Wilton, U.S.A. 80 B4 47 10N 100 47W
Wilton →, Australia 62 A1 14 45S 134 33 E
Wiltshire □, U.K. 11 F6 51 18N 1 53W
Wiltz, Lux. 15 E5 49 57N 5 55 E
Wiluna, Australia 61 E3 26 36S 120 14 E
Wimborne Minster, U.K. 11 G6 50 48N 1 59W
Wimmera →, Australia 63 F3 36 8S 141 56 E
Winam G., Kenya 54 C3 0 20S 34 15 E
Winburg, S. Africa 56 D4 28 30S 27 2 E
Winchendon, U.S.A. 79 D12 42 41N 72 3W
Winchester, U.K. 11 F6 51 4N 1 18W
Winchester, Conn., U.S.A. 79 E11 41 53N 73 9W
Winchester, Idaho, U.S.A. 82 C5 46 14N 116 38W
Winchester, Ind., U.S.A. 76 E3 40 10N 84 59W
Winchester, Ky., U.S.A. 76 G3 38 0N 84 11W
Winchester, N.H., U.S.A. 79 D12 42 46N 72 23W
Winchester, Nev., U.S.A. 85 J11 36 6N 115 10W
Winchester, Tenn., U.S.A. 77 H2 35 11N 86 7W
Winchester, Va., U.S.A. 76 F6 39 11N 78 10W
Wind →, U.S.A. 82 E9 43 12N 108 12W
Wind River Range, U.S.A. 82 E9 43 0N 109 30W
Windau = Ventspils, Latvia 9 H19 57 25N 21 32 E
Windber, U.S.A. 78 F6 40 14N 78 50W
Winder, U.S.A. 77 J4 34 0N 83 45W
Windermere, U.K. 10 C5 54 23N 2 55W
Windhoek, Namibia 56 C2 22 35S 17 4 E
Windom, U.S.A. 80 D7 43 52N 95 7W
Windorah, Australia 62 D3 25 24S 142 36 E
Window Rock, U.S.A. 83 J9 35 41N 109 3W
Windrush →, U.K. 11 F6 51 43N 1 24W
Windsor, Australia 63 E5 33 37S 150 50 E
Windsor, N.S., Canada 71 D7 44 59N 64 5W
Windsor, Ont., Canada 78 D2 42 18N 83 0W
Windsor, U.K. 11 F7 51 29N 0 36W
Windsor, Colo., U.S.A. 80 E2 40 29N 104 54W
Windsor, Conn., U.S.A. 79 E12 41 50N 72 39W
Windsor, Mo., U.S.A. 80 F8 38 32N 93 31W
Windsor, N.Y., U.S.A. 79 D9 42 5N 75 37W
Windsor, Vt., U.S.A. 79 C12 43 29N 72 24W
Windsor & Maidenhead □, U.K. 11 F7 51 29N 0 40W
Windsorton, S. Africa 56 D3 28 16S 24 44 E
Windward Is., W. Indies 89 D7 13 0N 61 0W
Windward Passage = Vientos, Paso de los, Caribbean 89 C5 20 0N 74 0W
Winefred L., Canada 73 B6 55 30N 110 30W
Winfield, U.S.A. 81 G6 37 15N 96 59W
Wingate Mts., Australia 60 B5 14 25S 130 40 E
Wingham, Australia 63 E5 31 48S 152 22 E
Wingham, Canada 78 C3 43 55N 81 20W
Winisk, Canada 70 A2 55 20N 85 15W
Winisk →, Canada 70 A2 55 17N 85 5W
Winisk L., Canada 70 B2 52 55N 87 22W
Wink, U.S.A. 81 K3 31 45N 103 9W
Winkler, Canada 73 D9 49 10N 97 56W
Winlock, U.S.A. 84 D4 46 30N 122 56W
Winnebago, L., U.S.A. 76 D1 44 0N 88 26W
Winnecke Cr. →, Australia 60 C5 18 35S 131 34 E
Winnemucca, U.S.A. 82 F5 40 58N 117 44W
Winnemucca L., U.S.A. 82 F4 40 7N 119 21W
Winnett, U.S.A. 82 C9 47 0N 108 21W
Winnfield, U.S.A. 81 K8 31 56N 92 38W
Winnibigoshish, L., U.S.A. 80 B7 47 27N 94 13W
Winnipeg, Canada 73 D9 49 54N 97 9W
Winnipeg →, Canada 73 C9 50 38N 96 19W
Winnipeg, L., Canada 73 C9 52 0N 97 0W
Winnipeg Beach, Canada 73 C9 50 30N 96 58W
Winnipegosis, Canada 73 C9 51 39N 99 55W
Winnipegosis L., Canada 73 C9 52 30N 100 0W
Winnipesaukee, L., U.S.A. 79 C13 43 38N 71 21W
Winnisquam L., U.S.A. 79 C13 43 33N 71 31W
Winnsboro, La., U.S.A. 81 J9 32 10N 91 43W
Winnsboro, S.C., U.S.A. 77 H5 34 23N 81 5W
Winnsboro, Tex., U.S.A. 81 J7 32 58N 95 17W
Winokapau, L., Canada 71 B7 53 15N 62 50W
Winona, Minn., U.S.A. 80 C9 44 3N 91 39W
Winona, Miss., U.S.A. 81 J10 33 29N 89 44W
Winooski, U.S.A. 79 B11 44 29N 73 11W
Winooski →, U.S.A. 79 B11 44 32N 73 17W
Winschoten, Neths. 15 A7 53 9N 7 3 E
Winsford, U.K. 10 D5 53 12N 2 31W
Winslow, Ariz., U.S.A. 83 J8 35 2N 110 42W
Winslow, Wash., U.S.A. 84 C4 47 38N 122 31W
Winsted, U.S.A. 79 E11 41 55N 73 4W
Winston-Salem, U.S.A. 77 G5 36 6N 80 15W
Winter Garden, U.S.A. 77 L5 28 34N 81 35W
Winter Haven, U.S.A. 77 M5 28 1N 81 44W
Winter Park, U.S.A. 77 L5 28 36N 81 20W
Winterhaven, U.S.A. 85 N12 32 47N 114 39W
Winters, U.S.A. 84 G5 38 32N 121 58W
Wintersville, U.S.A. 78 F4 40 23N 80 42W
Winterswijk, Neths. 15 C6 51 58N 6 43 E
Winterthur, Switz. 18 C8 47 30N 8 44 E
Winthrop, U.S.A. 82 B3 48 28N 120 10W
Winton, Australia 62 C3 22 24S 143 3 E
Winton, N.Z. 59 M2 46 8S 168 20 E
Wirrulla, Australia 63 E1 32 24S 134 31 E
Wisbech, U.K. 11 E8 52 41N 0 9 E
Wisconsin □, U.S.A. 80 C10 44 45N 89 30W
Wisconsin →, U.S.A. 80 D9 43 0N 91 15W
Wisconsin Rapids, U.S.A. 80 C10 44 23N 89 49W
Wisdom, U.S.A. 82 D7 45 37N 113 27W
Wishaw, U.K. 12 F5 55 46N 3 54W
Wishek, U.S.A. 80 B5 46 16N 99 33W
Wisła →, Poland 17 A10 54 22N 18 55 E
Wismar, Germany 16 B6 53 54N 11 29 E
Wisner, U.S.A. 80 E6 41 59N 96 55W
Witbank, S. Africa 57 D4 25 51S 29 14 E
Witdraai, S. Africa 56 D3 26 58S 20 48 E
Witham, U.K. 11 F8 51 48N 0 40 E
Witham →, U.K. 10 E7 52 59N 0 2W
Withernsea, U.K. 10 D8 53 44N 0 1 E
Witney, U.K. 11 F6 51 48N 1 28W
Witnossob →, Namibia 56 D3 23 55S 18 45 E
Wittenberge, Germany 16 B6 53 0N 11 45 E
Wittenoom, Australia 60 D2 22 15S 118 20 E
Witvlei, Namibia 56 C2 22 23S 18 32 E
Wkra →, Poland 17 B11 52 27N 20 44 E
Wlingi, Indonesia 37 H15 8 5S 112 25 E
Włocławek, Poland 17 B10 52 40N 19 3 E
Włodawa, Poland 17 C12 51 33N 23 31 E
Woburn, U.S.A. 79 D13 42 29N 71 9W
Wodian, China 34 H7 32 50N 112 35 E
Wodonga = Albury-Wodonga, Australia 63 F4 36 3S 146 56 E
Wokam, Indonesia 37 F8 5 45S 134 28 E
Woking, U.K. 11 F7 51 19N 0 34W
Wokingham □, U.K. 11 F7 51 25N 0 51W
Wolf →, Canada 72 A2 60 17N 132 33W
Wolf Creek, U.S.A. 82 C7 47 0N 112 4W
Wolf L., Canada 72 A2 60 24N 131 40W
Wolf Point, U.S.A. 80 A2 48 5N 105 39W
Wolfe I., Canada 79 B8 44 7N 76 20W
Wolfeboro, U.S.A. 79 C13 43 35N 71 13W
Wolfsberg, Austria 16 E8 46 50N 14 52 E
Wolfsburg, Germany 16 B6 52 25N 10 48 E
Wolin, Poland 16 B8 53 50N 14 37 E
Wollaston, Is., Chile 96 H3 55 40S 67 30W
Wollaston L., Canada 73 B8 58 7N 103 10W
Wollaston Lake, Canada 73 B8 58 3N 103 33W
Wollaston Pen., Canada 68 B8 69 30N 115 0W
Wollongong, Australia 63 E5 34 25S 150 54 E
Wolmaransstad, S. Africa 56 D4 27 12S 25 59 E
Wolseley, S. Africa 56 E2 33 26S 19 7 E
Wolsey, U.S.A. 80 C5 44 24N 98 28W
Wolstenholme, C., Canada 66 C12 62 35N 77 30W
Wolvega, Neths. 15 B6 52 52N 6 0 E
Wolverhampton, U.K. 11 E5 52 35N 2 7W
Wondai, Australia 63 D5 26 20S 151 49 E
Wongalarroo L., Australia 63 E3 31 32S 144 0 E
Wongan Hills, Australia 61 F2 30 51S 116 37 E
Wŏnju, S. Korea 35 F14 37 22N 127 58 E
Wonosari, Indonesia 37 G14 7 58S 110 36 E
Wonosobo, Indonesia 37 G13 7 22S 109 54 E
Wonowon, Canada 72 B4 56 44N 121 48W
Wŏnsan, N. Korea 35 E14 39 11N 127 27 E
Wonthaggi, Australia 63 F4 38 37S 145 37 E
Wood Buffalo Nat. Park, Canada 72 B6 59 0N 113 41W
Wood Is., Australia 60 C3 16 24S 123 19 E
Wood L., Canada 73 B8 55 17N 103 17W
Woodah I., Australia 62 A2 13 27S 136 10 E
Woodbourne, U.S.A. 79 E10 41 46N 74 36W
Woodbridge, Canada 78 C5 43 47N 79 36W
Woodbridge, U.K. 11 E9 52 6N 1 20 E
Woodburn, U.S.A. 82 D2 45 9N 122 51W
Woodenbong, Australia 63 D5 28 24S 152 39 E

Yetman

Woodend, Australia 63 F3 37 20S 144 33 E
Woodford, Australia 63 D5 26 58S 152 47 E
Woodfords, U.S.A. 84 G7 38 47N 119 50W
Woodlake, U.S.A. 84 J7 36 25N 119 6W
Woodland, Calif., U.S.A. 84 G5 38 41N 121 46W
Woodland, Maine, U.S.A. 77 C12 45 9N 67 25W
Woodland, Pa., U.S.A. 78 F6 40 59N 78 21W
Woodland, Wash., U.S.A. 84 E4 45 54N 122 45W
Woodland Caribou Prov. Park, Canada 73 C10 51 0N 94 45W
Woodridge, Canada 73 D9 49 20N 96 9W
Woodroffe, Mt., Australia 61 E5 26 20S 131 45 E
Woods, L., Australia 62 B1 17 50S 133 30 E
Woods, L. of the, Canada 73 D10 49 15N 94 45W
Woodstock, Australia 62 B4 19 35S 146 50 E
Woodstock, N.B., Canada 71 C6 46 11N 67 37W
Woodstock, Ont., Canada 78 C4 43 10N 80 45W
Woodstock, U.K. 11 F6 51 51N 1 20W
Woodstock, Ill., U.S.A. 80 D10 42 19N 88 27W
Woodstock, Vt., U.S.A. 79 C12 43 37N 72 31W
Woodsville, U.S.A. 79 B13 44 9N 72 2W
Woodville, N.Z. 59 J5 40 20S 175 53 E
Woodville, Miss., U.S.A. 81 K9 31 6N 91 18W
Woodville, Tex., U.S.A. 81 K7 30 47N 94 25W
Woodward, U.S.A. 81 G5 36 26N 99 24W
Woody, U.S.A. 85 K8 35 42N 118 50W
Woody →, Canada 73 C8 52 31N 100 51W
Woolamai, C., Australia 63 F4 38 30S 145 23 E
Wooler, U.K. 10 B5 55 33N 2 1W
Woolgoolga, Australia 63 E5 30 6S 153 11 E
Woomera, Australia 63 E2 31 5S 136 50 E
Woonsocket, R.I., U.S.A. 79 E13 42 0N 71 31W
Woonsocket, S. Dak., U.S.A. 80 C5 44 3N 98 17W
Wooramel →, Australia 61 E1 25 47S 114 10 E
Wooramel Roadhouse, Australia 61 E1 25 45S 114 17 E
Wooster, U.S.A. 78 F3 40 48N 81 56W
Worcester, S. Africa 56 E2 33 39S 19 27 E
Worcester, U.K. 11 E5 52 11N 2 12W
Worcester, Mass., U.S.A. 79 D13 42 16N 71 48W
Worcester, N.Y., U.S.A. 79 D10 42 36N 74 45W
Worcestershire □, U.K. 11 E5 52 13N 2 10W
Workington, U.K. 10 C4 54 39N 3 33W
Worksop, U.K. 10 D6 53 18N 1 7W
Workum, Neths. 15 B5 52 59N 5 26 E
Worland, U.S.A. 82 D10 44 1N 107 57W
Worms, Germany 16 D5 49 37N 8 21 E
Worsley, Canada 72 B5 56 31N 119 8W
Wortham, U.S.A. 81 K6 31 47N 96 28W
Worthing, U.K. 11 G7 50 49N 0 21W
Worthington, Minn., U.S.A. 80 D7 43 37N 95 36W
Worthington, Pa., U.S.A. 78 F5 40 50N 79 38W
Wosi, Indonesia 37 E7 0 15S 128 0 E
Wou-han = Wuhan, China 33 C6 30 31N 114 18 E
Wousi = Wuxi, China 33 C7 31 33N 120 18 E
Wowoni, Indonesia 37 E6 4 5S 123 5 E
Wrangel I. = Vrangelya, Ostrov, Russia 27 B19 71 0N 180 0 E
Wrangell, U.S.A. 72 B2 56 28N 132 23W
Wrangell Mts., U.S.A. 68 B5 61 30N 142 0W
Wrath, C., U.K. 12 C3 58 38N 5 1W
Wray, U.S.A. 80 E3 40 5N 102 13W
Wrekin, The, U.K. 11 E5 52 41N 2 32W
Wrens, U.S.A. 77 J4 33 12N 82 23W
Wrexham, U.K. 10 D4 53 3N 3 0W
Wrexham □, U.K. 10 D5 53 1N 2 58W
Wright, U.S.A. 80 D2 43 47N 105 30W
Wright Pt., Canada 78 C3 43 48N 81 44W
Wrightson Mt., U.S.A. 83 L8 31 42N 110 51W
Wrightwood, U.S.A. 85 L9 34 21N 117 38W
Wrigley, Canada 68 B7 63 16N 123 37W
Wrocław, Poland 17 C9 51 5N 17 5 E
Września, Poland 17 B9 52 21N 17 36 E
Wu Jiang →, China 32 D5 29 40N 107 20 E
Wu'an, China 34 F8 36 40N 114 15 E
Wubin, Australia 61 F2 30 6S 116 37 E
Wubu, China 34 F6 37 28N 110 42 E
Wuchang, China 35 B14 44 55N 127 5 E
Wucheng, China 34 F9 37 12N 116 20 E
Wuchuan, China 34 D6 41 5N 111 28 E
Wudi, China 35 F9 37 40N 117 35 E
Wuding He →, China 34 F6 37 2N 110 23 E
Wudinna, Australia 63 E2 33 0S 135 22 E
Wudu, China 34 H3 33 22N 104 54 E
Wuhan, China 33 C6 30 31N 114 18 E
Wuhe, China 35 H9 33 10N 117 50 E
Wuhsi = Wuxi, China 33 C7 31 33N 120 18 E
Wuhu, China 33 C6 31 22N 118 21 E
Wukari, Nigeria 50 G7 7 51N 9 42 E
Wulajie, China 35 B14 44 6N 126 33 E
Wulanbulang, China 34 D6 41 5N 110 55 E
Wular L., India 43 B6 34 20N 74 30 E
Wulian, China 35 G10 35 40N 119 12 E
Wuliaru, Indonesia 37 F8 7 27S 131 0 E
Wuluk'omushih Ling, China 32 C3 36 25N 87 25 E
Wulumuchi = Ürümqi, China 26 E9 43 45N 87 45 E
Wundowie, Australia 61 F2 31 47S 116 23 E
Wunnummin L., Canada 70 B2 52 55N 89 10W
Wuntho, Burma 41 H19 23 55N 95 45 E
Wuppertal, Germany 16 C4 51 16N 7 12 E
Wuppertal, S. Africa 56 E2 32 13S 19 12 E
Wuqing, China 35 E9 39 23N 117 4 E
Wurtsboro, U.S.A. 79 E10 41 35N 74 29W
Würzburg, Germany 16 D5 49 46N 9 55 E
Wushan, China 34 G3 34 43N 104 53 E
Wusuli Jiang = Ussuri →, Asia 30 A7 48 27N 135 0 E
Wutai, China 34 E7 38 40N 113 12 E
Wuting = Huimin, China 35 F9 37 27N 117 28 E
Wutonghaolai, China 35 C11 42 50N 120 5 E
Wutongqiao, China 32 D5 29 22N 103 50 E
Wuwei, China 32 C5 37 57N 102 34 E
Wuxi, China 33 C7 31 33N 120 18 E
Wuxiang, China 34 F7 36 49N 112 50 E
Wuyang, China 34 H7 33 25N 113 35 E
Wuyi, China 34 F8 37 46N 115 56 E
Wuyi Shan, China 33 D6 27 0N 117 0 E
Wuyuan, China 34 D5 41 2N 108 20 E
Wuzhai, China 34 E6 38 54N 111 48 E
Wuzhi Shan, China 38 C7 18 45N 109 45 E
Wuzhong, China 34 E4 38 2N 106 12 E
Wuzhou, China 33 D6 23 30N 111 18 E
Wyaaba Cr. →, Australia 62 B3 16 27S 141 35 E
Wyalkatchem, Australia 61 F2 31 8S 117 22 E
Wyalusing, U.S.A. 79 E8 41 40N 76 16W
Wyandotte, U.S.A. 76 D4 42 12N 83 9W
Wyandra, Australia 63 D4 27 12S 145 56 E
Wyangala, L., Australia 63 E4 33 54S 149 0 E
Wyara, L., Australia 63 D3 28 42S 144 14 E
Wycheproof, Australia 63 F3 36 5S 143 17 E
Wye →, U.K. 11 F5 51 38N 2 40W
Wyemandoo, Australia 61 E2 28 28S 118 29 E
Wymondham, U.K. 11 E9 52 35N 1 7 E
Wymore, U.S.A. 80 E6 40 7N 96 40W
Wyndham, Australia 60 C4 15 33S 128 3 E
Wyndham, N.Z. 59 M2 46 20S 168 51 E
Wynne, U.S.A. 81 H9 35 14N 90 47W
Wynyard, Australia 62 G4 41 5S 145 44 E
Wynyard, Canada 73 C8 51 45N 104 10W
Wyola L., Australia 61 E5 29 8S 130 17 E
Wyoming, Canada 78 D2 42 57N 82 7W
Wyoming □, U.S.A. 82 E10 43 0N 107 30W
Wyomissing, U.S.A. 79 F9 40 20N 75 59W
Wyong, Australia 63 E5 33 14S 151 24 E
Wytheville, U.S.A. 76 G5 36 57N 81 5W

X

Xaçmaz, Azerbaijan 25 F8 41 31N 48 42 E
Xai-Xai, Mozam. 57 D5 25 6S 33 31 E
Xainza, China 32 C3 30 58N 88 35 E
Xangongo, Angola 56 B2 16 45S 15 5 E
Xankändi, Azerbaijan 25 G8 39 52N 46 49 E
Xánthi, Greece 21 D11 41 10N 24 58 E
Xanxerê, Brazil 95 B5 26 53S 52 23W
Xapuri, Brazil 92 F5 10 35S 68 35W
Xar Moron He →, China 35 C11 43 25N 120 35 E
Xátiva, Spain 19 C5 38 59N 0 32W
Xau, L., Botswana 56 C3 21 15S 24 44 E
Xavantina, Brazil 95 A5 21 15S 52 48W
Xenia, U.S.A. 76 F4 39 41N 83 56W
Xeropotamos →, Cyprus 23 E11 34 42N 32 33 E
Xhora, S. Africa 57 E4 31 55S 28 38 E
Xhumo, Botswana 56 C3 21 7S 24 35 E
Xi Jiang →, China 33 D6 22 5N 113 20 E
Xi Xian, China 34 F6 36 41N 110 58 E
Xia Xian, China 34 G6 35 8N 111 12 E
Xiachengzi, China 35 B16 44 40N 130 18 E
Xiaguan, China 32 D5 25 32N 100 16 E
Xiajin, China 34 F9 36 56N 116 0 E
Xiamen, China 33 D6 24 25N 118 4 E
Xi'an, China 34 G5 34 15N 109 0 E
Xian Xian, China 34 E9 38 12N 116 6 E
Xiang Jiang →, China 33 D6 28 55N 112 50 E
Xiangcheng, Henan, China 34 H8 33 29N 114 52 E
Xiangcheng, Henan, China 34 H7 33 50N 113 27 E
Xiangfan, China 33 C6 32 2N 112 8 E
Xianggang = Hong Kong □, China 33 D6 22 11N 114 14 E
Xianghuang Qi, China 34 C7 42 2N 113 50 E
Xiangning, China 34 G6 35 58N 110 50 E
Xiangquan, China 34 F7 36 30N 113 1 E
Xiangquan He = Sutlej →, Pakistan 42 E4 29 23N 71 3 E
Xiangshui, China 35 G10 34 12N 119 33 E
Xiangtan, China 33 D6 27 51N 112 54 E
Xianyang, China 34 G5 34 20N 108 40 E
Xiao Hinggan Ling, China 33 B7 49 0N 127 0 E
Xiao Xian, China 34 G9 34 15N 116 55 E
Xiaoyi, China 34 F6 37 8N 111 48 E
Xiawa, China 35 C11 42 35N 120 38 E
Xiayi, China 34 G9 34 15N 116 10 E
Xichang, China 32 D5 27 51N 102 19 E
Xichuan, China 34 H6 33 0N 111 30 E
Xieng Khouang, Laos 38 C4 19 17N 103 25 E
Xifei He →, China 34 H9 32 45N 116 40 E
Xifeng, Gansu, China 34 G4 35 40N 107 40 E
Xifeng, Liaoning, China 35 C13 42 42N 124 45 E
Xifengzhen = Xifeng, China 34 G4 35 40N 107 40 E
Xigazê, China 32 D3 29 5N 88 45 E
Xihe, China 34 G3 34 2N 105 20 E
Xihua, China 34 H8 33 45N 114 30 E
Xiliao He →, China 35 C12 43 32N 123 35 E
Ximana, Mozam. 55 F3 19 24S 33 58 E
Xin Xian = Xinzhou, China 34 E7 38 22N 112 46 E
Xinavane, Mozam. 57 D5 25 2S 32 47 E
Xinbin, China 35 D13 41 40N 125 2 E
Xing Xian, China 34 E6 38 27N 111 7 E
Xing'an, China 33 D6 25 38N 110 40 E
Xingcheng, China 35 D11 40 40N 120 45 E
Xinghe, China 34 D7 40 55N 113 55 E
Xinghua, China 35 H10 32 58N 119 48 E
Xinglong, China 35 D9 40 25N 117 30 E
Xingping, China 34 G5 34 20N 108 28 E
Xingtai, China 34 F8 37 3N 114 32 E
Xingu →, Brazil 93 D8 1 30S 51 53W
Xingyang, China 34 G7 34 45N 112 52 E
Xinhe, China 34 F8 37 30N 115 15 E
Xining, China 32 C5 36 34N 101 40 E
Xinjiang, China 34 G6 35 34N 111 11 E
Xinjiang Uygur Zizhiqu □, China 32 C3 42 0N 86 0 E
Xinjin = Pulandian, China 35 E11 39 25N 121 58 E
Xinkai He →, China 35 C12 43 32N 123 35 E
Xinle, China 34 E8 38 25N 114 40 E
Xinlitun, China 35 D12 42 0N 122 8 E
Xinmin, China 35 D12 41 59N 122 50 E
Xintai, China 35 G9 35 55N 117 45 E
Xinxiang, China 34 G7 35 18N 113 50 E
Xinzhan, China 35 C14 43 50N 127 18 E
Xinzheng, China 34 G7 34 20N 113 45 E
Xinzhou, China 34 E7 38 22N 112 46 E
Xiongyuecheng, China 35 D12 40 12N 122 5 E
Xiping, Henan, China 34 H8 33 22N 114 5 E
Xiping, Henan, China 34 H6 33 25N 111 8 E
Xique-Xique, Brazil 93 F10 10 50S 42 40W
Xisha Qundao = Paracel Is., S. China Sea 36 A4 15 50N 112 0 E
Xiuyan, China 35 D12 40 18N 123 11 E
Xixabangma Feng, China 41 E14 28 20N 85 40 E
Xixia, China 34 H6 33 25N 111 29 E
Xixiang, China 34 H4 33 0N 107 44 E
Xiyang, China 34 F7 37 38N 113 38 E
Xizang Zizhiqu □, China 32 C3 32 0N 88 0 E
Xlendi, Malta 23 C1 36 1N 14 12 E
Xuan Loc, Vietnam 39 G6 10 56N 107 14 E
Xuanhua, China 34 D8 40 40N 115 2 E
Xuchang, China 34 G7 34 2N 113 48 E
Xun Xian, China 34 G8 35 42N 114 33 E
Xunyang, China 34 H5 32 48N 109 22 E
Xunyi, China 34 G5 35 8N 108 20 E
Xúquer →, Spain 19 C5 39 5N 0 10W
Xushui, China 34 E8 39 2N 115 40 E
Xuyen Moc, Vietnam 39 G6 10 34N 107 25 E
Xuzhou, China 35 G9 34 18N 117 10 E
Xylophagou, Cyprus 23 E12 34 54N 33 51 E

Y

Ya Xian, China 38 C7 18 14N 109 29 E
Yaamba, Australia 62 C5 23 8S 150 22 E
Yaapeet, Australia 63 F3 35 45S 142 3 E
Yablonovy Ra. = Yablonovyy Khrebet, Russia 27 D12 53 0N 114 0 E
Yablonovyy Khrebet, Russia 27 D12 53 0N 114 0 E
Yabrai Shan, China 34 E2 39 40N 103 0 E
Yabrūd, Syria 47 B5 33 58N 36 39 E
Yacheng, China 33 E5 18 22N 109 6 E
Yacuiba, Bolivia 94 A3 22 0S 63 43W
Yacuma →, Bolivia 92 F5 13 38S 65 23W
Yadgir, India 40 L10 16 45N 77 5 E
Yadkin →, U.S.A. 77 H5 35 29N 80 9W
Yaeyama-Rettō, Japan 31 M1 24 30N 123 40 E
Yagodnoye, Russia 27 C15 62 33N 149 40 E
Yahila, Dem. Rep. of the Congo 54 B1 0 13N 24 28 E
Yahk, Canada 72 D5 49 6N 116 10W
Yahuma, Dem. Rep. of the Congo 52 D4 1 0N 23 10 E
Yaita, Japan 31 F9 36 48N 139 56 E
Yaiza, Canary Is. 22 F6 28 57N 13 46W
Yakima, U.S.A. 82 C3 46 36N 120 31W
Yakima →, U.S.A. 82 C3 47 0N 120 30W
Yakobi I., U.S.A. 72 B1 58 0N 136 30W
Yakovlevka, Russia 30 B6 44 26N 133 28 E
Yaku-Shima, Japan 31 J5 30 20N 130 30 E
Yakumo, Japan 30 C10 42 15N 140 16 E
Yakutat, U.S.A. 68 C6 59 33N 139 44W
Yakutia = Sakha □, Russia 27 C13 66 0N 130 0 E
Yakutsk, Russia 27 C13 62 5N 129 50 E
Yala, Thailand 39 J3 6 33N 101 18 E
Yale, U.S.A. 78 C2 43 8N 82 48W
Yalgoo, Australia 61 E2 28 16S 116 39 E
Yalinga, C.A.R. 52 C4 6 33N 23 10 E
Yalkubul, Punta, Mexico 87 C7 21 32N 88 37W
Yalleroi, Australia 62 C4 24 3S 145 42 E
Yalobusha →, U.S.A. 81 J9 33 33N 90 10W
Yalong Jiang →, China 32 D5 26 40N 101 55 E
Yalova, Turkey 21 D13 40 41N 29 15 E
Yalta, Ukraine 25 F5 44 30N 34 10 E
Yalu Jiang →, China 35 E13 40 0N 124 22 E
Yam Ha Melaḥ = Dead Sea, Asia 47 D4 31 30N 35 30 E
Yam Kinneret, Israel 47 C4 32 45N 35 35 E
Yamada, Japan 31 H5 33 33N 130 49 E
Yamagata, Japan 30 E10 38 15N 140 15 E
Yamagata □, Japan 30 E10 38 30N 140 0 E
Yamaguchi, Japan 31 G5 34 10N 131 32 E
Yamaguchi □, Japan 31 G5 34 20N 131 40 E
Yamal, Poluostrov, Russia 26 B8 71 0N 70 0 E
Yamal Pen. = Yamal, Poluostrov, Russia 26 B8 71 0N 70 0 E
Yamanashi □, Japan 31 G9 35 40N 138 40 E
Yamantau, Gora, Russia 24 D10 54 15N 58 6 E
Yamba, Australia 63 D5 29 26S 153 23 E
Yambarran Ra., Australia 60 C5 15 10S 130 25 E
Yâmbiô, Sudan 51 H11 4 35N 28 16 E
Yambol, Bulgaria 21 C12 42 30N 26 30 E
Yamdena, Indonesia 37 F8 7 45S 131 20 E
Yame, Japan 31 H5 33 13N 130 35 E
Yamethin, Burma 41 J20 20 29N 96 18 E
Yamma-Yamma, L., Australia 63 D3 26 16S 141 20 E
Yamoussoukro, Ivory C. 50 G4 6 49N 5 17W
Yampa →, U.S.A. 82 F9 40 32N 108 59W
Yampi Sd., Australia 60 C3 16 8S 123 38 E
Yampil, Moldova 17 D15 48 15N 28 15 E
Yampol = Yampil, Moldova 17 D15 48 15N 28 15 E
Yamuna →, India 43 G9 25 30N 81 53 E
Yamunanagar, India 42 D7 30 7N 77 17 E
Yamzho Yumco, China 32 D4 28 48N 90 35 E
Yana →, Russia 27 B14 71 30N 136 0 E
Yanagawa, Japan 31 H5 33 10N 130 24 E
Yanai, Japan 31 H6 33 58N 132 7 E
Yan'an, China 34 F5 36 35N 109 26 E
Yanaul, Russia 24 C10 56 25N 55 0 E
Yanbu 'al Baḥr, Si. Arabia 46 C2 24 0N 38 5 E
Yanchang, China 34 F6 36 43N 110 1 E
Yancheng, Henan, China 34 H8 33 35N 114 0 E
Yancheng, Jiangsu, China 35 H11 33 23N 120 8 E
Yanchep Beach, Australia 61 F2 31 33S 115 37 E
Yanchi, China 34 F4 37 48N 107 20 E
Yanchuan, China 34 F6 36 51N 110 10 E
Yanco Cr. →, Australia 63 F4 35 14S 145 35 E
Yandoon, Burma 41 L19 17 0N 95 40 E
Yang Xian, China 34 H4 33 15N 107 30 E
Yangambi, Dem. Rep. of the Congo 54 B1 0 47N 24 24 E
Yangcheng, China 34 G7 35 28N 112 22 E
Yangch'ü = Taiyuan, China 34 F7 37 52N 112 33 E
Yanggao, China 34 D7 40 21N 113 55 E
Yanggu, China 34 F8 36 8N 115 43 E
Yangliuqing, China 35 E9 39 2N 117 5 E
Yangon = Rangoon, Burma 41 L20 16 45N 96 20 E
Yangpingguan, China 34 H4 32 58N 106 5 E
Yangquan, China 34 F7 37 58N 113 31 E
Yangtse = Chang Jiang →, China 33 C7 31 48N 121 10 E
Yangtze Kiang = Chang Jiang →, China 33 C7 31 48N 121 10 E
Yangyang, S. Korea 35 E15 38 4N 128 38 E
Yangyuan, China 34 D8 40 1N 114 10 E
Yangzhou, China 33 C6 32 21N 119 26 E
Yanji, China 35 C15 42 59N 129 30 E
Yankton, U.S.A. 80 D6 42 53N 97 23W
Yanonge, Dem. Rep. of the Congo 54 B1 0 35N 24 38 E
Yanqi, China 32 B3 42 5N 86 35 E
Yanqing, China 34 D8 40 30N 115 58 E
Yanshan, China 35 E9 38 4N 117 22 E
Yanshou, China 35 B15 45 28N 128 22 E
Yantabulla, Australia 63 D4 29 21S 145 0 E
Yantai, China 35 F11 37 34N 121 22 E
Yanzhou, China 34 G9 35 35N 116 49 E
Yao Xian, China 34 G5 34 55N 108 59 E
Yao Yai, Ko, Thailand 39 J2 8 0N 98 35 E
Yaoundé, Cameroon 52 D2 3 50N 11 35 E
Yaowan, China 35 G10 34 15N 118 3 E
Yap I., Pac. Oc. 64 G5 9 30N 138 10 E
Yapen, Indonesia 37 E9 1 50S 136 0 E
Yapen, Selat, Indonesia 37 E9 1 20S 136 10 E
Yapero, Indonesia 37 E9 4 59S 137 11 E
Yappar →, Australia 62 B3 18 22S 141 16 E
Yaqui →, Mexico 86 B2 27 37N 110 39W
Yar-Sale, Russia 26 C8 66 50N 70 50 E
Yaraka, Australia 62 C3 24 53S 144 3 E
Yaransk, Russia 24 C8 57 22N 47 49 E
Yare →, U.K. 11 E9 52 35N 1 38 E
Yaremcha, Ukraine 17 D13 48 27N 24 33 E
Yarensk, Russia 24 B8 62 11N 49 15 E
Yarí →, Colombia 92 D4 0 20S 72 20W
Yarkand = Shache, China 32 C2 38 20N 77 10 E
Yarker, Canada 79 B8 44 23N 76 46W
Yarkhun →, Pakistan 43 A5 36 17N 72 30 E
Yarmouth, Canada 71 D6 43 50N 66 7W
Yarmūk →, Syria 47 C4 32 42N 35 40 E
Yaroslavl, Russia 24 C6 57 35N 39 55 E
Yarqa, W. →, Egypt 47 F2 30 0N 33 49 E
Yarra Yarra Lakes, Australia 61 E2 29 40S 115 45 E
Yarram, Australia 63 F4 38 29S 146 39 E
Yarraman, Australia 63 D5 26 50S 152 0 E
Yarras, Australia 63 E5 31 25S 152 20 E
Yartsevo, Russia 27 C10 60 20N 90 0 E
Yarumal, Colombia 92 B3 6 58N 75 24W
Yasawa Group, Fiji 59 C7 17 0S 177 23 E
Yaselda, Belarus 17 B14 52 7N 26 28 E
Yasin, Pakistan 43 A5 36 24N 73 23 E
Yasinski, L., Canada 70 B4 53 16N 77 35W
Yasinya, Ukraine 17 D13 48 16N 24 21 E
Yasothon, Thailand 38 E5 15 50N 104 10 E
Yass, Australia 63 E4 34 49S 148 54 E
Yāsūj, Iran 45 D6 30 31N 51 31 E
Yatağan, Turkey 21 F13 37 20N 28 10 E
Yates Center, U.S.A. 81 G7 37 53N 95 44W
Yathkyed L., Canada 73 A9 62 40N 98 0W
Yatsushiro, Japan 31 H5 32 30N 130 40 E
Yatta Plateau, Kenya 54 C4 2 0S 38 0 E
Yavari →, Peru 92 D4 4 21S 70 2W
Yávaros, Mexico 86 B3 26 42N 109 31W
Yavatmal, India 40 J11 20 20N 78 15 E
Yavne, Israel 47 D3 31 52N 34 45 E
Yavoriv, Ukraine 17 D12 49 55N 23 20 E
Yavorov = Yavoriv, Ukraine 17 D12 49 55N 23 20 E
Yawatahama, Japan 31 H6 33 27N 132 24 E
Yazd, Iran 45 D7 31 55N 54 27 E
Yazd □, Iran 45 D7 32 0N 55 0 E
Yazd-e Khvāst, Iran 45 D7 31 31N 52 7 E
Yazman, Pakistan 42 E4 29 8N 71 45 E
Yazoo →, U.S.A. 81 J9 32 22N 90 54W
Yazoo City, U.S.A. 81 J9 32 51N 90 25W
Yding Skovhøj, Denmark 9 J13 55 59N 9 46 E
Ye Xian = Laizhou, China 35 F10 37 8N 119 57 E
Ye Xian, China 34 H7 33 35N 113 25 E
Yebyu, Burma 38 E2 14 15N 98 13 E
Yechŏn, S. Korea 35 F15 36 39N 128 27 E
Yecla, Spain 19 C5 38 35N 1 5W
Yécora, Mexico 86 B3 28 20N 108 58W
Yedintsy = Edineţ, Moldova 17 D14 48 9N 27 18 E
Yegros, Paraguay 94 B4 26 20S 56 25W
Yehuda, Midbar, Israel 47 D4 31 35N 35 15 E
Yei, Sudan 51 H12 4 9N 30 40 E
Yekaterinburg, Russia 26 D7 56 50N 60 30 E
Yekaterinodar = Krasnodar, Russia 25 E6 45 5N 39 0 E
Yelarbon, Australia 63 D5 28 33S 150 38 E
Yelets, Russia 24 D6 52 40N 38 30 E
Yelizavetgrad = Kirovohrad, Ukraine 25 E5 48 35N 32 20 E
Yell, U.K. 12 A7 60 35N 1 5W
Yell Sd., U.K. 12 A7 60 33N 1 15W
Yellow Sea, China 35 G12 35 0N 123 0 E
Yellowhead Pass, Canada 72 C5 52 53N 118 25W
Yellowknife, Canada 72 A6 62 27N 114 29W
Yellowknife →, Canada 72 A6 62 31N 114 19W
Yellowstone →, U.S.A. 80 B3 47 59N 103 59W
Yellowstone L., U.S.A. 82 D8 44 27N 110 22W
Yellowstone Nat. Park, U.S.A. 82 D9 44 40N 110 30W
Yelsk, Belarus 17 C15 51 50N 29 10 E
Yemen ■, Asia 46 E3 15 0N 44 0 E
Yen Bai, Vietnam 38 B5 21 42N 104 52 E
Yenangyaung, Burma 41 J19 20 30N 95 0 E
Yenbo = Yanbu 'al Baḥr, Si. Arabia 46 C2 24 0N 38 5 E
Yenda, Australia 63 E4 34 13S 146 14 E
Yenice, Turkey 21 E12 39 55N 27 17 E
Yenisey →, Russia 26 B9 71 50N 82 40 E
Yeniseysk, Russia 27 D10 58 27N 92 13 E
Yeniseyskiy Zaliv, Russia 26 B9 72 20N 81 0 E
Yennádhi, Greece 23 C9 36 2N 27 56 E
Yenyuka, Russia 27 D13 57 57N 121 15 E
Yeo →, U.K. 11 G5 51 2N 2 49W
Yeo, L., Australia 61 E3 28 0S 124 30 E
Yeo I., Canada 78 A3 45 24N 81 48W
Yeola, India 40 J9 20 2N 74 30 E
Yeoryioúpolis, Greece 23 D6 35 20N 24 15 E
Yeovil, U.K. 11 G5 50 57N 2 38W
Yeppoon, Australia 62 C5 23 5S 150 47 E
Yerbent, Turkmenistan 26 F6 39 30N 58 50 E
Yerbogachen, Russia 27 C11 61 16N 108 0 E
Yerevan, Armenia 25 F7 40 10N 44 31 E
Yerington, U.S.A. 82 G4 38 59N 119 10W
Yermak, Kazakstan 26 D8 52 2N 76 55 E
Yermo, U.S.A. 85 L10 34 54N 116 50W
Yerólakkos, Cyprus 23 D12 35 11N 33 15 E
Yeropol, Russia 27 C17 65 15N 168 40 E
Yeropótamos →, Greece 23 D6 35 3N 24 50 E
Yeroskipos, Cyprus 23 E11 34 46N 32 28 E
Yershov, Russia 25 D8 51 23N 48 27 E
Yerushalayim = Jerusalem, Israel 47 D4 31 47N 35 10 E
Yes Tor, U.K. 11 G4 50 41N 4 0W
Yesan, S. Korea 35 F14 36 41N 126 51 E
Yeso, U.S.A. 81 H2 34 26N 104 37W
Yessey, Russia 27 C11 68 29N 102 10 E
Yetman, Australia 63 D5 28 56S 150 48 E